EXPERIMENTAL METHODS in RF DESIGN

Wes Hayward, W7ZOI

Rick Campbell, KK7B

Bob Larkin, W7PUA

Editors:
Jan Carman, K5MA
Steve Ford, WB8IMY
Dana Reed, W1LC
Jim Talens, N3JT
Larry Wolfgang, WR1B

Technical Illustration:
David Pingree, N1NAS

Cover Design:
Sue Fagan
Bob Inderbitzen, NQ1R

Proofreaders:
Kathy Ford
Jayne Pratt Lovelace
Larry Joy, WN8P

Production:
Michelle Bloom, WB1ENT
Paul Lappen
Jodi Morin, KA1JPA

Copyright © 2003-2009 by
The American Radio Relay League

Copyright secured under the Pan-American Convention

International Copyright secured

This work is publication No. 288 of the Radio Amateur's Library, published by ARRL. All rights reserved. No part of this work may be reproduced in any form except by written permission of the publisher. All rights of translation are reserved.

Printed in USA

Quedan reservados todos los derechos

Revised First Edition

ISBN # 0-87259-923-9
ISBN13: 978-0-87259-923-9

CONTENTS

Contents

Preface

1 Getting Started
1.1 Experimenting, "Homebrewing," and the Pursuit of the New
1.2 Getting Started – Routes for the Beginning Experimenter
1.3 Some Guidelines for the Experimenter
1.4 Block Diagrams
1.5 An IC Based Direct Conversion Receiver
1.6 A Regenerative Receiver
1.7 An Audio Amplifier with Discrete Transistors
1.8 A Direct Conversion Receiver Using a Discrete Component Product Detector
1.9 Power Supplies
1.10 RF Measurements
1.11 A First Transmitter
1.12 A Bipolar Transistor Power Amplifier
1.13 An Output Low Pass Filter
1.14 About the Schematics in this Book

2 Amplifier Design Basics
2.1 Modeling Simple Solid State Devices
2.2 Amplifier Design Basics
2.3 Large Signal Amplifiers
2.4 Gain, Power, dB and Impedance Matching
2.5 Differential Amplifiers and the Op-Amp
2.6 Undesired Amplifier Characteristics
2.7 Feedback Amplifiers
2.8 Bypassing and Decoupling
2.9 Power Amplifier Basics
2.10 Practical Power Amplifiers
2.11 A 30-W – 7-MHz Power Amplifier

3 Filters and Impedance Matching Circuits
3.1 Filter Basics
3.2 The Low Pass Filter, Design and Extension
3.3 LC Bandpass Filters
3.4 Crystal Filters
3.5 Active Filters
3.6 Impedance Matching Networks

4 Oscillators and Frequency Synthesis
4.1 LC-Oscillator Basics
4.2 Practical Hartley Circuits and Oscillator Drift Compensation
4.3 The Colpitts and Some Other scillators
4.4 Noise in Oscillators
4.5 Crystal Oscillators and VXOs
4.6 Voltage Controlled Oscillators
4.7 Frequency Synthesis
4.8 The Ugly Weekender, MK-II, A 7-MHz VFO Transmitter
4.9 A Digital Dial
4.10 A General Purpose VXO-Extending Frequency Synthesizer

5 Mixers and Frequency Multipliers
- 5.1 Mixer Basics
- 5.2 Balanced Mixer Concepts
- 5.3 Some Practical Mixers
- 5.4 Frequency Multipliers
- 5.5 A VXO Transmitter Using a Digital Frequency Multiplier

6 Transmitters and Receivers
- 6.0 Signals and the Systems that Process Them
- 6.1 Receiver Fundamentals
- 6.2 IF Amplifiers and AGC
- 6.3 Large Signals in Receivers and Front End Design
- 6.4 Local Oscillator Systems
- 6.5 Receivers with Enhanced Dynamic Range
- 6.6 Transmitter and Transceiver Design
- 6.7 Frequency Shifts, Offsets and Incremental Tuning
- 6.8 Transmit-Receive Antenna Switching
- 6.9 The Lichen Transceiver: A Case Study
- 6.10 A Monoband SSB/CW Transceiver
- 6.11 A Portable DSB/CW 50 MHz Station

7 Measurement Equipment
- 7.0 Measurement Basics
- 7.1 DC Mesaurements
- 7.2 The Oscilloscope
- 7.3 RF Power Measurement
- 7.4 Attenuators
- 7.5 Measuring Frequency, Inductance, and Capacitance
- 7.6 Sources and Generators
- 7.7 Bridges and Impedance Measurement
- 7.8 Spectrum Analysis
- 7.9 Q Measurement of LC Resonators
- 7.10 Crystal Measurements
- 7.11 Noise and Noise Sources
- 7.12 Assorted Circuits

8 Direct Conversion Receivers
- 8.1 A Brief History
- 8.2 The Basic Direct Conversion Block Diagram
- 8.3 Peculiarities of Direct Conversion
- 8.4 Mixers For Direct Conversion Receivers
- 8.5 A Modular Direct Conversion Receiver
- 8.6 DC Receiver Advantages

9 Phasing Receivers and Transmitters
- 9.1 Block Diagrams
- 9.2 Introduction to the Math
- 9.3 From Mathematics to Practice
- 9.4 Sideband Suppression Design
- 9.5 Binaural Receivers
- 9.6 LO and RF Phase-Shift and In-Phase Splitter-Combiner Networks
- 9.7 Other Op-Amp Topologies, Polyphase Networks and DSP Phase Shifters
- 9.8 Intelligent Selectivity
- 9.9 A Next-Generation R2 Single-Signal Direct Conversion Receiver
- 9.10 A High Performance Phasing SSB Exciter
- 9.11 A Few Notes on Building Phasing Rigs
- 9.12 Conclusion

10 DSP Components
    10.1 The EZ-Kit Lite
    10.2 A Program Shell
    10.3 DSP Components
    10.4 Signal Generation
    10.5 Random Noise Generation
    10.6 Filtering Components
    10.7 DSP IF
    10.8 DSP Mixing
    10.9 Other DSP Components
    10.10 Discrete Fourier Transform
    10.11 Automatic Noise Blankers
    10.12 CW Signal Generation
    10.13 SSB Signal Generation

11 DSP Applications in Communications
    11.1 Program Structure
    11.2 Using a DSP Device as a Controller
    11.3 An Audio Generator Test Box
    11.4 An 18-MHz Transceiver
    11.5 DSP-10 2-Meter Transceiver

12 Field Operation, Portable Gear and Integrated Stations
    12.1 Simple Equipment for Portable Operation
    12.2 The "Unfinished," A 7-MHz CW Transceiver
    12.3 The S7C, A Simple 7-MHz Super-Heterodyne Receiver
    12.4 A Dual Band QRP CW Transceiver
    12.5 Weak-Signal Communications Using the DSP-10
    12.6 A 28-MHz QRP Module
    12.7 A General Purpose Receiver Module
    12.8 Direct Conversion Transceiver for 144-MHz SSB and CW
    12.9 52-MHz Tunable IF for VHF and UHF Transceivers
    12.10 Sleeping Bag Radio
    12.11 14-MHz CW Receiver

Contents of CD-ROM

Index

PREFACE

The predecessor for this book, *Solid State Design for the Radio Amateur* (*SSD*), was first published by ARRL in early 1977. The goal for that text was to present solid state circuit design methods to a community much more familiar with vacuum tube methods. But, another goal was integrated into the text, that of presenting the material in a way that would allow the reader to actually *design* his or her own circuits. Handbooks of the day presented only an encyclopedic overview of solid state devices with brief qualitative discussions about functionality. *SSD* described circuit elements in terms of models that could be used for analysis. Design consists of more than merely combining representative circuits from a catalog or handbook.

SSD succeeded with **design** becoming the key word in the title, especially in later years as the world became accustomed to all electronic equipment being predominantly solid state. What surprised many is that the book remained popular, even after many of the transistors used in the circuits were no longer available.

Experimental Methods in Radio Frequency Design (*EMRFD*) is the sequel to *SSD*, with **design** remaining as a central theme. Our goal is to present models and discussion that will allow the user to design equipment at both the circuit and the system level. Our own interests are dominated by radio frequencies, so the text discusses problems peculiar to radio communications equipment. A final emphasis in *EMRFD* is **experimentation**. A vital part of an experiment is measurement. We encourage the reader to not only build equipment, but to perform measurements on that gear as it is being built.

The word "experiment," often conjures memories of school exercises where a teacher has assembled equipment and we, as students, go through a prearranged set of steps to arrive at a conclusion, also predetermined. Although efficient, this is a poor representation of science. Rather, experimental science begins with a new idea. An experiment to test the idea is then generated, the experiment is built, measurements are made, and the results are pondered, which often results in new ideas to test. This can all be done by one person working alone. *EMRFD* encourages the participating reader to build equipment with an attitude of continually seeking to understand the equipment and to understand the primitive concepts that form the basis for the equipment and the circuits contained therein. Our greatest hope is that the text will illustrate the potential of amateur radio, and other personal science, as a training ground for the individual.

This text is aimed at a variety of readers: the radio amateur who designs and builds his own equipment; college students looking for design projects or wishing to garner practical experience with working hardware; young professionals wishing to apply their fresh engineering and physics coursework to kitchen table projects; non-engineers wanting to dabble in a technical field; engineering managers recapturing the fun of making things (instead of people) work; and technical explorers of all types.

The first chapter of *EMRFD* deals with the problems of getting started with experimentation. Numerous projects are presented, aimed at assisting the experimenter in beginning investigations in electronics. Chapters 2 through 5 then deal with specific circuit functions. Chapter 2 presents amplifiers while filters are discussed in Chapter 3. Oscillators emerge in Chapter 4, including the natural extension of frequency synthesis. Mixers, including frequency multipliers, appear in the fifth chapter. These chapters are laced with projects that can be constructed, but they also emphasize important basic concepts. Chapter 6 moves on to present communications equipment, predominantly using super-heterodyne methods. System design considerations are included, especially with regard to distortion and dynamic range. The chapter contains several projects including a high performance receiver. Chapter 7 deals with measurement methods and includes considerable test equipment that the experimenter can build. Chapter 8 then moves on to a fundamental discussion of direct conversion. This is followed by a thorough treatment of the phasing method of SSB in Chapter 9. Chapters 10 and 11 present fundamental concepts of digital signal processing and illustrate them with projects. The book concludes with Chapter 12 featuring a variety of experimental activities of special interest to the authors.

A Compact Disc is included with the book. This CD contains some design software, extensive listings for DSP firmware related to Chapters 10 and 11, and a sizeable collection of journal articles relating to material presented in the text. The design software is written for a personal computer using the Microsoft Windows operating system, while the journal papers are presented in Adobe Acrobat (PDF) format.

This book is a personal one in that we have only written about those things we have actually experienced. We specifically avoided an encyclopedic discussion of material that we had not actually experienced through experiments. Equipment of interest to the three of us dominates. The amateur bands up to 2 meters are considered, and are illustrated with CW and SSB gear. The book uses some mathematics where appropriate. It is, however, kept at a basic level.

The book contains numerous projects that are suitable for duplication. Printed circuit boards are not generally available for these, although boards may become available at a later time. Readers should keep an eye on the world wide web for PCB information and other matters related to the book. See **http://www.arrl.org/notes/8799**. We generally prefer that builders use the projects as starting points for their own designs and experiments rather than duplicating the projects presented.

Acknowledgments

The following experimenters have contributed to this book through experiments, direct correspondence, encouragement, and by example. We gratefully acknowledge their contributions.

Bill Amidon (sk); Tom Apel, K5TRA; Leif Åsbrink, SM5BSZ;
Kirk Bailey, N7CCB; Dave Benson, K1SWL; Byron Blanchard, N1EKV; Denton Bramwell, W7DB; Guy Brennert, K2EFB; Rod Brink, KQ6F; Kent Britain, WA5VJB; Wayne Burdick, N6KR;
Russ Carpenter, AA7QU; Dennis Criss; Bob Culter, N7FKI; George Daughters, K6GT; John Davis, KF6EDB; Paul Decker, KG7HF; Rev. George Dobbs, G3RJV;
Pete Eaton, WB9FLW; Gerry Edson, WA0KNW; Bill Evans, W3FB;

George Fare, G3OGQ; Johan Forrer, KC7WW; Dick Frey, K4XU;

Barrie Gilbert; Jack Glandon, WB4RNO; Joe Glass, WB2PJS; Dr. Dave Gordon-Smith, G3UUR; Mike Greaney, K3SRZ; Linley Gumm, K7HFD;

Nick Hamilton, G4TXG; Mark Hansen, KI7N; Markus Hansen, VE7CA; Neil Heckt; Ward Helms, W7SMX; Don Hilliard, W0PW; Fred Holler, W2EKB; Robert Hughson;

Pete Juliano, W6JFR;

Bill Kelsey, N8ET; Ed Kessler, AA3SJ; Paul Kiciak, N2PK; Don Knotts, W7HJS; O. K. Krienke;

Beb Larkin, W7SLB; John Lawson, K5IRK; Roy Lewallen, W7EL; John Liebenrood, K7RO; Larry Liljeqvist, W7SZ; B. F. Logan Jr., WB2NBD;

Stephen Maas, W5VHJ; Chuck MacCluer, W8MQW; Jacob Makhinson, N6NWP; Ernie Manly, W7LHL; Dr. Skip Marsh, W6TFQ (sk); Mike Michael, W3TS; Jim Miles, K5CX;

Dave Newkirk, W9VES;

Gary Oliver, WA7SHI;

Paul Pagel, N1FB;

Dave Roberts, G8KBB; Mike Reed, KD7TS; Don Reynolds, K7DBA (sk); Dr. Ulrich Rohde, KA2WEU; Dr. Dave Rutledge, KN6EK; Tom Rousseau, K7PJT;

Bill Sabin, W0IYH; Tom Scott, KD7DMH; Marty Singer, K7AYP; Derry Spittle, VE7QK;

Fred Telewski, WA7TZY;

Paul Wade, W1GHZ; Al Ward, W5LUA; Dr. Fred Weiss; Jim Wyckoff, K3BT;

Bob Zavrel, W7SX; Bob Zulinski, WA8MAM.

We have certainly missed some folks in our list. Please accept our apologies for our oversight and our thanks for your help with the book and related experiments.

Some folks have made special contributions and deserve special thanks. Colin Horrabin, G3SBI; Harold Johnson, W4ZCB; and Bill Carver, W7AAZ, collectively formed the "Triad," a group building the high performance transceiver partially described in Chapter 6. We sincerely appreciate their willingness to share their efforts and results with us. Thanks go to Roger Hayward, KA7EXM, for building some equipment described in the book as well as helping with field testing of numerous designs. Jeff Damm, WA7MLH, deserves special thanks for his efforts. He built equipment described in *SSD* and provided encouragement for this version. Special thanks to Merle Cox, W7YOZ, and Jim Davey, K8RZ, for several decades of bouncing around radio ideas, building the second prototypes, and manning the distant station for countless experiments. Very special thanks are extended to Terry White, K7TAU. Terry did high quality PC layouts for several of the designs presented in the text and in earlier *QST* articles. He also built some equipment shown in the book and provided measurement assistance on several occasions.

Special mention should be made of the efforts of the late Doug DeMaw, W1FB. As co-author of *SSD*, he provided interest and encouragement for this sequel. One of Doug's greatest qualities was his intense, sincere interest in radio communications. He designed and built radio equipment, used it on the air, and then clearly wrote about the efforts, establishing a standard for all to follow. We missed him often through the generation of this text.

Finally, we want to thank our families, and especially our wives: Charlene (Shon) Hayward, Sara Rankinen, and Janet Larkin. A book requires time and intense effort that often detracts from other activities. Our "better halves" have all tolerated these moments of distraction.

About the Cover Photograph

The cover photograph is an experimental 2.4 GHz IC direct conversion receiver front-end on a gallium arsenide die. The die is a little more than one millimeter wide, and less than one millimeter high. Gold-bond wires connect to the metal squares around the edge. The large spiral is a quadrature hybrid coupled inductor, and the matched inductors at the top are in a Wilkensen splitter. The passive circuitry is similar to Fig 9.39, and the photograph on page 9.43 shows this IC connected to baseband circuitry described in Chapter 9. Note the call signs on the die. "MAL," who was not licensed in 2001, is now K7MTL. Photograph by Dean Monthei.

About The Authors

All three of the authors share a similar early exposure to radio, obtaining an amateur license as a teen or earlier. They all started with the novice class license. Their early ham experiences expanded to become careers in science and electronics. All three are members of the IEEE Microwave Theory and Techniques Society and have published extensively in a wide variety of journals and books. All three writers contributed to all chapters of this text, but each author had a primary responsibility listed below.

Wes Hayward, W7ZOI

Wes received a BS in Physics from Washington State University in 1961 and an MSEE from Stanford University in 1966. He worked on electron device physics at Varian Associates, The Boeing Co., and Tektronix. He then did RF circuit design, first at Tektronix and then at TriQuint Semiconductor. Wes is now semi-retired, dividing his time between writing and consulting. Wes was the primary contributor to Chapters 1 through 7 and large parts of 12 and can be contacted at **w7zoi@arrl.net**.

Rick Campbell, KK7B

Rick received a BS in Physics from Seattle Pacific University in 1975, after two years active duty as a US Navy Radioman. He worked for 4 years in crystal physics basic research at Bell Labs in Murray Hill, NJ before returning to graduate school at the University of Washington. He completed the MSEE degree in 1981 and the PhD in EE in 1984. He served on the faculty at Michigan Tech University until 1996. Since 1996 he has been with the Advanced Development Group at TriQuint Semiconductor, designing microwave receiver circuitry. Rick had primary responsibility for chapters 8, 9, and large parts of 12. He can be contacted at **kk7b@arrl.net**.

Bob Larkin, W7PUA

Bob received a BS in EE from the University of Washington and a MS in EE from New York University. He worked for 12 years at Bell Labs in New Jersey in areas of circuit design and signal processing. In 1973 he and his wife Janet started Janel Labs where a variety of radio frequency products were manufactured. They moved the company to Corvallis Oregon in 1975 where it operated until being acquired by Celwave RF in 1991. He now works as a consultant specializing in microwave circuits. Bob was the primary contributor to Chapters 10 and 11 and wrote a section in Chapter 12. Readers can contact Bob at **w7pua@arrl.net**.

CHAPTER 1

Getting Started

1.1 EXPERIMENTING, "HOMEBREWING," AND THE PURSUIT OF THE NEW

Amateur Radio is a diverse and colorful avocation or hobby where the participants communicate with each other through the use of radio signals. The communications, which can encompass and extend beyond the planet, are often routine and predictable, but can at times be ethereal. The romance of communicating with the other side of the world blends with the joy of observing a complicated part of nature. For some of us, the wonder never disappears.

Although radio can be fun, our pragmatic society demands more than excitement when resources are used. The virtue that most often justifies our use of the radio spectrum is the growth of a proficient communications system that can be called upon in times of emergency. The examples of its use are numerous.

But, "ham" radio is more than this. It is a technical avocation of diverse educational potential. It has values that go well beyond that of a supplementary communications network.

Most radio amateurs have an interest in the technical details of the equipment they use. Historically, this was a requirement: The only way a radio amateur could assemble an operating station was to personally build his or her gear. Commercial equipment was rare, and was often prohibitively expensive. But today, high quality "ham" gear is readily available in most of the world, much of it at modest prices.

Although no longer necessary, it is still common for radio amateurs to build at least some of their own equipment. The reasons are varied and as numerous as the participants. A few purists consider building the equipment they use to be a non-optional, integral part of their hobby in the same way that a fly fishing enthusiast would *never* consider fishing with a fly that he or she had not fabricated. The majority take an intermediate path, building parts of their radio stations while purchasing others. For some, building is an exercise in craftsmanship, an opportunity to generate equipment with an individual imprint and personality.

Common to all of these, amateur radio presents an opportunity that is rare among avocations, a chance for individual, unrestrained investigations in fundamental science and technology. This is a rarity in an age when most research and design is performed by teams of investigators within large organizations, be they universities or the engineering arms of corporations. There, the subjects chosen for investigation are often those of corporate or national interest. It is increasingly rare that a study is initiated out of simple curiosity. Fortunately, we are not so constrained within our personal investigations of radio science.

Consider an example. An experimentally inclined radio amateur envisions a new scheme for a receiver. It might be a better front end circuit, a new block diagram, or a way to realize some receiver functions with a computer. The experimenter can analyze the scheme, design an example, build a prototype, build and assemble needed test equipment, measure the receiver performance, compare it with predicted results, and use the receiver on the air. Each part of the investigation can interact with the others. All of the activity can be done without interference from other sources. The program will never be cancelled by the changing goals of an organization. Nor will it be rushed by the economic pressures of a corporate program.

The inspiration for experiment varies. In rare cases, the experimenter may feel that his or her work could lead to a new twist in the state-of-the-art, a better receiver. But more often it will just be a casual thought that "Hey, I've never built one of these before and I'll learn something if I do." The most common is an effort spurred by a need; a ham wants a rig to take along on a hiking trip when no such thing can be purchased. No matter what the origin, the experimenter can enjoy the knowledge that he or she is learning more about the subject and about the research process.

In this book we encourage all levels of what has become known as radio "homebrewing," ranging from beginner projects to sophisticated multi-mode creations. We generally emphasize simple equipment described by primitive explanations. By *primitive*, we intend that the discussion relate to the most fundamental and basic circuit design concepts. The equipment and systems presented are themselves basic, often without the frills, bells, and whistles of commercial equipment. Some refinements will be discussed, allowing the experimenter to add those he or she needs.

This book emphasizes equipment design. Our interest is in basic circuit functions and the underlying concepts that allow them to be understood. This book is generally NOT a collection of projects for reproduction and construction. Although some of the equipment may be directly duplicated, we would prefer to have you adapt our results to fit your own needs.

This book is, in many ways, a sequel to an earlier effort, *Solid State Design for the Radio Amateur*.[1] That 1977 book, co-authored with the late Doug DeMaw,

W1FB, had goals similar to those outlined above, plus that of introducing solid-state methods to readers with experience limited to vacuum tube electronics. The later need has become arguable, for virtually all of our equipment is now based upon solid-state technology.

All of the circuits presented in this text have been constructed, tested, and used in practical, on-the-air situations. If there are exceptions where the authors have not actually built an example of what is discussed, we will so state in the related text.

We emphasize the traditional communications modes of CW, the original digital mode, and SSB phone. Building little rigs and radiating and receiving continuous waves are to a radio experimenter much like playing scales and folk tunes are to a musician. They are the first things we learn, are important parts of the daily practice routine throughout life, and we neglect them at our peril. The little rigs, and the concepts they represent, are at the core of wireless technology. It is not enough to play with them as a novice and then move on to other things; they need to be revisited over and over again at different stages of one's vocation, each time achieving a new level of mastery until finally one is probing the deepest mysteries of the art.

1.2 GETTING STARTED—ROUTES FOR THE BEGINNING EXPERIMENTER

What to build:

A frequent question asked by the prospective experimenter regards an initial project or subject for pursuit. A common choice for a first project comes from a desire to extend the capabilities of an existing station. The future experimenter already has experience with on-the-air activity and a working station. He or she then wants to extend that station to new bands, improved transceiver performance, or fabricate a rig offering portability. While these goals are all worthy, they can be difficult. They may be conceptually impossible for the beginner, and impractical for the seasoned experimenter with other life commitments. A better "first" experiment may well be something that is much simpler. Several simple projects are offered later in this chapter as suitable beginnings.

How to build it:

Another getting-started question regards the methods to use in building electronics. There are several options, all with their assets and weaknesses. A few are discussed below.

The primary construction scheme used in modern electronics is the printed circuit board (PCB). Here, pads or islands of metal are attached to an insulating material, usually epoxy-fiberglass. Wires on the parts are pushed through holes in the board and soldered to the pads, which are interconnected by printed metal runs, thus forming the circuit.

A PCB begins as a fiberglass sheet with copper laminated to one or both sides. The metal surfaces are then coated with a light sensitive "photo-resist" material. A pattern for the circuit is optically transferred to the surface and the unexposed material is washed away. The board is now placed in a solution that chemically etches some of the copper away, leaving only those regions needed to form the desired circuit. After etching, the board is washed and drilled. Pure copper is easily corroded, so it is common to plate boards with a tin coating, forming a more stable and solderable surface. Refined boards include copper on both sides, and even plating on the inside of the holes. Industrial boards will often incorporate many layers.

Modern practice features *surface mount technology*, SMT, using small components without wire leads. The leads have been replaced with metalized regions on the parts that are then soldered directly to the board. The soldering provides physical mounting as well as electrical connection. The SMT boards are cheaper to build and usually much more dense. SMT parts can be so small that they are hard to handle without a good microscope. SMT is an interesting way to build if there is a need for really small equipment. The small size of SMT circuits often results in improved high frequency performance.

Growing SMT popularity in manufacturing means that surface mounted is the only available form for a component. Many parts don't exist in leaded forms. In some cases they can be handled by the "Surfboards" by Capital Advanced Technologies which are found in DigiKey catalogs. These are small SMT boards with an interface that will adapt to other board forms.

Circuit boards have been built in a home environment by hams for generations. The reader should review the subject in *The ARRL Handbook* to find out more about the methods. A major problem with home etched boards is the disposal of the used etchant, usually a solution of ferric chloride. Disposal practices common in the past are now questioned in this era of enlightened recycling. Although some of the projects described in this text use etched boards, few of the boards were etched in our home labs.

, as applied to electronics, is a term from a time when early radio experimenters built their equipment on slabs of wood, often procured from the kitchen. The term remains as an industry-wide description of a preliminary experimental circuit. There are numerous modern methods that can be used to generate a one-of-a-kind circuit.

UGLY

A particularly simple method was outlined in an early   paper and is now know as "Ugly Construction."[2] Although certainly not unique, the scheme works well and continues as a recommended method. The scheme consist of the following:

1. A ground plane is established using an un-etched scrap of copper clad circuit board material.

2. Following the schematic for a circuit being built, grounded components are soldered directly to the ground foil with short leads.

3. Some non-grounded parts are soldered to and supported by the grounded components.

4. Other non-grounded components are supported with suitable "tie down points" consisting of high value resistors.

5. Once finished and working, the board can be mounted in a suitable box, hidden from view if desired, where it becomes a permanent application of the idea. Ugly construction is illustrated in **Fig 1.1**.

Casual circuit analysis allows the builder to pick the standoff resistor values. Any "high R" value resistors can be used. Usually, 1-MΩ resistors work well anywhere within RF circuits. The typical 1/4 W resistor of any value has a stray lead-to-lead parallel capacitance of about 0.3 to 0.4 pF, perhaps a little more with longer leads, and a series inductance of 3 to 5 nH.

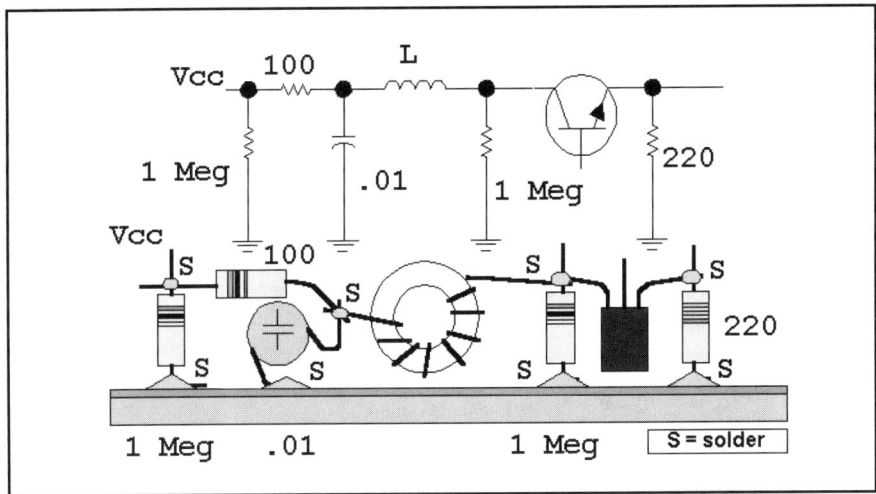

Fig 1.1—A partial circuit illustrating "ugly" construction.

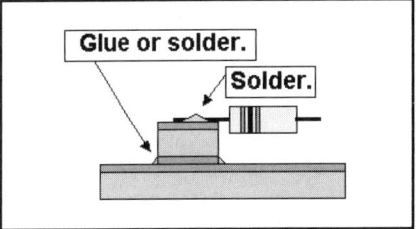

Fig 1.2—An example of "Manhattan" breadboarding.

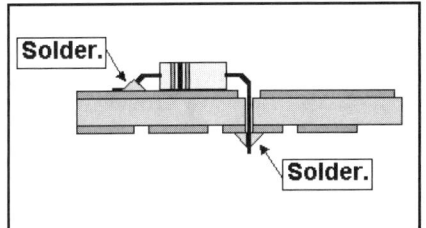

Fig 1.3—A "quasi-circuit board" scheme for breadboarding. The installed resistor here is soldered to ground and to a pad that connects to the rest of the circuitry.

Reactance is of little consequence for work up through 150 MHz or so. High R means that resistance is high with respect to the reactance of the inductance. We sometimes use R values as low as 10 kΩ. It is often surprising just how few standoff resistors are needed in an ugly breadboard.

The greatest virtue of the ugly method is low inductance grounding. Any construction scheme that preserves this grounding integrity will work as well. Picking a method is a choice that the builder has, a place where he or she can develop the methods that work best.

Integrated circuits can be placed on an ugly board with leads sticking up, "dead bug" style. There is little need to glue the chips down, for components and wires will eventually hold them in place. Grounded IC leads are bent and soldered directly to the foil.

Some builders prefer to maintain ICs with the IC label facing upward, allowing later inspection. They then bend all leads out in a "spread eagle" format.

We have never had a problem with ugly equipment being less than robust. Many of our ugly rigs have been hauled through the mountains of the Pacific Northwest in packs without incident. An outstanding example, the work of a friend, is the W7EL Optimized QRP Transceiver, a rig that has traveled around the world in suitcases and packs.[3] Few if any standoff resistors were used in that rig.

Several other construction schemes offer similar grounding fidelity, including those where small pads of circuit board material are glued or soldered to the ground foil. These pads then have components soldered to them. We have found this method to be especially useful for slightly massive components such as floating, non-grounded, trimmer capacitors. The specific glue type has little impact on circuit performance. Variations of this method have been called "Manhattan Construction," and can be mixed with other breadboarding schemes. The reader can find numerous examples on the Web on sites dealing with QRP experiments, as well as in **Fig 1.2**.

The proponents of Manhattan Construction often use small round pads that are glued to a ground foil with epoxy or similar glue. The pads are placed so that all components are parallel to board edges and close to the ground foil. This produces an attractive board resembling a commercial, PC board. This does not seem to compromise performance.

With traditional ugly construction, parts can be moved about to make room for another stage. In the extreme, an entire circuit can be lifted and moved, a stage at a time, to another board.

A primary virtue of a bread-boarding scheme is
, especially important when the primary purpose of building gear is information about circuit behavior.

Some folks prefer to rebuild a circuit after a breadboarding phase, replacing an ugly prototype with a more permanent, production-like version. These efforts take additional time and rarely produce performance superior to the original breadboards. Even looks can be deceptive when one hides ugly breadboards behind more attractive front panels.

Some experimenters prefer to build equipment that looks like a PCB, even when the board is not etched in a circuit-specific pattern. One method, called "checker-board," uses double sided circuit board with one side functioning as a ground foil. The other side consists of a matrix of small islands of copper. These regions are created either by etching or manually with a hack saw. Patterns of squares on 0.1-inch centers accommodate traditional ICs. Holes are drilled in the islands where components must reside. A large drill bit then removes ground foil around the hole without enlarging it. No holes are required where a ground connection is needed. Components usually reside on the ground side of the board. See **Fig 1.3**.

The double sided checker-board can also serve for breadboarding with surface mounted components. Parts then reside on the pattern side with holes drilled to reach ground. Small leaded components can also be surface mounted.

The checkerboard scheme, "Manhattan" variants, and even double-sided printed boards have fairly high capacitance from pads to ground. These are often poor quality capacitors with low Q, under 100 for epoxy fiberglass board material, and are subject to water absorption. A single sided format is preferred for critical sections of an LC oscillator application.

1.3 SOME GUIDELINES FOR THE EXPERIMENTER

With   came considerable interaction with the rest of the amateur radio community. A frequent question we heard was "How do I get started with experimenting?" Or, "I've read about and have even built some kits and published projects, but I want to go further. I want to do my own design. What is the next step?"

A set of guidelines is offered in an attempt to answer some of these questions. These are not firm, well established rules, but mere impressions and personal biases that we have generated, approaches that work for us. They are offered without guarantee.

•KISS: This British term is short for "Keep It Simple, Stupid." We often design equipment that is more complicated than needed. It is well worth some extra time during design to evaluate every part to see if it is really needed. The function of each part should be understood and justified. The circuit should function as intended. This does not imply that designs with the minimum number of parts are best. However, it is rarely justified to overdesign by adding extra components "because a problem might occur." For example, designs with a profusion of ferrite beads and "stability enhancing" resistors may be suspect.

•Avoid lore: Lore, in this case, refers to "knowledge" that is based upon experiences that are divorced from careful thought. A classic example in amateur radio regards the thermal stability of LC oscillators. Envision the amateur experimenter who built an oscillator using a toroid. The circuit drifted when he opened the window to the winter weather. The next evening he replaced the inductor with one wound on a ceramic coil form, noticing less drift when he opened the window. He concluded that ceramic forms are better than toroids, having never considered the specific coil forms that were used, the other components in the circuit, or the fact that the weather had improved. Poorly executed experiments like this often generate erroneous conclusions. The resulting lore, although interesting, should always be questioned. It is always better to do meaningful measurements.

•Plan your projects with block diagrams: Start with small diagrams where each block is a global element, perhaps containing several stages. Expand these to show greater detail. Block diagrams will be discussed further below.

•Generate modular equipment: A high performance receiver, for example, should consist of several sections, each designed so that it can be built, tested, modified, and redesigned as needed, with minimal change to the rest of the system. Even the simplest little rig should be built a stage at a time, turned on sequentially, tested, and modified as needed. Single board transceiver designs are popular in the QRP arena. But realize that the ones that work well are probably the result of several rebuilds, and even then, some don't work very well; others are superb.

• Avoid excessive miniaturization: It takes much more time to build small things than those where the circuitry can expand without bound. Even when building small portable QRP transceivers, it's often worthwhile to establish the design with a larger breadboard.

• Base projects on your own goals: Our central personal goal is learning through experimentation. Hence, we base projects on questions that need investigation rather than what we need or want for on-the-air operation. But your goals may be different. It is worthwhile to review and define them as a means of picking the best projects for you. Isolate primary goals from those that are serendipity.

• Be wary of "Creeping Features." The term "   " often describes the transceivers that we purchase for on-the-air communications. Appliances, even ones that we build ourselves, are usually expected to have many features, but these   can actually impede experimental progress. A single band, single mode transceiver can be as experimentally enlightening and informative as a multiple mode, general coverage transceiver.

• Use the literature. Peruse catalogs, data manuals, web sites, and even instruction manuals for circuit ideas. When a circuit method is not understood, it should be studied in texts appropriate to the technology. It is useful to build something with the part as a way to really understand that part.

• While planning is necessary, don't spend excessive time in the preliminary design phase of a project. Rather, outline preliminary ideas and goals, do initial calculations (on a computer only if they are really complicated), gather parts, and begin building. Enjoy the freedom that allows you to change your mind in the middle of an investigation. Refined calculations can occur during and after construction and are not just "design phase" activities.

•It's not about craftsmanship: A portion of the homebrewing community was schooled with the idea that "nice looking" circuit construction went along with good performance. But the two factors are generally isolated. This is illustrated in **Fig 1.4**. There is no relationship between having a nice looking, orderly circuit board and good performance from that board. Indeed, those saddled with the chore of designing a printed board to perform as well as an   breadboard may wonder if there might be an inverse relationship!

•Use breadboarding over a ground plane for communications circuits, especially when investigating new ideas. Use vector board or wire-wrap methods for slow digital circuits, but treat fast digital circuits as if they were RF functions. In general, build with those methods that will offer the best, low inductance, grounding while allowing circuits to be quickly designed, assembled, and tested. If you are concerned with aesthetic details, build a second version. Alternatively, an attractive panel can be used to hide ugly, but highly functional breadboards.

• Build what you use, and use what you build: Those of us in the homebrew end of amateur radio often kid our appliance operator friends, suggesting that a "real ham" should build instead of just operate. Some avid experimenters may take this too far; they build a rig, use it just long enough to confirm functionality, and go on to the next project, missing some exciting discoveries along the way. By using the equipment with tempered intensity, the experimenter will discover the strength and weakness of the rig, allowing the next project to be even more successful. The same arguments might be applied to software developments!

• Beware of the golden screwdriver: A good friend, WA7MLH, encountered a fellow on the air whose sole method for experimentation was to adjust all of his equipment for maximum output. He did this with a favorite screwdriver, which he treated as golden. After careful tweaking of all circuit elements that could be adjusted, he was almost always able to coax a 100-W transceiver into delivering 110 W of output. Unfortunately, what started as a good piece of equipment had become a distorted disaster. While we all tend to adjust circuits for "maximum smoke," linear circuitry should be confined to operate under linear conditions. It is important that the limits be recognized and adhered to. This is especially important when building SSB gear. Alignment means adjustment to the proper, measured level,

1.4   Chapter 1

Fig 1.4—"Nice looking" circuit construction does not always equate to good circuit performance.

which may differ from maximum.

• Always keep notebooks for experiments: Record those wild circuit ideas that come up while you cut the lawn or watch TV; record important data during experiments, including the temperature when you open the window; take notes on the circuits that you build, including changes that are made during building and "turn on". Date the notebook and place small dated labels inside the rigs so you can find the data when it's needed. Use bound or spiral notebooks rather than loose-leaf documents, for they are more permanent. A long term computer based index of notebooks is very useful.

• Find others with the same passion for experimenting: Although this guideline is pretty obvious, it's also easy for the experimenter to become isolated in his or her own world. Builder hams are rarely isolated. Finding the local ones will give you a place to communicate your ideas, hear about new thoughts, and to share junkbox parts as well as test equipment. Ask at local clubs to find out who is building. Listen to the appropriate nets and attend the specialty clubs. Write to fellows who author articles of interest, especially if they live nearby. Watch the chat sessions on the Internet or the Web. Amateur radio is about communications, so don't hesitate to communicate.

• Look toward the ordinary for explanations: When a design is not working as well as it should, we look for explanations that will explain the differences. All too often we consider the complicated answers, only to discover that the real answer is in the "obvious." It is always worthwhile to return to fundamentals.

• Strive to build equipment that does not pollute the already abused radio spectrum: Make an effort to generate equipment, meaning that it does not emit signals at frequencies other than the intended ones. While most of this concern is with transmitters, the ideas should also be applied to receivers. The difficult question is "How clean is clean enough?" The FCC has specifications for spurious emissions from US transmitters. These specifications depend upon transmitter output power. Even for equipment running full power, the specifications are generally easy to meet at HF. When power drops below 5-W output, they become even easier. Throughout this text we take the approach that even greater levels of cleanliness will be sought. This book includes a chapter on test equipment. One of the items featured there is a spectrum analyzer that will allow the builder to measure spectral purity.

A final "rule:" Don't let any of these rules get in the way of experimenting and building! It's OK if there are things that you don't understand even if that includes the project you are about to build, for you will understand much more when you are finished. The real goal of this pursuit, and of this book is to   The same can be said for other "rules" that may appear in the literature or on the web: Don't let them keep you from experimenting.

Getting Started   1.5

1.4 BLOCK DIAGRAMS

Fig 1.5 shows a collection of elements that can be used in a detailed block diagram of a radio. This short list is generally extensive enough to describe the non-digital designs in this book.

Schematic and block diagrams serve a variety of purposes in electronics. The purpose of the block diagram is to present the functions and their interconnection used in a piece of equipment. Schematic diagrams present the details.

A block diagram is a useful way to plan and describe the equipment we wish to build. The block diagram will serve as the starting point for mathematical analysis that we may apply to the overall system. It can also emphasize the functions required to complete the design. This is illustrated with **Fig 1.6** showing a direct conversion transceiver for the 40-meter band. Several filters are shown, illustrating the functions that are important for good performance. The low pass and the high pass between the mixer and audio amplifier are simple, consisting of one component each. There may be no components for the signal splitter, but the function remains.

Fig 1.7 shows a more elaborate circuit, a super-heterodyne SSB/CW transceiver for the 50-MHz band. The phasing method can also be used; such a 50-MHz transceiver is presented in **Fig 1.8**. Designing any of these systems begins by forming the block diagrams, which includes specifying each of the blocks. Once this is done, the individual circuits can be designed. Some elements are missing in the block diagram in the interests of clarity. It will be useful to add block detail during circuit design.

Some block details may differ from the final implementation, but functions remain. For example, the splitter and phase shifting functions are often combined in quadrature combiner circuits operating at RF. We sometimes show a 90-degree phase shift in one path with none in another where actual circuitry merely maintains a 90-degree difference.

These figures offer a glimpse of what the text will cover. The design of the block elements will each be discussed in individual chapters. Then, the blocks will be assembled in chapters related to filter, phasing, and digital signal processing systems.

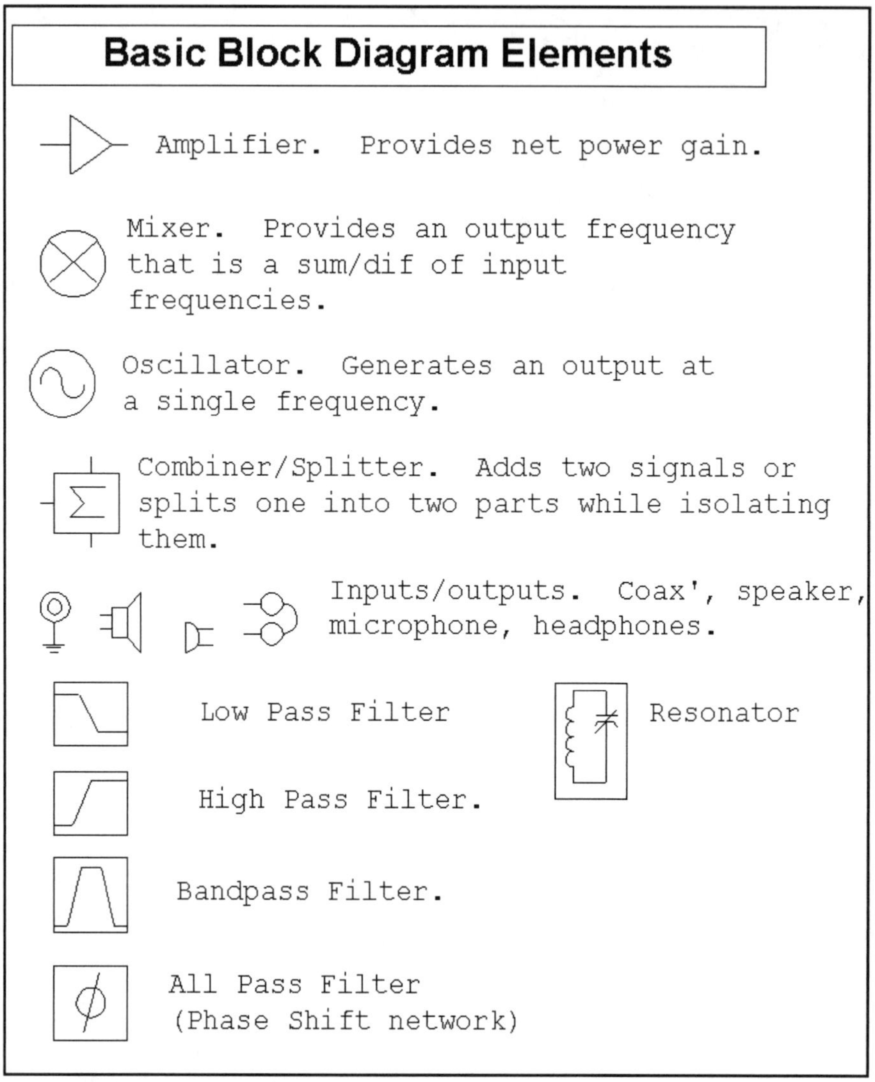

Fig 1.5—Common block diagram elements.

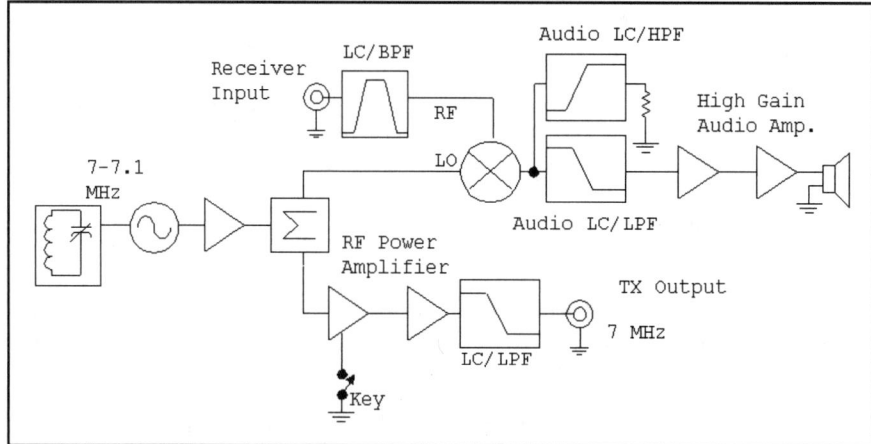

Fig 1.6—Block diagram of a direct conversion transceiver.

1.5 AN IC BASED DIRECT CONVERSION RECEIVER

This receiver design is one of the simplest possible that will allow CW and SSB signals to be received. It offers performance enough for on-the-air contacts while serving as an introductory construction effort.

The basis for this receiver is the NE602 (or NE612) integrated circuit. Originally introduced by Signetics in the late 1980s, the chip is easy to use and offers good performance among very low current receiver components. The NE602 contains a mixer and an oscillator, two essential blocks needed for a receiver. The mixer in a direct conversion receiver serves to heterodyne the incoming antenna signal directly down to audio. The oscillator provides mixer LO (local oscillator) injection for this conversion. The oscillator within the NE602 is a single transistor followed by a buffer amplifier of undisclosed complexity. The NE602 mixer is a doubly balanced circuit of a type known as the Gilbert Cell with operation outlined in a later chapter.

The LM386N audio amplifier following the NE602 completes the receiver. The LM386N will drive a small speaker, or headphones of high or low impedance. The ideal set of "cans" to use with this receiver is a light weight pair of the sort used with jogging receivers or similar consumer gear.

The receiver is shown schematically in **Fig 1.9**. Our version is built using the "ugly" methods outlined earlier. If you use a pre-etched and drilled circuit board, take the time to study the board layout in detail, and trace the circuit while studying the schematic diagram. Merely stuffing parts and soldering will provide you with no more than soldering practice.

The signal from the antenna connector is applied to a pot that serves as a gain control with output routed to a single tuned circuit using L1, a toroid inductor. This circuit drives the mixer input at NE602 pins 1 and 2. The load within the IC looks like a pair of 1.5-kΩ resistors from the input pins to a virtual ground.

The NE602 oscillator has a collector tied to the positive power supply. The base of that transistor is available at pin 6 while pin 7 goes to the emitter. Internal bias resistors set the voltage and establish a current of about 0.3 mA in the Colpitts oscillator. Feedback capacitors in our circuit run between pins 6 and 7 and from pin 7 to ground. A 270-pF capacitor then ties the base to the rest of the tuned circuit.

A simplified version of the oscillator circuit is shown in **Fig 1.10**. This illustrates the way a simplified circuit is used to calculate the resonant frequency. Fig 1.10A shows the complete oscillator. But, the two 680-pF feedback capacitors have a series equivalent of 340 pF, as shown in part B of the figure. In going from Fig 1.10B to Fig 1.10C, we resolve the 50-pF variable and 10-pF fixed into 8.3 pF; the 270 and 340 pF become 150 pF. We evaluated both variable capacitors at their maximum value. Fig 1.10C has nothing but parallel capacitors which add directly to

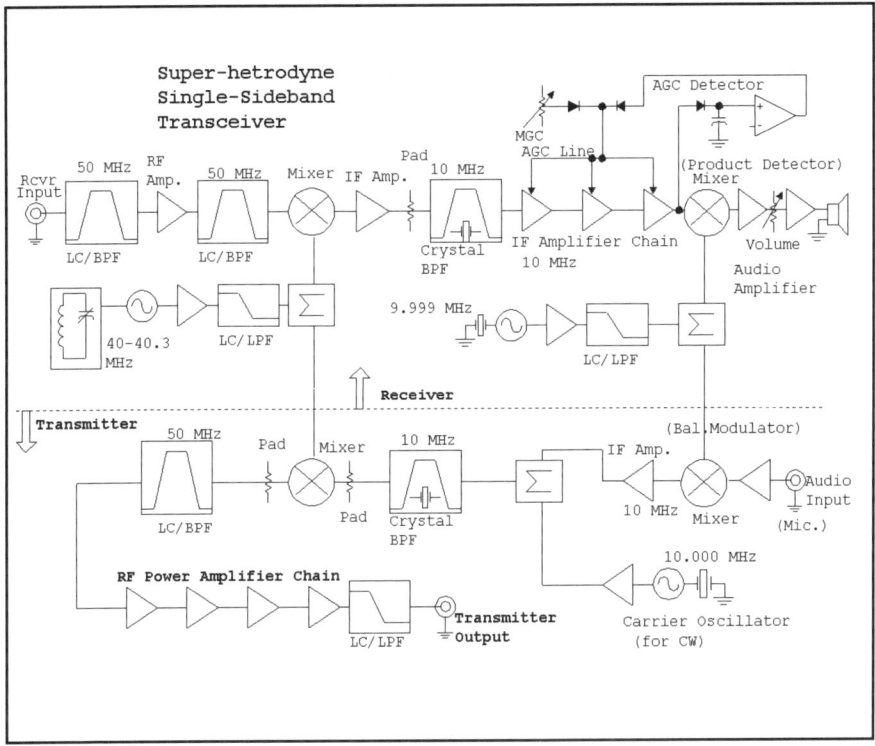

Fig 1.7—Block diagram of a super-heterodyne SSB transceiver.

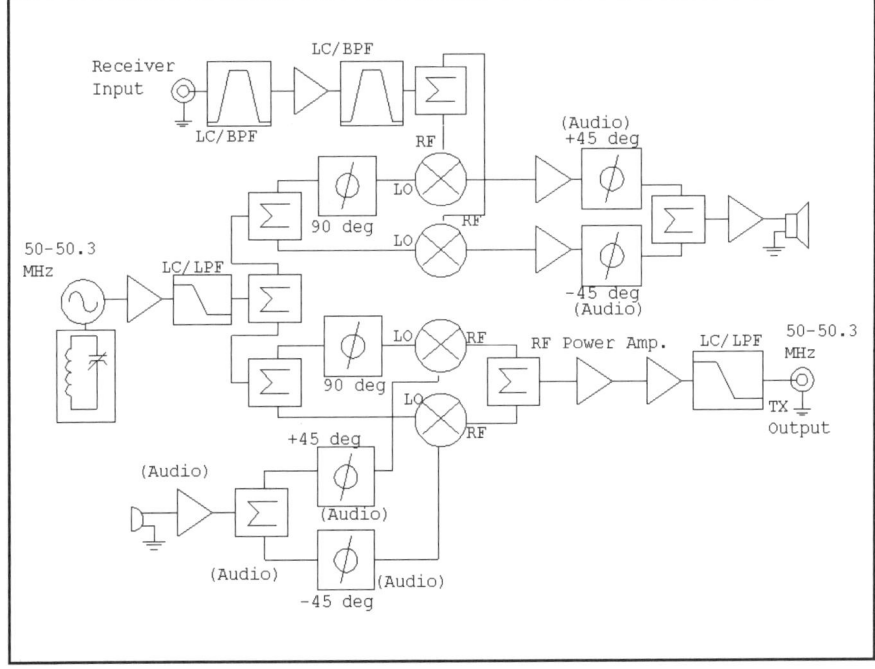

Fig 1.8—Block diagram of a phasing method SSB transceiver.

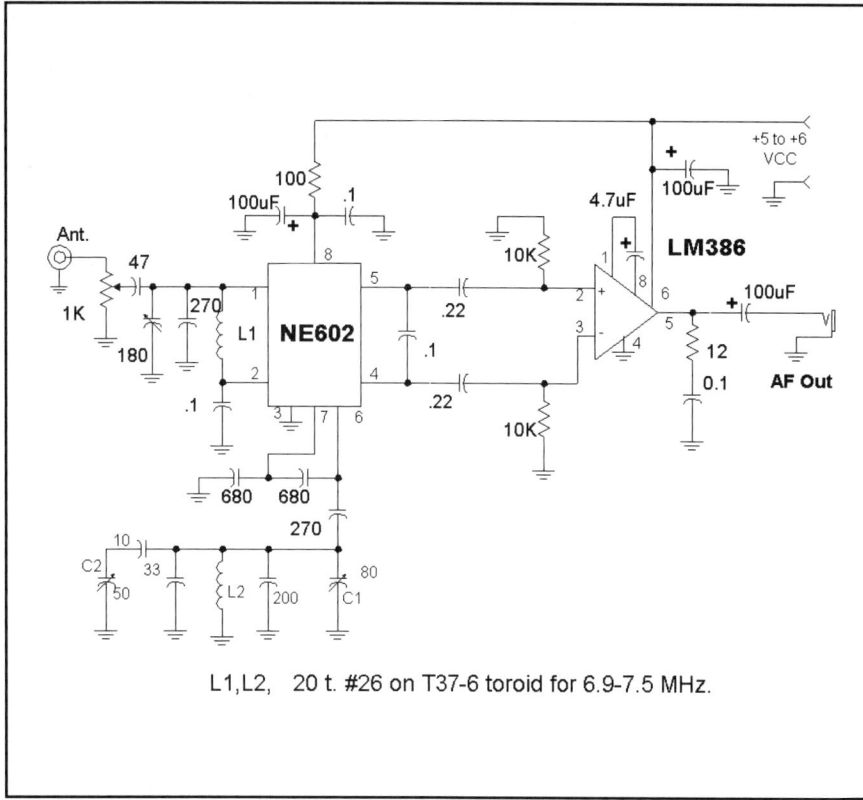

Fig 1.9—Direct conversion 7-MHz receiver using two integrated circuits.

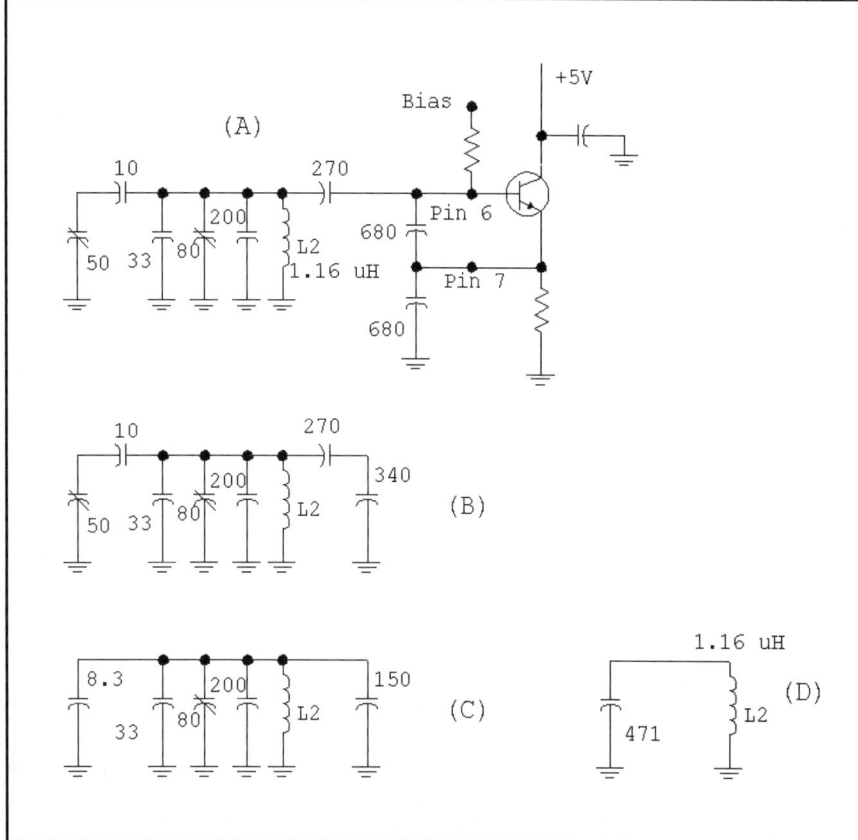

Fig 1.10—Simplified version of the oscillator in a NE602. See text for explanation.

form Fig 1.10D. A simple resonance calculation shows tuning to 6.9 MHz.

Two variable capacitors (C1 and C2) are used in our oscillator. They are nearly the same value. The larger, C1, directly parallels the inductor. A detailed analysis shows that it will tune over a wide range, the full 6.9 to 7.5-MHz span. C2 is "padded down" with a 10-pF series capacitor. C2 has a value ranging from 5 to 50 pF. The series capacitor then generates a composite C ranging from 3.3 to 8.3 pF, a 5-pF difference. Add capacitance in parallel with C2 to create even greater bandspread (resolution or low tuning rate).

All fixed capacitors should ideally be NP0 ceramic types, readily available from major mail order sources. But, don't hesitate to try other caps if you have them in your junk box. The worst that will happen is that the receiver will drift more than desired. New parts are easily substituted later.

These capacitor variations are doubly significant. First, you can adapt a tuned circuit to work with whatever you have on hand. For example, common 365-pF AM broadcast capacitors can be used in both positions with appropriate padding. Second, the use of two capacitors is a very practical means for building simple receivers while avoiding the mechanical complexity of a dial mechanism. We have used double cap tuning for transceivers in other parts of the book. Adapt the circuit to what you have available.

The mixer input network at L1 that injects antenna signals into the NE602 uses an inductor identical to that in the oscillator, tuned with a mica compression trimmer capacitor. Any variable can be used here. If a 365-pF panel mounted cap is used, the 270-pF capacitor could be reduced in value. If the only available variable capacitor is much smaller than 180 pF, you may have to resize L1, or add or subtract net capacitance a bit to hit resonance. The inductance can be reduced by spreading or removing turns, or increased by compressing turns. Both circuits are very tolerant of such changes.

Once the mixer has been wired, most of the receiver is finished. The LM386 is a low power part with no heat sink required. This receiver draws only 7 mA when signals are low, with more current with louder signals. A simple 5-V power supply works well. A 6-V battery pack will run the receiver for extended periods.

The NE602 mixer features excellent . This means that there is little LO energy appearing at the mixer RF port, and hence, the receiver antenna terminal. The presence of such energy can lead to a common problem of "tunable hum" with

Fig 1.11—Direct conversion receiver assembly.

some direct conversion receivers.

The receiver also has problems. Some, the audio images, are intrinsic to all simple direct conversion receivers. This is the price, but also the thrill of such a design. The selectivity is lacking. This can be remedied with audio filters that can be placed in the receiver. Examples of audio filters are found elsewhere in this book. These filters would go between the mixer and the audio amplifier. It is easy to add such things to a breadboarded receiver, but more difficult with a printed board.

The greatest performance deficiency is the poor strong signal handling capability of the receiver. Although helped a bit by placing the only gain control in the antenna lead, the problem is intrinsic to the NE602 mixer. The basic Gilbert Cell is capable of much more, but only when biased to draw considerably more current. The current is kept low in the NE602 by design, for it is intended for battery powered consumer equipment and not ham gear. Strong, high performance direct conversion receivers are described later in the book.

Initial turn-on and adjustment is straight forward. Apply power initially with a 100-Ω resistor in the power supply line. The resistor serves as a fuse if you have done something drastically wrong. Inserting the headphones when the output capacitor is uncharged will produce an audible pop. If the audio seems to be working, turn the receiver off, remove the extra resistor, and start again. Attach an antenna, advance the gain control and tune C1. Signals should be heard. Adjust the front-end tuned circuit for maximum signal. If you have a calibrated signal generator you can inject a signal and see if the operation is at the right frequency. If you have a general coverage receiver available, you can attach the antenna of this receiver to that of the general coverage receiver where you will be able to hear the LO signal. If an antenna is not available, you can throw 20 or 30 feet of wire out on the floor. While this is not going to compete with a good outdoor antenna, it will provide signals in abundance to listen to and confirm receiver operation.

The receiver in **Fig 1.11** was built for the 40-meter band. If you want to try a different band, all that is required is to change the two inductors. Increasing the 1.16-µH inductor to 4.5 µH will drop the receiver right into the 80 meter band. A band switching version would be practical.

The first popular receivers of this sort appeared in the USA in a paper by WA3RNC.[4] Variations of a similar sort were generated and published in Europe by George Dobbs, G3RJV. George used a double tuned circuit in the front end to improve signal handling properties.

1.6 A REGENERATIVE RECEIVER

There was a time when simple vacuum tube regenerative circuits were the only receivers available to the radio amateur. Even when super-heterodynes became possible, the regenerative design remained as the entry level radio.

Regenerative receivers have become popular again, but they now generally use semiconductors. Much of this popularity has been fueled by the work of Charles Kitchin, N1TEV.[5,6] People now build regenerative receivers for the sheer joy of listening to a receiver that is extremely simple, yet is capable of receiving signals from all over the world. The radio offered here tunes from 5.5 to 16 MHz, covering three amateur bands, 7, 10.1, and 14 MHz, as well as international short-wave broadcasts at 6, 7, 9.5, 12, 13.5, and 15 MHz.

The core of a regenerative receiver is the detector. **Fig 1.12** shows a JFET version of a classic regenerative detector using a "tickler coil." Signals from the antenna or a preceding radio frequency amplifier are applied to the tuned circuit, producing a voltage at the FET gate. This produces FET drain currents that vary at the RF rate. The RF drain current flows in the tickler coil which couples energy back to the original coil through inductive transformer action. If enough energy is coupled back, the circuit oscillates. Even when the coupling is weaker, insufficient for oscillation, the circuit can have very high gain. This makes the weakest signal large within the detector circuit. The presence of any large signal in a "square-law" device like a JFET will produce detection, which means that audio also appears within the circuit. It need only be coupled out and applied to headphones or an audio amplifier to complete the receiver.

Our receiver uses some slightly unusual circuits that simplify the design. The detector is based upon a little appreciated variation of a traditional Hartley oscillator, a variant transformer action. Instead, two series inductors, L1 and L2, serve as the traditional "tank," or resonator. Toroids were used, although Q is not critical and traditional cylindrical coils will also work. Indeed, low Q radio frequency chokes offer opportunity to the experimenter.

The detector, Q2, uses a junction field effect transistor. While we used a 2N5454, the detector worked well with any N-chan-

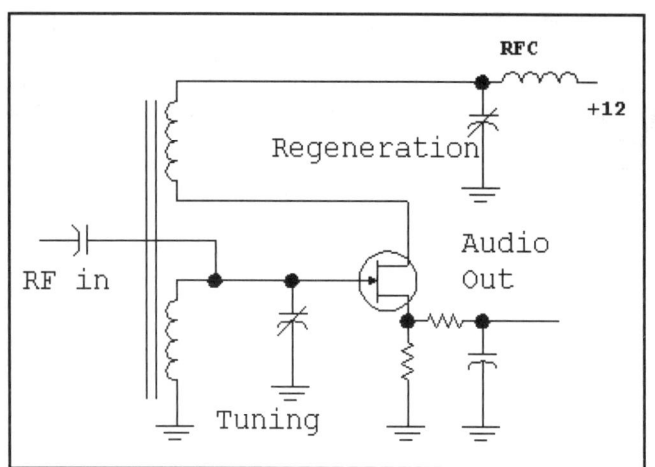

Fig 1.12—A classic regenerative detector.

nel depletion mode FET we could find in our junk box. This included the U309, J310, 2N4416, 2N3819, and MPF-102, as well as some even more obscure parts. We couldn't find an FET that would not work. Use what you have! The complete receiver schematic is shown in **Fig 1.13**, and a front panel photograph appears in **Fig 1.14**.

We wound our own 1-mH choke for L3 using a large ferrite bead. A 1-mH or 2.5 mH RFC will work well in this position. A 1-K resistor even functioned in place of L3, although the regeneration control was not as smooth as it was with an inductor.

The mechanical complications of a dial mechanism are avoided by tuning the receiver with two variable capacitors, C2 and C3, each with a large knob. C2 is a "bandset" while C3 is a higher resolution "bandspread" tuning, an action resulting from the series and parallel fixed capacitors around C3. Regeneration is controlled with another 365-pF variable capacitor. None of the variable capacitor values are terribly critical. If you find others at a flea market or hamfest, you can adapt the circuit to use them. That's part of the charm of a personalized regenerative receiver; it applies positive feedback to your imagination.

This circuit uses an RF amplifier, Q1. The gain is not really needed, or even desired. However, the amplifier provides a relatively stable driving impedance for the detector, and is a convenient way of varying the strength of the signals arriving at the detector. The RF amplifier is preceded by a 5th order low pass and 3rd order high pass filters. The high pass rejects signals from the AM broadcast band that could overload the receiver. The low pass attenuates FM and TV broadcast signals that could inter-modulate in the RF amplifier or detector, producing distortion within the receiver tuning range.

Audio gain is provided by Q3 driving

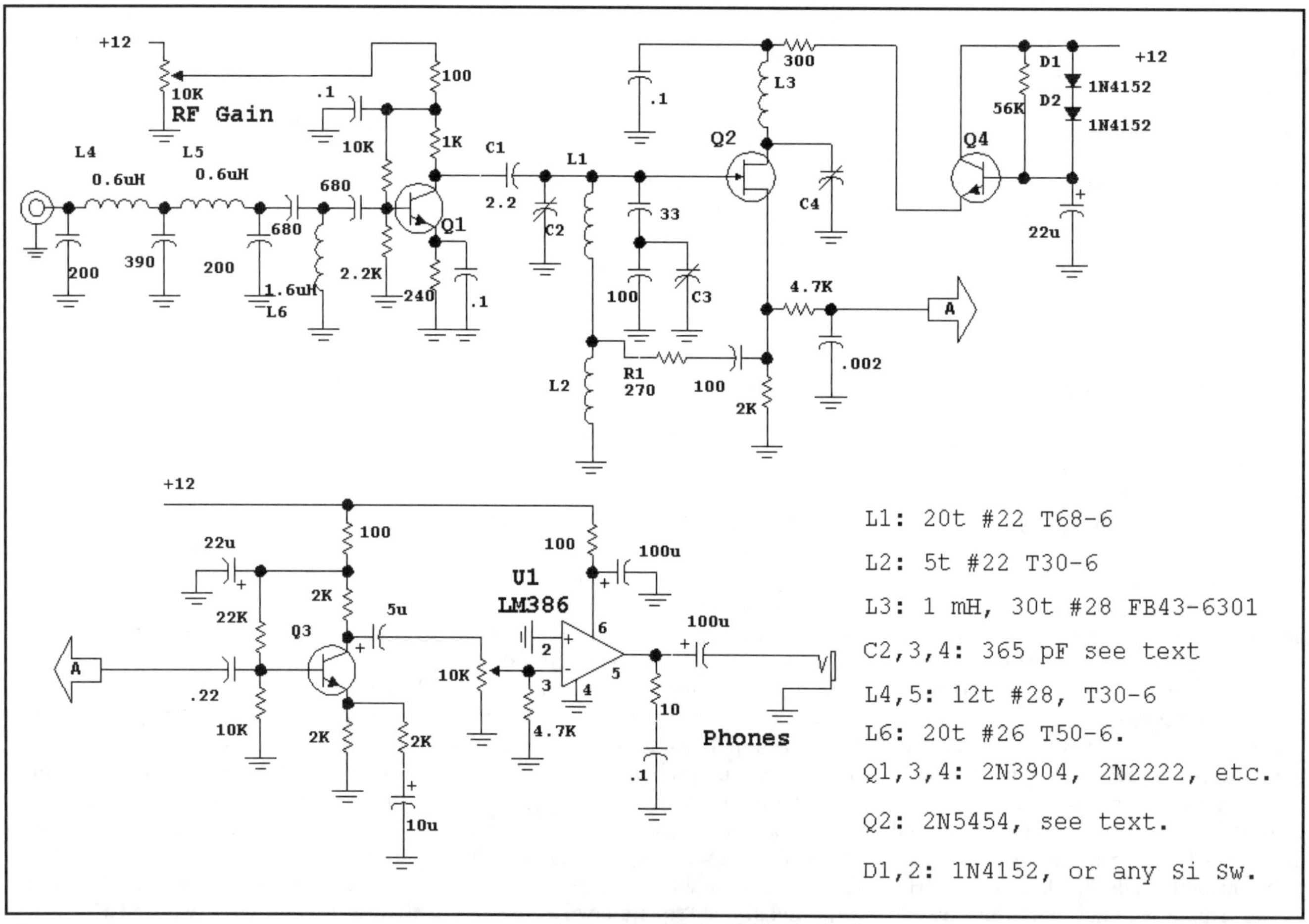

L1: 20t #22 T68-6
L2: 5t #22 T30-6
L3: 1 mH, 30t #28 FB43-6301
C2,3,4: 365 pF see text
L4,5: 12t #28, T30-6
L6: 20t #26 T50-6.
Q1,3,4: 2N3904, 2N2222, etc.
Q2: 2N5454, see text.
D1,2: 1N4152, or any Si Sw.

Fig 1.13—A regenerative receiver tuning from 5.5 to 16 MHz. See text for discussion of parts and construction.

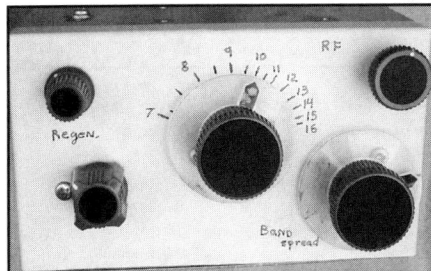

Fig 1.14—Front panel view of the regenerative receiver.

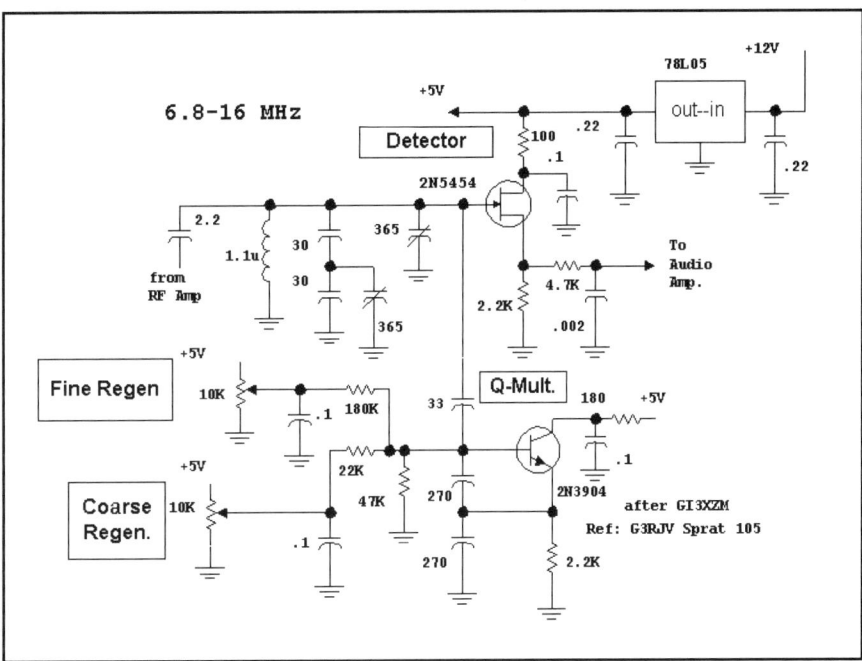

Fig 1.16—Alternative regenerative detector.

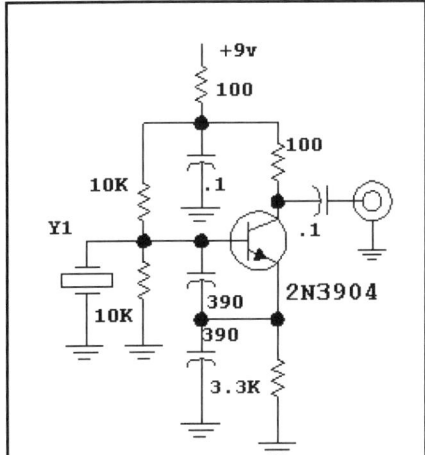

Fig 1.15—A simple crystal oscillator becomes a substitute for a signal generator.

U1, a common LM386N output amplifier. This will drive either low impedance "Walkman" type phones or a small speaker. Walkman is a Sony trademark. Q4 is an active decoupling filter that provides hum-free dc to the detector. Although the receiver of Fig 1.13 is shown with a 12-V power supply, it worked well with voltages as low as 6. Typical current is 20 mA at 12 V.

A signal generator with frequency counter is useful during initial experiments with the receiver. However, many builders may not have them available. **Fig 1.15** shows a suitable substitute, a crystal oscillator that will operate anywhere within the receiver range. Numerous inexpensive crystals are available from the popular mail order sources that will provide a starting point. For example, a 10-MHz crystal available for under $1 will mark the 10.1-MHz amateur and the 9.5 to 10-MHz SW broadcast bands.

The receiver can be built in any of many forms. A metal front panel is a must, affording shielding between circuitry and the operator's hands. However, the rest of the receiver could be as simple as a block of wood found in the garage. Our receiver was built "ugly" with scraps of circuit board material. One scrap will suffice, although our receiver used three, an indicator of earlier experiments. Other breadboards will work as well, but a printed circuit board should   be used for a regenerative receiver. Even if dozens are to be built, such as in a club effort, the project should emphasize open ended, flexible breadboarding to encourage experimentation.

Some experimentation may be required to set up the regeneration. Increasing L2 by a turn or decreasing R1 will both increase regeneration. However, too much inductance at L2 or too little resistance at R1 will produce such robust feedback that regeneration cannot be stopped or easily controlled.

Operation of this, or any regenerative receiver is a multiple control effort. Begin with the regeneration control, C4, at minimum capacitance, unmeshed, and set the two tuning controls at half. Set the RF gain for maximum gain, +12 V on the amplifier, with the audio gain in the middle and attach an antenna. Tuning C2 may produce a signal. Now slowly advance the regeneration, adding C at C4. It is normal for background noise to increase with a mild "plop" occurring in the headphones as the detector begins to oscillate. If the detector becomes overloaded, reduce the RF gain control. Tune the receiver until an AM signal is found. Then reduce regeneration until the "squeals" subside. CW and SSB are best received with the regeneration well advanced. While the receiver works best with an outside antenna, it will function with as little as a few feet of wire tacked to the wall. The signal generator of Fig 1.15 requires no more than a two foot piece of wire on its output, somewhere in the same room as the receiver.

There are numerous interactions between controls, features that offer challenge and intrigue for the experimenter who takes the time to enjoy them. Numerous circuit refinements are available to the experimenter who wishes to continue the quest. The experimenter will discover a great deal from his or her efforts in operating this receiver. The availability of very high gain through positive feedback can be used to great advantage. But operation can be a greater challenge than found with a more advanced receiver.

A more recent experiment used a different regenerative detector, shown in **Fig 1.16**. This circuit eliminates one of the variable capacitors used in the other circuit, replacing it with a pair of potentiometers. This circuit was featured in a recent issue of   by George Dobbs, G3RJV, although the circuit seems to be the brainchild of GI3XZM.[7] Performance of the two circuits is similar.

Getting Started 1.11

1.7 AN AUDIO AMPLIFIER WITH DISCRETE TRANSISTORS

The amateur literature is rich with older designs using high impedance headphones. These designs are often very battery efficient, a vital performance virtue for portable or emergency equipment. But high impedance headphones that can be used with the more efficient designs have become rare. The answer to this dilemma is a simple audio amplifier that will drive low impedance headphones while maintaining reasonable efficiency.

One solution to the problem is one of many integrated circuits. Throughout the book we used the LM386 or op-amps to drive headphones of the Sony "Walkman" variety. An alternative circuit is shown in **Fig 1.17**. This amplifier uses commonly available discrete transistors. The version of the circuit that we built used leaded parts, but could just as well be built with SMT components. Q1 functions as a gain stage. The 2.2-kΩ collector load (R8) with 100-Ω degeneration (R4) produce Q1 bias current of 2 mA for an approximate voltage gain of 20. Q2 functions as a floating voltage source that establishes bias for complementary emitter-follower output transistors Q3 and Q4. Negative feedback through R3 reduces gain and establishes overall bias. This circuit is similar to many of the simpler integrated circuits. This circuit functions well with power supplies from 5 to 15 V.

An IC is usually the preferred solution. However, the discrete solution is available when an IC is not. All of the transistors in this circuit are very inexpensive and usually found in the experimenter's junk-box.

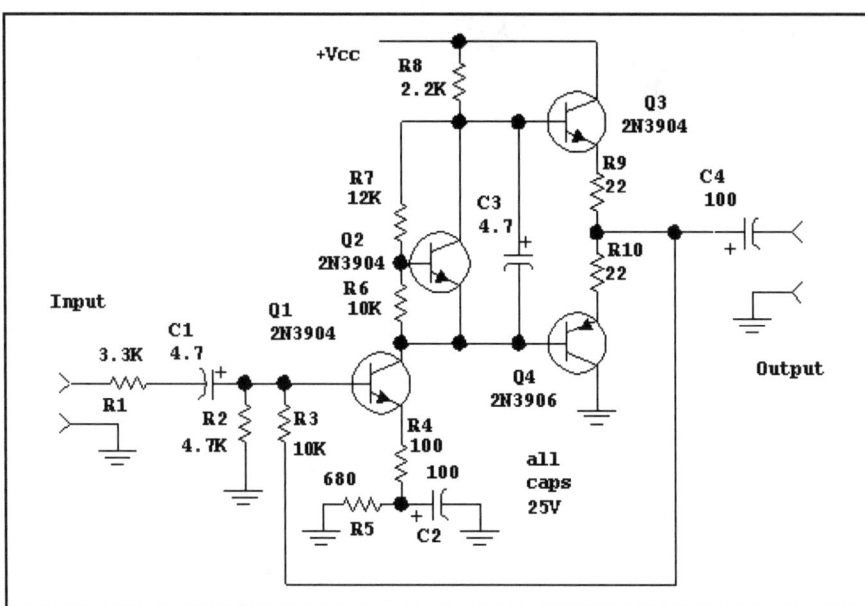

Fig 1.17—Simple audio amplifier using discrete components.

1.8 A DIRECT CONVERSION RECEIVER USING A DISCRETE COMPONENT PRODUCT DETECTOR

The direct conversion receiver described earlier used an NE-602 integrated circuit to fulfill both the detection and the local oscillator functions. Discrete (non-integrated) components can also be used in these applications. The receiver shown in **Fig 1.18** uses a differential amplifier as the product detector. This design, shown for operation in the 40-meter band, has been built with both traditional leaded components and with surface mounted technology (SMT) parts and appears in **Fig 1.19**.

Q1 functions as a local oscillator. Voltage control is used with any of several common tuning diodes. The Colpitts circuit uses small powder iron toroids for both leaded and SMT components. C1 is a combination of NP0 capacitors, selected during construction to resonate at the desired frequencies. With the parts shown, the receiver tunes over about a 50-kHz range in the 40-meter band. The range may be expanded by paralleling additional varactor diodes, increasing the value of the 82-pF blocking capacitor, decreasing the value of the 2.2-kΩ resistor in series with the tuning control, or combinations of these measures.

The oscillator is buffered with Q2, a common-emitter amplifier with emitter degeneration. This circuit, using negative feedback, uses a form found throughout the book, one where an added component reduces gain to improve performance. The output drives the mixing product detector consisting of Q3 and Q4. An RF signal is extracted from the antenna through a gain control, low pass filtered, and applied to the base of Q5 where it is amplified and converted to a current source feeding Q3 and Q4. The mixer collectors are bypassed for RF.

The detector output feeds a differential signal to a LM386 audio amplifier. De-coupling became important with this stage, owing to the internal resistance found with a normal 9V battery. An uncomfortable "howling" oscillation disappeared with high decoupling capacitance for the audio amplifier.

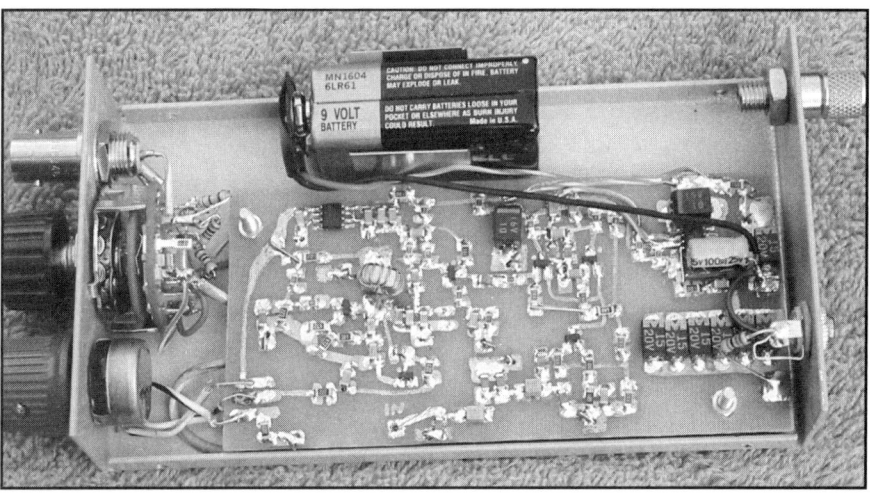

Fig 1.19—Inside view of SMT direct conversion receiver.

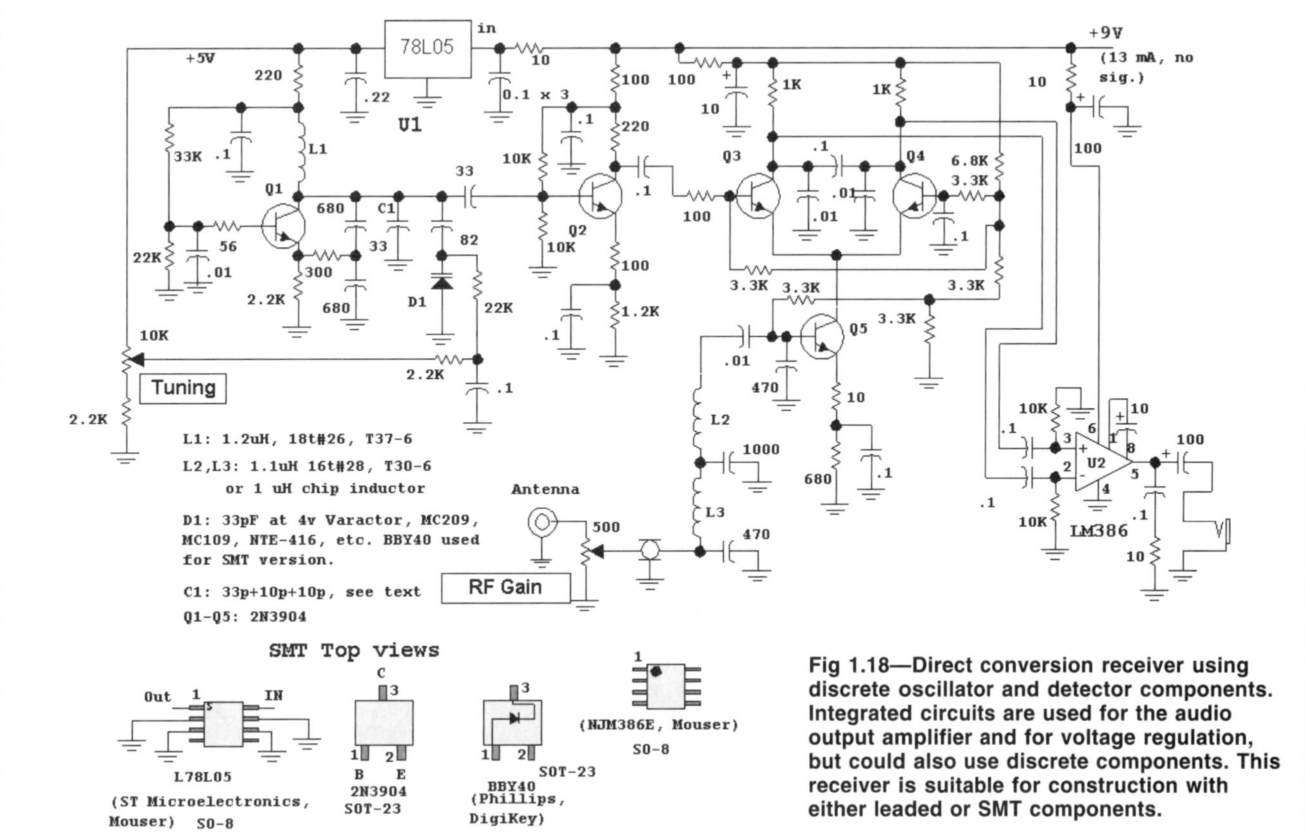

Fig 1.18—Direct conversion receiver using discrete oscillator and detector components. Integrated circuits are used for the audio output amplifier and for voltage regulation, but could also use discrete components. This receiver is suitable for construction with either leaded or SMT components.

1.9 POWER SUPPLIES

Among the many tools needed by the circuit experimenter, beginning or seasoned, is a power supply. Indeed, several are always useful. Batteries serve well for simple, low current applications. However, the more useful power supply extracts energy from the power mains. That ac voltage is applied to a transformer, is rectified, filtered with a large capacitor, and regulated with transistors and/or integrated circuits.

Two major design questions are presented to the beginner: What transformer should be selected and how large should the filter capacitor be? **Fig 1.20** shows an example 12-V, 0.5-A design we use to address these questions.

Transformers are rated for RMS output voltage with a load. The peak voltage will be higher by a factor of 1.414, so a 12.6-V transformer will have a peak output of 17.8 V. The transformer current rating should equal or exceed the maximum desired dc current, so a 0.5-A transformer is adequate for this application. This is shown in part A of Fig 1.20. A switch and protective slow-blow fuse is added to the transformer primary.

A bridge rectifier using four diodes is added to the circuit to generate a dc output. The bridge is preferred over circuits with just two diodes, for a center tapped transformer is then not required. Bridge rectifier diodes should have an average current rating above the maximum power supply current. 1-A diodes would be fine for this application.

Some waveforms are shown in **Fig 1.21**. The "before filtering" voltage is the result of rectification for the circuit of Fig 1.20A. The "V-cap" trace shows the voltage across the capacitor when it is added to the circuit, Fig 1.20B. The significant detail is the   , or variation in unregulated output voltage occurring at the filter capacitor. **Fig 1.22** shows ripple for two different capacitor values when the load current is 0.5 A.

A suitable regulator is the popular 7812. This three terminal regulator IC will provide the desired output with a   of about 2.5 V. Dropout is the minimum voltage difference between the regulated output and the higher unregulated input. With a 2.5-V dropout, the unregulated input must be 14.5 V or more over the entire cycle. Fig 1.22 shows that a 2000-μF capacitor will be adequate, but 500 μF will not. If we define ΔV as the difference between the peak rectified voltage and the minimum unregulated value, 17 − 14.5 = 2.5, I as the output current, and Δt as the time for a half cycle (.0083 second for 60 Hz), the minimum capacitor value in

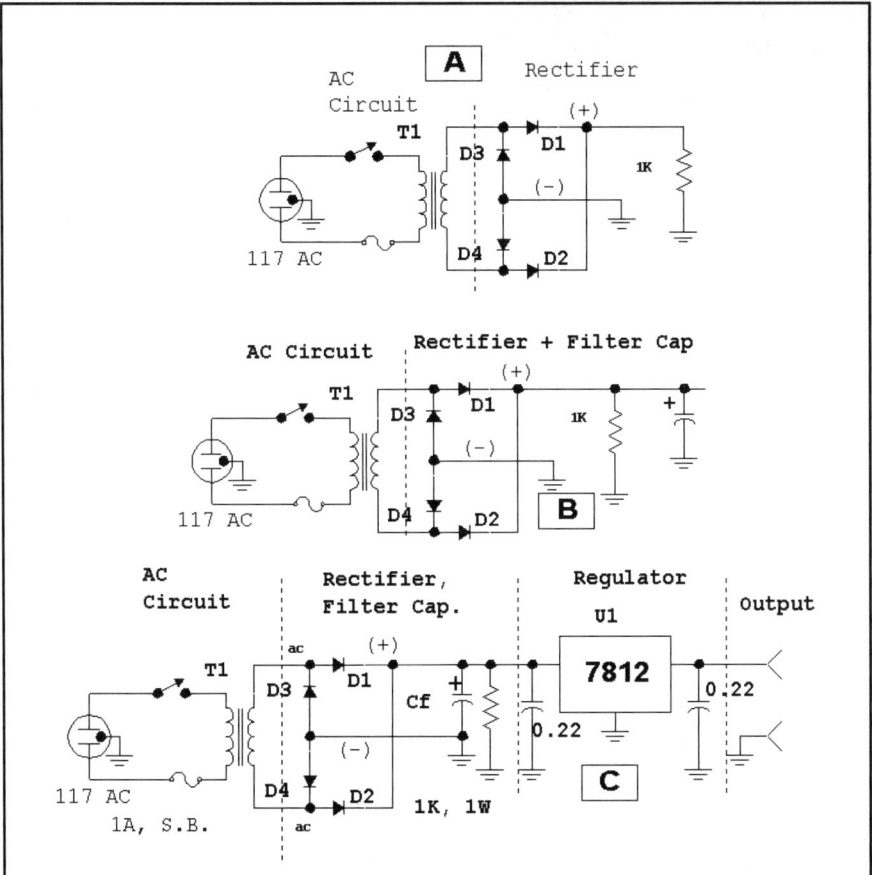

Fig 1.20—Fundamental power supply. Part A shows the transformer and rectifier, B adds the critical output filter capacitor, while C uses a 12-V regulator IC.

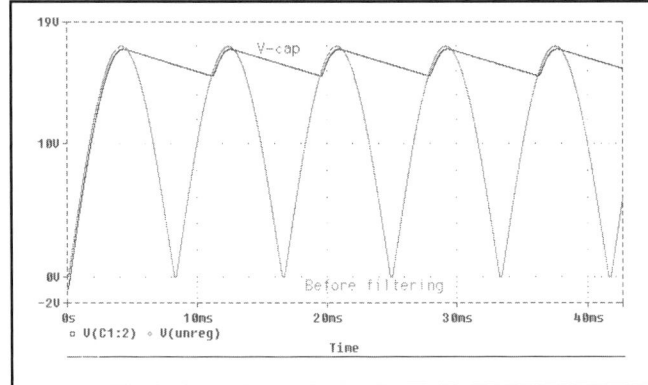

Fig 1.21— Waveforms for a simple power supply. The "before filtering" shows the raw rectified signal without any filter capacitor. The "V-cap" shows the voltage across the filter capacitor attached to the rectifier when loaded to a modest current.

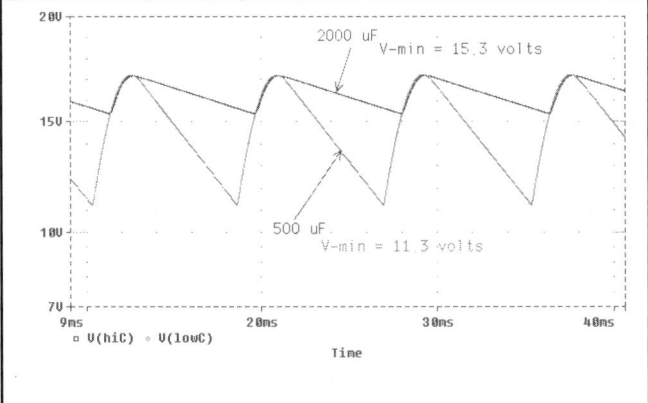

Fig 1.22—Waveforms showing the voltage across filter capacitors of two values when loaded with 0.5 A. See text discussion.

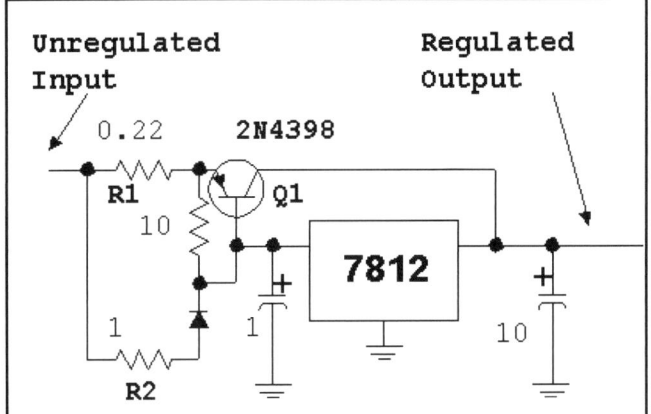

Fig 1.23—Extending the output current capability of a regulator with a "wrap-around" PNP transistor.

Farads is given by

$$C = \frac{I \cdot \Delta t}{\Delta V} \quad \text{(Eq 1.1)}$$

For this example, **Eq 1.1** predicts a minimum C of 1700 μF. A practical value of 2500 μF would be a good choice.

The complete circuit with the regulator is shown in Fig 1.20C. Extra capacitors, placed close to the regulator IC, serve to stabilize the IC. The user should check data sheets for the IC that he or she uses to evaluate stability. The 1-kΩ bleeder resistor consumes little current, but guarantees that the supply turns off soon after the switch is opened.

The 0.5-A rating of the 7812 becomes a problem when more current is needed. **Fig 1.23** shows a circuit that will extend the output current rating by adding a power transistor. Q1 now carries most of the current with the split being determined by the ratio of R2/R1. The dropout for the total circuit is now that of the IC plus a little more than a volt for the diode/transistor and R1 and R2.

Fig 1.24 shows a supply using an LM317. This is a programmable voltage part that can supply outputs from 1.2 up to 37 V, set with two resistors, for an output current of 1.5 A. The power supply we built, used extensively for developing many of the circuits in this book, was variable voltage and also included a 12-V regulator as a second output. An 18-V transformer was used, for we wanted regulated outputs up to 20 V.

Many other regulators are found in vendor catalogs, many with considerably higher output currents and lower dropouts. The experimenter is encouraged to build his own circuits using them. Switching mode regulators offer interesting performance virtues with equally interesting challenges.

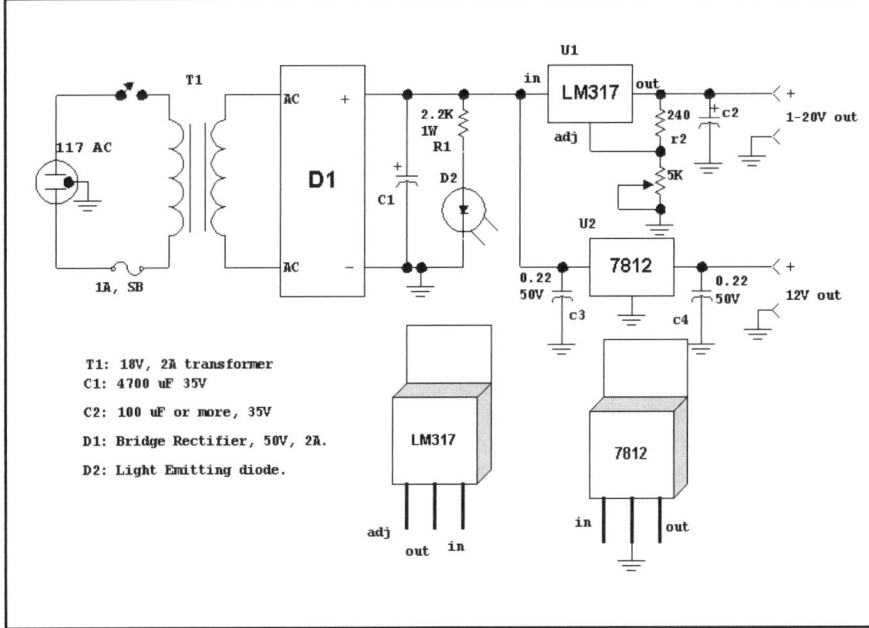

Fig 1.24—Practical dual output power supply featuring the LM-317 regulator.

1.10 RF POWER MEASUREMENTS

Before one can do any meaningful experiments with transmitters, you must be able to measure RF power. A basic scheme for doing this is shown in **Fig 1.25**. The RF is applied to the 50-Ω termination through a coaxial cable. It is necessary that a well defined impedance be available to absorb the transmitter power. The load must be capable of dissipating that power in the form of heat. So if the transmitter is capable of delivering, for example, 100 W, the 50-Ω load resistor must be capable of dissipating this power. The load must be a resistor that really appears as a resistor to

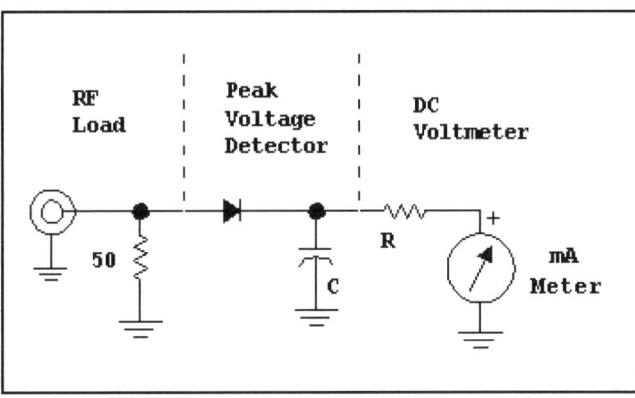

Fig 1.25—A basic RF power meter.

the radio frequency applied to it. This means that the usual power resistors sold by vendors, even if capable of dissipating 100 W, will not be suitable. They are usually built as a "wire wound" part, making them highly inductive for RF. It is sometimes possible to tune them, an interesting avenue for the advanced experimenter.

Suitable 50-Ω terminations, or "dummy loads" can be built with parallel combinations of 2-W carbon resistors, or similar 2 or 3-W metal oxide power resistors such as those manufactured by Yaego or Xicon. Some of these are used in power attenuators described in Chapter 7.

The RF power dissipated in the resistor will develop a corresponding RF voltage. That is rectified with a simple diode detector, providing a signal across the capacitor equaling the peak RF voltage, less 0.7 V for the diode turn-on voltage.

The power meter is completed with a suitable dc volt meter. It can be as simple as a 0-1-mA current meter and a resistor, an FET voltmeter, or even a digital voltmeter.

Fig 1.26 shows a dual range power meter. Essentially it is a pair of power meters sharing a single meter movement. The higher power part of the circuit starts with a 4-W load built from two parallel 100-Ω, 2-W resistors. These can be carbon or metal film resistors. If 2-W resistors are not available, four parallel 200-Ω 1-W parts will work as well. The resulting RF voltage is rectified with a silicon switching diode. This should be a 100-V part such as the 1N4148, 1N4152, or similar diode. The voltmeter part of the circuit is a 20-kΩ resistor driving a 0-1 mA meter.

Assume a transmitter is attached and keyed on to produce an indication of 0.6 mA. This represents a peak of 12 V, for the meter multiplier is the 20-kΩ resistor. The resulting power is then calculated from the formula given with the figure, 1613 mW, or 1.6 W.

The 50-mW input to the power meter uses a single 51-Ω, ¼-W, resistor with a more sensitive 1N34A rectifier diode. The meter multiplier is now just 1.5 kΩ. An approximate calibration curve is shown in **Fig 1.27**. The finished meter is shown in **Fig 1.28**.

Other schemes suitable for RF power measurement include terminated oscilloscopes, microwave power meters (usually using calorimeter measurement methods,) spectrum analyzers, and wideband logarithmic integrated circuits. Some of these will be covered in a later chapter.

Often we wish to examine an RF voltage to see if a circuit is "alive," and perhaps to adjust it. The classic method for doing this used an RF probe with a high impedance, usually vacuum tube or FET voltmeter. The method is still very useful, especially when instrumentation is limited. **Fig 1.29** shows a very simple RF probe. The photo in **Fig 1.30** shows an open breadboard version; it's the sort of circuit that one builds when a measurement must be done immediately. A long lasting version of the same circuit might better be built inside a cylinder at the end of the coaxial cable.

The probe may require calibration. This is best done with one of the other power meters and a small transmitter or similar RF source. The transmitter is attached to the power meter and the output is measured. The corresponding RF voltage is noted and the RF probe is attached to the power meter 50-Ω resistor, producing a result that can be compared.

Fig 1.31 shows a high impedance dc voltmeter suitable for use with this probe. It is also a good starting measurement tool for

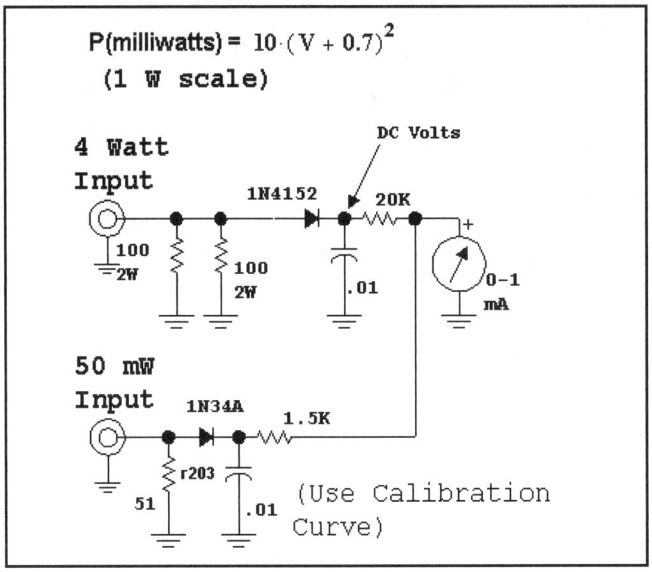

Fig 1.26—Dual range power meter. The 4-W input uses the formula to calculate power in milliwatts. The 50-mW range uses the curve of Fig 1.23.

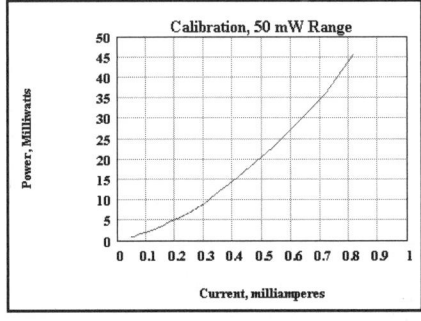

Fig 1.27—Calibration curve for the 50 mW range of the previous power meter.

Fig 1.28—The front panel of the dual-range QRP power meter.

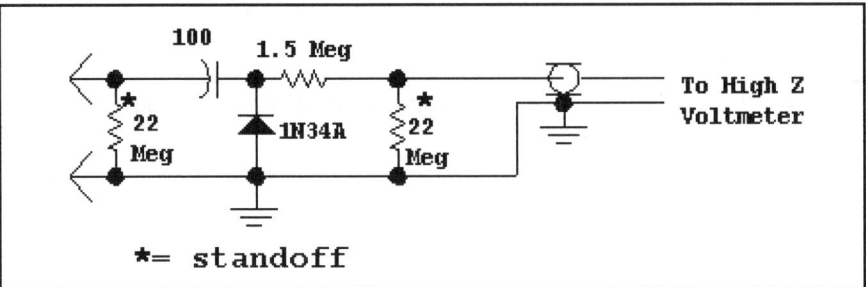

Fig 1.29—RF probe suitable for use with a VTVM, FET voltmeter, or even a DVM. Resistors marked with * are standoff resistors used for probe construction and have little impact on circuit operation.

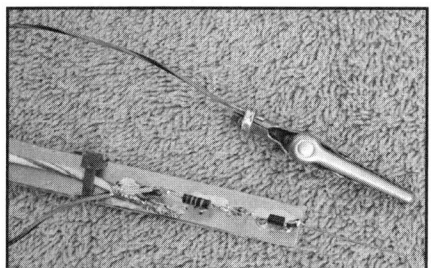

Fig 1.30—Close up view of an RF probe built on a strip of PC board material. The probe is a capacitor lead.

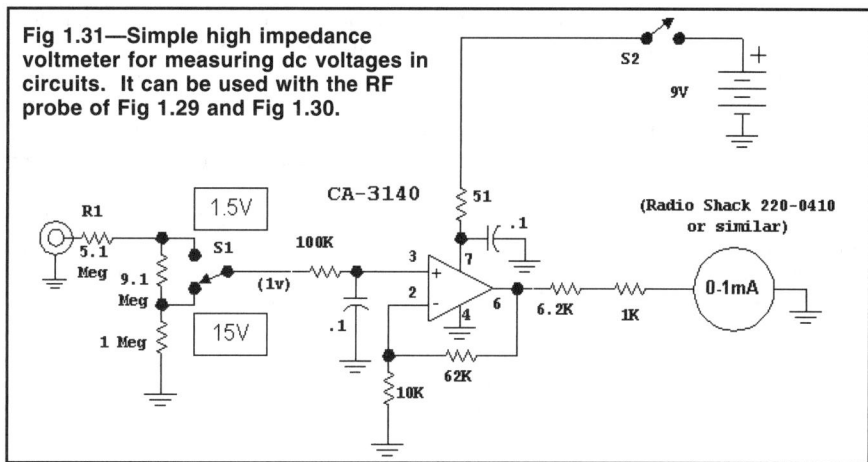

Fig 1.31—Simple high impedance voltmeter for measuring dc voltages in circuits. It can be used with the RF probe of Fig 1.29 and Fig 1.30.

use in the lab. For general utility, it is useful to have the 5.1-MΩ resistor at the tip end of a probe that is inserted into a circuit for measurements. This allows the dc to be measured without upsetting signals that may be present in the circuit. This circuit can be calibrated with a fresh 1.5-V battery; vary the 6.2-kΩ resistor if needed.

We will have more to say about RF power measurement in Chapter 7.

1.11 A FIRST TRANSMITTER

This section describes the design of a simple transmitter suitable as a first rig, a project for someone who has never built a transmitter. It uses robust circuits with few adjustments required during construction. It can be built with nothing more than a volt meter, a power meter, and power supply. We used an oscilloscope and a spectrum analyzer during the rig design phase and those results are presented. However, that equipment is not necessary for construction. The crystal controlled 2-W 40-meter transmitter is built with breadboard methods rather than with a printed circuit.

The circuit, shown in **Fig 1.32**, begins with Q1 functioning as a crystal controlled oscillator. Our crystal had a marked frequency of 7045 kHz. This was the specified frequency for operation with a 32-pF load capacitance. This Colpitts circuit uses a pair of series 390-pF feedback capacitors. The equivalent 195 pF parallels the crystal. Because this capacitance is much larger than the specified 32 pF, the operating frequency will be less than the marked 7045 kHz. If you want the frequency to be exact, a small trimmer capacitor can be placed in series with the crystal. We will eventually do this as a method of obtaining some slight tuning, but don't bother with this refinement in the beginning. The complexity of crystals is discussed in later chapters.

The oscillator is built on the end of a scrap of circuit board material. The crystal was held on the board with a piece of double sided foam tape (Tesa, 67601). The oscillator worked right off with several V peak-to-peak observed at both the base and the emitter with an oscilloscope and 10X probe. The RF probe described earlier could also be used. The oscillator functioned well with supply voltages as low as 2.5 V. A quick check with a receiver confirmed the frequency.

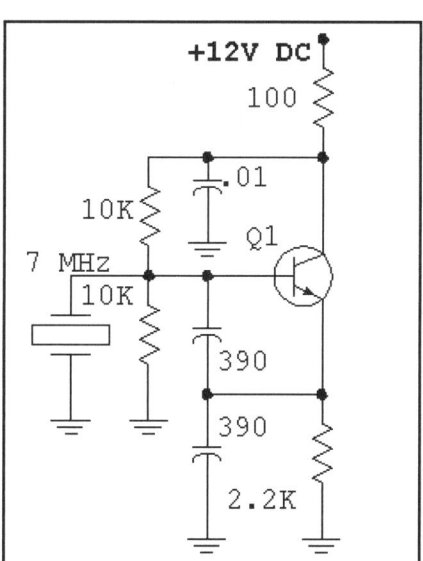

Fig 1.32—Crystal controlled oscillator that is the start of the beginner's transmitter.

The oscillator is followed by a buffer amplifier. A buffer is an amplifier that allows power to be extracted from an oscillator, or other stage, without adversely disturbing it. An ideal buffer often has a high input impedance so it can be attached without extracting any power. The best buffers have good reverse isolation, meaning that any signal present at the output is heavily attenuated at the input.

The first buffer tried was an emitter follower, a common choice to follow a crystal oscillator. Performance was poor. While the loading was light, the output was highly distorted. This problem behavior is discussed in detail in Chapter 2. The design was changed to the degenerated common emitter amplifier shown in **Fig 1.33**. We obtain the buffer input from the oscillator base instead of the more common emitter, for the waveform is cleaner, more sinewave-like, at that point.

The buffer is added to the crystal oscillator by soldering the required parts to the board or to other components. The board is not installed in a box at this time. Rather, it's loose where it is easiest to build and measure. We can tack solder small load resistors or coax connectors to the board to facilitate experimentation.

The buffer output transformer has a 4:1 turns ratio. The primary, the 12-turn winding on a FB43-2401 ferrite bead, or a FT37-43 toroid, which is virtually identical, has an inductance of about 50 uH. This has a 7-MHz reactance of 2.3-kΩ. The load on the output is transformed from 50 Ω up by the square of the turns ratio to 800 Ω, the approximate impedance presented to the collector of Q2. The inductive reactance is much higher, so it does not impact the circuit operation. The output is not tuned, allowing it to function well over a wide frequency range.

We measured the power from the 3-turn output link on T1 by attaching a small length of coax cable that ran to the 50-mW input of the power meter described earlier. The output was +10 dBm, 10 mW, with R1=270 Ω, and was up to +15 dBm with R1 of 150 Ω. Recall that the power meter has a 50-Ω impedance.

We want more than 10 mW from our transmitter and will eventually add a power amplifier to reach an output of two W. That amplifier will require modest drive of 200 to 300 mW. We could obtain more power by biasing the second stage for higher gain and output. A more conservative and stable, free from self-oscillation, approach adds a third stage.

The evolving design is shown in **Fig 1.34** with a class C amplifier for Q3. We want this third stage to provide a power gain of 10 and pick another 2N3904. With an F_t more than ten times the operating frequency, gain will be good. The 2N3904 also has a beta that holds up well at high currents, a useful characteristic for a power amplifier. While we wanted class C operation in the 3rd stage, stability was deemed vital, so the circuit is degenerated with a 10-Ω emitter resistor and a 100-Ω load is placed at the base. Class C operation is assured. Q3 current disappears when RF drive is removed from the amplifier.

The desired driver output power is 1/3 W. This can be realized by properly loading the stage. We must present a resistive load to the collector given by

$$R_L = \frac{(V_{CC} - V_e)^2}{2 \cdot P_{out}} \quad \text{(Eq. 1.2)}$$

where V_{cc} is the supply, V_e is the emitter voltage, and R_L is the load resistance in Ohms. $(V_{cc} - V_e)$ is about 11 V for this example, so the equation predicts a desired load of about 150 Ω. An L-network, L1 and the 200-pF capacitor, is designed to transform a 50-Ω load to "look like" 200 Ω at the collector. An RF choke provides collector bias for the transistor. While tunable components could have been used in the L-network to get the optimum output, we elected to use fixed values. The L-network design (see section 3.6) produced values of 1.97 μH and 197 pF. We wound an inductor on a toroid and then measured inductance, spreading or compressing turns slightly to produce the desired value. We then used a 5% value for the 200-pF capacitor. Variable elements are only needed in higher Q situations, or where it is not possible to find tight tolerance components.

Power output could be measured with the 4-W position of the watt meter. We used an alternative approach here. A 51-Ω 1/2-W resistor was tack soldered into the circuit at the output point shown in Fig 1.34 and the output voltage was measured with an oscilloscope and 10X probe. The Q3 output was 123 mW, 7 V peak-to-peak at the load, with R1=270 Ω in the buffer. Changing R1 to 150 Ω increased output to 314 mW. The DC current, 43 mA, was determined by measuring the voltage drop across the 10-Ω decoupling resistor. The calculated efficiency is then 62%, good for an amplifier which contains resistors in both the emitter and collector. The 2N3904 at Q3 is operating well within ratings. Generally, a TO-92 plastic transistor like the 2N3904 can dissipate a quarter of a watt for extended times, or half a watt for the shorter intermittent periods encountered in a CW transmitter. This "rule of thumb" can be stretched with heat-sinking, or easily violated in thermally isolated settings. Owing to the good efficiency, the dissipation is only 200 mW in Q3.

Q3 power output varied smoothly from very low levels up to the maximum 314 mW as V_{cc} was adjusted from 5 to 12 V. This is generally a useful method for ex-

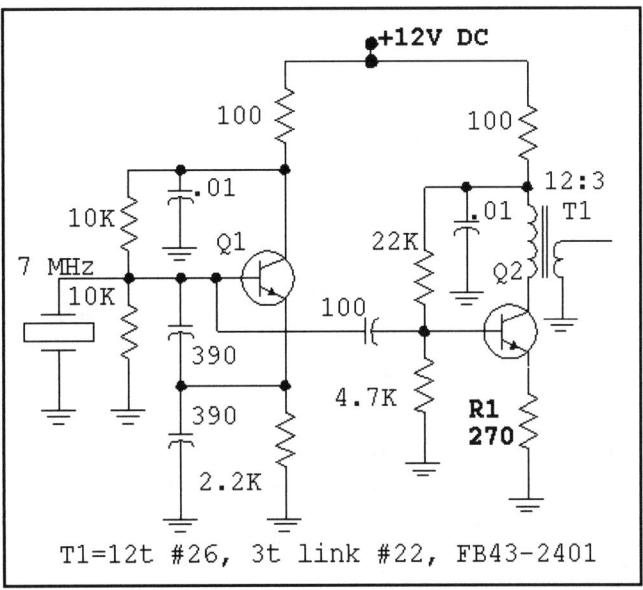

Fig 1.33—Evolving transmitter schematic showing the addition of a buffer amplifier, Q2.

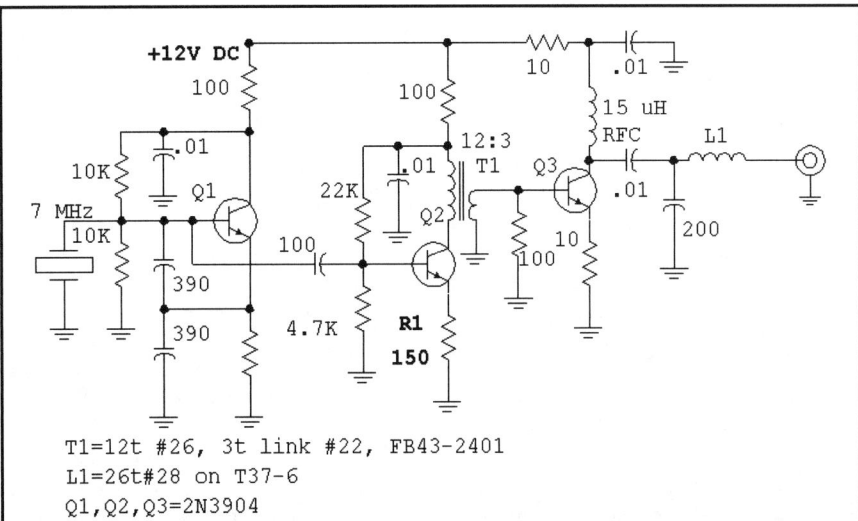

Fig 1.34—A Class C driver amplifier, Q3, is added to the transmitter.

amining stability. We will eventually add a "drive control" to the circuit.

Before continuing we need to address the issue of spectral purity. Some observed waveforms have departed from a sinewave. This means that these waveforms are harmonic-rich. This transmitter uses a crystal oscillator operating at the output frequency. The only signals that should be present anywhere within the transmitter are at 7 MHz or harmonics at 14, 21, 28, ... MHz. The only filtering needed is a low pass filter at the transmitter output. While the L-network that makes a 50-Ω load appear as 200 Ω at the Q3 collector has a low pass characteristic, it has only two components and is not very effective as a filter. If the driver amplifier is going to be used by itself as a transmitter, another low pass filter should be added to the output. There is, however, little value in adding a better low pass filter after the driver if it is to be used only to drive another stage which will also be creating harmonic distortion. Spectrum analyzer measurements showed spurious driver outputs at –27, –30, –43, and –49 dBc for the second through fifth harmonics when the driver was delivering full output. The harmonic suppression was actually worse at lower output levels. The term dBc refers to dB down with respect to the *carrier*.

1.12 A BIPOLAR TRANSISTOR POWER AMPLIFIER

The project now starts to get exciting as we begin to experiment with higher output powers. The transistor we have selected for a 2-W power amplifier (PA) is a 2N5321. This is a NPN device in a TO-39 case with a collector dissipation of 10 W in an infinite head sink, or 1 W in free air, 50-V breakdowns, the ability to switch a current of 2 A, and a 50-MHz F_T, all for less than $1. The low F_T restricts the device to the lower bands, but it also means that high frequency stability will not be an issue. The 2-W PA schematic is presented in **Fig 1.35**.

The first detail we must consider with the PA is a heat sink. Our intention was to increase power by about 10 dB to the 2 to 3-W level. If efficiency turns out to be 50%, we will have a collector dissipation that is the same as the RF output. The transistor can't support this power without a heat sink. We had a Thermalloy 2215A in the junk box which should be more than adequate. The transistor was mounted in the heat sink which was then bolted to a PC board scrap. Holes through the board made the leads available for soldering. Be careful to avoid any short circuits that are not intended. The transistor case is attached to the collector terminal in most TO-39 packaged devices.

It's always difficult to estimate heat sink sizes. While one can do thermodynamic calculations, it's generally adequate with small transmitters to experimentally treat the problem. Touch the heat sink often during initial measurements. If it's too hot to touch, the heat sink is not large enough. We always seem to err in the conservative area with more heat sink than is needed.

The formula presented in **Eq 1.2** shows that a 25-Ω load resistance presented to the collector will support the desired output. A simple pi-network was designed. The network Q was kept low, but was picked to generate a network with standard, and junkbox available, capacitors. A matching network design is presented in Chapter 3.

A 33-V Zener diode is attached from the collector to ground. The collector voltage will never reach these levels with normal Class-C operation, so the diode is transparent except for the sometimes substantial capacitance that it adds to the collector circuit. But, the diode conducts if the output load disappears, and prevents collector breakdown that might otherwise destroy the transistor. Care was taken to keep the emitter lead short when the amplifier was built, for even small amounts of inductance can alter the performance. This is *not always* bad.

Transmitter testing *always* begins by attaching a 50-Ω load to the output. This can be a power meter or a resistor of the proper rating. The PA should never be run without a load.

The first PA we built for this project used the simplified circuit of **Fig 1.36**. This circuit suffered from instabilities which became clear as we varied the drive from the earlier part of the transmitter. At one point, the RF output and the collector current both changed abruptly. The oscilloscope showed frequencies well below the desired 7 MHz. Changing the collector RF choke from the original 15 µH to a smaller 2.7-µH molded choke moved the frequency up, but the instability was still present. However, changing the base circuit to one with a lower drive impedance completely solved the problem. The output power and collector current now vary smoothly as the drive is varied. The base transformer is a 2:1 turns ratio stepdown that now drives the base from a

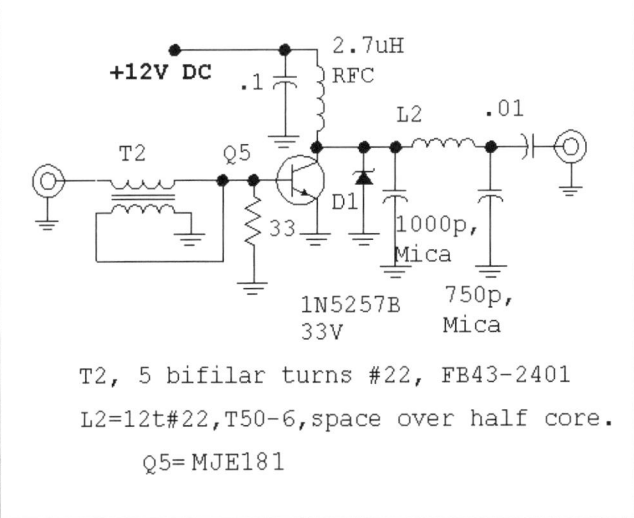

Fig 1.35—A 2 W power amplifier.

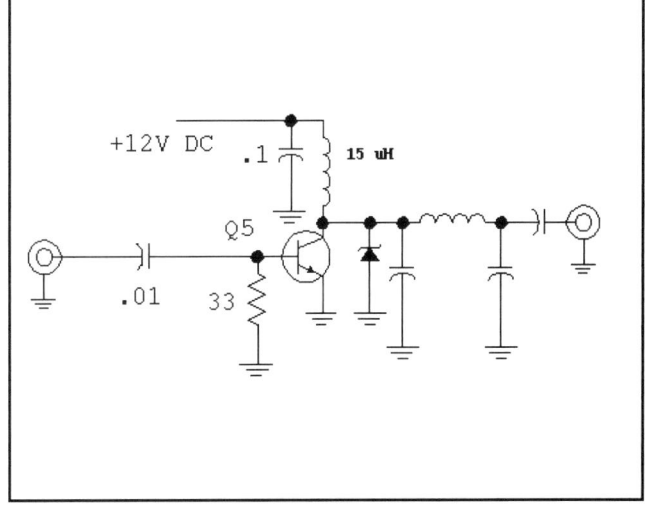

Fig 1.36—Earlier simplified PA design which suffered with stability problems. See text for discussion.

12.5-Ω source impedance. The 33-Ω base resistor absorbs some drive and tends to stabilize the amplifier. Changing this resistor is one of the experimental "hooks" available to the experimenter fighting instability.

The 2-W amplifier is installed in the transmitter. An output power of 2.25 W results from a drive of just over 100 mW. Increasing the drive produces higher output. But once the output gets much beyond 3 W, Q5 begins to heat. Although a higher power was observed with the oscilloscope when the key was first pressed, the power decreases over a period of a few seconds before stabilizing. We investigated this by looking at the collector waveform at differing drive levels. When driven to 2.25-W output, the collector voltage varied between 3 and 23 V. As drive increases, the bottom of the collector swing drops toward zero. But at this point the amplifier is fully loaded. Further excursions are not consistent with simple class C operation. More drive will cause higher current with little increases in output, allowing efficiency to decrease. This causes the heating. Changing both the matching network and drive power is needed for higher output.

1.13 AN OUTPUT LOW PASS FILTER

When the 2-W amplifier drive is adjusted for 2.25-W output, the measured efficiency was 47%. A spectrum analysis showed 2nd and 3rd harmonics at –36 dBc and –47 dBc. Addition of an outboard low pass filter removed all spurious responses to better than –75 dBc.

The outboard low pass filter is shown in **Fig 1.37**. This is a 7th-order Chebyshev design with a 7.5-MHz ripple cutoff frequency and a ripple of .07 dB. The rather obscure ripple was picked to fit standard value capacitors that were on hand. The inner capacitors are parallel combinations of 680 and 180 pF. The measured insertion loss for the filter was 0.11 dB at 7 MHz. The filter was built into a small aluminum box, **Fig 1.38**, as an outboard appendage so it could be used for other projects. Also, the performance is superior when the shielding around the filter is absolute. If the same filter was built into the transmitter, there is a greater chance that ground currents and radiation could provide paths for signals to leak around the filter.

This extreme filtering is probably redundant. A much simpler filter could be built into the transmitter, near the output coax connector, for adequate harmonic attenuation. Chapter 3 provides detail.

Practical Details

The modules built so far are mere scraps of circuit board material sitting on a bench with short pieces of wire to tie them together. They need to be refined and packaged to create a transmitter that we can put on the air. An almost complete schematic of the transmitter is shown in **Fig 1.39**.

The first refinement is a keying circuit. This function is performed by Q4, a PNP switching   . This is a favorite keying scheme of ours, allowing a grounded key to control the positive supply to a transmitter stage. Keying in the positive supply allows the grounded parts of the circuit to remain grounded without ever being disturbed by keying. Q4 serves the additional function of shaping the keying. When the key is pressed, current begins to flow in the 3.9-kΩ resistor. The current flows from Q4 base which "tries" to turn Q4 on. As the Q4 collector voltage begins to increase, the change is coupled back to the base through the capacitor. The positive going signal opposes the current extracted by the 3.9-kΩ resistor. Hence, the collector does not switch immediately to a high state. Rather, it ramps upward at an approximately steady rate until Q4 becomes saturated. Forcing the stage to turn on smoothly over a couple of milliseconds restricts the bandwidth of the modulation related to turning the carrier on. That bandwidth will extend a few hundred Hz on either side of the carrier. Beyond that, no clicks will be heard in a good receiver.

A power output control is added to the emitter of Q2. Owing to the class C nature of the following amplifiers, the output control will allow the transmitter to run from the maximum output down to virtually nothing. The control is a screwdriver adjusted pot mounted on the board.

A variable capacitor, C1, is added to the crystal oscillator. The capacitor used in our transmitter tuned from 5 to 80 pF and provided a tuning range of 3 to 4 kHz. Use whatever is in your junkbox. While certainly not a substitute for a VFO, it allows the user to dodge some interference. A "spot" switch, S2, allows the oscillator to function without placing a signal on the air.

Finally, a transmit-receive system is added. This function is performed with a multi-pole toggle switch, a simple but ad-

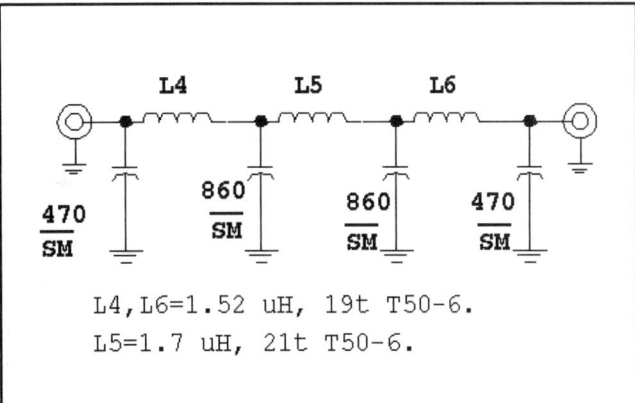

Fig 1.37—Low pass filter for use with the experimental transmitter.

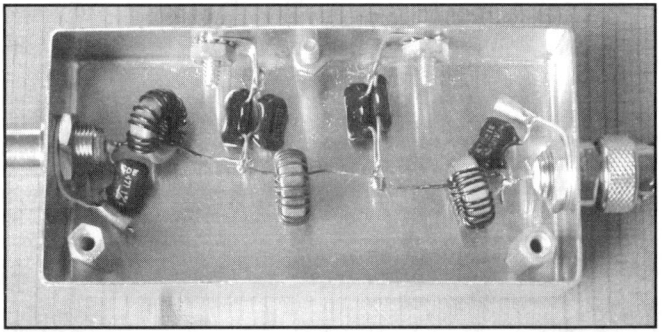

Fig 1.38—Inside view of the 7-element low pass filter built to go with the beginner's rig. The filter is also used with other equipment.

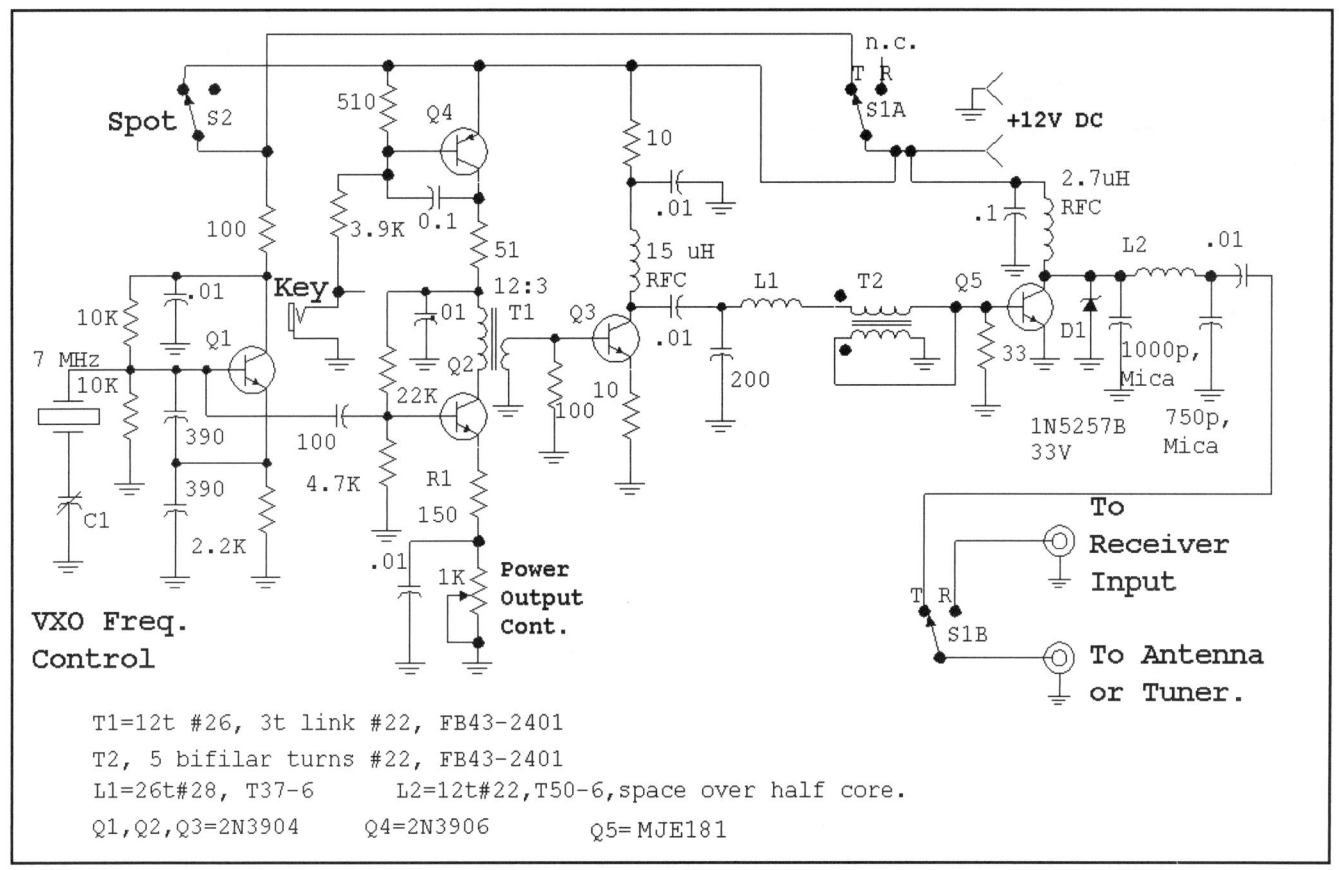

Fig 1.39—A nearly complete schematic of the transmitter. This version combines the PA with the earlier stages, adds shaped keying, power output adjust, T/R switching, and VXO action.

equate solution. S1A applies the +12 V supply to the oscillator during transmit periods. The supply is always available to Q3 and Q5 and does not need to be switched. The keying circuit, Q4, controls the supply reaching Q2. S1B switches the antenna from the receiver to the transmitter. The miniature toggle switch at S1 is suitable for powers up through a few watts. More refined T/R methods are presented elsewhere in the book.

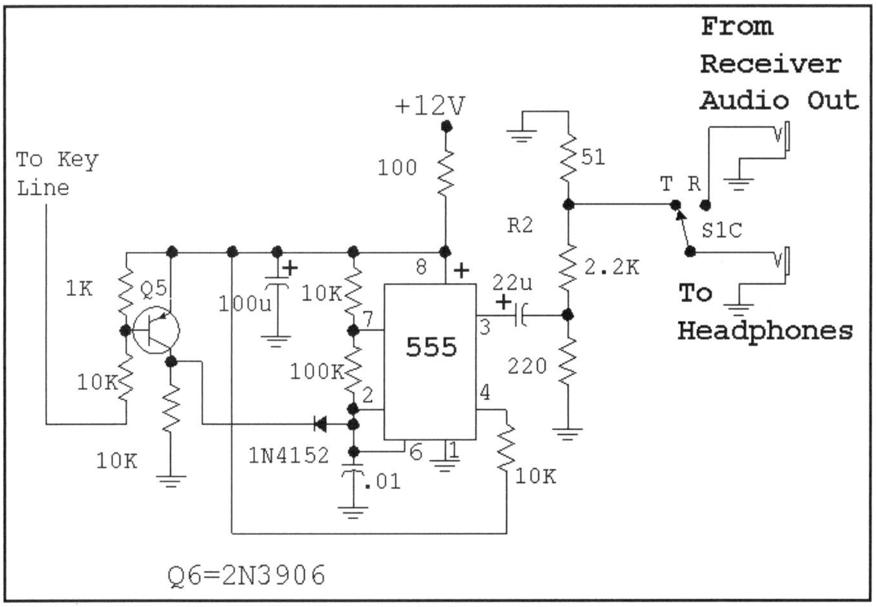

If this transmitter is to be used with a high quality modern receiver with a wide AGC range, a two pole switch is all that is needed at S1. The user can then listen to the transmitter in the receiver as the key is actuated. The more common scenario places this transmitter with a simple direct conversion receiver such as that described earlier in this chapter. It will then be impossible to turn the gain in that receiver down far enough to prevent overload. An answer to the problem is presented in **Fig 1.40** where a sidetone oscillator is added to the system. A 555-timer integrated circuit functions as the square wave oscillator which is keyed on and off with Q5. Q5 base current routes through a 10-kΩ resistor attached to the key in Fig 1.39. R2 must be adjusted for the headphones used with the transmitter. The headphones are disconnected from the receiver during transmit intervals, attached only to the sidetone oscillator. Two phone jacks are included on the transmitter. A short cable then routes the receiver audio output from the receiver to the transmitter where it is switched. This scheme does not prevent the receiver from being overloaded, but guarantees that you don't have to listen when it happens. The receiver won't be damaged by

Fig 1.40—Sidetone oscillator for the transmitter. This circuit is also suitable as a code practice oscillator.

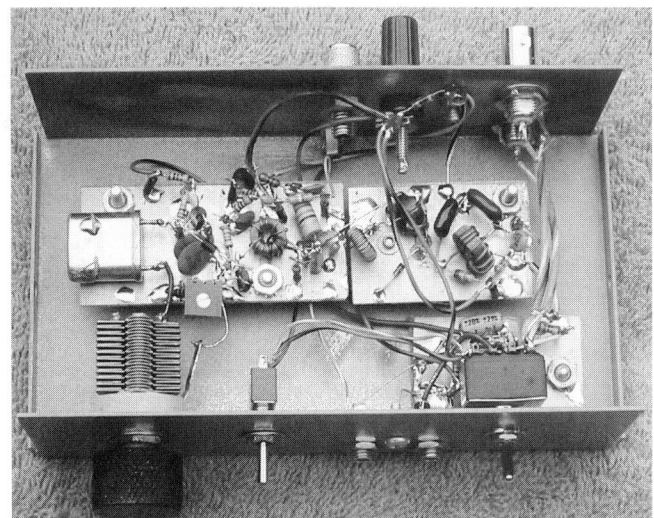

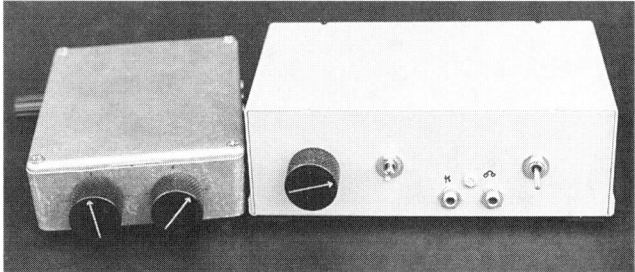

Fig 1.41—Overall view of the complete transmitter construction.

Fig 1.42—Outside view of the Beginner Station. At left is the beginner's direct conversion receiver with the transmitter at the right.

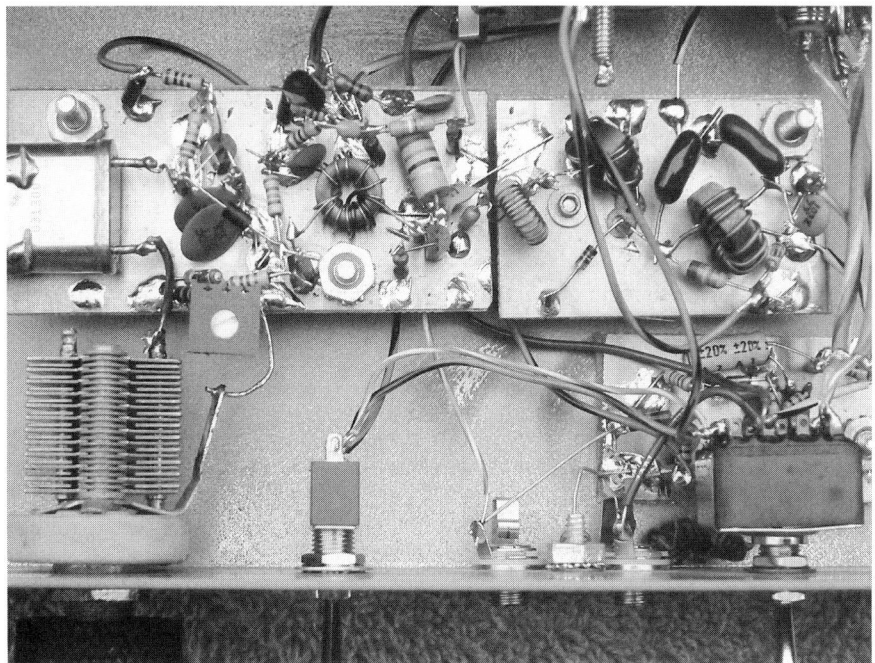

Fig 1.43—The inside view of the transmitter shows the capacitor and T/R switch mounted to the front panel with power and coaxial connectors on the rear. The left board contains the first three stages while the right board contains the 2-W power amplifier. A heat sink is under that board. A small board under the T/R switch contains the sidetone oscillator.

the overload. A third pole is needed on the switch for this refinement. Three pole double throw toggle switches are unusual, so we used one with four poles.

The complete transmitter is packaged in a standard box as shown in **Fig 1.41**. This one measured 2 x 3.5 x 6 inches, although whatever is available will work. Alternatively, you can build your own box. The outside of the box can be fixed to be as attractive as you would like it to be, consistent with personal tastes. The variable capacitor, C1, the spotting switch, S2, and the T/R switch are located on the front panel as shown on the right hand side of **Fig 1.42**. The key jack and a headphone jack are also located on the front. The rear panel contains power receptacles, a jack for the audio input from the receiver, and coaxial connectors for the antenna and a cable to the receiver input. The box we purchased for the transmitter had gray paint on it. Unfortunately, it had nearly as much paint on the inside as was on the outside. Inside paint was removed where components were grounded to the case. Details of the internal construction appear in **Fig 1.43**.

1.14 ABOUT THE SCHEMATICS IN THIS BOOK

The schematic diagrams used in this book differ slightly from other ARRL publications in that we use slightly different conventions. Not all details are presented in all schematics.

Capacitors are in microfarads if electrolytic or if they have decimal values less than 1. Values greater than unity are in picofarad if they are not electrolytic. Electrolytic caps always have a voltage rating greater than the V_{cc} or V_{dd} value used in the circuit with 25 V being typical. In some applications we will use C values in nF, which stands for nanofarad. 1000 pF = 1 nF.

RF transformers are specified by turns ratio rather than impedance ratio. Often this data is presented within the schematic diagram rather than as part of a caption. The same holds for inductance values. We strive to load the schematic with as much information as possible.

We generally label schematics with the parts that we used. But that does not mean that this is what you might want to use. An example is our frequent use the 1N4152 silicon switching diode. In all cases, virtually all of these can be replaced by the more common 1N4148 or 1N914. When there is a question about such details, look the part up and see if the parts you have on hand are similar. Then try the substitution.

REFERENCES

1. W. Hayward and D. DeMaw, , ARRL, 1977.
2. R. Hayward and W. Hayward, "The Ugly Weekender," *QST*, Aug, 1981, pp 18-21. See also G. Grammer, *Understanding Amateur Radio*, ARRL, 2nd Edition, 1976, p 144.
3. R. Lewallen, "An Optimized QRP Transceiver," , Aug, 1980, pp 14-19.
4. J. Dillon, "The Neophyte Receiver," , Feb, 1988, p 14-18.
5. C. Kitchin, "A Simple Regenerative Radio for Beginners," , Sep, 2000, p 61.
6. C. Kitchin, "An Ultra-Simple VHF Receiver for 6 Meters ," , Dec, 1997, p 39.
7. G. Dobbs, "A Stable Regenerative Receiver," , Issue 105, Dec, 2000, p 21.

CHAPTER 2

Amplifier Design Basics

2.1 MODELING SIMPLE SOLID STATE DEVICES

Small signal amplifiers are used in a receiver to bring weak signals up to the point that they can be heard in headphones. Large signal amplifiers in transmitters create even larger signals that, when applied to an antenna, propagate to be heard by the receivers. Clearly, the amplifier function is central to all that we do as radio experimenters.

Before we get into the details of the amplifier circuits, we examine devices that can amplify. A preliminary look at diodes soon evolves into a discussion of bipolar and field effect transistors. But, prior to that, we examine the modeling process.

Even the simplest electronic devices can be very complicated in their overall behavior, especially if all power levels and all frequencies are considered. Such a complete description can be overwhelming. Indeed, such a complete device picture would conceptually bury the salient behavior that the designer may seek when he or she uses a device. What is needed is something simpler, a *model* with enough complication to be useful in practical applications, but with no extra frills.

We use models for even the simplest of parts. A resistor, for example, is modeled as an ideal element, a part that obeys Ohm's Law, with no other characteristics. The real resistor is more complicated: even the smallest surface mounted part has capacitance and inductance. Wire leads only make the effects larger. The L and C alter circuit behavior, but can be described by more elaborate models.

The Junction Diode

The first device we model in detail is the junction diode. The diode is a device that has polarity dependant properties. Specifically, if we insert an ideal diode in a functioning dc circuit that carries a current, the circuit will be unchanged by the presence of the diode if the polarity is for "forward bias." But, current flow will cease if the diode is reverse biased. The schematic diagram of **Fig 2.1** illustrates a forward biased diode defined by this behavior. Reversing the diode leads eliminates current flow in the circuit.

The current in the circuit of Fig 2.1 is shown in **Fig 2.2**, a curve called an I-V characteristic. The current is that flowing through the diode and the voltage is that across the diode. Fig 2.2 plots a current that is completely determined by elements external to the diode. This particular part is called an "ideal" diode.

A real world diode departs from the ideal. First, a slight voltage drop appears across the forward biased diode. Current remains very small until that level is exceeded. Second, the flow of diode current causes a slight additional voltage drop. A refined model with these characteristics is shown in **Fig 2.3**. The model becomes an ideal diode, a 0.6-V battery, and a diode resistor, R_D, that is the ratio of a small increase in applied voltage, ΔV, and the resulting small change in current, ΔI. We sometimes refer to the threshold (0.6 V in the figure) as a *diode offset voltage*. The offset will vary with diode type. Silicon junction switching and rectifier diodes usually have an offset of 0.6 to 0.7 V. Germanium and hot-carrier silicon diodes will have lower values, while some compound semiconductor parts have

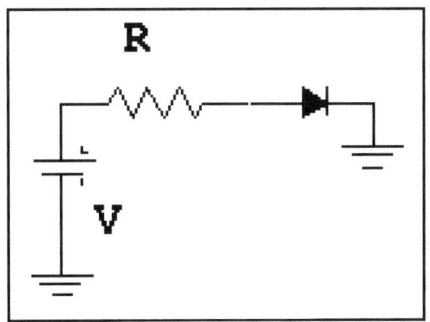

Fig 2.1—Forward biased junction diode.

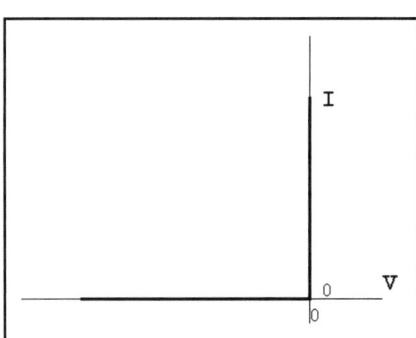

Fig 2.2—IV Characteristics for an ideal or perfect diode. The curve shows I for any possible V that might be applied to the ideal diode.

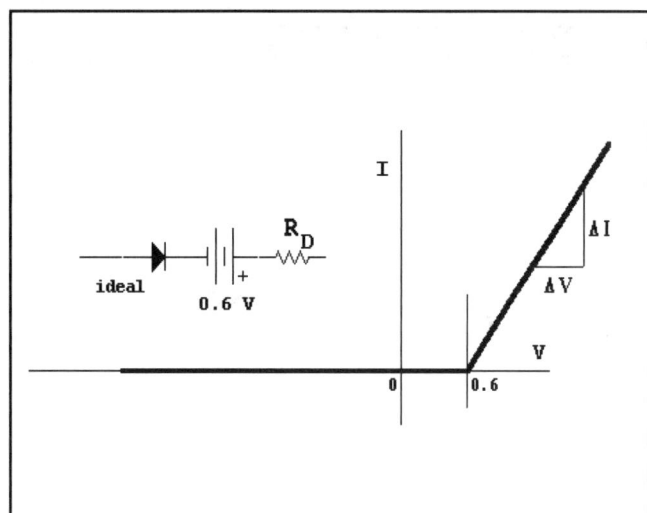

Fig 2.3—IV characteristic for a refined diode model.

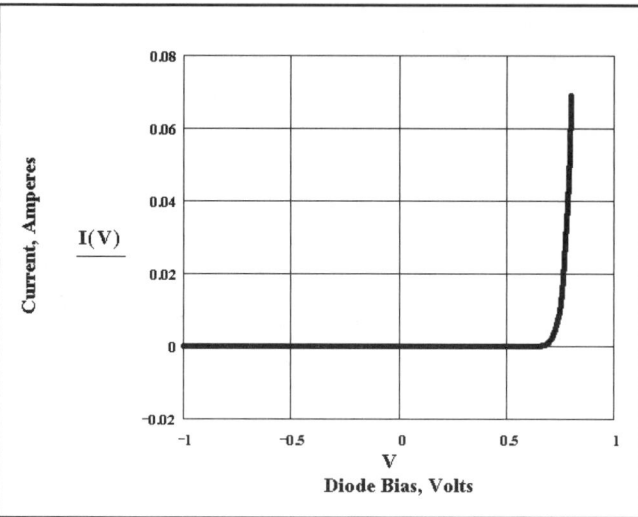

Fig 2.4—IV characteristic for a common junction diode. This follows the diode equation.

thresholds exceeding one volt.

The model of Fig 2.3 is more accurate than the ideal diode, but is still less than perfect in some situations. A much better diode representation is a mathematical model where current is given by an equation,

$$I = I_S \cdot \left(e^{qV/kT} - 1\right)$$
$$I \approx I_S \, e^{qV/kT} \qquad \text{Eq 2.1}$$

where I_S is called the saturation current in amperes, q is the charge on an electron, k is Boltzman's constant, and T is the diode temperature in degrees Kelvin. The second, approximate form is common. This model, known merely as *the diode equation*, is illustrated in **Fig 2.4** for the case of T=300 K (near room temperature) and $I_S = 3 \times 10^{-15}$ A, a value that we inferred from measurements for the popular 1N4148/1N4152 series of parts. Changing I_S generates new offset values. The diode equation is also significant because it originates as a description evolving from basic physics. Physics based models are generally preferred because they follow from fundamentals, even though they may not be as intuitive.

More refined diode models will include reverse breakdown, high frequency parameters (inductance and capacitance,) and even carrier lifetime. No matter what methods we use to analyze a circuit, the results of the analysis will only be as good as the models.

SMALL SIGNAL DIODE MODEL

The antenna signals that our receivers amplify are often in the microvolt region or less. We ask how the diode would behave if one microvolt was applied to it. The current flowing in the diode, **Eq 2.1**, would be essentially zero if a microvolt was applied directly. But, the diode might have a much different response if the diode already had a bias current flowing.

Fig 2.5 shows part of a diode IV curve. The point corresponding to 5 mA DC current flow is marked with a tangent line. The slope of this line defines a resistance, a change in current for an applied change in voltage that occurs when a small signal is applied to the biased diode. The diode has a resistance of about 5 Ω when the current is 5 mA, generally represented by

$$R_{IN} = \frac{26}{I(mA)} \qquad \text{Eq 2.2}$$

The factor 26 mV (or .026 V) comes from differentiation of Eq 2.1 and is a very common parameter in semiconductor electronics:

$$\frac{kT}{q} \approx .026 \qquad \text{Eq 2.3}$$

A small signal diode model is no more than a simple resistor. We will make extensive use of small signal models as we move on.

The Bipolar Transistor

The bipolar transistor is a three terminal device. If we use the same equipment that we used to examine diodes, we might conclude that the bipolar transistor is just a pair of diodes in one package, attached as shown in **Fig 2.6**. This is an incomplete, yet useful model.

Let's place this model in a test circuit, shown in **Fig 2.7**. A variable voltage bias source with a large base resistor is used, allowing us to control base current. A positive voltage is applied to the collector, *reverse* biasing the collector-base junction. The two-diode model would predict zero-collector current. But, collector current does flow in proportion to the current in the base. This is transistor action. The ratio of collector to base current is usually sig-

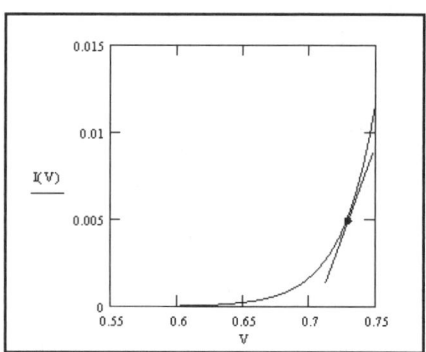

Fig 2.5—Small signal model for a junction diode represents it as a resistor with the slope shown. See text.

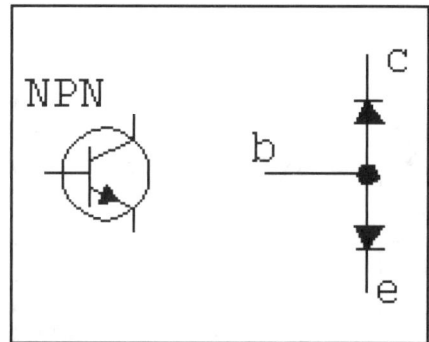

Fig 2.6—Apparent model of a bipolar transistor. This is what we would infer by examination with a VOM.

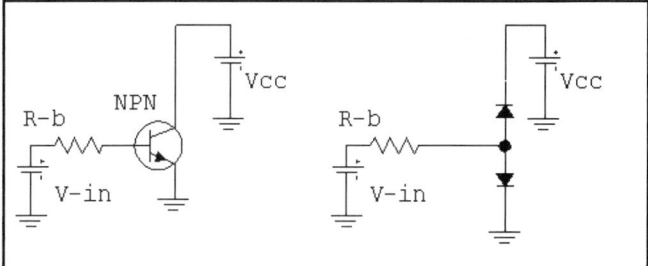

Fig 2.7—The circuit we used to bias a bipolar transistor for active operation. See text.

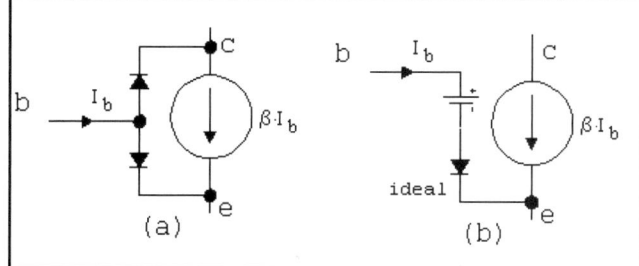

Fig 2.8—A current source is added to the diode pair to form a representative model. The diode is often ignored as in B.

nified by the greek letter beta, β. A typical value might be 100.

The simplified model on the right side of Fig 2.7 is clearly in error. The "collector" diode is reverse biased by V_{cc}, yet considerable current flows against the diode arrow. A better model is shown in **Fig 2.8A** where the original diode pair is supplemented by a current source proportional to the current in the base-emitter diode. The model in Fig 2.8B is the model we use for evaluation of biasing circuits. It neglects the collector-base diode and refines the base-emitter diode.

SMALL SIGNAL BIPOLAR TRANSISTOR MODEL

What happens with the bipolar transistor for small signals? How do we model it? The methods used with the diode are expanded to describe the transistor, as shown in **Fig 2.9**.

In Fig 2.9A, the input diode is replaced for small signals with a resistance. The resistance is exactly that used with the earlier diode, 26/I where I is now the DC current in milliamperes for the base-emitter diode. The current amplifying properties that we discovered earlier are preserved for small signals, so the small signal collector current remains at β×i_b. We use a lower case "i" to emphasize the small signal levels.

An alternative small signal model is shown in Fig 2.9B. Here the resistance in series with the base has been replaced with one in the emitter. This resistance, termed r_e, is given by

$$r_e = \frac{26}{I_e} \qquad \text{Eq 2.4}$$

where I_e is now the emitter current in milliamperes. The collector current exceeds that in the base by β, and the emitter current is the sum of the collector and base values, so the dc emitter current is greater than the base value by (β+1). Accordingly, the emitter resistor of Fig 2.9B is smaller than the resistor of Fig 2.9A by (β+1). Both models are equally valid, although that

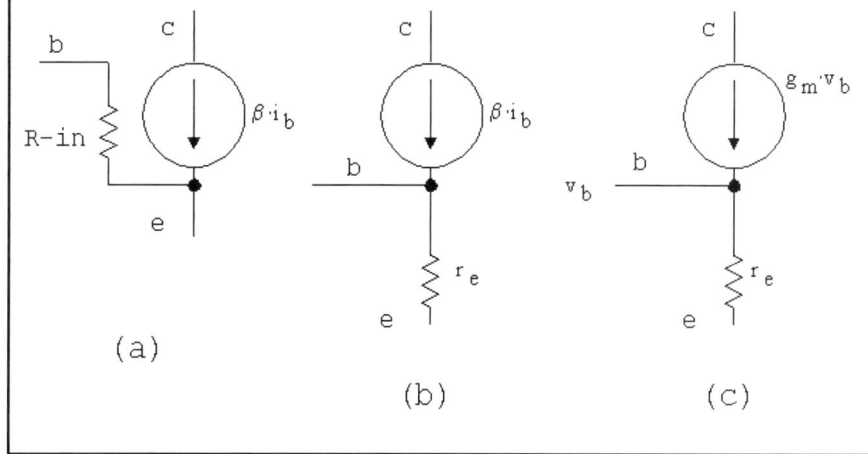

Fig 2.9 Evolution of a small signal transistor model.

using r_e is more common. Common emitter small signal amplifier input resistance is

$$R_{in} = (\beta + 1) \cdot r_e \qquad \text{Eq 2.5}$$

A traditional viewpoint emphasizes the bipolar transistor as a *current controlled* device with β representing current gain. But beta can vary considerably for a given transistor type, suggesting that the amplifier gain may differ for different transistors, which is not true. A preferred small signal model is shown in Fig 2.9C, where the part is viewed as a *voltage driven* component. The output current source is now specified by a transconductance, g_m:

$$i_c = g_m \cdot v_b \qquad \text{Eq 2.6}$$

The transconductance, g_m, is given by

$$g_m = \frac{I_E \, (mA)}{26} \qquad \text{Eq 2.7}$$

While β may vary among transistors, g_m is well defined by emitter current.

Another feature of the model is illustrated by a simple amplifier design, shown in **Fig 2.10A**. An NPN transistor is biased with a base resistor attached to a positive supply. A load resistor, R_c, is placed in the collector. The base resistor is adjusted until the emitter current is 1 mA. The small signal model shown in Fig 2.10B is used for analysis.

With 1 mA emitter current, the transconductance is g_m=1/26. Signal current is then $v_{in} \times g_m$. This current produces an output voltage because it flows in R_c, resulting in a voltage gain of $g_m \times R_c$, which is

$$G_v = R_c / r_e \qquad \text{Eq 2.8}$$

Knowing biasing details, voltage gain can be predicted "by inspection" as a resistor ratio, independent of beta. Current gain, or β, is still of significance, for it will alter the signal current that flows when drive voltages are applied, which defines input impedance.

Note that we have said nothing about transistor type. Our discussion has considered the NPN, but has said little else of a specific nature. This is not an oversimplification. Much of the utility of the bipolar transistor results from properties that

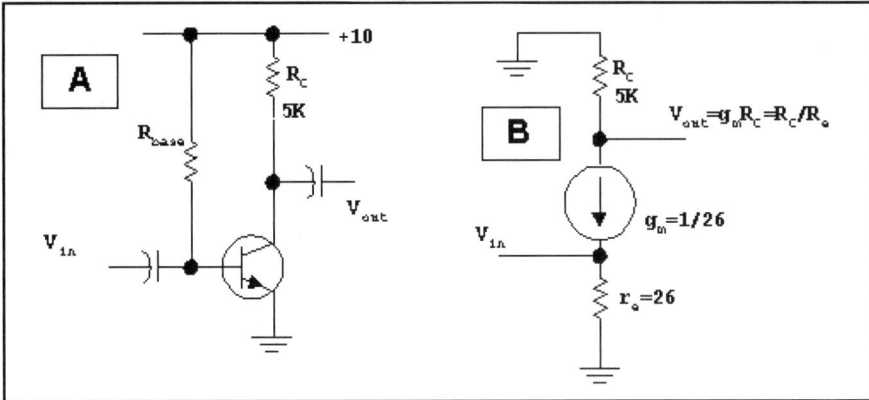

Fig 2.10—The simple amplifier at A is analyzed with the small signal model at B.

depend primarily upon the standing emitter current.

BIPOLAR TRANSISTOR BIASING

Accurate transistor current is vital to any design, because current determines small signal properties. The power dissipation, the power output capabilities, the distortion, and even frequency dependence are also determined by bias current and voltage. Biasing methods will be evaluated with the model of Fig 2.8B, where the base-emitter junction becomes an ideal diode with a 0.6-V battery. Collector current is then $\beta \times I_b$.

The first bias example we consider is that shown in **Fig 2.11**. A 1-kΩ load resistor appears in the collector, while the base is biased from the 12-V supply through a 100-kΩ resistor. The model assumes an offset of 0.6 V, so the base current is 11.4 V across 100 kΩ, or 114 µA. If transistor β=100, the collector current is 11.4 mA. But, the 1-kΩ collector resistor produces an IR drop of 11.4 V, leaving a collector voltage of only 0.6 V.

Repeating the calculation with slightly higher β predicts a negative collector voltage, impossible without a negative supply. Recall that earlier models included a collector-base diode that prevented the collector from being more than a diode drop below the base. Whenever the collector voltage equals or drops below that of the base, for an NPN, the transistor is said to be saturated.

The scheme of Fig 2.11 is, at best, a poor bias method. Slight changes in beta yield great uncertainty. Biasing is improved with negative feedback, with one form shown in **Fig 2.12**. The 100-kΩ resistor is biased from the collector rather than the 12-V supply. An intuitive examination shows that this is an improved method, even before we "crunch" any numbers. If the beta changes to drive the transistor toward saturation, the current through R1 will decrease from the reduced collector voltage. A lower than nominal beta will cause collector voltage to climb, forcing more base current to flow.

Application of the model and some algebra provides a general equation for Fig 2.12,

$$V_C = \frac{V_{cc} \cdot R_1 + V_{eb} \cdot \beta \cdot R_c}{\beta \cdot R_c + R_1} \quad \text{Eq 2.9}$$

An even better bias scheme is shown in **Fig 2.13A**, where the base is driven from the positive supply through a voltage divider, R_1 and R_2. The equivalent circuit for the divider is shown in Fig 2.13B. The base voltage with the transistor temporarily removed is found from divider action as

$$V_b' = V_{cc} \cdot \frac{R_2}{R_1 + R_2} \quad \text{Eq 2.10}$$

where the prime indicates that the base is open circuited, and absent from the calculation. The emitter voltage is below the base by the 0.6-V offset, placing the emitter voltage at 1.45 V. The emitter current is then determined by the 330-Ω emitter resistor as 4.39 mA. The collector current is almost the same as that in the emitter, and the drop across the collector load puts V_c at 7.61 V.

This analysis, although close, is in error. Base current flow produces an IR drop in the biasing resistor chain. This decreases the base voltage below the value shown in Fig 2.13 by about 0.25 V. There are two solutions to this problem. One would replace R_1 and R_2 with a "stiffer" voltage divider. Values of 3.3 kΩ and 680 Ω would work well, but at the price of greater power consumption. The other alternative is a more careful analysis. If this is performed, the emitter current is given by

$$I_e = \frac{(V_{cc} \cdot R_2 - V_{eb} \cdot (R_1 + R_2)) \cdot (\beta + 1)}{(R_2 + R_1) \cdot R_3 \cdot (\beta + 1) + R_1 \cdot R_2}$$

Eq 2.11

The I_e value for the components in Fig 2.13 is 3.759 mA.

PNP biasing is identical to that of the NPN, except that the voltages are measured with regard to the positive power supply, which may or may not be "ground." See **Fig 2.14**.

Fig 2.15 shows a natural refinement to the biasing scheme. Here another resistor is added, a normal part of a decoupling scheme. The added resistor provides negative feedback like that used earlier in Fig 2.12. This, in combination with the feedback from R_3 of Fig 2.13 further stabilizes bias.

A scheme useful for biasing an NPN transistor with a directly grounded emitter is shown in **Fig 2.16**. A PNP transistor emitter senses the dc collector voltage and compares it with the PNP base at a reference, V_r, established with voltage divider R_1 and R_2. The reference divider is usually designed to put most of the power supply on the NPN collector. The 0.1-µF capacitor stabilizes the negative feedback bias loop. With the values shown, the bias is defined by

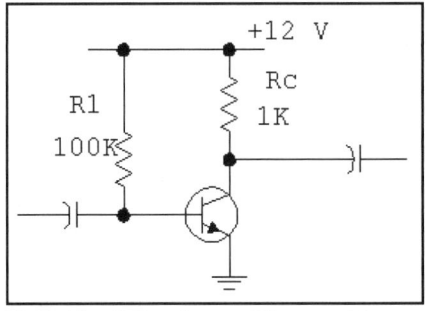

Fig 2.11—A simple amplifier used for bias analysis.

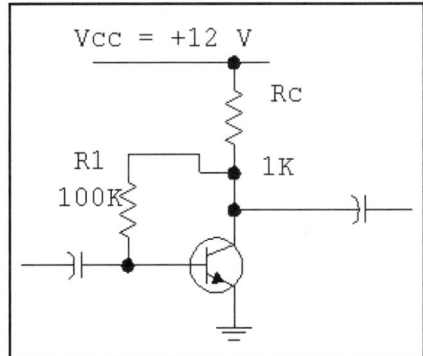

Fig 2.12—Improved bias is obtained from the collector.

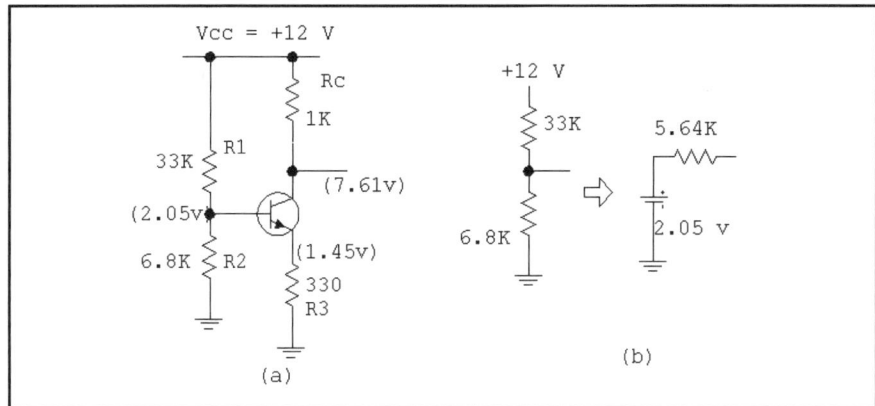

Fig 2.13—Evolution of base bias from a voltage divider.

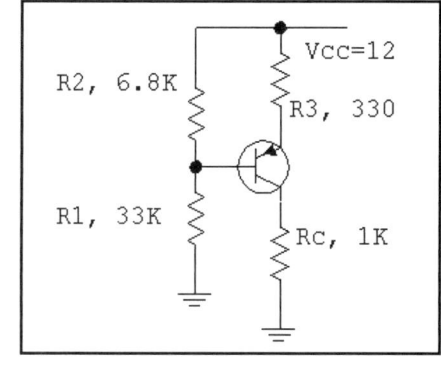

Fig 2.14—PNP biased to the same conditions as we established with the NPN example.

$$V_R = \frac{V_{CC} \cdot R_2}{R_1 + R_2}$$

$$I_C = \frac{V_{CC} - V_R - 0.6}{R_A} \quad \text{Eq 2.12}$$

The Field-Effect Transistor

Although the bipolar transistor is our work horse, various forms of *field effect transistor*, or FET, are close in popularity. Among FETs, one of the most common is the junction variant, the JFET. A JFET is much like vacuum tube triodes of the past and is easily biased and used in amplifier applications. FETs, including the JFET, generally lack the uniformity and predictability of a bipolar transistor. JFETs tend to be low noise devices. Not only is the noise figure low, but the low frequency flicker, or "1/F" noise is small. This combination makes the JFET especially useful for low noise oscillators.

Fig 2.17 presents the test setup that allows us to measure, and then model the JFET. The example is an N-channel Depletion mode JFET. A drain power supply, $+V_{dd}$, is applied. The gate voltage is then varied while examining the current that flows. **Fig 2.18** is a resulting plot of drain current vs gate-to-source voltage with constant drain voltage. The gate voltage is negative for most of the curve. The gate can be no more than 0.6 V positive, for the gate of a JFET is actually a diode junction. The metal oxide silicon field effect transistor, MOSFET, has similar properties, but uses an insulating gate. There is then no diode clamping action.

Once gate-to-source voltage drops to an adequate level, drain current goes to zero and the FET is said to be in "pinch-off." The pinch-off voltage, the gate-source V where current drops to (or nearly to) zero,

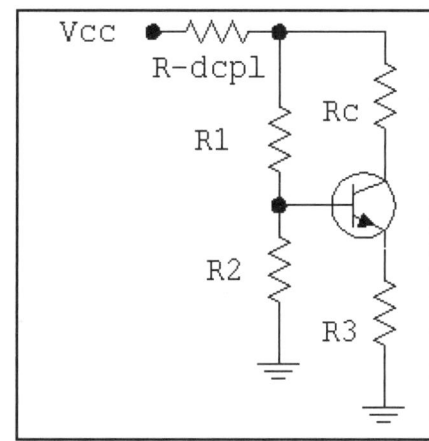

Fig 2.15—Decoupling resistor adds negative feedback to the biasing with an emitter resistor.

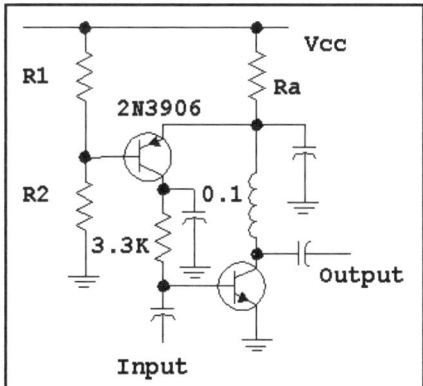

Fig 2.16—A "wrap-around" PNP biases an NPN with grounded emitter. The 0.1-µF capacitor stabilizes bias and is the dominant element in the bias loop.

is at -3 V for the example of Fig 2.18. These data are typical for the popular J310 JFET. A drain voltage higher than the magnitude of the pinch-off is usually required to ensure linear operation. This is

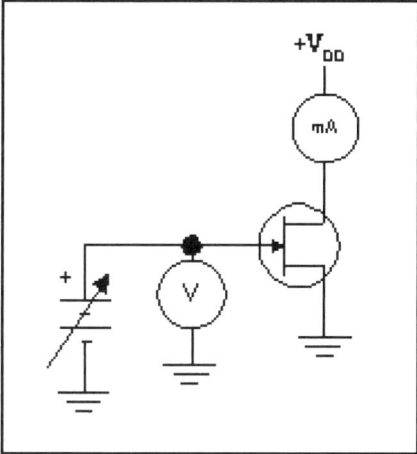

Fig 2.17—Test setup used to evaluate a JFET.

often called operation in the *saturation* region. *Saturation* is just the opposite condition in a FET from *saturation* in a bipolar transistor.

Fig 2.19 shows the usual source resistor method used for biasing an N-channel JET at a current below I_{dss}. The current flowing through the resistor establishes a positive source voltage. As current increases, the source voltage increases, causing the gate-to-source voltage difference to become more negative. This is the action needed to decrease current, eventually stabilizing the bias. The action of an external source R is a form of negative feedback, just as we used with an emitter resistor in the case of a bipolar transistor. Fig 2.19 includes some JFET equations.

SMALL SIGNAL JFET MODEL

Fig 2.18 showed a complete curve, describing large and small signal behavior as well as JFET biasing. The simplified small signal model is shown in **Fig 2.20**. Here an open gate terminal accepts an input voltage. That signal then controls an

Amplifier Design Basics 2.5

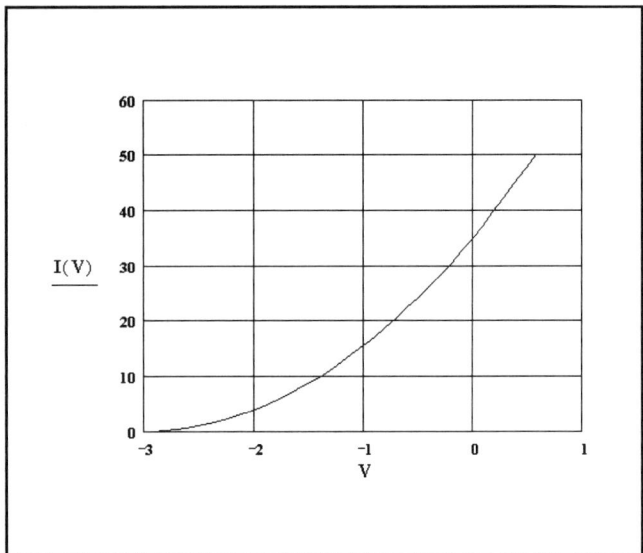

Fig 2.18—Drain Current vs Source-to-Gate Voltage for J310 type Junction Field Effect Transistor. I_{dss}=35 mA and V_p=-3 V. V_p is the voltage where drain current goes to zero. I_{dss} is the drain current when the gate and source are both at the same potential.

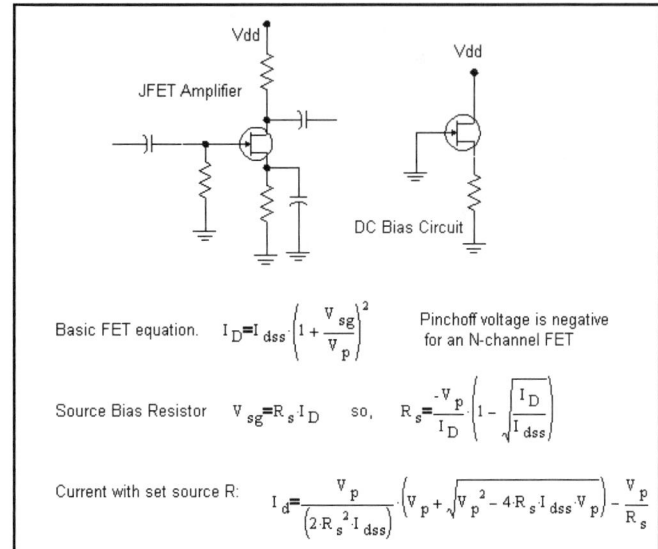

Fig 2.19—JFET bias circuit and equations. The left circuit is a practical amplifier, while that on the right is the bias equivalent. Pick a desired drain current, I_D (must be less than I_{DSS}), and use the middle equation to find the required source resistor. The resulting source voltage is given by Ohm's Law.

output current source related to the input by a transconductance, g_m, with

$$g_m = -2 \cdot \frac{I_{dss}}{V_p} \cdot \left(1 + \frac{V_{sg}}{V_p}\right) \quad \text{Eq 2.13}$$

For example, if we biased the FET for a gate voltage equaling half of the pinch-off value, with Idss=35 mA and Vp=–3 V, the small signal transconductance is 0.0117 S,

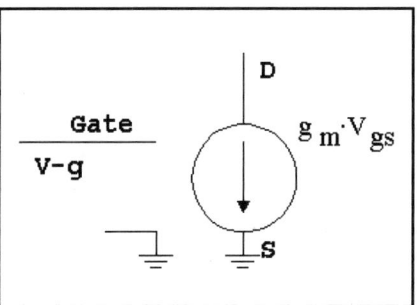

Fig 2.20—Simplified small signal JFET model.

or "amps per volt." From the equations in Fig 2.19, we see that the DC drain current is then 8.75 mA, which is realized with a source R of 171 Ω. The low frequency input resistance is essentially infinite.

2.2 AMPLIFIER DESIGN BASICS

Having examined basic device models and biasing, we now evaluate some basic amplifier designs, first with the bipolar junction transistor (BJT) and then the junction field effect transistor (JFET).

We begin with a single stage audio design, **Fig 2.21**. The circuit that we might build is presented in Fig 2.21a, while a biasing related part is shown in Fig 2.21b. The voltage divider, 10 kΩ and 3.3 kΩ, creates an equivalent source of 2.481 V at the base. This decreases by 0.6 V in moving through the transistor to produce an emitter voltage of 1.881 V. The emitter current is then 1.881 mA. With beta=100, base current is 19 µA, well below the 752 µA in the voltage divider. The collector voltage is then 10−1.881=8.119 V. The collector-to-emitter voltage, V_{ce}, is 6.238 and power dissipation is the product of this voltage with the standing current, 11.73 mW.

Small signal transistor characteristics are established by emitter current. The resulting small signal model is that in Fig 2.21c. The 1-kΩ emitter resistor has disappeared from the circuit for it is well bypassed by the 100-µF capacitor. The small signal r_e is $26/I_e(mA)=13.82$ Ω. The input resistance looking into the base is almost 1.4 kΩ = $r_e \times (\beta+1)$.

The input source is a 1-mV voltage generator in series with a resistance of 1 kΩ, which might represent a previous stage. The source is AC coupled to the base through a 10-µF capacitor which has a 1-kHz reactance of 16 Ω. Being very small compared with the amplifier input or the source, it may be neglected for a 1-kHz analysis. The same argument may be made for the output capacitor. The result is the small signal circuit of Fig 2.21d. The power supply is missing in the small signal models where V_{cc} is replaced by ground; the supply is fixed and does not change with audio signal current, so it is effectively a signal ground.

We characterized the BJT by a transconductance, g_m=0.0724 amp/volt. Also, we neglect any effect related to the base bias divider on the small signal model.

The 1-mV input signal is voltage

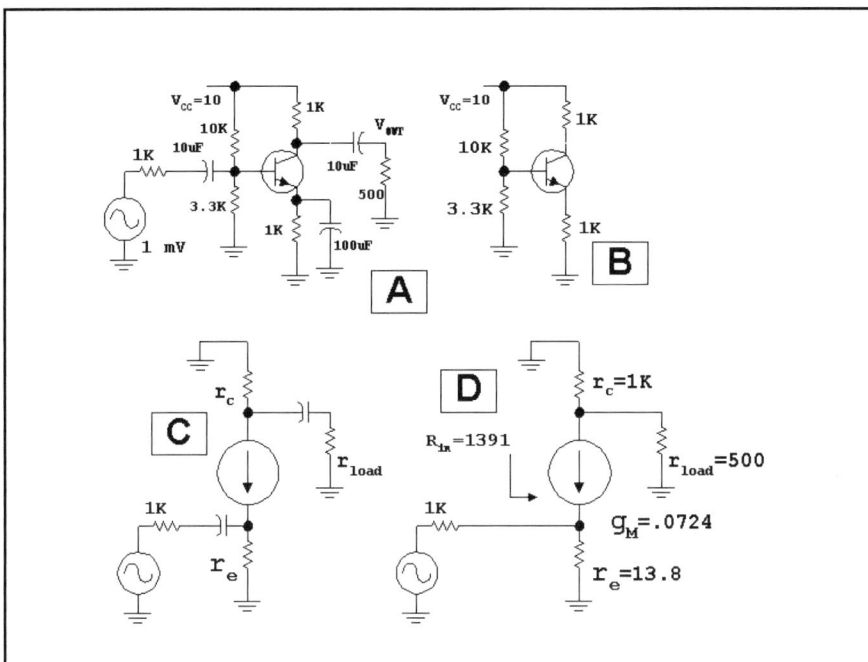

Fig 2.21—Single transistor audio amplifier design. See text for details.

divided between the 1-kΩ source resistance and the 1.39-kΩ input resistance. The base input voltage becomes 0.582 mV to produce a collector signal current of $i_c = g_m \times v_b = 0.0421$ mA. This current flows through a resistance of 333 Ω, the parallel equivalent of the 500-Ω load and the 1-kΩ collector resistance. The output voltage is then 0.0421 mA×333, or 14.02 mV for a circuit voltage gain of 24.1. Note that this is also exactly the ratio

$$G_V = \frac{R_L}{r_e} \qquad \text{Eq 2.14}$$

where the load is the total impedance *seen* by the collector.

The form of this equation is especially intuitive, emphasizing the role of r_e as a degeneration resistance. If we placed a 10-Ω resistor in series with the 100-μF emitter bypass capacitor, the net emitter resistance would be $10 + 13.8 = 23.8$ Ω and the voltage gain would become 14. The role of emitter current is clear; Increasing standing emitter current causes r_e to decrease, increasing voltage gain. Emitter degeneration is a common feedback scheme.

We have treated the bipolar transistor as a voltage controlled device. Beta was indirectly used in the calculation, but only to set transistor input resistance. This, in turn, established the fraction of the 1-mV input voltage that appeared at the base.

There is a counter intuitive nature to the modeling presented in Fig 2.21D. The schematic shows the input is tied to ground through r_e, the 13.8-Ω resistor, which would severely attenuate the signal. However, the current source representing the transistor is also attached to the input node, and that current moves in unison with the input voltage. This yields the results outlined.

We calculated a voltage gain. The gains of greater interest are power ratios. One of interest to the RF designer is, simply, *power gain*, the output power divided by input power. The output power is calculated (for Fig 2.21) as V^2/R where R is the 500-Ω load and V is the 14.02-mV output. Output power is then 3.93×10^{-7} W. The input power is the base voltage (0.582 mV) across the transistor input R of 1.4 kΩ, or 2.435×10^{-10} W. The power gain is the ratio of the two powers, 1614. Using a dB relationship, this becomes 32.1 dB. This is high but reasonable for a single transistor, for this amplifier operates at low frequencies. Such gain from a single transistor at radio frequencies is more difficult.

Power gain is fundamental but is not always the gain we measure. We usually measure *transducer power gain*, especially when working with RF circuits. Transducer gain is output power delivered to a load vs the maximum power *available* from the input generator. We have already calculated output power. The available power from the source is the power that would be delivered to a termination that was impedance matched to the generator. The generator was a 1-mV open circuit source behind a 1-kΩ resistor, so the load that would allow the maximum available power to be extracted would be a 1-kΩ resistor. The available input becomes 0.5 mV across 1 kΩ, or 2.5×10^{-10} watts, leaving a transducer gain of 1572, or 32.0 dB. This is nearly as high as the power gain. The gain difference is a consequence of the input impedance mismatch. We will have more to say about gains and dB later in this chapter.

A common practice converts a voltage gain to decibel form with the familiar $20*\text{Log}(G_v)$, 27.6 dB for this example. This is **not** a correct result, for the source impedance is not the same as the load impedance. The decibel construct is one that should only be applied to power ratios. It works with voltage ratios only when the related resistances are equal.

In the amplifier we analyzed, the input was applied to the base while the emitter was grounded through a large bypass capacitor. Hence, the input was applied between the base and the emitter. The output was extracted from the collector-emitter port. This is a common-emitter (CE) configuration, for the emitter is common to input and output. A common-collector (CC) amplifier is shown in **Fig 2.22**.

The complete amplifier circuit is shown in Fig 2.22A, while the small signal version is in Fig 2.22B. The open circuit dc base voltage is 5 V, so the emitter bias current is 4.4 mA, leading to $r_e = 5.91$ Ω.

The follower of Fig 2.22B is driven from a 1-kΩ source impedance. It is terminated in a pair of 1-kΩ resistors in parallel. The input resistance of a follower is given by

$$R_{IN} = (\beta + 1) \cdot (r_e + R_L) \qquad \text{Eq 2.15}$$

while the output impedance is

$$R_{OUT} = \frac{R_S}{(\beta + 1)} + r_e \qquad \text{Eq 2.16}$$

The voltage gain for the emitter follower is

$$G_V = \frac{R_L}{R_L + r_e} \qquad \text{Eq 2.17}$$

Substituting r_e into these equations shows that the follower has a gain of 0.988, essentially 1, accounting for the circuit name. Setting β to 100, the input resistance is 51 kΩ while the output resistance is 15.8 Ω. The input resistance and the voltage gain both grow if the follower is lightly loaded. The output resistance decreases as the source impedance drops.

It is very common to dc-couple a follower to a preceding amplifier; this is illustrated in **Fig 2.23**.

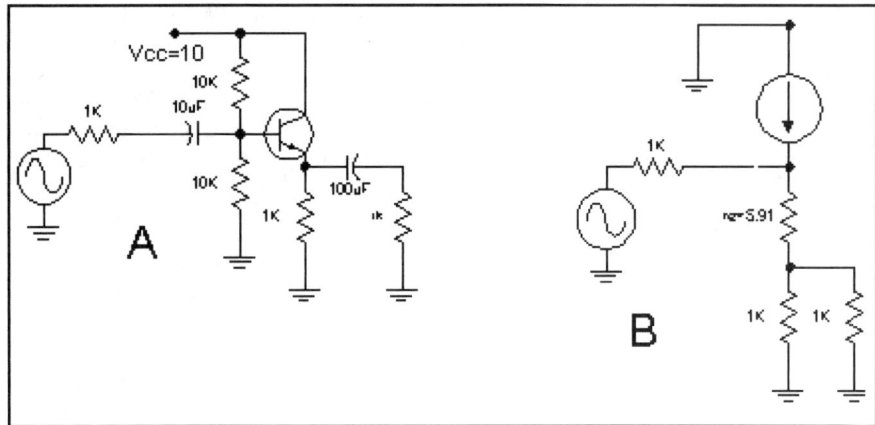

Fig 2.22—Common collector amplifier, also known as an emitter follower.

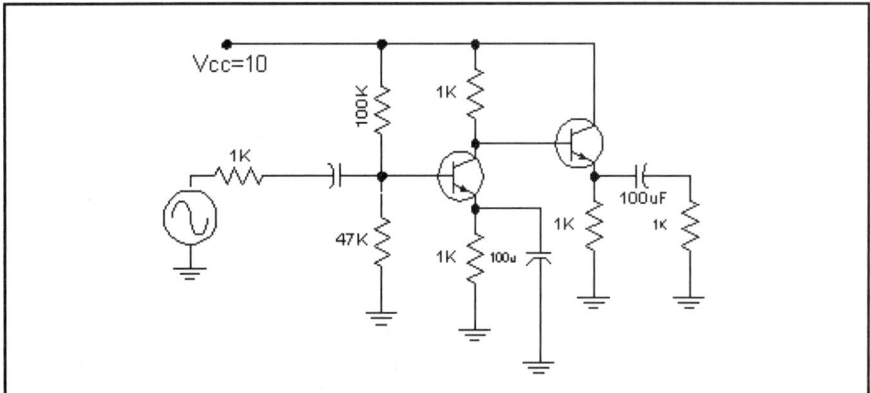

Fig 2.23—Voltage Amplifier with a DC coupled emitter follower.

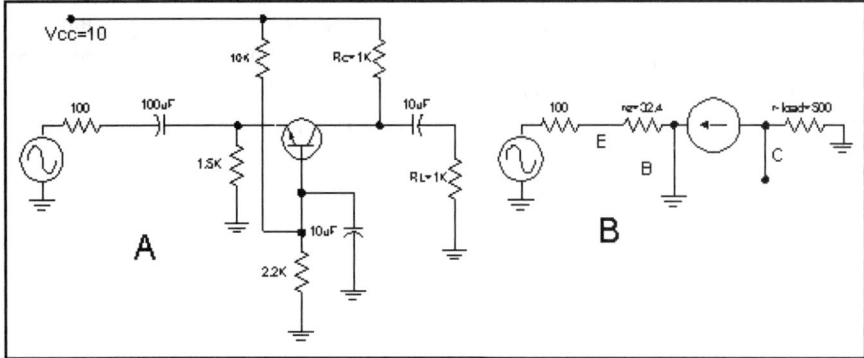

Fig 2.24—Common Base Amplifier with small-signal equivalent.

The third basic amplifier configuration is the common base (CB) amplifier of **Fig 2.24**.

The input resistance for the common base (CB) amplifier is

$$R_{IN} = r_e = \frac{1}{g_m} \quad \text{Eq 2.18}$$

The current gain for the CB amplifier is given by the parameter α,

$$\alpha = \frac{\beta}{\beta + 1} \quad \text{Eq 2.19}$$

which is normally very close to unity. We essentially assume that the current injected into the CB amplifier appears at the output. The voltage gain is then

$$G_V = \alpha \cdot R_L \quad \text{Eq 2.20}$$

The voltage gain for the CB amplifier can be very large. However, this is somewhat synthetic, for the input impedance is usually very low, making the amplifier difficult to drive. The common applications use a current source to drive the CB amplifier, realized by placing an extra resistance in series with the input.

The CB amplifier has the useful property that it offers excellent reverse isolation. That is, the input impedance of a CB amplifier is not affected by anything that happens to the output circuit. The example shown in Fig 2.24 is biased to a current of about 0.8 mA, producing an input resistance of 32 Ω.

The equations for the small signal properties of the various amplifiers are derived in *Introduction to RF Design*[1] and are discussed in *The Art of Electronics*.[2]

The CC amplifier had a low output impedance. Nothing was said about the common emitter and common base amplifier output resistance. Both are essentially infinite for the simple models considered where the BJT is modeled as an "ideal current source."

Most of the amplifier analysis we have done is based upon simple models, ones that have but one or two parameters. Beta has only minor impact on circuit performance. The dominant element in all of the models is r_e, the emitter resistance. This parameter is directly related to current, a parameter under the control of the circuit designer. This would suggest that *all* bipolar transistors are more alike than they are different and that the only major differences are in the frequency capability and size. This is generally an accurate view of the small-signal bipolar transistor.

Small-Signal FET Amplifiers

The field effect transistor families are similar to the BJT; as three terminal devices, they can be configured into three different forms. **Fig 2.25** shows the common source, common gate, and common drain (or source follower) configurations for an N Channel JFET.

There are many similarities between BJT and JFET circuits. The common gate FET amplifier (CG) has a low input impedance with a high output impedance. The topology offers excellent reverse isolation. The follower (CD) has a low output impedance with a very high input impedance.

JFET bias current is controlled by the designer, just as it was with the BJT. Resistor values may, however, have to be device specific, picked for a given FET to establish performance. Within a given JFET type, for example, a 3:1 variation in

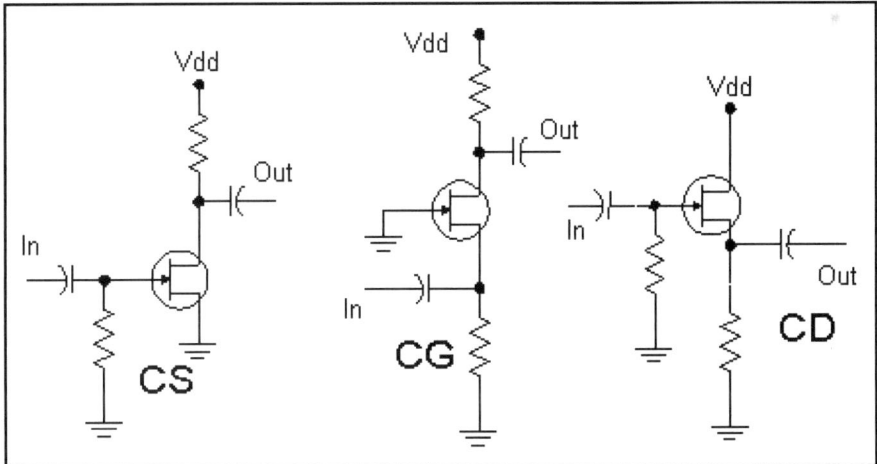

Fig 2.25—Common Source, Common Gate, and Common Drain JFET Amplifiers.

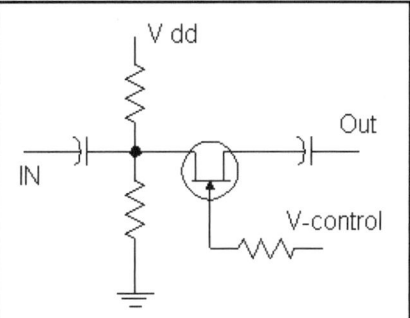

Fig 2.26—A JFET operating as a series switch.

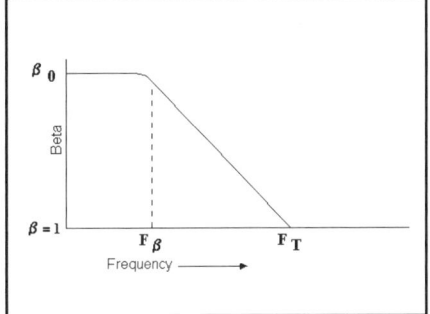

Fig 2.27—Current gain vs Frequency for a BJT.

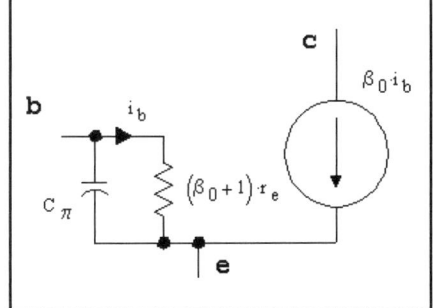

Fig 2.28—The hybrid-pi transistor model.

I_{dss} is common. A similar variation exists with pinchoff voltage. The combination of these two variables might lead one to feel that it would be nearly impossible to design with FETs. Fortunately, it's not that bad, for the variations are related to each other. That is, a given JFET in a family with a high I_{dss} will also tend to have a pinchoff with a more negative value, producing less variation in g_m, the dominant small signal characteristic.

There is good reason for the similarities between FET and BJT amplifiers. Many of the properties result from feedback that is added to a circuit by the configuration. For example, the follower has the load in series with the current source. The voltage developed across the load then generates a signal that contributes to the control of the current generator.

The JFET has an additional property not predicted by the preceding model, the switch action illustrated in **Fig 2.26**. The JFET functions here as a series SPST switch. An input ac signal is applied to the FET channel (the source-drain path) and is routed to the output when the control voltage is positive with regard to the channel. The channel is the current path between source and drain. The channel is biased above ground by the voltage divider. The switch is open circuited if the control voltage is more negative with regard to the channel than the FET pinchoff voltage. The switching FET may be modeled as a voltage controlled variable resistor in this application. Lowest R occurs when the control voltage is at or above the channel. The gate resistor is usually large, allowing the control to be several volts higher than the channel. Although the gate diode is then forward biased, current is small and of little consequence.

Virtually all FET types function well as switches. Enhancement mode MOSFETs offer the advantage of no gate diode to complicate the circuit. GaAs MOSFETs are useful in very high speed switching applications where they may be used for microwave signal control. JFETs and MOSFETs are useful audio switches in many applications.

The FETs may be used as voltage variable resistors. As such, they can function in gain control circuits.

High Frequency Effects

Little has been said about the effects of high frequency. Yet, much of our interest as radio experimenters is in the performance of transistor circuits at frequencies well beyond the range of our simple models.

The first thing that happens to the BJT as frequency increases is that β decreases over the dc and audio values. This is shown in the curve of **Fig 2.27** of β vs frequency. The low frequency β is shown as $β_0$. The frequency where β drops to a value of unity is called the current gain bandwidth product, or more often, just as F_t. Dropping to a frequency of $F_t/2$ will produce β=2. The frequency where β begins to depart from $β_0$ is called the "beta cutoff."

The role off of current gain with frequency is modeled with an added base capacitor, **Fig 2.28**. The other elements are generally unchanged, so the complete roll off may be attributed to the capacitor across the input. The circuit shown in Fig 2.28 is called the hybrid-π model.

At low frequencies an output signal from a transistor is either in phase (0 degrees) or out of phase (180 degrees) with the input signal. These simple phase relationships no longer hold above the β cutoff where the mathematics change, taking on a (formally) complex character.

A typical BJT is the 2N3904. This NPN transistor has a typical F_t of about 300 MHz and a low frequency $β_0$ of 100. This places the β cutoff at about 3 MHz. This device will have some phase shift effects at all frequencies within the HF spectra and higher.

Amplifier Design Basics 2.9

2.3 LARGE SIGNAL AMPLIFIERS

Our previous small signal viewpoint is now expanded. We will examine over-driven receiver circuits only intended for small signals. A more common large signal amplifier is a transmitter stage, a circuit intended to function at high levels.

Distortion is a consequence of large signal operation. Distortion in an amplifier merely means that the output is something different than a replica of the input. A distorting circuit driven by a sinewave will have non-sinewave outputs when viewed in the time domain, experimentally with an oscilloscope. In the frequency domain, the distortion appears as harmonics. A distorting circuit driven by two or more signals may contain outputs that are the result of intermodulation, frequencies that are sums and differences of input frequency multiples.

The BJT model of greatest popularity is an extension of the diode equation,

$$I_E = I_{ES} \cdot e^{\frac{q \cdot V}{k \cdot T}} \qquad \text{Eq 2.21}$$

where I_{ES} is called the emitter saturation current. V is the voltage on the base-emitter diode. The other parameters are the same as appeared with the diode equation in Section 2.1. This is part of the model known collectively as the Ebers-Moll equations. The non-linear exponential behavior is intrinsic to the bipolar transistor. Detailed use of this model takes us well outside the realm of this text, but is highly recommended for those with such interests.[3]

Many large signal properties of amplifiers are extensions of simple circuit analysis. Although the details are always buried within refined models, much can be discerned from careful analysis without analytic complexity. Some examples will be used to illustrate this.

Fig 2.29 shows a simple audio amplifier driven with a 1 kHz signal behind a 1-kΩ impedance. We observe an output voltage at the collector. The dc base voltage is approximately ¼ the power supply, so the emitter is at about 1.8 V. The emitter current is then 1.8 mA, producing a dc collector bias voltage of 8.2 V. The emitter current leads to a small signal r_e value of about 14 Ω. Voltage gain is 70 with the 1-kΩ collector load. The input resistance will be a little over 1 kΩ if β is 100. This means that the base signal voltage is just over half the generator value.

From the bias and small signal analysis, we predict that an input of 20 mV peak at the generator will produce a bit over 10 mV at the base. The voltage gain of 70 applied to this will give a peak collector signal of 0.7 V, or a peak-to-peak value of 1.4 V. The 8.2-V zero signal collector value will then move between 7.5 V and 8.9 V. This is still a long way from the +10-V supply or the 2.5-V base where saturation would be approached. We would expect a sinewave input to generate a sinewave output.

Fig 2.30 shows waveforms for three drive levels: .02 V, 0.1 V, and 0.5 V peak. The sinusoidal output is very close to the values we estimated. However, the other two cases are severely distorted. The 0.1-V drive case, five times stronger than the initial 20-mV input, is enough to cause the output to reach the 10-V positive power supply, causing collector current to drop to zero. The other part of the cycle is still well behaved with approximately sinusoidal outputs.

The most severely distorted output results from the largest input signal, 0.5 V peak, also shown in Fig 2.30. At the positive extreme, the transistor is cutoff with current having vanished. At the other end, the transistor current is well beyond the bias value. The collector has dropped below the base voltage and the transistor is saturated for the bottom, voltage-flat parts of the curve.

Simple models predict much of the nonlinear behavior, without formal analysis. The base-collector diode prevents collector voltages more than a diode-drop below the base. But, the collector current generator is capable of increasing "as needed" to supply larger currents, but only of the prescribed polarity. The larger drive examples would sound very distorted if this audio amplifier was part of a receiver.

The next example is a familiar emitter follower that might be on the output of an oscillator. A follower has a low output impedance, and should, we reason, be capable of delivering power to a low impedance such as a mixer. But this

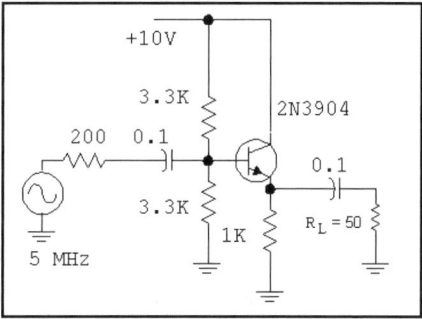

Fig 2.31—Emitter follower to drive a 50-Ω load. This circuit is not biased to deliver the needed output power.

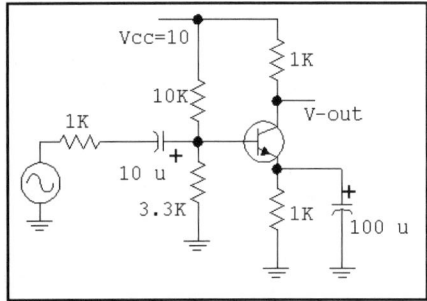

Fig 2.29—A simple audio amplifier examined for large signal performance.

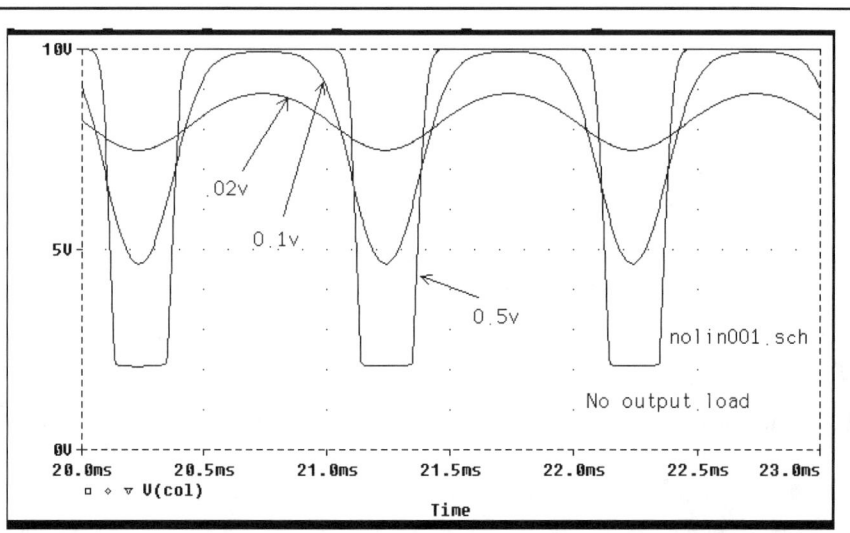

Fig 2.30—Output waveforms for the simple amplifier at several drive levels.

reasoning is flawed.

The emitter follower circuit is shown in **Fig 2.31**. A pair of 3.3-kΩ resistors bias the base at half the 10-V power supply, and the emitter is biased with a 1-kΩ resistor. I_e=4.4 mA, setting r_e to 5.9 Ω. The follower is driven from a 200-Ω source resistance for an output resistance of 7.9 Ω. If this circuit was going to be used to drive a 50-Ω filter, the 50-Ω resistance would be realized by adding a series 43-Ω resistor to the output.

This follower circuit is being driven by a signal source with a peak amplitude of 0.5 V. The input impedance is well above the 200-Ω driving source, so virtually all of the available generator signal is present at the base.

The modeling process is applied to capacitors with the same importance that it is to transistors. A capacitor accumulates charge through current flow, never allowing the voltage across the capacitor to instantaneously change. The capacitor could conceptually be replaced by a battery. In no-signal conditions the 4.4-mA transistor current flows in the 1-kΩ bias resistor with zero current in the 50-Ω load.

Applying a positive going signal to the base merely turns the transistor on harder. As the base voltage increases from the 5-V no-signal level to 5.5 V, the emitter will follow from 4.4 V to 4.9 V. We now have +0.5 V on the output load, forcing an output current of 10 mA to flow. The current in the 1-kΩ bias resistor has increased to 4.9 mA, so the total transistor current is 14.9 mA.

A negative-going base signal produces complications. A small negative base drive of 0.1 V to 4.9 V would drop the emitter to 4.3 V, which drops the output to –0.1 V. The current in the 50-Ω load becomes –2 mA.

With the emitter voltage at 4.3 V, we still have 4.3 mA flowing in the 1-kΩ resistor. The transistor current has now dropped to 2.3 mA. Because it is still positive, the transistor is still controlling the output and the follower continues to follow.

But what happens when the drive reaches the full negative value of –0.5 V? If the linear, small signal model applied, the base would drop to 4.5 V, leaving the emitter at 3.9 V with the output at –0.5 V, producing a current in the load of –10 mA. But the current flowing in the bias resistor would still be 3.9 mA, implying that the transistor current would be –6.1 mA. This is not possible! The transistor can supply current via the model current generator, but that current cannot be negative.

Fig 2.32 presents the waveforms. The negative going excursion is clipped at the point when the transistor emitter current drops to zero, leaving all output current to flow in the 1-kΩ resistor.

This simple circuit has illustrated the difference between small signal and large signal models. Currents of either polarity are allowed in a small signal model. The large signal behavior is restricted to that dictated by the model, in this case limited to the positive current flow predicted by the Ebers-Moll equation.

The low small signal output impedance of a follower was a consequence of negative feedback. The load in series with the output creates a voltage that is applied to the transistor in opposition to the signal driving it. If we allow the follower to "run out of current," the transistor is cut off with zero current flow. The low output impedance is no longer present during that part of the cycle when transistor current flow has ceased.

Fig 2.33 shows the output after the design was modified. The emitter bias resistor was changed from 1 kΩ to 330 Ω, increasing the emitter bias current to 12.6 mA. This is larger than the needed 10 mA, so the output remains clean. But, even a slight increase in drive could allow the distortion to return. The ultimate refinement might be a complementary output such as is found with many audio amplifiers.

The next example considered is a 10-MHz Class A amplifier intended to develop a few milliwatts of output power. The circuit is in **Fig 2.34**. The base is biased from a 10-V supply through a voltage divider of 10 kΩ and 3.3 kΩ, producing a DC emitter voltage of 1.64 V. The 200-Ω emitter resistor sets an emitter current of 8.2 mA, yielding a small signal r_e of 3.2 Ω. The 50-Ω output load sets the small signal voltage gain at 16.

A common approximation sets high

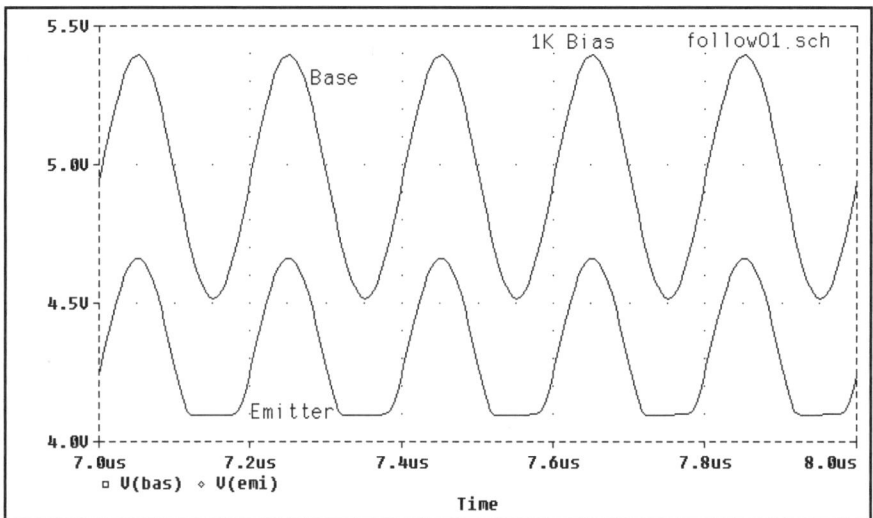

Fig 2.32—Follower waveforms.

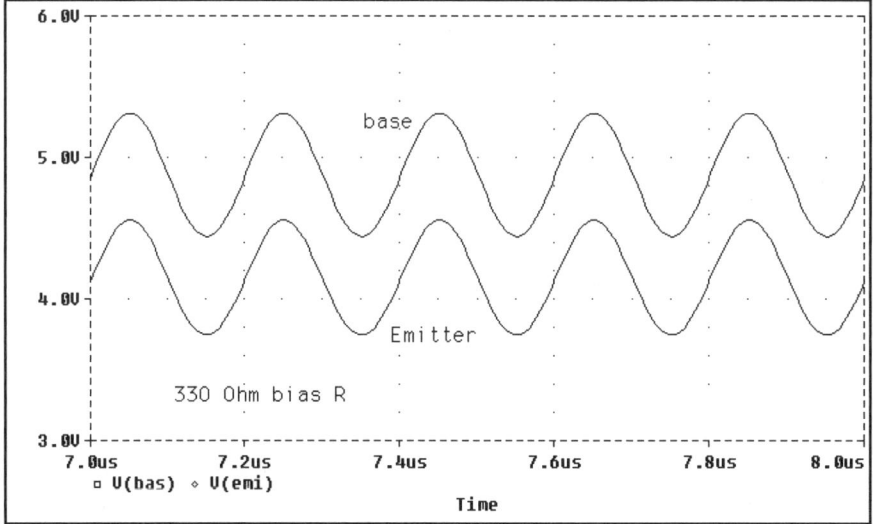

Fig 2.33—Follower output waveforms after increasing the standing bias current.

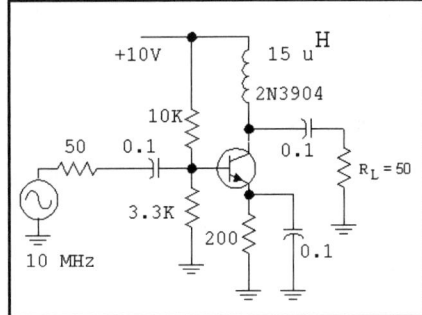

Fig 2.34—A class A amplifier.

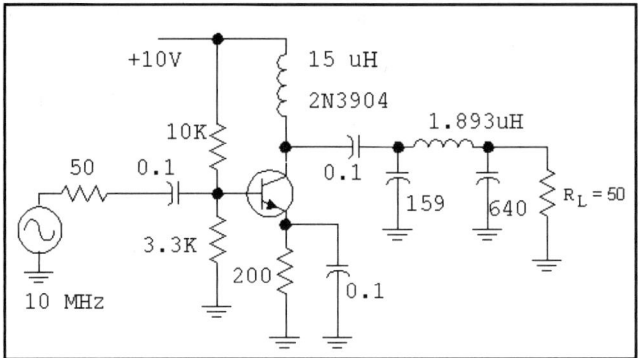

Fig 2.35—The class A amplifier is modified with output impedance transformation for higher output power.

frequency β at F_t/F, placing β at 30. This sets input resistance of about 100 Ω, which predicts that about 2/3 of the open circuit input voltage will appear at the base. An input signal of 10 mV peak produces about 6.7 mV on the base. Applying the small signal voltage gain, the output will be 105 mV peak. Perhaps of greater interest, the load current for this output is 2 mA peak. The transistor collector current varies from the quiescent (no-signal) value of 8.2 mA up to 10.2 mA and down to 6.2 mA. While small signal characteristics are preserved, the output current is already becoming a sizable fraction of the DC bias current.

A characteristic found with the present circuit that we did not see in earlier amplifiers results from the use of a collector RF choke. The inductor has the properties of a constant current source. As a dc current is established in the inductor, the action of the inductor "tries" to maintain that value. This allows the collector voltage to exceed V_{cc}, which never occurred when a collector resistor supplied bias current. This is shown in plots which follow.

We now increase the input drive to 50-mV peak. This is a five times increase over the 10-mV case, so we expect a similar increase in both the output voltage and current if small signal conditions are preserved. Measurements and computer simulations both confirm this general behavior, although the output signals depart considerably from sinusoids. Output voltage across the load is about 0.5 V peak. Collector current drops almost to zero at one point in the cycle but reaches a maximum of about 19 mA, about twice the bias value. Distortion is severe.

The amplifier with 0.5-V drive is current limited, for the current drops to zero at one point in the drive cycle. However, the voltage excursions are still small. The output power with a 50-Ω load is about 2.5 mW.

Consider changes in load resistance seen by the collector. If we maintain drive at 0.5 V peak, the collector signal current

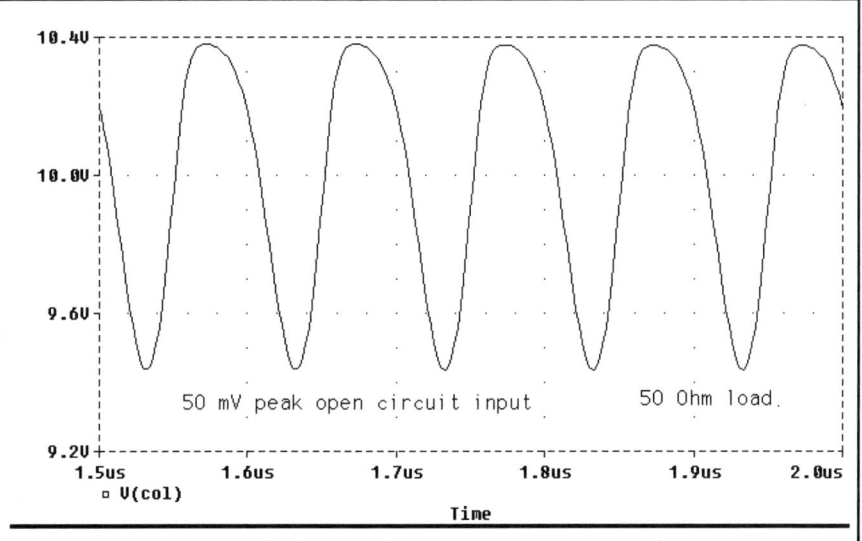

Fig 2.36—50-Ω termination on the class A amplifier.

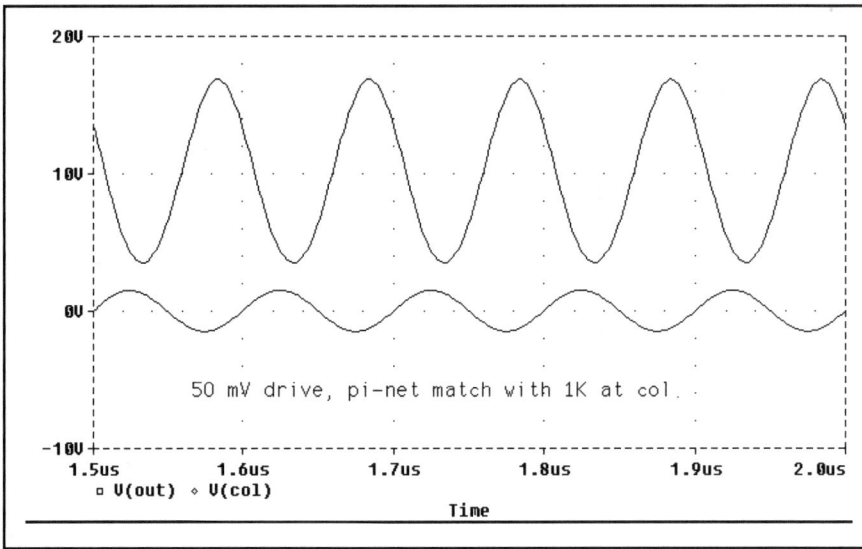

Fig 2.37—Collector (upper) and output load (lower) voltages with the pi network output circuitry.

will be the same. Output voltage can, however, increase as R_L grows. To obtain the maximum power output, we wish to pick a load that allows the collector voltage to drop nearly to the base value (saturation) while going an equal distance above V_{cc} at the opposite part of the cycle. This voltage excursion should occur as the current varies from twice the bias value down to zero. The load resistance that allows this is

$$R_L = \frac{(V_{CC} - V_B)}{I_E} \qquad \text{Eq 2.22}$$

where I_E is the dc bias value. A more familiar form expresses the load in terms of a desired output power,

$$R_L = \frac{(V_{CC} - V_B)^2}{2 \cdot P_{out}} \qquad \text{Eq 2.23}$$

where R_L is the load resistance in Ohms, V_{cc} is the power supply, V_B is the DC base voltage, and P_{out} is the output in Watts. This form applies to Class B and C amplifiers as well as the class A amplifier under discussion.

Application of **Eq 2.22** predicts a load resistance of just over 1000 Ω for maximum output. Changing the load to 1 kΩ in the circuit produces a 10-MHz output of 11 V peak-to-peak corresponding to a power of about 16 mW. Even larger resistance would have produced voltage limiting, so this is close to optimum.

More often than not, 1000 Ω is not the impedance that the designer wishes to use as a termination for the amplifier just designed. Rather, he or she wishes to measure the amplifier output with 50-Ω instrumentation and perhaps drive other circuits with a 50-Ω impedance. The solution is found in **Fig 2.35** where an impedance transforming π-network is inserted between the 50-Ω load and the collector. This network makes the termination "look like" 1000 Ω at 10 MHz. It also has low pass filtering characteristics, attenuating energy at 20 MHz, 30 MHz, and higher harmonic frequencies. **Fig 2.36** shows the collector waveform when the 50-Ω load is connected directly to the collector. The waveforms after matching are shown in **Fig 2.37**.

2.4 GAIN, POWER, DB AND IMPEDANCE MATCHING

Audio and other low frequency amplifiers are easily analyzed with the low frequency models used for biasing. But most of our interest is in higher frequencies where measurement difficulties persist. These encourage us to consider power instead of the voltages and currents that dominate the view of the circuit theorist. This emphasis is an integral part of RF design and forms the basis for this section.

The emphasis on power measurement goes back to early methods. Power at radio, microwave, and even optical frequencies was measured using a Bolometer. The Bolometer is based upon temperature measurements. A resistive load is embedded in a thermally well-insulated chamber. The application of RF power causes a temperature increase, which can be detected with a thermometer. But, the same increase in temperature can be produced with application of direct current. Measurement of the direct current and related voltage then provide a very fundamental determination of the RF power.

The other reason we are concerned with power is that it is power and not voltage or current that is more fundamental. Power is the rate that energy is transferred, whether it be a rate of dissipation, such as the power that becomes heat in a resistor, or the rate that energy may pass through a surface, such as the rate that a radio or light wave passes through a plane. That plane could well be the capture area of an antenna. The unit for power is the watt (W), or joules per second. We are more familiar with it being the product of current and voltage.

An amplifier application is presented in Fig 2.38 consisting of a voltage source with related source resistance, the amplifier, and an output load. While 50 Ω is common for both the source and load, this is certainly not necessary. But, if power is to be measured, we must have some resistance, for a voltage across an open circuit provides no power.

Consider the simple circuit of **Fig 2.39** consisting of a voltage source, V, and a source resistance, R_S. We will terminate this in a load R. Ohms Law provides the net current, while voltage divider action gives the voltage across the load, yielding the power

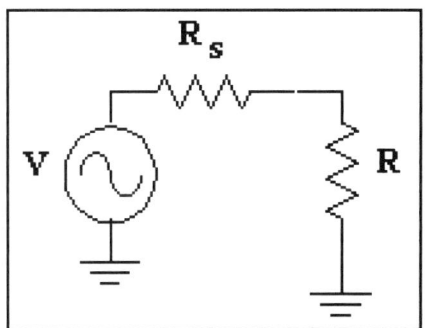

Fig 2.39—A voltage with a source resistance R_S delivers power to a load R.

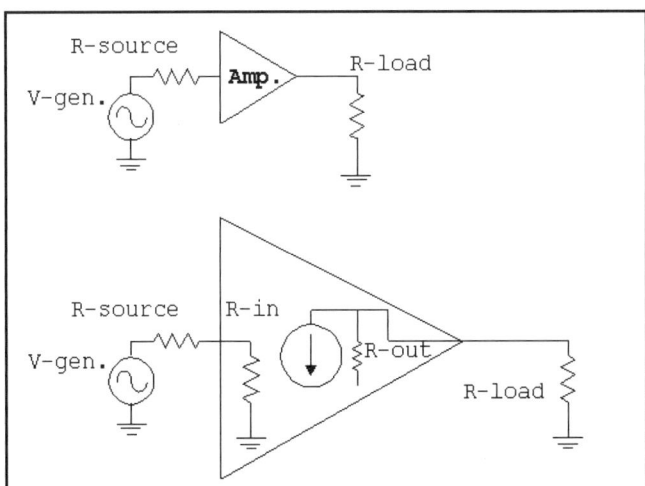

Fig 2.38—Basic amplifier with resistive input and output impedances.

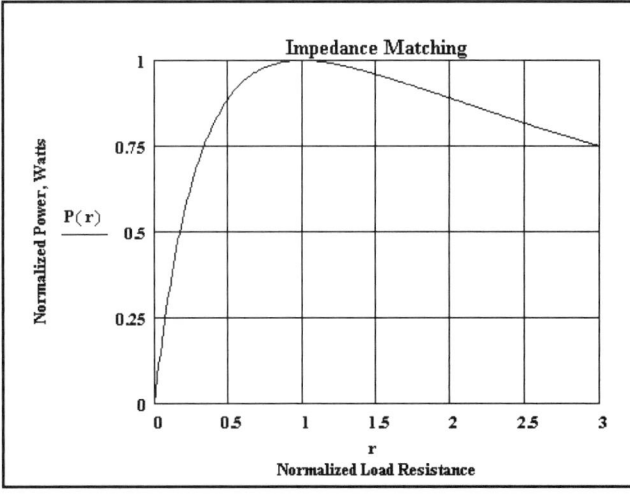

Fig 2.40—Power delivered to the load is maximum when the load resistance equals that of the source.

$$P = \frac{V^2 \cdot R}{(R_S + R)^2} \qquad \text{Eq 2.24}$$

A plot of power vs R is given in **Fig 2.40** where we have normalized the curve. The maximum power is shown as 1 and the normalized resistance, defined as $r = R/R_s$, is 1 when power is maximum. This is the familiar result that the maximum power transfer occurs when the load resistance, R, equals that of the source, R_s. We then say that the source is *matched* to the load. In the general case, the source impedance can have a reactive part. Then, maximum power transfer occurs when the load is a complex impedance with the same resistance as that in the source impedance.

When a generator voltage and the related source resistance are specified, the power extracted when the generator is terminated in a matched load is called the *available power*, for it is the maximum power that is available from that generator.

The amplifier of Fig 2.38 has an input resistance, R_{in}, and an output resistance, R_{out}. The rest of the amplifier is modeled with a controlled current generator. The amplifier will be matched at the input when $R_S = R_{in}$. The output is matched with a load $R_L = R_{out}$. Picking these source and load resistances will produce this perfectly matched amplifier. While it sounds easy enough, it can be very complicated in a practical RF application. In a practical amplifier R_{in} will depend upon the load, R_L, while R_{out} will depend on R_S. Eventually stability becomes a dominating issue. Circuits that are unconditionally stable can eventually be matched perfectly at both input and output.

Source and load resistances are not changed directly as a means of achieving matched conditions. Rather, a 50-Ω generator might be applied to an impedance transforming network that presents a different impedance to the amplifier input. These networks are discussed in greater detail in Chapter 3.

We always are interested in the "gain" of an amplifier. This usually means *power gain*, which is the ratio of two power levels. With a known source voltage, V, and source resistance, R_S, and the modeled input resistance R_{IN} (from Fig 2.38), we can calculate the input power. Output power can also be calculated when the amplifier is well modeled. Knowing the powers, the *power gain* is:

$$G_P = \frac{P_{OUT}}{P_{IN}} \qquad \text{Eq 2.25}$$

The maximum possible gain is that occurring when both input and output are matched.

The power gain of Eq 2.25 is rarely measured directly. Instead, we more often measure or calculate *transducer gain*, first mentioned in Section 2.2. Transducer gain is:

$$G_T = \frac{P_{OUT}}{P_{AV}} \qquad \text{Eq 2.26}$$

where P_{out} is the power delivered to the load and P_{AV} is the power *available* from the source. Power gain and transducer gain are equal in a perfectly matched amplifier. A variant of transducer gain is the *insertion power gain* obtained when a transmission line is broken, and an amplifier is inserted. This occurs when both R_S and R_L are identical, usually 50 Ω.

The Decibel, or dB.

Gain can be expressed as a numeric ratio, but is more often specified in decibels, given by

$$dB = 10 \cdot \text{Log}\left(\frac{P_1}{P_2}\right) \qquad \text{Eq 2.27}$$

where P_1 and P_2 are two different powers. If an amplifier has a 5 mW output and is being driven by a generator with an available power of 1 mW, the power ratio P_{out}/P_{AV} is 5, for a transducer power gain of 7 dB.

The dB construct was not invented to confuse the prospective designer. Rather, it is a natural consequence of the mathematics. Output power is calculated from an input power and a numeric gain by using multiplication. It is also calculated from a dB ratio, but now simpler addition is used.

The dB construct is useful for other comparisons. For example, we might examine the harmonic distortion in an amplifier and find that for a 3-mW drive at 7 MHz, output appears not only at 7 but at 14, 21 and 28 MHz. If the 14-MHz output is less than the 7-MHz output by a factor of 500, we say that the 2nd harmonic is 27 dB below the fundamental. The 7-MHz component is often regarded as a carrier and the 14-MHz component is then said to be at –27 dBc where the "c" indicates dB with regard to a *carrier* or reference power.

Another often used variation of the dB ideal occurs when a power is referenced against a standard of *one milliwatt*. We then say that the power is in dBm, meaning power referenced to *one mW*. This does *NOT* depend upon impedance. The dBm values will be positive or negative depending on their relationship to 1 mW. A one watt QRP transmitter has an output of 1000 mW or +30 dBm. But a strong received signal from the terminals of an antenna might be at one microwatt, 30 dB below the 1 mW, or at –30 dBm.

Many instruments are calibrated in dBm. The dBm output of a signal generator is a measure of the *available* output power of the generator. *Available power*, discussed above, was the power actually transferred for the single case when the load matched the source. It is common for the output to be specified in a 50-Ω system, a common RF standard. A signal generator set up for an output of +10 dBm will deliver that power to a 50-Ω load attached directly. It will also deliver that power to a 200-Ω load if an appropriate 2:1 turns ratio transformer is placed between the load and the generator.

RF detection instruments, such as RF power meters or spectrum analyzers, are also calibrated in dBm. These instruments usually have a 50-Ω input impedance. They behave like a 50-Ω resistive load when attached to a generator. A 50-Ω signal generator set for an output of –40 dBm should produce an indication of –40 dBm when attached to a spectrum analyzer.

Wideband instruments used for general purpose electronic measurements include wideband voltmeters and oscilloscopes. They usually have high input impedance, typically 1 MΩ. When used with a 10X probe, the input resistance becomes 10 MΩ. The measurement philosophy behind the design of these instruments is to present such a small load to a circuit being measured that the instrument can be ignored. The oscilloscope is usually used in an *in situ*, or in-place measurement. This contrasts with the measurement philosophy of many RF measurements, which use *substitution*. For example, we *substitute* a power meter for the antenna when we wish to measure transmitter output power.

The wideband oscilloscope can be used for measurements in a 50-Ω system, but it becomes vital to establish a well defined input impedance. This is done with a 50-Ω resistive termination. A form that can be built for the home lab is shown in **Fig 2.41**, while a photo in **Fig 2.42** shows a home-built version and a couple of commercial terminators. The commercial models are built with low inductance disk resistors that offer higher bandwidth than can be easily achieved with leaded parts in a homebuilt box.

Gain measurements in a 50-Ω environment are straightforward with the terminated oscilloscope and a signal generator. The generator is first attached directly to the terminated oscilloscope with a length of coaxial cable. The 'scope response is noted, and power is calculated to be sure

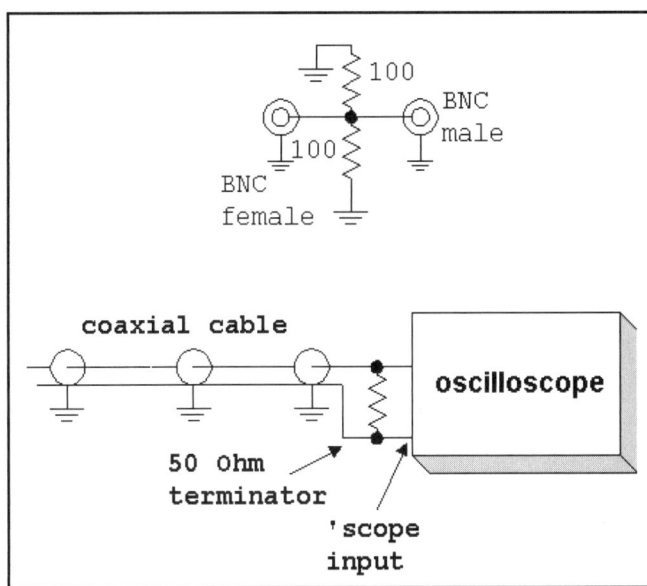

Fig 2.41—Terminators for oscilloscope input loading. See Chapter 7 for additional detail on power measurements.

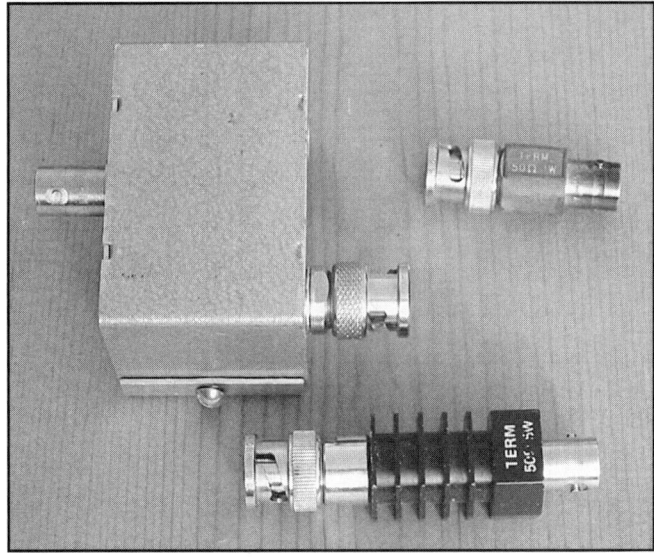

Fig 2.42—Homebrew and surplus terminators.

that this is not too large for the amplifier. The cable is then disconnected, the amplifier is attached, another section of cable is inserted to connect to the instrumentation, the amplifier is powered, and the new response is noted. The response will (hopefully) be larger than it was without the amplifier in place.

Several approaches can be used to determine gain. The first would be to measure the new voltage with the terminated oscilloscope and then calculate a new output power. The transducer gain then becomes 10 Log (P_{out}/P_{AV}). This scheme works well with a calibrated oscilloscope operating within it's bandwidth.

The alternative method removes all need for oscilloscope calibration and accurate response at the test frequency, but places a greater burden on the signal generator. The reference is first established with the signal generator attached directly to the oscilloscope. The response is noted, as is the output setting for the generator. The amplifier is then inserted in line, and the signal generator output is reduced until the 'scope response is exactly the same as noted earlier. The new generator output is examined and found to be lower than the original. The difference in generator settings in dB is then the transducer gain.

Gain can still be determined, even if the signal generator is not calibrated. A step attenuator is inserted in the generator output. Attenuation is increased when the amplifier is placed in the system until a reference 'scope response is duplicated. The attenuator difference is then the gain.

The oscilloscope can, of course, be used with a 10X probe to study the amplifier. Output power can be measured from a voltage determination at a load on the amplifier output. But amplifier input power is not defined when the input impedance is unknown. Although common, it is rarely valid to merely measure a voltage ratio to calculate a power or transducer gain.

Measures of impedance match and mismatch

In Fig 2.40 we saw that the power transferred from a source to a load depends upon the match between the two. This curve has a symmetry that is not immediately obvious. Although the power transferred from the source to the load is 100% only when the match is perfect, the degree of match depends only on the ratio of one resistor to the other without regard to which is larger. That is, if the source is 50 Ω, we see that power transfer is 88.9% effective for loads of either 25 Ω or 100 Ω. Similarly, 12.5-Ω or 200-Ω loads produce 64% power transfer and so forth. The ratio of these resistances to 50 Ω (always with the larger number taken) is called the *voltage standing wave ratio*, or VSWR.

The term VSWR arises from transmission line behavior and it relates to voltages measured along a transmission line that is not matched. While we can do this measurement with RF volt meters and suitable transmission lines, this is not the way we usually measure the degree of impedance match. (Actually, some microwave experimenters still do just this measurement.) Rather, we perform bridge measurements of a related term called *voltage reflection coefficient*, often signified by the Greek letter Gamma, Γ. Gamma is given for resistive loads,

$$\Gamma = \frac{R - R_0}{R + R_0} \qquad \text{Eq 2.28}$$

where R_0 is the reference resistance. In the examples we have discussed, R_0 would be the source resistance while R is the load. Gamma is related to VSWR through

$$\text{VSWR} = \frac{1 + |\Gamma|}{1 - |\Gamma|} \qquad \text{Eq 2.29}$$

where the bars around Γ indicate that only the magnitude of Γ is used. In the general case, Γ has both magnitude and angle, corresponding to complex impedance with both resistive and reactive parts. Also, the more general form of Eq 2.28 uses complex impedance to define Gamma, $\Gamma=(Z-Z_0)/(Z+Z_0)$.

Fig 2.40 showed power transfer effi-

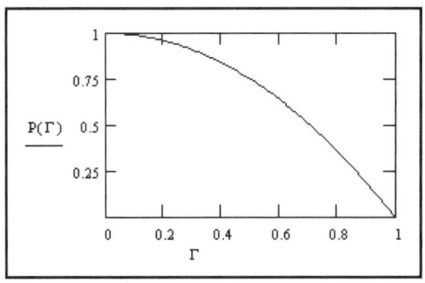

Fig 2.43—Power transfer related to reflection coefficient.

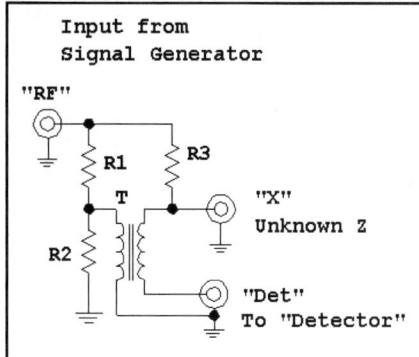

Fig 2.44—Return Loss Bridge. All resistors are normally 50 Ω.

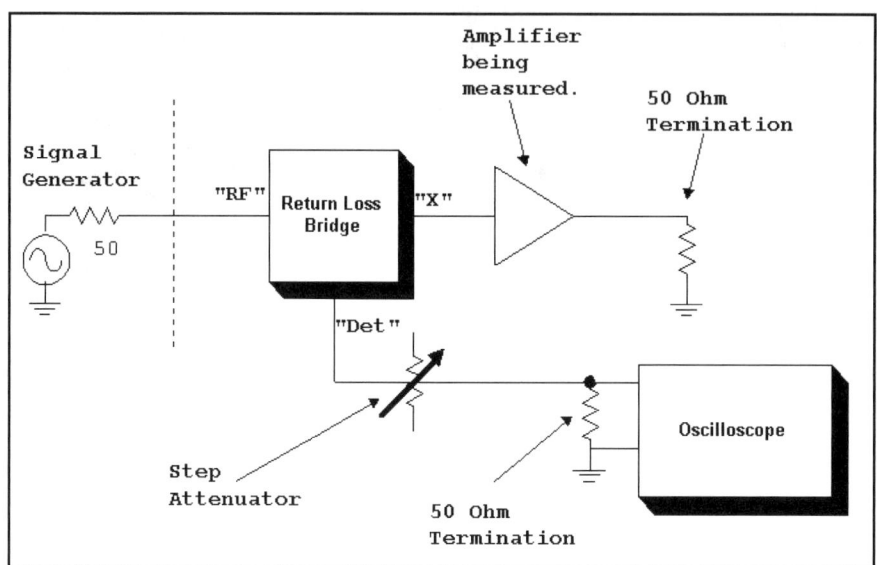

Fig 2.45—Using a return loss bridge with an amplifier.

ciency as a function of the terminating resistance. A similar plot is given in **Fig 2.43** where power is now plotted against reflection coefficient, Γ.

Although reflection coefficient, Γ, may seem like an esoteric impractical parameter, it is easily measured (in magnitude) using a simple apparatus that can be built in the home lab. This circuit, shown in **Fig 2.44**, is called a *return loss bridge*, or RLB. The three resistors in the bridge are 50 Ω when building a bridge for use in a 50-Ω system. The signal generator is assumed to then have a 50-Ω impedance as well. The transformer is a *common mode choke* (see Chapter 3.) Construction is discussed in Chapter 7.

The bridge action occurs because all resistors are 50 Ω. Assume that the "X" port, the unknown, is terminated in 50 Ω. Then half of the voltage applied at the "RF" port appears at the junction of R1 and R2. But half also appears at the "X" port. The voltages are equal on either side of the common mode transformer, so no signal appears at the detector. In contrast, a larger signal appears when the unknown "X" port is either open or short circuited.

Use of the return loss bridge is presented in **Fig 2.45**, where an amplifier input will be measured. The bridge is first open circuited at the "X" port, and the detector response is noted. Then, a 50-Ω terminator is placed on the "X" port. A large decrease in detector response should be noticed. This response is a measure of how well the RLB is functioning and is called the bridge *directivity*. An amplifier (power on) is now attached to the "X" port through a coaxial cable, and a terminator is attached to the amplifier output. The detector response will be lower than the level present with the "X" port open circuited by a ratio called the *return loss*, a dB value. The step attenuator in the detector can be adjusted to attenuate the reference to better measure return loss.

Return loss is related to Γ through

$$R.L. = -20 \cdot \text{Log}(|\Gamma|) \qquad \text{Eq 2.30}$$

The inverse form is

$$\Gamma = 10^{\frac{-R.L.}{20}} \qquad \text{Eq 2.31}$$

While we have illustrated the RLB with oscilloscope detection, a 50-Ω power meter or spectrum analyzer is preferred. Both are described in Chapter 7. These are 50-Ω instruments, so they do not require the external terminator so vital to the oscilloscope. The 'scope suffers from two problems that compromise this application. First, it is a wideband instrument, so noise limits the sensitivity, making it difficult to see the weak signals that are readily seen in a spectrum analyzer. Second, many of the terminations that we might measure are narrow bandwidth loads. As such, they will produce high return loss at one frequency, but not at the harmonics. The usual signal generator is harmonic rich. The harmonics are resolved and, hence, ignored in a spectrum analyzer measurement.

2.5 DIFFERENTIAL AMPLIFIERS AND THE OP-AMP

The differential amplifier, or *diff-amp*, is the foundation for most silicon analog integrated circuits in use today, making it a very important topology. Here we investigate differential amplifier fundamentals and examine a major derivative of it, the operational amplifier, or *op-amp*.

Following the name, the differential amplifier is a circuit intended to amplify a difference. The differential amplifier has two input terminals. The output, which can be between two collectors or from just one, is then proportional to the voltage (or current) difference between the inputs. The basic differential amplifier using NPN bipolar transistors is presented in **Fig 2.46**.

We start with two identical transistors biased at the same dc base voltage. The two emitters are attached and returned to ground through a common resistance, as in Fig 2.46A. Two identical collector resistors are attached, biased from a common supply. This circuit can have signals applied in two ways. If the two bases are driven together, the composite circuit would behave as one transistor. The two collector signals would then be identical. This operation is called *common-mode* drive or excitation. The large emitter resistor becomes a degeneration element, causing the common-mode gain to be low.

The other diff-amp drive is the *differential-mode*, where one base is driven in one direction while the other is driven by an opposite polarity. Assume that Q1 and Q2 are biased with a dc base voltage of 5. The voltage at the common emitter is then 4.4. Total current will be 4.4 mA for an emitter resistor of 1 kΩ. If the two transistors are identical, each will be biased to an emitter current of 2.2 mA. We now apply a differential signal causing V_{b1} to increase by 10 mV while V_{b2} drops by an equal 10 mV. The emitter voltage remains essentially constant. V_{c1} decreases while V_{c2} increases by an amount related to the gain. A useful property of this circuit is that total current does not change with differential drive.

Fig 2.46, part B shows the circuit variation found most often in integrated circuits where the emitter resistor is replaced by a third transistor. Set V_{b3} to 2 volts and pick the Q3 emitter resistor for the same 4.4 mA. This leaves bias conditions for Q1 and Q2 as they were, although the common mode gain is even lower.

Q3 is a constant current source, a circuit that acts as if the bias for Q1 and Q2 came from a very large negative power supply with an equally large resistor. The effect of this topology is to force the sum of the currents in Q1 and Q2 to remain constant. This has two important consequences.

First, a differential amplifier is very easy to decouple. With constant total current, signals are not injected onto the V_{cc} power supply, very important when the diff-amp is one of many such circuits within an IC.

The other consequence of the constant current source is that drive applied to just one input will result in differential output signals. This is shown in the amplifier of **Fig 2.47**. The two collector voltages have equal amplitudes and are out of phase with each other.

Although differential amplifiers are abundant in integrated circuits, they are also useful and practical in discrete form. **Fig 2.48** shows a diff-amp with readily available parts that might be used to provide balanced local oscillator drive to a mixer without transformers. This circuit is

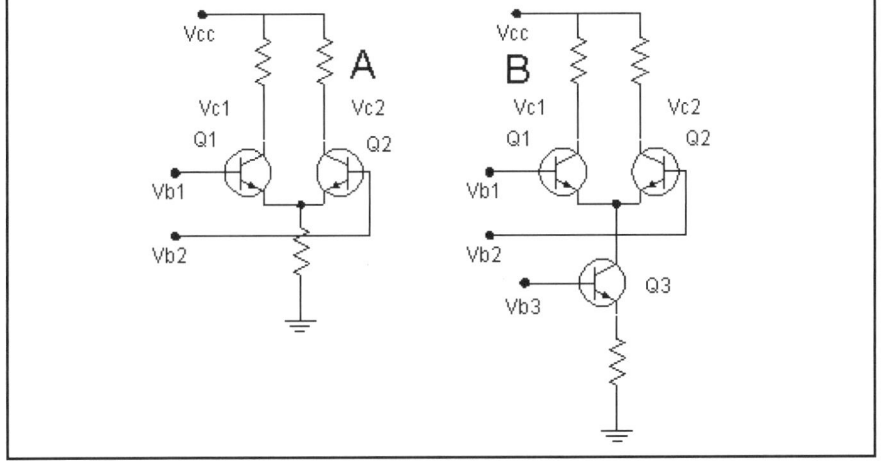

Fig 2.46—Differential Amplifiers.

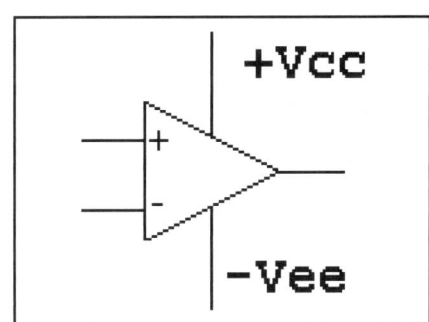

Fig 2.49—Schematic diagram for an operational amplifier.

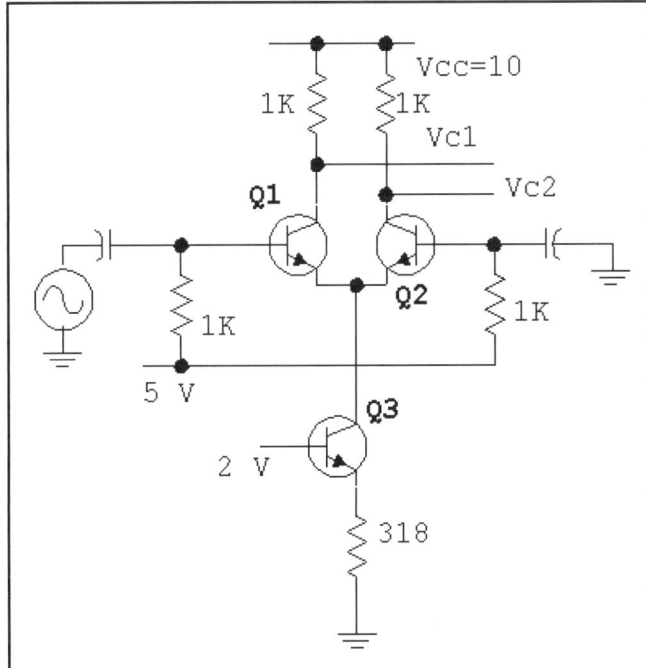

Fig 2.47—Differential Amplifier that converts a single ended signal into a differential one having two outputs with a differential relationship. The 2 and 5-V points are fixed voltage, usually generated within the IC containing this differential pair.

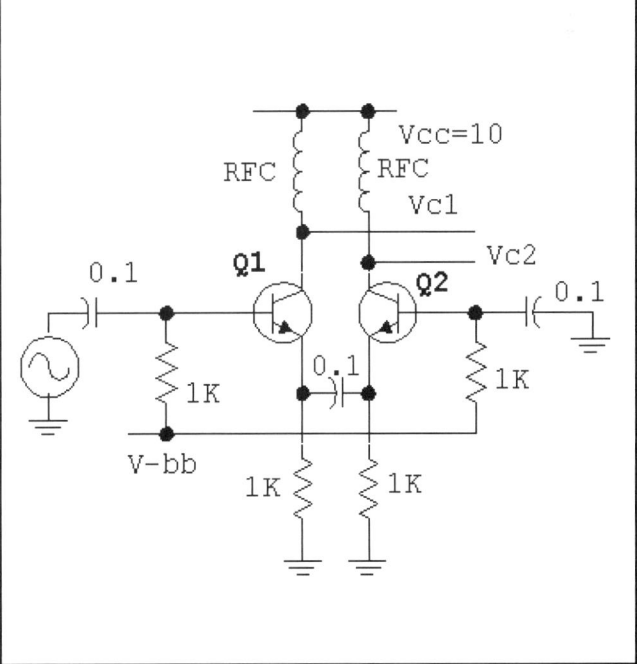

Fig 2.48—Differential Amplifier built with discrete components. The emitter resistors are adjusted for equal current in the two transistors. V_{bb} represents a base bias power supply, which could be a simple voltage divider from the higher supply.

Amplifier Design Basics   2.17

useful because it provides a balanced output with reduced even order harmonics as well as power gain. The use of two emitter resistors eases the need to have identical transistors.

Having examined properties of the diff-amp, we will now look at the "ultimate" diff-amp example, the operational amplifier. An op-amp is shown schematically in **Fig 2.49**. The internal circuitry can be rather complicated; familiar examples such as the 741 or 358, will include a dozen or more transistors while high performance variants will have many more.

The operational amplifier (Fig 2.49) is shown with two power supplies, although virtually all can be used with a single supply. The basic operation is, in some ways, exactly like the simple diff-amps discussed above. The op-amp has two inputs just as the diff-amp has two base inputs that affect their outputs. The usual op-amp, however, has just one, single ended output. Moreover, the output voltage can be either above or below the input voltages.

The usual op-amp has several gain stages, all cascaded with the output of one feeding the input of the next. As such, the low frequency voltage gain is often very high with values ranging from 50,000 up to over one million. While op-amp gains are often expressed in dB (using the familiar 20*LOG(Vout/Vin) formula), this is often incorrect. **The dB form only pertains to power ratios.** The equation relating voltage ratio is valid only when terminating impedances are equal.

A typical op-amp can provide output voltages from near the negative power supply up to within a volt or two of the positive supply. The inputs can also occur at a wide variety of voltages. A 741 op-amp will work with inputs that are from about V_{ee} +2 to V_{cc} −2. This span is called the *common mode input range*. Op-amps using PNP bipolar input transistors can have a common mode input range that extends all the way to the negative supply. Examples include the LM-324 and LM-358, which are especially useful with single power supplies.

Assume that the "−" input in Fig 2.49 is constant at ground with power supplies of +15 and -15 volts. Set the "+" input several volts negative. The output will then be very negative, as low as it can go. As the "+" input is increased, the output remains negative until the input gets close to ground. Then, the output will start to increase very quickly. The output goes above ground as the "+" input becomes just a few millivolts positive. The voltage gain may be evaluated from a curve of the output vs the input. With even modest inputs, the output reaches the positive power supply, or "rail." The "+" input is called the *non-inverting input* for the output polarity follows it in direction.

Circuit operation is similar if the non-inverting ("+") input is grounded and the positive going signal is applied to the "−" or inverting input, except that now the output moves in the opposite direction. That is, the output makes a transition from the positive power supply to the negative one. Repeating these experiments at reference voltages other than ground shows that the output depends only upon the **voltage difference** between inputs.

The input transistors for most op-amps are biased for low current operation, causing the input impedance to be quite high. We usually neglect R_{in} during the analysis of op-amp circuits.

Op-amps are rarely operated "open loop," as described above. Instead, they are used with negative feedback. This is illustrated in **Fig 2.50**. Power supplies are omitted in the op-amp circuits that follow, but are assumed to be + and − 15 volts.

Assume initially that the "+" input for Fig 2.50 is at ground. If the output was at a different voltage, the inverting input would then be at a level other than ground. This would then produce a difference voltage at the inverting input that forces the input toward ground.

Increase the non-inverting input to +1 volt. Similar arguments show that the output increases until the inverting input is also at +1 volt. The circuit of Fig 2.50 is a voltage follower with a gain of +1. The value of the feedback resistor is of no consequence for this circuit, for the input current is very small. (A practical unity gain follower normally has the output shorted to the inverting input.)

The modification in **Fig 2.51** adds an equal valued resistor from the inverting input to ground. Setting Vin to 0 forces the output to ground. However, when we set the input to +1 volt, we find that the output moves to +2 volts. Our circuit now has a non-inverting gain of 2. This is confirmed through voltage divider action. The voltage at the "−" input must be half of that at the output; a voltage other than +1 at the "−" input would produce an input difference that would move the output.

Fig 2.52 shows an inverting amplifier. The "+" input is grounded with an input applied to a resistor attached to the inverting input. We start with the amplifier input at ground. The output must then be at ground. Increasing the excitation to +1 volt causes the inverting input to "try" to go positive, an action that is inverted with gain in the op-amp. The system is in equilibrium when the output is −1 volt. The amplifier then has an inverting gain of 1.

A general behavior has emerged from this discussion, easing further analysis: *Negative feedback around an op-amp always has the effect of forcing the two inputs to have the same voltage.* This can be used to derive the usual formulas for gain of closed loop amplifiers. The char-

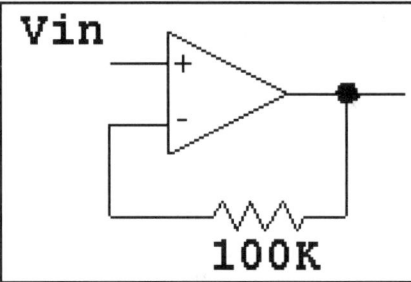

Fig 2.50—A unity gain follower.

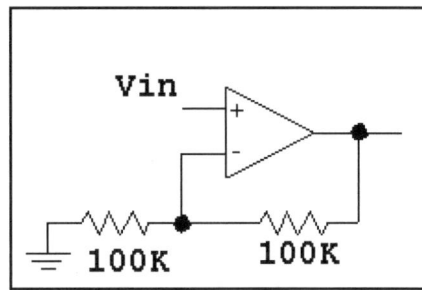

Fig 2.51—A follower with a gain greater than unity.

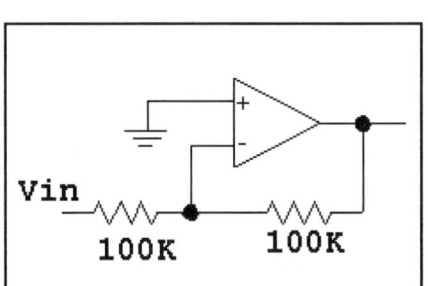

Fig 2.52—An inverting amplifier with unity gain.

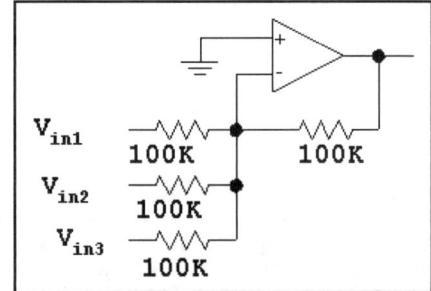

Fig 2.53—A summing amplifier with three inputs.

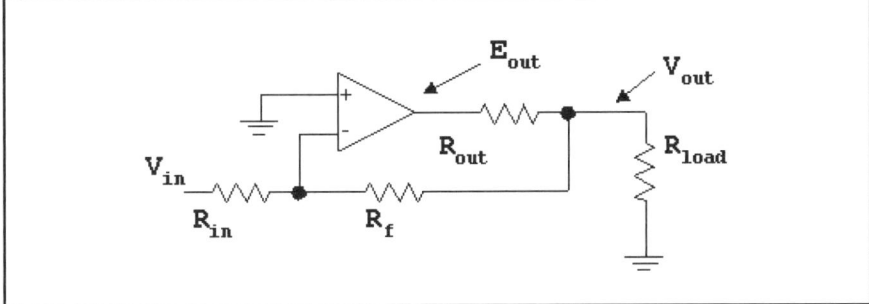

Fig 2.54—Feedback reduces an output resistance.

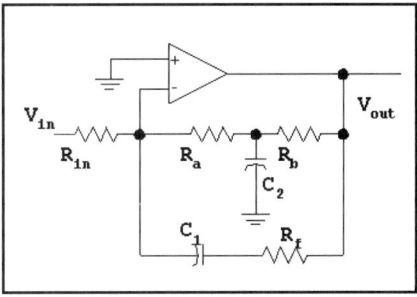

Fig 2.55—The Ra-Rb-C2 network establishes DC bias with little impact on AC gain. C1 and the related resistor then set AC gain. If C1 has a small reactance compared with its series resistor, the gain will grow with increasing frequency.

acteristic is maintained so long as all inputs and outputs are maintained within the allowed ranges.

The inverting input of a closed loop amplifier is often described as a "summing node," illustrated in **Fig 2.53** with three inputs. All three have the same input resistor values, so the gain for each input is the same at −1. This circuit is sometimes referred to as a "mixer" in audio circles, although the term *mixer* has a much different meaning for the RF experimenter. Analysis is direct. The feedback resistor maintains the two op-amp inputs at the same voltage, which is ground in this example. Any single input will change the output accordingly while feedback keeps the summing node at ground. We calculate the current entering the summing node for each input and note that the total current into the summing node, including that from the output via the feedback resistor, must be zero. This defines the output response.

A highly useful effect of negative feedback is that of altered impedance. The zero voltage difference at the inverting amplifier of Fig 2.52 tells us that the voltage at the "−" input is essentially zero. There is, however, signal current flowing into the node. The effect of the feedback is to reduce the impedance at that node to near zero.

Feedback also decreases output resistance. **Fig 2.54** shows an ideal op-amp with an added output resistance, R_{out}. Feedback is extracted from the output end of this resistor. Because V_{out} drives the feedback resistor, it is this point (V_{out}) that is controlled by the feedback element, R_f. Changing the load (R_{Load}) may have impact on E_{out}, the op amp direct output, but it has little effect on V_{out}; the output impedance at V_{out} is very low, a result of the feedback.

The effects of feedback from a parallel resistor are most dramatic with op-amps where the open loop gain (that gain without feedback) is very high. Negative feedback is also useful in single stage amplifiers using but one transistor. The effects are similar; parallel negative feedback reduces gain, making it depend primarily on resistor values, and reduces both input and output impedance. Not all forms of negative feedback reduce impedance. Emitter degeneration in a transistor amplifier increased amplifier input R as it reduces gain.

Placing capacitors (or inductors) in a feedback path will force the amplifier gain to depend upon frequency. An example is presented in **Fig 2.55** where C_1 causes gain to be lower at high frequencies. C_2 has the effect of allowing R_A and R_B to set DC conditions with little effect on gain for AC signals. But, this must done with care to avoid stability problems.

2.6 UNDESIRED AMPLIFIER CHARACTERISTICS

The ideal amplifier is linear with an output that is an exact replica of the input with the only difference being greater amplitude and a phase difference. There should be no other output frequencies. If two inputs are applied to an ideal linear amplifier, the result will be two outputs, each being just what would be seen if each input was applied alone, with nothing else added. Several phenomena compromise amplifiers from this ideal. They include noise, gain compression, harmonic distortion, and intermodulation distortion.

Noise in Amplifiers

Noise is a familiar corruption in an amplifier. The noise of concern is not what we most often hear coming from our HF receivers; that noise generally arises from thunder storms somewhere in the world, or power lines somewhere in our community. Rather, we are concerned with the noise that is generated within the circuitry. The dominant component of this noise, so called thermal noise, originates from random motion of the electrons within a conductor. This noise shows up as a voltage that appears between the two conductor ends. The available power present is kTB (in watts) where k is Boltzman's constant, T is absolute temperature in Kelvin, and B is the bandwidth we use to observe the noise. Although a power kTB is available from any conductor, the related voltage is very small if the conductor is a good one. A resistor, a conductor with larger resistance, allows a larger voltage to appear, but with the same available power. (*Available power* was discussed in an earlier section.)

Fig 2.56 shows a simple amplifier terminated in 50 Ω at both input and output.

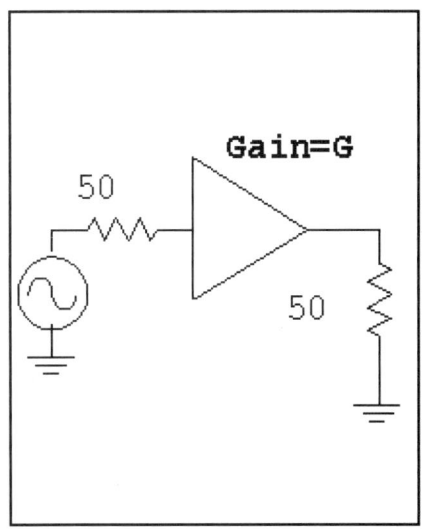

Fig 2.56—A terminated amplifier used for noise analysis.

Amplifier Design Basics 2.19

The source and load resistances generate noise. The noise generated by the output load is normally ignored during a noise analysis of the amplifier, for the circuit designer is primarily concerned with the *available* noise from the amplifier. The noise from the input source is increased by the amplifier gain, just as any signal would be increased. There is nothing that can be done to avoid this noise. If the amplifier available power gain is G and the available noise power from the input source is N_i, the output noise will be $G \times N_i$, even when the amplifier is perfect and noiseless.

A real world amplifier will have a noise output that is even higher than the amplified input noise. The output noise is greater by a ratio that we call the *noise factor* or *noise figure*, designated by F. The logarithmic form of *noise Figure* is NF(dB)=10*Log(F). The two forms, algebraic ratio or dB, are used interchangeably, although the algebraic ratio is used in all of the equations that follow. The extra noise is that generated within the active device and circuit components.

A formal treatment of noise[4] deals with noise power ratios. Noise factor is given by,

$$F = \frac{N_{out}}{G \cdot N_{in}} \qquad \text{Eq 2.32}$$

where N_{OUT} is the output noise power delivered to the load, N_{IN} is the noise power available from the input resistance, and G is the available power gain of the circuit. N_{IN} is the noise power available from the source resistance when it has a temperature of 290 K. NF is the ratio of two noise powers. The larger number (numerator) is the noise actually coming from the amplifier while the smaller (denominator = GN_{IN}) is the noise that would be coming from the amplifier if it generated no noise of its own. A perfect, noiseless amplifier would have F=1 from the equation, or converting to dB, NF=0 dB.

Gain, G, is the power gain we normally associate with an amplifier: output signal power delivered to the load, S_{OUT}, divided by S_i, an input signal power. If we insert this gain ratio into the noise figure defining equation, and rearrange the terms, we obtain

$$F = \frac{S_i / N_i}{S_{out} / N_{out}} \qquad \text{Eq 2.33}$$

This describes a combination of signal and noise. Essentially, noise figure can be interpreted to be a degradation in signal to noise ratio as we progress through the amplifier. This equation can be rearranged to

$$F = \frac{S_{in} \cdot N_{out}}{S_{out} \cdot N_{in}} = \frac{G_{NOISE}}{G_{SIGNAL}} \qquad \text{Eq 2.34}$$

where G_{NOISE} is the *noise gain*, the output noise power divided by the available input noise power. G_{SIGNAL} is the familiar signal gain used above. All forms of these equations are used in deriving some of the results we use with noise figure.

Typical NF values range from 1 to 10 dB for the amplifiers that we frequently use in RF systems. Mixers tend to have higher noise figures. Modern FET amplifiers are capable of NF as low as 0.1 to 0.2 dB at UHF with values under 1 dB even possible at 10 GHz.

We frequently ask for the noise factor of a cascade of two amplifiers. This result is

$$F_{NET} = F_1 + \frac{F_2 - 1}{G_1} \qquad \text{Eq 2.35}$$

where F_1 and F_2 are noise factors for stage 1 and 2, respectively, and G_1 is the available power gain for the first stage. While the noise from both stages contributes to the net noise factor, the 2nd stage noise contribution is reduced by the gain of the first stage. Clearly, if we can calculate NF for two stages, we can perform the calculations several times and obtain the result for any number of stages.

Noise figure is a vital amplifier and receiver characteristic at VHF where external noise (thunder storms, etc) is low. While a low noise figure is rarely needed at lower frequencies, it becomes more important when small antennas are used.

Noise figure is also a vital parameter within a receiver, for careful control of noise will allow the designer to use low gain, which keeps distortion low. Details are discussed in later chapters.

Recall that the noise power available from a resistor is kTB. A useful number to remember is that kT = –174 dBm at "room" temperature of 290 K. If the noise was observed in a receiver with a bandwidth of 3 kHz (a voice "channel"), B would be 3000 Hz and 10×LogB is 34.8 dB. The noise power available from the resistor would then be –174dBm + 34.8 dB = –139.2 dBm. A receiver can be thought of as a large amplifier. If the receiver had a 10 dB noise figure, the output noise would be the same as would appear if an input noise of –139.2 dBm + 10 dB = –129.2 dBm was applied to the input of a perfect, noiseless receiver.

The related noise voltage from a resistor is

$$V_N = \sqrt{4 \cdot k \cdot T \cdot B \cdot R} \qquad \text{Eq 2.36}$$

where k is again Boltzmann's constant (1.38×10⁻²³), T is the resistor temperature in K, B is bandwidth in Hz and R is the resistance in Ω. The available power, kT, is called a *spectral power density*, usually in W/Hz. The resulting voltage, V_n, is a *spectral voltage density* in volts-per-root-Hz. Op-amps often have noise specified in terms of an equivalent input *spectral voltage density* of noise. The same method is sometimes used for transistors, although noise figure is the more common parameter used to specify an RF design.

Amplifier noise figure is not always a

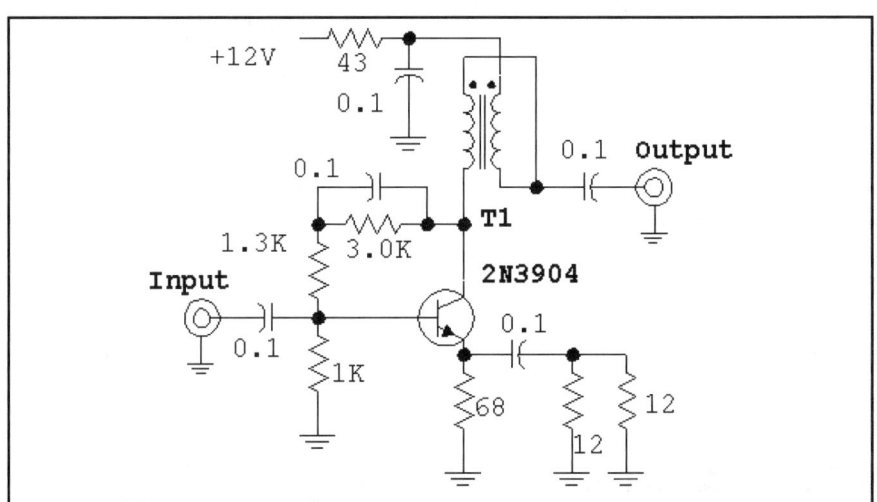

Fig 2.57—Feedback amplifier illustrating gain compression and distortion. This circuit has 20-mA I_c. T_1 consists of 10 bifilar turns on a FT-37-43 ferrite toroid core, although the specific core type is not critical. This circuit features a small signal gain of 20.5 dB and a good impedance match to 50 Ω at both input and output. See text for noise Figure, gain compression, and intercept results.

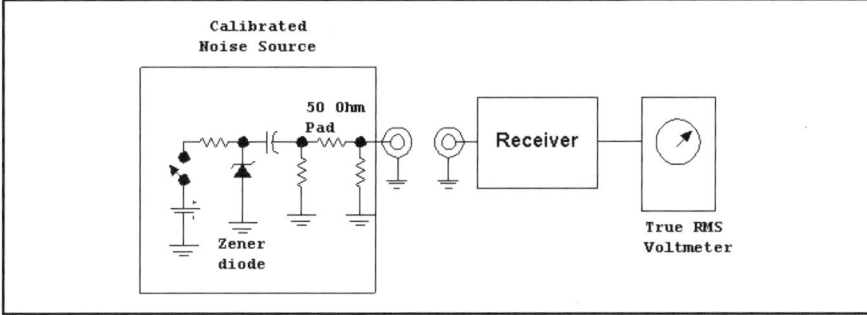

Fig 2.58—Scheme used to measure receiver noise Figure. Audio voltmeter examples are the HP3400A or the Fluke Model 89.

simple constant that may be extracted from a data sheet and applied to a design. Rather, data sheet noise figure is specific to a "typical" amplifier, or more often, is the best NF one can achieve. The noise figure of a specific design then depends upon device biasing and the impedance presented to the device input.

An example amplifier is shown in **Fig 2.57** in connection with our discussion of distortion. This amplifier was measured with an HP-8970 Noise Figure test set as 6 dB at 10 and 20 MHz. The circuit is discussed further as we investigate feedback amplifiers.

The most common method for noise-figure measurement is shown in **Fig 2.58**. This drawing deals with a receiver. However, the same source is used to measure an amplifier by following it with a receiver (or spectrum analyzer). After a measurement of the cascade is obtained, equation 2.35 is used to obtain the NF of the amplifier alone. The critical part of the measurement system is the noise source. The one used here is a Zener diode. When the switch is open, the diode is off. The pad attenuation, if large, forces the output impedance to be close to 50 Ω. When the diode is turned on by closing the switch, the noise increases by a large amount. The noise increase is called the *excess noise ratio*, ENR, and is about 22.5 dB for our noise source, which is described in Chapter 7.

With a 22.5 dB ENR, the noise output of a perfect, noiseless receiver would increase by 22.5 dB when the source is turned on. But the receiver is contributing noise of its own, so the noise *increase* will be less than 22.5 dB. The output increase is called the "Y-factor." Noise factor (a power ratio rather than dB) is related to the ENR and Y by

$$F = \frac{ENR}{Y-1} \qquad \text{Eq 2.37}$$

where both ENR and Y are power ratios rather than dB values. Consider an example:

A 22.5 dB ENR corresponds to ENR=178 as a power ratio. If we measure Y of 19 dB for a receiver, the corresponding power ratio is 79.4. F is then 2.27, or NF=3.6 dB.

Gain Compression

Most non-ideal amplifier behavior occurs at higher powers with a simple example being gain compression. Fig 2.57 showed a typical amplifier that illustrates gain compression and other problems. The circuit is a feedback amplifier with a 20 mA collector current. This circuit, which was built and measured, has migrated into numerous receiver and transmitter applications. No heat sink is needed in normal applications.

Small signal amplifier gain was 20.5 dB. Repeating the measurement at several input powers allows one to plot a graph of gain Vs power. Eventually a point is reached where the gain begins to drop. The output power where the gain is 1 dB below the small signal value is called the 1-dB compression point and occurred at an output of +16.5 dBm.

Harmonic Distortion

A familiar amplifier distortion appears in the form of harmonics. If an amplifier is driven at one frequency, amplifier non-linearity generates a distorted output. That output will contain the original input plus harmonic components. A harmonic is an integer multiple of the input frequency. The amplifier of Fig 2.57 was measured with a spectrum analyzer. The input was from a crystal controlled 14-MHz source followed by a 15-MHz low-pass filter, guaranteeing a drive free of harmonics. The measurement results are shown in **Table 2.1**.

The drive power was varied from −20 to +5 dBm with a step attenuator. The 14-MHz output, although increasing with drive, still showed gain compression, severe at the highest drive. At lower levels the harmonics (also shown in dBm) grow at a level proportional to the harmonic number. Hence a 10 dB drive change causes a change of about 20 dB in 2nd harmonic and about 30 dB in 3rd harmonic. This simple behavior disappears as the amplifier enters gain compression. Most linear circuits display harmonic amplitudes proportional to order with increasing drive.

It is common to specify harmonic (and other) distortions in terms of "dBc," which is dB with regard to the desired *carrier*. Hence, with a drive of −10 dBm, the desired output was +11 dBm, and the 2nd harmonic was −30 dBm, or −41 dBc.

Intermodulation Distortion, IMD

We next consider intermodulation distortion, IMD. Intermodulation describes the behavior of an amplifier when it is driven with two signals ("tones") that are generally close to each other in frequency. Second order IMD then creates undesired outputs at the sum and the difference frequencies. The *desired* output of a *mixer* is often a 2nd order IMD product between the RF and LO. Third order IMD from two tones at f_1 and f_2 generates products at $(2f_2-f_1)$ and $(2f_1-f_2)$. The *order* relates to the number of frequencies participating in a distortion process where $(2f_1-f_2)$ can be thought of as f_1, f_1 again and f_2. Order is also ambiguously related to the underlying mathematical description of the distortion.

Consider an example where two equal strength, −15 dBm tones at 14.0 and 14.2 MHz are applied to the amplifier of Fig 2.57. The desired outputs occur at the original frequencies at a level of +5 dBm, 20 dB above the drives. Also present are the third order IMD terms at 13.8 and 14.4 MHz. A sketch of the spectrum analyzer response is shown in **Fig 2.59** with the analyzer set for a +10 dBm reference level at the top of the display. The distortion

Table 2.1 All powers are in dBm, dB with regard to one mW.

Drive Power	14 MHz	28 MHz	42 MHz	56 MHz
−20 dBm	+1 dBm	−51 dBm	−72 dBm	—
−10	+11	−30	−46	—
0	+18	+3	−7	−35 dBm
+5	+21	+11	0	−1

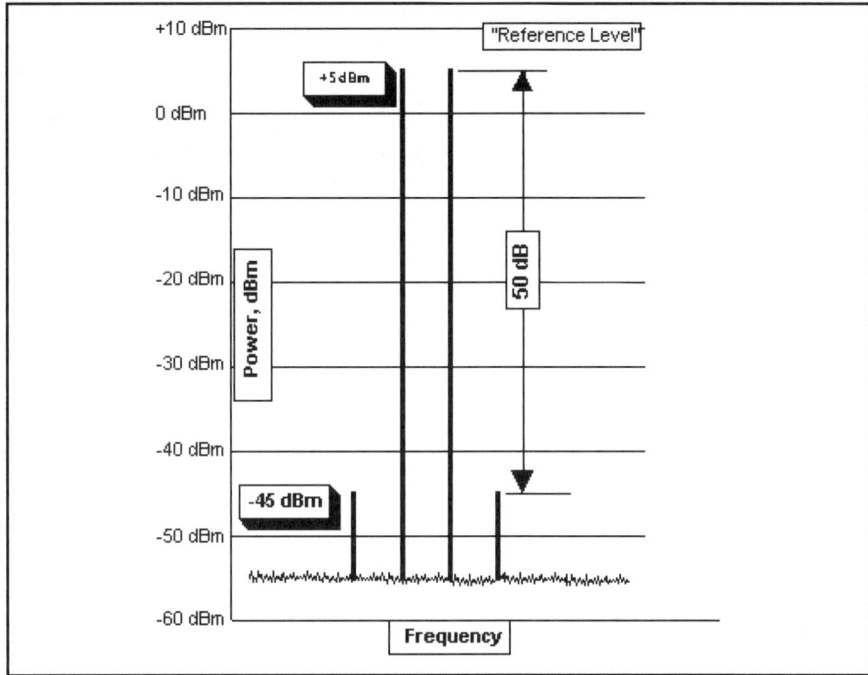

Fig 2.59—Spectrum from the feedback amplifier when driven with two tones. The smaller signals are third order intermodulation distortion. If this was the input to a receiver, all of these signals could be heard.

outputs have a power of –45 dBm. The IMD products are said to be 50 dB below one of two equal desired output tones.

Transmitters are sometimes described by an IMD that is below the desired output by a specified amount. But, implicit in such a specification is transmitter operation at rated output power. There is rarely a "rated output" for amplifiers like this one.

Amplifier intermodulation distortion generally depends upon drive level. Increasing drive by 1 dB will cause third order IMD powers to increase by 3 dB. This was readily confirmed during the tests to obtain the data of Fig 2.59. Continuing this procedure allows us to plot both desired output power for each tone and distortion power for each IMD product. This plot is shown in **Fig 2.60**. The curves are "log-log" form, with both x and y axis in dBm. The "desired output" plot is a linear straight line (slope=1) until gain compression is encountered. The third order distortion plot is a straight line following a steeper path.

It is useful to extend the two curves, each being straight lines on the log-log plot, until they intersect. The point where the desired and the third order curves cross is called the *third-order intercept point* or sometimes just the *intercept point*. There are two power values (input and output) associated with this point, with the values differing by the small signal amplifier gain. These values are very useful as a Figure-of-merit for the amplifier. The higher the *third order output intercept*, IP3out, the more immune that amplifier is to distortion problems. We sometimes see this called OIP3, with the "O" indicating that the number relates to the output. IIP3 is also popular to indicate *third order input intercept*. OIP3 and IIP3 differ by the stage gain.

Note that the intercept is mathematical; it is usually impossible to operate an amplifier with an output power as high as the output intercept. The amplifier intercept, IP3out or OIP3, is more than a mere figure of merit. If the operating output powers are known and if IP3out is specified, the distortion can then be calculated with

$$\text{IMDR} = 2 \cdot \left(\text{IP}_{3\text{OUT}} - P_{\text{OUT}} \right) \quad \text{Eq 2.38}$$

where IMDR is the *IMD Ratio* in dB, the difference between the desired signal and the distortion; IP_{3out} is the output intercept in dBm; and P_{out} is the output power in dBm. Both powers are "per tone," one of two identical values. For example, our test amplifier has IP_{3out} = +30 dBm. If we drive the amplifier with two tones to an output of –7 dBm per tone, the IMD ratio is

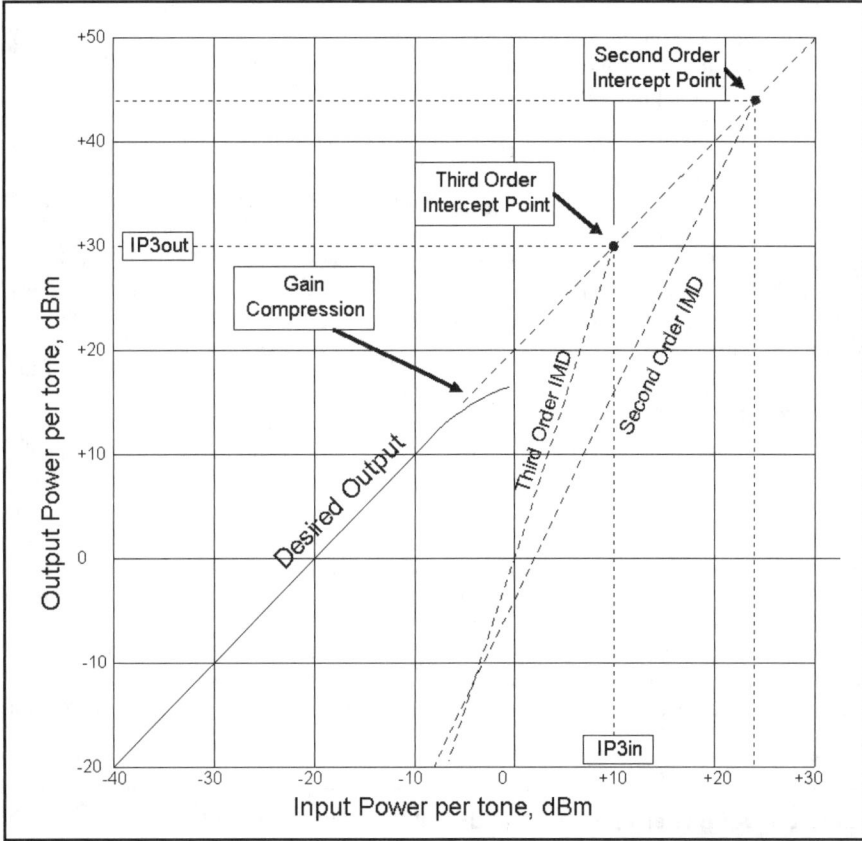

Fig 2.60—Plot of amplifier output vs input when two equal input tones are varied together. Both the desired output amplitude and the distortion product amplitudes are plotted, although only extrapolation distortion is shown. Gain compression is evident. The distortion products intersect the desired output at the intercept points.

74 dB, leaving the output distortion products at −81 dBm.

It is not necessary to actually draw the plot of Fig 2.60 to obtain the intercept. Rather, it can be inferred from a single distortion measurement with Eq 2.38; this is the usual practice.

Intercepts have another very important use. If the output intercepts of all stages in a cascade are known, a composite intercept can be calculated for the cascade. Consider the two-stage amplifier of **Fig 2.61**. Each stage has a gain of 12 dB, but the second stage has lower IMD than the first. The intercepts of each stage can be *normalized* to any desired point in the cascade. Picking the overall amplifier input as that point, the first stage (IP3out= +15 dBm) has IP3in=+3 dBm, while the second stage has an intercept at the cascade input of IP3cin= −4 dBm, 24 dB below that stage's output intercept. The second stage will dominate distortion, which becomes clear when they are compared at a single normalized plane within the chain. We can calculate the input intercept of the cascade with

$$IP_3(mW) = \left(\frac{1}{IP_{3A}(mW)} + \frac{1}{IP_{3B}(mW)} \right)^{-1}$$

Eq 2.39

where all powers are now mW rather than dBm. (See section 2.5 for the conversion.) Once we have the cascade input intercept, it can be moved to the output by adding the gain of the cascade. Eq 2.39, derived in *Introduction To Radio Frequency Design*,[5] describes coherent voltage addition of third order distortion products, so it represents a worst case. We have experimentally observed that this worst-case behavior is usually realistic.

Fig 2.60 also includes second order IMD. A second order intercept point, and values for IP2in and IP2out are defined in the same way as those of the third order products. If inputs occur at f_1 and f_2, second order IMD occurs at frequencies

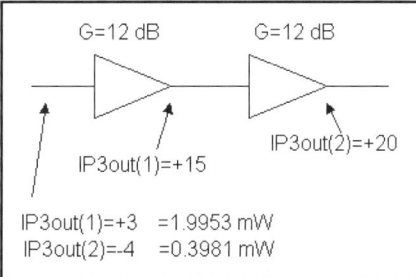

Fig 2.61—A cascade of two amplifiers, each well specified for gain and output intercept. The composite intercept is easily calculated. An extension of this allows an entire system to be analyzed for IMD.

($f_1 + f_2$) and ($f_1 - f_2$). These distortion frequencies are usually far removed from the inputs. Hence, they can be removed with a filter following the amplifier. This is not possible with third order products very close to the frequencies causing the distortion.

The test amplifier was found to have a second order output intercept of +44 dBm. Second order intercepts are generally numerically higher than the third order ones, although the second order distortion does not drop as quickly. Second order IMD can be a major difficulty in wide band designs, such as general coverage receivers or spectrum analyzers.

It is interesting to compare the 1 dB compression power with output intercepts. Our test amplifier had $P_{out}(-1\ dB)=+16.5$ dBm and $IP3_{out}=+30$ dBm, a difference of 13.5 dB. Differences of 13 to 16 dB are common for amplifiers with bipolar transistors. Smaller values (7 to 10 dB) are more common with silicon JFETs and with GaAsFETs. The difference is *not* intended to be a Figure-of-merit. Indeed, *smaller* numbers indicate that a device can be operated closer to it's intercept. Typically any of the devices we commonly use for amplifiers cannot operate at powers as high as their output intercepts.

A test set used to measure 2nd and 3rd order intercepts is show in **Fig 2.62**. The key to the scheme is the hybrid combiner that adds the output of two signal generators while preserving impedance match and isolating the two generators. A 6-dB hybrid is the preferred scheme owing to the excellent isolation afforded. But a 3-dB hybrid can be substituted if good quality signal generators are used. A 6-dB hybrid is a network with an output that is 6 dB lower per tone than each input. Note that the 6-dB hybrid has the same schematic as a return loss bridge. Hence, one instrument can be used to measure impedance match and to isolate signal sources. Every home lab needs at least one hybrid combiner.

The intercept formalization is generally restricted to circuits with constant, or nearly constant, bias current. A Class AB or B amplifier where current grows with applied drive is not generally described by an intercept. Rather, it is characterized with a simple IMD ratio, usually at full power output.

Further information on distortion and noise is found in *Introduction to Radio Frequency Design*.[6] The reader is also referred to Bill Sabin's presentation in the 1995 (and later) *ARRL Handbook*[7] concerning distortion, including that of 2nd order IMD.

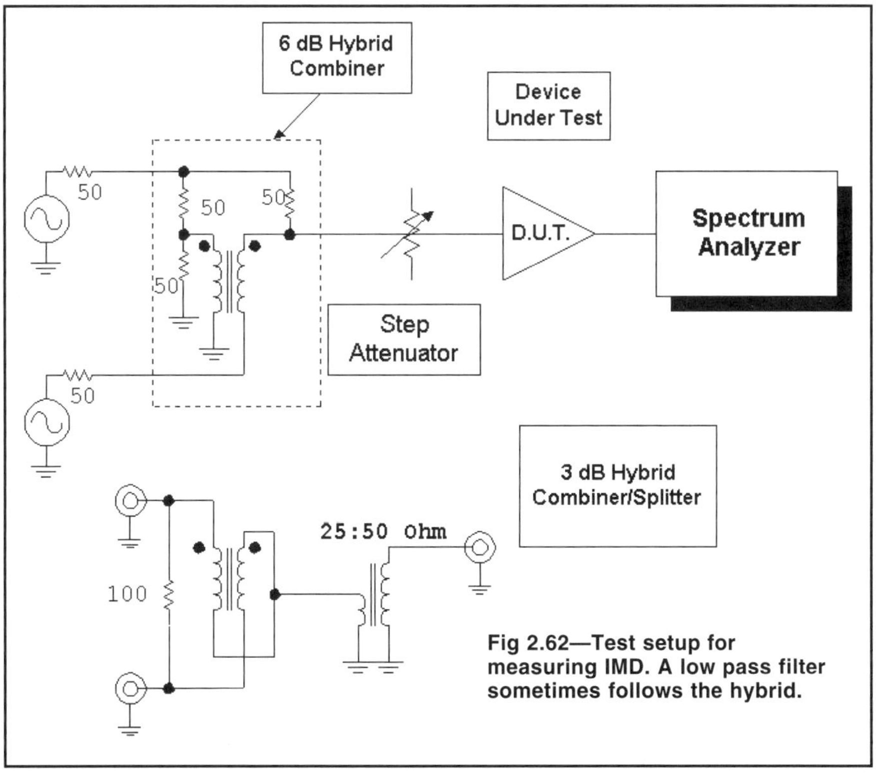

Fig 2.62—Test setup for measuring IMD. A low pass filter sometimes follows the hybrid.

2.7 FEEDBACK AMPLIFIERS

A circuit form appearing often in this book is the feedback amplifier. This is a circuit with two forms of negative feedback with (usually) a single transistor to obtain wide bandwidth, well controlled gain, and well controlled, stable input and output resistances. Several of these amplifiers can be cascaded to form a high gain circuit that is both stable and predictable.

The small-signal schematic for the feedback amplifier is shown in **Fig 2.63** without bias components or power supply details. The design begins with a NPN transistor biased to a stable dc current. Gain is reduced with emitter degeneration, increasing input resistance while decreasing gain. Additional feedback is then added with a parallel feedback resistor, R_f, between the collector and base. This is much like the resistor between an op-amp output and the inverting input which reduces gain and *decreases* input resistance.

Several additional circuits are presented showing practical forms of the feedback amplifier. That in **Fig 2.64** shows a complete circuit. The base is biased with a resistive divider from the collector. However, much of the resistor is bypassed, leaving only R_f active for actual signal feedback. Emitter degeneration is ac coupled to the emitter. The resistor R_E dominates the degeneration since R_E is normally much smaller than the emitter bias resistor. Components that are predominantly used for biasing are marked with "B." This amplifier would normally be terminated in 50 Ω at both the input and output. The transformer has the effect of making the 50-Ω load "look like" a larger load value, R_L=200 Ω to the collector. This is a common and useful value for many HF applications.

Fig 2.65 differs from Fig 2.64 in two places. First, the collector is biased through an RFC instead of a transformer. The collector circuit then "sees" 50 Ω when that load is connected. Second, the emitter degeneration is in series with the bias, instead of the earlier parallel connection. Either scheme works well, although the parallel configuration affords experimental flexibility with isolation between setting degeneration and biasing. Amplifiers without an output transformer are not constrained by degraded transformer performance and often offer flat gain to several GHz.

The variation of **Fig 2.66** may well be the most general. It uses an arbitrary transformer to match the collector. Biasing is traditional and does not interact with the feedback.

Feedback is obtained directly from the output tap in the circuit of **Fig 2.67**. While this scheme is common, it is less desirable than the others, for the transformer is part of the feedback loop. This could lead to instabilities. Normally, the parallel feedback tends to stabilize the amplifiers. The equations and curves presented below pertain to circuits with feedback taken directly from the collector.

The circuit of **Fig 2.68** has several fea-

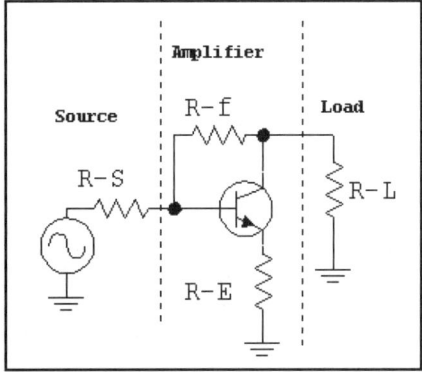

Fig 2.63—Small signal circuit for a feedback amplifier.

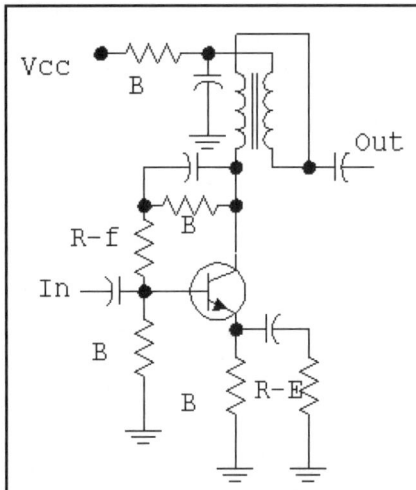

Fig 2.64—A practical feedback amplifier. Components marked with "B" are predominantly for biasing. The 50-Ω output termination is transformed to 200 Ω at the collector. A typical transformer is 10 bifilar turns of #28 on a FT-37-43 ferrite toroid. The inductance of one of the two windings should have a reactance of around 250 Ω at the lowest frequency of operation.

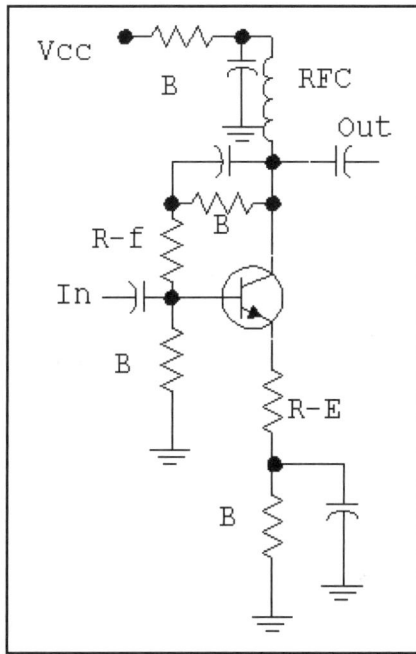

Fig 2.65—A variation of the feedback amplifier with a 50-Ω output termination at the collector.

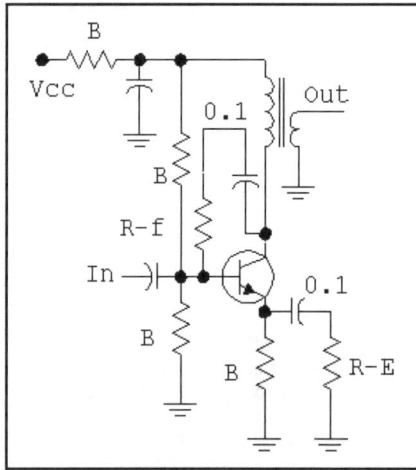

Fig 2.66—This form uses an arbitrary transformer. Feedback is isolated from bias components.

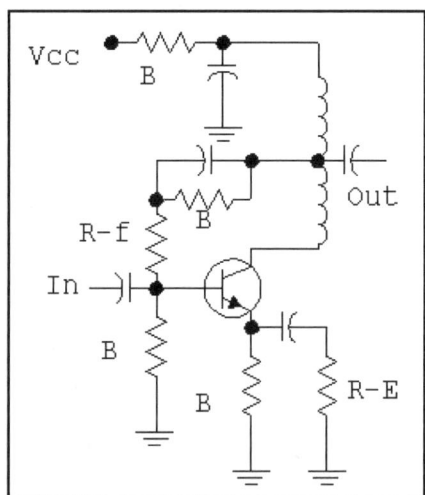

Fig 2.67—A feedback amplifier with feedback from the output transformer tap. This is common, but can produce unstable results.

tures. Two transistors are used, each with a separate emitter biasing resistor. However, ac coupling causes the pair to operate as a single device with degeneration set by R_E. The parallel feedback resistor, R_f, is both a signal feedback element and part of the bias divider. This constrains the values slightly. Finally, an arbitrary output load can be presented to the composite collector through a π-type matching network. This provides some low pass filtering, but constrains the amplifier bandwidth.

Design Procedure

Design begins by picking a bias current, usually dictated by output power and IMD requirements. Next the output load impedance presented to the collector (or drain) is chosen. A value of 200 Ω is probably the most common, for it affords good gain with reasonable current. With that load, the output power will be restricted to around 200 mW in 12-volt systems. Progressively lower impedances will allow higher output power. Most feedback amplifiers end up being designed for 50-Ω input resistance.

The emitter degeneration and feedback resistors are chosen next. A reasonable input and output impedance match occurs with

$$R_f \cdot R_e = R_S \cdot R_L \qquad \text{Eq 2.40}$$

where R_f is the parallel feedback and R_e is the net degeneration resistance, $r_e + R_E$. Here R_E is the external degeneration, and r_e is the current dependant value, $26/I_e(mA)$. For example, an amplifier driven by 50 Ω and terminated in 200 Ω might use 10-Ω external degeneration and 10-mA current for $R_e = 12.7$ Ohms. $R_f = 787$ Ω would produce $R_{in} \approx R_S$ and $R_o \approx R_L$, with R_{in} and R_o being the input and output resistances for source and load R_S and R_L. A practical choice would be $R_f = 820$ Ω, a standard value.

There is still a wide range of values that can be used for degeneration and feedback. The final choice is made on the basis of desired gain, which can be determined by the equations presented in **Fig 2.69**. The choice is eased by example data in **Table 2.2**. While the data in the table is for one current, 20 mA, it will provide an initial estimate.

The equations of Fig 2.69 appear long and messy, but are easily programmed for a calculator or computer.

Fig 2.70 shows the gain obtained when

Table 2.2

Simulated Gain vs Degeneration and Feedback Resistors for a 2N3904 biased with I_E=20 mA where r_e=1.3 Ω. Gain was calculated at 14 MHz, so β=300/14=21. Resistors were picked as standard values and to provide an input return loss better than 10 dB. The first example is the amplifier described in the previous section.

Load	R-degen	R-feedback	Gain
200 Ω	6 Ω	1.3 kΩ	20.3 dB
	3.9 Ω	3 kΩ	24.8 dB
	4.7 Ω	2.7 kΩ	23.9 dB
	5.6 Ω	2 kΩ	22.3 dB
	6.8 Ω	1.6 kΩ	20.7 dB
	10 Ω	910 Ω	16.8 dB
	12 Ω	750 Ω	15.1 dB
	15 Ω	560 Ω	12.6 dB
	18 Ω	430 Ω	10.3 dB
	22 Ω	330 Ω	7.7 dB
50 Ω	2.7 Ω	820 Ω	20.0 dB
	3.9 Ω	680 Ω	18.2 dB
	4.7 Ω	560 Ω	16.9 dB
	5.6 Ω	470 Ω	15.6 dB
	6.8 Ω	390 Ω	14.1 dB
	10 Ω	270 Ω	10.7 dB
	12 Ω	220 Ω	8.8 dB
	15 Ω	150 Ω	5.4 dB

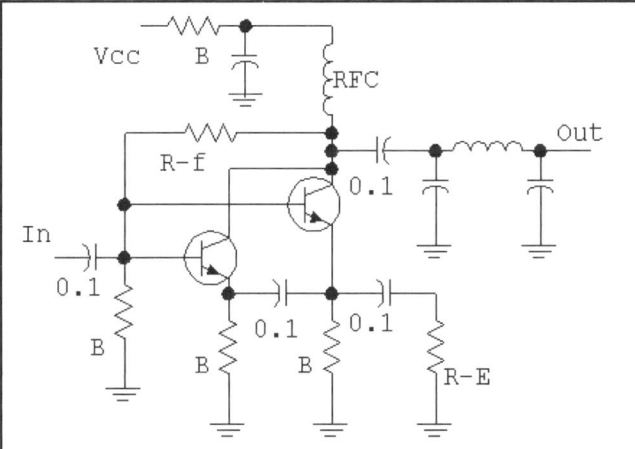

Fig 2.68—Feedback amplifier with two parallel transistors.

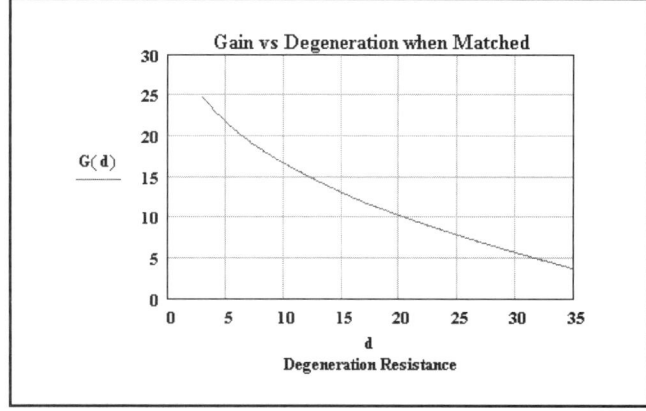

Fig 2.70—Gain Vs net degeneration resistance when the amplifier is matched. This evaluation occurred at 14 MHz with a 2N3904 biased to 20 mA with a 50-Ω source and 200-Ω load.

$$G := 10 \cdot \log \left[4 \cdot R_L \cdot R_s \cdot \frac{\left[(\beta+1)^2 \cdot R_e^2 - 2 \cdot \beta \cdot R_f (\beta+1) \cdot R_e \right] + \beta^2 \cdot R_f^2}{\left[\left[(1+\beta) \cdot R_e + R_s \right] \cdot R_f + \left(R_L + R_s + \beta \cdot R_s + \beta \cdot R_L \right) \cdot R_e + \beta \cdot R_s \cdot R_L + R_L \cdot R_s \right]^2} \right]$$

$$R_{in} := (1+\beta) \cdot (R_f + R_L) \cdot \frac{R_e}{\left[(1+\beta) \cdot R_e + \beta \cdot R_L + R_L + R_f \right]} \qquad R_o := \frac{\left[\left[(1+\beta) \cdot (R_f + R_s) \right] \cdot R_e + R_s \cdot R_f \right]}{\left[(1+\beta) \cdot R_e + R_s + \beta \cdot R_s \right]}$$

Fig 2.69—Transducer Gain G in dB, Input resistance, R_{in}, and Output resistance, R_o, both in Ohms for a feedback amplifier. The analysis is restricted to the case where parallel feedback is obtained from the collector. R_f is the parallel feedback and R_e is the total emitter degeneration (see text.) R_s and R_L are the source and load resistances, and are arbitrary for this analysis. β is the current gain and is approximated as a scalar value, $β=F_t/F$ where F_t is the current gain-bandwidth product and F is the operating frequency, both in MHz.

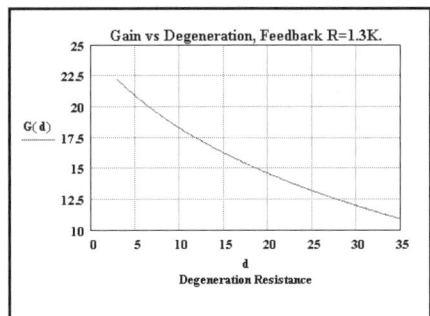

Fig 2.71—Gain Vs degeneration for fixed feedback R of 1.3 kΩ.

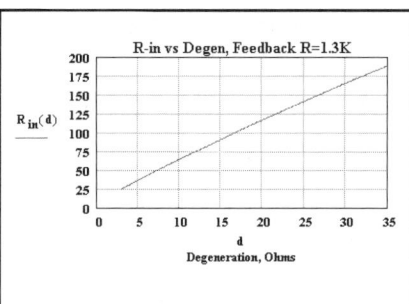

Fig 2.72—Input resistance Vs degeneration for fixed feedback resistance.

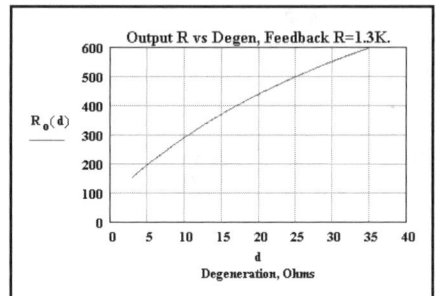

Fig 2.73—Output resistance Vs degeneration for a fixed 1.3-kΩ feedback resistance.

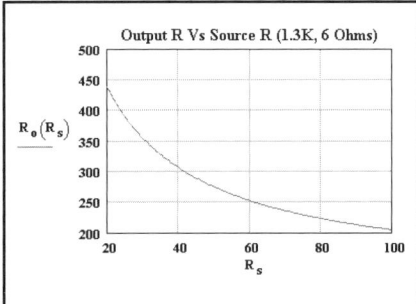

Fig 2.74—Output resistance depends on the source resistance.

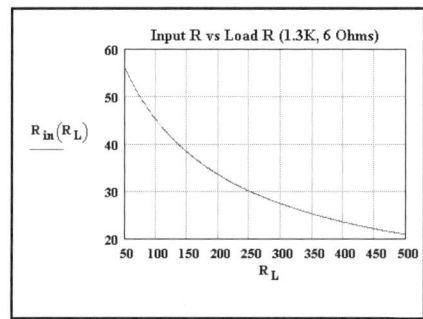

Fig 2.75—Input resistance as a function of load resistance.

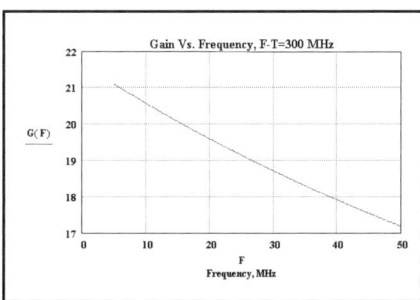

Fig 2.76—Feedback tends to flatten frequency response. This is even more dramatic with lower gain amplifiers.

Eq 2.40 is applied, forcing a reasonable input and output impedance match.

It is common to build an amplifier only to then find that the gain must be changed a little. The effect of changing the emitter resistor is presented in **Fig 2.71** for a fixed $R_f = 1.3$ kΩ. The same 14-MHz, 20-mA bias case is assumed. **Fig 2.72** and **Fig 2.73** show the related effect on terminal resistances.

A characteristic of feedback amplifiers (sometimes useful, sometimes frustrating) is that they are partially *transparent*. The input resistance becomes a strong function of the load while the output resistance depends upon the source. This is illustrated in **Fig 2.74** and **Fig 2.75**. Again, a 1.3-kΩ feedback R and 6-Ω external degeneration are used. The amplifier transparency is partially "fixed" with the addition of an attenuator at the amplifier output, especially useful when the amplifier must interface with filters and switching-mode mixers. Pads must be added with care, for they will decrease overall gain, available output power and output intercept.

Feedback extends the bandwidth of transformer terminated amplifiers. **Fig 2.76** shows gain vs F for the example amplifier with a 2N3904 at 20 mA, 6-Ω degeneration and 1.3-kΩ R_f, 50-Ω source and 200-Ω load. There is less than a 3-dB variation over the HF spectrum, and the amp is usable up to 50 MHz, even with a modest 2N3904. Higher F_t transistors can produce much greater bandwidth, especially when configured for low or modest gain without any transformers that might compromise frequency response.

While we usually think in terms of building feedback amplifiers with bipolar transistors, they are just as tenable with FETs. **Fig 2.77** shows a JFET version of the amplifier. This circuit uses no degeneration resistor. The FET is self-biased with a bypassed source resistor, and the biased FET transconductance is calculated using equations presented earlier. Having this value, we can then ask "what current (r_e) in a bipolar transistor would produce the same transconductance?" Finding that value, we then use the same equations for analysis that were applied to the bipolar, Fig 2.69.

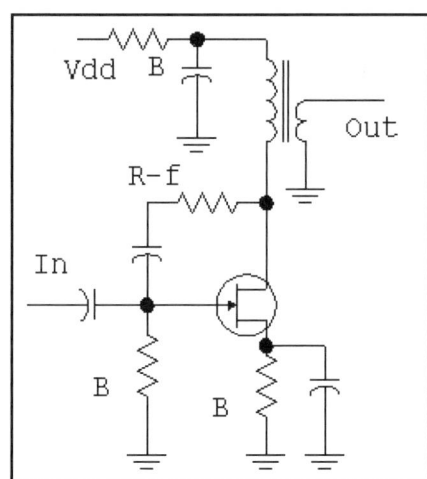

Fig 2.77—A feedback amplifier using a FET. See text for design details.

Feedback amplifier noise figure is usually greater than that from the same transistor without feedback. Noise available from the feedback resistors is injected into the circuit. A feedback amplifier was built using a 2SC1252 transistor ($F_t \approx 2$ GHz) with degeneration and feedback resistors of 5.1 Ω and 1.8 kΩ. Noise figure was measured with an HP8970B test set for differing standing currents. The noise figure was 1.8 dB in the HF spectrum for $I_e=10$ mA, increasing to 3.3 dB with 63 mA. Noise figure for the 2N3904 example amplifier featured in this section (20 mA, 6 Ω and 1.3 kΩ, 200-Ω load) was measured at 6 dB.

Fig 2.78 shows a feedback amplifier with two transistors in a Darlington configuration. This circuit is typical of several popular silicon monolithic integrated circuit amplifiers that are presently available. Those components within the dotted line are part of the IC. Q1 and Q2 usually have F_t above 5 GHz, so the amplifiers offer useful performance to 2 GHz and beyond with gain from 10 to nearly 20 dB.

These amplifiers are specified by their distributor for a voltage on the output pin with a specified current allowing the user to pick R_3 for an available V_{cc}. For example, the Minicircuits MAR-2 is specified for 25 mA at 5 V. Hence, for a 12-V power supply, 280 Ω would be needed for R_3. This IC should not be used without a dropping resistor. The power dissipation in the resistor should be checked. It's only 175 mW in this example, so a ¼-W resistor would suffice.

Fig 2.79 presents another two discrete transistor feedback amplifier. This is a buffer amplifier designed by W7EL. This circuit is similar to MAR circuits parts, but uses transformer output coupling for even greater available gain. The input resistor should be driven from a source at DC ground. Bandwidth depends on the output transformer with severe distortion possible at low frequencies if it does not have adequate reactance. A typical 7-MHz application uses a 20-turn primary on a FT-37-43 toroid with a 5-turn output link.

A common base amplifier with transformer output coupling is shown in **Fig 2.80**. This circuit uses no feedback other than the 47-Ω degeneration. This is presented as an evolutionary step toward a feedback amplifier, but it is very useful as shown. The common base topology features excellent reverse isolation, and, as such, it is an excellent VFO buffer. The amplifier is biased to about 4 mA collector current, so has an input resistance at the emitter of 6.5 Ω. Adding a series 47-Ω resistor creates a reasonable input match to a 50-Ω source. The power gain will be determined by the ratio of turns on the output auto-transformer.

An interesting variation of this circuit is presented in **Fig 2.81**. The 47-Ω input resistor has been replaced by a single turn link through the transformer core. The operation is easily understood if we think of driving the input with a current source. The low input impedance at the emitter has no impact on the current flowing. Essentially the same current flows in the collector (recall that the current gain of a common base amplifier is unity), but it now flows in the high impedance multiple turn transformer windings. This allows the circuit to provide power gain. We now "sample" the collector current with a winding, creating a voltage across the winding. The new "voltage" is placed in series with the low emitter input imped-

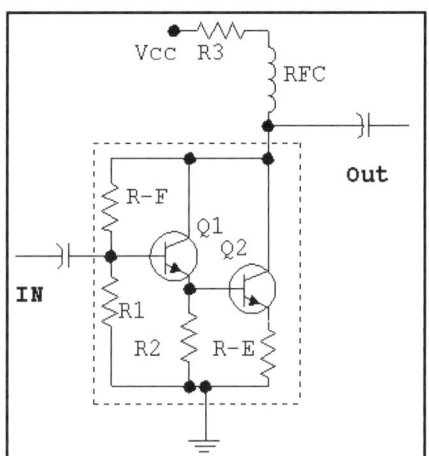

Fig 2.78—Feedback amplifier with a Darlington connection of transistors.

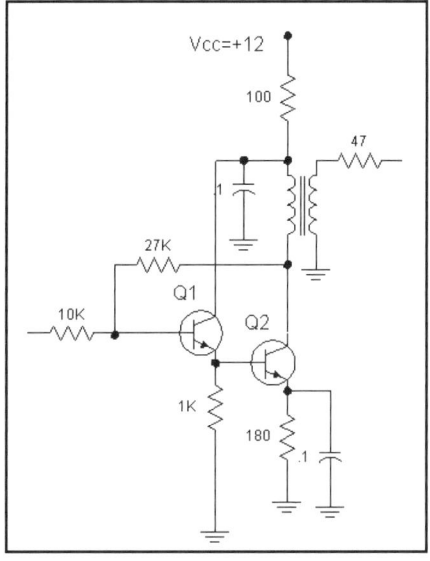

Fig 2.79—Feedback amplifier, the design of W7EL, often used as an oscillator buffer.

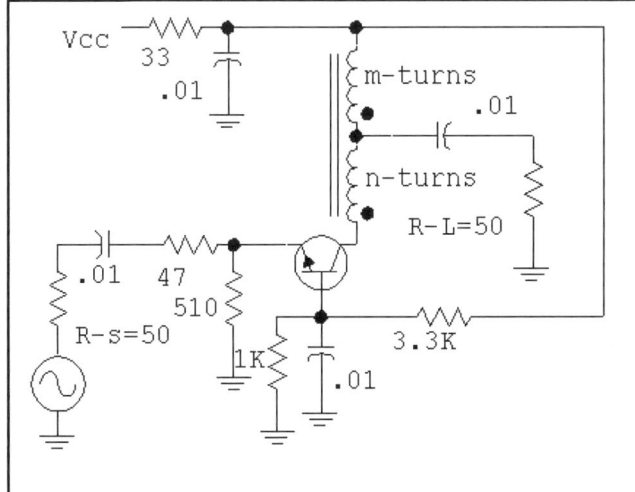

Fig 2.80—Common base amplifier with an input resistance. See text.

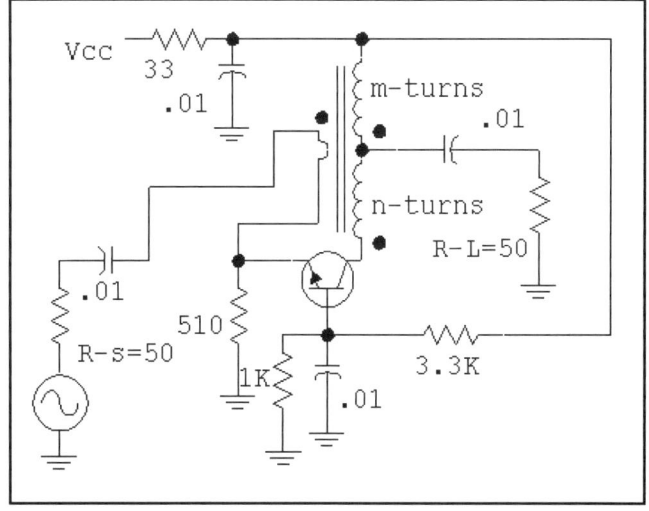

Fig 2.81—A transformer feedback amplifier designed by D. Norton of Anzac.

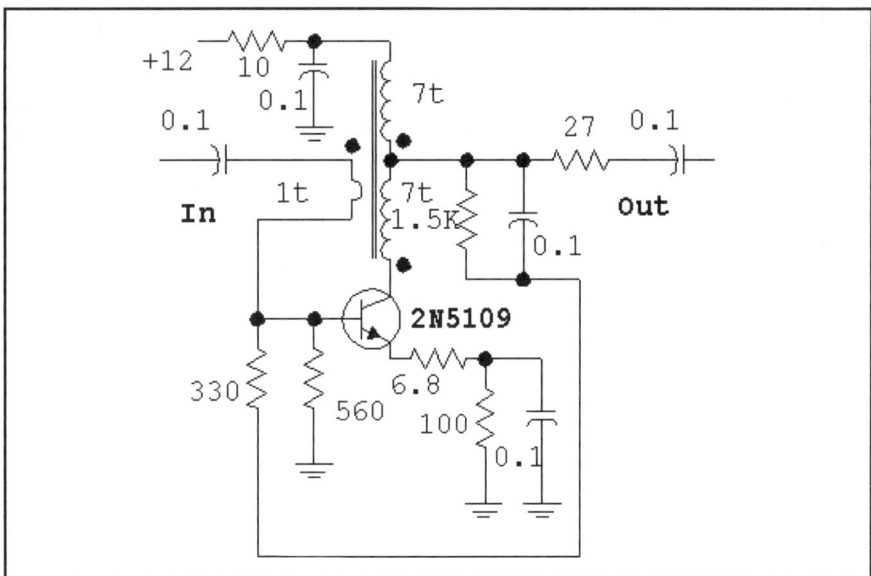

Fig 2.82—A modified feedback amplifier where transformer feedback increases input impedance.

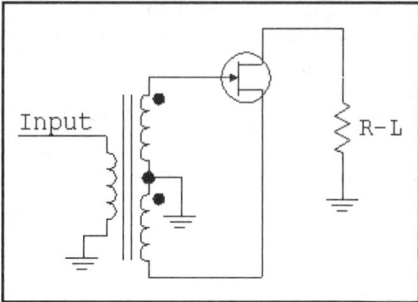

Fig 2.83—Small signal circuit of a transformer type feedback amplifier using a JFET.

Fig 2.84—A feedback amplifier example. This circuit supplements test equipment. With V_{cc}=12, I_c=65 mA and OIP3=+42 dBm, Gain=16 dB, and bandwidth exceeds 50 MHz.

ance to create a 50-Ω input termination. However, this is done without any resistors, so the noise figure is not compromised. This amplifier is the brainchild of David Norton of Anzac.[8]

The Fig 2.81 amplifier will be matched if

$$n = m^2 - m - 1 \qquad \text{Eq 2.41}$$

to produce a transducer power gain of 20Log(m) dB. For example, if m=3, n is then 5, and the power gain is 9.5 dB. The transformers for these amplifiers are often wound on a binocular-type balun core. A turn through such a ferrite core is counted as a single pass of wire through *both* holes. Polarity is vital to construction of the transformer. If wound wrong, the input impedance will be negative, almost guaranteed to create oscillation. In amplifiers of this kind that we have built, we measured excellent input impedance match (25-dB return loss) over a 5 to 100 MHz range with noise figure under 2 dB. This amplifier, however, suffers from a major problem; the terminal impedances depend strongly on the termination at the other port. The circuit is worse than resistive feedback amplifiers in this regard.

Transformers can be further applied to extend performance of amplifiers. **Fig 2.82** shows a generally traditional feedback amplifier that is modified by passing the input lead through the transformer core to alter input impedance. This topology is early work of Rohde.[9]

Fig 2.83 shows a FET amplifier (small signal circuit only) using an input transformer. A tapped transformer feeds signal to both the FET source and the gate. The winding driving the source sees a low impedance, so adjustment of turns ratio can ensure a perfect match. The gate winding, even though there is no signal current flowing, provides the gate voltage needed for gain and low noise performance. Design details are given in *Introduction to Radio Frequency Design*, p 216.[10] Bill Carver, W7AAZ, has built practical versions of this amplifier. See *QST*, May, 1996,[11] with further discussion in Chapter 6.

Transformer feedback amplifier design is a subject that continues to produce design activity. The reader can find more information starting with papers by Trask[12,13] and Koren.[14]

Fig 2.84 shows an example of a feedback amplifier.

2.8 BYPASSING AND DECOUPLING

Our amplifier designs have included grounded points that were not really at ground. Rather, those points are "signal grounded" through bypass capacitors. Obtaining an effective bypass can be difficult and is often the route to design difficulty.

The problem is parasitic inductance. Although we label and model parts as "capacitors," a more complete model is needed. The better model is a series LRC, shown in **Fig 2.85**. Capacitance is close to the marked value while inductance is a small value that grows with component lead length. Resistance is a loss term, usually controlled by the Q of the parasitic inductor. All components show some inductance, including a wire. Even a leadless SMT component will display inductance commensurate with the dimensions. A wire has an inductance of about 1 nH per mm of length.

Bypass capacitor characteristics can be measured in the home lab with the test setup of **Fig 2.86**. **Fig 2.87** shows a test fixture with an installed 470-pF leaded capacitor. The fixture is used with a signal generator and spectrum analyzer to evaluate capacitors. Relatively long capacitor leads were required to interface to the BNC connectors, even though the capacitor itself was small. The signal generator was tuned over its range while examining the spectrum analyzer response, which was minimum at the series resonant frequency. Parasitic inductance is calculated from this

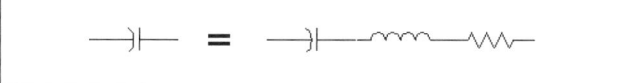

Fig 2.85—Model for a bypass capacitor.

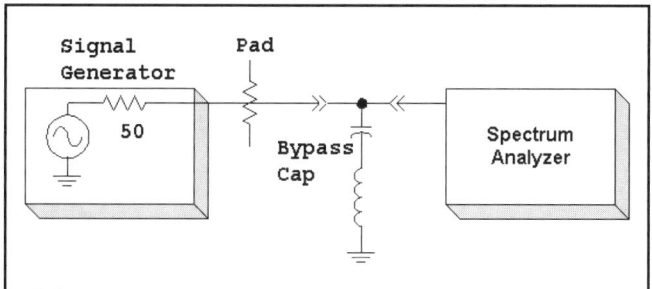

Fig 2.86—Test set for home lab measurement of a bypass capacitor.

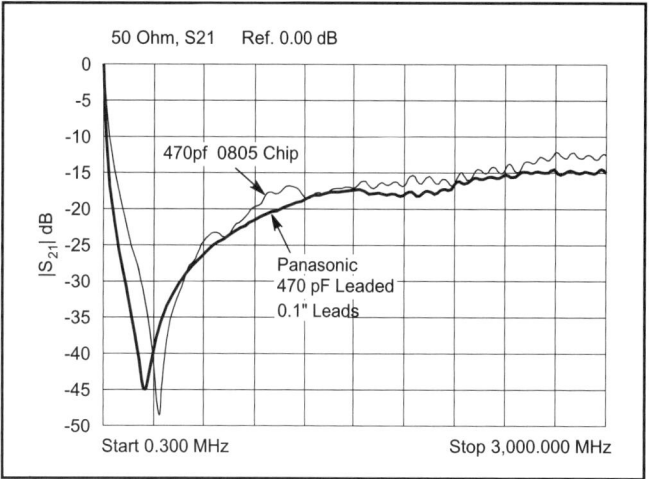

Fig 2.88—Network analyzer measurement of 470-pF shunt capacitors. Both SMT and leaded parts are studied.

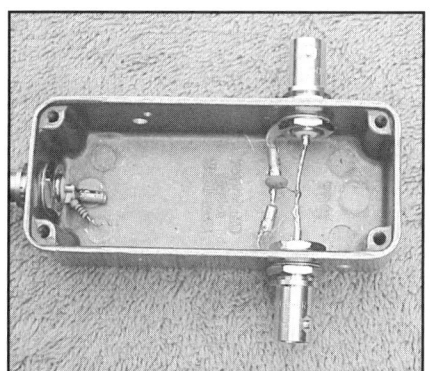

Fig 2.87—Test fixture for measuring self resonant frequency of capacitors.

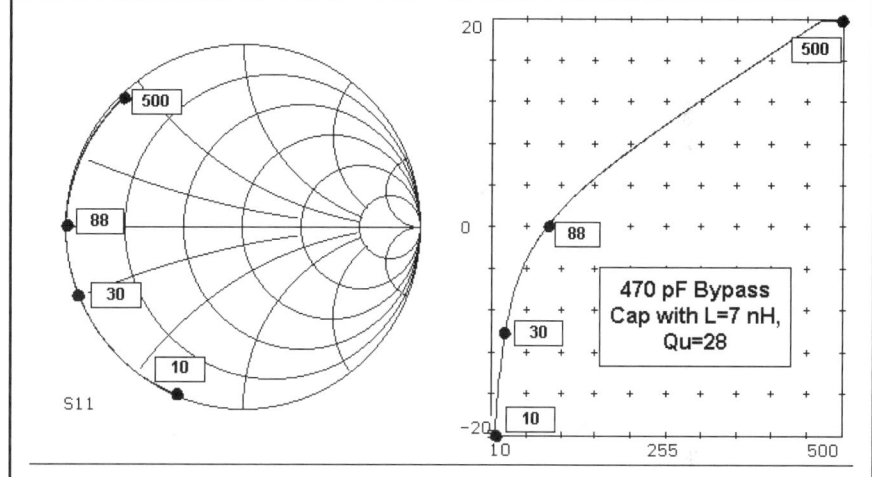

Fig 2.89—Impedance of a 470-pF bypass capacitor.

frequency. The C value was measured with a low frequency LC meter. Measurement gear is discussed in Chapter 7.

The measured 470-pF capacitor is modeled as 485 pF in series with an inductance of 7.7 nH. The L is larger than we would see with shorter leads. A 0.25-inch 470-pF ceramic disk capacitor with zero lead length will show a typical inductance closer to 3 nH. The measured capacitor Q was 28 at self-resonance of 82 MHz but is higher at lower frequency.

Data from a similar measurement, but with a network analyzer is shown in **Fig 2.88**. Two 470-pF capacitors are measured, one surface mounted and the other a leaded part with 0.1-inch leads.

Fig 2.89 shows two calculated plots for the 470-pF capacitor. The one on the left is a Smith Chart showing the behavior vs. frequency, while that on the right is a plot of component reactance vs. frequency. Reactance dominates, keeping the data on the edge of the Smith Chart, for the Q is moderate at 28. Bypassing is "perfect" at only one frequency, that of series resonance. An ideal (no inductance) capacitor would have a capacitive reactance of about 2 Ω at 150 MHz. The actual 150-MHz value is inductive with a magnitude of about 5 Ω.

Traditional lore tells us that the bandwidth for bypassing can be extended by paralleling a capacitor that works well at one frequency with another to accommodate a different part of the spectrum. Hence, paralleling the 470 pF with a .01-μF capacitor should extend the bypassing to lower frequencies. The calculations are shown in the plots of **Fig 2.90**. The results are terrible! While the low frequency bypassing is indeed improved, a high impedance response is created at 63 MHz. This complicated behavior is again the result of inductance.

Each capacitor was assumed to have a series inductance of 7 nH. A parallel resonance is approximately formed between the L of the larger capacitor and the C of the smaller. The Smith Chart plot shows us that the impedance is nearly 50 Ω at 63 MHz. Impedance would be even higher with greater capacitor Q. **This behavior is a dramatic example of lore that is generally wrong!**

Bypassing can be improved by paralleling. However, the capacitors should be approximately identical. **Fig 2.91** shows the result of paralleling two capacitors of about the same value. They differ slightly at 390 and 560 pF, creating a hint of resonance. This appears as a small "burble" in the reactance plot and a tiny loop on the Smith Chart. These anomalies disappear as the C values become equal. Generally, paralleling is the scheme that produces the best bypassing. The ideal solution is to

Amplifier Design Basics 2.29

place a chip cap on each side of a printed circuit run or wire at a point that is to be bypassed.

Additional capacitors were measured. A .01-µF disk (leaded, 50-V, 0.2-inch diameter) was resonant at 20 MHz in the test fixture shown, indicating an inductance of 6.5 nH. The Q was 5.7. Two different 0.1-µF leaded capacitors were investigated. Both had identical capacitance even though one was larger than the other. The inductance was about 4.5 nH with Q=5 for both.

Matched capacitor pairs form an effective bypass over a reasonable frequency range. Two of the .01-µF disks have a reactance magnitude less than 5 Ω from 2 to 265 MHz. A pair of the 0.1-µF capacitors was even better, producing the same bypassing impedance from 0.2 to 318 MHz. The 0.1-µF capacitors are chip components with attached wire leads. Even better results can be obtained with multi-layer ceramic chip capacitors. Construction with multiple layers creates an integrated paralleling. We have measured some 0.2-µF parts with an inductance of 2 nH. The multi-layer components are more expensive than the monolithic 0.1-µF parts investigated.

Some applications (e.g., IF amplifiers) require effective bypassing at even lower frequencies. Modern tantalum electrolytic capacitors are surprisingly effective through the RF spectrum while offering high enough C to be useful at audio. The parts should be evaluated for critical applications.

We have discussed the problem of bypassing, but have neglected the related problem of decoupling. The bypass capacitor usually serves a dual role, first creating the low impedance needed to generate a "signal" ground. It also becomes part of a decoupling low pass filter that passes dc while attenuating signals. The attenuation must function in both directions, suppressing information in the power supply that might reach an amplifier while keeping amplifier signals from reaching the power supply.

A low pass filter is formed with alternating series and parallel component connections. A parallel bypass is followed by a series impedance, ideally a resistor. Additional shunt elements can then be added, although this must be done with care. An inductor between shunt capacitors should have high inductance. It will resonate with the shunt capacitors to create high impedances just like those that came from parasitic L in the bypasses. This makes it desirable to have an inductance that is high enough that any resonance is below the band of interest. But series inductors have their own problems; they

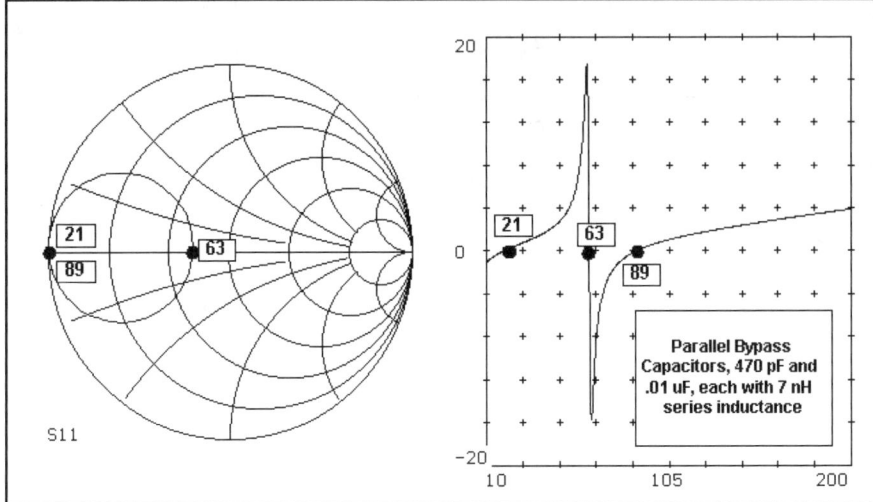

Fig 2.90—The classic technique of paralleling bypass capacitors of two values, here 470 pF and .01 µF. This is a terrible bypass! See text.

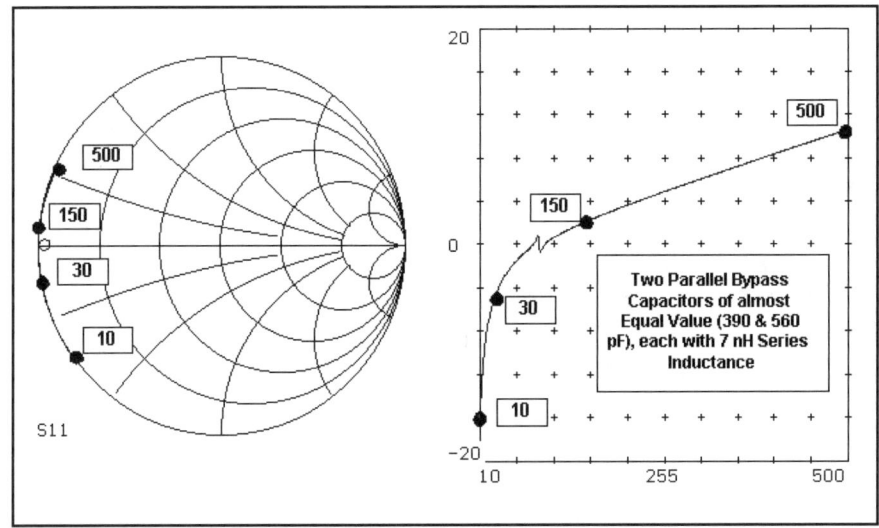

Fig 2.91—Paralleling bypass capacitors of nearly the same value. This results in improved bypassing without complicating resonances.

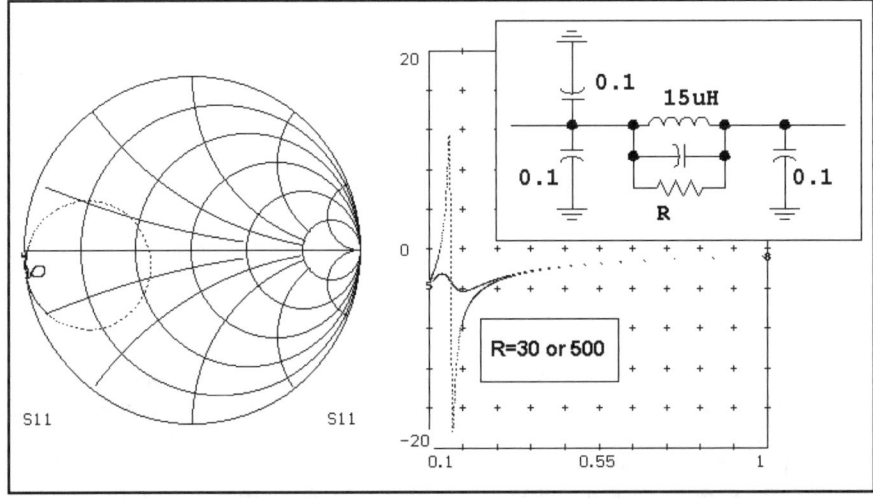

Fig 2.92—Two different resistor values parallel a decoupling choke. The lower, 30-Ω value is more effective. See text.

have parasitic capacitance that create their own self-resonance.

A couple of available RF chokes were measured (now as series elements) with the equipment described earlier. A 2.7-μH molded choke was parallel resonant at 200 MHz, indicating a parallel capacitance of 0.24 pF. The Q at 20 MHz was 52. A 15-μH molded choke was parallel resonant at 47 MHz, yielding a parallel C of 0.79 pF. This part had a Q of 44 at 8 MHz.

Large inductors can be fabricated from series connections of smaller ones. The best wideband performance will result only when all inductors in a chain have about the same value. The reasons for this (and the mathematics that describe the behavior) are identical with those for paralleling identical capacitors.

Low inductor Q is often useful, which encourages us to use inductors with ferrite cores. Inductors using the Fair-Rite (Amidon) -43 material have Q in the 4 to 10 region in the HF spectrum. One can also create low Q circuits by paralleling a series L of modest Q with a resistor.

Fig 2.92 shows a decoupling network and the resulting impedance when viewed from the "bypass" end. The 15-μH RFC resonates with a 0.1-μF capacitor to destroy the bypass effect just above 0.1 MHz. A low value parallel resistor fixes the problem.

A major reason for careful wideband bypassing and decoupling is the potential for amplifier oscillation. Instability that allows oscillations is usually suppressed by low impedance terminations. The base and collector (or gate and drain) should both "see" low impedances to ensure stability. But that must be true at all frequencies where the device can produce gain. It is never enough to merely consider the operating frequency for the amplifier. A parallel resonance can be a disaster. When the ultimate bypassing is not possible, negative feedback that enhances wideband stability is often used.

Capacitors also appear in circuits as blocking elements. A blocking capacitor, for example, appears between stages, creating a near short circuit for signals while accommodating different dc voltages on the two sides. A blocking capacitor is not as critical as a bypass, for the impedances on either side will usually be higher than that of the block.

Emitter bypassing is often a critical application. As we have seen, a few Ohms of emitter degeneration can drastically alter amplifier performance. A parallel resonant emitter bypass could be a profound difficulty while a series resonant one can be especially effective. Clearly, detailed modeling is the answer to component selection.

2.9 POWER AMPLIFIER BASICS

The remainder of this chapter deals with power amplifiers, a subject dear to the radio experimenter. The earliest tinker among us cut our teeth on attempts to extract more power from the already stressed amplifier devices of the day. We all recall stories of 6L6 receiving vacuum tubes being coaxed into providing high output power by immersion in an oil bath. The rest of us have tried to extract power from transistors, only to see the device disappear "in smoke." Experience of this sort is a "right of passage" for all RF experimenters; don't miss it!

Classes of Amplifier Operation

Many of the amplifiers considered so far have been "Class A." The class of operation of an amplifier is determined by the fraction of a drive cycle, or duty cycle where conduction occurs. The Class A amplifier conducts for 100% of the cycle. It is characterized by constant supply current, regardless of the strength of the driving signal. Most of the amplifiers we use for RF applications and many audio circuits in receivers operate in Class A.

A Class B amplifier conducts for 50 % of the cycle, which is 180 degrees if we examine the circuit with regard to a driving sinewave. A Class B amplifier draws no DC current when no input signal is applied. But current begins to flow with any input, growing with the input strength. A Class B amplifier can display good envelope linearity, meaning that the output amplitude at the drive frequency changes linearly with the input signal. The total absence of current flow for half of the drive cycle will create harmonics of the signal drive.

A Class C amplifier is one that conducts for less than half of a cycle. No current flows without drive. Application of a small drive produces no output and no current flow. Only after a threshold is reached does the device begin to conduct and provide output. A bipolar transistor with no source of bias for the base typically operates in Class C.

The large-signal models discussed earlier are suitable for the analysis of all amplifier classes. Small-signal models are generally reserved for Class A amplifiers.

The most common power amplifier class is a cross between Class A and B, the Class AB amplifier that conducts for more than half of each cycle. A Class AB amplifier at low drive levels is indistinguishable from a Class A design. However, increasing drive produces greater collector (or drain) current and greater output.

Amplifier class letter designators were augmented with a numeric subscript. A vacuum tube Class AB1 amplifier was one operating in AB, but with no grid current flowing. In the absence of grids, the numbers have disappeared.

While wide bandwidth Class A and Class B amplifiers are common, most circuits operating in Class C and higher are tuned at the output. The tuning accomplishes two things. First, it allows different terminations to exist for different frequencies. For example, a resistive load could be presented at the drive frequency while presenting a short circuit at some or all harmonics. The second consequence of tuning is that reactive loads can be created and presented to the amplifier collector or drain. This then provides independent control of current and voltage waveforms.

While not as common as A, B, and C, Class D and E amplifiers are of increasing interest. The Class D circuit is a balanced (two transistor) switching format where the input is driven hard enough to produce square wave collector waveforms. Class E amplifiers usually use a single device with output tuning that allows high current to flow in the device only when the impressed voltage is low.

Class A and AB amplifiers are capable of good envelope linearity, so they are the most common formats used in the output of SSB amplifiers. Class B and, predominantly, Class C amplifiers are used for CW and FM applications, but lack the envelope linearity needed for SSB. Recent work with a 4th method of SSB may change that, allowing distorting amplifiers to be used in SSB service.[15]

Efficiency varies considerably between amplifier class. The Class A amplifier can reach a collector efficiency of 50%, but no higher, with much lower values being

Amplifier Design Basics 2.31

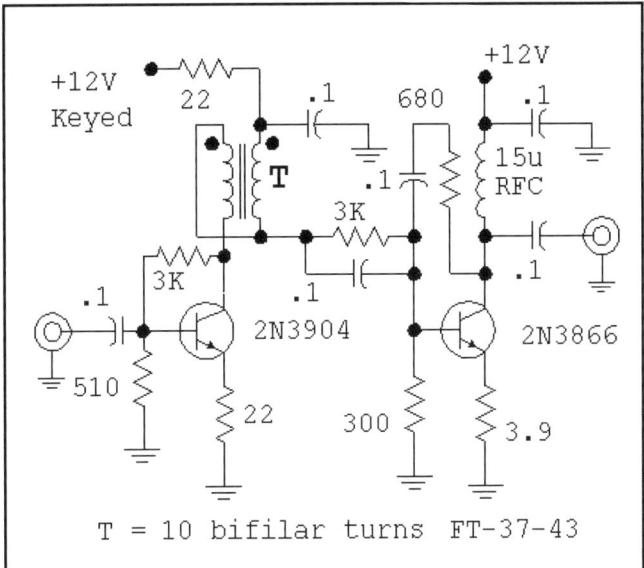

Fig 2.93—Class AB amplifier chain.

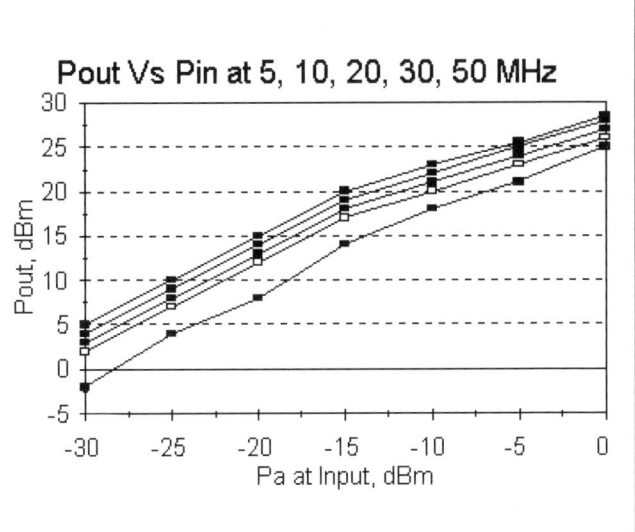

Fig 2.94—Gain compression characteristics for the simple power chain.

typical. Class AB amplifiers are capable of higher efficiency, although the wideband circuits popular in HF transceivers typically offer only 30% at full power. A Class C amplifier is capable of efficiencies approaching 100% as the conduction cycle becomes small, with common values of 50 to 75%. Both Class D and E are capable of 90% and higher efficiency.

An engineering text treating power amplifier details is Krauss, Bostian, and Raab's *Solid State Radio Engineering*.[16] A landmark paper targeted to the home experimenter was that presented by a group from Cal Tech in *QST* for May and June, 1997.[17]

A Two-Stage General Purpose Class AB Amplifier

The circuit of **Fig 2.93** operates in Class AB with an output of half a watt in the HF spectrum. This circuit was originally built as a general purpose gain block for CW transmitters. Total current is about 80 mA with no RF drive, reaching 200 mA or more when drive is increased with most of the increase occurring in the second stage. **Fig 2.94** shows P_{out} Vs. P_{in} at 5, 10, 20, 30, and 50 MHz for this amplifier when operating with a 12-V supply. The measurements were done with a signal generator and a spectrum analyzer. Low frequency gain is high at 35 dB, dropping to 28 dB at 50 MHz. Low frequency output power is over half a watt, with over a quarter of a watt available at 50 MHz. However, gain is severely compressed at this level. Higher output power is available with impedance matching.

A heat sink is used on the output transistor, for dissipation becomes high with large drive. The dissipation in the 2N3904 is 350 mW, safe for keyed (low duty cycle) CW applications, but marginal for SSB or digital modes.

The third order intermodulation distortion was measured at 14 MHz. With an output of +10 dBm per tone, the output intercept was +32 dBm. Increasing drive for +20 dBm per tone output (100 mW/tone or 400 mW PEP) yielded a higher value of IP3out=+35 dBm. This is expected, for total current is now higher at 180 mA.

The power supply for the input stage is normally keyed when used for CW transmission. The bias for the output stage is derived from the same supply resulting in a typical backwave 70 dB below full output. "Backwave" is the residual signal present from a CW transmitter during key-up periods.

This design, although lacking in efficiency, is otherwise very useful and has been used in over a dozen transmitters or transceivers in our stations. It can be driven by a crystal oscillator on any HF band to form an effective QRP transmitter. Preceding it with a feedback amplifier produces a DSB or SSB chain suitable for QRP use, or as a driver for a five watt PA.

2.10 PRACTICAL POWER AMPLIFIERS

This section presents several design examples for power amplifiers. A two watt bipolar power amplifier was presented in Chapter 1 with the "Beginner's Transmitter." Some simple power meter circuits were also included.

A CW-QRP Rig Amplifier

A familiar RF power amplifier encountered by the experimenter is that used with a low power (QRP) transmitter. The popular design provides about 1.5-W output from a 12-V supply. The load resistance the collector would "like to see" is then

$$R_L = \frac{V_{CC}^2}{2 \cdot P_{OUT}} \qquad \text{Eq. 2.42}$$

Evaluation yields R_L=48, so close to 50 Ω that no impedance matching network is required at the output. Only a low pass filter is required to attenuate the strong harmonics that are often created by the circuit. The amplifier circuit is shown in **Fig 2.95**. The 7-MHz design illustrates the design ideas, which are frequency invariant.

The amplifier input is to be driven from a 50-Ω source. While not required, it promotes convenient measurement. The builder can then test and adjust the driver stages alone, with the earlier transmitter stages, and without the complications of the output amplifier. This amplifier will usually require a drive power of 20 to 100 mW, depending upon the transistor type used in the amplifier. The 50-Ω drive is transformed downward to "look like" a 12.5-Ω source at the base. This transformation provides the high base current required for efficient operation. The 18-Ω base resistor serves as a wideband load for the input driver, even during the part of the drive cycle when the base is reverse biased. Decreasing this resistance can improve stability at the price of gain.

Base matching occurs with T1, a simple transmission line transformer consisting of a bifilar winding on a ferrite core. These transformers are discussed in the filter chapter. Other impedance transformation circuits can also be used, including tuned L, π, or Tee networks. The stage that must drive this will probably be loaded with a higher impedance, perhaps 200 Ω. Another bifilar transformer could be used, or a single ferrite transformer with a 4:1 turns ratio could make the transition from 200 to 12.5 Ω in one step.

It is important that the base drive be provided by a low impedance source. A higher source resistance might supply the needed base current, but then develop high voltage during the negative part of the drive cycle. This could lead to emitter base breakdown, a phenomenon that creates transmitted noise and a slow performance degradation in the output transistor. Emitter-base breakdown is easily observed with a wideband oscilloscope. A low driving impedance also helps stability.

A small heat sink is needed for a TO-39 transistor such as the 2N3866 or 2N3553. A clip-on heat-sink will suffice. The transistor can even be soldered into a hole in a circuit board. If the latter method is used, the hole must be isolated from circuit ground with extra capacitance absorbed into the design.

The amplifier includes extra components that are not always needed. One is the familiar Zener diode at the collector. This should have a breakdown value of about 3 times V_{cc} but less than the transistor breakdown. The diode's purpose is to load the amplifier if it loses an output termination. The diode conducts only if the collector voltage becomes too high, thus saving the more expensive output transistor from damage. The typical Zener diode will have a relatively high capacitance, even before breakdown, requiring that the input C in the low pass filter be reduced in value.

The virtue of this diode is open to debate. It is often seen in amateur applications, especially with transistors not intended for Class C RF applications. It is not so common in commercial applications using transistors intended for RF. The protection function is easily studied with a high-speed oscilloscope.

An RF choke routes bias to the collector. An extra inductor is placed in series with the supply, providing a series impedance for decoupling. A resistor then parallels the decoupling choke, as discussed in an earlier section. An optimum decoupling RFC uses large lossy ferrite beads.

A 7-MHz series tuned circuit is formed by the 50-pF, 10-µH combination. The back-to-back diodes provide a short circuit for large RF signals, generating a convenient electronic T/R system. This scheme, and similar T/R methods are discussed in Chapter 6.

A low ripple Chebyshev low pass filter with a cutoff frequency of about 7.5 MHz is recommended. Details appear in Chapter 3. The capacitance at the transistor end of the filter should be reduced to account for Zener diode capacitance and the 50 pF related to the T/R. No component values are shown for this example.

The ideal transmitter design will include variable RF drive. Besides being useful for communications, it is a very useful experimental tool.

Amplifier adjustment consists of nothing more than varying the drive power while watching the output to a 50-Ω load. Amplifier operation without a load should be avoided. The output power should change smoothly with drive, with any jumps suggesting instability.

It is interesting to monitor efficiency while drive is varied. Drive is adjusted, output power is measured, power supply current is noted, input power is calculated, and the resulting efficiency is calculated. Efficiency is usually low when the output is considerably less than the design level, but increases with drive. It will often be possible to drive the amplifier to an output greater than 1.5 W, usually at the price of efficiency. If you are interested in higher output, the output network should be re-designed accordingly.

It is useful to examine amplifier performance with a variety of loads. This is easily done with a transmatch. The dummy

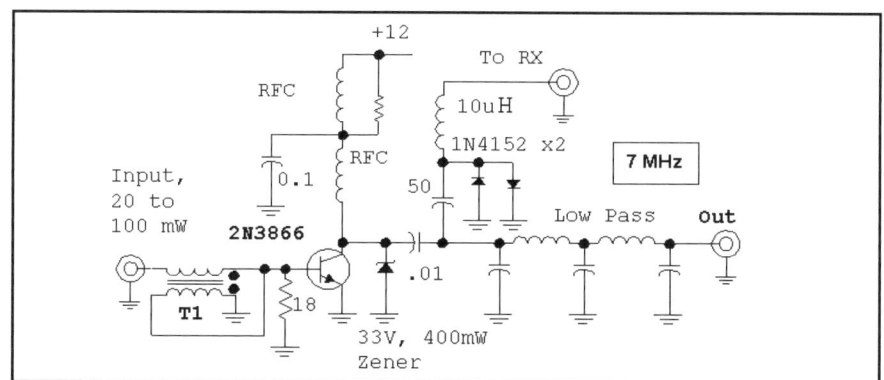

Fig 2.95—Typical output amplifier in a QRP transmitter.

Waveforms of a Class-C Amplifier

In an effort to garner intuition about the voltages in Class-C amplifiers, a low power experiment was performed with the circuit of **Fig A**. A signal generator provided base drive to the 2N3904 amplifier. The collector was biased at 5 V through a 4.5-μH high Q inductor. A variable capacitor allowed the inductor to be tuned to the drive frequency, or be detuned for an inductive collector termination. A Zener diode could be added to the circuit.

Test points are available at the transistor base and collector, allowing the voltages to be monitored with a high speed oscilloscope, a Tektronix 7704A in this case.

The first case examined was the reference for the experiment with results shown in **Fig B**. The low RF drive barely excites the base, but turns the transistor on at the peaks. The resulting current is a short spike, but still produces a very clean collector waveform, just reaching

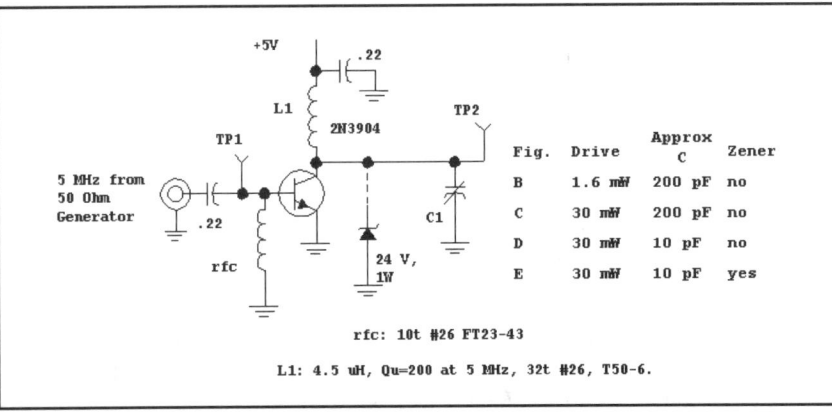

Fig A—RF Drive is applied to the base of a BJT while the un-terminated collector is biased through a tuned circuit. The data table relates results to operating conditions.

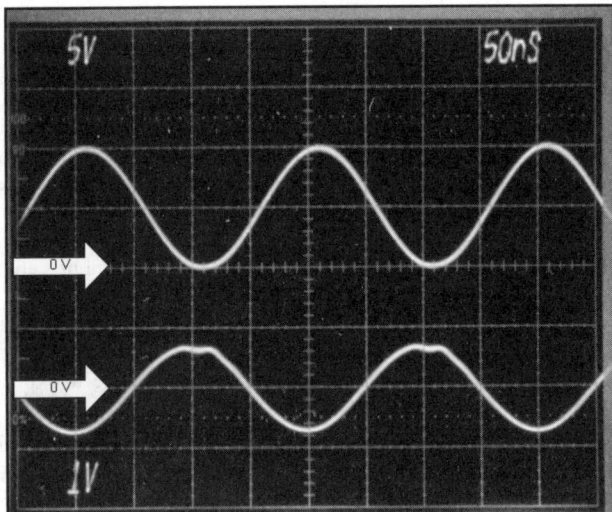

Fig B—Low drive produces a clean collector waveform in the upper trace. The lower trace shows the base voltage. In all cases, the vertical sensitivity is shown for each trace, and the 0-V line is marked at the left of the trace.

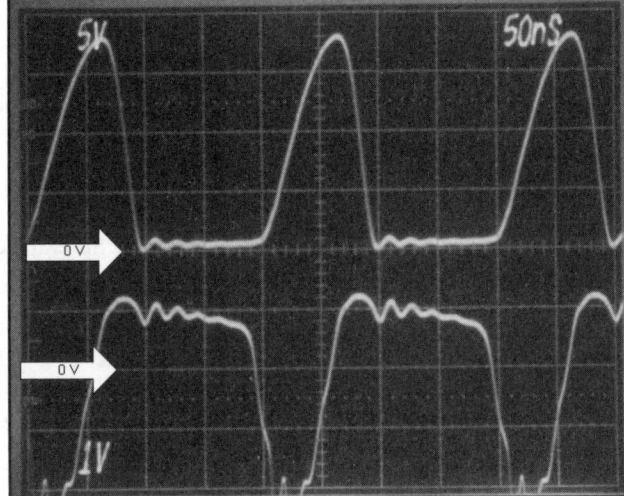

Fig C—Increased drive produces severe clipping in the base voltage and an 18-V peak collector signal.

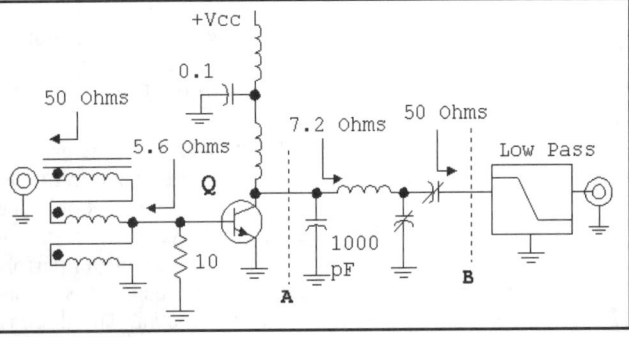

Fig 2.96—Schematic for a 10-W output Class C amplifier. The input autotransformer might consist of 3 turns through a binocular type balun transformer core. A Thomson 2SC1969 would be a good transistor choice, but try other parts as well. See text.

load is placed at the transmatch output, and the collector voltage is observed with an oscilloscope and 10X, 10-MΩ probe. The output power will be 1.5 W when the transmatch is properly adjusted. However, output power will drop considerably as the transmatch is "tweaked." The collector voltage will undergo major changes during this adjustment, with voltages sometimes going well beyond the expected 24-V value observed when operating in the usual class C mode with a

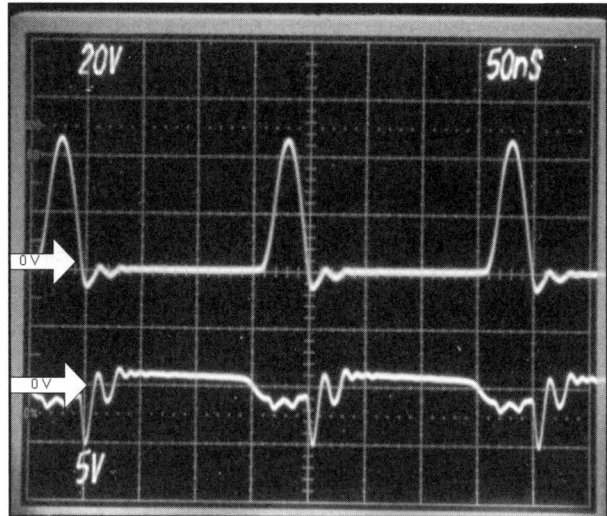

Fig D—Operation with an inductive load allows the collector voltage to ring up to over 40 V on positive peaks.

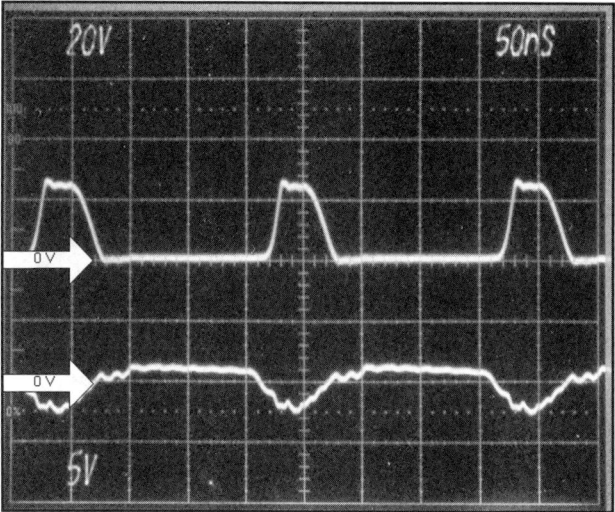

Fig E—The Zener diode is attached, effectively protecting the transistor from excess voltage.

zero at the bottom of the oscillation. The positive collector peak easily reaches twice the supply value. Just a hint of base conduction can be seen at the peak of the base waveform. The conduction must be occurring only over a small fraction of the applied waveform, for the base spends most of the cycle below 0.6 V. The Zener diode is disconnected for the first experiments.

The RF drive is now increased to 30 mW, more than we would normally use with this small transistor. The base voltage exceeds 1-V peak, which causes the collector voltage to drop to zero. The base voltage "tries" to stay on for more than half of the cycle, evidence of charge storage, a phenomenon intrinsic to the BJT. But when the base does stop conducting, the collector voltage "rings up" to 18 V, well beyond the 5-V supply. These results are in **Fig C**. Base voltage ringing at higher frequency is evident.

The collector resonance of the last example is eliminated by detuning the capacitor to a low value. The collector now sees a predominantly inductive impedance, resulting in the over 40-V peak signal of **Fig D**.

Note the change in vertical scale. The transistor is probably on the verge of damage at this point. Note also that the base voltage has changed, having been altered by the stressed collector.

The amplifier has no resistive load other than that represented by the unloaded resonator Q and provides no output power. The collector could be loaded by adding a resistor across the inductor, which would reduce the collector voltage. Even with loading, an inductive component in the collector impedance will allow high voltages to be generated.

The final experiment connects the Zener diode, producing the waveforms shown in **Fig E**. The collector voltage is now clipped at the 24-V breakdown of the Zener diode. The base conduction duty cycle is still high, a result of the high drive and charge storage. But the transistor is now saved from damage.

These experiments illustrate the effects of an inductive collector termination, Zener diode protection, and variable drive. The experiments could be extended with other devices, more aggressive applied stress, and loading that would allow DC collector current to increase.

"proper" termination. It is not unusual to see the amplifier go into oscillations during the severe mismatch that happens with this transmatch experiment. The oscillations should not be destructive at this power level, so long as the transistor has a modest heat sink and is protected against excessive collector voltage. It is a good idea to monitor the heat sink temperature (by touch is good enough) during these experiments. A current limited power supply is always useful, if not vital, during experiments of this sort.

Consider placing a pad between the transmitter and the transmatch. If we used, for example, a 1-dB pad, the worst-case return loss would be twice the attenuation, or 2 dB. The corresponding worst-case VSWR is 8.7:1 (see Eq 4.6.) If the amplifier can now withstand all possible adjustments of the transmatch, we say that the amplifier can withstand an 8.7:1 VSWR at all angles. The pad is, of course, removed after the test.

A 10-W CW Amplifier

While the 1.5-W amplifier is ideal for the seasoned QRP operator, others may want a bit more power. Outputs of 10 to 20 W are interesting. A few dB gain can make a big difference in results while still sporting and practical for portable operation.

There are numerous inexpensive bipolar transistors that will provide this power including many not normally used for RF. One should look for devices specified for a peak current that exceeds twice the anticipated level (1.5 to 2 A for this case), collector breakdown voltages well above the expected level (24 V here), and an F_t at least 3 to 5 times the expected operating frequency. Power dissipation should equal or exceed the planned output. A suggested 10-W amplifier is shown in **Fig 2.96**.

The input resistance is expected to be lower than for the 1-W amplifier, so we drive the circuit from a lower impedance source. This can be an auto-transformer, as shown in Fig 2.96, or a 3:1 or 4:1 turns ratio classic transformer. Binocular type ferrite balun cores are excellent in this application, noting that each turn now consists of one full pass through both holes in the core. Other wideband transformer configurations are listed in the transformer discussion in the Filter chapter. The input can also be driven from a low Q L-C-C Tee network like that used in the output, designed for an impedance of a few Ohms.

A 10-W output calls for a resistance of 7.2 Ω presented to the collector when Vcc=12. (See Eq 2.42) This amplifier uses tuned circuitry in the form an L-C-C type Tee network. This particular topology is excellent in that component values are usually practical, network Q can be kept low for low loss, and once designed, the network is easily "tweaked" for slightly different impedances. A good design value for Q is 2 to 3. The network between the dotted lines in Fig 2.96 is used for impedance transformation while the filter attenuates harmonics.

The normal Tee network is modified slightly; a fixed capacitor with a reactance magnitude near the load resistance value is placed at the collector. This kills high frequency gain, helping to ensure VHF stability. Silver mica capacitors are a good choice for network capacitors with ceramics for bypass and blocking elements.

A suitable test load is six paralleled 300-Ω, 2-W resistors. The drive is increased slowly while monitoring the RF output and collector current. The output Tee network capacitors are tuned for maximum output at each power level. An oscilloscope is especially useful during such experiments, allowing easy observation of oscillations, should they occur. More often than not, oscillations will occur at low frequencies, so a wideband 'scope is not mandatory. This amplifier will probably use no more than ¼-W of drive, so the builder may wish to add a pad if the driving transmitter delivers more than this.

The amplifier is set up for Class-C

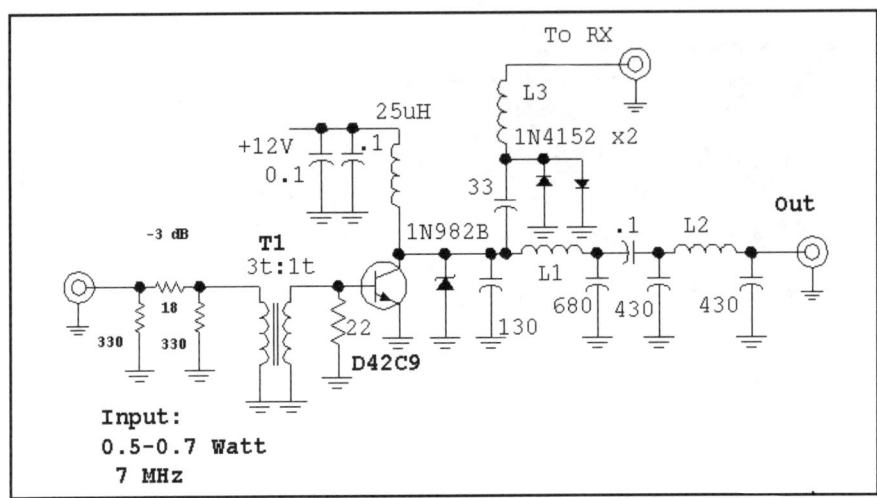

Fig 2.97—High efficiency amplifier after W7EL. T1=3-turn primary, 1-turn secondary, #30 wire, on Fair-Rite 2843002402 Balun core. Count one turn on a balun core as a pass through *both* holes. L1=0.71 µH = 13 t. on T44-6; L2= 1.05 µH = 19 t on T37-6, L3=15 mH molded RFC. Q is a GE D42C9 plastic power transistor.

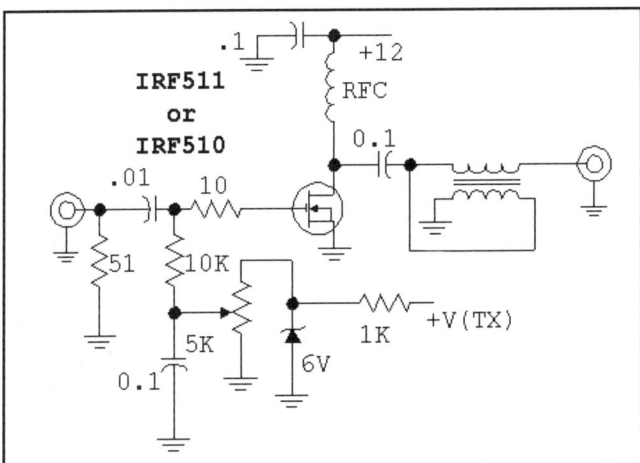

Fig 2.98—Simple HEXFET linear amplifier for QRP rigs.

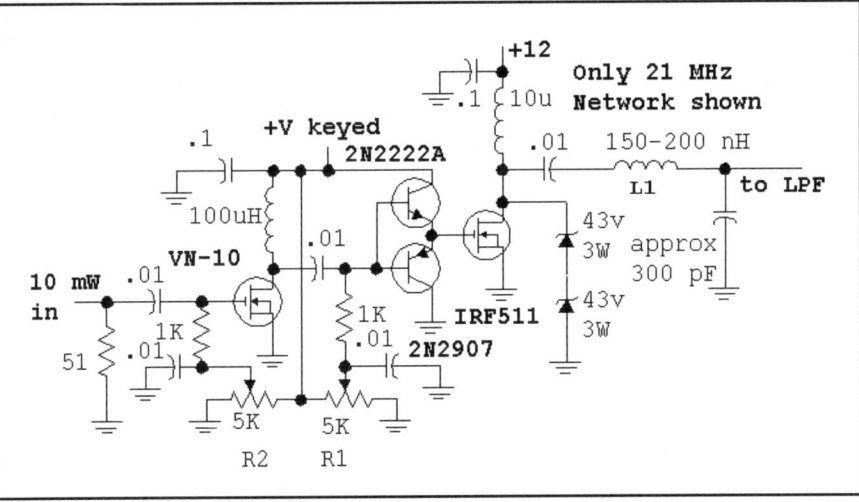

Fig 2.99—Dual band Direct Coupled HEXFET Amplifier after W7EL. This circuit operates at 14 and 21 MHz. L1 is 7 turns on a T37-6 and is the inductor for an L-Network at 21 MHz. The 1N5367 Zener diodes protecting the FET drain add about 140 pF to the circuit and are a vital part of the network. The band-switch adds more series inductance for a 14-MHz L-Network. Both impedance transforming networks are followed by low pass filters. R1, 5 kΩ, is adjusted for about 20-mA quiescent current in the IRF511, while R2, 5 kΩ, sets the quiescent current in the VN10 at 40 mA. The keyed driver power supply is less than +12 and is varied to establish output power.

operation. It is easily modified for class AB linear operation with the methods outlined below. Linear biasing is discussed below.

An Enhanced Efficiency Amplifier

An interesting and subtle amplifier from Roy Lewallen, W7EL, is presented in **Fig 2.97**. Dubbed the "Brickette," it was intended to follow a 1.5-W output, 7 MHz QRP transceiver.

This amplifier used an unusual transistor, a GE D42C9. The available drive is attenuated with a 3-dB pad, which was needed for stability. The original W7EL application used a 6-dB pad. The amplifier contains the usual Zener protection diode, but now with a 75-V breakdown. A peak collector voltage of 65 was measured with this circuit, even with $V_{cc}=12.0$ V. The circuit transforming the 50-Ω load to a lower value at the collector is a simple L-network. The resistance presented to the collector is higher than expected, and is inductive, allowing the high RF voltages. The net result is a collector efficiency of 85% or greater with an output of 7 to 9 W. What began as a Class C design probably now operates in Class E. The measurements have been repeated and confirmed with several versions of the circuit, all showing high efficiency.

The adjustment procedure was similar to that presented for the 10-W design. However, Roy kept increasing drive while adjusting the output network for increased power and efficiency.

The T/R series-tuned circuit is attached to the collector. Although the networks present an impedance less than 50 Ω to the receiver, the mismatch is not a problem at 7 MHz.

HEXFET Amplifiers

Power FETs became popular in the late 1970s. While some manufacturers introduced devices specified for RF, the market was dominated by switching applications. A major supplier is International Rectifier with a line of devices called HEXFETs.

The HEXFETs are available as both N and P channel enhancement mode parts with a gate threshold around 4 V. The transconductance of the typical N-channel device is very high, often rivaling that of a bipolar power transistor at comparable currents. While the input gate is a very high impedance at DC, high capacitance at all three terminals limits high frequency gain. HEXFETs are often high voltage devices, allowing a wide variety of supply voltages.

Fig 2.98 shows an RF amplifier using an IRF511 or the IRF510, preferred for higher breakdown. Either part has a low "on" resistance of 0.6 Ω, important for efficiency. This circuit is set up for an output of about 6 W from a 12-V supply. A 2:1 turns ratio transformer generates a 12-Ω drain load. This class AB circuit will function in either CW or linear SSB applications. The bias should be adjusted for a quiescent current of 100 mA or more for SSB while lower levels are suitable for CW. The output transformer is a bifilar winding on a ferrite core and is suitable for any of the HF bands. We have used this circuit up through 14 MHz. The FET should reside on a modest heat sink.

The HEXFET amplifier uses a 10-Ω gate resistor to preserve HF stability. A ferrite bead should **not** be substituted for the resistor.[18]

An interesting direct-coupled amplifier appears in **Fig 2.99**. This circuit, another creation of W7EL, uses a dc coupled IRF511 to generate an output of 5 W at either 14 or 21 MHz with a measured efficiency of about 75%.

Higher Powers

HEXFETs offer an inexpensive and interesting route to higher power. We have built single band CW amplifiers for output powers from 10 to 50 W on many of the HF bands. The inexpensive IRF530 HEXFET is an excellent choice for the bands up through 14 MHz. A 30-W 7-MHz CW amplifier is described later.

The IRFP440 and IRFP450 have been used in high efficiency CW amplifiers discussed later. These parts should also offer interesting opportunities for the experimenter. Although more expensive than HEXFETs, some vendors build parts especially for RF power applications. A search of the web can yield numerous data with suggested experiments. See, for example, an interesting paper by K4XU and the related Web site of Advanced Power Technology at **www.advancedpower.com**.[19]

SSB Amplifiers

The bipolar and FET amplifiers presented can be adapted for linear operation as shown in **Fig 2.100**. Bipolar transistor base bias should come from a voltage source. If the more typical current source is used, the DC current cannot easily increase with RF drive as is needed for Class AB operation. A voltage source bias uses a diode as a shunt "regulator," Fig 2-100A. The diode is biased with a resistor from the same supply that powers the amplifier. The silicon diode is in intimate thermal communications with the output transistor. Some designs us a stud-mounted diode bolted to the PA transistor heat sink. Others attach the diode to the transistor with epoxy.

The BJT amplifier is usually biased at the quiescent level recommended by the transistor manufacturer. A 10-W part might use an idling collector current of 20 to 30 mA. A larger current should flow through R-bias with the diode serving as a shunt regulator. Increasing the resistor current increases the standing current in the amplifier, one of the handles available to the experimenter for improved IMD performance from the amplifier.

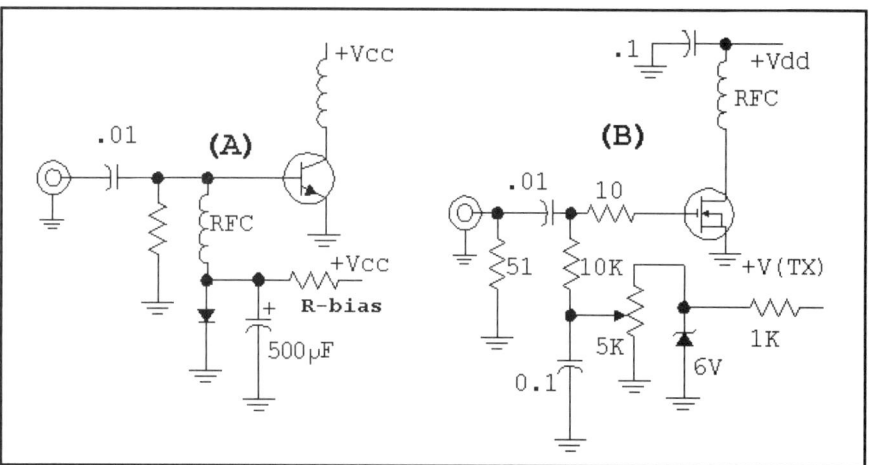

Fig 2.100—Biasing schemes for linear amplifier operation of (A) bipolar transistors and (B) power FETs. The base RFC used with the BJT can have small reactance, for the input impedance is low. The diode is bypassed with a 500-μF electrolytic capacitor. The base resistor may or may not be needed. R-bias in (A) should have moderate dissipation, for the current may be high.

Amplifier Design Basics 2.37

Fig 2.100B shows FET biasing for SSB. This is generally simpler than with a BJT, for bias current is low. The FET bias is easily controlled with small transistors, easing T/R switching problems. As with bipolar transistor amplifiers, the FET circuits present a compromise between efficiency and linearity. Amplifier IMD can be reduced with higher standing currents, although the heat sink requirements grow.

Amplifier biasing methods are discussed in more detail in the text by Dye and Granberg.[20] Included are schemes for temperature compensation.

Push-pull operation is common with both FET and bipolar linear amplifiers. There are several advantages to this. First, two devices are used instead of one, spreading the thermal load over a larger region. Second, transformer coupling between device inputs will prevent large reverse voltages on bipolar base-emitter junctions. One forward biased junction serves to clamp the reverse voltage on the other device. Finally, the balanced operation will reduce even order harmonic and intermodulation distortion.

Negative feedback is often used with Class AB amplifiers, usually in the form of an ac coupled resistor between base and collector, or gate and drain. Feedback stabilizes gain over frequency. The negative feedback is applied individually to each device in a push-pull pair. Negative feedback is sometimes extracted from a winding in an output transformer or bias element in a push pull pair.

Push pull bipolar transistors are essentially in parallel for biasing. For this reason, and to help maintain RF balance, RF power bipolar transistors are often sold in matched pairs. This has become so common that the price penalty is minimal.

The ease of FET biasing includes push pull amplifiers, which is illustrated in the practical circuit shown in **Fig 2.101**. This SSB linear amplifier, the work of AA3X (now K3BT), uses a pair of IRF511s in a push pull circuit to develop an output of 30 W PEP. The circuit uses a solid ferrite block for the output transformer. **Fig 2.102** shows a sketch for the output transformer, T3.

Separate bias lines set up a quiescent current for each FET. A DVM measuring total current during bias adjustment allows the two currents to be set equal to each other. While matched transistors might be

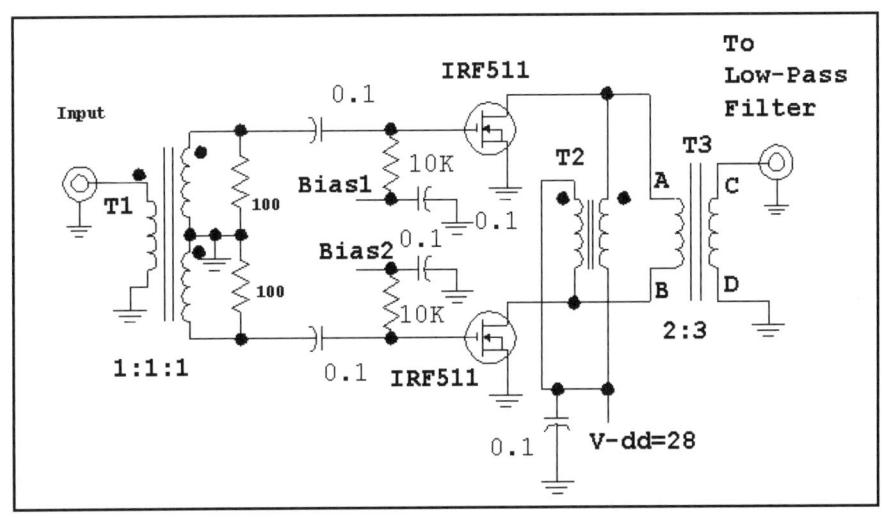

Fig 2.101—An amplifier using a push-pull pair of IRF511s. This circuit, the creation of AA3X, is capable of up to 30-W output with Vdd=28 V on the lower HF bands. Reduced output and gain are available at 14 and even 21 MHz. Input transformer T1 is 12 trifilar turns #26 on a FT50-43 ferrite toroid. T2 is 12 bifilar turns of #22 on a stack of two FT37-50 toroids. This amplifier was originally in *QST*, Hints and Kinks, for January, 1993, page 50.[21] See reference and text for practical details.

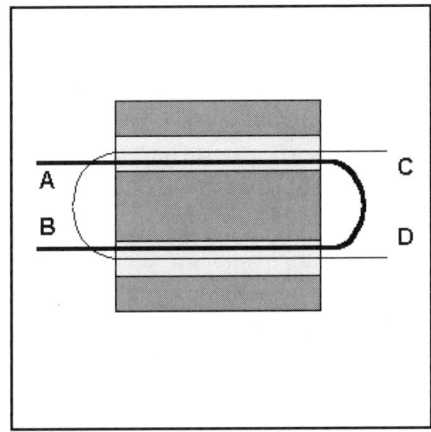

Fig 2.102—Transformer detail for T3 of the AA3X amplifier. The primary, A-B, shown here as a single turn, but actually uses two turns, two complete passes through the core. The secondary (also just shown as one turn) is 3 turns, three complete passes through the core. The windings end on opposite sides of the ferrite block, a BN-43-7051.

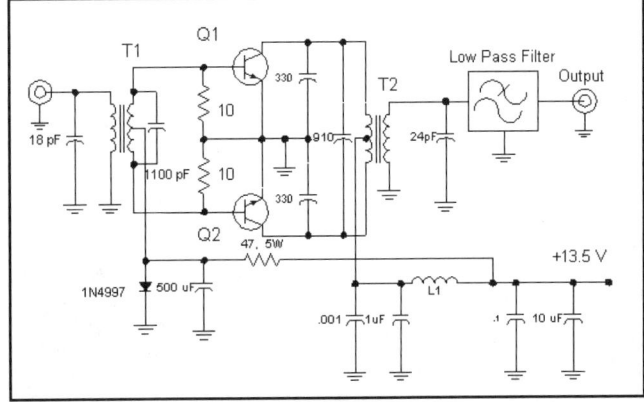

Fig 2.103—100-W BJT Amplifier. This circuit, originally described in *Motorola Engineering Bulletin*, EB63,[22] is capable of an output power of over 100 W from 3 to 30 MHz. Q1 and Q2 are matched MRF454s mounted to a large heat sink. L1 is a piece of #18 wire loaded with 9 ferrite beads. Both transformers have a 4:1 turns ratio with the winding, consisting of ferrite loaded brass pipe, attached to the transistors. The one-turn windings are center tapped. The 4-turn input and output windings are plastic covered wire wound through the center of the tubes. Similar transformers could be built with 3/16-inch diameter brass tubing (available at hobby stores) loaded with FB-77-63 Ferrite beads. T1 would use 4 while T2 would use 10 beads. A larger bead and tubing size would be better for T2. The transformers used in our amplifier were supplied with the kit from Communication Concepts, Inc. of Beavercreek, Ohio. See *QST* advertisements for a current address. CCI has several other kits for power amplifiers.

desirable, K3BT reports that he has had good results with devices with severely mismatched thresholds. Equal currents of about 20 mA per transistor are recommended. This amplifier has been used on the amateur bands from 3.5 to 21 MHz, although the available output power is less at the higher end.

The output transformer (3:2 turns ratio) presents a load of 22 Ω between the two drains. The resulting load is lower than might be desired for high efficiency, a common tradeoff with linear amplifiers favoring lower distortion. The K3BT amplifier should be built with a large heat sink, especially if experiments are planned with variable bias currents.

Careful low impedance termination of the HEXFET inputs provides stability. The power gain is still high enough to make the parts very useful, even with the reduced gain related to the low source impedance. The stability problem is largely the result of internal feedback within the FETs. While extremely difficult with bipolar transistors, it becomes possible with FETs to neutralize the circuits, canceling the destabilizing effects of internal feedback. These methods were common place with vacuum tubes, but have largely been ignored with semiconductors. A neutralized push-pull 18-MHz linear power amplifier using IRF-511s is included in Chapter 11.

A high power bipolar transistor amplifier is shown in **Fig 2.103**. This circuit was originally described in a Motorola engineering buildetin, EB63 (ref 22), and was offered in kit form from CCI. (**www.communication-concepts.com**) The amplifier is capable of over 100 W of output over the entire HF spectrum. A matched pair of MRF454s is used with a 13.5-V power supply.

This circuit is a classic, similar to many of the output amplifiers in typical transceivers. Brass pipe transformers are used at both the input and the output. Some negative feedback is used, along with capacitive loading to improve gain flatness. This version of the amplifier has been tested over the 2 to 30-MHz band and found to operate as described in the applications note, although we did not measure IMD. The circuit has been used extensively on the 40-M band. It performed well as a SSB amplifier, being easily driven by a 1.5-W QRP SSB transceiver. It has seen more service following a 1-W CW transmitter.

The original version of this amplifier included an RF actuated circuit to control a built-in T/R relay. The RF actuated scheme was found to be completely unsuitable for either CW or SSB use. When RF drive was initially applied, the relay was activated. But amplifier current started to grow before the output was properly terminated, causing the amplifier to draw excessive current. The power supply was current limited at 25 A. As the supply went into limiting, the voltage dropped to 7 V before starting to recover. The relay then dropped out and the cycle repeated. The relay chattered for about half a second before stabilizing. The RF actuated circuitry was eventually replaced with an electronic T/R system with diode switching.

T2, the output transformer, has a single turn between collectors with a 4-turn secondary. The 4:1 turns ratio transforms the 50-Ω load to appear as a 3.1-Ω load, collector-to-collector. The load applied to each collector is then 1.56 Ω. Rearrangement of Eq 2.42 shows that an output of 58 W should be available from each device at V_{cc}=13.5 V for a net output of 117 W.

In spite of the T/R problems, the amplifier is a recommended circuit. The MRF454 is very robust, and has provided us with classic power amplifier experience. We recommend modified bypassing to use parallel capacitors of **equal** value.

A Look at some High Efficiency Amplifiers

All of the power amplifiers presented are conceptually simple, many using the same or similar schematic diagrams, even though intended for differing applications. Class-C amplifiers are designed by picking a load resistance using Eq 2.42 and designing an output network to achieve that load at the operating frequency. The device is then biased for zero current without drive. With the usual threshold, application of an input sine wave produces Class-C operation.

Linear amplifier design is similar. An output network is designed for the peak envelope output, again with Eq 2.42. Moving toward even lower load resistance may enhance linearity at the price of efficiency. The linear amplifier is biased for class AB operation. This begins with class A bias, but usually allows device current to increase with applied RF drive. While efficiency at the peak envelope power is poor, the normal voice has an average power well below the peak, providing a useful compromise.

An amplifier discussed earlier (the Fig 2.97 circuit by W7EL) featured improved efficiency. It is interesting to examine the networks that produced this result.

Fig 2.104 shows a schematic and a Smith Chart impedance plot for the output matching network the Beginner's Transmitter of Chapter 1. Frequency sweeps from 3.5 to 21 MHz for this 7-MHz design. The impedance at 7 MHz is nearly real at about 25 Ω, providing the needed load for Class-C operation. The impedance is capacitive for all other frequencies. This

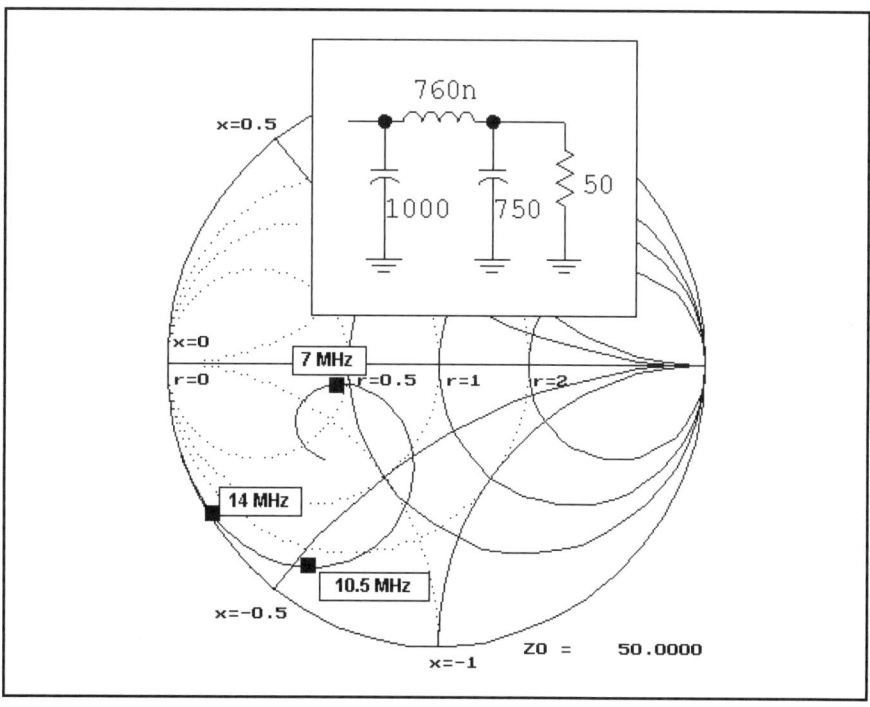

Fig 2.104—Smith chart plot of the impedance "seen" by the collector of the 2N5321 2-W "Beginner's Transmitter" from Chapter 1.

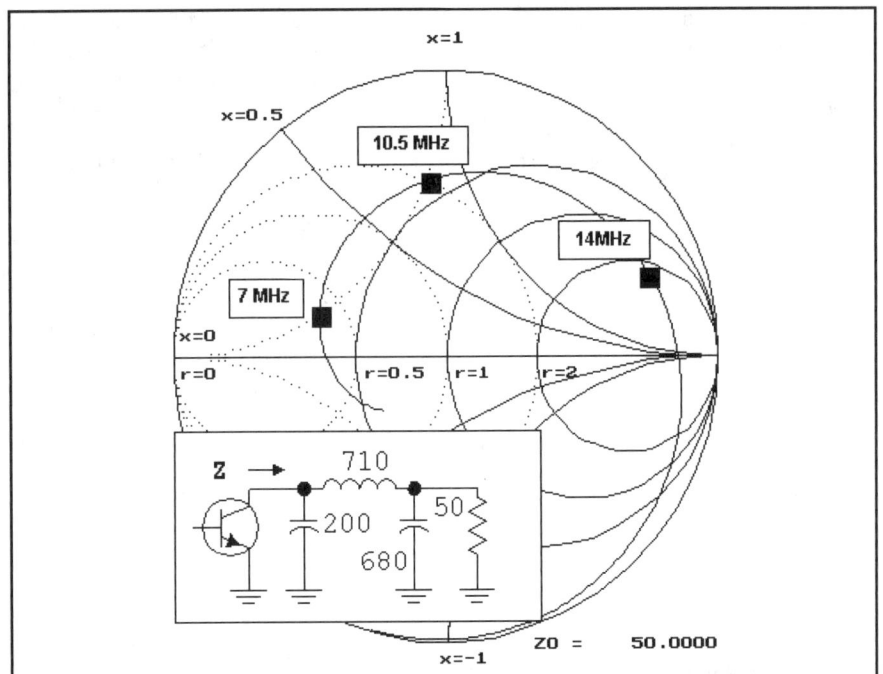

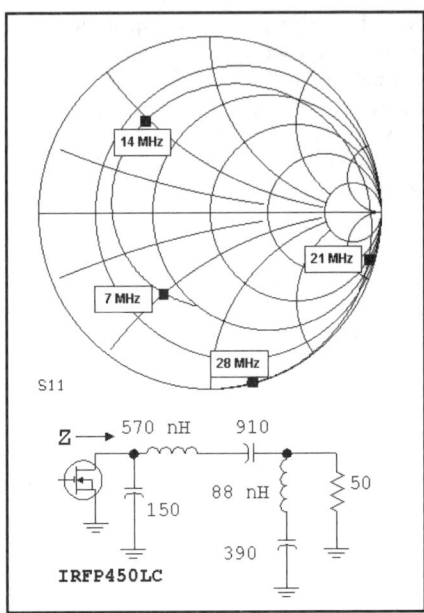

Fig 2.105—Smith Chart plot for the Brickette of W7EL, shown in Fig 2.97. The impedance is inductive until reaching the second harmonic. There is a slight change in the plot when additional C is added at the collector to account for the Zener diode.

Fig 2.106—50-Ω Smith chart display of impedance for a 400-W amplifier operating at 13.5 MHz. See text.

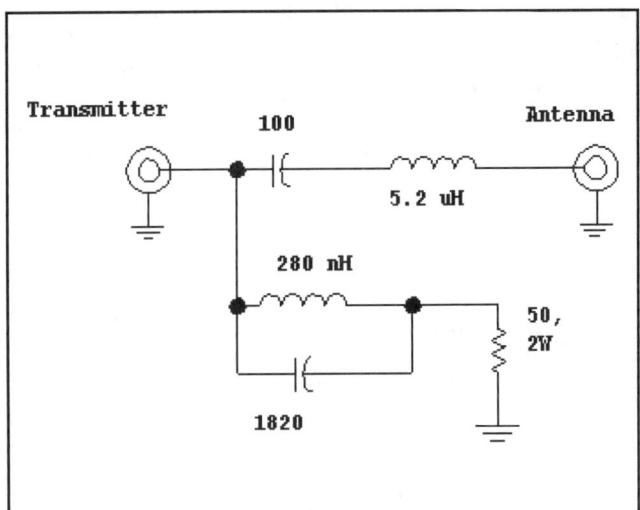

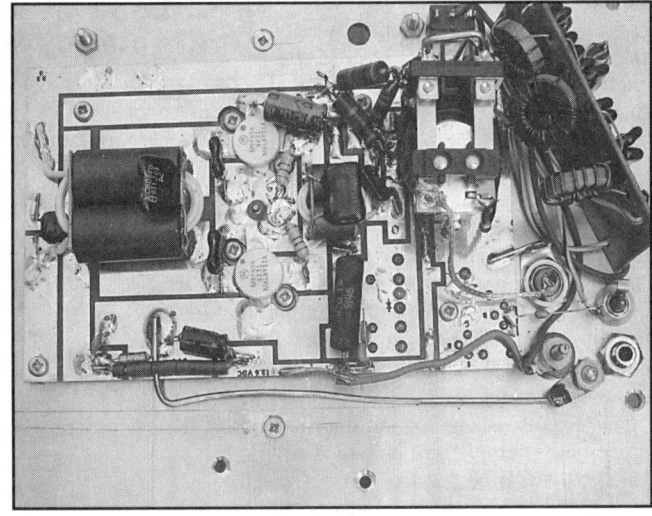

Fig 2.107—Diplexer, bandpass-bandstop type, used for harmonic attenuation from a 7-MHz transmitter. The reader should consult the original *QST* article[23] for details.

Fig 2.108—Top view of 100-W bipolar amplifier. The board is bolted to a large heat sink that is also the top of the module.

amplifier (7 MHz, 2.2-W output, 12-volt supply) was stable and reproducible, but had only 50% efficiency.

The contrasting amplifier was W7EL's "Brickette" of Fig 2.97. The output network is also a π-network, and the resulting impedance plot is shown in **Fig 2.105**. The plot differs from the simple Class-C circuit. The impedance has a real part of about 17 Ω near the design frequency, but is inductive for much of the sweep. R_L is about twice that we would use for a Class-C design. Z becomes capacitive only above the 2nd harmonic. This amplifier has excellent efficiency (85 to 90%) at 7 to 9-W output (7 MHz, 12-V supply) and has been stable.

Class-E amplifiers have become of increasing interest in the past few years. Recent HEXFET offerings from International Rectifier provide very high power capability at modest price. While the amplifiers are now used only for digital applications (including CW,) recent work has paved the way for SSB with non-linear high efficiency amplifiers.[24] The recent work of greatest interest to the experimenter evolves from the EE department at California Institute of Technology.[25]

Fig 2.106 shows an example of a high efficiency Class-E amplifier.[26] The partial schematic shows two modifications to the simple pi-network used in the other two circuits. First, the normal inductor is replaced by a series LC. This provides the same inductive reactance at the 13.5-MHz

Fig 2.109—A 1.5-W 7-MHz amplifier using a 2N3866.

Fig 2.110—An RF power amplifier using an IRF510 HEXFET. The output network is an LCC type Tee-network. Up to 10 W was obtained from this circuit.

Fig 2.111—A high efficiency 7-MHz amplifier (circuit of Fig 2.97).

design frequency, but greater inductive reactance at higher frequencies. This presents the needed load to the FET drain needed to allow the voltage to grow ("ring up") to values much larger than the supply and offer the phase control needed for efficiency. A Class-E amplifier is characterized by high current flowing only when the voltage across the device is close to zero.

The other modification is at the load end of the network. The usual parallel capacitor is replaced with a parallel-connected series tuned circuit (88 nH and 390 pF). This circuit is resonant at the 2nd harmonic of the 13.5-MHz drive frequency of this example. This amplifier provides an output of 400 W with a drain efficiency of 86%. This circuit, which uses a 120-V supply, could be adapted to the 20-meter amateur band. The load impedance is $13.5 + j19\ \Omega$ at the 13.5-MHz operating frequency, but is purely capacitive by the time the 2nd harmonic is reached. Eq 2.42 would predict an 18-Ω load for this output and V_{dd}. This circuit is very similar to the 7-MHz design presented in *QST* for May 1997.[27]

Spectral purity is an issue with these amplifiers. The resonant trap at twice the operating frequency included in the designs helps. One would normally insert additional low pass filters to attenuate harmonics. However, this normal low pass filter has an input impedance that is real and 50 Ω at the operating frequency, but is almost a short circuit at the harmonics. An improved harmonic reduction filter form is shown in **Fig 2.107**. This circuit is called a diplexer and has the characteristic that the input impedance is 50 Ω at all frequencies. Other diplexers are used elsewhere in the book.

Fig 2.108 through **Fig 2.111** show some of the design implementations described in this section.

2.11 A 30 W, 7 MHZ POWER AMPLIFIER

While QRP can be great fun, especially in a portable application, there are times when more power can make a large difference in station effectiveness. The amplifier shown in **Fig 2.112** is intended to boost the output of a QRP rig to the 30 to 40-W level with an inexpensive HEXFET. A moderate heat sink is used, allowing extended testing and operation.

The amplifier requires about 1 W of drive for full output. If more drive is available, it may be dissipated in an input attenuator. A 3.3-dB pad is shown in the figure. This is followed by T1, a bifilar wound ferrite transformer providing gate drive for the FET. The low impedance drive is needed to accommodate the high input C of the IRF530. A 10-Ω, 1-W resistor provides a wide band termination.

The drain circuit is supplied with a +25-V source through an RFC (L1) made with a large powdered iron toroid. The exact value is not critical. The RF resistance that should be presented to the drain for a 30-W output is 10 Ω. This is realized with T2, a bifilar wound ferrite transformer. This part of the circuit is open to considerable experimentation for those so inclined. T2 is followed by a low pass filter for harmonic attenuation. Inductor L5 is tuned for parallel resonance at 7 MHz. An attached resistor then provides a termination for the amplifier transistor at frequencies other than 7 MHz when a trans-match with a peaked high pass characteristic is used. The combination emulates the diplexer described earlier.

A T/R system is included to supply a signal to the receiver input. As shown, this system has a measured insertion loss of about 3 dB, the result of the low Q RF choke at L7 and the shunting effect of C1. This loss of no consequence at 7 MHz.

An adjustable bias is available for this amplifier, provided by a PNP switch circuit keyed with a signal from the driving transmitter. A grounding signal is applied at J1 to turn on the PNP switch. FET bias is adjusted at R1 (S1 open) for a few milliamperes of drain current with no RF drive during key-down periods. The switching

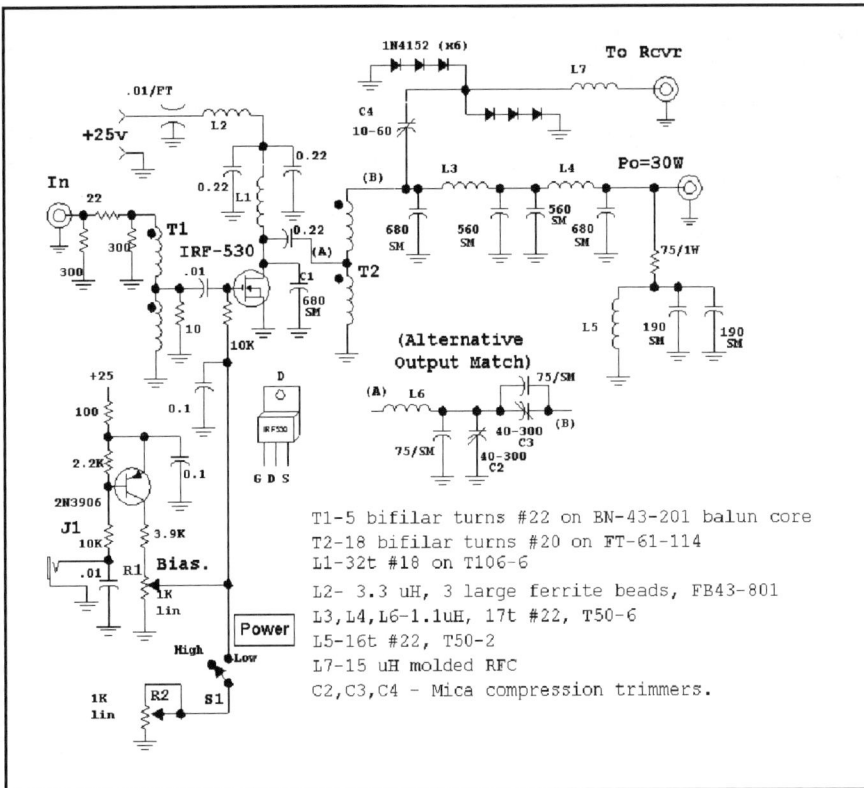

Fig 2.112—Schematic for the 30-W, 7-MHz power amplifier. See text for details.

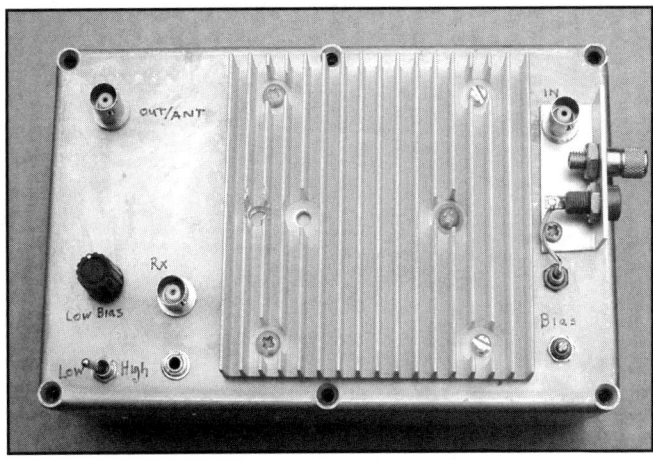

Fig 2.113 —The 30-W amplifier.

gests stability problems. We saw no such problems with this amplifier.

Monitoring drain voltage with an oscilloscope (60-MHz bandwidth) revealed some disturbing characteristics. When C1 is absent, the drain voltage contained extensive harmonic current, evident from the fine structure around the positive peaks. While these harmonics are blocked from the outside world by the low pass filter, they should be controlled or reduced at the FET where they can compromise efficiency. The low pass filter was temporarily removed from the system, allowing the wideband output load to appear at point "B" in the circuit. This immediately cleansed the signal at the drain, removing the high frequency spikes. The low pass filter appears as a large shunt capacitance at plane B in the figure. This load is reflected through T2, allowing the transformer leakage inductance to appear at the FET drain. This is the load that will allow the higher frequency currents to flow.

The ideal solution for this situation is a diplexed low pass output filter, mentioned above. Sabin studied diplexer filters and presented his work in *QEX* for July/August, 1999.[27] The amplifier used with these filters was described in the Nov/Dec 1999 *QEX*;[28] both papers are excellent and are included on the book CD.

We elected not to use a diplexer filter in this amplifier. Rather, C1 is included at the drain. With C1 in place, the drain voltage goes up to about 60 V, well within the FET ratings. Although there is still distortion in the drain waveform, harmonic currents are not excessive.

Several transformer structures were tried at T2. The most interesting variation replaced the wideband transformer with a narrow band LCC type Tee-network, also shown in the figure. This circuit was adjusted for maximum output while slowly advancing drive power. Over 45 W of output was available with this circuit. The drain waveform was very clean, reaching a peak of 75 V. C1 was still present at the FET drain during this experiment. The T-network was designed to provide 10 Ω to the drain with a Q of 5. Experiments with other networks will allow you to move over the ill-defined border between class B or C operation toward class E. FETs with higher voltage ratings should be considered for these experiments.

This circuit has been used in several variations for years and on several bands up through 14 MHz. Higher bands should also be possible with experimentation. We have always been impressed with the robust character of the devices. The typical power supply used is a surplus open-frame linear regulated type with 4-A

action removes bias during receive, preventing amplifier noise from overwhelming the receiver. The standing current for SSB operation can be adjusted to larger values, up to 1 A. Monitor heat sink temperature to be sure that it never becomes too hot during transmit periods.

Throwing switch S1 to the low power position allows the power output to be dropped to levels from well below a watt up to 5 W, controlled by a knob on R2. This scheme works well even with an output less than the input drive.

Initial turn-on begins by terminating the amplifier in a 50-Ω load with at least 30 W of dissipation capability. A current limited power supply is attached. RF drive well below the required level is applied while the output is monitored with an oscilloscope or RF detector. Drive is slowly increased while examining the output waveforms. Clean signals with smoothly varying levels should be seen with changes in drive. Any sudden change sug-

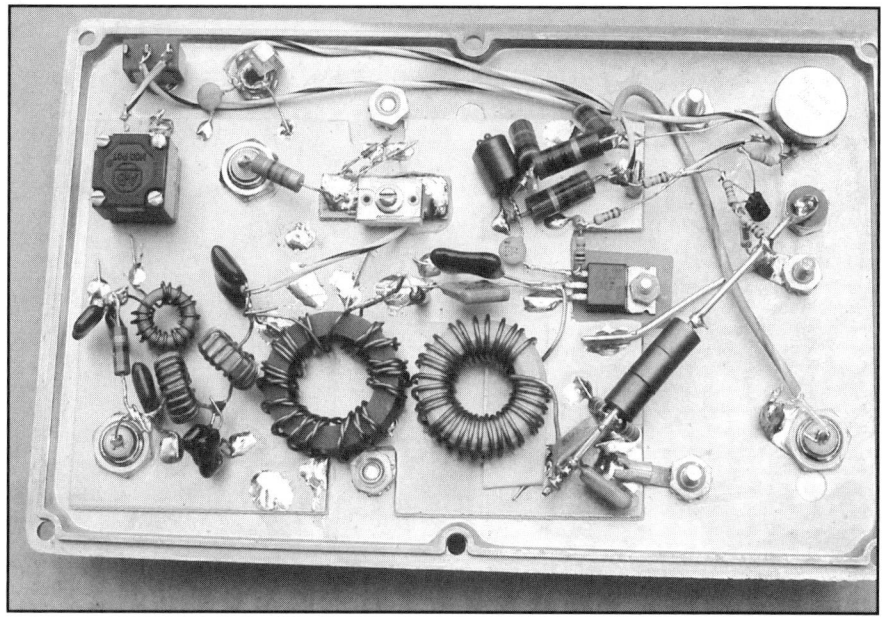

Fig 2.114—Inside the 30-W amplifier.

current limiting. Typical current is 2.5 A. The use of slight forward bias helps to guarantee stability.

The present interest in QRP operation is generally applauded as both fun and worthwhile. However, many folks miss some exciting experimental rewards by an overly strong adherence to a synthetic 5-W limit. This amplifier is a chance to examine the other side of the power switch. See **Fig 2.113** and **Fig 2.114** for two views of the 30-W amplifier.

REFERENCES

1. W. Hayward, *Introduction to Radio Frequency Design*, Prentice-Hall, 1982, and ARRL, 1994.

2. P. Horowitz and W. Hill, *The Art of Electronics*, Second Edition, Cambridge University Press, 1989.

3. P. Gray and R. Meyer, *Analysis and Design of Analog Integrated Circuits*, Second Edition, Wiley, 1984.

4. *IEEE Standard Dictionary of Electrical and Electronics Terms*, ANSI/IEEE Std 100/1984, Published by IEEE and Distributed by John Wiley, 1984.

5. See Reference 1.

6. See Reference 1.

7. *The ARRL Handbook for Radio Amateurs*, ARRL, 1995, pp 17.5-8, 17.10, 17.22-25.

8. D. Norton, "High Dynamic Range Transistor Amplifiers Using Lossless Feedback," *Microwave Journal*, May, 1976, pp 53-57.

9. U. Rohde, "Eight Ways to Better Radio Receiver Design", *Electronics*, Feb 20, 1975, p 87.

10. See Reference 1, p 216.

11. W. Carver, "A High-Performance AGC/IF Subsystem", *QST*, May, 1996, pp 39-44.

12. C. Trask, "Common Base Amplifier Linearization Using Augmentation," *RF Design*, Oct, 1999, pp 30-34.

13. C. Trask, "Distortion Improvement of Lossless Feedback Amplifiers Using Augmentation," *Proceedings of the 1999 IEEE Midwest Symposium on Circuits and Systems*, Las Cruces, NM, Aug, 1999, Vol 2, pp 951-954.

14. V. Koren, "A New Negative Feedback Amplifier," *RF Design*, Feb, 1989, pp 54-60.

15. R. Campbell, "A Novel High Frequency Single-Sideband Transmitter Using Constant-Envelope Modulation", *1998 IEEE MTT-S International Microwave Symposium Digest, 98.2*, (1998 Vol II [MWSYM]) pp 1121-1124.

16. H. Krauss, C. Bostian, and F. Raab, *Solid State Radio Engineering*, Wiley, 1980.

17. E. Lau, K. Chiu, J. Qin, J. Davis, K. Potter, and D. Rutledge, "High Efficiency Class-E Power Amplifiers" *QST*, May, 1997, pp 39-42 and Jun, 1997, pp 39-42.

18. Technical Correspondence, *QST*, Nov, 1989, p 61.

19. R. Frey, "A 300-W MOSFET Linear Amplifier for 50 MHz," *QEX*, May, 1999, pp 50-54 and "Letters to the Editor,"

QEX, Jul, 1999, p 63.

20. N. Dye and H. Granberg, *Radio Frequency Transistors: Principles and Practical Applications*, Butterworth-Heinemann, 1993.

21. J. Wyckoff, "Hints and Kinks", *QST*, Jan, 1993, p 50-51.

22. T. Bishop, "140W (PEP) Amateur Radio Linear Amplifier 2-30 MHz", *Communications Engineering Bulletin*, EB63, Motorola Semiconductor Products, Inc, Phoenix, AZ, Jul, 1978.

23. See Reference 17.

24. R. Campbell, "A Novel High Frequency Single-sideband Transmitter Using Constant-Envelope Modulation," *1998 MTT-S International Microwave Symposium*, Digest 98.2, (1998 Vol. II, [MWSYM]): pp 1121-1124.

25. See Reference 17.

26. J.F. Davis and D.B. Rutledge, "A Low-Cost Class-E Power Amplifier with Sine Wave Drive," *1998 MTT-S Inter-national Microwave Symposium*, Digest 98.2, (1998 Vol. II, [MWSYM]): pp 1113-1116.

27. W. Sabin, "Diplexer Filters for an HF MOSFET Power Amplifier," *QEX*, Jul/Aug, 1999, pp 20-26.

28. W. Sabin, "A 100-W MOSFET HF Amplifier", QEX, Nov/Dec, 1999, pp 31-40.

CHAPTER 3

Filters and Impedance Matching Circuits

Filters constitute one of the major blocks in a communications system and are especially important to the radio experimenter. The performance offered by a filter may well define the performance and/or cost of a project. The experimenter who can design and build his or her own filters has control over that performance and equipment cost.

There are several ways of segmenting filters into groups. The usual scheme segments filters according to frequency response, such as low pass vs high pass. Other methods segment by the kind of components used. In that regard, this chapter deals first with LC filters, and later with RC active and crystal filters. Filters can also be classified by the way they deal with impulses of energy. The filters presented in this chapter are generally "infinite impulse response" filters, or *IIR*. Finite impulse response filters (FIR) are detailed in a later chapter emphasizing digital signal processing (DSP).

3.1 FILTER BASICS

A filter is, in the most general sense, a circuit block that linearly modifies the nature of the signals applied to it. When we say linear, we mean that the output is a replica of the input, changed in amplitude and/or phase. However, no additional frequencies appear.

The term *domain* refers to our emphasis when describing and measuring a phenomenon. When a filter is examined in the frequency domain, we characterize the filter by the way it behaves with different frequencies. We may then change focus and examine the time domain response. For example, we may investigate the time delay imposed upon a signal as it passes through a filter. The DSP filter designer has the ability to simultaneously examine and often control both the time and frequency domain responses.

The response of a filter is measured by examining the transfer properties of the circuit. The voltage transfer function is the output voltage (usually across a termination) divided by the input voltage that caused the output. This is just the familiar voltage gain that we used with amplifiers. In the case of a filter, that "gain" is usually a loss, a number less than one, with a corresponding negative dB value.

Simple filters are built from mathemati-

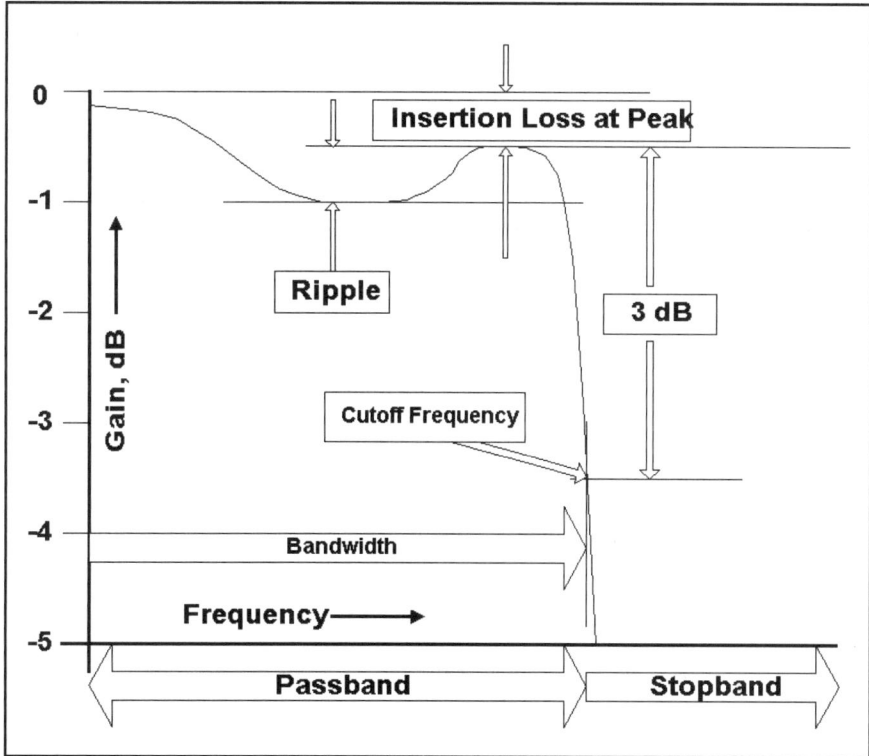

Fig 3.1—Low pass filter characteristics showing the passband and stopband, bandwidth, 3-dB cutoff, passband ripple, and insertion loss. This filter has approximately 0.5 dB IL at the frequency of peak response while passband ripple is also 0.5 dB. The vertical axis is the gain through the filter, output power Vs available input power when the filter is properly terminated. (Formally, the usual gain used is the forward scattering parameter, S21.) Horizontal axis is frequency.

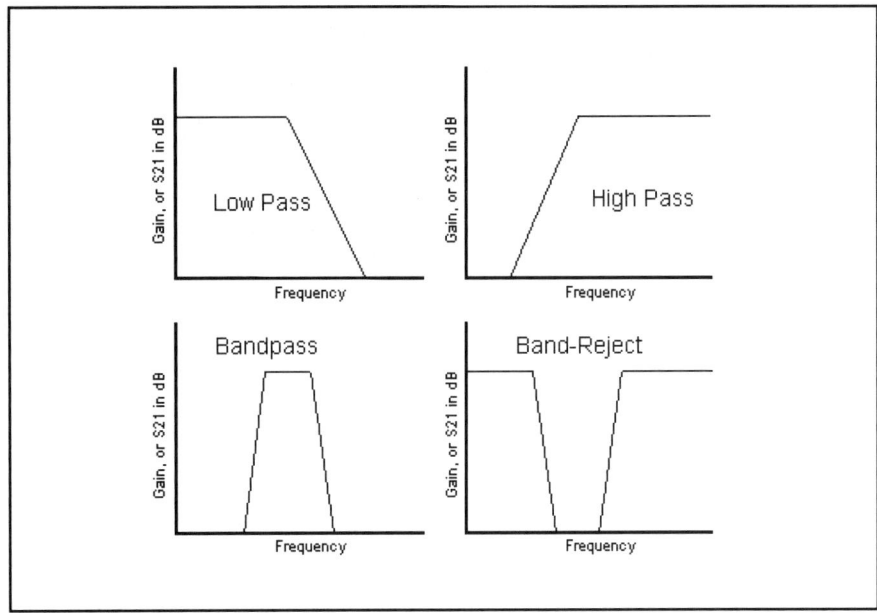

Fig 3.2—The frequency responses of various filter forms.

cally ideal inductors and capacitors. Such a filter, one without resistors, is called *lossless*. All of the power applied to a lossless filter is available at the output. Real filters containing resistive elements, desired or otherwise, will suffer from some loss. Loss in dB is a positive number, and loss as a power ratio is greater than 1.

The traditional filters we use are classified with regard to frequency domain response, illustrated with a low pass filter in **Fig 3.1**. This figure is a plot of filter gain vs frequency. We encountered several different kinds of power gain in Chapter 2. The one usually used with radio frequency filters is transducer gain.

A low-pass filter is one that transfers all input frequencies below a specified cutoff frequency. The spectrum below the cutoff is called the passband while the region of higher attenuation above the cutoff is called the stopband. A filter dissipates some of the available power applied, called insertion loss. The filter of Fig 3.1 has an insertion loss (IL) of about half a dB at the highest frequency peak. IL is about 0.1 dB at very low frequency. The cutoff frequency is usually defined as that frequency where the response is 3 dB less than the peak passband response. Additional variations in gain within the passband occur with some filters; these variations are termed passband ripple.

A high-pass filter is similar to the low pass except that the regions are interchanged; the passband, the region containing desired signals, is now above the stopband.

A bandpass filter is one that passes a given region, often narrow, while rejecting most frequencies. The bandwidth of a bandpass filter is the difference between two points 3 dB below a peak. A band-reject filter is the opposite, a filter that passes most of the spectrum while rejecting a specified region. Finally, an all-pass filter is one that passes all frequencies applied to its input. The all-pass filter is useful because it can alter the phase of signals passing through it without altering signal amplitude. The various types (except for the all-pass) are summarized with regard to frequency response in **Fig 3.2**.

Passive filters conserve energy; power flowing into the input must go somewhere. If input energy is at a frequency within the filter passband, that energy emerges at the filter output where it can be used. (A fraction of the energy is lost in any real, passive filter, being dissipated in the losses of the inductors and capacitors that form the circuit.) In contrast, energy in the filter stopband is reflected. That is, an impedance mismatch is created by the filter elements such that power is not efficiently delivered from the source, through the filter and to the output. Most LC filters display this property, allowing us to use input impedance match as another way to examine filter performance. The primary performance indicator remains the transfer function.

3.2 THE LOW-PASS FILTER—DESIGN AND EXTENSION

A low pass is a filter that passes frequencies below a specified cutoff frequency while attenuating those above. It is a vital component of almost any communications system. The low pass is also the basis for other filter forms. Once we have a low-pass filter designed, cataloged, and understood, the properties and the component values can be extended to generate any of the other basic filter types. One extension changes the low pass into a high-pass circuit. Another modification changes the low pass to a bandpass. A band-reject filter is a direct result of transforming a high-pass circuit, itself derived from a low-pass prototype. The practical application details of these methods will be presented, although many mathematical details will be ignored in this treatment. Analytic detail can be found in *Introduction to Radio Frequency Design*[1] or numerous other texts.

A simple three-element low-pass filter is given in **Fig 3.3**. This circuit consists of a series inductor and a pair of shunt capacitors. The filter is driven with a generator with a source resistance Rs, and is terminated in a load of RL. The source and load are a vital part of the circuit; the transfer function depends upon having both ends of the filter properly terminated. A filter that is terminated in resistive loads at each end, input and output, is called a doubly-terminated filter. Most of the LC filters that are interesting to us will be doubly terminated.

Figure 3.3B shows another three-element filter. This one uses two series inductors with one shunt capacitor. With proper design, this filter will have exactly the same transfer function as that of Fig 3.3A. This is a common detail of filters; they often have dual forms.

We can tell by inspection that both filters of Fig 3.3 are low-pass circuits. The series inductor is a short circuit at dc and has reactive impedance that grows with frequency. Hence, it will inhibit the flow of energy through the circuit more as frequency increases. The same argument can be made about the capacitors. They behave as an open circuit at dc. However, as frequency increases, they show lower and lower impedance, more effectively shunting the energy flowing in the circuit.

A low-pass filter will have a number of elements equaling the order. The filters of

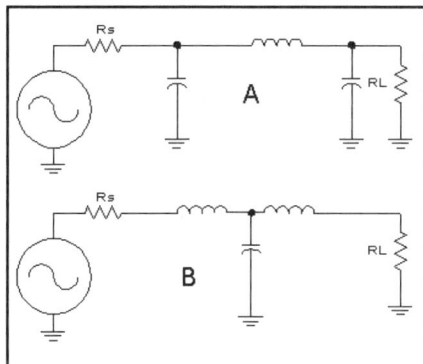

Fig 3.3—Three element, or 3rd-order low-pass filters.

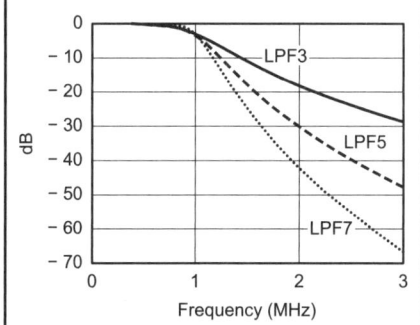

Fig 3.4—Transfer function for low-pass filters with order 3, 5 and 7. Adding sections will increase stopband attenuation.

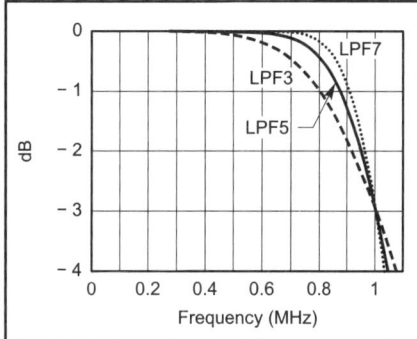

Fig 3.5—Butterworth filter transfer functions showing the passband details.

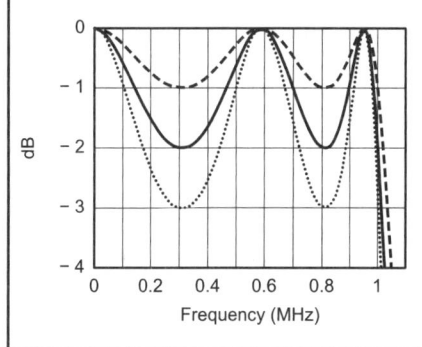

Fig 3.6—Chebyshev 5th-order low-pass filter transfer functions showing passband ripples of 1, 2, and 3 dB. These extreme ripple values are rarely used, but illustrate the concepts. Note that there is a half cycle of ripple for each filter element.

Fig 3.3 are 3rd-order filters. A low pass with 5 elements is a 5th-order circuit and offers greater attenuation in the stopband. The component type must alternate as we progress down the low-pass filter, going from series inductor to shunt capacitor and so forth. If there were, for example, two series inductors next to each other, they would behave as one single inductor. (The term "order" comes from the mathematics. A 5th-order low-pass filter has a transfer function where the denominator is a 5th-order polynomial, meaning that the frequency appears raised to the 5th power.)

Fig 3.4 shows response plots for three different low-pass filters. These circuits all have a 3-dB cutoff frequency of 1MHz, but differ in the number of components. These filters have order 3, 5 and 7. Odd-order pi filters are popular, offering maximum performance vs the number of inductors used.

Filter Shapes

All three of the filters analyzed in Fig 3.4 used a Butterworth design. This refers to the mathematical details that describe the filter; this one has a transfer function described as a Butterworth polynomial. Another popular shape is the Chebyshev. There are many more. The ideal is a brick wall low pass filter, an unattainable goal with an absolutely flat response throughout its passband, and infinite attenuation in the stopband. The responses of Fig 3.4 suggest that achieving the ideal is going to be difficult. Wanting to do as well as we can with minimum difficulty, we accept some compromise. By picking different compromises, we will end up with different filter shapes.

The Butterworth filter is one that is designed to be maximally flat within the passband. (The slope of the transfer function is to be zero at zero frequency.) This is illustrated in greater detail with **Fig 3.5**, a repeat of Fig 3.4 showing only passband details. All of the filters are flat at zero frequency. Although the curves are smooth throughout the passband, attenuation grows as we approach cutoff.

The Chebyshev filter allows a different kind of error. This filter type allows ripples of equal amplitude to occur within the passband. Three transfer functions for Chebyshev low-pass filters are shown in **Fig 3.6**. The three circuits are all 5-pole, or 5th-order low-pass filters, now using a 1 MHz ripple cutoff frequency. The circuits have passband ripples of 1, 2 and 3 dB. Even though the three filters show large ripples, they all show 0 dB loss at points through the passband. The frequencies are not a function of ripple value. These filters were designed for ripple cutoff frequency. That is, a filter with 1-dB passband ripple will have the last point of -1 dB response at the ripple cutoff frequency. Chebyshev filters can be designed for either a desired 3-dB cutoff, or a ripple cutoff. Odd ordered Chebyshev filters have zero attenuation at zero frequency while even ordered versions will have a dc attenuation equal to the ripple. Stopband attenuation is a strong function of passband ripple. The more ripple allowed within the passband, the greater the stopband is attenuated.

There are numerous other polynomial types that form useful and interesting low-pass filters. Some are of direct interest for low-pass filters while others are of greater utility as the beginnings of other filter types. For example, the Bessel filter, also know as the max flat delay filter, is often used as a starting point for bandpass filters with minimum ringing. This will be discussed later with LC and quartz crystal bandpass filter design.

Low-Pass Filter Design

The design of practical low-pass filters begins with tables of normalized values. These component values, $g(n)$, are either capacitor or inductor values for the n-th part in a low-pass filter with a 1 Ω termination and a cutoff frequency of $1/(2\pi)$ Hz. While this is rarely a filter that anyone would wish to build directly, it is a convenient form for scaling to practical filters. It's also a mathematical simplification.

Table 3.1 shows some $g(n)$ values for a few representative low-pass filters. The Butterworth part of the table gives data in terms of a 3 dB cutoff frequency, while the Chebyshev filter data are calculated on the basis of a ripple cutoff.

A practical low-pass filter is easily designed with data from Table 3.1. Design begins by picking a cutoff frequency in Hz and a resistive termination, in Ω, for each end of the filter. The filters that are designed from the table are doubly terminated in equal values. Having picked the critical parameters, a low-pass filter has inductor and capacitor values given by

$$L(n) = \frac{g(n) \cdot R_0}{2 \cdot \pi \cdot f} \quad \text{Eq 3.1}$$

$$C(n) = \frac{g(n)}{R_0 \cdot 2 \cdot \pi \cdot f} \quad \text{Eq 3.2}$$

where g(n) is the n-th normalized value from the table, R_0 is the terminating resistance in Ω, f is frequency in Hz, L(n) is the n-th inductor in Henries, and C(n) is the n-th capacitor in Farads.

The first part can be an L or C. If the first part is an inductor, the second one will be a capacitor, the third another inductor, and so forth. Both forms generate the same resulting transfer function.

Consider an example, a 4-th order Butterworth low-pass filter. The normalized values from the table are g(1)=0.7654, g(2)=1.85, g(3)=1.85, and g(4)=0.7654. Let's design this filter for a 3-dB cutoff of 10 MHz with a termination of 50 Ω at each end. The filter will begin with an inductor. Hence,

$$L(1) = \frac{0.7654 \cdot 50}{2 \cdot \pi \cdot 10 \cdot 10^6} = 0.609 \cdot 10^{-6}$$

$$C(2) = \frac{1.85}{50 \cdot 2 \cdot \pi \cdot 10 \cdot 10^6} = 5.889 \cdot 10^{-10}$$

$$L(3) = 1.472 \cdot 10^{-6}$$

$$C(4) = 2.436 \cdot 10^{-10}$$

The resulting filter is shown in **Fig 3.7A** while the dual form, the variation beginning with a shunt capacitor, is presented in Fig 3.7B.

The filter example picked for Fig 3.7 was a special case, an even ordered design. As such, the dual filter, which is the one starting with the alternative component type, is really the same filter, but with the input and output exchanged. If we had picked an odd order filter to illustrate the two filter types, we would have filter (A) with more capacitors than inductors while (B) would be dominated by inductors.

The denormalization equations are simple and easily programmed in a spreadsheet, a programmable calculator, or in any popular computer language.

What might be the obvious route to a filter design may not be the most practical. The logical sequence calculates the values, purchases and or builds the components, and then assembles the circuit. Inductors are not a problem, for the user can pick a number and position of turns as needed to realize a required value. But capacitors tend to come only in standard values. The non-standard values can be synthesized with parallel combinations of capacitors, although this often leads to bulkier and more expensive circuitry than desired, and parallel capacitors lead to additional resonances. An alternative route is:

• Design an initial low-pass filter.
• Analyze the filter to confirm that the desired response is realized. Computer programs such as *GPLA* or *ARRL Radio Designer*[2] work well. Other analysis programs are often found on the Web.
• Substitute available capacitors for those calculated in the design phase and analyze the results.
• Adjust inductor values to "fix" variations that might have occurred as a result of using practical capacitors.

Most low-pass filters, especially the simple Butterworth and Chebyshev designs, are insensitive to small component value changes. Slight adjustments toward practical values will often have so little impact that there will be no need for additional adjustments. If a refined program is used for design, it is easy to vary the filter order and ripple to obtain a desired response, especially in a low-pass filter.

The radio experimenter will often use a low-pass filter at a transmitter output, for a low pass will attenuate harmonics, the predominant distortions created in the output stages. An ideal low-pass filter, however, is not required. Rather, the need is for a filter that will attenuate harmonics and will pass a relatively narrow band of frequencies. The required passband is often no more than 10 or 20% in width. It is not necessary to do a good job at very low frequencies. Chebyshev or Butterworth filters may not be the best choices.

An interesting, and often practical filter type is the almost unknown ultra-spherical low-pass filter.[3,4] An ultra-spherical filter is like the Chebyshev to the extent that it has passband ripples. However, the

Half-Wave Filter

The popular half-wave filter is a very tolerant low-pass filter form. L and C have a reactance equal to the terminating resistance. The middle capacitor is twice that at the ends. This filter, a low pass, is designed at the operating frequency rather than a cutoff. This filter will have a 3-dB cutoff that is about 40% above the design frequency and only offers about 25-dB attenuation at the second harmonic. A 7-MHz half-wave filter will use L=1.1 μH and C=450 pF when designed for R=50 Ω. This filter will have a phase shift of 180 degrees at the operating frequency; hence, the circuit name.

$$R_S = R_{Load}$$

$$X_L = X_C = R_S$$

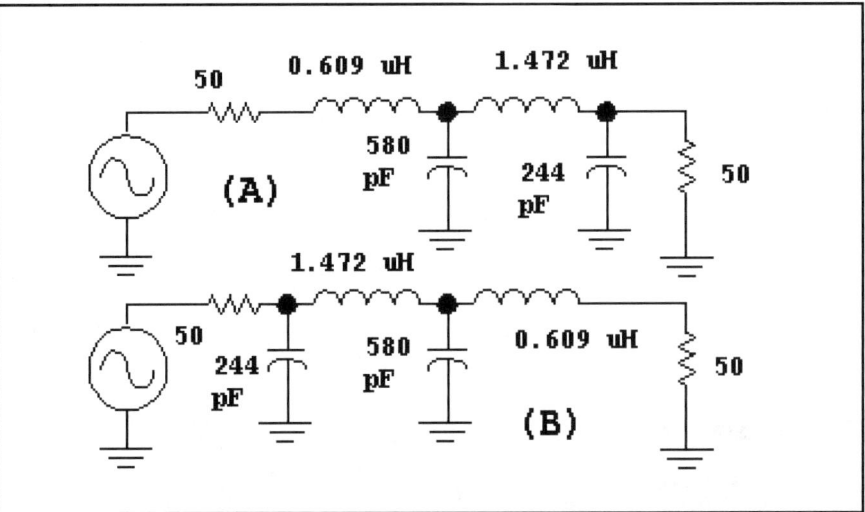

Fig 3.7—Two forms of a 4th-order, 50-Ω, doubly-terminated, 10-MHz cutoff Butterworth low-pass filter.

Table 3.1

Normalized Values for Butterworth and Chebyshev Low-Pass Filters. These are used with the Low Pass and High-Pass de-normalization equations. All of the data presented are for doubly terminated filters. Butterworth filters are designed on the basis of a 3-dB cutoff while a ripple cutoff is used for the Chebyshev filters.

Type	N	g(1)	g(2)	g(3)	g(4)	g(5)	g(6)	g(7)
Butterworth	2	1.414	1.414					
	3	1	2	1				
	4	0.7654	1.85	1.85	0.7654			
	5	0.618	1.618	2	1.618	0.618		
	6	0.5176	1.414	1.932	1.932	.1414	0.5176	
	7	0.445	1.247	1.802	2	1.802	1.247	0.445
.01 dB Chebyshev	3	0.6292	0.9703	0.6292				
	5	0.7563	1.305	1.577	1.305	0.7563		
	7	0.797	1.392	1.748	1.633	1.748	1.392	0.797
0.1 dB Chebyshev	3	1.032	1.147	1.032				
	5	1.147	1.371	1.975	1.371	1.147		
	7	1.18	1.423	2.097	1.573	2.097	1.423	1.18

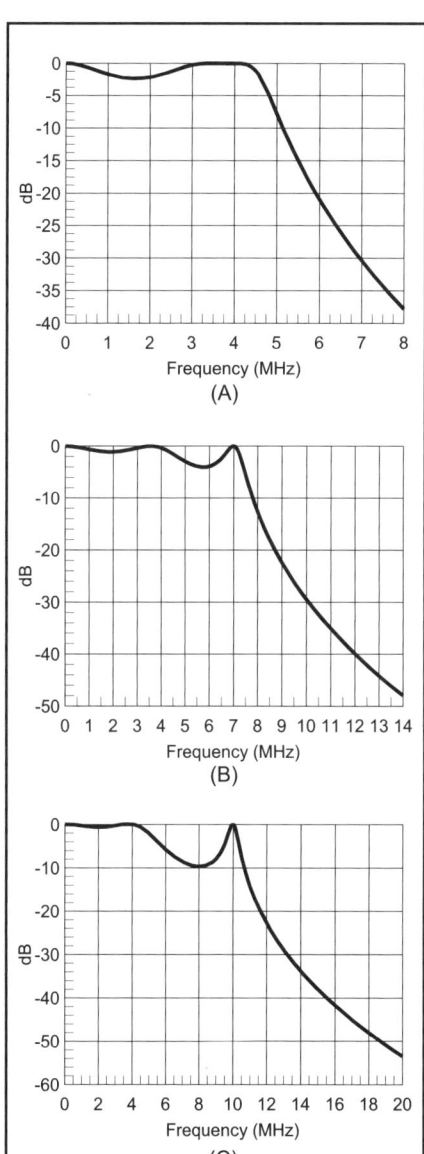

ripples are not necessarily of equal magnitude. The Chebyshev filter is a special case of the ultra-spherical. The transfer function for three variations of the ultra-spherical filter is shown in **Fig 3.8**. All of these 5th-order filters are designed at the highest peak frequency rather than at a cutoff frequency. **Eq 3.1** and **Eq 3.2** still apply. The g(n) values are shown in **Table 3.2**.

Fig 3.8A shows what we might call a wide ultra-spherical filter, a circuit with about a 20% bandwidth for 2.5-dB variation, yet having stopband characteristics like those of a very high ripple Chebyshev low pass. This example circuit was configured for complete coverage of the 3.5-4 MHz band.

Fig 3.8B shows a medium width ultra-spherical filter. The main virtue of this circuit is the extreme flexibility offered with regard to component value. The price of this is the need for an adjustable element in the middle of the filter. This is especially suited to junk box driven projects. The example is a filter for a 7-MHz transmitter. The end capacitors might, in practice, be 1200-pF silver mica while the middle capacitor could be a 1000-pF part paralleled with a 200-pF mica trimmer.

Fig 3.8C presents the result of a narrow ultra-spherical filter. This circuit has a peak 3-dB bandwidth of about 200 kHz at 10 MHz while offering 54-dB attenuation at the 2nd harmonic of the peak.

While the ultra-spherical filters offer band-pass filter like performance with low-pass stopband characteristics, they can also suffer from high loss with low-Q components. They should be analyzed or measured when applied to narrow band applications.

High Pass Filters

The low-pass filter is the basis for this section; it is the cornerstone that supports all other passive LC filters. Occasionally, a high-pass filter is required in a piece of equipment. A high pass has a passband that extends upward from a cutoff frequency. The stopband of a high pass is below the cutoff.

Once we have a set of normalized low pass tables, designing a high-pass filter is an easy extension. The conceptually easy approach is a two-step process: Having picked a cutoff frequency, a low pass of

Fig 3.8—(A) We might call this a wide ultra-spherical filter, a circuit with about a 20% bandwidth for 0.2-dB variation, yet having stopband characteristics like those of a very high-ripple Chebyshev low pass. This example circuit was configured for complete coverage of the 3.5-4 MHz band. (B) A medium width ultra-spherical filter. The main virtue of this circuit is the extreme flexibility offered with regard to component value. The price of this is the need for an adjustable element in the middle of the filter. This is especially suited to junk box driven projects. The example is a filter for a 7-MHz transmitter. The end capacitors might, in practice, be 1200-pF silver mica while the middle capacitor could be a 1000-pF part paralleled with a 200-pF mica trimmer. (C) The result of a narrow ultra-spherical filter. This circuit has a peak 3-dB bandwidth of about 200 kHz at 10 MHz while offering 54-dB attenuation at the 2nd harmonic of the peak.

Table 3.2
Normalized Ultra-spherical low-pass filter data.

Case	g(1)	g(2)	g(3)	g(4)	g(5)
Wide U.Sp.	1.759	0.704	3.352	0.704	1.759
Medium U.Sp.	2.717	1.087	2.56	1.087	2.717
Narrow U.Sp.	3.456	1.382	1.787	1.382	3.456

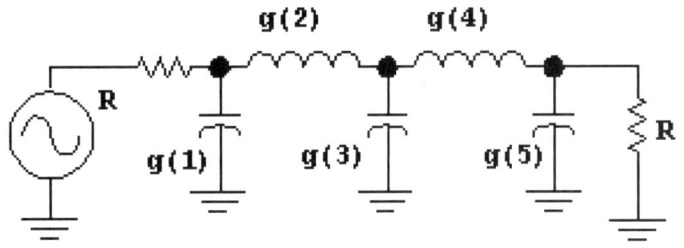

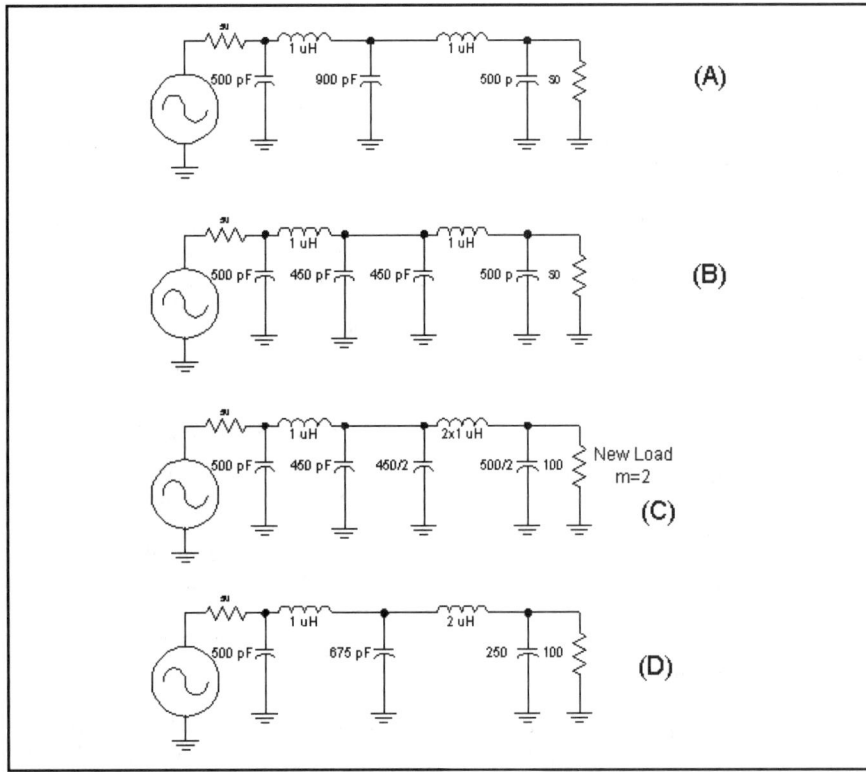

Fig 3.9—Low-pass filter illustrating Bartlett's Bisection Theorem that allows a termination to be changed to a new value.

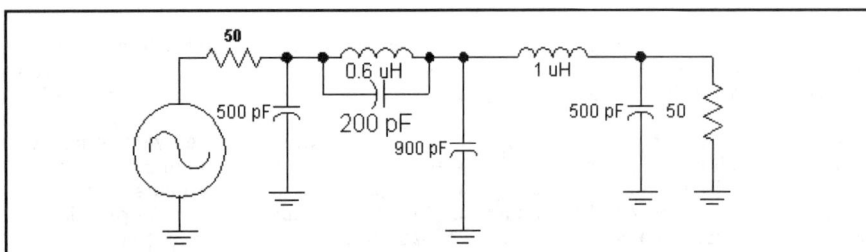

Fig 3.10—Changing an inductor to a "trap" creates a frequency of very high attenuation in the stopband.

the desired order and shape is designed. Then, each low-pass element is replaced with a high-pass one that has the same reactance at the cutoff. Series inductors are replaced with series capacitors; shunt capacitors become shunt inductors.

Alternatively, the tables of g(n) values may be used directly for high-pass filter design. The viable equations are then

$$C(n) = \frac{1}{g(n) \cdot R_0 \cdot 2 \cdot \pi \cdot f} \quad \text{Eq 3.3}$$

$$L(n) = \frac{R_0}{2 \cdot \pi \cdot f \cdot g(n)} \quad \text{Eq 3.4}$$

where g(n) are the normalized low-pass elements from Table 3.1, R_0 is the terminating resistance and f is frequency in Hz. The inductance value, L(n), is in Henrys and capacitance, C(n), is in Farads.

As with the low-pass filters, once a high-pass filter is designed, it should be confirmed with some appropriate calculations and, later, measured after construction.

Some Simple Transformations

There are several circuits that can be designed with relative ease once a low pass or high-pass filter is in place. Some will be discussed here, for they offer considerable flexibility and opportunity to the experimenter.

We often need different terminations at filter ends. A method for doing this is provided by the Bartlett's Bisection Theorem, illustrated in the low-pass filter shown in **Fig 3.9**.

The first filter, shown in Fig 3.9A, is a symmetric 50-Ω 5th-order low pass. The filter is a low pass with a 3-dB cutoff of about 10 MHz. This filter is redrawn in part B with the filter split in the mid point. The two half sections are identical. We wish to change the output termination to 100 Ω while preserving the same filtering characteristics. The ratio of the new termination, 100 Ω, to the original 50 Ω is 2. The filter is transformed by increasing series elements (the L) by m=2 in the right side. The shunt elements are decreased by the same factor of m. This is illustrated in Fig 3.9C with the final filter in Fig 3.9D. The multiplier m can be any value greater than 0.[5] This method is used later in the book in the design of some filters for a SSB transceiver.

The next filter modification that we consider adds capacitors or inductors to a filter. This scheme is used in the design of elliptic, or Cauer-Chebyshev low-pass filters where adding components that create

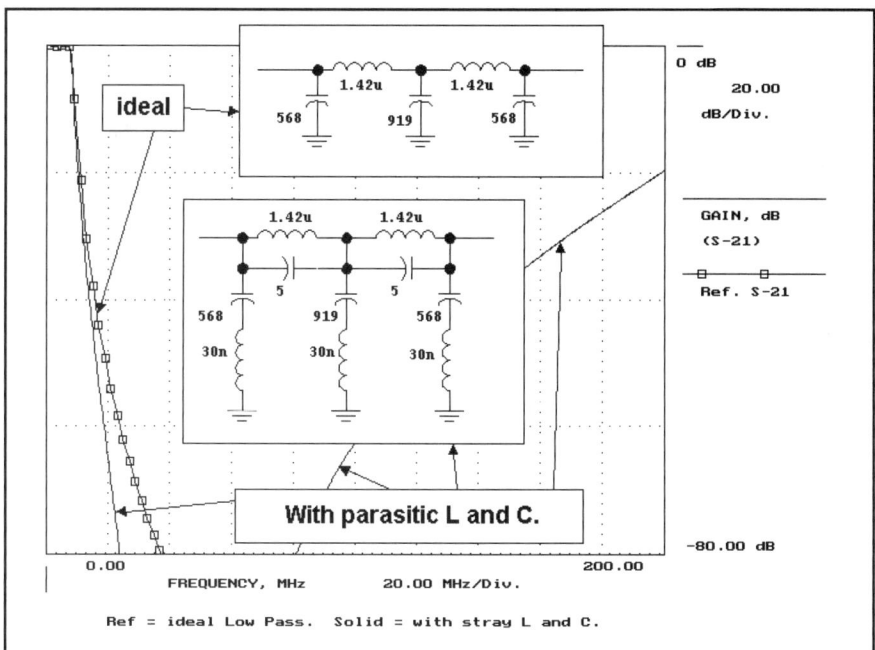

Fig 3.11—The VHF performance of HF low-pass filters is significantly altered by parasitic inductance and capacitance. The parasitic elements are modeled as being larger than normal to illustrate the effects.

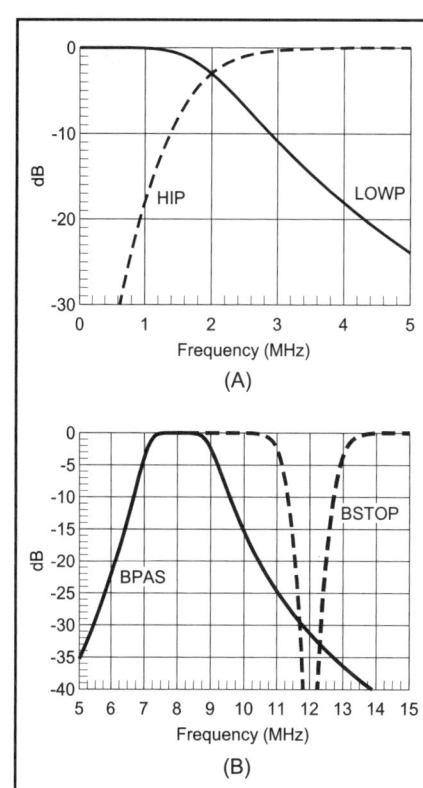

Fig 3.13—Transfer functions for the low pass and high pass (A) and the bandpass and bandstop filters (B).

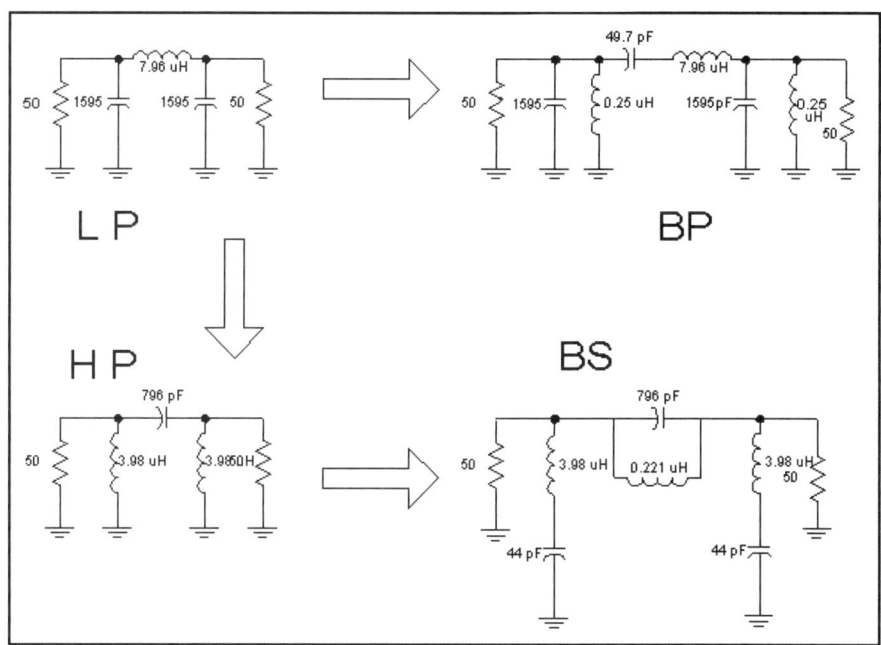

Fig 3.12—A low-pass filter (LP) is the prototype for the high pass (HP). The components in the low pass may be resonated to produce a bandpass (BP) filter with a bandwidth equaling the original low pass. Similarly, the high-pass elements are tuned to produce a bandstop filter (BS) with a 3-dB notch width equaling the bandwidth of the high pass.

"trap" frequencies alters the stopband of a filter. This is illustrated in **Fig 3.10** where a low-pass filter is modified. The first inductor, originally a 1-µH unit, is paralleled with a 200-pF capacitor. The inductor is reduced to 0.6 µH so the LC combination will have approximately the same reactance at the filter cutoff frequency.

This "elliptic" modification can be extended by converting both inductors to traps and by adding series inductance with any or all of the shunt capacitors. The modification shown leaves the passband almost unchanged, but increases the attenuation at 14 MHz. Unfortunately, the attenuation at the higher end of the stopband, above 20 MHz, is not as good as it was with the original low-pass filter; this is typical of elliptic filters. Another disadvantage of the method is that component losses have much greater impact than they did without the traps, especially near the cutoff frequency. All of these changes are easily modeled with computer analysis. Design tables are found in numerous standard filter texts such as Zverv.[6]

The trap characteristics we describe are always present to one extent or another, even when they are not featured. Assume we needed a low-pass filter to follow a 7-MHz transmitter. A 5th-order circuit was designed for a 0.2-dB ripple Chebyshev shape with a 7.5-MHz ripple cutoff frequency. The designed filter is the "ideal" circuit in **Fig 3.11** with response shown as the "reference." The analysis is extended out to 200 MHz. The other circuit in the figure includes the "accidental" effects of parallel capacitance across the inductors and inductance in series with the capacitors. Both improve the steepness of the rolloff. But they both contribute to a severely degraded VHF stopband attenuation.

The next transformation we consider resonates the elements of low pass and high-pass filters. We begin by designing a

Filters and Impedance Matching Circuits 3.7

3rd-order low pass with a cutoff of 2 MHz. A similar 2-MHz high pass is designed; the filters are shown in **Fig 3.12**. Once the low and high-pass circuits are in place, each element is resonated. The three-element low pass maps into a 6-component bandpass filter. The new filter is centered at the resonance frequency, here 8 MHz, with the 2-MHz bandwidth of the parent low pass. This method is generally limited to wide bandwidths, perhaps 20% or more. Impractical component values are sometimes avoided by terminating the filter in resistances greater than 50 Ω.

A similar transformation is applied to the high-pass filter, resulting in a bandstop filter. A frequency of 12 MHz was picked for this example. The same restrictions that accompanied the wide bandpass filter apply to this design.

The transfer function for the low pass and high pass are given in **Fig 3.13** along with the response for the bandpass and band stop.

3.3 LC BANDPASS FILTERS

The LC bandpass filter is a critical function in determining the performance of a typical RF system such as a receiver. An input filter, usually a bandpass, restricts the frequency range that the receiver must process. A later IF filter determines the overall receiver bandwidth. This filter often uses crystals, although LC filters were popular in older receivers. Audio filters often use LC elements, although RC active circuits, or the computational abilities of digital signal processing add further selectivity and confine the noise to a desired spectrum.

The LC filters we discuss in this section are narrow with a bandwidth from 1 to 20 % of the center frequency. Even narrower filters are built with resonators with higher Q; the quartz crystal is an example that will be discussed later where bandwidths of less than a part per thousand are possible. The basic concepts that we examine with LC circuits will transfer to the crystal filter.

Losses in Filters and Q

The key elements in narrow filters are tuned circuits made from inductor-capacitor pairs, quartz crystals, or transmission line sections. These resonators share the properties that they store energy, but they have losses. A chime is an example. Striking the chime with a hammer produces the waveform of **Fig 3.14**. A parameter called Q, for quality factor, describes the rate that the amplitude decreases with time after the hammer strike. The higher the Q, the longer it takes for the sound to disappear. The oscillator amplitude would not decrease if it were not for the losses that expend energy stored in the resonator. The mere act of observing the oscillation will cause some energy to be dissipated.

The chime was an acoustic resonator, but the same behavior occurs in electric resonators. A pulse to an LC causes it to ring; losses cause the amplitude to diminish. The most obvious loss in an LC

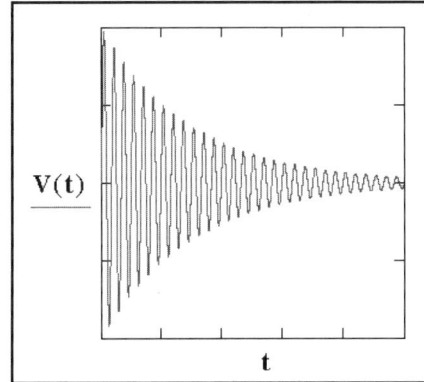

Fig 3.14—The amplitude of a chime's ring after being struck by a hammer. Units are arbitrary.

circuit is conductor resistance, including that in the inductor wire. This resistance is higher than the dc value owing to the skin effect, which forces high frequency current toward the conductor surface. Other losses might result from the motion of magnetic regions in an inductor core or the movement of dielectric parts of a capacitor.

An inductor is modeled as an ideal part with a series or a parallel resistance. The resistance will depend on the Q if the inductor was part of a resonator with that quality. The two resistances are shown in **Fig 3.15**.

$$R_{Series} = \frac{\omega \cdot L}{Q}$$
$$R_{Parallel} = Q \cdot \omega \cdot L$$

Eq 3.5

The higher the inductor Q, the smaller the series resistance, or the larger the parallel resistance is needed to model that Q. It really does not matter which component is used.

The Q of a resonator is related to the bandwidth of the tuned circuit by

$$BW = \frac{f_C}{Q}$$

Eq 3.6

where f_C is the tuned circuit center frequency. This Q is also that of the inductor in a tuned circuit if the capacitor is lossless.

The single tuned circuit is presented in two different forms in **Fig 3.16**. In the top, a parallel tuned circuit consisting of L and C has loss modeled by three resistors. The one labeled by Rp is the parallel loss resistance representing the non-ideal nature of the inductor. (Another might be included to represent capacitor losses.) But the LC is here paralleled by three resistors: the source, the load, and the loss element. Rp would disappear if the tuned circuit was built from perfect components. The source and load remain; they represent the RF world where a source resistance must be present if power is available and a load resistance must be included if power is to be extracted.

Eq 3.5 and **Eq 3.6** can be applied in several ways. If the resonator is evaluated with only the intrinsic loss resistance (in either series or parallel form) the resulting Q is called the unloaded Q, or Qu. If, however, the net resistance is used, which is the parallel combination of the load, the source, and the loss in the parallel tuned circuit, the resulting Q is called the loaded value, Q_L. If we were working with the series tuned circuit form, the loaded Q would be related to the total series R.

Consider an example, a parallel tuned circuit (Fig 3.16 top) with a 2-µH inductor tuned to 5 MHz with a 507-pF capacitor. Assume the parallel loss resistor was 12.57 kΩ. The unloaded Q calculated from Equation 3.5 is 200. The unloaded bandwidth would be 5 MHz /200 = 25 kHz.

Assume that the source and load resistors were equal, each 2 kΩ. The net resistance paralleling the LC would then be the combination of the three resistors, 926 Ω. The loaded Q becomes 14.7 with a loaded bandwidth of 339 kHz. The loaded Q is

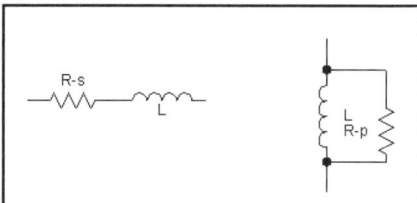

Fig 3.15—Inductor Q may be modeled with either a series or a parallel resistance.

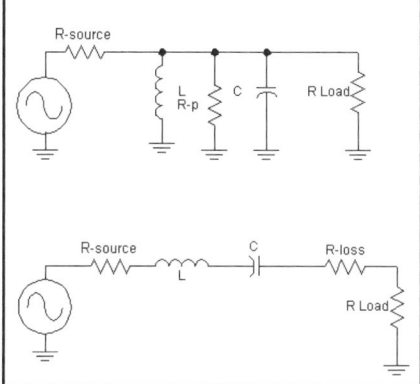

Fig 3.16—Two simple forms of the single tuned circuit.

$$\text{IL}(\text{dB}) = -20 \cdot \text{Log}\left(1 - \frac{Q_L}{Q_U}\right)$$

Eq 3.7

The Q of a tuned LC circuit is easily measured with a signal generator of known output impedance, R_0, and a sensitive detector, again with a known impedance level, often equaling the generator R_0 at 50 Ω. The test-set is shown in **Fig 3.18**. The test setup of Fig 3.18 uses equal loads of value R_0 and equal capacitors to couple from the terminations to the resonator. Equal capacitors, C1 and C2 guarantees that each termination contributes equally to the resonator parallel load resistance. The voltmeter across the load is calibrated in dB.

To begin measurement we remove the tuned circuit and replace it with a direct connection from generator to load. The available power delivered to R_0 is calculated after the voltage is measured. The resonator is then inserted between the generator and load, and the generator is tuned for a peak. The measured power is less than that available from the source, with the difference being the insertion loss for the simple filter. Capacitors C1 and C2 are adjusted until the loss is 30 dB or more. With loss this high, the intrinsic loss resistance of the resonator will dominate the loss.

The generator is now tuned first to one side of the peak, and then to the other, noting the frequencies where the response is down from the peak by 3 dB. The unloaded bandwidth, δF, is the difference between the two 3 dB frequencies. The unloaded Q is calculated as

$$Q_U = \frac{F}{\delta F}$$

Eq 3.8

This method for Q measurement is quite universal, being effective for audio tuned circuits, simple LC RF circuits, VHF helical resonators, or microwave resonators. The form of the variable capacitors, C1 and C2, may be different for the various parts of the spectrum, but the concepts are general. Indeed, it is not even important how the coupling occurs. The Q measurement normally determines an unloaded value, but loaded values are also of interest when testing filters.

Coupling

Coupling refers to the sharing of energy between resonators. Two resonators in a filter are generally tuned to exactly the same frequency. However, when an element (L or a C) is attached to cause energy in one to be shared with the other, two re-

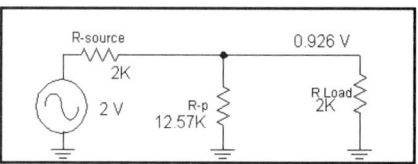

Fig 3.17—Simplified parallel tuned circuit at resonance. The effect of loss is illustrated.

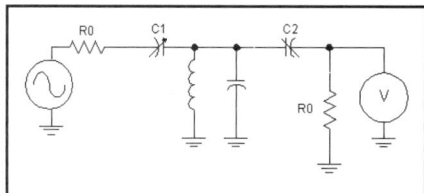

Fig 3.18—Test setup for measuring the Q of a resonator. The source and load are assumed identical. The two coupling capacitors are adjusted to be equal to each other. The output signal is measured with an appropriate ac volt meter, a high impedance oscilloscope, or a spectrum analyzer.

sponse peaks often appear with frequency separation becoming a measure of the coupling. This is illustrated with the circuit of **Fig 3.19**, which results in the curves of **Fig 3.20**.

The frequency separation between peaks is a measure of the coupling between the resonators. The utility of this parameter is in the measurements that become possible. The filter designer needs only to generate a method for coupling to produce a desired frequency difference in order to realize a given filter. Such measurements (or calculations) are a vital part of building filters with unusual tuned circuits, such as UHF helical resonators. A natural extension of this measurement is a collection known as the Dishal Method.[7] The Dishal method is extremely useful in the adjustment of multiple resonator filters. The method is discussed further in *Introduction to Radio Frequency Design* and in Chapter 9 of Zverev's text.[6]

Multiple Resonator Bandpass Filters

Bandpass filters with several tuned circuits are designed with relative ease with careful application of some basic steps:

The resonators must have an unloaded Q that is higher, usually by a factor of 3 or more, than the desired filter Q, which is $f_C/\Delta f$ where f_C is center frequency and Δf is bandwidth.

A filter shape (e.g., Butterworth or

also called the filter Q, for it describes the bandwidth of the single tuned circuit, the simplest of bandpass filters.

This filter has an insertion loss. This is illustrated in **Fig 3.17**, which shows the filter without the L and C, effects that cancel at resonance. We use an arbitrary open circuit source voltage of 2. The available power to a load is then 1 V across a resistance equaling the 2-kΩ source. If the resonator had no internal losses, this available power would be delivered to the 2-kΩ load. However, the loss R parallels the load, causing the output voltage to be 0.926 V, a bit less than the ideal 1 V. Calculation of the output power into the 2-kΩ load resistance and the available power shows that the insertion loss is 0.67 dB.

This exercise illustrates two vital points that are general for all bandpass filters. First, the bandwidth of any filter must always be larger than the unloaded bandwidth of the resonators used to build the filter. Second, any filter built from real world components will have an insertion loss. The closer the Q of the filter approaches the unloaded resonator Q, the greater the insertion loss becomes.

A parallel tuned circuit illustrated these ideas; the series tuned filter would have produced identical results. Generally, the insertion loss of a single tuned circuit relates to loaded and unloaded Q by

Filters and Impedance Matching Circuits 3.9

0.2- dB Chebyshev, etc) is defined by the loaded Q of end resonators and by coupling between resonators.

These end Q values and coupling values between resonators are obtained from normalized tables of k and q. Some values for double and triple tuned filters are given in **Table 3.3**.

Bandpass filter design with normalized coupling and loading uses k and q tables. These are directly related to the normalized g_K values used for low-pass filter design. The g_K data is useful for quickly estimating the insertion loss of virtually any bandpass filter we might design. The loss in dB is

$$\text{Loss (dB)} = 4.34 \cdot \frac{F}{Q_U \cdot B} \cdot \left(\sum_k g_k \right)$$

where F, B, and Q_U were defined above. The g_K values are the normalized low-pass elements for the shape in question. Assume that we wish to build a 4th order bandpass filter with a 0.1-dB Chebyshev shape. The low pass parameters are g1=1.109, g2=1.306, g3=1.77, and g4=0.818. The sum of the elements is then 5.003. If we were going to build this filter at 144 MHz with a bandwidth of 5 MHz and we had managed to build resonators with Q_U=500, we would then expect an insertion loss of 1.25 dB. This formula is attributed to Cohn.[8,9]

The sidebar equations (see page 3.14) may be used to write a computer or calculator program for designing these circuits. This has been done to form **Table 3A**. The inductors used are all wound on toroid cores; the inductance values shown are very close to actual values when the toroids are wound with a single, evenly spaced winding. The Q_U values are approximate, although they are typical of measured data. Larger wire size will increase Q slightly. The data in the table are calculated values, but are typical of those we have built and confirmed on numerous occasions.

Double-Tuned Circuits

The double tuned circuit (DTC) can take on many forms, all showing the same basic shape around the passband so long as they develop the same end section Q values and the same coupling between resonators. A familiar "top coupled" DTC uses a series capacitor to couple terminations to parallel tuned circuits to set end section Q. Coupling between resonators is established with a small valued capacitor between the "hot" ends of the tuned circuits. The DTC in this form is presented, with design equations, in the sidebar on page 3.14.

Filter shape options are available in the sidebar DTC procedure. The Butterworth is generally a good starting point, for it is easily realized with practical components.

The Triple-Tuned Filter

While the ever-popular double-tuned circuit is often adequate, there are many

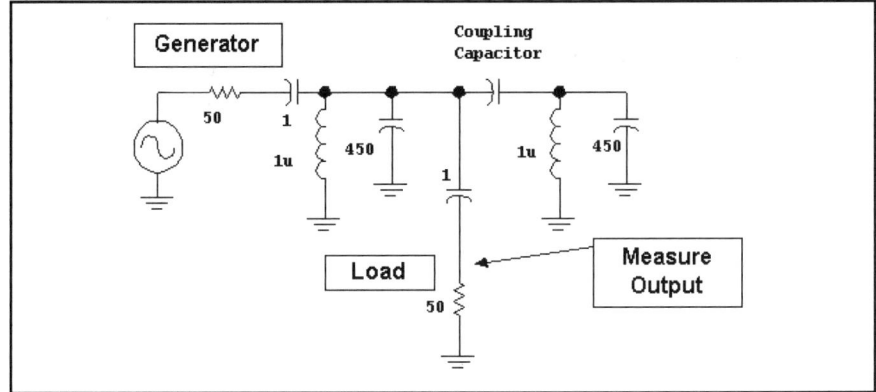

Fig 3.19—Scheme for measuring and defining coupling between two tuned circuits. C12 is either 10 or 20 pF while the resonators are both 1μH paralleled with 450 pF. "Probe" capacitors are 1 pF.

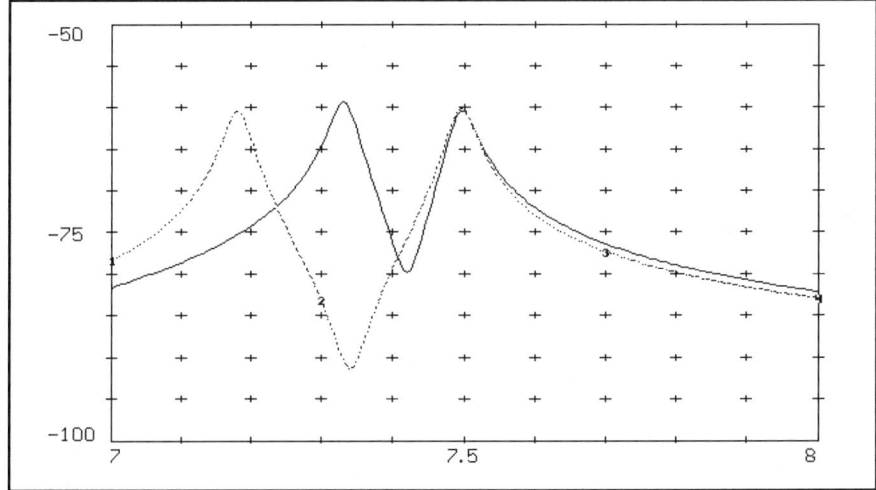

Fig 3.20—Separation of response peaks indicating coupling between two resonators. The solid line uses a 10-pF coupling capacitor while the dotted line uses 20 pF.

Table 3.3
k and q Values for Two- and Three-Pole Filters

Passband Ripple, dB	n	k	q
Butterworth	2	0.7071	1.414
0.1 dB	2	0.7107	1.638
0.25	2	0.7154	1.779
0.5	2	0.7225	1.9497
0.75	2	0.7290	2.091
1.0	2	0.7351	2.3167
1.5	2	0.7466	2.452
Butterworth	3	0.7071	1.000
0.1	3	0.6617	1.4328
0.25	3	0.6530	1.6330
0.5	3	0.6474	1.8640
0.75	3	0.6450	2.0498
1.0	3	0.6439	2.2156
1.5	3	0.6437	2.5169

cases where more performance is needed. The third-order bandpass is a special case, easily designed with the same equation (and hence, software) used for a double-tuned circuit. This possibility emerges if you compare a double-tuned circuit with the example triple-tuned circuit shown in **Fig 3.21**. This particular filter is centered at 16.2 MHz with a design bandwidth of 0.5 MHz. **Fig 3.22** shows the response of the triple-tuned filter, along with that of a double-tuned circuit built with the same inductors.

The triple-tuned filter is designed with different k and q values than used for a double-tuned circuit. Set $q=1$ and $k=0.707$ for a triple tuned Butterworth filter. Then, the coupling capacitors and the end matching capacitors are the values provided by the sidebar equations. The last equation in that series provides the tuning capacitor values for the end sections. The middle tuning capacitor is given by

$$C_M = C_0 - 2 \cdot C_{12} \qquad \text{Eq 3.9}$$

Building a triple-tuned filter is no more difficult than one with two resonators. If it is designed for a slightly wider bandwidth than might be used with a 2-pole design, the filter is often easier to align, has similar insertion loss, and offers improved stopband attenuation, the usual primary goal of bandpass filtering.

The design of higher order (N>3) bandpass filters is similar to the DTC. Coupling between resonators (numbered m and n) is described by a normalized coupling coefficient, k_{mn}. The values will generally differ for each pair of resonators. End loading, perhaps different for the two ends, is described by normalized end section q values, q_1 and q_n for a filter with n resonators. Denormalization establishes loaded end Q values that are then established as with the DTC. The individual parallel-tuned circuits are individually tuned to the filter center frequency with all other parallel resonators short-circuited. A calculator or computer program written for the design of double-tuned circuits may often be used, without modification, for the design of higher-order filters.

The bandpass filters examined so far used parallel tuned circuits. Series resonators may also be used. This variation is shown in **Fig 3.23** with the design procedure given in the literature.

With either form, values for normalized k and q are obtained from a table of values such as those published in the classic book by Zverev. The values may also be calculated in computer programs. Sometimes one encounters tables of predistorted k and q values. Predistortion is a process to retain a desired filter shape, even with losses present.[10,11,12]

Some filters are mixtures between the forms presented. An example is presented in **Fig 3.24** where the familiar small coupling capacitor is replaced with a shunt capacitor, usually large in value. A small value shunt inductor could also be used.

Filters at VHF and Higher

Bandpass filters are sometimes easier to realize at VHF and above than at lower frequency, the result of higher available resonator Qu at VHF. Building an air-core coil with a Q of even 200 at 2 MHz requires a considerable volume. However, one with such a Q at 200 MHz can be very small. This results from skin effect changing with frequency.

The book CD includes a tutorial paper on the DTC.[13] That article outlines methods for experimentally realizing simple bandpass filters at any frequency. The methods outlined there are easily applied to VHF and microwave filters, including those using transmission-line resonators.

Resonators can take on much different forms at higher frequency. One common and popular form is the quarter-wavelength long resonator. This is built by forming a section of transmission line that is just slightly less than 0.25 wavelength. One end is then short circuited while the other is open circuited. The resonator Q will depend upon frequency, geometry, and dielectric material. Air (or vacuum) dielectrics offer highest Q. The conductivity of the surface metal will significantly affect Q. Copper surfaces are excellent, with silver being even better.

Fig 3.25 shows a method for evaluating a transmission line resonator. This is a schematic, yet practical scheme for building filter elements with, for example,

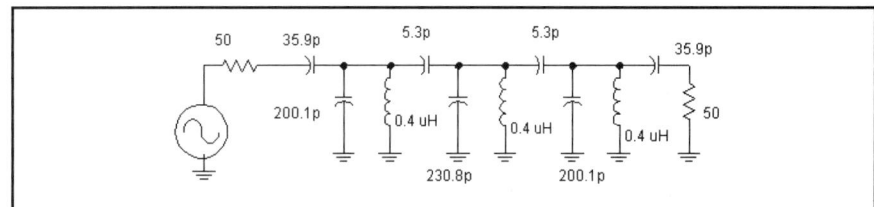

Fig 3.21—A triple-tuned circuit centered at 16.2 MHz with a bandwidth of 0.5 MHz.

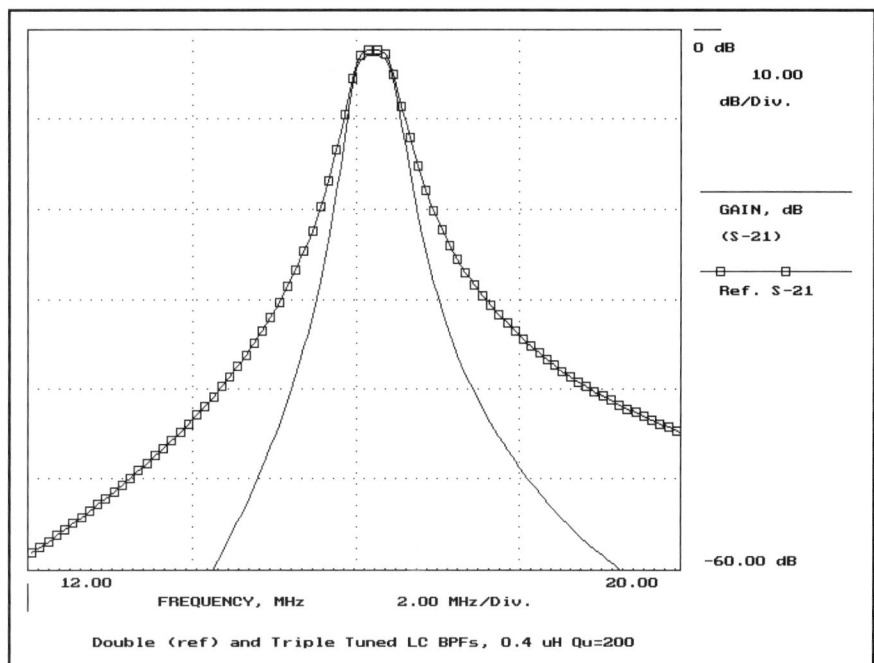

Fig 3.22—Response of triple and double-tuned circuits built with 0.4 µH inductors with Q_U=200.

Stopband Attenuation of Bandpass Filters

A 9-MHz bandpass filter required for a mixer experiment was built with available components. A triple-tuned circuit was fabricated from top-coupled parallel tuned circuits. The filter was examined in greater detail after the experiment was finished. While the filter satisfied the immediate need, the performance was far from ideal. A deep notch appeared in the stopband at about 11 MHz. Then what should have been an ideal filter became a disaster with a stopband attenuation of only 40 dB at 40 MHz.

This behavior had been observed earlier in a 7-MHz bandpass filter, shown in **Fig 3A**. The circuit was built on a scrap of circuit board material that was then bolted into an aluminum box. The BNC connectors at each end were "grounded" to the board with short wires from solder lugs under the connector nut. The filter was excited with a signal generator while examining the other end with a spectrum analyzer. We observed that the stopband attenuation improved slightly when a screw driver blade short circuited various spots on the circuit board edge to the aluminum box. This pointed toward grounding as a major problem with this filter.

A new 9-MHz bandpass filter was then built. The components used in the original, which was built like the 7-MHz filter "bad filter," were lifted and used in the new one. But the new circuit was fabricated in a box built from circuit board material (**Fig 3B**). The walls were soldered to the box floor, creating a cleaner ground. One of the long walls was initially left off, easing the filter construction. Filter performance was improved even before the 4th wall was added. The wall was added and the circuit was measured, revealing a stopband null at 43 MHz. The depth was at –110 dBc, near the limits of our measurement capability. The response at 70 MHz, the top of the spectrum analyzer range, was –83 dBc.

A single shield was added to the filter that removed the null and dropped the 70-MHz response to –96 dBc. The filter is shown in the photo "good filter."

The behavior observed is easily modeled with the circuit of **Fig 3C**. The stray coupling, related to ground currents, is modeled by lifting all ground connections in the filter and

Fig 3A—Bad filter—This bandpass filter performed well around the 7-MHz passband but had poor stopband attenuation. A very deep attenuation notch appeared at about 15 MHz.

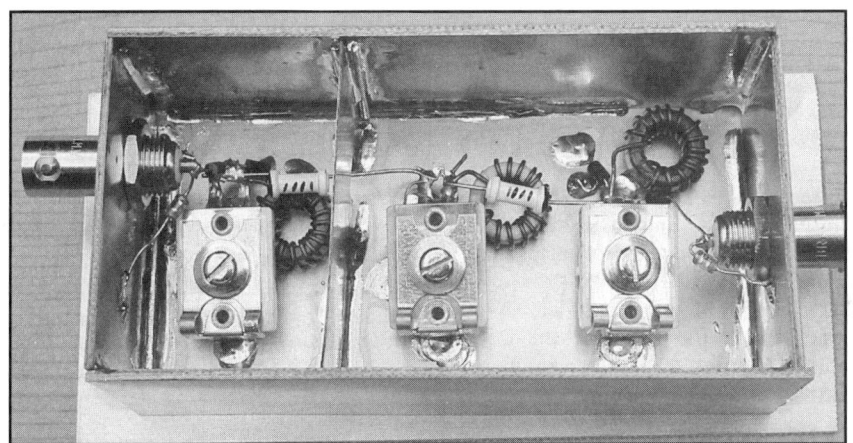

Fig 3B—Good filter—A box built from scraps of circuit board material produced a response with good stopband attenuation.

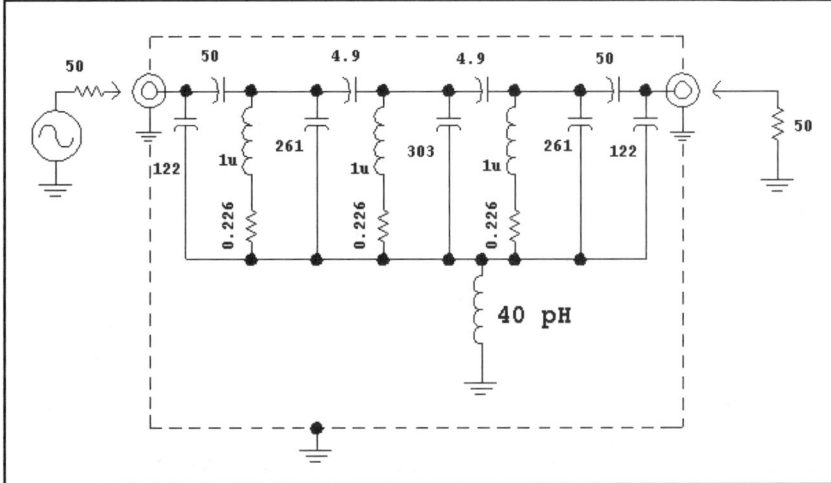

Fig 3C—The traditional bandpass filter is modified with a mutual inductor, raising the bandpass filter above ground. The resistance in series with the 1-µH inductors represents Qu of 250 at 9 MHz.

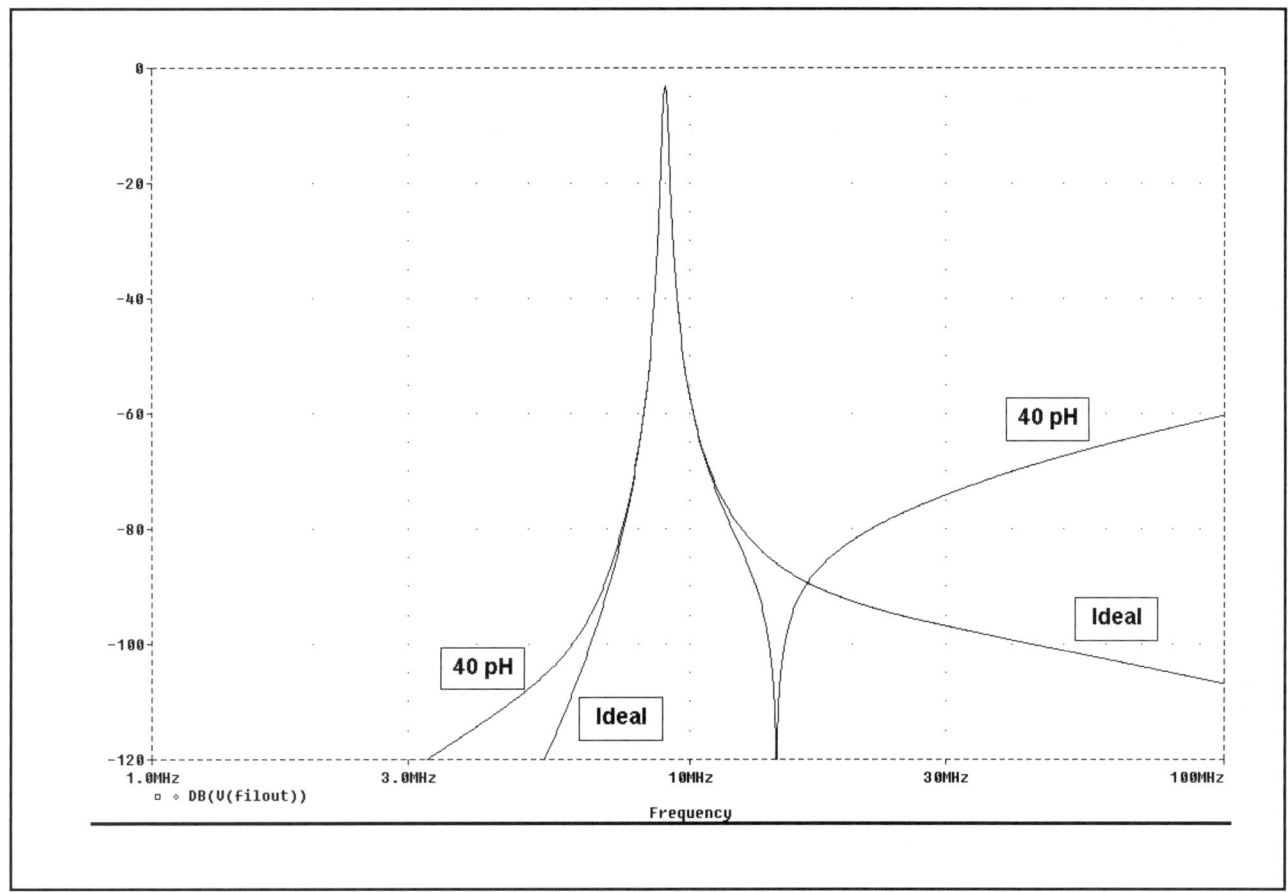

Fig 3D—The response of the ideal filter and that of the mutual coupling inductor are compared. The ideal response was realized in measurement when one shield was added to the filter.

attaching them to a common inductor. An inductance of only 40 pico Henry (yes; pH and not even nH) produced coupling that matched the measured performance. The "before and after" transfer responses are shown in **Fig 3D**.

Clearly, ground integrity is a vital part of an RF circuit, especially a bandpass filter using high Q resonators. Enclosures fabricated from soldered scraps of circuit board material or similar solid conductor are ideal, often far superior to aluminum boxes, especially following oxidation. Painted aluminum boxes are even worse. Clearly, measurements should always be performed.

0.141-inch outside diameter semi-rigid coaxial cable like that used in microwave systems. The center conductor is made available at both ends. It is shorted with as little inductance as possible at one end. Then, a 50-Ω generator and a 50-Ω load with detector are loosely coupled to the "hot" end of the resonator. The coupling capacitors may be nothing more than small pieces of wire spaced a small distance from the high impedance end of the resonator. The couplings from the generator and to the detector should be on opposite sides of the line to reduce direct interaction. The coupling is adjusted for a high insertion loss and the frequency is swept until the center frequency is found. The unloaded Q is measured by determining the 3-dB bandwidth. Center frequency may be adjusted by adjusting line length.

If a bandpass filter is to be built with the lines, the end section loading may be realized with the scheme shown in **Fig 3.26**. The "grounded" end of the resonator is attached to a coaxial connector in a ground plane. The center wire is attached to the connector and a short is created with a small inductor consisting of nothing more than a very short wire. The wire length is adjusted to set end section Q. The line shield should be carefully grounded very close to the coaxial connector.

Once proper end section Q is established and resonators are tuned to the proper center frequency, a working filter can be built by placing the two close enough to each other that the "hot" ends are in close proximity. This scheme works well for filters for the 432 and 1296-MHz bands. The line sections may be bent to fit available space.

The transmission-line double-tuned circuit just described used semi-rigid coaxial cable. Another common transmission line filter uses so-called hairpin circuits. Micro-strip transmission lines are printed on circuit board material in this filter. The lines are each a half wavelength long and are bent into a "U", or hairpin shape. An example of a hairpin filter with three resonators is shown in **Fig 3.27**.

The design of these filters is a straight-

DTC Design

Pick a center frequency, F, and a bandwidth, B, both in Hz. Pick an inductor; it can be of essentially arbitrary value, although a good "starting value" would be L=10/F where L is in Henry and F is still in Hz. The unloaded inductor Qu should be approximately known. One must also pick normalized k and q values. For a Butterworth shape, k=0.707 and q=1.414. For a filter with some passband ripple, but steeper skirts, use 0.25 dB Chebyshev values of k=0.7154 and q= 1.779. The design equations are:

$$\omega = 2 \cdot \pi \cdot F$$

$$C_0 = 1/(\omega^2 \cdot L)$$

$$C_{12} = C_0 \cdot \frac{k \cdot B}{F}$$

$$Q_E = \frac{q \cdot F \cdot Q_U}{B \cdot Q_U - q \cdot F}$$

$$C_E = \frac{1}{\omega} \cdot \frac{1}{\sqrt{R_0 \cdot Q_E \cdot \omega \cdot L - R_0^2}}$$

$$C_T = C_0 - C_E - C_{12}$$

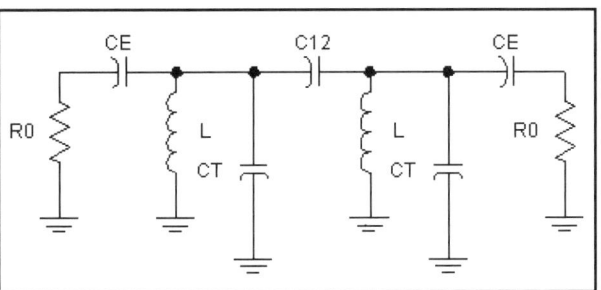

Table 3A

Double Tuned Circuits using the sidebar circuit. All filters are doubly terminated in 50 Ω at each end. The core designators use the copyrighted numbering scheme of Micrometals, Inc.

F-MHz	BW-MHz	Core	Turns	L-µH	Q-u	C-end	C-12	C-tune
1.85	0.1	T68-2	35	6.98	200	250 pF	41 pF	775 pF
3.55	0.1	T68-2	35	6.98	200	62	5.7	220
3.6	0.2	T68-2	35	6.98	200	93	11	177
3.9	0.2	T68-2	35	6.98	200	79	8.7	152
7.1	0.2	T50-6	17	1.156	250	56	8.7	371
7.05	0.1	T50-6	17	1.156	250	35	4.4	402
7.05	0.1	T50-6	20	1.6	250	30	3.2	286
10.1	0.1	T50-6	17	1.156	250	14	1.5	199
10.1	0.1	T50-6	10	0.4	250	20	4.4	597
14.1	0.2	T50-6	10	0.4	250	21	3.2	295
14.2	0.2	T50-6	10	0.4	250	34	6.3	271
18.1	0.2	T50-6	10	0.4	200	10	1.5	182
21.1	0.2	T50-6	10	0.4	200	6.1	1.0	135
21.25	0.5	T50-6	10	0.4	200	16	2.3	122
25	0.2	T50-6	10	0.4	200	2.9	0.57	98
28.2	0.4	T50-6	10	0.4	150	5.6	0.8	73
28.35	0.7	T50-6	10	0.4	150	9.8	1.4	68
50.2	1.0	T50-6	10	0.4	150	3.5	0.4	21
14.1	0.2	T50-6	5	0.1	200	38.7	12.8	1224
14.1	0.2	T50-6	7	0.196	200	27	6.5	617
14.1	0.2	T50-6	10	0.4	200	19	3.2	296
14.1	0.2	T50-6	15	0.9	200	13	1.4	127
14.1	0.2	T50-6	20	1.6	200	9.5	0.8	69
14.1	0.2	T50-6	25	2.5	200	7.6	0.5	43
14.1	0.2	T50-6	30	3.6	200	6.4	0.36	28.7
14.1	0.2	T50-6	35	4.9	200	5.4	0.26	20.3

Note: Only a couple of core types are needed to cover the entire spectrum from 1.8 to 50 MHz. The last eight table entries describe the same filter, a 14.1-MHz circuit with a 200-kHz bandwidth. The number of turns is allowed to vary, illustrating the freedom available to the filter designer. The builder with a computer program set up for design can vary inductance and bandwidth to realize a desired filter with standard (and junk-box available) component values.

Small Numeric Value Capacitors

Top coupled LC bandpass filters often use capacitors with small numeric value. These are becoming increasingly difficult to obtain. However, a simple substitution will provide the same coupling, but with larger more convenient values, picked with the equations shown. For example, assume a filter design calls for a capacitor with C_{JK}=1.2 pF. The substitute network can use any value of C_{SER} that is greater than 2.4 pF. Assume we use series capacitors of 10-pF value. The parallel capacitor is then C_{PAR}=63.3 pF. A practical value would be either 56 or 68 pF. The new network will have an equivalent parallel component at each end; you must reduce the capacitance that tunes the resonators accordingly.

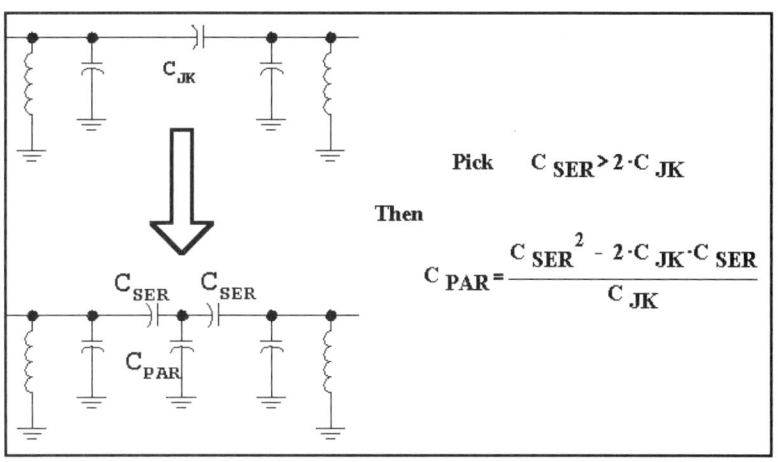

Pick $C_{SER} > 2 \cdot C_{JK}$

Then

$$C_{PAR} = \frac{C_{SER}^2 - 2 \cdot C_{JK} \cdot C_{SER}}{C_{JK}}$$

forward chore with a modern computer, although it's a job for professional-level microwave simulation software.

The total length of each section is 0.5 wavelength for proper tuning. The two end sections are usually identical. The lengths of the end sections are 2(X4) + X5 while that for the middle section is 2(X4) + X3. End section loading is determined by X2, essentially the spacing from the center of the end resonators, a virtual ground point. Coupling between resonators is established across the "gap" shown in Fig 3-27, analyzed by considering the overlapping sections as directional couplers. It is important for the computer analysis to include the junctions to the 50-Ω lines (Tee junctions) and a proper model for the open line ends. The designer must also have good information about the board material including loss, dielectric constant, and thickness between the pattern layer and the ground foil below.

The hairpin filter is generally a lossy structure when built on conventional circuit board materials used by amateurs. This material generally has a loss tangent of .02, producing resonator Q of 50. As such, narrow filters are not possible. Hairpin filters generally have 10 to 20 % bandwidth unless built on some of the more exotic materials.

Hairpin filters have responses at harmonic frequencies. A half wave resonator is resonant at frequencies where the line is 1, 2, 3, etc wavelengths long.

Another popular structure for higher frequencies is the helical resonator. These were very popular for UHF FM mobile radios of just a few years ago. A helical resonator is a section (usually one quarter wavelength) of line using a helical transmission line. A helical line is a solenoid coil-like structure placed inside a shielded enclosure. We can think of a wave as propagating along the wire at the speed of light. Hence, the propagation velocity parallel to the axis is much less than that of light. This is a slow wave structure. Cutting a quarter wavelength section, grounding one end with the other open circuited, forms a resonator. The usual helical resonator is just under a quarter-wavelength long. The extra length required for resonance is compensated by adding a small adjustable capacitor to the end, often nothing more than a grounded metal screw close to the "hot" end of the center conductor.

Numerous review articles have appeared describing the helical resonator and filters using them. Equations are often given for resonator dimensions, an implication that they must conform to a well-defined structure. Generally, there is much greater freedom available to the builder. A helical filter may still work well if built in a volume that is "too small."

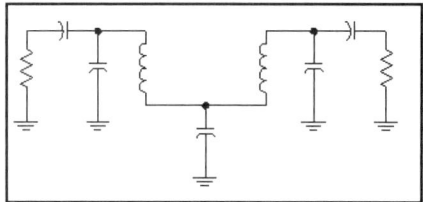

Fig 3.24—Double-tuned circuit with a shunt capacitor for coupling between resonators. This illustrates one of numerous bandpass filter topologies that are mixtures of the two methods presented.

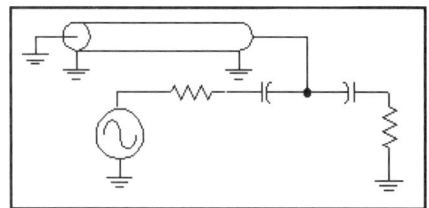

Fig 3.25—A quarter wavelength of transmission line forms a resonant tuned circuit.

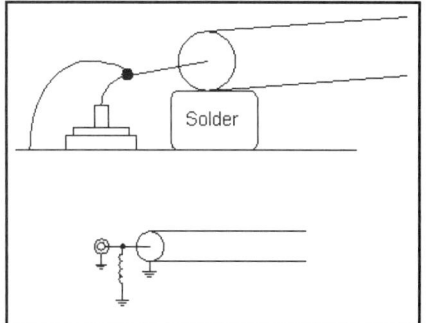

Fig 3.26—Loading (coupling to the "outside world") can be controlled with small wire inductors.

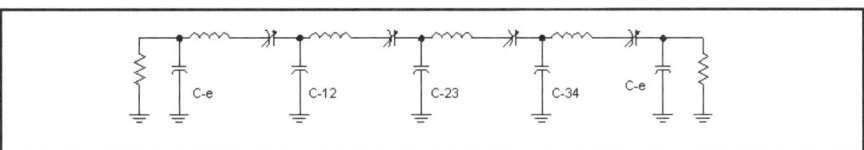

Fig 3.23—Bandpass filter using series tuned circuits. In this example, N=4.

A casual glance may not reveal a true identity. That is, a helical resonator with a tuning capacitor looks like a shielded LC resonator. However, the difference becomes clear if wideband measurements are done with loosely coupled probes like the ones that have been described for Q measurement. Such measurements will show a high Q at the fundamental frequency and additional responses (also having high Q) at 3, 5, and other odd harmonics of the fundamental. In contrast, a pure LC resonator will not show these departures. If capacitance is added to a helical resonator to decrease fundamental frequency, the higher frequencies will not move as fast. Slight capacitive loading might move the first "spurious response" to 4 F0 with greater departure as loading grows. Q remains high and excellent filters can still be built.

Helical resonators are coupled to each other with a variety of methods, although the most popular is through apertures, or holes in the walls between adjacent resonators. As with other filter types, the coupling can be related to the frequency spread between peaks when the resonators are unloaded. End section loading is realized in a variety of ways with helical resonators. A small line from a coaxial connector can be tapped onto the helix. The usual tap point is very close to the grounded end, often a small fraction of one turn. Again, the loading may be adjusted to establish an end section loaded Q.

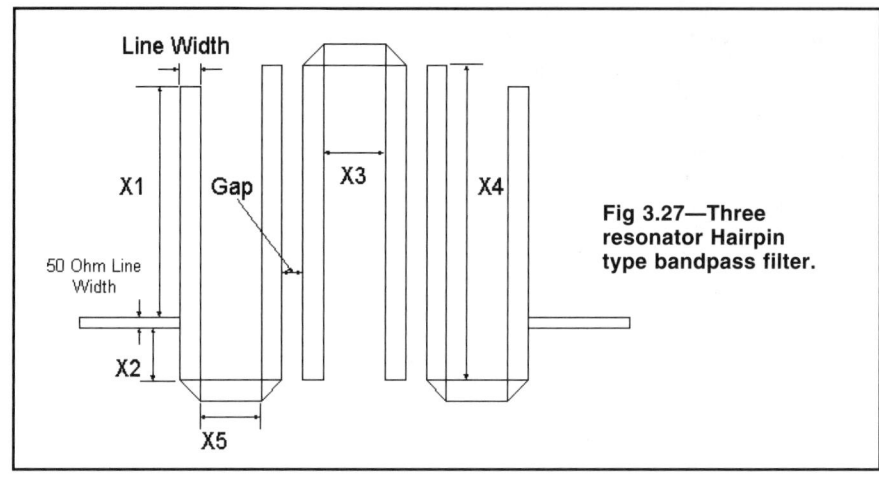

Fig 3.27—Three resonator Hairpin type bandpass filter.

We have only scratched the surface with some filter types we have built. A detailed review of the literature will reveal numerous other filter topologies of interest. The bandpass filters presented here are transformed from simple low-pass filters, the so-called all-pole low-pass circuits with nothing more than series inductors and shunt capacitors. Other low-pass filters such as the Elliptic can be transformed to bandpass form to generate bandpass circuits with transmission zeros next to the passband.

Another variation injects a transmission zero in a passband with no additional inductors. This is realized by an additional coupling capacitor that couples energy between non-adjacent resonators. This method was used in a 144 MHz transceiver discussed later in the book.[14] There is a great deal of work available to be done by the curious experimenter.

3.4 CRYSTAL FILTERS

No element is more intimately related to radio receivers than the quartz crystals used in filters. The early superheterodynes of the 1930s obtained single-signal selectivity with a crystal filter using but one crystal, a practice that continued through the 1970s. The use of high quality filters using a multiplicity of crystals became popular in the 1950s as SSB replaced classic AM as the radiotelephone method of choice.

Crystal Fundamentals

A modern quartz crystal is usually a round disc of single crystalline quartz with metalization on each side. The metal films serve to create (and sense) an electric field within the quartz. The basic structure is shown in **Fig 3.28**.

The basis for the interesting circuit properties of a quartz crystal is the piezoelectric effect. This effect is a material characteristic where an electric field causes a mechanical displacement. The mechanical motion is at right angles to the electric field in the quartz crystal. An electric field occurs when a voltage is placed between the two metalization layers attached to the crystal. The opposite effect also occurs; a mechanical motion generates an electric field.

The action of a quartz crystal when subjected to an electrical impulse is analogous to striking a bell or chime with a hammer: the energy of the impulse causes an oscillation to occur, a ringing that dies out in time. The resonant frequency of the chime is related to mechanical dimensions. In the same way, the resonant frequency of a quartz crystal is related to the crystal thickness. The Q of a quartz crystal can be very high, from 10,000 to over one million. The motions of a quartz crystal are transverse with the crystal vibrating parallel to the surface. This allows the Q and resonant frequency to be altered by surface effects. The reader with an interest in the physics of quartz crystals is referred to the classic text by Virgil Bottom.[15]

The quartz crystal is modeled as the LC tuned circuit shown in **Fig 3.29**. L_m and C_m are termed "motional" parameters for they relate to the mechanical motion of the crystal. The equivalent series resistance, ESR, is an element representing losses; it is related to the crystal Q. The final element, C_0, is the parallel, or holder capacitance. This C is a simple consequence of the crystal construction as a parallel-plate capacitor. This value is the sum of the parallel plate C (the dominant element) and some stray C related to the package housing the crystal. The parallel and the motional capacitance are related in the usual AT cut crystal. (AT cut refers to the crystallographic orientation of the crystal. Many of the crystals we deal with in radio are AT cut.) The relation between capacitors is approximately

$$C_0 = 220 \cdot C_M$$

Table 3.4 shows some measured representative values for some junk-box crystals. A crystal placed between a 50-Ω signal generator and 50-Ω load shows a response like that of **Fig 3.30**. If the crystal was a simple series tuned circuit without the parallel capacitor, C_0, the response would be a simple peak.

A crystal filter can be built with a single crystal with the scheme of **Fig 3.31**. L-networks at each end transform 50 Ω to present 500 Ω at the crystal. Transformer T1 provides an out-of-phase voltage to drive a phasing capacitor. This signal combines with the energy flowing through the crystal parallel capacitance to control the position of the notch. The 10-pF capacitor increases the effective parallel C of the crystal, moving the notch closer to the peak while the 25-pF capacitor resonates the ferrite transformer. **Fig 3.32** and **3.33** show the result of tuning the phasing capacitor.

Changing the terminating L-networks can alter the filter response. The bandwidth will decrease if the terminating impedance is dropped. A link could be used on T1 to replace the input L network while an output could be terminated with another wideband transformer. The modified circuit would then function well with a wide variety of crystals. Bandwidth will, of course, vary considerably as the com-

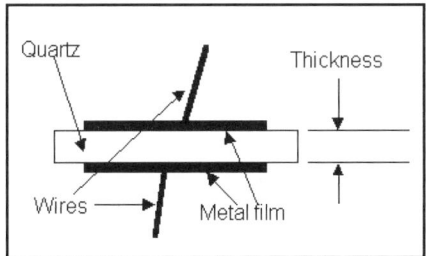

Fig 3.28—Cross section of a quartz crystal.

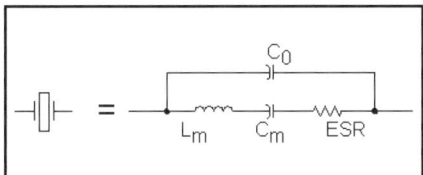

Fig 3.29—Symbol and circuit model for a quartz crystal.

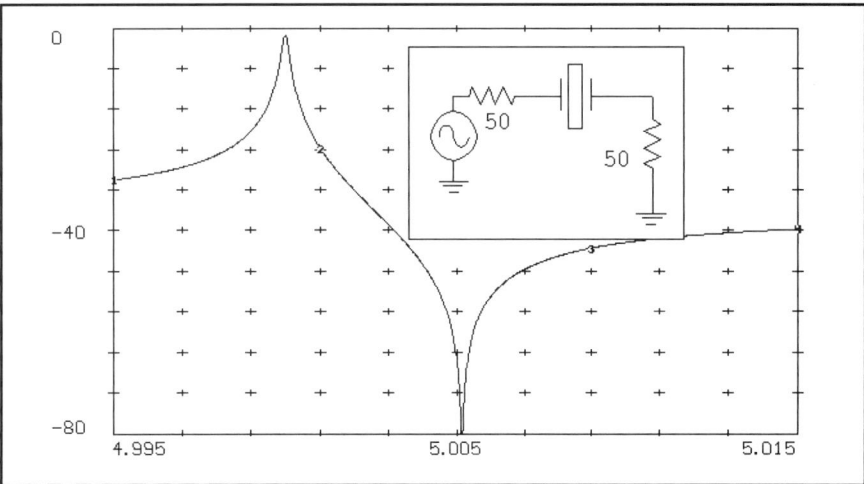

Fig 3.30—Crystal in a 50-Ω system with response. This crystal has a 5-MHz series resonant frequency, L_m=.098 H, Q=240,000, and C_0=5 pF.

Table 3.4

Freq. MHz	L_m, H	C_m, pF	C_0, pF	Q	ESR, Ω
3.58	0.13	.0152	3.35	50,000	58
5.0	.098	.0134	2.275	240,000	12.8
10.0	.020	.01267	2.8	200,000	6.3

ponents are changed. This filter type could even be used ahead of a receiver.

Crystal Measurement and Characterization

Earlier we swept an LC tuned circuit that was loosely coupled to a generator and a detector. A bandwidth measurement produced a Q_U. Loose coupling to a parallel tuned circuit occurred with a high impedance source and load. The crystal is a series tuned circuit and needs a low impedance environment for the loose coupling required for measurements. We can measure a crystal in the 50-Ω system shown in **Fig 3.34**.

The signal generator should be well buffered and extremely stable. The input of the circuit shown begins with a 20-dB pad, compensating for mismatch. The load can be a 50-Ω terminated oscilloscope, a spectrum analyzer, or a sensitive power meter. (See Chapter 7 or *QST*, June, 2001.) A 50-Ω, switched, 3-dB step attenuator is a useful aid in determining bandwidth.

A crystal is inserted in the test set (Fig 3.34) and the generator is tuned for a peak output. Note the peak response amplitude and the frequency F0 where it occurs.

Having measured peak response, remove 3-dB attenuation from the system, increasing the response. Tune the generator upward until the response drops to the level of the previous peak and record the frequency. This is one of the –3 dB frequencies. Repeat this step by finding the lower -3 dB point. The frequency difference, ΔF, is the 3 dB-loaded bandwidth in Hz for this test setup, which will be greater than the unloaded crystal bandwidth.

Knowing ΔF, return the generator to the frequency of peak response. Remove the crystal and plug the 100-Ω pot into the test set. Adjust the pot for the same meter reading; remove the pot from the test setup and measure its resistance with a digital multimeter. This is approximately the ESR of the crystal.

Some experimenters have mounted the pot in a panel and switched it into the circuit as needed. This may give inaccurate results owing to stray inductance. The pot should be mounted to a suitable "dummy crystal" with short leads.

A detailed analysis of the method reveals errors. These can be reduced substantially by shifting to lower measurement impedance.

The test set of Fig 3.34 is complete, providing both motional parameters and Q information. However, measurements with this apparatus become tedious. A simple crystal oscillator can provide the motional parameters. This circuit, **Fig 3.35**, includes a series capacitor that may be switched into the circuit to produce a frequency shift. Related equations are included with the figure.

The required Q_U for filter applications will depend upon the filter bandwidth and center frequency as well as on the filter shape and the number of resonators. A reasonable rule of thumb for most filters (LC and crystal) is that the "normalized Q" must exceed twice the number of resonators. Normalized q, q_0, is defined as Q_U

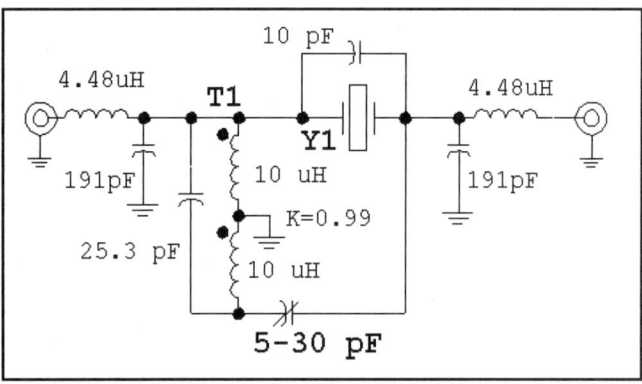

Fig 3.31—A single crystal filter using the crystal of Fig 3-30. T1 is 12 bifilar turns #26 on a FT-50-61 ferrite toroid. This filter has a 3-dB bandwidth of 1.4 kHz.

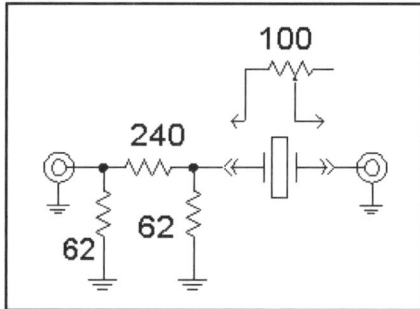

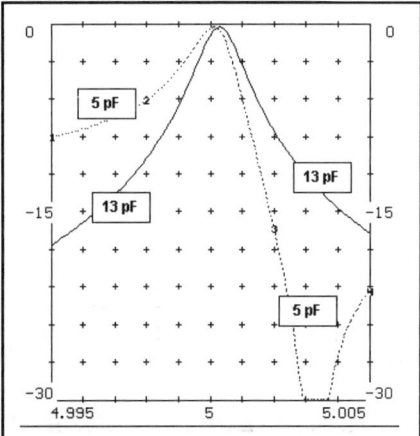

Fig 3.32—Response of the single crystal filter of Fig 3.31 when the phasing capacitor is at minimum value of 5 pF. The solid line represents the case of exact balance when the phasing capacitor equals the crystal C_0.

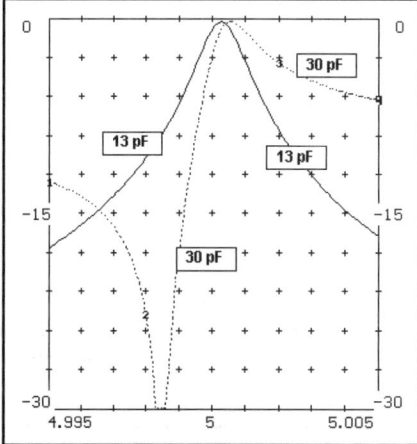

Fig 3.33—Response of the single crystal filter of Fig 3.31 when the phasing capacitor is at maximum value of 30 pF. The solid line represents the case of exact balance when the phasing capacitor equals the crystal C_0.

Fig 3.34—Simple test set for crystal measurement. The pad is a 20-dB, 50-Ω circuit. The output should be terminated in 50 Ω. A maximum input power from the generator would be about –10 dBm, resulting in a maximum to the crystal of –30 dBm. The 100-Ω pot is substituted for the crystal for ESR measurement. See text. Approximate equations for motional parameters are:

$$Q_U = \frac{1.2 \cdot 10^8 \cdot F}{\Delta F \cdot R_S}$$

$$C_M = 1.326 \cdot 10^{-15} \cdot \frac{\Delta F}{F_0^2}$$

$$L_M = \frac{19.1}{\Delta F}$$

F= Crystal Freq in MHz, ΔF=BW in test fixture in Hz, R_s= ESR, equivalent series resistance.

divided by the filter Q, or

$$q_0 = \frac{Q_U \cdot B}{F} \qquad \text{Eq 3.10}$$

A 500 Hz bandwidth filter at 5 MHz would have filter Q of 10,000. If crystal Q_U=100,000, q_0=10 and the filter would be practical with 5 crystals.

Generally, the most practical way to build crystal filters in the home lab begins with a large number of essentially identical crystals. These can sometimes be found at local surplus houses, often for very low prices. Equally good sources are mail order catalogs selling microprocessor crystals. Measurements (by W7AAZ) confirmed that many crystal brands offer good Q_U with a minimal frequency spread. But this is changing, even at this writing. The experimenter might consider ordering a small lot (perhaps 10) of a given crystal type. He or she can then measure them for Q and frequency distribution. If results are suitable, another order can be placed for a larger number. Typical cost for these crystals is around $1 each, so a batch of 10 crystals is still much less expensive than ordering even one special crystal.

Crystals should be matched to within 5 to 10% of the filter bandwidth to build effective filters. Hence, crystals for a 500-Hz wide CW filter should be matched within 25 to 50 Hz of a nominal frequency.

The recommended measurement procedure begins by numbering and marking all crystals in a set with stick-on labels. The crystals are measured for oscillation frequency in the same oscillator. If the "G3UUR" oscillator is used, be sure you specify which switch position is used, and record it in the notes. Measure motional parameters for several crystals to guarantee that there is small spread between crystals. It is also worthwhile to measure a few crystals for Q_U. The data is then entered into a computer spreadsheet where it is sorted according to frequency, making it easy to select matched crystals for a filter.

How many crystals should be purchased to make one filter? The answer is difficult, for it could vary a great deal with the crystal manufacturer. Generally, the purchase of 2 or 3 times as many crystals as the number of filter resonators is a good start. More is always useful. A larger lot, perhaps 100, almost guarantees a large selection of filters using most of the crystals. Left over crystals will be used in oscillators. It is rarely practical to build homebrew filters for already existing equipment.

Designing Simple Crystal Filters

Having characterized a set of crystals, we can now consider a filter design. The procedure will depend on the quality of the filter to be built. Some filters are easy, while others may require extensive and very careful measurement as well as computer simulation. Both extremes will be discussed.

Most of the filters we will discuss use the lower sideband ladder topology. An example is presented in **Fig 3.36**. The crystals are series elements in a ladder. Shunt capacitors couple energy between adjacent crystals. A mesh is one loop of a ladder, one crystal and the two shunt coupling capacitors on either side of it. A mesh could also be a load, a matching capacitor, a crystal, and one coupling capacitor. Some meshes include a series capacitor to tune the mesh to the same frequency as the other meshes in the filter.

The first method presented ignores the parallel crystal capacitance, treating the crystal as a simple series LC circuit. This scheme is suitable for simple CW filters. (Although we think of narrow filters as being more exotic than wide ones, it is generally easier to build narrow crystal filters.) This will be illustrated with an example, a 4th-order filter at 5 MHz with a 400 Hz bandwidth and a Butterworth shape. The $n=4$ Butterworth is a symmetrical filter with $q_1=q_4$=0.7654, k_{12}=0.8409, k_{23}=0.5412, and k_{34}=0.8409. The crystals have a 5-MHz center frequency, a motional inductance of 0.098 H, parallel C of 3 pF, and Q_U of 240,000. Normalized Q is q_0=19.2, so this is a realizable filter. Calculating the motional C from resonance at 5 MHz, we find C_m=0.010339 pF. We calculate the cou-

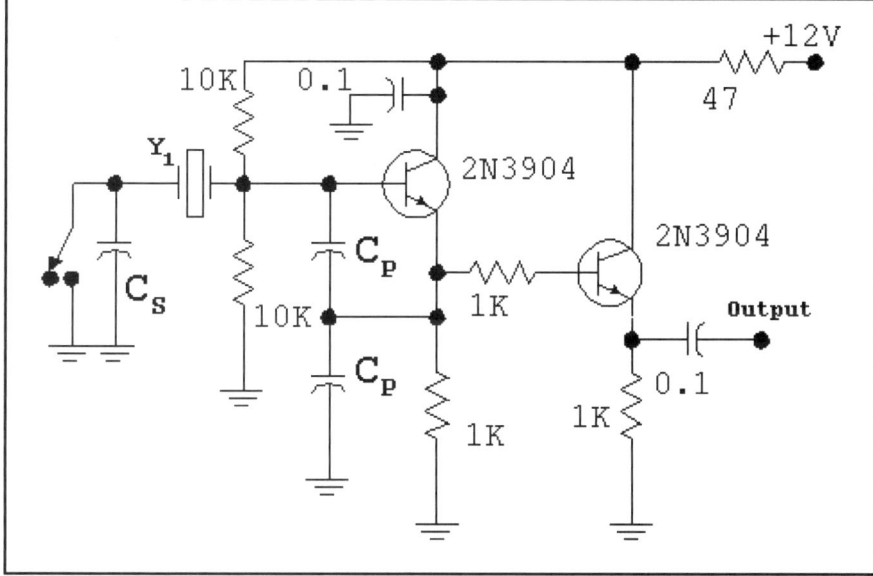

Fig 3.35—The G3UUR method for measuring quartz crystal motional parameters. A simple circuit to measure the motional parameters of fundamental mode quartz crystals. A crystal to be evaluated is placed in the circuit at Y1 and oscillation is confirmed. The frequency is measured. Then the switch is thrown and the frequency is measured again. Typical values are C_p=470 pF and C_s=33 pF. C_m will have same units as C_s. Be sure that C_s includes the stray capacitance of the switch as well the circuit part. Then:

If

$$C_S \ll C_P$$

then

$$C_M \approx 2 \cdot C_S \cdot \frac{\Delta F}{F}$$

and

$$L_m = \frac{1}{\omega^2 \cdot C_M}$$

where $\omega=2\pi F$ with F now in Hz. ΔF is the F difference observed when the switch is activated. Example: Use capacitors mentioned above, 10 MHz crystal; $F=1\times10^7$, ΔF=1609 Hz, to yield L_m=.0239H and C_m=10.6 fF. (1000 fF = 1 pF.)

pling capacitors with

$$C_{jk} = \frac{C_m \cdot F}{k_{jk} \cdot B} = \frac{1}{4 \cdot k_{jk} \cdot B \cdot \pi^2 \cdot F \cdot L_m}$$

Eq 3.11

where B is the bandwidth; F and B are both in Hz. Substituting, we find $C_{12}=C_{34}=154$ pF and $C_{23}=286$ pF. The end terminating resistance is given by

$$R_E = \omega \cdot L \cdot \left[\frac{B}{q \cdot F} - \frac{1}{Q_U} \right] \approx \frac{2 \cdot \pi \cdot B \cdot L_m}{q}$$

Eq 3.12

The end resistance is 309 Ω, yielding the preliminary filter as shown in Fig 3.37A. The filter has yet to be tuned. The filter would, otherwise, be finished if we wanted to terminate in this resistance. To illustrate the general case, we will terminate in a larger value, 450 Ω.

A termination R_0 will "look like" a smaller value R_E if it is shunted with a parallel capacitance, C_E where

$$C_E = \sqrt{\frac{(R_0 - R_E)}{R_E \cdot \omega^2 \cdot R_0^2}}$$

Eq 3.13

Using the values from above, we obtain an end capacitor of 47 pF, producing the next version of the filter as shown in Fig 3.37B. Only filter tuning remains.

The end meshes, 1 and 4, are terminated in a parallel RC circuit. The equivalent series RC consists of the original end resistance, R_E, and a capacitance C' where

$$C' = \frac{C_E^2 \cdot \omega^2 \cdot R_0^2 + 1}{C_E \cdot \omega^2 \cdot R_0^2}$$

Eq 3.14

C' is 153 pF, R_0 is 450 Ω, and R_E is 309 Ω for this example.

The end meshes are shown, isolated from the other meshes, in Fig 3.37C while the interior meshes are shown in isolation in Fig 3.37D. The end meshes have a net series C of 76.7 pF while the interior ones have a net series C of 100.1 pF. Both will be detuned from the nominal crystal 5 MHz, but the meshes with the smallest capacitance will be detuned by the largest amount. The lower meshes can be properly tuned by added series C so that they have the same net series C as the highest frequency one. This will occur with a tuning C of

$$C_T = \frac{C_{High} \cdot C_{Mesh}}{C_{Mesh} - C_{High}}$$

Eq 3.15

Using $C_{Mesh}=100.1$ pF and $C_{High}=76.7$ pF, a proper tuning capacitor is 328 pF. The final filter circuit is shown in Fig 3.37E.

The computer generated response for this filter is shown in **Fig 3.38**.

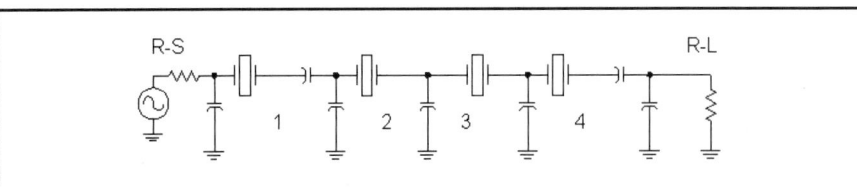

Fig 3.36—Lower sideband ladder filter with four crystals. The four meshes are labeled for reference in the discussion.

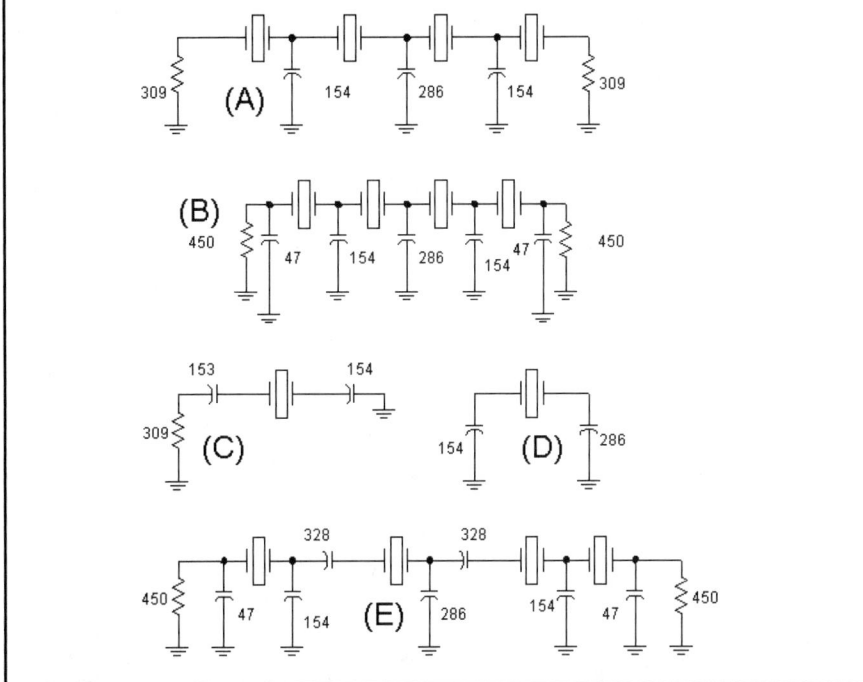

Fig 3.37—Evolution of a bandpass filter showing the steps in the design. See text for details.

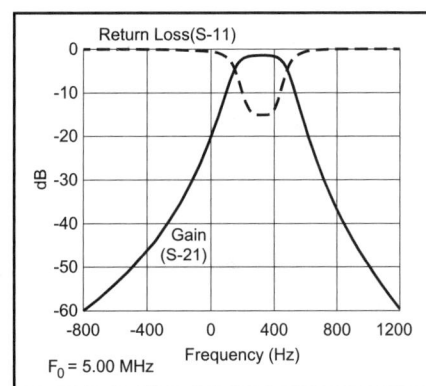

Fig 3.38—Response for the crystal filter designed in Fig 3.37.

Accounting for Parallel Crystal Capacitance

The quartz crystal model of Fig 3.29 is generally an accurate one. C_0 has little effect in filters that are sufficiently narrow, so was ignored in the previous design. The 5-MHz CW filter just presented was designed for a 400-Hz bandwidth with a Butterworth shape. The shape is very close to an ideal Butterworth.

Problems increase as filter bandwidths grow. This is illustrated with **Fig 3.39** which shows the response of two different 3-kHz bandwidth filters using 3.58-MHz TV color burst crystals. The

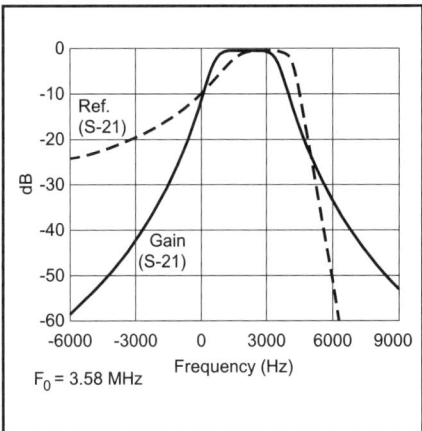

Fig 3.39—The response of two crystal filters built from 3.58-MHz color burst crystals. One uses ideal crystals with zero C0 to produce a symmetrical shape. The other (with dashed line) uses C0=4 pF crystals.

solid curve is the response we would like, designed with ideal crystals with zero parallel capacitance. $C_0=4$ pF produces the other response. The filter bandwidth is too narrow and the attenuation is markedly increased. It is for this reason that this circuit is named the lower sideband ladder filter.

Response distortion results because the parallel C_0 makes the series resonators behave as if they had a larger motional L than is measured. This effect is plotted in **Fig 3.40** for the 5-MHz crystals used in the earlier CW filter design. The lower curve shows the effect of a 2-pF parallel capacitance while the upper curve is for $C_0 = 5$ pF. Here, X is the ratio of L_{eff} to L_m. The horizontal axis in the curve is δF, the offset from the series resonant frequency. These effects were discussed in greater detail in *QEX* for June, 1995, where detailed design equations are given. The corrections related to the effective inductance are included in the program *XLAD.exe*. Both the program and the 1995 *QEX* paper are included on the book CD.

The effective inductance is larger than the normal motional L by a factor of 2 or more. This reduces the effective motional capacitance by the same factor. Accordingly, the coupling capacitors must be reduced by the same factor. The change also alters the calculation of end resistance. The new terminations and reduced coupling capacitors will then alter the filter tuning.

One can build symmetric filters if the effect of parallel capacitance is eliminated. One way to do this parallels each crystal with a large inductance. The value required is one that resonates with C_0, forming a parallel trap that is then bridged by the series resonant portion of the crystal. An experimental filter was built to examine this idea. The inductance used was smaller than required for resonance, so small trimmer capacitors were added. The filter, built with 3.58-MHz color burst crystals for a 3.5-kHz bandwidth, is shown in **Fig 3.41**. The measured response is presented in **Fig 3.42**.

Crystal filters built with paralleled inductors suffer from degraded stopband response. Although the performance around the filter center is as designed, it degrades a few hundred kHz away from center, necessitating the crystal filter be supplemented with an LC bandpass.

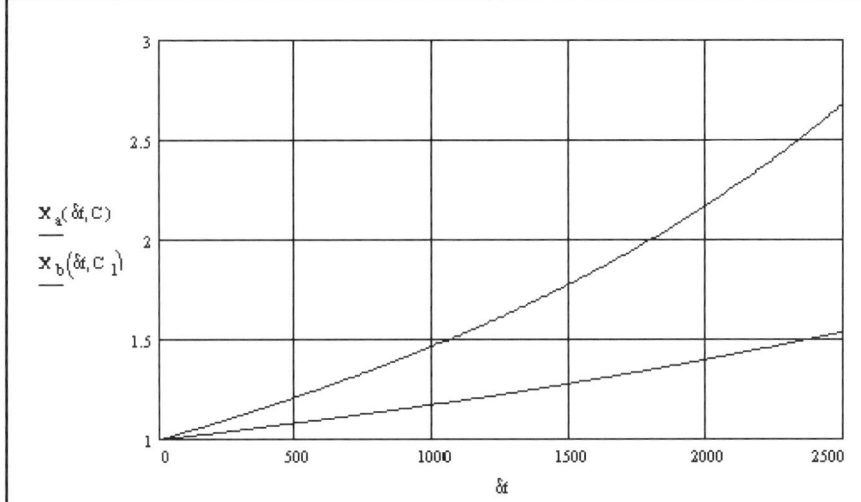

Fig 3.40—X, defined as L_{eff}/L_m, is plotted for frequency offset, δf, above crystal series resonance in Hz. These 5-MHz crystals had parallel C of 2 and 5 pF.

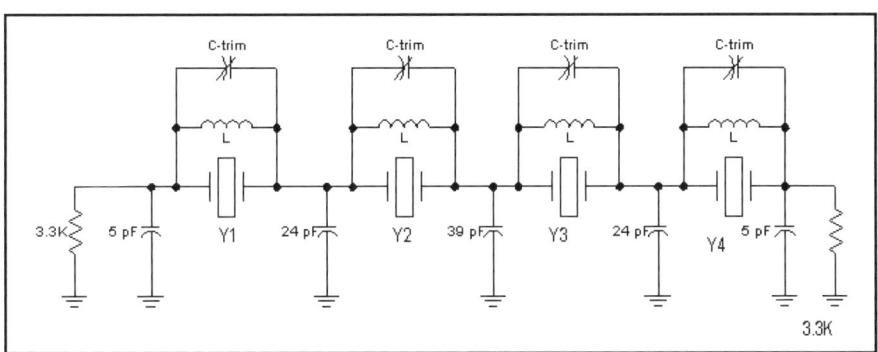

Fig 3.41—Experimental crystal filter.
Y1,2,3,4 = 3.58-MHz surplus color burst crystals. (L_m=0.117H, C_0=4 pF)
L = 151 µH, 48 turns #30 on FT-50-61 Ferrite toroid.(Amidon)
C-trim = 3-12 pF ceramic trimmer. See the referenced *QEX* paper for adjustment procedure.

The Min-Loss Filter of Cohn and other Simplified Forms

A simplified non-mathematical scheme for building crystal filters uses the Min-Loss circuit. This circuit is the result of fundamental work by S. B. Cohn where he described a family of coupled resonator filters that achieved very low insertion loss while maintaining good stopband attenuation.[16] A really interesting property of these filters was the fact that they used identical resonators that were coupled to each other with equal values of coupling. This means that all shunt coupling capacitors in a Min-Loss crystal filter are equal. If the filters are designed without shunt end loading capacitors, tuning is greatly simplified. A Min-Loss type crystal filter is properly tuned if

• all crystals have the same frequency,
• all coupling capacitors are of the same value, C,
• series capacitors having the same capacitance as the coupling C are placed in series

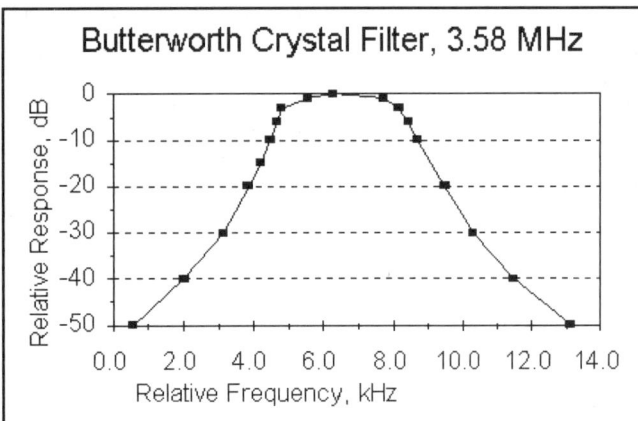

Fig 3.42—Measured response for the filter shown in Fig 3.41.

A three element crystal filter at 10 MHz. The metal can crystals have small wires soldered to them that are then grounded to the foil.

Three experimental crystal filters. The top circuit uses 10 crystals in a circuit with equal coupling between resonators (Cohn). The bottom filter is that from Fig 3.41.

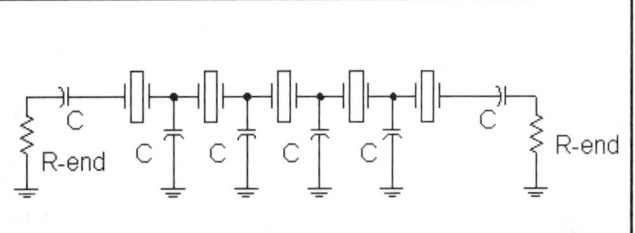

Fig 3.43—Min-Loss type crystal filter with equal coupling and simplified tuning.

with both end crystals
• both terminations are equal and properly related to coupling.

A crystal filter of this type, with five resonators, is shown in **Fig 3.43**.[17]

This filter topology often appears with the name "Cohn Filter," titled for the original circuit theorist who contributed so extensively to our design methods. Other filters have also appeared with the Cohn name. Here we have divorced the name from this simple crystal filter, for it is but one example from Cohn's body of work, a collection that is much richer and more extensive than has been presented in the amateur literature.

While most of the Min-Loss crystal filters we build are fabricated without design (i.e., without any mathematical analysis), they may certainly be studied and designed on the computer. The normalized coupling coefficients and end section Q for this filter type are approximately given by

$$k_{jk} = \frac{1}{2} \cdot \exp\left(\frac{\operatorname{Ln}(2)}{N}\right) \quad \text{Eq 3.16}$$

$$q = \frac{1}{k_{jk}} \quad \text{Eq 3.17}$$

where N is the number of resonators. These values are tabulated for n from 2 to 10 in **Table 3.5**. (The first few points appeared in the original Cohn paper, while k and q for N>5 are extrapolations via our above equations.)

Shown in **Fig 3.44A** are transfer function plots for two different filters of this type. The wider, lower loss one has 3 resonators while the other has 8 crystals. Both circuits were designed for 5 MHz with a 500-Hz bandwidth using high Q crystals with L_m=0.098 H. Part A of the figure shows close-in details while Fig 3.44B shows the response to the −80 dB level. Part C of the figure shows the group delay for the filter with 8 resonators. (More will be said about group delay shortly.) All three plots are computer generated re-

Table 3.5

N	k	q
2	0.707	1.414
3	0.63	1.587
4	0.595	1.683
5	0.574	1.741
6	0.561	1.782
7	0.552	1.811
8	0.545	1.834
9	0.54	1.852
10	0.536	1.866

sponses, although they are in good agreement with measurements on similar filters. We have built Min-Loss crystal filters up to 10th order.

The data of Fig 3.44 illustrate the salient properties of the Cohn filter. The passband shape is smooth with minimal ripple for the low order filters (N=3), but becomes distorted as the number of resonators grows beyond five. The ripples on the passband edges near the skirts become extreme with wider bandwidth filters. The N=8 data of Fig 3.44B illustrate the excellent shape afforded by the Min-Loss filter. However, the time domain performance as depicted in the group delay plot suggests that this filter may have severe ringing if built for narrow (CW) bandwidths.

Although the two filters (N=3 and N=8) described in Fig 3.44 have different responses, they are remarkably similar in component values. The N=3 filter used 146-pF capacitors and 181-Ω terminations while the N=8 filter used 168 pF and 155 Ω. A filter designed with two or three crystals can be extended with the same capacitor values and terminations. This becomes extremely useful for the experimenter.

The Min-Loss crystal filter has virtues of low insertion loss and good skirts, but at the price of poor passband shape with higher orders. Some other filters offer similar non-mathematical simplicity and better passband performance, with a group of crystals all at the same frequency. **Fig 3.45** shows such a filter. This design is a Butterworth design at 10 MHz with normalized parameters of $q=0.765$, $k_{12}=k_{34}=0.841$, and $k_{23}=0.541$. This filter is designed with a pure resistive termination at the ends (no shunt end capacitors.) The equations predict the end resistance and the shunt capacitors. The series tuning capacitors are yet to be established. However, the values are clear from inspection. If the end capacitors are set to the value of the center capacitor (85 pF,) each mesh has the same capacitors in the related loop.

Design with the equations does not take the parallel crystal capacitance effects into account. This is done with curves like those of Fig 3.40 that establish an increased effective inductance value that can then be applied with the equations. Approximate designs without the curves will still result in practical filters at the higher frequencies (8 MHz and up) although the bandwidth will be a bit narrower than the design values.

Ringing, Group Delay and Filter Passband Shape

All serious receiver experimenters have their favorite efforts, receivers with specifications differing little from others, but with a "crisp sound" that sets them apart from the ordinary. There are numerous phenomenon that tend to degraded performance and remove "crispness." One that can ruin an otherwise excellent receiver is an IF filter with excessive group delay. All filters have time delay, a truth that cannot be avoided. The filters that "sound" the best are those that have small delay for a given bandwidth and, of greater import, behave like a transmission line with little variation in group delay over the passband.

The group delay of an eighth order Min-Loss filter was presented in Fig 3.44C. The delay was high, exceeding 10 milliseconds in part of the passband. The group delay variation over the passband was also severe. This filter, although very selective, would probably not sound good, especially with noise pulses.

Two 5-MHz filters were designed for a bandwidth of 500 Hz, each with five crystals. One filter used a 0.1-dB ripple Chebyshev response while the other used a linear phase response. The Chebyshev results are shown in **Fig 3.46** while the linear phase response is given in **Fig 3.47**. Both plots overlay group delay and gain. The "ears" of the Chebyshev group delay plot line up with the 3-dB edges of the passband, so all delay variations are heard. In contrast, the region of low group delay in the linear phase filter extends well beyond the filter bandwidth edges. Both of these filters have been built and tried in an experimental CW receiver. The linear phase filter was more difficult to build, but sounded much better. The skirts were steep in the Chebyshev, so it presented adequate selectivity. We found the linear phase filter in need of more skirt selectivity. Although not shown in the figures, the Chebyshev filter group delay was 2.5 times as large as the linear phase filter delay.

We have also had good results with an intermediate filter shape, the Gaussian-to-6 dB response. This is a filter with a rounded peak shape for the top 6 dB, but with steep Chebyshev-like skirts. Transitional filters (Gaussian-to-6 dB, Gaussian-to-12 dB, linear phase, and maximum flat delay) are slightly more difficult to build than the Min-Loss, Butterworth, or Chebyshev filters, for they lack the sym-

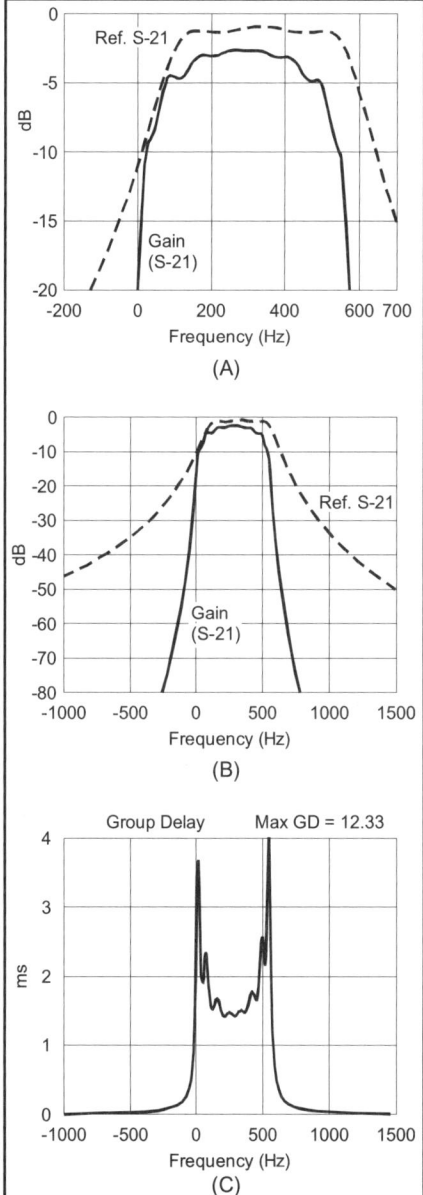

Fig 3.44—Min-Loss crystal filter responses. A and B compare 3rd and 8th order filters in responses to –20 and –80 dB. C shows the group delay for the 8th order filter.

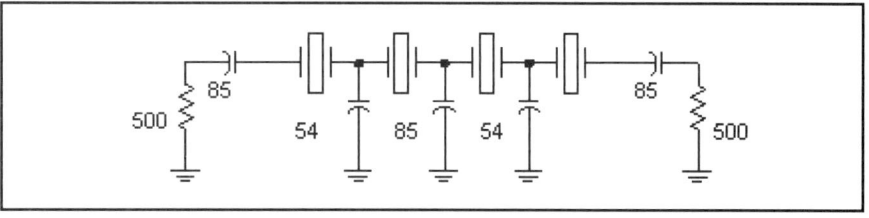

Fig 3.45—10-MHz SSB bandwidth filter using crystals with identical frequencies and "easy" tuning. This filter has a Butterworth shape; the simplified tuning method often works well with N=4 Chebyshev filters.

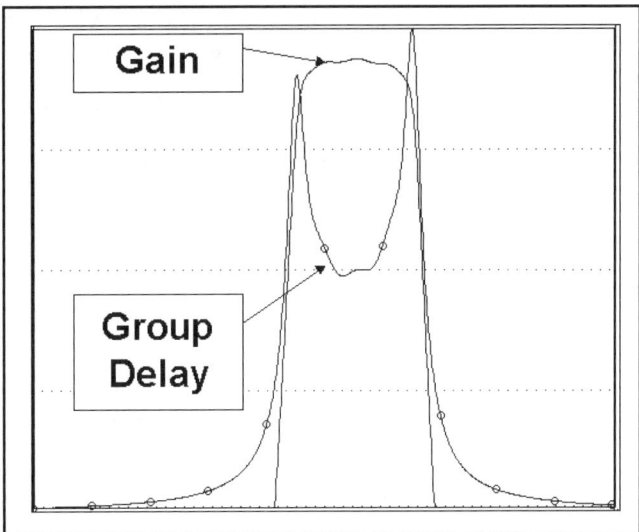

Fig 3.46—Group delay and gain for a Chebyshev crystal filter. The gain is plotted over a 20-dB range.

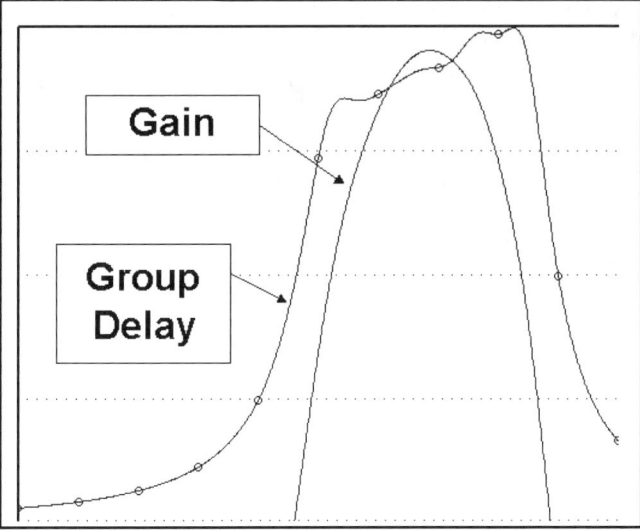

Fig 3.47—Group delay and gain for a linear phase crystal filter. The gain is plotted over a 20-dB range.

metry of the traditional types. If the transitional filters were commercially available, they would probably be very expensive. On the other hand, they offer a challenge that is well worth the effort for the advanced experimenter. The reader should review the work of Carver[18].

Intuition would suggest that a FIR (finite impulse response) filter, usually realized with DSP, would have significantly reduced ringing. Some do, but some others still show significant ringing. Extreme selectivity always seems to bring some ringing. Generally, it is the less selective schemes with smooth peak shapes that always sound the best, without regard to the method used to achieve it, traditional hardware or digital signal processing.

3.5. ACTIVE FILTERS

While most receivers are super-heterodyne designs with an IF, some simple superhets as well as virtually all direct conversion receivers obtain much of their selectivity from audio filtering. Audio frequency inductors have become available in recent years, making traditional LC designs viable at low frequencies. Even prior to the arrival of those parts, some builders had built audio filters with surplus telephone toroids. Still, the most common method for audio filtering uses RC active circuits. An RC active filter combines gain with resistors and capacitors to synthesize inductor behavior.

The Low Pass Filter

Figure 3.48 shows an active low pass filter form known as the voltage controlled voltage source (VCVS). It uses an operational amplifier configured as a non-inverting amplifier, usually with a gain of one. Two resistors and two capacitors complete the circuit. For example, with R=10 kΩ, C1=.01 μF, C2 = .02 μF, the cutoff is 1125 Hz. Eq 3.18 can be solved for R for any cutoff or capacitor, but is restricted to A = 2, which is the Butterworth filter. A peak appears in the response as A exceeds 2. The circuit provides a voltage gain of 1.7 when A=10.

The filter has a two-pole Butterworth response when A=2. For A = 2 and for equal R, the 3 dB cutoff frequency is given by

$$F_{C.O.} = \frac{\sqrt{2}}{4 \cdot \pi \cdot R \cdot C}$$ Eq 3.18

Fig 3.48 is a special version with two equal resistors and a capacitor that is A times the value of the first. **Fig 3.49** shows representative responses for R = 10k, C = .01 μF and A = 1, 2, 5 and 10.

If A exceeds 2 the filter takes on a peaked response. It is then more convenient to work with the peak frequency as a function of R, C, and A, the capacitor ratio. If A>2, the peak frequency is given by

$$F_{PEAK} = \frac{\sqrt{A-2}}{2 \cdot \pi \cdot A \cdot C \cdot R}$$ Eq 3.19

Some values of low pass voltage gain at

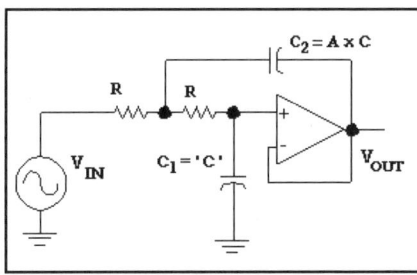

Fig 3.48—RC active low-pass filter. The op-amp is assumed to be powered from dual supplies around ground. Other biasing schemes are presented later. The operational amplifier is configured for a non-inverting gain of 1. C2, the feedback capacitor, is A x C1 where A is a value greater than 1.

Table 3.6

A	Voltage Gain	A	Voltage Gain
2.2	1.004	6.8	1.41
2.4	1.014	10	1.67
3.3	1.088	22	2.4
3.6	1.12	33	2.9
3.9	1.14	47	3.46
4.7	1.22		

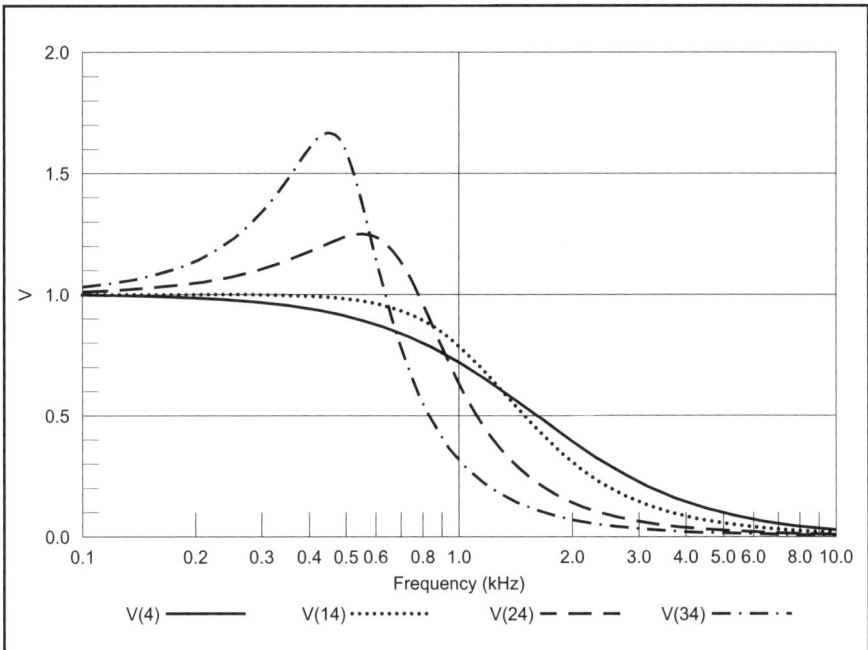

Fig 3.49—Response of the filter shown in Fig 3.48 with A=1, 2, 5, and 10. These curves, and several others in this section, were generated with *Super Spice* from Compact Software. The solid line corresponds to A=1 while the highest peak is for A=10.

High-Pass Filters

Figure 3.52 shows a VCVS type high-pass filter. This circuit is the dual of the low pass just discussed. It is designed with equal valued capacitors. The resistors now differ by a factor "A". The usual filters have the grounded resistor as the one with larger value. **Fig 3.53** shows the response for four different filters, all with

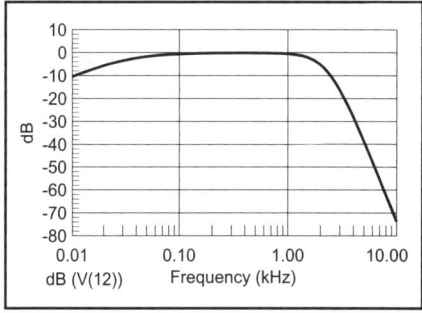

Fig 3.51—Response for the cascade of identical low-pass sections presented in Fig 3-50. This is a calculated result, although we have built several similar designs.

the response peak are tabulated vs A, the capacitor ratio, in **Table 3.6**.

There are numerous ways to design practical low-pass filters with the equations. A cascade of sections like those in Fig 3.48 would form Butterworth or Chebyshev filters of high order. Each capacitor corresponds to one pole in the response, one L or C in the traditional filter. Generally, each two-pole low-pass section will differ from the others in higher order Butterworth or Chebyshev filters. For details, see the text by Johnson, et al.[19]

Alternatively, several identical low-pass sections can be cascaded to form a useful circuit. These filters are easy to analyze and design, and offer excellent performance, especially with simple direct conversion receivers. An example of a filter of this type is shown in **Fig 3.50**.

Three two-pole sections with A=2 are cascaded to form a 6-pole filter suitable for SSB reception. The response for this filter is shown in **Fig 3.51**. The dip at low frequency results from the 1-μF input coupling capacitor.

Cascades of peaked low-pass filters (A>2) can be very useful. The gain can be considerable when several stages are cascaded. These filters take on a bandpass like shape, offering an attractive response for direct conversion receivers intended for CW use.

The filter shown in Fig 3.50 is biased for single power supply operation. This scheme is especially attractive with the low-pass filter, for an entire chain of filter sections may be biased with only one divider. If LM-358 or LM-324 op-amps are used, a pull down resistor should be connected from the amplifier output to ground. The resistor should pass a standing current of about 1 mA. Severe crossover distortion will result without this loading.

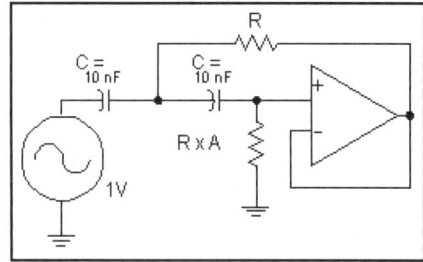

Fig 3.52—Voltage controlled voltage source high-pass filter. The operational amplifier is again set for a closed loop gain of +1.

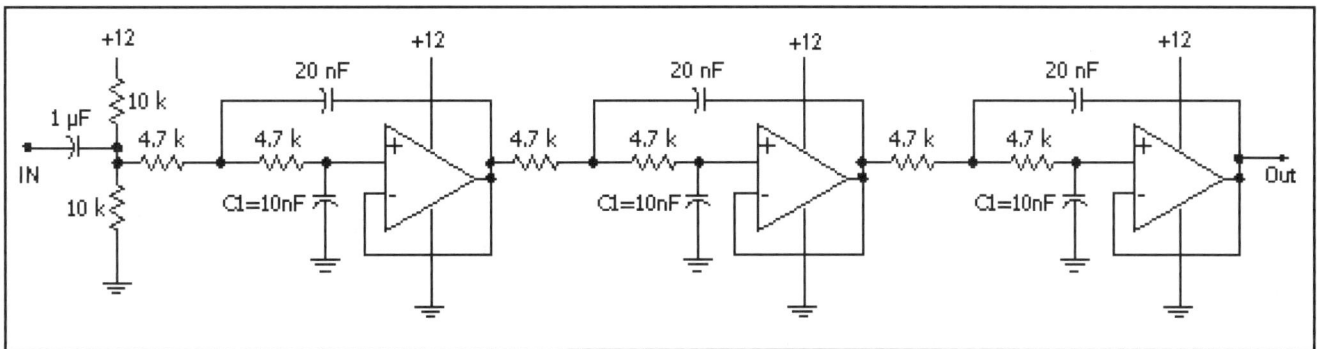

Fig 3.50—Practical low-pass filter that can be built with common op-amps, such as the 741, 1458, 358, 324, 5532.

Filters and Impedance Matching Circuits **3.25**

The VCVS low-pass filter with equal resistors has a transfer function of

$$H(s) = \frac{1}{1 + 2 \cdot s \cdot C \cdot R + s^2 \cdot C^2 \cdot R^2 \cdot A} \qquad \text{Eq 3.20}$$

where s is now the complex (LaPlace) frequency, s=jω in the frequency domain. C is the shunt capacitor while A×C is the feedback capacitor. The corresponding frequency domain response is

$$\left| \frac{V_{out}}{V_{in}} \right| = \sqrt{\frac{1}{\left(1 - 8 \cdot \pi^2 \cdot f^2 \cdot R^2 \cdot C^2 \cdot A + 16 \cdot \pi^4 \cdot f^4 \cdot R^4 \cdot C^4 \cdot A^2 + 16 \cdot \pi^2 \cdot f^2 \cdot R^2 \cdot C^2 \right)}} \qquad \text{Eq 3.21}$$

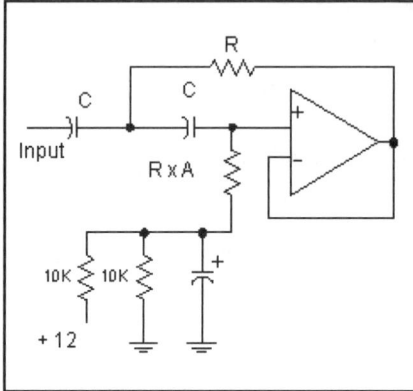

Fig 3.54—Biasing method for high-pass filter sections. A voltage divider creates a synthetic ground at half of the single supply.

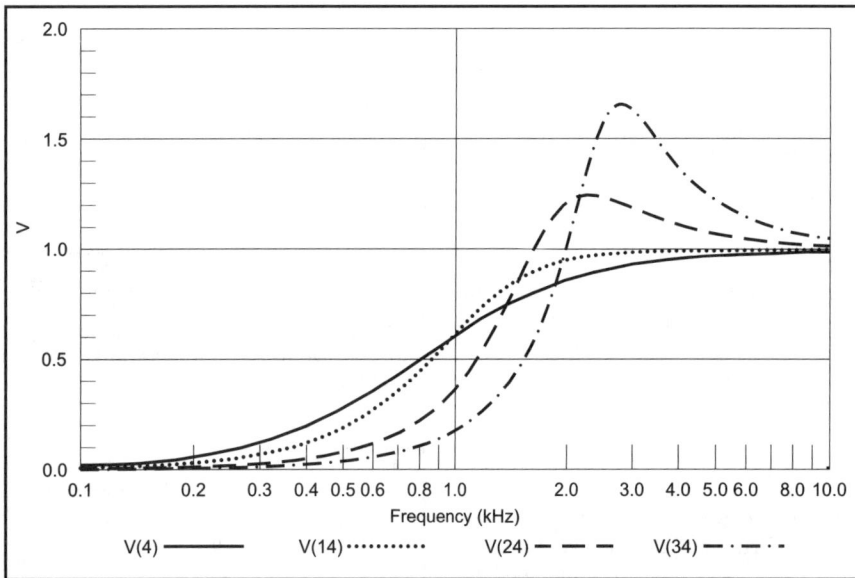

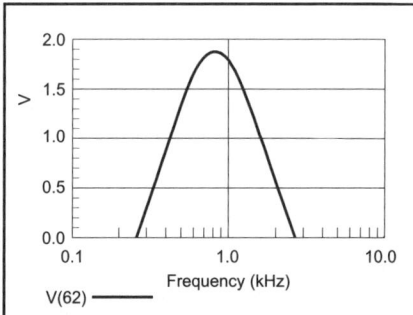

Fig 3.55—The 4x4 filter, a cascade of four peaked low-pass sections (6.8 kΩ, 10 nF, and 50 nF) followed by four peaked high-pass sections (20 nF, 27 kΩ, and 5.6 kΩ)

Fig 3.53—Transfer functions for four versions of the high pass section of Fig 3.52. The resistor ratio varies, taking on values of A=1, 2, 5, and 10. The solid line corresponds to A=1 while the highest peak is for A=10.

Active Bandpass Filters

A bandpass-filter section is shown in **Fig 3.56** using an operational amplifier in an infinite gain multiple feedback circuit. The IGMFB circuit is practical with common op-amps such as the 741, 1458, and 5532. The topology is represented with two equal valued capacitors and three resistors. One of the resistors allows the user to specify circuit gain as well as center frequency and Q or bandwidth. The design begins by picking these values for voltage gain K (a dimensionless ratio), Q, f_0 in Hz, and C in Farads. The required resistors are then

10-nF capacitors and a 20-kΩ grounded resistor. The ungrounded resistor varies to set gain and peaking. The values used are 20 kΩ, 10 kΩ, 4 kΩ, and 2 kΩ.

The characteristics of the high-pass section are much like those of the low pass. The circuit begins to take on a peaked response when A exceeds 2. A peaked high pass will have a peak frequency given by

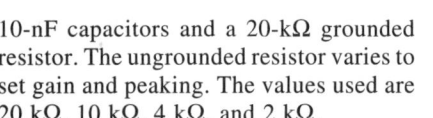

There is no peak if A<2. The pure N = 2 Butterworth then has a 3 dB cutoff frequency given by

$$f_{c.o.} = \frac{\sqrt{2}}{4 \cdot \pi \cdot R \cdot C} \qquad \text{Eq 3.23}$$

The VCVS high-pass sections do not have a dc path through them that allows the easy biasing afforded by the low pass. A high-pass section may be biased with the methods shown in **Fig 3.54** when dual power supplies are not available.

The high pass and low-pass forms may be combined in a cascade to form bandpass filters with excellent stopband attenuation. An example response is shown in **Fig 3.55** where four peaked low-pass sections are cascaded with four peaked high-pass sections.

$$R_1 = \frac{Q}{K \cdot \omega_0 \cdot C} \qquad \text{Eq 3.24}$$

$$R_2 = \frac{Q}{(2 \cdot Q^2 - K) \cdot \omega_0 \cdot C} \qquad \text{Eq 3.25}$$

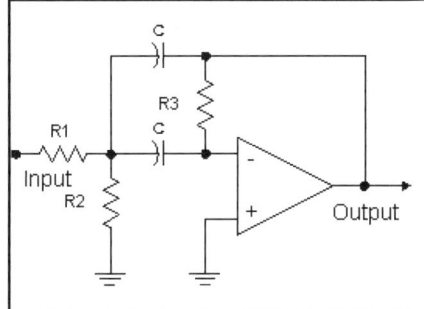

Fig 3.56—Infinite gain, multiple-feedback (IGMFB) bandpass filter. This topology is capable of moderately high Q and gain with practical components.

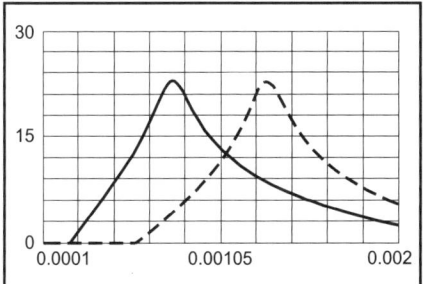

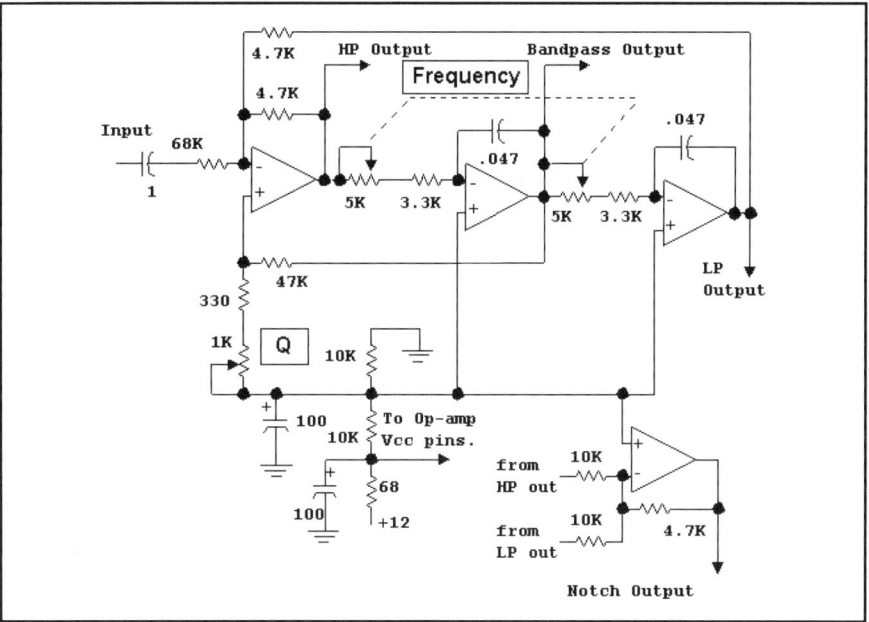

Fig 3.58—State-variable audio filter for CW receiver applications. All op-amps are 741 or 1458. The op-amp pin numbers are not shown. The builder must also connect the power supply line to the V_{cc} point on the op-amps. This circuit was inserted between the audio gain control and the output amplifier in a high performance CW receiver.

Fig 3.57—Calculated gain in dB for the IGMFB bandpass filter shown in Fig 3-56. This version used the resistor and capacitor values calculated in the text for Q=5 at 800 Hz with a gain at resonance of 2. The solid curve represents the nominal response while the dashed curve shows the result of tuning R2 to a lower value. Changing R2 to a 1 kΩ-variable in series with a 560-Ω fixed resistor would produce a tunable bandpass characteristic with essentially constant gain and bandwidth. This tuning scheme works well only when R1>R2. This sweep was generated with *Super-Star Professional* from Eagle Software.

$$R_3 = \frac{2 \cdot Q}{\omega_0 \cdot C} \qquad \text{Eq 3.26}$$

where $\omega_0 = 2 \times \pi \times f_0$. We see from **Equation 3.25** that the gain should be less than $2Q^2$. For example, a filter using 22-nF capacitors with a center frequency of 800 Hz, a Q of 5, and a gain at resonance of 2 is built with R1=22,600 Ω, R2=942 Ω, and R3=90.4 kΩ. The transfer function for this filter is shown in **Fig 3.57**.

The IGMFB bandpass filter must be biased with the method shown earlier for a high pass filter if a single power supply is to be used. This filter form is ideal if several sections are to be cascaded. It is sometimes useful to provide a rotary switch allowing the user the ability to select one of several outputs in a cascade. Each section of an IGMFB filter can have a Q as high as 10 or 20.

Other bandpass circuit forms are also suitable. An especially interesting one is the so called state-variable filter, which uses three operational amplifiers. The one circuit will simultaneously provide low pass, high pass, and bandpass outputs. Adding one more op-amp will even allow a notch filter function. An example is shown in **Fig 3.58**. This circuit is tunable over the normal range used for CW notes and has variable Q. The notch is not included in the version that was built, but could be added with the circuitry shown.

The reader interested in more information on the state-variable filter should examine the article by Howard Berlin.[20] The state-variable filter is an especially interesting circuit for those with a mathematical inclination, for the circuitry is an exact replication of the equations.

The All-Pass Filter

An especially interesting, but very simple RC active filter circuit is the all-pass of **Fig 3.59**. This circuit uses an op-amp, a single section RC low pass filter, and a pair of resistors. Although we analyze the circuit with mathematics, much of the behavior is clear from inspection. At very low frequency, the capacitor is an open circuit. The op-amp input impedance is very high, so the input voltage is also that appearing at the point marked "E." The negative feedback action forces the inverting op-amp input to also be E. The only way for this to happen is for the output to also equal E. At low frequency the output is in phase with the input and has the same magnitude for unity gain. In contrast, at very high frequency, the capacitor is a short circuit. The op-amp

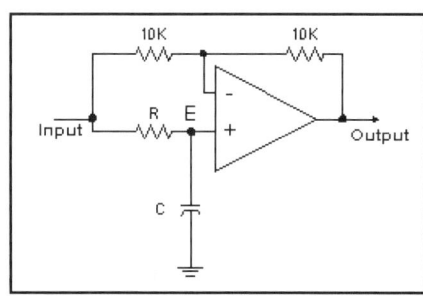

Fig 3.59—Basic, single section all-pass filter. This circuit has unity gain at all frequencies, but has a continually changing phase response. It is useful for phase shift networks such as those used for the phasing method of single sideband.

Filters and Impedance Matching Circuits 3.27

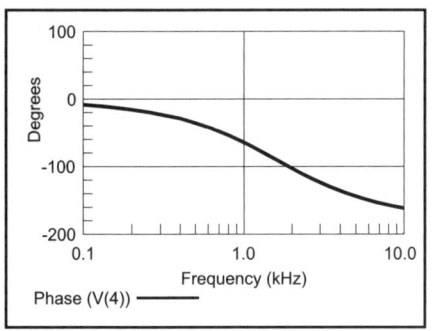

Fig 3.60—Phase response for an all-pass filter.

then behaves as the familiar inverting amplifier (180 degrees of phase shift) with unity gain.

The transfer function for this circuit is

$$H(j\omega) = \frac{1 - j \cdot \omega \cdot R \cdot C}{1 + j \cdot \omega \cdot R \cdot C} \qquad \text{Eq 3.27}$$

where $\omega = 2 \times \pi \times f$ with f in Hz. This circuit has an amplitude response of unity at all all frequencies and a phase shift given by

$$\theta = \cos^{-1}\left[\frac{1 - \Omega^2}{1 + \Omega^2}\right] \qquad \text{Eq 3.28}$$

where $\Omega = f/f_0$ with f_0 being the frequency where the network has a 90 degree phase. f_0 is given by

$$f_0 = \frac{1}{2 \cdot \pi \cdot R \cdot C} \qquad \text{Eq 3.29}$$

The phase response of the network is presented in **Fig 3.60** for the case of R= 10 kΩ and C=10 nF.

A common application for the all-pass network is to generate the audio phase shift needed in a phasing type SSB receiver or transmitter. Examples are found in Chapters 8 and 9.

A FIR Bandpass Filter

The all-pass filter serves as a frequency dependent delay element for a variety of applications. An unusual one is in a special bandpass filter, one with a finite impulse response. The basic, repeated element in this filter is a delay element, shown in **Fig 3.61**. The delay arises from a cascade of two all-pass networks. The RC in the all-pass is picked for 90 degrees of phase shift at 800 Hz. Hence, the cascade of two has 180° shift at 800 Hz. The shift is less at lower frequency, but more at higher frequency. The circuit of Fig 3-61 behaves like a transmission line with length of one half-wave at 800 Hz.

The halfwave lines are repeated and cascaded to form a line that is, in this example, 4.5 wavelengths long at 800 Hz, shown in **Fig 3.62**. The line is tapped at each half wave point. Because the line is built from several operational amplifiers, the tap points are low impedance and can be loaded without interaction or other adverse consequence, difficult with a real transmission line.

A sinusoidal audio signal at 800 Hz is applied to the input. The signal looks the same at all points along the line except for changes in phase. If we extract two signals from two taps on the line that are separated by one full wavelength, the two signals will be in phase. If the two signals are added, they will produce a signal that is twice the original. If, however, the two taps are one (or three, or five,...) half wavelengths apart, the result is complete cancellation, for the two components are then equal in magnitude, but out of phase. The cancellation can be turned into positive reinforcement if we add 180 degrees of phase shift to one before addition; this results from an inverter.

Fig 3.62 shows a complete filter. All taps with even numbers are summed together in a summing amplifier U1. U2 serves a similar role for signals from odd numbered taps. U3 inverts one resultant signal with the final output extracted from U4 as the sum of the two. An output response is presented in **Fig 3.63**.

This filter has a characteristic that differs from the typical audio filter, the finite nature of the impulse response. The usual bandpass audio filter, such as described earlier, will ring virtually forever when subjected to a noise impulse. The long ringing is evident from the mathematics; it is also evident from listening to such a filter. In contrast, the FIR filter has an impulse response that is limited to the total delay of the all pass structure. A filter like this one will still "color" noise, but that noise will not bring about the sometimes terrible ringing that would occur with a cascade of high Q resonators. Note the rounded peak shape; it's similar to that found with filters with the better time domain responses.

The filter circuit shown in Fig. 3.62 is not completely impractical, although it is not recommended as a construction project. One of the authors built several FIR audio bandpass filters in the late 1970s. In some, the signals from the taps had unequal weighting, accomplished by changing the summing resistors from each tap. The number of taps grew to impractical extremes. (Don't ask!) Taps can be added as the delay length grows. The results were mixed with the eventual conclusion that a filter of this type was not practical in simple analog form. The experiments were, nonetheless, among the most enlightening that we have ever experienced!

A large number of taps is possible and completely practical today in FIR filters based upon digital signal processing. It is informative to continue the analogy.

- A DSP audio filter begins by sampling the incoming signal. The incoming signal is merely a voltage that changes with time. Sampling means that the signal is captured at one instant in time. This must occur quickly and often, at least twice for every cycle for the highest frequency that our audio system will process.
- Each sample is applied to an analog-to-digital converter. The A to D provides a stream of data that can be processed. It can be done in a high speed general purpose computer or in special circuitry designed specifically for this task. The digitized data is stored in computer memory.
- Computer memory also contains data that was stored from earlier moments. Remember that we are sampling the signal at least twice per cycle for the incoming data we wish to process. The memory has the data just sampled, that from one sample period back, from two periods back, and so forth, extending into the past by a number of "taps" commensurate with our ability to store and process.
- At each interval in the process, we will multiply each of the stored numbers by a constant, weighting the samples in the same way that they are weighted by the summing resistors in the analog filter. They are then added together to obtain a

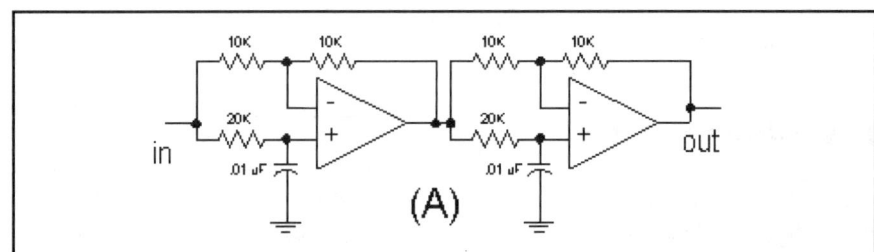

Fig 3.61—Half wave transmission line emulator.

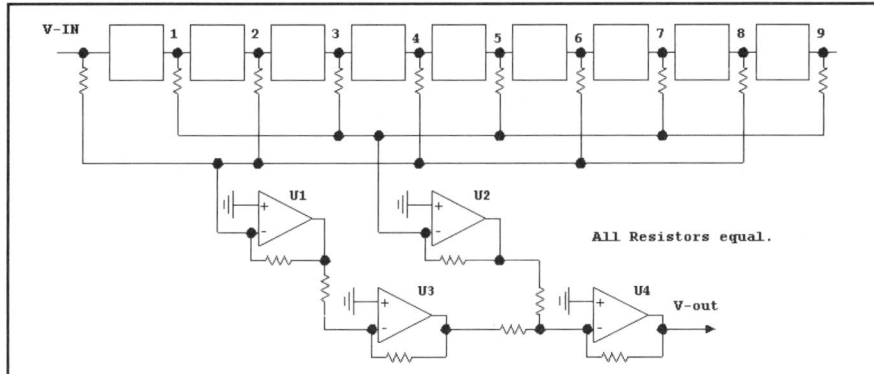
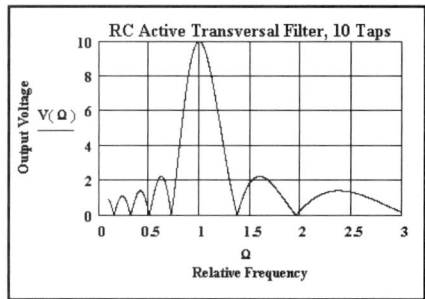

Fig 3.63—Transfer function of a 10-tap FIR filter.

Fig 3.62—A Finite Impulse Response, or FIR bandpass filter built from a cascade of all-pass filters. This filter has 9 taps. Op-amps U1 through U4 serve to add signals from the various taps.

final result.
- The digital output "word" is applied to a DAC, a digital-to-analog converter that provides a signal that can be injected into an audio amplifier and, eventually, headphones.
- Data is eliminated from memory at each step in the process. We only go as far back in time as our computing power will allow.

Among the significant lessons that emerge from a study of FIR filters is the realization that filtering is a comparative process; a signal is compared with a replica from an earlier point in time. The nature of the comparison is direct and clear in the FIR filter. It is present in the simpler filters, be it a single LC resonator or crystal, or an active version with an identical function. The signal components from earlier times vanish from the resonator as they dissipate in the tuned circuit losses.

3.6 IMPEDANCE MATCHING NETWORKS

Most filters built from inductors and capacitors were designed to achieve a desired frequency domain result: They accepted an input consisting of many frequencies, but allowed only a few to emerge at the output. Other LC circuits are designed for impedance transformation. An impedance transforming or matching network is one that accepts power from a generator with one characteristic impedance, the source, and delivers virtually all of that power to a different impedance, the load. Both source and load may be complex with both real and imaginary (reactive) parts. Simple designs are performed at only one frequency. More refined methods can encompass a wide band of frequencies.

Impedance transforming networks generally have filtering properties, even if they are not designed for that characteristic. We found earlier, for example, that a modified low-pass filter could be terminated in an impedance that differed from the original design value, serving a wideband matching role.

Directional Impedances

Consider point A in the circuit of **Fig 3.64**. A frequent question we hear is, "What is the impedance at point A?" This question does not have a good answer, for we did not ask the right question. Impedances are directional. A better question would have been, "What is the impedance looking into the amplifier from the plane marked by A."

The circuit in the figure is a simple amplifier operating at, for example, 50 MHz. The input impedance looking into the base is $20 - j10\ \Omega$. This value would be reasonable for an RF transistor biased to a few mA and operating at $F_T/10$. Wishing to transfer as much power into this amplifier from the source as possible, we will strive for a conjugate input match by designing a suitable input network. One of many possible networks that will realize such a transformation is the L-network shown, transforming from 50 down to 20 Ω. If we then add an inductance with 10-Ω reactance in series with the inductor of the L-network, we will have transformed the 50-Ω source to look like the desired $20 + j10$ needed by the amplifier.

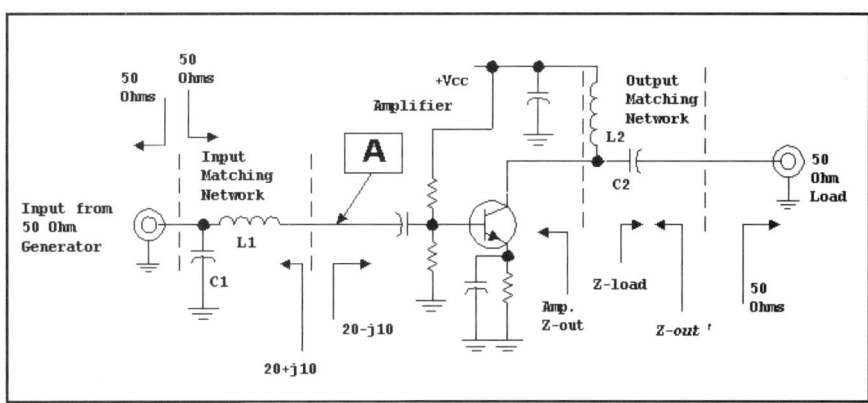

Fig 3.64—An amplifier with matching networks at input and output illustrating directional impedances. See text.

Filters and Impedance Matching Circuits 3.29

We were careful to match the input, but will not seek a conjugate match at the output. This often occurs with, for example, power amplifiers where we present a specific load, Z_{LOAD}, to the collector in order to realize a well defined output power. But this load will generally be different than a conjugate match to the amplifier output impedance, Z_{OUT}. Although a conjugate output match may well provide the highest gain and the maximum output power for small signal conditions, that output load could produce limiting that constrains large signal output power.

Input matching resulted from a low-pass type L-network. An input blocking capacitor is an integral part of the amplifier. Output matching is performed with a high-pass type L-network, which serves double duty by providing a route for V_{cc} to reach the transistor. There is no "perfect" match anywhere through the output. Recall also that changing the load presented to the amplifier will probably alter the input impedance.

We often build transforming networks that will present impedances for reasons other than matching. Output loading for power was mentioned. We sometimes present impedances at the input of low noise amplifiers that will optimize noise figure, usually different than those that provide best gain. We must be clear in defining our goals when designing matching circuits, and exercise similar clarity when talking about such circuits.

The L, π and Tee-Networks

Perhaps the most common LC impedance transforming network is the L, so named because it uses two elements, one as a series element with the other as a parallel one, resembling the capital L on it's side. Both L-network forms are shown in **Fig 3.65**. The lower value resistor, R_1, is transformed by adding a series reactance. The higher value, reactive impedance, is resonated at one frequency with a parallel reactance, yielding a load that looks like a real impedance of value R_2.

The same equations apply if we wish to transform a higher resistance, R_2, to "look like" a lower one, R_1. This bilateral nature is a general characteristic of all lossless networks. The derivation of these equations is outlined in Chapter 4 of *Introduction to RF Design*.[21]

We can define a network Q as the ratio of the parallel resistance, R_2 in this example, to the reactance of the parallel element. That is, we treat the network as if it was a parallel tuned circuit. Network Q is related to the voltage transformation of the network, but is not always a good indicator of network bandwidth.

$$X_S = \sqrt{R_1 \cdot R_2 - R_1^2} \quad \text{Eq 3.30}$$

$$X_P = \frac{R_1^2 + X_S^2}{X_S} \quad \text{Eq 3.31}$$

$$Q = \sqrt{\frac{R_2}{R_1} - 1} \quad \text{Eq 3.32}$$

Consider an example: We wish to transform a 10-Ω resistance to look like 50 Ω at 7 MHz. The series reactance, from the equations, is 20 Ω and the parallel one is 25 Ω. The low-pass form, the L-network with a series inductor, would use L= 0.455 µH and 909 pF. The high-pass form would use 0.568 µH and 1137 pF. Both networks offer essentially identical performance at the design frequency, but differ in their filtering properties. The Q of this L-network is 2. Q is a characteristic of the L-network that is established by the transformed impedances.

Another popular network is the pi, named because its three elements resemble the Greek π. This network is shown in low pass form in **Fig 3.66**. Again, R_1 is restricted.

Q is now a network parameter that the designer must pick. It can take on a wide variety of values, although they are bounded. The lowest Q allowed is defined by **Eq 3.32**, presented above for the L-network. If you used this value, the

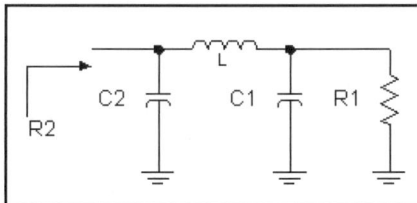

Fig 3.66—Schematic and corresponding design equations for the popular π-network.

pi-network equations collapse to those for the L. Low Q values are generally preferred with the low impedances usually found with solid-state circuits, offering more practical component values and lower network loss. Higher Q tends to restrict bandwidth, just as it would in a simple tuned circuit. It also exacerbates the effects of loss in the network L and C parts.

As an example, we examine the same 10-Ω load that must be transformed to 50 Ω; we pick a network Q of 5. The results are X_{C2}=10 Ω, X_{C1}=4.88 Ω, and X_L= 13.56 Ω. At 7 MHz, the respective component values are 2274 pF, 4660 pF, and 0.308 µH.

A high-pass variant of the pi network is also possible. The pi-network component values may not be as practical as those in some other circuits, especially when Q is high.

$$R_2 \geq R_1$$

$$X_{C_2} = \frac{R_2}{Q} \quad \text{Eq 3.33}$$

$$X_{C_1} = R_1 \cdot \sqrt{\frac{R_2/R_1}{Q^2 + 1 - R_2/R_1}} \quad \text{Eq 3.34}$$

$$X_L = \frac{Q \cdot R_2 + R_1 \cdot R_2 / X_{C_1}}{Q^2 + 1} \quad \text{Eq 3.35}$$

Although less common, a very practical and useful network is the Tee using two capacitors and one inductor. Component values are practical and loss is low, especially for the low impedances found with solid state circuits. The design begins by picking a network Q.

The T-network has the same minimum Q as the pi network, which is the Q of the L-network given by Eq 3.32. The Tee circuit is shown in **Fig 3.67**. Intermediate variables, A and B, are used in these calculations.

We pick the same example used before

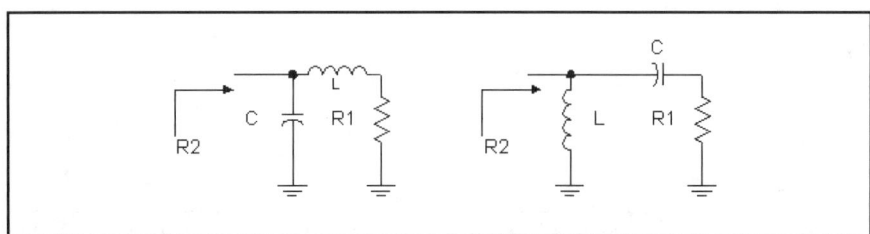

Fig 3.65—L-Network with design equations when $R_1 < R_2$.

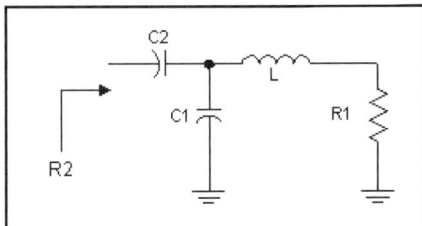

Fig 3.67—LCC type Tee-network and design equations.

with $R_1 = 10$, $R_2 = 50$, and $Q = 5$.

The resulting reactance values become $X_{C_1} = 88.12$, $X_{C_2} = 102.5$, and $X_L = 50$, all in Ω. At 7 MHz, these values correspond to 258 pF, 222 pF, and 1.137 μH, respectively. These components are especially practical for both input and output networks of RF power amplifiers if mica compression variable capacitors are used.

$R_2 > R_1$

$$B = R_1 \cdot (Q^2 + 1) \qquad \text{Eq 3.36}$$

$$A = \sqrt{\frac{B}{R_2} - 1} \qquad \text{Eq 3.37}$$

$$X_L = Q \cdot R_1 \qquad \text{Eq 3.38}$$

$$X_{C_1} = \frac{B}{Q - A} \qquad \text{Eq 3.39}$$

$$X_{C_2} = A \cdot R_2 \qquad \text{Eq 3.40}$$

Increasing the inductor, then adding a series capacitor that cancels the added inductive reactance, may modify all the networks described. The modified networks are more easily adjusted and can provide narrower bandwidth.

We often view π or T-networks as back to back L-networks, transforming from a nominal impedance to another, and then back. This has the effect of increasing overall circuit Q or selectivity. Cascaded L-networks can have the opposite effect of decreasing selectivity, an extremely powerful tool when building circuits to function over wide bandwidth.[22]

The Transmission Line as a Transformer

Transmission lines have well known impedance transforming properties. A termination of value R_1 is transformed to a new value, R_2, by a transmission line that is a quarter of a wavelength long with a characteristic impedance Z_0 given by

$$Z_0 = \sqrt{R_1 \cdot R_2} \qquad \text{Eq 3.41}$$

If, for example, we wished to transform a 10-Ω load to appear as 50 Ω at 7 MHz, we would use a line with a characteristic impedance of 22.4 Ω. The length would be $\lambda/4$ at 7 MHz, about 25 ft in cable with a velocity factor of about 0.7. This characteristic impedance is impractical, but could be approximated with parallel sections of higher impedance lines. (Line with $Z_0 = 25$ Ω can be purchased.) Transmission line transformers are sometimes practical at this low frequency, especially in antenna systems where the lines are needed anyway. Coaxial transmission lines can be coiled with virtually no impact on their behavior so far as the fields within the line. The quarter wavelength lines are often called "Q-Sections." A transmission line need not have a $\lambda/4$ to serve as a transformer. A Smith Chart is often used for the design of these elements.

Transmission lines become more practical circuit elements at higher frequencies. One printed line form is microstrip, shown in **Fig 3.68**. The lower conductor is a ground plane on the back of a circuit board while the upper conductor is a printed run. Electric field lines between the conductors are found in the dielectric as well as in air. Hence, these transmission lines have a velocity factor part way between that of air and that of the higher dielectric constant insulator.

Microstrip is versatile, for it can be designed for about any characteristic impedance in the 10 to 100-Ω region, or more. The wider lines have lower Z_0. Robert Wilson, KL7ISA and Hal Silverman, W3HWC, in "Wire Line—A New and Easy Method of Microwave Circuit Construction," described a wonderful

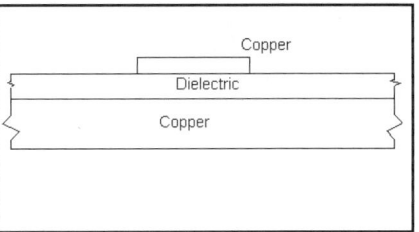

Fig 3.68—Microstrip transmission line shown in cross section. The dielectric material is the insulated portion of a printed circuit board. The lower conductor is usually a solid ground plane. The drawing is not to scale.

variation that the experimenter can build without etching in the July 1981 *QST*.[23]

Another practical transmission line form is a simple twisted pair of insulated wires. Wire insulated with plastic often produces lines with a characteristic impedance around 100 Ω. Enameled #24 wire will produce line with an impedance near 50 Ω when tightly twisted.

A variation on the quarter-wave line matching uses synthetic transmission lines. Here, a transmission line is replaced by a pi-network using inductors and capacitors. A sidebar earlier in this chapter discussed the half-wave filter, a variation of this circuit. **Fig 3.69** shows a synthetic quarter-wave example, the same case considered earlier at 7 MHz. Transforming from 10 to 50 Ω occurs with a 22.4-Ω line.

Powdered Iron Toroid Inductors and Transformers

Inductors are realized with many structures, ranging from straight wire pieces to solenoid and toroid coils. The solenoid is easy to wind and can exhibit high Q, especially at VHF. However, the magnetic field of a solenoid extends well outside the coil

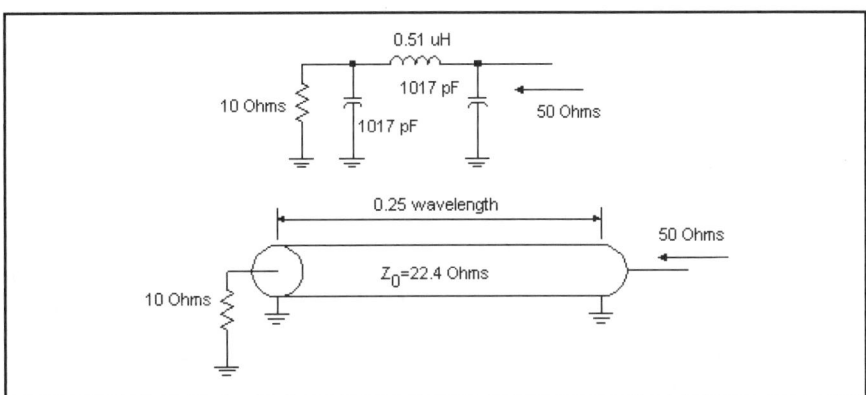

Fig 3.69—A synthetic quarter wavelength line is formed at 7 MHz with three equal reactance values of Z_0 of a Q section.

dimensions, leaving it free to couple to other circuit elements in close proximity, including conductive walls that can alter Q. In contrast, the toroid inductor has most (but not quite all) of its magnetic field confined to the core interior, allowing a toroid to be mounted directly against a ground plane with minimal change in inductance or Q. The Q available for low volume coils is generally much higher for toroids up through 30 MHz.

Toroids are more difficult than solenoids to wind, creating apprehension among beginning experimenters. It is, however, straight forward, even if time consuming.

Toroid inductance is almost exactly proportional to the square of the number of turns,

$$L = K \cdot n^2 \qquad \text{Eq 3.42}$$

A common core is the T30-6 from Micrometals with inductance constant, K, of 3.6 nH/t² (nano-henry per turn squared.) Various manufacturers use other units that can be related directly to the K we find convenient for RF parts. A coil with 15 turns evenly wound around most of this core has a predicted inductance of 810 nH, or 0.81 µH. Generally, the highest Q will result when the cores use the largest wire that will fit in one layer. It is important for Q, and especially for temperature stability, that the wire be tightly wound against the core. A more temperature-stable coil can often be built with a wire size smaller than that producing the highest Q.

Micrometals, Inc copyrights the usual toroid numbering scheme, illustrated here with T30-6. The –6 indicates a specific core material or "mix," while the 30 indicates an outside diameter of 0.30 inch. A manufacturer or vendor catalog might list the inductance constant for the T30-6 as 36 µH per 100 turns. The user can convert these constants to whatever form he or she prefers.

A toroid is wound by counting the number of passes through the center hole. While solenoids can have a fractional number of turns, this does not happen with toroids. A single turn on a toroid consists of the wire passing through the hole just one time.

We built the inductor mentioned by winding 15 turns of #28 wire over about 90% of a T30-6 core. Using an Almost All Digital Electronics L/C Meter IIB, the inductance was measured as 872 nH, 8% above the prediction. Part of the difference was probably the result of slight bunching of some of the turns. The permeability tolerance normally associated with these

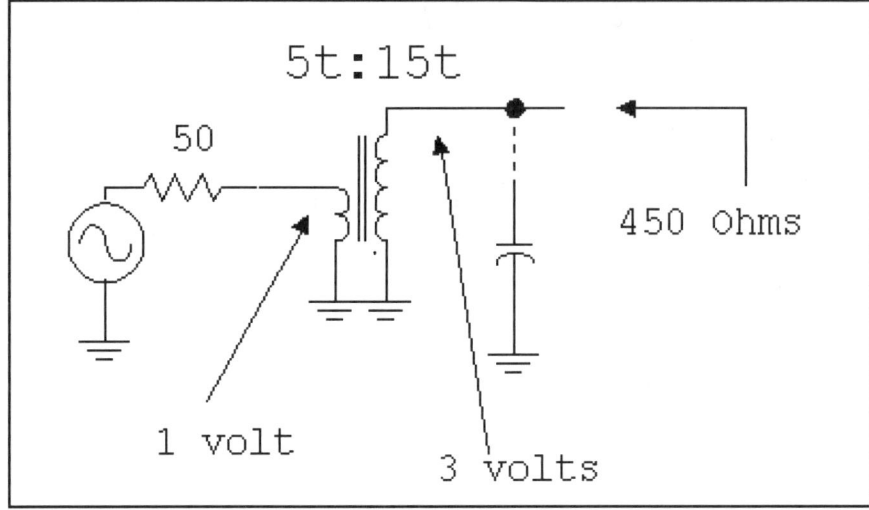

Fig 3.70—Circuit illustrating the transfer characteristics of an ideal transformer.

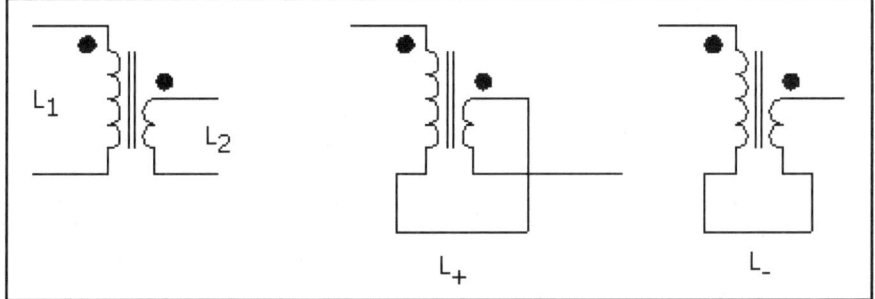

Fig 3.71—Method for connecting windings that allows coupling coefficient to be calculated. This method is general and can be applied with powdered iron or ferrite core transformers. The results become less accurate when coupling is strong, and it is not unusual to calculate $k>1$. This is usually an indication of capacitance.

cores is +/-5%. The accuracy is usually better as inductance and core size grow.

The windings were then compressed to cover only 60% of the core, increasing inductance to 1.039 µH. This 15 to 20% increase is typical and offers a convenient means for adjustment.

This inductor can be used directly in impedance matching networks, or as part of an L/C filter. The reader should consult the extensive data available from Amidon Inc. This is found at an excellent Web site, **www.amidon-inductive.com/**.

A common impedance matching network uses a powered iron inductor with a second winding, forming a transformer. The inductor we just described was modified by adding a 5 turn link of #26 wire on the remaining bare portion of the core. The measured inductance was 206nH. This is much higher than the 90 nH the formula would predict, but the coil is severely compressed. (Even with the 5 turns spread over the complete core, L=121 nH.) The 15 turn winding L was unchanged at 1039 nH.

We expect RF voltage to increase in proportion to the turns ratio and impedance to transform with the square of the turns ratio in an ideal transformer. Hence, a 50-Ω generator attached to the 5-turn link should provide three times the voltage across the 15-turn winding with the combination looking like a 450-Ω source to the following circuitry, as shown in **Fig 3.70**. If it was terminated in a 450-Ω load, the impedance match looking into the link should be perfect. This transformer might be used to match between a 50-Ω amplifier and a 450-Ω, 10-MHz crystal filter.

But, these ideals are not realized. First, the impedances are highly reactive. This is remedied by tuning the secondary with a parallel capacitor, 244 pF at 10 MHz. This brings the voltage gain nearly up to the predicted 3 when the output is terminated, but impedance match is still poor. This is a result of less than ideal coupling.

The coupling coefficient is easily measured with the same instruments used to measure inductance. This is shown in **Fig 3.71**. L_1 and L_2 are the 5 and 15 turn windings and are measured with the other winding open circuited. The two windings are then connected as shown in Fig 3.71 and the composite inductance values are measured as L_+ and L_-. The coupling coefficient is then given:

$$k = \frac{(L_+ - L_-)}{4 \cdot \sqrt{L_1 \cdot L_2}} \qquad \text{Eq 3.43}$$

This method was presented by Bill Carver, W7AAZ, in the January, 1998 issue of the *QRP Quarterly*.[24] When the method was applied to the test transformer, we measured L_+=1533 nH and L_-=872 nH, leading to a coupling coefficient of k=0.357. The input VSWR exceeds 2:1 for this transformer, even when tuned and properly terminated.

Ideally, all inductors should be measured after they are wound. While the traditional tuned transformer is still a practical component, it may require more design effort than an impedance transforming network built from discrete elements.

The Ferrite Transformer

The powered iron core transformer discussed above had to be resonated to function as desired. Even after tuning, it suffered for a lack of coupling. Both problems are overcome with higher inductance, which occurs with the much higher permeability found in ferrite cores. The toroid is the most common form, but balun cores, with their binocular shape, are also popular. Most of the powered iron cores we use have initial permeability under 10 while typical ferrites show μ_i values between 40 and 5000.

Recall the classic inductor, a component that "tries" to maintain whatever current is flowing at any instant. It is the dual of the capacitor, which does not allow voltage to change instantly. Consider a switch that connects a battery to an inductor. The inductor current is zero before the switch closed, so it must be zero immediately afterward. There is no restriction on the voltage. The voltage impressed on L changes quickly, soon reaching the battery value. The current conserving characteristic of the inductor is a result of the magnetic field. When the switch is closed, current begins to flow. But as soon as the field starts to build up, the changing magnetic field generates an electric field (hence, a voltage) that opposes the electric effect that caused the current in the first place. This is a non-rigorous statement of Faraday's Law, one of Maxwell's equations. The inductor is shown with curves illustrating the behavior in **Fig 3.72**.

Inductor current increases without bound in the ideal, lossless case. Losses, resistance within the wire and the battery, would limit the current to a finite, but large level in a practical circuit.

Consider now a modified structure. The single winding inductor is replaced with a pair of windings, shown in **Fig 3.73**, that are very close together. The wires, although isolated from each other, occupy virtually the same space and see essentially the same magnetic field. If we left the second winding (BB') open circuited, voltage from A to A' builds up in the same way that it did with the simple inductor. Measurement across either winding will show the same voltage profile. But, no current flows in BB' when it is open circuited.

The behavior changes when we repeat the experiment with a load at BB'. As the voltage builds, load current will begin to flow. Transformer action begins. The current in the second winding will generate a magnetic field, just as that in the primary winding did. But the field from the secondary is in a direction opposite to that from the first winding. Because the net magnetic field has been reduced (nearly) to zero, current flow is determined by R, the external load.

The transformer described (Fig 3.73), with the two wires in close proximity, is said to be bifilar. Bifilar windings are often twisted. One manufacturer supplies Multifilar® wire with strands of differing colors, simplifying transformer construction. (Multifilar® parallel banded magnet wire from MWS Wire Industries.)

The dots on the transformer schematic are useful. An increasing voltage at one dot produces an increasing voltage at the other. Current entering the A dot equals that leaving the B dot. This behavior arises because the magnetic field vanishes within the core. If the primary (AA') had N_P turns while the secondary (BB') had N_S turns, the currents would obey the more general boundary condition that

$$N_P \cdot I_P = N_S \cdot I_S \qquad \text{Eq 3.44}$$

Bifilar winding and the use of a high permeability magnetic material produce tight coupling, approaching k=1. Coupling is measured for a ferrite transformer with the same method outlined for a powdered iron design, Fig 3.71. Strong coupling means that all of the magnetic field lines created by the primary also couple into the secondary. In a practical transformer, some of the primary field loops out from the core, only to return without communicating with the secondary.

The transformer is often modeled as an ideal one with added components, shown in **Fig 3.74**. The ideal transformer has a voltage ratio proportional to the turns ratio and a current ratio defined by **Eq 3.44**. L_P is the primary inductance, the value we would measure if the primary was examined without a secondary termina-

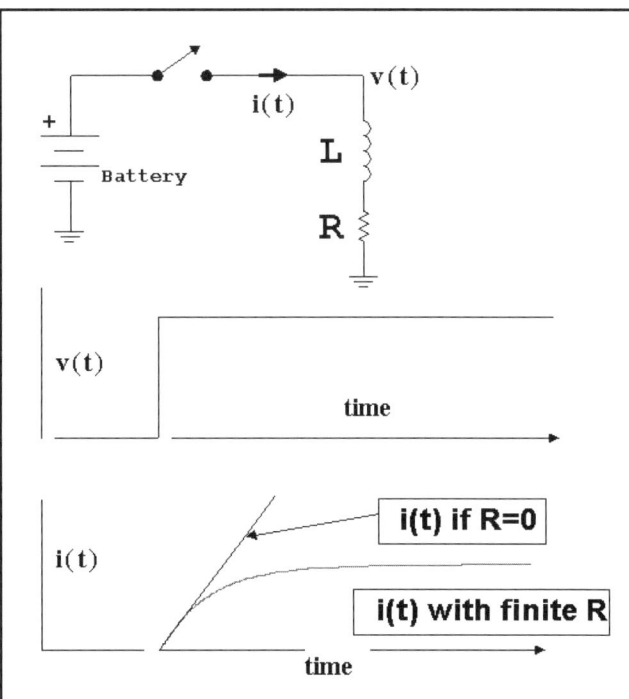

Fig 3.72—Principles of an ideal inductor, with waveforms. The current would grow linearly forever in an ideal component. Resistance establishes an ultimate value.

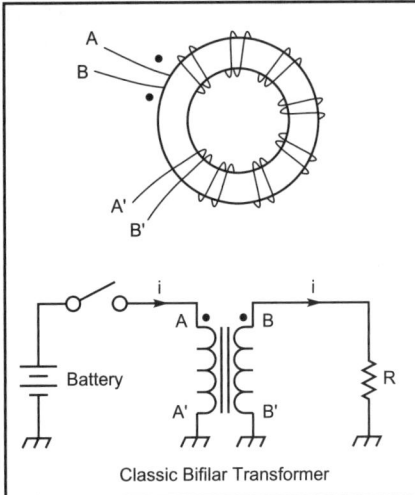

Fig 3.73—Current flow in a bifilar wound transformer.

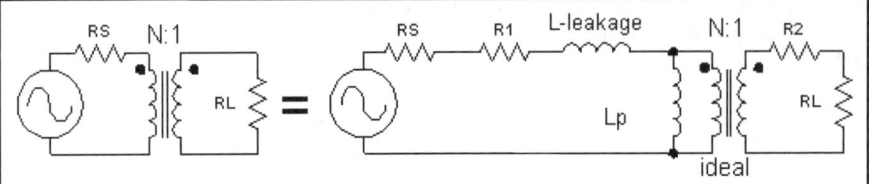

Fig 3.74—A transformer model.

RF transformers can be built by placing ferrite beads over brass tubing that forms a single turn winding. Circuit board material connects the tubing ends with a short at one end. A multiple wire winding is then threaded through the middle of the tubing, guaranteeing tight coupling.

tion. The L-leakage is the inductance accounting for the magnetic flux that does not pass through both windings. R1 and R2 account for losses. The transformer is a bandpass circuit with L_p presenting a short at dc and very low frequency; L-leakage, a series element, presents a high impedance at high frequency.

A practical transformer will have a primary inductance with a reactance at least 5 times the terminating resistance at the low frequency limit and a leakage inductance reactance less than 1/5 the resistance at the highest frequency, and loss resistances small with respect to the source and load. Inductance of windings on ferrite cores is proportional to the square of the turns, although the higher permeability of ferrite produces dramatically higher "k" constants for use with **Eq 3.42**. For example, the popular FT37-43 ferrite toroid has k of about 360 nHt^{-2}. Core loss can be modeled as a parallel resistance, which is also proportional to the square of n, although this formulation is not in general use.

Examples of practical transformers are found throughout the text. A wonderful treatment of the modeling of this "simple" component is presented by Clarke and Hess.[25] A more complete review of transformer modeling is presented by Chris Trask.[26] We generally use powdered iron toroid cores for high-Q inductors with good temperature characteristics while ferrites are relegated to low-Q wideband transformer application. However, this distinction is not required. Some powdered iron cores are suitable for wideband transformers while some ferrites have excellent Q at HF. A good example of the latter is –63 material from Fair-Rite Products Corp (**www.fair-rite.com**), often producing Q values of several hundred at HF.

Ferrite Transmission Line Transformers

The example presented above to illustrate basic transformer action used a bifilar winding, with one wire as primary and the other as a secondary. A pair of wires also forms a transmission line. As such, it can operate as a transmission line transformer such as a Q-section according to Eq 3.41. Even if it is not a proper λ/4 length, it will still transform the impedance seen at one end from that presented at the other. The transmission line properties persist if the line is wound in the shape of a coil, including a toroid. But the structure then assumes a different extended behavior, summarized in a classic paper by Ruthroff.[27]

The simplest ferrite transmission line transformer is that shown in **Fig 3.75**. This structure, formed with a bifilar winding on a toroid was at one time called a balun. A balun is a structure that generates a balanced voltage from one that is single ended. This connection does not force such balance and is, hence, not strictly a balun, even though it does perform some of the isolation chores that we might ask of a balun. Perhaps a better name is *isolation transformer*. Transformer action, described above, does force equal currents in the two windings, so this circuit is sometimes also called a *current balun*.

The isolation transformer is labeled AB at one end of the winding while the other end is A'B'. Wires A and B are not attached to each other, a useful detail to keep in mind when winding such transformers without wires of differing color. Viewing this structure as a transmission line, cur-

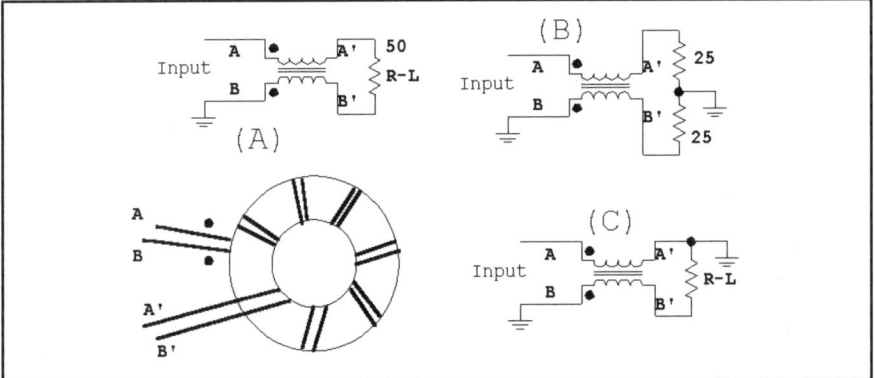

Fig 3.75—Part A: Basic isolation transformer using a transmission line on a ferrite toroid. This structure has some balun like properties. Part B shows a balanced load connected to a single-ended drive while C shows polarity inversion.

rent at point A' is delayed from that at A. However, the ferrite core and traditional transformer behavior would force equal current through a winding, and indeed, in the other winding.

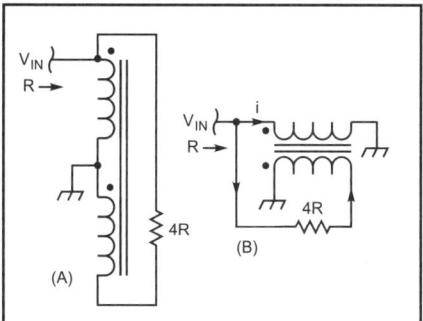

Fig 3.76—A 4:1 step-up balun transformer.

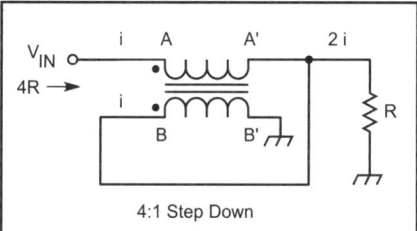

Fig 3.77—A single ended impedance step down transformer.

The isolation transformer of Fig 3.75 has a single ended input. The single ended drive will appear as a balanced output on a balanced load such as that in part B. In this sense, it is a balun structure. However, if the load becomes unbalanced, as in Fig 3.75C, the input may still be applied to the termination.

It is instructive to mentally connect the two wires at one end (A and B) together, doing the same thing at the other (A' and B') end. The result is an inductor. Several turns on a high permeability ferrite would produce considerable inductance. This is termed a common mode inductance. Separating the wires, a load placed across one end, A'B', is then seen differentially (between A and B) at the other end. This structure is often called a common mode choke for common mode signals at one end are isolated from the other by the large inductance, while differential signals are not impeded.

The isolation properties of this structure allow us to drive one end while treating the other end as if it were a separate generator. An isolation transformer (Fig 3.75C) can produce a polarity reversal.

It is useful to connect the output of an isolation transformer in series or parallel with the input. An interesting example is shown in **Fig 3.76** where a load is connected between the input and the inverted output. The composite input will carry twice the current that one transformer winding carries, resulting in a true balun, for it forces equal, but out of phase voltages to appear between the ends. This is a 4:1 impedance transforming balun.

The same structure is reapplied in **Fig 3.77**. The transformer forces twice the current to flow in the output as at the input. The isolation properties of the transmission line transformer are used to parallel an output with a "direct connection" to the input. This circuit now serves an unbalanced-to-unbalanced role. This circuit is used for transforming from 50 Ω down to the 12.5-Ω input on a RF power amplifier. We also saw it used extensively to cause a 50-Ω load to look like 200 Ω at the collector of a feedback amplifier.

These wideband transformers may be viewed as either transmission line circuits or as conventional transformers. Their operation is consistent with either set of boundary conditions. The transformers are designed with about λ/8 to λ/4 of transmission line at the upper frequency of the circuit. The characteristic impedance of the line is consistent with line behavior for the terminations considered. If, for example, we built a 4:1 step down from 50 to 12.5 Ω using Fig 3.76, Z_0 should be 25 Ω. This could be realized by paralleling two 50-Ω windings on the core. A 50-Ω winding consists of a tightly twisted pair of #24 enamel wires.

The transformer of **Fig 3.78** is a true 1:1 balun. The termination impedance is that seen at the input, but the circuit creates two voltages that are equal in magnitude, but out of phase.

A useful step down circuit for high power single ended amplifiers is the 9:1 circuit of **Fig 3.79**. This transformer uses two cores to drop from 50 Ω down to about 6 Ω. Series connections at the input side drive parallel ones at the output. A similar series/parallel circuit is presented in **Fig 3.80** where two cores form a balanced to balanced 1:4 impedance ratio step up transformer.

Numerous other kinds of transmission line transformer can be built, some almost diabolic in their cleverness. The reader is referred to Motorola Applications Note AN-593[28] for further interesting examples.

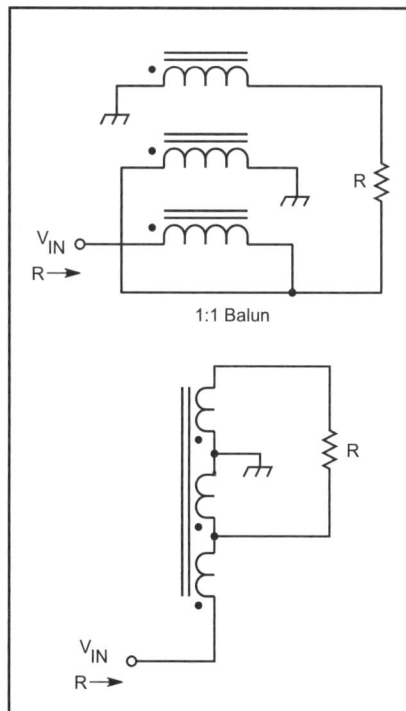

Fig 3.78—A 1:1 impedance ratio true balun transformer.

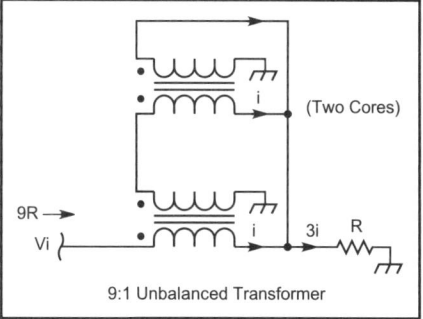

Fig 3.79—Illustration of a 9:1 unbalanced transformer.

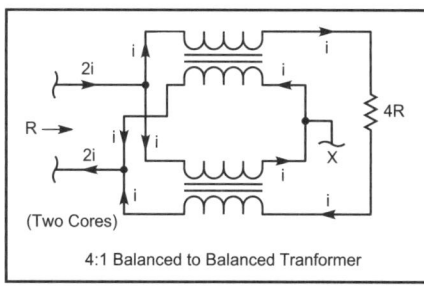

Fig 3.80—A 4:1 balanced-to-balanced transformer.

Some Multiple Port Networks

All of the networks presented in this section have used but two ports, an input and an output. There are, however, several multiport networks that are of special interest to the radio amateur. The first is the so called "Splitter/Combiner" shown

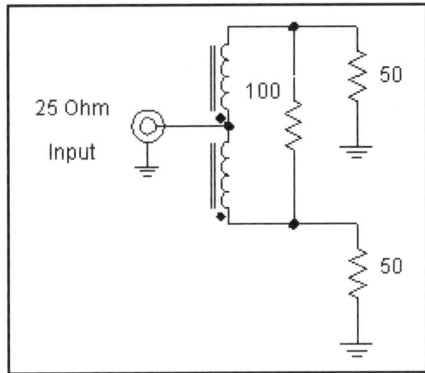

Fig 3.81—An in-phase splitter/combiner network. Use 10 bifilar turns on a FT-37-43 ferrite toroid for the HF spectrum.

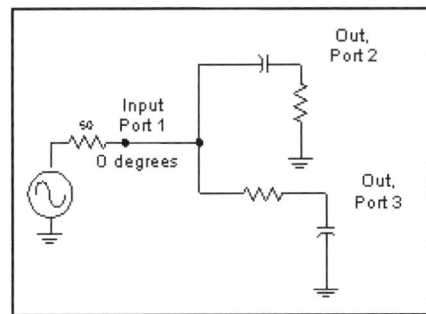

Fig 3.83—Phase shift network for RF phasing in simple SSB equipment.

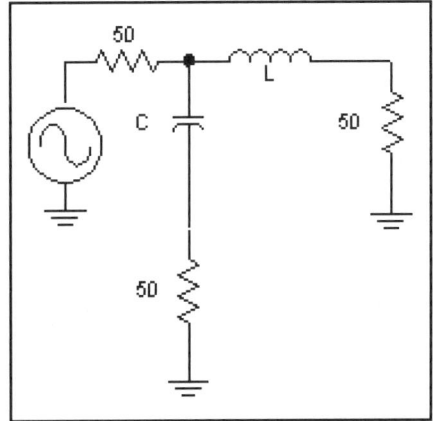

Fig 3.82—First-order low-pass/high-pass diplexer.

in **Fig 3.81**. This circuit, using nothing more than a bifilar winding on a ferrite toroid, accepts energy from a single generator with a 25-Ω characteristic impedance and supplies that energy to two outputs, each with a 50-Ω impedance. A 50-Ω input can be transformed down to 25 Ω with any of the matching schemes presented above. Variations of this network use transmission lines or L-Networks. The 100-Ω resistor absorbs excess power that becomes available when one of the two output ports is miss-terminated. A common application splits the output of a local oscillator chain to drive two mixers. The circuit isolates the two outputs. This circuit is called a 3-dB hybrid transformer, for the power in each output, neglecting losses, is 3 dB below the input, while Hybrid refers to transformer-like circuits that provide isolation between two of three ports. Hybrids were used in early telephones to isolate the microphone from the earphone.

Fig 3.82 shows a three port circuit where two outputs receive drive from a single input. This circuit, a diplexer, is similar to a crossover network used in audio systems. Frequencies below a cutoff pass through the inductor and are dissipated in the related termination. Signals above cutoff pass through the capacitor to the related resistor. The L and C are picked with regard to the source impedance such that there is always a perfect impedance match presented to the generator. If the cutoff frequency is f, then the related angular frequency is $\omega_c = 2\pi f$. Then, the L and C for a perfect match are

$$L = \frac{R}{\omega_c} \qquad C = \frac{1}{\omega_c \cdot R} \qquad \textbf{Eq 3.45}$$

The diplexer is applied where mixers (e.g., diode rings) must be terminated in a wideband 50 Ω to minimize distortion. The diplexer shown is an especially simple one where each arm is a one pole low pass or high pass filter. Nic Hamilton, G4TXG, has described high order low pass high pass diplexers.[29] A third-order example of this design is shown in the diplexer sidebar. Diplexers can also be built with combinations of band-pass and bandstop networks, also summarized in the sidebar.

An interesting, yet simple phase shift network is shown in **Fig 3.83**. A generator drives two one pole filters that are terminated at their output in open circuits. The two capacitors, equal in value, are picked to have a reactance at one frequency equal to R, the resistor value used in each arm. The phase difference for this network is 90 degrees at all frequencies. However, the two output amplitudes are equal only at the design frequency.

An especially interesting four-port circuit form is the directional coupler. The coupler has an input and output, usually with low loss between them. A third is called the "forward" coupled port, for the energy available is proportional to the energy flowing from the "input" to the "output." A fourth is the "reflected" coupled port with energy proportional to that flowing from the "output" to the "input." **Fig 3.85** shows a schematic representation of a directional coupler, which is also a practical topology in microstrip form. Part B of Fig 3.85 shows a wideband variation using ferrite transformers.[30] A practical version of the wideband coupler using three transformers was designed by Roy Lewallen[31] and is included on the book CD.

The directional coupler is extremely useful for a variety of applications. When used with a power meter or spectrum analyzer, reflected energy is a measure of the impedance at the output port, leading to popular in-line power meters such as the W7EL design. But the coupler can also be used to inject signals on a line. The coupling value is the power ratio between the output and the coupled ports and is $1/N^2$ for the ferrite version. Most directional couplers have coupled energy that is in phase with the output. The microwave literature abounds with interesting couplers.

A coupler is also characterized by directivity. Assume that the thru path is terminated in an open (or short) circuit and a power P1 is measured in the reflected port. If the main path is now loaded with a perfect match, the reflected power will drop to P2. The ratio of P1 to P2 is called the directivity. We consider directivity with a number of bridge circuits in Chapter 7.

Directional couplers can be built with lumped components, even at VHF. A lumped element example with −28 dB coupling with 20-dB directivity at 144 MHz is included in a design discussed later in the book and included on the book CD. That design is a quadrature coupler, discussed below.[32] There are numerous references in the literature to directional couplers. See, for example, Andre Boulouard.[33]

The twisted-wire quadrature hybrid directional coupler is a very useful variation. This circuit was described by Reed Fisher, W2CQH.[34,35] Fisher's *QST* article is included on the book CD-ROM. Also see.[36,37] For information on distributed couplers, see.[38,39] This is a 3-dB coupler, for the coupled output is below the input by 3 dB, producing two outputs of equal strength. The circuit is called a quadrature coupler because there is a 90-degree phase difference between the two output ports. An HF variation, built for the 7-MHz band, is shown in **Fig 3.84**.

The design equations for the coupler are identical to those presented for the diplexer, **Eq 3.45**. However, in this case, the capacitance is the total C in the circuit.

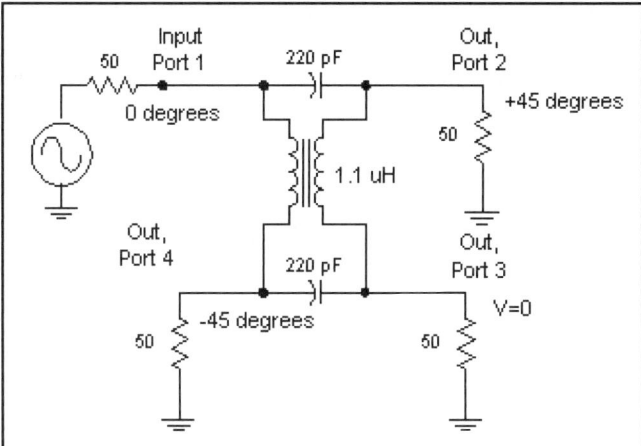

Fig 3.84—Quadrature coupler for 7 MHz.

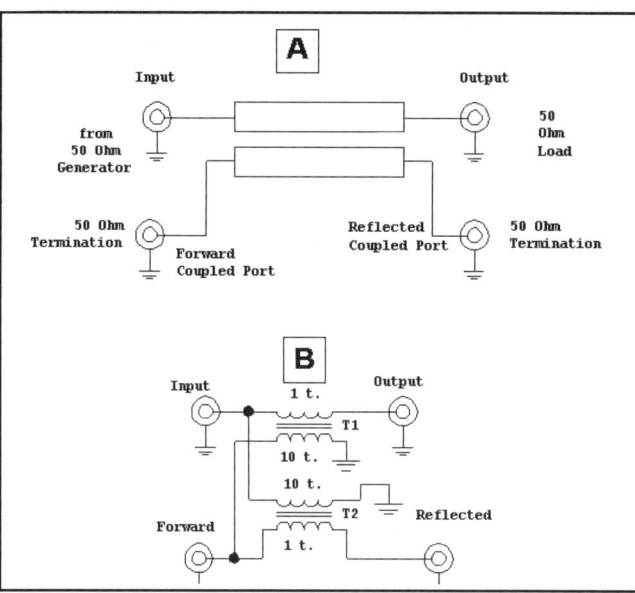

Fig 3.85—Part A shows a general schematic for a directional coupler while B presents a wideband version using ferrite core transformers. The coupling on B is 20 dB owing to the 10:1 turns ratio used. This is a practical circuit if wound with FT37-43 or FT37-75 cores. A single binocular core can be used for both transformers.

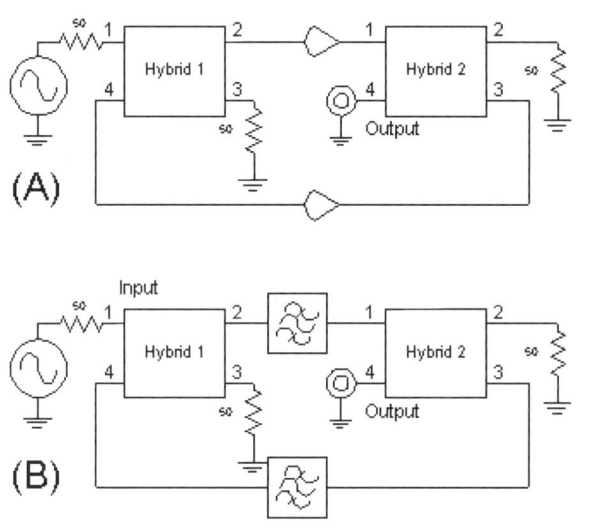

Fig 3.86—Some applications for quadrature hybrids. Identical amplifiers (A) or filters (B) are combined to form termination insensitive linear circuits. The extra terminations required are shown in the circuits.

This must be halved to build the circuit. As Fisher points out, the capacitance of the tightly wound bifilar pair (12 pF in his example) is measured and removed from the calculated C before construction. The inductance is that of the two windings in parallel, essentially the same as that of a single winding on the core of interest. Fisher used a low-permeability ferrite core, while we have generally used powdered iron cores, owing primarily to availability. Small powder iron cores such as the T25 in the -6, -12, or -17 materials are suitable through 150 MHz.

At the design frequency, the circuit is a 3-dB coupler, providing equal power at port 2 and 4. However, the coupling is different at other frequencies. The very interesting properties of the quadrature hybrid are summarized:

1. There is power transfer from port 1 to 2.

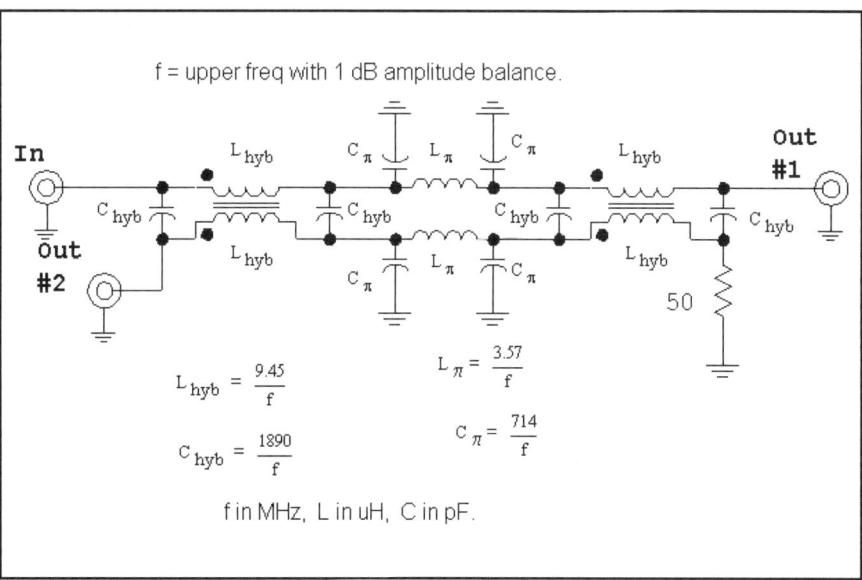

Fig 3.87—Extended bandwidth quadrature hybrid network.

Filters and Impedance Matching Circuits   3.37

Typical diplexer configurations and equations.

Third order Low Pass High Pass Diplexer

$Z_{IN} = 50$ at all F

L and C values shown are reactance at the cutoff frequency.

Bandpass-Bandstop Diplexer

1. Pick cutoff frequency F and Q (from 1 to 10)
2. $\omega = 2 \cdot \pi \cdot F$
3. $L = \dfrac{50 \cdot Q}{\omega}$
4. $C = \dfrac{1}{\omega^2 \cdot L}$
5. $C_p = \dfrac{L}{R^2}$
6. $L_p = \dfrac{1}{\omega^2 \cdot C_p}$

---

2. Power is transferred from port 1 to 4.
3. There is *no* power transfer from port 1 to 3 when all ports are properly terminated.
4. There is *no* reflected power back out of port 1, again with proper terminations.
5. The phase difference between ports 2 and 4 is 90 degrees.

The characteristic of greatest interest will depend upon the application. The phase difference is important in the construction of phasing-method SSB equipment. However, it is the isolation from reflection problems, item 4, that leads to some of the more subtle applications. Two examples, each using a pair of couplers, are shown in **Fig 3.86**. In part A, two amplifiers are combined, while in B, two filters are combined. In both cases, the two elements must be identical. However, the networks to be combined need not be impedance matched for a good match to exist at the input. For example, the two amplifiers could be FET circuits that have an L network at the input. Such a circuit produces a very poor input impedance match, but an excellent noise figure. Alternatively, two conditionally-stable amplifiers can become an unconditionally stable circuit when imbedded in quadrature hybrids. This balanced scheme is attributed to Engelbrecht and Kurokawa.[40,41,42] A termination insensitive crystal filter is described in Chapter 6 where quadrature couplers are applied.

The circuit of Fig 3.86 is narrow bandwidth with identical output amplitudes at only one frequency. However, the bandwidth can be extended to an octave by cascading two identical quadrature hybrids with a pair of pi-networks between. This topology, with related design equations, is shown in **Fig 3.87**.

REFERENCES

1. W. Hayward, *Introduction to Radio Frequency Design*, Prentice-Hall, 1982; ARRL, 1984. Also see *The ARRL Handbook*, 1995 or later editions.

2. *GPLA* accompanies *Introduction to Radio Frequency Design* (see Ref. 1) as a DOS program. *GPLA 2002* is a Windows version included on the book CD. *ARRL Radio Designer* was formerly available from ARRL.

3. W. Hayward, *Ham Radio Magazine*, Jun, 1984, p. 96.

4. D. Johnson and J. Johnson, "Low Pass Filters Using Ultraspherical Polynomials," *IEEE Transactions on Circuit Theory*, Vol CT-13, No. 4, Dec, 1966, pp 364-369.

5. Tortorella, *RF Design*, Mar/Apr, 1983.

6. Zverev, *Handbook of Filter Synthesis*, Wiley, 1967.

7. M. Dishal, "Alignment and Adjustment of Synchronously Tuned Multiple-Resonant-Circuit Filters," *Elect. Commun.*, Jun, 1952, pp 154-164.

8. S. B. Cohn, "Dissipation Loss in Multiple-Coupled-Resonant Filters," *Proc. IRE*, Aug, 1959, pp 1342-1348.

9. G. Matthaei, L. Young, E. M. T. Jones, *Microwave Filters, Impedance-Matching Networks and Coupling Structures*, McGraw-Hill, 1964.

10. See Reference 6.

11. A. B. Williams, *Electronic Filter Design Handbook*, McGraw-Hill, 1981.

12. W. Hayward, *Introduction to Radio Frequency Design*, ARRL, 1994, Ch 3.

13. W. Hayward, "The Double-Tuned Circuit: An Experimenters Tutorial," *QST*, Dec, 1991, pp 29-34.

14. R. Larkin, "The DSP-10: An All-Mode

2-Meter Transceiver Using a DSP IF and PC-Controlled Front Panel, Part 1," *QST*, Sep, 1999, pp 33-41.

15. V. Bottom, *Introduction to Quartz Crystal Unit Design*, Van Nostrand Reinhold, 1982.

16. S. B. Cohn, "Dissipation Loss in Multiple Coupled Resonators", *Proc IRE*, Aug, 1959.

17. W. Hayward, "Designing and Building Simple Crystal Filters", *QST*, Jul, 1987, pp 24-29.

18. Carver, K6OLG, "High-Performance Crystal Filter Design," *Communications Quarterly*, Winter, 1993.

19. D. E. Johnson, J. R. Johnson, and H. P. Moore, *A Handbook of Active Filters*, Prentice-Hall, 1980.

20. H. Berlin, "The State-Variable Filter," *QST*, Apr, 1978, pp 14-16.

21. W. Hayward, *Introduction to Radio Frequency Design*, ARRL, 1994, Ch 4.

22. G. L. Matthaei, "Tables of Chebyshev Impedance-Transforming Networks of Low-pass Filter Form," *Proc IEEE*, Aug, 1964, pp 939-961.

23. R. Wilson and H. Silverman, "Wire Line - A New and Easy Method of Microwave Circuit Construction," *QST*, Jul, 1981, pp 21-23.

24. W. Carver, "Measuring Capacitors and Inductors," *QRP Quarterly*, Jan, 1998, p 37.

25. Clarke and Hess, *Communications Circuits: Analysis and Design*, Addison-Wesley, 1971.

26. C. Trask, "Wideband Transformers: An Intuitive Approach to Models, Characterization and Design," *Applied Microwave and Wireless*, Nov, 2001.

27. Ruthroff, "Some Broad-Band Transformers", *Proc. IRE*, Aug, 1959.

28. N. Dye and H. Granberg, *Radio Frequency Transistors: Principles and Practical Applications*, Butterworth-Heinemann, 1993, Ch 10.

29. Hamilton, "Improved Direct Conversion Receiver Design", *Radio Communications*, Apr, 1991, Appendix.

30. W. Hayward, *Introduction to Radio Frequency Design*, ARRL, 1994, Ch 4.

31. R. Lewallen, "A Simple and Accurate QRP Directional Wattmeter," *QST*, Feb, 1990, pp 19-23, 36.

32. R. Larkin, "An 8-Watt, 2-Meter 'Brickette'," *QST*, Jun, 2000, pp 43-47.

33. A. Boulouard, "Lumped-Element Quadrature Couplers," *RF Design*, Jul, 1989.

34. R. Fisher, "Broadband Twisted-Wire Quadrature Hybrids," *IEEE Transactions on Microwave Theory and Techniques*, Vol. MTT-21, No. 5, May, 1973, pp 355-357.

35. R. Fisher, "Twisted-Wire Quadrature Hybrid Directional Couplers," *QST*, Jan, 1978, pp 21-23.

36. J. D. Cappucci and H. Seidel, US Patent 3,452,300, *Four Port Directive Coupler Having Electrical Symmetry with respect to Both Axes*, issued Jun 24, 1969.

37. J. D. Cappucci and H. Seidel, US Patent 3,452,301, *Lumped Parameter Directional Coupler*, issued Jun 24, 1969.

38. B. M. Oliver, "Directive Electro-Magnetic Couplers," *Proc. IRE*, Oct, 1954.

39. S. B. Cohn, "Shielded Coupled Strip Transmission Line," *MTT*, Oct, 1955.

40. K. Kurokawa, "Design Theory of Balanced Transistor Amplifiers," *Bell System Technical Journal*, Vol. 44, No. 10, Oct, 1965, pp 1675-1698.

41. R. S. Engelbrecht and K. Kurokawa, "A Wideband, Low Noise, L-band Balanced Transistor Amplifier," *Proc. IEEE*, Vol 53, Mar, 1963, pp 237-246.

42. R. S. Engelbrecht, US Patent 3,371,284, *High Frequency Balanced Amplifier*, Feb 27, 1968.

CHAPTER 4

Oscillators and Frequency Synthesis

Almost all of the Amateur Radio equipment we build will contain at least one oscillator. It may be a simple crystal controlled circuit, a tuned LC variable frequency oscillator, or even a direct-digital synthesizer, a circuit that provides an output similar to what we might expect from a simpler circuit. A basic oscillator might be a simple one tuned by a mechanical variable capacitor. Alternatively, it might be voltage controlled. Combinations of all of these are possible and are common in modern communications equipment.

The *local oscillator* (LO) is a critical part of any communications system. Modern transceiver performance is often compromised by LO systems that suffer from excess phase noise, effectively limiting the receiver dynamic range. While *quiet* oscillators, those with low phase noise, can be built using traditional methods, these circuits often lack the thermal stability of a synthesizer.

Beyond their practical importance, oscillators are extremely interesting circuits. An effective oscillator can be built with a single transistor. Yet, this simple, primitive circuit will include both positive feedback, causing oscillation to start at the desired frequency, and negative feedback that maintains operating amplitude constant with time.

A frequency synthesizer offers outstanding thermal stability and frequency accuracy. A synthesizer using a handful of integrated circuits, each containing hundreds of transistors, is less expensive to manufacture than a high quality mechanically tuned LO system. It is more reliable, owing to a reduced number of moving parts. Frequency synthesis is not, however, the answer to all of the LO problems presented to the experimenter. Some PLL synthesizers are burdened by excessive phase noise. Those using DDS, while quieter, emit spurious outputs, often in profusion. Both use an excess of digital circuitry that can often corrupt a receiver environment.

4.1 LC-OSCILLATOR BASICS

Oscillators may be classified in a number of ways. One categorizes the circuit by the devices used for the active element and the resonator, such as the *bipolar transistor, crystal controlled oscillator* and the *JFET LC oscillator*. One can also classify oscillators according to a historic circuit form, such as the Colpitts or Hartley. An oscillator can be classified by the active device configuration, such as *common-emitter*. Finally, it can be classified according to the method used during design, such as a *negative resistance* oscillator.

The first question we ask (or should ask) is if an oscillator will indeed oscillate when power is applied.

Fig 4.1 shows a block diagram of an oscillator. The circuit is segmented into two elements: a resonator or tuned circuit, and an amplifier. The tuned circuit output is applied to the amplifier input. But, the amplifier output is routed back to the input of the tuned circuit.

Assume that the circuit has a power supply attached, but through some means or another the resonator is short-circuited with a switch or otherwise altered so that the circuit is not oscillating. The switch is then opened, restoring resonator functionality. The amplifier is operational with normal operating bias applied; hence, it generates noise. The noise present at the

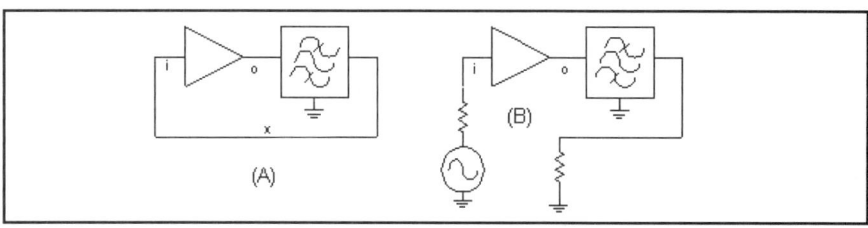

Fig 4.1—Block diagram of an oscillator. Part A shows the basic oscillator while part B illustrates the method used for analysis. This analysis can be applied to either LC or crystal oscillators, or even circuits using RC filters to replace the resonator. Amplifier input and output is labeled with "i" and "o."

input is amplified to appear at the output with greater amplitude. This noise is spread more or less evenly over a wide bandwidth. The amplifier output is applied to the tuned circuit where it is filtered and phase shifted. The resulting signal emerges where it is again applied to the amplifier input. For each frequency, the signal that has traversed the amplifier-resonator loop emerges with a new amplitude and new phase. If the amplifier has a net gain at the resonator center frequency, the signal at that frequency is larger after having traversed around the circuit. It will continue to grow with each round trip.

There will be one unique frequency where there is no net phase shift as energy at that frequency traverses the loop. This eventually establishes the oscillator operating frequency. Energy at frequencies above and below the center *carrier* frequency will be shifted further in phase with each trip around the loop, eventually emerging 90 degrees away where it no longer contributes to the power.

We have just described oscillator *starting*. Oscillation will begin if the signal grows in amplitude with each pass around the loop and if the phase is the same as it was in the beginning. These are the so-called Barkhausen criterion. They are measured or analyzed with the system in Figure. The loop has been broken at "X" in part "A" of the Figure. A signal source and a load are inserted that allow the gain to be measured, shown in part "B."[1]

The amplitude cannot continue to grow without bound. Something must occur within the circuit that will reduce the overall gain to the level just needed to maintain a stable amplitude. This usually occurs through current or voltage limiting, with current limiting generally preferred. (Automatic gain control can also be used.) Biasing details usually establish limiting and set oscillator operating level. A high operating level is generally desired.

We rarely analyze starting in an HF oscillator we wish to build for a project. Rather, we merely build and examine the oscillator to see if there is an output.

The Colpitts and Hartley Circuits

While there are numerous named LC oscillators, they can generally be categorized as Colpitts or Hartley variations with both circuits named for their inventors, early radio pioneers from the Bell Labs of the 1920s and 1930s era. The basic forms are shown in **Fig 4.2**, A and B. The only difference between the two is in the means for feedback. The Hartley (B) uses a tapped inductor while the Colpitts (A) uses capacitors.

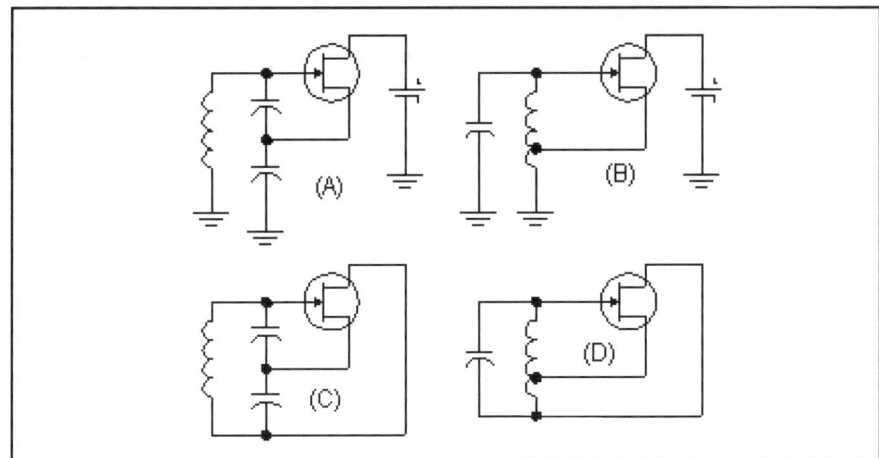

Fig 4.2—Colpitts (A) and Hartley (B) oscillators. The versions at (C) and (D) have the ground removed, allowing any of the three FET terminals to be grounded. The bias is eliminated from the last two circuits. Although illustrated with FETs, bipolar transistors are often used.

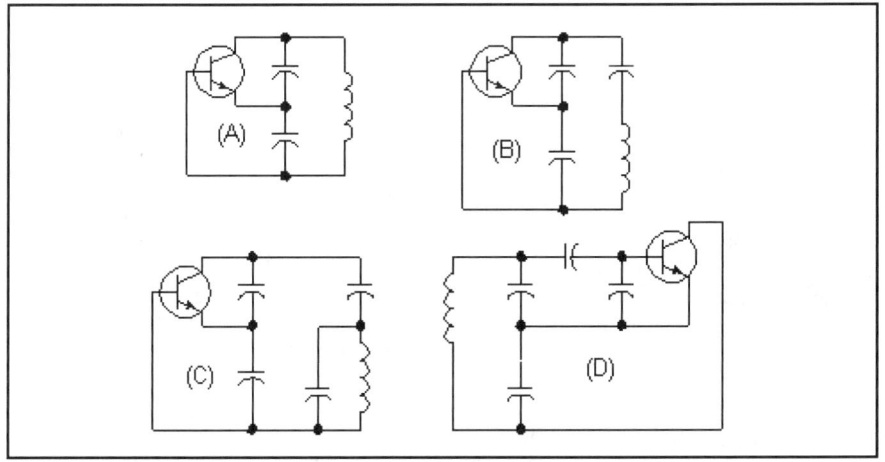

Fig 4.3—The Colpitts (A) evolves into the Clapp (B) and then the Seiler (C). The Vackar oscillator at (D) is yet another variation on the Colpitts where the base is driven from a lower impedance, achieved with a capacitor tap across one of the usual "Colpitts Capacitors." These oscillators can be designed with either FETs or bipolar transistors.

The Hartley and the Colpitts oscillators of Fig 4.2 A and B use a source follower amplifier. This distinction is an arbitrary one, as is illustrated with the two variations of Fig 4.2 C and D, which are drawn without a ground. The ground and biasing can then be inserted as needed by the designer.

The operation of the Hartley is often explained with transformer action. The source follower of Fig 4.2B has a high input and relatively low output impedance, and a voltage gain close to 1. The amplifier output signal is applied to the tap on the tuned circuit. Transformer action then increases the voltage that appears at the gate. Breaking the loop at either the FET gate or source will show the required greater-than-unity, zero phase shift starting gain.

The Colpitts circuit (Fig 4.2A) may not be as intuitive. Detailed circuit analysis will show that driving the capacitive tap with a low impedance source will produce the required voltage step up in the composite tuned circuit. Indeed, a similar analysis shows that the same action occurs in the Hartley oscillator even if there is no magnetic coupling between the two inductor sections. Transformer action is not required! A Hartley is easily built with two separate coils, an occasionally useful variation.

The Hartley oscillator with positive feedback resulting from inductors can have an advantage over the Colpitts: If it is tuned with a variable capacitor with mini-

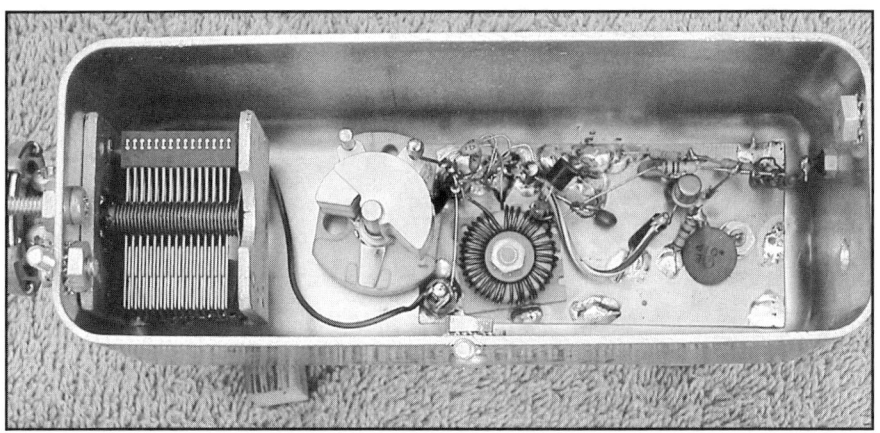

This Hartley Oscillator is mounted in a stamped box. A vernier drive is attached to the capacitor shaft and is fixed to the box with a single bolt that prevents rotation. Spade lugs allow a lid to be attached to the box.

mal fixed capacitance, it will produce a wider tuning range than is easily realized with a Colpitts. There is no other fundamental advantage of one over the other.

The Colpitts oscillator has several popular variations shown in **Fig 4.3**. The first circuit (A) is the basic Colpitts, now shown with a bipolar transistor. Part B shows the Clapp oscillator, also called a *series tuned* Colpitts. The Clapp starts with a Colpitts circuit, but replaces the usual inductor with a larger one. Then, the extra inductive reactance is removed with a series capacitive reactance. Part C shows yet another variation, the Seiler, where a Clapp is modified. The Clapp inductor is replaced by a smaller one paralleled with a capacitor. The Clapp is capable of greater energy storage than a similar Colpitts while the Seiler allows the active device to be well decoupled from the resonator. These three are analyzed in greater detail in *Introduction to Radio Frequency Design*, Chapter 7.

A final variation shown in Fig 4.3D is the Vackar. In this circuit, the Colpitts capacitor attached to the base is expanded, allowing the base to be driven from a lower source impedance. This would provide excellent decoupling between the active transistor and the resonator. The Vackar is discussed later in greater detail.

4.2 PRACTICAL HARTLEY CIRCUITS AND OSCILLATOR DRIFT COMPENSATION

A good oscillator is thermally and mechanically stable in frequency and has low noise. We'll look at the stability issues in this section, leaving noise for later, and will illustrate the ideas with practical circuits suitable for duplication.

The first circuit we examine is a simple LC Hartley oscillator suitable as a LO in the HF spectrum. We have used this circuit in applications from 1 to 50 MHz, and have breadboard variations that extend from audio to 3 GHz. The 7-MHz circuit presented in **Fig 4.4** uses a JFET.

Generally, an inductor with reactance of around 100 Ω offers a good starting point in design, although this is very noncritical. The tap position is similarly uncritical; start with a tap up from ground by about 20% of the number of turns.

If this oscillator is built with no fixed capacitance other than stray values, a frequency range approaching 4:1 can be expected. Much of the capacitance in the tank is fixed for narrow tuning ranges. All fixed capacitors should be NP0 types. NP0 is an abbreviation for *negative positive zero*, a capacitor type with a capacitance that does not change with temperature. The capacitor between the hot end of the resonator and the FET gate should have a small C value. The input C of the FET is typically

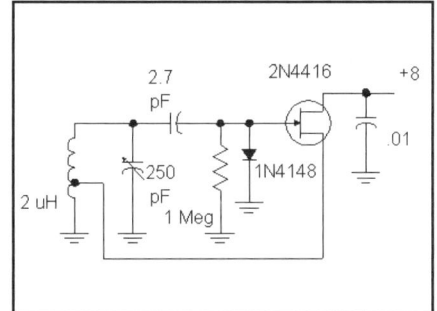

Fig 4.4—Practical 7-MHz Hartley oscillator.

around 1 pF, so any series capacitor with a similar or slightly larger size will do.

The oscillator of Fig 4.4 uses one large variable capacitor for tuning. A typical circuit will use combinations of fixed and variable capacitors, configured to tune a narrow range with the variable element. The equations are shown in a sidebar.

The gate diode is often described as a "clamping element," for it does not allow the gate to become more positive than about 0.6 V. However, the primary function is a detector to supply the FET with negative bias. A signal voltage present on the tank circuit causes diode current when the anode is positive by 0.6 V. The current through the 2.7-pF blocking capacitor charges it. The average dc voltage on the tank side of the capacitor must be zero, for the coil is at dc ground. Hence, the charged capacitor causes an average negative voltage to appear at the FET gate. This negative bias builds toward FET pinchoff as oscillator amplitude increases. If the oscillator operating level changes during tuning, the negative bias will change, allowing FET gain to change as needed to maintain a nearly constant output. This automatic gain control (AGC) action is much like the limiting that also occurs in the Hartley. Limiting will occur on a cycle-to-cycle basis while the AGC responds to an average level. The AGC offers a coarse control, leaving the limiting to set the final level.

The voltages described are easily observed with a high-speed oscilloscope with a 10X probe. Even a high quality probe will load the HF oscillator tank, compromising accuracy, but qualitative details can still be seen.

This oscillator normally operates with a 5 to 20-V peak-to-peak signal on the tank. It can be even higher if an extra shunt capacitor is used at the gate, mimicking that design feature in the Vackar oscillator. The phase noise capabilities of the Hartley oscillator of Fig 4.4 are good,

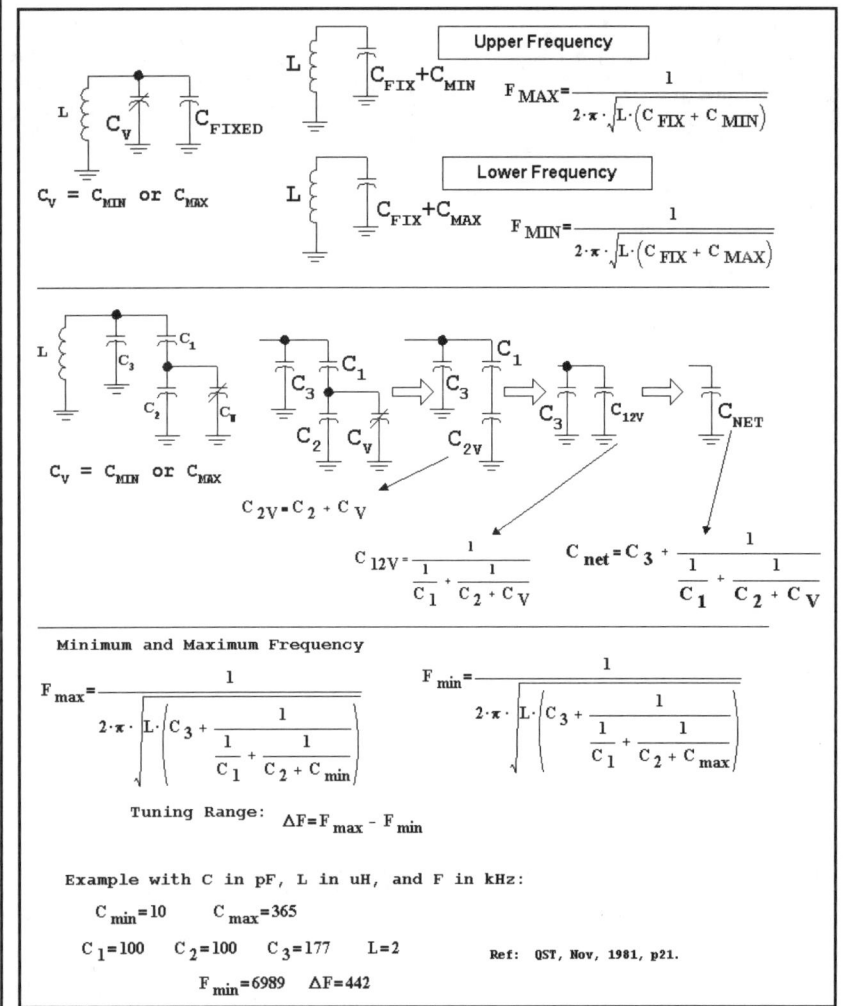

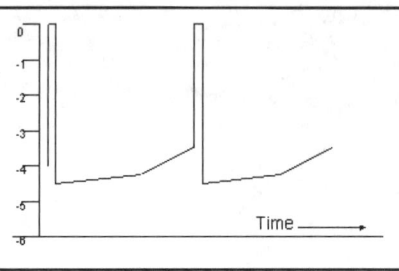

Fig 4.5—Squeeging in a Hartley oscillator, an on-and-off mode where the oscillator is not functioning except during short periods. The vertical scale shows the gate voltage. Extreme values of blocking capacitor and bias resistor are required to produce this behavior in the FET Hartley oscillators.

A simple resonant circuit is tuned with parallel capacitors as shown in the top section. The tuning range is controlled by the ratio of the variable capacitance to the fixed one.

Often an available variable capacitor has greater capacitance than required for a desired frequency range. While plates can sometimes be removed, a better solution embeds the variable capacitor in a network of fixed capacitors. The evolution of this network is shown in the middle section. The variable, C_v and C_2 are paralleled to form the equivalent C_{2v}. This is then placed in series with C_1 for the equivalent C_{12v}. This is paralleled by C_3 to form the total capacitor, C_{NET}. The overall frequency is calculated from the usual resonance relationship. The equations are shown, with capacitors in Farads, inductance in Henrys and frequency in Hz.

There is considerable flexibility available to the designer, afforded by picking C_1 and C_2 values. Some combinations with C_1 much smaller than the variable capacitor can produce highly nonlinear tuning.

although not the ultimate. (Phase noise is discussed later in this chapter.)

The 1-MΩ resistor represents a load on the tank. It also discharges the series blocking capacitor. If a smaller resistance is used, the blocking capacitor will discharge more quickly. The energy to maintain bias comes from the RF envelope, further loading the resonator. Resistor values around 1 MΩ are generally optimum.

Experiments were performed to examine the effect of resistor and blocking capacitor values, and unloaded resonator Q. If extreme values (long time constant) were used with degraded tank Q, the oscillator could become amplitude unstable, producing a phenomenon called *squegging*. A sketch of the observed gate voltage is shown in **Fig 4.5** for an oscillator using a 2N4416 FET. This unusual behavior was observed when tank $Q_U = 30$, the gate resistor was increased to values much larger than 1 MΩ, and blocking capacitors of 200 pF or more were used.[2]

The supply voltage used with this oscillator should be larger than the magnitude of the FET pinchoff. A supply of +5 is high enough for a 2N4416 with pinchoff of –3 V. The supply should be regulated and come from a moderately low dc impedance. In one experiment, we built this oscillator with a 6-V Zener diode with a 3.9-kΩ resistor fed from a 12-V supply. The high resistance value was picked for overall efficiency. The oscillator would not start. DC voltmeter measurements showed that the FET only had 1 V on the drain. The FET was *trying* to draw a current of I_{dss}, leading to excessive drop across the 3.9-kΩ resistor. A smaller (470-Ω) dropping resistor solved the problem, but at the cost of higher power consumption. A better solution is a 100-Ω drain-decoupling resistor supplied by a dc emitter follower with the base referenced to a Zener diode paralleled by a large electrolytic capacitor. A small charging current can then be used, maintaining efficiency. Three terminal regulator ICs also work well in this application. This is one of many examples where *extra* circuitry *improves* efficiency.

Temperature Compensation

Generally, the most important characteristic of oscillators built for radio application is frequency stability. Stability relates to a change in frequency other than the desired ones that occur with tuning. This change, or *drift*, occurs in two forms. One is the warm up drift occurring when an oscillator is first turned on and allowed to operate at constant temperature. The sec-

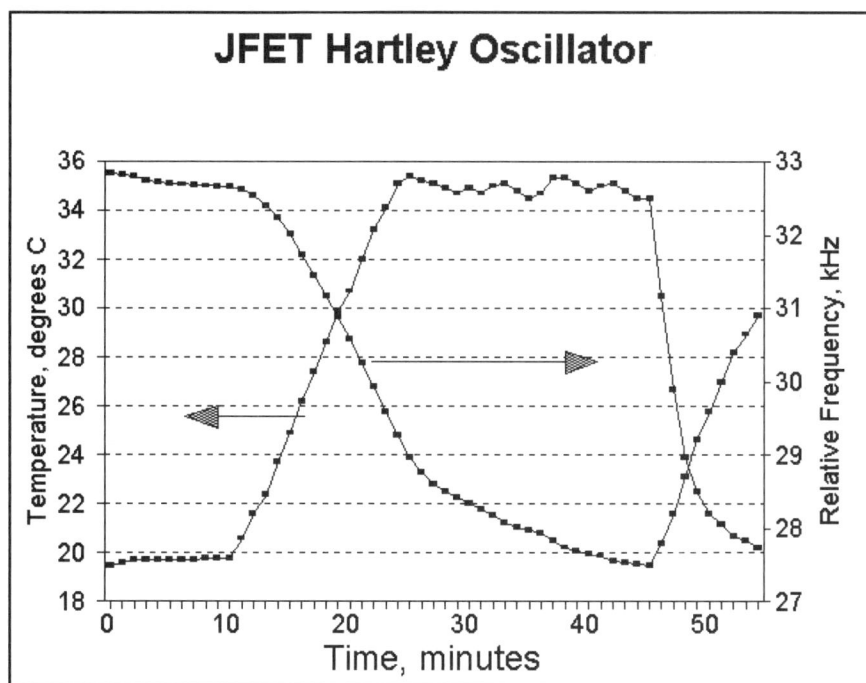

Fig 4.6—Temperature and frequency vs. time for a Hartley oscillator operating in a simple environmental chamber. The heat was turned on at 10 minutes. It was cycled off and on after 25 minutes to maintain an approximately constant temperature. The chamber lid was removed and a cooling fan was turned on at 46 minutes.

ond is the drift with changing temperature. Both effects are thermal in origin, but the warm up drift is caused by temperature changes in individual components resulting from heating by the circulating currents within the circuit. Warm up drift is normally small compared with the drifts that occur when an oscillator is subjected to even a modest temperature change.

Thermal drift may be of little consequence when equipment is built and used in a typical home environment where room temperatures are stable. But the oscillator that was "rock solid" during home operation may become a very poor performer when subjected to portable environments. The most extreme examples we have encountered occurred when we took equipment on mountaineering trips. The temperature at the summit of a glacier clad, cloud covered mountain can be below freezing, even in mid summer. But the temperature can quickly shoot up when the clouds blow away for a few minutes, only to plummet downward as soon as the clouds return. It's important to design for these extremes if they might be encountered. While not as severe, drift problems are common even when we are on the flatlands.

Oscillator temperature compensation is surprisingly easy, requiring little equipment beyond the simple frequency counter and DVM that most experimenters already possess. All that is needed is a simple environmental chamber with a thermometer. The chamber is built from an inexpensive Styrofoam box. A light bulb is placed inside the box along with the circuit being tested. A small fan stirs the inside air to complete the chamber. Temperature is measured with an integrated circuit intended for this purpose. Leads supply power to the IC and route a dc signal out of the chamber for measurement with a DVM. An oscillator to be tested is placed in the chamber with cables routed to the outside for power and for frequency measurement. The oscillator is turned on for a while before the heat source is applied, providing a measure of warm-up drift. Heat is then applied, causing the temperature to increase.

Data for a 7-MHz Hartley oscillator is shown in **Fig 4.6** where frequency and chamber temperature are plotted vs time. The oscillator was operated for 10 minutes before applying the 60-W heat source, producing a typical 150-Hz warm-up drift. Chamber temperature immediately started to increase when the heat source was turned on. The frequency did not respond immediately, for the oscillator was housed in a moderately tight container. When frequency began to drop, it moved about 5 kHz for a 15°C temperature increase. The external heating induced drift was over 30 times the warm up drift! The heat source was only operated intermittently after the 25-minute mark to maintain chamber temperature. Oscillator drift continued as the internal components came up to temperature.

Measurements are simpler when the tested oscillator is only a small board with low thermal mass, capable of quicker temperature changes.

Thermal frequency stability depends on the resonator coil and all related capacitors. Most oscillators we built use toroid inductors wound on SF (–6) material. A newer material with a –7 designation is reported to be slightly more stable. The –6 material has a permeability of about 10 and a temperature coefficient of inductance (*TCL*) of +35 parts per million per degree Celsius (C). This means that an inductor of 1 micro-henry will increase by 35 pH (i.e., 0.000035 µH) when the temperature increases by 1 degree C. Temperature coefficients are generally specified in normalized, dimensionless form, (parts per million) allowing convenient scaling. The normalized rate of change of frequency, TCF, is related to all of the components in the oscillator resonator. If, for example, a tank consisted of two parallel capacitors and an inductor, the temperature coefficient of frequency is related to that of the components by

$$TCF = \frac{\Delta F}{F} = -\frac{1}{2} \cdot \left[TCL + TC_{C1} \cdot \frac{C_1}{C_{TOT}} + TC_{C2} \cdot \frac{C_2}{C_{TOT}} \right]$$

Eq 4.1

where C_1 and C_2 are the capacitors with temperature coefficients TC_{C1} and TC_{C2}, TCL is the temperature coefficient of the inductor, and TCF is the temperature coefficient of frequency of the oscillator in normalized parts. C_{TOT} is the total capacitance, C_1+C_2. The negative sign arises because an increase in L or C leads to decreasing frequency. The factor of one half comes from the square root relationship of frequency to L and C.

Consider a 7-MHz example, using a 2-µH inductor carefully wound on a T50-6 toroid. Assume TCL is +50 ppm/°C, slightly worse than the quoted material performance, which will be explained later. Initially assume that the inductor is paralleled with 250 pF of perfectly non-drifting NP0 capacitors. The only part that will drift will be the inductor. From **Eq 4.1**, the 50 ppm/°C will produce a TCF of –25 ppm/°C, or –25 Hz per MHz. The 15-degree shift of Fig 4-6 would then produce a frequency change of –2.6 kHz.

We now replace the single capacitor with two, a 150-pF NP0 ceramic and a 100-pF polystyrene. The nominal frequency remains 7.118 MHz. Assume that the NP0 capacitor is not perfect, having a TC of +5 ppm/°C. The poly cap has TC = -150 ppm/°C. The TCF for the circuit is

$$TCF = -\frac{1}{2} \cdot \left[50 + 5 \cdot \frac{150}{250} - 150 \cdot \frac{100}{250} \right]$$

Eq 4.2

This oscillator has a much improved TCF of +3.5 ppm per degree C. This is 3.5 Hz drift per MHz of observed frequency per °C. A 10-degree C temperature rise would produce a 245-Hz frequency increase, a very stable VFO. The stability results from the use of a combination of parts with temperature coefficients that cancel each other.

The temperature coefficient of frequency, TCF, is reduced from that of the compensating capacitor to half the ratio of the compensating capacitor to the total resonator C. Capacitors with a temperature coefficient of –750 ppm/°C are readily available. They can be placed directly across a resonator or in series with a NP0 capacitor for compensation. If capacitor C1 has a known TC, but is placed in series with an NP0 capacitor, C2, the resulting TC of capacitance is given by

$$TC_{net} = \frac{C_2}{(C_1 + C_2)} \cdot TC_{C1} \quad \textbf{Eq 4.3}$$

For example, if we place a 47-pF capacitor with TC of –750 ppm/°C in series with a 10-pF NP0 capacitor, the result is 8.2 pF with a TC of –132 ppm/°C.

Although polystyrene capacitors can be used for compensation, they are not ideal. The TC of –150 ppm/°C is not a precise number. The TC itself has a tolerance of +/–50 ppm/°C, allowing a polystyrene capacitor to have a TC ranging from –100 to –200 ppm/°C. This variability is common, even among NP0 capacitors. For example, one of the best commonly available NP0 capacitor types is one with a so-called C0G characteristic, where the G designates a TC tolerance of +/–30 ppm/°C.

Our example used an inductor TC that differed from the published value for the powdered iron core. The difference relates to the way the core is wound. If a large wire is hand wound on a toroid, with the wire size picked to fill the core to produce highest possible Q, there is a good chance that the wire will gap away from the core for part of each turn. This leaves unsupported loops that can expand or contract with heat, producing ill-defined characteristics. A more temperature stable coil is produced with a wire size that is smaller than that producing maximum Q. The Q degradation is usually not large.

Temperature coefficients are themselves temperature dependent. An oscillator that has been compensated at one temperature may not be as stable at temperature extremes.

Another subtle problem has to do with stress built into the wire during the winding process. We first observed this while temperature testing bandpass filters built from toroids. The filter frequency would change as temperature increased, but would not come back to the original frequency when the circuit returned to room temperature. However, a second excursion to high temperature and back would produce the expected return. Evidently, the first excursion to high temperature (85 °C) and back relieves the stresses left in the metal during winding. W7EL has dropped coils into boiling water after winding; subsequent cooling produces a more stable inductor.

None of the temperature stability and compensation arguments relate to oscillator topology. There is nothing that will make one type more stable than another so long as the circuit does not degrade tank Q from improper limiting. The compensation methods described here for the Hartley apply equally to other circuits presented later. Capacitor variability makes it difficult to predict and control stability, encouraging the serious builder to measure his or her VFO.

Powdered iron toroid cores (–6 and –7 material from Micro-Metals) produce stable and reproducible inductors if carefully wound. Some other coil forms may produce stable coils, although the reader should not trust poorly documented testimonials (lore) regarding slug tuned forms or other schemes that are not easily duplicated and quantified.

The most stable oscillators are built from collections of components that *all* have low drift. A really bad component can be compensated, but only over a narrow temperature range.

Drift measurements in a measured, variable temperature environment are much more meaningful than mere warm up drift measurements. A suitable chamber can be built at very low cost in an evening. The chamber is described in a paper included on the book CD.[3]

Variations on the Simple Hartley

The oscillator described has been a long

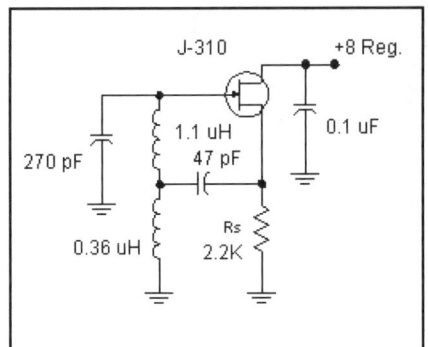

Fig 4.7—A Hartley oscillator using source bias and two inductors. The larger inductor is 17 turns on a T50-6 toroid. The smaller one is 10 turns on a T30-6. Output can be extracted from the source or directly from the resonator with a capacitive tap and appropriate buffering.

time favorite among QRP experimenters. There are, however, some variations that should also be considered. **Fig 4.7** shows an oscillator without ac coupling into the gate, removing the AGC action of earlier oscillators. The amplitude is determined by more traditional current limiting. The FET in the example has a pinchoff voltage of –3 V. The source resistor places the source at a positive potential, even before oscillation has started. As oscillation builds, follower action causes the source voltage to reach large positive values. The gate also reaches positive values, but is always offset below the source. During part of the cycle, the gate-source voltage drops to or below pinchoff; the greater the fraction of each cycle spent in this condition, the greater will be the gain reduction, which establishes the final operating level. With a 2.2-kΩ source resistor, the gate signal was 11 V peak-to-peak. This dropped significantly when the source R was increased to 10 kΩ.

The oscillator of Fig 4.7 has an additional unusual feature: The usual tapped coil is replaced with two isolated coils. This has the advantage that the circuit is easily band-switched, a sometimes-messy problem with tapped inductors.

The "Huff 'n Puff"

Frequency counter circuitry can be used to stabilize a moderately good oscillator, achieving nearly the stability of a synthesized oscillator.[4]

This scheme uses normal frequency counter circuitry such as that in **Fig 4.8**. A stable crystal oscillator is the foundation. The result is divided with a large counter, a straightforward operation with CMOS circuits such as the 4060 or similar indus-

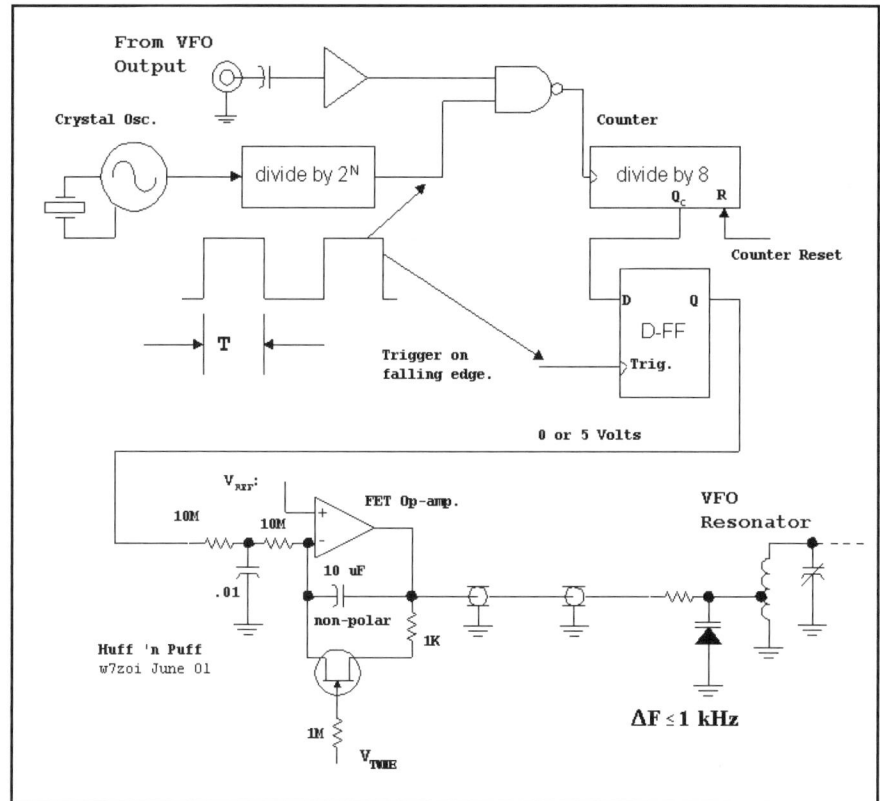

Fig 4.8—This scheme uses normal frequency counter circuitry. A stable crystal oscillator is the foundation.

riod with a logic 1 in the output digit, indicating a count of 4, 5, 6, or 7. The negative edge of **T** is detected and used to trigger a D-Flip-flop that memorizes the result. The saved digital 1 causes Q of the FF to be at 5 V. This signal is applied to the input of an op-amp integrator circuit which generates an output that ramps downward, but at a very low rate. This slowly changing voltage causes the VFO frequency to decrease.

The frequency goes down slightly as a result of the applied signal. Finally, after a few cycles of counting, it will have dropped enough that the signal held in memory becomes a logical zero, resulting in an integrator input of 0 V. This now causes the op-amp output to again ramp upward, slowly increasing the frequency. The overall effect of the added circuit elements is to force the oscillator to never be at a fixed, exact frequency, but to move (huffing and puffing) between two frequencies. These two references are 40-Hz apart for our example, so changes are not noticed in normal applications. Greater resolution is available with a shorter count or longer sample period.

We now allow a slow thermal drift to occur. This has the effect of altering the time when we reach one of the transition frequencies. However, the drift will be cancelled so long it is well under 40 Hz in a 0.2-second window.

A FET switch is placed across the integrator timing capacitor. This FET is turned on when the oscillator is tuned.

The Huff 'n Puff scheme can be extremely useful for adding stability to a circuit that is already reasonably solid. It is a wonderful tool for the experimenter, for it can be added to an already existing design. Several experimenters have expanded the basic system in recent times.[5]

trial timer parts. The division is extended to produce a square wave with a positive half period of length **T**. Assume **T**=0.1 second. A well-buffered sample of the VFO is applied to a conditioning amplifier followed by a gate controlled by the timing signal **T**. This allows timing data to reach a counter for 0.1 second. Let's assume the oscillator to be stabilized has a frequency 300 Hz above 5.0 MHz, and is thermally stable with no drift of it's own. In a 0.1 second period the 8 bit counter input will see 500,030 transitions, so it will overflow again and again. When the gate signal terminates at the end of the period **T**, the 8 bit counter will have overflowed a total of 62,503 times and will end the pe-

4.3 THE COLPITTS AND OTHER OSCILLATORS

One of the most popular oscillator circuits among radio experimenters has been Colpitts in one of its many forms. The basic circuit, along with several of its derivative forms, was presented at the beginning of the chapter. Some practical variations are presented here.

Fig 4.9 shows a simple Colpitts oscillator using a junction FET. Although very simple, this circuit is capable of excellent performance. The variation shown operates at approximately 7.5 MHz with a common drain JFET. Addition of a variable capacitor and trimmer to this circuit

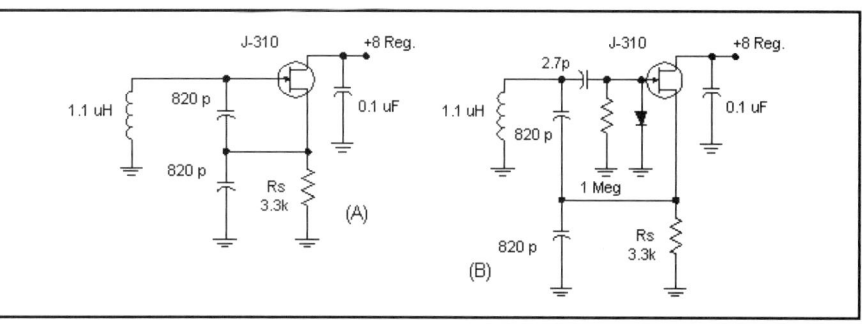

Fig 4.9—Two versions of a Colpitts oscillator. The variation at B is more tolerant of FET variations. The lower noise versions of this oscillator have larger C with reduced L values.

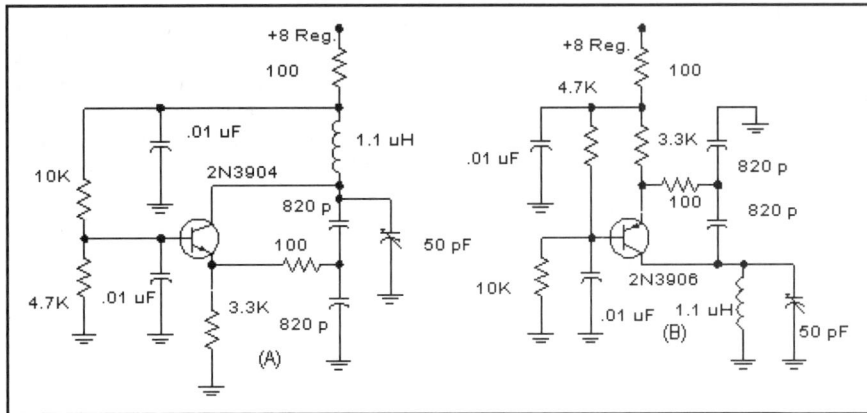

Fig 4.10—Colpitts oscillators using bipolar transistors. Although these circuits were designed around the 2N3904 (NPN) and 2N3906 (PNP), transistor type is not critical for general-purpose applications. The 2N5179 is a good general-purpose choice for VHF applications. The PNP has the advantage that the tank is at ground, removing the bypass capacitor of the NPN tank from the frequency-determining loop. The PNP is also handy when varactor diode tuning is planned.

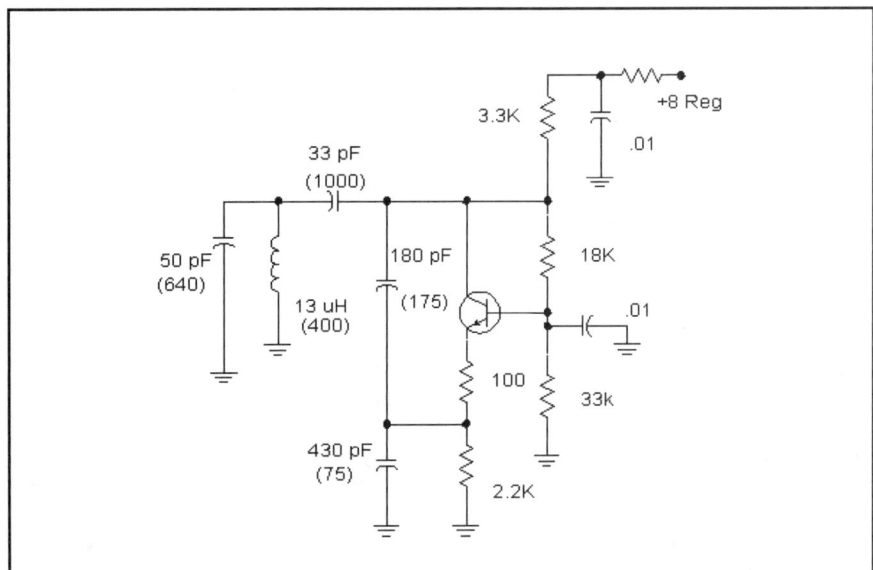

Fig 4.11—A Seiler oscillator for 5-MHz operation. The values shown in parenthesis are reactances, allowing the circuit to be scaled to other frequencies. Transistor type is not critical, although the circuit works well with a 2N3904.

will drop it down into the 40-meter band. The circuit uses a source resistor to set operating level. In the variant with diode clamping, the source resistor may be replaced with a choke, although the negative feedback at low frequency from the resistor is believed to improve phase noise close to the carrier. While shown with a J-310 FET, FET type is not critical. The J310 used for the measurements on this oscillator had a pinchoff voltage of –3.1 V and I_{dss} of 37.5 mA. The circuit draws just over 1 mA during operation. The R_s value may require adjustment if built with a low gain JFET.

While the preferred device for HF Colpitts oscillators and variations is usually the JFET (owing to reduced low frequency noise), bipolar versions are still popular and effective. Bipolar Colpitts oscillators are shown in **Fig 4.10**. The familiar form is that in A using an NPN transistor. The PNP version (Fig 4.10B) is convenient, for the dc grounded collector removes the need for a good bypass capacitor that becomes part of the frequency-determining resonator.

The two oscillators presented in Fig 4.10 are designed for operation near 7 MHz. Like any of the circuits presented, they can be scaled to any frequency within the HF and low VHF spectrum, and even down to audio. The frequency stability will depend upon the criterion outlined earlier. That is, if quality NP0 capacitors and –6 or –7 toroid inductors are used, reasonable stability is predictable. Temperature compensation can be applied to further improve the performance.

A subtlety haunts the bipolar Colpitts circuits of Fig 4.10 in the form of ill-defined limiting. The circuit will nearly always oscillate. However, if the 3.3-kΩ emitter bias resistor is reduced, the transistor will go into saturation at the negative extreme of the collector voltage waveform. This action extracts energy from the tank and dissipates it in the transistor saturation resistance. This can severely degrade the loaded tank Q, compromising phase noise and thermal stability. The emitter degeneration decreases starting gain and helps to establish current limiting as the mechanism determining operating level. Transistor saturation is easily detected with a high-speed oscilloscope.

A simple Colpitts should be built with high capacitance and low inductance, storing the greatest energy in the resonator. But there is a practical limit to this trend. Eventually, stray inductance of the capacitors and the wiring in the tank, including bypass capacitors, will all contribute to the overall L in greater proportion. The stray inductance generally has a considerably lower Q and poorer stability than that of a powdered iron toroid inductor.

Fig 4.11 shows a Seiler oscillator using a bipolar transistor. The values shown are for 5-MHz operation, with reactance at the operating frequency shown in parentheses, allowing scaling. As mentioned earlier, the Seiler can be analyzed as a variation of the Clapp, which is the familiar "series tuned" version of the Colpitts. This circuit has some very useful characteristics. First, the *Colpitts capacitors* (the 180 and 430-pF capacitors providing the in-phase feedback from collector to emitter) are large compared with the 33-pF coupling capacitor to the inductor. This decouples the active device, including parasitic capacitance, from the rest of the tank. Second, current limiting is well established with this circuit. (Computer analysis shows that the transistor stays well away from saturation when the 100-Ω degeneration is used.) Even though the current is small in this circuit, about 1 mA, the signal voltages can be quite high. We measured over 10 V pk-pk across the inductor. The collector signal is much smaller at 2.5 V peak-to-peak. Output can be obtained from the junction of the Colpitts capacitors.

The Colpitts oscillators presented have all operated at the lower end of the HF

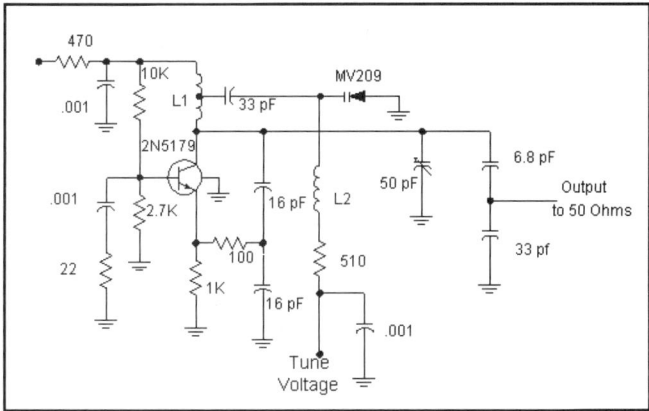

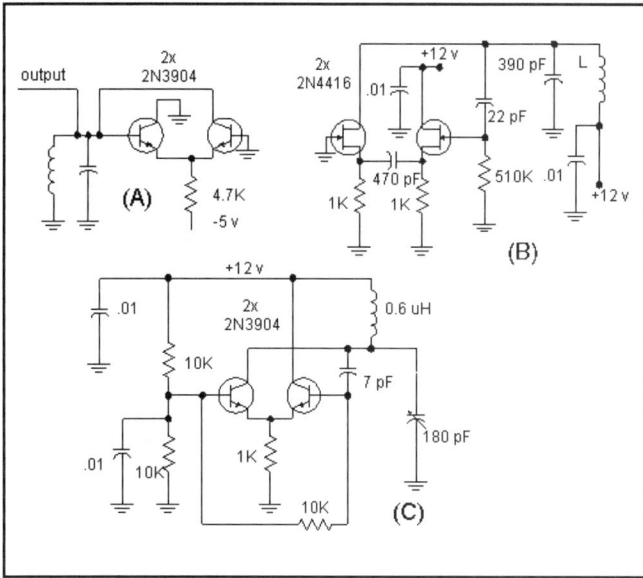

Fig 4.12—A Colpitts VHF oscillator. L1 is 50 nH, 3 turns of #22 bare wire. It is initially wound on a 1/4-20 machine screw as a former. The bolt is then removed. The varactor diode is attached to a tap (approximately center) on the coil in order to reduce the tuning sensitivity. The diode tunes the oscillator by 4 MHz around 134 MHz with a voltage from 5 to 12. L2 is a 2.7 µH RFC. The trimmer capacitor allows the circuit to tune from 71 to 153 MHz. Power output is -2 dBm to a 50-Ω termination.

Fig 4.13—Negative resistance one-port oscillators for application at HF and VHF. See text for discussion.

spectrum. The Colpitts and Hartley can both be scaled for operation at much higher frequencies. Shown in **Fig 4.12** is a VHF Colpitts oscillator. This circuit was originally set up as a voltage controlled local oscillator in an SSB transceiver at 144 MHz. It can, however, be set up for a wide frequency range by spreading or compressing the turns on the coil, which uses an air dielectric.

Numerous other oscillator forms are available for wide frequency range applications. Three are shown in **Fig 4.13**. The first bipolar circuit (Fig 4.13A) is a primitive variation of the scheme used in the Motorola MC-1648. The version shown uses NPN transistors with a negative supply. The same circuit will work with a single positive power supply with PNP transistors such as the 2N3906. The oscillator is a one-port type where two non-inverting amplifiers, an emitter follower and a common-base, are cascaded. The output is returned to the input with a shunt-tuned circuit attached at the common point. This scheme can be built on the bench and made to function over an extremely wide frequency range. Low Q tank circuits are favored. This circuit suffers from very low stored tank energy, the result of voltage clipping by the transistors.

The second circuit uses J-FETs in a variation of the same topology. This circuit, similar to one used in the HP-8662 synthesized generator[6], does not suffer from the voltage limiting found with the simple bipolar version. The circuit shown in Fig 4.13B is one

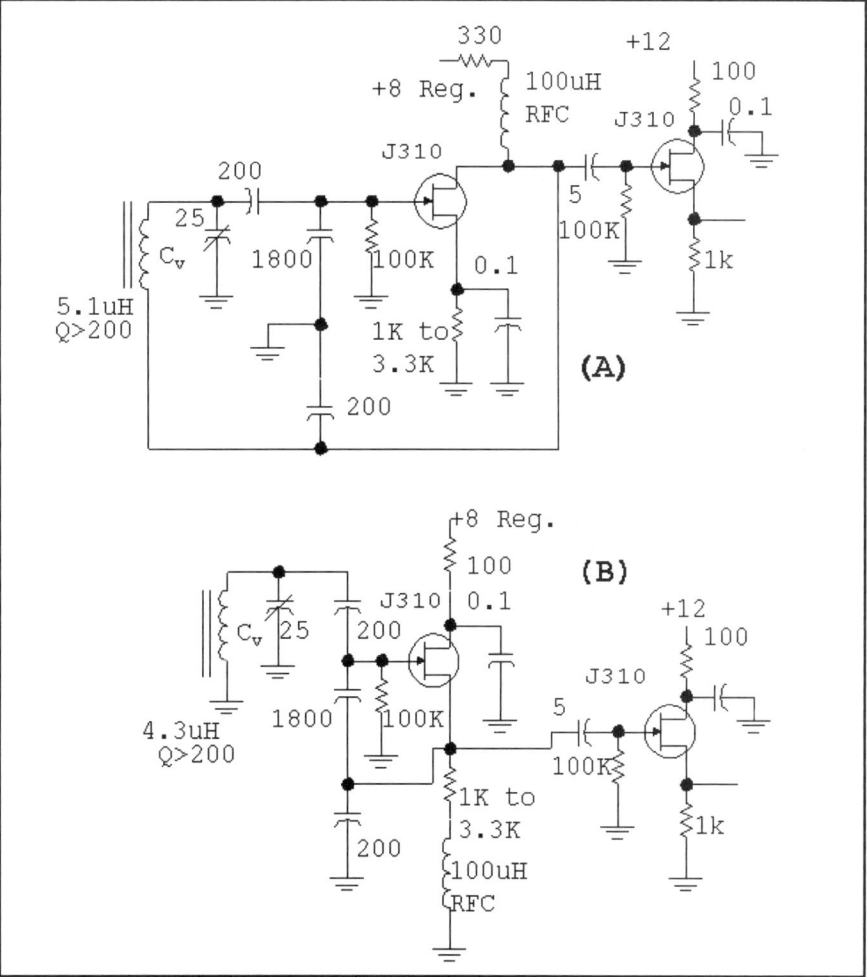

Fig 4.14—The Vackar circuit shown is identical to the Seiler circuit presented earlier except for the choice of component values.

Oscillators and Frequency Synthesizers 4.9

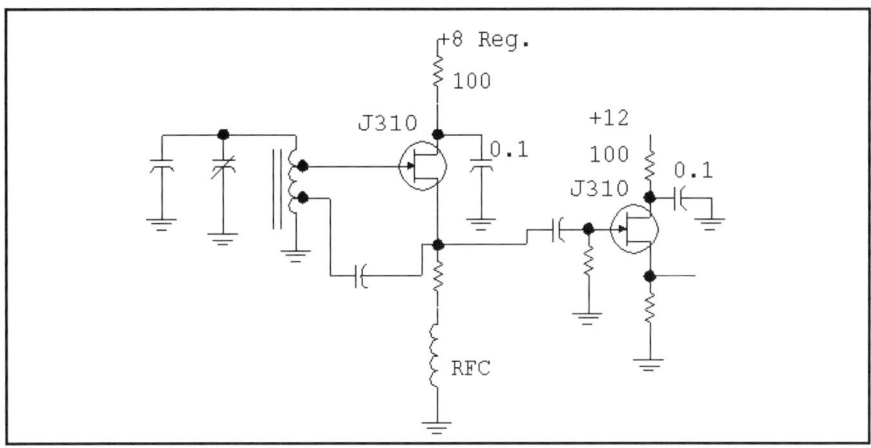

Fig 4.15—This figure shows a variant of the Vackar oscillator with a Hartley theme. The source and gate are both tapped down on the resonator as a means of isolating the tank from the resonator.

that was breadboarded from available components. With an inductor consisting of 20 turns on a T50-2 toroid, the circuit operated at 5.34 MHz with 20-V peak-to-peak on the tank. Changing the resonator allowed operation up to 200 MHz.

Figure 4.13C shows a third version of this oscillator that was built, this time using 2N3904 bipolar transistors. Again, the signal was 20-V peak-to-peak across the resonator.

Fig 4.14 shows the Vackar oscillator. Part A is a JFET adaptation of a vacuum tube design appearing in the 5th edition of the *RSGB Radio Communications Handbook* with components chosen for 7-MHz operation.[7] Output is extracted with a high input impedance buffer attached to the oscillator drain. Part B of the Figure is essentially the same circuit with the ground point shifted from the source to the drain. The inductance value is slightly lower in B than in A, for variable capacitor C_V connects to ground in B. If the capacitor had been returned to the FET source in B, the L value would be the same as at A for 7-MHz resonance.

The Vackar circuit in Fig 4.14B is identical to the Seiler circuit presented earlier except for the choice of component values. The unique component in the Vackar is the large capacitor across the FET gate-source. This component is critical; increasing the value will drop the starting gain to the point that oscillation will not commence. A decrease in inductor Q will have a similar effect. The decoupling between resonator and FET is near optimum in the Vackar. Passive component temperature coefficients will still dominate thermal stability.

Fig 4.15 shows a variant of the Vackar oscillator with a Hartley theme. The source and gate are both tapped down on the resonator as a means of isolating the tank from the active device. This circuit is a direct transformation of that of Fig 4.14B and is often used at VHF for low noise oscillators.[8]

4.4 NOISE IN OSCILLATORS

Some mention has already been made regarding oscillator noise. We don't traditionally think of noise when discussing oscillators. However, noise is present in any practical electronic circuit; the oscillator is certainly no exception. Indeed, excess LO noise is typically the dominant phenomenon limiting the performance of most transceivers in the late 1990s time frame.

Before discussing oscillator noise, we should consider some RF measurements. A spectrum analyzer (SA) is the instrument normally used to examine radio frequency signals. The SA is essentially a calibrated, swept receiver, usually without audio output. Signal strengths are displayed on a CRT or similar screen. When a sinusoidal carrier is applied to a spectrum analyzer, a response is noted at the frequency of that carrier. Changing the analyzer bandwidth will have little impact as we observe the carrier. The amplitude is unchanged. It is specified as a power in dBm. (See Chapter 2 for a discussion of dBm.)

Noise is different. If strong, wideband noise is applied to a spectrum analyzer, it will cause the baseline to rise. If we increase the spectrum analyzer bandwidth by a factor of 10, the baseline will further increase by 10 dB. We cannot describe the noise with a simple "dBm level." Rather, noise is specified as a power density, the power that would appear in a 1-Hz bandwidth. If we apply a wide band noise source to a spectrum analyzer set to a resolution bandwidth of 10 kHz and the response comes up to the –60 dBm line, we say that the spectral density of noise is –100 dBm/Hz; the 10-kHz bandwidth is "40 dB wider" than a 1-Hz wide filter. Recall that $10 \cdot \mathrm{Log}(10{,}000) = 40$.

If a carrier was also present in the noisy display described, we might make reference to a *carrier to noise ratio* (CNR.) (We use the term "ratio," for we are examining the ratio of power. However, we calculate this with a simple subtraction, for the power values are already in a dBm format.) If the carrier was –15 dBm with the noise at –60 dBm with a 10-kHz bandwidth, which corresponded to –100 dBm/Hz, we would say the CNR was 85 dBc/Hz, with dBc standing for dB with respect to a *carrier*. (We usually talk of CNR, carrier to noise ratio, rather than NCR, noise to carrier ratio, for the carrier is much stronger than the noise and is the louder. There is often a sign discrepancy in these discussions, requiring care on the part of the reader.)

Recall the earlier discussion of oscillator starting. (Fig 4.1) Wideband noise at the amplifier input port was amplified, but was then filtered in a resonator. The "signal" within the bandwidth of the resonator is transferred with little attenuation and is again applied to the amplifier input. With a few "trips" around the loop, the signal has grown to the point that limiting begins. As limiting occurs, the net gain around the loop diminishes, eventually stabilizing at unity, the level needed to sustain amplitude-stable oscillation, but no more. Unity gain occurs at the resonator center frequency (or very close to it) where the net phase shift is zero degrees.

Consider the gain characteristics at frequencies close to but slightly removed from the carrier. For example, suppose we build an LC oscillator operating in the amateur 20-meter band with a *loaded* tank Q of 100. The 3-dB bandwidth will then be 1% of 14 MHz, or 140 kHz. Signals 70 kHz on either side of the carrier are attenuated by 3 dB and shifted in phase by

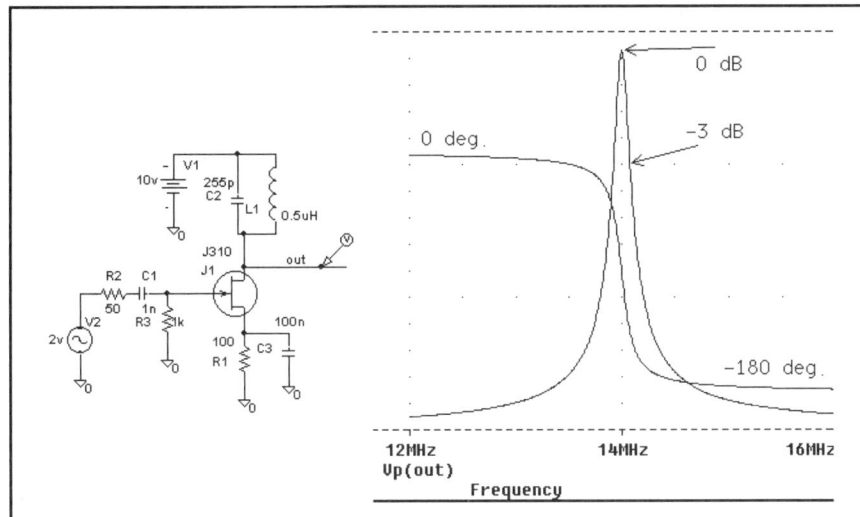

Fig 4.16—An example circuit of an amplifier followed by a resonator. The amplitude and phase responses are shown vs frequency.

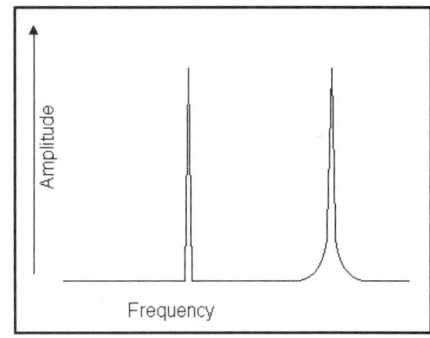

Fig 4.17—A spectrum analyzer output showing two signals with identical amplitude. The peak at the left is "perfect," having a vertical spike shape. The width represents the spectrum analyzer bandwidth. The right hand signal has noise, which appears as a modulation on either side of the carrier. The flat horizontal line is the background noise level of the spectrum analyzer.

+ or – 45 degrees. Signals closer to the carrier have less phase shift and less than 3-dB attenuation. This behavior is illustrated with the amplifier and resonator of **Fig 4.16**.

Although amplifier gain in an oscillator is limited, noise is still present. That noise will still be amplified and filtered in the resonator. Each time a burst of noise energy passes through the resonator, it is shifted in phase and attenuated. Noise very close to the center must travel around the loop several times before it is phase shifted and attenuated enough to disappear. Signals further from the carrier will disappear with fewer passes around the loop.

The noise arises from two sources. One is the wideband noise of the transistor. The other noise starts at a lower frequency. This *baseband* signal modulates the carrier to generate sidebands in the same way that a low frequency sine wave might modulate a carrier to generate discrete sidebands. The modulation happens within the circuit *nonlinear* amplifier, a nonlinearity that is always present in a self limited oscillator.

Noise associated with an oscillator is usually *phase noise*, a variation in frequency or phase. Amplitude noise is also present, but it is usually much less than the phase variation, a result of limiting. Also, oscillators are often used with mixers with limiting characteristics with regard to LO power, further reducing the impact of amplitude noise.

A sketched spectra of an oscillator observed in a spectrum analyzer is shown in **Fig 4.17**. The left peak represents a perfect signal, one without noise. The right peak contains excess noise sidebands typical of that found in a noisy oscillator or synthesizer. If the SA bandwidth is increased, the noise will increase. The response to the carrier peak, however, will not change. A photographed spectral display is also shown.

The spectrum of an oscillator with noise is shown in greater detail in a sidebar figure. A wideband noise floor exists within the oscillator feedback path. The noise then grows at frequencies within the loaded bandwidth of the oscillator resonator.

The phase noise of an oscillator can be predicted with the equations given in the sidebar.[9]

Consider a typical example, an average 14-MHz oscillator. It uses a loaded resonator Q of 100, tank capacitance of 100 pF, transistor noise figure of 10 dB, and a peak tank voltage of 4 V. Analysis with the sidebar equations shows a wideband phase noise floor of –162 dBc/Hz and, at 10 kHz, noise of –146 dBc/Hz.

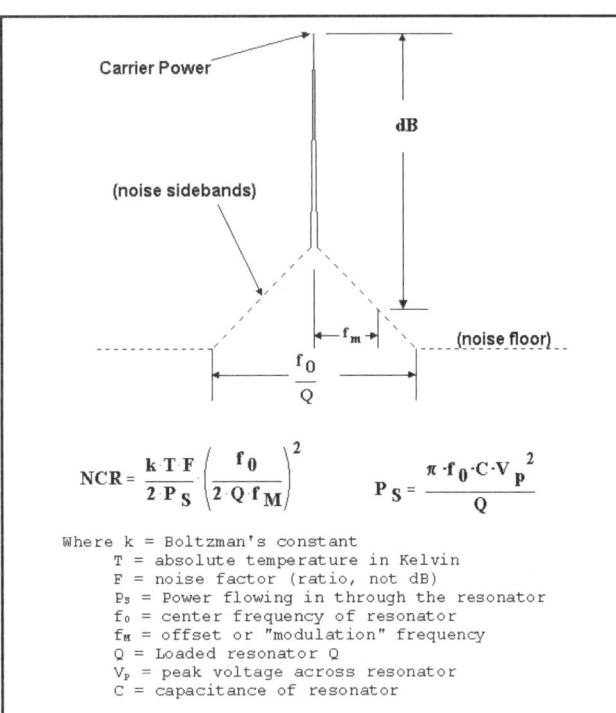

Noise spectrum of an oscillator based upon the work of D.B. Leeson.

$$NCR = \frac{k \cdot T \cdot F}{2 \cdot P_S} \left(\frac{f_0}{2 \cdot Q \cdot f_M}\right)^2 \qquad P_S = \frac{\pi \cdot f_0 \cdot C \cdot V_p^2}{Q}$$

Where k = Boltzman's constant
T = absolute temperature in Kelvin
F = noise factor (ratio, not dB)
P_S = Power flowing in through the resonator
f_0 = center frequency of resonator
f_M = offset or "modulation" frequency
Q = Loaded resonator Q
V_P = peak voltage across resonator
C = capacitance of resonator

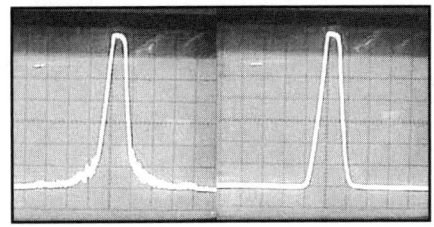

Spectrum analyzer plots from two oscillators. The left is especially noisy, producing noise sidebands where the signal merges into the noise floor. The quiet oscillator (right) lacks these excess sidebands, allowing the signal to go all the way down to the noise floor set by the spectrum analyzer. The left trace was produced with an Epson SG-8002 Programmable Oscillator (4.26 MHz) while the right trace came from a 7-MHz crystal controlled oscillator.

The Effects of Phase Noise

At first glance, phase noise sounds like an esoteric detail that probably has little impact on practical communications. This is generally true. Few oscillators are so noisy that they hamper normal communications in a band occupied with weak to average signals. But things change dramatically when a local station shows up on a band or when a contest starts with attendant stronger signals.

Assume that a receiver uses an ideal filter (perfect skirts) with a bandwidth of 500 Hz. The receiver uses noiseless oscillators. Even if a very strong noiseless carrier is applied to the receiver, a listener will hear a strong response when the receiver is tuned to it, but nothing as soon as the receiver is tuned away.

Consider now a carrier with noise, perhaps keyed with "CQ" so we can recognize it. As the receiver tunes toward the keyed carrier, we first hear some keyed noise. The noise grows in strength as we get closer to it, until finally the carrier is within the receiver passband, producing a clean, crisp note. The noise re-appears on the other side, symmetrical with the first side.

We can't always put the blame on "the other guy." Assume that the keyed carrier applied to the receiver is noiseless, but that we now use a noisy oscillator as the LO in our receiver. The perceived result is exactly the same as we heard before with the noisy CW signal. The effect that we hear is called "reciprocal mixing."

This result is expected. The IF response is the difference (or sum) frequency of the LO and the RF signal. Any frequency change in either one will cause the IF to contain the same change, the same phase or frequency noise. The phase noise is just an instantaneous change in frequency of one of the oscillators.

While our illustrations have presented oscillator noise as viewed in a spectrum analyzer, few analyzers are good enough to actually do this measurement for the local oscillators we need in our HF and VHF transceivers. Like receivers, spectrum analyzers have limited dynamic range. Consider the oscillator mentioned earlier with a phase noise density of –146 dBc/Hz 10 kHz from the carrier. 146 dB is the difference between the carrier and the noise if analyzer bandwidth is 1 Hz. If we used a more practical bandwidth of 1 kHz, the carrier to noise ratio is still 116 dB. An analyzer capable of looking at this carrier and the noise at the same time would need a dynamic range greater than 116 dB. This is close to the present state of the art. Oscillator noise measurements for typical oscillators (at HF) must use modified methods. An example will be given later.

Designing Quiet Oscillators

Many of the methods used to design good LO systems are implicit in the Leeson design equations presented in the earlier sidebar. Some rules are:

- Use moderately low noise transistors in low noise circuits.
- Use a high Q resonator so that the noise sideband width is low. It is *loaded* Q that is important. A high unloaded Q that is degraded by the circuit does little good. If an oscillator is built with a loaded Q close to the unloaded Q, the insertion loss through the resonator will be high, which increases operating gain and increases noise. (This effect was treated in the filter chapter.) This degrades the wideband noise floor.
- The goal is a high carrier-to-noise ratio, which is enhanced with a high carrier. Hence, the best oscillators are those operating with high stored energy in the resonator. This means high power. Even with 8 or 10-V power supplies, it is not unusual to find oscillators with over 50-V peak-to-peak across resonator components. High energy also results from high capacitance in simple resonators.
- Limiting characteristics are critical in an oscillator, with current limiting being preferred. The circuit should operate in a way that allows the transistor current to drop to zero over part of the cycle to limit gain. Less desirable voltage limiting occurs when a low impedance is created over part of an operating cycle; that low impedance then loads the resonator, degrading Q.
- The transistors used in an oscillator should have low noise at both the operating frequency and at baseband. This is important because low frequency noise is heterodyned up to the operating frequency in a working oscillator to modulate the carrier. For this reason, MOSFETs and GaAsFETS, normally perceived as low noise parts, are not as desirable in oscillators as quiet bipolar transistors or JFETs.
- The better oscillators are often those without excessively large starting gain. This places less demand on limiting within the oscillator. The operating circuit is closer to a linear amplifier which has less tendency to mix low frequency noise up to modulate the carrier. Emitter or source degeneration is often a useful modification.

An excellent example of a low noise oscillator is shown in **Fig 4.18**. This oscillator was originally designed by Linley Gumm, K7HFD, and is a good example of a simple circuit that functions well. It features excellent phase noise performance and high output power.

The circuit was designed specifically for high stored resonator energy and high power. Total emitter current is 28 mA, or 14 mA per transistor. The emitter RF choke converts the 47-Ω emitter R into an constant current sink.

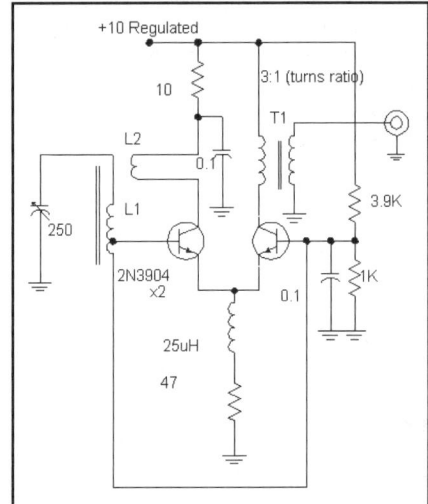

Fig 4.18—Low Noise 10-MHz Oscillator designed by K7HFD. L1 is 1.2 μH, consisting of 17 turns on a T68-6 toroid core. The tap is at 1 turn from the grounded end while the link is 2 turns wound over L1. The link must be properly phased for oscillation. Although not shown, ferrite beads were used on both bases and collectors.

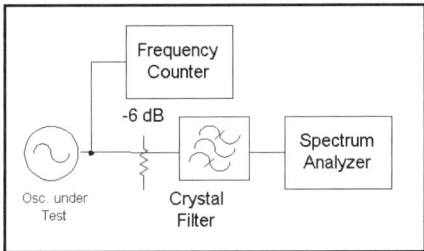

Fig 4.19—System used to measure phase noise in the K7HFD oscillator.

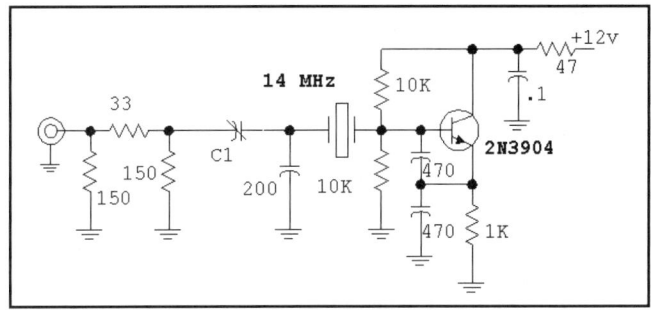

Fig 4.20—Crystal oscillator used for receiver reciprocal mixing measurements. C1 is adjusted for a power output of -10 to -20 dBm.

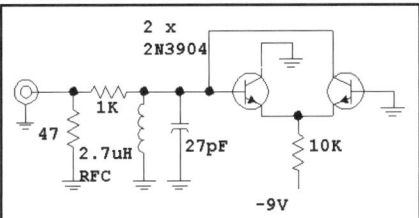

Fig 4.21—Easily built example of a noisy oscillator that the reader can construct to observe phase noise. It is instructive to evaluate this circuit with the design guidelines offered earlier to see just why this is such a poor oscillator.

The circuit of Fig 4.21 is especially bad for phase noise. This can be built as a simple experiment that will allow you to hear the results in a station receiver.

A differential amplifier with heavy base drive will behave as a limiting switch. The total current will oscillate between the two transistors with one collector, and then the other conducting the total current. The high standing current is further increased with an output autotransformer, yielding a measured 10-MHz output power of +17 dBm.

The peak current in the T1 primary also appears in L2, the 2-turn "tickler" link coil over L1. The load presented to the transistor by the link comes from the transistor base and the intrinsic loss of L1. Neglecting the transistor for the moment, the unloaded resonator Q is about 250 for a T68-6 core wound with heavy wire. At 10 MHz, the effective parallel resistance across L1 is about 18 kΩ. This value is diminished by the square of the turns ratio to present a 250-Ω load to the collector. The signal current through this load produces a peak collector signal of 3.3 V. This transforms to a base signal of 1.6 peak V; the signal across L1 is similarly calculated as 56 V peak-to-peak. These values are all significant. The low collector impedance establishes current limiting with no chance of voltage limiting. The restricted base drive guarantees that emitter-base breakdown will not occur.

A crystal filter, shown in the system of **Fig 4.19**, was used to evaluate the oscillator noise. The outboard filter had a 3-kHz bandwidth and skirts that were steep enough to provide over 50-dB rejection to signals 10 kHz away from the filter center. The oscillator was tuned to the filter center and the power reaching the analyzer was measured. The LO was then tuned 10 kHz away. The attenuation in the analyzer could then be reduced enough to measure the noise response. The K7HFD circuit produced phase noise that was below the carrier by 156 dBc/Hz. Even though this circuit was originally built and tested in the early 1970s timeframe, it still holds its own with modern equivalents.

Other oscillator circuits, many of them relatively simple, also offer good phase noise performance. For example, the simple Hartley circuit of Fig 4.4, has been measured several times. Versions operating at 5 MHz often indicate phase noise of –150 dBc/Hz at 10 kHz spacing. Rohde reports that computer simulations suggest this Hartley topology will have degraded performance closer to the carrier.[10]

The Hartley oscillator results were measured indirectly by measuring a crystal oscillator with a receiver using the Hartley. A typical circuit used for the testing is shown in **Fig 4.20**. This circuit can be used with a crystal filter 10 kHz away from the oscillator, or with a crystal notch filter at the oscillator frequency. Assuming the crystal oscillator to be *perfect*, all phase noise observed is attributed to the receiver LO. Even without the assumption, observed results will bound the LO performance. The crystal filter is required because of the limited dynamic range of the typical receiver. The loaded Q of a crystal, the "tank" in a crystal oscillator, can be a thousand times higher than that of a typical LC tank. The resulting phase noise is often quite low, in line with Leeson's equation.

Fig 4.21 shows an oscillator at the other extreme. This 15-MHz circuit is rich in phase noise. It is well worth building and applying to a general coverage receiver to observe first hand just what a noisy oscillator will sound like in a receiver.

4.5 CRYSTAL OSCILLATORS AND VXOS

One of the most common oscillator forms is that using a quartz crystal as the resonator. They may be ordered from a number of sources for modest cost with only a short manufacturing delay. A crystal cross-section, symbol, and an equivalent circuit are shown in **Fig 4.22**. Crystals were also discussed in the filter chapter.

A typical crystal oscillator circuit is the Colpitts shown in **Fig 4.23**. It is the series LC of the crystal model, Fig 4.22, which now serves as the "inductor" in this circuit. Owing to the series motional C, this circuit is actually a Clapp oscillator variant. With the components shown, the circuit will function with fundamental mode crystals from about 2 to 20 MHz or more. Transistor type is not critical with the ubiquitous 2N3904 being a good choice. If the crystal is specified for a "load capacitance" of 32 pF, the oscillator can be adjusted to the exact frequency with C1. This will occur when the total loop capacitance is 32 pF, which is approximately the series equivalent of the two 470-pF capacitors and C1. In many applications C1 can merely be eliminated.

Output can be extracted with an emitter follower driven by Q1's emitter. The signal on the base of Q1 is often about the same magnitude, but is spectrally cleaner. It is also possible to insert a small resistor (100 Ω or so) in the Q1 collector and to use the developed signal voltage as an output. While well isolated from the resonator, the collector signal is usually very rich in harmonics.

Fig 4.24 shows another scheme for extracting an output signal. Here, C1 becomes a selected, fixed capacitor in series with the crystal. It is no longer convenient to adjust the frequency with C1, for the capacitance will vary both F and output voltage. However, an output obtained in this manner can be extremely clean with all harmonics being over 50 dB below the desired output. Phase noise is also low with this topology.

A popular and especially simple crystal oscillator is the Pierce circuit shown in **Fig 4.25**. If the circuit is redrawn with the ground shifted to either the base or the collector, we see that this is yet another version of the Colpitts. This circuit functions well with a wide variety of crystals from 2 to 20 MHz or even higher. The circuit generally operates at the crystal fundamental. Output is easily obtained with a follower from either the collector or base. If C_1 is lifted from ground, a direct output of a few milliwatts is available.

Another Colpitts variation is presented in **Fig 4.26**. This oscillator is capable of 10 to 25-milliwatts output and can function at either fundamental or overtone frequencies (explained below). The circuit uses the relatively high base-emitter capacitance of the transistor as part of the capacitive feedback needed for oscillation, again as a Colpitts variation. External C3 vanishes except for the 1.8 and 3.5-MHz bands where values of 330 and 200 pF can be used, respectively. C2 varies from 100 pF at 3.5 and 7 MHz to 22 pF at 28 MHz and 10 pF at 50 MHz. L1 uses a toroid with a reactance of about 250 Ω. The output link is 10 to 20% the number of turns on L1. This is a very robust oscillator that takes little experimentation to get going.

A crystal overtone is a different operating mode for an AT-cut quartz crystal. Any crystal will display a fundamental resonance as well as overtone responses. Sometimes the crystals are manufactured in a way that will substantially enhance one mode over another. A general model for a quartz crystal including overtones is shown in **Fig 4.27**. The model presented so far included only the fundamental mode, related to N=1 in the figure. But

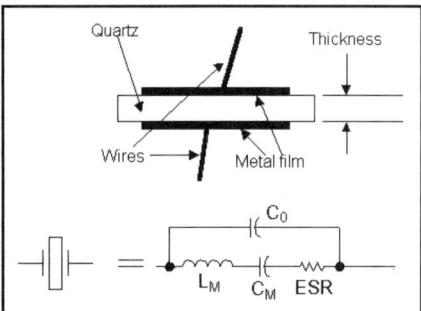

Fig 4.22—Cross-section, symbol and model for a quartz crystal.

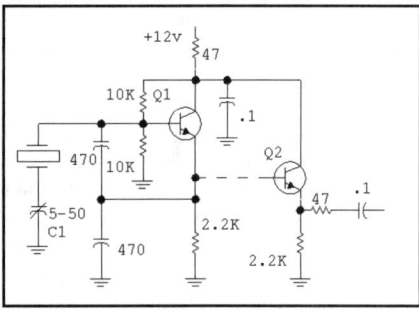

Fig 4.23—Typical Colpitts crystal oscillator. Power output is low. Extra amplifiers are usually used to increase power to the level needed to drive a ring mixer or function in simple transmitters.

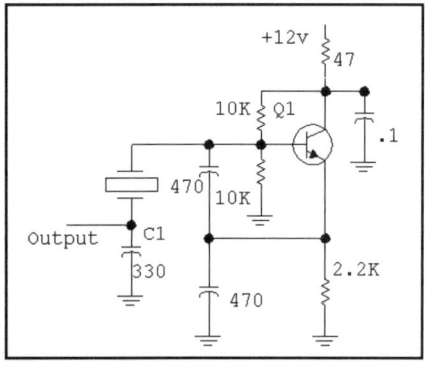

Fig 4.24—Method for extracting low noise, low distortion output from a crystal oscillator.

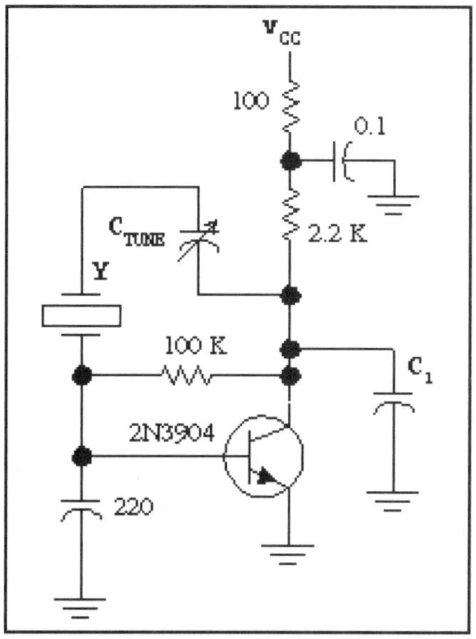

Fig 4.25—Pierce type crystal oscillator. C_1 can be as little as 10 to 20 pF. Vcc can be from about 3 up to 15 V. C_{TUNE}, often omitted, is a trimmer with a maximum of 50 or 100 pF.

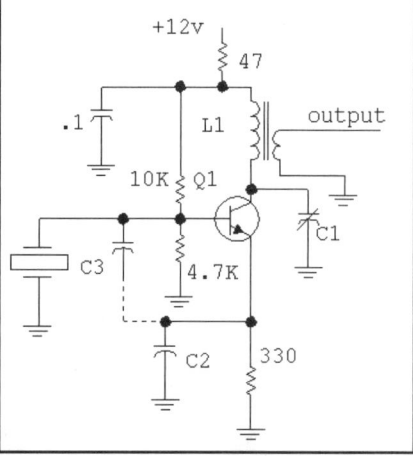

Fig 4.26—General purpose power oscillator for use from 2 to 70 MHz. Q1 is a 2N3904 or similar medium F_T device. See text for component value discussion.

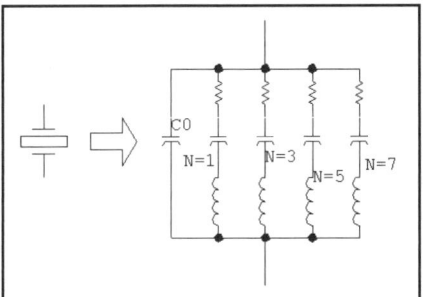

Fig 4.27—More detailed model for a quartz crystal. All motional inductance values are identical, with motional capacitance scaling with frequency. See text.

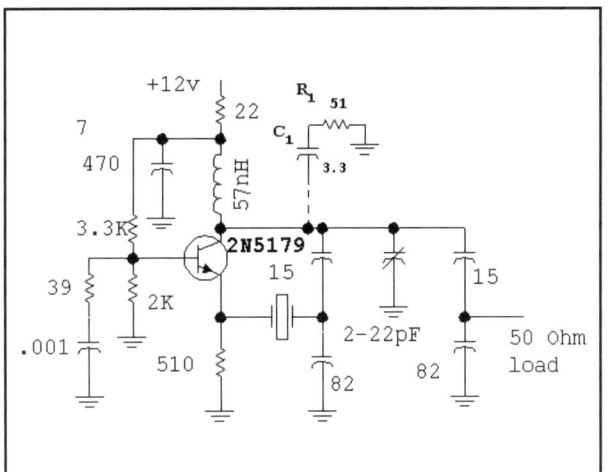

Fig 4.28—Butler oscillator for 100 MHz. L=57 nH. This is formed with a 1.7 inch piece of #22 enameled wire wound in the threads of a 6-32 machine screw.(3.3 mm dia, 12.6 turns/cm) The wire ends are stripped and 3 turns are wound on the screw, which is then removed. C_1 and R_1 form a network to suppress UHF oscillations at 500 to 1000 MHz. The suppression circuit generates a UHF load that is largely absent at the operating frequency.

other *odd* harmonic modes are also possible. (Even order harmonics are not consistent with the mechanical boundary conditions needed so support oscillation.)

An oscillator operating at an overtone must include additional frequency dependant circuits that will select the desired overtone. Simple fundamental mode circuits, such as those presented, will emphasize the lower frequencies where starting gain is higher. The circuit of Fig 4.26 included a tuned circuit peaked at the operating frequency.

Fig 4.28 shows a popular and effective overtone circuit, the Butler oscillator. This circuit is essentially an LC Colpitts oscillator with a quartz crystal inserted in the feedback path. The LC tank should have a loaded Q from 10 to 20. A Q that is too low could allow oscillation at the wrong overtone, while a Q too high will make tuning difficult. An excellent method to align this circuit replaces the crystal with a resistor equaling the equivalent series resistance (ESR) of the crystal. If ESR is unknown, use a 33-Ω resistor in place of third overtone crystals and a 56-Ω for 5th overtone crystals. The oscillator is adjusted for the proper operating frequency with the resistor in place. The resistor is then replaced with the crystal with no additional adjustment needed. Most overtone circuits, including the Butler, can be used for fundamental mode operation by proper adjustment of the tuned circuit.

This circuit is sometimes "neutralized" by placing an inductance in parallel with the crystal. The value resonates with C0, the crystal parallel capacitance. If C0=3 pF for the 100-MHz crystal of Fig 4.28, the inductance would be 0.84 mH. Be sure that the inductor used has a self-resonance well above 100 MHz. We have generally found that this inductor can be eliminated from the circuit.

The Butler oscillator shown in Fig 4.28

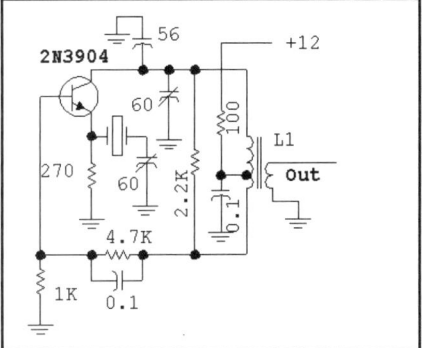

Fig 4.29—An oscillator designed by inserting a crystal in series with the feedback path of a Hartley LC oscillator. The ground point is then shifted to the tap on the coil. The version shown is set up for 10 MHz operation, but tuning can be shifted to other frequencies. Eliminating the tuning capacitors and replacing the transformer with one using a ferrite core also works well.
Y1=10 MHz fundamental;
L1=30t T 50-6, tapped at 7 turns and 6 turns for the link.

will provide an output of 10 mW to 50 Ω. The load is part of the design; if the load is ill defined, use a 50-Ω pad at the oscillator output. Never try to adjust the oscillator without the load in place. The Butler oscillator generally exhibits excellent phase noise. Although a trimmer capacitor in series with the crystal will allow some frequency adjustment, it is much less effective with overtone crystals than with fundamental mode parts. Never try to adjust oscillator frequency with the crystal by changing collector tuning, for that could cause the circuit not to start when power is first applied.

The Butler used a Colpitts as the basis. **Fig 4.29** presents a useful variation of this circuit that begins as a Hartley with the crystal in the feedback path from the coil tap to the emitter. The ground point is then

A Butler oscillator. The circuit of Fig 4.28 is breadboarded here without the crystal. Instead, a 51-Ω resistor is placed in the crystal position. This is a useful way to test the oscillator.

shifted, placing ground at the coil tap. This puts one end of the crystal at ground, or connected to a trimmer. This circuit functions well as either an overtone or fundamental mode oscillator with low phase noise and moderate output. The circuit functions well (fundamental mode only) if the tuned output transformer is replaced with a ferrite transformer.[11]

The VXO

The crystal oscillators shown so far have often included a trimmer capacitor for fine frequency adjustment. If the tuning range can be made larger, the circuit can be used as a high stability substitute for a variable frequency LC oscillator, taking on the descriptor *VXO*. A typical VXO circuit is shown in **Fig 4.30**.

The circuit of Fig 4.30 was built and tested with numerous crystals from our junk box. Crystals at and above 14 MHz could typically be tuned by 0.1% of the marked frequency when L=0, with the bottom frequency being close to the marked crystal frequency. For example, a crystal

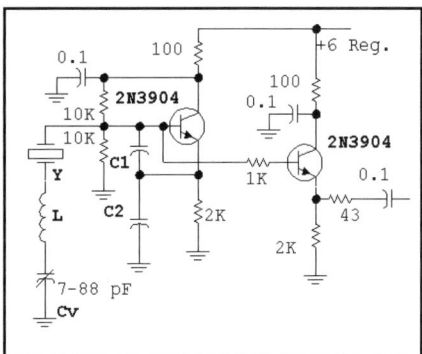

Fig 4.30—Basic VXO circuit. C2 is typically twice C1, which is 100 pF at 10 MHz and higher, doubling for 7 MHz. L is determined by experiment. C_V can be about any variable capacitor, but should be one with small minimum capacitance. L may = 0, 2.7 µH or 5.4 µH.

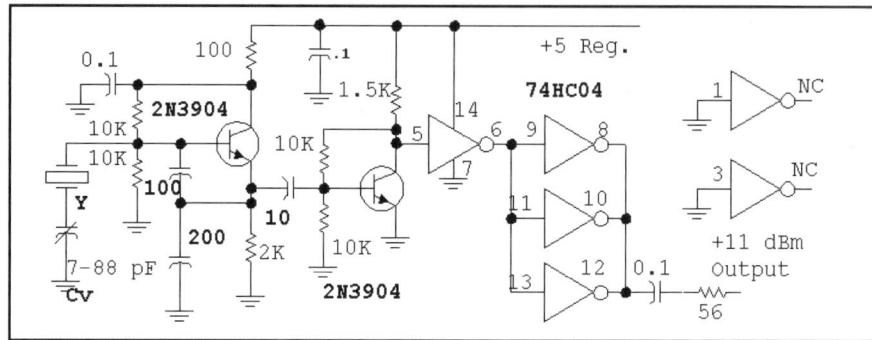

Fig 4.31—Adding HCMOS inverters can substantially flatten the output of a VXO. Output filtering will be required.

marked 14060 kHz tuned from 14059.0 to 14070.4 kHz (11.4-kHz shift) with C1 and C2 of 100 and 200 pF. Adding inductance moved the bottom of the range downward with a much smaller change in the upper edge. L=5.4 µH produced 14053.0 to 14068.4 kHz (15.4-kHz shift.) In another example, an 18-MHz crystal shifted 13.3 kHz with no inductance, but shifted over 25 kHz when 3.7 µH was added. 5.4 µH in that circuit produced unstable operation, emphasizing the need for experimentation.

In some cases a variety of crystals were available from different manufacturers, all at approximately the same frequency. Results varied only slightly. Larger values for C1 and C2 were required for oscillation at 7 MHz and lower.

With even greater added inductance, the lower frequency drops further and the range expands. However, stability also degrades. Eventually, if oscillation is maintained, it may not be crystal controlled. Experimentation and careful analysis can both pay large dividends. With zero or only modest added inductance, the frequency stability of a VXO is nearly as good as the original oscillator. This makes the circuit especially attractive for narrow tuning range equipment such as VHF/UHF CW and SSB rigs.

Extreme tuning nonlinearity is common with most VXO circuits. Most of the frequency shift tends to be compressed at the high frequency (low C) end of the range. This effect is so extreme that it is very difficult to implement a predictable shift for use in, for example, a direct conversion transceiver.

The typical VXO suffers from considerable variation (unflatness) in output power with tuning. The VXO of Fig 4.30 can vary by nearly 10 dB. This is relieved with the circuit shown in **Fig 4.31** where a CMOS inverter is added as an output buffer. The circuit shown provides an output of +11 to +12 dBm for a total current of around 35 mA. The output is very rich in harmonics, so low pass filtering will often be required. Different numbers of parallel inverters may be used to control output power. The square waveform at the inverter output can also be useful for frequency multiplication.

Two VXO circuits are shown in **Fig 4.32** that are of special interest to the experimenter. One adds a second crystal, producing almost double the tuning range of the same circuit with one. The crystals should be close in frequency, but need not be an exact match. We encountered this circuit in the worldwide web where it is known as the "super VXO."[12] The two elements in parallel behave like one crystal, but with twice the motional and fixed capacitances and half the motional inductance. This is the direction needed for greater "tunability."

The second VXO of Fig 4.32 uses a quar-

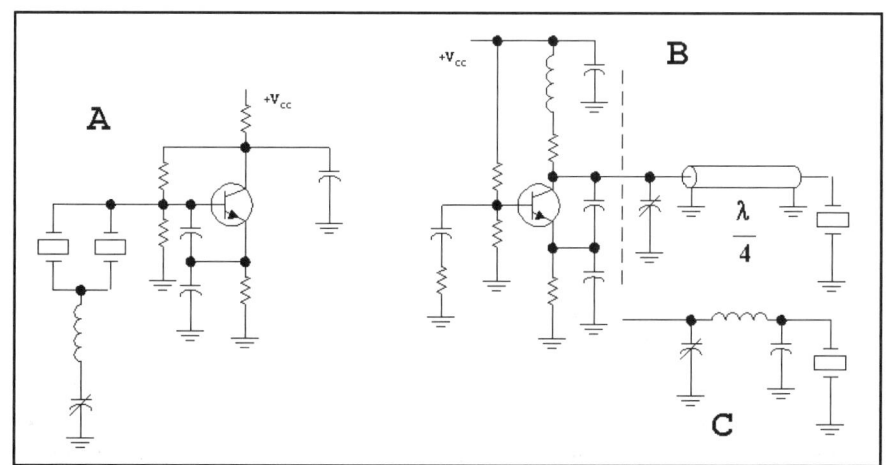

Fig 4.32—Two VXO circuits of interest to the experimenter. That at A is known in Japan as the Super VXO, and is the creation of JA0AS and JH1FCZ. The circuit at B uses a quarter wavelength of transmission line while that at C is the lumped element equivalent.

ter wavelength of transmission line to convert a crystal series resonance to appear at the collector as a parallel resonance. The alternative version of this circuit uses a lumped element equivalent for the transmission line. The real virtue of this scheme is that the troublesome crystal parallel capacitance is absorbed into the "line." The performance of this circuit can be truly outstanding, although the circuit can be difficult to adjust. In one experiment we were able to tune a 7-MHz crystal by a range of over 100 kHz. The circuit has problems that present challenge to the designer/ builder. The Q of the equivalent parallel resonator varies dramatically over the tuning range, making it difficult to maintain clean limiting in the transistor or to obtain an output with a stable amplitude.[13]

The Hartley theme circuit presented earlier (Fig 4.29) is especially well suited to VXO applications, especially when built with ferrite transformers. This topology is used in a 28-MHz VXO transmitter presented in Chapter 12.

4.6 VOLTAGE CONTROLLED OSCILLATORS

The oscillators presented so far have used mechanical variable capacitors for tuning. The other traditional tuning scheme is inductive, the *permeability-tuned oscillators* of Collins fame. Both depend on well-engineered mechanical designs, a desirable, but disappearing characteristic. The voltage-controlled oscillator is replacing the "simple" mechanically tuned oscillator of the past. That oscillator is then used as part of a frequency synthesizer. In a few cases, the VCO is used "open loop," without synthesis.

The dominant component used for voltage control of oscillators of concern in this text is the varactor diode. Any diode will exhibit a capacitance. When the diode is reverse biased, the capacitance will vary inversely with the applied voltage. The reverse biased diode is inserted in a VCO circuit to become the tuning element in that oscillator.

Figure 4.33 shows a 7-MHz voltage tuned oscillator. This circuit was designed to serve as the main control for a direct conversion transceiver. (Described later as the *Western Mountaineer*.) Q1 functions as a high C Colpitts oscillator. Inductor L1 is resonated with the 470-pF Colpitts capacitors and C1, a fixed capacitor of over 600 pF. The value was hand picked for resonance, with only a small, 10-pF trimmer for final adjustment.

Earlier measurements with a small environmental chamber had established the tuning diode temperature coefficient at 5 V as +442 ppm/°C. This is generally quite severe, over ten times worse than NP0 oscillator components.

This oscillator was initially built without the diode, stable operation was confirmed, the diode was added, and environmental chamber measurements were done. The tuning diode D1, a Motorola MV209, was temperature compensated with a second diode, D2. The sense diode is placed in the same thermal environment as the tuning diode. The complete oscillator and its buffer are shielded from the rest of the circuitry, for the oscillator runs at the same frequency as the transmitter PA in this rig. The diode standing current is adjusted by picking R1, generating the needed voltage change with temperature. R1=10 kΩ worked well in our circuit, but should be picked with the environmental chamber for individual applications. This compensation scheme was suggested to us by WA7TZY.

The oscillator supply is regulated with U1, a 78L05 three-terminal regulator. The original Zener regulation was unstable with temperature, adding extra complication. The regulated voltage also provides

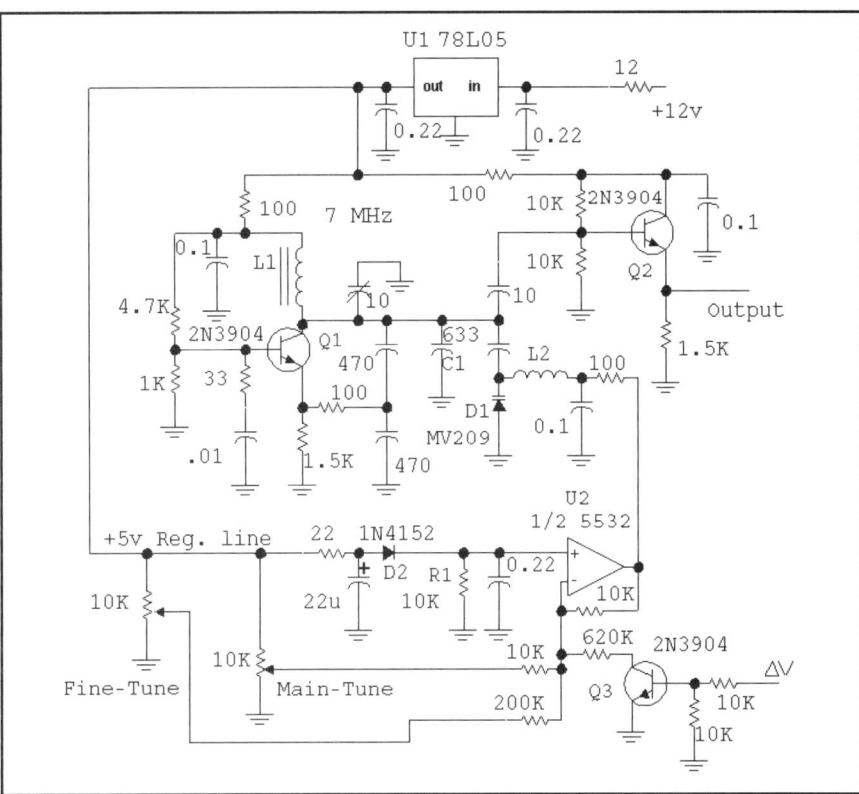

Fig 4.33—A varactor tuned 7-MHz oscillator with a restricted tuning range of about 60 kHz. Temperature compensation is provided with D2, a sense diode. L1=12 turns #26 on a T30-6 toroid. L2 is a 15-µH RF choke.

Inside view of the 14-MHz VCO.

sense-diode biasing and serves as the supply for the tuning controls.

The op-amp, U2, combines two tuning controls and an offset voltage while providing a regulated tuning voltage. The circuit is configured to maintain at least 4.3 V on the tuning diode. In many varactor-tuned oscillators, RF voltage will be rectified by the diode, allowing conduction during part of the cycle, degrading stability, phase noise, and tuning linearity. This occurs with low tuning voltage and is usually detected as a decrease in VCO output.

The final temperature coefficient realized with this oscillator was about 2 ppm/°-C. The transceiver has appeared to be "rock solid" during field operation, including winter snowshoeing treks.

A 14-MHz VCO is shown in **Fig 4.34**.

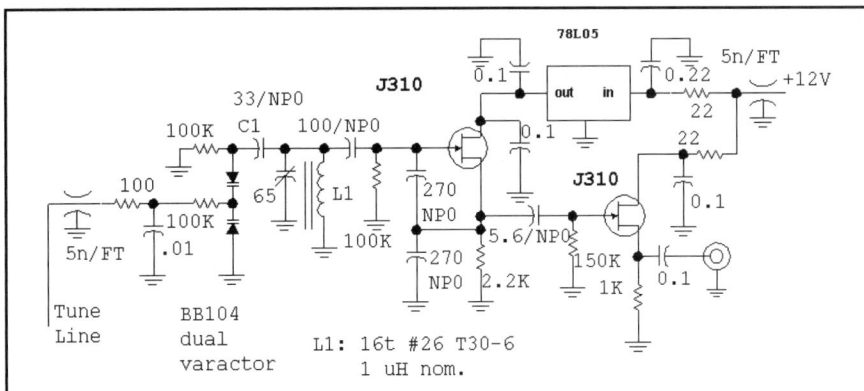

Fig 4.34—14-MHz VCO for use in synthesizer experiments. L1=16 turns on a T30-6 toroid coated with Q-dope to reduce micro-phonic effects.

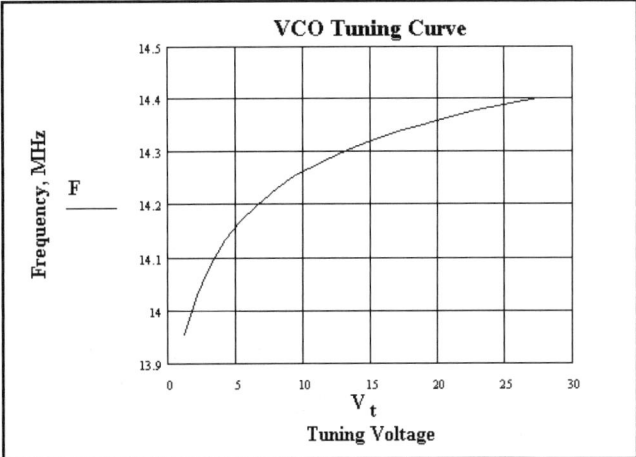

Fig 4.35—Frequency vs control voltage for 14-MHz VCO. An average sensitivity for this circuit over the 2 to 10 V range is 30 kHz/V.

A J310 FET was used with source resistor biasing. The varactor diode was a surplus BB104, similar to the Motorola MV104. The Toshiba 1SV103, used in some imported equipment, might be a suitable substitute. Two individual varactor diodes can also be used. This oscillator can be set up for a wider frequency range by picking C1. Over 1 MHz of tuning was available with C1=100 pF. C1 was dropped to 33 pF for a reduced range. The tuning characteristics for this oscillator are shown in **Fig 4.35**. The circuit is built in a Hammond 1590A enclosure with coaxial output and feedthrough capacitors for power and tuning. A 65-pF plastic trimmer provides coarse tuning.

The use of back-to-back varactor diodes is common in VCOs, for it reduces the effects of rectification of the oscillator signal. It is also common to see many diodes operated in parallel. This topology shows lower noise than a smaller number of higher-capacitance diodes.

The phase noise of this oscillator was measured using a 14-MHz single conversion superhet receiver with extensive crystal filtering. The VCO was battery powered with the battery also biasing the varactor, which was filtered further with a 100-μF capacitor. The signal was attenuated to −31 dBm and applied to the receiver input through a step attenuator. Audio output was monitored with an HP3400A true-RMS voltmeter with receiver AGC turned off. The audio noise output in the meter was noted 5 kHz away from the carrier. The receiver was then tuned to the carrier and the step attenuator was increased until the response was the same as observed with the noise. Additional attenuation of 110 dB was required to reach this response. The noise bandwidth was 500 Hz, producing a measured CNR of −137 dBc/Hz. It is not clear if this noise comes from the VCO or from the receiver VFO, but this value is a useful "worst case" limit. No phase noise could be detected at 10 kHz offset. No outboard crystal filter was used for this measurement, placing us at the limit of what we can measure with this setup.

Voltage tuning with diodes tends to compromise noise and stability performance. However, reasonable results are available if the tuning range is kept small. An attractive scheme uses varactor tuning over a small range with PIN diode switching in larger frequency steps. PIN diode capacitor switching is illustrated in a transceiver (*The Lichen*) offered in Chapter 6.

The reader working on a synthesizer for a high performance (wide dynamic range) receiver should review the extensive literature on voltage controlled oscillators. Numerous methods are available to design these circuits. It is often the varactor diodes that ultimately limit noise performance. Noise supplied to the diode on tuning lines can also compromise performance.[14]

4.7 FREQUENCY SYNTHESIS

Virtually all of the local oscillator systems used in modern communications equipment now use frequency synthesis in one form or another. Two circuit types dominate synthesis: the phase-locked loop (PLL) and direct-digital synthesis (DDS). The two schemes are often used together. The Huff 'n Puff scheme described earlier is a frequency lock method and is not usually the basis for synthesis. The reason is that frequency lock allows frequency errors, which are absent in PLL or DDS synthesizers.

A PLL for frequency synthesis in its simplest form is shown in **Fig 4.36**. The first component is a voltage-controlled oscillator characterized by a tuning sensitivity in Hz/V. This sensitivity usually varies over the tuning range. The next component is the phase, or phase difference detector, a circuit that provides a dc output proportional to the phase difference between two RF inputs. The third element is a "loop filter." In its simplest form, this is (for a second order loop) a single pole RC filter with a couple of resistors and one capacitor. More often it's an operational amplifier offering low frequency gain as well as filtering properties. The low pass filtering is needed to remove signal components coming from the phase detector. The dc from the detector and loop filter must be of the proper magnitude to drive the VCO tuning line. Because this is a negative feedback system (a type of *servo loop*,) the phase of the feedback signal as it moves through the loop to eventually reach the VCO must be tailored for loop response.

A PLL that is "locked" forces the VCO to be at exactly the same frequency as the reference. If the reference is tuned, the VCO will follow, maintaining not only the same frequency but a phase relationship that depends on the characteristics of the detector. If the loop dynamics are "wrong," the VCO may not respond smoothly to a change in the reference frequency. In the extreme, the loop can oscillate.

We begin our discussion of the PLL with an experiment to evaluate a Mini-Circuits SBL-1 mixer operating as a phase detector. Most of us have no easy way to accurately measure phase, but we can do things to infer it. In this vein, we first characterize a piece of coaxial cable, a 25-foot length available in a home lab. A "half wave" balun was fabricated from the cable, shown in **Fig 4.37A**. The two balanced output points were attached to 100-Ω resistors with the junction attached to an RF spectrum analyzer. The signal generator was tuned until a null was found at 12.88 MHz. This occurs when the cable is a half wavelength long, producing 180 degrees of phase shift between the two ends. A half wavelength in free space at this frequency is 38.2 feet, so the velocity factor of our coax is 0.65, which is about what we would expect. The phase delay in the coaxial cable will be directly proportional to cable length and to frequency. We know the length and frequency that yield a phase shift of 180 degrees, so we can calculate the phase for any arbitrary frequency.

The characterized coaxial cable is now used in the test set of Fig 4.37B. The signal generator output is divided in a power splitter consisting of three 51-Ω resistors. This preserves a 50-Ω environment while equally splitting the input power. One signal is applied directly to the SBL-1 LO port. The other is attenuated by 10 dB, phase shifted with the cable, and applied to the mixer RF port. The output was low pass filtered with a simple RC filter and measured with a digital voltmeter. The signal generator amplitude was adjusted to produce the specified +7 dBm LO drive level. This overall circuit is familiar as a delay-line discriminator.

A quick tuning of the signal generator showed that the output was zero at 6.4 MHz where coax phase shift is 90 degrees. Data was taken over the 3 to 10-MHz spectrum to generate a plot (**Fig 4.38**) showing output voltage as a function of phase. This is close to a straight line over a wide phase range, with the departure at low angles resulting from a signal generator output decrease near 3 MHz. (We used a modest drive at the mixer RF port; the mixer is approximately linear to RF drive at this level.) Examination of the data in Fig

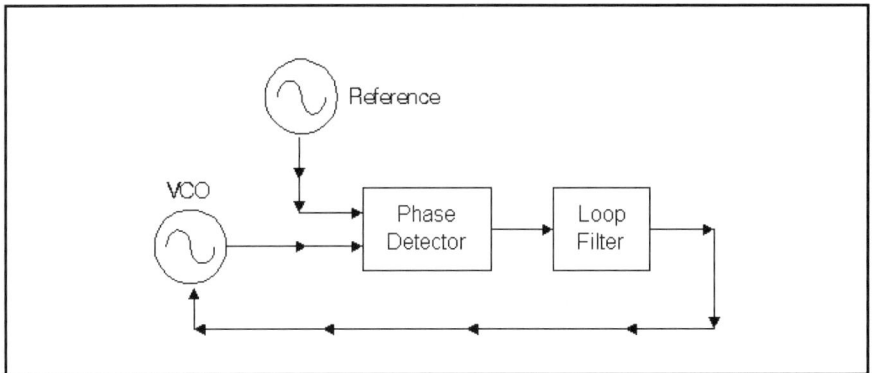

Fig 4.36—Basic Phase Locked Loop.

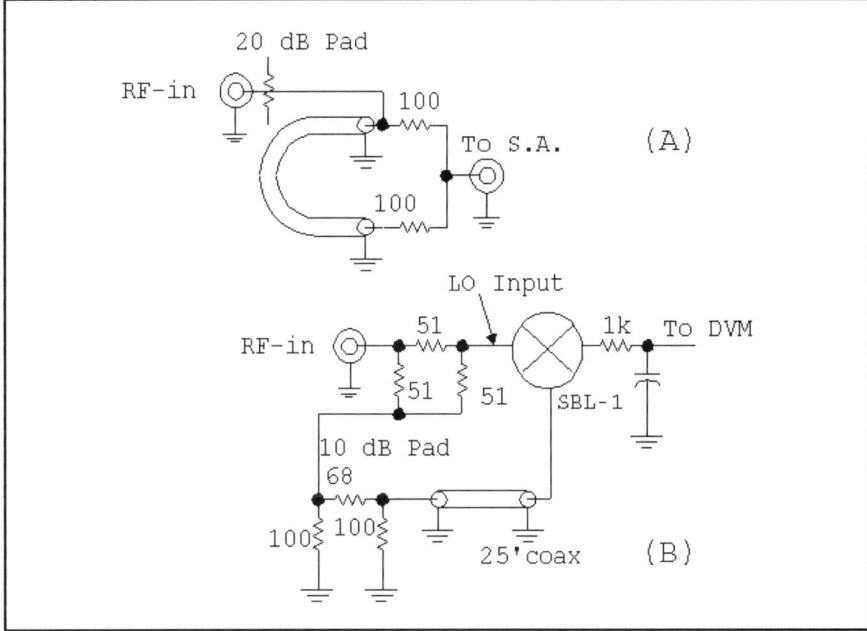

Fig 4.37—Part A characterizes the phase shift in a section of coax cable that is then used in part B to evaluate a SBL-1 as a phase detector.

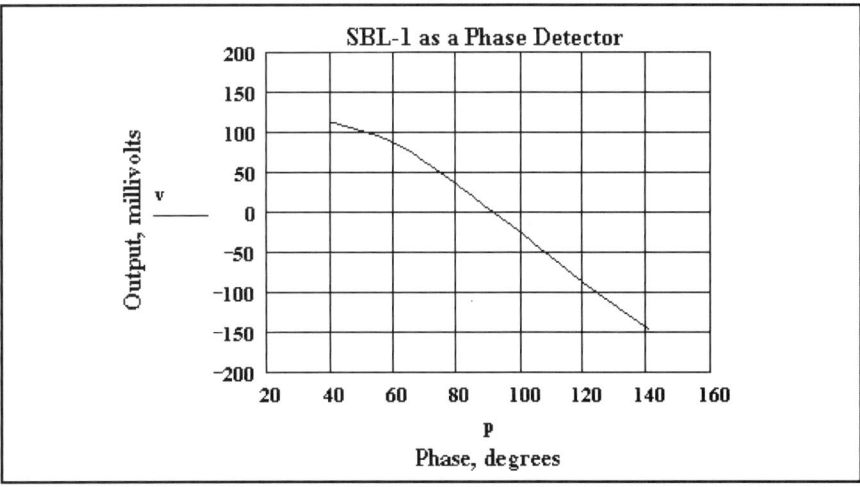

Fig 4.38—Dc output vs phase for a SBL-1 operated as a phase detector.

4.38 shows that the slope (phase gain) is –2.96 millivolt per degree, or –0.17 V/radian. Repeating these experiments with other cable lengths show that this circuit responds to phase rather than frequency.

Having characterized the phase detector, we can now build a phase locked loop. We will use the 14-MHz VCO described earlier (Fig 4.34), an oscillator with an average tuning sensitivity of 30 kHz/V with the available voltages when we use a 12-V bench supply. A general-purpose signal generator is the "reference" in the loop shown in **Fig 4.39**. The SBL-1 details are shown to emphasize the dc isolation properties of the ring and transformer windings. An operational amplifier increases the relatively low output of the detector to drive the VCO tune port. The LM358 used was available for the experiment; a better choice would be an OP-27 or similar low noise part.

The loop was originally tested while running the phase detector at the low RF port level used for measurements. Although phase lock was possible, performance was poor. Increasing the levels to +7 dBm at both mixer ports produced more robust behavior. The circuit is initially turned on without seeing any indication of "lock." An oscilloscope was used to monitor the op-amp output, which came up to about 4 V, the level set by the 3.9-kΩ/2.2-kΩ voltage divider. The signal generator was then tuned. Lock was achieved when it passed through the VCO resting frequency. The VCO will then track the reference over the full op-amp output range.

Intuition suggests that achieving lock would be difficult, that both signals would have to be at the same frequency before phase lock can ever be realized. But lock does occur, even with a slight frequency difference. Consider two input signals, a reference and a VCO, separated by 1 kHz and applied to the *phase detector*, which is the same topology as a mixer. The mixer will produce 1-kHz currents. This low frequency component will generate sidebands about both the reference and the VCO. These components appear in the mixer output. One of the VCO sidebands is now directly on top of the reference, producing a dc component that will pass through the loop filter where it can be amplified and move the VCO toward a locked condition. A similar sideband is on top of the VCO frequency.

Analysis like this offers some explanation, albeit sketchy, of a related phenomenon called *injection locking*. This occurs when an external signal is applied to an operating oscillator. If the signal is strong enough, it can cause the oscillator to move frequency until it becomes locked to the injected frequency. The same modulation sidebands are created within the oscillator and operate in much the same way.

Although these modulation processes are powerful, they are restricted. A simple PLL will have a well-defined *pull-in range* where capture is possible.[15]

This experimental loop was designed for a closed loop bandwidth (open loop unity gain frequency) of 2 kHz with a damping factor of 5, parameters determined by the choice of the resistor and capacitor values of the loop filter. Although we pick *loop filter* components, the parameters describe the overall PLL and not just the op-amp and related parts.

This seemingly simple circuit is useful, not only as an illustration of the concept, but as a way to obtain two signals that have a well-defined phase relationship to each other. With a diode ring phase detector, the locked oscillator will differ from the reference by 90 degrees. A sidebar shows a practical PLL with a diode ring phase detector.

Other mixers, including the popular Gilbert cell, work well as a phase detector. The most popular phase detectors use digital circuits. **Fig 4.40** shows a common circuit, a so-called phase-frequency detector. This digital circuit is fed with digital voltages to the clock inputs of two *data flip-flops*. The D-FF is a topology that transfers the level on the Data input to the Q output when a clock transition occurs. The data, in this circuit, is just a logic 1, for the D input is tied to the positive power supply. A NAND gate resets both D-FFs when both have a high Q output. If the two inputs are signals at the same frequency and are in phase, the output will be a very narrow spike, defined by the logic speed. If, however, there is a phase difference, the Q related to the first FF triggered will stay positive for a short period, producing an output with a net dc component.

This circuit will also compare frequencies. If one frequency is higher than the other, the dc average of the two outputs will, after filtering, cause the VCO to sweep toward equal frequencies. Even if this detector is not the primary phase detector in a PLL, it can still serve to compare two frequencies, a handy feature in some applications.

The digital phase frequency detector uses digital logic. However, the simple loops discussed so far have dealt with analog signals. An analog signal is easily converted to digital form with the circuit shown in **Fig 4.41**. The 10-kΩ and 4.3-kΩ resistors form a voltage divider with a voltage gain of about 1/3. But, to be active, the

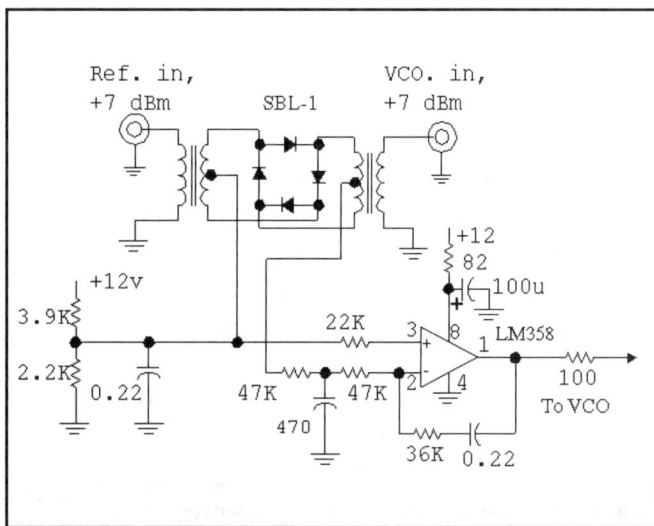

Fig 4.39—Phase locked loop using the phase detector.

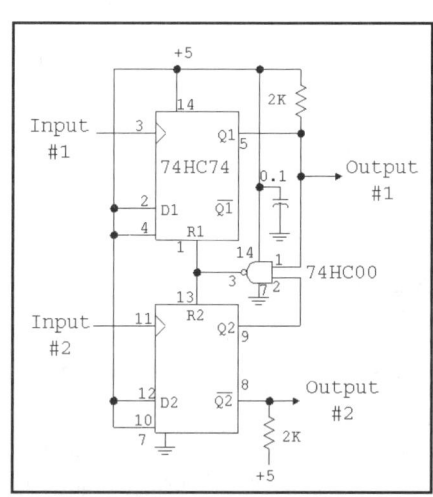

Fig 4.40—Phase frequency detector using digital integrated circuits.

A Practical Frequency Multiplying PLL LO System without a Loop Filter

The phase locked loops we have described are second order loops, ones with a capacitor in the loop filter that alters loop response. A simpler form for loops is possible, a first order circuit. This occurs when we take the dc output from a phase detector, perhaps with some amplification, and apply it directly to a VCO. This is exactly the sort of negative feedback used when we control the gain of a simple op-amp by connecting the output to the input through a resistor. The circuit is stable so long as the gain before feedback is inverting. The second order loop, with its additional capacitor, introduces the possibility of a delay between an output error and the signal reaching the amplifier input to correct that error.

An analogy may be appropriate: A rider proceeding down a hill on a bicycle controls direction with a first order feedback loop. The VCO represents the bicycle handlebars; a direction error is corrected with immediate feedback applied to the handgrips. The second order loop places springs between the rider's controlling hands and the handlebars, effecting a delay in the feedback. The system with springs might be smoother on a gentle hill, but clearly needs much more effort on the part of the designer. The consequences of failure are dramatic.

We had built a VHF transceiver (described later in the book) tuning from 52 to 53 MHz that receives most modes. While normally used with microwave transverters, we wanted to also use this for casual HF reception. We needed a stable LO that would operate in the 48 to 70-MHz area that could drive a mixer to convert HF signals to VHF. The needed LO could take on frequencies that were multiples of 1 MHz. This was done with a first order phase locked loop, shown in **Fig 4A**. The basis for the LO is a pair of off-the-shelf modules from Mini-Circuits: a POS-100 voltage controlled oscillator tuning from 50 to 100 MHz and a SBL-1 serving as a phase detector.

The VCO output is split with most of the energy routed to a coaxial output for mixer use. A sample is applied to a common gate amplifier, Q1, and then to the SBL-1 phase detector with a level of about +7 dBm. The RF input to the phase detector, the "reference" for the loop, is a harmonic of a lower frequency crystal controlled oscillator. The harmonic signal should be between –40 and 0 dBm at the desired frequency.

A dual op-amp provides the rest of the control for the system. U1A is a unity gain voltage follower driven by a 10-turn 2-kΩ pot. The output signal, from 0.3 to 6 V, is applied to the diode ring in a way that this level also reaches the VCO. Note that this is not easily realized with all ring mixers. Phase detection occurs in the diode ring, creating a dc signal that is added to the applied dc bias. This is then differentially amplified with a voltage gain of +11 in U1B and routed to the VCO.

The system is generally very easy to use. The 10-turn pot is merely tuned until a lock is obtained, producing stable output signals in the receiver. A chart of the various frequencies vs the setting of the 10-turn control is kept, allowing an easy return. The capture range (how close you must tune the 10 turn control to achieve lock) is about 100 kHz if the corresponding input is at –10 dBm, but drops to 10 kHz for a –30 dBm input. The reference spurious responses in the output at plus and minus 1 MHz were at –60 dBc when the loop was locked.

This circuit should be built over ground plane with relatively short leads in the RF areas. The U310 common gate amplifier is critical. While the gain is low, the reverse isolation is very good, needed to prevent 1-MHz energy from reaching the VCO where it can create sidebands. The amplifier is built by drilling a hole in the ground foil for the FET and soldering it in place. This is possible with the U310, for the gate is attached to the metal can. A J310 could be substituted if caution is devoted to keeping the circuit stable. Such circuits are discussed in Chapter 6.

Capacitor C1 is a VHF bypass that filters the dc coming from the phase detector. The value is small enough that it does not impact loop performance.

The greatest virtue of this circuit is its tolerance to experimental changes. Because there are no loop filter components to pick, there is little design to be done. Yet the resulting performance can be excellent.

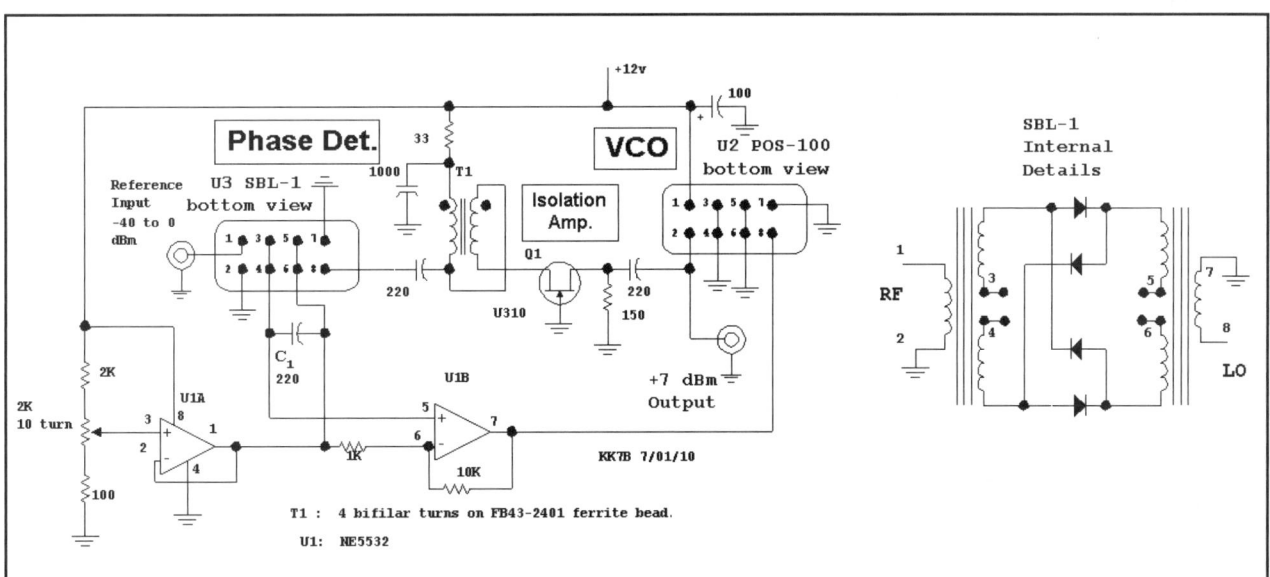

Fig 4A—A first order PLL allowing a VHF VCO to lock to harmonics of a 1-MHz input.

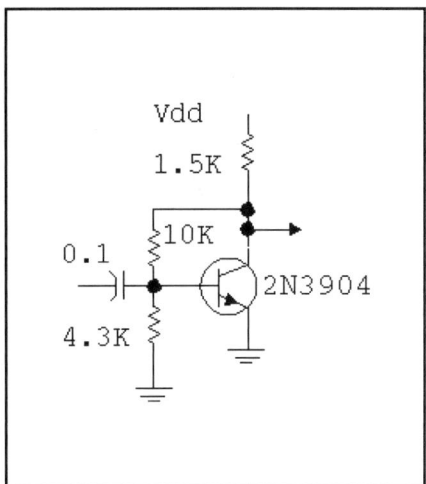

Fig 4.41—An analog signal is easily converted to digital form with the circuit shown here. The 10-kΩ and 4.3-kΩ resistors form a voltage divider with a voltage gain of about 1/3. But, to be active, the transistor base must be biased at about 0.7 V.

A One-on-one Tracking Phase-locked Loop

The PLL scheme becomes more tractable when a mixer is added to the system, shown in **Fig 4.42**. The frequencies are those used in a practical VFO, a circuit designed for a two-band output. A 14-MHz VCO is mixed with a 12.5-MHz crystal oscillator and the down-converted output is selected with a low pass filter. The result is applied to a phase-frequency detector. The reference for the detector comes from a stable, free running 1.5-MHz oscillator. The detector output is filtered in the "loop filter" with the dc output controlling the VCO.

The most obvious virtue of this system is stability; the VCO has the frequency stability of the two oscillators in the system. The 12.5-MHz oscillator is crystal controlled and quite stable. The free running 1.5-MHz VFO operates at a low frequency and is also very stable.

Good long-term stability measured over periods of seconds to minutes is but one virtue. Another is short-term stability, the cycle-to-cycle behavior that we have characterized by phase noise content. The noise of the 1.5-MHz reference oscillator is transferred to the VCO within the bandwidth of the phase locked loop. Outside the loop bandwidth, the phase noise is dominated by the intrinsic performance of the VCO.

The astute reader is certainly posing a question at this point: Why a PLL? Why not merely mix the 1.5-MHz VFO with the 12.5-MHz crystal oscillator to directly generate the desired 14-MHz signal? The question is a good one, as is the method. A direct heterodyne approach, which will be discussed in a later chapter, is ideal. However, if the output is to be spectrally clean, the filter at 14 MHz must be a good one. The up-conversion process will generate an image at 12.5–1.5 = 11 MHz. This must be well suppressed. There are other higher order mixing products that can also compromise the performance.

Another virtue is low cost. The LC filter is a relatively expensive circuit. A direct heterodyne system would be even more difficult and expensive if the frequencies were changed to, for example, a 13.5-MHz crystal oscillator and a 0.5-MHz VFO. But, the PLL for the new scheme would be virtually unchanged. Notice in Fig 4.42 that there are no bandpass filters in the system, not even simple ones. A very simple low pass filter picks the down-converted product.

transistor base must be biased at about 0.7 V. Hence, the feedback loop holds the collector close to 2 V, which is between a logic 0 and 1 for TTL and for CMOS running at 5 V. This circuit will function with RF signals of –30 dBm from a 50-Ω generator, or even less, depending on frequency.

The normal phase-frequency detector outputs come from Q1 and Q2*. (Q2*=Not Q2.) Q2* is shown as Q2 with a bar above it in the schematic shown in Fig 4.40. During phase locked operation, Q1 and Q2 are both low between clock pulses. So, Q2* will be high. When Q1 and Q2* voltages are analog added with an op-amp, the result is a signal at half of the digital supply. Even when both make transitions together, the net result is the same as the resting state so long as there is no phase difference. This balance helps to suppress spurious pulses from the detector. This detector suffers from gain that drops with zero phase difference. More refined phase-frequency detectors use logic schemes that generate a gain that is constant at all phase differences.

The phase frequency detector output is sampled data. A sample occurs once per cycle and then disappears. The averaged dc component is extracted and applied to the op-amp that follows. The primary function of the loop filter is to attenuate the high frequency part of these pulses. In contrast, the output of a diode ring phase detector is continuously present, so long as there is sine wave excitation. But when clipping occurs at both inputs, which is common, the data begins to take on sampled characteristics.

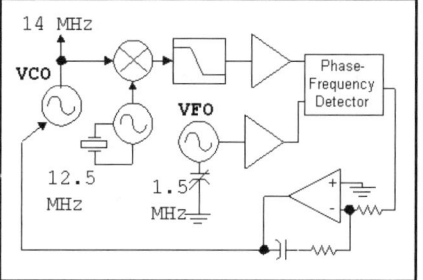

Fig 4.42—A practical one-on-one or offset tracking PLL.

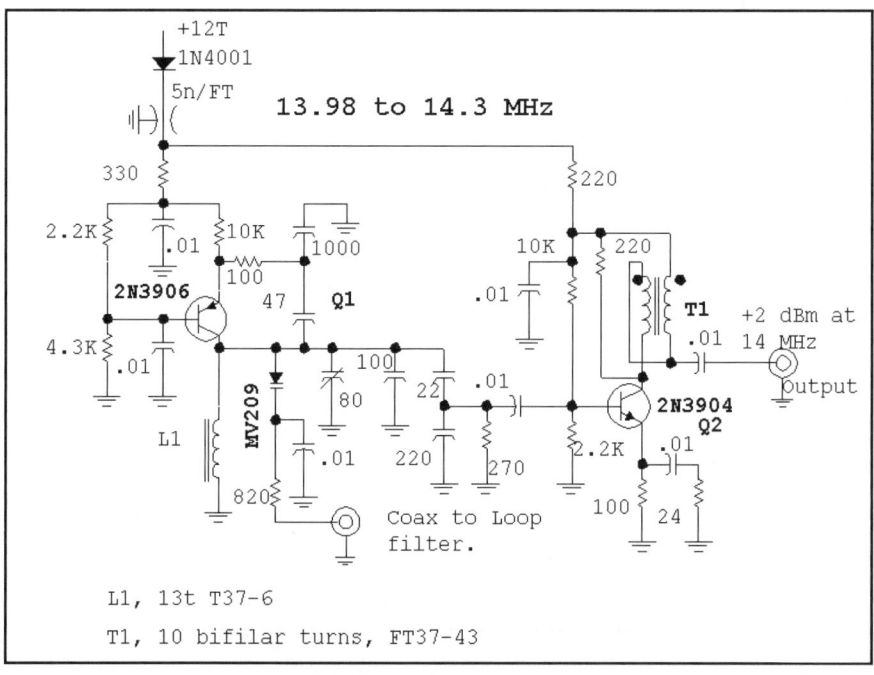

Fig 4.43—VCO for the 14-MHz tracking loop.

The PLL still has filtering properties. A detailed analysis will show that the loop behaves like a single tuned circuit at the VCO frequency with a bandwidth equaling the loop bandwidth. This *tracking filter* moves along with the output, transferring the characteristics of the reference to the VCO output. This filtering characteristic is not available to one building the more conventional heterodyne system.

Schematics are presented for a practical implementation of the system of Fig 4.42, a design we used for a 10-year period. Two output frequency bands were available: 7 to 7.1 and 14 to 14.2 MHz. The 14-MHz output was also frequency doubled to produce a 28-MHz signal. The basic circuit is a 14-MHz PLL, but the output is digitally divided to produce the 7-MHz component.

The 14-MHz VCO is shown in **Fig 4.43**. A 2N3906 PNP (Q1) oscillator is tuned with a MV209 abrupt junction varactor diode. The grounded collector facilitates diode biasing. The emitter current in the PNP guarantees an operating level that never forward biases the tuning diode. A buffer increases the output to +2 dBm. There are no large bypass capacitors within the shielded VCO, for the +12-V supply is keyed.

The VCO output drives a passive power splitter where the two applications are isolated, shown in **Fig 4.44**. One path routes 14-MHz energy to Q3 where it is amplified to a 2.5-V pk-pk level to serve as the LO for Q4, a dual gate MOSFET mixer. The 12.5-MHz signal is generated with Q5. The level reaching the mixer is adjusted to prevent overdriving the mixer. The mixer output is filtered in a 1.7-MHz low pass filter.

The other splitter output is applied to Q6, a stage providing 14-MHz output. Some energy is "stolen" at the emitter to drive Q7 and U1, a D-flip-flop operating as a divider. The resulting square wave is further buffered in Q8 and is low pass filtered to produce a clean 7-MHz signal. The low pass is a peaked (ultra-spherical) design, offering greater than normal harmonic attenuation. A band-switch selects the appropriate output. Even though the 7-MHz circuitry continues to operate when the 14-MHz band is in use, the 40-meter output is still 70 dB below the desired output. The 0-dBm output was used to drive a two stage, 1-W power amplifier. This was low pass filtered and used on the air for QRP activity, or applied to a FET power amplifier for more aggressive efforts.

The 1.5-MHz output from Fig 4.44 is applied to the phase frequency detector, shown in **Fig 4.45**. This then drives a loop filter using an LM301 op-amp. The loop was designed for an 10-kHz loop bandwidth. The reference VFO (not shown) for the phase detector was a JFET Hartley buffered with a MOSFET.

Keying and timing details, although not shown, are critical in this system. The VCO was keyed with a "+12T" voltage that started as soon as the key was pressed.

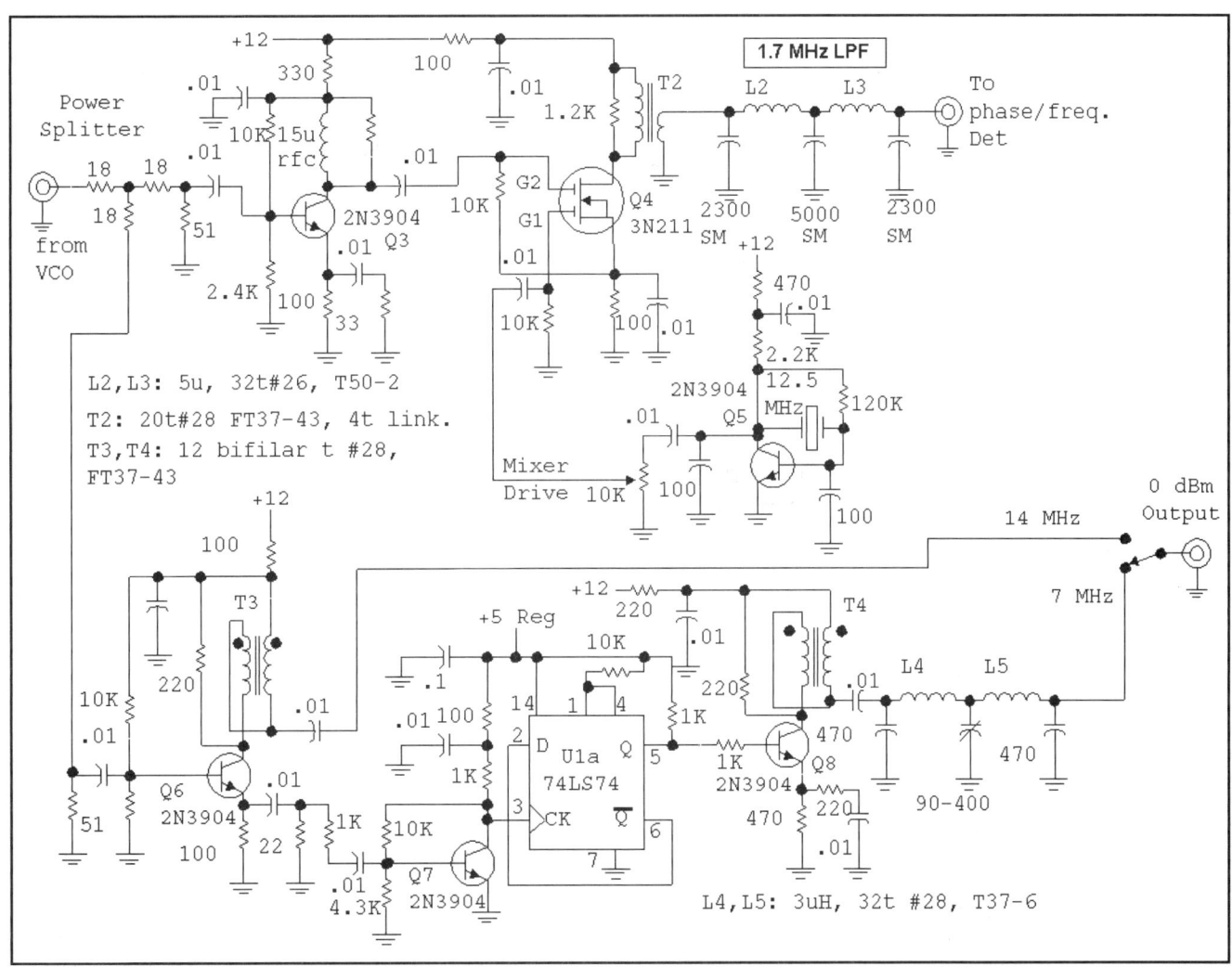

Fig 4.44—Mixer section for the tracking PLL.

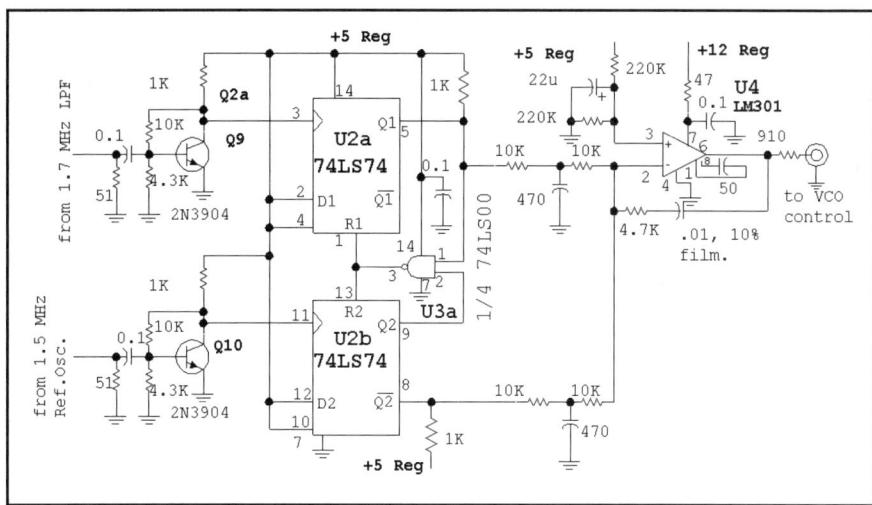

Fig 4.45—Phase-frequency detector and loop filter for the tracking PLL.

Phase-frequency detector using LS-TTL logic. This circuit is shown in Fig 4.45.

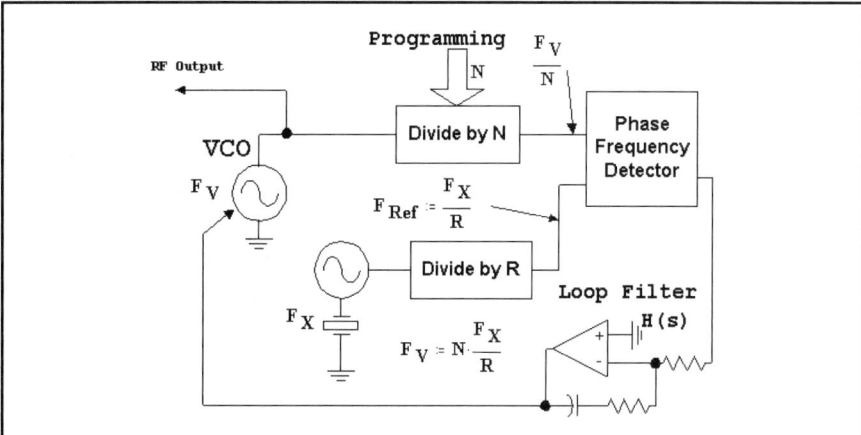

Fig 4.46—A single loop Divide-by-N PLL.

Careful listening and examination with an oscilloscope showed that phase lock was fast and always occurred before a signal was applied to the antenna. A "hold-off" circuit was included that prevented the keying voltage from reaching the power amplifier until the key had been down for 5 milliseconds. This was applied only on initial key closures. VOX-like circuitry then maintained the system in transmit mode (VCO on and T/R relay closed) for half a second or so. This system would be compromised if the VCO locking had not been quick.

A number of changes would be implemented if this system was rebuilt today. The dual-gate mixer would be replaced with a balanced circuit. The op-amp would become an up-to-date alternative, such as the OPA-27. High speed CMOS would replace the LS-TTL used. Finally, the VCO would run continuously without keying, but would operate at a different frequency. This could be 28 or 56 MHz where direct division would produce the desired outputs.

Divide-by-N Phase Locked Synthesis

The most common scheme for frequency synthesis is the divide-by-N PLL **Fig 4.46**. A crystal oscillator at F_X is divided by a (usually) fixed integer R, producing a reference signal at the phase frequency detector at F_X/R. The VCO is divided by a programmable integer, N. The divided VCO must also appear at F_X/R, so $F_V = NF_X/R$. Consider an example: We wish to build a synthesizer for the 9 to 9.5-MHz range. We divide a 2-MHz crystal oscillator by R=200 to generate a 10-kHz reference. N must be set to 900 to produce a 9-MHz signal. Increasing N causes F_V to increase in 10-kHz steps, reaching 9.5 MHz with N=950.

This system would work well as a transceiver local oscillator (LO) in an environment where signals were spaced at 10-kHz intervals. It would not, as shown, be very useful as a general purpose LO.

Modifications could improve resolution. For example, increasing R to 2000 produces a 1-kHz reference. N ranging from 9000 to 9500 would then cover the desired range in 1-kHz steps. (A means of pulling the 2-MHz crystal oscillator by a mere 222 Hz would then generate all LO frequencies within the desired range.)

Generally, 100-Hz resolution produces understandable SSB while 10-Hz steps yield natural sounding voices. But dividers of 90,000 or 900,000 are impractical, even though they are easily achieved with digital logic.

Consider the 1-kHz step system with N=9000 to 9500. The detector reference frequency is 1-kHz, the step value. The loop filtering (plus balance effects) must produce considerable attenuation at 1 kHz. Generally, a system with a 1-kHz step would use a loop bandwidth of 100 Hz or less. The dc from the loop filter includes a small 1-kHz component that frequency modulates the VCO carrier at 1 kHz. The spectrum is a carrier with 1-kHz sidebands. These would be transmitted if the LO was part of a transmitter. If part of a receiver, the contaminated LO would cause a strong signal to be received in a couple of extra frequencies, albeit at reduced strength.

Timing problems occur when N is incremented to tune such a synthesizer. While the N change is instantaneous, the result is not. A filter with 100-Hz bandwidth is capable of change in a time commensurate with 1/B where B is the loop bandwidth, here 10 milliseconds. The effect can be a "chirpy" sound with tuning.

There is yet another problem, a degradation in phase noise. The PLL with a division-by-N is a frequency multiplier. Assume that the reference at the phase detector changes by 1H2. With N=9000, that 1-Hz shift becomes a 9-kHz shift in the VCO. If we think of the 1-Hz reference shift as a noise, the result after frequency multiplication by N is a noise increase by 20Log(N) dB, 79 dB for this case. Clearly, PLL synthesizers with large N should be avoided.

PLL synthesizers are still practical. With large frequency steps, perhaps 10 kHz or more, tuning seems instantaneous while keeping reference sidebands well suppressed. Gaps between steps can be filled in with schemes using additional PLLs, VXO tuning of the reference, or direct digital synthesis—a method that we will discuss later.

Numerous schemes are available for programmable frequency division, limited by the experience of the designers. One is shown in **Fig 4.47**. The incoming signal is *digitized* and applied to the down counting clock input of an Up/Down counter, a 74HC193. The state of the counter decrements by 1 with each clock pulse. When it reaches 0, the "borrow" line goes low. This is fed to the data input of a D-FF. When the Q of that part goes low, the "load" command on the '193 is executed, causing the data on the "jam" inputs, J_a to J_d, to be loaded in the counter, beginning the cycle anew. This overall circuit will divide by the number loaded at J_A to J_D (0 to 15) plus 2. Several 74HC193s can be cascaded to realize large divisors. The 74HC74 forces the output to be synchronous with the input clock.

Many PLL frequency synthesizers use a prescaler, a divider that divides by a fixed amount before reaching programmable circuitry. This reduces the complexity of the programmable parts, but has the disadvantage of multiplying the synthesizer step size by the pre-scale value.

This difficulty is eliminated with a variable modulus prescaler, a chip that divides by one of two different values, depending on the status of a control pin. For example, the Motorola MC12015 is a divide by 32/33: it divides the incoming frequency by either 32 or 33. Extra circuitry is required in the programmable part of the synthesizer to accommodate prescaler programming, but the programmable circuitry is relatively slow, easing design and reducing power.

Numerous commercially manufactured LSI (large-scale integration) chips are available for phase locked loops. One example is the Motorola MC145170, which includes programmable N and R dividers, phase-frequency detector, crystal oscillator, and digital control and memory circuits.[16] This IC functions up to 160 MHz, receiving instructions as a 16-bit serial word. While the use of this chip simplifies a synthesizer, it often means that a microprocessor or computer must be present in equipment using such a synthesizer. The MC145170 and the National LMX1501A are used in a synthesizer on the book CD, the DSP-10 transceiver.

The frequency multiplication and the resulting phase noise degradation between the reference and the VCO is a fundamental property of a divide-by-N synthesizer that cannot be avoided with "improved" design. For this reason, it is becoming common for manufacturers of PLL integrated circuits to specify the phase noise of their ICs at the phase detector. Spectral noise density in the –160 dBc/Hz region is common. The final system design is then degraded by 20Log(N). It will be even worse if other noise sources come into play, such as a poor VCO.

A VXO Extending Synthesizer

A simple PLL synthesizer with a single loop can be used in conjunction with a VXO for numerous special applications. This could be a divide-by-N design like that of Fig 4.46, or a modified design that includes a mixer, shown in **Fig 4.48**. The crystal oscillator (VXO) now serves as the LO for a mixer and as a divided programmable clock for the phase detector. The step size is no longer uniform, a consequence of the variable reference divider. However, the scheme is capable of producing very small steps with a relatively high reference frequency.

Consider an example: A 6.892-MHz oscillator is placed in the circuit of Fig 4.48 with N ranging from 32 to 64. Some (but not all) output frequencies, step sizes, and reference frequencies are given in **Table 4.1**.

The reference frequency varies according to the crystal frequency divided by N while the step size varies with F_X/N^2. Con-

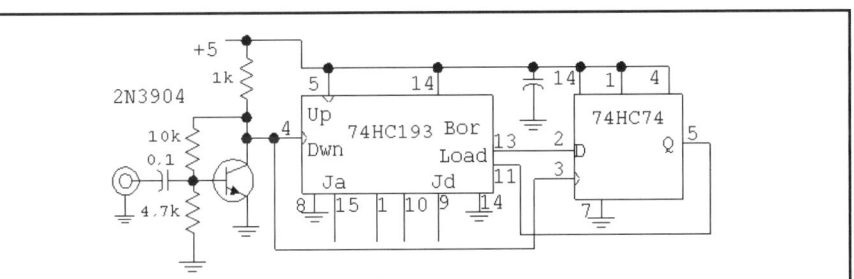

Fig 4.47—A simple programmable divider. See text.

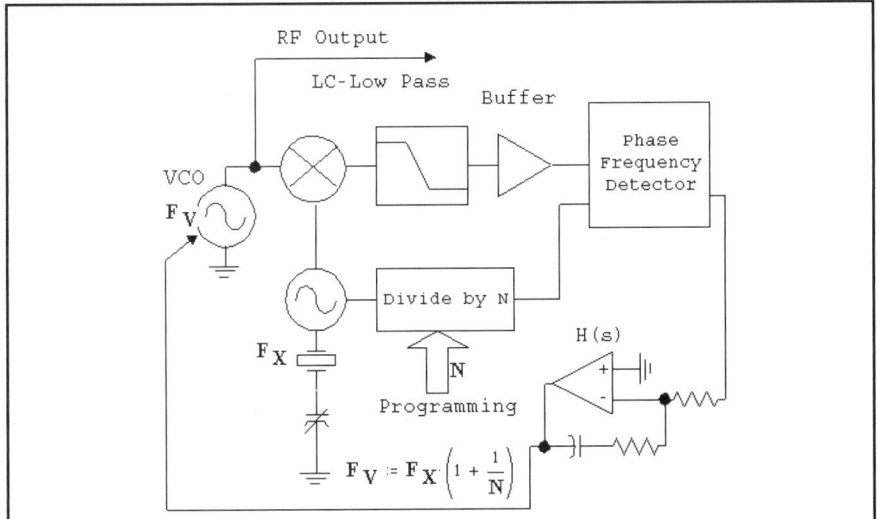

Fig 4.48—A simple PLL synthesizer featuring frequency steps much smaller than the reference frequency.

Table 4.1

N	VCO Output	Step Size	Ref. Freq.
32	7107.7 kHz	6.5 kHz	215.4 kHz
33	7101.2	6.3	208.9
63	7001.7	1.74	109.4
64	7000.0	1.68	107.7

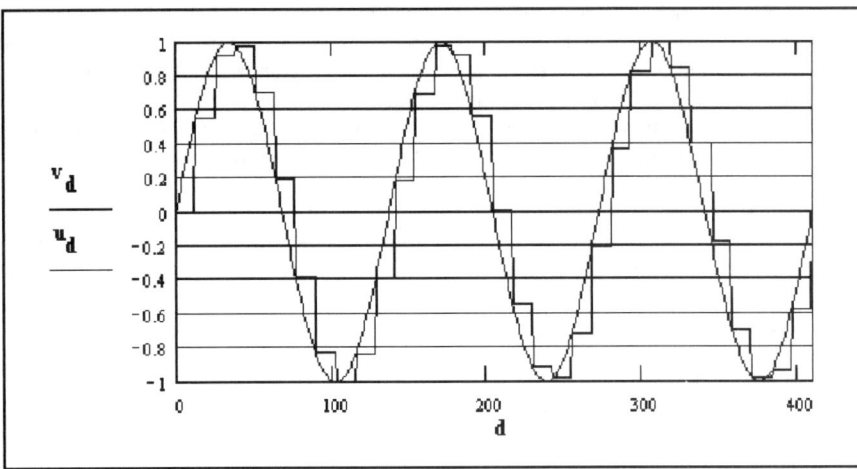

Fig 4.49—A sine wave is generated in DDS with a stepped approximation. Both the stepped, or "sampled" waveform and the desired sine wave result are shown.

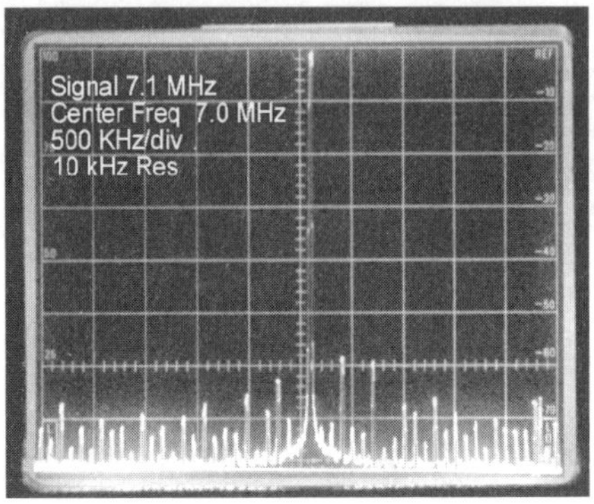

Fig 4.50—Measured output of a direct digital synthesizer using the Analog Devices 9831. Measurements were performed with a Tektronix 494A spectrum analyzer set for a center frequency of 7.0 MHz. The signal is at 7.1 MHz. This DDS device uses a 10-bit D-to-A converter and the manufacturer reports similar spurious responses.

verting the crystal oscillator to a VXO fills the gaps. When this is done, it may not be necessary to use all possible N numbers. Synthesizers of this kind are useful as a means of extending the range of a VXO to cover a larger band. However, they are best used with an independent frequency counter that provides readout. A practical project using this scheme is given elsewhere in this chapter. A practical, general-purpose counter is also presented.[17]

Direct Digital Synthesis

DDS, or *direct digital synthesis* is very powerful and is easily implemented with special, large-scale integrated circuits. The concept is deceptively simple: Digital approximations to values for a synthesized sine wave are calculated or looked-up from memory. These values are loaded into a digital-to-analog converter (DAC) with a new value being periodically generated after a fixed sample time.

A typical DDS IC might be clocked with a 40-MHz crystal oscillator. This signal serves as a clock for updating the output with a new sample that will persist for 25 nanoseconds (1/40 MHz) until the next update arrives. To illustrate the operation, assume we want to generate a 3-MHz sine wave with a 1 V amplitude. This is given as

$$V = \sin(2 \times \pi \times f \times t) \quad f = 3 \text{ MHz}$$

Eq 4.4

At time zero, the desired, output sine wave will have zero amplitude. But 25 nS later, it will have an amplitude calculated by inserting 25 nS into the equation, 0.454 V. At 50 nS, the signal will be 0.809 V, and so forth.

One could plot these values against n to obtain the usual sine wave. However, this is *not* what you would see when examining the DAC with a high-speed oscilloscope. Rather, you would see a line that is flat and level for 25 nS. It would then jump almost instantaneously to 454 millivolts and remain there for another 25 nS. At 50 nS it would jump to 0.809 V, and so on. This behavior is shown in **Fig 4.49** where a sine wave is sampled about 10.7 times per cycle.

If we had used an even 10 samples for each cycle of the sine wave being generated, the lowest frequency in the overall signal would be that of the output. The only distortion would be harmonics. Consider a slightly different case, one where we use 10.333 samples for each cycle of the final oscillation. Three cycles of the output waveform would then be generated with 31 samples. There is a longer periodic character to the overall waveform that would create spurious outputs at one-third the output frequency. All harmonics of the low frequency are also available. The spurs become more numerous as the periods become longer.

Fig 4.50 shows the measured output of an Analog Devices AD-9831 residing on a demo-board from Analog Devices. The part used a 25-MHz clock. An output of 7.1 MHz was synthesized for this example, producing spurious outputs over a wide spectrum. Other examples produced spurs confined to limited regions. There are even some "sweet spots," output frequencies that are virtually free of spurs!

Limited DAC accuracy is a common reason given to explain spurs in a DDS synthesizer. While this is usually dominant, it is not the only source of spurs. The analysis presented above assumed a perfect DAC and still generated spurs. The very stair-step waveform of Fig 4.49 is an approximation to a more ideal sampling waveform reconstruction.[18]

The wideband phase noise in the output of a DDS synthesizer is often very low, comparable with the best Divide-by-N PLL systems. However, this is of little consequence if the noise is merely replaced by a family of coherent spurious responses.

Most current commercial transceivers use a combination of PLL and DDS technology. Unfortunately, it is very difficult to gain even a basic understanding of these systems from the sketchy manuals. Rohde described an excellent example of a dual technology synthesizer.[19] That design used DDS to generate a 10.7-MHz signal that was tunable in small steps. The result was bandpass filtered with a 10-kHz wide crystal filter and then frequency divided to 100 kHz where it served as the reference for a PLL controlling a 75 to 105-MHz VCO.

4.8 THE UGLY WEEKENDER, MK-II, A 7-MHZ VFO TRANSMITTER

The "Ugly Weekender" is a viable project for both the beginner and the seasoned builder. The major feature, and the source of the name, is the construction method outlined in Chapter 1. This section describes a version of that transmitter that uses frequency doubling to achieve improved oscillator isolation.

The transmitter (**Fig 4.51**) begins with a 3.5-MHz variable frequency oscillator. The familiar Hartley topology is used, although others would work as well. The oscillator, Q1, runs continuously to avoid repeated warm-up drift, oscillating a few kHz above the normal frequency, but is shifted to the desired frequency during transmit intervals. The VFO is temperature compensated with a combination of NP0 and polystyrene capacitors in the 3.5-MHz tank circuit. The combination was picked and confirmed with repeated temperature runs in a home-built environmental chamber.

The VFO is buffered with a keyed dual-gate MOSFET amplifier, Q2. A JFET source follower driving a feedback amplifier would also provide the needed 10-milliwatt output needed to drive the frequency doubler.

The 2X-frequency multiplication occurs with a pair of diodes, as discussed in greater detail in Chapter 5. The doubler output is selected with a single tuned circuit. A 10% bandwidth double tuned circuit would be a better choice in this position. The power lost in the passive frequency multiplication is regained with a buffer amplifier using Q6 and Q7.

The 7-MHz output from Q7 is applied to a 500-Ω drive control with output to a keyed feedback amplifier, Q8, shown in **Fig 4.52**. The keying voltage is derived from Q4, an integrating waveform shaping circuit.

A feed-through capacitor in the two box version of this circuit routes the Q4 collector voltage between modules. This component was eliminated in the single compartment version.

The output power amplifier, Q9, an ever-reliable 2N3866 with a small heat sink, is shown in Fig 4.52. Numerous other

Outside view of "Ugly Weekender" transmitters for 7 (left) and 3.5 MHz.

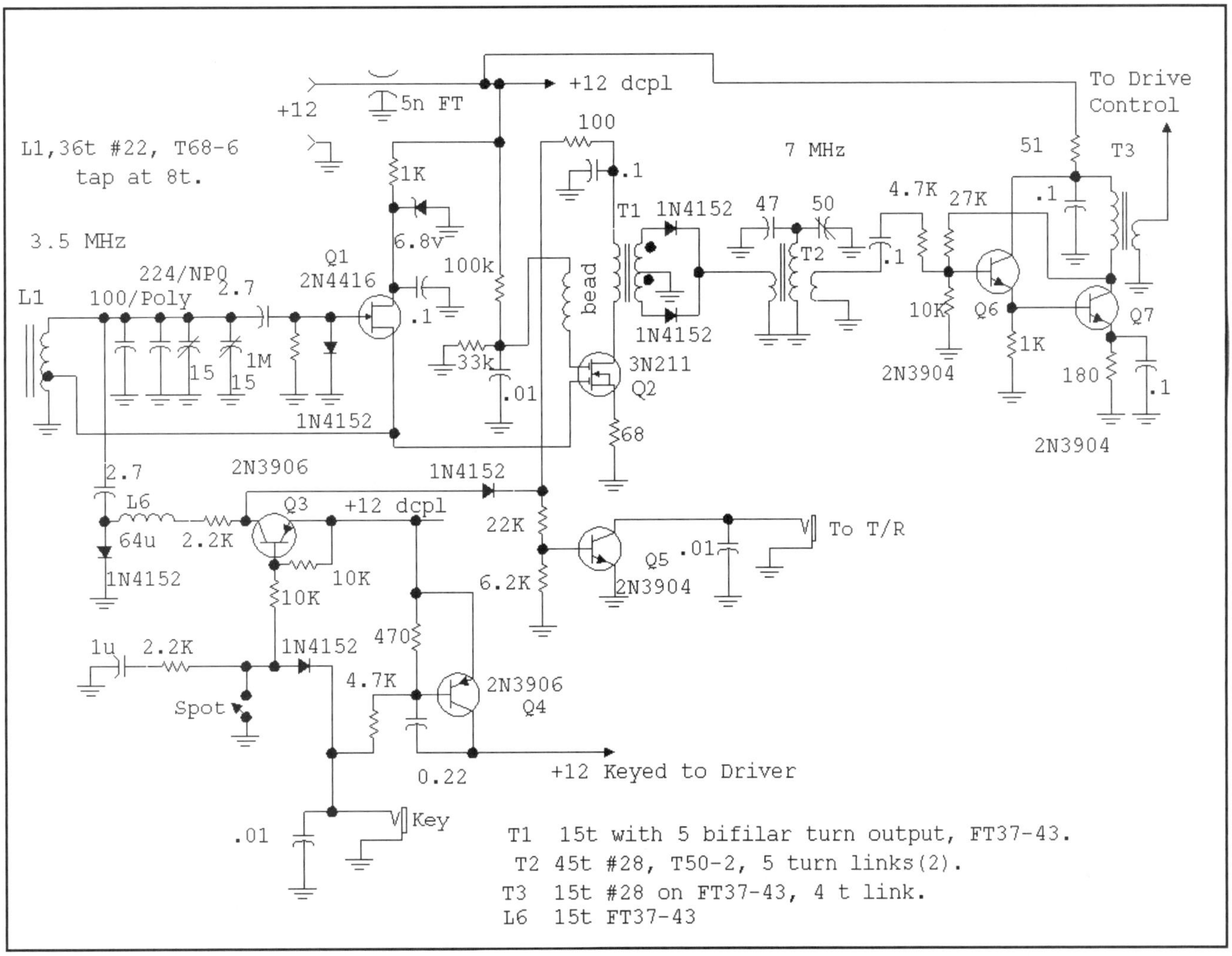

Fig 4.51—VFO and frequency multiplier for the Mk II Ugly Weekender.

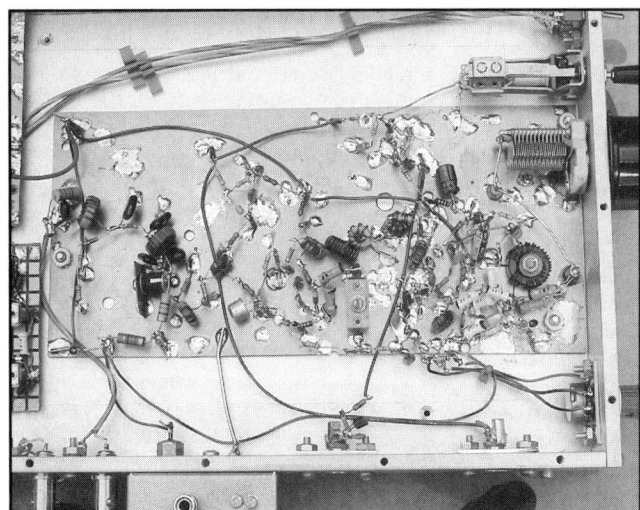

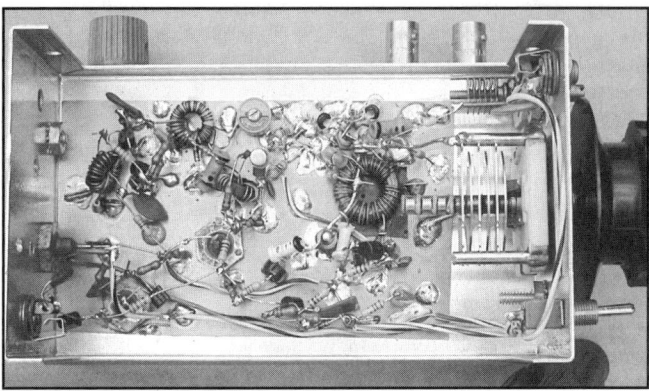

Inside view of a single board version of the 7-MHz transmitter. A receiving converter is at the rear (left) of the box.

The VFO portion of the transmitter, including diode frequency doubler.

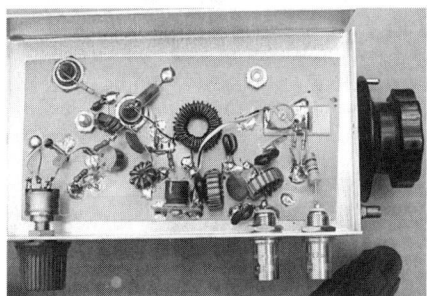

The power amplifier for the 7-MHz version.

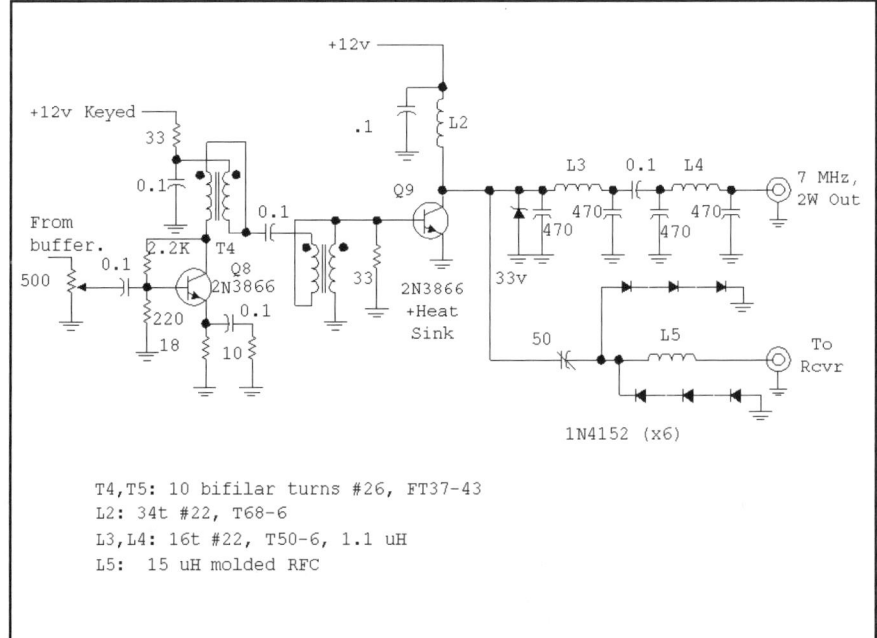

T4,T5: 10 bifilar turns #26, FT37-43
L2: 34t #22, T68-6
L3,L4: 16t #22, T50-6, 1.1 uH
L5: 15 uH molded RFC

Fig 4.52—Driver and power amplifier portion of the Mk II Ugly Weekender.

parts will function in this position with circuit details discussed in Chapter 2. Output power is just over two watts with the drive control at maximum. A T/R system is included for QRP applications.

Q5 is a transistor switch that generates a grounded line when the key is pressed. This signal is timed to hold for a short period after the key is opened to control an electronic transmit-receive switch with a 100-W power amplifier sometimes used with this exciter.

4.9 A DIGITAL DIAL

The frequency counters we see in the amateur literature are either general-purpose test instruments or special designs, intended as a readout for a receiver or transceiver. This unit falls into the later category, but it could be expanded to serve general applications.

We wanted this design to use standard parts. Excellent special purpose counter chips are available, but they are often expensive and difficult to find. Micro-processors, such as the popular PIC and Basic Stamp Series, can be configured as counters, while serving all related display chores. But a simpler solution was sought, one that was usable without special programming skills.

This circuit uses a small number of readily available, inexpensive integrated circuits, including the four-LED displays. The design was intended to be cheap enough for repetitive use in a variety of projects. The approximate $10 parts cost included the time base crystal, but did not include a PC board.[20]

This counter avoids multiplex methods, which are prone to RF noise generation. Frequency resolution is 100 Hz.

Figure 4.53 shows a functional block diagram for the frequency counter. Signals to be counted are applied to a single transistor conditioner that drives a gate controlled by the counter time base. For 100 Hz resolution, the gate must be "open" for 10 milliseconds. However, this design has an extra divide-by-10 to suppress last digit flicker, so a 100-mS count window is used. After the counting is finished and the gate is closed, a "strobe" signal is applied to ICs that *remember* the counted result and decode it to a format suitable to drive the 7-segment light emitting diode displays. This is followed by a pulse that resets the counters to zero, ready for the next cycle.

The time base, shown in **Fig 4.54**, begins with a crystal controlled bipolar tran-

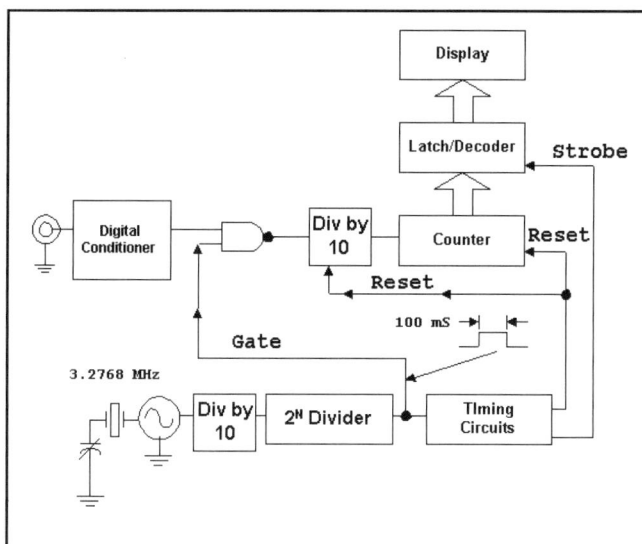

Fig 4.53—Block diagram for counter.

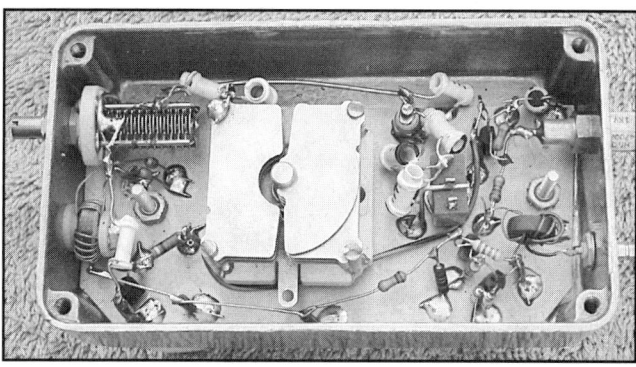

A clean way to fabricate an LC oscillator uses a Hammond 1590B box, offering excellent shielding. DC enters through a feedthrough capacitor and RF leaves on coaxial cable. This oscillator used a differential capacitor, but only one side is connected.

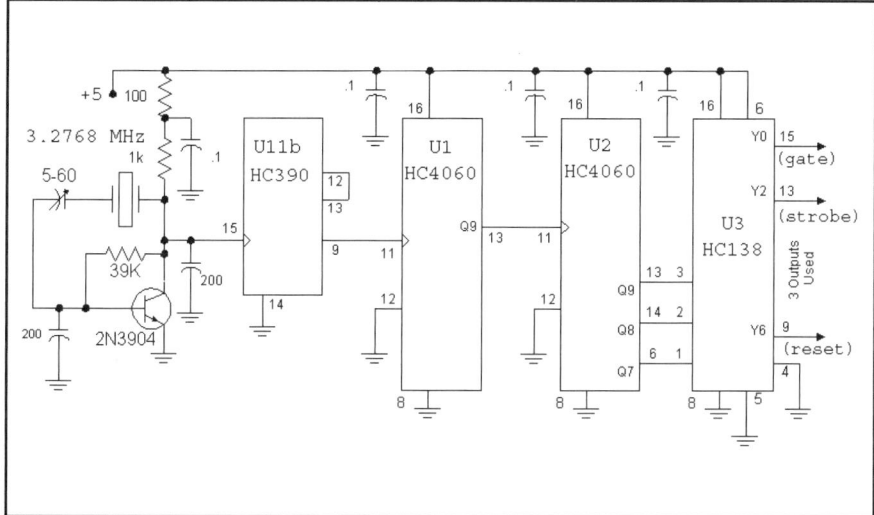

Fig 4.54—Time base portion of frequency counter.

sistor oscillator operating at 3.2768 MHz. The crystal is a readily available, off-the-shelf part. The oscillator is divided by 2^{15} in U1 and U2, a pair of 74HC4060 "timer" ICs, resulting in the desired 100-millisecond gate window. Further division in U2 provides a chain of additional 100 mS windows. These are decoded in U3 to generate strobe and reset pulses.

The rest of the counter is shown in **Fig 4.55**. The signal to be counted is conditioned with Q1 with the resulting logic applied to U4A, part of a quad NAND gate with other sections serving as inverters. The output is then counted by U11a, U5, and U6, 74HC390 dual decade counters. These drive the decoder drivers, U7 through U10, using 4511B decoder-driver ICs. This configuration will display kHz to the left of a decimal place and tenths of a kHz to the right of the decimal place.

We have used ICs from two families in this design. Most of the system uses "HC" *high-speed CMOS* parts. This allows the circuit to function to 50 MHz or beyond. However, there is no need for high speed in the display function, so the decoder drivers use the slower standard CMOS parts. Using slower parts here should help to minimize RF noise and current consumption. We used common cathode, seven segment LEDs, type MAN4740.

Early versions of this counter used only two digits of display, showing only 0 to 99 kHz. While this worked well as a digital substitute for a mechanical dial, it became frustrating in some applications. We found ourselves wanting more resolution, including a digit to the right of the kHz decimal place. A more complete display with digits to the left allows complete elimination of mechanical dials in many systems. The lower current two-digit format is available by eliminating the related 4511 drivers and LEDs in the design presented.

Total current depends upon the digits being displayed. With 5-MHz input signals, current was about 80 mA when the display read "888.8", dropping to 30 mA with "111.1." The sensitivity was excellent with a 5-MHz input, counting reliably with an input of less than –40 dBm from a 50-Ω generator. The counter continues to function to over 50 MHz, but requiring higher RF drive power.

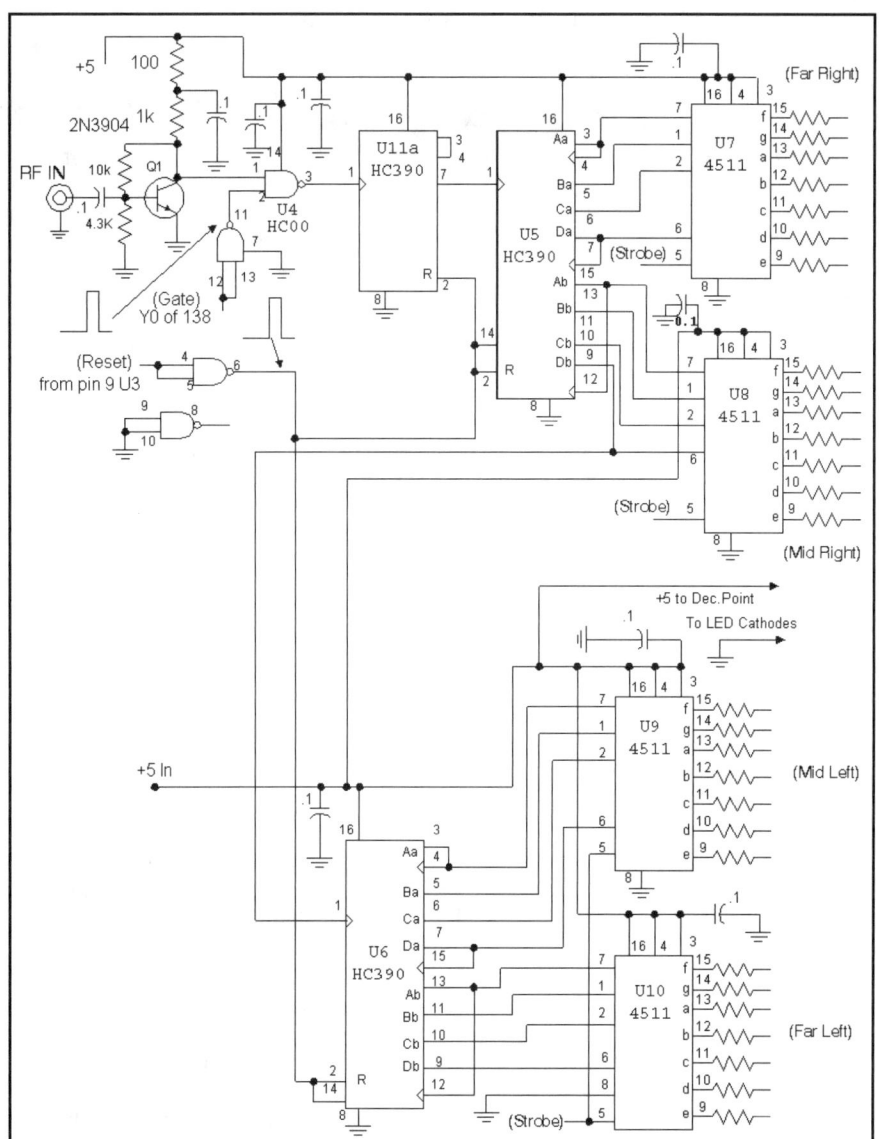

Fig 4.55—Input circuit, counter detail, and display portion of frequency counter.

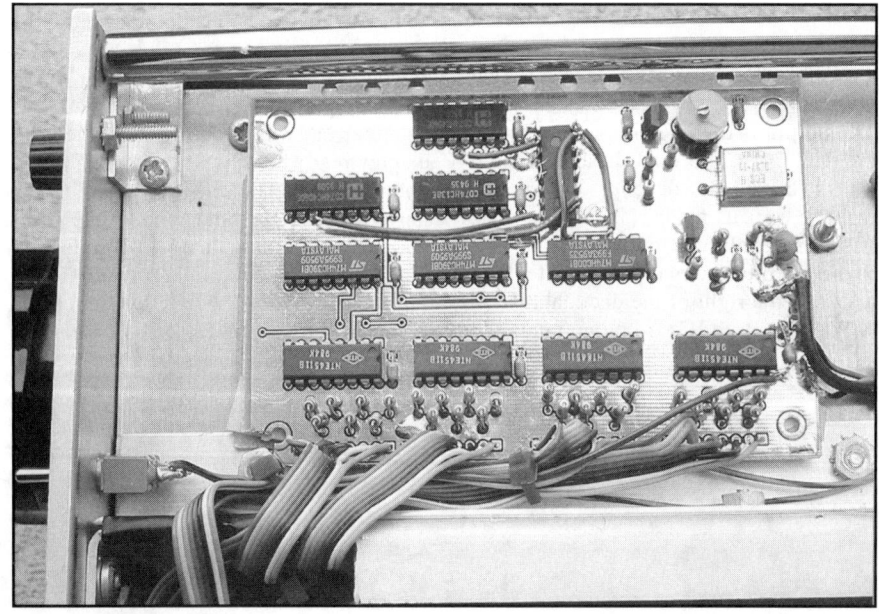

Frequency counter installed in a receiver. U11 was added "dead bug style" to eliminate flicker.

Regarding Counter Accuracy

The simple counter described above is capable of good accuracy so long as the crystal and the oscillator components are stable. The capacitor in series with the crystal should be adjusted to produce the proper count when a known frequency is applied to the counter input.

The counter as shown is suitable for use with simple direct conversion transceivers or superhet systems where the intermediate frequency is an even multiple of 100 kHz. The "dial" then functions accurately when the LO alone is counted, except for the left most digit. If a "less friendly" IF is used, other schemes must be applied. The usual transceiver might have several intermediate frequencies, all of them with uneven values. The corresponding oscillators, including BFOs or carrier oscillators, could all be counted. A mixture of up and down counting might be required with the various oscillators, depending upon the way the final frequency is calculated or measured. Clearly, this would be a good application for a microprocessor.

A simple counter that would still be accurate over a wide frequency range could be built with circuitry much like that in Fig 4.55, even if the IF is "unfriendly." The simple up counters would be replaced with presetable up-down counters. Instead of resetting the counters to zero at the end of each cycle, the counter would be *loaded* with an appropriate digital word that causes the LO counting to produce the right readout.

It is possible in some applications to obtain reasonable results over a narrow tuning range merely by changing the crystal frequency. This counter uses a clock oscillator of 3276.8 kHz. That value is divided by a fixed value to produce a time window that drives the counting gate. The final count is the number of cycles that pass through the gate during the time interval. The display is a number that is a constant multiplied by the ratio of the two frequencies. If the crystal frequency is changed, the "dial" can still be exactly right for one frequency. It might not be too far off at others that are close.

Consider an example, a 7-MHz transceiver using a crystal filter at 1.98 MHz. The VFO will then be tuned to 5.02 MHz when the transceiver is at 7.000 MHz. Using the counter with the standard 3.2768-MHz crystal would produce a count of "20.0" instead of the desired "00.0." If the clock crystal was changed to 3.2899 MHz, a 13.1-kHz difference, the count would be proper at 7 MHz. The error at 7.1 MHz would be 0.4 kHz. This may be tolerable for some applications.

There are several options available to the builder wanting to use a microprocessor controlled counter. Simple units are available in kit form, ready for installation in QRP rigs and the like, with references found on the web. Some examples are also included on the book CD.[21]

4.10 A GENERAL PURPOSE VXO-EXTENDING FREQUENCY SYNTHESIZER

Fig 4.56 shows the block diagram for a unique frequency synthesizer. Although this example was built for 14 MHz using an off-the-shelf TV color-burst crystal in the VXO at 14.318 MHz, the system can be adapted for many other applications. VHF examples are given later. This example used the VXO design presented in Fig 4.30. The VCO used with the synthesizer is that of Fig 4.34, which can be scaled to other frequencies.

We only discuss the synthesizer circuits in detail here. The VCO provides the needed output. It will usually be split in a hybrid with one component used in an intended output while the other drives the synthesis circuitry. A level of −6 dBm is needed by the synthesizer at both the VCO and the VXO inputs.

The programmable frequency divider is a version of the circuit shown in Fig 4.47 using two 74HC193 chips, allowing division by up to 258. The detailed circuit is shown in **Figs 4.57** and 4.57A. The division ratio is derived from two more 74HC193 chips, now operated as an up-down counter. Pulses to the "up" or the "down" inputs increment or decrement the frequency by one step. The user must establish the division range, controlled by four hard wired points below U2, marked A, B, C, and D in Fig 4.57. The four inputs are connected to logic 0 (ground), logic 1 (+5 V), or to the outputs from U4. Some possible variations are shown in **Table 4.2**.

The frequency determining up-down counter, U3 and U4, may also be loaded with an often-used setting, such as a recognized calling frequency. Each line must be hard-wired by the user to establish this frequency.

The Up/Down commands are buffered with U6. Grounding an input line (P9 or P10) will cause an up or down pulse to appear at U3. A ground command on J8 also causes the "calling frequency" to be

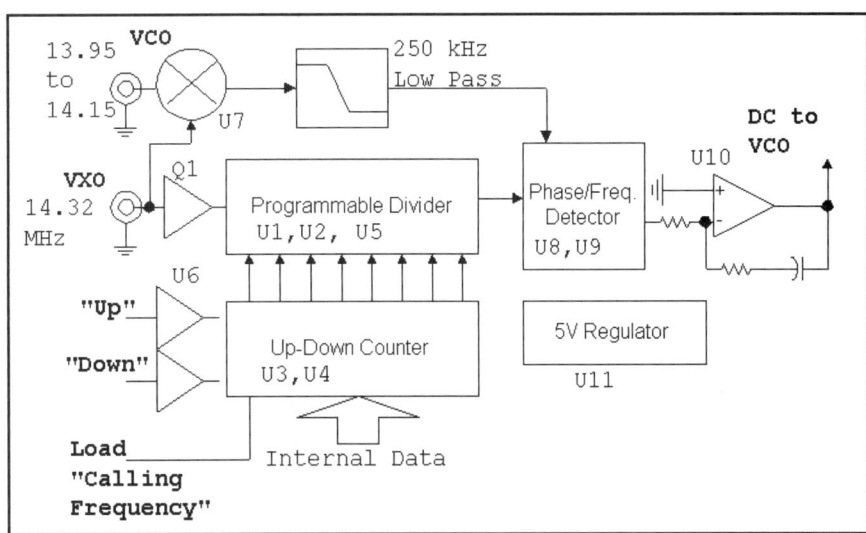

Fig 4.56—Block diagram for the VXO extending synthesizer.

Table 4.2

Available States	A	B	C	D
2 to 258	U4	U4	U4	U4
2 to 130	U4	U4	U4	0
66 to 130	U4	U4	1	0
34 to 66	U4	1	0	0

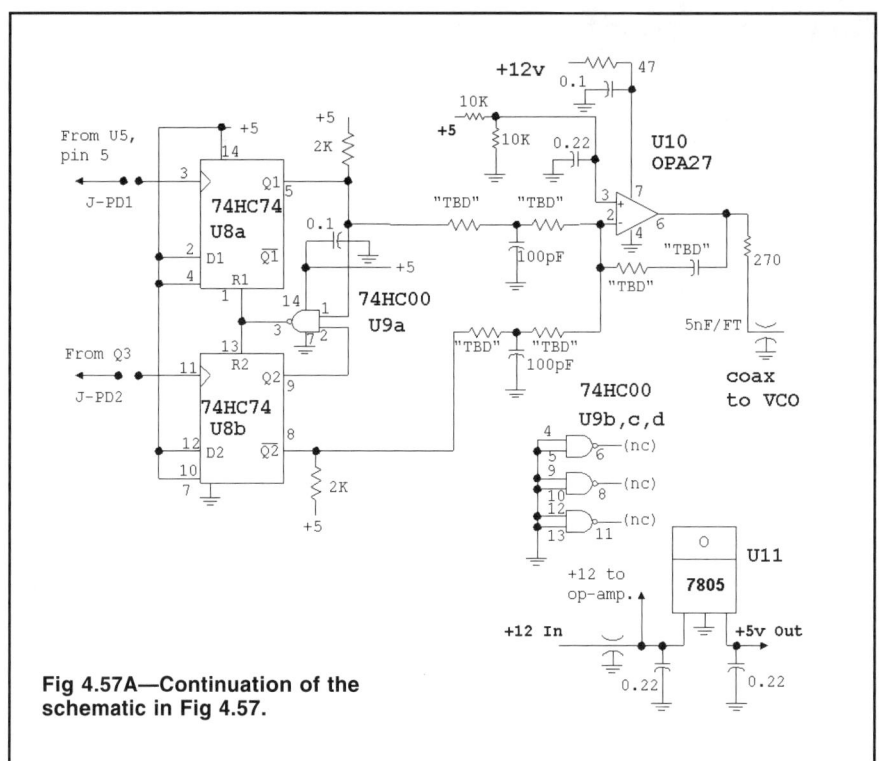

Fig 4.57A—Continuation of the schematic in Fig 4.57.

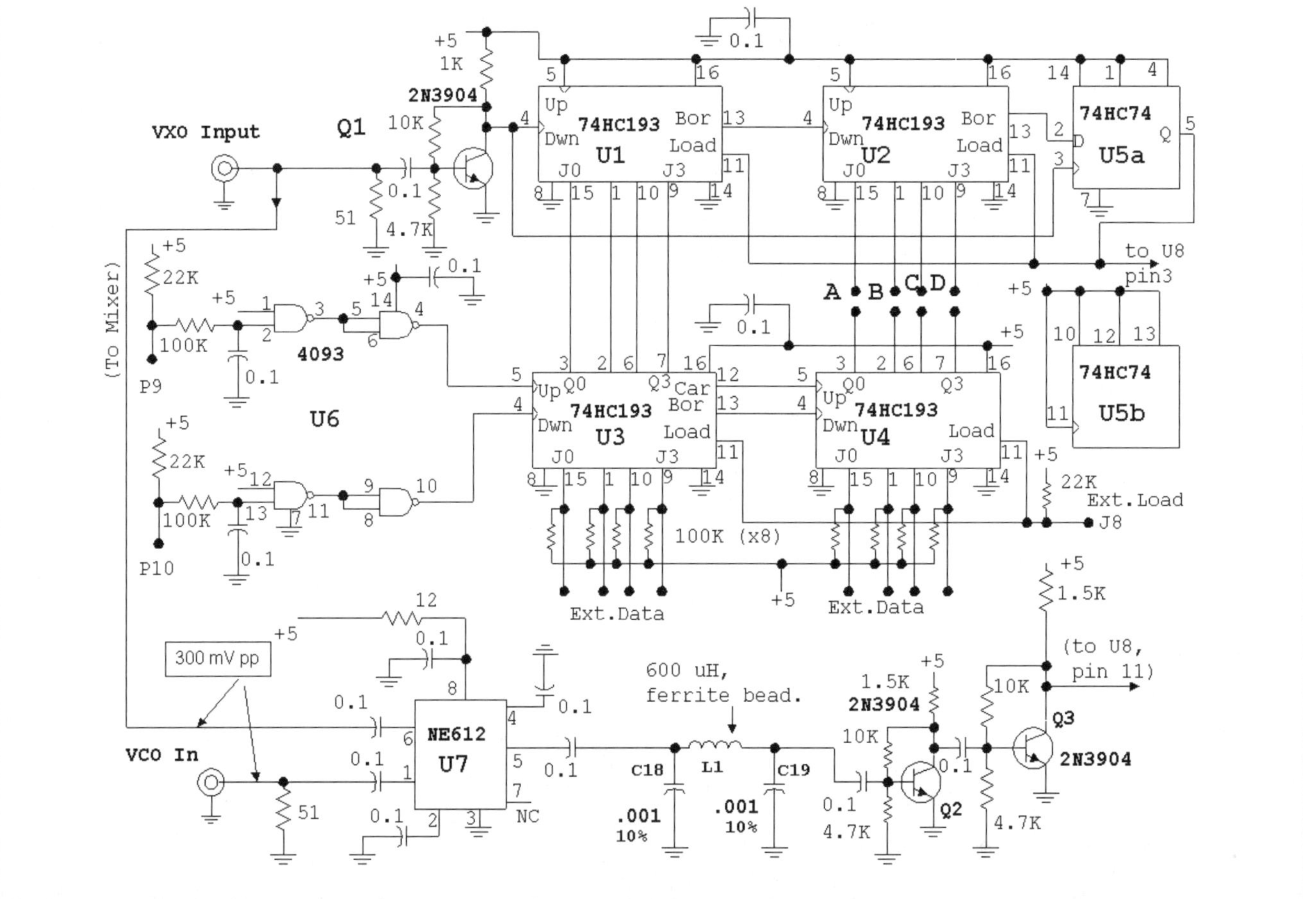

loaded. The user may wish to add more interface circuitry to the Up/Down lines; standard CW keyer circuits work well, as does a keyer paddle or a computer mouse as an input device.

The VXO and VCO are both applied to mixer U7, an NE-612. The low frequency output is low pass filtered and impedance transformed with a pi-network using L1. In the example, a 200-kHz signal is transformed from 1.5 kΩ to 500 Ω with the pi network formed by L1, C18, and C19. The 600-μH inductor consists of 22t #26 on a FB43-6301 ferrite bead. The low pass filter components will change with other applications. The low pass filter output is

Fig 4.57—Schematic for the experimental synthesizer. See text for details.

amplified and conditioned for digital levels with Q2 and Q3.

Two programmable jumpers are provided at J-PD1 and J-PD2. While pin 3 of U8 is normally driven from U5 in applications with the crystal below the VCO frequency, it may change to drive from Q3 in other systems. The frequency scheme shown has the crystal above the VCO. A VCO tuning polarity may also require a change.

The loop filter uses a premium op-amp, the OPA-27. This fast, low noise part is ideal for this application. The four input resistors are all 47 kΩ while the feedback elements are 10 kΩ and 1.0 µF for the 14-MHz example. All of these components are subject to change with other applications and are marked TBD in the schematic for "to be determined." They are picked with the PLL computer program that accompanies *Introduction to Radio Frequency Design*. Phase lock loops must be designed with some care and component values are *not* well suited to casual selection.

The 14-MHz version of this design is summarized in the equation sheet of **Fig 4.58**. The programming sets N for values from 34 to 66 with some frequencies listed

Frequency synthesizer installed in a Hammond 1590BB box. Coaxial inputs are from the VCO and reference VXO. All input/output lines are attached to feedthrough capacitors.

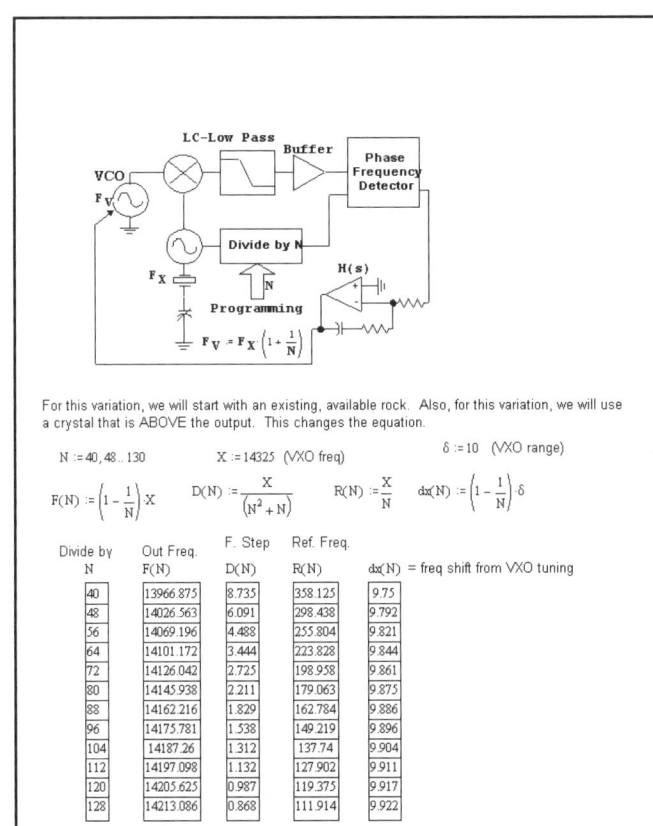

Fig 4.58—Summary of available frequencies and characteristics of the 14-MHz "VXO extender." This data was generated with *MathCad 7.0*.

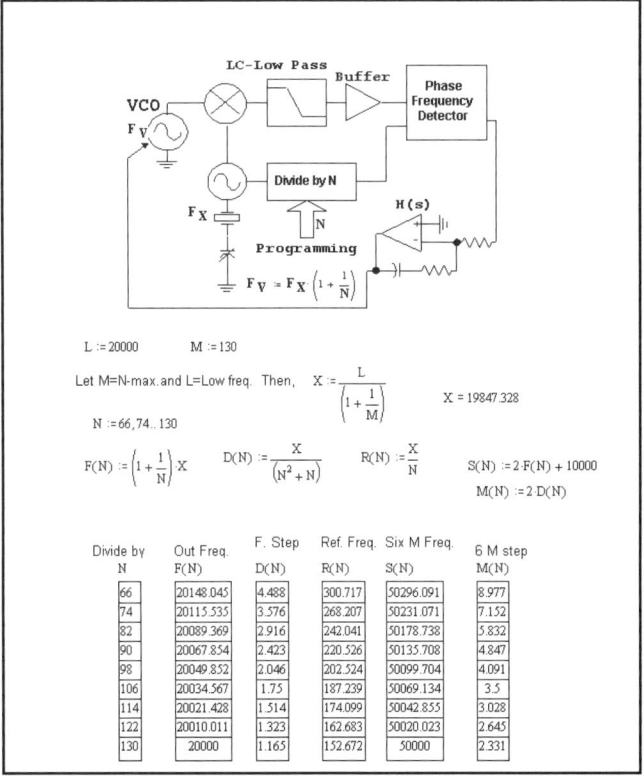

Fig 4.59—Summary of available frequencies and characteristics for a 20-MHz "VXO extender." The result will be frequency doubled where it then serves as the LO for a 50-MHz transceiver based upon a 10.0-MHz IF. This data was generated with MathCad 7.0.

in the table. The design equations use a minus sign for this case, for the crystal is above the VCO.

The synthesizer board is housed in a milled aluminum box (Hammond 1590BB) with either coaxial cables or feedthrough capacitors for all interfaces. The VXO and the VCO are each housed in individual milled boxes (Hammond 1590A.) While it is possible to include both digital and RF/analog circuitry on a single board, the isolated and shielded approach is less prone to spurious responses and is recommended.

Once the boards are functioning and checked out, the system is turned on with relative ease. An oscilloscope senses the dc on the control line while the VCO course tuning is adjusted.

Fig 4.59 shows a design for the 6-meter band. It is intended to be used in a mono-band super-heterodyne transceiver with a 10.0-MHz IF. The synthesizer operates in the 20-MHz range with a 19.847-MHz VXO. It is then frequency doubled and filtered to provide a 300 kHz range at 40 MHz. The circuit could also be adapted for 25-MHz operation; frequency doubling would then allow use with a 6-meter phasing transceiver.

A similar version could be built for the two-meter band where an injection frequency of 144–10=134 MHz is needed. An especially useful scheme would use a synthesizer operating at a tenth of this frequency, 13.4 MHz. If N varies from 66 to 130, the required VXO would operate at 13.298 MHz. The synthesizer output would be multiplied by 5 with a 74HC04 and bandpass filtering, followed by a X2 diode multiplier and 134-MHz filter. The 10X scheme leads to simple frequency counting. The system can also be adapted for direct phasing at 144 MHz. Nearly one full MHz of range is available at the 2-meter band.

The "VXO Extender" is an experimental synthesizer, something of a departure from the normal schemes in use. The method is one that provides relatively small step sizes with much higher reference frequency, but at the price of uneven step size.

Single loop synthesizers can be configured in a more traditional format with modest step size while still being used for general-purpose applications. For example, the Elecraft K2 CW/SSB transceiver uses a single loop synthesizer with 10-kHz steps. The "clock" is a voltage controlled crystal oscillator that is then driven by a DAC, allowing all gaps to be filled in with small steps. Clever firmware on the part of the designers remove tuning ambiguities.

REFERENCES

1. W. Hayward, *Introduction to Radio Frequency Design*, Chapter 7, Prentice-Hall, 1982; R. Rhea, *Oscillator Design and Computer Simulation, Second Edition*, Noble Publishing, 1995.

2. For a discussion of the squeeging problem, see Clarke, *IEEE Transactions on Circuit Theory*, Vol CT-13, No. 1, Mar, 1966.

3. W. Hayward, "Measuring and Compensating Oscillator Frequency Drift," *QST*, Dec, 1993, pp 37-41.

4. K. Spaargarem, "Crystal Stabilized VFO," *RadCom*, Jul, 1973, pp 472-473.

5. J. Makhinson, "A Drift-Free VFO," *QST*, Dec, 1966, pp 32-36; K. Spaargaren, "Frequency Stabilization of LC Oscillators," *QEX*, Feb, 1996, pp 19-23.

6. U. Rohde, *Digital PLL Frequency Synthesizers Theory and Design*, Prentice-Hall, 1982.

7. "The RF Oscillator", *Radio Communications Handbook*, Sixth Edition, RSGB, 1994, p 6.36.

8. U. Rohde, *Digital PLL Frequency Synthesizers Theory and Design*, Prentice-Hall, 1982; U. Rohde, "Designing Low-Phase-Noise Oscillators," *QEX*, Oct, 1994, Fig 15, p 10; H. Johnson, personal correspondence with author; "Demphano - A Device for Measuring Phase Noise," J. Makhinson, *Communications Quarterly*, Spring, 1999, pp 9-17.

9. D. B. Leeson, "A Simple Model of Feedback Oscillator Noise Spectra," *Proc. IEEE*, Vol 54, Feb, 1966, pp 329-330.

10. U. Rohde, personal correspondence with author.

11. W. Hayward, "Variations in a Single-Loop Frequency Synthesizer," *QST*, Sep, 1981, pp 24-26.

12. http://www.qsl.net/7n3wvm/supervxo.html

13. W.S. Mortley, "Frequency-Modulated Quartz Oscillators for Broadcasting Equipment," *IEEE Proceedings*, Part B, May, 1957, pp 239-249; W.S. Mortley, "Circuit Giving Linear Frequency Modulation of Quartz Crystal Oscillator," *Wireless World*, Oct, 1951, pp 399-403; V. Manassewitsch, *Frequency Synthesizers: Theory and Design*, Third Edition, John Wiley & Sons, 1987, pp 401-405.

14. See, eg, U. Rohde, *Digital PLL Frequency Synthesizers: Theory and Design*, Prentice-Hall, 1983; U. Rohde and D. P. Newkirk, *RF/Microwave Circuit Design for Wireless Applications*, Chapter 5, John Wiley & Sons, Inc., 2000.

15. F.M. Gardner, *Phaselock Techniques*, Second Edition, Wiley, Apr, 1979; V. Manassewitsch, *Frequency Synthesizers: Theory and Design*, John Wiley & Sons, 1976.

16. *CMOS Application-Specific Standard ICs*, Motorola Inc, Publication DL130/D, 1991, pp 5-101. Data sheet has a good set of references. See also design equation page.

17. W. Hayward, "Variations in a Single-Loop Synthesizer," *QST*, Sep, 1981, pp 24-26; Talbot, "N-over-M Frequency Synthesis," *RF Design*, Sep, 1997.

18. E.O. Brigham, *The Fast Fourier Transform*, Section 5.4, "Sampling Theorm," Prentice-Hall, 1974, pp 504-510.

19. U. Rohde, "A High-Performance Hybrid Frequency Synthesizer," *QST*, Mar, 1995, pp 30-38.

20. This circuit is similar to one described by G. Adcock, G4EUK, "A Simple Frequency Counter for DC Receivers," *Sprat 73*, Winter, 1992/93, p 10.

21. For the ultimate, high performance circuit, see W. Carver, "The Modular Dial," *Communications Quarterly*, Spring, 1998, pp 35-44. See also N. Heckt, "A PIC-Based Digital Frequency Display," *QST*, May, 1997, pp 36-38; and D. Benson, "Freq-Mite—A programmable Morse Code Frequency Readout," *QST*, Dec, 1998, pp 34-36.

CHAPTER 5

Mixers and Frequency Multipliers

5.1 MIXER BASICS

Nearly all of the equipment we build uses at least one mixer. Even the simplest direct conversion receiver uses a product detector, which is one form of mixer. **Fig 5.1** shows the block-diagram symbol for a mixer. A mixer is a three-port circuit with two input signals and one output occurring at a frequency that is the sum and/or difference of the two input frequencies. One input, the *local oscillator* (or *conversion oscillator*) is usually much stronger than the other, the *RF input*. The output in typical receiver applications is called an *intermediate frequency*, or *IF*, for it is often part way between a higher input frequency and baseband. While this historic relationship does not always apply to modern systems, the *IF* term remains

We begin our examination of mixers with an experiment designed to analyze a simple mixer with the goal of extracted understanding. What are the device characteristics that allow mixing (difference and sum frequencies) and what are the resulting signal levels? Are there undesired output signals?

Our experimental mixer is the single JFET circuit of **Fig 5.2**. Both local oscillator and RF are applied at the gate. While this may not be the most common scheme, it lends itself to analysis.

Examination begins with the *bias* circuit of **Fig 5.3**. Our goal is to model the FET and to then bias it half way between pinchoff and full drain current. The Fig 5.3 circuit is built without a "test" resistor, producing a source voltage of 3.74 V. (These are actual measured results with a J310 FET.) The FET current is very low owing to the high value source resistor, so the FET pinchoff voltage will be close to -3.74 V. Test resistors from 10 kΩ down to 15 Ω were then

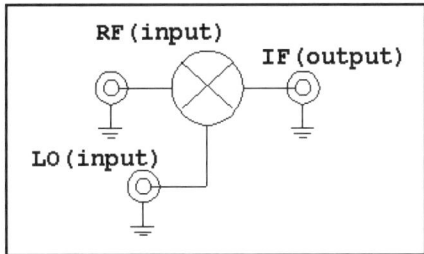

Fig 5.1—Block diagram element for a mixer.

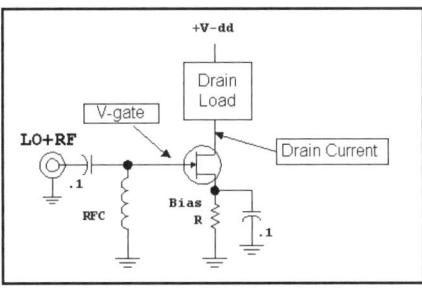

Fig 5.2—Basic JFET mixer with LO and RF applied at the gate. The drain will then have all available outputs. It can be tuned to emphasize one mixer product.

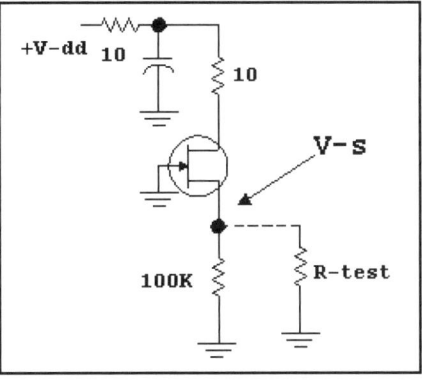

Fig 5.3—Biasing setup for FET modeling.

placed in the circuit, measuring source voltage for each. This allowed us to form a curve of drain current vs gate-source voltage, **Fig 5.4**. The data scatter (the bumps) resulted from thermal effects at higher current levels. The smooth curve is calculated for an ideal JFET with a – 4.2 V pinchoff and I_{DSS}=45 mA. These parameters produced a good fit to the measured data over most of the range.

This exercise provides a mathematical model, something to use to study the mixing process. A 150-Ω resistor provides the desired bias that sets the source voltage at 2 V, about half way between full current and pinchoff.

Fig 5.5 is a modification of the smooth, modeled data. The zero voltage point has been shifted to the middle of the graph, the bias point chosen with the 150-Ω source resistor. The voltage is the actual value appearing at the gate in Fig 5.2. The total current has been split into three segments. The first is a constant, the bias current with no signals present. The second is the linear term, a straight line. The third is a parabola. The three components add to form the previous curve.

We now consider each of the three curve segments by themselves as signals are applied to the mixer input. The bias is a fixed value; the fixed current does not depend on any applied signal. This is evident in the bias curve in Fig 5.5, which is flat.

The linear term becomes more useful. If we apply a sine wave to the gate that causes the voltage to oscillate between –0.5 and 0.5 V, a 1 V peak-to-peak swing, the current will vary by about 11 mA peak-peak. A high impedance in the drain allows the signal current to develop an output voltage. This is the characteristic we seek when we use the JFET

Mixers and Frequency Multipliers   5.1

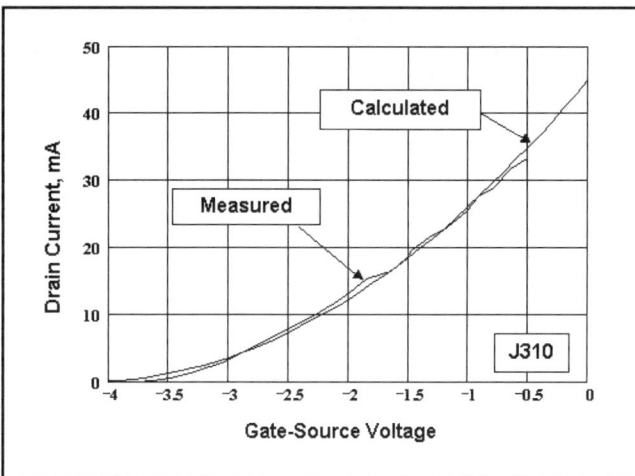

Fig 5.4—Curve fit of data for FET modeling. The bumps are the result of thermal effects in data, while the smooth curve is calculated.

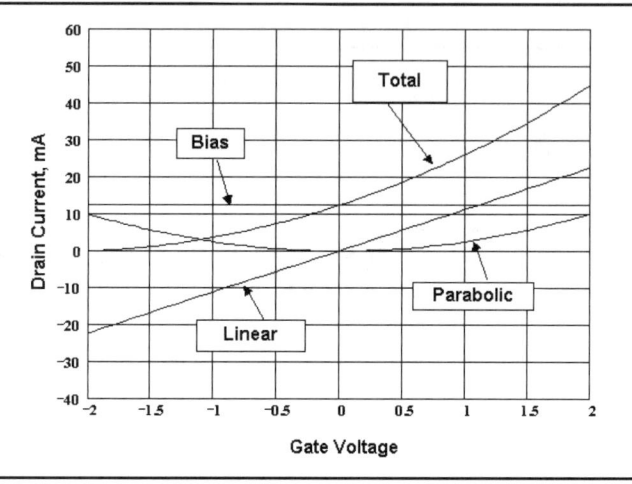

Fig 5.5—The FET current is split into three components: a fixed bias, a linear term and a parabola.

as an amplifier.

Consider the linear curve when two signals are applied to the input: Two sine wave voltages at the gate produce two sine wave currents, but nothing more; no mixing occurs as a result of the linear term. There is also no distortion. This is the behavior we intend when we speak of linearity.

It is the parabola that becomes interesting, taking us beyond amplifier behavior. A low amplitude gate signal causes no current, for the parabola is zero everywhere near 0 V. But current flows as the signal grows. Moreover, it is distorted. This is evident; for a positive excursion will produce the same positive current that is generated by a negative excursion. A large amplitude sine wave will produce two output current pulses per cycle as the signal swings both positive and negative about the bias point. We have built a frequency doubler.

We now apply the sum of two signal voltages to the gate. Again, the bias curve produces nothing. The linear curve will generate two response currents, each a replica of the input, but nothing more. No mixing occurs from the linear response. But the parabolic curve generates interesting results. Not only do we see each input frequency doubled, but we now see sum and difference products. This is not evident directly from the curves, but follows directly from the related mathematics. This is available on the book CD as a *Mathcad* file, *mixer_jfet1.mcd*. A file is also available (*mixmath.pdf*) that can be viewed even if the reader does not own *Mathcad*.

The two-component input uses one part, the "local oscillator," at a higher level than the other, the "RF." When this term is applied to the parabolic curve, the result is a product of two sine waves. Multiplication is the reason our mixer symbol, Fig 5.1, uses a large multiply sign. High school trigonometry identities convert the product of two sine waves into sine waves at the sum and difference frequencies, the mixing result that we seek. The sum is often called the *upper sideband*, while the difference is the *lower sideband*, terminology left over from modulation theory. Most of the circuits that we call modulators are actually mixers. The power amplifier in a classic amplitude modulated (AM) transmitter operates as a power mixer. The circuit traditionally called a "modulator" is really just an audio power amplifier.

Fig 5.6 shows a practical version of the circuit we have designed. We use a 1-V local oscillator signal at 10 MHz with RF amplitude of 0.2 V at 14 MHz. The drain is terminated in 50 Ω by way of a wideband transformer with a 5:1 turns ratio, resulting in a drain load of 1.25 kΩ. The calculated output powers for all frequencies appear in **Table 5.1**. These are very close to those measured when we built the circuit with the FET we had characterized. The calculations are in the *Mathcad* file mentioned earlier.

The two converted, or mixed outputs at 4 and 24 MHz have equal amplitudes, which are much less than the amplified RF output. The amplified LO is a large signal, close to the maximum possible from a J310 FET with a 12-V supply with the drain impedance used. This mixer topology is normally built with a tuned output. Tuning would eliminate the large drain voltage at the 10-MHz LO frequency. This would then allow a larger LO power to be used, which would increase conversion gain.

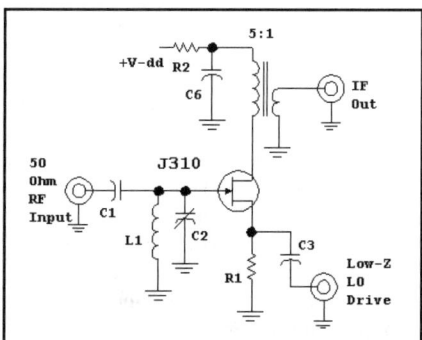

Fig 5.6—JFET mixer with a wideband output termination using a 5:1 turns ratio transformer. LO power is applied to the source, but this still results in LO between the source and drain, making this circuit the equivalent of Fig 5.2.

Table 5.1

4	−8 dBm	Lower Sideband mixed (down converted) output
10	+18.9	Amplified LO (feedthrough)
14	5	Amplified RF (feedthrough)
20	−0.1	Frequency doubled LO
24	−8	Upper Sideband mixed (up-converted) output
28	−28	Frequency doubled RF

Generally, FET mixers (including those using MOSFETs) will have an optimum conversion gain that is below the amplifier gain by 12 dB when the same terminating impedances are used.

The JFET example presented is but one of many devices that will produce mixing action. Mixing usually arises from *nonlin-*

ear device behavior. Mixing can also be produced in a system with time-dependent parameters. But, an ideal linear amplifier will never produce mixing. Even-order curvature in a device characteristic is the nonlinearity needed for mixing.

The simple single ended JFET mixer of Fig 5.6 becomes a practical circuit when the drain is tuned. But, it suffers from the wide spread in FET characteristics, making it difficult to use in a "plug-and-play" mode. A builder really needs to examine the FET to determine pinchoff and I_{DSS}, to establish bias, and to pick the right LO level. The following procedure may be used:

(1) Build the mixer with a 100-kΩ source resistor. Measure the source voltage to approximately establish the pinchoff.
(2) Place a small resistor or even a short circuit across the source resistor to infer I_{DSS}. (optional)
(3) Find (mathematically or experimentally) a source resistor that sets the dc source voltage at half the magnitude of the pinchoff.
(4) Apply LO power from a low Z source and increase LO amplitude until the peak voltage approaches the dc bias value. In the J310 example, the optimum LO signal would be nearly 2-V peak, or 4-V peak-to-peak. A high-speed oscilloscope is required.

The low impedance LO drive allows the FET to "look like" the source is grounded for RF input signals. Similarly, the RF tuned circuit should be one where the gate looks back into a low impedance at the LO frequency.

The single JFET mixer, when carefully done, is capable of excellent performance. We have measured 4 to 6-dB NF with input intercepts (third order) from 0 to +10 dBm with a 2N4416. The J310 is more difficult to drive owing to the increased I_{DSS}, but is capable of higher IIP3.

A bipolar transistor can be operated as a single-ended active mixer, shown in **Fig 5.7**. Lowest distortion will result from higher standing current, but this produces very low input impedances presented to the local oscillator, making drive difficult. Emitter degeneration reduces drive power, but can compromise noise figure. We have not performed careful measurements on this mixer.

Fig 5.8 shows a mixer using a single diode. Such mixers were once very common, especially for microwave applications. They have largely disappeared in modern times.

The usual diode mixer has no bias applied, but the LO signal is large enough that it causes the diode to conduct. When the diode conducts, it looks like a small resistance, allowing current to flow as the result of the applied RF. We envision the diode as a *switch* that is controlled by the LO. The switch is "on" for half of the LO cycle, and off for the rest. When on, virtually all of the RF power available can be delivered to a load at the IF port. But when the switch is off, none of the power can reach the load. With the RF reaching the IF load only half of the time, the voltage developed across the load from the RF generator is only half as high as it would be if present all of the time. Accordingly, the mixer has a 6-dB loss. **Fig 5.9** shows waveforms for a single diode switching mode mixer.

Switching mode mixers are extremely common, with most of the mixers we use in communications operating in this way. These mixers are typically passive and use no power supply; they offer no gain. The diode mixer of Fig 5.8 uses a series switch,

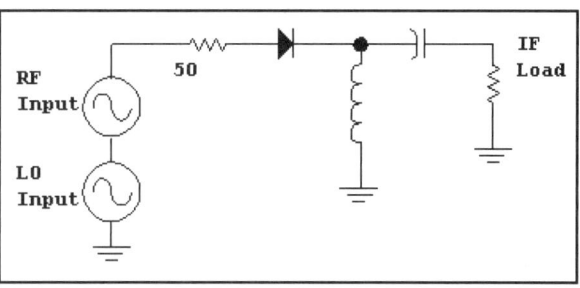

Fig 5.8—A simple diode mixer. RF and LO inputs generate an IF output, but the output is rich in signal feedthrough.

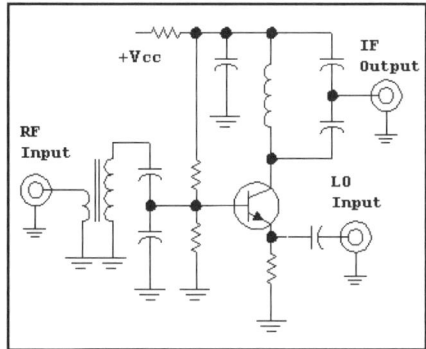

Fig 5.7—Simple bipolar mixer.

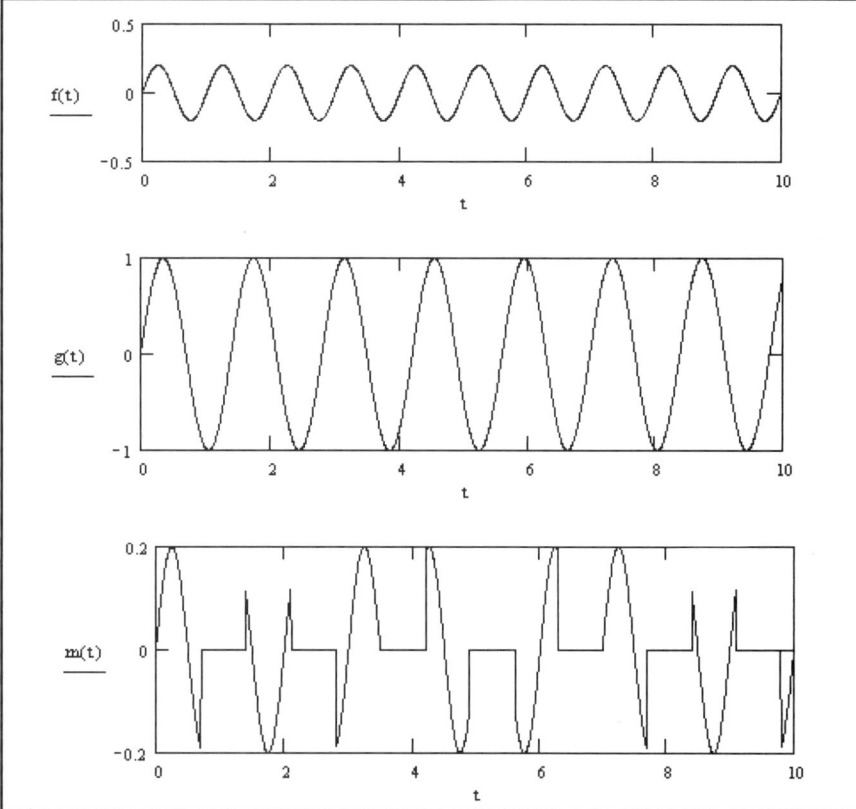

Fig 5.9—Time domain waveforms for a single diode switching mode mixer. The IF output at any instant is the RF input if the LO voltage is positive, but 0 when the LO is 0 or negative.

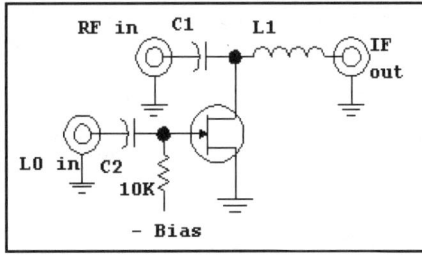

Fig 5.10—Switching mode mixer using a single FET. Although a JFET is shown, the mixer can also be implemented with a bipolar transistor, a MOSFET, or a GaAs FET. This circuit typically has a conversion loss of 6 dB. Input intercept (third order) can be from 0 to +20 dBm, depending on the FET type. LO energy at the RF port is typically reduced by 10 to 15 dB. Operating frequency will dictate the components in the diplexer filter, C1 and L1. See text.

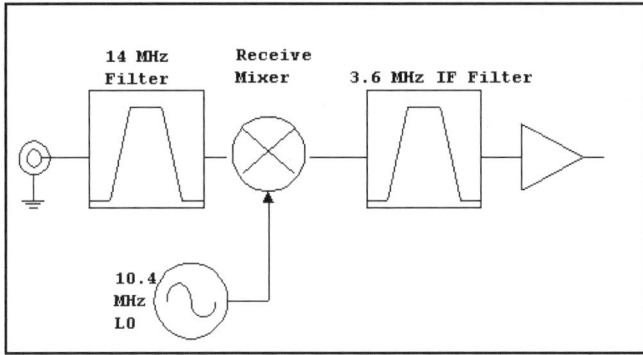

Fig 5.11—Partial block diagram of a 14-MHz receiver. The IF is 3.6 MHz, produced with a 10.4-MHz local oscillator.

but shunt switches also work well. FETs and bipolar transistors can be used in switching mode mixers.

Fig 5.10 shows a single FET as a shunt switch mixer. Steve Maas presented this circuit in detail in a 1987 paper.[1] We have used this mixer extensively in integrated form in GaAs integrated circuits.[2] The FET often has a bias applied to the gate, a negative voltage equaling the FET pinchoff. The LO is typically a sine wave with a peak value equal to or just over the pinchoff. All three ports are terminated in 50 Ω, but the LO presents a severe mismatch. The configuration shown is a down-converter with an IF below the RF and LO. Up-converters exchange the RF and IF ports.

The diplexer filter, C1 and L1 in Fig 5.10, isolates the IF from the RF port. The capacitor is a single element high pass filter while the inductor is a low pass circuit. A common application might use an IF much lower than the RF. One can then calculate a "crossover" frequency that is the geometric average of the IF and RF. L1 and C1 are then picked to have a reactance at the crossover equal to the terminations. Higher order diplexer filters will be needed if the IF and RF are closer. A bandpass/bandstop diplexer can also be used.

Mixer Specification and Measurement

We now examine mixers in more detail, seeking the properties needed to specify and understand mixers for use in a communications system.

Chapter 2 included some vital, yet less common specifications for amplifiers including noise figure and IMD. These phenomenon, which also occur in mixer circuits, are illustrated by the system of **Fig 5.11**, a CW receiver for 14 MHz with 10.4-MHz LO and 3.6-MHz IF.

IMAGES, SIDEBANDS, SUMS AND DIFFERENCES

The example receiver mixer is preceded by a 14-MHz bandpass filter that ideally passes only frequencies close to the 20-meter band. The 10.4-MHz LO drives the mixer to produce an IF output at the 3.6-MHz difference between the RF and LO frequency, 14 –10.4.

Temporarily remove the input bandpass filter and attach a wide range signal generator at the receive mixer RF input. There is now also a response at 6.8 MHz, for 10.4 – 6.8 = 3.6. The response to a 6.8-MHz input is called the image response. We evaluate the receiver, now with the bandpass filter reconnected, by attaching a signal generator to the input. Tune the generator to 14 MHz, deactivate receiver AGC, and measure the receiver output signal. This measurement works best with a modest input signal, perhaps -100 dBm. Note the audio output, then tune the generator to 6.8 MHz. Increase the generator level until the receiver output is identical to the original. The ratio of generator power levels is the receiver *image suppression*.

It is straightforward to build a bandpass filter at 14 MHz that will suppress 6.8-MHz signals by 100 dB or more. Early receivers, the old instruments now sought by collectors, used intermediate frequencies near 500 kHz, allowing 14 MHz to be received with a 13.5-MHz LO. The image response would then be at 13.0 MHz. It was difficult to obtain significant (by modern standards) suppression of 13 MHz in a 14-MHz filter.

The receive mixer example has two inputs: 10.4 and 14 MHz. We use the 3.6-MHz *difference* output response. But the mixer output will also contain a *sum* response, 10.4 + 14 = 24.4 MHz. The 3.6-MHz response is terminated in the usually reasonable impedance match of the 3.6-MHz bandpass filter. But all 24.4-MHz energy is generally reflected by the IF filter. That energy can get back into the mixer "output" where it might be reconverted back to 14 MHz, but in a different phase than the original signal where it can alter conversion gain and distortion performance. These problems are especially insidious with the popular diode ring mixers. It is for this reason that we often see extra resistive pads used with such mixers. They are often used in all three ports. Active mixers such as the FET discussed earlier are much less prone to this problem.

Assume that the incoming 14-MHz signal is modulated, containing a single upper sideband at 14.002 MHz. We analyze the behavior of the sideband by considering it to be an independent signal. It will be mixed down to IF without any disturbance from the original carrier. The sideband ends up at 3.602 MHz, still above the 3.600-MHz carrier appearing at the IF; it is still a USB signal.

Our receiving mixer would function just as well if we used a 17.6-MHz LO, 3.6 MHz above the input. An upper-sideband at 14.002 MHz applied to such a receiver would produce an IF response at 3.598 MHz, now below the 3.6-MHz carrier. *Sideband inversion* has occurred. This possibility should be investigated in any SSB system. The analysis is equally valid when a carrier is suppressed. Sideband inversion is often a practical advantage to the builder/designer. For example, a popular crystal filter form is the lower sideband ladder with greater stopband attenuation on one side than the other.

ISOLATION

We are always concerned about the output at one port of a mixer as signals are applied to the others. For example, we might ask how much LO signal appears at a mixer's RF port. This would be important in a receiver; we don't want a large LO signal to be radiated, for the mixer RF port may be attached to the antenna with minimal filtering. Even without radiation considerations, isolation can be important. If excessive LO was present, it could be reflected by a filter to re-appear at the mixer RF port where it would be converted to produce a dc output component. This could, in some mixers, alter the bias to change the mixer properties.

Isolation is easily measured for a mixer that is not already imbedded within a piece of equipment. If you are concerned with, for example, LO to RF port isolation, apply LO at a known level while examining the output at the RF port by attaching it to a spectrum analyzer or measurement receiver. The LO power at the RF port will be lower (we hope!) than that available from the LO source. The difference is the suppression. This will depend on mixer tuning in circuits such as the JFET described earlier. Often we hear folks talking about "mixer balance" in dB. Usually, they are concerned with port-to-port isolation, which can be enhanced with balanced circuits, a method discussed later.

SPURIOUS RESPONSES

Consider the transmitter application shown in **Fig 5.12**. In this example, we want to build a 7.1-MHz transmitter that works with an existing receiver using a 5-MHz IF. This will be accomplished by mixing the signal from a 2.1-MHz LO with that from a 5-MHz crystal oscillator. The output is filtered with a bandpass filter to produce the desired output.

The ideal output response from this mixer, assuming that the output filter is removed, is that shown in **Fig 5.13**. The desired sum product at 7.1 MHz is accompanied by a difference response at 2.9 MHz.

The ideal is rarely realized. **Fig 5.14** shows what we might actually see. This is a result of harmonic responses. Specifically, the output of a mixer excited by an LO at L MHz and RF at R MHz will be at F MHz,

$$F = n \cdot L \pm m \cdot R \qquad \text{Eq 5.1}$$

where n and m are integers. This spurious response, or *spur* generation relates to harmonics *created within the mixer*, even when the inputs are free of harmonics. The upper part of Fig 5.14 presents what we would see if n and m were allowed to take on values from 0 to 7 with the bandpass filter missing. The lower display is even more extreme, allowing values of n and m up through 15. (These data were generated with *Spurtune.exe*, a program distributed with *Introduction to Radio Frequency Design*.)

These uncalibrated displays are discouraging. Undesired outputs in such abundance would discourage anyone from ever using a mixer in a transmitter! Fortunately, not all spurious responses are of equal magnitude. The spurs tend to get weaker as the total order (n+m) increases. Further suppression can occur with some spurs as a consequence of balance that might be used in the mixer.

Spurs are also less with some system architectures over others. For example, if the transmitter considered here used a 12.1-MHz LO instead of 2.1, the outputs of **Fig 5.15** result.

A spur related to order "m" for the RF

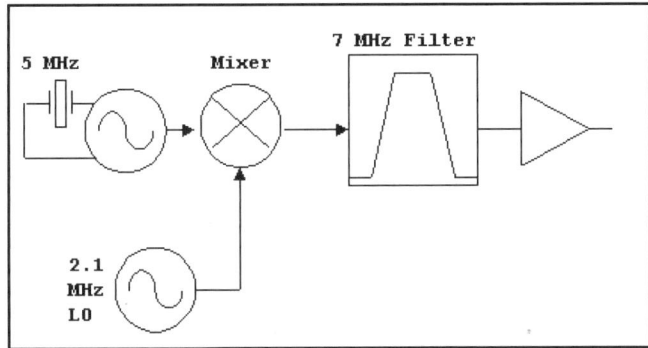

Fig 5.12—Mixer section of a 7-MHz transmitter with a 2-MHz LO and a 5-MHz crystal "carrier" oscillator.

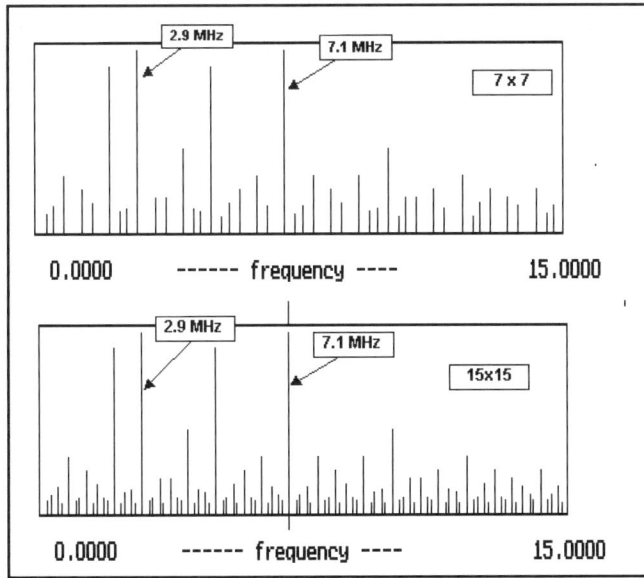

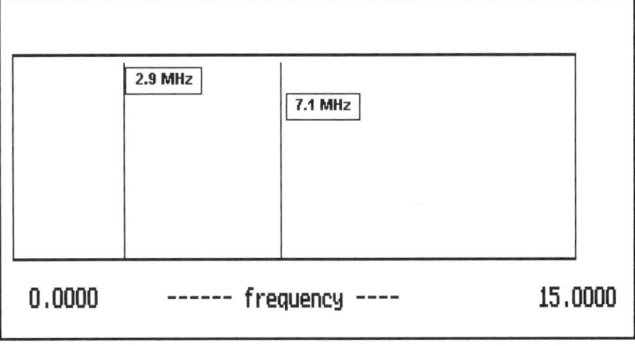

Fig 5.13—Idealized mixer output for the circuit of Fig 5.12 without the output filter.

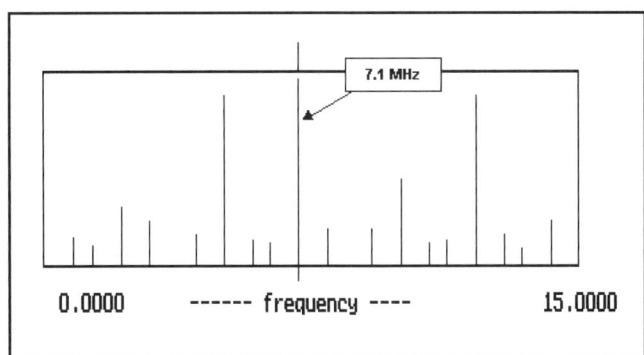

Fig 5.14—Mixer outputs with a variety of orders allowed, n and m to 7 in the upper curve and 15 in the lower.

Fig 5.15—Spur spectrum for the same transmitter, but with a 12.1-MHz LO. Spur orders through 7 are shown.

will generally have a strength proportional to the "m th" power of the input at the R mixer port. Hence, decreasing the RF input by 1 dB will drop an m-order spur by m dB. Mixer overdrive should be judiciously avoided. The worst possible cases are those where the IF is related to the output by a small integer, IF = k × RF, or IF = RF/k.

LO DRIVE LEVEL

Most commercial mixers are specified with regard to LO drive level. For example, the typical diode ring mixer is specified for +7 dBm. This is *not* the power that is actually delivered to the mixer port. Rather, it is the power available to a 50-Ω termination from the source that will eventually drive the mixer. Oscilloscope examination of the LO drive to a diode ring shows a severely distorted signal with less amplitude than the original sine wave driving a pure 50-Ω load. Many of the measurements we do with RF applications are substitutions rather than the familiar in-situ measurements of analog electronics.

Various mixers behave differently as LO power is varied. A small change in LO power makes almost no detectable difference with the typical diode ring. In contrast, the JFET studied earlier will show output decreasing almost linearly as LO drive drops.

CONVERSION GAIN (OR LOSS)

Mixers are all characterized by a conversion gain, meaning that we examine the converted output power vs that available to the RF port. The method of specifying the gain will vary slightly. A diode ring mixer, a passive circuit, might be specified with a loss, with 6 dB being a typical value. Active mixers such as the JFET considered earlier will be specified by power gain in a well-defined circuit, or perhaps by a conversion transconductance.

Terminal impedance is specified for a mixer. Most passive mixers show an RF input impedance that equals the IF termination while the JFET mixer at the beginning of this chapter shows a nearly open circuit as the input impedance at the gate, or a low impedance at the source like that of a common gate amplifier. Output (IF) port impedances are usually high with active mixers, but related to other port terminations with switching mixers. That is, the impedance seen at the IF port equals the value presented to the RF port.

NOISE FIGURE

Mixers all exhibit noise that can be characterized by noise figure. The measurement is similar to that of an amplifier. A wideband resistive termination at 290 K is first presented to a mixer input and the noise output is noted. Then, a stronger but known noise source is applied to the input, again while observing output noise. The "noise gain" is compared with normal available power gain to infer a noise figure.

The procedures, both for definition and for measurement, are nearly identical to those used with an amplifier. Two different mixer noise figures are available during any given measurement, as shown in **Fig 5.16**, with the difference being the image-stripping filter. (An image-stripping filter is one that prevents an image from reaching the input of a mixer.) Single sideband noise figure is the desired parameter, for most systems use filters to eliminate the image. Care is required to guarantee that SSB NF is measured, for noise figure is defined only for a single signal case.

Passive mixers usually have a noise figure equaling the numeric value of the loss. Hence, the usual diode ring with a 6-dB conversion loss will have a noise figure of 6 dB, or just a bit more.

INTERMODULATION DISTORTION AND GAIN COMPRESSION

While noise figure limits the weakest signal a mixer can process, intermodulation distortion and gain compression usually define strong signal behavior. IMD measurement is the same as is used with an amplifier, except that the output signals are observed at the converted frequency. Two RF signals or tones are combined in a suitable hybrid circuit with the result applied to the mixer being tested. The output tones are then observed at the mixer output frequency, along with the distortion products. An intermodulation ratio is established by the measurement, allowing an

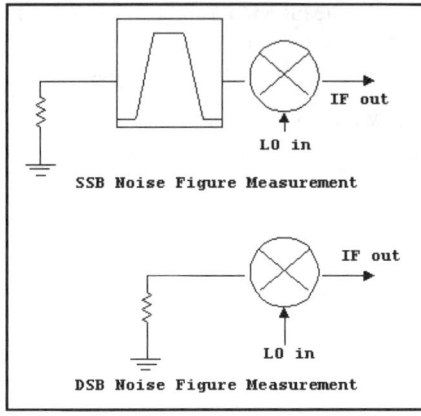

Fig 5.16—Scheme for measuring mixer noise figure. The upper circuit determines the usual single sideband NF. The lower applies noise at two frequencies and establishes what is often called double sideband noise figure. The bandpass filter eliminates any image response from the mixer input. DSB noise figure is typically 3 dB higher than the desired SSB noise figure.

input or output intercept to be calculated.

Gain is a constant for small signals, but eventually decreases as the RF level increases. A useful parameter is the available RF input power where the gain is below the small signal value by 1 dB.

Most mixer manufacturers specify their mixers by an input intercept value. This is in direct contrast to the amplifier folks who focus on the output. Both forms are fine, so long as the reader understands what is being specified.

Implicit in a mixer input intercept specification is an impedance. The usual specification uses 50-Ω terminations at all ports, and those terminations are wideband ones. This usually implies that the mixer was driving the input of a spectrum analyzer during the measurement, an instrument with a good 50-Ω input impedance at all frequencies. This occurs when the analyzer is set for at least 10 dB of input attenuation. This becomes very important with switching mode mixers where a poor output termination can destroy otherwise excellent IMD performance.

5.2 BALANCED MIXER CONCEPTS

Some intrinsic mixer problems can be reduced or eliminated when circuits are modified by adding balance. Consider **Fig 5.17**, part A, where we start with the familiar JFET active mixer. Local oscillator energy is applied at the source. FET gate-source capacitance couples the source voltage to the gate, degrading LO to RF isolation. Connecting a spectrum analyzer to the RF port reveals considerable LO energy at the RF port.

The term balance implies symmetry, a circuit with two sides or parts. A circuit becomes a balanced mixer through duplication, shown in Fig 5.17. The duplication presented in part B did not improve LO to RF suppression, but that in C does. The sources in C are in parallel, but the two gates are differentially driven. LO energy transferred to the gate of the first FET is exactly duplicated by that at the second FET, resulting in gate voltages that are in phase. But the transformer gate connection results in no net current, and no LO frequency signal at the transformer primary. The LO to RF port isolation is now excellent. Practically, one might expect a 30-dB improvement with balance.

The reverse, RF to LO isolation is also improved. A signal applied at the RF port results in gate voltages that are out of phase. But the sources are paralleled, resulting in reduced output at the LO port. RF to IF isolation is similarly improved, for the drains are paralleled. However, LO to IF isolation is not altered. LO is applied as an unbalanced or single-ended signal, with IF extracted from a similar single-ended connection. There are no balanced currents that can produce any cancellation. Generally, balance improves isolation between ports that have differing termination forms, differential vs single ended.

Although suppression has been improved in the circuit shown in section C of Fig 5.17, the topology has cancelled the desired IF output. The designer/builder must exercise care to guarantee that the addition of balance will not eliminate the mixing action.

A working variation of the previous mixer might use a drain transformer at the IF port, shown in **Fig 5.18**. A basic mixer, Q1, is duplicated in Q2, with a differential output connection through the transformer. The LO is still single ended, but is now a current from the drain of Q3 applied to the sources of Q1 and Q2. Although RF is applied only to the Q1 gate, this is a differential excitation, for Q1 and Q2 are a differential pair. As such, RF at the Q1 gate causes RF signal currents in Q1 and Q2 that are equal, but out of phase. Balance in this mixer improves LO to IF suppression (single ended to differential ports), but does not help RF to IF isolation.

The active balanced mixers presented are all assumed to be built from identical transistors. Although best when the circuits are fabricated in integrated form, they can still be practical with discrete devices.

Fig 5.19 shows balanced diode mixers. Part A presents a simple, yet very useful two-diode mixer circuit. LO is applied to a transformer and causes the diodes, now behaving as switches, to turn on during the positive half of the LO cycle. The diodes are off for the other half cycle. This mixer is configured as a down-converter; a higher frequency RF signal is applied to the diode junction through C, while lower frequency IF energy moves from the junction to the IF port.

It is instructive to examine the transformer action in greater detail. LO power causes, at one instant, a positive voltage at a dot on the transformer. But a positive voltage on one dot causes a positive signal on the other. The windings are wired to generate the polarities shown, one positive and the other negative at one instant in time. The diodes are identical, with

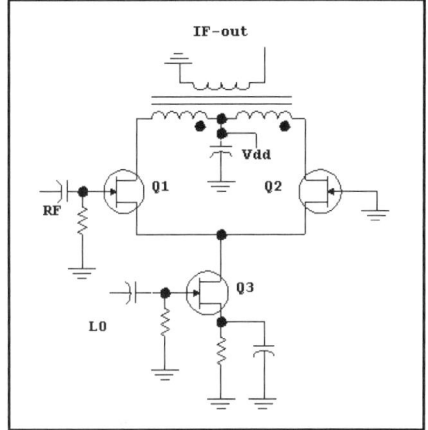

Fig 5.18—A JFET balanced mixer with single ended LO and differential IF ports. This mixer is similar to a bipolar classic, the RCA CA3028A. The RF and LO ports can be interchanged with little performance difference.

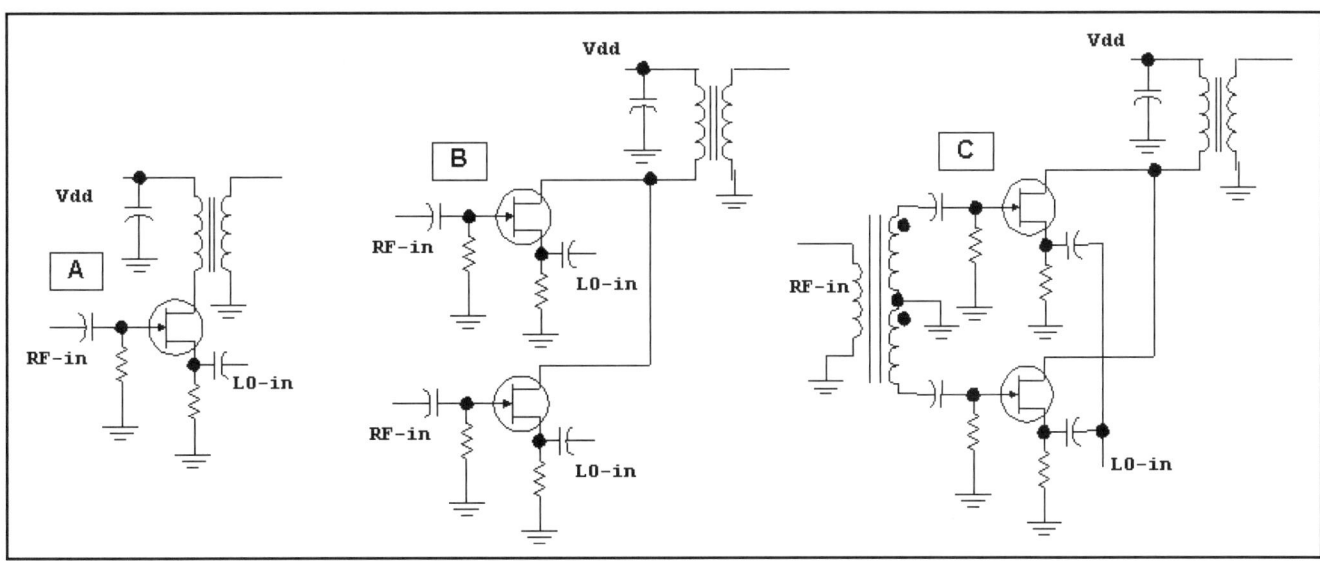

Fig 5.17—Evolution of balanced JFET mixer. The circuit in C offers good suppression, but poor gain.

matched on-resistance. Voltage divider action then causes the junction to be at ground, or zero LO voltage. Even when the LO polarity reverses, the identical diode reverse capacitance values generate zero LO voltage at the junction. LO to RF and LO to IF suppression are both enhanced.

The L and C values form a diplexer filter (see Chapter 3) in Fig 5.19A. The usual crossover frequency used is the geometric mean of the RF and IF, the square root of $(f_{RF} - f_{IF})$. Then, if the RF and IF impedances are 50 Ω, L and C are picked to have 50 Ω of reactance at the crossover frequency. More complicated diplexer filters may be needed if the IF is not small with regard to the RF.

Diode LO current is established by the diode characteristics and the source impedance provided by the LO system. The open circuit voltage must be high enough to cause the diodes to *turn on*. Greater available LO power produces higher diode current, which means that the diode on resistance is lower and conversion loss is lower. Hot carrier diodes are normally used in mixers of this sort, for they usually turn on with less voltage than a silicon junction type. The absence of a junction eliminates charge storage effects, allowing quicker diode turn-off, improving UHF performance. This mixer is still very practical at HF with silicon switching diodes such as the 1N4148. The diodes in a mixer should all be matched for voltage drop when forward biased to a few mA.

The local oscillator essentially causes the diodes to switch on and off. This combines with the transformer behavior to generate low impedance between the transformer center tap and the diode junction when the diodes are conducting. The impedance is high when the diodes are *off*. This behavior is extended to form a wideband mixer with the circuit of Fig 5.19B.

The mixers in parts A and B of Fig 5.19 present a poor load to the LO generator, for LO current only flows on half of each cycle. The addition of two more diodes, Fig 5.19C, provides a load on both halves of the LO waveform. With this connection, the LO action can be thought of as a square wave.

These three mixers (Fig 5.19, parts A, B, and C) are singly balanced with differential connections only at the LO port. But they evolve into a doubly balanced mixer in Fig 5.19D, which is labeled with LO polarity. During the polarity shown, diodes d1 and d2 conduct while diodes d3 and d4 are open circuit. The diode roles interchange when the LO polarity changes.

The switching action is further illustrated in **Fig 5.20** showing the two LO polarities. Diodes d1 and d2 conduct with d3 and d4 off in part A. Transformer action generates a low impedance connection between the diode junction and the T1 center tap. Bold lines in Fig 5.20 emphasize the current that now flows as a result of applied RF. Part B of the figure is the same, except for an opposite LO polarity. The diode ring mixer essentially creates a direct connection between the RF input, through the RF transformer T2, to the IF load. However, the polarity of the connection changes in synchronism with the applied LO. This process is called commutation; the diode ring is the classic example of a commutation mixer.

Fig 5.20 reveals another interesting property of this circuit: The RF transformer, T2, communicates the IF termination through to the RF port *without* impedance transformation. The transformer used at T2 is often thought of as having a 4:1 impedance ratio, and it can certainly function this way in some applications. But this is not consistent with the figure. Rather, one half of the center-tapped secondary carries current for each polarity of the LO. The inactive side has voltage across it from transformer action, but no current other than that needed to charge stray capacitance. (Care must be exercised whenever transformers with more than two windings are used with nonlinear devices.)

Time domain waveforms for a commutation mixer are shown in **Fig 5.21**. The LO does no more than to commute polarity of the RF signal appearing at the IF port.

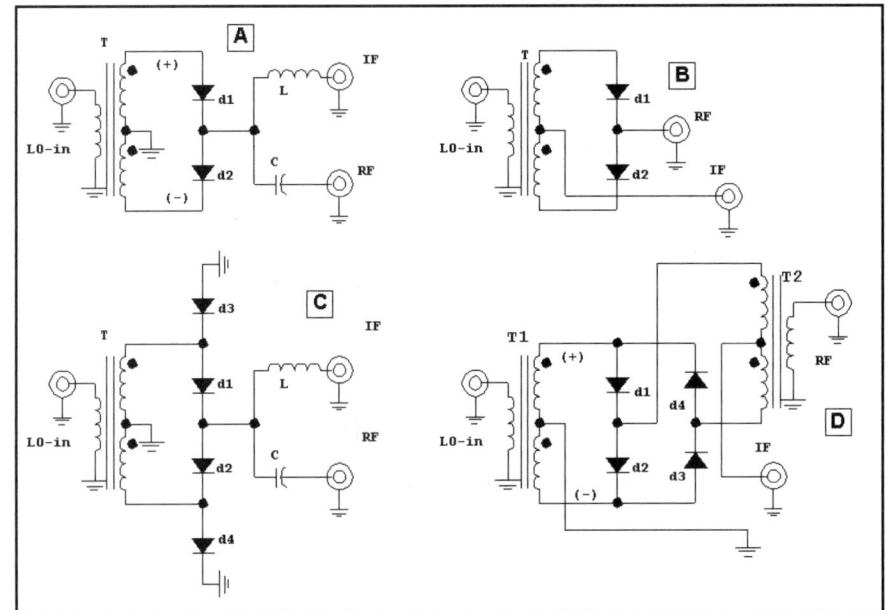

Fig 5.19—Evolution of diode mixers. Parts A and B show narrow and wideband versions of a two-diode mixer. The mixer is expanded to 4 diodes in part C, a circuit offering a better termination for the LO generator. These evolve into a diode ring, doubly balanced mixer in part D.

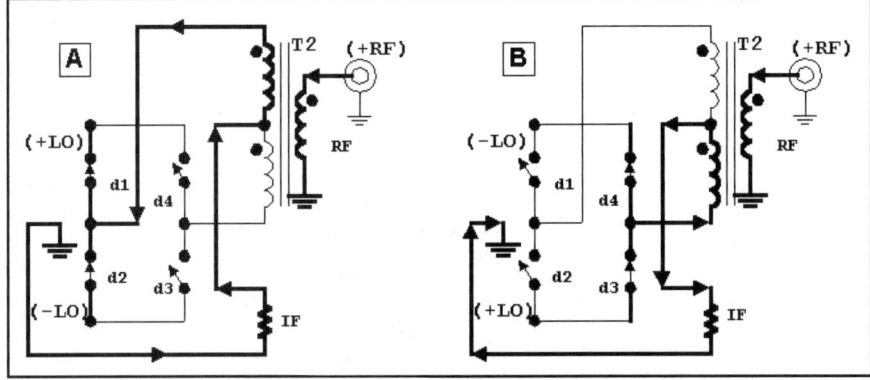

Fig 5.20—Diode ring commutating balanced mixers. See text for discussion.

5.8 Chapter 5

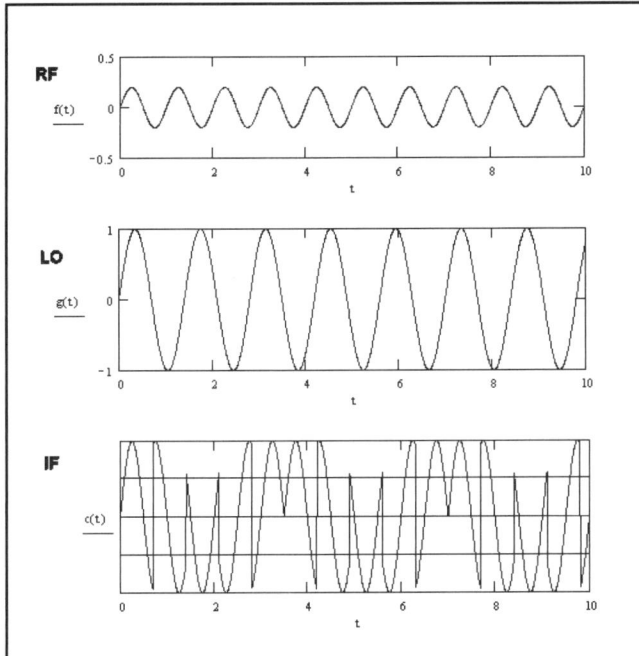

Fig 5.21—Waveforms for a diode ring commutation mixer. The RF and LO signals are those seen when the sources are examined into resistive loads. The IF signal is merely the RF waveform, except that the polarity is reversed when the LO is negative.

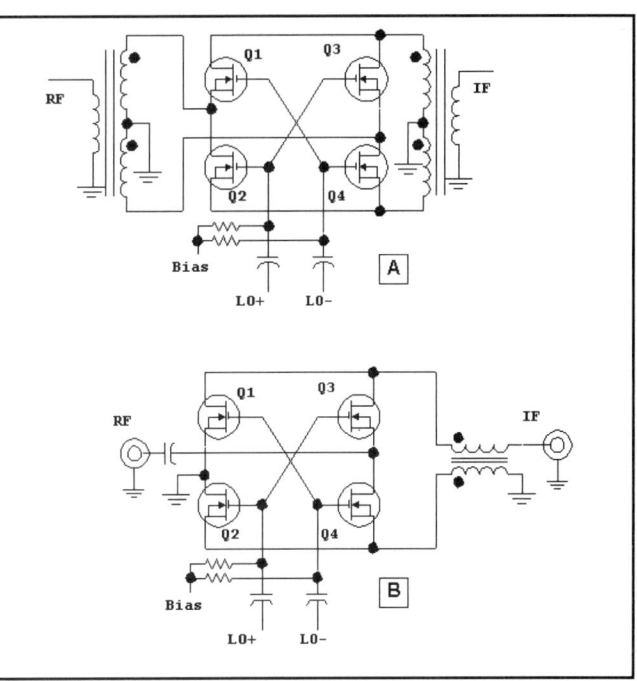

Fig 5.22—FET ring mixers using MOSFETs. The circuit at A is that originally describe by Oxner while that at B is a minimum transformer topology.

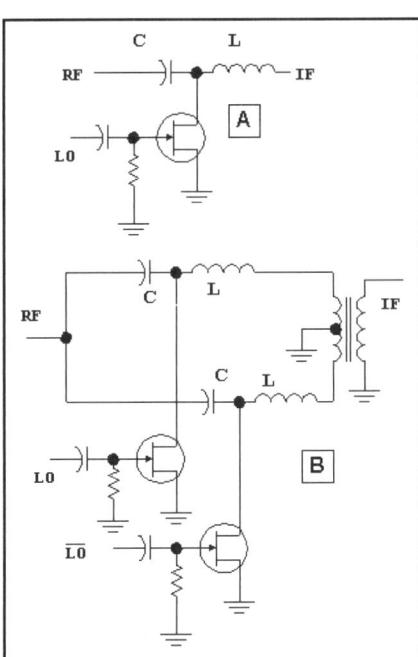

Fig 5.23—Evolution of the Maas mixer where balance improves LO to RF isolation.

Field effect transistors can also be used in switching mode commutation mixers as shown in **Fig 5.22**. Part A is a doubly balanced FET ring described by Ed Oxner of Siliconix.[3] Oxner's mixer originally used an integrated array of MOSFETs, the Siliconix SD8901. Many quad analog switches are also suitable in this application, although one should use those featuring low on-resistance MOSFETs. Discrete MOSFETs will also function in this circuit. A detailed analysis shows that exactly the same commutation action occurs in this mixer as we saw with the diode ring.

Oxner's mixer is an excellent performer, offering third order input intercepts in excess of +30 dBm. This low IMD occurred with a conversion loss of about 8 to 9 dB. The mixer functions well at HF, but degrades significantly at VHF. The FET ring mixer can be extended to higher frequencies with other technologies. In some measurements we saw conversion loss under 6 dB with large area monolithic GaAsFETs, but IMD was not as low as observed with the MOSFETs.[4]

The variation in Fig 5.22 part B uses only one transformer. Performance is similar to the other ring, although the intercepts are usually not quite as high.

The passive FET mixer using shunt FETs, **Fig 5.23**A, can also be extended with balance. Duplicating the circuit with differential LO and IF, but a single ended RF results in a singly balanced mixer, Fig 5.23B. Typical LO to RF isolation is 40 dB, even at low microwave frequencies.

Balance is an extremely powerful and general design tool that can often be applied to enhance port-to-port isolation. If any mixer is lacking in, for example, LO-to-RF isolation, placing two of them in a balanced pair will often enhance isolation by another 30 dB, with a bonus of a 3 dB increase in IIP3.[5]

Mixers and Frequency Multipliers   5.9

5.3 SOME PRACTICAL MIXERS

The Gilbert Cell

By far the most popular integrated mixer circuit available is the Gilbert Cell, named for Barrie Gilbert of Analog Devices. Gilbert developed a "four quadrant" multiplier circuit as an extension of a circuit presented earlier by Jones in US Patent 3,421,078 issued in 1966. The revised circuit is described in more detail in the text by Gray and Meyer.[6]

The Gilbert Cell is based upon the simpler mixer circuit shown in **Fig 5.24**. RF drives the base of Q1 to produce the combined dc and RF current that is then applied to the common emitters of a differential amplifier, Q2 and Q3. LO energy applied differentially to the dif-amp bases causes the RF to be toggled from one collector to the other. The IF termination is a balanced load, usually created with a transformer. This topology improves RF to IF and LO to RF isolation, for the RF input is single ended while the IF output and LO input are differential. This circuit was available from RCA in IC form as the CA3028A. This mixer suffers from poor LO to IF isolation, for differential drive at the bases of Q2 and Q3 produce directly amplified responses at the differential collectors.

The Gilbert Cell in rudimentary form, shown in **Fig 5.25**, contains a pair of these differential amplifier mixers. RF is applied to the lower differential amplifier, Q1 and Q4, producing two currents containing dc bias and the RF signal. These drive the emitters of identical differential pairs that are switched by the same LO signal. The Q3 and Q5 collector currents are in phase with each other with regard to LO drive; Q2 and Q6 share the other phase. However, one of the two output collector connections is "twisted" before attachment, producing a connection that cancels LO appearing at the IF. Port to port isolation is now excellent for all combinations.

Most Gilbert Cell mixers are integrated. The popular MC1496 and similar devices have been replaced with ICs that include internal biasing resistors. The most popular of these is the NE-602 shown in **Fig 5.26**. This version includes load resistors as well as input biasing. One can actually measure the collector resistors with an Ohmmeter; the RF input resistors do not really appear to be there, although network analyzer measurements show the resistors to represent a good model. The test circuit of **Fig 5.27** was fabricated to evaluate the NE602.

The conversion gain for this mixer was 20 dB with LO drive of 0 dBm (632 mV pk-pk at pin 6) with the test circuit of Fig 5.27. Early Signetics data recommends a minimum LO of 200 mV peak-peak, −10 dBm in our test circuit. Conversion gain dropped to 14 dB at this level in our measurements.

Both the RF and IF ports were floating in the test circuit, allowing balanced drive

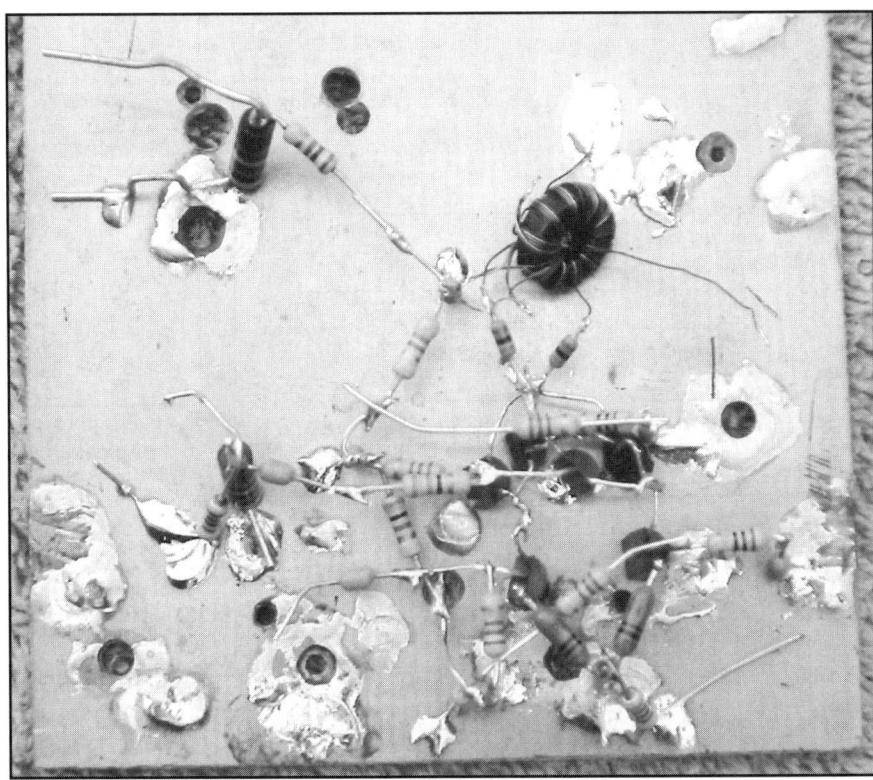

Experimental discrete transistor version of a Gilbert Cell Mixer.

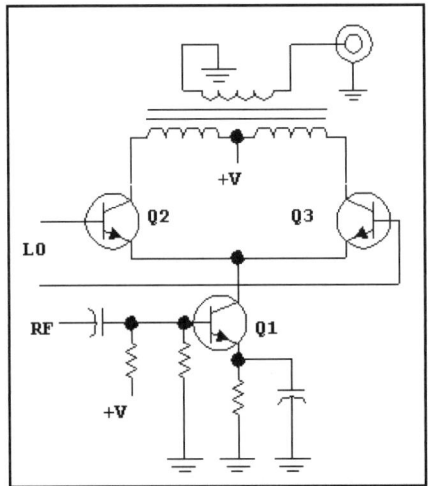

Fig 5.24—The basic bipolar differential amplifier mixer that is the basis for the Gilbert Cell. This mixer can be built with a CA3028A, or fabricated from discrete transistors. The 2N3904 would be suitable for HF applications. Biasing resistors (not shown) set the Q2 and Q3 bases at approximately mid supply.

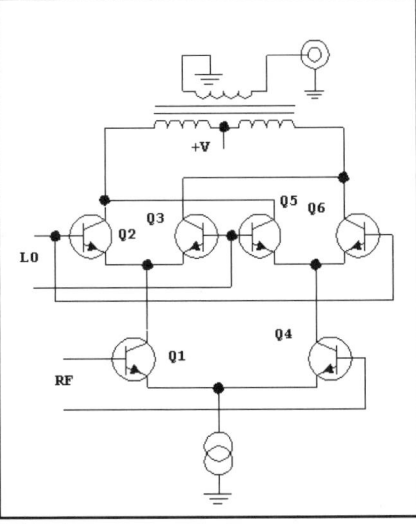

Fig 5.25—Fundamental Gilbert Cell mixer. The collector load is sometimes realized with resistors, although this will degrade intercepts, for internal load resistors absorb power that would otherwise be available to an external load.

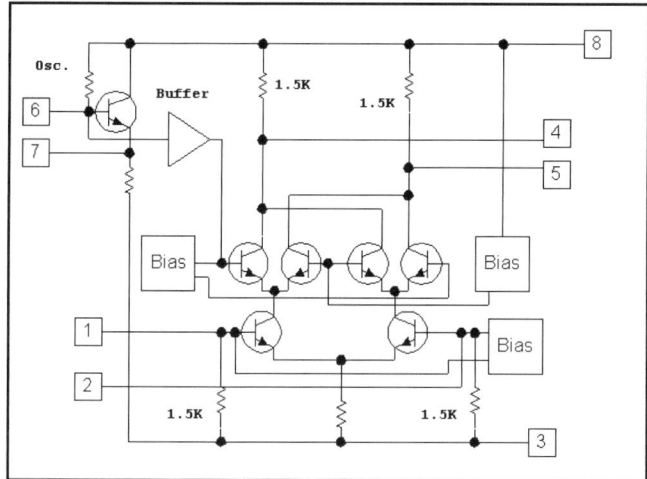

Fig 5.26—Equivalent circuit of the Philips NE602/NE612.[8]

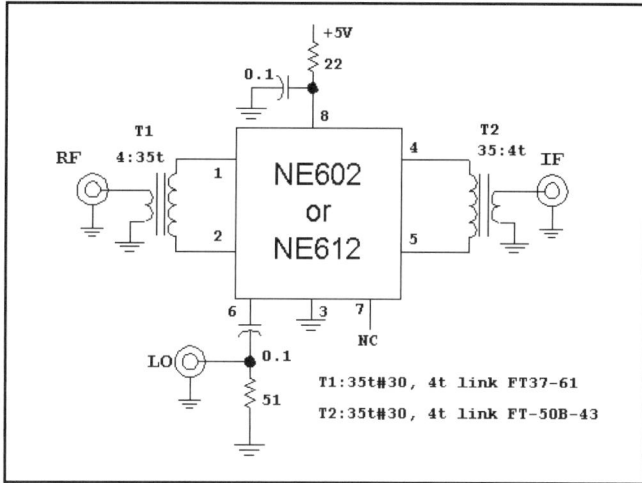

Fig 5.27—Test circuit used to evaluate the performance of the NE602. Most measurements used a 14-MHz RF, 19-MHz 0-dBm LO, and an IF of 5 MHz. The output 1 dB bandwidth extended from 0.5 to 10 MHz with the transformer shown. The RF port impedance match was a return loss of 19 dB while that at the IF was 15 dB. The internal oscillator was not used in these experiments.

to balanced loads. This balance could be altered experimentally by bypassing one end of the transformer. Bypassing pin 2 reduced gain by 2 dB and degraded the input impedance match. A similar exercise at the output (pin 5) degraded gain by 4 dB. Of greater import, unbalanced termination at either port degraded port-to-port isolation. Balanced RF drive will also alter product detector performance.

Our best IMD performance resulted with a single ended RF drive. IP3in was then –17.5 dBm with conversion gain of 18 dB and 0 dBm LO drive.

Single sideband noise figure was measured at 7 dB for this test circuit. This measurement was realized with a 15-MHz low pass RF filter and a 19-MHz LO.

We usually think of the Gilbert Cell as an integrated circuit. However, there is nothing fundamental to preclude building these mixers in discrete form. A discrete Gilbert Cell mixer built from 2N3904 transistors is shown in **Fig 5.28**. No special transistor matching was used, although all transistors came from the same bag with identical manufacturer and date codes. The chance is reasonable that they came from the same silicon wafer.

The circuit presented some VHF oscillation difficulty when power was initially applied. Although the problems occurred at VHF, LO harmonics mixed with the VHF signal to produce a low frequency output that moved in frequency as our hand was moved close to the circuit. The frequency could also be tuned with changing supply voltage. The oscillations were suppressed with the 10- and 36-Ω resistors included in Fig 5.28.

The mixer was biased to either 5 or 15 mA with most experiments performed at the higher level. Single-ended drive is used for both RF and LO inputs, slightly compromising port-to-port isolation. **Fig 5.29** shows the IF port output spectra. Conversion transducer gain for this circuit was 18 dB (15 mA, P-LO = 0 dBm, F-LO = 10.4 MHz and RF = 14.3 MHz.) Increasing LO drive by 10 dB made no difference in gain, but a drop to –10 dBm produced a 1-dB gain decrease. RF and LO signal appear in the wideband IF output with both about 14 dB below the respective input levels. Numerous other spurious outputs are present, all expected mixer spurious responses. Most would be lower in magnitude if the circuit was actually integrated. This circuit had a third-order input intercept of +11 dBm with 15-mA bias and 0-dBm LO power.

Decreasing the standing current to 5 mA produced an IP3in=–2 dBm, with 16-dB gain, still dramatically better than the

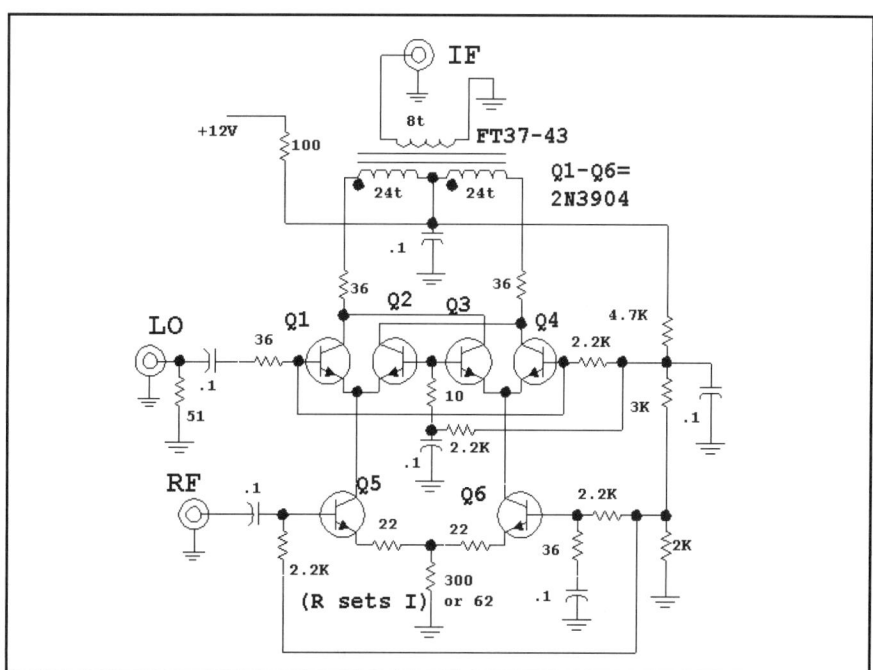

Fig 5.28—Gilbert Cell mixer built with discrete transistors. A resistor (300 or 62 Ω) at the bottom sets the bias current for the overall circuit.

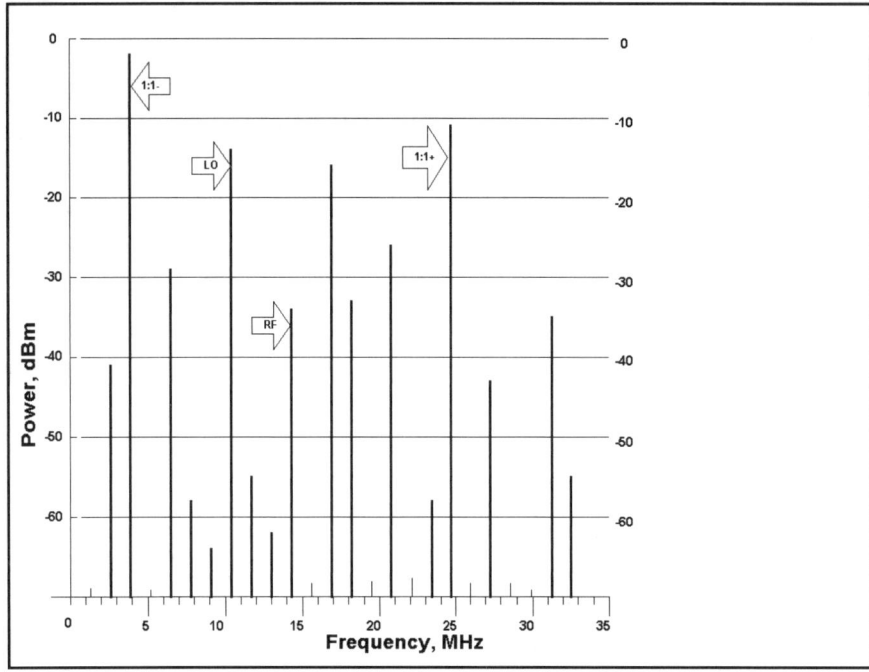

Fig 5.29—Output spectrum observed with the mixer of Fig 5.28. See text for details.

NE602. A diode noise source was used to measure DSB noise figure of 10.8 dB.

Degeneration (22-Ω resistors in the emitters of Q5 and Q6) was needed in the RF input stage to reduce IMD. However, this degraded the noise figure.

Although the main tool used to improve IMD performance in a Gilbert Cell is to increase current, feedback can also be applied. The experimenter should examine the work of Trask.[7]

Some of the integrated Gilbert Cell mixers that were once popular (e.g., MC1496, NE602) are becoming difficult to find. The topology remains popular and is often found as part of a larger, multiple function IC. Some Gilbert Cells are available internationally, although design data is sometimes difficult to obtain. One example is the SN16913P, from Texas Instruments Japan. This device is slated for discontinuation at this writing. It appears similar to another discontinued TI part, the TL442. The Toshiba TA7358P is still in production and could be a viable replacement in new designs. (Thanks to JG1EAD and JA3FR for information on Japanese parts.) There is ample challenge available to the experimenter.

Dual Gate MOSFET Mixers

JFET mixers were discussed earlier. A related device is the *metal oxide silicon field effect transistor*, or MOSFET. While the usual JFET is a depletion mode device, the typical MOSFET is an enhancement mode part. See the References chapter of any recent issue of *The ARRL Handbook* for definitions and further information. MOSFETs were, at one time, often built with two gates with that closest to the source termed "gate 1." When one of the gates is forward (positive) biased with respect to the source, the device behaves much like a JFET with the remaining gate as the controlling element. These devices are often modeled as a cascode connection of single gate FETs. Mixers can, of course, be built with MOSFETs, for they exhibit the same quadratic transfer characteristic seen with the JFET.

Fig 5.30A shows a mixer type that was very popular from the mid 1960s until about 1990. This circuit uses a dual gate MOSFET, an insulated gate topology with two parallel gates. A rule-of-thumb is that a dual gate FET will display a narrow band conversion transconductance of $^1/_4$ the gm expected for an amplifier biased at a similar current with similar terminating impedances. (This guideline is consistent with more refined analysis.) Traditional dual gate MOSFETs required an LO drive of about 5 V pk-pk at gate 2 to realize optimum gain.

Dual gate MOSFETs, although still available, are not as abundant as they once were. The alternative mixer of Fig 5.30B uses a cascode-connected pair of JFETs in a similar circuit. This connection was evaluated for noise figure, gain, and intercept. The 2N5454 FETs from our junk box are similar to the popular 2N4416, TIS-88, MPF-102, 2N5485, 2N5486, and many other components; any of these parts should perform well in this topology. Our initial attempt with this circuit presented a stability problem with an oscillation occurring at the resonant frequency of the input circuit. This was observed with a power meter attached to the IF output. The oscillation was eliminated when R1 was inserted across the transformer primary. A broadband IF output transformer is wound on relatively low loss type 61 ferrite core with a turns ratio to present a good output match to 50 Ω. An alternative winding

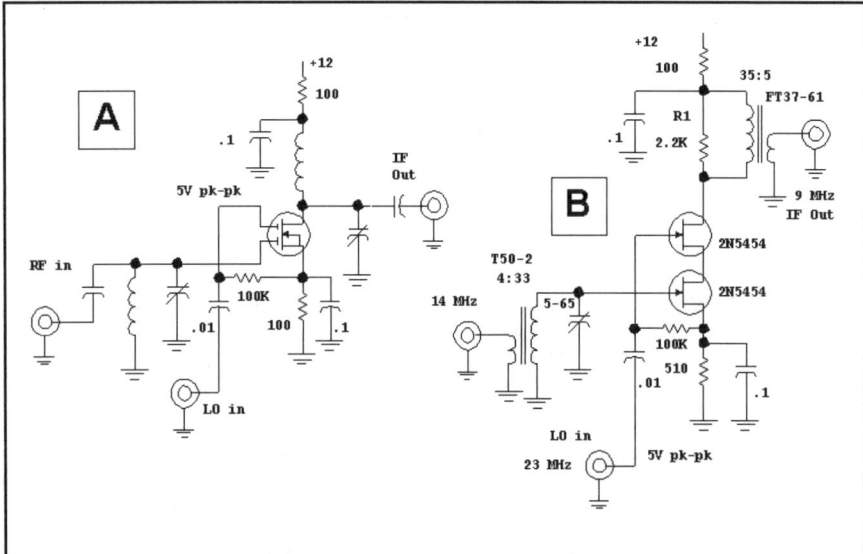

Fig 5.30—Part A shows a mixer using a dual gate MOSFET. Best gain occurs with around 5 V pk-pk at gate 2 for LO injection. The mixer at B uses a pair of JFETs in a cascode connection. This mixer is easily fabricated with nearly any available JFET type. See text.

5.12   Chapter 5

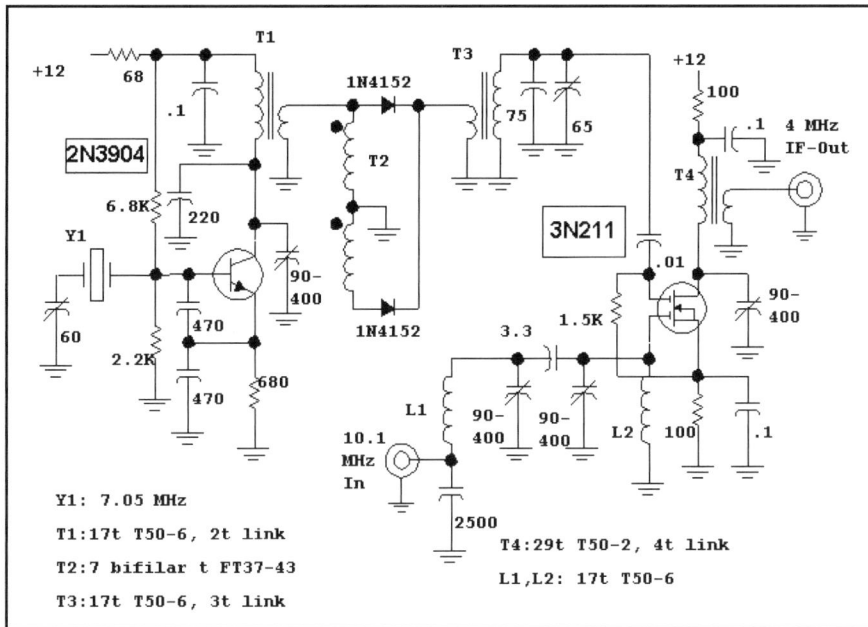

Fig 5.31—Schematic for a low noise 10.1-MHz converter.

would allow matching to a crystal filter. The mixer shown, biased for 3.4 mA at 12 V, has a measured conversion gain of 8 dB with a noise figure of 10 dB and IIP3 of +5 dBm. There is no balance in this circuit, so LO and RF energy is available at the IF port. This mixer is used in a simple superhet receiver appearing later in the book.

Many dual gate MOSFETs show very low *amplifier* noise figure with values of 1 dB being common. They can also function well in mixer applications. **Fig 5.31** shows a receiving converter with a measured NF of 6.6 dB and a conversion gain of 22 dB. This circuit needed an LO of 14.1 MHz to convert 10.1 MHz to 4 MHz. An available 7.05-MHz junk box crystal was used with a frequency doubler. The oscillator provides 10 mW to drive the passive diode doubler. The single tuned circuit then increases the voltage to the required level. This mixer has a low noise figure because gate 2 "sees" a low impedance at all frequencies other than that of the LO injection. Hence, noise energy within the LO system at the 4-MHz IF and at the 10.1-MHz RF does not reach the mixer output. The same mixer with a wideband LO drive circuit will usually have a noise figure closer to 10 to 12 dB. We did not measure IMD with this circuit.

The traditional dual gate MOSFET mixer biased for 5 mA at about 10 V will have OIP3 of around +20 dBm. The input intercept will be this value reduced by the conversion gain. The best dynamic range for mixers of this sort will occur when the impedance presented to gate 1 (RF input) produces lower gain. Lower impedances will also alter noise figure. The advanced experimenter (the one willing to measure and optimize results) can expect outstanding performance from either mixer in Fig 5.30.

Diode Ring Mixers and Related Circuits

The diode ring has become the workhorse for the communications industry. Although the mixer has loss, noise figure is low and intercepts are generally high, making it the best choice when dynamic range is critical. The lack of gain is not, in itself, a problem. It is important to use the ring with care if best performance is to be realized.

Probably the most critical characteristic of a diode ring, and most other switching mode mixers, is the need to carefully terminate the IF port. A proper termination (usually 50 Ω) means that output energy available from the mixer is absorbed. If power is reflected from the IF, it then impinges back upon the mixer IF port where it can be reconverted back to the RF, or to image frequencies. Reconverted components can then exit the mixer RF port where they are yet again available for absorption or another reflection. With each reflection can come phase shift and distortion.

Fig 5.32 illustrates the termination problem. A diode ring is used in a 14-MHz receiver where a 10-MHz LO converts the desired signal to a 4-MHz IF. But the mixer output also contains a 24-MHz signal. The mixer is terminated in an IF amplifier with the first selectivity appearing *after* the amplifier. Typical amplifiers have an input impedance that varies with frequency. Even if the amplifier input is close to 50 Ω at 4 MHz, it probably will not be 50 Ω at 24 MHz as well. The 24-MHz component will then be scattered from the amplifier input back to the mixer output where it can participate in further conversions, all undesired.

The mixer needs to be properly terminated for any and all signals that emanate from it. Assume the receiver is tuned to 14.00 MHz, but a strong signal appears at 14.01 MHz. That signal, once translated to the IF, is probably out of the crystal filter passband. It will then be reflected by the filter and returned to the amplifier output, possibly creating excess distortion there. If the amplifier uses negative feedback, the poor output termination for the 14.01-MHz signal will be reflected back to the amplifier input, creating an improper termination for the mixer.

The obvious question that arises when a *good* impedance match is specified is "How good?" Generally, we look for an IF termination that is better than a 2:1 VSWR, or a 10-dB return loss. This match is easily measured in the home lab with a return loss bridge, signal generator, and sensitive detector. The detector could be a special receiver, a spectrum analyzer, power meter, or even an oscilloscope (see Chapter 7). The match should be examined over a wide frequency range, and with a signal level low enough to guarantee that the terminating circuitry is not overdriven.

In many situations the IF port termina-

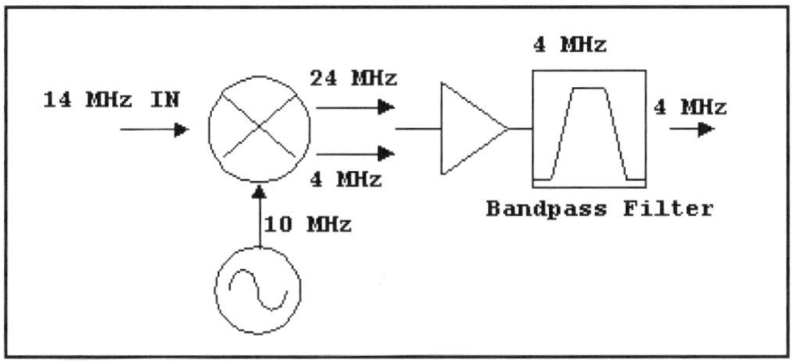

Fig 5.32—A 14-MHz receiver front end illustrating the problems of terminating a diode ring mixer.

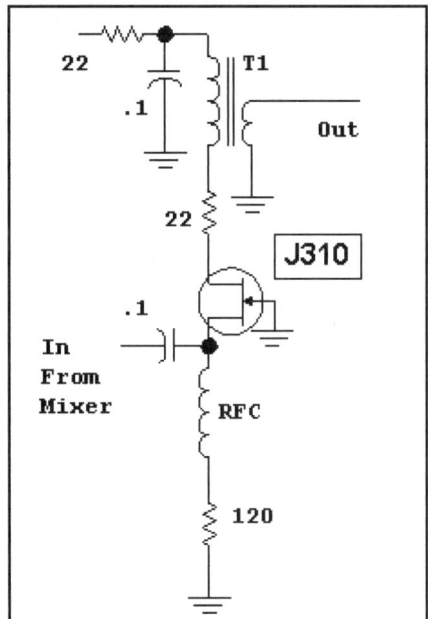

Fig 5.33—A post mixer amplifier using a junction FET. A high I_{dss} FET is required such as the J310. See the text for transformer discussion.

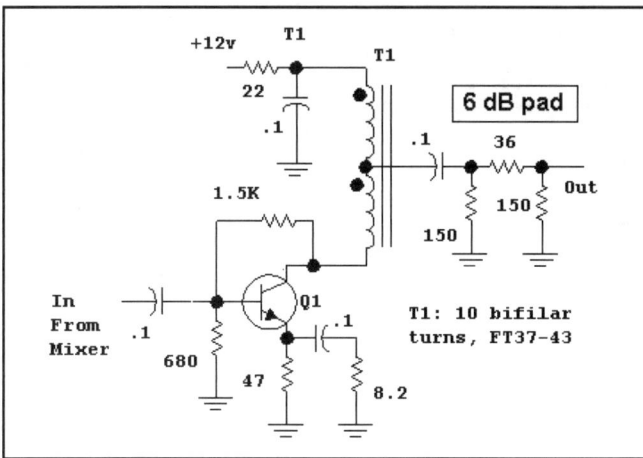

Fig 5.34—Post mixer amplifier using a medium power, high F-t bipolar transistor. See text.

tion requirements may be relaxed if the match is improved at the RF port. Generally, distortion and gain measurements will reveal the problems. The aggressive experimenter can build the instrumentation needed for these measurements.

Ideally, the best amplifier for terminating a switching mode mixer is one with excellent reverse isolation and a frequency invariant ("flat") input impedance. The amplifier must have good distortion properties, for it is often subjected to an entire band full of signals. The noise figure should be low, for it will add directly to the mixer loss to set the noise figure looking into the mixer. Finally, the gain should be high enough to compensate for mixer loss and loss in the filter that will follow, but not a lot more. Excess gain means that the signals become too large, stressing the following filter (crystal filters can be damaged by excessive signals, and can generate their own IMD) and stressing the distortion properties of the amplifier.

A grounded gate J310 JFET amplifier suitable for post mixer applications is shown in **Fig 5.33**. This circuit has good reverse isolation, so a crystal filter may be driven directly. The output transformer determines gain. A drain impedance of about 1200 Ω yields a gain of about 10 dB. We measured a third-order output intercept of +28 dBm for this amplifier when biased for $I_d = 14$ mA. A noise figure of less than 3 dB is possible with a slight input mismatch. The amplifier will normally yield an input match (return loss) better than 10 dB. Good input match and modest intercepts are found only with high current, which happens only with fairly high I_{DSS} FETs.

A favorite amplifier of ours (**Fig 5.34**) for terminating a switching mixer is a bipolar transistor feedback amplifier followed by a 6-dB pad. Negative feedback is used to set the gain and to stabilize the input and output impedances. This circuit was discussed in detail in the amplifier chapter. The output termination on a feedback amp will strongly influence the input impedance. As such, one should avoid driving a crystal filter directly with such an amplifier. The filter impedance changes rapidly with frequency, especially in the region at the passband edges. What may be a fine termination in the passband becomes an open or short circuit in the skirts and stop band. The resulting mixer termination may cause severe IMD problems.

These problems are largely avoided by placing a 6 dB pad in the amplifier output. This then guarantees an amplifier with a stable, frequency independent input impedance to terminate the mixer. It also guarantees a good source impedance for the crystal filter, another vital consideration.

The amplifier of Fig 5.34 uses a transistor usually specified for RF power or Community TV service. They are bipolar devices with a 1 W or better output capability and with an F_T that is at least 10 times the highest frequency IF where they will be used. The 2N3866 and 2N5109 are both available at this writing and work well in this service. Many other parts are suitable. Paralleled 2N3904s or similar plastic cased devices are also suitable and are shown later. The amplifier in the figure uses a bias emitter current of 50 mA and a collector termination of 200 Ω, provided with a bifilar transformer. The input impedance is very close to 50 Ω and is fairly flat through the HF spectrum. Typical OIP3 is +41 dBm if the attenuator is not part of the measured circuit. The 6-dB attenuator decreases the overall output intercept to +35 dBm. The gain is 21 dB, dropping to 15 dB with the 6-dB pad.

This particular amplifier uses the feedback resistor for transistor biasing, so changing circuit elements will alter biasing as well as feedback. Altering feedback with constant bias current will maintain the output intercept while changing the gain. Input intercept will change accordingly.

Noise figure for the amplifier of Fig 5.34 will vary with transistor type and bias, but values of 5 dB are typical. Careful measurements on one version of this circuit showed lower NF with reduced current, offering some DR optimization.

An attenuator at the input of a feedback amplifier will generate stable port impedances as well as good output intercept. However, the input pad degrades noise figure.

Some receiver designs (with high level mixers) demand amplifiers with higher intercepts. This is possible with higher current. However, the output pad compromises efficiency. A better solution uses two feedback amplifier stages with attenuation between. The impedances are stable and noise figure and intercepts are maintained.

There are some situations where no amplifier is required. It is still important to maintain the proper mixer terminations. An example might be the front end of a spectrum analyzer, shown in **Fig 5.35**. The first mixer is preselected with a low pass filter and produces a first IF of 1.5 GHz. The pad in the mixer output stabilizes impedance in both directions, ensuring mixer and filter performance. The second mixer produces a 50-MHz IF where an amplifier with a pad is now used. This topology has a much higher noise figure than the usual receiver, but is capable of excellent IMD performance, the parameter of greater interest for measurements.

Fig 5.36 shows a different approach to the problem. Here, a mixer is followed by a diplexer filter that then drives a post mixer amplifier using a dual gate MOSFET. (40673, or 3N211 used.) The 2.2-kΩ gate resistor is transformed to look like 50 Ω to the mixer through an L-network, L1 and C1. This only provides a termination at the IF, 1.9 MHz in this example. Sum products are terminated with a high pass filter paralleling the L-network. The preselector filter was a triple tuned circuit in this example with about 3-dB loss while the MOSFET amplifier has a noise figure of about 3 dB, for a net NF of 12 dB. Overall gain is 9 dB. Measured input intercept for the system was +15 dBm. This two-decade-old scheme is not as strong as others, but can be an efficient one for battery operation. The broadband impedance match is marginal.[9]

Perhaps the ultimate IF termination for the switching mixer is a special crystal filter that presents a proper impedance at all frequencies. This filter, and similar amplifiers result from a now classic method described by Kurokawa, et al.[10] Such a filter is discussed in the next chapter.

Parts like the MiniCircuits SBL-1, TUF-1, and ADE-1, a SMT part, represent the standard diode rings. There are, of course, many more listed in their catalogs. These mixers are specified for a LO drive power of +7 dBm. (Recall that this is *available power from the LO source*.) The mixer is usually well saturated at this +7 dBm and LO drive changes do not alter gain. The "+7-dBm" mixers will continue to function with LO drives as low as 0 to +3 dBm, with reduced gain and degraded intercepts. Some Mini-Circuits parts are available for LO power as low as 0 dBm.

Mini-Circuits +7 dBm mixers are specified for an input 1 dB compression power of +1 dBm. A rule of thumb states that the input intercept of a diode mixer is 10 to 15 dB above P_{-1dB}, placing IIP3 at +11 to +16 dBm. These values are in line with our measurements for the TUF-1 and SBL-1.

Most mixer manufacturers also build mixers specified for LO power of +17 dBm. These mixers usually use two series connected diodes in each leg of an otherwise conventional ring. One example, the TUF-1H, has a +14 dBm value for P_{-1dB}, placing IP3in at +24 dBm or higher. Even higher power mixers are available, including some "level 27-dBm" devices with P_{-1dB} = +24 dBm.

A recent *QEX* paper examines the termination of high-level mixers to improve IMD.[11] That paper considers diplexer filters at both the IF and RF ports, as well as some modified LC filters. It strikes us that the Engelbrecht-Kurokawa methods may also be suitable for RF port terminations. The excellent paper by Stephensen is included on the book CD.

High Level FET Mixers

Very wide dynamic range receivers and low noise transmitters both demand high-level mixers. While some diode-based designs are suitable, they demand high LO power, a practical difficulty. Several workers have examined other devices as switches. The notable example mentioned earlier was the MOSFET ring described by Ed Oxner.

Perhaps the most exciting work published in the past decade in this area was a note appearing in Pat Hawker's ever popular and consistently informative Technical Topics column in Radio Communications.[12] Hawker presented previously unreported work on a new mixer topology by Colin Horrabin, G3SBI. This four-FET mixer, shown in **Fig 5.37**, differed from earlier circuits. Oxner's design used FETs as series switches while Horrabin used the FETs as grounded switches. This is still a commutating mixer, but transformer action now generates the needed signals. Horrabin's circuit used a monolithic quad

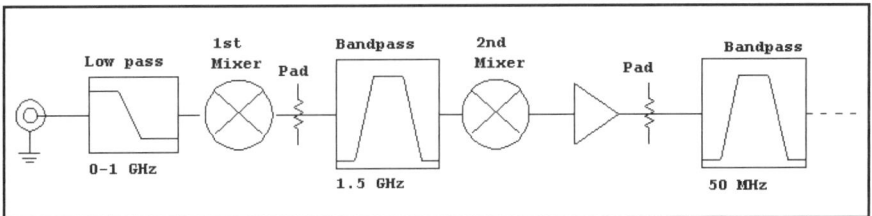

Fig 5.35—Front end of a spectrum analyzer showing ring mixers without amplifiers.

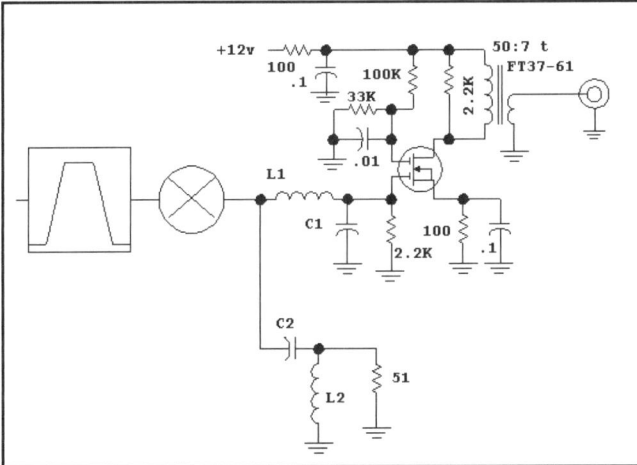

Fig 5.36—A mixer-terminating amplifier using a diplexer filter. This is a combination of a low pass and a high pass filter in this example, but could also be a bandpass and bandstop filter. This example uses a considerable impedance transformation at the amplifier input.

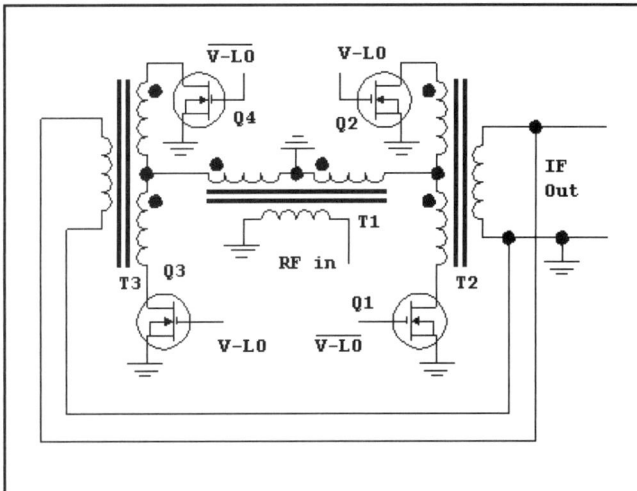

Fig 5.37—H-mode mixer using grounded FETs. This mixer, the work of Colin Horrabin, G3SBI, has produced third order input intercepts as high as +55 dBm. The circuit takes its name from the "H" shape presented by the transformers.

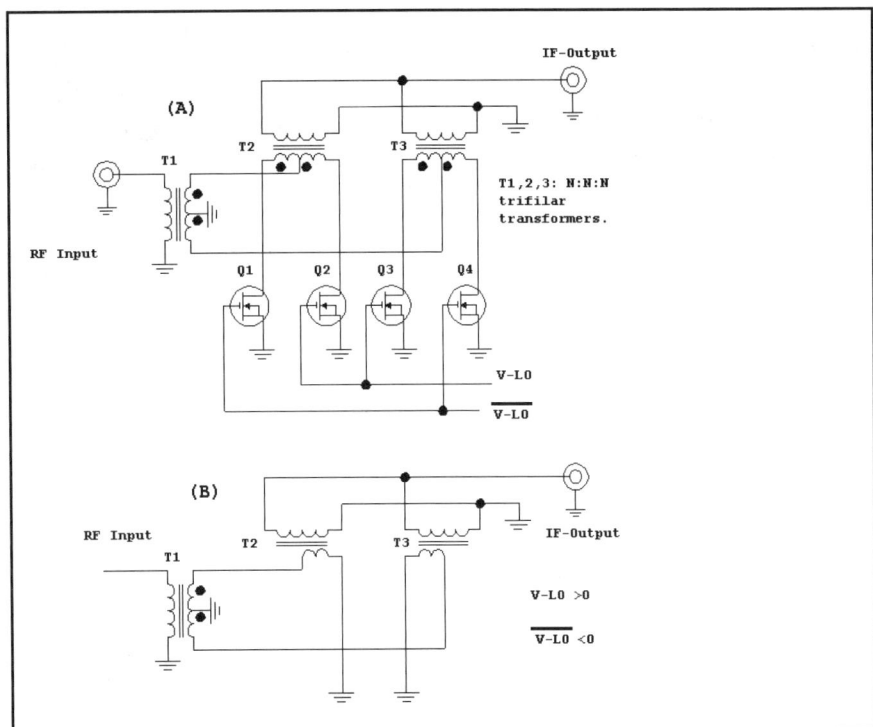

Fig 5.38—The H-mode mixer is redrawn to clarify operation. See text for explanation.

of MOSFETs, the Phillips SD5000, which is essentially the same MOSFET as used in Oxner's Si8901.

The operation of the H-mode mixer is understood with the redrawn circuit of **Fig 5.38**. Part A of the figure shows the basic circuit. Assume that at one point in time V-LO is positive. This causes FETs Q2 and Q3 to be on, creating a low impedance to ground. The other two FET switches are off, now modeled as open circuits. The resulting circuit is shown in part B of the figure. Transformer T1 is one with essentially three identical windings, with two configured as a larger center tapped secondary. Each secondary winding is now connected to separate output transformers T2 and T3. Part of the transformers are not shown, for they are connected to open circuits at this point in time. The currents in T2 and T3 add at the IF output.

The polarity changes as we advance one half of a LO cycle. Q1 and Q4 are now on with Q2 and Q3 off. The other two secondary half-windings are now connected. Although not shown in the figure, detailed examination confirms commutation.

Horrabin has measured values as high as +55 dBm for IIP3. It becomes challenging to build low IMD amplifiers to accompany this robust mixer. It is difficult to measure intercepts this high, and considerable effort has been expended by Horrabin and his colleagues in this pursuit. They attribute the excellent performance to a removal of RF input signals from the gate-source switch-on path. The configuration with grounded FET sources makes it much more difficult to modulate the LO action with applied RF. Practical front-end examples using this mixer are presented in Chapter 6.

5.4 FREQUENCY MULTIPLIERS

Closely related to the mixer is a commonly used circuit, the frequency multiplier. This is a circuit with the predominant output occurring at a frequency that is an integer multiple of the input. We saw frequency multiplication when a local oscillator was first applied to a mixer; the action was a natural consequence of the circuit nonlinearity.

The simplest frequency multipliers resemble a simple amplifier with a single device (bipolar or FET). If the output is tuned to a multiple of the input frequency and if the circuit is driven harder than it would normally be driven for amplifier service, efficient frequency multiplication can occur. Example circuits are shown in **Fig 5.39**.

While these circuits are simple and easy to implement, they often suffer from poor spectral purity. If the circuit is tuned to operate as a frequency tripler, the dominant output will certainly be at 3 times the input. However, there is a good chance that

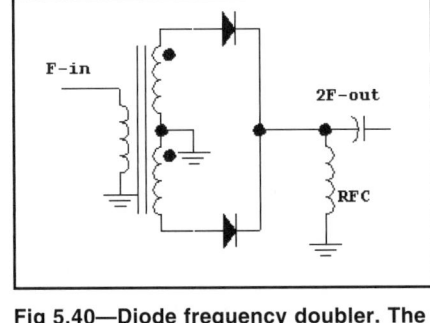

Fig 5.40—Diode frequency doubler. The diodes, ideally identical, can be silicon-switching types, such as the 1N4152 or 1N918 for use at HF and low VHF. Hot carrier diodes are recommended for UHF applications, or for critical, low phase noise HF applications. The transformer can be the familiar 10 trifilar turns on a FT37-43 core for HF applications. Often, this doubler drives a link on a single tuned circuit, eliminating the need for the RFC.

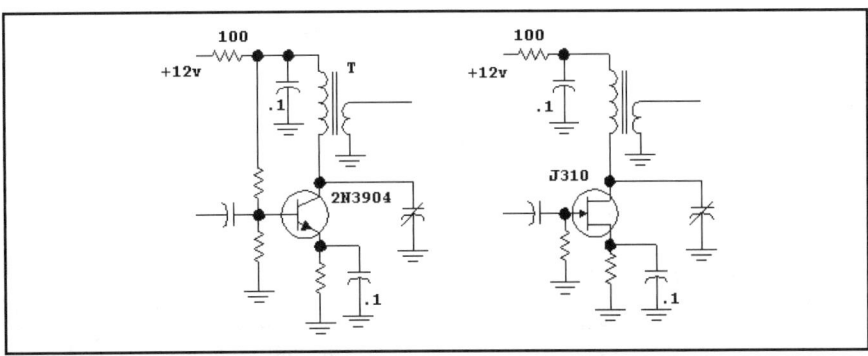

Fig 5.39—Simple, single-ended frequency multipliers using a bipolar transistor and a JFET. These classic circuits can still be useful in modern designs, but only if built with careful measurements.

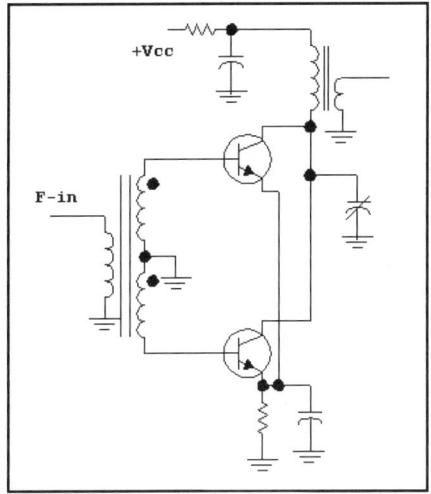

Fig 5.41—Basic push-push frequency doubler using balanced bipolar transistors.

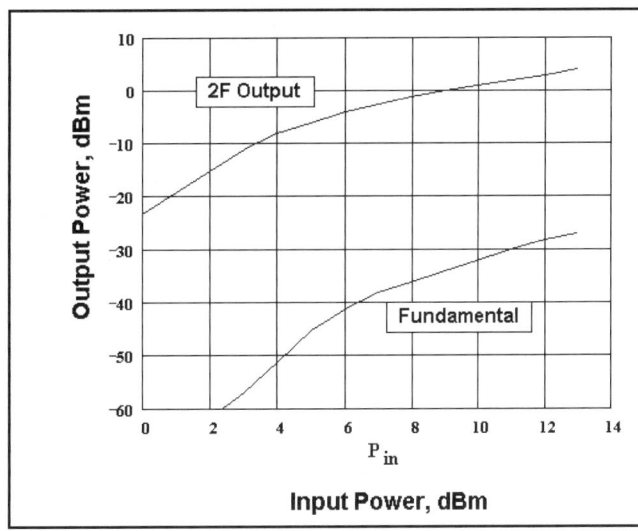

Fig 5.42—Output power and fundamental feed-through for a diode doubler using the circuit of Fig 5.40. The diodes were 1N4152 that had been matched with a DVM.

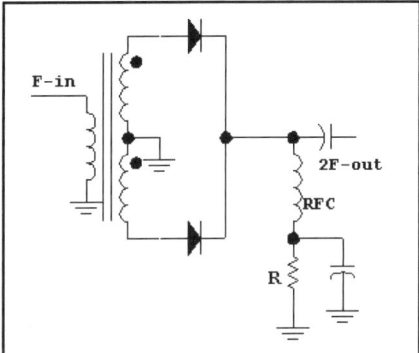

Fig 5.43—Improved balanced diode frequency doubler. Typical resistor values are from 10 to 220 Ω. See text.

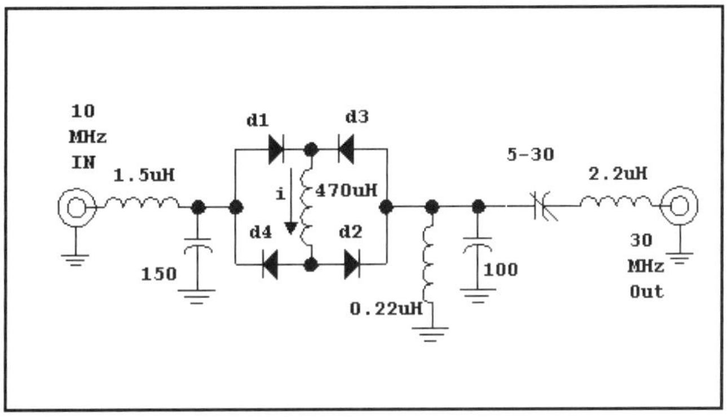

Fig 5.44—Frequency tripler using four diodes and a large inductance choke to generate a square wave. The output circuits are tuned to the 3rd harmonic of the input drive.

there can be considerable energy at the fundament frequency (the input), the 2nd, and the 4th harmonics of the input. The only way to improve the performance is through more filtering.

Not all output components occur at harmonics. As with Class C amplifiers, nonlinear C_{cb} of a bipolar transistor can result in non-harmonic spectral components.

As with mixers, we reduce the occurrence of spurious outputs with balanced circuits. A balanced frequency doubler is shown in **Fig 5.40** where two diodes operate in a circuit that is more familiar to us as a full-wave power supply rectifier. However, we now short circuit the dc output with a radio frequency choke, extracting only the 2F output. If the input transformer is well balanced and if the diodes are matched, it is common for the fundamental feedthrough for this circuit to be 30 to 40 dB below the 2F output. This circuit is passive and has no gain.

The diode frequency doubler idea is often extended to form the push-push doubler shown in **Fig 5.41**. This circuit is capable of gain and higher output power than is possible with the diodes. A passive doubler followed by an amplifier to regain the power lost in the diodes has similar power consumption and spectral purity.

The output power from the classic diode doubler (Fig 5.40) is typically around +2 dBm with a +10-dBm drive. A curve is shown in **Fig 5.42**. Although output grows with drive, gain drops. Gain tends to be more constant with the modified circuit of **Fig 5.43** where a bypassed resistor is added to "terminate" the dc component. The dc signal also provides a convenient tuning indicator. The added resistor decreases multiplication gain at drive levels below +10 dBm. However, gain is higher at the highest drive levels of +20 dBm where an output of +12 dBm has been measured. At a drive of +20 dBm, the 4× output is –1 dBm.

The drive to a balanced frequency doubler should be relatively free of even order harmonics. A distorted drive can destroy balance, which compromises the suppression of fundamental feed-through.

Odd order frequency multiplication is also common. Although possible with the single device circuits presented earlier, it is generally done with a balanced circuit that generates a square wave. Mathematics reveals that a square wave contains no even order harmonics. **Fig 5.44** shows a frequency tripler using a diode bridge tuned for a 10-MHz input with output at 30 MHz. The input circuit provides some impedance transformation from a 50-Ω source as well as some low pass filtering that helps to preserve a sine wave drive. Diodes d1 and d2 conduct on the positive drive polarity while d3/d4 conduct on the negative half of the cycle. Note that the current flowing in the intermediate inductor, shown with an arrow, is the same for both polarities. The multiplication gain for this circuit can be around -9 dB, but is level dependent. The circuit can also be tuned for ×5 multiplication with reduced gain. This circuit originated from Charles Wenzel.[13] The Web site in this reference is a wonderfully useful site with many other applications listed.

A slightly simpler odd order multiplier is presented in **Fig 5.45**. This circuit,

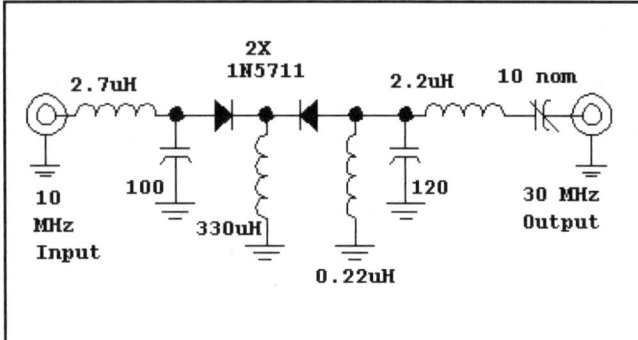

Fig 5.45—A simplified tripler circuit using only two diodes. This circuit is described in the Web site from Wenzel Associates. See text.

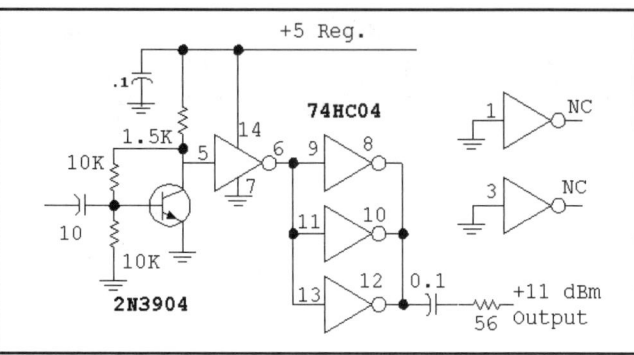

Fig 5.47—Simple limiting amplifier using a digital IC. Here, a HEX inverter generates an output with over 10 mW at the fundamental drive frequency. The inputs to unused sections should never be left floating.

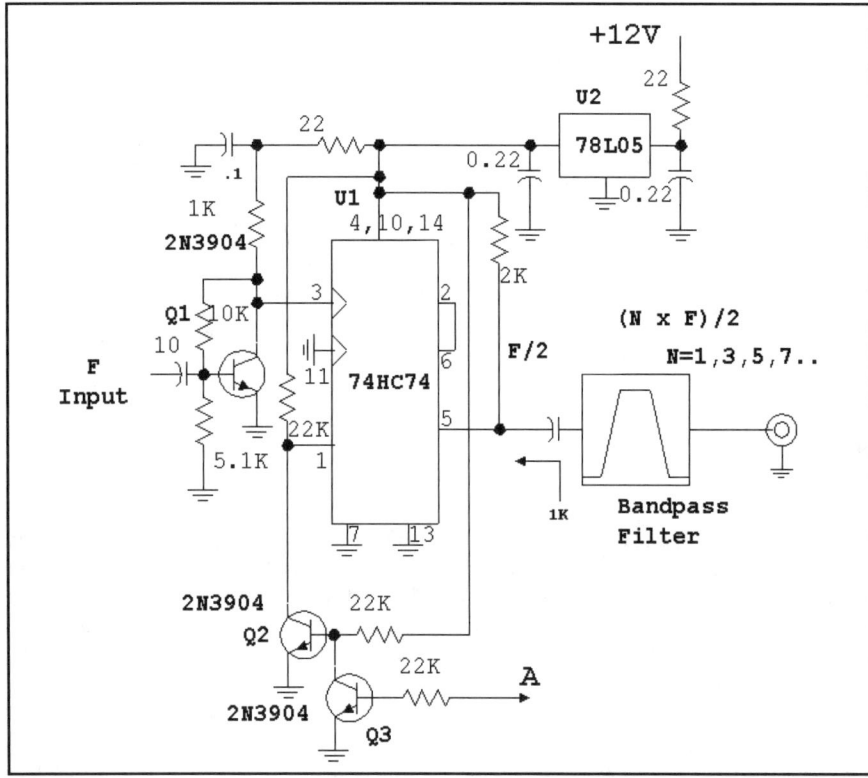

Fig 5.46—This frequency multiplier begins with a frequency division by 2 in a digital integrated circuit. The result, after division, is a very precise square wave. Odd harmonics can then be selected with a suitable bandpass filter. The output from the filter is typically –5 dBm when n=3. The bandpass should be designed for a termination of 1 kΩ at the IC end.

which uses only two diodes, can also be tuned for x5 operation. While we have not yet done the experiment, it would be very interesting to examine the insertion of resistance in series with the large inductance. The tripler circuits from Wenzel work well with either junction diodes or hot carrier devices, although the hot carrier diodes are preferred for low noise applications. The Wenzel web site discusses diode selection.

Square waves are easily created and processed with digital integrated circuits. This provides design opportunities for many interesting applications. **Fig 5.46** shows a scheme we have used for numerous VXO based transmitters. A signal is injected at the input to Q1 where it is converted to a logic friendly format. Levels from -10 to 0 dBm are suitable. The signal is then frequency divided with a 74HC74 D-flip-flop, resulting in an accurate square wave. This output is then applied to a bandpass filter where the appropriate harmonic is selected. Transmitters using this scheme are presented later. One example might use a 14-MHz crystal in a VXO. The divider output is a 7-MHz square wave, but one rich in 21-MHz energy. A 5% bandwidth triple tuned circuit bandpass filter selects the desired 21-MHz output while providing over 60 dB suppression of 7, 14 and 28-MHz components. This scheme offers two additional advantages: First, the oscillator operates at a frequency that is well isolated from the output, so buffering is extremely effective. Second, the output is easily turned on or off with the digital input at "A", allowing keying without disturbing the operating oscillator. Shaping to remove clicks must be applied to later amplifiers.

Other digital schemes that generate square waves are useful for odd-order frequency multiplication. The buffer of **Fig 5.47** can serve this function. For example, this circuit could be driven by a VXO at 14.4 MHz and followed by a triple tuned bandpass filter at 72 MHz. The signal would then be amplified to a level of +10 dBm or so where it can be used to drive a two diode frequency doubler with a double tuned circuit at 144 MHz, resulting in 0 dBm at 2 m, ready for use with simple transmitters or transceivers.

The example of Fig 5.47 used a Hex inverter, but other digital parts are also useful. For example, an exclusive-OR gate can be used as a digital balanced mixer, offering 40 dB or greater suppression of both "LO" and "RF" input signals before bandpass filtering.

The frequency multipliers designed by Wenzel featured low phase noise. While the multiplied output has higher noise than the driving source, that noise is worse only by the normal 20×Log(N) factor for an ideal multiplier. The multipliers using digital logic elements may well be worse than this. We have not performed the measurements needed to establish this performance.

5.5 A VXO TRANSMITTER USING A DIGITAL FREQUENCY MULTIPLIER

The original goal for this project was a transmitter that would function on the 21-MHz amateur band while using an available 14-MHz crystal. The single band transmitter described here develops an output in the 14-MHz band. 28-MHz and 50-MHz designs are presented elsewhere in the book.

The basis for the transmitter is shown in the block diagram of **Fig 5.48**. A crystal oscillator drives a digital divide-by-2 circuit to generate a square wave at half the oscillator frequency. This waveform is rich in odd-order harmonics while nearly devoid of even ones. A bandpass filter is fabricated to extract the harmonic of interest while suppressing the rest. The resulting signal is then amplified to the desired power.

There are several advantages to this scheme when applied to a transmitter design. First, the digital divider and related circuitry form a high gain buffer, providing excellent isolation from the output. While a common problem with a VXO is

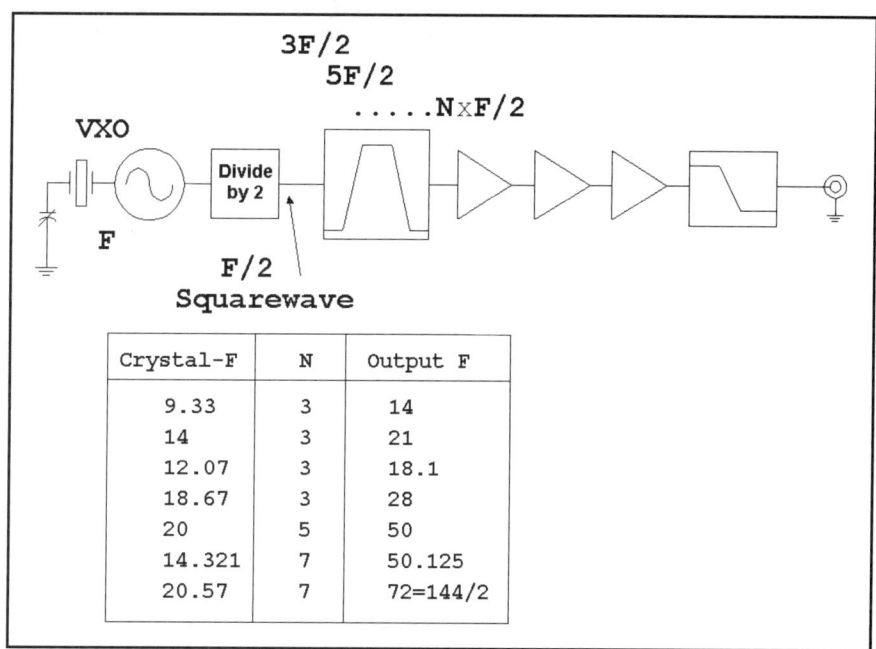

Crystal-F	N	Output F
9.33	3	14
14	3	21
12.07	3	18.1
18.67	3	28
20	5	50
14.321	7	50.125
20.57	7	72=144/2

Fig 5.48—Block diagram showing the transmitter concept. The table shows some possible applications.

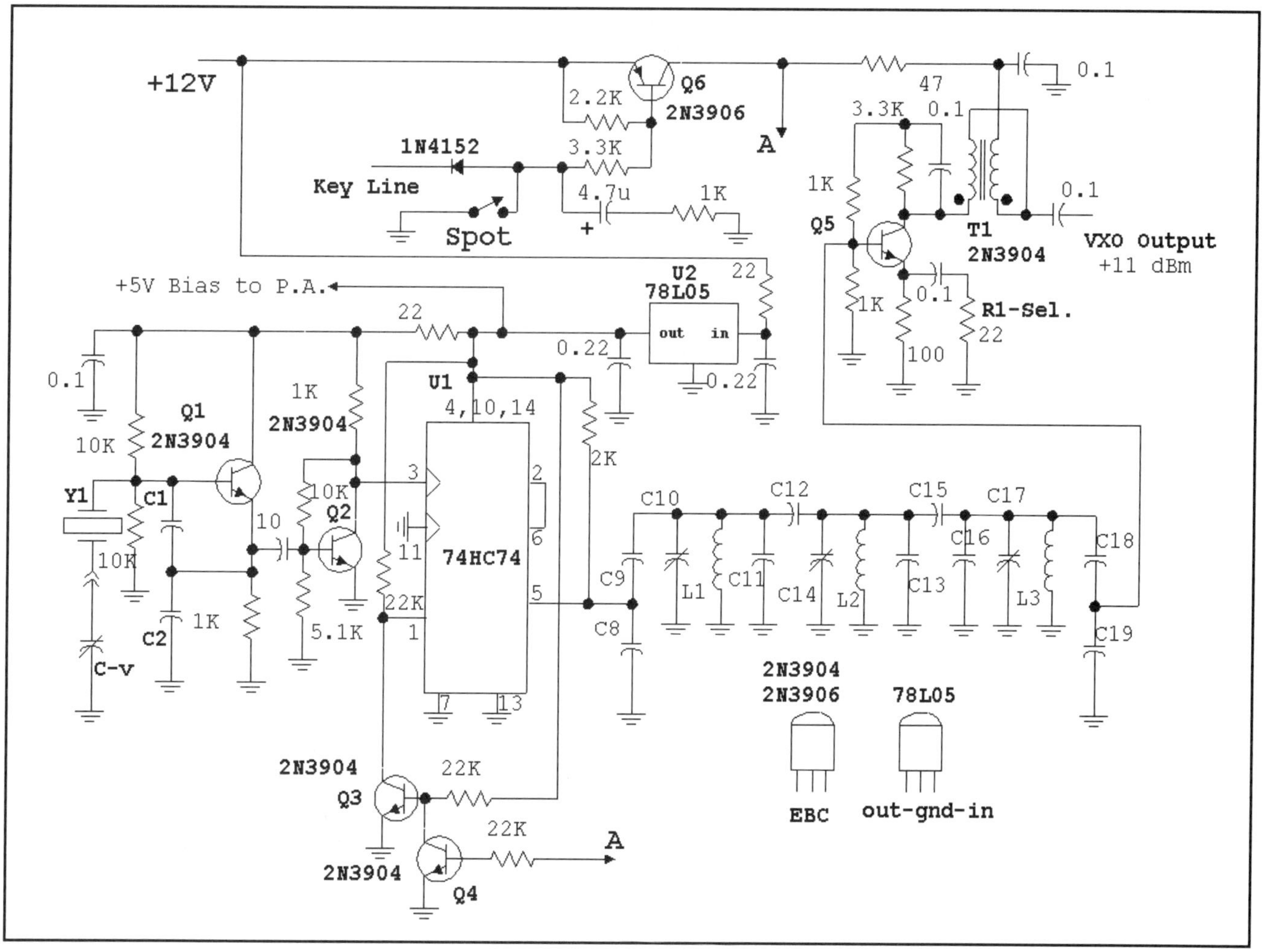

Fig 5.49—Schematic for the oscillator, divider, 14-MHz bandpass filter and buffer amplifier for the VXO transmitter.

Mixers and Frequency Multipliers

output variation with tuning, this output is constant for the total tuning range. The oscillator frequency is not directly related to the transmitter output frequency, so there are few problems relating to stray power amplifier energy in the oscillator circuitry. Finally, the output can be turned off and on by controlling a digital reset line in the divider. As such, there is a perfect method for keying without ever changing the oscillator operating frequency. The oscillator runs continuously and does not change frequency during a transmit interval. The usual mechanisms for generating chirp are absent.

The oscillator, divider, and filter portion of the 20-m transmitter is shown in **Fig 5.49**. A crystal at 9.373 MHz (HC-49, 20-pF load) was chosen to provide about 10 kHz of tuning around the desired output frequency of 14.06 MHz. The range is obtained without any crystal series inductance. However, the builder may wish to add inductance to extend the tuning range. The output from oscillator Q1 drives Q2, conditioning the signal for logic compatibility. This then drives a 74HC74 divide-by-2 chip. During normal key-up conditions, pin 1 is held low by Q3. This "reset" prevents any output from appearing from the IC. When the key or spot switch are pressed, pin 1 goes high, and the divider generates the desired 4.687-MHz output. The 5 V bias for U1 is obtained from U2, a low power regulator.

A 2-kΩ pull up resistor on U1's Q output helps to ensure that the output goes all the way to 5 V during operation, establishing the logic level, and hence, the RF output level. The combination of the resistor and the chip circuitry generate a load of approximately 1 kΩ to provide filter loading at the input end, establishing the values for C8 and C9. The filter is designed for a 50-Ω output load. The available power at the third harmonic is about 0 dBm. This filter is designed for a bandwidth of 400 kHz at 14 MHz. With the inductors used, the filter insertion loss is about 3 dB.

A buffer amplifier, Q5, increases the output from the filter to a comfortable +11 dBm. Q5 is only powered on key-down intervals, controlled by a delayed switch, Q6, which also provides the needed control signal "A" for U1. A 4.7-µF capacitor keeps this switch "on" for a short interval after key down. The 1-kΩ resistor in series with the 4.7-µF capacitor allows the "A" signal to immediately change with the initial application of the key while the transmitter output is still shaped with the circuitry around Q9. This creates a "time sequence" keying scheme, similar to one applied to vacuum tube transmitters of the 1950's era.

The buffer output is applied to a 100-Ω pot functioning as a Drive control, and then to a keyed driver, Q7. This stage and the output power amplifier are shown in **Fig 5.50**. These components are on a separate board from the earlier circuitry, further isolating the circuits. The driver, a medium power bipolar feedback amplifier, is capable of an output of up to 300 mW. The keying is done with Q9, a shaping integrator-switch.

The output amplifier uses an inexpensive HEX FET. Some regulated 5-V energy is stolen from the other board and applied to a pot that generates bias for the FET PA. The bias is adjusted by monitoring the FET drain current with a sensitive meter and is set for a current of close to 1 mA. This amplifier will run in Class B, off during key up conditions, allowing the use of electronic T/R switching. However, forward FET bias enhances both gain and stability. The FET output is matched with a modified LCC type T-network consisting of L5 and a pair of mica compression trimmer capacitors. This is followed by additional low pass filtering. The output is set to 4 W by adjusting the drive and tun-

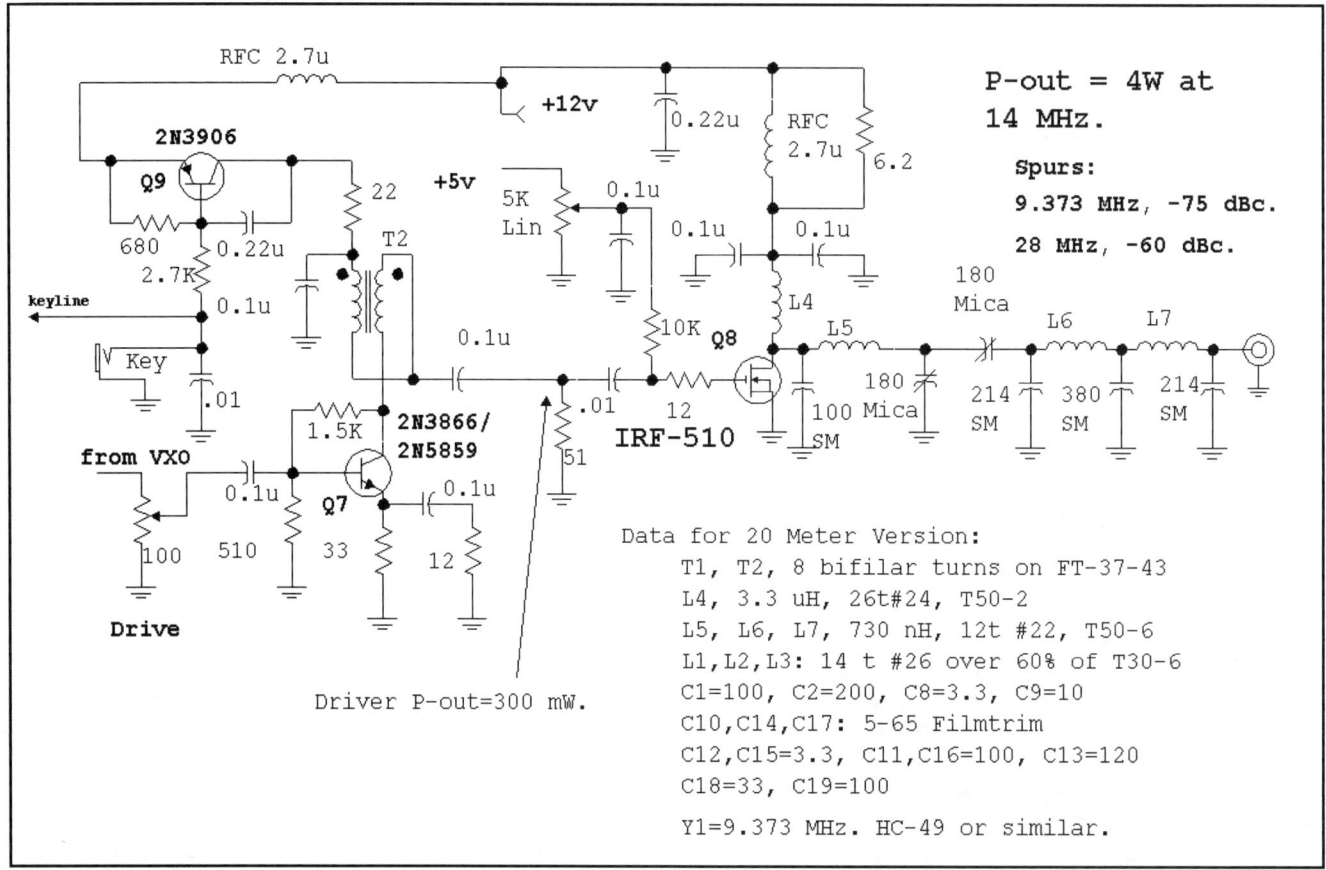

Fig 5.50—Keyed driver and power amplifier for the transmitter.

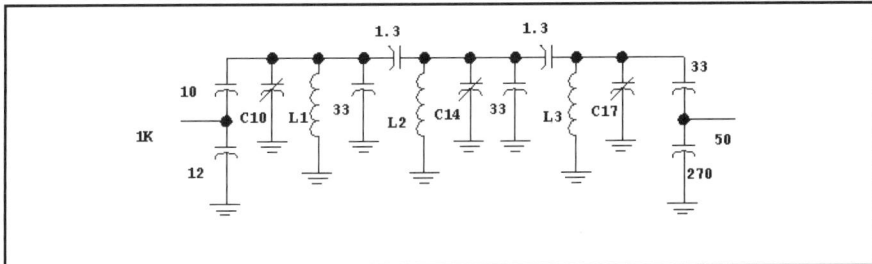

Fig 5.51—A 21-MHz bandpass filter. The inductors and the variable capacitors are identical to those used in the 14-MHz design.

The 4-W output power amplifier is shown at the top of the photo. The board includes the keyed driver, drive control pot, and bias pot. The box housing this rig also includes a 20-meter receiver (The "Easy 90-14") described in Chapter 6.

ing the T-network capacitors for maximum output.

A subtle instability was noted during the transmitter turn-on process. In an effort to make the transmitter as clean as possible, an extra 2.7-µH RFC had been included in the drain line. But a low level oscillation was noted in the PA. An oscilloscope examination revealed a frequency of 300 kHz. This turned out to be the result of a resonance between the 2.7 µH inductor and the bypass capacitors. A 6.2-Ω resistor was paralleled across the RFC and the oscillation was eliminated. This illustrates the subtlety of wideband bypassing of power stages in a transmitter. See the information on decoupling in Chapter 2.

The only spurious responses noted in the output were at the crystal oscillator frequency and at the transmitter 2nd harmonic, but they were below the desired output by 75 and 60 dB, respectively. Yet the transmitter is built with no internal shielding or other complexities.

A 21-MHz version of this design would be especially practical, for it could use an existing 14-MHz crystal. A 21-MHz bandpass filter is shown in **Fig 5.51** to aid the designer/builder in realizing a rig for that band.

Although the digital divider was originally implemented for use with simple low power transmitters, it lends itself well to general-purpose applications with LC oscillators as well as crystal-based designs.

REFERENCES

1. S. Maas, "A GaAs MESFET Mixer with Very Low Intermodulation," *IEEE MTT-35*, No. 4, April, 1987.

2. W. Hayward, "Experiments with Primitive FET Mixers," *RF Design*, Nov, 1990.

3. E. Oxner, "A Commutation Double Balanced Mixer of High Dynamic Range," *Proceedings of RF Technology Expo '86*, Anaheim, CA, pp 309-323. See also *RF Design*, Feb, 1986.

4. W. Hayward, "Experiments with Primitive FET Mixers," *RF Design*, Nov, 1990.

5. Li and Corsetto, *Microwave Journal*, Oct, 1997.

6. Gray and Meyer, *Analysis and Design of Analog Integrated Circuits*, 2nd Edition, Wiley, 1984.

7. B. Zavrel, W7SX, "Feedback Technique Improves Active Mixer Performance," *RF Design*, Sep, 1997.

8. B. Zavrel, W7SX, "Double Balanced Mixer and Oscillator", *Signetics NE/SA602*, Nov 9, 1987.

9. W. Hayward, "CERverters," *QST*, June, 1976, pp 31-35.

10. K. Kurokawa, "Design Theory of Balanced Transistor Amplifiers," *Bell System Technical Journal*, Vol. 44, No. 10, Oct, 1965, pp 1675-1698. See also R. S. Engelbrecht, US Patent 3,371,284, "High Frequency Balanced Amplifier," Feb 27, 1968; and Kurokawa and Englebrecht, "A Wideband Low Noise L-Band Balanced Transistor Amplifier," *Proc IEEE*, Mar, 1965.

11. J. B. Stephensen,"Reducing IMD in High-Level Mixers," *QEX*, May/June, 2001, pp 45-50.

12. P. Hawker, "G3SBI's High Performance Mixer", Technical Topics, *Radio Communications*, Sep/Oct, 1993, pp 55-56.

13. C. Wenzel, "New Topology Multiplier Generates Odd Harmonics," *RF Design*, July, 1987. See also **www.Wenzel.com/documents/2diomult.html**.

CHAPTER 6

Transmitters and Receivers

6.0 SIGNALS AND THE SYSTEMS THAT PROCESS THEM

The basic building blocks of amplifiers, filters, oscillators, mixers, and frequency multipliers have been discussed. We now begin to combine these components to build the equipment that provides communications. We begin the chapter with a look at CW, AM, DSB, SSB, and FM signals. Block diagrams are then shown for the equipment we build to deal with these signals. Later sections will present detailed design methods and examples.

Signals are presented as equations. We then show graphs in the time and frequency domains, the results we would observe with either an oscilloscope or spectrum analyzer. This discussion is not intended to be complete, but is merely a sketch of signal forms. A complete treatment is found in communications texts.[1]

The first signal we consider is the audio, or *baseband* representation. This might represent the output of a receiver or a voice signal that we apply to a transmitter microphone input. A receiver output from a CW signal is generally a rather pure sine wave, perhaps at a frequency of 1000 Hz. Mathematically this is

$$v(t) = \sin(2\pi f t) \quad \text{Eq 6.1}$$

where v(t) indicates that the voltage is a function of time, f is the frequency in Hz, and t is time in seconds. Graphed in the time domain, the tone is the familiar sine wave, **Fig 6.1**. The energy is confined to a single frequency, so the spectrum, or frequency domain representation is a single line, **Fig 6.2**. The 1-V amplitude has a spectrum with a height of 1 V. It is more common within the radio frequency design arena to see spectra calibrated in terms of power.

The human voice is not a sine wave, but a combination of tones forming complicated patterns in both time and frequency. The actual signals are difficult to handle with simple equations and are different for every voice. So, we approximate a voice signal with several sine waves. The baseband example we use (**Figs 6.3** and **6.4**) has three tones of $f_1 = 1000$, $f_2 = 2500$, and $f_3 = 400$ Hz with respective amplitudes of 0.6, 1, and 0.5 V. The total baseband signal is

$$v_b(t) = 0.5 \sin(2\pi f_3 t) \\ + 0.6 \sin(2\pi f_1 t) \\ + 1 \sin(2\pi f_2 t) \quad \text{Eq 6.2}$$

Traditional amplitude modulation is familiar as an AM broadcast signal. This is generated by changing—or modulating at an audio rate—the amplitude of a *carrier*. The carrier is merely a single sinusoid. A frequency of 100 kHz is used in our

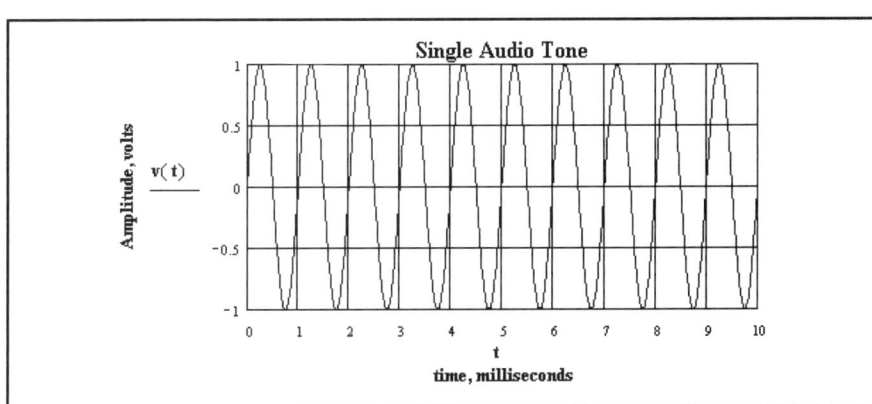

Fig 6.1—A single audio tone as a function of time.

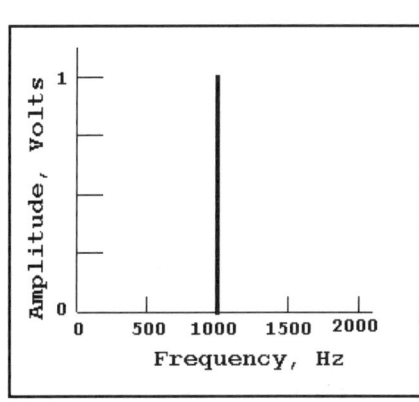

Fig 6.2—The 1000 Hz audio tone in the frequency domain.

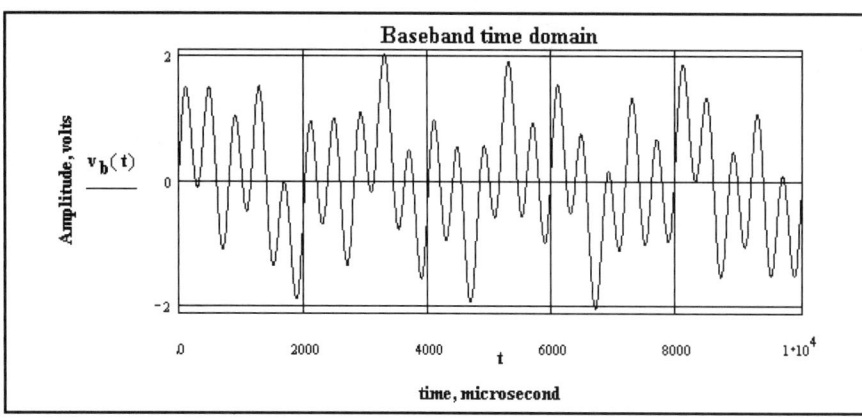

Fig 6.3—The time-domain graph of the three audio tones.

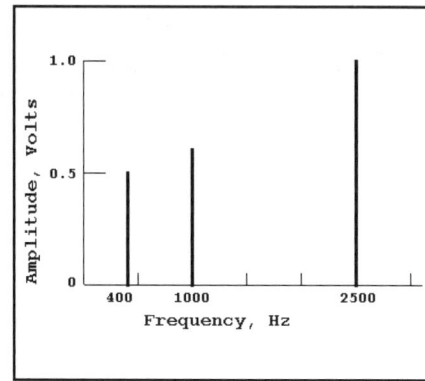

Fig 6.4—The frequency-domain graph of the three audio tones.

examples. The graphs and equations are the same as the earlier single-tone audio signal, except that the frequency is higher.

The carrier amplitude is modulated to generate the AM signal of **Eq 6.3**.

$$v_a(t) = \left(1 + 0.3 \sin\left(2\pi f_{aud} t\right)\right) \times \sin\left(2\pi f_c t\right)$$

Eq 6.3

where f_c is the carrier frequency of 100 kHz and f_{aud} is the audio frequency of 1 kHz. The 0.3 factor is a modulation index and indicates 30% modulation. The time domain signal is shown in **Fig 6.5** with a spectrum in **Fig 6.6**. The two curves are related through appropriate mathematics, which follow from the trig identity shown in the *Trig Identities for Signal Analysis* sidebar. A detailed mathematical analysis will always tie the two domains together. Modulations that are simple in one domain are often complicated and messy in the other.

The time domain waveform shows that the amplitude of the RF sine wave varies, exceeding the original carrier amplitude for part of the cycle. The frequency domain graphs show that extra energy to be contained in the frequency domain sidebands while the carrier remains constant with no audio variation. This is easily confirmed by observation with a spectrum analyzer or receiver that will resolve the carrier from the sidebands.

A 100-kHz carrier modulated by the three-tone baseband signal is shown in **Fig 6.7** and **Fig 6.8**.

The multi-tone amplitude modulation is described by

$$v_{am}(t) = \left(1 + 0.3 v_b(t)\right) \sin\left(2\pi f_c t\right)$$

Eq 6.4

where the sine term represents the carrier and $v_b(t)$ is the baseband signal from **Eq 6.2**. The first set of parentheses on the right side of the equal sign in **Eq 6.4** contains the unity term, which leads to the carrier in the final result, and the complex audio signal $v_b(t)$ that generates the sidebands.

A double sideband signal results when audio is applied to a *balanced* modulator driven by a local oscillator. The resulting output for a single modulating audio tone is

$$v_d(t) = \sin\left(2\pi f_{aud} t\right) \sin\left(2\pi f_c t\right)$$

Eq 6.5

where the first term is the audio while the second is the carrier. The term with unity in Eq 6.4 is missing from Eq 6.5, indicating that the carrier is no longer present. The waveforms are shown in **Fig 6.9** and **Fig 6.10**.

The result of a double-sideband generator driven with the multiple-tone audio is then

$$\begin{aligned} v_{dsb}(t) = &\sin\left(2\pi f_{U2} t\right) + \sin\left(2\pi f_{L2} t\right) \\ &+ 0.6 \sin\left(2\pi f_{U1} t\right) + 0.6 \sin\left(2\pi f_{L1} t\right) \\ &+ 0.5 \sin\left(2\pi f_{U3} t\right) + 0.5 \sin\left(2\pi f_{L3} t\right) \end{aligned}$$

Eq 6.6

where the frequencies shown represent the

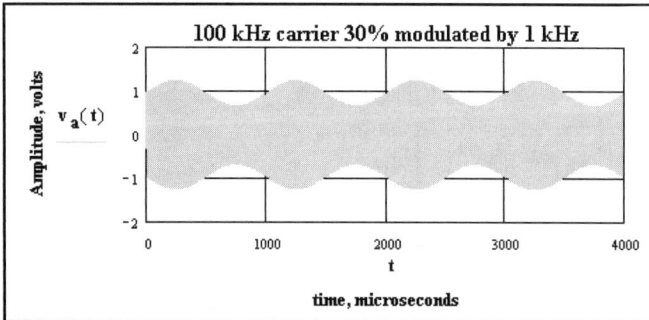

Fig 6.5—The carrier amplitude here is 1 V. Modulation causes the amplitude to depart from this value. The energy appears in the figure to be a solid mass of energy, but if we zoom in, plotting only a small fraction of the curve shown, we will see the details of the RF oscillation. This could be done experimentally with an oscilloscope triggered from the RF waveform.

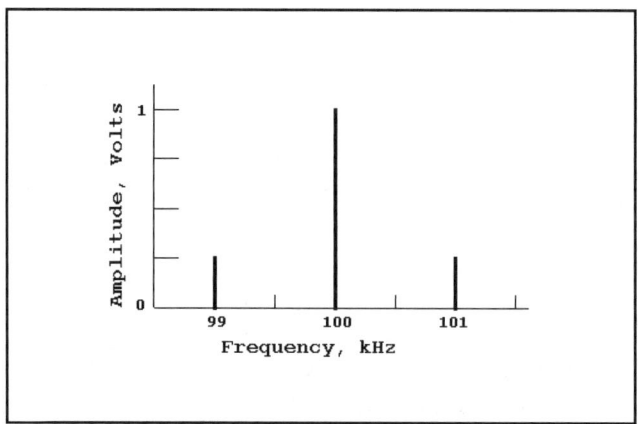

Fig 6.6—Frequency-domain representation of an AM signal. The carrier at 100 kHz is modulated at 1 kHz to generate two sidebands below and above the carrier.

Trig Identities for Signal Analysis

In high school trigonometry class you may have learned some useful identities. One of them relates the product of two sine functions:

$$\sin(a)\sin(b) = \frac{1}{2}\cos(a-b) - \frac{1}{2}\cos(a+b)$$

Our analysis of amplitude modulation started with a carrier of amplitude A:

$$A\sin(\omega_c t)$$

where $\omega_c = 2\pi f_c$ is a carrier frequency expressed in radians/sec, with f_c in Hz. The amplitude is allowed to vary about a base value.

$$A = A_0\left(1 + m\sin(\omega_a t)\right)$$

where ω_a is an audio frequency in radians/sec and m is a modulation index. The modulated wave becomes:

$$v(t) = A_0\left(1 + m\sin(\omega_a t)\right)\sin(\omega_c t) \text{ which expands to:}$$
$$v(t) = A_0\sin(\omega_c t) + A_0\sin(\omega_c t)m\sin(\omega_a t)$$

The first term is the carrier, which varies only with time at the carrier rate, ω_c. The second term is the product of audio and RF carrier sine waves. Expansion with the identity yields:

$$A_0\, m\sin(\omega_a t)\sin(\omega_c t) = A_0\, m\left[\frac{1}{2}\cos\left[(\omega_c - \omega_a)t\right] - \frac{1}{2}\cos\left[(\omega_c + \omega_a)t\right]\right]$$

and then:

$$A_0\, m\sin(\omega_a t)\sin(\omega_c t) = A_0\, m\left[\frac{1}{2}\cos\left[2\pi(f_c - f_a)t\right] - \frac{1}{2}\cos\left[2\pi(f_c + f_a)t\right]\right]$$

The two cosine waves on the right are the lower and upper sidebands of the AM signal.

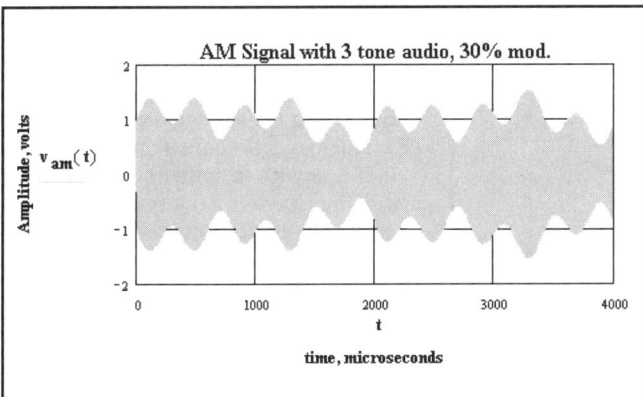

Fig 6.7—A three-tone baseband signal modulates a 100-kHz audio tone.

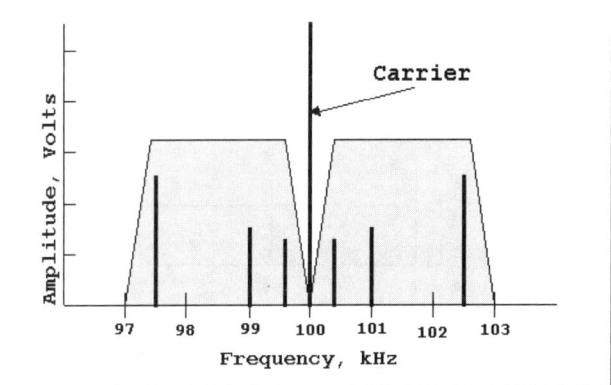

Fig 6.8—Frequency-domain view of amplitude modulation with a three-tone baseband signal. The two sideband regions are now shaded.

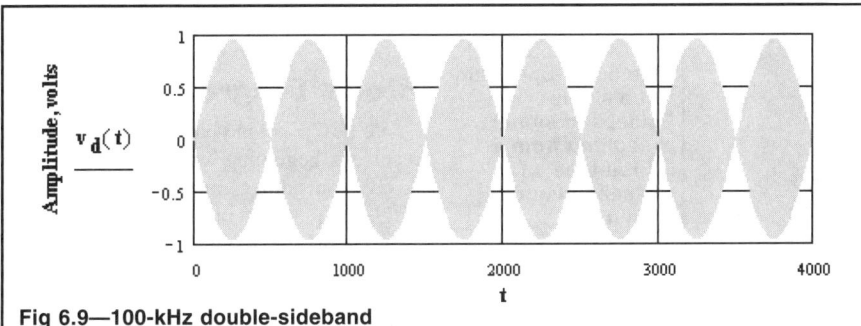

Fig 6.9—100-kHz double-sideband output with 1-kHz audio.

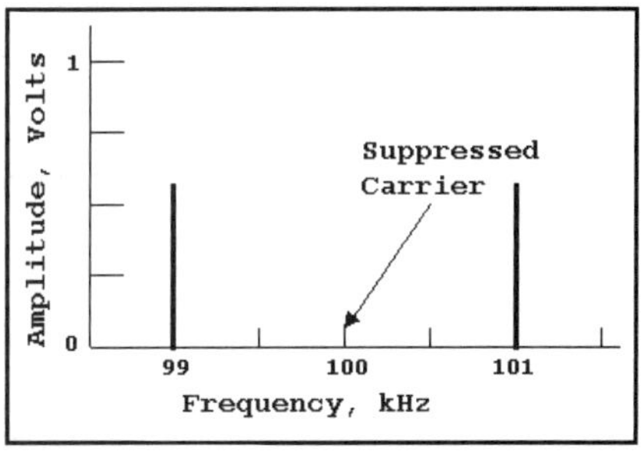

Fig 6.10—Frequency-domain view of a DSB signal with a single audio tone. Two output frequencies are created.

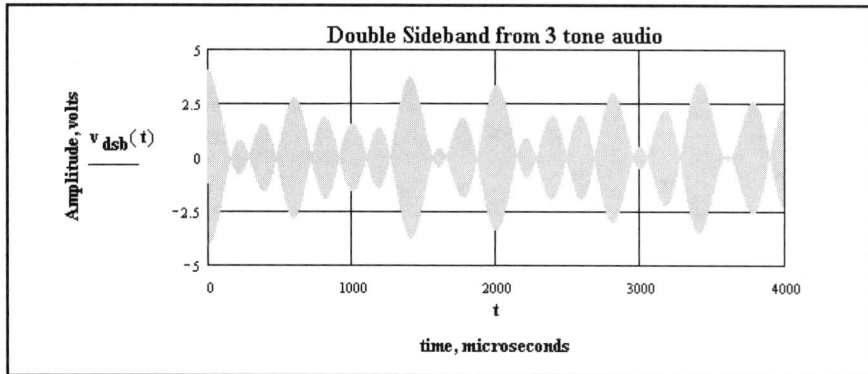

Fig 6.11—Double sideband with a multi-tone audio, time domain.

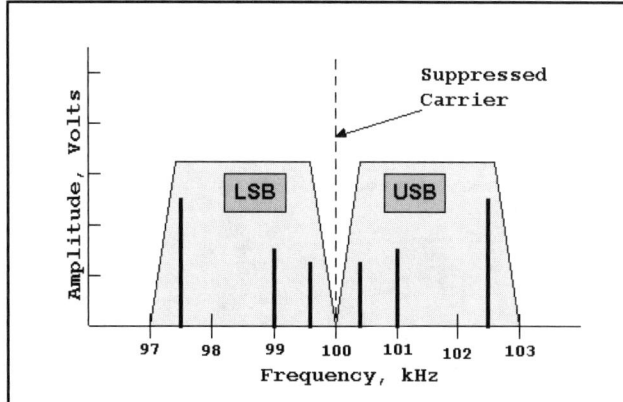

Fig 6.12—Frequency-domain representation of DSB with multiple-tone audio. The upper and lower sideband parts of the spectrum are highlighted.

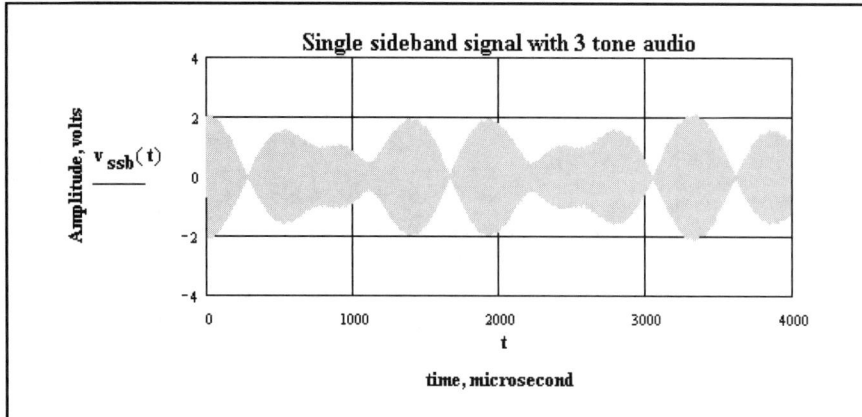

Fig 6.13—Single-sideband signal from a three-tone baseband input.

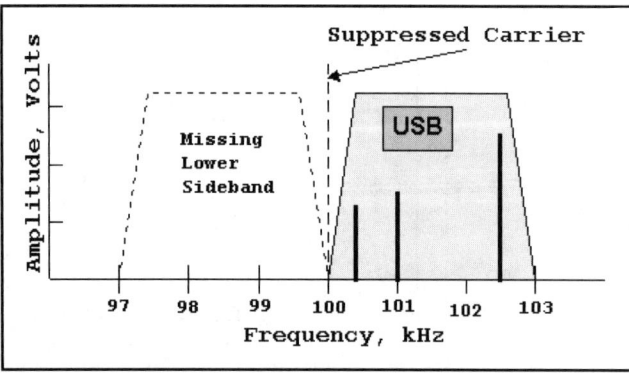

Fig 6.14—Spectrum of a single-sideband signal resulting from a three-tone baseband audio input.

upper and lower sideband components resulting from audio components at f_1, f_2, and f_3. The DSB signals are shown in **Fig 6.11** and **Fig 6.12**.

A single sideband (SSB) signal is described by eliminating one of the sidebands. For this example, we retain the upper sideband, resulting in

$$v_{ssb}(t) = \sin(2\pi f_{U2} t) + 0.6 \sin(2\pi f_{U1} t) + 0.5 \sin(2\pi f_{U3} t) \quad \text{Eq 6.7}$$

The corresponding graphs are **Fig 6.13** and **Fig 6.14**.

The SSB signal, when viewed in the frequency domain, is really nothing more than an exact replica of the original baseband signal, except that it is now translated linearly to a higher frequency. If a lower sideband signal had been generated, it would have been a replica of the original with an inversion. That is, what had started as a high audio frequency of 2500 Hz now appears as the lowest frequency.

A frequency-modulated signal is described by

$$v_{fm}(t) = \sin[2\pi f_c (1 + m \sin(2\pi f_a t))t] \quad \text{Eq 6.8}$$

If we pick a 10-kHz carrier and modulate it with a 1-kHz audio signal, we see the time domain signal of **Fig 6.16**. The amplitude is constant, but the frequency varies.

Extracting the spectrum for this signal is mathematically much more difficult than it was with the other signals, for the audio sine wave is now *inside* the argument for the basic signal before modulation, as seen in **Eq 6.8**. Signals appear about the carrier, spaced by the audio frequency. However, several sets appear. A 1 kHz audio tone produces signals at +/–1, +/–2 kHz and so on, as shown in **Fig 6.17**. The strength of the sidebands *and the carrier* depend on m, still a modulation index, and are described by Bessel functions.[2] No FM equipment is described in this book, but the equations are included for completeness.

Block Diagrams

We now examine basic transmitters and receivers, beginning with simple CW gear. A CW transmitter generates a carrier at a single frequency with no modulation other than the off-on keying that imposes the familiar encoding. A simple CW transmitter is shown in **Fig 6.18**. The circuit begins with an oscillator operating at the final output frequency. Typical oscillators are usually fol-

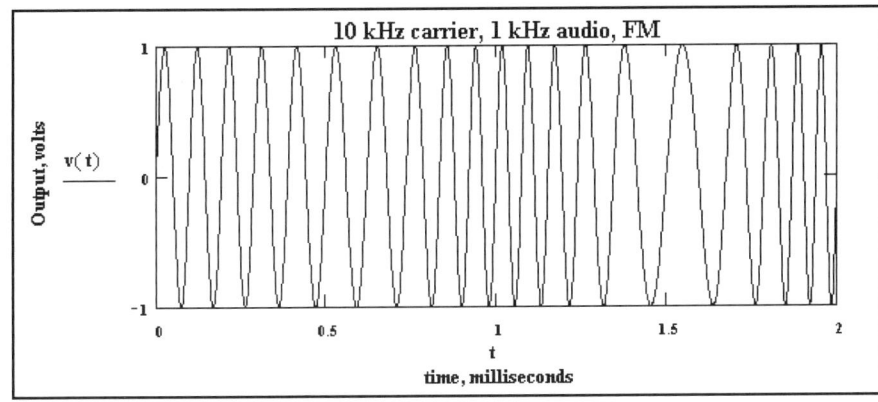

Fig 6.16—Time domain representation of an FM signal.

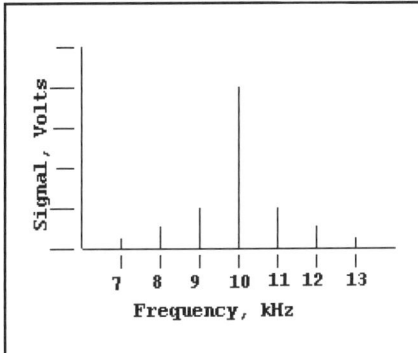

Fig 6.17—Spectrum of an FM signal, 10-kHz carrier with 1-kHz audio. This graph represents what we might observe with a typical spectrum analyzer. We often see plots like this with some components below the frequency axis, indicating a sign change when frequency is modulated rather than amplitude.

the frequency to change (pulling) when the amplifiers are keyed on. The output frequency then differs from that when the amplifier is off.

The modified circuit of Fig 6.18B uses a frequency multiplier between the oscillator and the power amplifiers. The buffering action of a frequency multiplier is profound. Signals travelling from the output backward in a buffer remain at the output frequency. The buffer input, including the oscillator, is not usually sensitive to this. With the transmitter output at a multiple of the oscillator frequency, it no longer has components within the bandwidth of the oscillator tank, so is not susceptible to the pulling mentioned. Indeed, it is often practical to build transmitters with no inter-stage shielding whatsoever if multipliers are used. A bandpass filter is used at the multiplier output to suppress direct feed-through from the oscillator and harmonics— other than the desired one— that are often present. The filter can often be as simple as a single resonator if the multiplier is just a balanced frequency doubler. More often, we use double or triple tuning at the output of multipliers.

A mixer is often used within a CW transmitter with a bandpass filter to select the desired frequency, shown in **Fig 6.19**. This example has a 2-MHz variable-frequency oscillator, a 5-MHz crystal-controlled oscillator, and an output at 7 MHz. The VFO tunes a 150-kHz range to cover the CW portion of the 7-MHz band. The bandpass filter must be wide enough to pass the entire range, but should not be a lot wider, for spurious mixer products must also be suppressed by the filter. The 5-MHz component will be suppressed by balance

lowed by amplifiers (perhaps several) to increase output power. The final block is a low-pass filter to remove harmonics.

The amplifiers serve the additional function of buffering the oscillator. Buffers may have low gain, but have much more gain in the normal forward direction than in the reverse one. A typical 20-dB gain design might have a gain of –30 dB in the reverse direction. This serves to prevent large transmitter output signals from reaching the oscillator. Common-base (gate) amplifiers usually feature excellent reverse isolation.

A crystal or an LC resonator determines the oscillator frequency. The oscillator should be shielded from the rest of the transmitter to prevent transmitter output components from reaching it. An oscillator is most sensitive to signals at frequencies within the loaded bandwidth of the resonator controlling the oscillator. Hence, shielding is especially important for the simple transmitter of Fig 6.18. Poor shielding or inadequate buffering allows

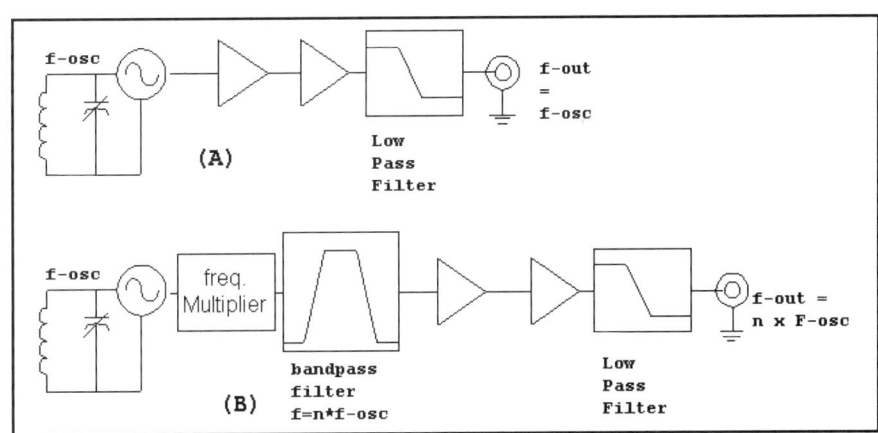

Fig 6.18—Simple CW transmitters with a *master oscillator* and a *power amplifier* are traditionally called a MOPA design. Design "A" has the oscillator and amplifier operating at the same frequency while that at "B" uses frequency multiplication.

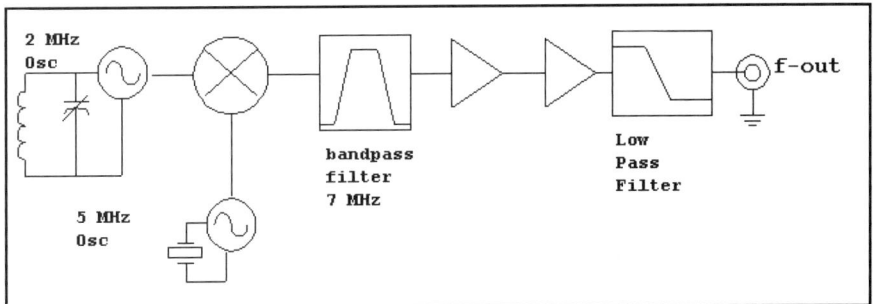

Fig 6.19—A CW transmitter using a mixer. Frequency stability is improved owing to use of a lower frequency for the variable-frequency oscillator. Careful bandpass filtering is required at the mixer output to preserve spectral purity.

in the mixer, but may often need to be further attenuated by the bandpass filter. A typical circuit would often use a triple-tuned filter if intended to meet modern standards.

These methods are not restricted to simple CW transmitters. Heterodyne methods are also useful when building local oscillator systems for SSB or similar equipment.

A CW signal is received by heterodyning the radio frequency energy down to baseband where it can be heard. This may occur in one step in a direct-conversion (including regenerative) receiver or in several steps in a conventional superheterodyne. The key element in a direct-conversion receiver is the mixer, or as it is usually called in applications with an audio output, the product detector. The input signal, usually relatively weak, is applied to the RF port of a mixer driven by a strong local oscillator. Two mixer outputs will appear, but only the audio difference frequency is used. The signal is usually amplified further and is applied to headphones. A block diagram is shown in **Fig 6.20**. The input preselector filter protects the receiver from strong signals at frequencies far removed from those being received. The low-pass filter routes audio to the amplifiers while preventing other mixer products or mixer feed-through components from reaching the amplifier. Direct conversion receivers are covered in much greater detail in Chapter 8.

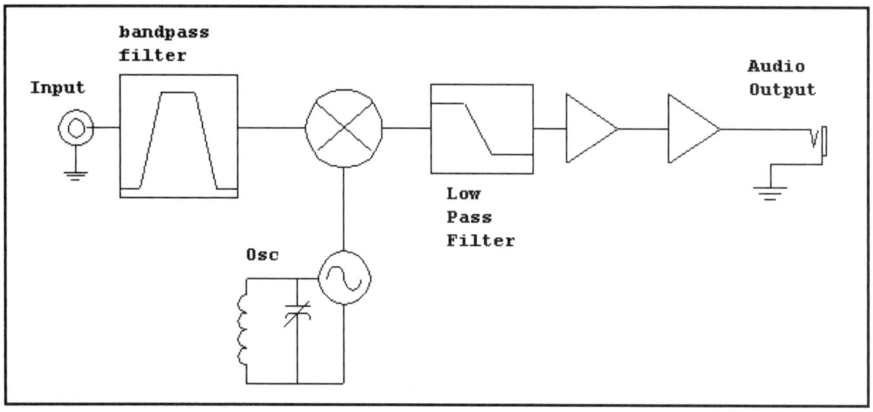

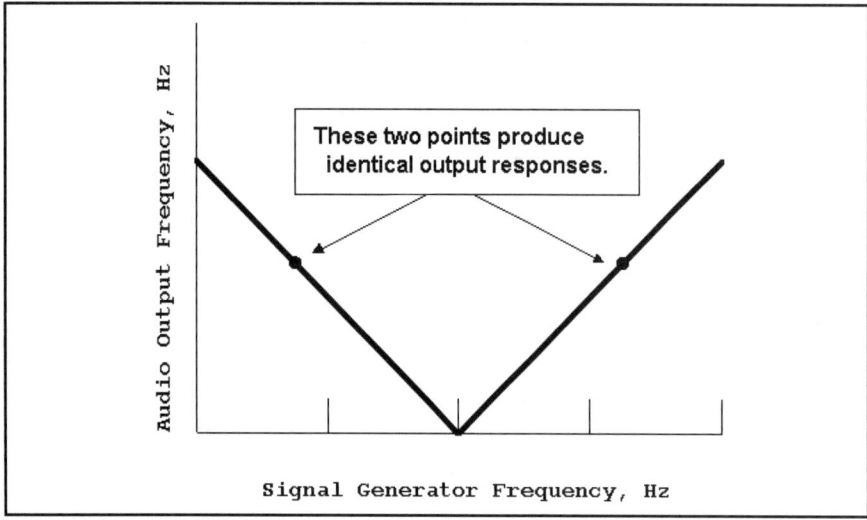

Fig 6.20—Direct-conversion receiver. The incoming signal is applied to a mixer where it is converted directly to audio without intermediate processing.

Fig 6.21—Tuning response of a fixed-tuned DC receiver while varying a signal generator applied to the input. A 1000-Hz beat note is available from the generator at two different generator frequencies. One response is the audio image of the other.

An instructive experiment tunes the frequency of a signal generator attached to a direct conversion receiver. One will then hear an audio beat note, the difference frequency between the generator and the receiver local oscillator. The output frequency is shown in **Fig 6.21** as a function of generator frequency. Tuning the receiver with a fixed generator produces an identical result. The response is double sided; for every tuning of a simple direct conversion receiver, there are two different input frequencies that can produce the same output signal. One response is called the audio image of the other. This makes it challenging to use such a receiver in severely congested bands. But the simplicity and other good qualities of a direct conversion receiver will often compensate for this problem.

The traditional solution to the audio image problem is the single-signal superheterodyne receiver shown in the block diagram of **Fig 6.22**. The incoming signal is processed in a preselector filter and then applied to a mixer. The output is still at a radio frequency, but one that is different from the incoming signal, an intermediate frequency, or IF. This 7-MHz receiver uses a 1-MHz IF with an LO in the 6-MHz region. The 1-MHz signal from the mixer is filtered with a narrow bandwidth circuit. It is further amplified and applied to a second mixer, now functioning as a product detector to produce an audio output. After some audio gain, headphones are driven. The LO for the product detector is called a beat frequency oscillator, or BFO.

Assume that the 1-MHz IF filter has a bandwidth of 500 Hz, centered exactly at 1 MHz. The receiver LO will be tuned to 6.040 MHz. This means that the incoming signals that will produce an output are cen-

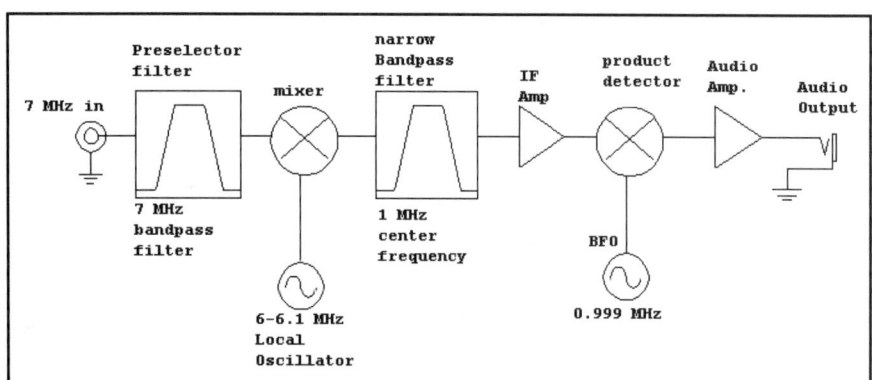

Fig 6.22—A simple single-conversion superheterodyne receiver featuring a "single-signal response." A narrow filter, usually using a quartz crystal, follows the mixer.

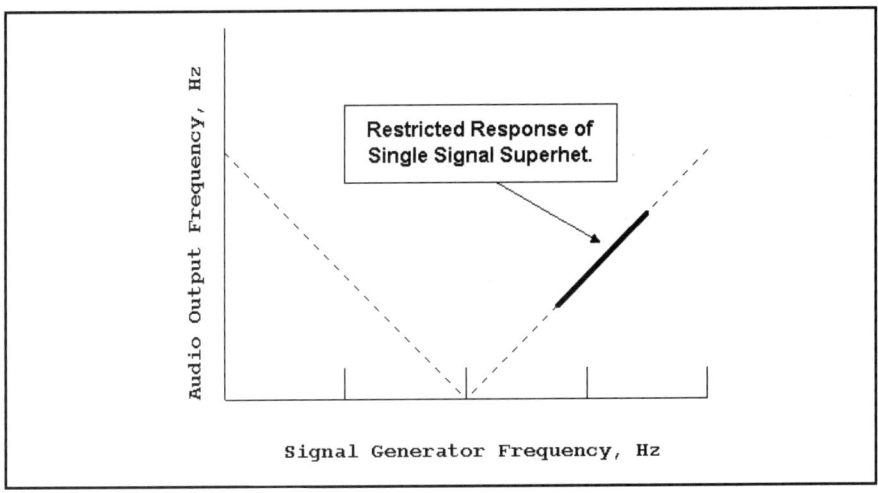

Fig 6.23—Tuning response to the single-signal superhet. The output from a single source occurs in a single area on the dial.

tered at 7.04 MHz and occur in a 500-Hz band, 250 Hz on either side of 7.04 MHz. Signals within that band are the only ones that will produce an IF output. Set the BFO to 0.999 MHz, 1 kHz away from the IF center. An IF signal at 1 MHz will then produce a 1-kHz beat note. But the only beat notes that are possible for this BFO setting are in a 500-Hz wide span from 750 to 1250 Hz. Repeating the earlier experiment performed with the direct conversion receiver yields the result of **Fig 6.23**.

A single-signal response can also be obtained with phasing methods, and related schemes. These are covered in detail in Chapter 9.

DSB

Let's return to the transmitter problem, but now consider the generation of a double sideband signal. (Double-sideband, full-carrier amplitude modulation is of great historic interest, especially to collectors, but is not the most-used method of voice communications today. We won't treat the method in this book.) The key element needed to generate DSB is a balanced mixer. It will be driven with a suitable RF local oscillator and low level audio from an amplified microphone. The output, shown earlier in Fig 6.10, contains the two sidebands symmetrically spaced about a suppressed carrier. Further amplification and low-pass filtering completes the transmitter. A simple DSB transmitter is shown in **Fig 6.24**. A typical simple DSB transmitter will have a carrier that is suppressed by 30 to 40 dB with respect to either sideband. Although simple and compatible with existing SSB equipment, DSB transmitters are rarely used today, largely due to the excess spectrum used.

Audio intelligence is impressed on the signal in DSB and SSB transmitters with a block traditionally shown as a balanced modulator. The modulator is really just a mixer with a particular application. It is usually a balanced circuit, for that is the mechanism used to suppress the carrier output. See balance in Chapter 5.

The direct-conversion receiver shown earlier (Fig 6.20) will allow DSB signals to be received. Each of the two sidebands will be heterodyned down to baseband where they will add to produce an audio output. It is *vital* that the BFO be *exactly*

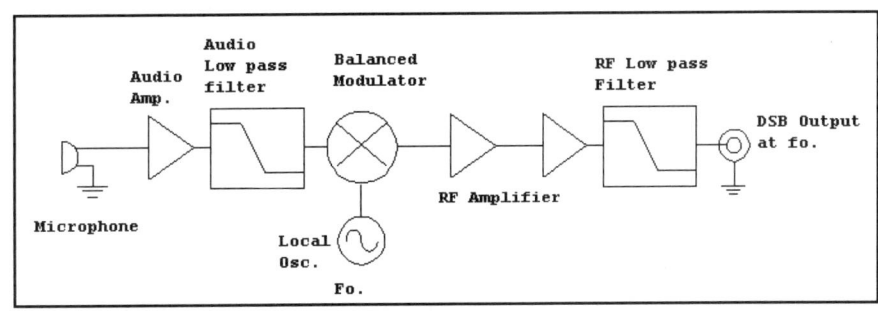

Fig 6.24—A double sideband transmitter.

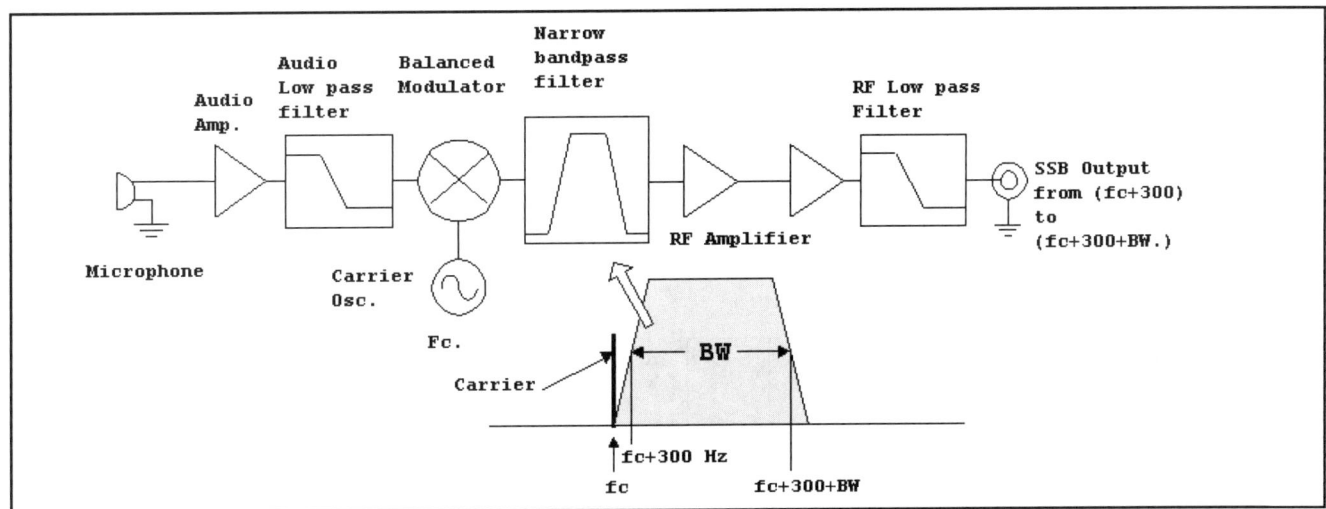

Fig 6.25—A traditional SSB transmitter using the filter method. A narrow filter follows a balanced modulator to remove one of two sidebands present on the DSB output of the modulator.

on the frequency of the suppressed carrier. This is so difficult in practice that a DC receiver is normally not suitable for DSB applications.

The most popular method used to generate SSB is shown in **Fig 6.25**. This is traditionally called the *filter method*, for a narrow bandpass filter is used to select one of two sidebands generated by a balanced modulator. See Figs 6.12 and 6.14. The other dominant way to get SSB is the phasing method, treated in great detail in Chapter 9. The phasing method is based upon mathematics following from the *Trig Identities for Signal Analysis* sidebar earlier in this chapter where multiplication of two sine waves is performed with a doubly balanced mixer.

The SSB transmitter shown in Fig 6.25 has a severe difficulty—it operates at only a single frequency, that of the filter used to generate the sideband. A practical filter-type SSB transmitter topology is presented in **Fig 6.26** where an SSB signal is generated at an intermediate frequency. The resulting SSB is then heterodyned to a desired output frequency where it is bandpass filtered, amplified, low-pass filtered, and applied to an antenna.

Assume the narrow filter used to create the SSB signal at IF is configured to create an upper sideband. For example, let the carrier frequency be 9.000 MHz with a filter extending from 9.0003 to 9.003 MHz, a bandwidth of 2.7 kHz. Set the LO to 37.4 MHz and design the LC bandpass filter to cover 28 to 29 MHz. The resulting signal is then at 28.4 MHz. The transmit mixer has both sum and difference frequency outputs and the LC bandpass has selected the difference, producing a carrier output of $(F_{LO} - F_C)$ for the suppressed carrier. The sideband frequency within the IF will be $F_C+\delta$ where δ is a small positive difference frequency. This value is greater than the carrier, so this is an upper sideband. Because the LC bandpass is configured for a difference output, the signal output will be $(F_{LO} - (F_C+\delta))$, which expands to $(F_{LO} - F_C - \delta)$. This is less than the suppressed and translated carrier at $(F_{LO} - F_C)$, so we now have a lower sideband signal. A designer must always be aware of such inversions. They can be useful for the designer, for crystal filters without ideal symmetry (lower sideband ladder of Chapter 3) are easily built.

The simple direct-conversion receiver in Fig 6.20 is effective in receiving an SSB signal. The difficulty that we encountered with DSB is no longer present, for there is no coherent information in the spectrum formerly occupied by the suppressed sideband to be heterodyned to baseband, eliminating the need for extreme stability. If the BFO is in error by 100 Hz, the received voice may sound unusual, but will still be intelligible.

Even though there is negligible opposite-sideband energy transmitted by a properly designed and adjusted SSB transmitter, that does not mean that the spectrum where that opposite sideband would have been is not used. That spectrum is usually occupied by another SSB station. If a direct-conversion receiver was tuned to a desired signal, the undesired signal would produce completely garbled audio, making simple direct-conversion receivers unsuitable in a densely populated band.

A superheterodyne receiver like that in **Fig 6.27** is usually used to receive SSB. The incoming signal is filtered in a preselector, heterodyned to an IF, and is passed through a bandpass filter. The bandwidth of that filter, usually built with quartz crystals, is wide enough to pass all of the speech spectrum that is transmitted, but little more. A typical SSB receiver will have a bandwidth from 2 to 3 kHz. The filter shape is fairly flat over the passband, but then has steep skirts so that energy in an adjacent "channel" will not interfere with the signal being received. The nar-

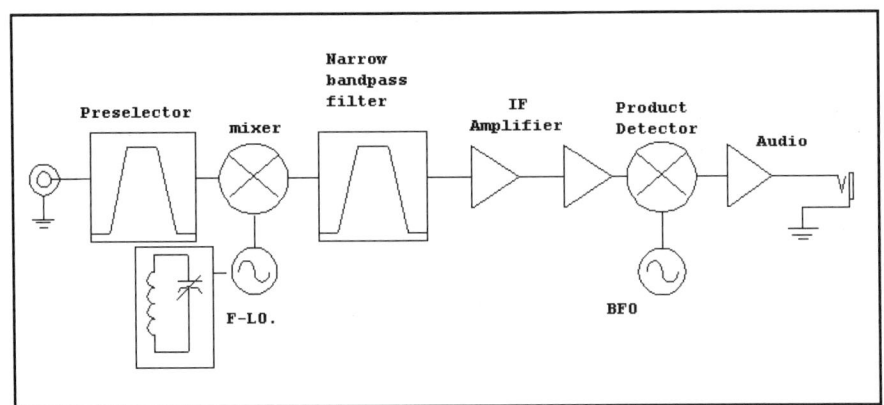

Fig 6.27—A traditional superhet SSB receiver. The response from only one sideband is allowed owing to the narrow-bandwidth crystal filter and the relationship of the BFO frequency to that filter.

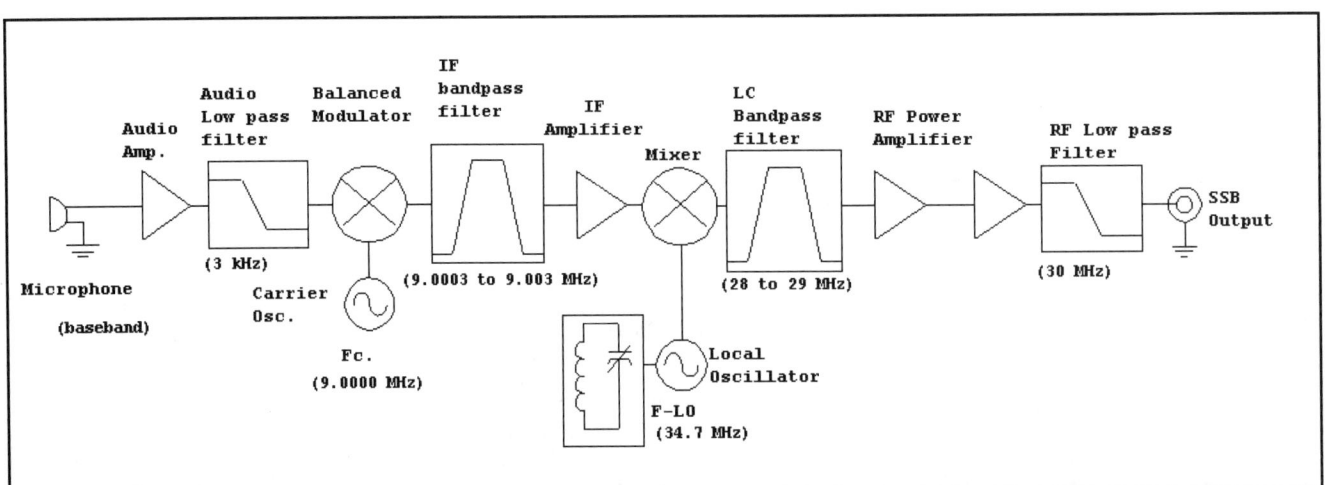

Fig 6.26—A practical filter type SSB transmitter where a mixer translates the output of a fixed-frequency SSB generator to a variety of outputs.

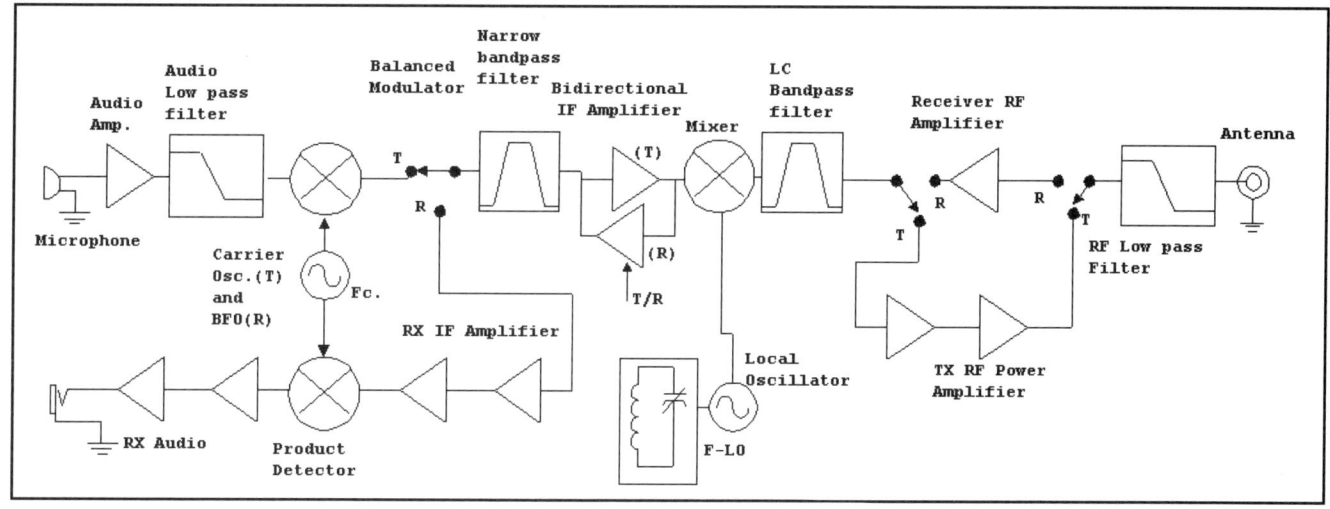

Fig 6.28—An SSB transceiver, a system for both receiving and transmitting an SSB signal. Economy and operating convenience are gained by sharing elements between functions. It is most common to share oscillators and a crystal filter, which is done here. This circuit also shares a mixer between the receiver and transmitter, and uses a bidirectional IF amplifier, a circuit that, with dc switching, will amplify signals moving in either direction. The amplifier circuits are presented later in the text.

row bandpass filter in the SSB receiver is followed by IF amplifiers, a product detector with BFO, and an audio amplifier.

The BFO must be carefully set in the SSB receiver. It should be fixed so that one edge of the filter (a –6 dB point) corresponds to an audio note of about 300 Hz. The other edge will be determined by the filter bandwidth. Typically the BFO is at a point on the filter response that is 20 or 30 dB below the nominal, flat response. The same constraints are used in setting up the carrier oscillator in the filter method transmitter.

The SSB receiver can produce sideband inversion just as we illustrated in the transmitter. The builder/designer should go through the numbers to confirm the behavior. Using popular vernacular, "You do the math."

The SSB receiver, although designed to receive SSB, is also well suited to CW. So long as the filter has good stopband attenuation, the response will also be single signal, as can be confirmed by repeating the experiment we have done with both the direct conversion and the CW superheterodyne. Readjustment of the BFO can compromise the single signal characteristic. An SSB filter is often considered too wide for optimum CW performance, especially in a heavily used band.

The SSB receiver is also well suited for reception of DSB signals. The filter in the receiver rejects one of the sidebands present at the receiver antenna terminal.

Finally, we see that combining Figs 6.26 and 6.27 will result in a transceiver where many circuit elements can be shared between transmit and receive functions. Most transceivers share all oscillators and the crystal filter between the two functions. **Fig 6.28** shows a typical block diagram, here with a design that also shares a mixer between functions, and uses a bidirectional amplifier. No matter what schemes the designer may elect to use, he or she should take care to preserve performance in both transmit and receive functions.

6.1 RECEIVER FUNDAMENTALS

A receiver is characterized by numerous parameters. It must have considerable gain, for the signals we wish to hear are weak. The receiver must also be selective, allowing signals with only slightly differing frequencies to be isolated, received, with useful information processed. The receiver must also include detection in one form or another, producing an output frequency that we can hear. The detection may consist of a rectifier that extracts information about amplitude variations of the radio frequency signal, a discriminator that evaluates signal frequency, or a mixer excited by an LO with a frequency at or very close to the incoming one.

All functions must be executed in a way that does not compromise the information from an original signal. Hence, local oscillators must be stable with respect to the stability of the signals being processed. Filters that provide selectivity must be wide enough to pass the desired information related to the received signals. The gain must be generated without adding excessive noise. Receiver performance specifications generally relate to how well the various required jobs are done.

We begin our receiver investigation with a primitive experiment, an examination of headphones, the generally preferred transducer for converting an electrical signal into sound. (Although we all tend to assume that headphones are optimum, some will argue that a speaker is preferred for weak signals. Individual experiments are required.) The experiment uses a 50-Ω audio-signal source with known output power. See Chapter 7.

A large collection of monaural and stereophonic headphones were examined, old and new. The two ear-pieces were usually operated in series. The typical phones were low (4 Ω) to medium impedance (20 to 35 Ω per side), often representing a reasonable impedance match to the 50-Ω generator. The signal source was adjusted with each headphone set until a signal was just detectable in a quiet room.

The most sensitive headphones were obsolete, inexpensive types consisting of little more than 2-inch diameter speakers mounted next to each ear. Two pair from our collection were capable of producing a detectable output with an available input of –85 dBm. That is, the applied signal was 85 dB below one milliwatt from a 50-Ω audio source.

Several of the phones were nearly as sensitive including some newer Koss TD/65 (90 Ω per side) used for routine communications. The Koss sensitivity was –80 dBm, with better clarity than provided by many others. Several lightweight inexpensive phones (Sony Walkman class) had sensitivity from –60 to –70 dBm. Very old high impedance phones had similar sensitivity, but only after being impedance matched.

A typical listening level will be significantly higher than our *threshold*, but still well below a milliwatt. From these experiments, we will assume that a minimum receiver must be capable of producing an output of –50 dBm for the weakest signal to be encountered. The weakest signals that we normally encounter in HF CW communications are –130 to –140 dBm, indicating a needed gain of around 90 dB. Although this is a subjective result, it represents a design beginning.

Our first simple receiver is shown in **Fig 6.29**. A high-gain audio amplifier with low input and output impedance was built with a gain of 87 dB. The amplifier is combined with an external diode ring mixer, 7-MHz local oscillator and input 7.5-MHz low-pass filter to form a complete direct-conversion receiver. An antenna was connected, producing numerous signals in the 40-m band. The receiver had the usual bright response that we expect from direct-conversion designs. (DC receivers are discussed in much greater detail in Chapter 8.)

The amplifier did more than make the signals louder. It generated noise, apparent when power was first applied. While the noise was not so loud as to be objectionable, it would obscure some weak signals we expected to hear. When a signal generator was attached and adjusted, the best we could hear was about –130 dBm, well away from the –140 dBm expected with many simple direct-conversion receivers.

Why is this receiver so noisy? Little noise is generated in the first element in the system, the diode ring mixer, a passive element without gain. Rather, the noise in this design is generated in the amplifier that follows the mixer.

This noise is not the result of a poor op-amp choice, but a poor design with respect to noise. Negative feedback in an amplifier reduces input impedance. The impedance looking into the inverting amplifier input of a 5532, *with a 5.6-kΩ feedback resistor*, is about 1 Ω. We modify this with an added series 56-Ω resistor to generate a 57-Ω impedance to approximately match the mixer, a requirement for low mixer distortion. The available signals from the mixer are all absorbed, but only the fraction of the power delivered to the 1-Ω input is amplified. The remaining power is merely converted to heat. All of the available noise current from the input resistor flows in the op-amp input. The result is poor noise figure, a degradation in the input signal-to-noise ratio in the process of amplification. This amplifier is contrasted with the popular design where the first audio amplifier is a common-base bipolar transistor. In that design, almost all of the available power is presented to the active device.

The fundamental receiver parameter used to characterize the noise that limits sensitivity is noise figure (NF), introduced in section 2.6. NF is a measure of the degradation of signal-to-noise ratio by a processing element, be it a complete receiver or a single stage.

Let's assume that we wish to infer receiver noise figure by driving the receiver with a signal generator. The input *signal* power is established by the *available* power from the generator. (This may differ from the actual power delivered to the load.)

Input available noise power is that available from whatever resistor might be attached to the input, given by

$$P_n = k\,T\,B \qquad \text{Eq 6.9}$$

where k is Boltzmann's constant, T is temperature in kelvins, and B is the bandwidth in Hz in which the noise is observed. The standard temperature used for noise determinations is 290 K, close to a normal room temperature. This noise power is independent of the resistance. The noise power is distributed uniformly over all frequencies. If receiver bandwidth is increased, the noise power increases accordingly.

Attaching a room temperature resistor to the input of a receiver provides a source of noise. The signal generator, with its output resistance, will also serve this function. If the generator level is changed by attenuation, output resistance seen by the receiver remains constant to maintain a constant available noise power.

The output signal and noise are measured by attaching a load (usually a speaker or earphones) monitored by an ac voltmeter, ideally one that provides a true rms response. Noise output can be monitored alone by momentarily turning the generator off. When the signal is again applied, along with the input noise, the

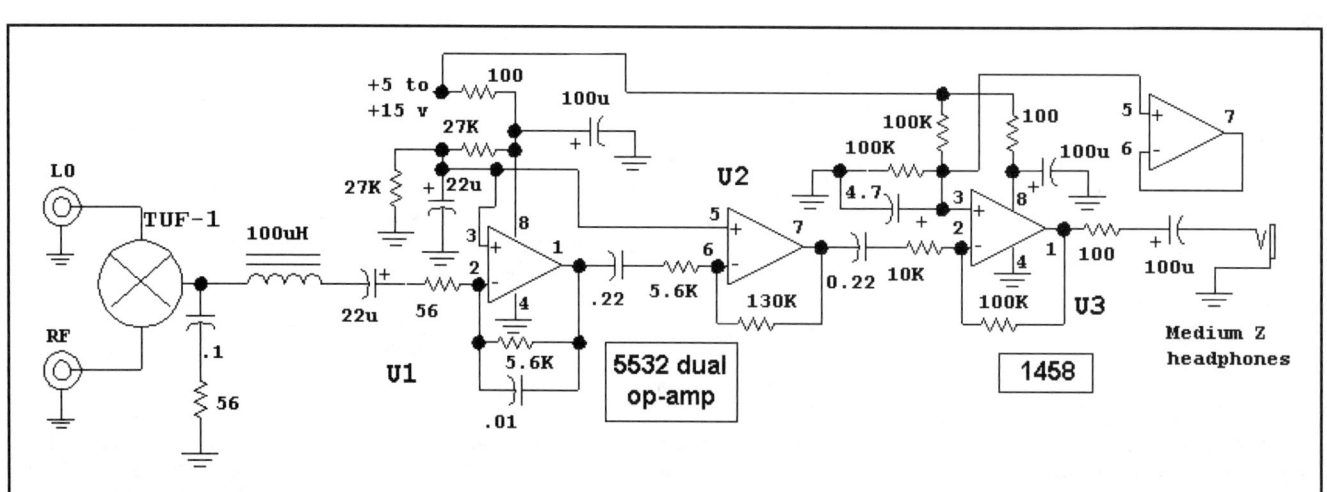

Fig 6.29—A basic direct-conversion receiver. An audio amplifier with a gain of 87 dB follows the diode ring. See text for discussion.

output will be an output signal + noise power. An output signal-to-noise ratio can then be calculated. Noise figure can then be calculated.

Noise figure is usually measured with a noise source of known power, usually well above the noise power available from a 290-K resistor. See Section 2.6 and noise measurements in Chapter 7.

The greatest virtue of noise figure as a receiver parameter is that it is bandwidth invariant. If we increase the bandwidth during an NF measurement, we will process more noise in the receiver. But the output will also increase in proportion, leaving the noise gain, the ratio of output noise to input noise, a constant.

Another measure of receiver sensitivity is *minimum discernable signal*, or MDS. This is the available input signal from a generator that will cause the output power to increase by 3 dB over what is present without the applied signal. In this condition the signal and the noise have equal output powers.

MDS is directly related to room temperature NF by

MDS (dBm) = −174 (dBm/Hz) + 10log B (dB × Hz) + NF(dB)

Eq 6.10

We measured the noise figure of one of our receivers to be 7 dB with a nominal bandwidth of 500 Hz. **Eq 6.10** then predicts MDS of −140 dBm. A direct measurement of MDS where we look for a 3-dB increase in output above the noise floor as we apply signal produced an almost identical result of −141 dBm.

It is interesting to listen to this receiver with the signal generator attached. We find that we can hear the MDS, but not much further into the noise.

We now increase the receiver bandwidth to 2.4 kHz by switching in a new crystal filter, increasing the bandwidth factor in Eq 6.10 to 33.8 dB. MDS becomes −133.2 dBm with a 7-dB noise figure. A measurement will usually confirm this number. Noise measurement in a wider bandwidth is generally easier than it is with narrow band systems owing to less fluctuation in the meter movement. But major errors can and often do occur as a result of slight gain variations with frequency in either the IF or the receiver audio circuitry— errors that generate a narrower noise bandwidth than expected. A direct NF measurement is generally preferred over one of MDS, where only a ratio of two noise powers must be determined.

An ideal receiver with measured MDS commensurate with the filter BW will often let a listener hear signals that are much weaker than indicated by the MDS. Why? The human ear and brain are a vital part of the communications system and they are capable of acting like a filter of considerably narrower bandwidth than the voice bandwidth of the receiver. This effect is observed with both wide bandwidth superheterodyne designs and direct conversion receivers. Indeed, many seasoned weak-signal VHF enthusiasts including moonbounce specialists normally use wider SSB-bandwidth filters.

Many argue that noise figure is rarely a significant receiver parameter, especially for HF reception. An NF of 10 or 12 dB at 28 MHz, with much higher numbers at lower frequencies will usually provide as much sensitivity as one can use. A practical receiver test is very simple: While listening to background noise on a band, disconnect the antenna. If the noise drops significantly, the receiver NF is as good as it needs to be.

NF is much more important as a design parameter. The essence of modern receiver design is a quest for dynamic range, and NF specifies the lower end of such a range.

Equation 6.10 relates NF to MDS, suggesting that little is to be gained with extremely low noise figures. Consider, for example, a receiver with a 200-Hz bandwidth and 3-dB NF. Equation 6.10 predicts MDS of −148 dBm. Dropping noise figure to a spectacular 0.5 dB results in only a 2.5 dB sensitivity improvement to −150.5 dBm. This is what a careful MDS measurement would demonstrate. But in reality, the practical improvement could be much more than this. The dilemma comes about when we pick a noise temperature of 290 K for our standard. This choice defined the "input" noise in Eq 6.9. But if the input noise resulted not from the 290 K resistor related to our measurement, but from an antenna pointed at a quiet part of the sky, the input noise might well relate to a resistor with a temperature as low as 20 K. A more refined calculation would show that MDS could be as low as −158 dBm for this example. A related concept of *noise temperature* was used to obtain this result.[3]

The noise factor of a two-stage cascade is

$$F = F_1 + \frac{(F_2 - 1)}{G_1}$$ **Eq 6.11**

where F is the net noise factor, F_1 and F_2 are the noise factors for the first and second stage, and G_1 is the available power gain for the first stage. All numbers are power ratios and not dB values.

Consider an example shown in **Fig 6.30**. The first amplifier has a gain of 12 dB and

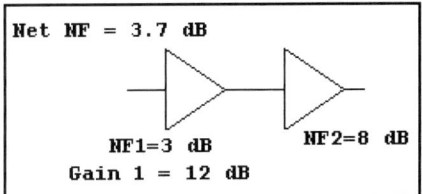

Fig 6.30—Example calculation for noise figure of a cascade of two stages.

a 3-dB NF while the second stage has an 8-dB NF. Related power ratios are $F_1 = 2$, $F_2 = 6.3$, and $G_1 = 15.8$, yielding F = 2.34, or NF_{NET} = 3.7 dB. The first stage noise performance dominates in this example. Once we know how to evaluate a cascade of two stages, we can apply the process in steps to evaluate an arbitrary cascade, including an entire receiver front end.

Many of the circuit blocks that we used in receivers and transmitters are room temperature passive parts with no gain elements. These include not only the popular passive switching-mode mixers, but attenuators and filters. Generally, the NF of a passive circuit equals the insertion loss of that circuit. Hence, a diode ring mixer with a 6 dB conversion loss (gain = −6 dB) will have a 6-dB NF. A bandpass filter with an insertion loss of 2 dB will, similarly, have NF = 2 dB and Gain = −2 dB.

Fig 6.31 illustrates a receiver front end where several elements contribute to the noise figure. This circuit will include an RF amplifier, for we are interested in relatively low noise figure. Two bandpass filters are used. The first is a single resonator ahead of the RF amplifier while the second is a double tuned circuit. A diode-ring mixer is followed by a feedback amplifier that uses a bipolar transistor with high dc emitter current. The overall cascade has net gain of 15 dB and a net noise figure of 7.1 dB.

Front-end bandpass filters usually do not impact overall noise figure. In the receiver example just presented the system bandwidth is determined by a crystal filter that follows the attenuator. This filter is usually narrow (3 kHz or less) and the two L/C bandpass filters shown as the first and third elements in the cascade are wide (a few hundred kHz). The crystal filter then sets the overall response. The bandpass filters in the cascade have no more impact on noise figure than an attenuator would. The situation would be considerably different if the crystal filter was replaced with a wide L/C filter with equal or wider bandwidth than those in the front end.[4]

Some RF Amplifiers and Attenuators

Many modern HF receivers use no RF amplifier, for adequate noise figure can be obtained without it. Most commercial gear has a NF of 10 to 12 dB at and below 30 MHz. A practical sensitivity test was outlined above. There are some situations where an RF amplifier can be useful, even at HF. This is especially true at 21 and 28 MHz during periods of marginal propagation. It is then useful to switch a low noise amplifier into the signal path. Such an amplifier is not normally needed and should not be used merely to make signals louder. We will illustrate a few circuits that we have built, used, and measured.

A favorite RF amplifier is a common gate JFET circuit. A J310 is used for HF applications, while a U310 is preferred for VHF and UHF. (The surface mounted version of the J310 should be excellent for both!) The basic amplifier is shown in **Fig 6.32**. The FET is biased for a current of 12 to 14 mA, determined by FET I_{DSS} and source resistor. The gain is only about 2 dB with this amplifier if the drain load resistor, R, is set at 680 Ω. In spite of the low gain, the amplifier is still very useful. It has a good input and output impedance match, so offers a good interface to filters and mixers. It is most useful for the excellent reverse isolation. The reverse gain (S12) was measured as –43 dB. This is an excellent amplifier for use with direct conversion receivers when attempting to reduce *tunable hum*, discussed in Chapter 8. The circuit is turned on with $V_{CONTROL}$ = +5 or so. The gain is reduced by 40 dB when turned off.

Gain goes up to 6.5 dB in this circuit when the drain load resistor is eliminated. In that configuration, the third order output intercept was +28 dBm, measured at 14 MHz with fairly flat gain up to 50 MHz. (Intercepts were introduced in section 2.6.) Lower frequency performance is improved with a larger inductance RF choke.

Higher gain is available if the output is tuned, shown in **Fig 6.33**. The output drain resistance for this amplifier is close to 10 kΩ, allowing it to form one termination of a bandpass filter. The variation shown with a single tuned output circuit has a typical gain of 12 to 13 dB with a 50-Ω load. The 50-Ω input match is a 15-dB return loss. Noise figure was 5.0 dB at 21 MHz.

This amplifier has no tuning at the input, for C1 and L1 are both large. Lower noise figure is often obtained with a suitable input network, one that usually degrades input impedance match. The designer can generally design an input network that will present a needed impedance to the input if the value for optimum NF is known. We didn't have that data for the J310, but were able to find hints. Specifically, Chip Angle, N6CA, has built amplifiers with the U310 for several VHF bands. The U310 is the same chip, but is packaged in a metal can with the gate attached to the can. We were able to analyze his circuits and scale his input networks to lower frequency. The result was an amplifier with a measured 1.5-dB NF, but with a poor input match and gain of only 12 dB. This occurred at 21 MHz with L1 = 1.26 µH and C1 = 39 pF. The noise match point that we inferred was $\Gamma_{OPT} = 0.89$ at 7°.[5]

A common source JFET should be capable of low noise performance. The practical difficulty in building such a circuit is often stability. Cascode connected JFETs should be considered. Neutralization is also practical, although rarely used.

The humble source follower should not be discounted as a low-noise amplifier. A suitable circuit is shown in **Fig 6.34**. A link-coupled input drives the gate through

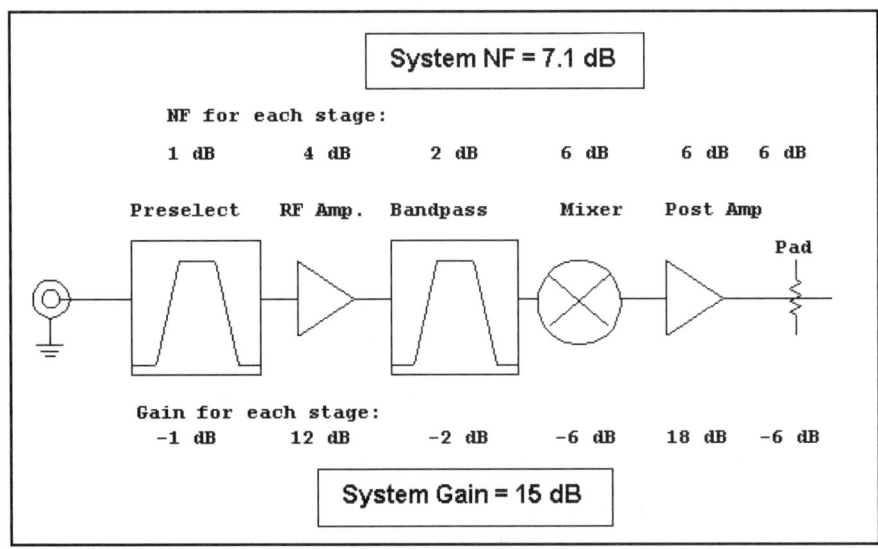

Fig 6.31—A six-stage cascade showing a typical receiver front end. The stages consist of a wide filter, an RF amplifier, a steeper skirted bandpass filter, a diode ring mixer, a post-mixer amplifier, and finally, a 6-dB attenuator.

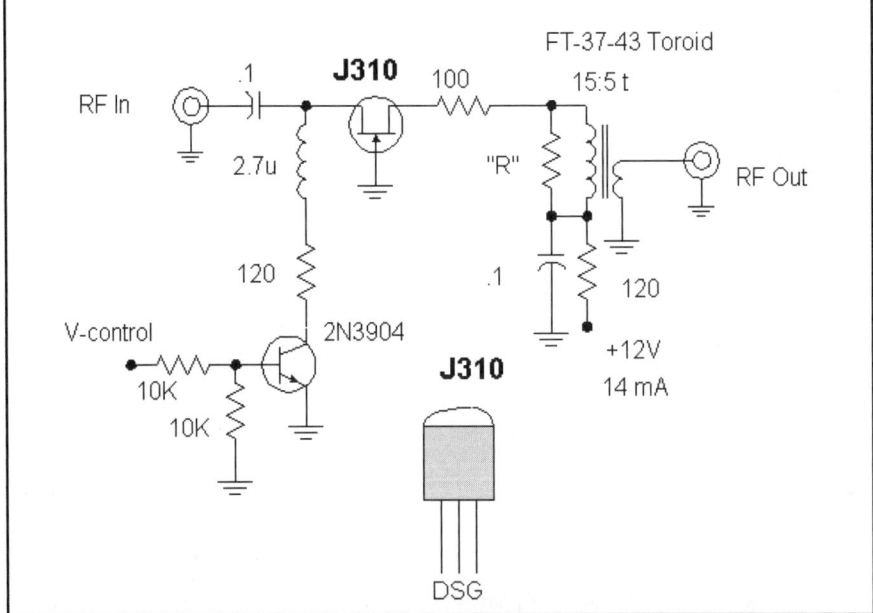

Fig 6.32—A common-gate amplifier using a JFET. The 100-Ω resistor at the drain suppresses UHF oscillations. See text regarding the drain load resistor, "R."

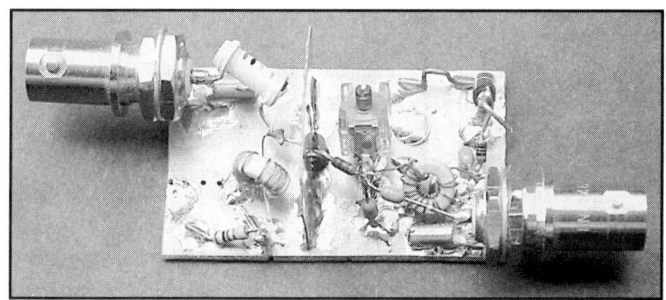

Close up of common-gate low-noise amplifier using a J310.

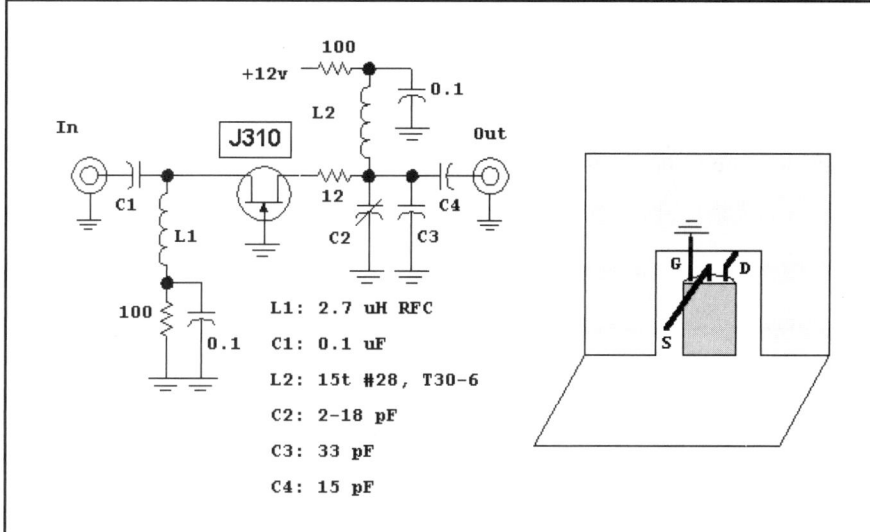

Fig 6.33—A 21-MHz RF amplifier. This circuit, with the values shown, provides a gain of 14 dB with a 5-dB noise figure. Redesign of the input network produced an NF of 1.5 dB, but with reduced gain of 12 dB. A shield between the source input circuit and the output drain circuit is advised, especially if high-Q solenoid coils are used. It is generally not required when using toroids, although the gate should be grounded with short lead length.

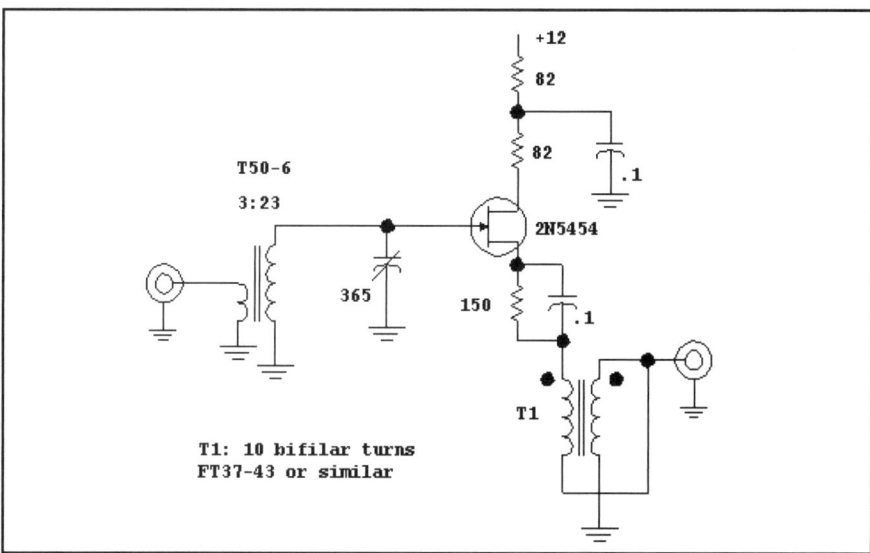

Fig 6.34—Source follower functioning as a low-noise amplifier. The drain resistor serves to suppress UHF parasitic oscillations. The components shown will tune from 6 to 22 MHz.

a tuned circuit with a sizable impedance transformation. The output is then extracted from the source with a ferrite transformer. An example amplifier measured gain of 11 dB with NF = 1.9 dB. No stability problems were noted. The output match was good, although the input is severely mismatched.

Dual gate MOSFETs make excellent RF amplifiers as shown in **Fig 6.35**. This circuit was tuned for both the 21 and the 14 MHz bands with similar results obtain with each. The 14-MHz circuit is shown. A pi-network transforms the 50-Ω source to "look like" an impedance of 2000 Ω at gate-1 of the FET. The network was designed for a Q of 10 and used an existing 2.7-µH RFC. The drain is matched with a ferrite transformer followed by a 6-dB pad. This amplifier provides a gain of 16.5 dB (including the loss of the pad) with a 3.6-dB noise figure. The circuit had an output intercept of +12.5 dBm.

The gain is often excessive with dual-gate MOSFETs. Better overall receiver dynamic range is afforded by reduced gain. The pad helps, but it compromises the amplifier intercept performance, for the amplifier must have a 6 dB higher intercept to get the quoted value. Even the 1200-Ω drain load resistor compromises IMD performance. Source degeneration provides an alternative, achieved by disconnecting the source bypass capacitor. Gain dropped to 9 dB for the circuit shown (with pad), and the noise figure increased slightly to 4.1 dB with OIP3 = +14 dBm.

The low-Q inductor used in the input pi-network compromises the noise figure. Replacing it with a toroid dropped the 3.6-dB NF to 2.5 dB. Even lower values are available if a higher impedance is chosen for the pi network. The input match is very poor with all variations of this amplifier.

Many of the feedback amplifiers described throughout this text are suitable for RF amplifier application. The noise figures can be in the 3 dB area with some transistors. For example, we have measured a 3-dB NF with a 2SC1252 operating with 20-mA emitter current.

The modern trend in amateur receivers is to include an RF amplifier that can be switched into the circuit if needed. That switching is best done with relays, although PIN diodes can also be used if done with extreme care to avoid second-order intermodulation. It is also common to include one or two attenuators that can be switched ahead of a receiver. An attenuator equally decreases the strength of all signals reaching the front end. Often the signals we are trying to copy are strong enough that an attenuation of 10 dB will not cause a sensitivity problem. The real

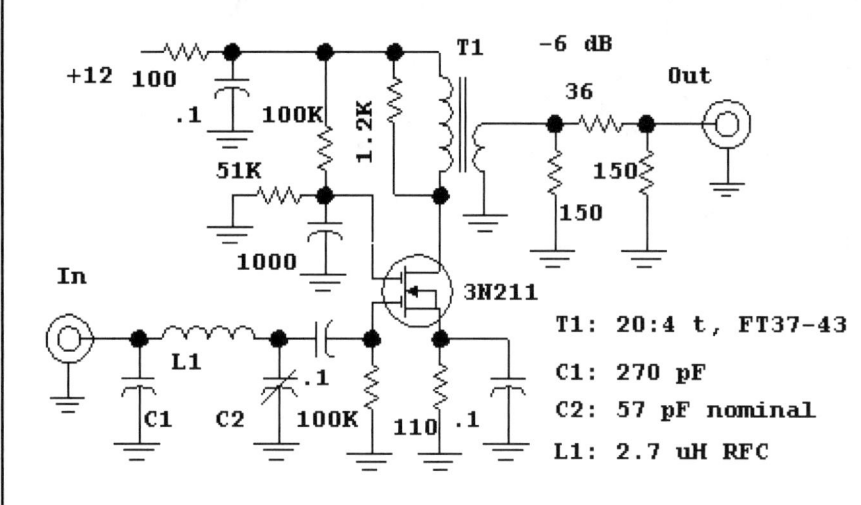

Fig 6.35—Dual-gate MOSFET RF amplifier. This version used an RF choke at L1 with Qu = 50. A higher Q inductor will drop the amplifier noise figure. See text.

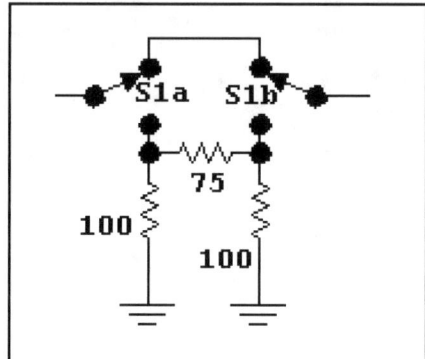

Fig 6.36—A 50-Ω, 10-dB pad using standard resistors and a toggle switch. Short lead lengths should be used to provide good performance over the HF region. Relay switching could also be used.

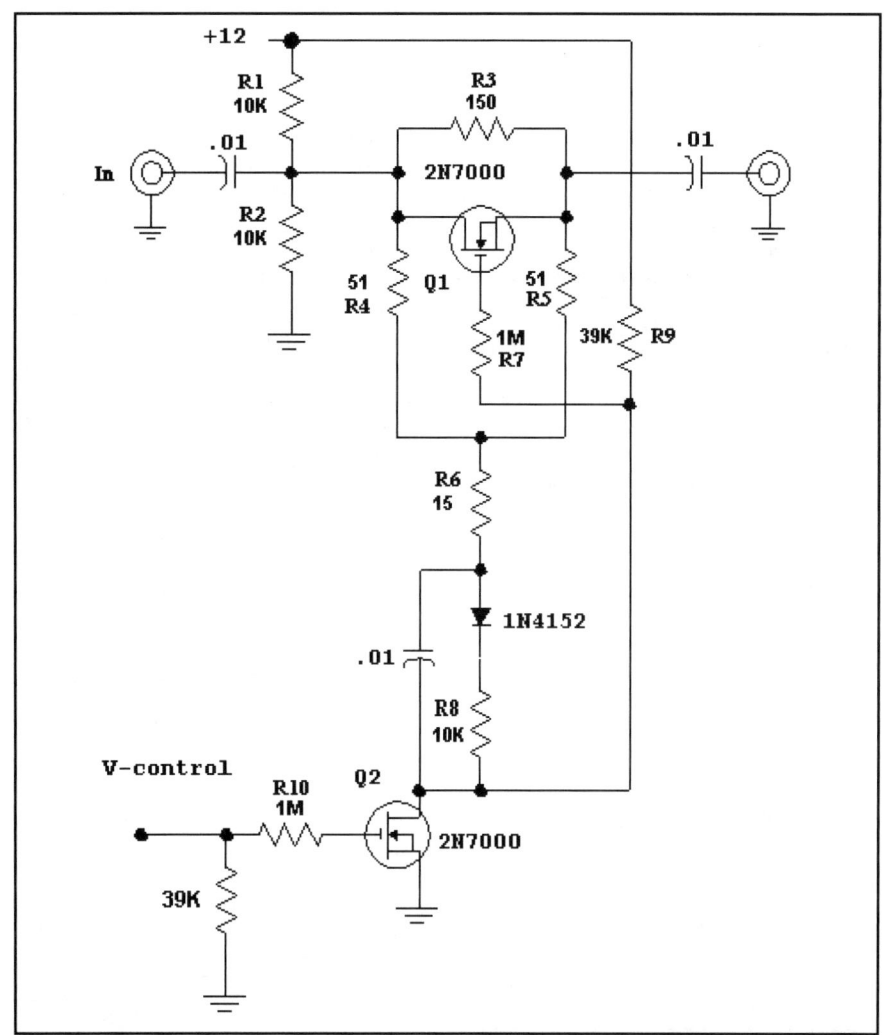

Fig 6.37—A 10-dB pad using electronic switching. A bridged-Tee pad (R3, 4, 5, 6) is switched with low-cost MOSFETs. During thru operation, Q1 is on while Q2 is off. Q2 comes on during attenuated operation. Current consumption is about 1 mA.

Dual-Gate MOSFET Availability

The dual gate MOSFET was a very popular consumer device from 1970 to 1980 and was readily available from a number of sources. The part provides low noise, moderate to high amplifier intercepts, and reasonable power consumption. They also offer good AGC performance. They are now more difficult to obtain than they were in the past.

But Dual-Gate MOSFETs are still available. Several suppliers in Japan continue to manufacture a variety of components. The NEC 3SK131 is an excellent part, but it is available only in a surface-mount form.

Phillips manufactures a large variety of dual-gate devices. These are often listed in some US catalogs. Again, these devices appear predominantly in SMT format.

Generally, it is quite straightforward to substitute one MOSFET in a circuit designed for another. There may be a few different biasing details, but these can be extracted from data sheets, which are generally available on the World Wide Web. Experiments may be required if data is not available.

Finally, most circuits using dual-gate MOSFETs can be built with N-channel JFETs in a cascode configuration. This is illustrated in the IF amplifier part of this chapter.

utility of an attenuator is that most distortions drop faster with signal strength than the signals themselves. Hence, if strong signals within a band are causing gain compression or intermodulation distortion, a small decrease in the strength of the offending signals can completely eliminate the problems.

A passive attenuator is shown in **Fig 6.36**. The typical miniature toggle switch works well for pads of this sort with 10 to 20 dB attenuation.

A scheme is shown in **Fig 6.37** where 2N7000 MOSFETs replace a mechanical switch. The FETs are both RF and dc switches in this application. A pair of resistors, R1 and R2, create a 6-V supply. R9 will bias Q1 into conduction in the low attenuation position with the Q2 gate low. The Q1 channel is then held at 6 V. But when Q2 is turned on, R6 is switched to RF ground. The dc potentials also change to turn Q1 off. We measured an insertion loss of 0.38 dB with this circuit, with a 10 dB gain step. The 14-MHz IIP3 exceeded +35 dBm during low attenuation, and was +26.5 dBm in the attenuation position.

6.2 IF AMPLIFIERS AND AGC

A superheterodyne receiver uses an intermediate frequency between an initial mixer and detector, primarily as a means for obtaining selectivity. It is this selectivity that selects the sideband received, or provides single-signal CW reception. The IF is the usual place for adding and controlling receiver gain through voltage control.

Voltage-controlled gain is usually realized with integrated circuits. But the most popular parts are slowly, but surely disappearing as the consumer markets evolve toward larger scales of integration. Accordingly, this section contains two goals. First, we hope to illustrate some IF amplifier methods that can be applied before the semiconductors disappear. And of greater import, we hope to illustrate some methods that others can use to develop their own IF circuits.

Early superhets used tuned IF amplifiers, providing selectivity throughout the amplifier while modern designs usually use local filtering. Signals exit a mixer, pass through a filter (usually built from quartz crystals) to reach the IF amplifier. As such, the IF amplifiers are protected from strong out of band signals, the sources of performance-compromising distortions. Reasonable linearity is still useful to preserve low *in-band* distortion.

The importance of IF noise figure is illustrated in **Fig 6.38** where we calculate receiver noise figure for a system with the front end treated earlier. The front end had a 7.1-dB NF with total gain of 15 dB. We start with a lossy crystal filter with 10-dB insertion loss and find that overall system noise figure is always above 10 dB, even if the IF NF is as low as 3 dB. A more realistic filter loss of 3 dB provides an overall NF in the 8 to 9 dB region, even with fairly noisy IF amplifiers. IF Amplifier noise figure, including the loss of any filter ahead of it, can have a major impact on system performance!

The distortion properties of IF amplifiers will become more important in emerging receiver topologies. These receivers, largely based upon digital signal processing, use wide IF filters followed by an IF amplifier driving an analog-to-digital converter. The receiver is then completed through digital calculations. Distortion within the IF amplifier and the A-to-D converter become vital.

In the following pages we will consider a number of IF amplifier circuits. We will examine them for noise figure, gain, gain variation, and IMD. Some complete IF systems will be shown.

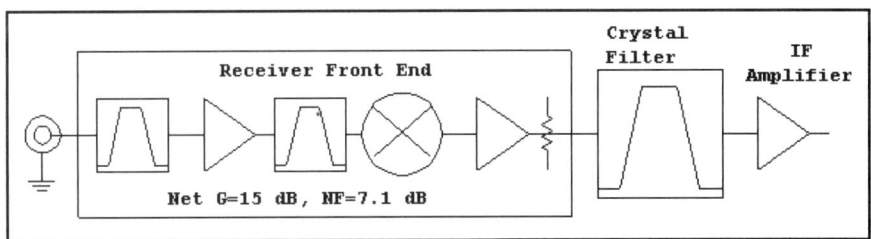

Fig 6.38—The front end presented earlier in Fig 6.31 is combined with a crystal filter of known insertion loss, followed by an IF amplifier. If the filter has a 10-dB IL, a 7-dB IF noise figure will produce a system NF of 10.6 dB.

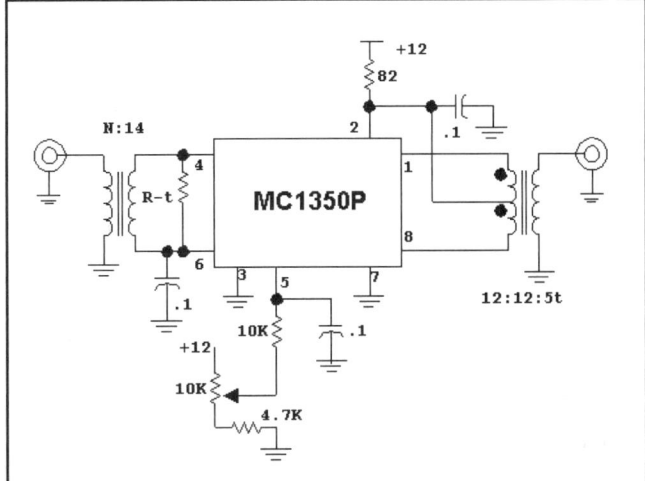

Fig 6.39—Amplifier for examination of the MC1350P. Gain is reduced by over 60 dB by increasing the dc current into pin 5.

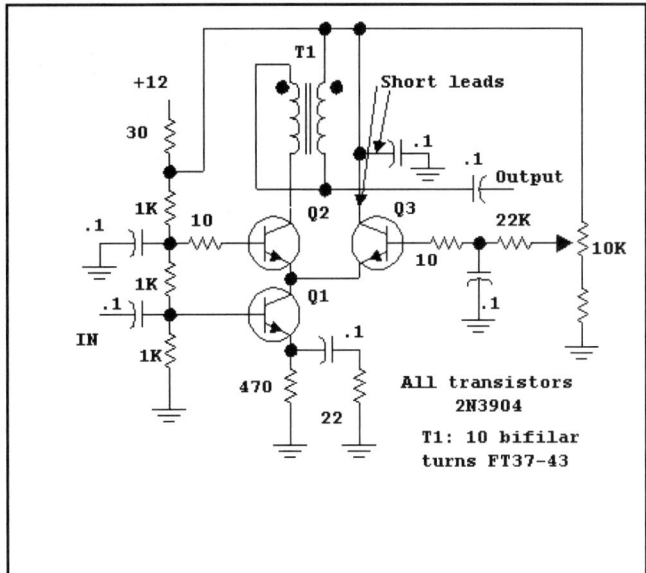

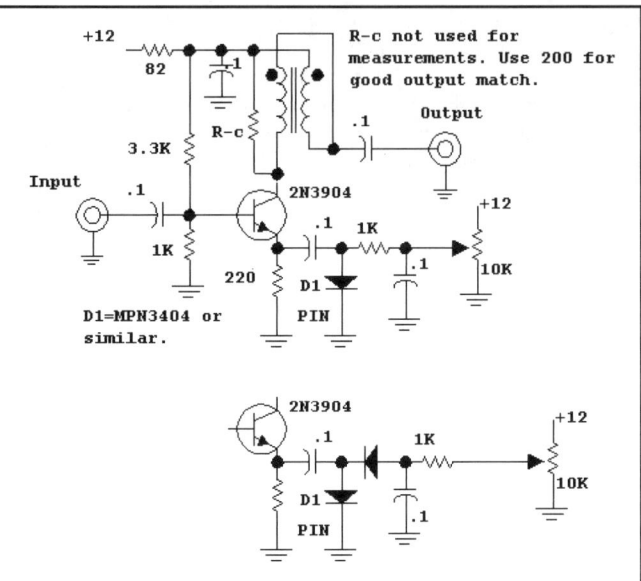

Fig 6.40—Bipolar transistor discrete IF amplifier with gain reduction using the same mechanism as used in the MC1350P. Control range was 70 dB, experimentally controlled with a 10-kΩ manual IF gain.

Fig 6.41—Simple gain-controlled amplifier. The inset shows the use of two PIN diodes to increase the control range slightly with the same control current. Many diode types work with this circuit; see text. The 10-kΩ pot establishes manual IF gain.

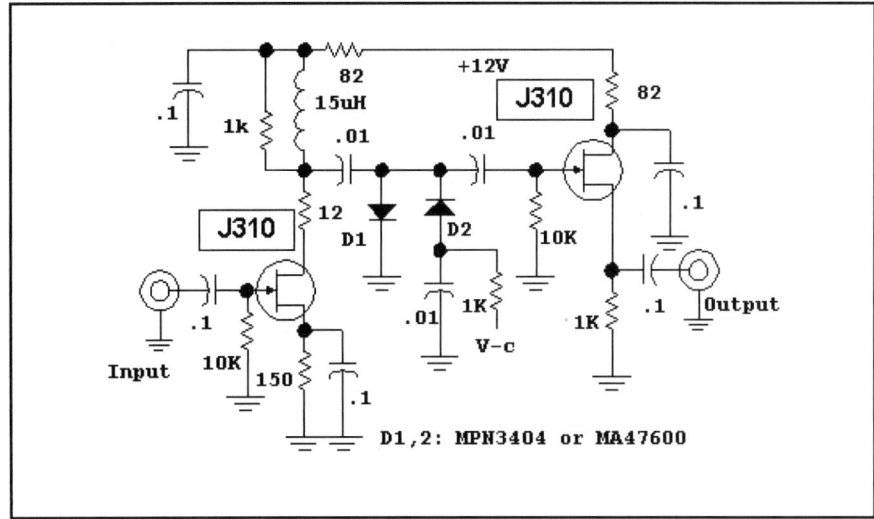

Fig 6.42—AGC amplifier with FETs and PIN diodes.

The first amplifier presented uses the popular Motorola MC1350P. Although this device is, at this writing, slated to be discontinued, it will probably be available for a while from distributors, or from surplus. The methods used in the 1350 can also be realized with discrete components. The MC1350P test circuit is shown in **Fig 6.39**.

The input between pins 4 and 6 (the input differential pair) looks like a 2700-Ω resistance paralleled by 8 pF at 10 MHz. This was approximately matched with a 2:14 turn ferrite transformer with no R_T used. The output, consisting of open collectors of a differential transistor pair, was terminated with a ferrite transformer, producing a 10-MHz gain of 47 dB. The gain-control range was over 65 dB. The noise figure was 5.1 dB, but degraded to 10.3 dB when the gain was reduced by 10 dB.

The relatively high input impedance is rarely suitable for termination of crystal filters. Extra resistance, R_T, is often paralleled with the input to achieve a needed impedance. $R_T = 620$ Ω produced a net impedance near 500 Ω, a common value needed to terminate crystal filters. This was matched to 50 Ω with a 4:14 turn ratio ferrite transformer. Gain dropped to 39 dB, as expected. Full gain noise figure was 6.6 dB, increasing to 14.1 dB with 10-dB gain reduction. Changing R_T to 220 Ω with a new matching transformer produced further degradation.

Fig 6.40 shows a breadboard circuit with internal workings similar to the '1350, although the IC has additional differential input and output buffering. The Q1 collector current passes through Q2 that operates as a common base amplifier. Gain is reduced by increasing the base bias on Q3 so that emitter current and signal current are both *robbed* from Q2. This circuit provided measured gain of 16.5 dB, 70-dB gain-control range, and good IMD performance. Noise figure was 7 dB at maximum gain, but degrading to 19 dB with 10-dB gain reduction. We noted a noise peak when Q2 and Q3 conducted equal currents. Careful examination revealed the same effect with the MC1350.

A bipolar transistor circuit using PIN-diode emitter degeneration is shown in **Fig 6.41**. Although simple, this circuit offers promise. Gain at 10 MHz was measured at 30 dB with a MPN3404 PIN diode. Gain control range was also 30 dB. A builder may wish to load the collector with a resistor to produce slightly less gain per stage with a better output impedance match. Noise figure was 5.2 dB and hardly changed with a 10-dB gain reduction. Several diode types were evaluated in this circuit. Power rectifiers such as the 1N4006 or 1N647 worked well with low distortion, although large diode capacitance reduced gain control range. While a 1N4152 worked, IMD was severe at some currents.

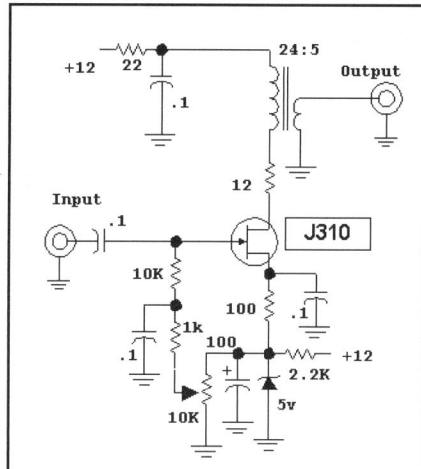

Fig 6.43—A single JFET is biased toward pinchoff with the reverse bias developed across the Zener diode. This amplifier offers 13.5 dB gain and a 37-dB gain range. The transformer, wound on an FT37-43, was available on the bench at the time of testing. The 10-kΩ pot sets gain.

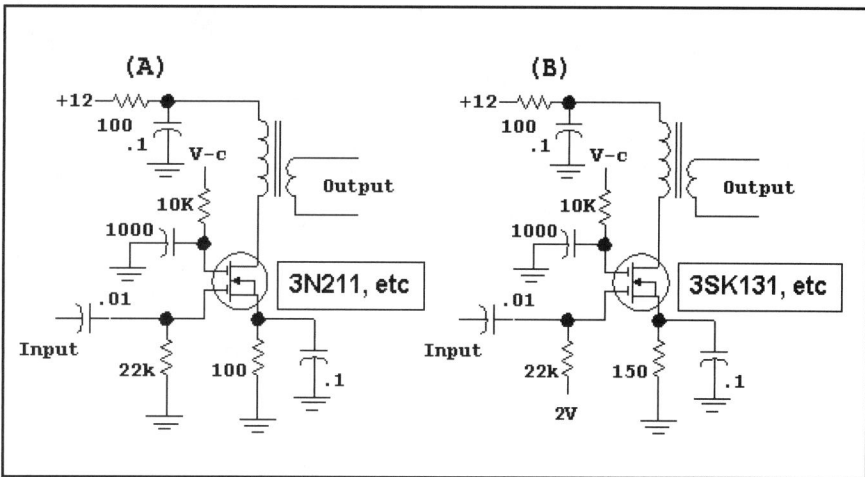

Fig 6.44—Two variations of a basic dual-gate MOSFET amplifier with variable gain. The circuit at (B) has the larger gain variation. The labeling of FETs is arbitrary, for these circuits are intended to be generic. The 3SK131, an SMT device from NEC is popular and is recommended.

PIN diodes can be combined with FETs for interesting IF amplifiers. **Fig 6.42** shows an amplifier where an FET serves as a common-source amplifier, followed by shunt PIN diodes driving a source-follower output. Output could also be obtained from the first FET drain through a transformer. This topology has many possibilities. Gain was 13 dB with a 60-dB gain range when the FET was driven from 50 Ω. NF was poor in this topology, but became very good when the first FET was driven from a higher impedance via an L-network. Gain also increased.

The performance of this amplifier is critically dependent on diode type. IMD was very low with MA47600 diodes from Microwave Associates. Experiments with devices from HP are recommended using the 5082-3080, or HSMP-3814. We observed some gain compression in this circuit with the MPN3404.

A very simple JFET IF amplifier is shown in **Fig 6.43** where gain is reduced as gate bias moves toward pinchoff. This circuit is configured (with a Zener diode) for a single power supply, although a negative supply for the biasing would be preferred. The circuit shown barely has adequate power supply voltage, but basic performance is excellent. Initial gain is 13.5 dB (at 10 MHz) with a smooth control range of 37 dB. Noise figure at maximum gain was 4.6 dB, increasing to 7.6 dB with 10 dB of gain reduction. Input intercept was +10 dBm at maximum gain, dropping eventually to −7 dBm as gain drops. However, intercept degrades slower than gain, so IMD products are always decreasing with gain reduction. The measurements were done with 50-Ω input drive. An input network presenting a higher impedance to the gate will increase gain and drop noise figure.

A popular IF device is the dual-gate MOSFET. See the earlier sidebar regarding part availability. With two basic

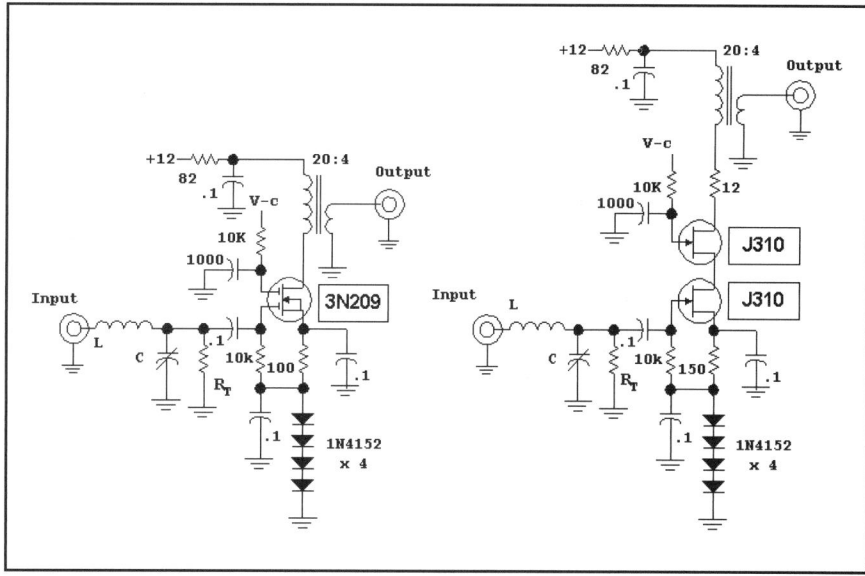

Fig 6.45—An IF amplifier using either a dual-gate MOSFET or a cascode connection of JFETs. These amplifiers use diode strings in series with the FETs for biasing, allowing substantial gain reduction with reduced control voltage. Transformers use #28 wire on an FT-37-43 ferrite toroid. Measurements were done at 10 or 14 MHz.

configurations in **Fig 6.44**, that at (A) is the more fundamental. The FET is self-biased with a source resistor while gate 1 is at dc ground. Gate 2 is normally biased at about 1/3 of V_{dd} to produce maximum gain. Moving the voltage on gate 2 in *either* direction will reduce gain. This topology has a limited gain reduction (less than 10 dB) available unless gate 2 is

Transmitters and Receivers 6.17

extended to negative voltages.

Fig 6.44B shows a popular variation used in many imported transceivers. Here, gate 1 is positively biased to about 2 V. With this biasing on gate 1, stage gain variation exceeds 30 dB with positive gate 2 voltages.

Fig 6.45 shows additional variations we examined. One uses a 3N209. The biasing is similar to that in the previous figure, part A, but uses a string of diodes in the source lead with gate 1 biased at the top of the diodes. With the 3N209 circuit shown and without R_T, maximum gain was 28 dB and gain variation was nearly 60 dB. The noise figure was 2.5 dB with the L network designed to present an impedance of 2.3 kΩ to gate 1. Inserting a 3-kΩ resistor for R_T generates a proper termination for the L-network, causing gain to drop to 20 dB and NF to increase to 6.6 dB, but now with a well-matched input. Noise figure degrades only slightly with gain reduction. Very careful gate-2 bypassing is required with all circuits using dual-gate MOSFETs to prevent UHF oscillation. The bypass capacitor should have fairly small C (1000 pF) so AGC dynamics are not altered, and capacitor lead length should be short. A drain resistor (10 to 100 Ω) will also help stability.

IMD performance was modest with a typical IIP3 being –11 dBm. However, intercepts improved as gain was reduced. This means that distortion products always drop faster with gain reduction than the desired signals.

The circuit on the right side of Fig 6.45 uses a cascode connection of J310 JFETs. A slightly larger source resistor was used to obtain similar stage current, typically 8 mA at full gain. This amplifier produced a maximum gain of 28 dB with a 34-dB gain variation (R_T absent.) The 3-dB NF degraded little with 10 dB gain reduction. A typical input intercept was –3 dBm with IMD products dropping faster than the desired output signals.

IF Systems

As we begin to assemble a complete IF system, the first question we ask is "How much gain is needed?" Often, the required gain is very small. In such a case, one can still realize AGC in the IF with a voltage-controlled attenuator. Such a circuit is shown in **Fig 6.46** where PIN diodes are arranged in a ladder of series and shunt elements. Diode current is controlled with a bipolar differential pair. Q2 is completely "on" at maximum gain, conducting all the current offered by Q3. This current flows through series elements with no current flowing in the shunt parts. Some of the current is shifted from Q2 to Q1 as gain is reduced, increasing I in shunt elements and removing it from the series ones. This circuit has a gain range of about 50 dB. Performance is better (lower insertion loss at max. gain) with premium PIN diodes, but is surprisingly good with 1N4006 rectifier diodes. Rectifiers often use a PIN structure to secure high breakdown voltage, but may still have high capacitance when compared with "RF parts."

The total IF gain needed in a traditional AM receiver can be relatively high, for the usual AM detector requires high drive for reasonable fidelity. The product detectors used in CW and SSB receivers are linear to low levels. IF gain is then picked for good sensitivity with the weakest signals and is

IF amplifier using a cascode JFET pair.

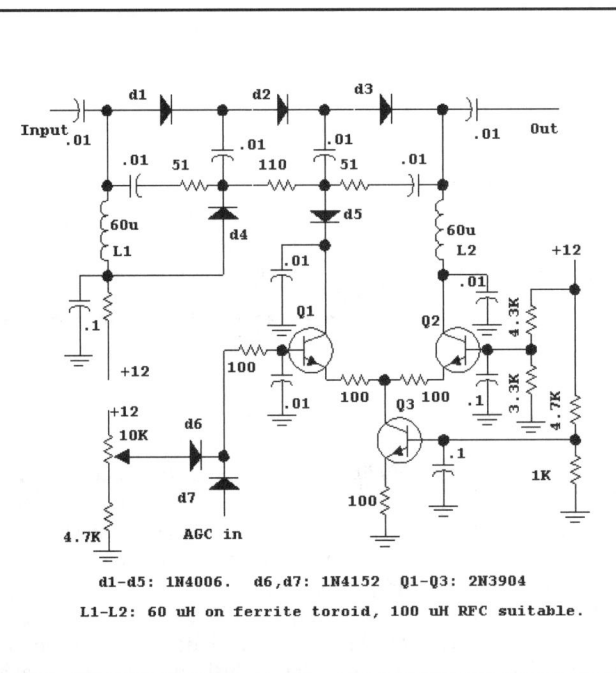

Fig 6.46—IF attenuator circuit offering a 50-dB gain-control range with an insertion loss of about 2 dB at 5 MHz. The 10-kΩ pot is a manual gain control.

reduced as signals get larger. The IF in a digital receiver (one where an IF signal is applied to an A-to-D converter) may have more severe requirements related to matching the input signal requirements of the A-to-D.

The usual IF system provides two outputs. One drives the signal detector while the other is applied to an AGC detector, a circuit providing dc output in proportion to the RF input voltage. Some AGC detectors are shown in **Fig 6.47**. The two outputs must be well isolated. It is especially important that BFO energy from the product detector not reach the AGC detector where it can be detected to reduce IF gain. Noise on the BFO (see the oscillator chapter discussion of noise) that reaches the IF can also inter-modulate with signals to compromise performance.

A dc signal emerges from the AGC detector. It is usually amplified and processed with op-amps for application to the controlled stages. The detector may have a threshold with no output until a minimum input signal is applied. This dc threshold must be exceeded before any gain reduction occurs, resulting in a threshold for RF detection. Once the signals are strong enough to exceed the detector threshold, the AGC holds the output nearly constant with only a slight increase with louder applied signals. **Fig 6.48** shows a plot for one of our receivers, showing output signal vs available input power. The threshold was adjustable and was set to occur with an input signal of –97 dBm. MDS for this CW receiver was under –140 dBm, so there is a moderate range of signals available before any AGC action occurs. This is an "ear-saver" design, one that protects the user from loud signals, but produces a receiver sound not compromised by AGC. Most commercial transceivers use AGC systems designed to make all signals sound nearly the same. This is clearly an open area for the individual designer/builder.

Diodes are often used to combine two control signals applied to an IF amplifier, shown in **Fig 6.49**. The two signals can come from a manual gain control and an AGC detector, or they may originate from two parts of an AGC system. Similar methods are used to mute receive IF amplifiers during transmit periods.

Fig 6.50 shows a system with two stages of gain with cascode connected J310s followed by a fixed gain differential amplifier. A 1:1 turns ratio ferrite transformer couples the signal from the cascode to the dif-pair. IF output is extracted from one collector of the pair while the AGC detector is driven by the other isolated output.

The experimental development of this circuit started with the first stage, Q1 and Q2. The gain control range was only 30 dB with three diodes in the chain, but increased with 5 diodes. Single stage current was 10 mA at maximum gain, but dropped to about 1 mA at minimum gain. A second stage, Q3 and Q4, was added, sharing the

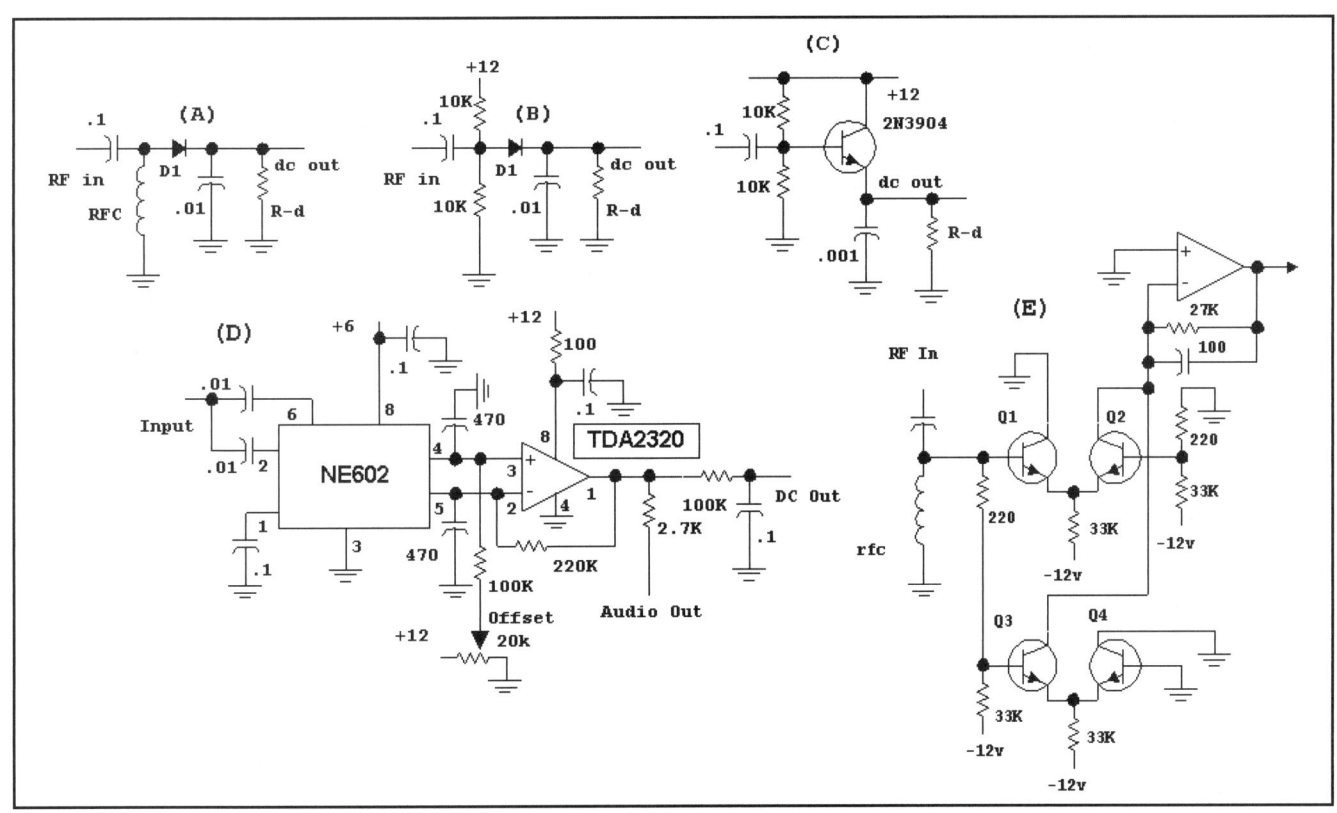

Fig 6.47— Several RF detectors suitable for examining the output of an IF amplifier. (A) shows a traditional diode detector with fast signal diodes. (B) is similar although the diode anode is now biased for a small direct current. (C) shows an emitter follower functioning as a detector. As the input voltage becomes more positive, causing the normal rectification in the e-b diode, collector current flows to charge the capacitor. (D) shows a sensitive detector, suitable for AM demodulation as well as level detection. The Gilbert cell mixer now functions as a multiplier, for both input ports are driven by the same signal. A 10-mV input yields several volts of dc output. If that input is 40% modulated, the audio output will be several volts peak-to-peak. This circuit was designed by W7AAZ. Many op-amps are suitable including the TL074 and NE5532. (E) uses a pair of differential amplifiers, each with an 80-mV input offset, causing each to operate as a detector. Cross coupling of the outputs cancels ac in the output through balance, producing a current input to an op-amp. A dual supply is usually required for this circuit. This detector was used by Carver (W7AAZ) in his high-performance IF system.[6]

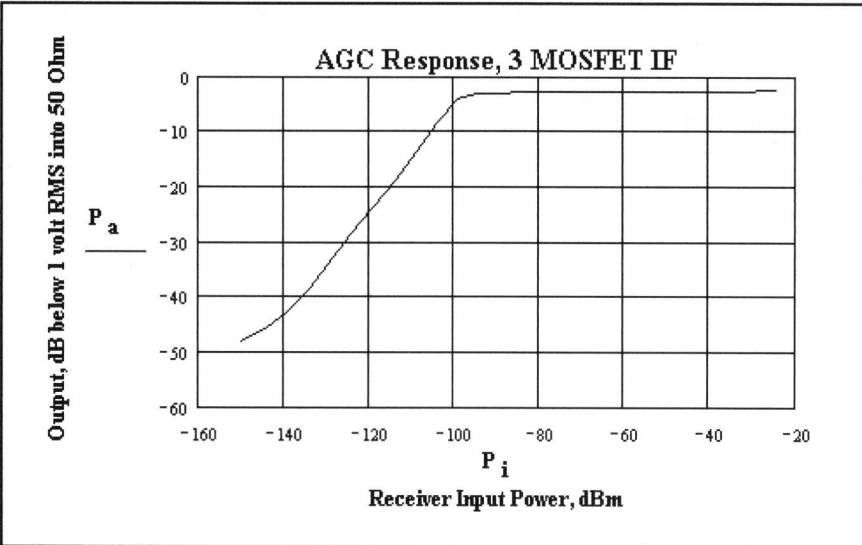

Fig 6.48—Receiver output vs input for a CW receiver. The threshold was specifically set in accord with operator preferences. The IF amplifier is shown later in Fig 6.56.

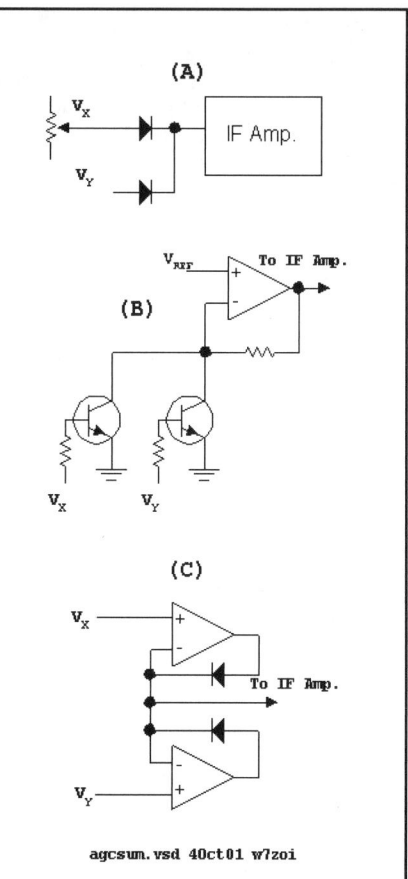

Fig 6.49—Diodes combine signals applied to an AGC amplifier. At A, an AGC signal and one from a manual gain control are selected with the more positive one setting the voltage applied to the amplifier. In B, two signals applied to transistor bases establish currents that are summed in an op-amp. Both inputs contribute in this case. The version in C uses diodes within feedback loops of op-amps to form "perfect rectifiers," which establish a very sharp transition between active inputs. This scheme was elegantly used in Carver's IF amplifier.[7]

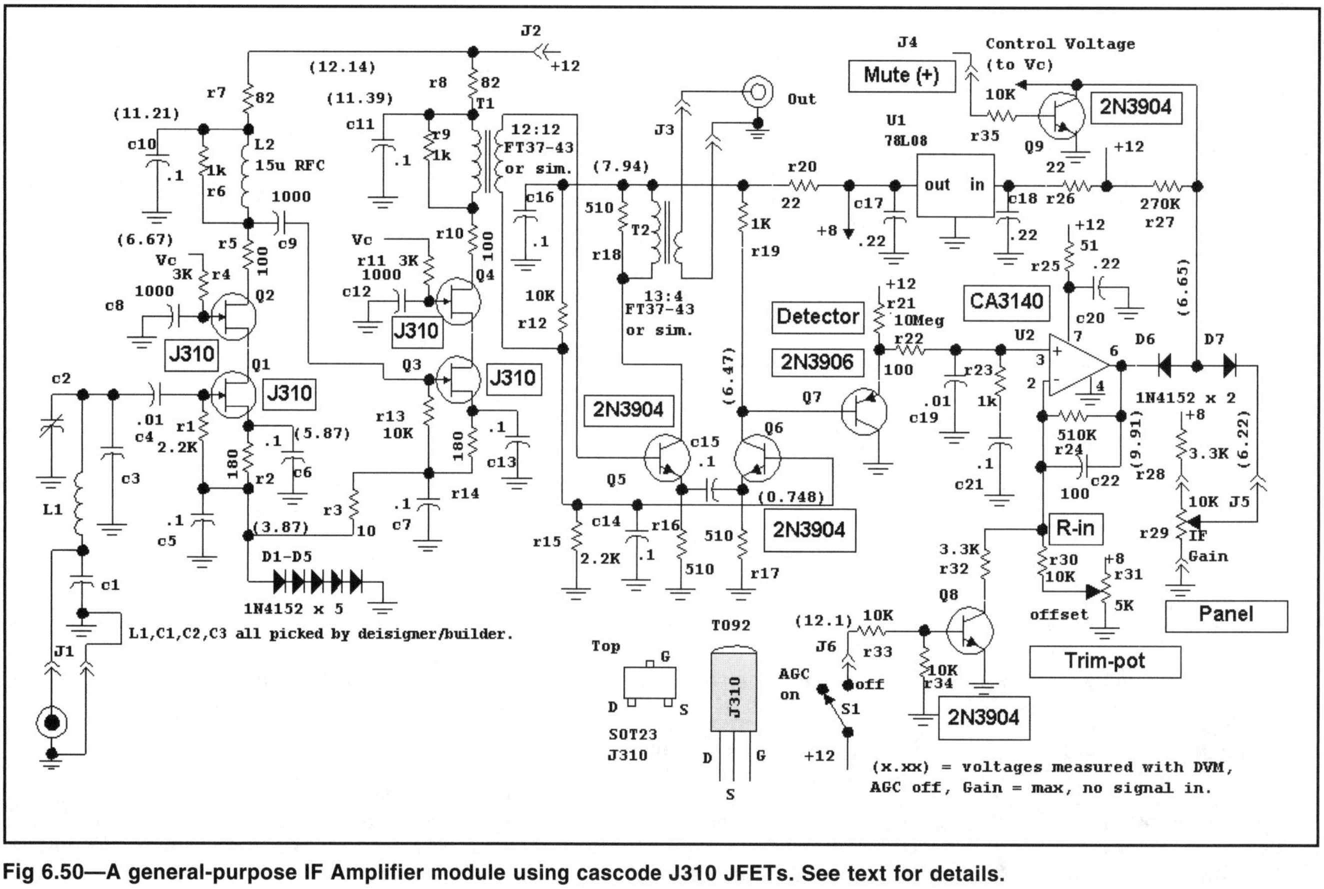

Fig 6.50—A general-purpose IF Amplifier module using cascode J310 JFETs. See text for details.

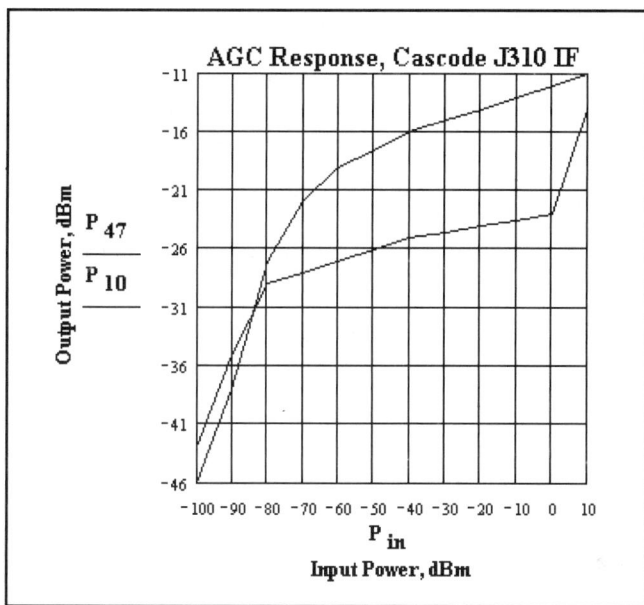

Fig 6.51—IF system output vs input for the IF system using two cascode-connected J310 stages. The two curves are for two different values of "input resistor" in the op-amp, which alters system dc gain. See text for details.

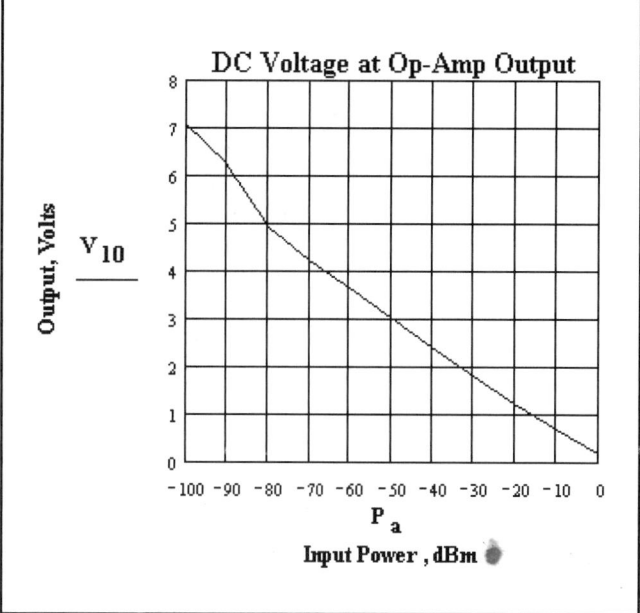

Fig 6.52—DC level at the op-amp output. This voltage may be used directly to drive an "S meter," driven with an op-amp dc follower.

diode chain with the first pair. A J310 source follower was temporarily added to provide an output. The gain variation was now 93 dB at 10 MHz, increasing to 108 dB at 5 MHz. There was a high pass gain characteristic, a result of the 15 µH RFC. Larger values should be used at lower frequency. The gain control voltage should be between 0 and 6 V. Values above 6 V produced a slight gain *decrease*, so that region should not be used.

The 9-MHz gain was 28 dB with no input network other than a blocking capacitor. NF was then 7 dB with R1 at 10 kΩ. A 9-MHz pi network was then added to present a 2-kΩ impedance to the first gate, causing gain to jump to 44 dB while NF dropped to an impressive 1 dB. The NF was maintained with 10-dB gain reduction. We then replaced R1 with a 2.2-kΩ resistor, so the network now causes a good 50-Ω impedance match to appear at the input. NF was now up to 5 dB, increasing to 6 dB with a 20-dB gain reduction. The designer/builder needs to design his or her own networks to apply this circuit to the filters used.

The rest of the circuit was now built, initially using 47 kΩ for R30, R_{IN} at U2. The no-signal dc voltage at the detector output (emitter of Q7) was 6.8 V, so the arm of "offset" pot R31 was set initially to this value. The Op-amp, U2, buffers the control voltage appearing across the timing capacitors, C19 and C21. The loop is closed, generating AGC action, when the op-amp output is connected to the controlled stages through diode D6. The response is shown in the upper curve of **Fig 6.51**. Although the loop is well behaved, it is not very *tight*, allowing considerable output variation between threshold and the upper input-signal limit. Input resistor R_{IN} was dropped to 10 kΩ (increasing loop gain) to produce the preferred response in the lower curve. But the

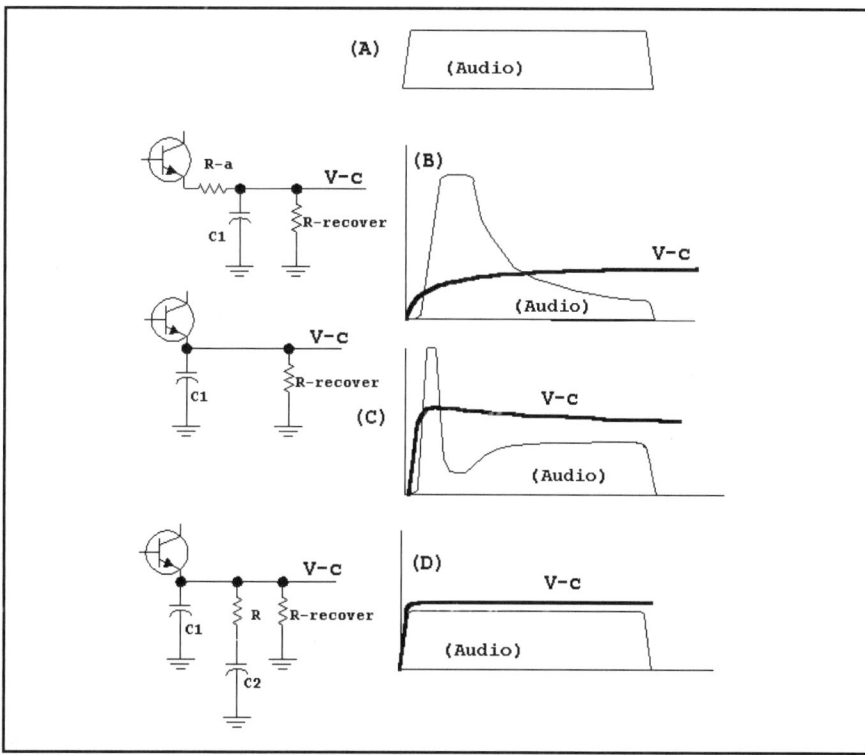

Fig 6.53—Audio envelopes and timing capacitor values vs time. See text for details.

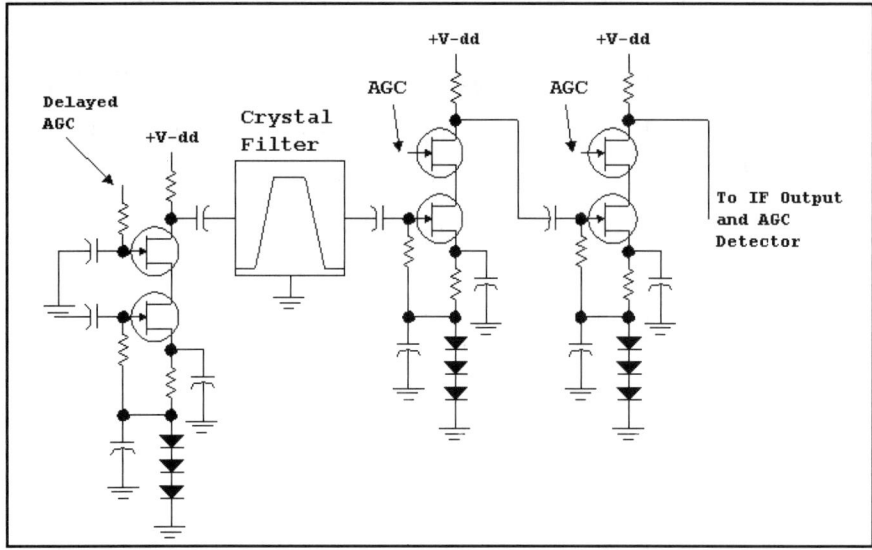

Fig 6.54—System with a crystal filter within the AGC loop. See text for discussion.

system is now ineffective at input levels above 0 dBm. The reason for this becomes clear if we examine the curve of **Fig 6.52** showing dc voltage at the U2 output. The dc voltage has reached 0 by the time the input gets to 0 dBm, so no further gain reduction is possible. Adjustment of the offset pot, R31, will probably fix this anomaly, if it becomes a problem. Such levels would rarely be encountered in most receivers.

The relatively clean dc variation in Fig 6.52 suggests that a signal-strength meter could be driven directly by the op-amp. If this is done, additional circuitry should be added for any "calibration" that might be desired with the S meter. The offset pot is not intended for this purpose, but only to set AGC threshold.

The attack time in the circuit of Fig 6.50 is determined by the detector (Q7) output impedance, by timing capacitors C19 and C21, and by R23. R21 and the capacitors establish recovery characteristics. The values shown were approximate and may require later changes.

A PNP detector was used in the previous circuit. Consider a more general case with an NPN (or a diode) detector charging memory capacitors. **Fig 6.53** shows some audio envelopes and related capacitor values. The input to the receiver (or IF system) is a chain of Morse dots (dits). Even if the receiver is to be used only for SSB, this represents a good test method. Set the strength of the dits to be low, AGC to "off," and the manual gain control to drop the IF gain to produce the response shown in Fig 6.53A. This is an ideal audio envelope with a well-defined rise and fall time.

Having observed the ideal system without AGC, we now increase the strength of the dit chain and activate AGC. Generally we wish to have a near instantaneous fast attack, with a slow decay, yielding the same audio response we saw with the ideal case. But that does not always occur.

Fig 6.53B shows a single timing capacitor, C1, with a modest detector output impedance, R_A. The resulting slow attack allows the audio to climb to high levels, and then drop over the course of the dit as the capacitor voltage stabilizes. Decreasing the attack time, realized by reducing R_A, reduces this distorting behavior. But in the extreme this generates the behavior shown in Fig 6.53C where the timing capacitor charges very fast before the gain is reduced. The audio drops to a level below the proper one, but grows to the right value after the loop "catches up." In the extreme, there is no audio for a period until the timing capacitor discharges enough to allow the IF gain to increase to a value that produces a stable result. This is the well known "pop" occurring with some AGC systems.

A solution is found with two (or more) timing capacitors, C1 and C2. C1 is smaller than before and can be charged quickly with the detector output impedance. This may reduce the gain, but for only a short time. Much of the charge on C1 discharges through R to be deposited on C2, increasing that voltage and the resulting V_C value. The process repeats with each cycle of the IF system. This behavior, closer to the ideal, is presented in Fig 6.53D.

The process is more complicated than the simple picture we have painted, for there are delays within all IF amplifiers.

For example, the control gates of the JFET cascode circuits are connected to bypass capacitors with series decoupling resistors. The bypassing is a necessary part of the cascode connection. The related RC forms a low pass filter that causes the signal at the controlled gate to arrive *after* an input is applied. The delay is short with the values we used, but can be much larger. Signals arriving at the IF input are delayed through a narrow bandwidth filter, generating an output that grows at a finite rate, allowing a fast AGC system to keep up.

In some applications we wish to apply AGC to an RF or IF amplifier preceding a narrow filter, shown in the example of **Fig 6.54**. The filter delay is now within the loop. That is, we detect after the delay of the filter, allowing the signal to grow too large to avoid overloading early stages. The delay can cause severe overshoot or popping if gain reduction is applied directly to the first stage. The preferred solution is to purposefully *delay* the control signal applied to the early stage with a long time constant. Good system dynamics result only when the controlled elements after the narrow filter have enough range and speed to reduce the gain far enough to restrict the output for a short while, only to recover, allowing the delayed stage time to assume part of the overall gain reduction.

We used a string of Morse code dots as a means for evaluation and adjustment of an AGC system. This is not a mere illustration, but a useful experimental method. A simple PIN diode modulator called the *Ditter* is presented in the measurement chapter for just this purpose. The dits are created with a 555 timer IC, but could be generated with a function generator, now offering adjustment ability. The Ditter includes an output to drive the external triggering input of a dual trace oscilloscope. One 'scope channel then shows the control voltage while the other monitors audio or IF output. Ideally, an AGC loop needs to be tested over a wide range of signals, for stability can vary with level.[8]

Audio Derived AGC

Simple equipment sometimes uses audio derived AGC where a detector samples the audio signal to charge a timing capacitor. That voltage is processed and applied to IF amplifiers for AGC. The attraction of this is that audio amplitudes are large, for most of the receiver gain has been realized. Little more gain is required to complete the AGC system. But there is a major difficulty with audio derived AGC. This relates to the sampling nature of the detection process. The detectors we

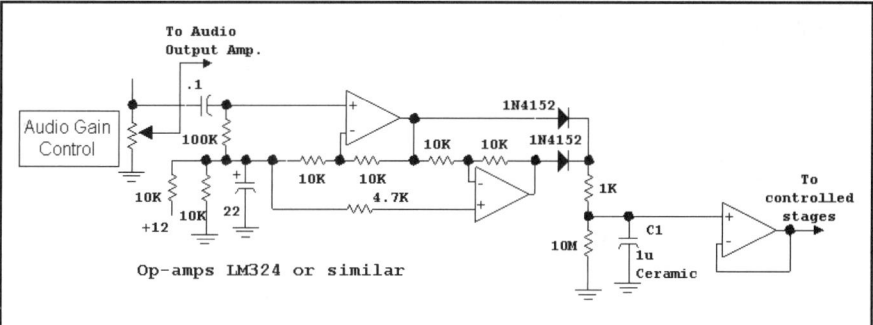

Fig 6.55—Full wave audio detector for use in simple AGC systems.

have examined obtain one sample for each peak of the waveform being detected. Audio waveforms have fewer peaks, especially if the signal is a low-pitched CW carrier. This allows the receiver to be overwhelmed in the period between peaks.

A partial solution to the low frequency difficulty lies in audio filtering. A high-pass filter (with several elements) ahead of both the AGC detector and audio output will prevent very low beat notes from reaching either. A cutoff of around 300 Hz is suggested.

A typical full wave detector for use in an audio derived AGC is shown in **Fig 6.55** with both positive and negative audio peaks contributing to the output. A slow recovery is set by the 10-MΩ resistor across C1, which can be made faster with a smaller resistor. Shorting C1 will turn the AGC off. The system shown is suitable for IF amplifiers like the MC1350P. Level shifting or inversion may be required for other controlled circuits.

Mention was made earlier of difficulties with filters within an AGC loop. This problem can be especially severe when audio filters are included within a loop. Audio filtering is better applied after detection for the AGC loop.

Although audio derived systems present major design challenges, good performance is still possible. This becomes evident when high-end professional-level audio-recording equipment is studied.

Practical FET IF System Examples

The Cascode JFET amplifier presented earlier was developed as a complete, practical module, Fig 6.50, for use in a Monoband SSB/CW Transceiver. This circuit can be built with other FET types, with appropriate circuit changes. The JFETs should be roughly matched for I_{DSS} (+/–10%) and should all be of the same type.

The initial adjustment of the IF amplifier starts by removing one end of R30 from the board. The AGC is turned on with no signals present and the voltage on pin 6 of U2 is measured and recorded in the notebook. The voltage on the arm of R31 is then set for the same value. R30 is again installed in the circuit. R31 can be readjusted later to alter AGC threshold.

A similar MOSFET IF amplifier is shown in **Fig 6.56**. This circuit uses three gain stages using 3N209 MOSFETs, a type available in our junkbox. Those wishing to duplicate this circuit should consider the 3SK131 or similar available SMT parts. After three gain stages, the signal is applied to a differential PNP amplifier. One side is terminated in a 510-Ω resistor, providing a properly matched drive for a "tail end" crystal filter. This filter serves to eliminate noise generated within the IF amplifier at frequencies other than that of the main filter. It also distributes the selectivity improving the stopband attenuation of the overall system.

The main IF input selectivity is provided by a 10th order filter with a 500-Hz bandwidth, designed for a Gaussian-to-12-dB response. (This filter, a KVG XL-10M, is regrettably no longer available. They are sometimes found on the surplus market, but few were manufactured.) The IF system was breadboarded without printed boards in a multiple-section surplus milling. One section contains the main filter input while another has the output and the first IF amplifier. Another houses the 2nd and 3rd IF stages while yet another holds the differential amplifier and an NPN detector. Feedthrough capacitors route the signal through the milling where the dc parts of the AGC loop reside.

The input circuitry is critical to the components used. A ferrite transformer matches the 50-Ω drive to the main crystal filter impedance of about 300 Ω. The filter output is then transformed up to 2200 Ω with a low Q pi-network where a 2.2-kΩ input resistor at Q1 terminates the filter. This topology guarantees a reasonable noise figure with a proper impedance match for the crystal filter, vital in preserving the specified performance. The pi-network used an existing RF choke, although a toroid with higher Q_U would be preferred.

This IF has a bandwidth just under 500 Hz with a measured system sideband suppression in excess of 120 dB. The dc AGC response was presented earlier in Fig 6.48. The threshold may be adjusted with R-th (2.5 kΩ) shown in the schematic.

The attack and recovery are determined by the components in the *Timing* section of the circuit. An NPN detector, Q6, charges a feedthrough capacitor that feeds a signal out of the milled enclosure to a CA3140 op-amp that then drives inverter Q8. The Q8 collector then drives the timing capacitors. The primary one is a .01 µF, which is tied to a 0.1µF/10 kΩ combination paralleled by a 1µF/100 kΩ pair. These values were established with the ditter mentioned earlier. The voltage on the timing capacitor and the audio signal are shown in a photo.

A Hang AGC System

Fig 6.57 shows an AGC system with the unusual characteristic of using two timing systems. One is driven by IF signals, so it has the advantages of quick attack. The other comes from the audio.

During receiver operation, signals within the IF cause C2 to charge, which reduces receiver gain. If the signal is a short lived one or even a noise burst, C2 will quickly discharge through Q10, a low pinchoff JFET switch. However, if the signal is present for a reasonable period (around a hundred milliseconds), audio will have been amplified by Q11 to charge C1 positively. This drives Q10 into pinchoff, which disconnects it from C2. The only discharge path for C2 is now a 22-MΩ resistor, so recovery is slow, causing the gain to *hang* at a nearly constant level. But if the audio disappears for a short period, C1 discharges, Q10 is no longer pinched off, and C2 quickly discharges, returning the receiver to full gain.[9] The audio detector is an "open-loop" process that modifies the basic closed-loop IF AGC system, so does not alter system dynamics.

The hang scheme can be adapted to our FET IF systems with relative ease, as shown in **Fig 6.58**. The partial circuit in (A) is set up for NPN detectors while (B) accommodates PNP detectors. The builder

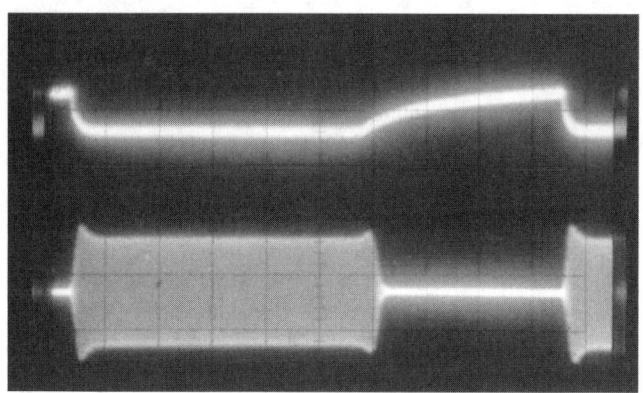

Timing signals for the MOSFET IF Amplifier during AGC testing.

Perhaps the most impressive IF design we have seen is that presented by Bill Carver, W7AAZ, in *QST*, May, 1996.[10] This circuit is based upon the AD600 series of integrated circuits from Analog Devices. Although expensive, these parts offer outstanding performance. They feature a wide AGC range that is extremely *dB linear* (the gain in dB is directly proportional to the control voltage). Bill's complete paper is included on the CD included with this book.

Carver's IF amplifier included a number of outstanding features not found in other circuits. His circuit used three amplifier blocks where gain reduction occurred, just as one of the previous circuits shown (Fig 6.56) used three stages. Our simple circuit had gain reduction applied to all stages at once. But Carver's IF used a sequential gain reduction. The last stage had gain reduced by 40 dB before any other reduction occurred.

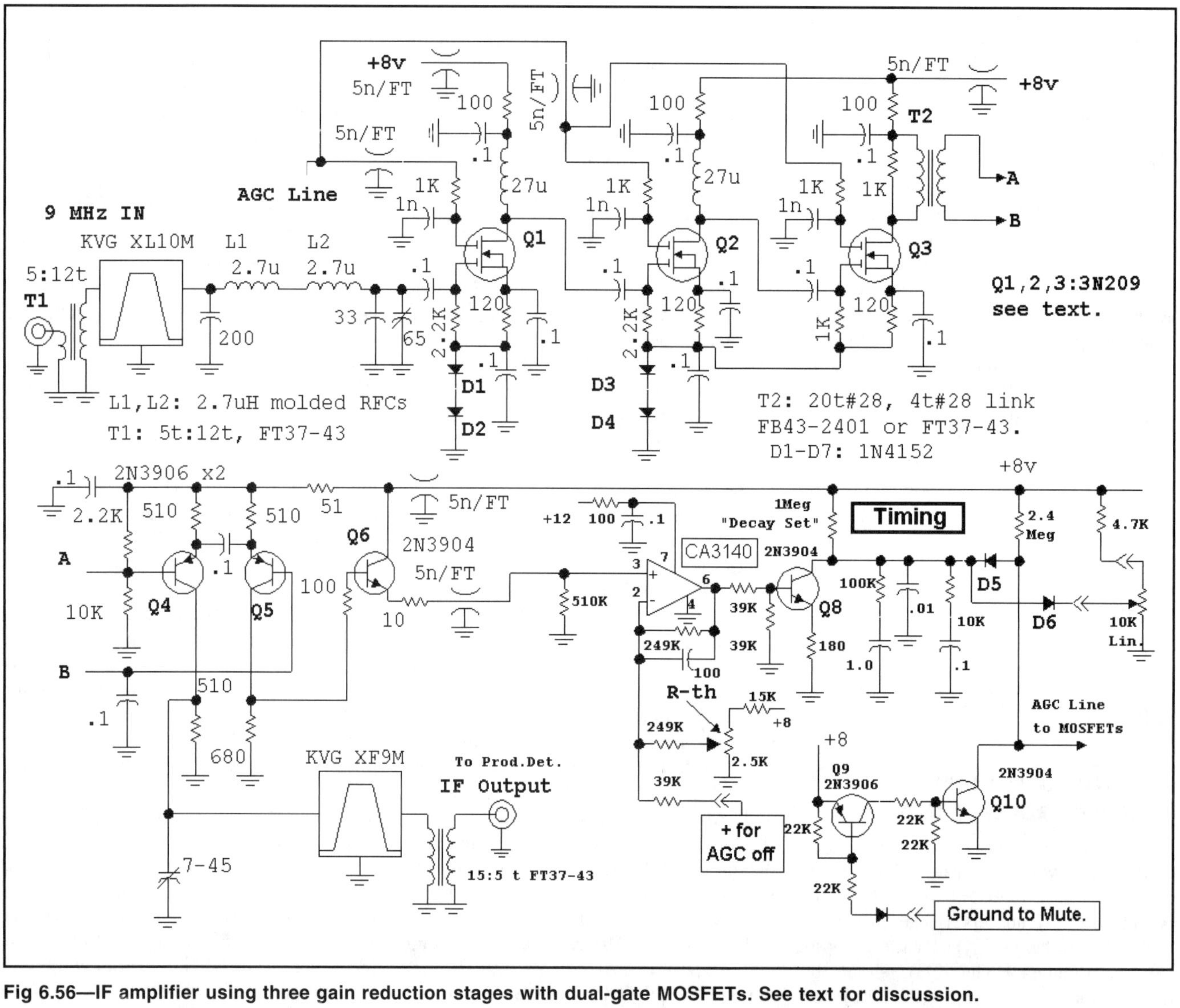

must provide many design details.

Evolving Designs

Clearly, many of the methods can be combined. For example, W7AAZ built an IF amplifier using a dual-gate MOSFET input stage followed by an MC1350P. By applying AGC to both FET gates, he was able to obtain a very wide AGC range in a relatively simple design. PIN diodes can also be added to existing systems to stretch the range of FET or bipolar amplifiers, integrated or not.

Fig 6.56—IF amplifier using three gain reduction stages with dual-gate MOSFETs. See text for discussion.

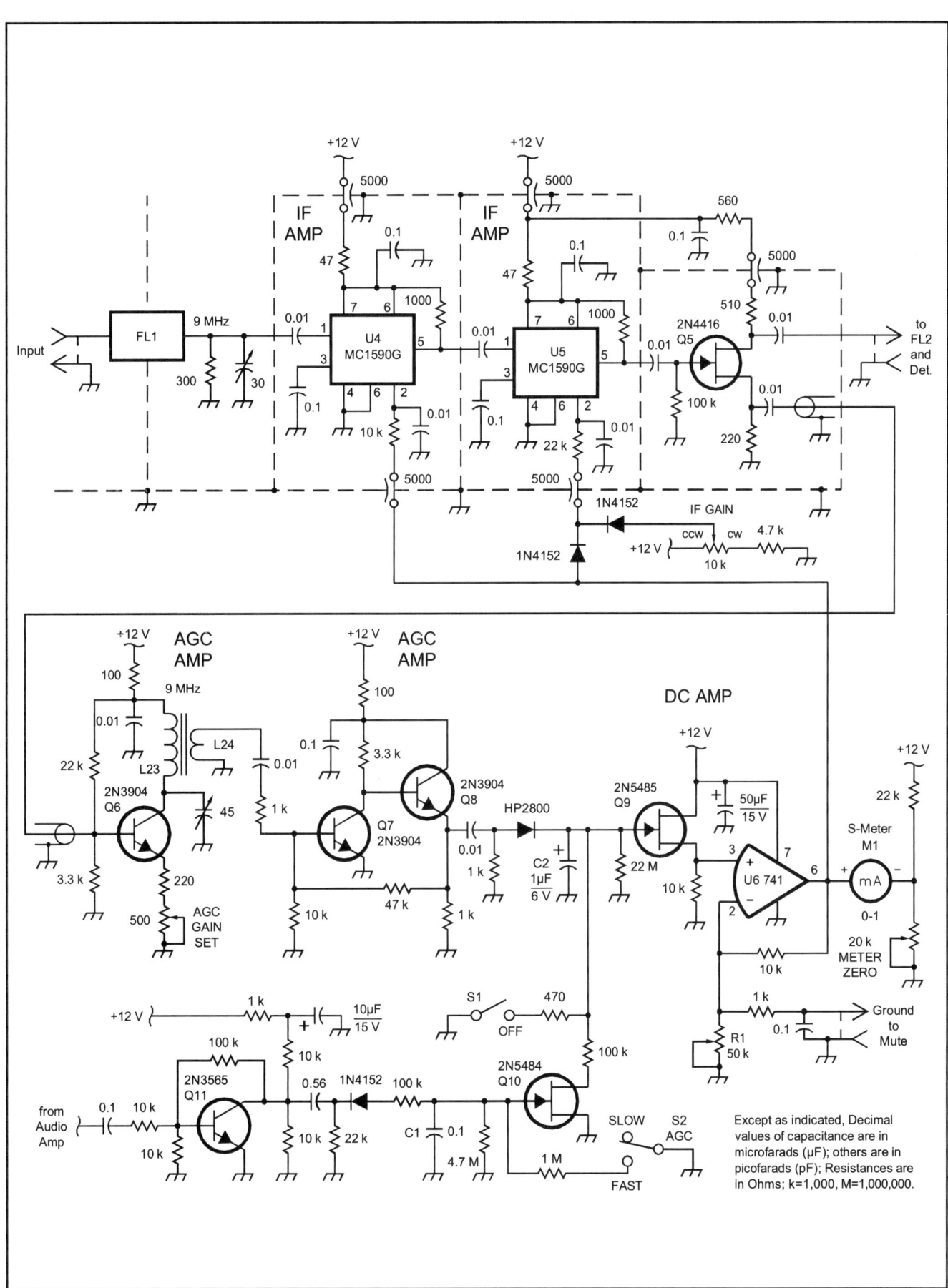

Fig 6.57—A full hang-type AGC system with two timing systems. The IF-derived AGC offers quick attack while "hang time" is established by the audio.

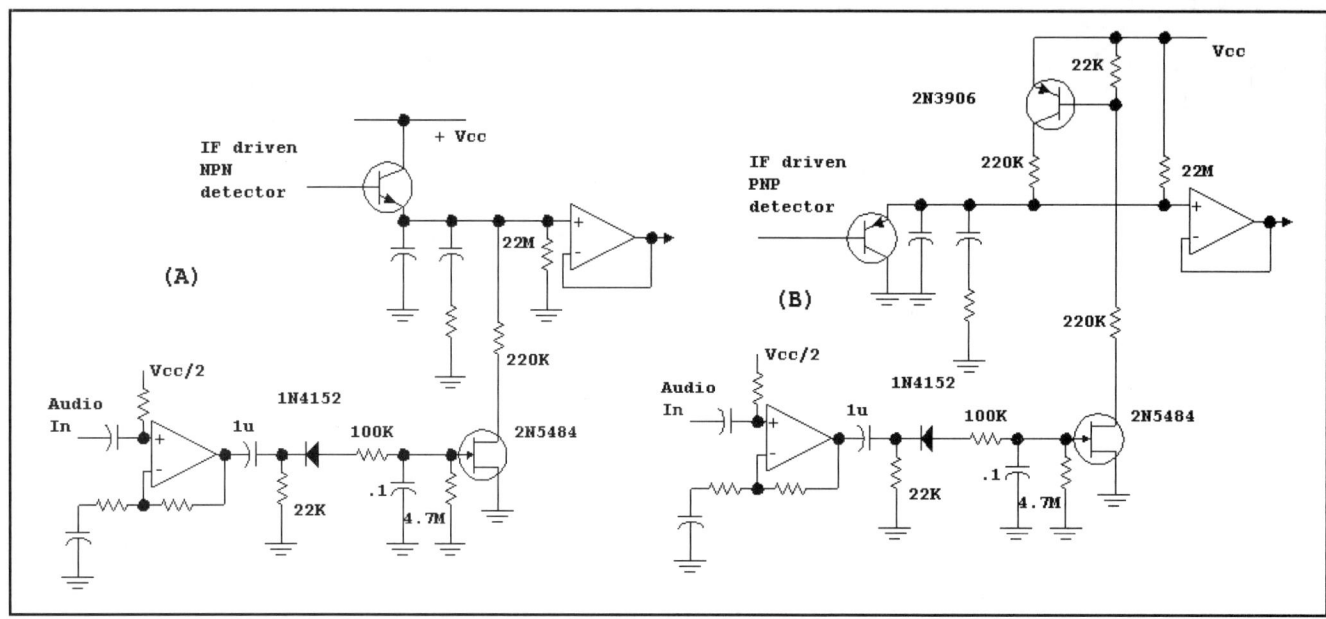

Fig 6.58—Adapting a hang AGC to IF amplifiers with NPN or PNP detectors. See text discussion.

Further reduction was applied then to the middle stage, and after a total of 80-dB reduction, to the input stage. This was possible because of the buffering used within the AD600 and the use of "perfect rectifiers" in the control circuits. The Carver system also used a second gain reduction loop with a bandpass filter between stages, optimizing dynamic behavior while keeping noise low.

The Carver paper included another unusual feature that will become more common with emerging receivers: He used a feed-forward scheme where the AGC detector not only controlled the gain of stages ahead of the detector, but altered the gain in stages following detection. In principle, one could carry these methods to the extreme where an accurate detector establishes gain in later stages without a need for negative feedback. This could be realized with hardware (a log amplifier and detector with variable gain IF amplifiers and stepped gain audio amplifiers) or software with a DSP system. Delay in filters or amplifiers presented a problem with traditional negative feedback systems, but now becomes an asset, providing time for calculations in a DSP based system. These DSP methods have already, at this writing, been used for a few years in some very high-performance military equipment from Rohde and Schwarz, and will be described for use in DSP transceivers described in this book.[11]

6.3 LARGE SIGNALS IN RECEIVERS AND FRONT END DESIGN

The range of signals available to our receivers can be very wide indeed. The weakest signals we can hear are limited by noise, and drop to typical levels of –140 dBm or less in a CW bandwidth. These are rare at HF, but common at VHF. But signals can also be very strong. The strongest sky-wave propagated signals we encounter will depend on our antenna, but can sometimes be as strong as a microwatt (–30 dBm,) or even more with high gain antennas.

Most of our concern for large signal performance relates to the receiver *front end,* the part of a receiver between the antenna connector and the place where receiver bandwidth determining selectivity is obtained, usually the first crystal filter. The front end usually consists of much more than the "first stage."

We have two concerns when dealing with large signals. First, "How loud can the signals be that we try to copy with our receivers?" This problem relates to both front ends and to gain control. Second, "What is the range of signals that can be present within the receiver front end without causing problems when we attempt to receive average or weak signals?" This is the more complicated and subtle problem with the more interesting challenge.

Fig 6.59 shows a partial receiver block diagram for a 14-MHz single-conversion superhet with a 2-MHz IF. The calculated front-end filter response is shown in **Fig 6.60**. The center frequency response is normalized to 0 dB, so the response at 10 MHz can be used to evaluate worst-case image rejection, 76 dB for this example. The front-end bandwidth, over 400 kHz, is wide enough to not require any adjustment during receiver use. These filters contribute little to the receiver signal selectivity and do *not* impact noise figure and sensitivity. But they are vital in protecting the receiver from other responses.

The narrow crystal filter in the IF determines the receiver selectivity. The response of two crystal filters are shown in **Fig 6.61**. Both filters were designed for a bandwidth of 2500 Hz, but one filter uses four crystals while the more selective one uses eight. The beat frequency oscillator (BFO) is normally placed 300 Hz below the lower passband edge for an upper sideband response. The voice frequencies then recovered by this 2500-Hz bandwidth filter extend from 300 to 2800 Hz. Opposite sideband response is then well defined. Owing to the filter skirt shape, sideband suppression is critically dependent on position within the passband. For the four-element filter, sideband suppression extends from only 14 dB at the low audio end to 43 dB at the high end. The 8-element filter offers much better sideband suppression, but is still only 27 dB at the low audio end. It grows to 87 dB at the high audio extreme. Similar response can be expected in a filter method SSB transmitter. The improved response of the phasing method is dramatic for sideband suppression *at low audio frequency*. This suggests that combinations of a superhet and the phasing method may offer spectacular performance, an old, but still viable option.

Several undesired phenomena occur in

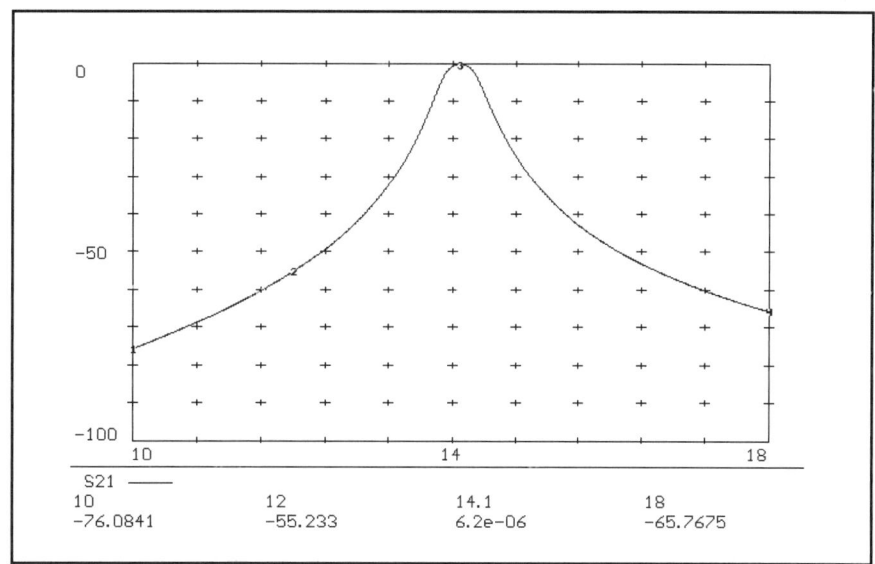

Fig 6.60—The response of the front end from 10 to 18 MHz. The image rejection at 10 MHz is 76 dB. This is a computer generated ideal plot. The 3-dB bandwidth is 0.41 MHz, centered at 14.1 MHz. This response results from a single- tuned circuit at the antenna and a double-tuned circuit between the RF amplifier and the mixer.

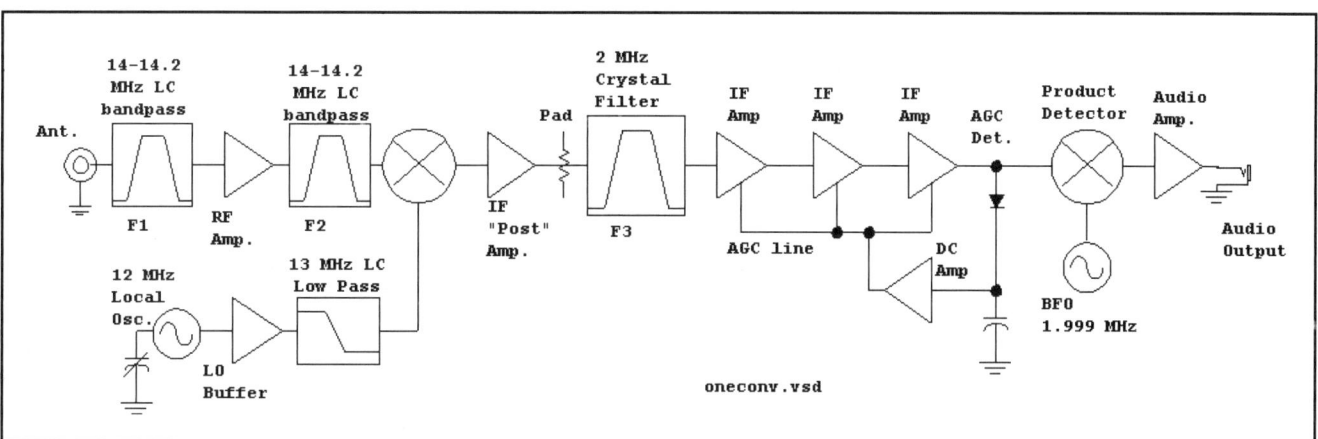

Fig 6.59—14-MHz receiver with a 2-MHz IF. The LO tunes from 12 to 12.2 MHz, so the image extends from 9.8 to 10 MHz.

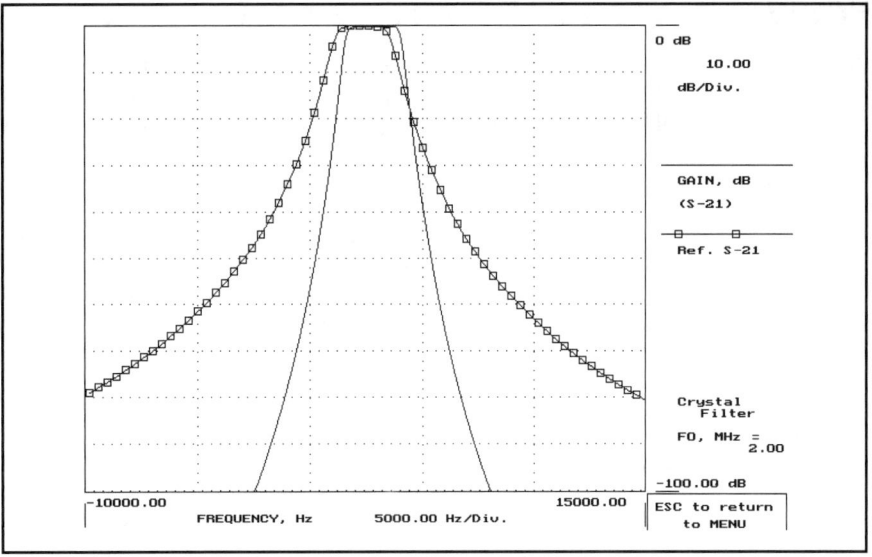

Fig 6.61—Response of two crystal filters. While both have a bandwidth of 2.5 kHz, one uses only 4 crystals (trace marked with small squares) while the other uses 8. Both were designed for a Butterworth response. Steeper skirts are afforded by a Chebyshev response. See text for discussion.

a receiver front end to compromise performance. These include:

- **Gain compression:** If we examine the front end as a module and measure gain, we find a constant value for most signals. However, as the signals grow, we eventually find a level where the gain is reduced over the small signal value. We usually specify the 1 dB compression point, that available input power in dBm where gain is reduced from the small signal value by 1 dB. A simple way to measure gain compression uses two signals or "tones." One is of weak to average strength and is the one tuned by the receiver during the test. The other is much stronger and is placed within the front-end bandwidth, but well outside the receiver bandwidth. A typical spacing for an SSB receiver might be 20 to 50 kHz. The strong signal is increased until the weaker one drops by 1 dB. This can be a difficult measurement to perform. The IF filter must have enough stopband attenuation to keep the strong signal from creeping into the IF where undesired AGC detection might occur. Further, the measurement is often compromised by reciprocal mixing, or noise blocking, which is described below. Gain compression is easily defined, but rarely a great problem.

- **Cross modulation:** This was a common specification when AM was the dominant modulation mode. It is measured with two input signals. The first is an *average* strength carrier with no modulation of it's own. The second is a much stronger modulated carrier spaced away from the weak carrier by several receiver bandwidths. It is often 30% amplitude modulated by an audio sine wave. We increase the strength of the modulated carrier while the receiver is tuned to the weaker one, waiting until the modulation of the stronger appears on the weaker one.

- **Phase noise blocking, or reciprocal mixing:** This problem was described in the oscillator chapter. Phase noise blocking occurs when a strong signal is applied to the receiver at a frequency slightly away from the receiver's tuned frequency. Noise sidebands on the receiver LO will mix with the incoming signal to produce an IF response. The offending energy is a noise rather than a carrier, so the response is proportional to receiver bandwidth. For this reason, the response, when measured, is usually normalized to a 1-Hz bandwidth. Measurement is complicated by noise on a generator that might be used to measure it. It is difficult to differentiate between the two, justifying the term *reciprocal mixing*. Noise blocking shows up as a problem on the air when a strong local signal appears. If the offending signal is on CW, the noise shows up as a keyed hiss that becomes stronger as the receiver is tuned toward the signal. It is a fundamental problem that is "fixed" only with careful LO design. Reciprocal mixing is a major problem with frequency synthesized radios and offers *the single most fundamental challenge* to the design of advanced communications equipment. An integral part of this challenge is that of eliminating spurious responses in frequency synthesizers, sometimes quite significant when DDS is used.

- **Second-order intermodulation distortion:** Generally, intermodulation distortion (IMD) occurs when two or more signals are applied to the input of a receiver, creating distortion products at frequencies other than the input. Second-order IMD produces sum and difference frequencies. The sample receiver of Fig 6.59 used a 2-MHz IF, so two inputs that were separated by 2 MHz could generate an output at the IF. Inputs at, for example, 13 and 15 MHz could generate the distortion products. However, this is unlikely, for our receiver is preceded with considerable filtering. Signals at these frequencies are attenuated before reaching the later parts of the front end. Second-order IMD is characterized by an intercept, as outlined in Chapter 2.

- **Harmonic distortion:** This is a distortion created within the receiver where the output is a harmonic of an input. For example, second-order harmonic distortion would occur if a strong 1 MHz signal was applied to the front end. A second-harmonic signal would then be created within the receiver and produce a signal in the 2-MHz IF. A more common distortion might be generated from a strong 7-MHz signal. The 14-MHz second harmonic created in the receiver front end is available for subsequent conversion. But the example front end filtering is extreme enough that little 1 or 7 MHz energy would ever reach the front end. Direct harmonic distortion is rarely a problem in a well pre-selected receiver, one with good input filters. But most commercial receivers, today, are not well pre-selected.

- **Third-order intermodulation distortion:** Like second-order IMD discussed above, this distortion is the result of two input tones. This product is perhaps the most difficult distortion to eliminate, for it occurs close to a pair of incoming frequencies. It is a *third-order* product because there are essentially three frequencies that create the product. If two input frequencies, f_1 and f_2, are applied to a receiver, the distortion occurs at $(2 f_1 - f_2)$ and $(2 f_2 - f_1)$. In the first example, f_1 is used twice, so the 3 inputs are f_1, f_1, and f_2. (Note that *order* can also be related to the exponent on a dominant term in a power series description of the distorting device, but that relationship is often ambiguous.[12]) Consider two example inputs of 14.04 and 14.05 MHz, directly within our input filters. The distortion products now

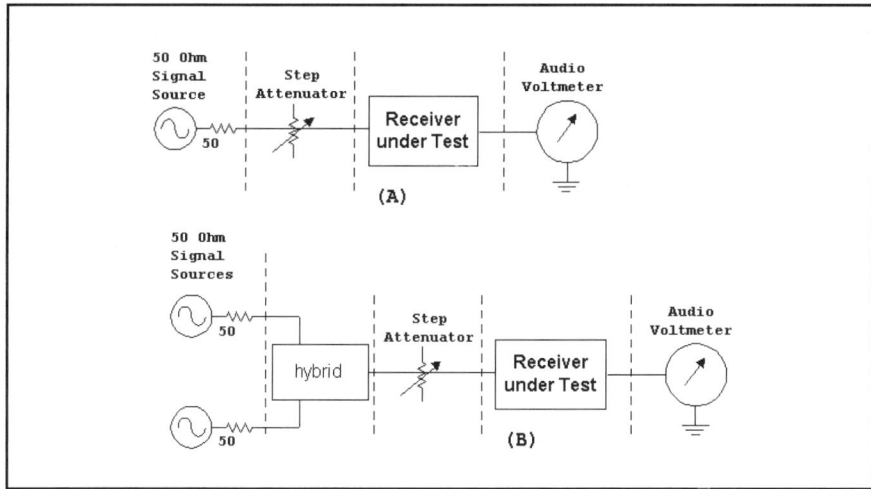

Fig 6.62—Setup for measurement of receiver dynamic range. See text for discussion.

measure MDS is shown in **Fig 6.62A**. The signal in dBm available to the receiver is the generator output less the attenuation value in dB.

After measuring MDS, a second signal source is added to the test set, as shown in Fig 6.62B. The sources are adjusted to have equal outputs. The *hybrid* in that figure is a circuit element that combines the outputs of two 50-Ω generators to form one 50-Ω source while isolating the two generators from each other. (See Chapter 7 under Return Loss Bridge.) The combined output is adjusted as needed in the step attenuator. The level available to the receiver input is adjusted until the response on the meter is exactly the same 3-dB-above-the-noise response that we saw when measuring MDS.

Consider an example. First, turn AGC off for all DR and intercept measurements. With no input signals, the audio output from our receiver is 5 mV, RMS. This is the result of receiver noise. We now inject a 14.010-MHz signal from a generator and adjust the level and receiver tuning until the audio output is 7.1 mV, 3 dB above the noise level. This happened with a generator output of –130 dBm, which becomes the MDS. Next, we set up the signal generators at 14.03 and 14.05 MHz, leaving the receiver tuned to 14.01 MHz. We increase the level of the two tones until we get the same output that we saw with the MDS measurement. This occurs with a signal at the input of –44 dBm per tone. Each tone is 86 dB above MDS, so our two-tone dynamic range is 86 dB.

We can measure the receiver input third-order intercept directly with the same

appear at 14.03 and 14.06 MHz. The front end filtering does nothing to attenuate the original signals that cause the distortion, nor does it attenuate the products once they have been generated. First impressions suggest that this distortion would ruin all communications, but things are not that severe. The detail that saves our receivers is the characteristic that a third-order distortion product will increase or decrease in proportion to the *cube* of the input signals. So, if input signals become 1 dB weaker, the resulting distortion decreases by 3 dB. Third-order IMD in a receiver is characterized by a third-order input intercept. Although third-order IMD is an insidious problem, it is easy to measure. Generally, anything we do to a front-end design to improve IMD will also improve gain compression and second-order IMD. For these reasons, the third-order input intercept becomes a central design consideration for receivers.

Dynamic Range and Intercepts

We often hear folks talking about *dynamic range* of an amplifier or receiver, but the term is often ill defined. When asked about it, the person will say it is the difference in dB between the largest signal that a circuit can handle and the smallest. But what is the weakest signal and what defines it? How large can the largest be and how do we define that?

We use the following *receiver* definition: Two-tone dynamic range is the dB difference between two signal levels; The weakest signal that a receiver can deal with is the *minimum discernable signal*, or MDS while the strongest signal is one of two signals of equal strength that produce a third-order distortion product with a re-

sponse equal to that of the MDS.

MDS was defined earlier and is the available power from a room temperature signal source that will cause the output to increase by 3 dB above the background noise. MDS is related to receiver noise figure and bandwidth by

$$\text{MDS (dBm)} = -174 \text{ dB} + 10 \log(\text{BW}) + \text{NF} \qquad \text{Eq 6.12}$$

where BW is the receiver *noise* bandwidth in Hz and NF is the noise figure in dB. Noise bandwidth is usually close to signal bandwidth at the –6 dB points.[13] For example, a receiver with a 2.5-kHz bandwidth and a 10-dB noise figure has a –130-dBm MDS. The test setup used to

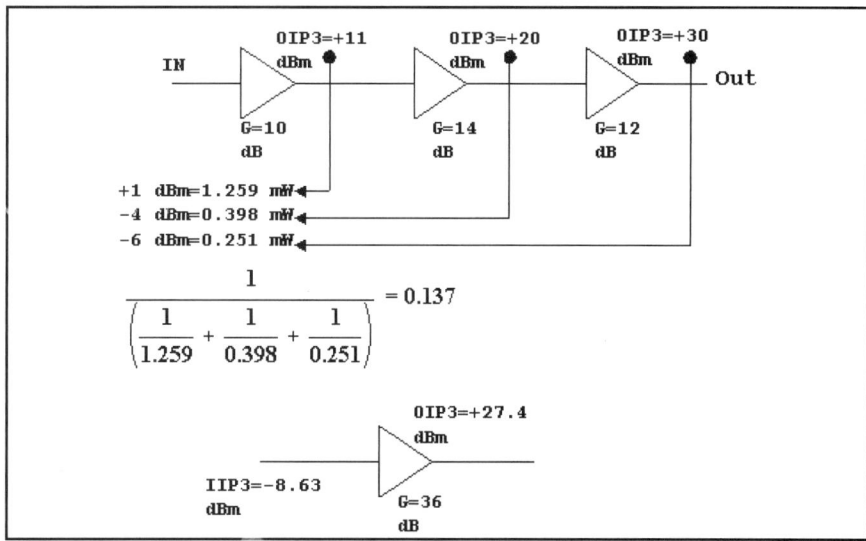

Fig 6.63—Three amplifier stages are cascaded. The intercept for the cascade is calculated by normalizing the intercepts to one plane in the system, converting values from dBm to mW, combining values in the way that resistors in parallel are combined, and then converting back to dBm. See text for details.

equipment. (See Chapter 2, section 6, to see how intercept is defined and measured.) Set the attenuator output for a larger output per tone than was used in the direct DR measurement. Tune the receiver to 14.01 MHz and note an output of 100 mV in the audio voltmeter. We note that the available signals at 14.03 MHz and 14.05 MHz is –31 dBm per tone or per signal. We now tune the receiver to 14.03 MHz where we encounter a very loud signal. The attenuator is increased until the output level is again at 100 mV, finding that this happened when we had added 60 dB of attenuation. Hence, the distortion products are 60 dB below the desired response. This is the IMD Ratio, or IMDR. Rewriting an equation from section 2.6

$$IP_{3in} = P_{in} + \frac{IMDR}{2} \qquad \text{Eq 6.13}$$

allowing us to calculate the input intercept for the receiver as –1 dBm. While doing this measurement, it is instructive to change the input from –31 to –29 dBm, or a similar small amount. With 2-dB-larger input signals, we see IMD products that are 6 dB stronger. The IMDR becomes 56 dB, still leaving an input intercept of –1 dBm. If IP_{3in} remains a constant, the front end is said to be *well behaved*.

Two formats are used to indicate intercepts. The one we have used for an input intercept is IP_{3in}. The IP_3 part indicates that it is a third-order intercept while *in* signifies an input rather than output intercept. An equally valid designation is IIP3 where the first *I* denotes input. The second format relates to the *output* intercept, symbolized by IP_{3out} or OIP3. Avoid associating the term *intercept point* with a number, for it is only confusing when the plane of definition is not specified. Strictly speaking, intercept point is the intersection of two curves.

Intercepts are not mere esoteric curiosities or receiver figures-of-merit. Rather, they are tools, useful parameters available to the designer. Intercepts offer two major capabilities:
- If the input intercept of a receiver (or any system) is known, the intermodulation distortion is well defined for all input levels.
- If the intercepts and gains for all stages in a system are known, they can be combined to calculate the intercept for the complete system. Input and output intercepts for a single stage differ by the small-signal stage gain.

Equation 6.13 lets us calculate distortion for any input level.

The intercept of a cascade was treated earlier and is illustrated here with an example; a three-stage amplifier shown in **Fig 6.63**. This cascade might be part of a wideband amplifier to be used in an SSB transmitter. The output intercepts of the three stages are known: +11, +20, and +30 dBm. The respective gains are 10, 14 and 12 dB. Recall that the input intercept of an amplifier is related to the output intercept through the stage gain. This difference is not restricted to a single stage. The output intercepts for each stage can be normalized, or "moved" to the input of the overall system, becoming +1, –4, and –6 dBm. The individual intercepts are merely adjusted by the gains in the movement process. The normalized values are converted from dBm to power in milliwatts. The values are then combined in the same way that *resistors-in-parallel* are combined, producing a net input intercept of 0.137 mW, or –8.6 dBm. The parallel resistor analogy has no significance other than being an easily remembered formula.

This can also be presented in a generalized equation

$$IP_3 = -10 \log \left(\sum_{i=1}^{N} 10^{-\frac{IP_i}{10}} \right) \text{ (General case)}$$

$$= -10 \log \left(10^{-\frac{IP_1}{10}} + 10^{-\frac{IP_2}{10}} + 10^{-\frac{IP_3}{10}} \right)$$
(N=3, 3-stage example)

Eq 6.14

where IP_3 now represents the intercept of the cascade and IP_i is the intercept of the i-th stage with all intercepts being normalized to a single plane in the amplifier. In our example, we normalized all intercepts to the system input. However, we could have picked the output, or any interface between stages. (The equation is derived in *Introduction to Radio Frequency Design*.) This method is a worst-case analysis where the intermodulation *voltages* from each stage add in phase. Our measurements indicate that this analysis works well in practical systems, so long as the individual stages are well-behaved, as defined earlier.

Receiver dynamic range is related to intercept and MDS by a simple equation. MDS is further related to bandwidth and noise figure, offering a more general equation.

$$DR(dB) = \left(\frac{2}{3}\right)(IIP3 - MDS)$$

$$= \left(\frac{2}{3}\right)(IIP3 + 174 - NF - 10\log(BW))$$

Eq 6.15

where IIP3 is the input third order intercept, NF is system noise figure, BW is the system bandwidth. Recall that kT = –174 dBm at 290 K, explaining that term in the equation.

Some Front-End Design Examples

We are now in a position to evaluate some receiver front-end designs. A few examples will be presented using data obtained from measurements we have performed.

The first example is a popular one among the QRP clan, a receiver front end based upon the Phillips NE602 or NE612. Our evaluation data was presented in Chapter 5. A front-end block diagram, **Fig 6.64**, includes gains, intercepts, and noise figures for the stages. The result of applying the dynamic range analysis is also included. This is a simple design with only one active block, the mixer. The dynamic range is modest at 83 dB, although

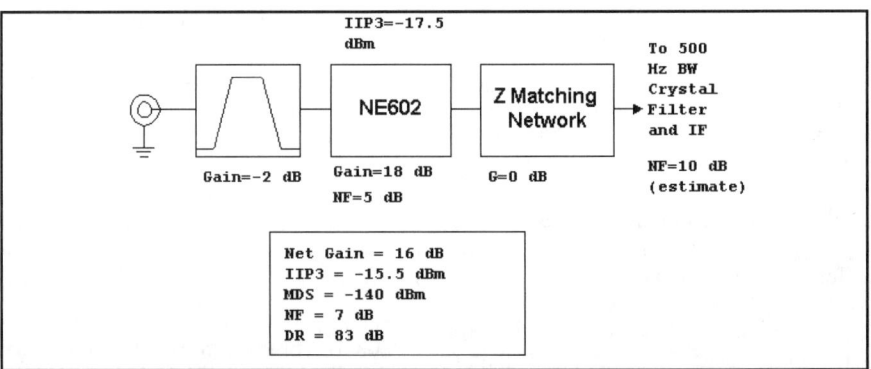

Fig 6.64—A simple receiver front end using the NE602. The IF system is estimated to have a noise figure of 10 dB.

sensitivity is quite good. The noise figure is essentially that of the IC plus the insertion loss of the bandpass filter preceding it. Care must be exercised in implementing this design if this DR is to be realized. For example, chip intercept could be altered if output is extracted only from one output terminal. On the other hand, careful mismatch at the input may decrease gain to actually increase input intercept with only a modest noise figure change. Some builders claim a 90-dB dynamic range with NE602 front ends with this bandwidth. Clearly, careful measurements are always worthwhile.

In spite of the good MDS obtained from the NE602, some builders are tempted to add an RF amplifier. In other situations, an NE602 is used as a second mixer in a receiver, having been preceded with gain. The trade-off is illustrated in **Fig 6.65**. A bandpass filter with a 1-dB loss is followed by a low gain RF amplifier. The signal then passes through the original 2-dB-loss filter before arriving at the mixer. This design offers a 2-dB improvement in sensitivity, but at the price of a 5-dB decrease in dynamic range.

The next sample front end, **Fig 6.66**, is the opposite extreme. Here we use a diode ring mixer as the first element, followed by a post mixer amplifier with high current. This is the sort of front end we recommend for the 160, 80 or 40-m amateur bands where low noise figure is rarely needed. Although MDS is 8 to 10 dB higher than the previous designs, dynamic range is 98 dB. The mixer in this design is a +7-dBm-type ring such as the Mini-Circuits SBL-1, TUF-1 or TUF-3. If an even stronger TUF-1H was substituted, DR over 100 dB is easily within reach in a simple design. The post mixer feedback amplifier would ideally use a part specified just for this application, such as the 2N5109 with 40 or 50 mA. However, a parallel pair of 2N3904s will do a surprisingly good job, again with 40 mA of total current.

Many builders question the use of a passive mixer with no gain. But it is exactly this lack of gain that leads to the low noise. The passive nature of the circuit eliminates the noise-generating elements that compromise some other mixers. There is no substitution for actual design.

The high noise figure of the bare-ring-mixer front end is usually not suitable for the higher bands. The designer will often want to add an RF amplifier to obtain lower NF. This modification is illustrated in **Fig 6.67**. The modest RF amplifier improves sensitivity by several dB while only reducing dynamic range by 2 dB. Too much RF gain could severely compromise performance.

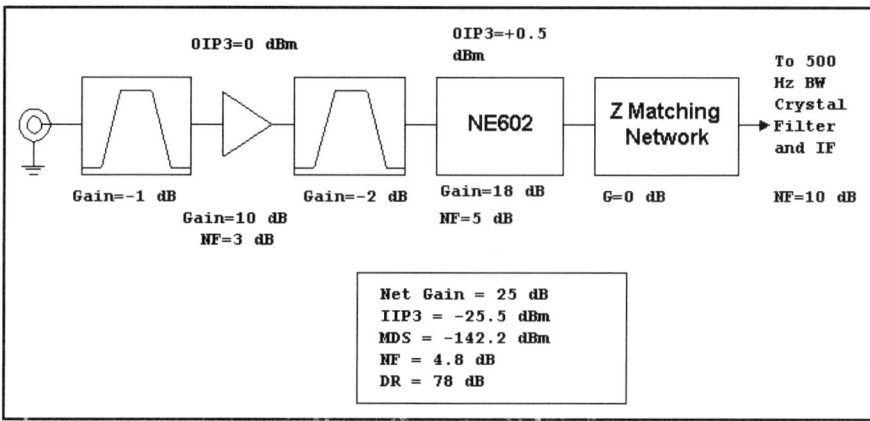

Fig 6.65—An RF amplifier is added to the previous design, offering slightly improved MDS at the cost of degraded dynamic range.

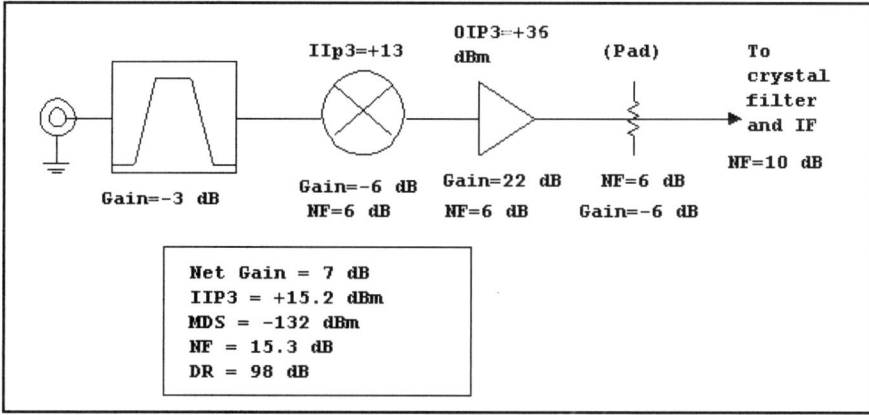

Fig 6.66—Basic front end with a diode-ring mixer followed by a high-current bipolar feedback amplifier.

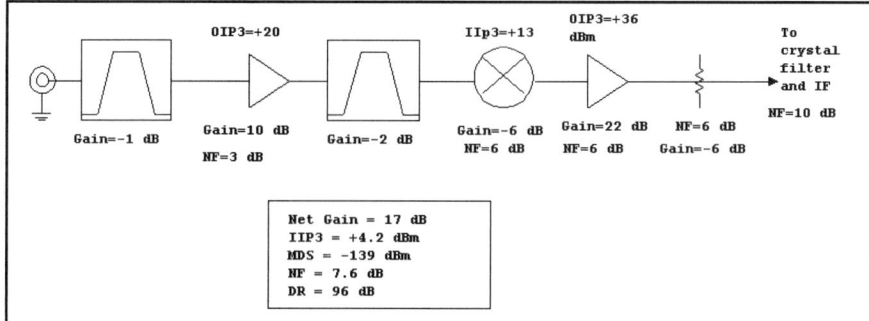

Fig 6.67—An RF amplifier is added to the basic diode-ring front end, significantly improving noise figure while compromising DR by only 2 dB.

The Receiver Factor

The two-tone dynamic range presented above has a major disadvantage as a receiver figure-of-merit: DR is a strong function of bandwidth. This is a direct result of MDS used in the DR equation. A CW receiver with a 500 Hz bandwidth will produce a higher DR than an SSB design with much wider bandwidth. Measurements of MDS are difficult, often complicated by un-planned filtering in the receiver audio section. While this filtering may or may not have much impact on the way a receiver sounds, the measured results are altered.

Both input intercept and noise figure for a receiver are generally bandwidth invariant parameters. The first is a measure of strong signal performance while the other defines weak signal behavior. They can be combined by taking the difference. We call this the receiver factor, **R = IIP3-NF**. The receiver using a diode ring front end without RF amplifier, Fig 6.66, had R=0 dBm while the NE602 receiver with an RF amplifier, Fig 6.65, provided R = –30.3 dBm. While both sample receivers used a CW bandwidth, the R-values would be the same if they were built with SSB filters. Later in this chapter we will describe a receiver with an astounding R = +35 dBm!

The noise figure, and hence, the receiver factor may change slightly with bandwidth with some receivers. This is usually the result of differing filter insertion loss as bandwidth is switched.[14]

A General Purpose Monoband Receiver Front End

Although there are numerous routes to the construction of a high performance front end, a dependable robust topology consists of the following cascade:
- A simple bandpass filter;
- A low-gain RF amplifier;
- A bandpass filter with two or more resonators;
- A diode-ring mixer;
- A post-mixer amplifier using a low-noise bipolar transistor with negative feedback;
- An attenuator that creates a stable impedance at both the output and, through the behavior of the feedback circuitry, the input of the post mixer amplifier;
- A crystal filter;
- And finally, an IF amplifier.

Generally, receivers designed with this front end have produced dynamic range within a couple of dB of the values pre-

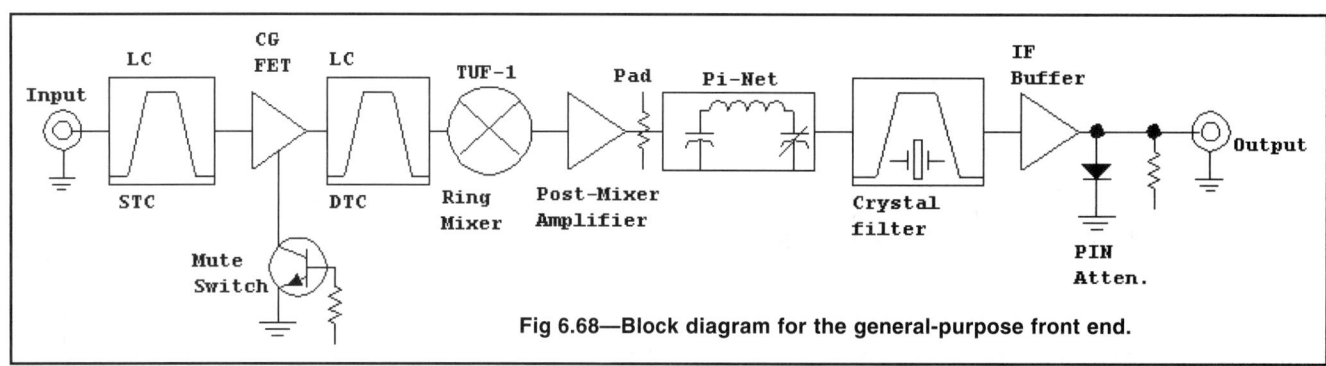

Fig 6.68—Block diagram for the general-purpose front end.

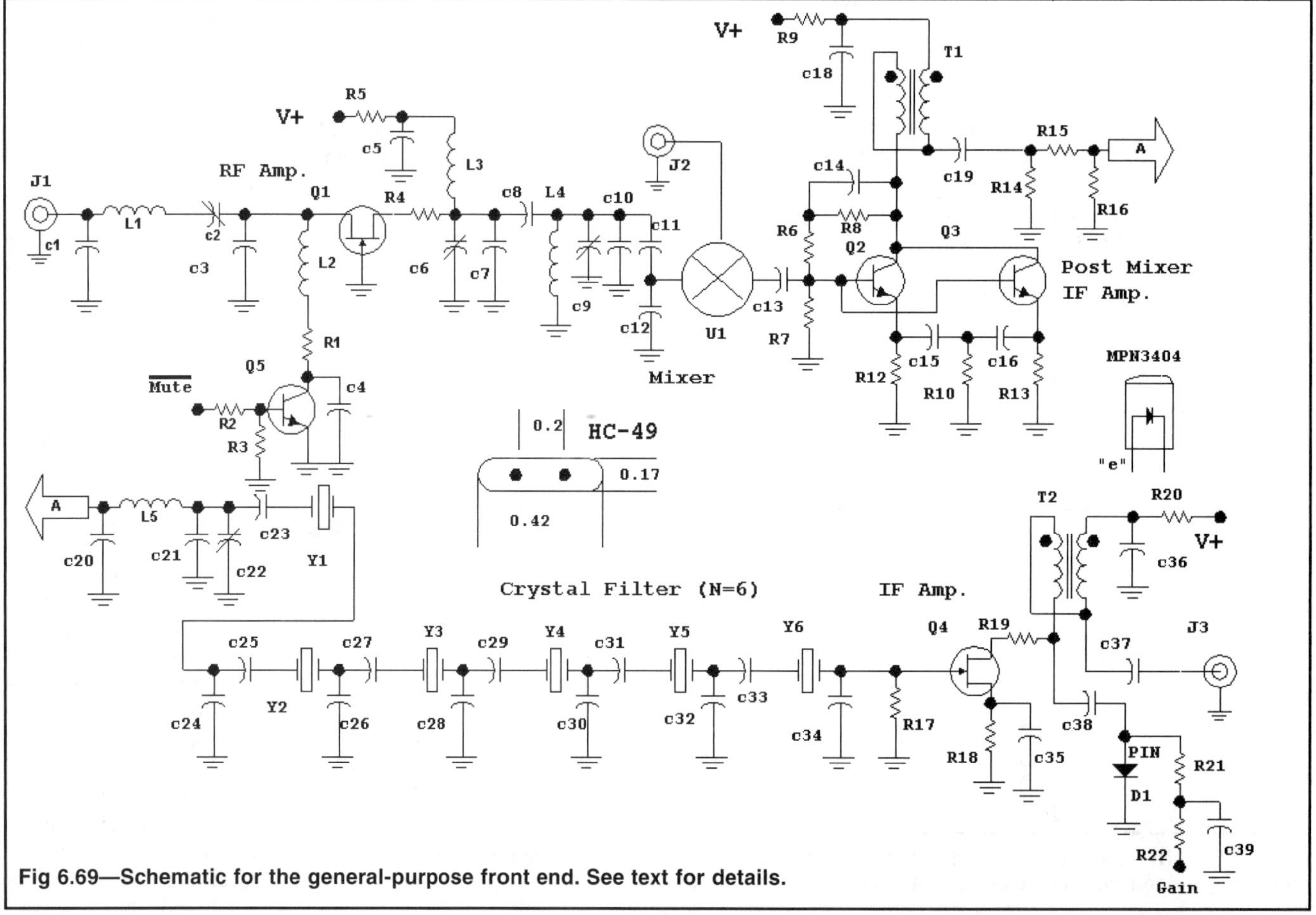

Fig 6.69—Schematic for the general-purpose front end. See text for details.

dicted by the analysis presented when using measured data for the individual stages. The block diagram for this front end is shown in **Fig 6.68**.

A small circuit board was designed and fabricated for this front end and includes a crystal filter of up to 6 crystals. The 50-Ω impedance of the pad is increased with a pi-network to whatever value needed by the filter. The other end of the crystal ladder is terminated in the proper resistor and a common source JFET amplifier. A PIN diode attenuator is also included in the IF amplifier output for those applications where no other IF gain control is available. A muting switch for the RF amplifier is also included. The complete schematic is given in **Fig 6.69**.

The input pre-selector filter is a single tuned circuit. It begins as a 3-element low-pass filter, but the usual inductor is replaced with a series tuned circuit. This simple topology degenerates into a low pass filter in the VHF stopband, a useful attribute when trying to avoid spurious responses related to stray VHF signals.

The second bandpass filter, a double-tuned circuit, appears after the RF amplifier where noise figure has been established. Insertion loss is not as critical as it might be without the amplifier. This means that the filter bandwidth can be narrow enough to ensure very good image rejection. It also allows us to use small toroid cores, if desired.

Two bandpass filters should be used in designs that include an RF amplifier. An RF amplifier that is not preceded by a filter is subject to overload from local signals, particularly the strong VHF broadcasts that most of us experience. A filter should also appear after the RF amplifier, immediately preceding the mixer. This circuit, often termed the image-stripping filter, establishes image rejection. If it was only present ahead of the RF amplifier, it would not suppress noise at the image frequency that is created by the RF amplifier.

The RF amplifier we chose is a common-gate JFET design. It is capable of very low noise figure while offering good intermodulation distortion and high power output when needed. It also can have very good reverse isolation, serving to suppress signals at the mixer that would otherwise find their way to the antenna terminal. But it can also be challenging, for the common gate FET amplifier can tend to oscillate. The spurious oscillations, which usually occur at a few hundred MHz, occur when the layout is poor or leads are too long. Generally, too much fuss is propagated in much of the electronics literature regarding long leads in solid-state circuitry, but this is a place where it really does matter, particularly with the FET gate lead.

A cure for the instability is resistance in series with the drain. This is not a mere experimental band-aid, but a circuit detail justified with analytic evaluation. Greater resistance generates even better stability. We have used 100 Ω in this application, for it provides margin without altering the low frequency (HF and low VHF) gain. The resistor should be placed as close to

Gain of a JFET Amplifier

The IF amplifier used in the output of the general-purpose receiver front end is a common-source configuration with a transformer output presenting a 200-Ω load to the FET drain. Amplifier gain depends on the impedance presented to the input.

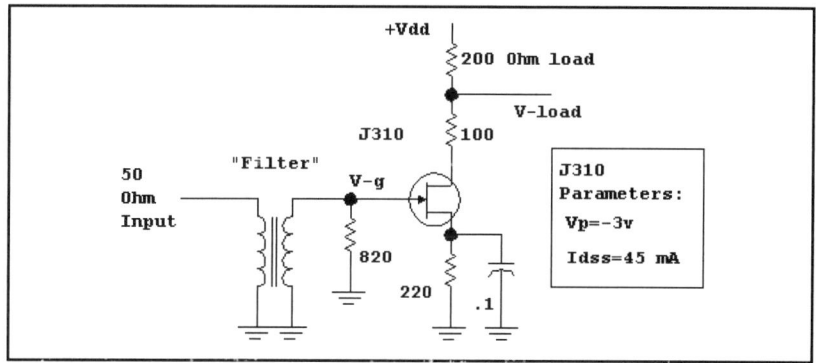

The "filter" is the combination of an impedance-transforming network and a crystal filter in this instance. The 50-Ω source is transformed to match a higher resistance, 820 Ω in the schematic above, with the composite "filter." If 1 mV is presented to the input, the voltage at the gate will be increased by the square root of the impedance ratio, here a factor of 4.05. So, Vg = 4.05 mV. The FET bias current is 7.92 mA in this instance, so the transconductance is gm = 0.0126 S, using equations presented in Chapter 2. The drain signal current is then

GM•VG = 0.051 milliampere.

This current develops an output voltage across the 200 Ω load, Vout = 10.21 mV. (The 100-Ω resistor is significant only in reducing the effective supply voltage. It is included to suppress parasitic oscillations.)

Output power is $V^2/200$ = 5.21 × 10^{-7} W. But the available input power is 1 mV across 50 Ω, or 2 × 10^{-8} W, so transducer power gain is 26, or 14.2 dB. The important detail here is that power gain is a strong function of the impedance terminating the filter, shown in the curve below.

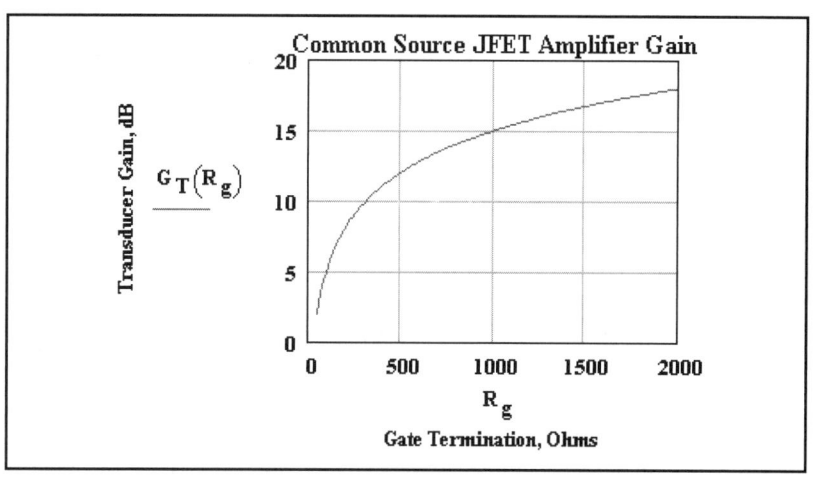

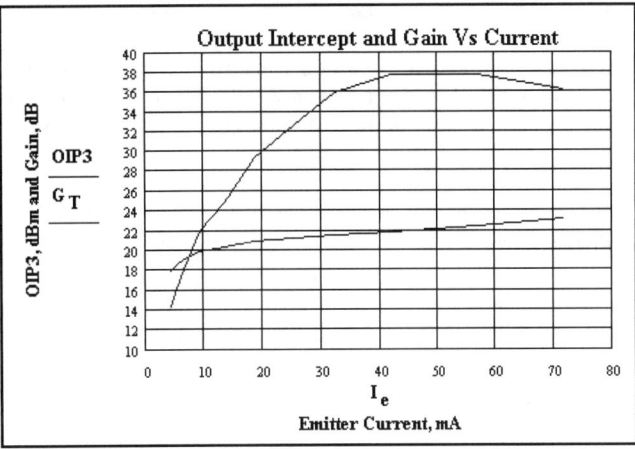

Fig 6.70—Gain (lower curve) and output intercept for one or two 2N3904s in parallel. Two devices should be used for currents above 20 mA, while total current over 40 mA is not recommended except as an experiment.
T1 = 10 bifilar turns on an FT37-43.
R9 = 47, R8 = 1kΩ, R6 = 1.5 kΩ, R7 = 680 Ω, R10 = 6.8 Ω,
R12 = R13, which are picked to set the dc emitter current.
R12 = R13 = 100 Ω for 30 mA total current.
C13,14,15,16,18,19 = 0.1μF.

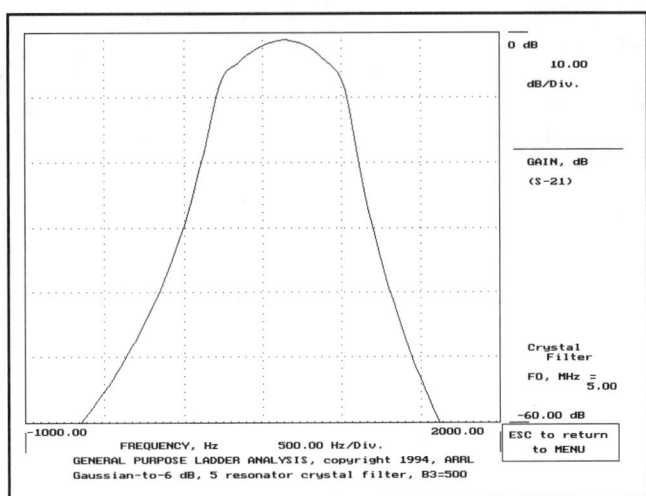

Fig 6.71—Calculated response for the Gaussian-to-6-dB crystal filter. The shape is Gaussian for the top 6 dB, but then reverts to a Chebyshev-like skirt response. The k and q data for this filter were obtained from Zverev's *Handbook of Filter Synthesis*, Wiley, 1967.

the FET as the board layout or breadboard allows. A simple shielding method for a J310 RF amplifier was shown earlier in this chapter. The shield was not needed on this circuit board.

The RF amplifier output resistance is around 10,000 Ω. That value was used while designing the input termination for the double tuned circuit while the output is set for a 50-Ω termination.

The RF amplifier FET is biased on when the NPN switch is saturated. The builder should design control circuitry to apply a positive voltage to the control input during receive intervals.

This module uses mixers in the TUF family from Mini-Circuits. Either the TUF-1 or TUF-3 should work well with +7 dBm of LO power. A high level mixer (TUF-1H or TUF-3H with +17-dBm LO power) will also fit in the board and will provide even higher dynamic range, but only when followed by an adequately strong post-mixer amplifier. The mixer is generally the DR defining element within the system.

The post-mixer amplifier is a critical element. Enough current should be used to guarantee the desired dynamic range. However, too much current can also be wasteful, especially in applications where batteries are used. The layout used in the general-purpose board is for two paralleled 2N3904s, shown in Fig 6.69. Resistors R12 and R13 determine the total current, which should be equal. Only one transistor is required if total current is 20 mA or less. Gain and output intercept are presented vs total amplifier current in Fig 6.70. A home station design where power is abundant might use 30 or 40 mA while 10 mA may be enough for a portable application. No heat sink has been needed for a pair of 2N3904s at 40 mA total current. Larger transistors with higher power dissipation ratings can, of course, be used.

The designer/builder must design the crystal filter for the desired bandwidth. While the board will accommodate up to 6 crystals, fewer may suffice. In one application using a 5-crystal CW bandwidth filter, we found that stopband attenuation was less than indicated by calculations. Two measures restored performance: First, all crystal metal cases were grounded to a wire bus. Second, a shield was soldered to the ground foil between the crystal filter and the post mixer amplifier.

The builder/designer has considerable flexibility available when choosing the terminating resistance for a crystal filter. This choice impacts the design of the IF amplifier. The design procedure is summarized in the *Gain of a JFET Amplifier* sidebar. Higher gain is available with higher impedance values.

The PIN diode will provide up to 30-dB attenuation. This is especially handy for applications where no additional IF gain is used.

The Easy-90 Receiver

The general-purpose front end was used to build a simple receiver for the 20-m CW band, dubbed the *EZ90-14C*. The 90 indicates a two-tone dynamic range in excess of 90 dB, which is achieved with ease with this receiver. The receiver architecture is one without an IF/AGC amplifier. Front-end parts are tabulated in the following list.

The 5-element 5-MHz crystal filter for this receiver was designed for a 3-dB bandwidth of 500 Hz and a Gaussian-to-6-dB shape. This shape has the virtue of a good time-domain characteristic, keeping ringing to a minimum in a narrow filter. The stopband attenuation is still reasonable. An added virtue of transitional filters, including this Gaussian-to-6-dB, is a relative insensitivity to exact component value, allowing a minor degree of "slop" when being constructed. On the down side, this filter lacks the familiar circuit symmetry of Butterworth and Chebyshev designs. We built this 5-MHz filter with available crystals that had good Q, often over 200,000. Crystal frequencies were matched to within 10 Hz. Design details are presented in Chapter 3. A calculated response for this crystal filter is shown in **Fig 6.71**.

Several different filter designs were tried in this receiver. While a Cohn design worked, it used a terminating resistance under 200 Ω. This severely impacted the IF amplifier gain as outlined in the IF sidebar. (A Cohn type crystal filter is, of course, possible with a higher terminating impedance, but the simple design method presented in Chapter 3 is then invalid.) A Gaussian-to-6 dB filter with a 250-Hz bandwidth and 500-Ω terminations worked well, but was too narrow for the intended application

The front-end board output is routed directly to the product detector, shown in the detector-audio board in **Fig 6.72**. This

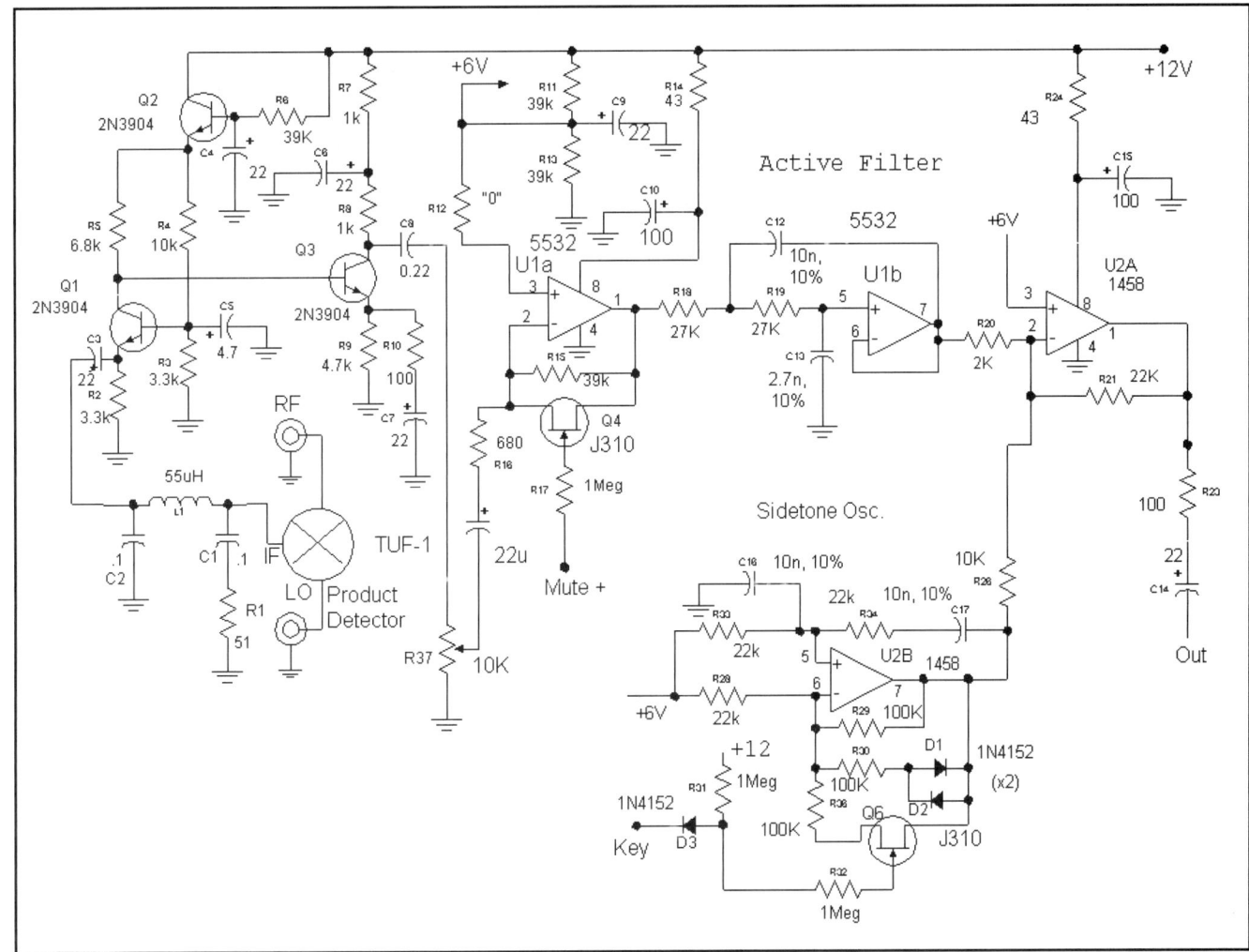

Fig 6.72—Audio amplifiers, product detector, and sidetone oscillator for the EZ-90C receiver.

EZ90-14C
Parts List for the 20-Meter "Easy 90" Receiver

C1,C3: 470 pF SM or NP0 ceramic
C2,C6,C9,C22: 65 pF, 10 mm air variable (Sprague Goodman GYC65000)
C4,5,13,14,15,16,18,19,35,36,37,39: 0.1 µF
C7: 82 pF
C8: 2.2 pF
C10: 56 pF
C11: 22 pF
C12: 200 pF
C20: 820 pF
C21: 220 pF
C23: 470 pF
C24: 68 pF
C25: short circuit
C26: 100 pF
C27: 150 pF
C28: 100 pF
C29: 100 pF
C30: 150 pF
C31: 100 pF
C32: 82 pF
C33: short circuit
C34: not used
Q1, Q4: J310
Q2, Q3, Q5: 2N3904
D1: MPN3404 or similar PIN diode
L1: 27t #28 on T30-6
L2, L5: 4.7 µH molded RFC, Q>=50
L3, L4: 1.04 µH, 16 t #28, T30-6
T1:T2 10 bifilar turns #28, FT37-43
R1: 180
R2, R3: 10 kΩ
R4: 100
R5: 47
R6: 1.5 kΩ
R7: 680
R8: 1 kΩ
R9: 47
R10: 6.8
R12, R13: 100
R14, R16: 150
R15: 36
R17: 820
R18: 220
R19: 100
R20: 47
R21: 1 kΩ
R22: 680
U1: TUF-1 or TUF-2 or TUF-3
Y1, 2, 3, 4, 5: HC49 crystals, 5 MHz, Lm=98 mH, C0=3 pF (see text)
Y6: not used; add short circuit

module design has been used in several projects. A TUF-1 provides the detector function. Bipolar audio amplifiers drive an audio gain control, followed by an op-amp providing gain and an RC active low pass filter with a peak at 700 Hz. The Q is kept low in this version. The audio is muted with a shunt FET switch.

The BFO for the product detector is shown in **Fig 6.73**. This is breadboarded on a small scrap of circuit board material.

Fig 6.74 shows a 9-MHz VFO for the EZ90-14C. The oscillator is a voltage-tuned Colpitts circuit purposefully configured for low inductance. The high fixed-tank capacitance is desirable for low phase noise. This LO produces a narrow tuning range of about 20 kHz with the available tuning diode. This receiver is used with a transmitter with restricted tuning range, so the narrow range is acceptable. The builder/designer may wish to use a combination of varactor tuning and a traditional variable capacitor to achieve a wider tuning range. Alternatively, higher L could be used to cover the entire CW band with a varactor diode.

The VCO output is extracted from an FET follower that then drives a power amplifier to provide the +7 dBm LO power needed by the ring mixer. Power amplifier degeneration is adjusted to set output level. An 8-V regulator supplies the VCO. It also provides a stable bias for the tune pot and a stable 4 V for an op-amp reference. The gain and offset in the op-amp are set up to supply a 5 to 10 V swing on the varactor diode.

A receiver noise figure measurement produced NF = 6.6 dB. If a noise bandwidth of 800 Hz is used with this, MDS of –138 dBm is suggested. However, a direct measurement of MDS produced –141 dBm. The difference is attributed to the narrow audio filter that restricts overall noise bandwidth. DR measurement produced a value of 95 dB, for IIP3 = –1.5 dBm. Using this value for IIP3, receiver factor is R = –8.1 dBm.

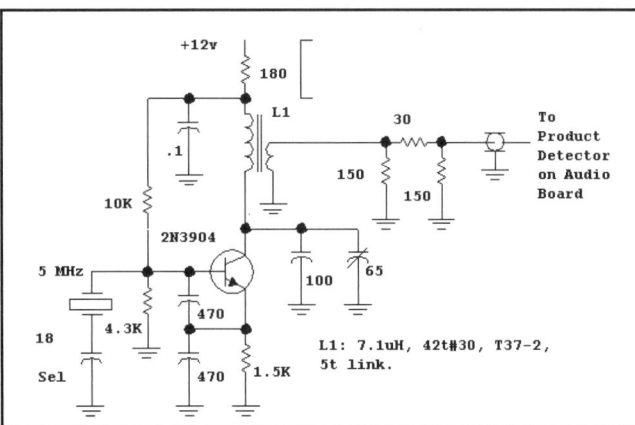

Fig 6.73—BFO for the EZ90-14C. A variable capacitor can be used in series with the crystal for final adjustment. It was replaced with a fixed capacitor in our receiver.

General-purpose receiver front end board used in the EZ90-14.

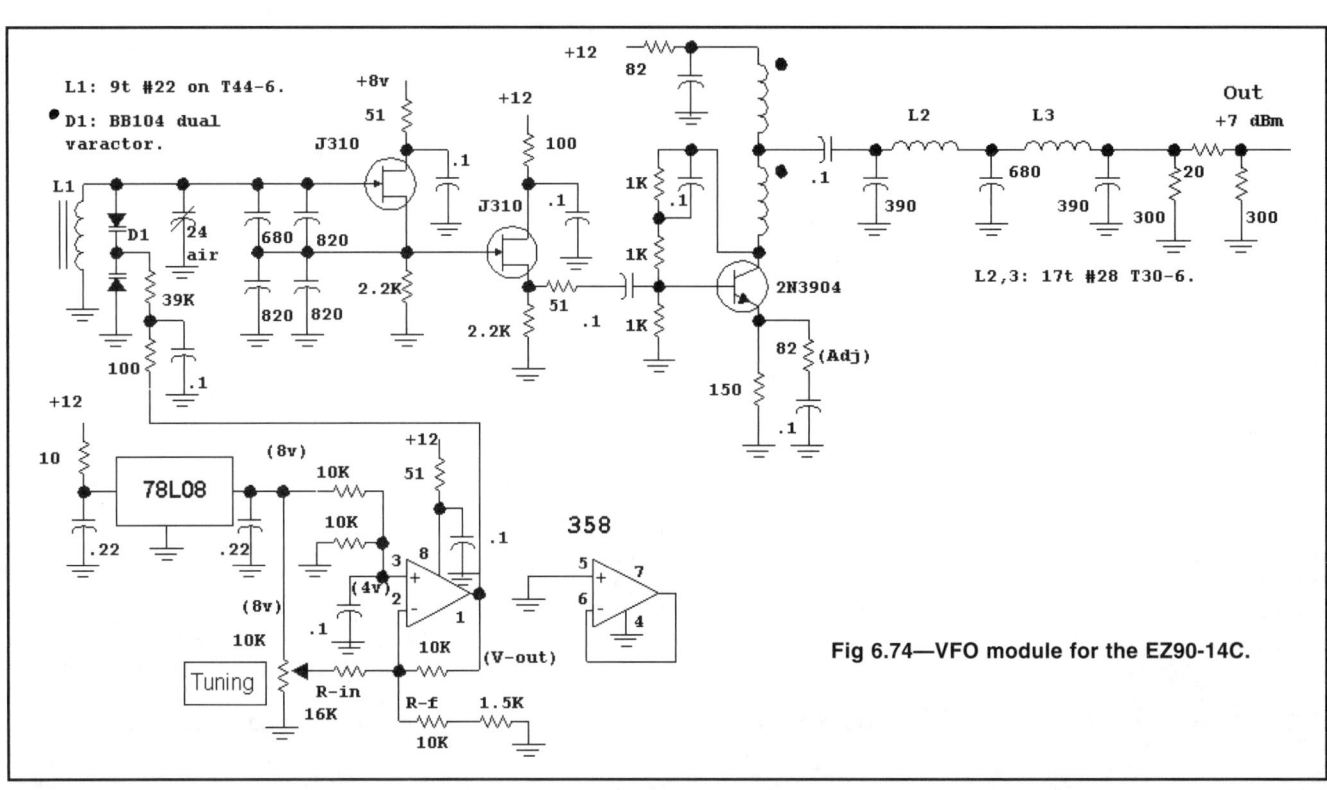

Fig 6.74—VFO module for the EZ90-14C.

The receiver is packaged with a 14 MHz VXO transmitter described in Chapter 5. The narrow receiver tuning range eliminates most birdies from being a problem. In spite of this, one was encountered in the form of a feedthrough of 15-MHz WWV energy. This signal got into the enclosure on the antenna connector where it then found it's way onto the grounds that reached the product detector. There, the normal third harmonic response of the diode ring allowed the 15-MHz component to be directly converted, to produce baseband audio. The problem was eliminated with a 5-MHz low-pass filter inserted in the line between the front end and the detector audio board. The problem would never have occurred if the receiver had not been built with completely unshielded boards.

Generally this receiver will hold up well in a contest environment, although we find it in need of some AGC for those moments when a really strong signal is encountered. Limiting in the audio output op-amp produces a clipped response when the strong signals appear, saving the operator's ears. The very "hot" receiver (low MDS) was designed for portable situations where noise levels are much lower than we find in a home environment.

A 14-MHz Receiver

This receiver is an updated version of two earlier designs.[15] The changes include repackaging (smaller size) with improved shielding, a new frequency counter with lower power requirements, and a reduced noise IF system. This receiver is similar to the EZ90, but features the shielding needed for high dynamic range.

The receiver is a CW only design using filters with reasonable time domain characteristics. While these filters are no longer available, it should be possible for the aggressive builder to build viable substitutes. The 9-MHz IF system was described earlier in detail in Fig 6.56. The design features three stages of gain using dual-gate MOSFETs and crystal filters at both the IF input and output. The IF circuitry is built with breadboards into a multiple section milled aluminum enclosure.

The front end (**Fig 6.75**) begins with a bipolar RF amplifier biased to $I_e = 12$ mA, which produces low noise figure while maintaining an intercept that is high enough to not degrade overall receiver IIP3. The amplifier is preceded by a single resonator preselector and followed by a double tuned image-stripping filter.

The mixer uses a TUF-1 with +7 dBm LO drive. A higher LO level is applied to a 3 dB hybrid that splits the signal into two isolated components. One drives the mixer while the other is attenuated and available for transceive applications. The mixer has two inputs, selected by a small relay. One is the normal 14 MHz signal from the double tuned circuit while the other comes from other equipment at either 4 or 14 MHz. The mixer output is applied to the familiar feedback amplifier and pad combination. The front end is housed in a 4 × 4 × 1 inch milled aluminum box.

The BFO and Product Detector, shown

General-purpose receiver front end board installed in the EZ90-14 Receiver.

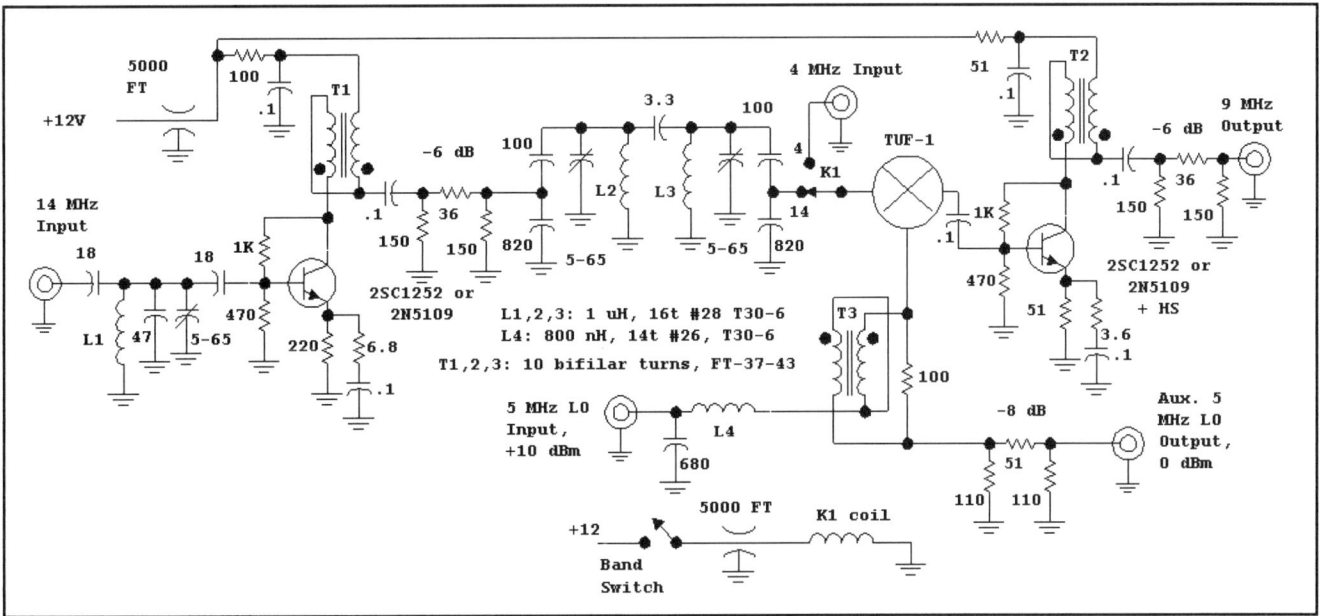

Fig 6.75—Front end for the 14-MHz receiver. The circuit is built largely with breadboarding methods.

Front panel view of receiver.

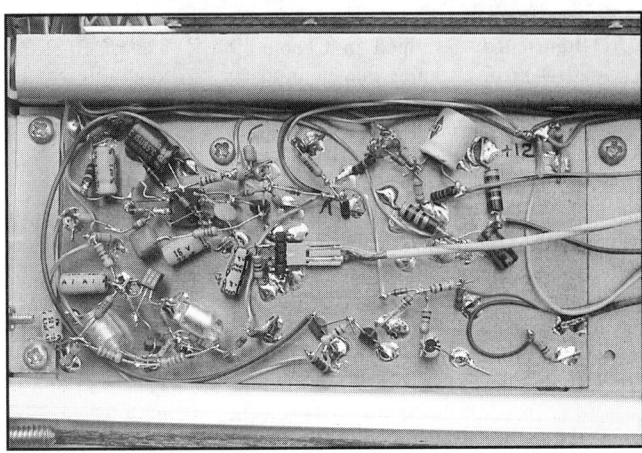
Close up view of audio amplifiers.

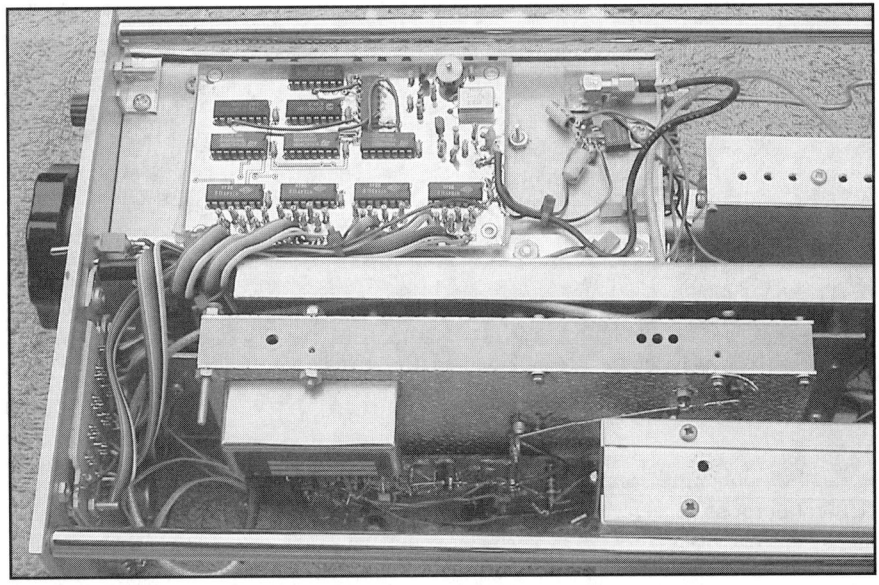
Inside of 14-MHz receiver. Upper left is the frequency counter, upper right is the front end, middle is IF chain, and lower right is product detector/BFO.

in **Fig 6.76**, is traditional. A diode ring moves the 9-MHz IF signal to base band while a bipolar transistor serves the BFO function.

The 5-MHz local oscillator is shown in **Fig 6.77**. The design uses a Colpitts VFO with a JFET. A JFET buffer drives a feedback amplifier output stage. The output power is large enough to drive the hybrid splitter and mixer in the front-end module. Varactor diode tuning will eventually be added to provide an RIT function. The related CMOS frequency counter was described in Chapter 4.

The receiver audio system is shown in **Fig 6.78**. U1 provides audio gain, muting, and a convenient place to inject a sidetone signal. This drives an audio gain control and the output stage, U2 and Q2. The output operates as a class A amplifier with a standing current of about 90 mA. This will drive a small speaker or headphones of virtually any impedance. The high current is not a prob-

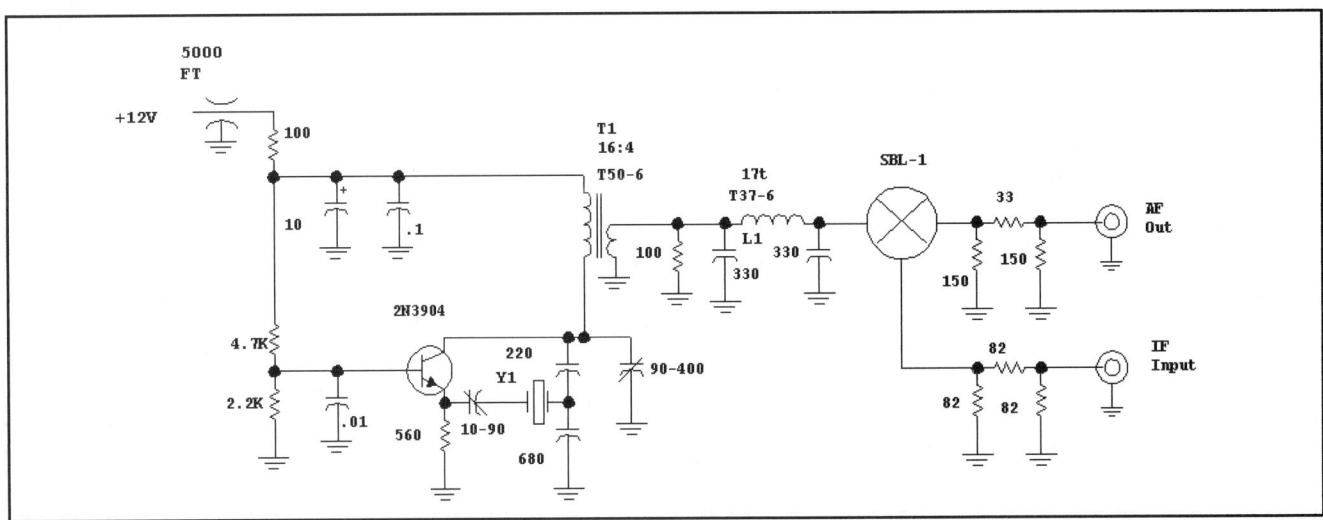

Fig 6.76—BFO and Detector for the 14-MHz receiver.

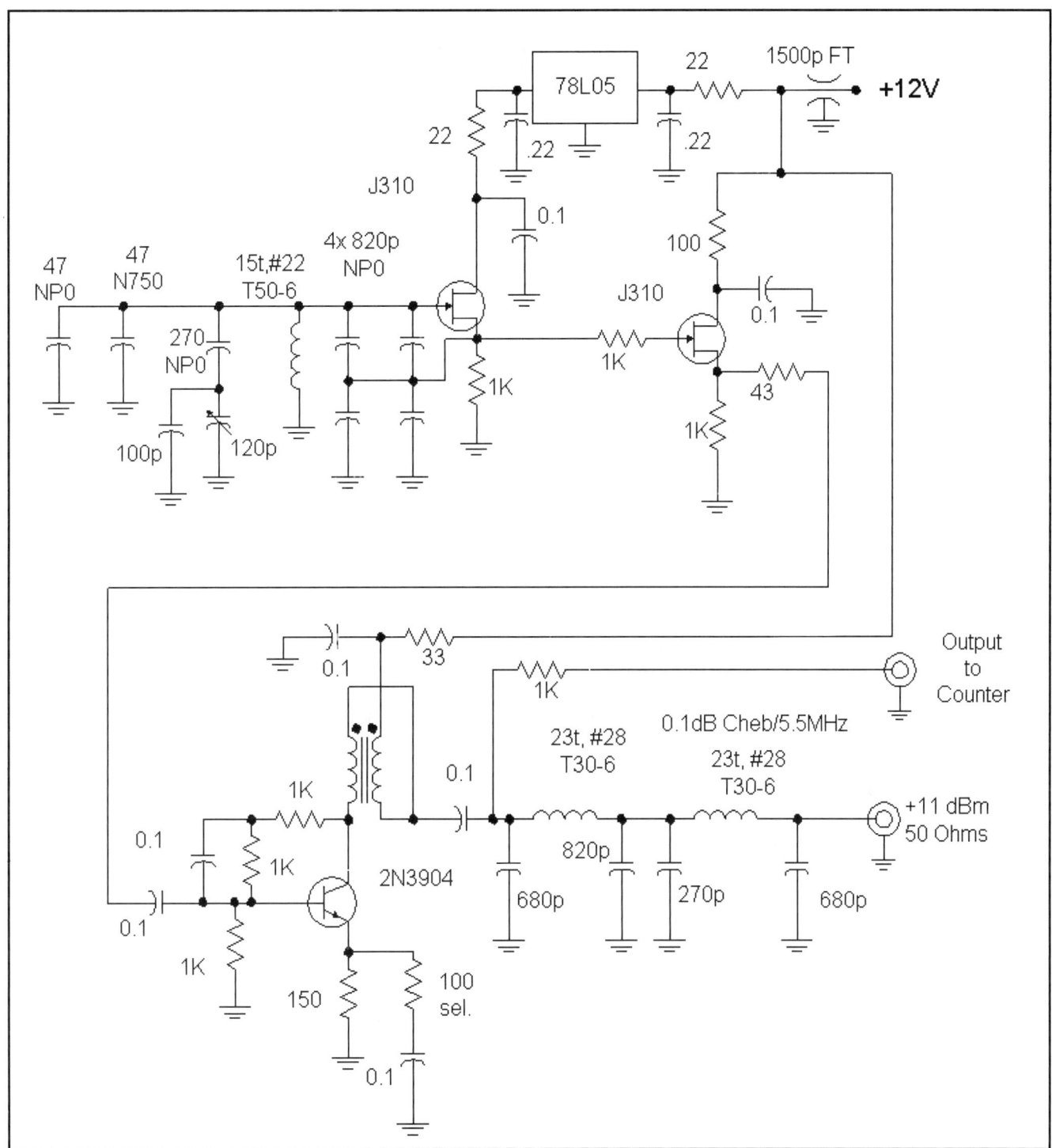

Fig 6.77—LO system for the 14-MHz receiver. The N750 capacitor provides temperature compensation as measured with a small homebuilt thermal chamber. All other capacitors in the oscillator have an NP0 temperature coefficient.

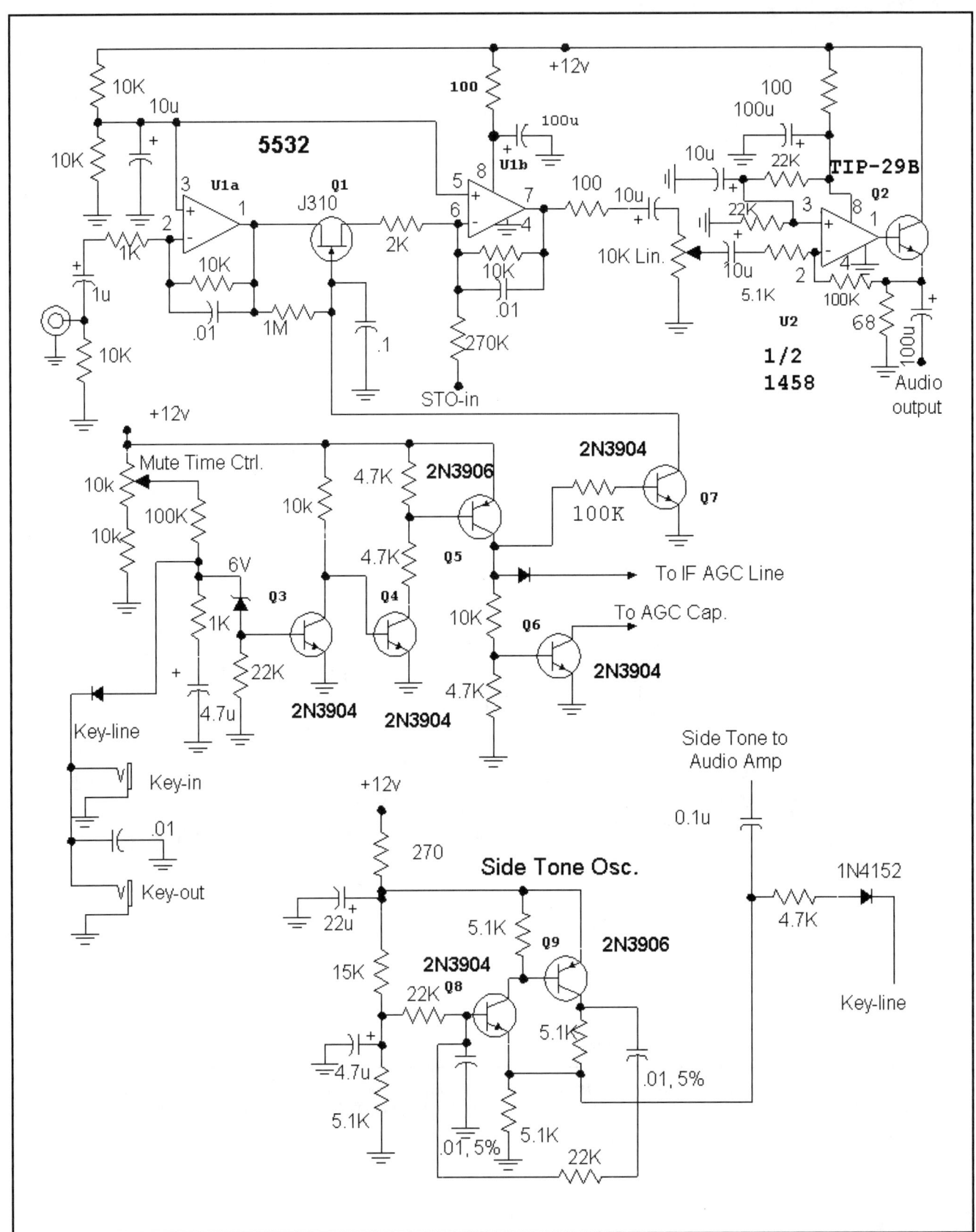

Fig 6.78—Audio and control system for the receiver. See text for details.

lem, for the receiver is used only in a home environment. Q3 and related components generate a time delay, establishing the time the receiver is muted following a key closure. Placing the function in the receiver allows use with many transmitters that may not include interface circuits. The key line loops in and out of the receiver.

Q8 and Q9 form an unusual Weinbridge sidetone oscillator. In key-up conditions the two transistors and the two 5.1-kΩ emitter resistors form an amplifier with a non-inverting gain of two. This is not high enough to support oscillation. But when the key is pressed, the 4.7-kΩ resistor causes the voltage gain to exceed 3, allowing oscillation to begin. The frequency is determined by the 5% capacitors and 22-kΩ resistors. Oscillator output is obtained from the emitter of Q8. This point does not change dc value as the circuit is keyed, preventing a keyed voltage spike in the audio.

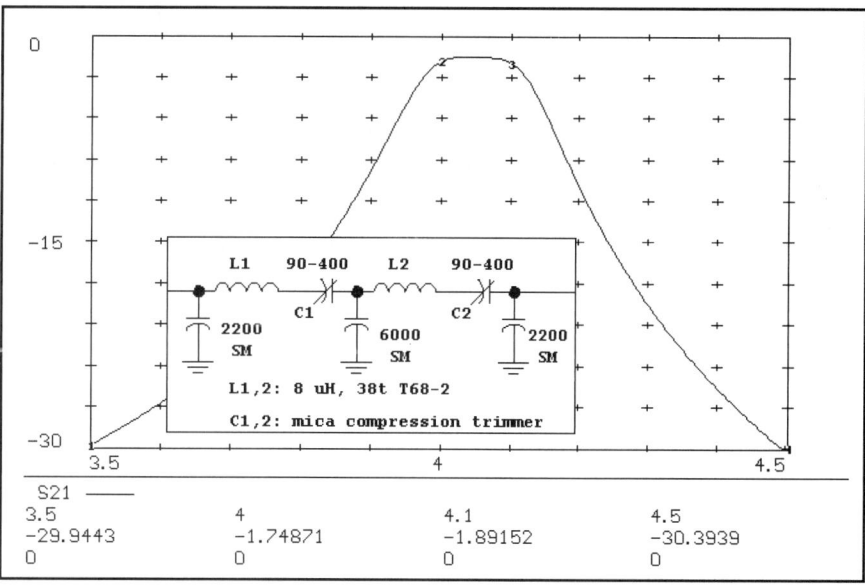

Fig 6.79—Filter for use at the output of crystal controlled converters to be used with the 4-MHz input in the 14-MHz receiver.

Overall Results

This receiver is a design that has evolved for several years, so the performance is fairly stable. Prior to a major rebuild in 1998, the receiver used an IF based upon MC-1350P integrated circuits. While adequate, the noise performance was marginal. Receiver noise figure is now maintained as IF gain is reduced, producing a receiver that continues to sound "bright," when used for weak or strong signals.

Noise figure was measured as 7 dB. The measured MDS was around –141 dBm while IIP3 was +1.5 dBm for DR of 95 dB. The LO system, although difficult to evaluate, seems to have phase noise less than –140 dBc/Hz at a 5 kHz carrier offset. Thermal stability is excellent, although this occurred only after a minor struggle. Examination showed that an RF choke in the oscillator FET source had poor temperature characteristics. Removal of that component and further compensation produced a stable oscillator, illustrating the virtue of careful testing and response to test results. The LO, although lacking the control features of a synthesized system, is completely free of spurious responses. The receiver is just as much fun to use as the original was in 1974.

Converters

The receiver has been used with crystal-controlled converters for numerous bands. Although a traditional dual conversion system does not offer the dynamic range of a single conversion design, it can be close if converter gain is kept low. The typical converters consist of a preselector filter, a diode ring mixer with crystal controlled oscillator, a post mixer amplifier, and pad. An RF amplifier is used for the higher bands. Some sort of 4-MHz bandpass filter is then required to guard against any second conversion images. One filter we have used is shown in **Fig 6.79** with calculated response. The filter may reside with the converter or with the basic receiver. All of our converters use a crystal 4 MHz above the incoming band, preserving the frequency counter accuracy.

6.4 LOCAL OSCILLATOR SYSTEMS

Fig 6.80 shows a number of traditional LO configurations found in receivers and transceivers. Not shown are the common synthesized schemes found in "modern" commercial equipment. Frequency synthesis was discussed in Chapter 4. Many considerations presented here apply to synthesizers as well as simpler systems.

The simplest system is that of Fig 6.80A. A free running LC oscillator operates at the desired output frequency. It is buffered, sometimes with more than one amplifier if higher power is required. Low pass or bandpass filtering is included to remove harmonics. The signal will eventually drive a mixer, with many types requiring LO drive that is free of even-order harmonics. Odd harmonics are

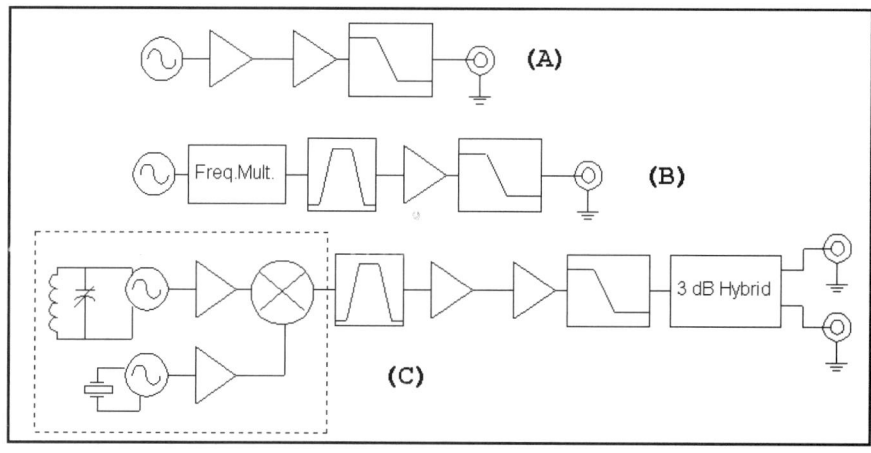

Fig 6.80—Local-oscillator systems for use with communications systems. See text for details.

allowed with the familiar diode rings, for they produce a symmetrical signal, a square wave in the extreme. Even-order harmonics can upset the balance needed for good port-to-port isolation. Details are discussed further in Chapter 8.

Frequency multiplication is often used, Fig 6.80B, for the buffering offered is excellent. In some cases a multiplier is needed to increase the frequency of a fundamental-mode VXO to the VHF region. While crystal-controlled oscillators may be possible at the needed frequency, overtone modes are usually used at VHF, which cannot be pulled with the ease of a fundamental mode oscillator. A bandpass filter follows the frequency multiplier. This is needed to select the desired harmonic while suppressing all other components. Balanced frequency multipliers are recommended when possible, for they ease the level of filtering and shielding required. The frequency multiplication process is often a lossy one, so more amplifiers may be required. More than one gain stage may be required. Finally, a low pass filter reduces the harmonics generated by the amplifiers.

A frequency multiplier system like that of Fig 6.80B need not alter stability. Any drift in the oscillator will be multiplied with the carrier signal. So a 1-kHz drift in an oscillator that is frequency tripled will produce a 3-kHz shift in the output, leaving the fractional change constant. This drift is still low with multiplied crystal oscillators.

The premix scheme of Fig 6.80C is popular, using a mixer to produce an output resulting from two oscillators. One input is usually from a free running LC circuit while the other is crystal controlled. For example, a 25-MHz transceiver with an IF of 6 MHz might use a 31-MHz LO system. This could be realized with a 4.5-MHz free running VFO and a 26.5-MHz crystal-controlled oscillator. The frequency drift is dominated by the LC circuit, which can be fairly stable owing to the low frequency.

Assume this example system is to tune a 300 kHz range from 30.9 to 31.2 MHz. The VFO will then tune from 4.4 to 4.7 MHz. Before construction begins, or a crystal is ordered, a spur analysis should be performed. This was discussed in the mixer chapter. There are no severe problems with the frequencies used in this example.

Spurious responses, when present, can be reduced with careful attention devoted to LO mixer drive levels. A normal diode ring should be driven with a LO signal of +7 dBm, the 26.5-MHz signal in our example. The "RF" input should be confined to a maximum level of –10 dBm. The "specifications" for the mixer list a much higher level, around 0 dBm. This is the level allowed without damage to the mixer. But spurious responses grow dramatically with drive level. It is important to actually measure levels. An available RF power of –10 dBm should be established with a suitable substitutional measurement with a power meter or 50 Ω terminated oscilloscope, discussed further in Chapter 7.

The example mixer will have –17 dBm outputs at 22 and 31 MHz. A bandpass filter will select the higher. Either a double or triple tuned circuit is suitable. This application requires at least a 300-kHz bandwidth. A wider filter may be preferred, for a 1% bandwidth LC filter is lossy with typical toroid coils. But a 1-MHz bandwidth at a 31-MHz center would be an easy filter to design, build, and tune. A typical filter insertion loss might be 3 dB, resulting in a filter output of –20 dBm. If the eventual system output must be +10 dBm, a net gain of 30 dB is required. This is difficult with one gain stage, but easily realized with two. Feedback amplifiers with general-purpose transistors such as the 2N3904 or MPSH10 are suggested. Again, measurements are required. Avoid input overdrive as a means of obtaining the desired mixer output.

Layout can be critical with the mixer system. The filtered mixer output is low at –20 dBm. Yet there are two very strong signals present: an *RF* input (the VFO) at 4.4 to 4.7 MHz, and a crystal generated *LO* at a robust +7 dBm at 26.5 MHz. Spurious mixer outputs should be at least 50 or 60 dB below the desired level of –20 dBm, or at –80 dBm. The crystal oscillator output reaches +7 dBm. It is reasonable to obtain 50 to 60 dB of suppression between points on a circuit board. But 87-dB suppression presents a greater challenge.

Fig 6.81 shows one way we might build this LO system. The block diagram is in part A while part B shows a typical single board layout. This might be either a breadboard or a printed circuit board, either using a nearly solid metal top foil. When this layout is built and measured, we see the spurious outputs mentioned earlier. The crystal oscillator signal (26.5 MHz) is present in the output, as is a weaker VFO component at 4.5 MHz. But spurious outputs may not just indicate an inadequate bandpass filter. Even when that filter is improved, the spurs may persist, a result of poor layout.

A number of problems are present with this layout. Large RF currents flow in the oscillators, often larger than indicated by the output levels. Those currents flow in the ground plane. If a solid ground plane is used, attenuated oscillator current will be found in the ground foil around and beyond the bandpass filter, now free to feed into the output. The amplifier after the filter has a wide bandwidth and increases the spurious level.

Radiated oscillator signals reach the output coaxial connector. The center wire and the ground connection between the box wall and the circuit board foil form an open loop. That loop is now free to intercept some of the radiated energy. A better connection to the outside world would extend coaxial cable on a bulkhead connection until the board is reached. A twisted-wire pair also works well.

Single-point grounds for each stage are common in audio systems and are appropriate for RF designs. Similar regional grounding can confine oscillator ground current to a small part of the overall board. This would also prevent coupling between the individual oscillators.

The scheme that produces much better performance is shown in part C of Fig 6.81. The board ends with the mixer, situated very close to an output connector. The loop area related to the output connection is kept small. A coaxial environment is maintained through the bandpass filter with the following amplifiers then built on an open board. Examples are shown later in photographs. A 5-element low-pass filter follows, attenuating harmonics created in the amplifiers. The final element in most systems is a splitter-combiner, allowing two 50-Ω loads to be driven. This circuit usually has a 25-Ω input impedance, provided by a modification to a 50-Ω low-pass filter.

Active mixers with lower LO power requirements may be preferred for premixed LO applications. While the NE602 is suitable, higher-level Gilbert Cells like the MC1496 or the Texas Instruments Japan SN16913P are preferred. The later part is soon due to be discontinued with no similar replacement on the horizon. The AD-831 or AD-8343 from Analog Devices should be investigated.

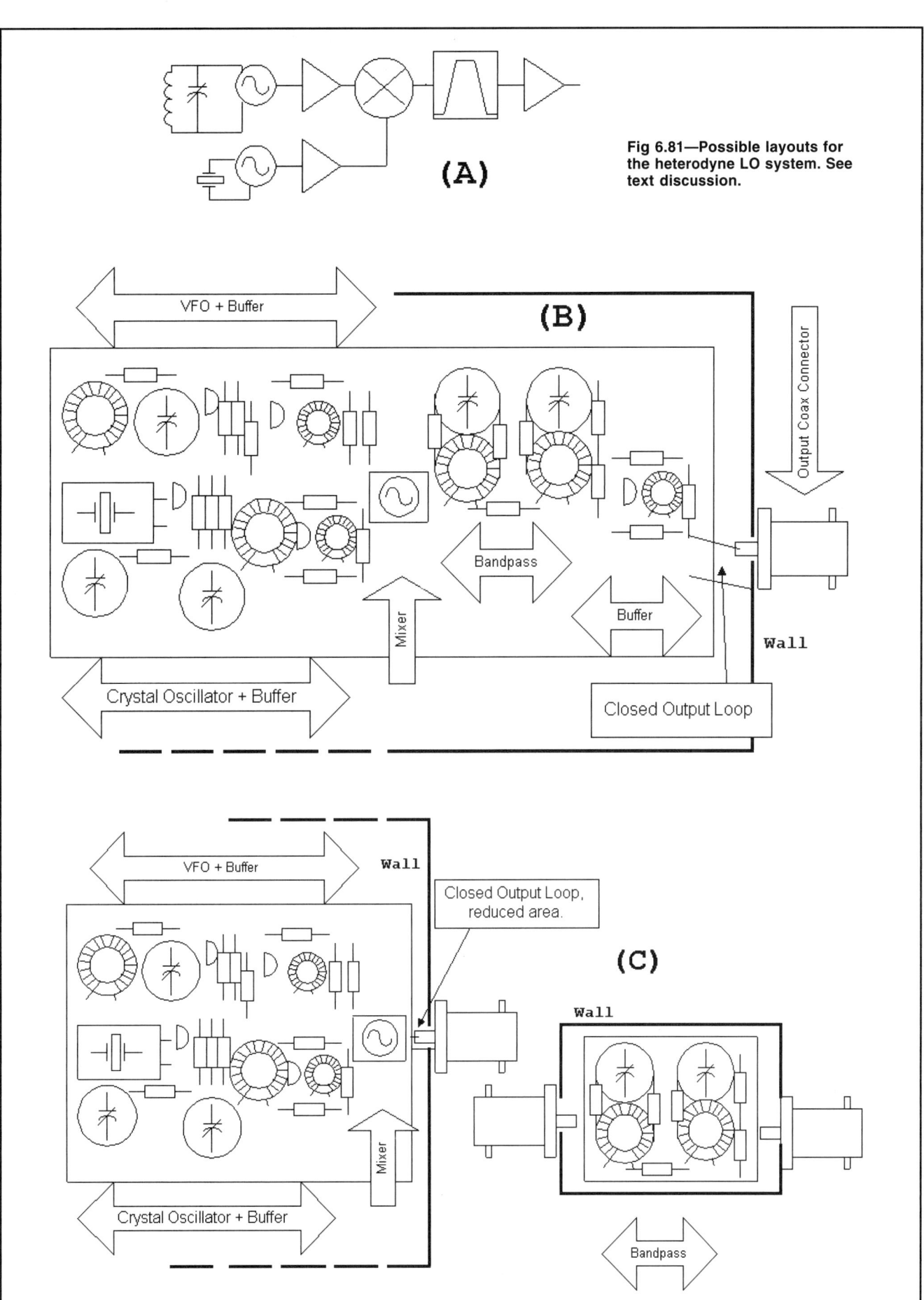

Fig 6.81—Possible layouts for the heterodyne LO system. See text discussion.

6.5 RECEIVERS WITH ENHANCED DYNAMIC RANGE

All of the elements within the front end must be enhanced when striving for high dynamic range (DR). It is usually the mixer (or mixers) that are the critical elements, the parts to be upgraded. However, as soon as we improve a mixer in a typical receiver, the amplifiers become stressed. It is mandatory that we examine all components up to and including the selective filters.

Intermodulation intercept and noise figure are both vital elements in a wide DR receiver. Any NF improvement will allow reduced gain in critical areas, thus relaxing intercept requirements.

A major change in receiver architecture can sometimes make a large difference. We will show a front end later that eliminates *all* gain ahead of the initial selectivity, thus achieving stellar intercept performance while maintaining an adequately low noise figure.

In the last chapter we saw that the input intercept (IIP3) for a +7 dBm LO type diode ring mixer could be +11 to +16 dBm. This is the value that we might measure with a 50-Ω, wideband termination. A high level mixer with +17 dBm LO drive will show IIP3 values 10 dB higher, with typical values in the vicinity +24 dBm.

Fig 6.82 illustrates these design concepts with a front-end block diagram. The first element is a single tuned circuit preselector filter. The wide bandwidth of 1.5 MHz keeps the insertion loss (IL) below 0.5 dB so long as inductor Q_u exceeds 250. Decreasing bandwidth to 350 kHz would cause IL to increase to 1.6 dB, again using inductor with $Q_u = 250$.

The next element is an RF amplifier. A bipolar feedback amplifier with a pad is used here, shown in **Fig 6.83**, which includes the input preselector schematic.

The next system element, **Fig 6.84**, is the main preselector filter, the one that establishes image rejection and protects the mixer from spurious responses. The circuit begins with a 9-element low-pass filter, followed by a 3-resonator bandpass with a bandwidth of 300 kHz. The receiver using this filter was a committed CW design that tuned only the bottom 150 kHz of the band, so the narrow preselector was not a limitation. The circuit ends in a 3-dB pad that establishes filter termination and helps preserve mixer performance. The low-pass filter guarantees stellar suppression of VHF signals, a problem in a metropolitan environment.

One might argue that this preselector is more extensive than needed. Our goal was to realize a "100 dB" receiver. That meant not only that the two-tone dynamic range should exceed 100 dB, but that all spurious responses should be suppressed by the same amount. One such spur occurred with 16-MHz input signals that reached the IF via a 5:1 spur. The image and spur rejection plus the 9-MHz IF feedthrough rejection could only be guaranteed with an extensive preselector. Such filters have high insertion loss, 6.5 dB here when the pad is included. It is this high loss that made the RF amplifier necessary.

Two preselector networks are required whenever an RF amplifier is used. Some initial selectivity protects the system from out of band energy. A single network at the input is generally insufficient, for it would allow image noise generated in the RF amplifier to be converted to the mixer IF.

The next element is the mixer, an SRA-1H using +17-dBm LO injection. The mixer is driven from a 5-MHz LO system. A design with fewer spurious responses would move the LO to 23 MHz. A heterodyne approach shown earlier (Fig 6.81A)

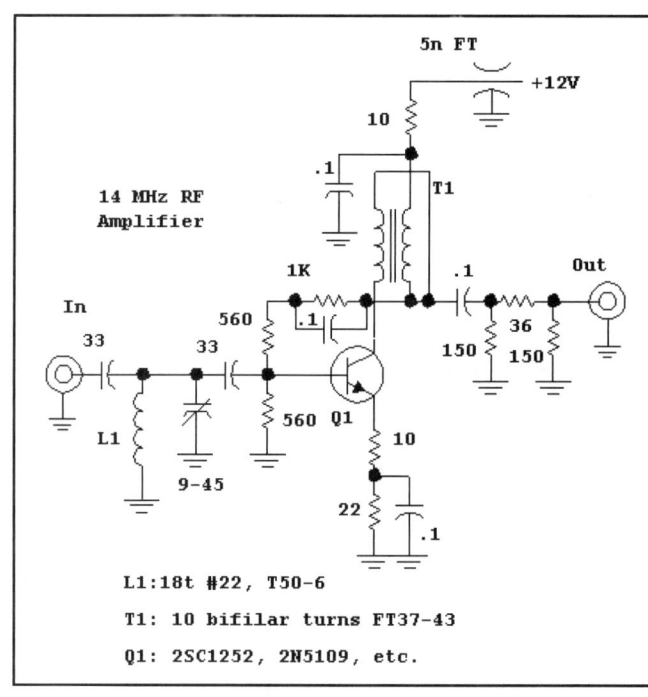

Fig 6.83—RF amplifier with preselector network. This amplifier used parallel feedback from the output tap. Feedback directly from the collector is preferred.

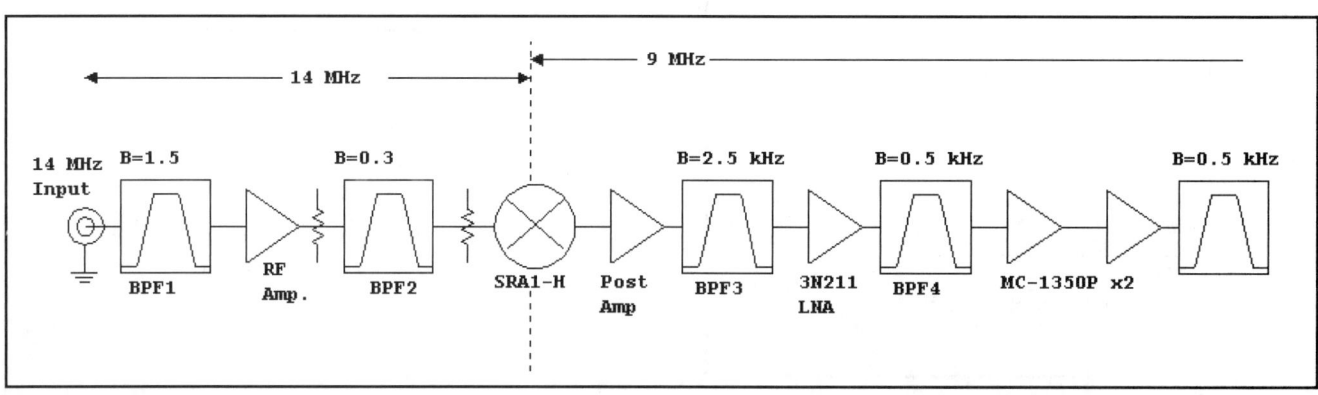

Fig 6.82—Block diagram of an early high-dynamic-range receiver. The various elements are shown in schematics. See text for stage-by-stage discussion.

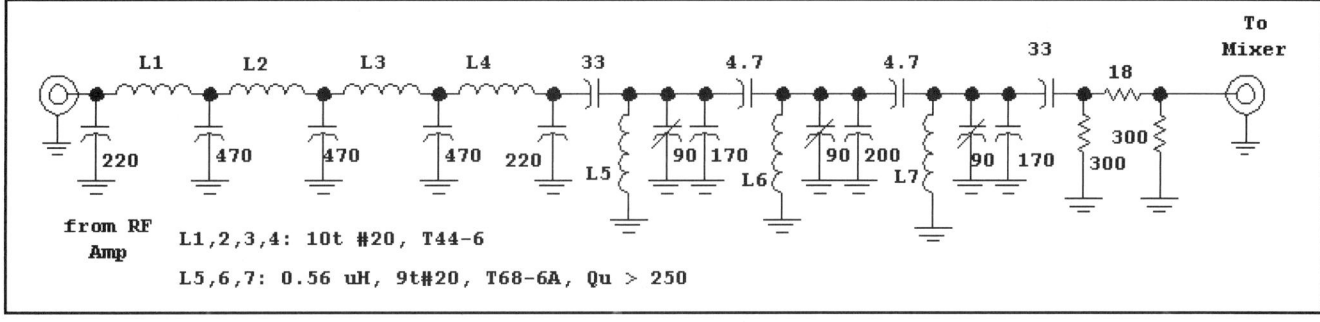

Fig 6.84—Image-stripping preselector filter used with the receiver. This filter provides over 100-dB suppression of images and other spurious responses.

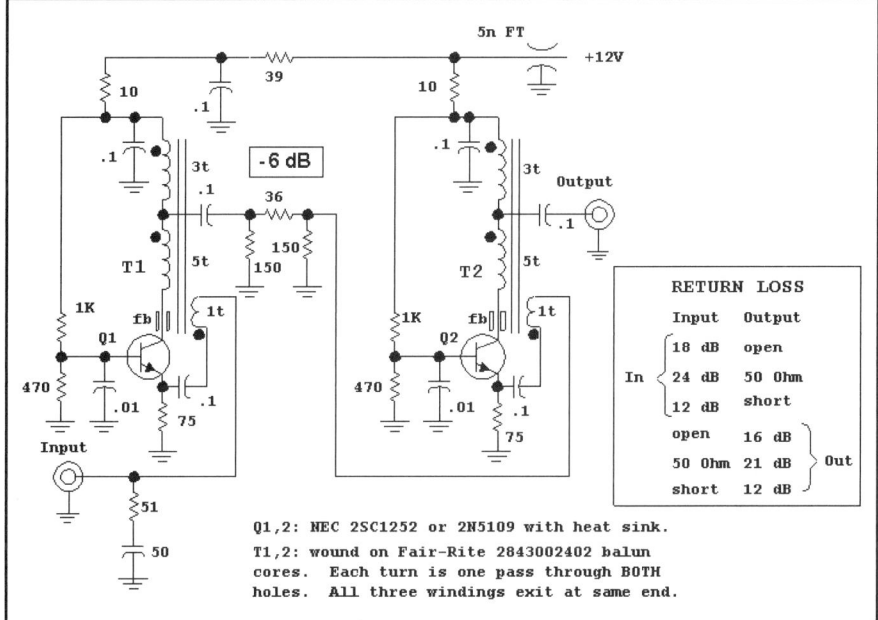

Fig 6.85—Two-stage Norton amplifier used in the CW receiver.

Two-stage Norton Amplifier.

is suitable. This would allow a wider preselector bandwidth with reduced loss, allowing less gain to be used in the RF amplifier, extending dynamic range.

The next front-end element is a post mixer amplifier, shown in **Fig 6.85**. This circuit uses the transformer feedback Norton amplifier topology presented in Chapter 2. That circuit has good noise figure and low IMD, but poor port-to-port isolation. Moreover, the terminal impedances are strongly dependent on the load at the opposite ports. This means that the strongly varying crystal filter input impedances would appear at the mixer output, degrading IMD performance. Placing a pad between two Norton amplifier stages solved the problem here. Overall amplifier gain was 11.5 dB with OIP3 = +42 dBm and NF = 5.7 dB. The individual stages had a 4.1-dB NF. The figure includes measured return loss for the input when terminated in a variety of outputs, and similar results for the output.

Overall front-end gain is low in this receiver. The main crystal filter that this design used was a 10-element circuit with 500-Hz bandwidth, which had a 10-dB insertion loss. The high IL was an acceptable price for the spectacular performance. But receiver NF would be compromised if the IF was driven from the low gain front end. So, a "roofing filter" was used to follow the front end. This lower loss filter with a 2.5-kHz bandwidth was followed by a fairly low noise amplifier that then drove the narrow CW filter. This topology compromises dynamic range for very close tone spacing, but is an otherwise useful technique.

Evaluation of this receiver produced an 8-dB noise figure (MDS = –139 dBm) with IIP3 = +13 dBm for dynamic range = 102 dB and Receiver factor R = +5 dBm. The receiver served as a self-test vehicle during development. The IF system was built and used with an earlier receiver. It then provided the narrow bandwidth needed for IMD measurements. This allowed direct evaluation of mixers, amplifiers, and filters. A key to the

Transmitters and Receivers   6.45

development was the ability to insert attenuators between stages. This then allows the designer/builder to pinpoint the distortion source.

Some interesting details emerged from this investigation. Our first attempts to use the 2.5-kHz roofing filter were frustrated by IMD in the filter, confirmed with the insertion of pads in the system. A new filter from a different manufacture eliminated this difficulty, leaving the mixer as the critical element. The mixer was not *well behaved*, showing better IIP3 when operated at higher levels than it did when IMD products were close to the receiver MDS. Lower level data is quoted.

The receiver was built with the front end segmented into several modules, each in a shielded box and interconnected with coaxial cable. The shielding continues through the IF, BFO, and Product Detector. Power is supplied to the modules via feedthrough capacitors. The 50-Ω interface allows easy measurement of individual modules and quick changes in gain distribution. It also prevents the sorts of interactions and instabilities that can (and usually do) arise when such systems are built in the open. Finally, it provides shielding against radiated and conducted energy from digital circuitry that might be used in other parts of the receiver. Shielding "by the stage" is generally much more important and useful than shielding afforded by one metal box around equipment. This is an old design and duplication is not encouraged.

Fast Forward—Modern Receivers

A more up-to-date front end is shown in **Fig 6.86**, where the incoming signal is converted to a VHF first IF. The design shown is not an example we have built, but one that should be possible with existing technology. It has features not found in earlier designs, but also introduces problems. Up-conversion is typical with most modern gear.

The first IF in this example is 70 MHz with the LO running above the IF. These up-converted designs are usually general coverage receivers, tuning from 50 kHz to 30 MHz. The example receiver uses a 70 to 100-MHz LO injection, generated by frequency synthesis. The input low-pass filter has a cutoff at 30 MHz and establishes image rejection. The image for this example is at the sum of the LO and the IF, 140 to 170 MHz. Images are no longer an issue so long as the low pass filter works as designed.

A bandpass preselector filter is still used in the front end of Fig 6.86. If none were used, the receiver would be subject to overload by signals far removed from the input. On the other hand, it is now practical to keep the preselector bandwidth wide enough that IL is low, which helps to maintain a low noise figure. Common practice uses half octave filters with two bandpass filters for each frequency doubling. This is often approximated with filters of around 5-MHz bandwidth. Narrower filter bandwidth could be useful.

Gains, noise figures, and intercepts are given with critical stages in Fig 6.86. The passive high-level (+17-dBm LO) mixer resembles that of the last receiver with 6-dB NF and conversion loss with an input intercept of +25 dBm. The post mixer amplifier has 12 dB gain, a low noise figure of 2 dB, and IIP3 of +25 dBm (OIP3 = +37 dBm.) Note that this amplifier is actually weaker (lower intercepts) than the post-amp used in the earlier receiver. This is practical, for signals are smaller, a result of using no RF amplifier. (Also, the post-amp in the previous receiver was stronger than necessary!)

Some design rules emerge from these studies: If the output intercept of one stage equals the input intercept of the following stage, each will contribute equally. If one of the two stages is to be dominant, it should have an intercept at the common plane that is 6 dB above the other. Note that these are not "rules-of-thumb," but results of analysis.

Data is included in the figure for crystal filter IL and IF noise figure. The result for this receiver is an overall noise figure of 10.5 dB with IIP3 of +26 dBm and R = +15.5 dBm. In a 500 Hz bandwidth this would generate a two tone DR of 108 dB. The mixer is the critical, performance-determining element defining system IMD.

Although the numbers appear good in this design, there are a couple of details that can severely degrade them. The first is the 70-MHz crystal filter. This element has a bandwidth of 20 kHz, easily realized with today's technology. But with such a wide bandwidth, a tone separation of 50 to 100 kHz would be required to achieve the calculated intercept. The same measurements done at 10 or 20-kHz separation would produce lower IIP3 values.

A second major problem relates to the bandpass filters used in the design. They are typically switched with PIN diodes at the filter *input* and output. Diodes at the input are not protected by the bandpass filters and are then subjected to a wide frequency spectrum. Both second and third-order intermodulation distortion can then generate products that severely compromise performance. Performance can sometimes be improved by increasing the bias current for conducting diodes. The better solution is substitution of improved diodes. The HP-8052-3081 is recommended.[16] The Siemens BAR17 or M1204 are also recommended.[17]

A vital diode parameter is *carrier lifetime*, which should be greater than 2 ms in this application. (Carrier lifetime is a measure of the life of carriers within the diode when reverse biased after a period of conduction.) Some high voltage rectifiers display long enough lifetimes, but tend to be lossy. PIN diodes built specifically for RF switching display lower loss, but only some have the long lifetimes needed for switching at HF and especially MF. The popular MPN3404 and similar devices used in this text are not generally suitable for high DR applications. Diodes need to be measured and characterized for RF performance so they can be included in a system analysis.

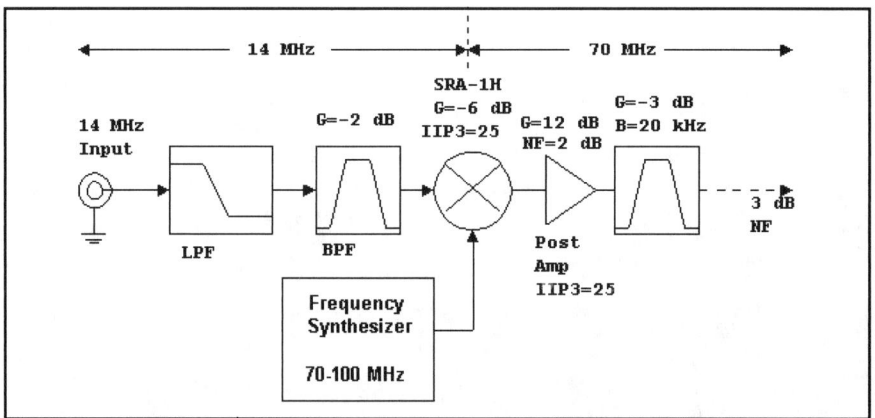

Fig 6.86—Front end typical of modern equipment, although this example is designed for performance beyond the norm. The bandpass filter, shown for 14 MHz center frequency, will have a bandwidth of several MHz and will be switched with relays or PIN diodes. See text.

Another flaw with the up conversion block diagram arises with the VHF crystal filter. IMD in these filters is often worse than seen with lower frequency filters. It should be characterized and considered in system analysis. The filter should have enough selectivity to allow the VHF IF signal to be converted down to a lower frequency IF where additional processing occurs. The conversion should be relatively spur and image free. It is common in current designs to amplify and heterodyne the signal to a low enough frequency that it can be applied to an analog to digital converter (ADC), producing a digital data stream suitable for digital signal processing (DSP.)

Additional distortion sources are found in the low-pass and, more often, in the bandpass filters ahead of the mixer. Filter intercepts depend primarily on the magnetic properties of the inductors used in the filters. They will also depend on the peak energy stored in the component during operation. Running 1 mW of power through a low-pass filter usually results in relatively low current flowing in the inductors used in that filter, so small cores are suitable. But the same 0 dBm applied to a narrow bandpass filter may produce much higher inductor current, producing intermodulation distortion. For example, we have observed in-band IIP3 of approximately +30 dBm for a three-resonator 10-MHz filter with 300-kHz bandwidth. Changing from T37-6 to larger T50-6 cores increased IIP3 to about +50 dBm. We have also observed severe IMD with inexpensive slug tuned coils. As with all things related to high DR equipment, meticulous measurements should replace lore.

Moving toward higher Dynamic Range

The front ends described can be extended to provide even better performance by substitution of improved circuit elements. Primarily, the high-level mixer can be improved. Higher-level diode rings are available, some using up to $1/2$ W (+27 dBm) LO power. With another 10 dB of LO comes a similar increase in IIP3.

Perhaps the more appealing mixers are those using FETs. They are capable of very high intercepts, have IL similar to the highest-level diode mixers, but require little LO power. This does not imply though that LO drive can be treated with casual abandon. Passive FET mixers usually have LO signals applied to the gates. They must be driven hard to ensure fast switching; symmetry is critical to preserve balance. The FET ring popularized by Oxner is capable of IIP3 up to +40 dBm or a bit higher with conversion loss around 8 dB. Makhinson,

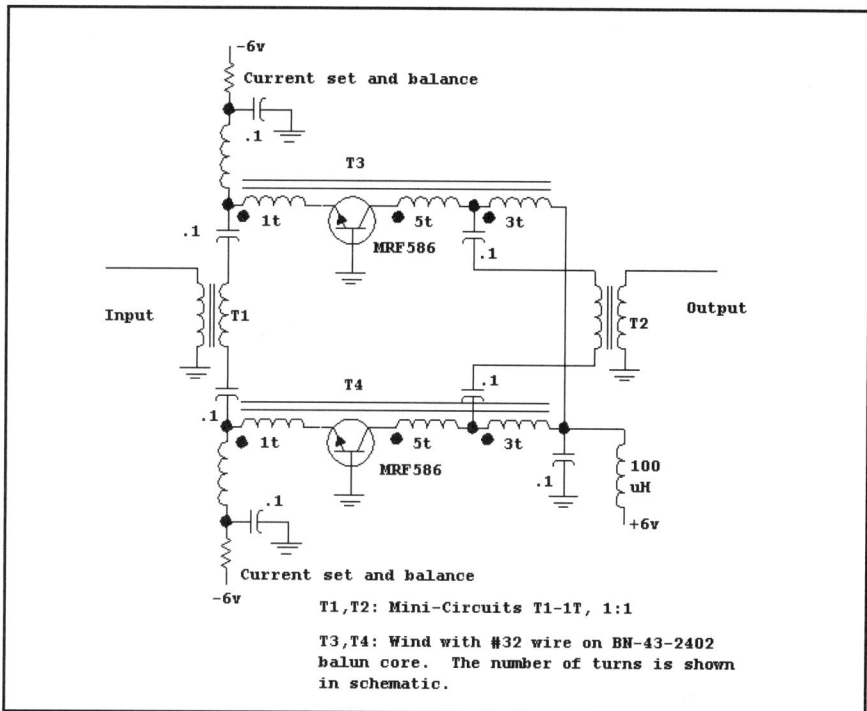

Fig 6.87—High-performance post-mixer amplifier. The transistors were biased to 40 mA each for OIP3 = +48 dBm. Dual power supplies are used for amplifier bias. This amplifier represents very good performance that we have not duplicated.

N6NWP, reported 7 dB loss with square wave LO drive.[18] The mixer of greater interest is the H-mode mixer generated by Horrabin, G3SBI.[19] IIP3 of +55 dBm was reported with a conversion loss from 8 to 9 dB when using the same FETs as applied with the Oxner mixer. A simplified version will be described later featuring IIP3 > +40 dBm with loss at 5 dB.

One of these high intercept mixers may well have OIP3 of +35 dBm or higher. To be dominant, post mixer amplifiers should have IIP3 of +40 dBm or higher. This moves the output intercepts into the +48 to +52 dBm range. Such amplifiers are possible with very high currents, or with modest currents and careful design. **Fig 6.87** shows the amplifier used by Makhinson in his receiver. Two Norton-type transformer-feedback amplifiers are used in push-pull to achieve a gain of 8 dB with OIP3 of +48 dBm and NF = 2.5 dB.

Colin Horrabin built a version of this amplifier with improved performance. He shifted to a single ended power supply, but increased current to 60 mA per transistor. He changed transistor type to the MRF-580A and added ferrite beads to the collectors for stability considerations. Transformers were hand wound on balun cores and the transistors were heat sunk to a copper substrate. He obtained the spectacular results of OIP3 = +56 dBm with Gain = 8.8 dB and noise figure under 1 dB! The amplifier was at the limits of his NF measurement capability, and IMD determination was also stressed. He also reported that transformers had to be selected for lowest IMD. Nothing is casual at this performance level.[20]

In a later variation of his earlier amplifiers, Makhinson used a push-pull pair of Norton feedback amplifiers that drove a differential pair of common base amplifiers. The second stage common-base circuit provided good reverse isolation while the input transformer feedback design afforded low noise. The lower second-stage reverse isolation generated an input impedance independent of output termination for the two-stage design.[21]

Another approach to balanced amplifier design is that of Engelbrecht, shown in **Fig 6.88** (see Chapter 3.)[22,23] The incoming signal is split in a 3-dB quadrature coupler. The two hybrid outputs are then 90° out of phase with each other as they are applied to the amplifier inputs. If the impedance match at amplifier #1 is less than perfect, there will be a power reflection. The action at the input to amplifier #2 will be identical, for the amplifiers are identical. Each reflected component undergoes another 90° reflection as it progressed back to the input. The two reflected components are 180° out of phase with each

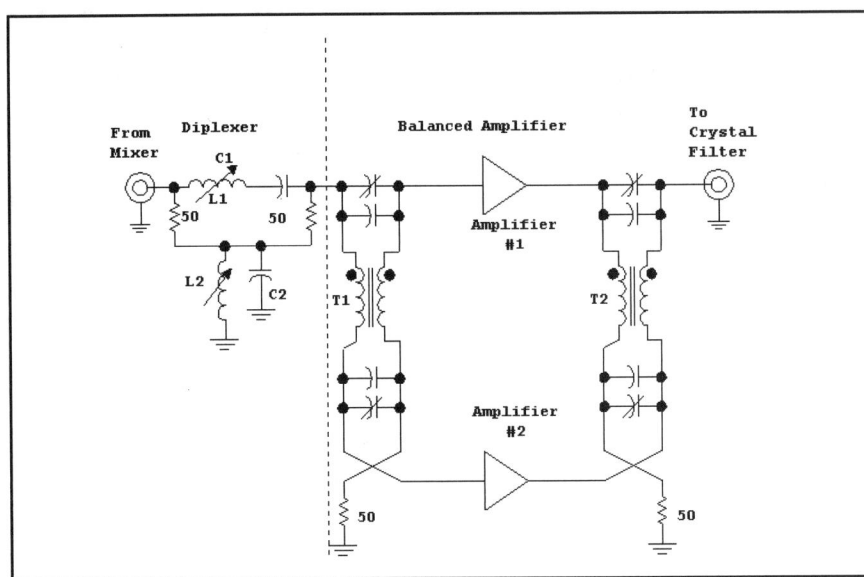

Fig 6.88—Balanced amplifier method of Engelbrecht. See text for discussion.

other by the time they reach the input, so the input impedance is always 50 Ω.

The coupler of Fig 6.88 generates a 90° phase shift at all frequencies, but equal output amplitudes at only one crossover point. A bandpass/bandstop diplexer provides a termination at all frequencies far from the design center. Depending on the nature of the crystal filter, a diplexer may be useful at the output port as well.

Front Ends Without Early Amplifiers—The Triad Receiver

The up-conversion system of Fig 6.86 is a child of compromise, illustrating the tradeoffs often taken to achieve general coverage. The ability to tune the entire HF spectrum was once considered a performance virtue. It is now, since the advent of WARC bands, merely an economic ploy. The aggressive designer/builder need not adhere to such guidelines. He or she can configure a system that will offer high performance on a few selected bands. The IF can be at HF where crystal filters can be narrow without severe loss and with low IMD. Preselector filters with only modest loss can be used with the best available mixers.

The problems with post mixer amplifiers remain. The ideal solution is to merely eliminate them. This can be done with a switching-mode mixer if a crystal filter with constant, frequency flat input impedance can be applied. Such a block diagram is shown in **Fig 6.89**. The circuit is the result of several years of collaborative effort on the part of Bill Carver, W7AAZ, Harold Johnson, W4ZCB, and Colin Horrabin, G3SBI—collectively referred to here as the *Triad*.[24]

The Mixer

The key element in this receiver is the H-mode mixer shown in **Fig 6.90**. The basic mixer was presented in Chapter 5. This example uses a readily available and inexpensive quad-MOSFET-Bus Switch, the Fairchild FST3125M. The device is also available from other vendors. (This part was suggested to the Triad by Giancarlo Moda, I7SWX.) The H-mode mixer is one with RF applied to a transformer, T1, which generates a balanced source of RF. The two resulting signals are then applied to the center taps of transformers T2 and T3. Four FETs connect windings to ground in pairs. Two IF outputs are generated on the secondary windings of T2 and T3.

The FST3125M uses a 5-V bias, required by the quad logic inverters included in the IC. The FETs and related transformers are biased at half this supply with a resistive divider. Symmetry is emphasized in the construction method. A sandwich of two circuit boards contains the mixer, diplexer and following crystal filter described below. The mixer chip is on the lower board while the diplexer and filter are on the upper one. The three transformers actually reside between the two boards, serving as the routes from one to the other and back.

The digital portion of the mixer circuitry dealing with the LO is shown in **Fig 6.91**. A signal of +10 dBm is applied to the mixer board at *twice* the desired LO frequency. It is converted to a digital form with two NAND gates (74AC00) and is then routed to a divide-by-two circuit using a 74AC109 J-K flip-flop. The flip-flop contains an inhibit input, which is driven by the remaining NAND gates, providing a convenient means for turning the LO off. This may be used during receiver mute periods or as a noise blanking input. This method of blanking is especially effective, for it is out of the main signal path and has few of the distorting effects usually related to the blanking function, other than that intrinsic to modulation.

An alternative logic section is presented in **Fig 6.92**. This scheme uses a VHF local oscillator that is then divided by any even integer from 4 to 18. This method is used in the W7AAZ version of the receiver.

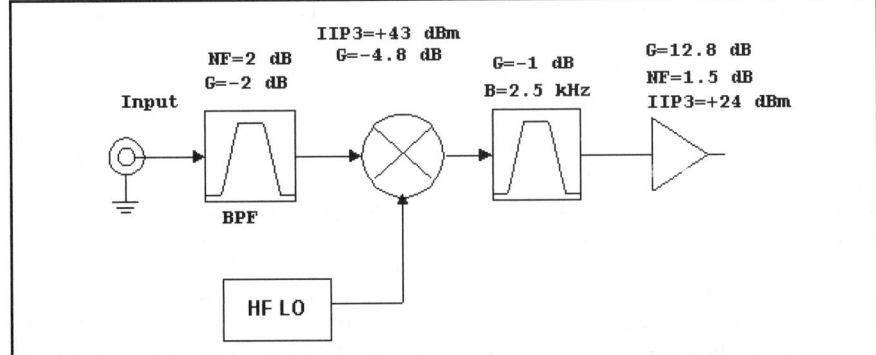

Fig 6.89—Receiver front end using no amplifiers before initial selectivity is obtained. This is the basis of the W7AAZ/W4ZCB/G3SBI receiver described below.

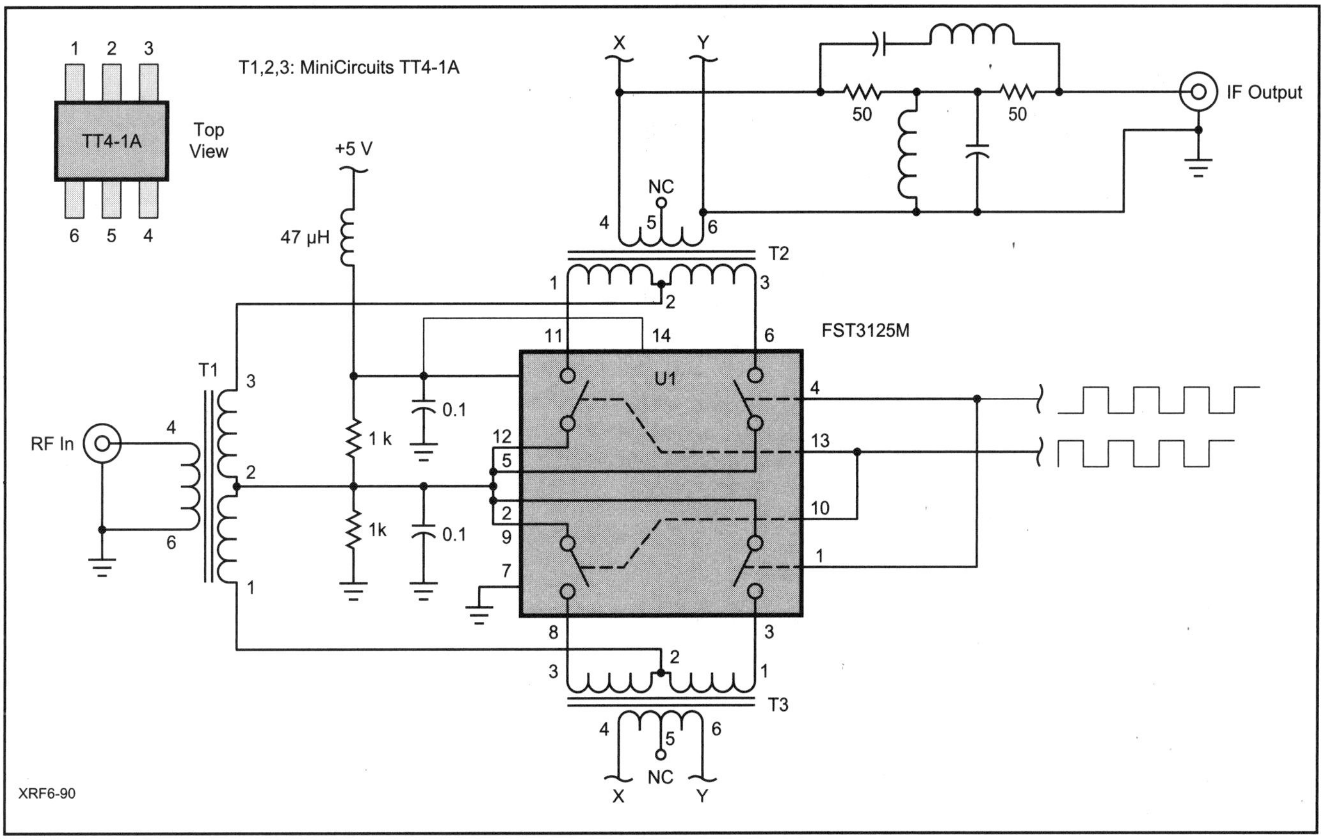

Fig 6.90—Mixer portion of the high-level front end. Commercially available transformers are used in this design. U1 consists of four MOSFET switches controlled by lines 1, 4, 10 and 13, linked with the dotted lines in the figure. See Chapter 3 for design of the Q = 1 diplexer at the IF port for compatibility with the chosen IF.

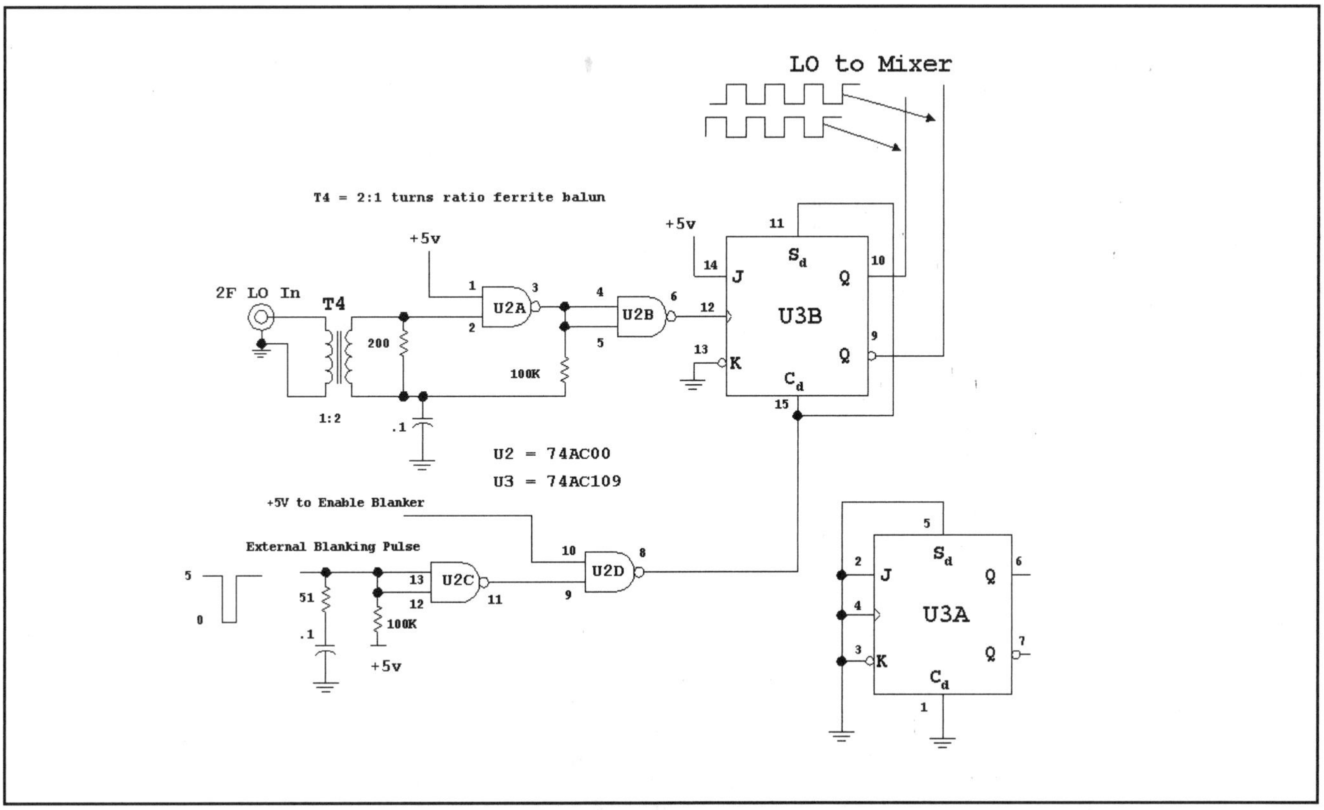

Fig 6.91—Logic circuits provide high-frequency LO drive for the H-mode mixer. Input is at twice the needed LO frequency. The designer/builder must add power supply connections to the ICs.

Transmitters and Receivers 6.49

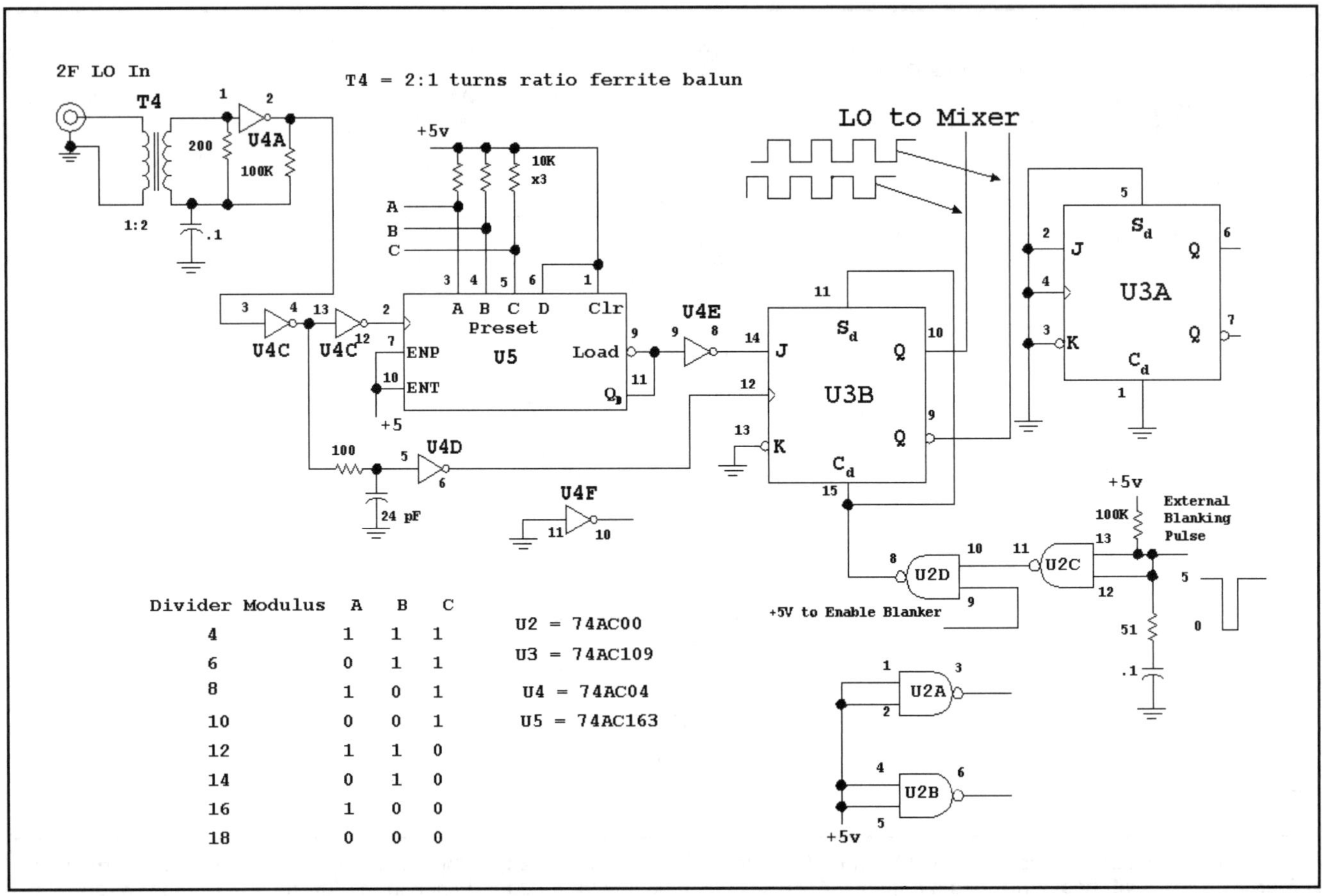

Fig 6.92—Logic circuits accept an input from a VHF synthesizer. The output is then divided by an even number between 4 and 18 before reaching the high-level mixer. The designer/builder must add power supply connections to the ICs.

The Roofing Crystal Filter

A poor mixer termination will severely degrade IIP3. A filter with a 50-Ω input impedance at all frequencies, inside and outside the passband, is shown in **Fig 6.93**.

The crystal filter is a critical element in the overall front end and requires careful design and adjustment by the designer/ builder. The crystal frequencies are picked to produce a passband that overlaps that of the dominant filter in the receiver IF system, measured before this filter is built. The crystals will then be ordered from a reliable supplier. High crystal Q should be sought, for it will directly impact filter IL. The builders saw their best filters with loss under 1 dB with others under 2 dB. Even if the receiver is to be used mainly on CW, a wider design filter bandwidth is used in the interest of low loss.

Careful measurements are required to adjust this filter. A spectrum analyzer with a tracking generator is ideal, but should have stability commensurate with narrow crystal filters. Sweeps measuring input and output impedance match should, however, extended from near dc to VHF.

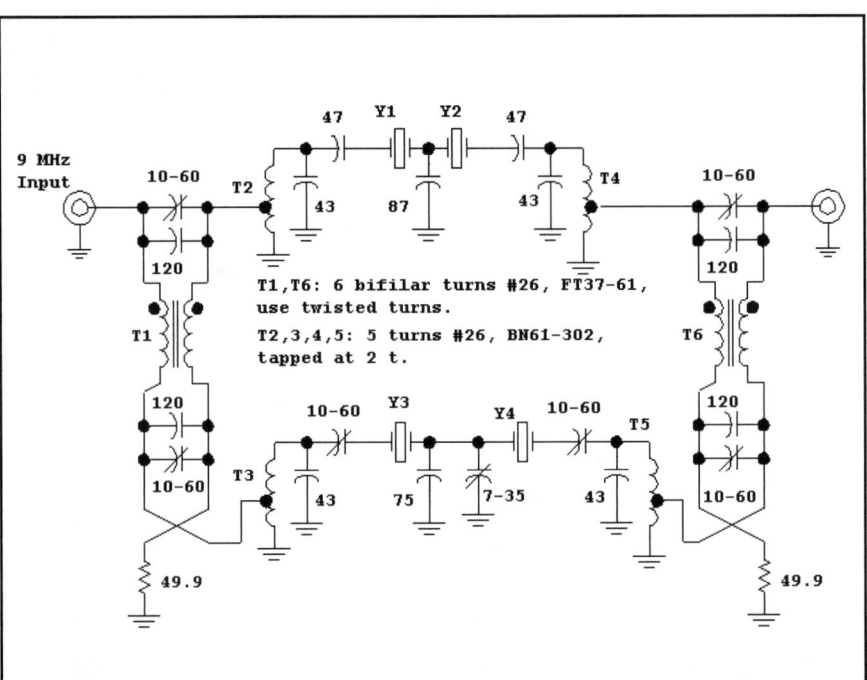

Fig 6.93—Crystal filter serving a "roofing" function. This circuit operates at 9 MHz, but can be redesigned for other frequencies within the HF spectrum. The variable capacitors with Y3 and Y4 are adjusted to match that filter to the one using Y1 and Y2. The quadrature hybrids are adjusted for optimum impedance match at both ports. See text.

An Amplifier to follow the Roofing Filter

Fig 6.94 shows the amplifier that follows the mixer. This circuit must have reasonable performance, although not as stellar as would be needed without the filter. With only two crystals per side, the roofing crystal filter has limited skirt selectivity, allowing some large signals to appear beyond the filter.

The amplifier is a feedback circuit with four parallel JFETs. The total current is high at 85 to 100 mA, so the circuit has good distortion performance. The circuit began conceptually as a transformer matched common-gate amplifier; a topology with a well-defined, low input impedance.[25] A winding is added to the transformer to apply some signal to the gate. The result is a circuit that has neither terminal as common, yet has a well-defined 50-Ω input impedance while featuring low noise figure. This circuit has a typical NF of 1.5 dB with some versions measuring 1.2 dB. The output is transformer coupled with a drain load resistor to ensure a good output match.

Bill Carver, W7AAZ, modified the bifilar output auto-transformer with another winding that drives an adjustable capacitor, C-N, to couple energy back to the gate. This capacitor is adjusted for low reverse coupling. The result is a neutralized amplifier featuring low noise, high IIP3, excellent input and output impedance match, and good reverse isolation.

This circuit can be adjusted for an input return loss greater than 30 dB in the 3 to 30-MHz region. Typical gain is 12.8 dB with IIP3 = +24 dBm. A heat sink is built for the four FETs by drilling four holes in a piece of 1/8-inch-thick aluminum. The FETs are pushed into the holes, which are then filled with epoxy. Carver has also built similar amplifiers with six FETs, but the same 100-mA total current. These circuits require no heatsink.

The Preselector

The final element in the front end is the preselector filter. The basic form is shown in **Fig 6.95**, a top coupled set of parallel resonators. Reed relays are used at each end for band switching. Extensive decoupling (not shown) is used with the relays. The filters were designed to have a maximum insertion loss of 2 dB. A 5-resonator filter was used for 160 m while 3 or 4 were sufficient for the other bands. Toroids were used for all inductors with emphasis on larger sizes for high unloaded Q and low IMD. A 6 mix was used for the lower bands with 10 for the upper ones. Most capacitors were 1% silver mica types. The only variable capacitors were some trimmers used for coupling on the highest bands. Components were carefully measured prior to installation and inductor turns were spread or compressed slightly for fine-tuning. This was suffi-

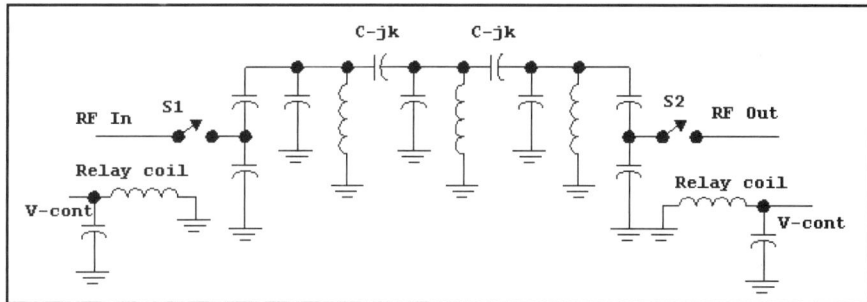

Fig 6.95—General form of preselector filters used for the high-performance receiver. While a 3-element filter is shown, some bands used up to 5 resonators.

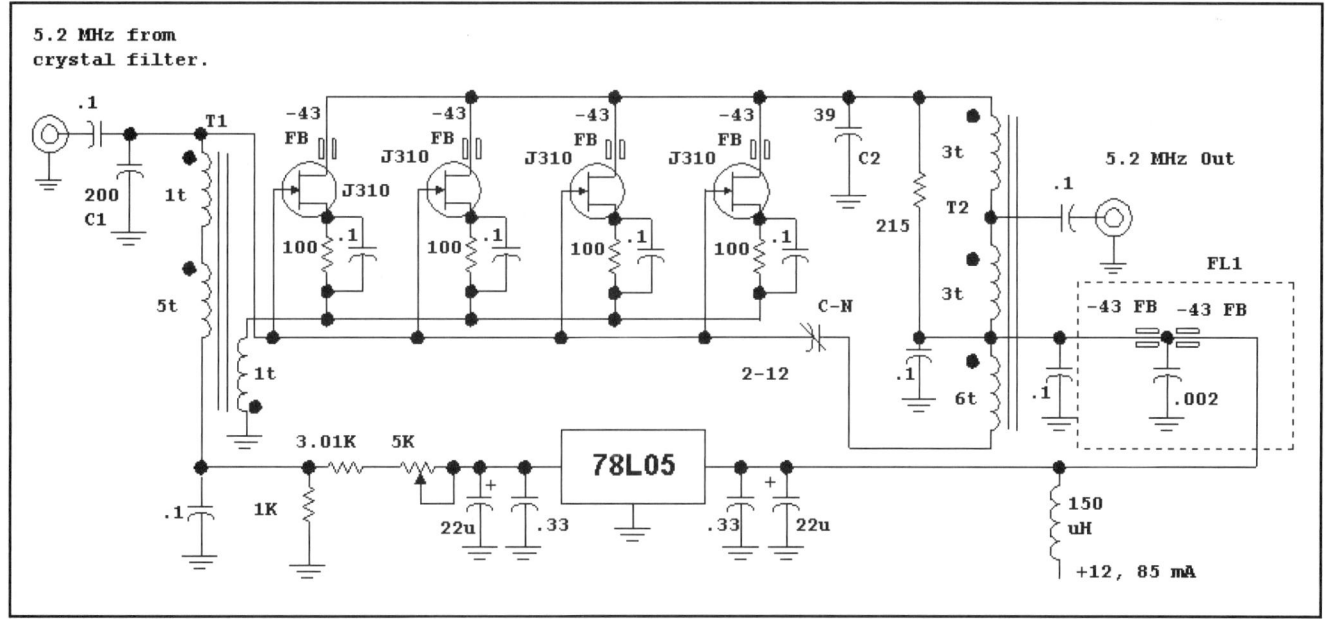

Fig 6.94—Amplifier that follows the roofing crystal filter. This particular version operates at 5.2 MHz, but can be optimized for any frequency in the HF spectrum. T1 is wound on a BN61-202 two-hole balun (binocular) core. The primary (grounded winding) is made from small copper or brass tubing through the balun holes. Alternatively, braid from RG174 coaxial cable may be used. The 5-turn and 1-turn windings are then wound with #28 or smaller wire. T2 consists of a pair of bifilar windings on a BN43-202 two-hole balun core. One bifilar winding forms the two 3-turn windings while the other bifilar pair is connected to form the 6-turn winding. Remember that one turn on a two-hole balun core is a pass through both holes. C1 and C2 are approximately resonant with transformers T1 and T2. FL-1 is a three wire monolithic element, but can be built with discrete components. C-N is adjusted for best reverse isolation (lowest S12.) All resistors are 1% metal film, 1/4 W.

cient for the lower bands while the Dishal method was applied for the upper frequencies.[26,27]

The design goal for the preselector filters was a stopband attenuation of 90 dB or more. This was realized, but it required considerably more effort than anticipated. The filters were all built on boards with components in a long narrow line for best input to output isolation. The stopband performance was only realized after the on-board grounds were isolated. Each resonator was grounded directly to the large metal plate that supported the boards. It was also important to carefully place the various filters in the stack. A situation to avoid was an adjacent filter that operated at an image. For example, if the receiver used a 5-MHz IF with LO at 9 MHz, the 4-MHz image is 14, so the 80 and 20-m filters should not be next to each other. Details of construction are shown in the photograph. This is yet another place where detailed measurements are required.

21 MHz bandpass filter used in W7AAZ version of the Triad Receiver. No variable tuning capacitors are used. The trimmers adjust coupling.

An Oscillator

A voltage-controlled oscillator developed by Harold Johnson, W4ZCB, is presented in **Fig 6.96**. It has been applied in a number of ways including acting as the controlled oscillator in experimental synthesizers and one-on-one phase-lock loops. The circuit operates in the 80 to 110 MHz region and is then divided from VHF in the circuit shown earlier in Fig 6.92.

The heart of the VCO is a helical resonator. This element offers an unloaded Q of 700, performance difficult to obtain at HF. A metal lathe is needed for the construction. The resonator is housed in a section of 1.5-inch-diameter copper tubing with copper-pipe ends. The helix consists of 9 turns of #12 wire wound on a 0.75-inch diameter tubular form that was machined from RF grade polystyrene rod. After the rod was machined to 0.8 inch outside diameter, threads were cut at an 8-turn-per-inch pitch. The inside of the rod was then removed with a large drill bit, leaving a wall thickness of approximately $1/8$ inch. Material was retained at one end for mounting. The #12 wire was wound and approximately spaced before being threaded onto the form.

The helix has two taps. Output is extracted from one $1/4$ turn up from ground while the drain is attached at $1/2$ turn from ground. The outputs are buffered with a quad buffer. One output drives the mixer while the other is for synthesizer use. Detailed information regarding tap placement and resonator construction is given in a note from W4ZCB included on the CD that accompanies this book.

Two different methods were used for phase noise measurement. In one, the VCO under test was phase locked to an HP-8640B signal generator. The baseband output was filtered, amplified, and analyzed with an HP-312 selective voltmeter. The other system uses the HP8640 as a local oscillator with a high level mixer. The output is applied to a narrow crystal filter. The signal is amplified and further filtered, and is then detected. The two systems offer good agreement.

This oscillator, after division by 8, provided phase noise of –155 dBc/Hz at a 20 kHz spacing. The noise dropped to –163 dBc/Hz at 50 to 75 kHz; at 100 kHz it was beyond the range of the measurement equipment. A one-on-one PLL will provide some close in clean-up. Thermal stability was good enough to allow direct use without any stabilization, although this is not common and should not be expected with similar designs.

The Overall Triad Receiver

We have described the receiver front end, the portion that generates the wide dynamic range. The four-FET amplifier (Fig 6.94) is normally followed by the major crystal filter used in the receiver. The bandwidth and performance vary with the members of the Triad. The main IF system is the design offered by Carver in *QST* for May, 1996, a circuit based upon the Analog Devices AD600. The rest of the receiver is standard, although DSP en-

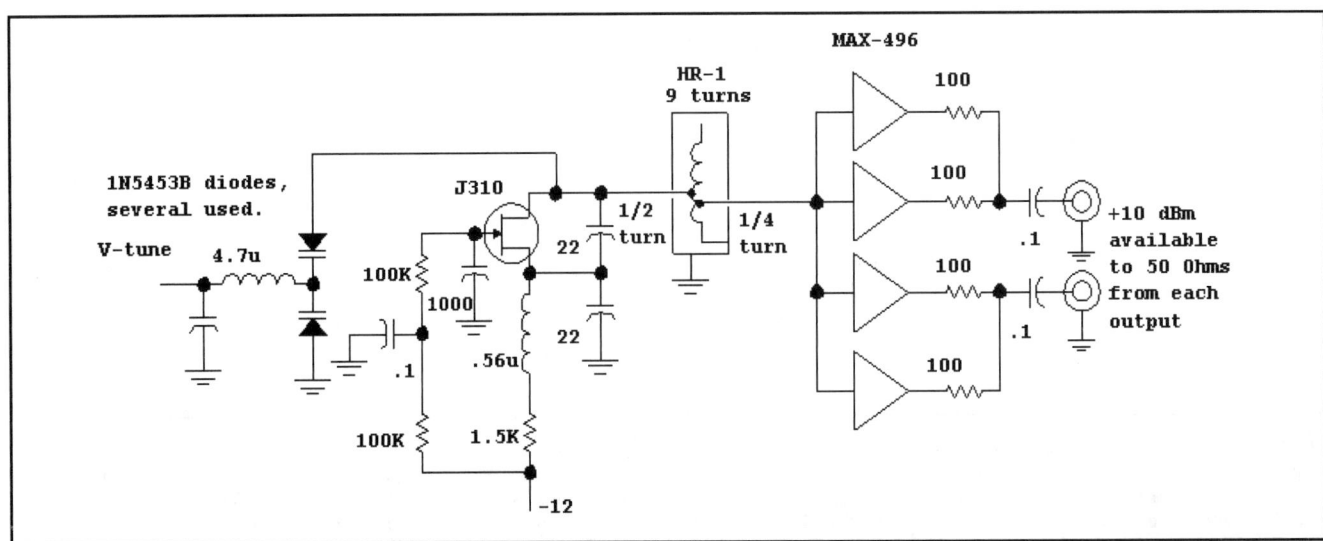

Fig 6.96—VHF helical-resonator voltage-controlled oscillator. See text for additional detail. Although a back-to-back pair of varactor diodes is shown, more may be required. It may also be useful to switch extra capacitance into the circuit with relays or PIN diode switches.

A working version of the Triad built in the UK. (TNX to George Fare, G3OGQ.)

hancements are planned. The plans also call for full transceive capability.

The receiver performance has been outstanding with different triad members having obtained slightly varying results. With careful adjustment of the preselector and post filter amplifier, slightly under 10-dB noise figure has been measured in a receiver also showing an input intercept of +45 dBm. This is slightly under the early goal of achieving a 120-dB DR in an SSB bandwidth, but the ease of duplication of the FST3125 mixer makes it preferable over one using the Si8901. That part had a 3-dB higher conversion loss, making it impossible to achieve a 10-dB noise figure without an amplifier in the "wide open" part of the front end. The present system with +45 dBm IIP3 and 10 dB NF (R = +35 dBm) will yield DR of 121.3 dB in 500-Hz BW.

There are some dramatic implications embedded within this work, ones that may well alter the way we design the next generation of receivers. It is clear that a lossy mixer can be followed directly by a narrow filter without compromising large signal performance. Use of the Engelbrecht technique is not new with filters, but it has not been routinely applied for experimenter equipment. The methods will work just as well with diode mixers as with FET mixers.

The typical high dynamic range receiver of recent vintage has consumed considerable power. This was generally accepted as the price one must pay for such performance. FET mixer based designs can, however, provide very high intercepts without high power. The oscillator powers are low, and with no early amplifiers, there is no compelling reason to use a high power amplifier anywhere in the system, especially if higher order, low loss roofing filters can be designed. Low loss and simplified matching should be possible with monolithic filter technology. We can now envision a very high dynamic range receiver that is as sensitive as we will ever need on the HF bands that operates efficiently with batteries.

But adequate challenge remains. The frequency synthesis problem continues to plague us. We certainly want new transceivers to include all of the refinements found in the older ones, and most of these features depend on frequency agility. The high phase noise of casual PLL synthesizers will drastically limit the performance. While somewhat better wideband phase noise is available from DDS, this is of little consolation when the noise is merely replaced by numerous coherent spurious responses. Some experimenters expect exciting things to happen in synthesis in the near future, which will help.[28]

But synthesis is not the major problem we face. Rather, it is the compromised nature of the transmitters that we usually encounter. It does little good to build a receiver that is so free of distortion that we become concerned about receiver damage when we measure it, only to find that the on the air signals we encounter are distorted.

Modern communications systems have been engineered with a sense of balance, using compatible transmitters and receivers. The receives have kept pace with the transmitters, but with little extra margin. The radio amateur service has not, however, grown in this way. Early stations had separate equipment for each function. We have had a DX based fetish for receivers, traditionally dealing with the classic axiom that "if you can't hear 'em, you can't work 'em." This left us ignoring our transmitters.

Many solutions to transmitter problems are found in the receiver design details. Improved receiver synthesizers will benefit our transmitter. High-level mixers, low-distortion amplifiers, and clean filters are elements common to both. The problem unique to the transmitter is in the higher power stages where distortion usually occurs. Even here, there is new technology that offers solutions. *Feed forward* methods offer one route to reduced IMD.[29,30,31] Feedback and predistortion offer alternative routes.[32,33] Predistortion is discussed, with references, in Chapter 10.

6.6 TRANSMITTER AND TRANSCEIVER DESIGN

System Considerations; Transmitters with Mixers

A block diagram for a simple CW transmitter was presented at the beginning of this chapter, Fig 6.18. In the simplest form an oscillator is amplified, low pass filtered and applied to an antenna. The more elaborate scheme uses a frequency multiplier, allowing the use of a lower frequency oscillator, isolated from the higher power amplifiers later in the system. These represented the simple equipment that many of us used as we began our experimental efforts in radio. It remains a good design. Even with frequency multiplication, the only spurious responses are either harmonics of the output, or harmonics of the lower frequency oscillator. The former are substantially reduced with suitable low pass filtering while the latter are reduced through bandpass filtering immediately after the frequency multiplier.

The best frequency multipliers are those with balanced circuitry. Appropriate circuit symmetry will suppress the fundamental and some undesired harmonics. For example, a push-push doubler, a balanced circuit with two diodes, will suppress the fundamental drive component in the output by 30 to 40 dB. Selective circuits afford additional suppression. Multiple resonator filters are recommended over single tuned circuits.

We can calculate the performance of low pass filters that might appear in a transmitter output. **Table 6.1** shows the suppression at the second and third harmonics of a carrier that is passed through a low-pass filter with a cutoff frequency 10% above the input frequency. The fil-

Table 6.1

N	Attenuation at 2f	3f
3	10 dB	21 dB
5	30	50
7	51	79
9	72	108

ters were designed for a 0.1-dB-ripple Chebyshev response. Filters with 3, 5, 7 and 9 components are considered.

The simpler filters are poor performers. The N = 3 low pass with two capacitors and one inductor offers surprisingly little harmonic attenuation. Other passband ripples may enhance performance slightly, but the dominant effect is just the number of components.

The more common transmitter block diagram, Fig 6.19, uses two oscillators heterodyned together in a mixer to produce the desired output. A bandpass filter is again needed to select the desired output component while suppressing the image as well as various spurious products. While frequency multiplier balance enhanced performance, a balanced mixer does nothing to suppress an image. The filter must now do all of the work. Frequencies should be chosen wisely.

Although we occasionally see a heterodyne transmitter using nothing more than a single tuned circuit, two or three resonator filters offer much better performance with only slight added complexity. Intuition suggests that the added insertion loss of a third order filter would complicate design. But one can increase bandwidth with a triple tuned filter to realize the same loss with greater stability, better stopband attenuation, and ease-of-alignment. Some special cases, such as VHF applications demand even higher order filters.

An often abused, sensitive parameter is mixer drive level. A normal diode ring (+7 dBm LO) should generally be driven with an RF input less than –10 dBm. Third-order IMD is not excessive at this level (important in SSB transmitters) and high order mixer spurious products are low. However, spurious products grow at an alarming rate with greater RF drive.

Mixer drive level should be established through careful measurement. Even if the builder does not have a high frequency oscilloscope or spectrum analyzer, he or she can always build and use a low-level power meter, often used with a step attenuator. See the measurement chapter.

A high level (+17 dBm LO) diode ring functions well with an RF drive of 0 dBm. Higher-level mixers are capable of even greater drive. Diode mixers are usually 50-Ω parts and are aligned with substitutional measurements, outlined in the measurement chapter. A Gilbert Cell mixer (NE602, MC1496) is usually a high-input-impedance circuit. It operates with a single-ended local oscillator level of 0.3 to 0.6 V, peak-to-peak, usually established with an in-situ (*in place* within the circuit) measurement. This is measured with a 10× 'scope probe attached to the LO or RF input of the mixer IC. The measurement may also be done with an RF probe and high impedance dc voltmeter, although this measurement is rarely as accurate owing to levels that crowd diode thresholds. The allowed RF drive can be 0.3 V peak-to-peak for a Gilbert Cell used in a CW transmitter, also established with an in-situ measurement.

Transmit mixers are best driven with harmonically clean sources. It is often worthwhile to low pass filter the LO input to a diode ring mixer, mainly for reasons of waveform symmetry. Excess even-order harmonic distortion may unbalance the mixer. The clipping action of the mixer diodes will convert a sine wave drive into a square wave, rich in odd-order harmonics. The RF input signal should be low in harmonics, for they can mix to generate spurious outputs. The usual diode mixer does not generate these harmonics in the same abundance that it does odd-order LO products. Similar arguments apply to Gilbert Cell mixers.

The levels recommended are derived from our observations, and could vary with different mixers. Mixers in SSB equipment are driven at an RF level dictated by IMD requirements while mixers in CW rigs are only constrained by spurious outputs far from the desired output. These spurious products can and should be reduced with filtering, but that is not possible with the closely spaced IMD products in SSB. The levels given are conservative results based on our results. Clearly, spectrum analyzer measurements are always preferred over simpler power level determinations.

Linear Power Amplifier Chains

Design begins with a pair of equal IF signals, or two tones. Recall that the *peak envelope power* (PEP) of two identical signals or tones is 6 dB above one of the tones. The output from a normal (+7-dBm LO) diode ring mixer driven with RF = –16 dBm per tone is –23 dBm per tone, or –17 dBm PEP. A typical bandpass filter might have a 3-dB insertion loss, producing a –20 dBm PEP output. Assume this will be used in a transmitter with a 10 W PEP output (+40 dBm PEP or +34 dBm/tone). The output low pass filter usually has negligible insertion loss, so a net gain of 60 dB is required. This can be obtained with three stages, although four, each using negative feedback, would be preferred, especially if wide bandwidth was needed.

Design of the amplifier chain is based upon cascade intercept calculations if SSB or other linear modes are planned. Assume our design goal is IMD at least 40 dB below each output tone (46 dB below PEP) during two-tone transmitter testing. Each output tone will be 6 dB below PEP, or 2.5 W (+34 dBm) per tone. The related IMD must then be over 40 dB lower at –6 dBm per tone. The required output intercept must then be half of this ratio, or 20 dB above the output, +54 dBm. Such levels are obtainable with high-level class-A amplifiers. The block diagram for this amplifier chain is shown in **Fig 6.97**. We have assigned the gain-per-stage values shown across the top of the figure. The intercept values for the individual stages were then adjusted to meet the specification. The final calculated result of OIP3 = +54.2 dBm is less than the value

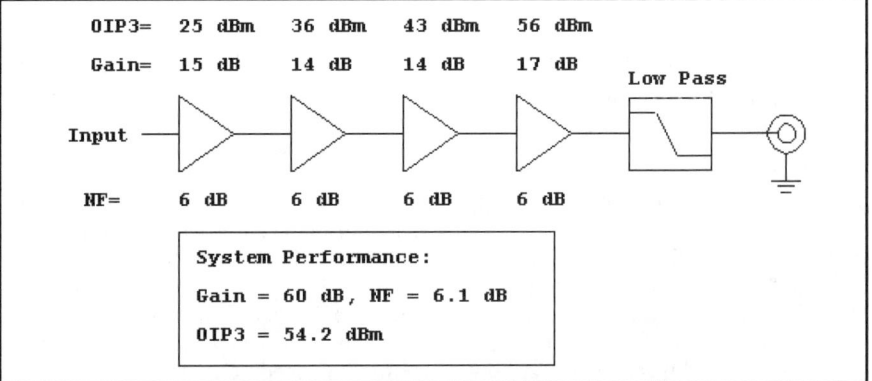

Fig 6.97—Individual stage parameters are combined for a cascade of four stages in an amplifier.

for the output stage itself of +56 dBm, allowing some of the distortion to occur in earlier stages. Increased output stage gain would relax the required earlier stage performance, but would reduce the margin for applying feedback in that stage. As in any practical design, this one is a collection of trade-off factors.

Noise figure is also calculated for the cascade, 6.1 dB based upon an assumed NF of 6 dB for each stage. If we assume a moderately low noise IF followed by a 10 dB loss in the mixer and bandpass filter, the output noise is essentially that of a resistor attached to the amplifier input. That noise is –174 dBm in a 1 Hz bandwidth. Adding 6.1 dB for the NF and 60 dB for gain, the wide band output noise density is –107.9 dBm/Hz. If this noise was to be sampled in a receiver with a 500-Hz bandwidth, total power would be –80.9 dBm. This is a very low power and would probably not be a problem for others using the same frequency. However, if another 20 dB of gain was added, bringing the output to 1000 W, the noise would be at –61 dBm. This noise would drop into the background at a distance, but could be troublesome for other stations in close proximity. This is a common difficulty with many stations in close proximity.

Transmitted phase noise is usually (much) greater than broadband amplifier noise. Consider a poorly designed transmitter with a synthesized LO generating phase noise of –120 dBc/Hz spaced 20 kHz from the carrier. If the carrier is amplified to a level of 1000 W (+60 dBm), the transmitted phase noise has a density 120 dB lower, or –60 dBm/Hz. If received with a 500-Hz-wide receiver, the noise is –33 dBm, or 0.5 µW. A low power transmitter of this level would probably not be heard at any distance, but can be copied by stations within a mile. The noise closer to the carrier will be much more evident.

The individual stages in the cascade of Fig 6.97 could be simple feedback amplifiers, biased to a high enough current that the individual stage intercepts are realized. The stages should present input and output impedances that match the adjacent stages, especially when wide bandwidth is desired. One may be more cavalier for a single-band CW design, although matched feedback amplifiers are still preferred, for they tend to preserve wideband stability. The emitter degeneration may be adjusted in a single band CW design to alter stage gain as needed for the desired output power. This practice should be used with more care when dealing with SSB.

A Class-A RF power chain can generally be built on a single board, for gain is modest. However, the board should end in a stage of around 1 to 10 W output. Higher-powered amplifiers should have separate power supply lines and an isolated thermal environment. A straight-line layout is recommended, separated from the bandpass filter that would normally follow the transmit mixer.

Fig 6.98 shows a two-stage class-A amplifier first presented over two decades ago. The design (like aging designers) is useful and robust in spite its age. The first stage uses a single TO-39 transistor biased to about 50 mA. Emitter degeneration and parallel feedback create low input and output impedance, presenting a good match at both ports. The second stage uses a parallel pair of TO-39 or similar transistors biased to about 250 mA. This circuit has a gain of 36 dB below 4 MHz, dropping to 29 dB at 29 MHz. The saturated output is a little over 1 W. IMD measurements at 14 MHz produced OIP3 of +43.5 dBm, making this a good starting point for low power SSB equipment. This circuit can also be used in CW applications by keying the positive supply to both stages with a robust PNP switch such as a 2N5322 or TIP-32.

A single-ended Class-A power amplifier is shown in **Fig 6.99**. This was built to investigate the performance of a variety of FETs as low distortion circuits. A 2N5947 bipolar feedback amplifier with measured OIP3 of +42 dBm preceded the circuit.

The first experiments used an IRF-510 HEXFET for Q1. With R2 = 1 Ω, an input network consisting of R1 = 47 with no input transformer, and with a 15 V power supply and bias adjusted for 0.5 A I_D, we measured OIP3 = +48 dBm. Increasing the

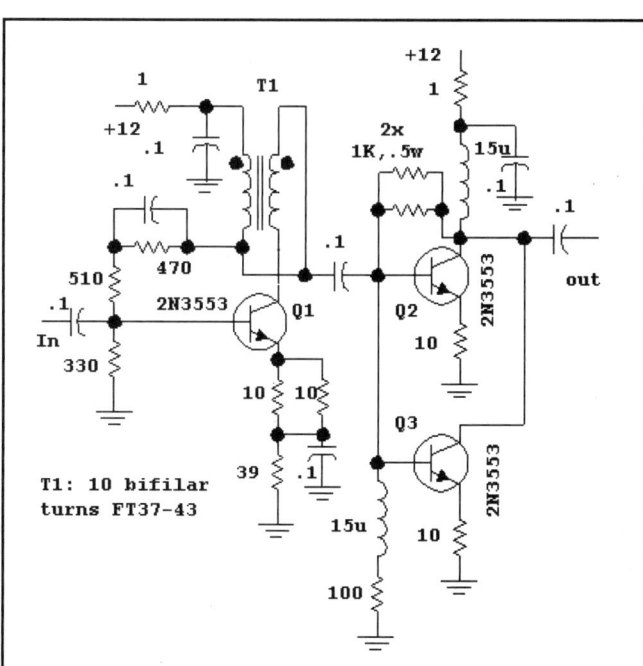

Fig 6.98—1-W power amplifier. Q2 and Q3 should have robust heat sinks if long operating periods are planned. If the 2N3553 is difficult to find, a Panasonic 2SC2988 can be considered for substitution. A single 2SC1969 might be a good substitute for the Q2 and Q3 pair.

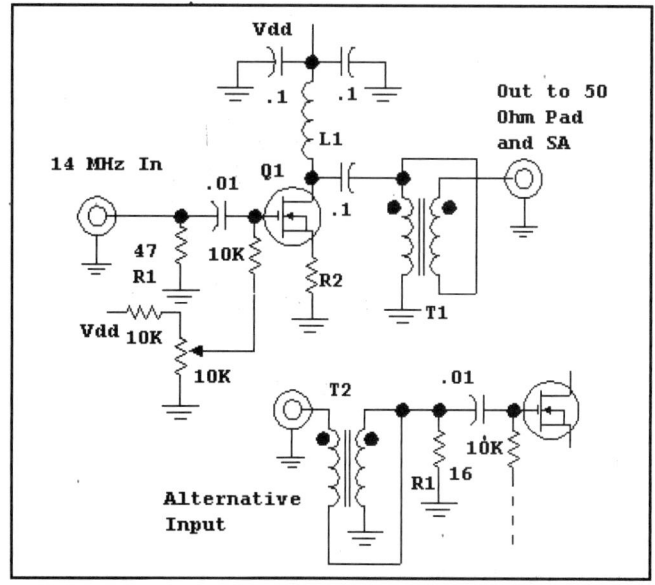

Fig 6.99—Class-A power amplifier experiment. Several MOSFET types were tried at Q1 while seeking high output intercept. L1 is 4 µH of #22 wound on a T68-2 toroid. T1 is 10 bifilar turns #18 on an FT-82-43 ferrite toroid. T2 is 8 bifilar turns #22 on an FT-37-43. R1 should have a 1-W power rating. Class-A amplifiers like this should be mounted on a large heat sink, for efficiency is not a feature of the design. See text for details.

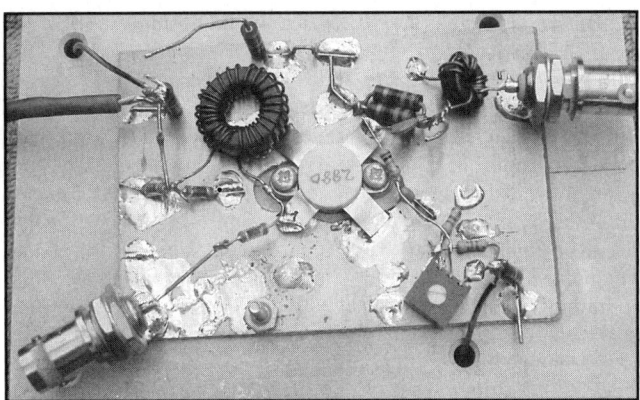

Experimental Class-A FET RF power amplifier.

One-watt output Class-A bipolar-transistor power amplifier.

power supply to 25 V with $I_D = 0.75$ A yielded OIP3 = +51 dBm with 19-dB gain. The HEXFET seemed to *want* high drain voltage and did not provide low distortion performance with a 12-V supply. Experiments with the larger IRF-530 and the alternative input network produced similar results. The HEXFETs were thermally unstable at high drain current without the source degeneration resistance.

The next tests used an FET specified for RF performance, a now obsolete Siliconix DV-2880T. The alternative input network provided a lower driving impedance for the gate. High drain voltage was again required to obtain low distortion. With $V_{dd} = 25$ V and $I_D = 0.8$ A, this device produced OIP3 = +57 dBm with 21-dB gain. The measurements were performed with outputs of +30 dBm per tone, or 4-W PEP. Slightly higher standing current should be used for a full 10-W PEP output.

The designer/builder could investigate other available FETs or power bipolar transistors. It appears that intercepts around +60 dBm will be available with moderately priced devices, allowing construction of Class-A power chains offering stellar performance at the 10-W PEP output level when compared with that offered by commercial transceivers. The experimental methods presented can certainly be extended to higher power levels.

Class-A power amplifiers are very inefficient with values of 25 or 30% being the best one can expect with reasonable distortion. Indeed, 50% is the theoretical maximum. Solid-state Class-AB amplifiers are also inefficient with values of 30% being typical. But the numbers obtained with two-tone testing are only part of the story. The Class-AB amplifier uses only enough bias to turn the devices on, perhaps to a maximum of 10% of the peak current used. With typical speech containing low average power compared to the peak value, average current is low. The average to peak power ratio is usually increased with speech processing, but net current is still far below Class A values.

An outstanding example of a medium power Class-AB FET amplifier was offered by Sabin.[34] That design is on the book CD.

Balanced Modulators

The voice signal from a microphone is amplified and converted to an intermediate radio frequency with a mixer. After up-conversion, it is usually processed with a crystal filter to eliminate one sideband. A balanced mixer is virtually always used in this application, a requirement to eliminate the local oscillator feedthrough. The mixer used in this application is usually described as a *balanced modulator*; the local oscillator that drives it is *the carrier*. All of the considerations presented earlier for mixers continue to apply. The popular diode ring mixers perform well in this application, often needing no adjustments for carrier suppression. The newer (physically smaller) TUF series parts from Mini-Circuits are preferred over the older and larger SBL-1, both for size and carrier suppression.

Fig 6.100 shows a simple balanced modulator design using two diodes. This is suitable for simple transmitters where the expense of a packaged mixer is to be avoided. The LO should be high enough to produce output that does not vary with LO drive, usually +7 to +10 dBm. Diode type is not critical. Silicon switching diodes such as the 1N4148 or similar will work well through the HF spectrum. Diodes should be matched for forward voltage drop with a current of a couple of mA.

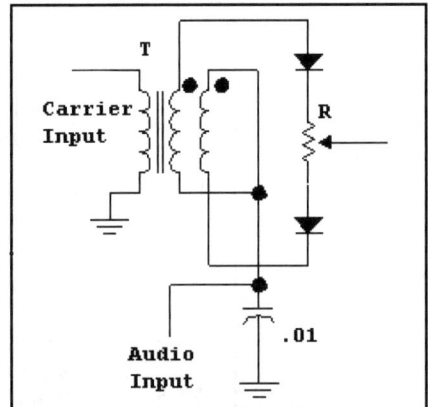

Fig 6.100—Simple balanced modulator for use in simple transmitters. R can be a small trim pot with R from 100 Ω to 2 kΩ. T is 10 bifilar turns on an FT-37-43 for HF applications.

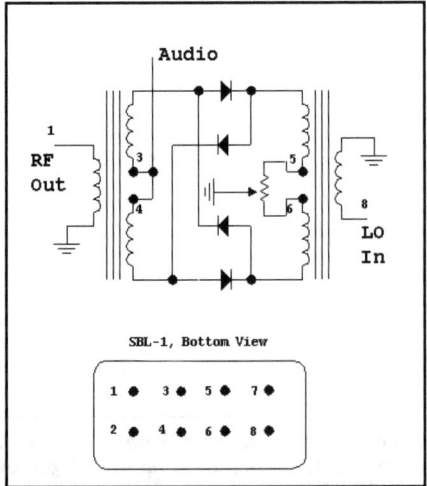

Fig 6.101—Adding balance adjustment to a balanced modulator using the SBL-1.

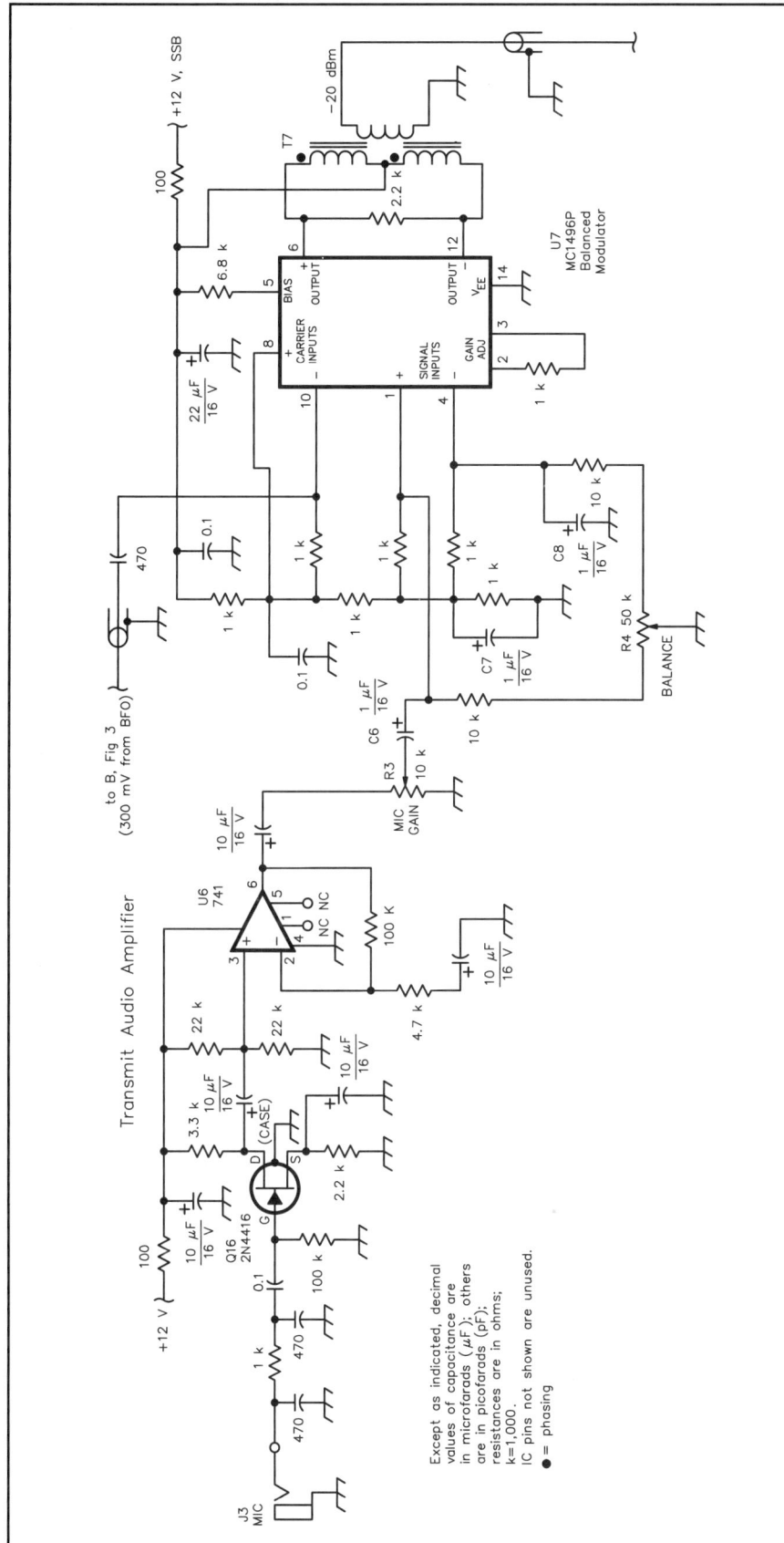

Fig 6.102—Speech amplifier and balanced modulator using an MC1496P. The transformer is 10 bifilar turns #28 on an FT37-43 with a 3-turn output link, used at 9 MHz. The carrier-balance pot is adjusted for minimum output at the carrier frequency. The dual in line version of the MC1496 is used here. Builders should consult manufacturer's data when using other variants.

Some builders have built very effective balanced modulators with the SBL-1 and similar Mini-Circuits mixers. But the topology is modified slightly from the expected where audio would be applied to pins 5 and 6, which were short circuited to each other. A modification used by W6JFR, shown in **Fig 6.101**, opens the short and inserts a low resistance (50 to 200 Ω) pot between pins. Adjustment of the pot allows the carrier to be nulled. Drive level considerations are still important.

The Gilbert Cell is an effective and popular balanced modulator. **Fig 6.102** shows a simple speech amplifier and balanced modulator using the Motorola MC1496P. The internal circuitry for the MC1496 is found in the manufacturer's data, with fundamentals presented in Chapter 5. This circuit is capable of a carrier suppression exceeding 50 dB. Indeed, one can probably adjust it to even greater suppression, although it may be difficult to maintain this performance over time and temperature variations. The output with audio drive should be kept to about –20 dBm with this circuit. LO drive is 300 to 500 mV peak-to-peak, usually measured (in-situ) with an oscilloscope with a ×10 probe.

The speech amplifier used in Fig 6.102 will accommodate both high and low impedance microphones. FET type is not critical. Most of the gain is provided by the op-amp. The builder may wish to use a dual op-amp with the other section configured as an active low pass filter. A project elsewhere in the book used this topology with a diode ring balanced modulator.

Transmitter IF Systems

The modulator output is routed to an IF amplifier. With a level of –20 dBm from the modulator and a requirement for only –10 dBm for a typical transmit mixer, little IF gain is needed. Indeed, most of the function of a transmit IF amplifier is that of signal conditioning and level control rather than gain. **Fig 6.103** shows an IF system. The first stage uses a common base amplifier, which provides good isolation between the modulator and crystal filter that follows. The amplifier also sets the termination impedance for the crystal filter. The amplifier and follower after the filter will establish the proper output level and gain. The follower provides a 50-Ω output impedance to drive a ring mixer while a 10-mA bias current sets low distortion.

A commercial crystal filter was used in the IF shown, part of an early transceiver.[35] The filter can be as simple as a 4th order Butterworth design. However, we have been disappointed with these simple filters. Filters with 6 to 8 crystals

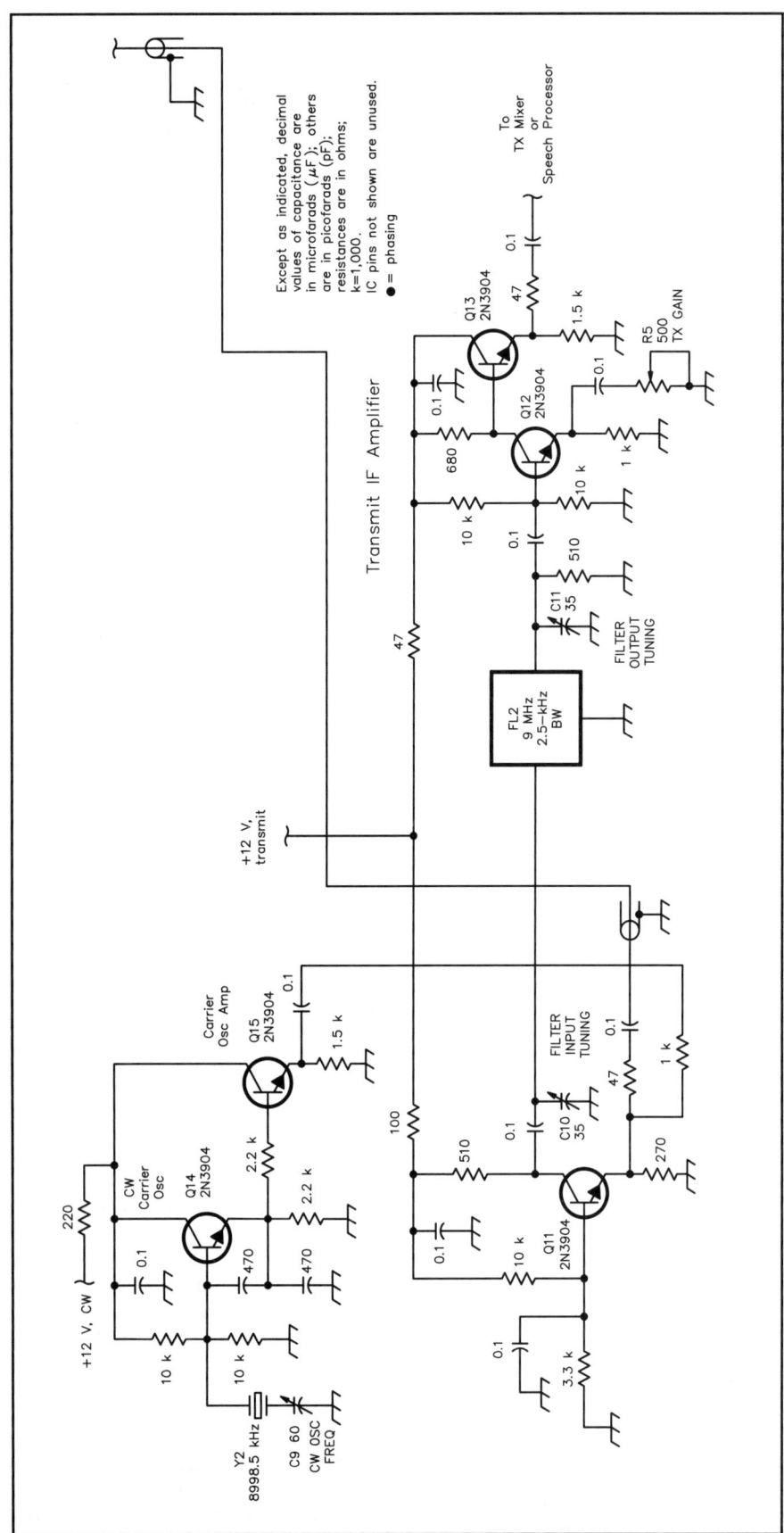

Fig 6.103—IF amplifier for an SSB transmitter. Very little IF gain is usually needed for this application. The trimmer capacitors were needed to terminate the crystal filter used on a transceiver using this amplifier, but may not be needed for other applications.

are little more complicated than a 4-pole circuit once the builder has been through the crystal characterization exercise needed when building filters. (See Chapter 3 for design details.) Yet the sideband suppression is dramatically better. Suppression is illustrated in **Fig 6.104** where overlapping 4 pole Chebyshev filter responses are presented. The level 6 dB down from the filter tops is marked, indicating the filter "passbands". The worst-case sideband suppression is about 30 dB, occurring for a 300-Hz audio note. Suppression approaches 60 dB at the highest audio input.

A Chebyshev filter shape is recommended for SSB applications over the simpler Cohn filter, which often suffers from poor passband shape. A comparison is made in **Fig 6.105**. The Cohn response, however, does have steep skirt attenuation, comparable to a 1.0-dB-ripple Chebyshev filter. Further, Cohn (equal coupling) filters built with lower Q_u crystals tend to have a smoother passband shape.

It is interesting also to compare available sideband suppressions with the responses of a phasing transmitter. The phasing system has the virtue of offering good suppression over the entire passband including the region close to the carrier. Hybrid systems with a phasing exciter followed by a filter could offer spectacular performance. (The same can be said for SSB receivers. See Chapter 9.)

CW Carrier Generation

The IF amplifier of Fig 6.103 includes a crystal-controlled carrier oscillator needed for CW generation. The oscillator and follower are relatively rich in harmonic energy, which might normally constitute a problem. However, the harmonics are removed by passing the signal through the crystal filter. The carrier is injected into the IF strip at the common base stage. The 1-kΩ resistor can be adjusted so the CW level is the same as the peak SSB power. An even simpler IF system is clearly in order for designs intended exclusively for CW. The important criterion is to provide the right level for the transmit mixer, but no more.

The CW carrier oscillator shown in Fig 6.103 functioned well in this application. This oscillator was turned off and on only at the relatively slow T/R rate. A faster rate is needed in many higher speed applications. But keyed crystal oscillators are subject to chirp, a change in frequency occurring as oscillation builds in the circuit. The problem often gets worse at lower frequency. There are several solutions to the problem. The crystal oscillator can be configured for lower loaded crystal

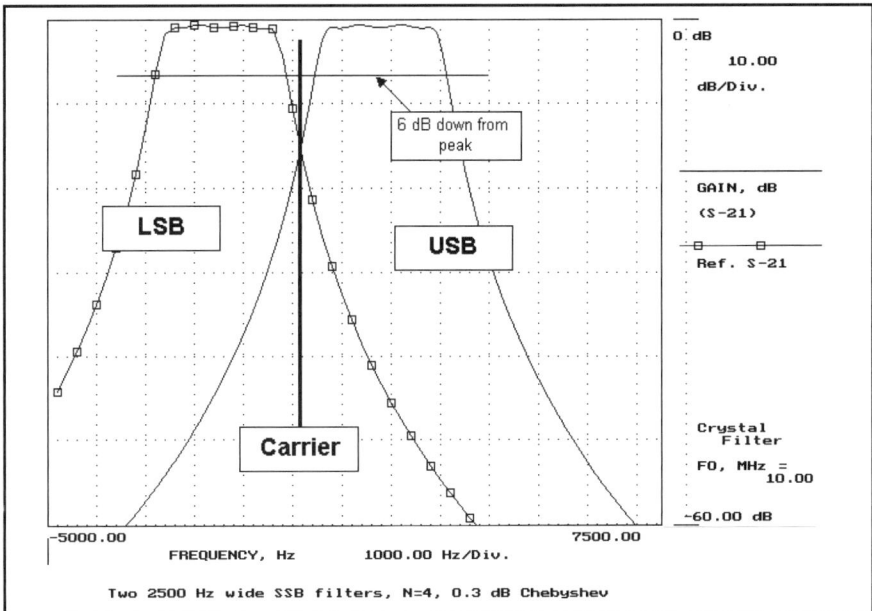

Fig 6.104—Two overlapping filters illustrating sideband suppression. See text. In a practical application, the filter response is measured and recorded in the builder/designer's notebook. The lower frequency 6-dB point is noted (for USB generation) and the carrier is placed 300 Hz below this point. The carrier is so marked in the figure.

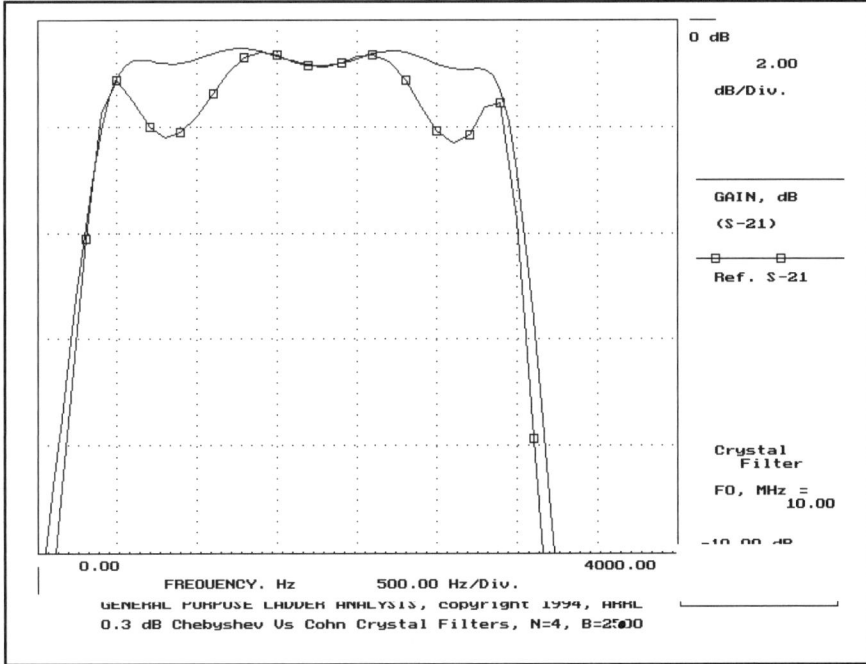

Fig 6.105—Two 4-element crystal filters are compared. The shape marked with small squares represents the Cohn filter while the other was designed for a 0.3 dB Chebyshev response. The two filters have similar skirt response, which is much better than a Butterworth shape, but much worse than a higher-order filter.

Q, often a difficult design task. A better solution uses an oscillator that is not keyed. The receiver BFO usually found in a transceiver is such an oscillator, but it is offset, operating at the wrong frequency. This slight change can be compensated with a suitable offset in the VFO. This is often a convenient solution, for RIT circuitry is already present in the transceiver.

Another alternative is a non-keyed crystal oscillator other than the BFO. But one can't normally use one within the receiver IF bandwidth, for it would be heard unless monumental efforts were taken to shield and isolate it from the receiver. Oscillator operation at a harmonic is often a convenient option. The signal is then divided with a digital divider during key down periods. One of our designs used a 5-MHz IF, but slight chirp was encountered when a 5-MHz crystal oscillator was keyed. The solution to the problem is shown in **Fig 6.106**.

Even though the free running oscillator in this scheme does not operate within the receiver IF, shielding is still required. A steady tone was heard when the 10-MHz oscillator was physically near the 5-MHz IF, a result of BFO second harmonic energy mixing with the higher frequency signal. Shielding and use of feedthrough capacitors for power and control eliminated the problem.

The non-integer frequency multiplication scheme described in Chapter 4 would also be well suited to generation of a CW carrier. That scheme divides a free running oscillator by 2, then uses one of the robust odd harmonics present in the square wave. In the prior example with a 5 MHz IF, a crystal oscillator at 3.3333 MHz could be used. It would be divided by 2 to produce a 1.667 MHz square wave that has a strong harmonic at 5 MHz. This could be filtered in a 5 MHz crystal or LC filter.

IF Speech Processor

The –10-dBm signal developed by the transmitter IF (Fig 6.103) is ready to drive a transmit mixer. Alternatively, it can be applied to an IF speech processor, shown in **Fig 6.107**.

The voltage related to a –10-dBm signal in a 50-Ω cable is only 0.1 V peak. This is not enough to turn on a diode. However, it can be increased with a transformer until diode clipping occurs. After the signal has been clipped, it is amplified and filtered. The filtering from the second crystal filter is necessary; without the filtering, intermodulation distortion products generated by the clipping circuitry would appear outside the IF bandwidth. Clipping cannot be done prior to initial filtering, for that clipping of the double sideband signal would create some distortion products within the eventual IF passband that would not otherwise occur.

The IF speech processor has the effect of increasing the average power within the speech sideband without increasing the peak. This higher average power increases intelligibility without excess distortion out of the normal passband. This processor, with the levels shown, increases the average to peak power by about 10 or 12 dB, readily observed with an oscilloscope.

The IF processor has a second advantage: It confines the IF level to prevent overdriving the transmit mixer. Without the processing, it would be desirable to add ALC, or "Automatic Level Control." This is an AGC loop in the transmitter that maintains the level through the overall power chain.

Intermodulation distortion is rarely a factor in a transmitter IF system. With so little gain required, the IF system can be simple. But the builder/designer should be careful to be sure that distortion is not an issue. It would be folly to design an extremely low distortion RF power chain only to feed it with the output of a mixer driven by a distorted IF signal.

Bidirectional Amplifiers

One view of an SSB transmitter says that it is nothing more than a superheterodyne SSB receiver with signals moving in the

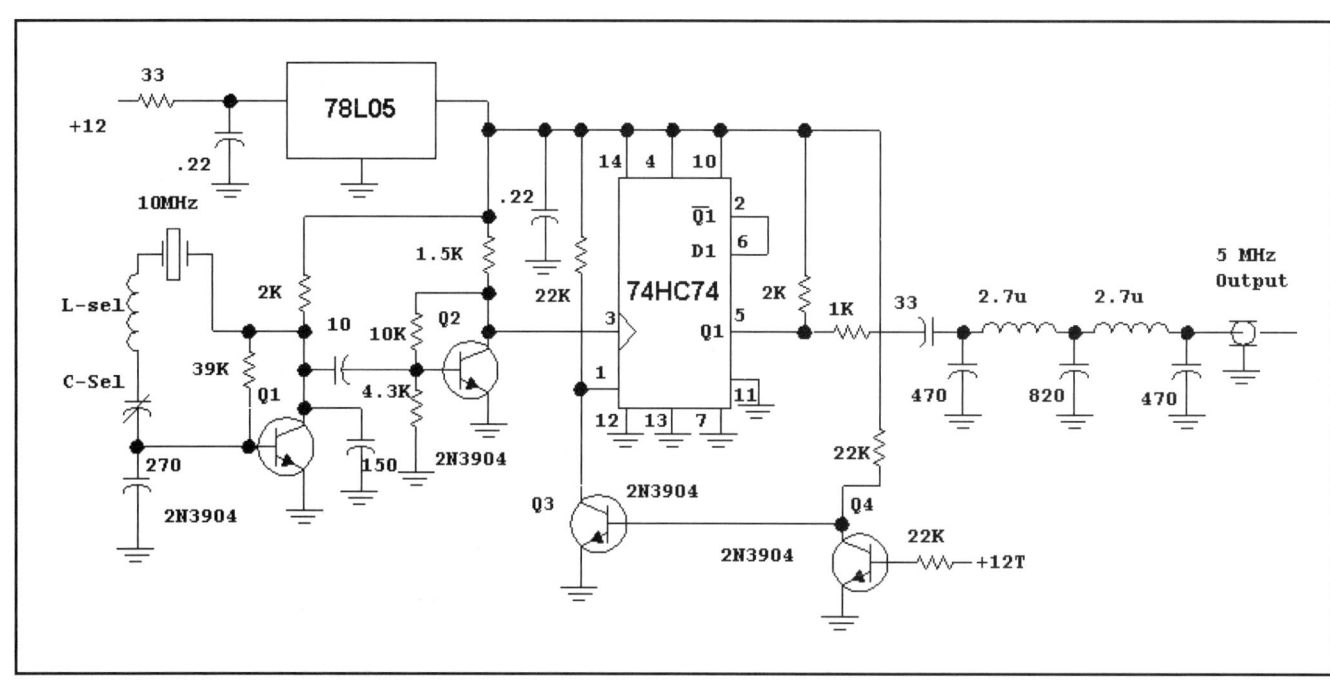

Fig 6.106—Alternative carrier-oscillator system for CW generation. A free-running 10-MHz crystal oscillator is divided with a digital divider to generate 5 MHz when needed. The divide-by-2 circuit is controlled with an IC reset line. See text.

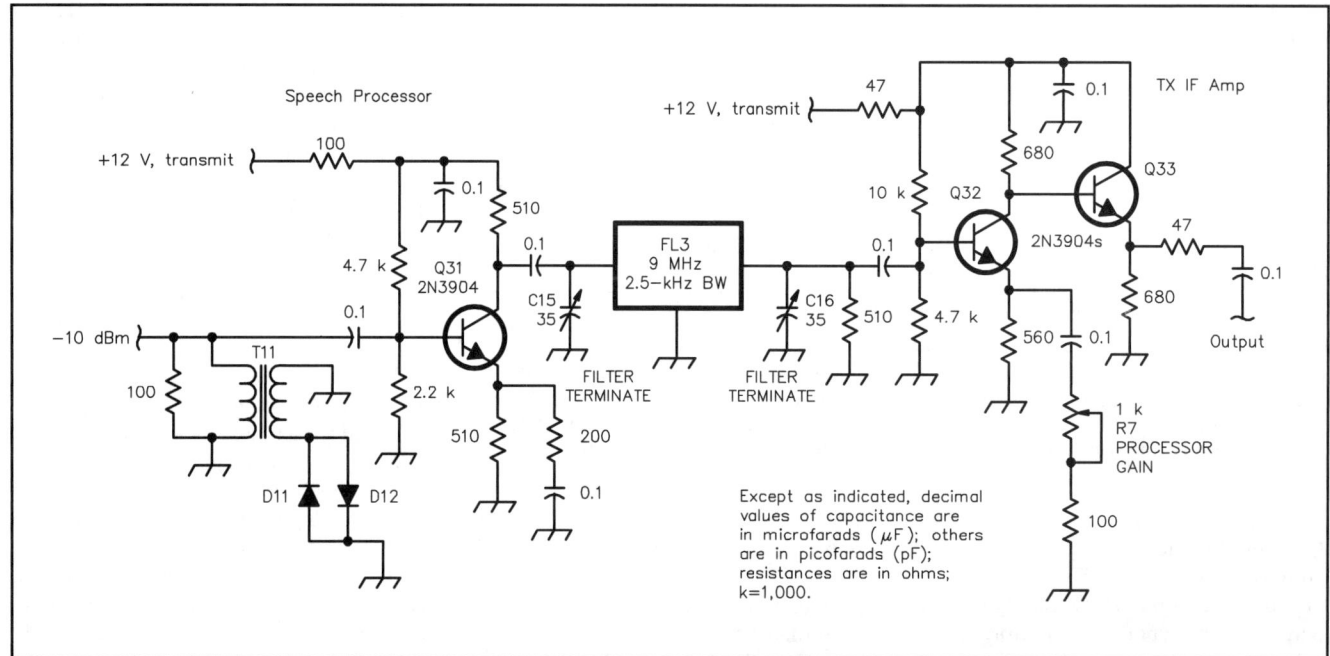

Fig 6.107—IF speech processor. Back-to-back diodes clip the IF signal. The resulting voltage is amplified and filtered in a crystal filter. It is then amplified and set to provide the desired −10 dBm to drive the transmit mixer. Schottky diodes are used in the clipper circuit. The diodes are driven by a 16-turn winding on an FT-37-43 ferrite toroid. The link on the 50-Ω line is 3 turns.

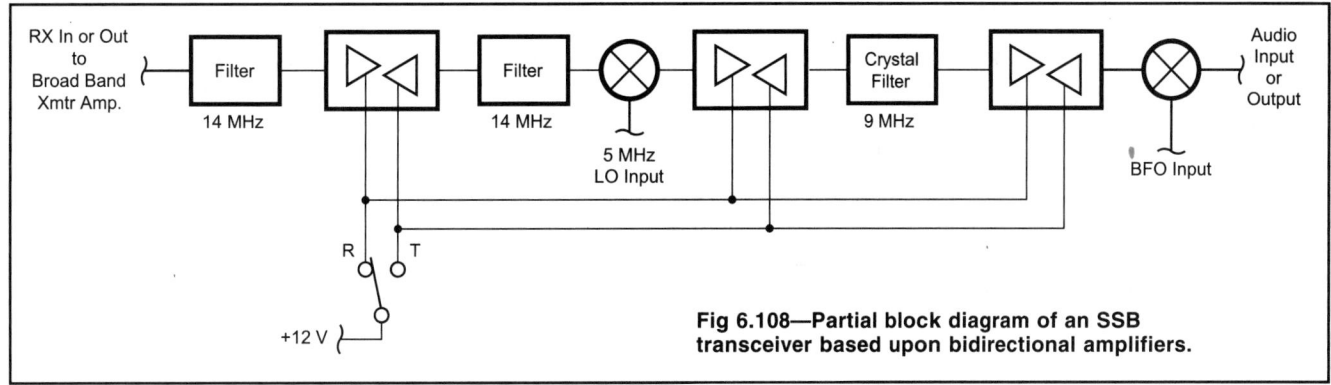

Fig 6.108—Partial block diagram of an SSB transceiver based upon bidirectional amplifiers.

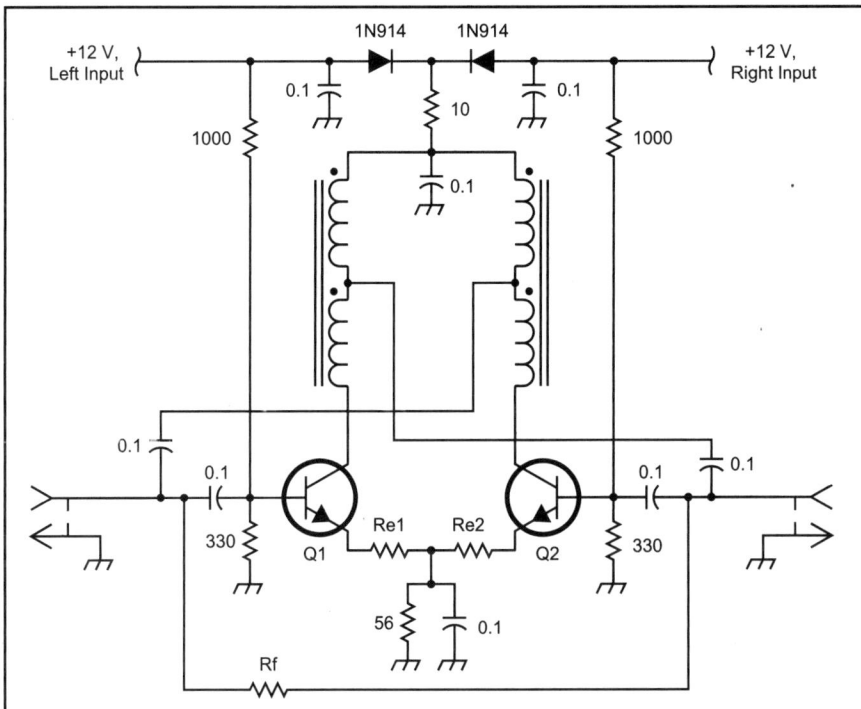

Fig 6.109—Bidirectional amplifier with bipolar transistors. Q1 and Q2 can be 2N5109s or similar parts. The input and output impedances are 50 Ω in both directions.

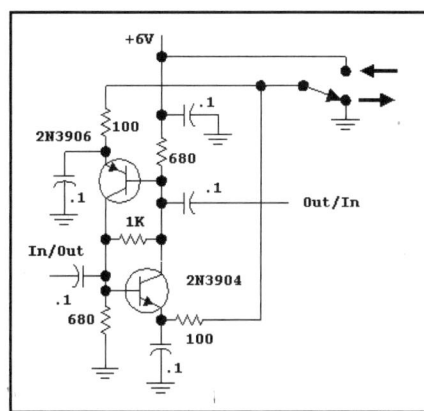

Fig 6.110—Bidirectional amplifier with complementary transistors. Only one transistor is on for each direction. Operation is clear if one of the transistors is mentally removed and the remaining circuitry is analyzed. See text for details.

opposite direction. The transmitter needs the same filters and oscillators as used in the receiver to create an SSB signal. Many transceiver designs have used this concept. A block diagram is shown in **Fig 6.108**. All of the RF and IF chain amplifiers are bidirectional; they provide gain to signals going in either direction when a dc control signal is changed. Diode-ring mixers are also bidirectional circuits, as are both LC and crystal filters. Audio signals can be switched with ease with integrated or discrete FET switches.

Fig 6.109 shows a circuit designed by the late Mike Metcalf, W7UDM. This circuit uses high F-t transistors biased to high current in the feedback amplifier circuit used throughout this book. The direction of operation is selected by applying Vcc to one of the two control inputs.

Very few of the components in the amplifier of Fig 6.109 are shared with switched directions. W3TS brought our attention to a simple bidirectional amplifier used in some "Manpack" transceivers built by Plessey.[36] We adapted this to the 50-Ω feedback circuit shown in **Fig 6.110**. The amplifier shown should be operated from a low V_{cc} to ensure that the emitter-base breakdown of either transistor is not exceeded. No emitter degeneration is used in the transistors, for each transistor is only biased to about 3.5 mA. Degeneration can be added for reduced gain or improved IMD. This amplifier will provide about 17 dB gain up to about 40 MHz. If redesigned for higher current, the 680-Ω resistors are replaced with smaller resistors in series with suitable inductors.

The junction field effect transistor is ideally suited to bidirectional amplifiers, owing to the usual symmetry of the physical device where the source and drain regions are identical. The drain only assumes drain-like properties when it is positively biased. A bidirectional amplifier using this is presented in **Fig 6.111**. A single-ended variation ("A" in the figure) shows the resonant drain network needed to generate high gain. This circuit appears twice in the bidirectional version ("B") of the circuit. A PIN diode short-circuits C-t when that portion of the circuit is used as an input. The low impedance then effectively short-circuits much of the tuned network. Input tuning can be implemented, if needed, by replacement of the RFC with small inductors. This circuit uses the metal can U-310 rather than the more common J-310, allowing a grounded gate with extremely low inductance, important for UHF stability.[37]

Transmitters and Receivers 6.61

Fig 6.111—Bidirectional amplifier using a junction FET in a common-gate topology. Part A shows a single-ended amplifier where L, C-v, and C-t form a resonant network that presents a high impedance to the drain. Part B shows the bidirectional variation. See text.

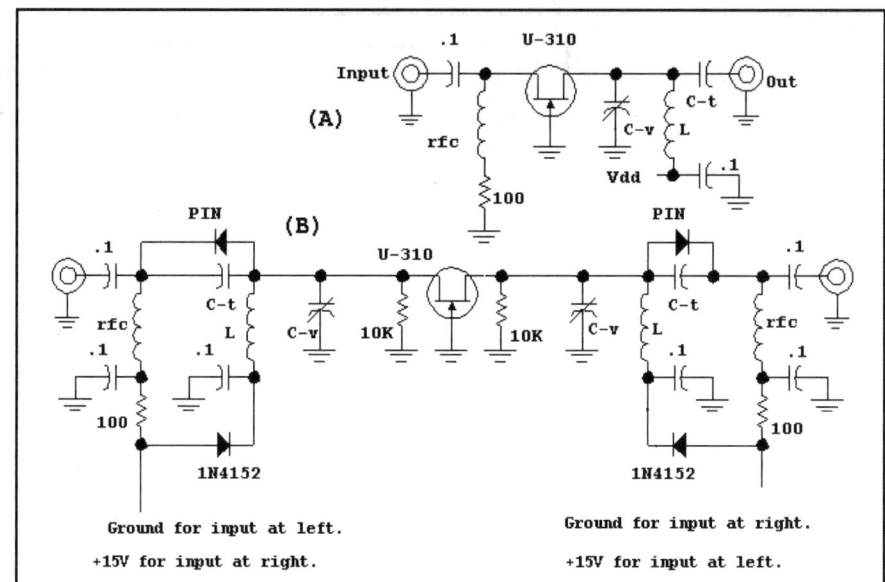

Bidirectional Crystal Filter Circuits

Fig 6.112 shows a system with diode switching, allowing a crystal filter to be shared between receive and transmit functions. Diode D1 routes the signal to the filter input during receive while D2 connects to transmit circuits. R1 and R2 set D1 current during receive. The positive voltage developed across R1 serves to reverse bias the diode in the off path.

Part B of Fig 6.112 shows an option with an added transistor, Q1, in the receive path. Q1 helps to reverse bias the D1 anode and creates a low impedance to ground during transmit, both increasing the switch on to off ratio. Typical switch performance at 10 MHz will be a 45 dB on to off ratio with a 1 dB insertion loss.

While the diode switching looks simple enough, it is a critical transceiver circuit. The switching and the interfacing circuits should present the same impedance to the filter with switching to preserve filter performance. All components must be examined and, if needed, characterized for IIP3 as well as switching performance.

The best diodes to use in this application are PIN types. Lower cost high voltage rectifier diodes are often suitable, although they have higher off capacitance. We have measured IIP3 higher than +50 dBm for 1N647 and 1N4007 diodes. Less robust, but lower capacitance switching diodes are often used when crystal filters with a 500-Ω impedance are used. Careful experiments are then required to maintain IMD performance.

A scheme using a shared filter is shown in **Fig 6.113**. This method using NE602 Gilbert Cell mixers is the brainchild of K7RO.[38] Part A of the figure shows a partial schematic for a NE602. This part has good isolation between ports, a result of balance and the virtual cascode internal topology. This allows two mixers to be tied together to present a constant composite impedance to a filter, shown in part B of Fig 6.113. The mixer output impedance is 1.5 kΩ and remains even when the part is biased off. The input impedance is 3 kΩ, but is present only when the mixer is biased into operation. The output of U1, a receiver front-end mixer, and U2, a transmitter output mixer, are paralleled, pre-

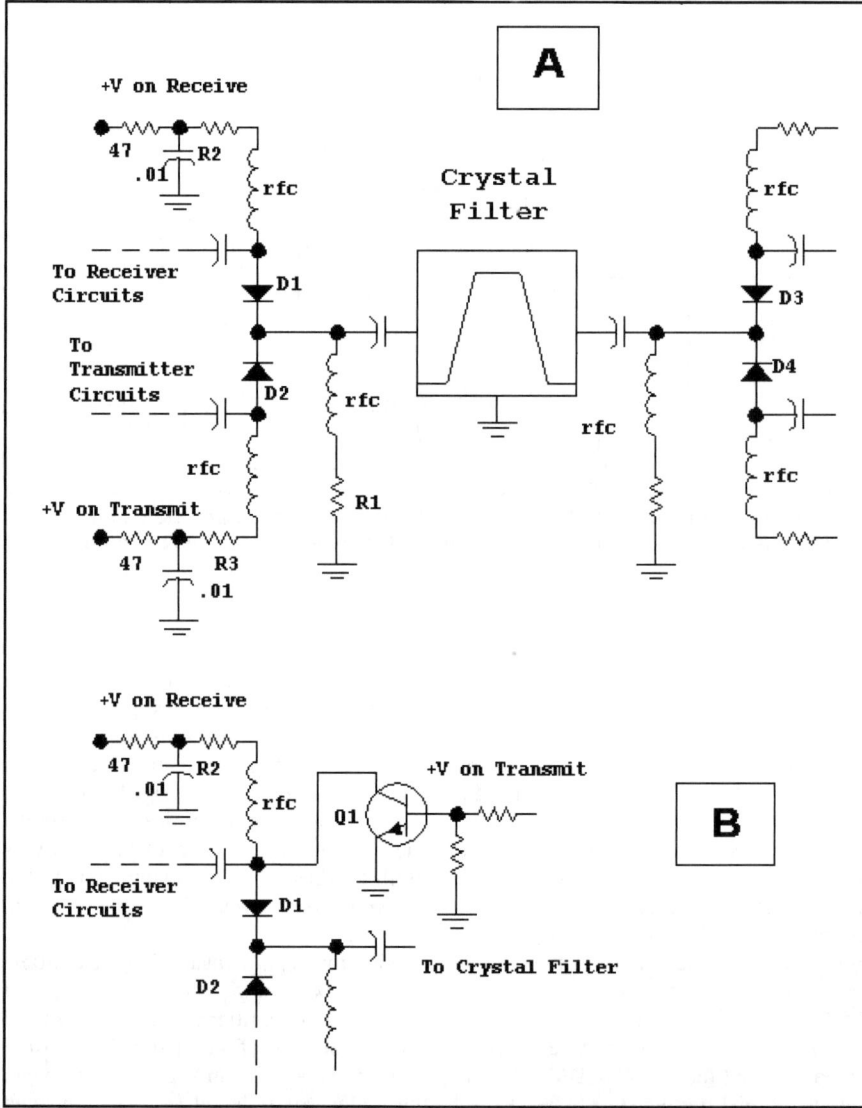

Fig 6.112—Diode switching of a crystal filter between transmit and receive functions. See text for details.

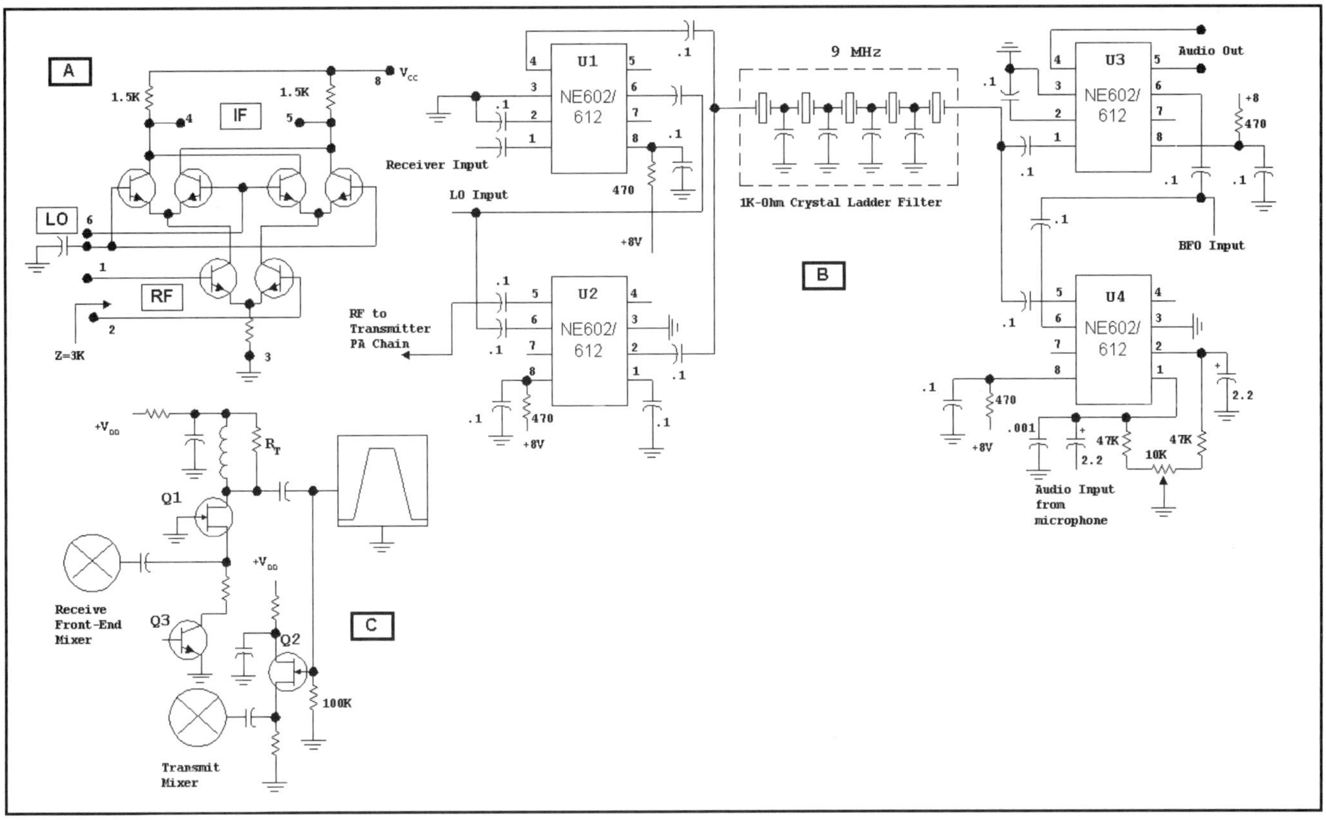

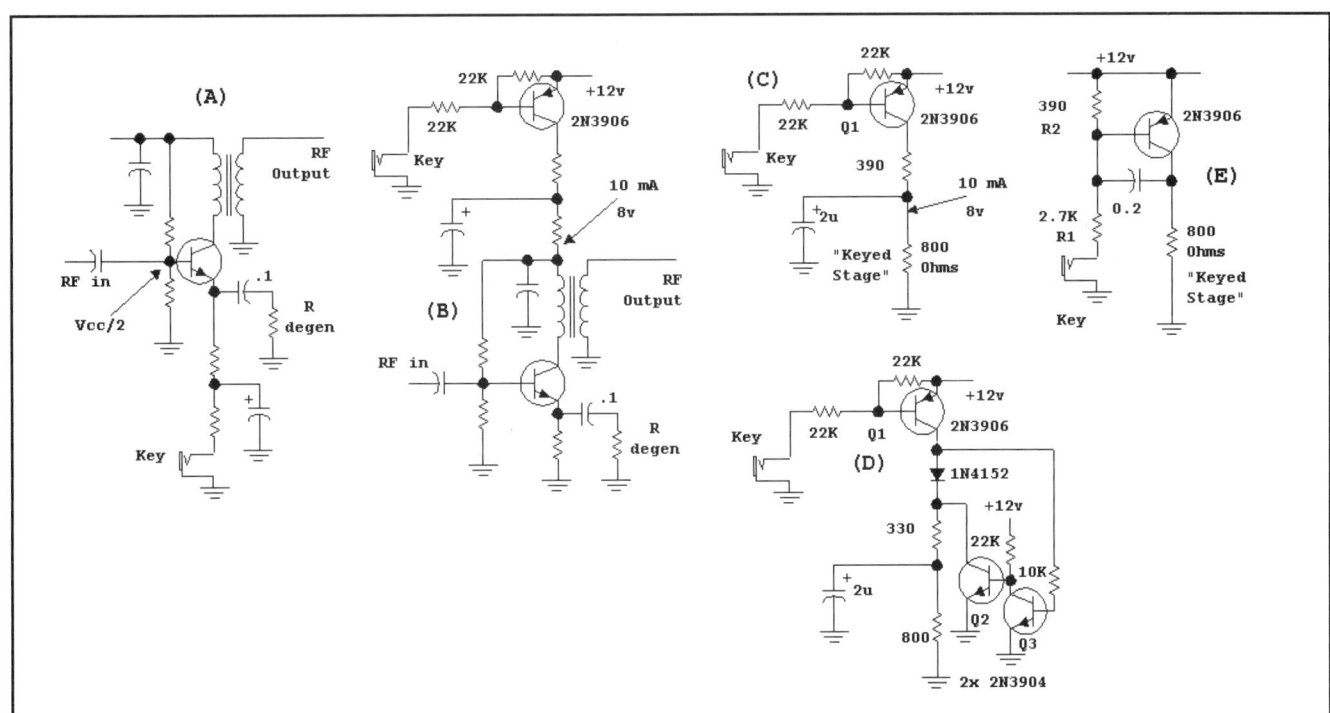

Fig 6.113—A scheme for sharing a crystal filter between functions. Part A shows a partial schematic for an NE602. Part B presents the basic scheme generated by K7RO while C shows FET buffers that allow other mixers and filters of many different impedances. The scheme in C has not been tried. See text for details.

Fig 6.114—Circuits used to shape keying of a transmitter amplifier stage. Part A is a general case of switching an emitter current to ground. Part B uses a PNP switch to apply a keyed waveform to an NPN amplifier. If that stage draws 10 mA with 8 V applied, it is modeled as an 800-Ω resistor, as in Part C. Analysis of C shows an asymmetry. The rise is controlled by the equivalent of 390 Ω in parallel with 800 Ω while the fall is the result of the 800-Ω value alone. Part D provides nearly identical rise and fall times. E shows a modified switch where the PNP now functions not only as a dc switch, but as an integrator that shapes the rise and fall. See text for discussion.

Transmitters and Receivers 6.63

senting a 1.0-kΩ impedance to the crystal filter. Local oscillator energy is simultaneously applied to both mixers.

Two more NE602 mixers are used with a similar connection to serve as a product detector (U3) and transmit balanced modulator (U4). Biasing is slightly altered in U4 to adjust balance.

One would ideally switch the mixers off and on to match their application. However, turning a mixer off that has an input that is shared with the output of another part will change the terminating impedance. The experimenter may wish to insert appropriate buffer amplifiers in the system to solve these problems.

The transceiver designed by K7RO used a crystal filter designed to have the impedance required by the mixers. Greater flexibility is afforded by the system in part C of the figure. Q1 functions as a common gate buffer amplifier, presenting a low input impedance such as might be needed for a diode ring receiver mixer. Q2 is a simple JFET follower to drive a variety of mixer types for the transmit function. Q3 is a dc switch that allows Q1 to be shut down during transmit intervals. Resistor R_T is the dominant element terminating the crystal filter.

Keying

Keying is the on-off control that is applied to a transmitter stage to generate RF in the pattern of International Morse Code. The keying circuitry can also control stages in a SSB transmitter when we wish to eliminate power consumption during receive periods. In principle, keying can be applied nearly anywhere in a transmitter. It is usually applied at an intermediate level and more than one stage is often keyed, especially when the following stages use linear amplifiers. It is acceptable to key just one stage when the following stages are nonlinear where bias is derived from RF input. The behavior we seek is a low *backwave*, meaning that the transmitted RF is low when the key is open. Backwave levels of –80 dBc are easily achieved.

Fig 6.114 shows several schemes for keying. Part A switches the emitter current, while the base is biased at about half the power supply. The electrolytic capacitor, the related stage current, and the resistor values time the rise and fall of the amplifier current. Both the rise and fall times should occur in a period of one or two milliseconds. Much shorter times allow key clicks to be created. Testing is normally done by examining the RF envelope with a high-speed oscilloscope, ideally while triggering the oscilloscope from the controlling dc.

The various parts of Fig 6.114 show a variety of shaping circuits, outlined in the caption. But the most popular is the simple integrator popularized by W7EL shown in part E.[39] The PNP transistor serves a dual role. The dc is switched, creating the basic function. But the transistor is also an amplifier that, in combination with the capacitor between base and collector forms an integrator circuit. No current flows when the key is up, bringing both base and emitter to +12 V, with the collector at ground. As soon as the key is pressed, current begins to flow in R1, causing the base voltage to begin to drop below +12 V. As soon as it gets to 11.3, base current begins to flow, forcing collector current to also flow which increases collector voltage. But the increasing collector voltage is coupled back to the base through the capacitor in a direction that "tries" to *reduce* the base current. This negative feedback does not let the collector voltage increase quickly, but forces it to ramp up at an approximately linear rate until the transistor begins to saturate.

The action is similar when the key is opened. The open R1 tries to reduce base current, which will let the collector voltage drop. But as that happens, base current will continue to flow through the capacitor as the collector voltage drops, again linearly, until the transistor finally turns off. R1 and C set the rising characteristic while R2 and C determine the fall. The traditional shapes of **Fig 6.115** approximate the linear ramp. Indeed it is the ramping part that is more effective in reducing clicks than is the rounded corners at the end of the shaping.

There are many methods that may be used to shape keying. In another W7EL creation (unpublished), a diode detector monitored the output of a transmitter. That signal was then compared with an ideal rise and fall in a dc-only circuit with an op-amp output controlling the gain of an amplifier. Shaping can even be done with DSP firmware, as presented in later chapters.

One sometimes sees simple transmitter circuits where a crystal oscillator is keyed. The result is often better than expected. This results from a general characteristic of oscillators—oscillation cannot start immediately, but must overcome the delay related to the bandpass filter intrinsic to all oscillator resonators. The resonator is the high Q crystal in this case. This behavior is usually not planned and should not be confused with design.

Although we emphasize shaping to reduce key clicks, some parts of the keying function must happen quickly. If an oscillator is keyed, it should occur quickly using circuitry isolated from shaping. The requirement for quick starting often precludes keying crystal oscillators. But keyed oscillators often suffer stability problems, adding challenge.

Generally, the following events must occur in sequence when a transceiver is keyed:

1. The receiver is operating normally.
2. The key is pressed to start a character.
3. The receiver is muted, preventing further audio from exiting.
4. Additional receiver muting is activated, preventing overload by strong transmitter signals.
5. The antenna is disconnected from the receiver input and is attached to the transmitter output. (In some cases, the transmitter output is already connected.)
6. Bias is established on important transmitter stages.
7. Oscillators are started and/or a frequency synthesizer is shifted and/or an RIT (detailed later) is shifted into transmit mode to establish the transmitted

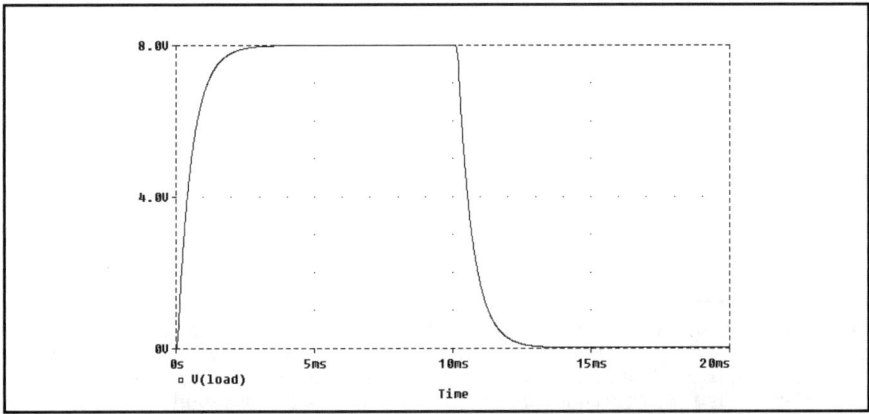

Fig 6.115—Desired waveform that should be applied to a keyed stage.

frequency.
8. The keyed stages are supplied with the shaped dc that causes the desired waveform to be generated.
9. The dot or dash continues to be sent for the desired length.
10. The key is opened.

The sequence outlined is reversed, with the final event being the unmuting of the receiver, allowing the receiver function to return to normal.

Although not listed, it may be desirable to activate circuitry that "remembers" the gain state of a receiver at the exact beginning of a keyed interval so the receiver can immediately return to that state after the transmit interval is finished.

Muting a receiver can be a major challenge, especially if very high speed is desired. The high-speed operation is especially useful for QSK, or break-in CW operation where ideally a CW operator can hear other stations between high-speed dots. This facility is considered an advantage in competitive operations, but is also useful while exchanging routine or emergency traffic messages.

The simple way to mute a stage in a receiver is to remove the power supply. Unfortunately, this does not allow the gain to diminish or grow immediately, for bypass capacitors within the stage must charge and/or discharge with the switching. This process can often create transients that are as troubling as the presence of signal. The better method of muting a stage applies a gain altering bias that reduces gain without changing other dc parameters.

Even the "simple" circuit task of injecting an audio sidetone can be a challenge. Often a sidetone oscillator is keyed on or off in a way that creates a dc transient. That is, the "key down" waveform has an average value that differs from the value when the key is up. A better sidetone oscillator is one that has no change in dc level as it is turned on and off, and the best ones have shaping applied to the keyed waveforms.

6.7 FREQUENCY SHIFTS, OFFSETS AND INCREMENTAL TUNING

Oscillator Modifications

Both direct conversion and superhet transceivers usually include a provision to shift the frequency of the main oscillator when the key or push-to-talk button is pressed, causing the rig to shift from a receive to a transmit mode. There are various reasons for this shift, depending on the application.

Fig 6.116 shows several partial oscillator schematics that allow the frequency to be shifted in a discrete step as a control voltage is changed. The voltage changes between two well-defined levels producing two closely spaced output frequencies. The circuit in Fig 6.116A is an LC tuned VFO. The frequency is changed when a small variable capacitor, C_{var}, is shifted into the circuit with a diode switch. When the "control" signal is positive, dc current flows in the diode and C-var is part of the frequency determination. However, when the control voltage is set at 0, very little current flows in the diode switch, so C-var is removed from the circuit. The same coil tap used for oscillator feedback is used for offset. Additional capacitance, Cx, paralleling the diode will reduce shift, providing an adjustment.

A crystal-controlled oscillator with a diode switch is shown in Fig 6.116B. This circuit is ideal for shifts of only a few hundred hertz. The shift will depend upon the crystal parameters and the circuit design, so experimentation with C_{delta} is required.

A transistor is used as a switch in Fig 6.116C. The transistor saturates when the switch has base current applied, creating effectively a RF short circuit. When

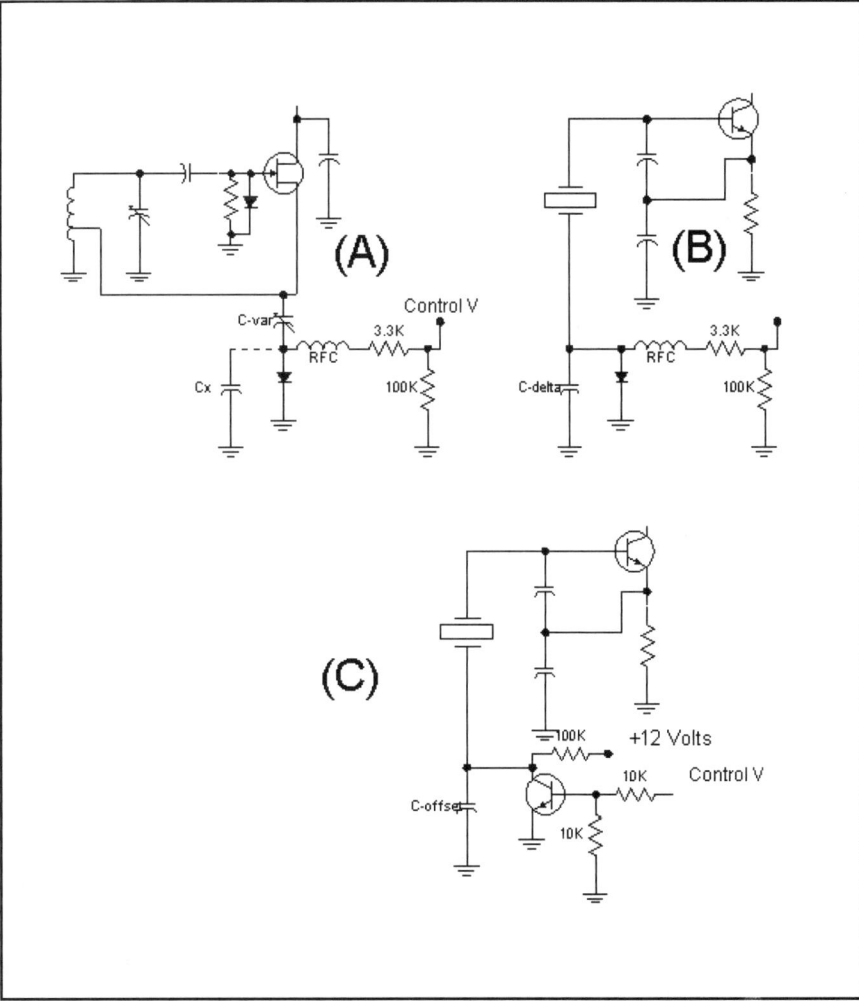

Fig 6.116—Oscillator circuits, including a means for frequency shifting.

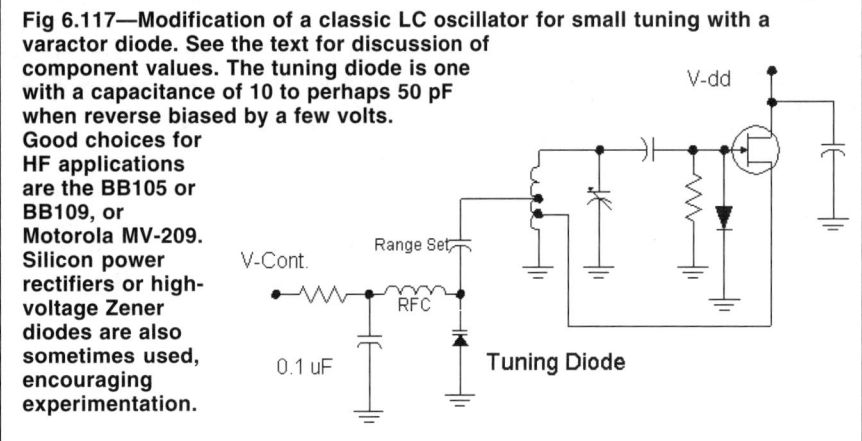

Fig 6.117—Modification of a classic LC oscillator for small tuning with a varactor diode. See the text for discussion of component values. The tuning diode is one with a capacitance of 10 to perhaps 50 pF when reverse biased by a few volts. Good choices for HF applications are the BB105 or BB109, or Motorola MV-209. Silicon power rectifiers or high-voltage Zener diodes are also sometimes used, encouraging experimentation.

base current is removed, the 100-kΩ collector resistor causes the collector voltage to rise, placing a reverse bias on the collector. The switch is then a small capacitor (a picofarad or two) that has less impact on the circuit.

A VFO example is shown in **Fig 6.117** where a traditional oscillator is modified with the addition of a varactor diode. For best stability, the "range set" capacitor is kept small, producing no more frequency shift than needed. Also, the voltage tuning range is picked to always reverse bias the tuning diode, even in the presence of large RF voltages. A typical circuit might have control voltage V_C that varies between 5 and 10 V dc. If the control drops close to zero, the RF will be rectified in the diode, altering V_C. This will often alter the Q of the oscillator tank and, in extreme cases, can cause oscillation to cease.

The bypass capacitor related to the tuning diode is shown as a 0.1 µF. A smaller value may be sufficient to decouple the RF. Values that are too large will slow the rate that frequency can change when the control voltage is altered, producing CW chirps or missed SSB syllables.

Superhet RIT

The most familiar application for the variable offset is *receiver incremental tuning*, or *RIT*, featured in most commercial transceivers. RIT is a simple function: During transmit periods, the transceiver frequency is determined by the main tuning system. But incremental tuning can become active during receive, allowing the user to adjust the received frequency by a small amount around the nominal transmit frequency. A typical range is +/– 3 kHz. Usual transceivers have a provision to turn the RIT function off, forcing the frequency of both transmitter and the receiver to be identical.

The RIT function might be controlled with the circuit in **Fig 6.118** where an operational amplifier determines the VCO control voltage. A 5-V regulator provides a stable voltage to drive the tuning pots and to power the oscillator. This is divided to provide 3 V for the noninverting op-amp input. A logic signal that is high during transmit periods is applied to the NPN, Q2. This saturates Q2 and cuts Q1 off, disconnecting the 10-kΩ summing resistor from the RIT pot, forcing the control to +7.5. The same result occurs when the "RIT-off" switch is closed.

The usual superhet transceiver generates the transmitted carrier by mixing the VFO output with a crystal controlled oscillator residing in the middle of a narrow IF bandwidth. During transceiver construction and alignment, the crystal oscillator is turned on and adjusted for a frequency that provides a desired beat note in the receiver, usually about 800 Hz. Then, during operation, the transceiver is tuned until an 800-Hz note is heard. Pressing the key then generates a signal that is exactly in zero beat with the received one.

Note that this operation and alignment is slightly different than that when SSB is generated in a superhet. In that case, the same circuit (usually crystal controlled) serves as the receiver beat frequency oscillator and the transmit suppressed carrier. It is important that an experimenter understand the frequency scheme used in his or her transceiver and the resulting operating mode. Also be careful to know when the RIT is active.

Offsets with Direct Conversion Transceivers

These basic superhet schemes will also work with direct conversion rigs. Consider a very simple 7-MHz direct-conversion CW transceiver using a VFO without off-

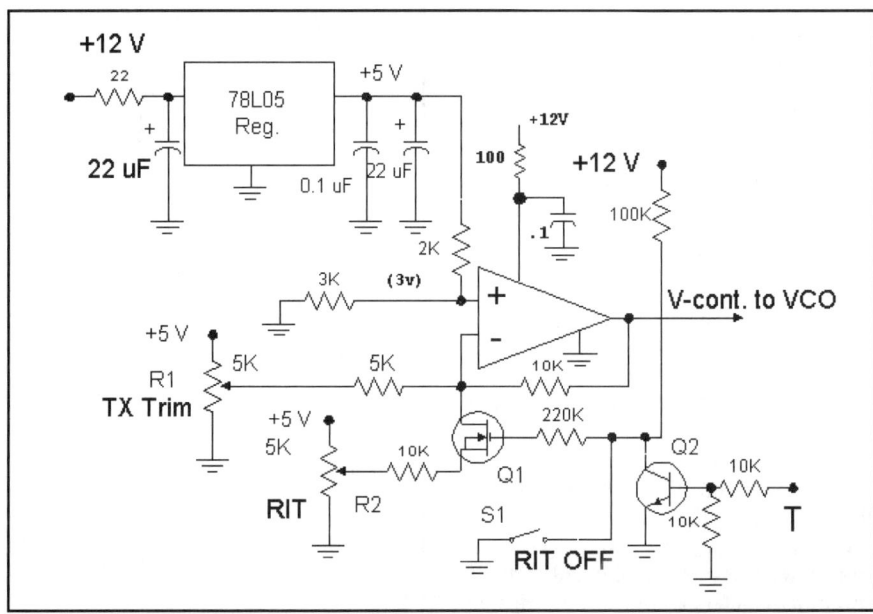

Fig 6.118—Circuitry to control RIT. Q1 is a TO-92 N-Channel MOSFET such as a 2N7000 or VN-10 or Zetex ZVNL-110A. Q2 = 2N3904 or similar. R1 sets the control voltage during transmit. The SPST switch is closed when the RIT is off. In this state, the control voltage should be approximately 7.5 V. The control voltage should vary between 4 and 10 with RIT on. Op-amp type is not critical; it could be a 741, half of a 5532 or 358, or similar.

set or RIT circuitry. A simple switch transfers the antenna between transmit and receive functions, as needed. The transceiver is turned on and attached to a suitable antenna. The VFO is tuned, producing the expected collection of signals. A station is found calling CQ on 7040 kHz. Assume that you had been slowly tuning *up* the band when you heard this station. If you stopped tuning and listened to an audio note of 1 kHz, your VFO will be at 7039 kHz. If you tried to answer him, there is a high likelihood that he would miss you and would merely call CQ again. He will probably listened most intensely on his transmitter frequency of 7040 kHz.

A similar situation would have occurred if you had been tuning *down* the band. You would have stopped with your VFO at 7041 kHz to listen to a similar 1-kHz audio note, again transmitting off frequency.

Clearly, you must do something so that you transmit on the right frequency. One simple answer uses an offset generating circuit like that shown in Fig 6.116A. This circuit shifts the VFO *downward* by a fixed amount when the control is switched positive. The exact shift can be adjusted with a frequency counter, or by ear by listening to strong signals. The schematic is duplicated in **Fig 6.119**, which now includes needed control circuitry.

The system shown in Fig 6.119 is common for D-C transceivers. Pressing the key causes immediate PNP base current to flow. The collector goes up to +12 V, shifting the VFO frequency downward. When the key is let up, the frequency remains shifted for a short period controlled by the 10-µF capacitor and related resistors.

One tuning method emphasizes the SPOT switch. When a station is heard that you wish to call, the SPOT switch is closed and the station is tuned to zero beat (zero audio frequency). This switch action is the same as pushing the key with the frequency shifted to the transmit state. Once the station is tuned to zero beat, the **SPOT** switch is opened. The station should then be heard with a 1-kHz note.

A second method is faster. When tuning and looking for stations to call, be sure that you are always tuning *down* the band, taking care not to tune through zero beat. It may be useful to mark the front panel with a small arrow next to the tuning knob, indicating the *proper* tuning direction. An error in picking the right tuning direction will now produce a 2-kHz error.

Extended use of a D-C transceiver reveals a subtlety: there is often interference when the VFO is on one side of the desired signal, but the other side is clear. It would be useful to be able to reverse the role of the offset. This leads to a modification of the usual scheme called "Almost Incremental Tuning," or AIT, shown in **Fig 6.120**.

Like the simpler system, the system with AIT is easy to use with a spot switch. Upon finding a station that you wish to work, tune to zero beat. Then throw the AIT switch. If there is interference, tune to zero beat and toggle the switch again.

RIT with Direct Conversion

An RIT system is often included with a D-C transceiver. The utility of the feature helps immensely to overcome the deficiencies of the double-sided response. RIT can be accomplished at two different levels. W7EL popularized the simple scheme shown in **Fig 6.121**.[40]

A varactor diode is coupled to the oscillator through a small capacitor. During transmit or "zero" intervals, the bias on the diode is maximum at the level of the 9 V regulated supply. The voltage applied to the tuning diode during receive is less than the regulated supply, causing a downward shift in VFO frequency. The amount of the offset is tunable via the 20-kΩ RIT control. This scheme works well, providing all the adjustment needed for normal operation.

The complete superhet system can also be applied to a D-C rig.

One often encounters articles in the literature where VFO *offset* in direct conversion transceivers is discussed. The variety of offset options presented are sometimes referred to as having to do with "sideband selection." This term is not correct. The usual direct-conversion receivers using

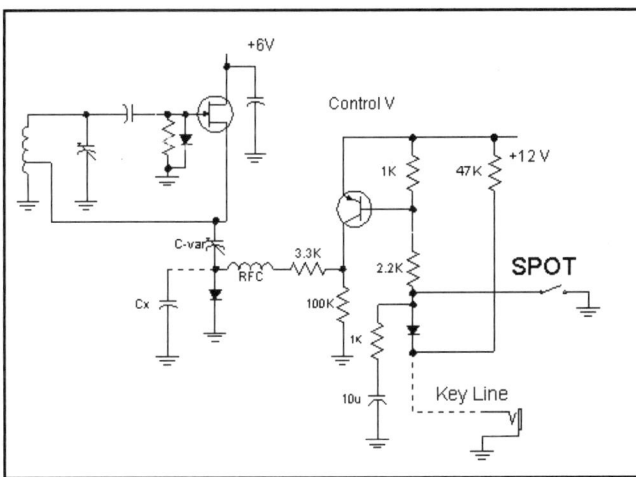

Fig 6.119—Offset system for a simple direct-conversion transceiver.

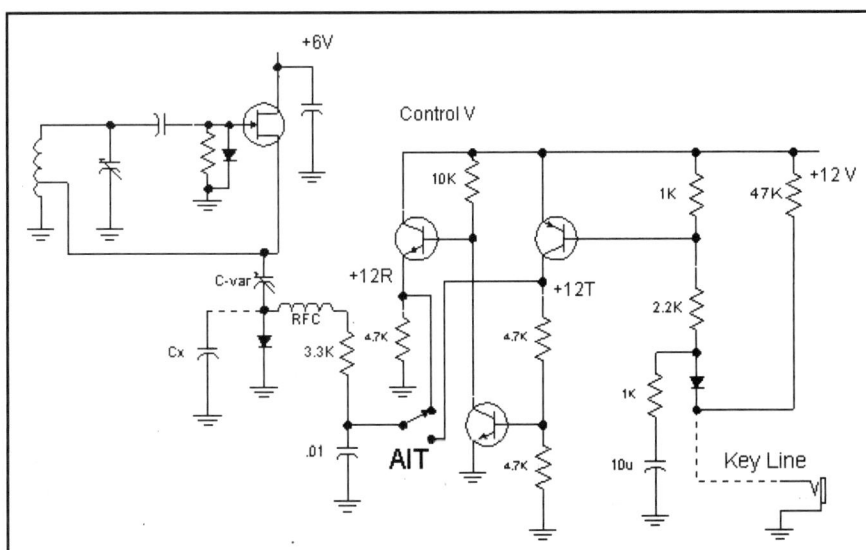

Fig 6.120—A VFO with offset capabilities and "AIT," Almost Incremental Tuning. This scheme allows the downward frequency shift in the VFO to occur on either transmit or receive, providing greater flexibility to avoid interference.

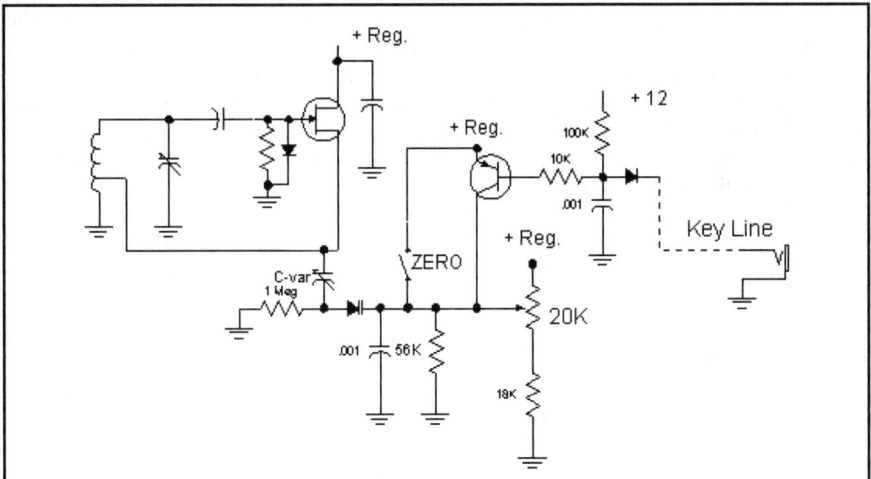

Fig 6.121—Simple RIT system developed by W7EL. This is a single-sided design where incremental tuning moves the VFO downward in frequency during receive periods, but only on one side of the transmit frequency. The general flexibility for effective RIT is retained. The tuning diode used by W7EL was actually a medium-voltage Zener diode, illustrating the simplifications that can be realized when one understands the behavior of the components. The system built by W7EL used a fixed capacitor where C-var is shown.

6.8 TRANSMIT-RECEIVE ANTENNA SWITCHING

An interesting design detail for a transceiver, and generally for any station is the way the antenna is switched between the receiver and transmitter. Something as simple as a manual switch will work and is used in some equipment in other chapters. However, the more common route uses either a relay or electronic switching methods. A traditional relay switch is shown in **Fig 6.122**. The RF part of the circuitry is presented in part A. The relay can be placed directly at the antenna terminal, but is shown here on the transmitter side of the usual low pass filter. Generally this scheme is preferred because the filtering is useful in both receive and transmit functions.

The example circuit in Fig 6.122B uses a 500-Ω relay coil. The relay current is switched with Q1, a saturated switch. Generally, the base current should be the collector value diminished by 10 to 20, so R1 is about 20 times the relay coil value. (The factor 20 is called a "forced beta" in this example.) Diode D1 serves to "catch" the voltage spike that will always occur when Q1 is turned off. Without the diode, the current that had been flowing in the inductive relay coil would "try" to continue flowing, generating the large spike as it charges the collector capacitance of Q1. This voltage surge can easily be large enough to destroy Q1.

If Q2 was not present, Q1 and the relay would be on. The base current in Q1 is shunted to ground through the collector of Q2 to control the relay. The Q2 base current is reduced by another factor of 20 over that in Q1. In the receive mode without the relay energized Q2 base current flows through R2 and the Zener diode, D2. The Zener voltage level is not critical, but should be near half the supply. The base current flows from the pot, R3, which provides a voltage from 6 to 12 V.

When the key is pressed, or a push-to-talk or VOX line goes low, the base current in Q2 is diverted away from the base. Q2 then stops conducting, causing Q1 and the relay to switch on. Pressing the key, etc,

but one balanced mixer are not single sideband receivers (even though they can be used to receive SSB.) Moreover, they are usually used to listen to CW signals that do not include sidebands other than the closely spaced key clicks.

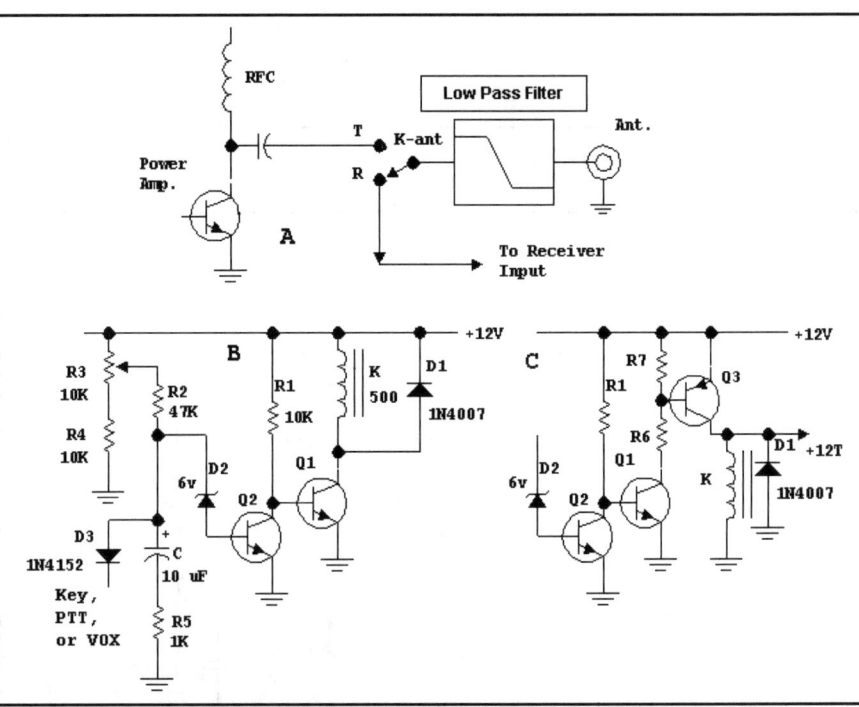

Fig 6.122—Relay T/R switching. The RF portion of the T/R switch is in part A while B shows a simple means for relay control. An expanded version is shown at C where higher relay current is allowed. Experimenters might wish to replace some of the transistors with some using built-in resistors found in parts catalogs, manufactured by Panasonic and others.

also discharges capacitor C. Resistor R5 in series with C restricts the current that must be conducted in the key when switched. The circuit does not change states immediately when the key is released. Rather, switching is delayed by the time interval required for C to charge through R2 until it reaches the Zener voltage.

Plastic switching transistors such as the 2N3904 are fine for Q1 and Q2. Fig 6.122C shows a scheme with a PNP that can be used when the relay current is much higher, or when additional current must be supplied for other transmit circuit functions. R6 is picked to provide a Q3 base current of about 5 to 10% of the current that must be supplied by the Q3 collector. General purpose PNPs for this application are the 2N5322 or the TIP32.

Fig 6.123 shows a common transmitter topology where the power amplifier (PA) is always attached to the antenna. The PA is cut off during receive periods, so it is essentially an open circuit with some parallel capacitance. Antenna energy is extracted through switch S1 to the receiver. This scheme is common, but it must be applied with care. The PA must not be conducting during receive; if it was, the collector resistance would absorb some of the signal that would otherwise reach the receiver. Also, conduction would generate excess noise that would compromise the receiver. It is also important to tap the receiver signal from a point in the low pass filter where the response will be maintained. For example, replacing the broadband transformer with a tuned network might lead to a shunt tuned circuit that would short some of the receiver energy to ground.

In some designs, a transmitter matching network might present an impedance lower than 50 Ω to the PA. This occurs when the output power is more than a watt or so from a 12-V supply. It is often tempting to tap the receiver signal from the PA collector. This may work, although if the impedance is much less than the receiver input impedance, the resulting mismatch can compromise performance. A matching network may be needed at the receiver input to increase the impedance back to 50 Ω.

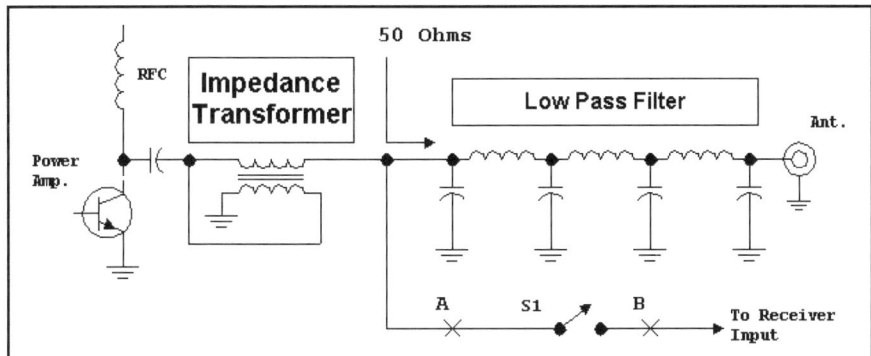

Fig 6.123—The RF portion of a T/R switch using a single switch. The transmitter is always connected to the antenna.

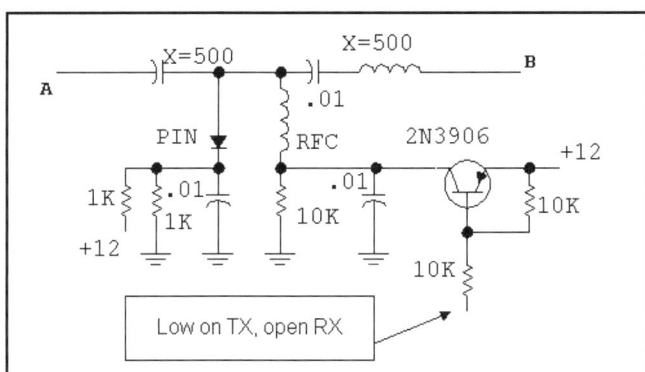

Fig 6.124—T/R switch with a shunt PIN diode.

The two sides of S1 are marked with A and B. A variety of switch circuits may be applied to generate the desired function. One is shown in **Fig 6.124**. Here, the switch is not a series element, but a shunt one realized with a PIN diode. The PIN diode is a common type used for RF switching. It departs from a normal PN switching diode with an intermediate region of *intrinsic* silicon. This has the effect of reducing switching speed, now a feature rather than a deficiency. The diode appears as a low valued resistor to radio frequency signals, but still as a diode for the dc controls. A PIN diode is capable of switching an RF current that is much larger than the dc current flowing. In contrast, a normal switching diode must be biased to a direct current that exceeds the peak RF current that is to be switched. The circuit in the figure biases the diode to 6 mA during transmit periods.

The shunt switch is effective in switching because it occurs within a tuned circuit. The usual capacitor at the end of a 50-Ω low-pass filter will have a reactance

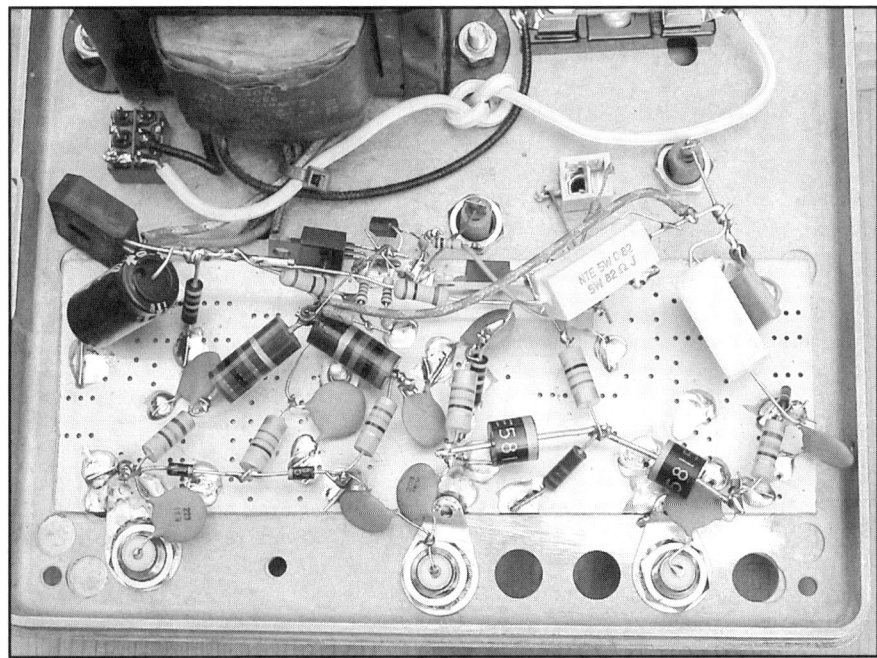

Inside view of 100-W T/R switch using inexpensive diodes.

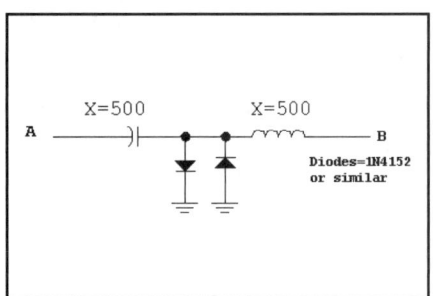

Fig 6.125—T/R switch with shunt PN diodes.

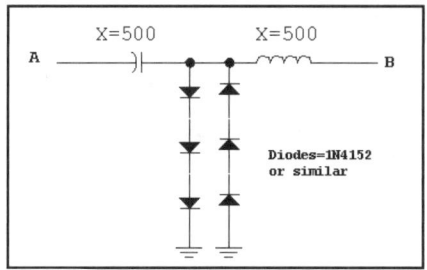

Fig 6.126—T/R switch with multiple PN diodes in each arm. This circuit features improved IMD. See text.

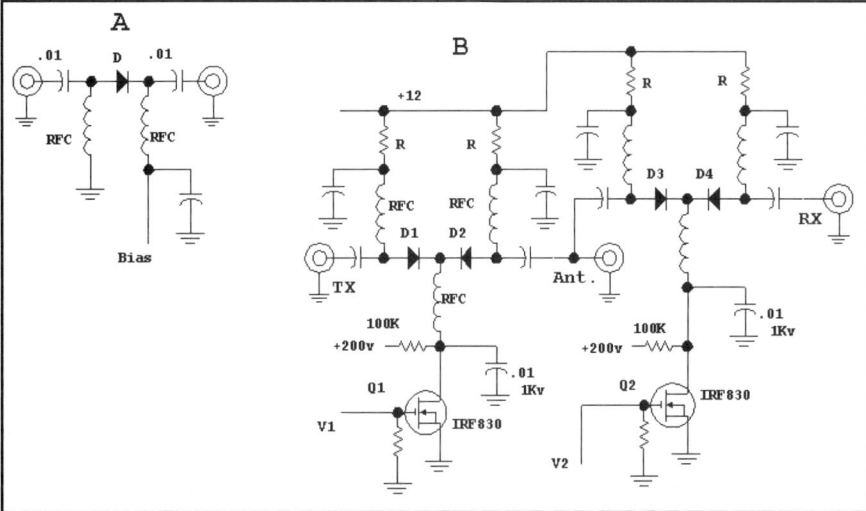

Fig 6.127—Part A shows the evaluation circuit. Poor "off" performance dictates the use of two series-connected diodes in each leg of the circuit in part B. Pick R to set the "on" current in the diodes.

around 50 Ω. The antenna signal is extracted from the low pass filter through a relatively small valued capacitor, one with a reactance of about 500 Ω. There is minimal receive loss, for it is tuned with a series inductor also with a 500-Ω reactance. When the junction of the two is switched to ground during transmit, the capacitor is merely paralleled with that in the end of the low pass filter, which will have little impact on transmitter performance. The inductance now in series with the receiver is useful in attenuating transmitter energy that might otherwise get to the receiver input.

A T/R switch of this sort is easily tested *before* a receiver is attached to guarantee that the power available to the receiver is low. The receiver end (B) of the switch is merely attached to a power meter and compared with the safe value for the receiver front end. A typical receiver with a diode ring as the first active element can usually tolerate 10 mW without damage.

The most common variation of the shunt T/R switch is shown in **Fig 6.125**.[41] Two common switching diodes (1N4152 typical) are placed in opposition. There is no controlling dc. Rather, when the transmitter is turned on, the RF causes the diodes to conduct, forming a relatively low impedance path to ground. We have measured this topology often (every time one is built) with the same result: The available output power at the receiver terminal is typically –10 dBm, easily within safe ratings for virtually any receiver. This power is independent of transmitter power.

The shunt diodes in Fig 6.125 can compromise the receiver dynamic range. Measurements with a 14-MHz example produced IIP3 of –3 dBm for the T/R switch, clearly a potential problem with high DR receivers. A solution is found in **Fig 6.126** where the single diodes are replaced by several series diodes. Two diodes per leg produced IIP3 of +7 dBm while three diodes per leg, the topology shown, yielded IIP3 = +13.5 dBm. The signals available at the receiver input increased to –4 and –1 dBm for the two and three diode per leg circuits. These levels will not cause damage to a receiver front end, but severe overload may occur.

Care is also required if these simple schemes are to be used at higher power. We have been able to extend the methods to the 100-W level, although only with circuit modification. The primary parameter to consider is the maximum current capability of the switching diodes. The 1N4152 that we have used in many circuits has a maximum current rating of 100 mA. The extended designs are discussed in a *QEX* paper.[42] This article is included in the CD that accompanies this book.

Another subtle, but significant problem occurs with this T/R scheme. The series-tuned LC is a tuned circuit that can interact with the tuned circuit(s) that follow to create a multiple-tuned circuit not in the designer's plans. The direct connection at (B) often leads to severe over coupling. The coupling can usually be adjusted to a proper level by inserting a suitable shunt capacitor at (B). Careful analysis is required.

Although the shunt diode switches presented are very useful for low power transceivers, they suffer from both IMD and power limitations, and are restricted to a single band. A wideband SPDT switch design with series diodes in the transmitter and receiver path would be more general. Our investigation of this topology begins with a simple single pole switch, shown in **Fig 6.127**, part A. This circuit is used to measure insertion loss and IMD with both forward and reverse diode bias. The IMD measurements should be done for both receiving conditions and at transmitter power levels when SSB use is planned.

High-power RF switching PIN diodes are available and discussed in the professional literature.[43] However, they are expensive and sometimes difficult to purchase. Our investigation, encouraged by K5CX, was directed toward inexpensive solutions. Many rectifier diodes are actually PIN structures, for this device topology tends to increase reverse voltage breakdown. The best inexpensive PIN diodes we found are the Motorola 6A6, a power supply rectifier specified for 6-A forward current and 600-V reverse breakdown. Diodes Inc manufactures similar parts. A forward bias current of 200 mA is enough for reliable operation at the 100-W level. We found identical performance with an NTE8515. We also had good results with the 1N4006, a 1-A, 800-V part.

While the forward biased performance was outstanding, the diode capacitance with reverse bias was relatively high, much higher than found with devices specified for RF switching. This made it necessary to put two diodes in series to obtain adequate reverse isolation. The SPDT topology used with a 100-W amplifier is shown in Fig 6.127B. It was necessary to go to 150 to 200 V of reverse bias to reduce capacitance of "off" diodes.

The reverse capacitance for the 6A6

diode was still 30 pF at 80-V reverse bias. The 1N4006 dropped to 3.6 pF at the same bias. We also investigated a Motorola 1N4007, a 1-A, 1000-V part and measured 2.1 pF at 80-V bias. In our final design we used the NTE8515 for D1 and D2 of Fig 6.127B, while 1N4006 diodes were used at D3 and D4. The 1N4006 was also satisfactory at D1 and D2 at the 100-W level, although this was not used for prolonged operation. The details of the T/R switch are shown in the *QEX* paper mentioned earlier. We used high-voltage HEXFETs for the bias switching. The switch insertion loss was so low that we could not measure it. Isolation was 56 dB between the TX and RX ports when the ANT port was 50-Ω terminated. IIP3 was greater than +40 dBm in the receive path. The IMD measurement was limited by the spectrum analyzer used and IIP3 may be even better.

We often wish to use a power amplifier driven by a transceiver. A suitable switching topology for this chore is shown in **Fig 6.128**. Three switches are shown. Only that at the PA output, SW3, would require the higher current diodes. SW1 and SW2 could use the less expensive 1N4006 or 1N4007.

Fig 6.129 shows a single band T/R switch using shunt PIN diodes, suitable for VHF as well as HF application. Quarter wavelength transmission lines interconnect the ports and switches. The diodes have reverse or zero bias during receive, but are forward biased during transmit. D1, behaving as an open circuit during receive, causes a short circuit to appear at the transmitter output. But open circuit D2 allows the nominal 50-Ω input of the receiver to appear at the antenna port. Switching to transmit forward biases both diodes. D1, now a short, reflects as an open circuit at the transmitter output. D2, also a short circuit, protects the receiver and presents an open circuit at the antenna port. The antenna impedance now appears at the transmitter output. This circuit can be implemented with true transmission lines or with pi networks as shown in Fig 6.129. The pi-network that behaves like a quarter wave 50-Ω line has L and C each with a 50-Ω reactance at the operating frequency. This circuit is used in a 17-m DSP-based transceiver presented later in the book.

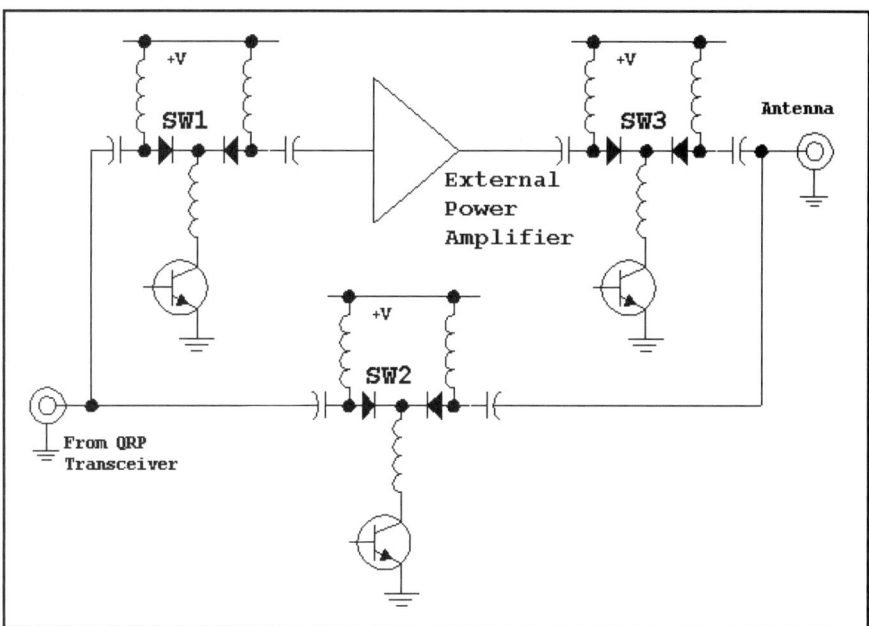

Fig 6.128—A T/R switch topology suitable for use following a low- power transceiver. We have not built this circuit.

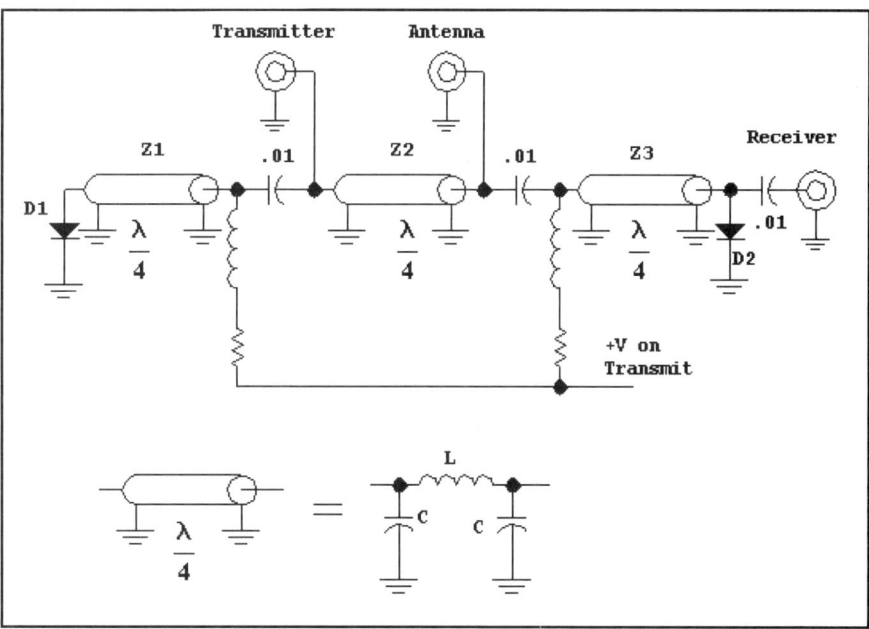

Fig 6.129—A T/R switch with shunt diodes using the impedance-reflection properties of quarter-wavelength transmission lines.

6.9 THE LICHEN TRANSCEIVER: A CASE STUDY

There are several suitable block diagrams for single sideband transceivers. The one we prefer shares only the oscillators, allowing receiver and transmitter optimization without compromise of interaction.[44] Although that scheme uses more parts, all basic functions are isolated with minimal interaction.

This transceiver, which is more efficient in its utilization of components, is an outgrowth of an architecture used by VE7QK in several versions of his *Epiphyte*.[45,46,47]

This format, used in some early military SSB gear, shares many of the circuit elements between modes with signals flowing in the *same* direction in transmit and receive. The transceiver is presented here to illustrate design ideas and to present some of the steps needed to build such a transceiver.

Block diagram

The system with two mixers is shown in **Fig 6.130**. The first serves as the front end mixer during receive and as a transmit balanced modulator. The second is a receiver product detector and an IF-to-RF converter during transmit.

The original Epiphyte used NE602 mixers with no IF gain. The rig was intended for field use in the rugged mountains of the British Columbia Coast Range. The Lichen uses diode-ring mixers and includes IF gain. The 75-m-band Lichen can be adapted to many other bands.

The price of simplified signal flow is complex LO and carrier oscillator switching. The NE602 mixers used in the Epiphyte required little power in the 3-MHz LO, allowing switching with CMOS parts. The Lichen performs the switching with diodes, a scheme selected for compatibility with higher frequencies.

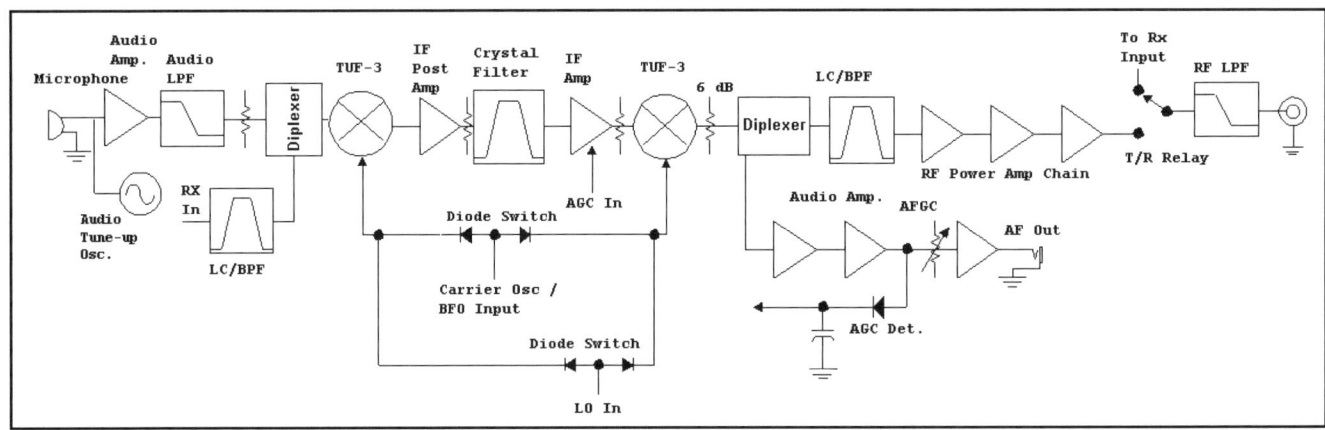

Fig 6.130—Block diagram for the Lichen transceiver.

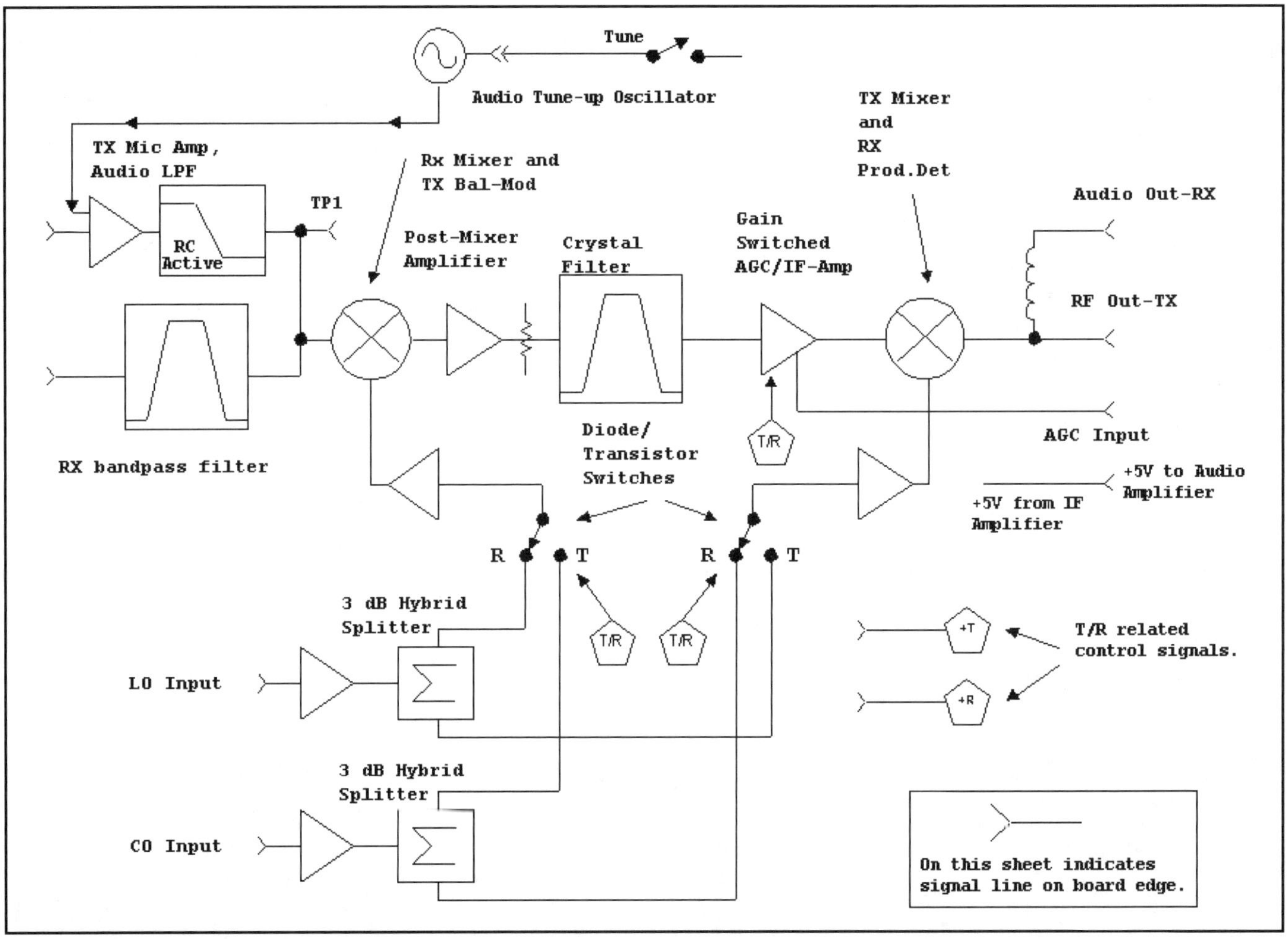

Fig 6.131—Block diagram for the transceiver main board.

6.72   Chapter 6

Signal flow in the "Main Board"

The transceiver is broken into several boards, a definite aid to the tedium of detailed measurements. The "main" board contains the receiver input preselector, a microphone amplifier, the two mixers, the IF system including crystal filter, and LO buffers and switching. The board includes an audio oscillator to facilitate testing. A block diagram is shown in **Fig 6.131**. The complete schematic is in **Fig 6.132**.

The main board begins at J2 where a signal enters the receiver input. ("J" numbers designate pads at the edge of a board.) The receiver preselector is a double tuned circuit using series resonators formed from molded RF chokes. The filter output is applied directly to the first mixer, U2. Bandpass filters for other bands are listed in **Fig 6.133**. The 160 and 80-m filters use $Q_U = 50$ RF-choke inductors while the higher bands use toroid inductors with $Q_U = 200$.

The microphone input is amplified and low pass filtered in U1. An RFC in the mixer line with capacitors in the receiver input filter form a diplexer to combine audio and receiver RF signals for the mixer. The microphone-amp is adjusted for a (lower than normal) signal of –20 dBm applied to the mixer.

The prototype transceiver used a commercial crystal filter while another (Fig 6.132) used a homemade 9.2-MHz crystal filter. The filter input is driven by a 2N3904 post mixer amplifier. *Post-amp* gain is 19 dB, reduced to 13 dB by the 6-dB pad, and has a 50-Ω input and output impedance. The sixth-order crystal filter is designed using the methods presented in Chapter 3.

An L-network (L4, C36) transforms the post-amp 50 Ω to the needed filter source impedance. Transformer T3 matches the relatively low filter impedance to the 2.2-kΩ input resistance of the following IF amplifier. T3 uses a 61-material ferrite core to keep the loss low. The filter should be built and measured before incorporation in the transceiver. The exact –6-dB filter frequencies should be recorded for later use. The designer/builder will have to design matching networks and transformers as well as the crystal filter.

Front panel view. Switch between tuning and audio gain is a sub-band switch. The push-button injects an audio tone for tuning.

JFETs Q6 and Q8 provide IF gain. These stages are gain switched by Q7 and Q9 with higher gain during receive. Reasonable IMD performance is vital, for the amplifier is in the transmit signal path. This system (Q6 and Q8) has a small signal receive gain of 27 dB with 70 to 80 dB of available gain reduction. Gain drops to 12 dB in transmit. IMD performance is good at OIP3 = +18.5 dBm, dropping to +14 dBm in transmit mode. IMD degrades with gain reduction, but the intercepts do not degrade as fast as the gain, a requirement to preserve output cleanliness. Receiver AGC is disconnected during transmit; R58 is switched in to establish a transmit level.

Transmit mixer, U3, should see maximum drive of –10 dBm for a spur free output, as discussed earlier. The post-amp, Q5, including pad has a gain of 13 dB while typical crystal filter loss is 4 dB. With a balanced modulator input of –20 dBm, the signal at the input to T3, just past the crystal filter, is –17 dBm. Transmit gain of 12 dB in the IF brings the level at U3 to –5 dBm. A slight IF gain reduction and a 3 dB pad in the IF output sets the –10 dBm level. If the balanced modulator had been driven at its nominal level of –10, the IF would be overdriven, resulting in overdrive for the second mixer.

This gain distribution degrades carrier suppression to 30 dB. If the post-amp gain could be reduced by 10 dB during transmit, the carrier suppression would be improved by a like amount.

With a U3 mixer drive of –10 dBm, the 6 dB conversion loss produces an output of –16 dBm. A 6-dB pad after the mixer and a bandpass filter (described later) with a 2-dB loss produce an eventual output of –24 dBm, established by R58.

The audio tune-up oscillator included in Fig 6.132 can be used during normal operation to generate a carrier for transmatch tuning. It is also available for

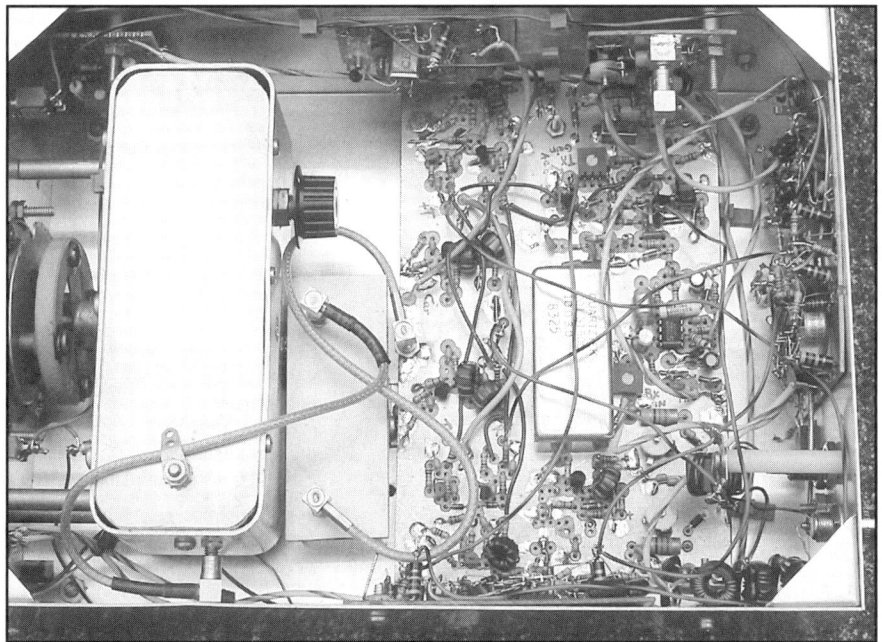

Top view showing LO module with "main board" to the right. The small box built from scrap circuit board material contains the 14-MHz-LO bandpass filter.

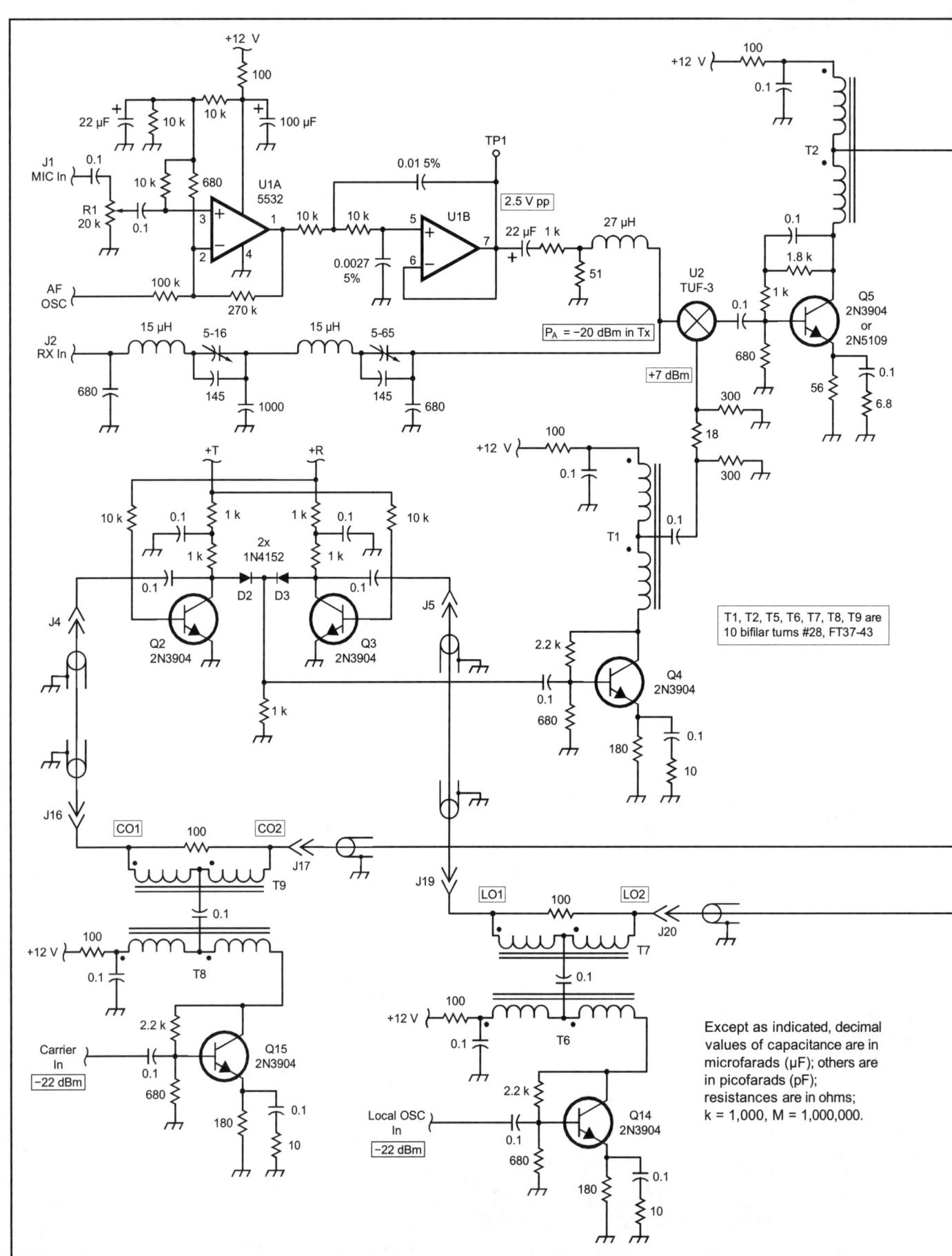

Fig 6.132—Schematic for the main board. See text for detailed discussion.

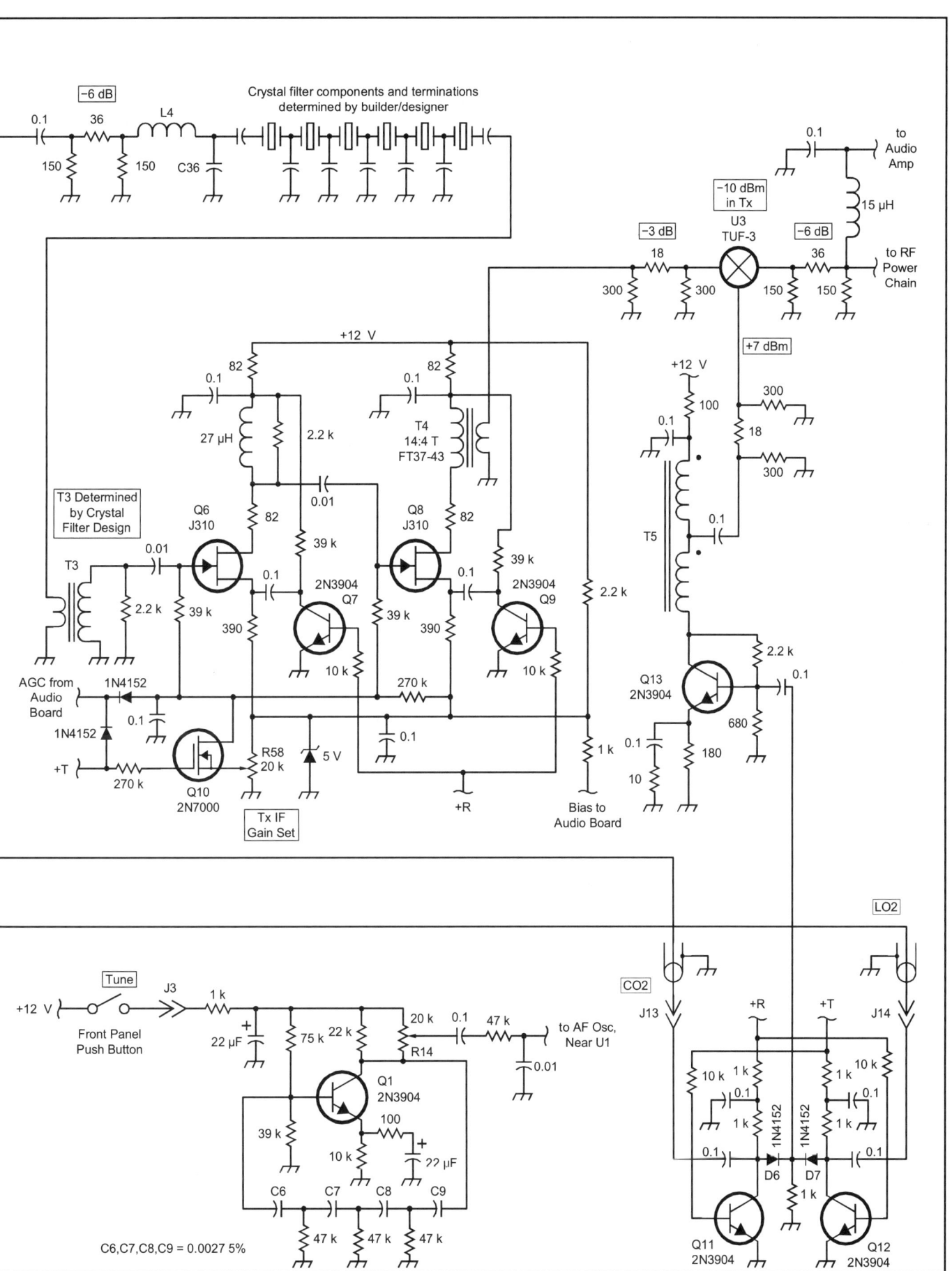

testing during board development. The microphone is attached at the amplifier input, J1, and the level at test point TP1 is observed. Audio gain (R1) is adjusted for 2.5 V peak-to-peak at TP1 on voice peaks with a normal voice into the microphone. The tune-up oscillator level, R14, is then set to produce the same level.

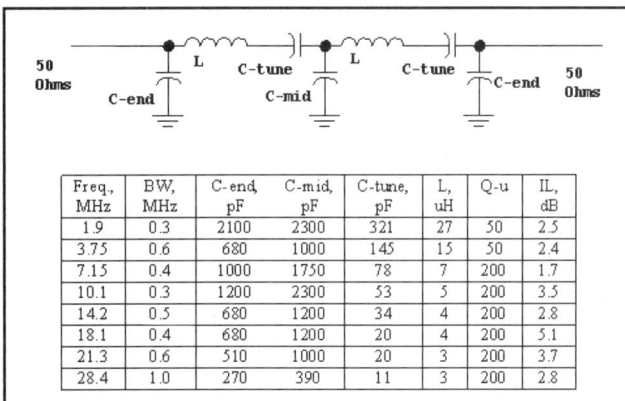

Fig 6.133—Receiver bandpass filters using series resonators.

Freq., MHz	BW, MHz	C-end, pF	C-mid, pF	C-tune, pF	L, uH	Q-u	IL, dB
1.9	0.3	2100	2300	321	27	50	2.5
3.75	0.6	680	1000	145	15	50	2.4
7.15	0.4	1000	1750	78	7	200	1.7
10.1	0.3	1200	2300	53	5	200	3.5
14.2	0.5	680	1200	34	4	200	2.8
18.1	0.4	680	1200	20	4	200	5.1
21.3	0.6	510	1000	20	3	200	3.7
28.4	1.0	270	390	11	3	200	2.8

Close up of the main board.

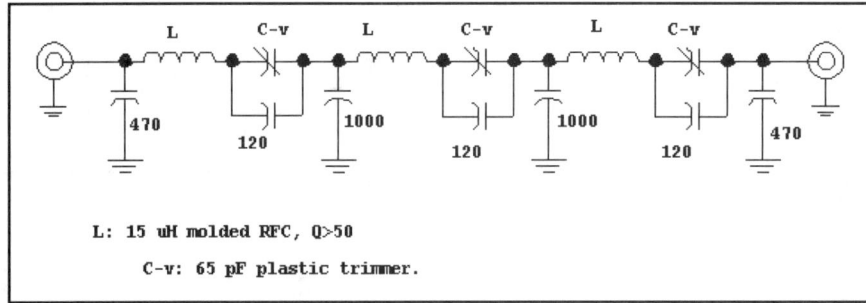

L: 15 uH molded RFC, Q>50
C-v: 65 pF plastic trimmer.

Fig 6.134—Triple-tuned 3.5 to 4-MHz bandpass filter for the output of the transmit mixer.

Mixer Injection Switching

The 10-MHz IF version uses a 13.5 to 14-MHz LO and a 10-MHz carrier oscillator (CO). The LO must be applied to U2 in receive while the CO drives U3. Roles are then reversed in transmit with the LO driving U3 and the CO driving U2.

Each ring mixer requires nominal LO power of +7 dBm. But lower power levels are switched. Drive amplifiers Q4 and Q13 reduce the switched power to –9 dBm, easily controlled with normal silicon diodes biased for modest current. Diodes D2 and D3 switch the signals going to U2 while D6 and D7 route energy to U3. These switches are controlled by signals labeled with T or R, indicating positive bias on either transmit or receive. These signals, appearing often throughout the transceiver, are generated on the RF power amplifier board. The diode switches route a desired signal to an intended load, but do not present as much attenuation of the *off* path as we would like. Shunt transistor switches Q2, Q3, Q11, and Q12 were added to provide about 50 dB reduction in the *off* paths.

Although the shunt transistor switches improve performance, they add a complication: Each input (LO and CO) is amplified and buffered in an amplifier, Q14 and Q15. If those amplifier outputs were routed directly to the composite diode/transistor switches, they would always be short circuited. Isolation results from transformers T7 and T9 which function as a splitter-combiner, described in Chapter 3. These switching methods can be extended to UHF. LO and CO signals are required at the board inputs with a power of –22 dBm.

The circuit board contains short lengths of coaxial cable to route the LO and CO signals. The two LO components, LO1 and LO2, move respectively from J19 to J5 and from J20 to J14 on cable. The CO signals CO1 and CO2 move respectively from J16 to J4 and J17 to J13. The best place to measure LO chain power is just before the mixers. Lift C29 or C59 at the pad ends and measure the power coming from the LO system. Those powers should both be close to +10 dBm. The LO amplifiers use 2N3904s, but the less robust MPS3904 is *not* suitable. The MPSH10 (Fairchild and Philips) is also an excellent choice.

Transmit Bandpass Filter

The Main board RF output at 3.5 to 4 MHz has a 23.4 to 24 MHz image. The lower range is selected with the filter shown in **Fig 6.134**. This circuit is best

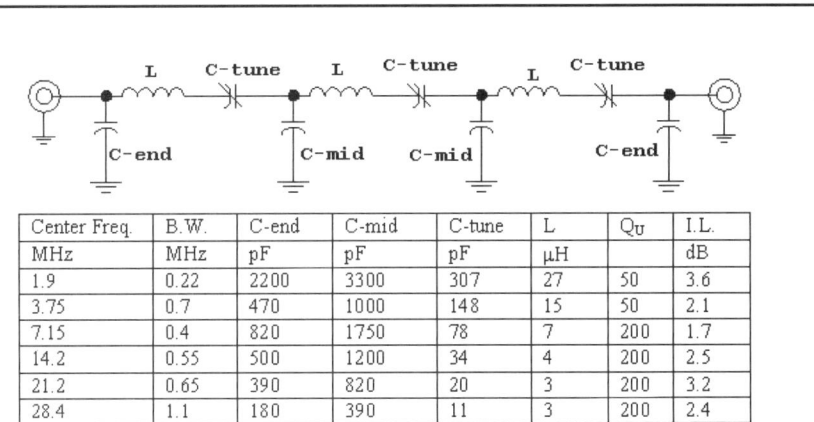

Center Freq. MHz	B.W. MHz	C-end pF	C-mid pF	C-tune pF	L µH	Qu	I.L. dB
1.9	0.22	2200	3300	307	27	50	3.6
3.75	0.7	470	1000	148	15	50	2.1
7.15	0.4	820	1750	78	7	200	1.7
14.2	0.55	500	1200	34	4	200	2.5
21.2	0.65	390	820	20	3	200	3.2
28.4	1.1	180	390	11	3	200	2.4

Fig 6.135—Triple-tuned bandpass filters for several HF bands. The required unloaded Q (vital) is also given.

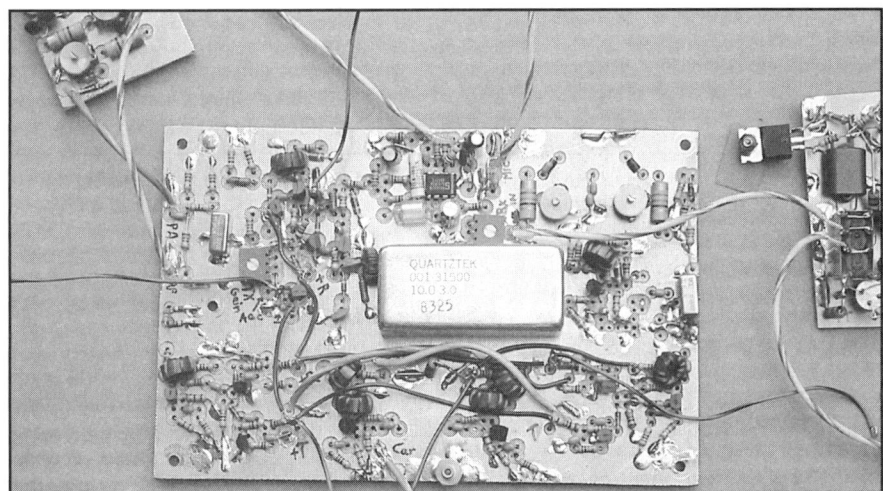

Main board removed from cabinet. Circuitry below crystal filter is for the LO and carrier oscillator buffers and switches. Upper right corner contains RF Input bandpass filter.

assembled and tested in a 50-Ω environment prior to use in the transmitter. A table of computer generated values is given in **Fig 6.135** for several additional bands.

The Local Oscillator

The LO tunes from 13.5 to 14 MHz with the heterodyne system of **Fig 6.136**. Q402 is a 2.5 to 3-MHz Colpitts oscillator buffered with a common-base amplifier, Q405. Output is kept low, for only –10 dBm is needed by diode ring mixer U402. The output is established with the pad driving the RF port. This level, and that at the mixer LO port should be measured during construction.

A 365-pF variable capacitor tunes only half of the range. The other half is tuned by switching in an additional capacitor, C403. The switching is performed with a pair of PIN diodes, D401 and D402. When a positive voltage is applied to J401, Q401 is saturated, causing both PIN diodes to conduct.

A crystal controlled 11-MHz oscillator provides the drive for the diode-ring mixer. The two oscillators are both placed inside the shielded LO enclosure, along with the ring mixer. The output is then routed through coaxial cable to a triple-tuned LC bandpass filter, **Fig 6.137**.

A change in IF from 10.0 MHz will result in the need for a new LO frequency on the part of the designer/builder.

The Carrier Oscillator

A carrier oscillator (CO) drives the balanced modulator in transmit and the BFO in receive. The CO must have the same –22 dBm level as the LO when applied to the Main board. The CO circuit is shown in

View of LO.

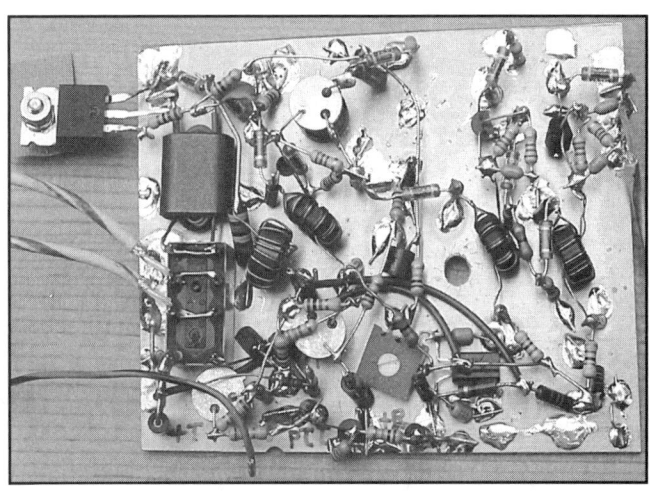

RF Power chain. The HEX-FET PA is normally attached to the cabinet that serves as a heat sink.

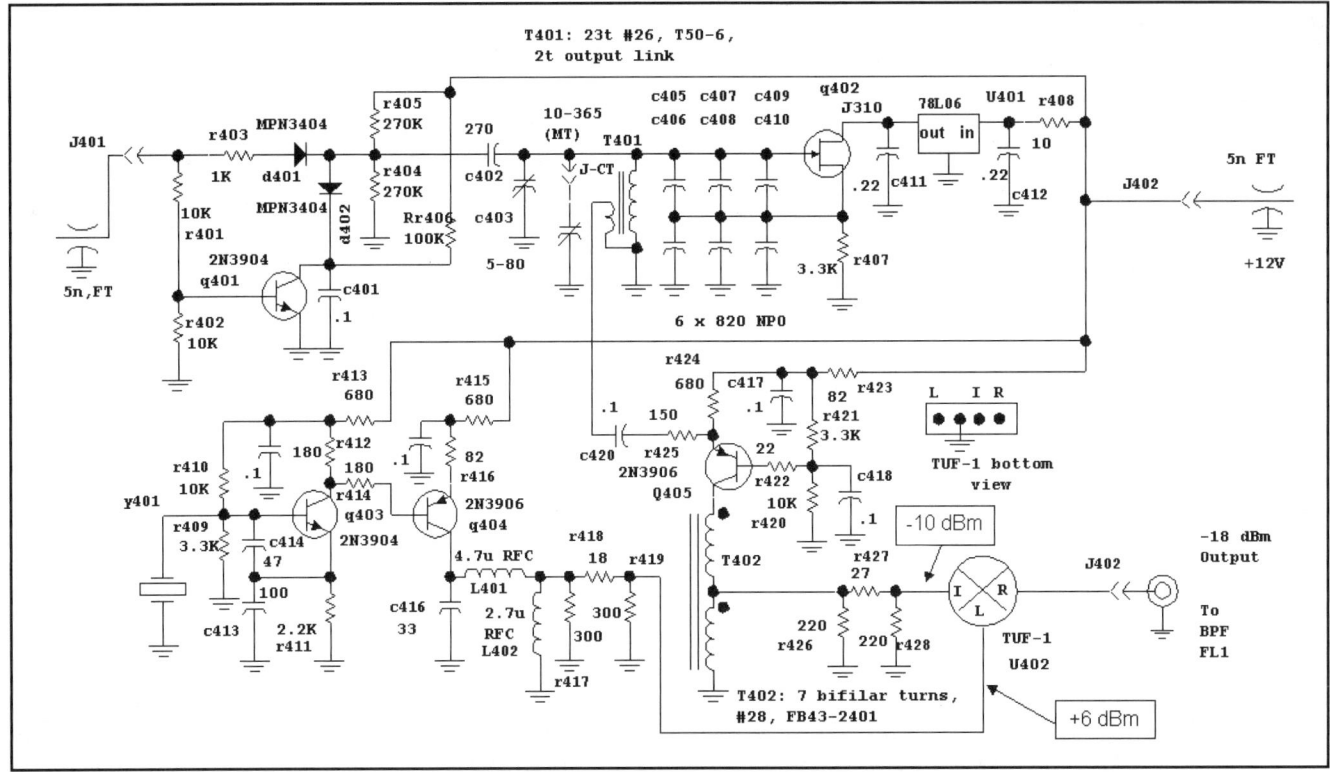

Fig 6.136—Transceiver LO system produces output at 13.5 to 14 MHz. The bandpass circuit of Fig 6.137 filters the mixer output. L401 and 402 are labeled as RF chokes because the network is formed with molded RF chokes.

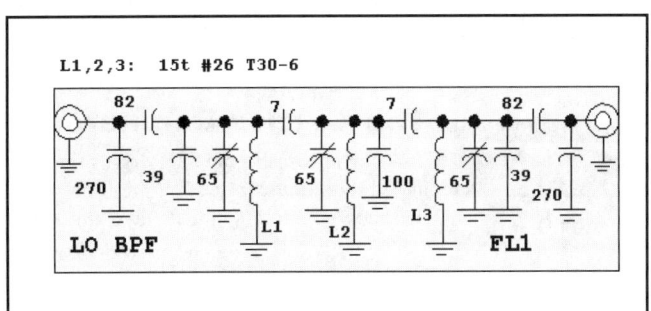

Fig 6.137—LO bandpass filter.

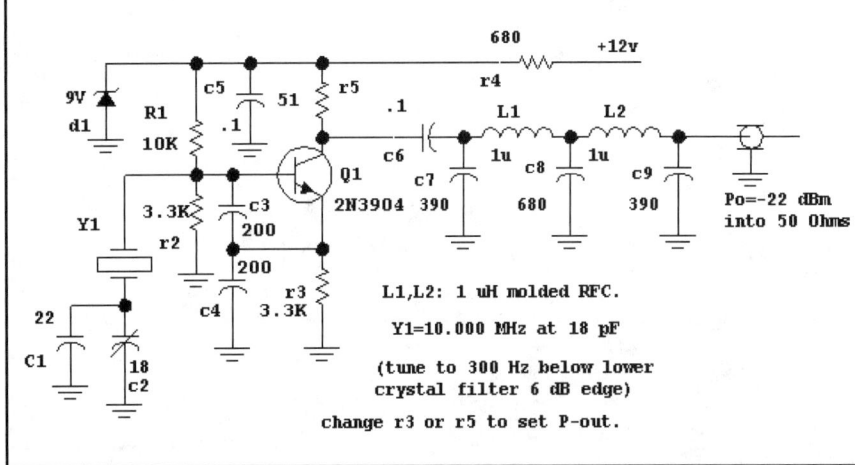

Fig 6.138—Carrier oscillator.

Fig 6.138. The output power is set at –22 dBm by adjustment of R5 in the oscillator collector. The power supply is regulated more as a means to stabilize amplitude than frequency.

We measured the crystal-filter response during circuit development. Knowing the exact lower 6 dB passband edge, we placed the carrier oscillator at a frequency 300 Hz below that edge. The resulting 10-MHz USB signal is inverted to become an LSB output at 3.8 MHz. Slight frequency adjustment may be done to optimize signals.

The Receiver Audio System

Fig 6.139 shows the audio system. The product detector output reaches the board via coaxial cable where it is amplified by Q301 and Q303, and applied to an off board audio gain control. The result is then amplified in two op-amp stages, U301, and applied to headphones.

The signal at the gain control is sampled and routed to op-amps U302 for full wave rectification. This charges the AGC sampling capacitor, C315, a 1 µF stacked metal film type (Panasonic V-series or similar.) R325 controls attack time while R324 sets recovery. U303A is a follower to drive the IF system with dc. Normal audio muting is not required. AGC was disconnected from

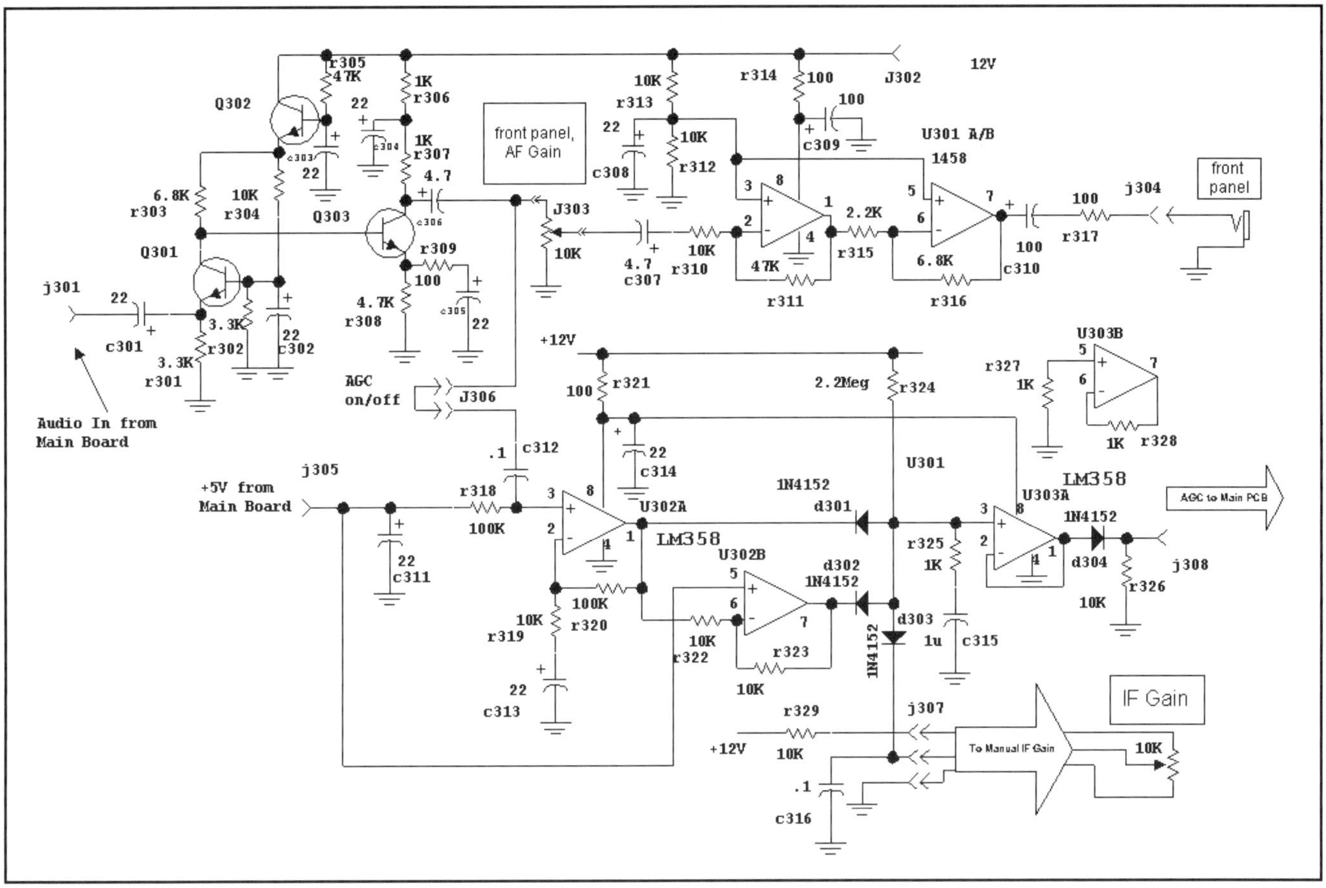

Fig 6.139—Audio system and AGC detector.

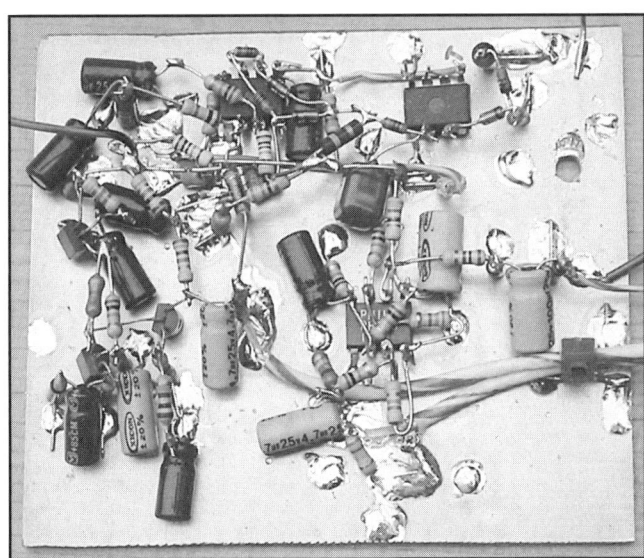

Audio Amplifier.

Carrier Oscillator.

the IF during transmit with D4, D8, and Q10 on the Main board.

The RF Power Chain

A four-stage RF power chain, **Fig 6.140**, completes the transceiver. Three bipolar transistors drive a HEXFET PA for a 5-W output.

The first two stages use a 2N3904 while the third uses a 2N3866 with a small heat sink. The three are respectively biased at 10, 17 and 50 mA. A 6-dB pad is placed after the first stage, providing a convenient place to alter gain for use on other bands. **Fig 6.141** shows gain vs frequency for the three stage bipolar driver. Although gain is dropping, the driver chain is useful through the entire HF spectrum. We realized another 3-dB gain at 50 MHz when Q101 and Q102 were changed to MPSH10s. IMD was measured at 14 MHz for the driver chain, producing OIP3 = +39 dBm with either transistor type in the first two stages. The nominal output for Q103 is +10 dBm per tone with a two-tone test, or +16 dBm (40 mW) PEP.

The PA, an IRF-510 HEXFET, is biased

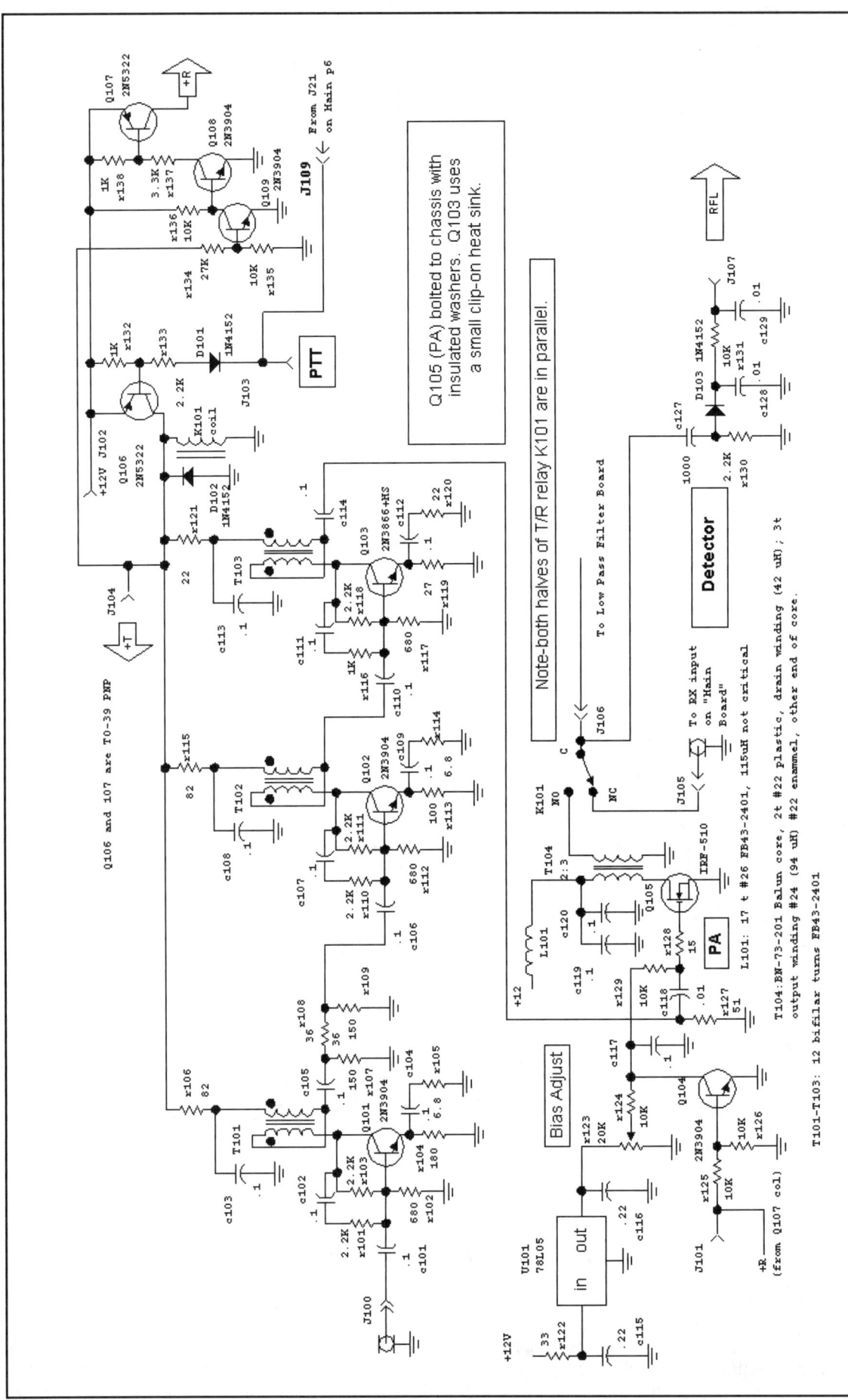

Fig 6.140—RF driver chain for the Lichen transceiver uses 4 stages for an output of 5 W. The T/R relay is a Nais DS2Y-S-DC12V or similar.

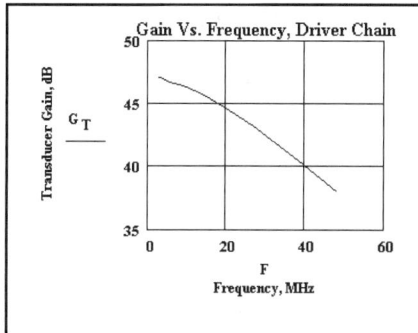

Fig 6.141—Small signal gain vs frequency for the three-stage bipolar driver chain.

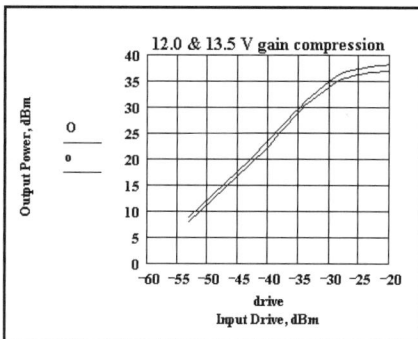

Fig 6.143—Gain compression measurement for complete RF power chain.

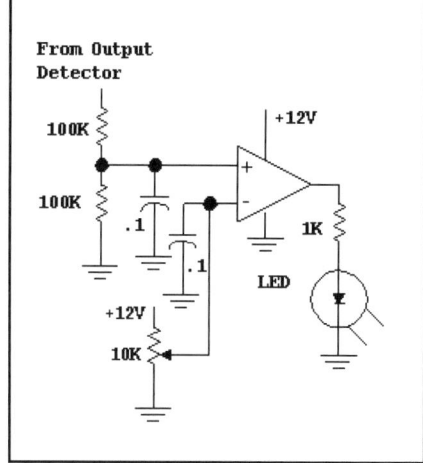

Fig 6.144—LED driver circuit that can be driven by the output peak detector. Op-amp is a 741, 1458, LM358, LM324, or similar part.

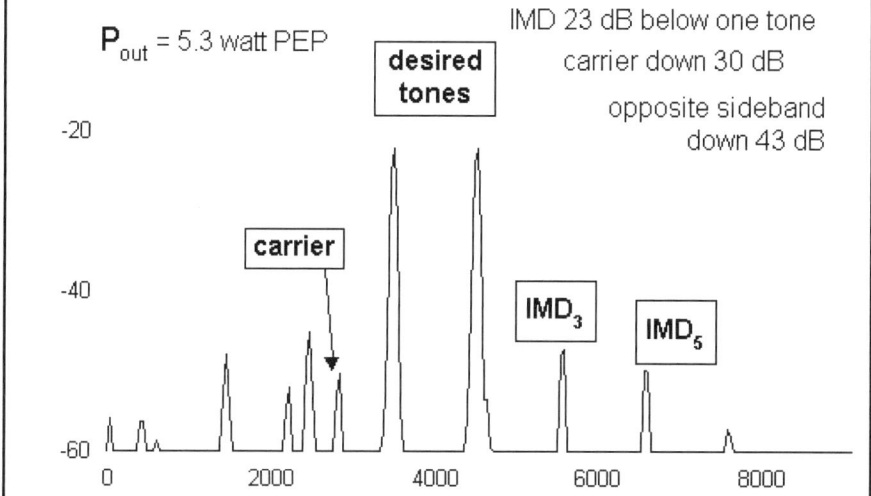

Fig 6.142—Spectrum analyzer view of transmitter output under two-tone testing. For software, see www.n0ss.net.

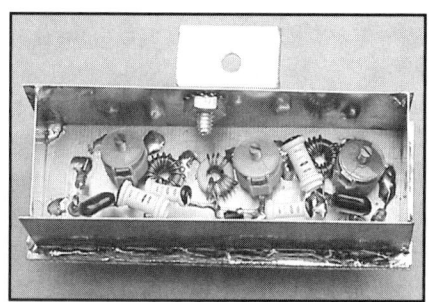

A view of the 14-MHz bandpass filter used for LO in transceiver.

from a pot driven by U101, a 78L05. Bias current with no drive is set for about 40 mA, a level producing excellent gain and distortion acceptable for QRP efforts. Transmitter output is shown in the two-tone test spectrum of **Fig 6.142**. This was obtained with a FFT spectrum analysis program, Spectrogram, running on a laptop computer, augmented with a converter. (See spectrum analysis discussion in Chapter 7.) Third order IMD is only 23 dB down from each tone, or 29 dB below PEP. The 30-dB carrier suppression is also shown. Opposite sideband suppression was 43 dB for a 1700-Hz single audio tone. Earlier driver chain measurements confirm the FET PA as the distortion source.

Fig 6.143 shows power chain output power as a function of drive power. This gain compression measurement was done with single-tone drive. The amplifier is relatively linear up to the +33 to +35 dBm output. This is a measurement that can be performed in the home lab that has yet to include a spectrum analyzer.

A peak detector is included at J107, useful during transmitter setup. It can also be used to drive a front-panel LED through a circuit like that shown in **Fig 6.144** where an op-amp serves as a comparator. Alternatively, the detector could drive an auto level control (ALC) circuit to provide negative feedback to the IF.

An IF speech processor was described in an earlier section where limiting within the IF constrained the output level. That scheme had the added advantage of preventing excessive levels in the transmit mixer and following amplifiers, eliminat-

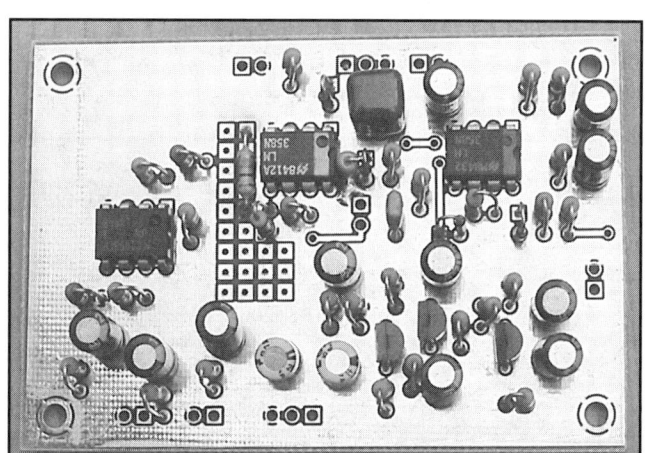

Printed circuit audio amplifier. (TNX to K7TAU)

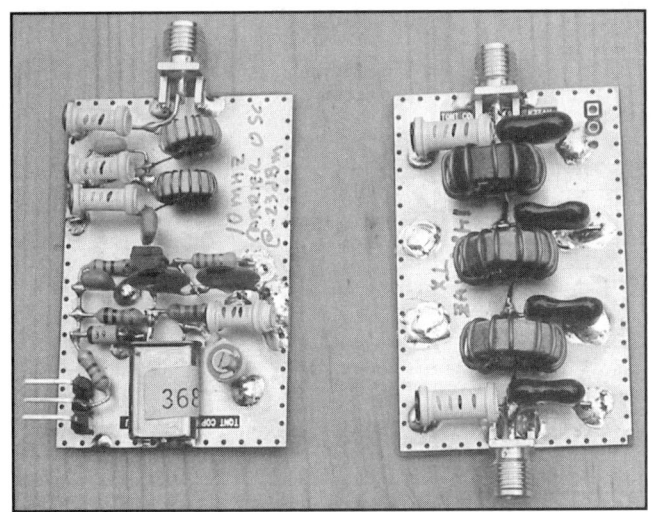

Breadboarded carrier oscillator and TX low-pass filter for a 14-MHz version of the transceiver by K7TAU.

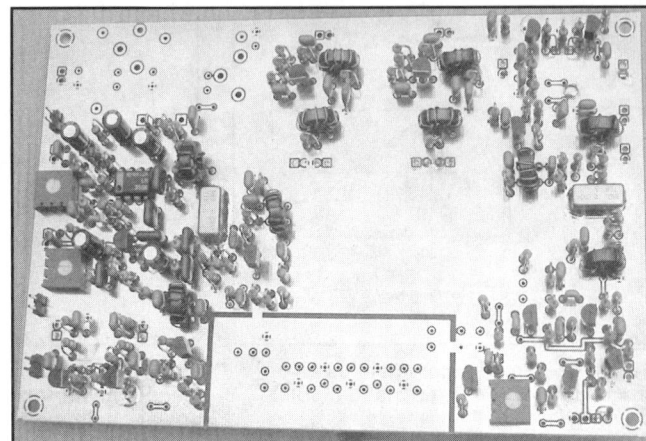

Partially built printed main board.

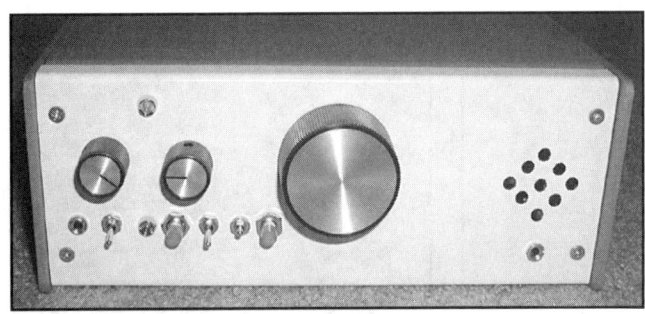

Front panel of 75-meter version built by AA7QU. One of the buttons activates a "Freq-Mite" frequency keyer that then reads the frequency and presents it in morse code.

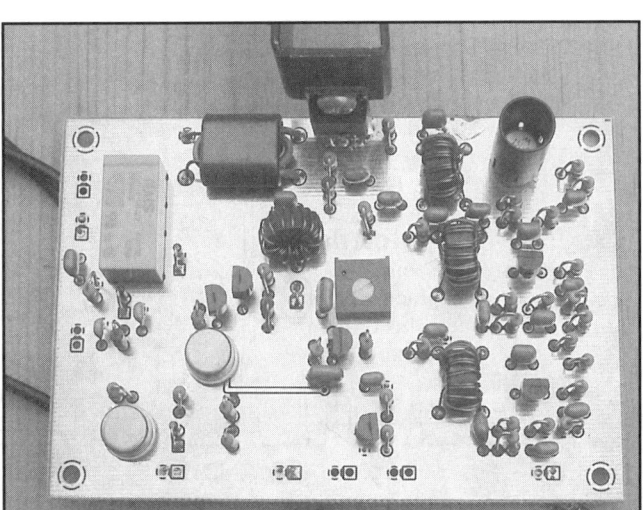

Printed Circuit Version of RF Power Chain. (TNX to K7TAU)

ing the need for ALC. The IF limiter has the minor disadvantage of requiring another crystal filter. However, it would be a dramatic virtue in this transceiver. Not only would it enhance transmitter performance, but it would generate excellent receiver skirt selectivity.

A seventh-order low pass filter follows the FET power amplifier, as shown in **Fig 6.145**. The filter is built on a separate board, isolated from the rest of the PA.

Control Circuits

The transceiver uses push-to-talk (PTT) operation, realized with the control circuitry included in Fig 6.140. When the microphone PTT button is pushed, a line goes low at J103 to saturate PNP switch Q106. That transistor powers antenna relay, K1, and feeds a +12V-T signal to the many places in the transceiver marked with "T." Q107, 108, and 109 then provide a similar +12V-R to control the receive function. PA bias is shorted with Q104 during receive periods. Both sections of the DIP antenna relay are paralleled for the T/R switching.

Extensions and Results

Once the boards are built and measured, they can be assembled and combined.

The system using a 10-MHz IF is reasonably clean with the second harmonic at −57 dBc as the dominant spur. Three non-harmonic spurs were found with strength from −67 to −62 dBc. A 9.2-MHz IF version (built by AA7QU) had similar performance. We were disappointed in the IMD performance offered by the HEXFET PA.

Receiver performance was adequate for the 75-m band. The relatively high noise figure of 18 dB is not a problem for this frequency. Measured IIP3 was +16 dBm and two-tone DR was 92.7 dB. The dynamic window is skewed to favor high intercept rather than low noise. A low-

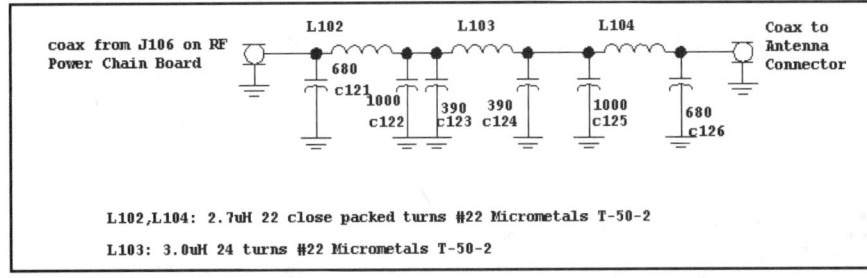

Fig 6.145—Low-pass filter for the 75-meter Lichen. Capacitors can be silver mica or ceramic.

noise RF amplifier with modest gain would substantially improve noise figure with little DR penalty, making this general topology useful at higher frequency.

Several boards were used in favor of a few, allowing the designer/builder to measure those parameters so critical to success. If the Main board was built without the input preselector filter, it would contain no band-specific components. The RF power chain and audio board are also band-independent, suggesting a multi-band design. Relay switching is recommended in the receiver front-end over PIN diodes to avoid second-order distortion problems.

6.10 A MONOBAND SSB/CW TRANSCEIVER

Although this transceiver was designed for operation on any single band within the HF spectrum, there is no fundamental reason it will not also function at VHF. Like the Lichen presented earlier, it is based upon homebrew crystal filters fabricated by the designer/builder.

This radio was designed for flexibility and performance. A common local oscillator system and common BFO/Carrier Oscillator are shared between the transmit and receive functions. The other functions are independent, allowing each to be optimized to meet the needs of the designer/builder/user. This seemingly inefficient approach becomes practical and inexpensive when one builds his or her own crystal filters. Although more extensive, the project is often less tedious than other sideband transceivers, for the receiver can be finished and made operational before dealing with the transmitter.

A collection of small circuit boards was used. Some were etched while others were merely breadboarded. The use of many small boards rather than just a few large ones provides improved isolation between functions and enhanced testability. A transceiver block diagram is shown in **Fig 6.146**.

The block diagram includes some shaded areas where circuit modules already presented are applied. The receiver begins with the "General Purpose Monoband Receiver Front-End" of Fig 6.68. That board includes a crystal ladder filter with up to 6 resonators. The next block is an IF amplifier. The recommended design here is that presented in Fig 6.50 using cascode connected J310 JFETs. Designs using some of the more up-to-date integrated circuits from Analog Devices should also be considered. Neither the front-end nor the IF will be discussed here.

The RF power chain is also shaded in the block diagram of Fig 6.146. A similar module developed for the Lichen transceiver would be suitable. Substitution of a different PA is recommended if the system is built for bands at the high end of the HF range, or for VHF. The poor IMD performance of the IRF510 would also be justification for a new PA design.

The monoband transceiver version

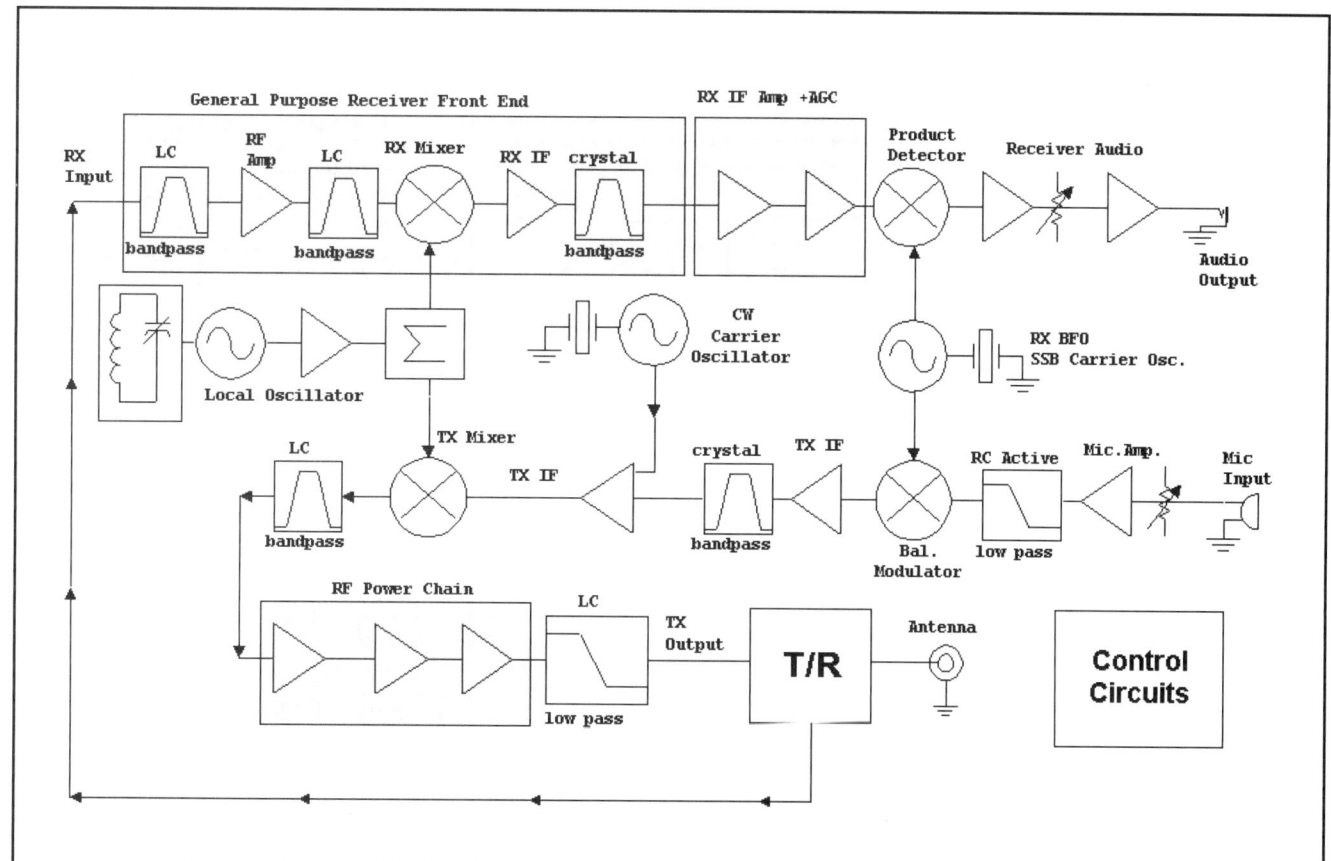

Fig 6.146—Block diagram for the SSB/CW transceiver. The version we built is for the 6-m band, but can be adapted to any band from 1.8 to 144 MHz. The system shown in the block diagram uses a non-heterodyne VFO system.

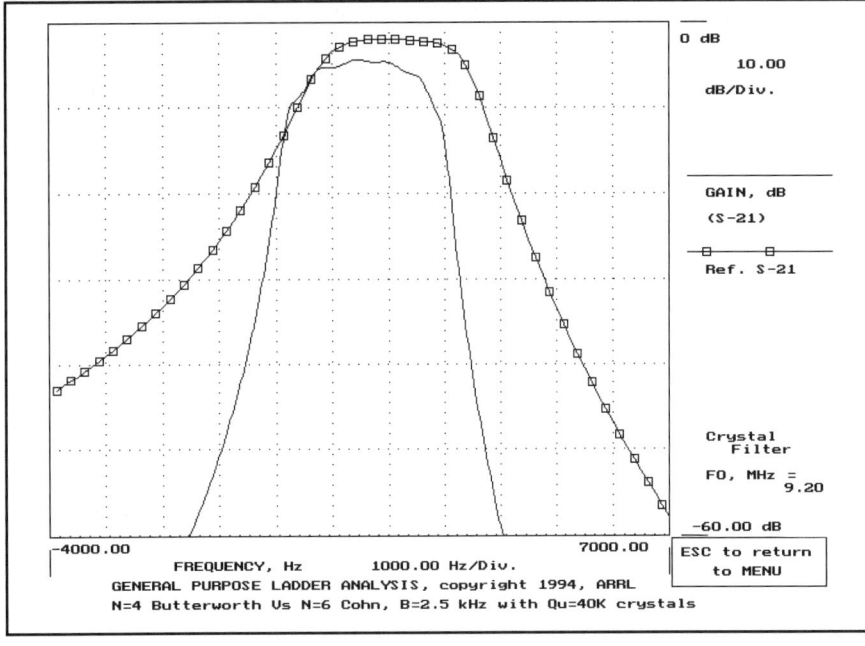

Fig 6.147—Crystal filter responses for two crystal filters. The Cohn is the preferred design for this transceiver even though the low crystal Q_u rounds the passband corners. See text.

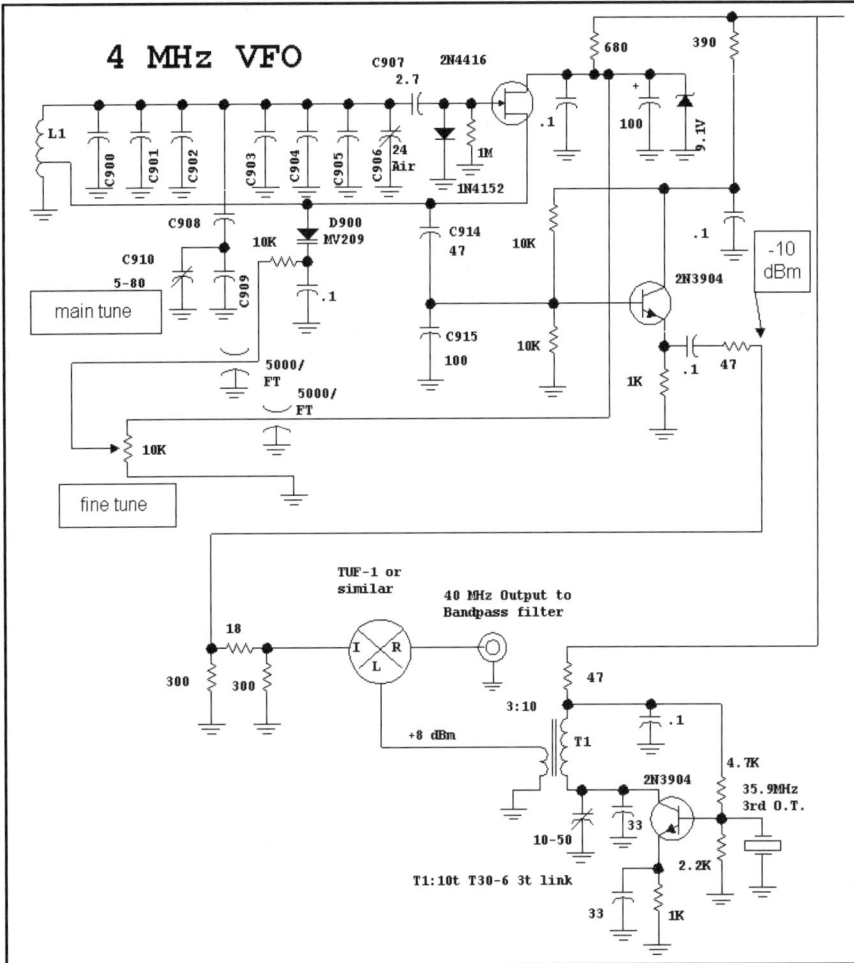

Fig 6.148—VFO for the 6-m transceiver. L1 is unspecified, but will generally be around 5 µH. The many resonator capacitors allow flexibility in setting the frequency. Details are set by the designer/builder.

described here was built for the 6-m VHF band using a 10-MHz IF. However, there is nothing special about that frequency. 10.7 MHz is a good general purpose IF suitable for both HF and VHF. 4.915 MHz has been used in several HF QRP transceivers with good success, based upon available computer crystals.

Our 6-m transceiver initially used only 4-pole crystal filters. They were cut for a 2.5-kHz bandwidth with 500-Ω terminations and a Butterworth shape. While the filters performed well, we often wished for better stopband attenuation in both functions. The original thought, that a casual 4-pole filter would be suitable for VHF applications, was clearly not valid when the 6-m band opened in the spring months! **Fig 6.147** shows the calculated response of a 9.2-MHz sixth-order Cohn filter with a 2.5-kHz bandwidth. This is an easy filter to build and duplicate for both functions. The plot also includes a plot for a Butterworth filter with four crystals. The aggressive designer/builder might expand his or her filter efforts to include extra filters to enhance receiver performance and for transmit IF speech processing.

LO System

The local oscillator system for the 6-m transceiver is shown in **Fig 6.148**, beginning with a conventional 4-MHz Hartley VFO. An emitter follower buffers the output to a diode ring mixer. A capacitor (C915) is selected to establish a follower output of –10 dBm. The VFO uses a 9-V regulated power supply established with a Zener diode. That regulated voltage is routed out of the shielded enclosure on a feedthrough capacitor to a front panel pot. The voltage generated is run back inside the shield where it controls bias on a varactor diode, D900. The diode tuning range is set up to be about 10 kHz. The main tuning cap, C910, uses a large knob with no vernier drive, offering mechanical simplification. This scheme has been surprisingly effective, even with a tuning range of 350 kHz, a direct result of a large tuning knob on a smooth capacitor. Digital readout provides the needed resetability.

The diode ring mixer and a 35.9-MHz third-overtone crystal oscillator occupy the same enclosure with the VFO. The mixer output is then applied to a coaxial connector through a short run of coax cable. The LO box output is routed on coaxial cable to a 40 MHz bandpass filter, shown in **Fig 6.149**. A triple tuned filter is used to enhance spectral purity. We measured 80-dB rejection of the 35.9-MHz component and the 32-MHz image.

The filtered LO signal is relatively weak

The audio amplifier and product detector board for the Universal Monoband Transceiver.

Front panel of the 6-meter transceiver. The very large tuning knob allows surprisingly smooth tuning without a vernier drive. The knob below the main tuning controls a varactor fine tune function.

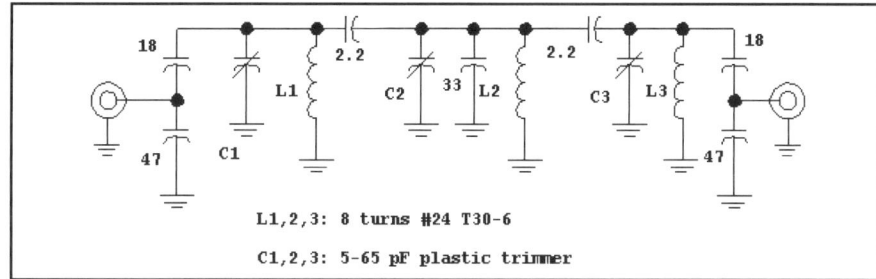

Fig 6.149—Triple-tuned 40-MHz bandpass filter. This circuit was built on a small scrap of circuit board material (approximately 1 × 3 inches) with coaxial connectors mounted at each end. After the filter was tested, a wall was built from ¾-inch brass sheet and soldered to the board. A lid was soldered to the brass walls after filter tuning. The filter was designed for a 2-MHz bandwidth. The inductors had an unloaded Q of 130 at 40 MHz.

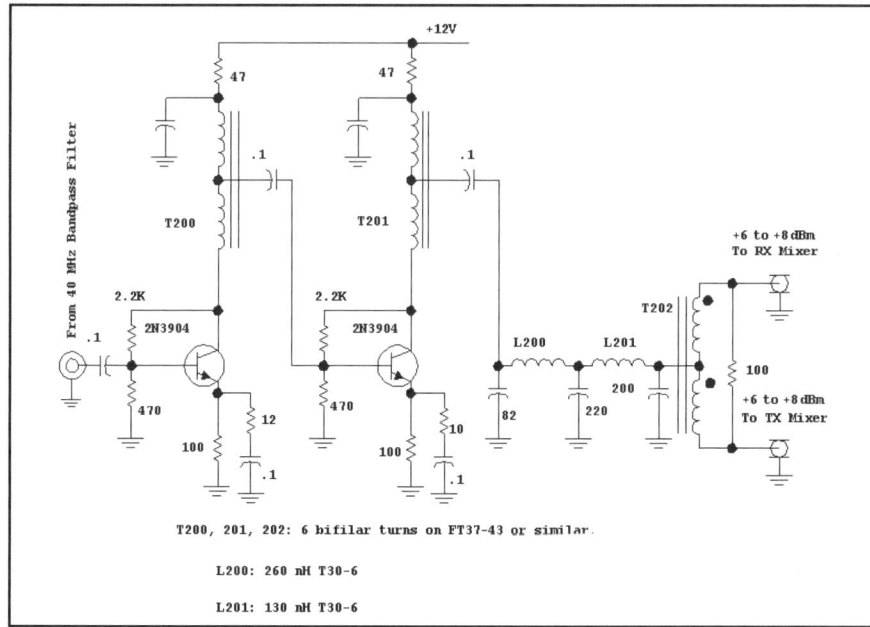

Fig 6.150—LO amplifier feeding 40-MHz energy to the two ring mixers used for the receiver front end and the transmit mixer. T200, 201, and 202 are all 10 bifilar turns #28 on a FT-37-43 toroid. L200 is 8 turns of #24 on a T30-6 core. L201 is 6 turns of #24 on a T30-6.

(about –20 dBm) as it exits the ring mixer and bandpass filter. The level is increased with the two-stage feedback amplifier shown in **Fig 6.150**. The second-stage output is low-pass filtered and applied to a hybrid splitter that delivers two isolated signals, each with a power of +7 to +8 dBm. The hybrid input impedance terminated in a pair of 50-Ω loads is 25 Ω. A low pass filter, initially designed for 50-Ω terminations, was then modified for a 25-Ω load using the procedure of Chapter 3.

BFO/Carrier Oscillator

A traditional Colpitts crystal controlled oscillator generates 10-MHz energy, shown in **Fig 6.151**. The oscillator was modified with inductor L300 allowing oscillation below crystal resonance. Two buffered outputs are available, providing +7 dBm to the product detector and the transmitter balanced modulator. A +12 T supply is applied to only one buffer during transmit periods.

SSB Generator

The SSB Generator board, **Fig 6.152**, begins with an op-amp speech amplifier followed by an RC active low pass filter. A test point allows the audio signal to be monitored to prevent overdrive of the balanced modulator. The peak-to-peak audio signal at TP600 should be 0.4 V for –10 dBm available at the balanced modulator input, which uses a TUF-1 or SBL-1 mixer.

Q600 amplifies the DSB signal from U600 and also sets the driving impedance for the crystal filter. R617 is picked to have the same value as R615, which is the desired termination value for the crystal fil-

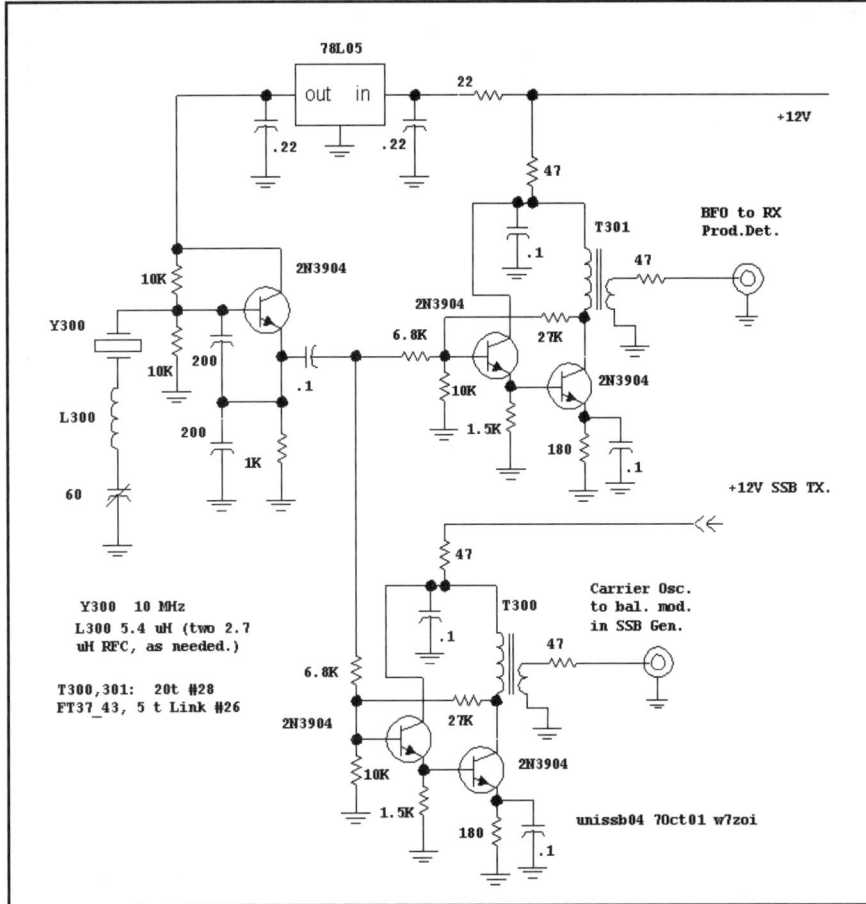

Fig 6.151—BFO and carrier generator. T301 and T300 each have a 20-turn primary with a 5-turn secondary on FT-37-43 cores. The amplifier input resistors, now 6.8 kΩ, can be changed to set the output power.

The RF power chain through Q3. See Fig 6.154.

ter. Further gain is obtained with Q603, 604, and 605. R635 allows a level to be picked that will not overdrive the transmit mixer, U601.

The mixer output drives a 50-MHz LC bandpass filter shown in **Fig 6.153**. This triple tuned filter is built in an isolated box with the same methods used for the LO filter of Fig 6.149 and has a bandwidth of 2.5 MHz.

Transmitter Power Chain

Fig 6.154 shows the driver stages for the RF power chain. This is a class-A design with increasing current in each stage through the chain. A heat sink is needed for the second and third stages. Gain for the chain is 47 dB with an output of 300 mW. The output low pass filter was included for QRP use before a "brick" was added. The low pass could be eliminated (or abbreviated) if a higher power amplifier is planned to follow Q3. A 2SC2988 might be a suitable substitute for Q3 operating at 50 MHz.

The power amplifier used with this transceiver is based upon the Mitsubishi M57735 hybrid integrated circuit, **Fig 6.155**. The hybrid (obtained from Down East Microwave) is an especially convenient part to use, providing 21 dB of small signal gain from a two-stage class-AB circuit. Power output is 14 W for the IC. The chip, which includes a built in low pass filter, is built on a flange that bolts directly to a grounded heat sink. A strip of scrap circuit board material is bolted next to the IC, offering a convenient place for additional circuitry.

Three terminals on the RF module require a power bias. Two use 12 V and feed the two collectors while the third provides base bias networks with 9 V. The 9-V supply should be regulated. In the process of setting up a LM-317T regulator, we realized that it could also function as a programmable circuit. This modification is included in **Fig 6.155** for complete power control over the amplifier. The bias on pin 3 of the IC module is 9.1 V in transmit, dropping to 1.27 V during receive.

The decoupling capacitors used are those suggested by the manufacturer. We measured these networks, finding that the 22-µF electrolytic capacitors we used are modeled with an inductance of 65 nH with very low Q. A better wideband bypass might be several parallel 0.01 µF.

Although the M57735 is ideal for general-purpose applications, it is an expensive part. **Fig 6.156** shows a QRP power amplifier that can be used in place of the hybrid. The output from this stage is 3 W

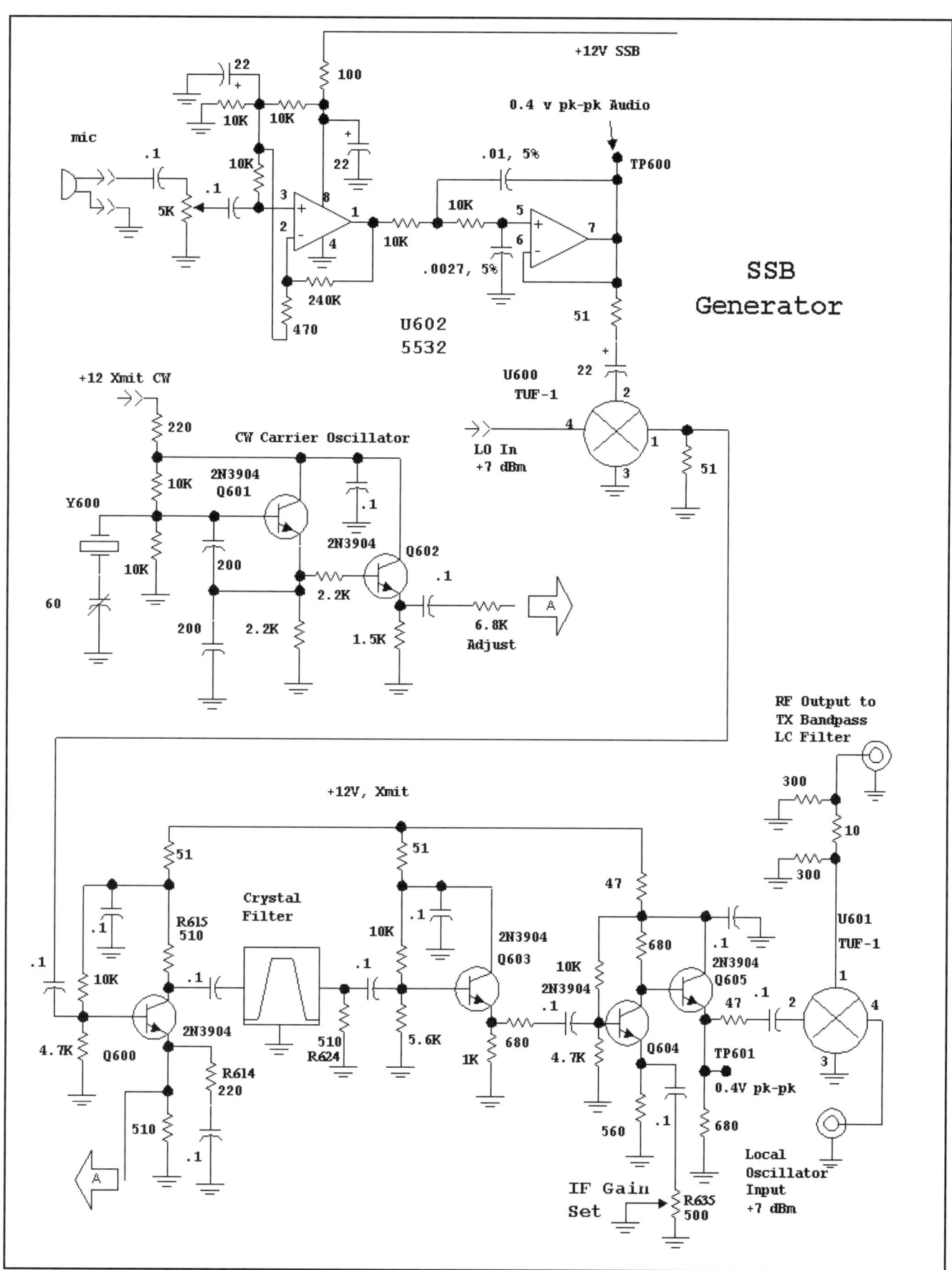

Fig 6.152—SSB generator. R615 and R624 should be picked to equal the desired terminating resistance for the crystal filter, which is a designer/builder-determined element. R614 can be varied to change gain, if needed. R635 is adjusted for 0.4 V peak to peak at TP601 during transmit. That level should be identical in CW and SSB.

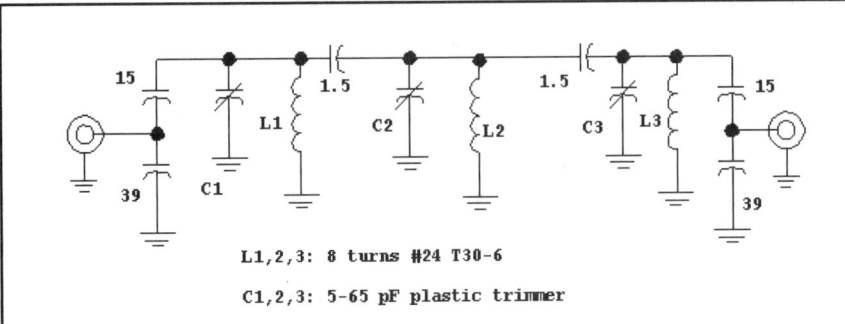

Fig 6.153—Triple-tuned 50-MHz bandpass filter.

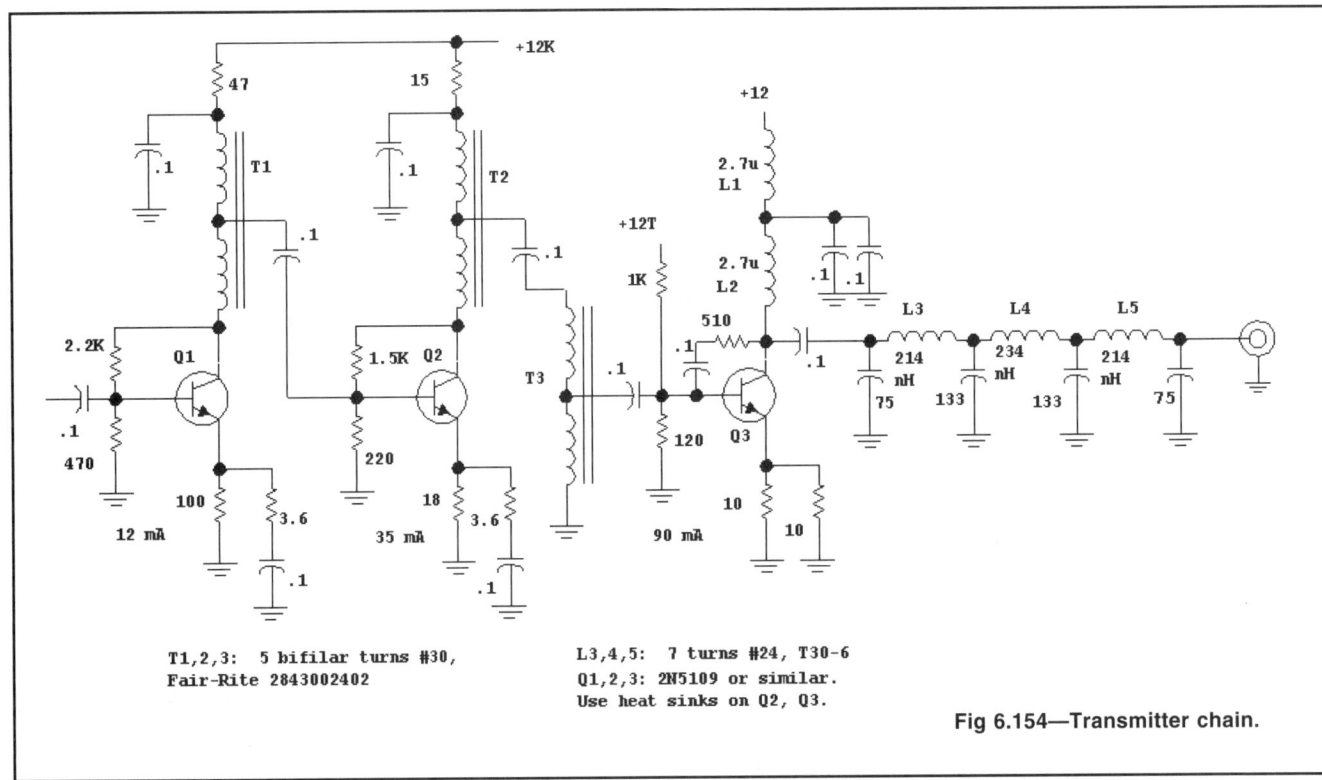

Fig 6.154—Transmitter chain.

The large board is the SSB generator and transmit mixer. This version used SBL-1 mixers. The transmit bandpass filter is in the box fabricated from scrap circuit board material. The control board is above the bandpass filter.

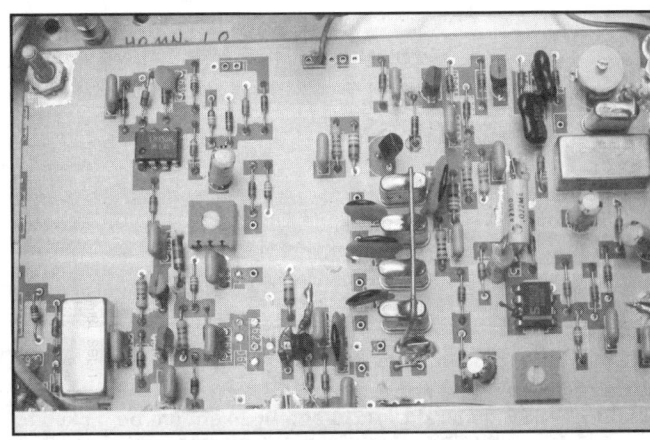

Close up view of SSB generator.

6.88   Chapter 6

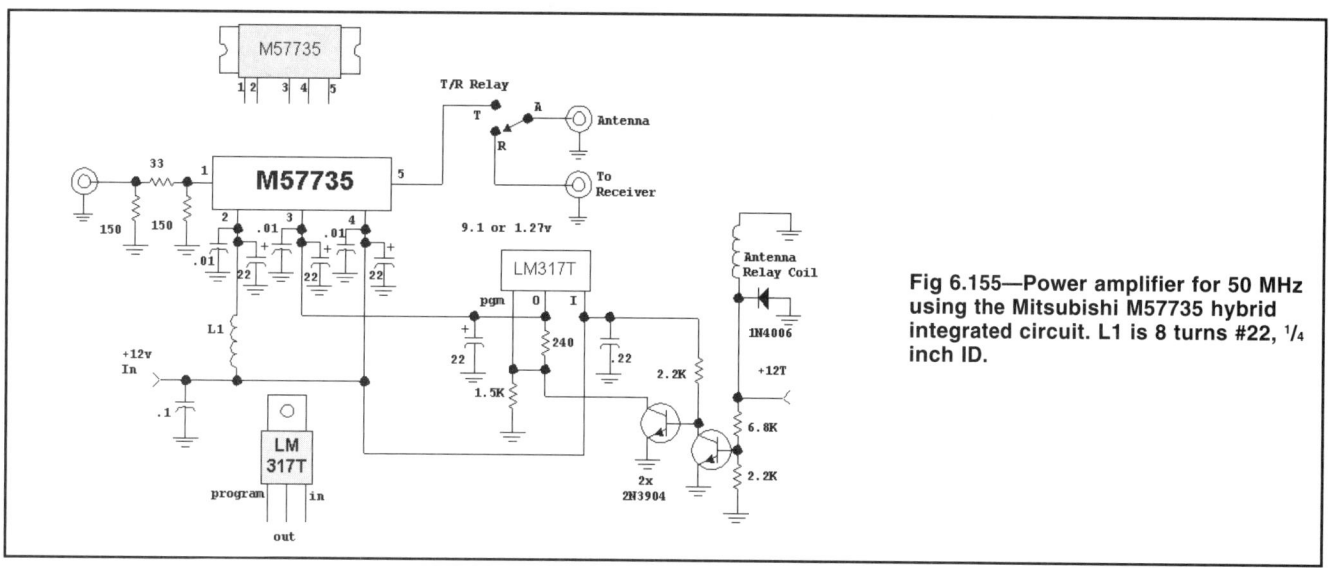

Fig 6.155—Power amplifier for 50 MHz using the Mitsubishi M57735 hybrid integrated circuit. L1 is 8 turns #22, 1/4 inch ID.

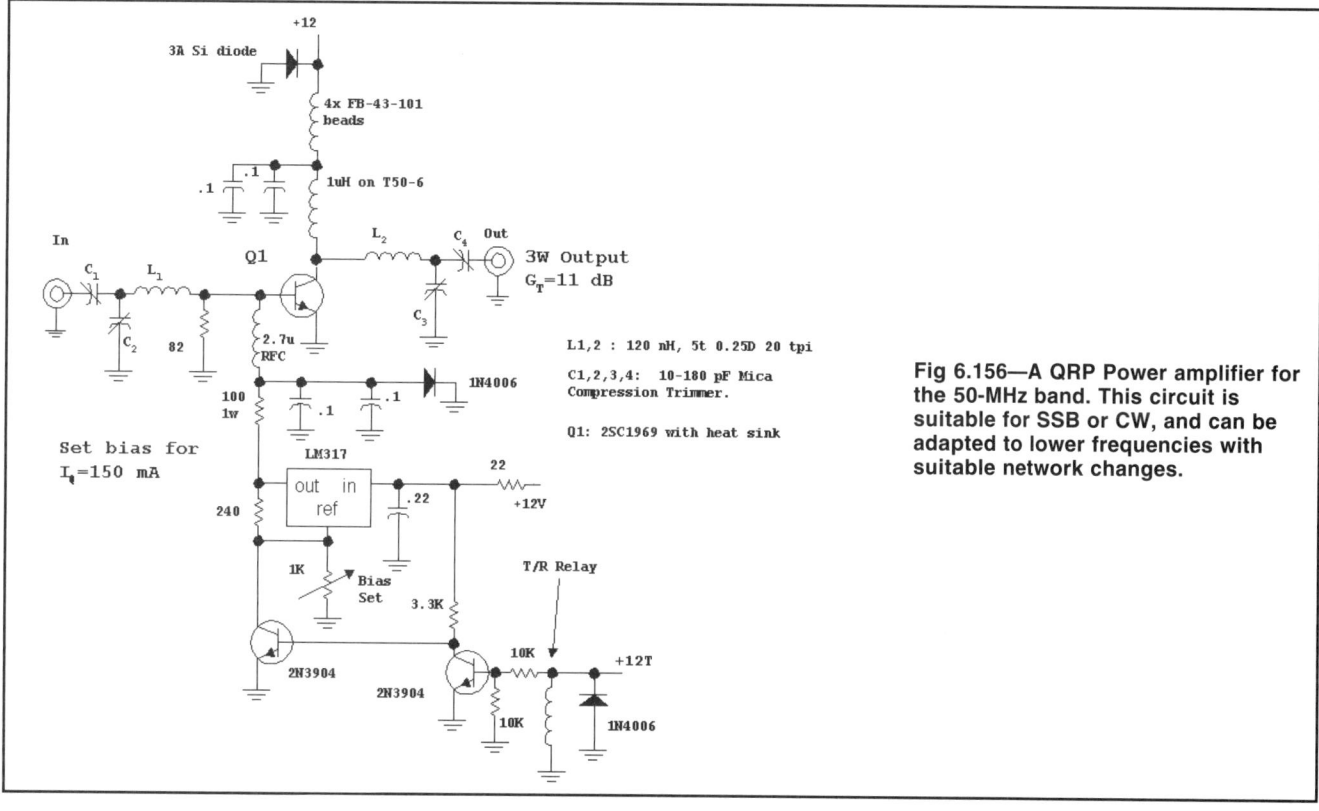

Fig 6.156—A QRP Power amplifier for the 50-MHz band. This circuit is suitable for SSB or CW, and can be adapted to lower frequencies with suitable network changes.

Receiver RF amplifier and preselector filter for the 50-MHz portable station. The variable capacitor tunes the transmitter VXO.

View of RF power amplifier using the Mitsubishi Hybrid. Output is up to 14 W.

with a power gain of 11 dB. This circuit can be adapted to any of the lower-frequency bands, with higher power gain expected. The 2SC1969 transistor is very robust, modestly priced, and available from Mouser.

Receiver Circuits

The receiver circuits resemble others used in this chapter and will not be repeated here. This transceiver uses a low gain RF amplifier, which would not be required for the lower HF bands. We used a shielded double tuned circuit built as a small, measurable filter module as the preselector ahead of the diode ring mixer. The post-mixer amplifier was a 2N5109 with 30-mA bias.

Control Circuits

Fig 6.157 shows the control circuitry used with this transceiver. The design is quite general and is suitable for any transceiver with a relay for T/R. With some modification, it should also be suitable for use with PIN diode antenna switching.

The board generates three outputs: +12 relay, +12 transmit, and +12 keyed. These are produced by TO-39 PNP transistors. We have used 2N5401 and 2N5322 in this application. About any PNP capable of switching about 500 mA (often less) will do as well. The TIP-32 should work. Q403, which provides the +12-V keyed

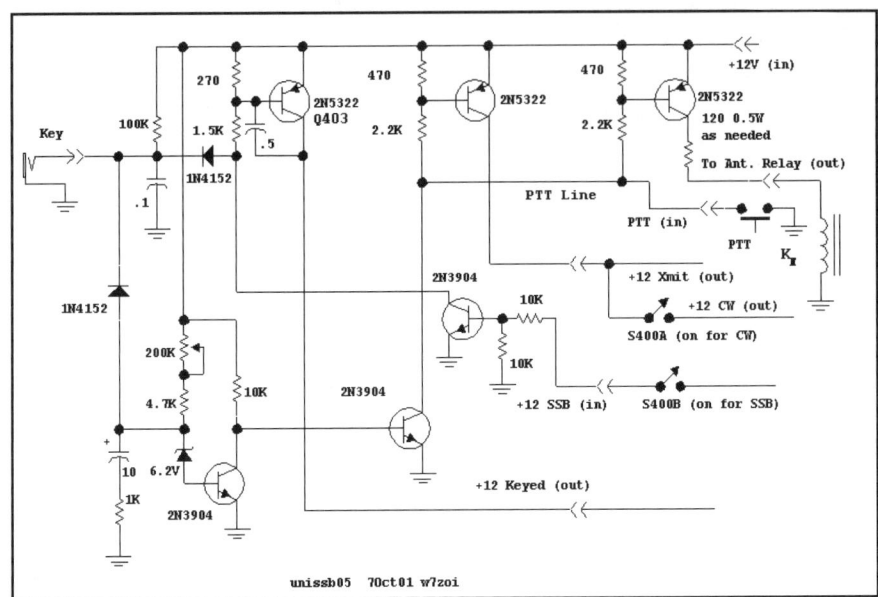

Fig 6.157—Control circuits for the SSB transceiver.

signal, generates the shaping required to suppress clicks.

Most of the signals available at the board are inputs. These include a +12 V supply, a ground-active key line, a similar ground-active push-to-talk (PTT) line, and a +12 SSB line. S400B is a DPDT front panel switch that provides +12 SSB during receive and transmit while in SSB, and +12 CW while in transmit mode in CW.

Results

This transceiver has generally been a useful and enjoyable addition, having provided an enjoyable sampling of "The Magic Band." But it is an evolving design that we plan to modify with better crystal filters and a different receiver IF amplifier. The circuit is suitable for operation from a battery, allowing some portable activity.

6.11 A PORTABLE DSB/CW 50 MHZ STATION

A favorite activity for all three of us is VHF operation from interesting locations, usually areas inaccessible to all but one traveling on foot or kayak. Equipment must be fairly light weight. This 6-m transceiver weighs 3 pounds and has an output of 0.3 W.

The rig uses a VXO-controlled DSB and CW transmitter. An 8-MHz direct-conversion receiver is coupled with a simple converter. The transmitter VXO, shown in **Fig 6.158**, uses an off-the-shelf 14.318 MHz color burst crystal. This oscillator is on at all times, but no output is present at 50 MHz until the key or push-to-talk (PTT) switch is closed. U1 then divides the signal by two, producing a 7-MHz square wave from circuitry presented in Chapter 5. The seventh harmonic, occurring in the desired part of the 6-m band, is selected with a double-tuned circuit, amplified with a Mini-Circuits MAR-2 amplifier and further filtered in a second bandpass. The filter output is –3 dBm with the worst spurious response at –64 dBc.

The VXO output is now routed to the transmitter circuit (**Fig 6.159**) where it is increased to +8 dBm with U4, a MAR-3 amplifier, and applied to a TUF-1 operating as a balanced modulator. U4 is driven with either audio from a microphone or dc to provide a CW signal. The –16 dBm modulator output is increased to +14 dBm through MAR-3 and MAV-11 amplifiers, U6 and U7. This then drives a 2N5947 class A amplifier. Suitable substitute transistors would include a 2N5109. The output is about 0.3 W in CW or DSB. The PTT switch on the microphone will ground the key line that also activates the antenna relay circuitry.

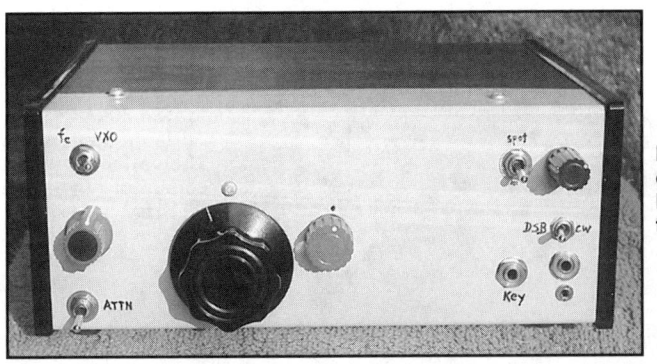

Front panel view of the portable DSB/CW transceiver.

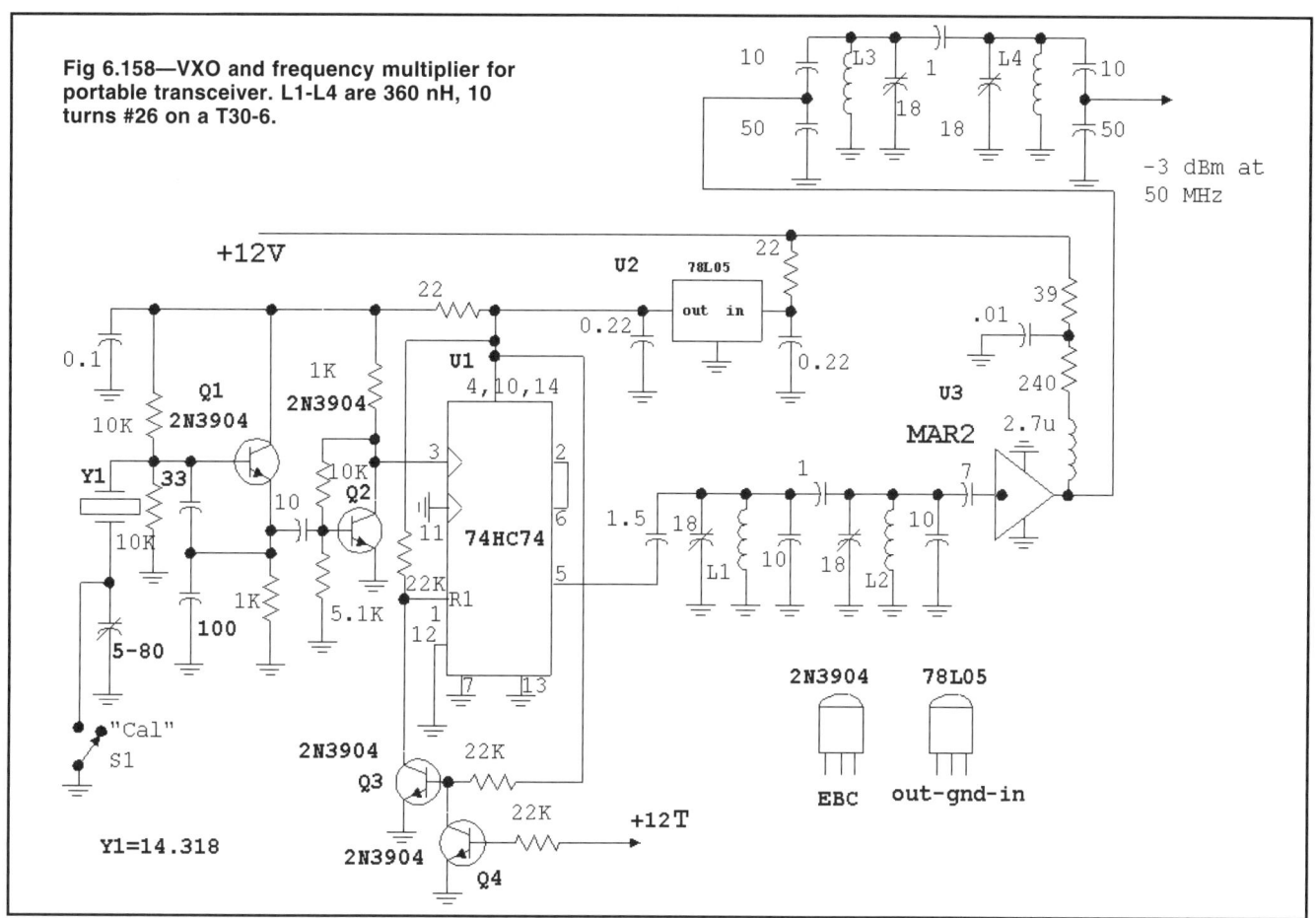

Fig 6.158—VXO and frequency multiplier for portable transceiver. L1-L4 are 360 nH, 10 turns #26 on a T30-6.

The receiving converter, shown in **Fig 6.160**, begins with a single tuned circuit driving a MAR-2 RF amplifier with a gain of about 12 dB. A double tuned circuit then preselects the signal before it is applied to a TUF-1 mixer followed by a 2N5109 post mixer amplifier. A switched 20-dB pad can reduce the signal before the product detector. A PIN diode at the mixer offers additional attenuation.

The converter output is 8 MHz, used merely because a 42-MHz crystal was available in the junk box. A better choice would be 43 MHz. The D-C receiver could then function on the 7-MHz band. The MAR-2 RF amplifier with its input filter could also be eliminated for typical applications, keeping only the double tuned circuit preselector.

Fig 6.161 shows the 8-MHz VFO used with the receiver. This circuit drives a fully shielded board containing the product detector, audio amplifier with switched attenuator, and sidetone oscillator. This module is described in Chapter 12.

Double sideband offers a very simple way to get a phone signal on the VHF bands, one that is compatible with SSB. If we were building this station anew, the minimalist phasing SSB transceiver described in Chapter 9 would probably be used. The VXO used with this rig would provide the needed 50-MHz injection.

REFERENCES

1. Krauss, Bostian, and Raab, *Solid State Radio Engineering*, Wiley, 1980.

2. An excellent summary of modulation is given in Krauss, Bostian, and Raab, *Solid State Radio Engineering*, Wiley, 1980, Chapter 8.

3. W. Hayward, *Introduction to Radio Frequency Design*, ARRL, 1994, pp 205 and 349.

4. H. T. Friis, "Noise Figures of Radio Receivers," *Proceedings of the IRE*, 32, 7 (Jul, 1944), pp 419-422, or R. Pettai, *Noise in Receiving Systems*, John Wiley & Sons, 1984.

5. www.ham-radio.com/n6ca/50MHz/50appnotes/U310.html; See also Gonzalez, *Microwave Transistor Amplifiers, Analysis and Design*, Prentice-Hall, 1984 for designing for lowest noise.

6. W. Carver, "A High-Performance AGC/IF Subsystem", *QST*, May, 1996, pp 39-44.

7. Ibid.

8. For further discussion of AGC loop dynamics, see U. Rohde and T. Bucher, Chapter 5, *Communications Receivers: Principles and Design*, McGraw-Hill, 1988.

9. W. Hayward, "A Competition-Grade CW Receiver," *QST*, Mar, 1974, pp 16-20, 37 and Apr, 1974, pp 34-39. Also see W. Hayward and J. Lawson, "A Progressive Communications Receiver," *QST*, Nov, 1981, pp 11-21.

10. W. Carver, "A High-Performance AGC/IF Subsystem", *QST*, May, 1996, pp 39-44.

11. Personal correspondence between the author and Ulrich Rohde, 1997.

12. W. Hayward, *Introduction to Radio Frequency Design*, pp 219-232. Also see K. Simons, "The Decibel Relationship Between Amplifier Distortion Products," *Proceedings of the IEEE*, 58, 7 (Jul, 1970), pp 1071-1086.

13. W. Hayward, *Introduction to Radio*

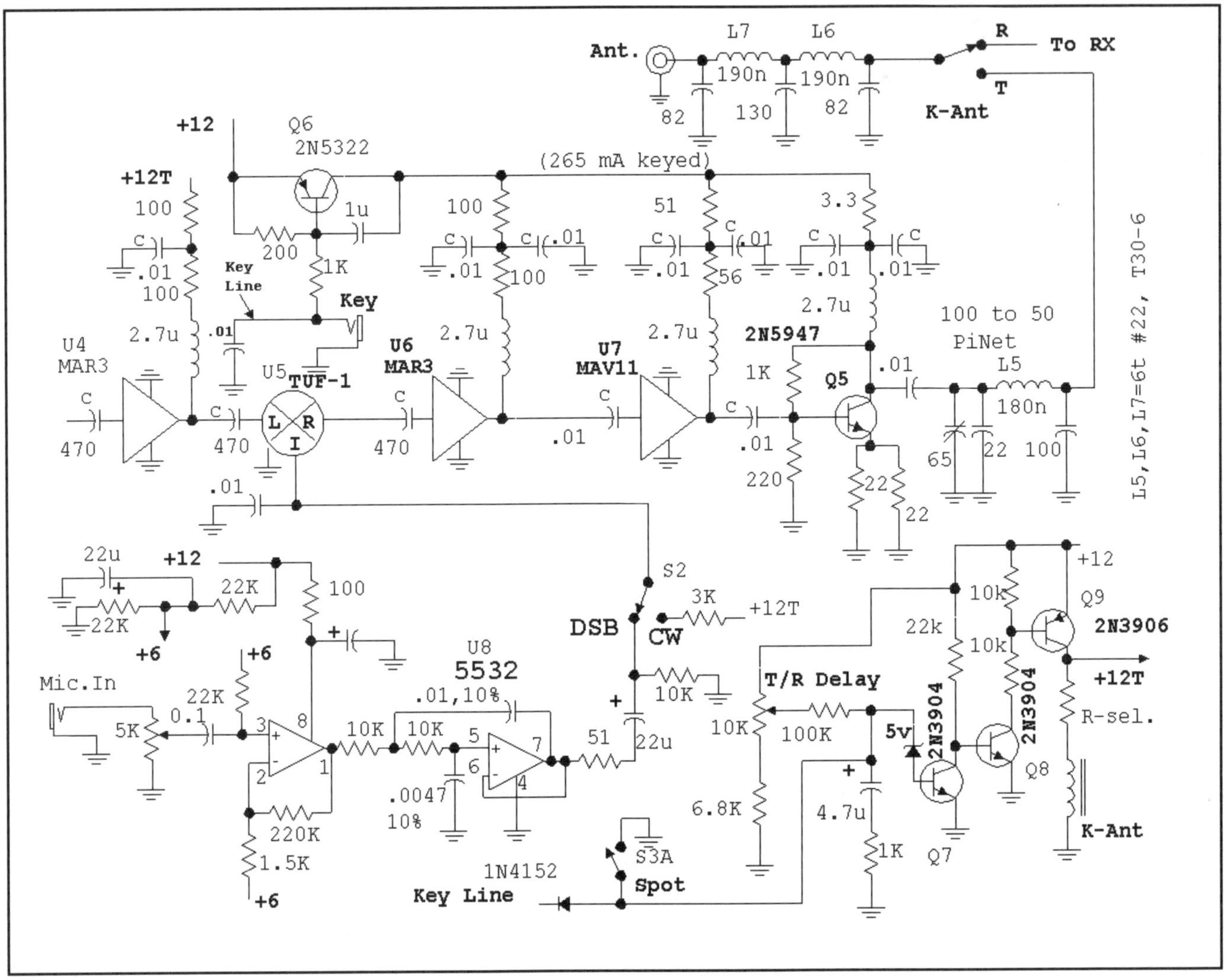

Fig 6.159—Transmitter portion of the 6-meter station.

Transmitter chain for portable rig. Audio microphone amplifier is on the other side of the board.

The audio amplifier and product detector for 8-MHz direct-conversion IF system are all in a Hammond 1590B box with coax and feedthrough capacitor interface connections. The 42-MHz crystal oscillator and 8-MHz low pass filter are on the small boards. The long board across the bottom of the figure is the VXO and ×3.5 frequency multiplier chain.

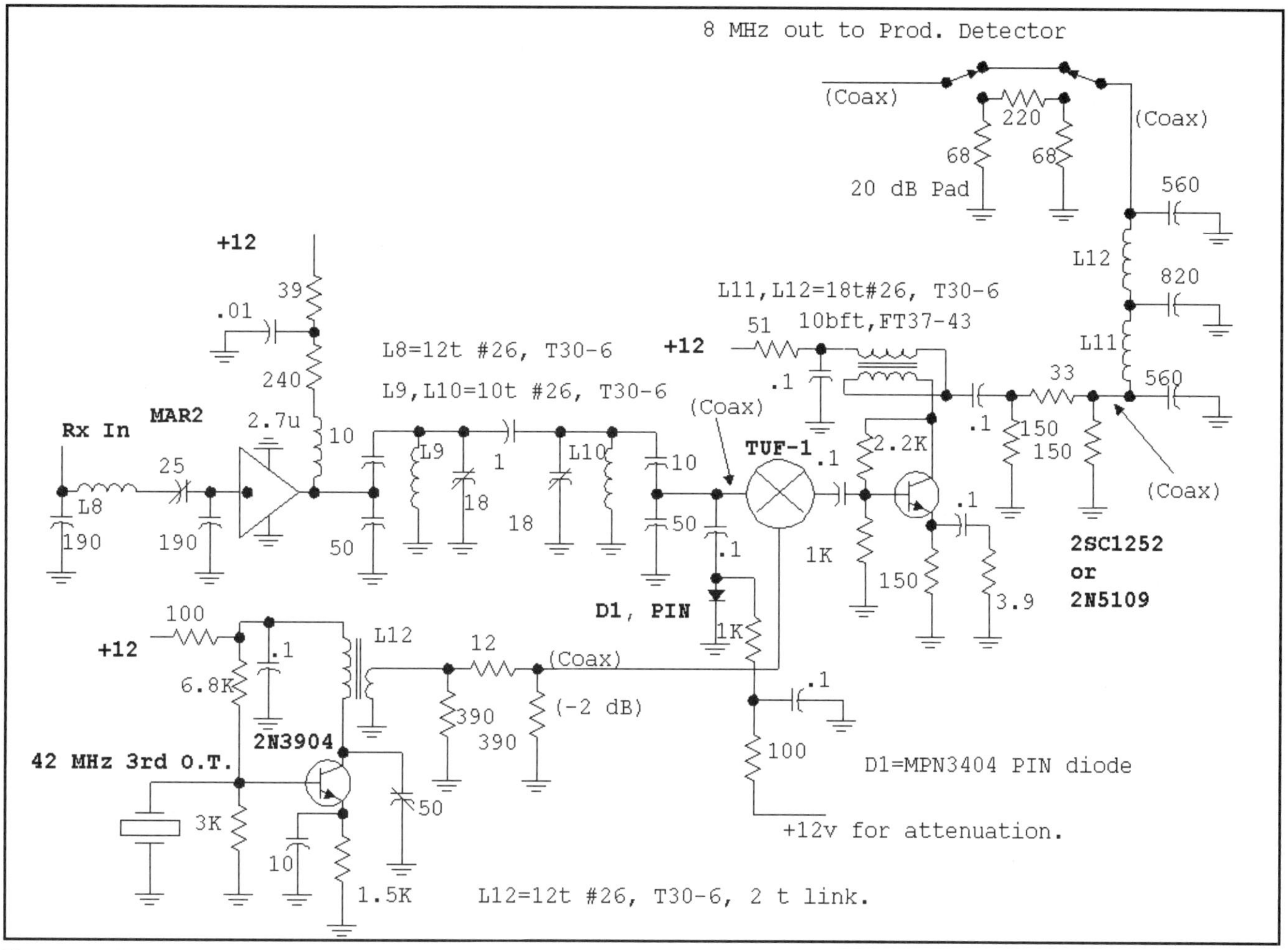

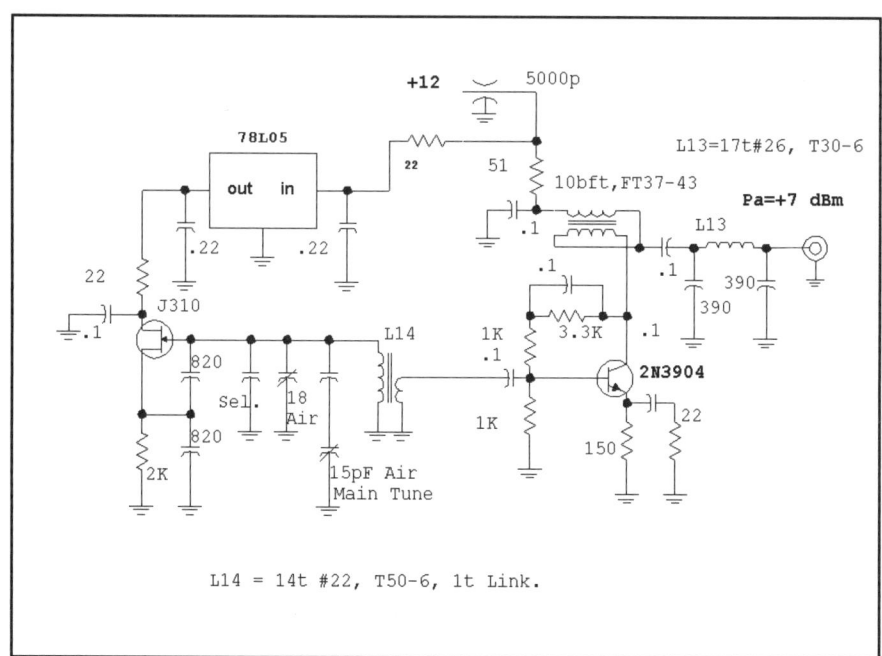

Fig 6.160—Receiving converter used with the 6-m portable station.

Fig 6.161—Eight megahertz VFO for the 6-m station receiver. Tank capacitors are selected to establish resonance at the desired operating frequency

Frequency Design, p 209.

14. W. Hayward, "Further Thoughts on Receiver Specification," "Technical Correspondence," *QST*, Nov, 1979, pp 48-49.

15. W. Hayward, "A Competition-Grade CW Receiver," *QST*, Mar, 1974, pp 16-20, 37 and Apr, 1974, pp 34-39. Also see W. Hayward and J. Lawson, "A Progressive Communications Receiver," *QST*, Nov, 1981, pp 11-21.

16. U. Rohde, "Key Components of Modern Receiver Design," *QST*, May, 1994, pp 29-32, Jun, 1994, pp 21-37 and Jul, 1994, pp 42- 45.

17. P. Hawker, "Technical Topics," *Radio Communications*, Dec, 1995, pp 70-73.

18. J. Makhinson, "A High-Dynamic-Range MF/HF Receiver Front End," *QST*, Feb, 1993, pp 23-28.

19. C. Horrabin in P. Hawker's "Technical Topics," *Radio Communications*, Oct, 1993, pp 55-56.

20. C. Horrabin in P. Hawker's "Technical Topics," *Radio Communications*, Sep, 1993, pp 54-56. Also personal correspondence between W. Hayward and C. Horrabin, Nov 1995 and Oct 2000.

21. J. Makhinson, "A Termination Insensitive Amplifier," *QEX*, Jul, 1995, pp 21-29.

22. R.S. Engelbrecht, US Patent 3,371,284, "High Frequency Balanced Amplifier," Feb 27, 1968.

23. Kurokawa and Engelbrecht, "A Wideband Low Noise L-Band Balanced Transistor Amplifier," *Proceedings of the IEEE*, Mar, 1965, pp 237-244.

24. C. Horrabin, D. Roberts and G. Fare, "The CDG2000 HF Transceiver," *Radio Communications*, Jun, 2002, pp 19-22.

25. U. Rohde, "High Dynamic Range Two-Meter Converter," *Ham Radio*, Jul 1977, pp 55-57. Also see W. Hayward, *Introduction to Radio Frequency Design*, p 216.

26. M. Dishal, "Alignment and Adjustment of Synchronously Tuned Multiple-Resonant-Circuit Filters," *Proceedings of the IRE*, Nov, 1951, pp 1448-1455.

27. A. Zverev, *Handbook of Filter Synthesis*, Chapter 9, Wiley, 1967.

28. B. Goldberg, "Frequency Synthesis Technology and Applications: A Review and Update," *QEX*, Sep/Oct, 2000, pp 3-12.

29. W. Sabin and E. Schoenike, Chapter 13 by E. Silagi, "Ultra-Low-Distortion Power Amplifiers," *Single Sideband Systems and Circuits*, Second Edition, McGraw-Hill, 1995.

30. H. Seidel, "A Microwave Feed-Forward Experiment," *Bell System Technology Journal*, Nov, 1971.

31. R. Meyer, R. Eschenbach and W. Edgerley, "A Wide-Band Feed Forward Amplifier," *IEEE Journal of Solid-State Circuits*, Vol SC-9, No. 6, Dec, 1974, pp 422-428.

32. M. Johansson and T. Mattsson, "Transmitter Linearization Using Cartesian Feedback for Linear TDMA Modulation," *IEEE Vehicular Technology Conference*, 1991, pp 439-444.

33. E. Pappenfus, W. Bruene and E, Schoenike, Chapter13, *Single Sideband Principles and Circuits*, McGraw-Hill, 1964.

34. W. Sabin, "A 100-W MOSFET HF Amplifier," *QEX*, Nov/Dec, 1999, pp 31-40.

35. W. Hayward, "A QRP SSB/CW Tranceiver for 14 MHz," *QST*, Dec, 1989, pp 18-21 and Jan, 1990, pp 28-31.

36. D. Holman, "Receivers and Transceivers" reprinted by P. Hawker, "Technical Topics," *Radio Communications*, Sep, 1986, p 638.

37. M. Thompson, "A Bidirectional Amplifier for SSB Transceivers," *RF Design*, Jun, 1990, pp 71-72.

38. J. Liebenrood, "The Cascade: A 20/75 M SSB Transceiver," *QRPp*, Dec, 1995. *QRPp* is the quarterly journal of NORCAL, the Northern California QRP Club.

39. R. Lewallen, "An Optimized QRP Transceiver," *QST*, Aug, 1980, pp 14-19.

40. Ibid.

41. Ibid.

42. W. Hayward, "Electronic Antenna Switching," *QEX*, May, 1995, pp 3-7.

43. W. Doherty and R. Joos, "PIN Diodes Offer High Power HF-Band Switching," *Microwaves and RF*, Dec, 1993, pp 119-128.

44. W. Hayward, "A QRP SSB/CW Tranceiver for 14 MHz," *QST*, Dec, 1989, pp 18-21 and Jan, 1990, pp 28-31.

45. *QRPp*, quarterly journal of the Northern California QRP Club, Sep, 1994.

46. *SPRAT*, Summer 2001.

47. S. Price, "Sideband Can Be Simple," *Radio Communications*, Sep, 1991, pp 41-45.

CHAPTER 7

Measurement Equipment

7.0 MEASUREMENT BASICS

Measurements are fundamental to all that we do as radio experimenters. The beginner needs a voltmeter to debug the kit he or she has just built, a simple power meter to evaluate it, and a bridge to use in setting up an antenna to use with it. At the other extreme is the designer/experimenter who lives with the equipment needed for the design efforts.

There was a time when the test equipment used by the radio amateur was no more than the indicator level gear needed to build basic gear (VOM and dipper) with the "high end" consisting of service equipment. Today's expectations demand more. Not only do we wish to build some of the equipment that we use, but we want to understand the performance. Our questions probe further as we seek to design that equipment, placing greater demands on measurements. Traditional service gear is usually inadequate, lacking range and accuracy. But it is impractical to purchase the laboratory equipment we would really like to have. The experimenter's measurement gear is often specialized, aimed at performing a few fundamental measurements, but doing so with meaningful accuracy.

This represents a research attitude, emulating the way we might examine a new field where no instrumentation exists, but where the questions must still be answered. The researcher expects to develop new skills as he attacks his or her work. The usual engineer is only expected to possess the skills at the beginning of a project, willing to deal with technology, without an expectation to develop it.

This chapter addresses measurement needs by describing some fundamental test equipment. We begin with some of the equipment needed by the beginner mentioned above, but expand to include the gear needed by the hard-core experimenter. This equipment is based upon some specific guidelines:

1. The experimenter should measure everything that he or she can. Even if you do not have the "right tool," you can often perform an approximate determination. The most casual measurement is still more informative than none.
2. Test equipment need not be refined. That is, simple equipment is still adequate if you can perform a calibration that provides information.
3. The equipment in this chapter is designed for the RF experimenter with a primary interest in building radio equipment. It is easy to become a "test equipment junky" by building and purchasing a great collection of good test gear, with no resources left for the original experiments. Individual goals must be the guideline.

In Situ vs Substitution Measurements

Measurements usually fall into two classes. The in situ or *in place* measurement is one where instruments are attached to a working system. A goal is to extract as much information as possible without disturbing the system any more than is absolutely necessary. Most of the measurements we do with an oscilloscope or a voltmeter occur in situ. Such measurements are the basis of analog electronics.

The contrasting measurement uses a substitution. In this case, part of a system is examined in isolation from the rest, with test equipment substituted for some components. Clearly, this is a major disturbance; the studied system ceases to function during the measurement. However, things can be evaluated that cannot be measured in situ. An example of a substitution measurement would be determination of receiver sensitivity. An oscilloscope or voltmeter can't measure the sub micro-volt signals that are applied to the antenna terminal of the receiver. So, we examine the receiver output while applying a calibrated signal source to the input. Substitution measurements provide the basis for radio frequency electronics.

We *will often describe the measurements we discuss as being substitution* or *in situ*. It is important to isolate the two, for a piece of equipment suited to one mode may be useless for the other. Some equipment can move into both worlds so long as it is applied with care.

Using This Chapter

We will describe a variety of test equipment in the following pages. Some is simple while some is more complex. The order of presentation does not generally coincide with complexity or utility, leaving the beginner searching for the suitable starting point.

The novice experimenter should begin with the simplest gear such as a voltmeter for kit building. Add an instrument for measuring inductor and capacitor values as you progress beyond these beginnings. If you are building any RF communications gear you will want a power meter or some other means for power determination.

As your commitment to experimentation deepens, you will want more test equipment. An inexpensive oscilloscope is probably one of the most useful tools one could acquire. It is useful for the classic in situ analog measurements, the substitution measurements of RF, and even the timing measurements of digital electronics. The oscilloscope then becomes the foundation for numerous other measurement tools.

But, no matter what equipment is being used, simple or sophisticated, keep your goals in mind. Our goal is to understand: Does the gear we build perform as fundamental concepts tell us that it should? This means that the test equipment is in constant use during construction of a project. Each stage in a complicated system is evaluated and confirmed as the system grows. The user should divorce himself from the oversimplified idea that test equipment is merely a tool for final evaluation.

7.1 DC MEASUREMENTS

The most basic instrument of electronics is the galvanometer of fundamental physics. Current flows in a coil to produce a magnetic field, interacting with another field to cause force against a spring. The resulting motion has an attached scale to indicate current.

The simple 0 to 1 mA meter movement is a modern equivalent. This meter usually has a very low internal resistance of 25 to 100 Ω. Larger currents are measured with meter "shunt" resistors while voltage is measured with a series "multiplier" resistor. A 1 mA meter movement would need a 10-kΩ resistor to measure 10 V. Hence, a voltmeter so built would load the circuit being measured as if a 10K resistor was attached to ground. See **Fig 7.1**.

The loading problems are significantly reduced when active circuits append the meter movements. The traditional active instrument is the classic VTVM, or *vacuum tube voltmeter*. A modern equivalent is a voltmeter using an op-amp with an example shown in **Fig 7.2**. The input signal is applied to a very high impedance voltage divider, resulting in a signal to the non-inverting input of an op-amp. The 1kΩ in series with the meter, R_{CAL}, can become a 2-kΩ pot if calibration is required.

Most experimenters tend to purchase general-purpose meters rather than build them from scratch. The typical unit is a digital voltmeter, or DVM that will measure dc and ac voltage and current and dc resistance. Some have become so good and so inexpensive that it is justified to purchase a general-purpose instrument to build into a special application.[1] The typical DVM will have an input resistance of 10 MΩ when measuring dc voltage. Some traditional VTVMs also had a 10-MΩ input resistance, but also had a resistance (1 MΩ or more) built into the tip of the probe used with the instrument. This allowed the probing of sensitive circuits with little loading, even at high frequencies. While the modern DVM will not cause problems with dc loading, the long test lead can certainly cause problems for circuits containing signals at audio or higher frequencies.

While the resolution and accuracy of a modern DVM is outstanding, many users still prefer an analog indication when a circuit is being adjusted. Some DVMs approximate an analog meter movement with a digital bar graph.

In spite of their justified popularity, the user should be careful when using DVMs, for they create some unique problems. Probably the greatest is the assumption that they are as accurate as their resolution. We should not assume that a meter reading a voltage to 1 mV or better is accurate to that level. See the meter's manual. Another often-overlooked problem is the "burden" of these meters when measuring current. Burden is the voltage drop across the meter when measuring current. This can often be several tenths of a volt for high currents, a departure from the classic multimeters of the past.

We often wish to measure audio signals from the output of receivers. This is best done with a true RMS responding voltmeter. Some of the newer DVMs from Fluke and other vendors include this highly useful feature. The user with older meters can still perform true RMS audio measurements by building an appropriate adapter.[2] This paper is included on the CD that accompanies this book.

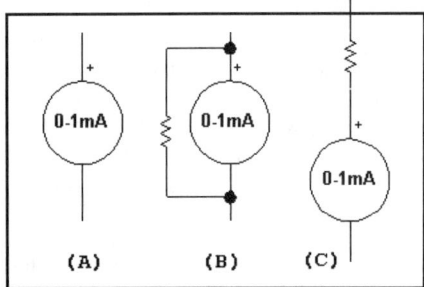

Fig 7.1—A basic 0-1 mA meter (A); measures higher current (B), or voltage (C) with the addition of resistors. Resistance can be measured with these through application of Ohm's Law.

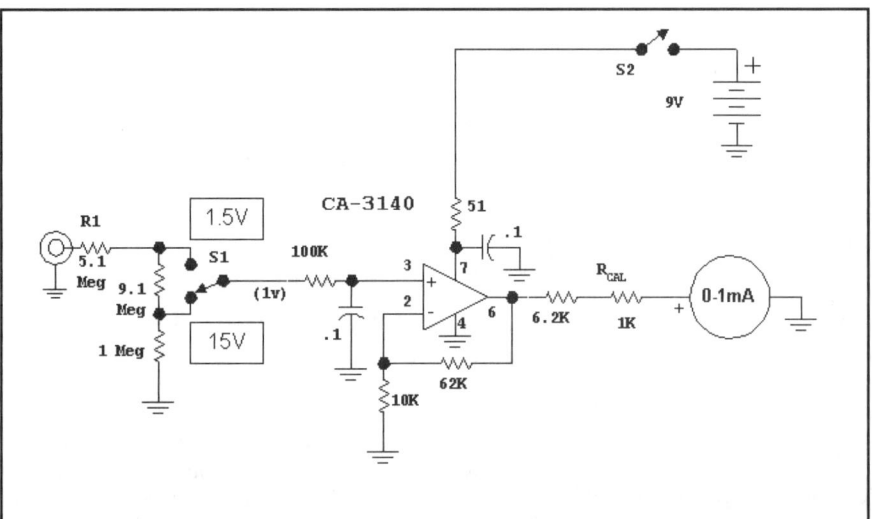

Fig 7.2—A simple op-amp based voltmeter. The meter is one normally intended for use as a 0-15 V meter where a 0-1 mA movement is used with an external 15-kΩ multiplier. The 0 to 15 indication on the meter is now used to register 0 to 1.5 or 15 V, but with a 15-MΩ input resistance. This circuit operates with an op-amp voltage gain of about 7, generating an output of 7 V for a full scale response. With a 9-V supply it becomes virtually impossible to damage the meter movement with excess voltage.

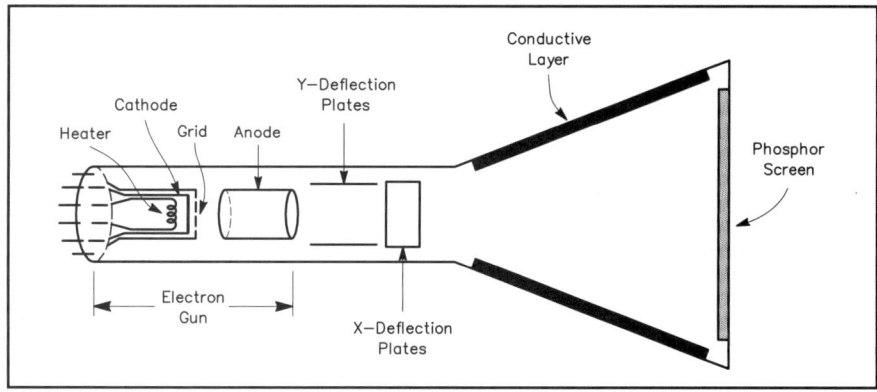

Fig 7.3—Cross section view of a cathode ray tube.

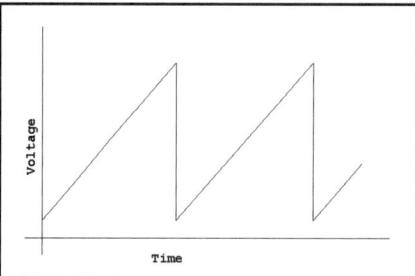

Fig 7.4—Linear ramp applied to the X axis of a CRT. A repeated ramp is called a *saw tooth* waveform.

7.2 THE OSCILLOSCOPE

The ultimate measurement tool for the time domain (explained later) is the cathode ray oscilloscope, or just oscilloscope or scope. This is an instrument that usually measures a voltage that varies as a function of time and displays the result as a time graph. Other measurements are also possible and will be outlined.

The basis for a traditional oscilloscope is the cathode ray tube, shown in **Fig 7.3**. This device begins at the left with a heater and a cathode, the electron-emitting element in the structure. Those unfamiliar with the basics of vacuum tubes can examine their construction and operation in the *ARRL Handbook*. The CRT cathode is much like that in any other vacuum tube, although it is usually a flat or planar surface. Directly to the right of the cathode is a grid. Normal bias slightly negative with respect to the cathode prevents the electrons from leaving the region close to the cathode. Changing the grid bias slightly in a positive direction allows some electrons to escape. They are then accelerated toward an element called an anode. This, plus other electrodes not shown, causes the electrons to be formed into a beam, or *ray*, in accordance with the classic name. The part of the CRT described is called the *electron gun*.

The region after the electron gun contains the deflection electrodes. These will alter the beam direction and allow it to eventually strike the faceplate where it will impinge on a phosphor, a material that gives off light when struck by energetic particles.

Most of the electron gun is biased negatively at a potential of −500 to −2000 V while the deflection region is close to ground. The rest of the CRT is also near ground potential for simple 'scopes. Higher performance instruments often include a high voltage post deflection acceleration (PDA) region for greater brightness.

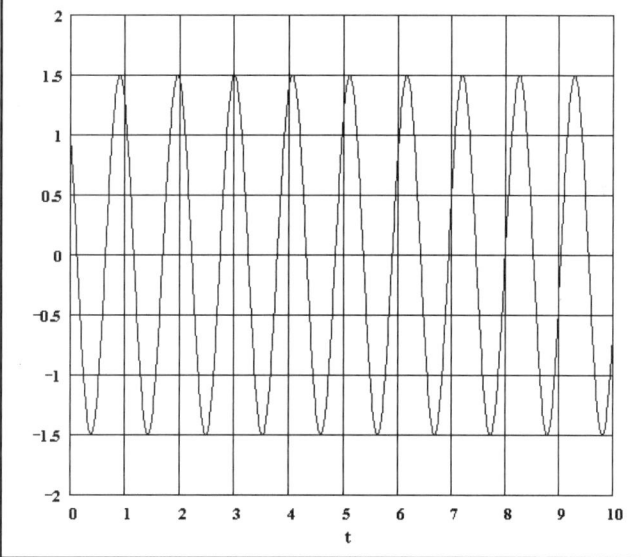

Fig 7.5—The appearance of an oscilloscope faceplate while examining a sinusoid. This ac voltage moves from zero to 1.5 V, back to zero, to −1.5, and again to zero, with the sequence repeating for a long time. This signal is measured as 3 V peak-to-peak. The display shown has a vertical sensitivity setting of 0.5 V per division.

The electron beam leaving the gun passes between deflection plates, often no more than parallel sheets of metal. The beam passes first through the vertical, or Y plates, and then enters the horizontal or X deflector. A voltage applied between the plates generates an electric field causing the electrons to move toward the more positive plate. The electrons are moving quite fast as they enter the deflection region, so the change in direction brought about by the deflectors may be slight. But a few volts across the horizontal plates will cause a beam originally headed for the faceplate center to strike at the edge.

The voltage applied to the X plates will cause the beam position to vary with the applied voltage. If we apply a voltage that is a linear ramp with time, shown in **Fig 7.4**, the result is a horizontal line across the faceplate.

The electrons move predominantly along what is usually referred to as the "z" axis. A signal applied to the grid next to the cathode is called a z axis or intensity modulation.

There are numerous applications for this versatile configuration. For example, if a fast ramp is repeatedly applied to the X axis (called a raster) while a slow one drives the vertical, the entire faceplate area is scanned. Modulation applied to the intensity controlling grid then allows television to be displayed.

Oscilloscope measurements usually begin with a ramp, a voltage that grows linearly in time, applied to the X axis. A signal being studied then drives the Y axis. If that signal, for example, is a simple sine wave, the user sees a sine pattern on the face of the CRT. This result is shown in **Fig 7.5**.

The operation just described would work well if the CRT was very bright and just one sweep occurred. The sinusoid would be seen right after it occurred, but would then decrease in intensity as the phosphor decays in time. Most of the signals we study are repeated in time and we use a low intensity beam that appears again and again. If we did this without doing something special to force the horizontal sweep and the vertical excursion to synchronize, we would have a display like that of **Fig 7.6** where no information is conveyed.

The elements that cause this synchronization are called trigger circuits, critical parts of an oscilloscope now shown in greater detail in the block diagram of **Fig 7.7**. The trigger is a circuit that looks at the signals present in the vertical channel. Once a predetermined level set by a front panel control (trigger level) is reached, a pulse is generated that is sent to two parts of the system. The pulse reaching the sweep circuit where the sawtooth wave is generated starts the ramp. The pulse reaching the Z-axis system *un-blanks* the electron gun, turning on the electron beam. Once just one sweep is finished, it terminates, but starts again when a new trigger pulse is generated.

Most 'scopes have an automatic trigger mode that causes a continuous sequence of sweeps to occur. However, as soon as a valid trigger pulse is generated by a vertical signal, that action dominates. While the vertical signal is the most obvious and useful source for triggering, others can also be used. An external trigger terminal is useful for sources that have a well defined associated signal. It is also useful to trigger from the 60 Hz line, allowing related (hum) signals to be examined.

The scope vertical input drives a resistive attenuator that establishes vertical sensitivity. The most sensitive position is typically 10 mV per division, increasing to 10 V per division in a 1-2-5 sequence. All modern scopes are dc coupled, although the user has the option of ac coupling. That is, applying a dc voltage will produce change in the sweep position that remains as long as the dc is present.

The availability of two or more vertical channels is also common. A variety of schemes are used to share one electron gun with the two.

The horizontal sweep is usually calibrated with a wide range of sweeps. One of the instruments used for much of our work is a Tektronix 453 with sweep rates of 0.5 second to 0.1 microsecond per division. Both the vertical and the time base can be operated in un-calibrated modes in most scopes. Further, both X and Y channels have related position controls, allowing the display to be moved to fit the incoming data.

The input impedance of the typical vertical channel is 1 MΩ paralleled by about

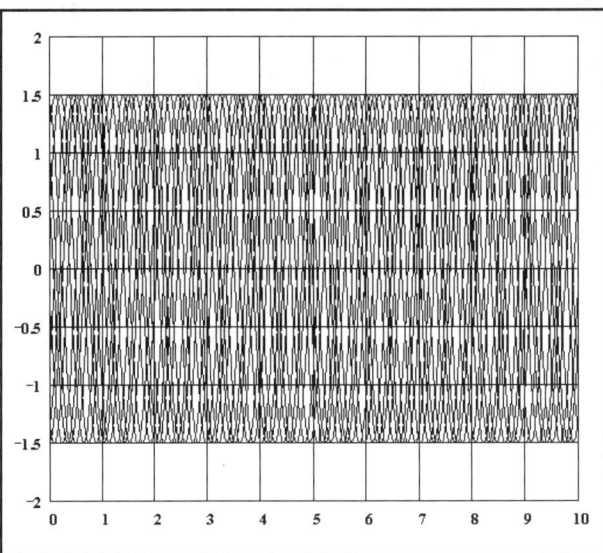

Fig 7.6—The sine wave of Fig 7.5 viewed without triggering. See text.

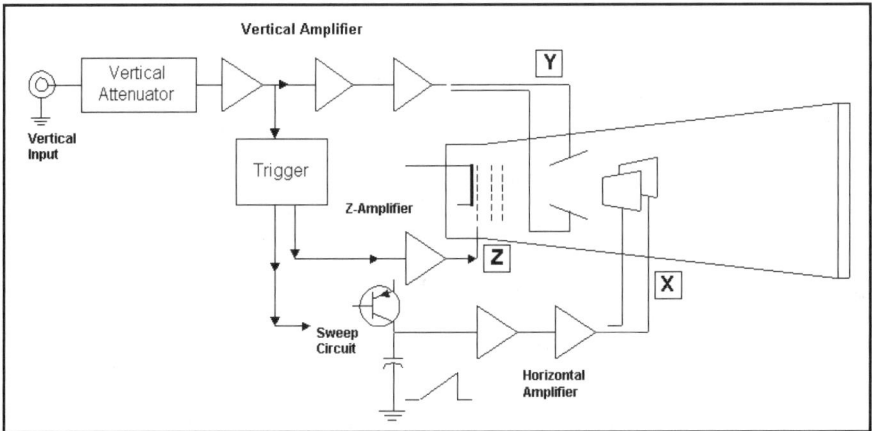

Fig 7.7—Partial block diagram for an oscilloscope. See text for details.

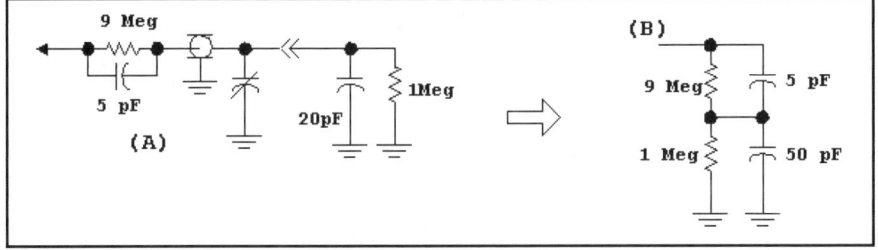

Fig 7.8—A 10X oscilloscope probe. Part A shows the probe and the input to the attached scope while B shows an equivalent circuit. See text.

20 pF. As such, the loading imposed by the 'scope is not severe. However, it can still be substantial, often dominated by the capacitance of the cable needed to connect the instrument to a circuit being tested.

A typical oscilloscope accessory is a 10X probe, used to reduce the capacitance seen by a circuit being tested. A 10X probe circuit is shown in **Fig 7.8**. A fixed capacitance parallels a 9-MΩ resistor to drive the cable and a variable capacitor. The combination drives the input RC of the scope. The capacitor is adjusted to produce clean, sharp edges when driving the probe from a 1-kHz square wave, the usual calibrator built into most oscilloscopes. Without the 10X probe, the scope input has a low pass characteristic formed by the circuit resistance and the scope input capacitance. The two capacitors of Fig 7.8B form a low pass – high pass combination with effects that cancel (an all pass filter), extending performance to the probe tip.

It is common to find beginners who acquire a new oscilloscope, but do not get the probes to go with it. Don't! The 'scope without the 10X probes is an invitation to misleading measurement attempts resulting from the loading from high oscilloscope input capacitance. Almost all high frequency measurements done with a 'scope are performed with the 10X probe. Even this loading is extreme in many applications.

Most oscilloscopes also have an X-Y mode where one vertical channel drives the Y axis, but the other is attached to the X axis. If you use this setup with two sine waves, you can infer something about the phase relationship between them. Two sine wave signals of the same frequency will produce a slanted, 45 degree line if they are in phase with each other. But a 90 degree phase difference will produce a circle when both have the same amplitude. These are called Lissajous patterns. The X-Y mode is also useful with other instruments that include their own time basis (sweep,) such as a homebuilt spectrum analyzer discussed later.

The up-to-date oscilloscopes offered for industrial and research applications differ from the traditional picture we have painted. While many of the changes relate to extended features, others deal with the very nature of the products. Modern scopes rarely feature the high performance CRTs of earlier times. Rather, the input connectors drive amplifiers that then drive high speed Analog to Digital converters, producing a digital version of a picture that is eventually presented for viewing on an inexpensive display. The performance is often impressive, as are the prices.

As you become accustomed to a new oscilloscope, you will find numerous ways to apply it. It is effective in measuring dc levels as well as the ac signals within a circuit. Careful triggering and setting of horizontal position will allow surprisingly accurate frequency measurements, although not up to counter standards. We will comment on various applications throughout the rest of this chapter.

A good general purpose reference on traditional oscilloscope measurements is the paper by K7OWJ, which is included on the CD that accompanies this book.[3]

7.3 RF POWER MEASUREMENT

One of the first things the beginning communications experimenter wishes to measure is radio frequency power, usually from a transmitter. Although not hard in concept, it can be a difficult measurement to perform with good accuracy.

The simplest way to measure RF power uses a termination with a dissipation exceeding the highest power to be measured, a diode, and a capacitor in a peak detector, shown in **Fig 7.9**. A transmitter to be tested is attached to the load and the signal is rectified by the diode, which then charges the capacitor. The capacitor will reach a voltage nearly equaling the peak ac value. Although virtually any meter can be used, one with a high dc impedance is preferred. A DVM works well, although if adjustments are being done, analog action is still useful.

Assuming a diode drop of 0.6 V, the RF power is given by **Eq 7.1** where R is usually 50 Ω. The breakdown voltage for the 1N4152 diode is 100 V, so dc levels of 50 V can be measured, corresponding to a little over 25 W. One can use higher breakdown diodes or tap the diode part way down the resistor to measure higher power, shown in Fig 7.9B. One must, however, alter the equation to reflect the voltage division.

$$P_{watts} = \frac{(V_{dc} + 0.6)^2}{2 \cdot R} \qquad \text{Eq 7.1}$$

R1 can be a parallel or series combination of resistors to reach the needed dissipation. Two or three watt resistors can be stacked between parallel sheets of circuit board material to reach the 100-W level. If the resistors are spaced from each other, and open to the air, they can be stressed beyond their normal rating for short intervals. One termination we use for 100-W measurements consists of 30, 1.5-kΩ 2-W resistors. These methods are generally confined to 50 MHz and lower.

We can add a voltmeter to the circuits of Fig 7.9 for a stand alone instrument requiring no external meter. Two versions are shown in **Fig 7.10**. The one at (A) uses a 1-mA meter movement with a 15-kΩ resistor to form a voltmeter with a maximum of 15 V. Using Eq 7.1, the maximum power would then be 2.43 W, so the 50-Ω load resistor should have this dissipation rating or greater. A valid choice would be two parallel 100-Ω, 2-W resistors. In practice, 1-W resistors would work well for short tests. The circuit at (B) is actually two power meters with one meter movement. This scheme functions because the typical milliampere meter has a low internal resistance.

The two ranges of the meter at Fig 7.10 are quite different. The one at the right hand input is much like the others discussed while the left input has a 50 mW full-scale reading (+17 dBm). This range is best calibrated against a calibrated signal generator. Alternatively, a higher power meter can be used to measure a

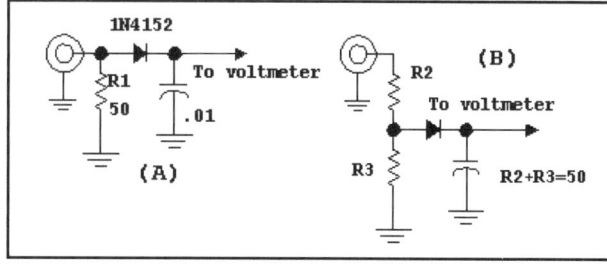

Fig 7.9—A peak detector (A) measures the peak RF voltage across a load, allowing calculation of RF power. The scheme at (B) allows higher powers to be determined without taxing diode breakdown voltages.

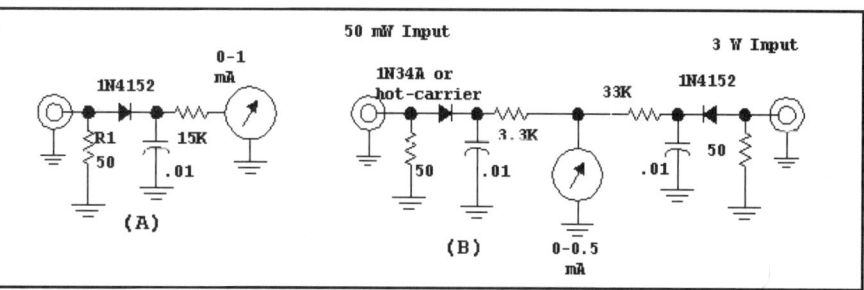

Fig 7.10—(A) shows an instrument with built in meter while the version at (B) has two RF inputs available. See text for details.

dB Arithmetic

Two RF powers are compared as a ratio, or in dB form with

$$dB = 10 \, \text{Log} \left(\frac{P_1}{P_2} \right)$$

...where the powers P_1 and P_2 are both in the same units of W, mW or μW. The dB, as well as other logarithmic forms is useful because a change in power ratio is analyzed with addition or subtraction. dB is defined only when *two* powers are considered.

We often specify a power in dB terms with respect to some reference. dBW is dB with respect to 1 W. The familiar dBm is power referred to one mW. These are both ratios, with the 1 (mW) understood. While many power measurements we perform that read out in mW happen in 50-Ω systems, this is certainly not necessary. There is nothing to preclude us from referring to 1.5 peak V across a 150-Ω resistance (7.5 mW) as +8.75 dBm, even though this is not the result we would read if the related power source was applied to a 50-Ω power meter.

With most measurements, an increment from one value to another occurs with a step value of the same units. For example, we change the length of a 50-inch antenna by one inch to become 51 inches. The inch unit is used in all cases. But this is not the case with dB and dBm. An absolute power of 20 mW (+13 dBm) is increased with an amplifier by a factor of 5 (7 dB) to 100 mW (+20 dBm.) A dBm value is altered by adding a dB value to become a new dBm value. The ratio of two powers is obtained by taking the difference of their dBm values to get a power ratio in dB.

It is usually not correct to "increase a +27 dBm power by *3 dBm*," which would literally mean increasing 500 mW by 2 mW. What was probably intended was to double (3 dB increase) the power of a +27 dBm (one half watt) source (500 mW) to 1000 mW (+30 dBm or one watt.)

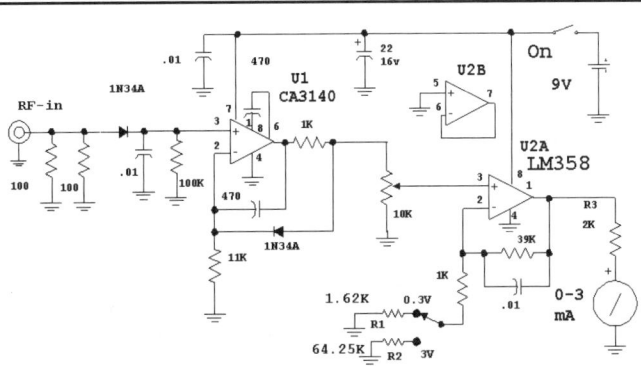

Fig 7.11—This power meter, based on the work of W7EL, has full scale readings of 0.3 and 3 volts RMS with sensitivity of less than –10 dBm. The circuit can be adapted to other ranges. R3 can be changed to 6 kΩ if a 0-1 mA movement is used. See text for details.

50-Ω power meter using the compensation method of W7EL.

Inside view of the W7EL type power meter.

Thirty parallel 2-W, 1.5-kΩ resistors sandwiched between postcard-sized pieces of circuit board material form a medium power termination. Although the rating is only 60 watts, the wide spacing between resistors allows 100 watts to be dissipated for modest times. The wire hooks are convenient places to attach an oscilloscope 10X probe.

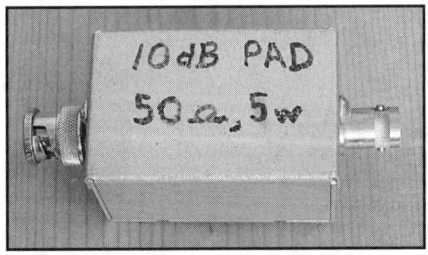

A 10-dB pad built into a small box is a valuable piece of test equipment as well as a station accessory suitable for reduced power experiments.

suitable source such as a QRP transmitter. A step attenuator is then used to decrease the power in known steps to calibrate the 50-mW input. The more sensitive meter can detect powers as low as 1 or 2 mW.

The intended purpose of power meters with small maximum power is not to test very small transmitters. Rather, it is to measure RF power in the early stages of transmitters or in receiver LO systems. A very common example is when setting up a diode ring mixer using hot carrier diodes for LO power of +7 dBm (5 mW.) This is a substitution measurement where a source is set for an available power of 5 mW into 50 Ω, even though it is attached in practice to a less ideal termination.

Microwatt Meter Circuits

Several methods can extend the sensitivity of power measurements, allowing lower levels to be read. One uses an op-amp to follow the RF detector. This guarantees a high impedance load for the detector. Then a matching diode is placed in the op-amp feedback path, which essentially removes the effects of diode

One box contains three power meters with full scale responses of 100 mW, 2 W, and 20 W.

Nine parallel 470-Ω resistors form the RF load for the 20-W power meter. The diode detector and meter multiplier hang on one side. The BNC connector mounts the board to a wall.

offset. This method was presented by Grebenkemper in 1987 and then applied to an in-line QRP power meter by Lewallen in 1990. Both papers are outstanding and are included on the book CD.[4,5] Both instruments included built-in directional couplers that allowed them to be used for in-line power and VSWR measurement.

The simple power meter shown in **Fig 7.11** was adapted from Lewallen's design. The input is a 50-Ω termination followed by the detector. The following op-amp includes a diode within the feedback path. The major effect of this diode is to cancel the effect of the voltage drop across the detector diode, forcing the meter to generate a reading closer to the RF value. The panel meter available when this was built had a 0-3 mA movement, so the instrument was set up for full scale readings of 0.3 and 3 V, RMS. This does not mean that a true RMS voltage is being read. It's still essentially a peak reading circuit, but is calibrated with regard to the related RMS value. Resistors were selected at R1 and R2 to establish the ranges. Lewallen used pots in his meter. The circuit in the figure easily responds to signals less than –10 dBm.

Fig 7.12 shows a power meter using two other methods to obtain greater sensitivity. The first is bias: The diodes are biased at about 20 μA in this system. Two diodes are used in a differential arrangement to reduce temperature drift. The bias

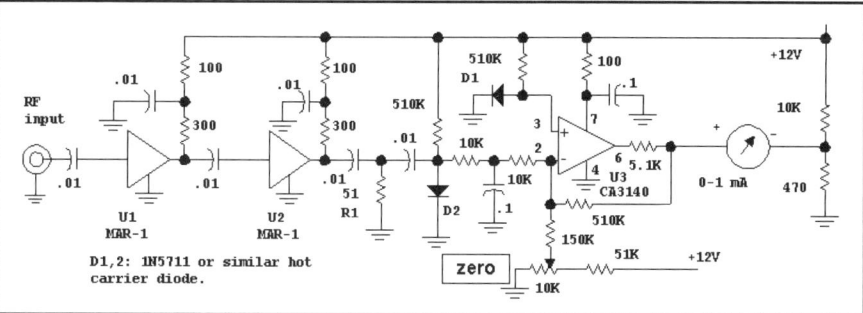

Fig 7.12—Low-level power meter capable of well under 1 μW full scale. This circuit is calibrated against a calibrated signal generator, or against an attenuated QRP transmitter that has been measured with a simple power meter.

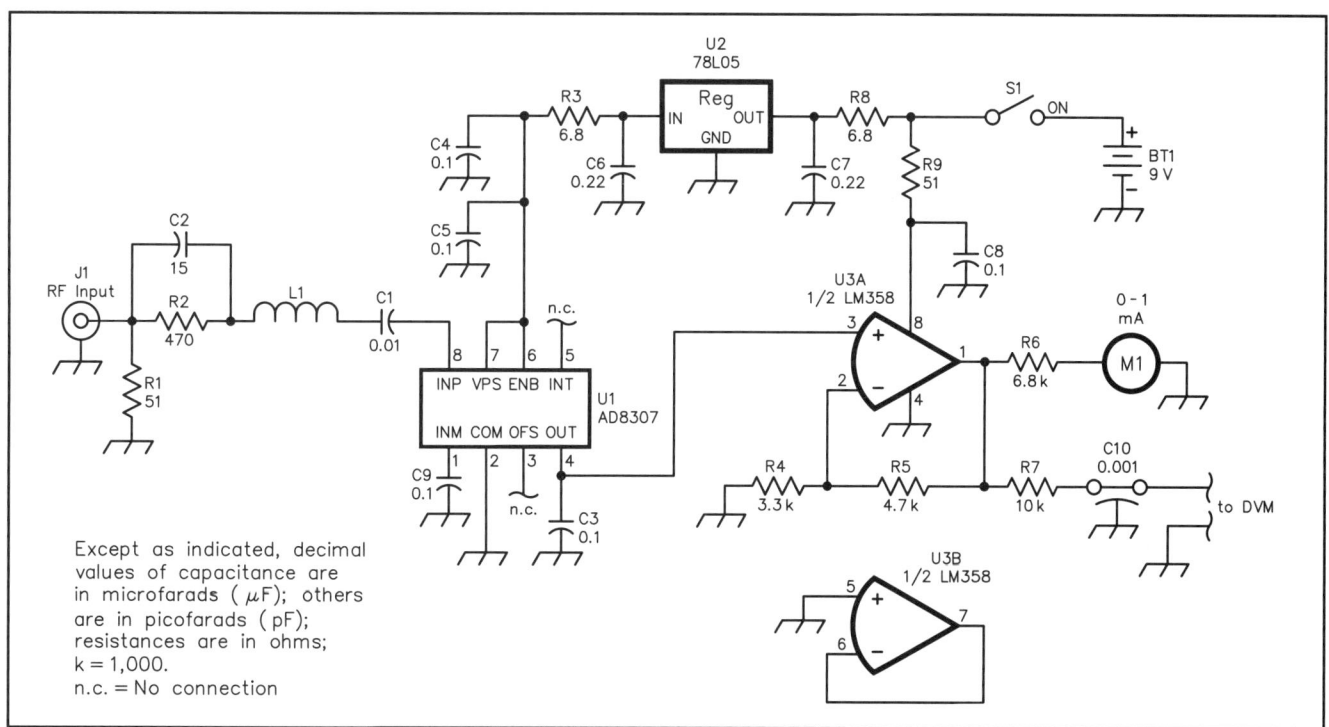

Fig 7.13—Logarithmic power meter capable of reading signals from –80 to +13 dBm.

allows us to see signals of –10 dBm or better at R1. Leaded or surface mounted hot carrier diodes are used. This circuit worked with 1N4152 diodes, although the sensitivity was reduced by a couple of dB. This detector functions well to over 1 GHz. An op-amp provides an interface between the diodes and the meter, and protects the meter against damage from overdrive.

Second, we enhance sensitivity with amplifiers before detection. Here, we use some of the inexpensive monolithic microwave integrated circuits (MMICs) from Mini-Circuits. Discrete feedback amplifiers could also be used.

This power meter will detect signals as low as –40 dBm full scale. This circuit displays about 10 dB of change in the meter motion, making it ideal for careful adjustment of filter circuits. The simpler peak detector power meters (Fig 7.9) typically had 18 dB or higher scale range.

Even greater sensitivity is available from the circuit of **Fig 7.13**. This power meter is based on a logarithmic amplifier integrated circuit from Analog Devices, the AD8307. This circuit functions as a logarithmic detector, accepting signals from audio up to 500 MHz over a power range from around –80 dBm up to over +10 dBm. The output is then a dc signal that tracks with spectacular accuracy, changing by 25 mV for each dB input change. The chip has a sensitivity that drops with frequency, but the circuit shown is compensated to be flat to beyond 500 MHz. This power meter is described in detail in a paper on the CD that accompanies this book.[6]

Any of the low level power meters described can be extended to higher levels with a variety of methods. One is a power attenuator, described later. Another is the 40 dB "tap" shown in **Fig 7.14**. This is essentially a small metal box with a wire connection through to an output attached to a high power termination, or *dummy load*. But the path is sampled with a large value resistor that then drives a 50-Ω terminated connector leading to the power meter. The power available at the tap is, in this example, 40 dB below that flowing in the main path. The *wire* between J1 and J2 is actually a piece of metal, approximately 1 x 1.5 inches, trimmed to fit the box, a Hammond 1590A. With the compensated power meter of Fig 7.13 with a maximum power of +13 dBm, signals beyond +50 dBm, or 100 W can be measured with the tap. The designer/builder should run the circuit only for short periods at full power, for the resistors used in the tap are otherwise taxed.

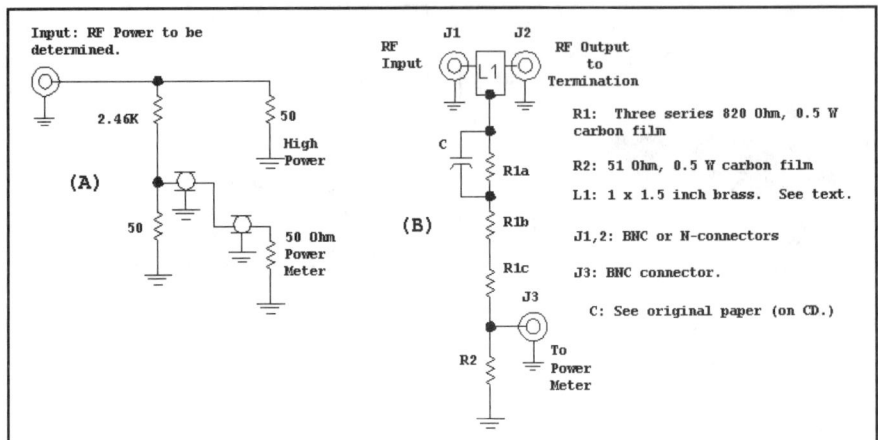

Fig 7.14—Power tap with 40-dB attenuation. Part A shows the basic concept while B shows the version built. See text and original paper on the book CD.

The power meter using the AD8307 was originally described in a *QST* article that is included on the CD. The tap information is in that paper.[6]

The in-line power meter referenced earlier by Grebenkemper used two simultaneous detectors attached to the forward and reflected ports of a directional coupler. This allowed both components to be displayed at once. Further, calculations could be performed on the resulting data. (Op-amps would probably be used.) N2PK has used a pair of AD8307 ICs to obtain similar performance with reduced powers.

RF Power Measurement with an Oscilloscope

Fig 7.15 shows how RF power is measured with an oscilloscope. A key element is the 50-Ω terminator. This is a 50-Ω resistance that can be paralleled with the oscilloscope input connector. The usual 'scope vertical input is 1 MΩ paralleled by 20 pF, essentially an open circuit for low impedance RF. The terminator is effective in setting impedance to 50 Ω. A terminator used for power measurement should *always* appear at the scope end of the coax cable and never at the transmitter end.

This method is limited to the power dissipation of the terminator used and by the vertical input limits. Higher powers can be measured by adding a 50-Ω attenuator in the line. Much higher power can be measured by routing a transmitter output to a 50-Ω load through a directional coupler or tap (described earlier) in the interconnecting cable.

A 10X probe forms the second recommended method for RF power measurement, shown in **Fig 7.16**. A power termination (dummy load) is connected to the transmitter with a coaxial cable. The voltage across the load is then measured with the probe. This method is generally suitable for powers up to 100 W at HF, 3 to 30 MHz. The ground lead should be clipped to the ground part of the load.

Voltages exceeding around 300 V can damage the usual oscilloscope probe, and additional de-rating is required above 10 MHz or so. For example, a 10-X probe may well present an impedance of only 5 kΩ by the time you reach 10 MHz, even though the resulting voltage measurement is accurate.

An often used, but generally inaccurate measurement is shown in **Fig 7.17**. An external dummy load is used, but the interconnect is realized with sections of 50-Ω cable. The difficulty results from transmission line behavior. We wish to examine the voltage across the 50-Ω termination while configuring the lines so that a 50-Ω load is presented to the transmitter under test. A 50-Ω load at one end of a coaxial cable with 50-Ω characteristic impedance presents 50 Ω at the other end. These measurement requirements are satisfied by the setup of Fig 7.15, but not with that of Fig 7.17.

Once a voltage measurement has been performed, it is easily converted to power with one of several equations, shown in **Fig 7.18**.

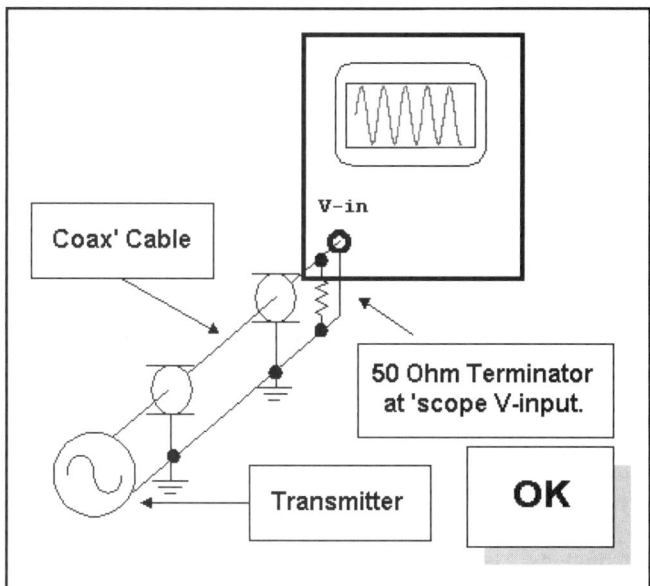

Fig 7.15—Power is measured with an oscilloscope and a 50-Ω terminator at the scope input connector.

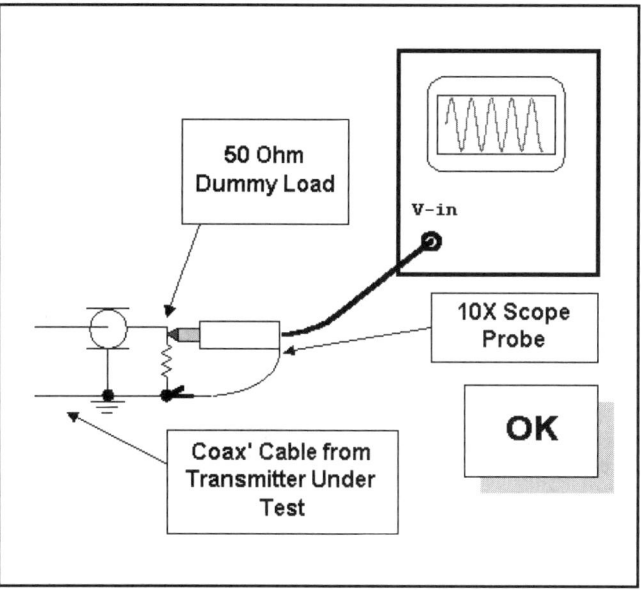

Fig 7.16—A 10X probe is used with an oscilloscope for power measurement.

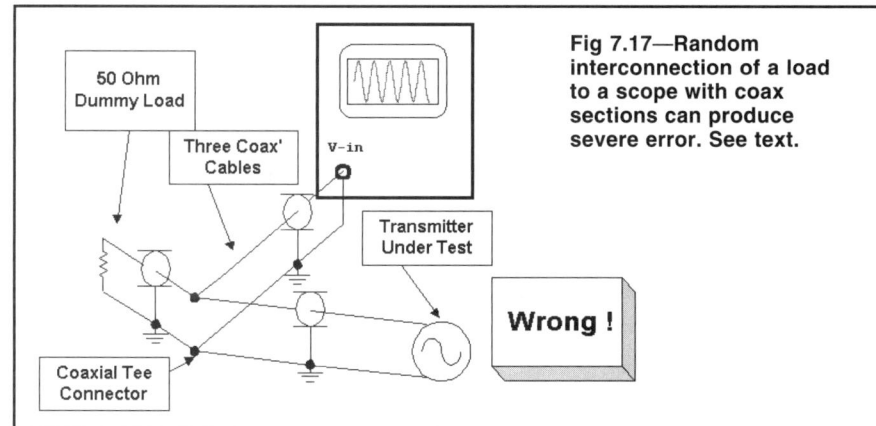

Fig 7.17—Random interconnection of a load to a scope with coax sections can produce severe error. See text.

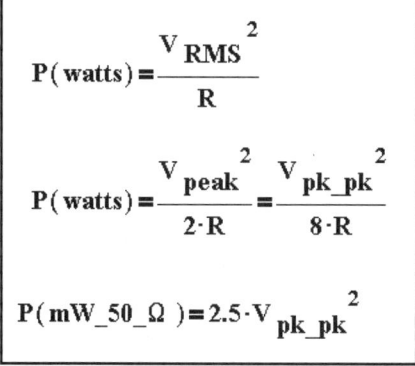

$$P(\text{watts}) = \frac{V_{RMS}^2}{R}$$

$$P(\text{watts}) = \frac{V_{peak}^2}{2 \cdot R} = \frac{V_{pk\_pk}^2}{8 \cdot R}$$

$$P(mW\_50\_\Omega) = 2.5 \cdot V_{pk\_pk}^2$$

Fig 7.18—Equations used to calculate power from oscilloscope readings.

7.4 ATTENUATORS

Attenuators form one of the most important and useful components in any RF measurement laboratory. They become especially useful in a home lab, for they are easily constructed and calibrated with dc. Once available, they can be used to extend numerous measurements to lower or higher levels.

Three attenuator network forms are shown in **Fig 7.19**. The series resistors have value S and the parallel ones a resistance P. When terminated in R (usually 50 Ω) at the right, the input resistance looking in at the left will also be R. This condition leads to a mathematical relationship between the series and the parallel resistors. Setting the attenuation, which established the output voltage V for a 1 V input, allows another equation for each type to be derived. Solv-

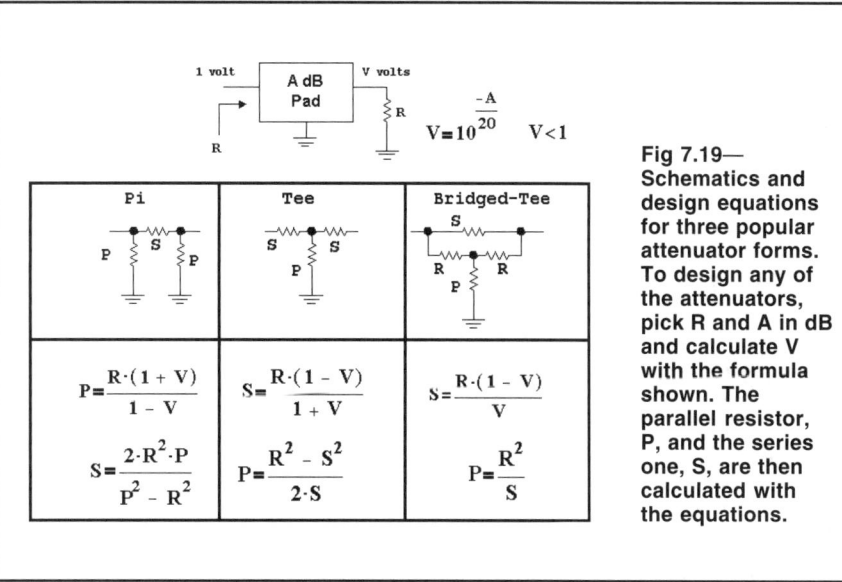

Fig 7.19—Schematics and design equations for three popular attenuator forms. To design any of the attenuators, pick R and A in dB and calculate V with the formula shown. The parallel resistor, P, and the series one, S, are then calculated with the equations.

Measurement Equipment 7.9

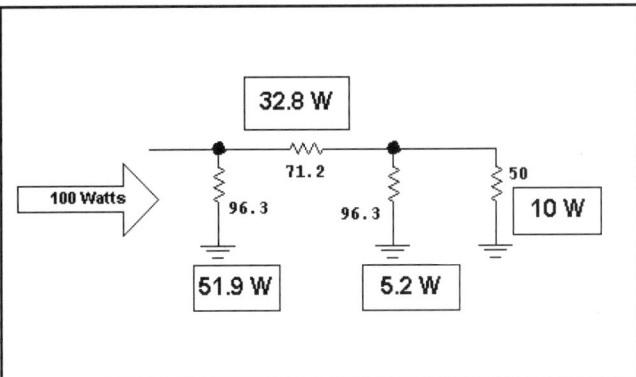

Fig 7.20—Power dissipated in each resistor is shown for a 10-dB pad with 100 W applied. The numbers are also percentages.

Fig 7.21—Power π attenuators built by Fred, W2EKB. The resistors were purchased from a catalog of electronic components. The 262-xxx numbers are from a Mouser catalog.

ing these two produces design equations included in Fig 7.19. If we pick A=4 dB as an example, V will be 0.631, resulting in P=221 Ω and S=24 Ω for the pi, P=105 Ω and S=11.3 Ω for the Tee, with P=85.5 Ω and S=29 Ω for the Bridged-Tee.

The pi and Tee both use three resistors and are equally useful. The pi may fit better with switches (described below.) The bridged-Tee uses 4 resistors, but only two need changing for different attenuation, so it tends to be a good topology for further design of adjustable circuits.

The dB attenuation value is a weak function of the actual resistance values, allowing one to use close 5% values to build practical circuits. For example, building the 4-dB Tee pad mentioned earlier with 12-Ω series resistors and a 100-Ω shunt would produce a 4.2 dB attenuation with input resistance of 50.3 Ω.

One must use care when designing attenuators for use with transmitters delivering modest to high power. **Fig 7.20** shows a 10-dB Pi-pad with 100 W applied to the input. The powers dissipated in the output and the three resistors are shown. The numbers are also the percent of the input power dissipated in each element. Clearly, for example, over half of the applied power appears in the first resistor. Analysis of this sort will allow one to design higher power attenuators. Two high power pads built by W2EKB are shown in **Fig 7.21**. When asymmetric pads are built, the input should be carefully labeled.

Care must be exercised when picking resistors for attenuator applications. Many power resistors use wire wound construction, often hidden in ceramic, making them too inductive for RF use. Carbon composition and the various types of film resistors are generally suitable for RF through UHF.

Fixed attenuators have two significant applications for the experimenter. The ob-

This photo shows some typical terminators. The smaller two are surplus with power dissipation of 2 and 5 W. The box is a homebrew terminator containing four paralleled 200-Ω, 2-W resistors.

Power Resistors at Radio Frequency

Several resistors were evaluated with an HP-8714 network analyzer to establish suitability for use as RF terminations or as elements in attenuators. The results are shown in the attached figure. The RF measurements were performed at the listed measurement frequency, establishing RF resistance and inductance. A maximum frequency was then calculated as that where the inductive reactance goes up to half of the RF resistance. Clearly, traditional wire-wound power resistors are not suitable as RF loads.

Part	Spec. R	DC R	RF R	L at RF (μH)	Freq. for RF Measurements (MHz)	Maximum Frequency (MHz)
A	50	52.2	51.5	6.4	3.5	0.64
B	100	99.6	99.4	0.194	30	40.8
C	50	56.2	59	0.24	30	19.6
D	47	47.2	49	0.0099	250	395
E	47	46	47	0.0095	250	394

Parts Key
A: Lectrohm 10W Wirewound
B: Tru-Ohm 20W Non-Inductive
C: Sprague KookOhm 5W
D: Xicon 3W Metal Oxide
E: Allen Bradley 2W Carbon Composition

A step attenuator for the HF spectrum is easily built with slide switches and 1/4-W resistors. This design used a brass box with the switches soldered in place. This was hard on the plastic parts of the switches, making hardware mounting preferred.

vious one is that of reducing power by a known amount. The other, often just as important, is that they serve to establish impedance level. Assume you have a receiver that you wish to use for measurements in a 50-Ω system. The input impedance of the typical receiver is rarely well matched to 50 Ω, even if it was designed for use with a 50-Ω antenna. However, inserting a suitable pad alleviates the problem. If, for example, we used a 10-dB pad, the return loss we would measure looking into that pad would be 20 dB when the output was left open, and would improve with any termination. A 20 dB return loss corresponds to VSWR=1.2. The receiver with the pad is now a good impedance match. We often use pads in the output of signal generators to force a clean output impedance.

The Step Attenuator

The core of many basement RF laboratories is a step attenuator. Although simple and even relatively inexpensive, such an instrument allows measurements performed at a modest level where they are easy to be extended to other powers where they are difficult. A step attenuator consists of fixed pads that are attached to a switch. Each pad is then switched in or out of a signal path, allowing a total attenuation to be established by adding the individual values.

Several switch types can be used. Most of our experience is with inexpensive DPDT slide switches (eg, CW Industries G and GF series) found in component catalogs. Use those with mounting flanges. The attenuator is built in a trough-like enclosure fabricated from scraps of PC board material. Rectangular holes are cut for the switch handles and the switches are mounted in a line. The resistors are then mounted with very short leads. Short wires are attached to extend one switch section to the next. WB6AIG and WA6RDZ described this circuit in a classic paper and found that vhf performance was improved by adding shields across the center of each switch section.[7] Shriner and Pagel built a similar design, using shields between sections. Bramwell did a more recent version of this classic where careful attention was devoted to maintaining the 50-Ω characteristic impedance within the trough structure.[8] The last two papers are included on the CD that accompanies this book.

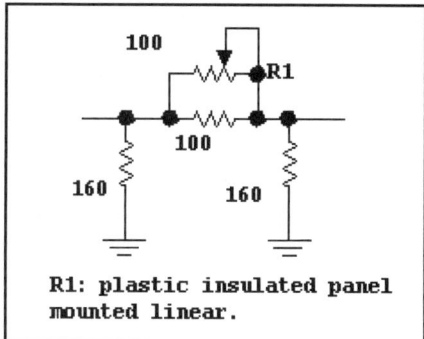

Fig 7.22—Continuously variable attenuator with about a 4-dB range.

It is sometimes useful to have a continuously variable attenuator. **Fig 7.22** shows an attenuator that we have used in the output of homebrew signal sources. This design has an attenuation ranging from 2.5 to 6.7 dB. The exact range obtained will depend on the surrounding impedances. This design will certainly be compromised at higher frequency.

7.5 MEASURING FREQUENCY, INDUCTANCE AND CAPACITANCE

Frequency Determination

The frequency counter is now the most practical instrument for measurement of frequency up to a few GHz. The ICs that form the basis for such measurements are available in virtually all digital formats and are all relatively easy to use in this application. We are not going to say much about counters in this chapter, but note that a simple and inexpensive counter was described in Chapter 4. That circuit could be adapted for general purpose counting with little additional effort. We have built versions with 2, 3, and 4 digits, but would recommend 6 or 8 for a general purpose lab instrument.

Counters are available in all price and frequency ranges, often at less than $100 for a unit that will count to beyond 1 GHz. Resolution at low frequency is typically 10 Hz, although some units are found that will count to 1 Hz. The higher resolution is easy to build if one is brewing an instrument for the home lab and is well worth the extra effort for those cases when it is needed. We find that 1 Hz or better resolution is especially useful when measuring parts for use in crystal filters.

Battery operation is also a useful feature. A battery operated counter will let one build numerous simple instruments that can then be carried into the field for antenna measurements.

It has become popular to build counters from single chip microprocessor of the PIC or BASIC Stamp variety. This offers some hardware simplification and a useful task to use as a mechanism to learn more about the use of these processors. It also offers some unusual possibilities. For example, one kit vendor (Small Wonder Labs) offers a frequency counter designed for use with low power transceivers where the counter uses no visual frequency display. Rather, when a button is pushed to start the circuit, the frequency is counted with the value sent to the user in Morse code. In another design, a single digit display is used sequentially to read up to 8 digits, offering economy and simplicity.[9]

Some inexpensive counters only have high (1 Hz) resolution when digital circuits are investigated. An example is from RadioShack, catalog no. 22-306. A simple interface can be built that will accept a low level RF input while providing a TTL or CMOS compatible output, shown in **Fig 7.23**. This circuit will usually function

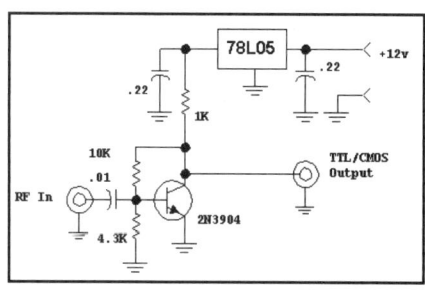

Fig 7.23—Low-level RF to TTL/CMOS converter for simple counting applications. The 10kΩ/4.3kΩ resistive divider sets the collector voltage at about 3 times the 0.7 V emitter-base offset, guaranteeing bias in the active region.

with inputs of −20 dBm at 10 MHz or −10 dBm at 30 MHz (substitution measurements from a 50-Ω signal generator).

Using counters is not difficult, although it is always useful to read the manual. The longer gate times, sometimes controlled by the user, will provide greater resolution, but with longer time between readings. Many counters have a 50-Ω input impedance, but also have a maximum input power. Don't over drive them for it will damage the counter. Instead use an attenuator after you have used a power meter to examine the source you plan on counting. Often a 10X 1-MΩ oscilloscope probe works very well at the input to a counter, even with 50-Ω inputs.

Some users will attach a small link to a piece of coax driving the counter. The link is then used to sniff the circuit under test. This may work, although the power to the counter is not well defined. Moreover, if the source is rich in harmonics, you can end up counting a harmonic instead of the fundamental. Don't try to use the counter as a spectrum analyzer; it may be an interesting measurement anomaly, but it is not a good method.

L and C measurement

The traditional experimenter measured inductance or capacitance by finding a resonant frequency with a dip meter. An unknown C was paralleled by a known inductor, the combination was "dipped," and the value was calculated. An identical process measured an unknown L. But the frequency measurement was poor, leaving the experimenter wondering about his or her results.

The same general method can be applied today, but the dipper is completely eliminated from the measurement. A stable LC oscillator is built in its place with a buffer to drive the frequency counter. Unknown components are then attached to the oscillator to alter its frequency. This produces the data needed to obtain the L or C. This method was the basis for a simple instrument built by Bill Carver.[10] This instrument is shown in **Fig 7.24**.

The instrument is ruggedly built with three binding posts labeled *L, C* and *Ground*. Operation always begins by placing a wire between the L and the C terminals and measuring frequency. Calibration can then be

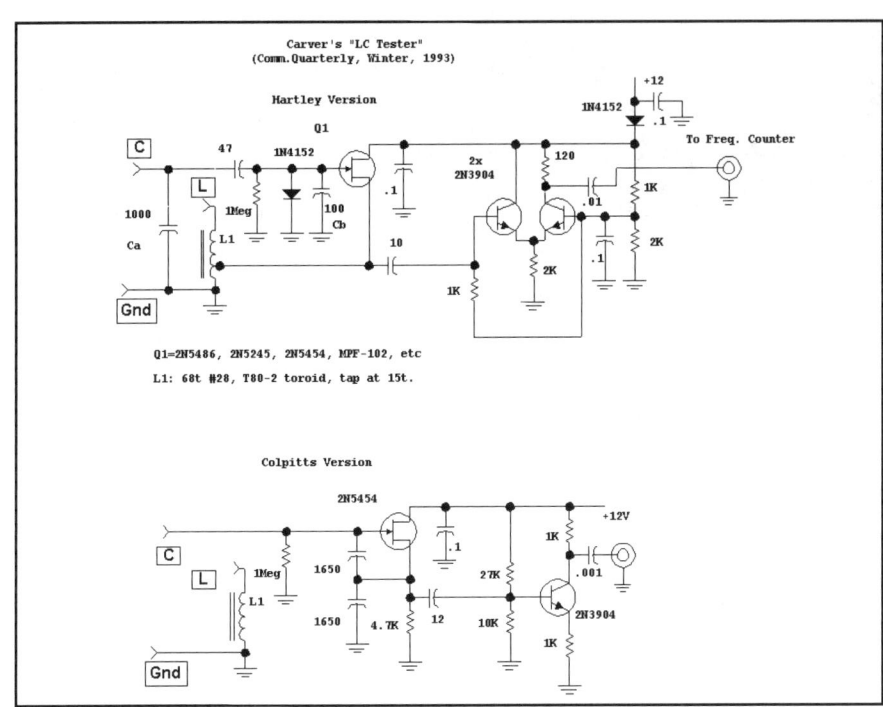

Fig 7.24—"The LC Tester" offered by Bill Carver, W7AAZ, in *Communications Quarterly*, Winter, 1993. **The two modes essentially offer identical performance. See text.**

performed, (not necessary with every measurement) by placing a known capacitor between the C and the ground posts with L and C still shorted. A good calibration value would be a 1000 pF 1% capacitor. A new frequency is measured with the CAL cap in place. From the two frequencies and the known CAL capacitor value, the net fixed capacitance and the inductance value can be calculated, C_o and L_o.

Measurements are now performed by parallel or series connections of the unknown components. The instrument is turned on and an initial frequency, F_1, is counted. An unknown inductor is then attached either between C and ground, or between L and C. The new frequency, F_2, is measured. Knowing Co, a new inductance can be calculated. If a series connection was used, $F_2 < F_1$ and L is found by subtracting L_o from the measured value. If a parallel connection was used, $F_2 > F_1$, and the measured L will be less than that of the one connected. The same resonance concepts give capacitance results.

Carver's original circuit used the Hartley circuit shown. When we breadboarded the circuits, we also tried a Colpitts variation that allowed larger capacitor values to be determined. Either large C or small L between the C and ground terminals can cause oscillation to cease. The two topologies are otherwise identical.

Once the instrument is built and in use, a computer or calculator program can be written to expedite calculations. Carver includes such a program in his paper.

Carver's paper also mentioned a preliminary version of the instrument that used a PIC microprocessor, performing the counting function as well as the calculations. Since that paper was published, a similar instrument has arrived on the market by Almost All Digital Electronics, which is offered as an easily constructed kit. (**www.aade.com/**)

The experimenter has a choice of building his or her own LC Tester or purchasing the kit from AADE. Whatever the choice, the modern experimenter cannot afford not to have this measurement capability. This instrument essentially replaces the classic grid dipper for the electronics experimenter of the 21st century!

7.6 SOURCES AND GENERATORS

A signal source or generator is needed to align and adjust most projects, or for most fundamental circuit experiments. Two or more are required for many other experiments. In this section we present a wide variety of sources

The one instrument that would do most of what we need is a "lab quality RF signal generator." But there is more to the name than suspected. A traditional signal generator used for servicing consumer radio and TV receivers consisted of a wide tuning range oscillator covering all input and intermediate frequencies that the service person might encounter. These boxes usually had modulation capability, allowing the user to align AM receivers. However, they did not qualify as the lab quality instrument we really want. A good signal generator will have the mentioned characteristics plus accurate frequency readout, a 50 Ω output impedance, low phase noise, low spurious outputs close to the carrier frequency, excellent buffering, good isolation from the power supply, and un-compromised shielding. Long term stability and low harmonic content are also useful, but are not dominant specifications.

Many instruments presented as *signal generators* don't qualify because they can't be made weak enough to test a receiver that is useful for communications. When you disconnect the generator, but perhaps attach an antenna to a receiver under test, the generator is still heard. The problem may be poor shielding, signal conduction through the power supply, or both.

The sources we describe in this chapter will not result in a lab quality instrument. Rather, we will describe specialized sources that will satisfy some of these needs, but not in one instrument. The surplus market is full of good equipment that will fulfill many of the experimenter's needs. Having one of these is useful as a means to calibrate home built sources.

Audio sources

A whistle or a few words spoken into a microphone may serve as a first functionality test for a phone transmitter. However, we need something more when testing a transmitter. A simple generator is shown in **Fig 7.25**. This circuit is battery operated from a 9-V cell, a very convenient feature when seeking good isolation from other sources. This topology is called a phase shift oscillator. The transistor is biased as an inverting amplifier (180 degree phase shift) with a voltage gain of just under 50, established with feedback and biasing. The output is routed back to the input through

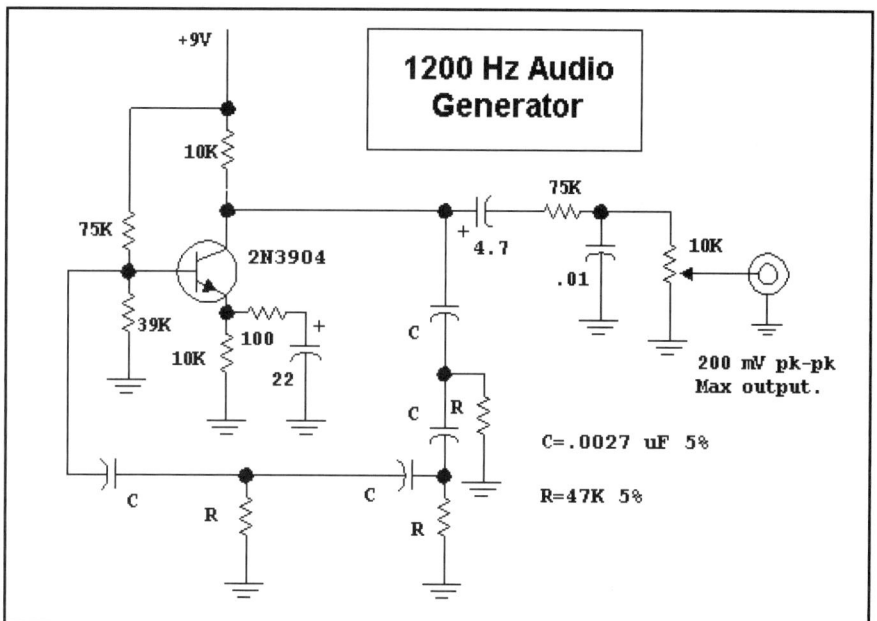

Fig 7.25—A simple audio generator for transmitter testing.

an RC high pass filter. Oscillation occurs at the frequency where the total phase shift is 360 degrees, half provided by the frequency dependant feedback network. Output is extracted from the collector, attenuated, low pass filtered, and applied to an output level control. This oscillator operates at 1200 Hz. There is nothing special about the exact component values. This one was based upon a handful of 0.0027 µF capacitors on hand. The measured 2nd harmonic was 40 dB below the desired output.

The circuit is built on a small scrap of circuit board material. Another board scrap is mounted to the original to hold a BNC output connector and a level control.

The maximum output from this circuit is about 200 mV peak-to-peak, more than that supplied by most microphones. Use begins by attaching a microphone to a speech amplifier in a transmitter. A few words into the microphone while looking at the amplifier output with an oscilloscope allows us to set audio gain. The microphone is then replaced with the audio oscillator with the level set to establish the same maximum level. This can then be used for extended bench testing.

Fig 7.26 shows a two tone generator useful for testing SSB transmitters. One generator operates at about 650 Hz while the other is at 1650, a non-harmonic higher frequency. A Wien Bridge circuit, shown in the inset, is used for each source. Each oscillator had a measured third harmonic that was only suppressed by about 30 dB, so

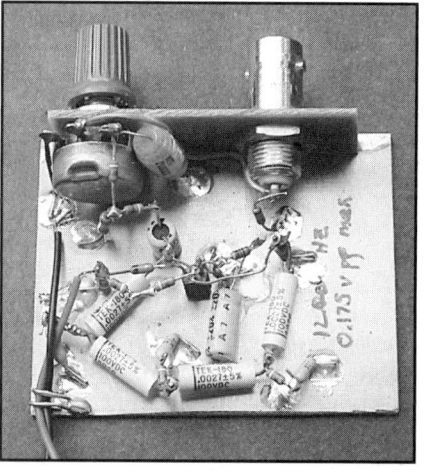

A simple audio oscillator for transmitter testing.

suitable active low pass filters are added. The two signals of about 3 V peak-to-peak are added and attenuated in U3A while U3B provides a 600-Ω output impedance.

There are many other ways to build audio sources including some special function generator ICs. These are circuits intended to generate triangle and square waves, but with modifications to also approximate a sine wave. The Exar XR-2206 and the Maxim MAX038 are examples. A DSP-based solution is also presented in Chapter 11.

The two-tone generator is attached to a transmitter mic input and the level is adjusted for the desired output. One tone can

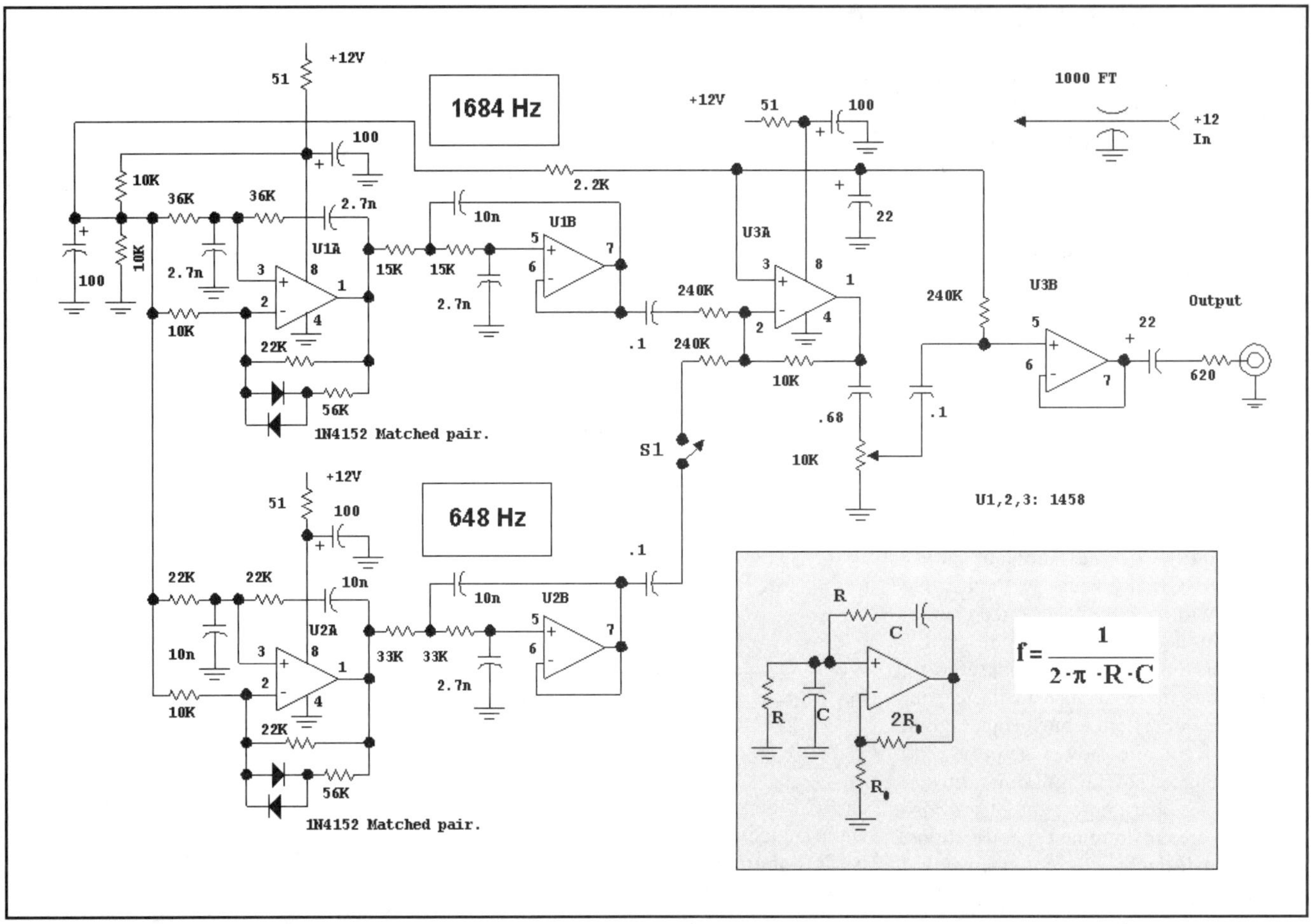

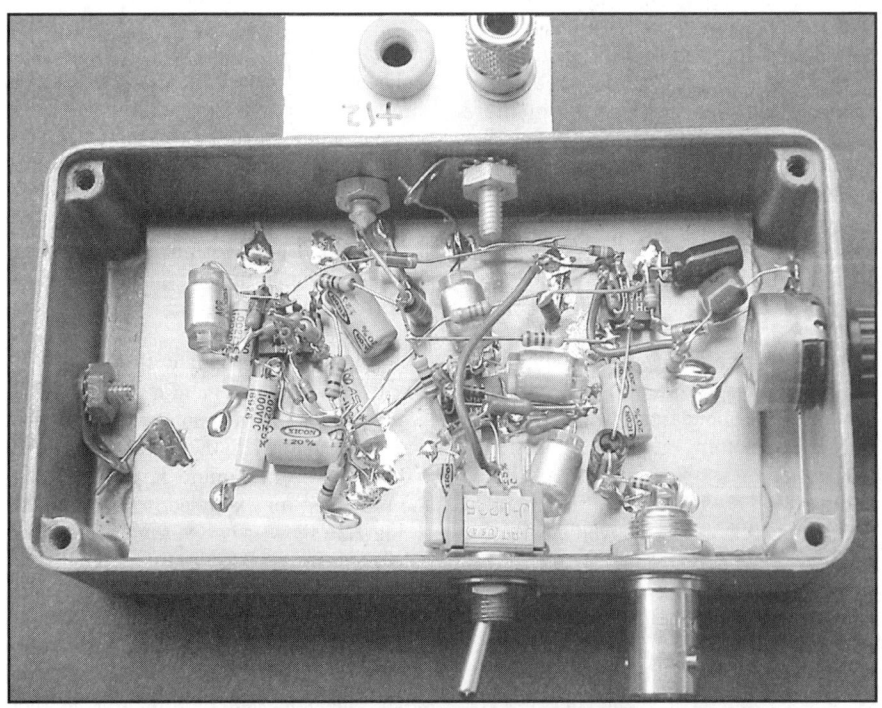

Fig 7.26—Two tone audio source. Each oscillator uses a matched pair of diodes with matching done with a DVM in the diode test position. Matching was done to 10 mV.

be turned off with S1 so single tone power can be measured. With two tones present, the composite signal moves through all stages of the SSB transmitter to produce a two tone output that can be observed with an oscilloscope or spectrum analyzer, or ideally, both. The intermodulation distortion products (or flat topping in a 'scope display) are then the result of distortion in the transmitter. It is vital that the source be free of these products.

General Purpose RF Sources

No lab is complete without a general purpose RF generator. Like power supplies and step attenuators, one more is always useful. The early sources we built consisted of an LC oscillator, link coupled to a feedback amplifier and pad to provide an output power of +5 dBm or more, enough to drive a diode mixer. Although the design was useful, the buffering was sometimes inadequate, especially for crystal filter testing. The addition of a com-

Two-tone audio generator for SSB transmitter IMD measurements.

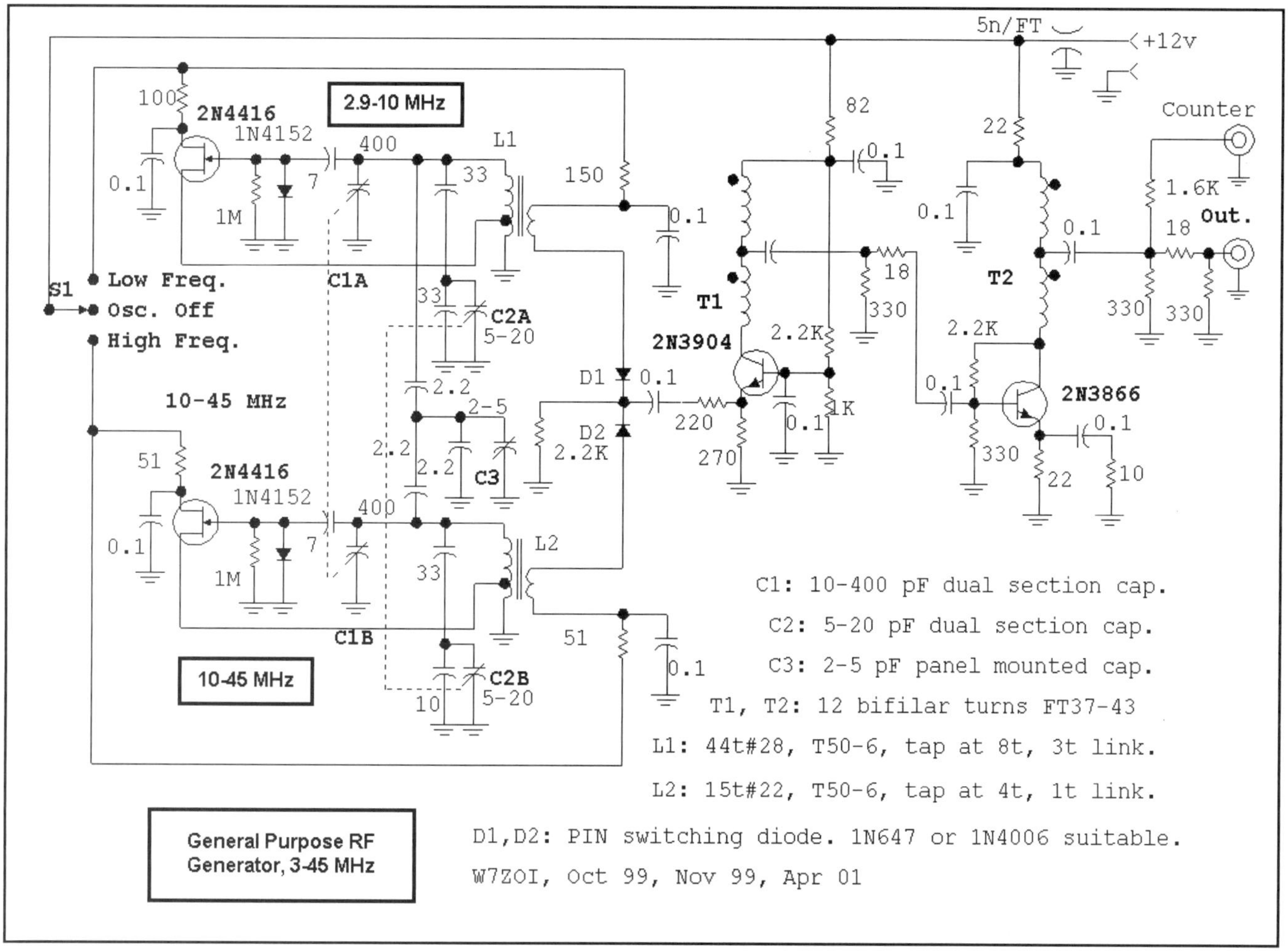

Fig 7.27—General purpose oscillator tuning the range from 3 to 45 MHz in two ranges. See text for details.

General purpose RF source tuning from 3 to 45 MHz.

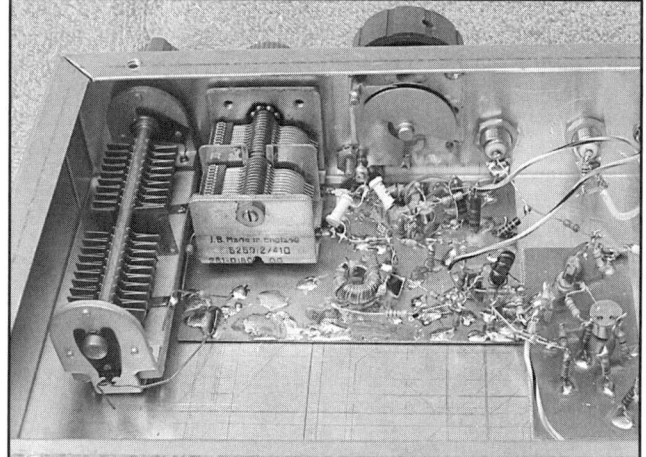

Inside view of 3-45 MHz RF Generator.

mon-base buffer amplifier has solved these problems.

A wide tuning range oscillator is shown in **Fig 7.27**. Two Hartley oscillators are tuned by dual section capacitors, C1 and C2. The Hartley topology is optimum, for it uses an inductor tap to obtain feedback. As such, all resonator capacitance can be variable, providing the widest possible tuning range. This circuit achieves 2.9 to 10 MHz in one of the oscillators with the other tuning 10 to over 45 MHz. C1 is the main tuning while C2 provides bandspread. Even greater bandspread is provided by C3, now a single section capacitor. C3 is coupled to both resonators in such a way that the inoperative oscillator does not disturb the other. The bandspread afforded by C3 allows the generator to be set accurately, even at the high end.

Another scheme that could provide bandspread would add a variable capacitor from the cathode of the PIN diode switches to ground. This capacitor would then be switched between oscillators with the diodes. But because it reaches the resonator through a link, it tunes over a proportionally smaller range.

Band switching is performed with a SPDT toggle switch with a center-off

Measurement Equipment 7.15

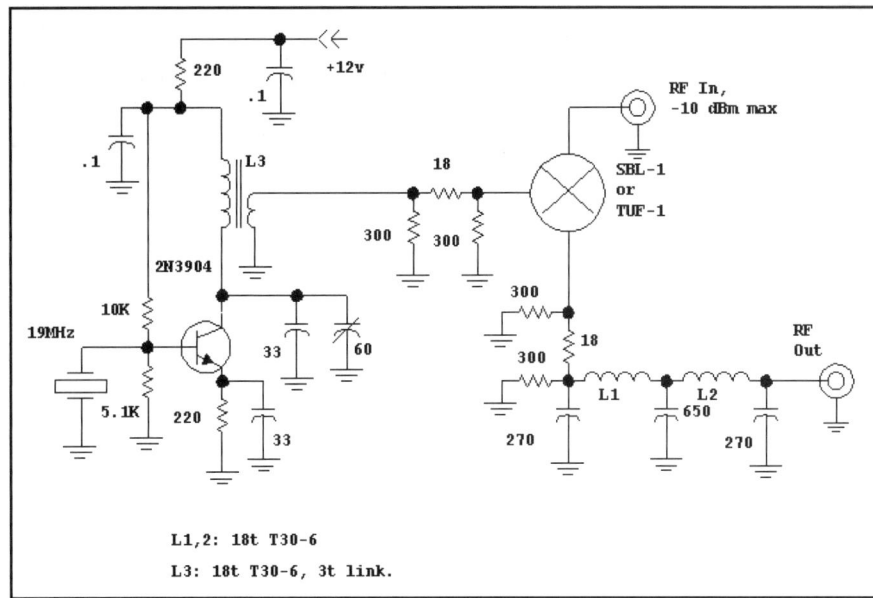

Fig 7.28—Signal Generator Extender.

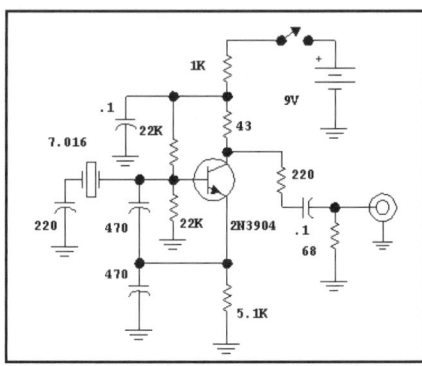

Fig 7.29—Crystal controlled oscillator used for receiver testing. This unit doubles as a spectrum analyzer calibration source with a 7-MHz output of −20 dBm.

position. The "off" mode has been useful to completely extinguish a signal without changing other settings. The toggle switch applies power to one of the two oscillator circuits and biases a PIN diode that routes the output to the buffer amplifiers. A high speed switching diode (1N4152, etc) should *not* be substituted here, although many rectifier diodes work well. The diode switch output is applied to the common base buffer amplifier, preferred over a common emitter amplifier or an emitter follower. The output stage is a 2N3866 common emitter feedback amplifier with a 3-dB pad. A bit of the output energy is tapped and supplied to an auxiliary output feeding a frequency counter. The output power from this source is around +10 dBm on both bands, although it is not as flat (constant amplitude with frequency) as we would like. But this is also the case with many very good signal generators, such as the classic HP-608 series and the surplus URM-25 line. A PIN diode leveling loop could be added to solve this problem, but should be done with considerable care, for such loops can generate additional problems.

Single band variations of the oscillator of Fig 7.27 have been built, all with a virtually identical circuit. One version was built into the remains of a surplus BC-221 frequency meter. The tuning range was purposefully restricted to about 30 kHz around 5 MHz. The oscillator is then used for crystal and crystal filter measurements.

These RF generators do not lend themselves to easy duplication owing to the unique components used. The junk box is the basis for much of our test gear. If dual section capacitors are not available, single range versions of this oscillator may be built. The circuitry is generally simple, tolerant of component value changes, and inexpensive except for the variable capacitors. These oscillators are running at moderately high power with over 10-V peak-to-peak across each resonator. While this is ideal for low phase noise, it means that one *cannot* casually substitute a varactor diode in these circuits.

The dual range source has been used for numerous applications, ranging from antenna measurements to IMD testing.

There are many generators found on the surplus market that cover ranges from 10 MHz upward. Examples include the HP-608 and HP-8654. A useful lower range may be added with the "extender" shown in **Fig 7.28**. An available 19 MHz junk box crystal was used in a crystal controlled oscillator driving a diode ring mixer. The signal generator is applied at the input above the crystal frequency and at a level of −10 dBm or less. The mixer output is attenuated in a pad and low pass filtered. This unit is especially useful, for the original generator amplitude calibration is retained with a 9-dB offset. We have also used this same box as an audio source. A 19-MHz VXO can then be used in place of a signal generator. The low pass filter following the mixer has a cutoff just above 10 MHz, the maximum output frequency for this box.

A useful variation of this instrument would use a high level (+17 dBm LO) mixer. More 19 MHz LO energy would be required. This would then allow operation at 10 dB higher levels, needed for some IMD measurements.

Outside view of matching crystal controlled RF sources used for receiver testing. The outboard amplifiers provide the higher signals needed for testing mixers and high-level amplifiers.

Close up view of outboard amplifiers for IMD testing.

An off-the-shelf 14.318 MHz color burst crystal becomes a convenient RF source for the 50-MHz band. Built by KA7EXM.

Crystal controlled sources

Most of the careful receiver measurements we do require good stability in both the receiver and the equipment used to test it. The ideal (affordable) solution uses crystal controlled test oscillators. **Fig 7.29** shows a general purpose source that was originally built as a spectrum analyzer calibration source. The crystal chosen lies within the 7 MHz amateur band, so it serves well as a general alignment tool. The harmonics at 14, 21, and 28 MHz are also useful. The 7 MHz output is –20 dBm. This unit is built into a Hammond 1590B box with a battery contained on the inside, providing the ultimate power supply filtering.

VHF experimenters are always in need of a source to test their equipment, and a crystal controlled oscillator will often serve this need. **Fig 7.30** shows a source using an inexpensive, standard "color burst" crystal to generate signals at 7.16 MHz and at 50.125 MHz. The marked crystal frequency is 14.318 MHz. This is frequency divided in a 74HC74 divider circuit to produce a squarewave at 7.16 MHz. Some low pass filtering strips most of the harmonic energy away for use at 7 MHz. The 7th harmonic of the square wave is extracted with a double-tuned circuit to provide the needed source for the 6-m band. This source was built by KA7EXM.

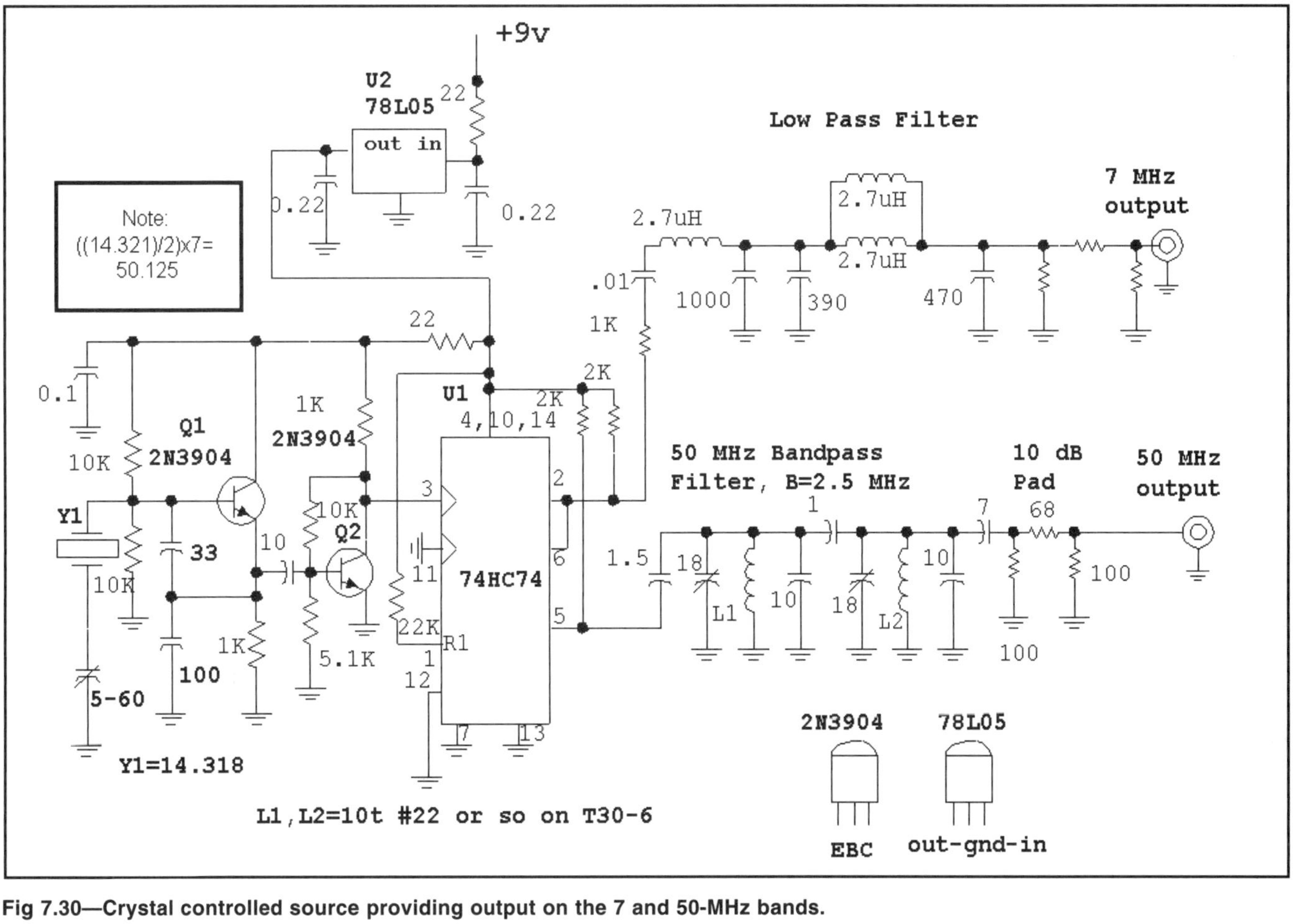

Fig 7.30—Crystal controlled source providing output on the 7 and 50-MHz bands.

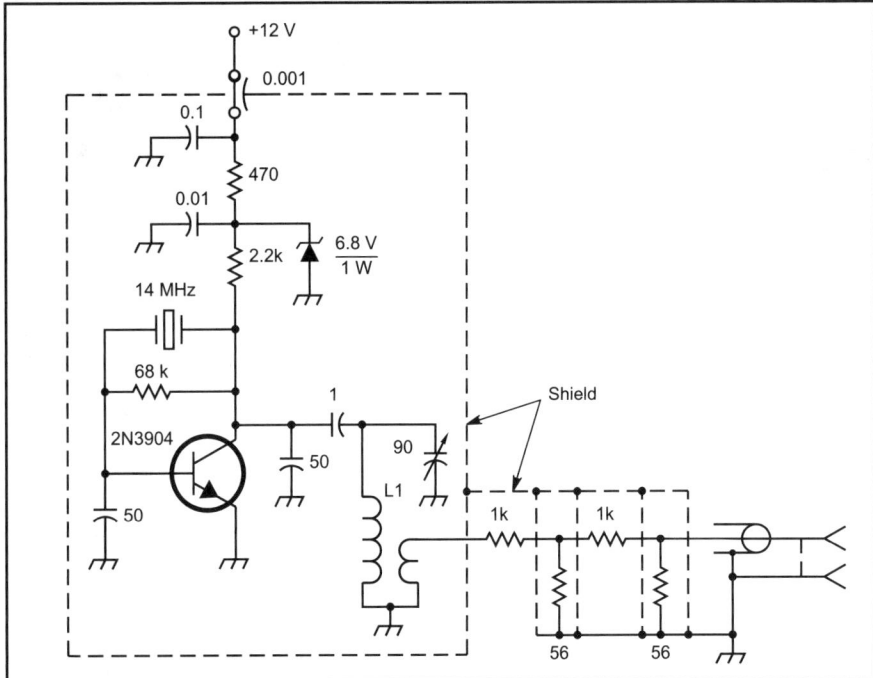

Fig 7.31—Crystal controlled oscillator for receiver MDS measurements. The output is set for about −100 dBm. A builder may wish to add a small resistor or an inductor between the feedthrough capacitor and the 0.1 μF capacitor. A few turns on a ferrite bead should work well. L1 is chosen for resonance at the crystal frequency—the one or two-turn link provides output.

A Weak Signal Source for MDS measurement

The source shown in **Fig 7.31** is similar, but has considerable attenuation included within the box. This unit is predominantly used as a weak signal source for receiver *minimum detectable signal* (MDS) measurements. The oscillator is built at one end of a narrow box fabricated from scrap PC board. Shields are then added with sections of attenuation between. The attenuation is then set to establish the desired output. Levels around −110 to −100 dBm are good, for they are easily attenuated further in a step attenuator to drop to the MDS levels often found with HF receivers. After the output is set, a shield lid is soldered to the box. If double sided board is used, be sure that the inside and outside are attached to each other at the lid.

The unit is calibrated with a CW receiver and another signal generator. The crystal oscillator is tuned with the receiver (AGC off) and the output is measured with an audio voltmeter. The signal generator is then tuned to the same frequency and the amplitude is adjusted until the same output response is observed. The level is noted in your notebook and is marked on the outside of the MDS generator.

MDS can then be measured with the oscillator and a step attenuator. The source is attached to the receiver (AGC still off) and the receiver is tuned to the generator frequency. Attenuation is then added to weaken the source. The source is momentarily turned off and the noise level is noted in the audio meter. The source is turned on again and the attenuation is adjusted until the meter response is 3 dB above the noise. The strength of the source less the added attenuation is then the MDS.

It's worthwhile to listen to the receiver as a means for growing a "calibrated ear." Although this signal is weak, it is clearly audible above the noise, even if the bandwidth is a kHz or more. As receiver bandwidth drops, the MDS will become smaller but there is less difference between the measured MDS and that perceivable by ear. When running a relatively wide SSB bandwidth, a signal at measured MDS sounds rather loud. It is not surprising that many weak signal VHF enthusiasts including EME aficionados will use the wider bandwidths when QRM is not an issue.

Crystal Oscillators for Intercept Measurements

Having measured receiver MDS, we now need "loud" generators that can be used to measure the strong signal performance, the receiver input intercept, IIP3. The measurement was described in detail for an amplifier or mixer in Chapter 2 and then applied to a receiver in Chapter 6. The basic source we use for receiver testing is shown in **Fig 7.32**. The crystal oscillator is carefully tailored to operate with current limiting, avoiding the Q degrading voltage limiting. The following buffer has an input impedance dominated by a single resistor, but then operates as a limiter, developing an output substantially independent of drive level. That output is low pass filtered and attenuated in a 6-dB pad and then applied to a common base output amplifier, picked for good reverse isolation.

We use two identical versions of the source of Fig 7.32, usually separated by about 20 kHz. The sources are always checked ahead of each use, confirming power and match between units. The output level chosen is 0 dBm for each source. These are usually applied to 6 dB-pads and then to a 6-dB hybrid combiner. The combiner, described later, is a return loss bridge used in a different way. The hybrid output is attached to a 15 MHz low pass filter and then to a step attenuator. This setup, shown in **Fig 7.33**, provides signals of −12 dBm per tone and lower. The role of the hybrid is to add the two signals while preventing the output of one source from reaching the other. If the output from one oscillator reached the other, inter-

Inside one of the crystal controlled RF sources.

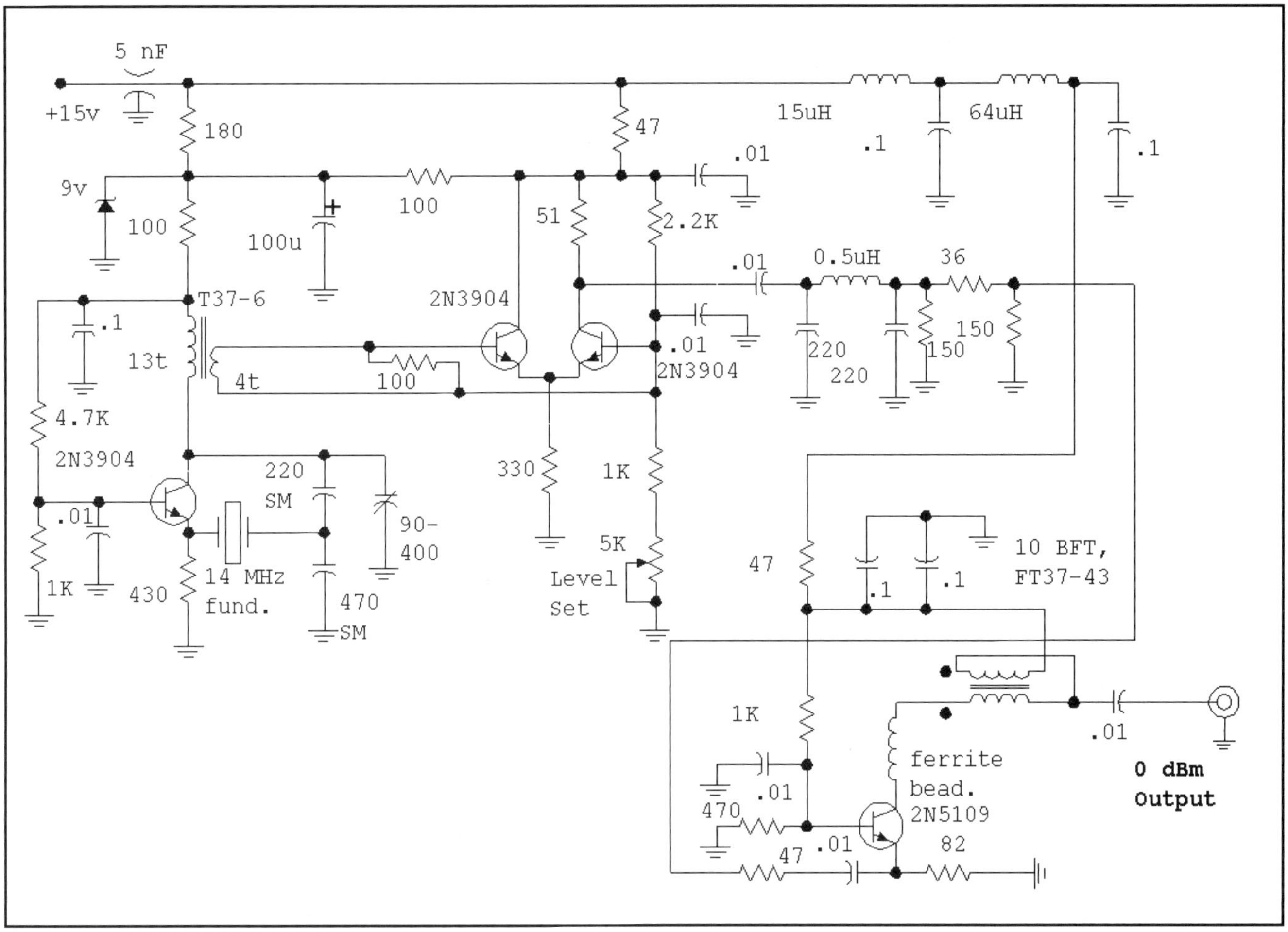

Fig 7.32—A source with an output of 0 dBm suitable for receiver testing. See text for discussion.

modulation could occur, creating spurious signals at the same frequencies as produced by the third order IMD that is usually measured with this system.

There are alternatives to the 6-dB hybrid. A 3-dB Splitter-Combiner is sometimes used and can offer excellent performance. Some experimenters will even use a 50-Ω power divider, which preserves impedances but provides no isolation. A 50-Ω power divider consists of three 50-Ω resistors in a "Δ" configuration, or three 18-Ω resistors in a "Y." The 6-dB hybrid is recommended.

Assume that the two generators have crystals to put their frequencies at 14.03 and 14.05 MHz. Tuning to either of these signals produces a large meter response. These signals impinging on the receiver front end will intermodulate, generating distortion products above and below the two desired signals, at 14.01 or 14.07 MHz. These products are created within the receiver, usually in the circuitry ahead of the main IF filter. With the two test signals separated by 20 kHz, the distortion signal will be 20 kHz above the upper desired signal and 20 kHz below the lower one.

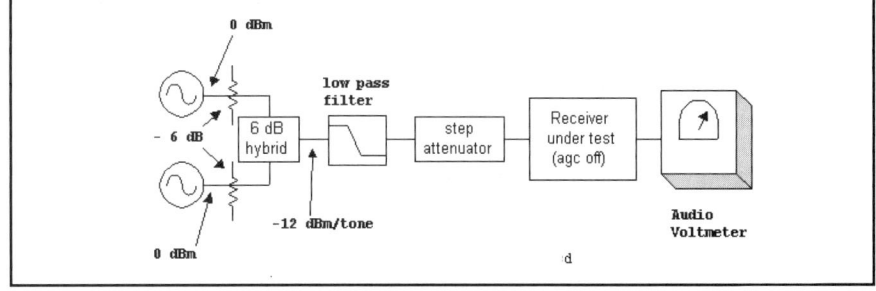

Fig 7.33—Test setup for determining a receiver IIP3, or "input intercept." See details in Chapters 2 and 6.

We tune to either of these IMD responses to measure them, seeing a loud, but still manageable response. Assume an audio signal of 50 mV when tuned to one of the distortion frequencies and that this occurs with the step attenuator set at 30 dB. The signals are then –42 dBm/tone at the receiver antenna terminal. But how strong is this response compared with the input signals? We find an answer by tuning the receiver to one of the main tones and increasing attenuation. When the net attenuation inserted is 110 dB, the audio output is again 50 mV. We have increased the attenuation by 80 dB to depress the main signals to the point where they produce the same response as was seen from intermodulation. The intermodulation distortion ratio, IMDR, is then 80 dB. The input intercept is then given by

$$IIP_3 \text{(dBm)} = P_{in} \text{(dBm)} + \frac{IMDR \text{(dB)}}{2}$$

Eq 7.2

Measurement Equipment 7.19

For this example, $P_{in} = -42$ dBm and IMDR=80 dB, so IIP3= −2 dBm.

Let's repeat the experiment, but start with less attenuation. Instead of 30 dB in the beginning, start with 24-dB attenuation to apply signals that are 6 dB stronger. The response at the distortion frequencies is now much larger, significantly more than the 6 dB increase in the main tones. Assume that it's about 400 mV in the audio voltmeter. We record this level and then tune the receiver to one of the main signals and increase the attenuation. After adding 68-dB attenuation, for a net attenuator setting of 92 dB, we observe 400 mV of audio. The applied power is −36 dBm/tone and IMDR=68 dB, so **Eq 7.2** predicts IIP3= −2 dBm.

This example illustrates the utility of the intercept concept. If we know the input intercept for the receiver, we know what the response will be to any input signals. If we allow the mathematics to get a little more complex, we can even predict the response to input signals of unequal amplitude.[11]

Let's say that this receiver had MDS of −139 dBm, a reasonable sensitivity for a CW receiver with a bandwidth of perhaps 500 Hz (NF=8 dB). The two-tone DR would then be

$$\mathrm{DR}\,(\mathrm{dB}) = \frac{2}{3}\cdot\left(\mathrm{IIP_3}\,(\mathrm{dBm}) - \mathrm{MDS}\,(\mathrm{dBm})\right)$$

Eq 7.3

or, 91.3 dB in this example. But what does this mean?

The meaning of two-tone DR is clarified with a more direct measurement, still using the example receiver we have been examining. First, we use our weak signal source with the step attenuator to measure MDS. Assume that the receiver gains are set to produce an output of 10 mV with the weak signal source. When we turn the source off, the level drops by 3 dB to 7 mV. Receiver AGC is still off and we don't touch any of the gain controls.

We now replace the weak source with the two tone generator setup of Fig 7.33. We tune the receiver to one of the distortion product frequencies and adjust the attenuator until we get the same response we saw with the MDS measurement, 10 mV on the meter. We tune the receiver to one side and the other of the distortion product to be sure that the response drops to the noise floor of 7 mV. This happens in our example with the attenuator at 36 dB, which places a strong signal of −48 dBm at the receiver input. We record these levels in our notebook and then retune the receiver to one of the strong tones. (It's a good idea to *not* have the headphones on during these experiments!) We now add attenuation until the response from a strong tone is again 10 mV. This occurs with a total attenuation of 127 dB. This is 91 dB lower than the signals that produced the distortion responses.

This experiment has illustrated the real meaning of receiver two-tone dynamic range: DR is the difference between the weakest signal we can hear with that receiver and the strength of one of a pair of signals that will produce intermodulation distortion at the same level as that minimum. This is a severe test, but it is measurable with carefully built test equipment.

The high attenuation levels needed for DR measurements, especially the direct one, may be intimidating. It's hard to obtain over 100 dB of attenuation, especially in casual homebuilt designs. For this reason, an indirect measurement is often easier. That is, measure IIP3 with two moderately well shielded strong sources with levels that can be confirmed with a power meter, a spectrum analyzer, or terminated oscilloscope measurement. Perform an independent measurement of MDS with a special generator you have built for just that purpose. Then calculate DR from **Eq 7.3**. It is, however, best to work with weaker "strong" signals, for most receiver mixers will then be "well behaved," as defined in Chapter 6.

The procedure we recommend eliminates the MDS measurement, replacing it with a noise figure determination. This will be discussed later.

Component Intercept Measurements

While the receiver builder may wish to perform IIP3 and MDS measurement to obtain DR, the designer is equally interested in evaluation of component parts of a receiver or transmitter. The two tone source is again used, driving the component, followed by a spectrum analyzer. (Analyzers and their design are described later.) The test setup is given in **Fig 7.34**. Frequency spacing is adjusted as needed for the component being investigated.

The test setup is more illuminating than the receiver evaluation, for it is a swept measurement showing the main signals and the distortion products on a calibrated screen, all at the same instant. A step that should always be done is to apply the signal from the step attenuator directly to the spectrum analyzer, prior to inserting the component. Any distortion seen would then be occurring in the analyzer or in the generators. Once a distortion-free test setup is confirmed, the amplifier is inserted, the analyzer input attenuation is readjusted to keep the main signals on the screen, and the data is recorded. The gain of the amplifier (or whatever) is now observed, equal to the change in spectrum analyzer sensitivity needed to keep the main signals in the same position on the screen. We know the input levels, for we measured them before inserting the amplifier, and the IMD ratio can be observed directly on the screen, so the input intercept, IIP3, can be calculated from Eq 7.2. The corresponding output intercept, OIP3, is just IIP3 plus the amplifier gain.

It is very informative at this time to vary the strength of the input tones used to test the amplifier, achieved by adjusting the step attenuator. The desired output signals should change on a dB-for-dB basis with the inputs. However, the distortion products above and below the desired two signals will move on a 3 dB per one dB input change rate. It is not necessary to collect all of the data to actually plot traditional intercept curves, such as were shown in Chapter 2 of this book.

Measurements normally performed with a spectrum analyzer can also be done with a receiver. It will be necessary to put an attenuator ahead of the receiver to control the levels reaching it, always tak-

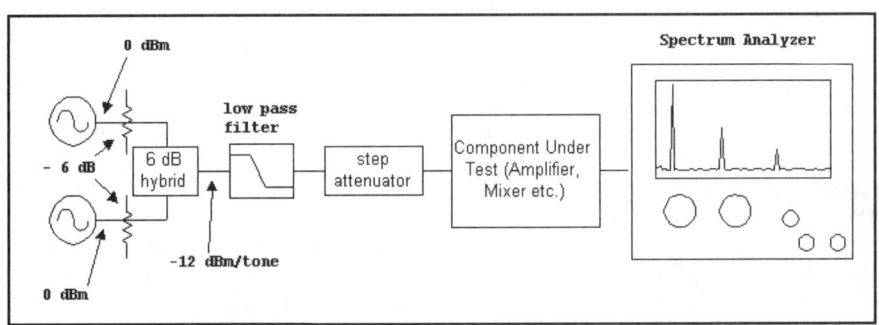

Fig 7.34—Test setup for testing components.

ing care that IMD in the receiver is not dominant. One then proceeds to add an amplifier, followed by further attenuation to maintain signal levels at the receiver input. If a receiver is to serve this function, it must have much better shielding and decoupling than it would for normal use, for we don't want signals from our generators to enter the receiver via any path other than the antenna terminal.

It is even possible to test receiver components (mixers, amplifiers, etc) that are part of a receiver while using that receiver for the measurements. Essentially one does intercept measurements as described, followed by a repeat measurement with a fixed attenuator added between stages. If the IMDR does not change when the pad is added, the distortion is occurring before the pad location.

Some components may require larger signals for testing, a prime example being high level switching mode mixers. Such circuits may have IIP3 of +30 dBm or more. To examine such circuits, we place an amplifier after each generator. **Fig 7.35** shows a sample feedback amplifier while the application is shown in **Fig 7.36**.

Even greater power may be obtained with another stage or by eliminating the output pad. Eventually the point is reached where IMD in other elements may come into play. W7AAZ and the other members of the "Triad" (see Chapter 6) reported seeing IMD in hybrid combiners.

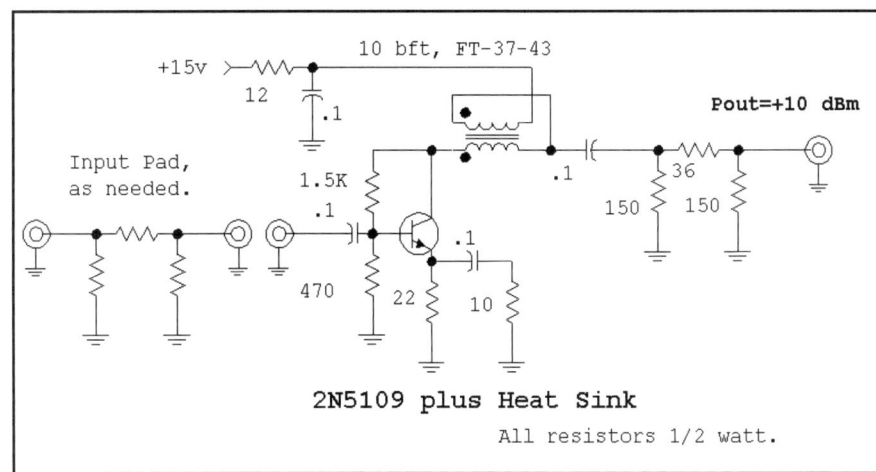

Fig 7.35—Feedback amplifier used following *each* IMD generator to increase the power to +10 dBm per tone. Amplifier gain is 22 dB at 14 MHz, which is reduced to 16 dB with the output pad.

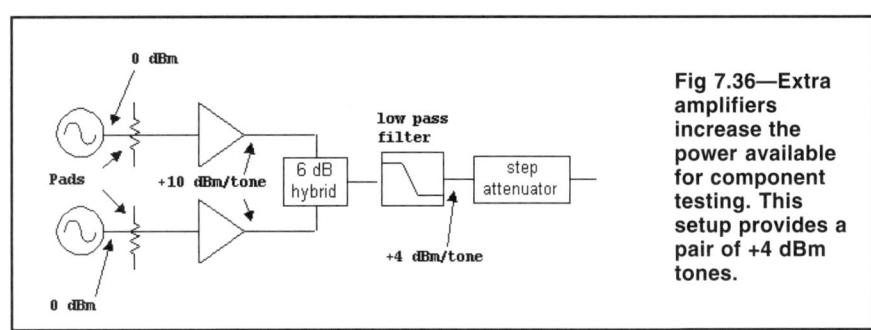

Fig 7.36—Extra amplifiers increase the power available for component testing. This setup provides a pair of +4 dBm tones.

7.7 BRIDGES AND IMPEDANCE MEASUREMENT

We are always interested in measuring impedance, be it for antenna experiments or to set up a termination for a filter. These measurements are difficult with homebuilt equipment, but they are becoming less so with the changing technology we enjoy. Traditional bridge circuits included built-in diode detectors, a restriction that is no longer necessary or even desired.

Shown in **Fig 7.37** is the circuit for a basic Wheatstone bridge. Assume that 1 V is applied to the RF input. If R1 and R2 are equal, point "x" will be at 0.5 V. Point "y" will also be at 0.5 V if the unknown impedance is 50 Ω resistive. A detector between x and y will show no output and a null is detected. If the unknown departs from 50+j0 in either the real or imaginary part, the null is not complete and an error appears at the detector port.

There are two ways that the bridge circuit can be used. The traditional examines the "detector" port between x and y as a place to seek a null. The bridge elements

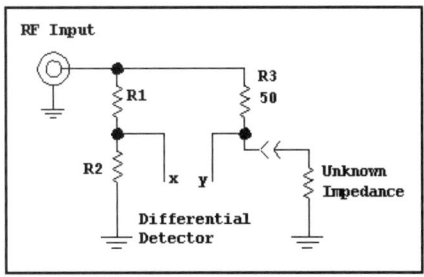

Fig 7.37—A basic bridge circuit.

are adjusted to produce the desired perfect null. The alternative places meaning on the indication at the detector port. We will examine both applications here.

We can form simple bridges with the circuit shown in Fig 7.37. (This one even works with dc.) When all three resistors are 50 Ω (use 51 if building one), the input will appear as 50 Ω to the RF source when the unknown becomes 50 Ω. The voltage between points x and y is roughly the voltage reflection coefficient, which goes to zero for a perfectly matched 50-Ω unknown Z. Such a bridge can be used to tune an antenna or transmatch. We will show some practical examples later.

A useful variation is adjustable. In this form, R1 and R2 are replaced by a 100-Ω pot with the arm serving as "x." Assume the bridge is loaded with 25 Ω as the unknown and the pot is tuned until a null is produced. Analysis shows this to occur when the pot arm is 1/3 of the way up from the ground end.

RF bridges with variable resistors have long been popular with the experimenter. The traditional instruments included a built-in diode detector and meter as the null indicator. They suffer a common problem: the sensitivity suffers with low RF drive owing to the threshold voltage presented by most diodes. Measurements that do not rely upon diode detection of a low level RF signal are preferred.

Fig 7.38 shows an RF resistance bridge with an external detector. This circuit was designed to measure RF resistance while using a sensitive power meter, spectrum analyzer, or 50-Ω terminated oscilloscope as the detector. An unknown resistive impedance is attached to the bridge and R1 is adjusted for a minimum response. The bridge is normally driven with a low level source of around 0 dBm. Less power is used when the termination will be an active circuit; more may be appropriate for antenna measurements. When working with antennas, it is useful to alternately tune the signal frequency and pot R1 to get the deepest null.

The instrument was calibrated at 14 MHz with resistors from 10 to 1000 Ω. T1 is wound on a low permeability, low loss core. Primary inductance was about 50 µH, allowing operation down to 2 MHz or less. Transformer T2 is a common mode choke with about 20 µH per winding that isolates the T1 secondary from ground. This bridge had over 30 dB directivity over the HF re-

Exterior view of RF Resistance bridge after calibration.

Exterior view of return loss bridge.

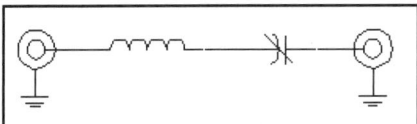

Fig 7.39—Tuned circuits can be added to the bridge to extract complex impedance information.

gion. Directivity is the change between the open circuit response and that when the unknown-Z port is terminated in 50 Ω.

Performance was flat throughout the lower part of the HF spectrum. However, as the frequency moved toward 30 MHz and higher, the 50-Ω point on the scale moved toward the high R end. Further refinement is required.

A series-tuned LC circuit can be cascaded with the unknown port for the measurement of reactive impedances, shown in **Fig 7.39**. The capacitor (or inductor) is then adjusted to deepen the dip. Repeated R1 adjustment may be necessary. A traditional instrument would have suitable scales, but that is not necessary. Rather, after adjustment of a trimmer capacitor, it could be measured with an instrument like the W7AAZ LC tester or the similar instrument from AADE. See Fig 7.24.

If the resistance bridge is used without the auxiliary tuned circuit, complex terminations will produce shallow dips. It's common to look at the meter scale and erroneously conclude that the impedance has a magnitude close to the value shown. This is rarely a valid interpretation, further justifying the reactance measuring options.

The bridge of Fig 7.38 was calibrated at 14 MHz with a handful of carbon resistors with the values then marked on the panel. While this is handy, it may not be necessary. Consider the variation shown in **Fig 7.40**. This is equivalent to the other bridge at RF where the capacitors are virtual short circuits. However, the design with capacitors can be measured with a digital voltmeter attached to the "unknown" port. The dc measurement tells the user the status of the pot, allowing the RF resistance to be inferred.

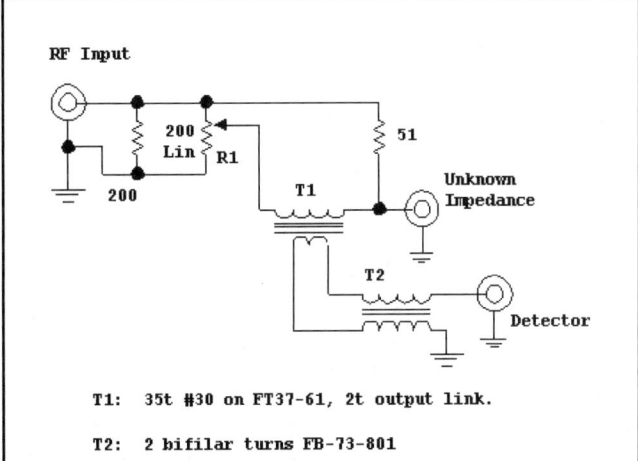

Fig 7.38—RF bridge for HF measurements. R1 is ideally a 100-Ω linear pot, but all we had was 200 Ω. The Clarostat 1/2-inch diameter conductive plastic parts should offer reasonable performance, although we have not used them in this application.

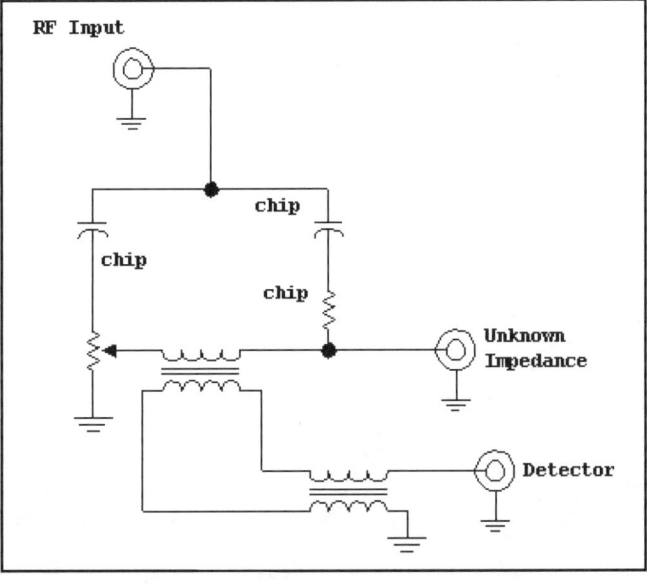

Fig 7.40—Optional variation of the resistance bridge.

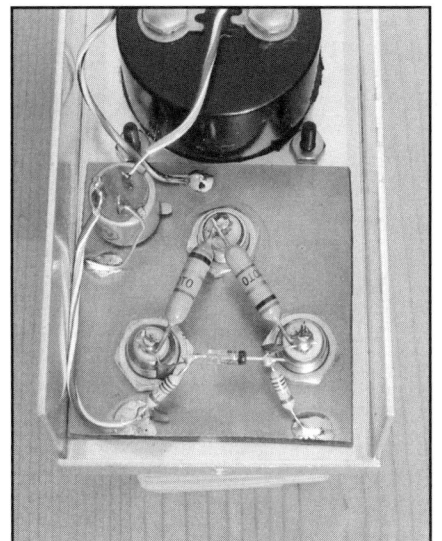

Interior of the RF impedance bridge. Symmetry and short lead lengths are maintained during construction. Long leads are okay with the dc parts of the circuit.

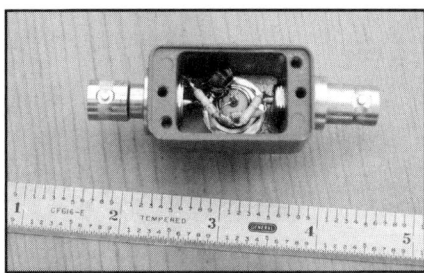

Interior view of return loss bridge. This one is built with 49.9 Ω, 0.1 W, 1% resistors.

Return loss bridge for HF range.

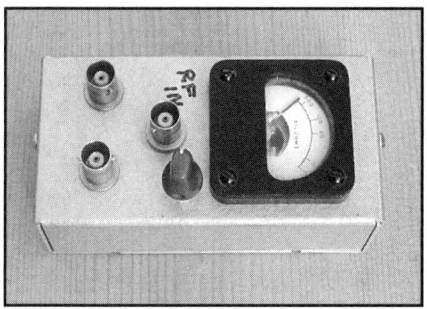

RF impedance bridge with built in meter. A reference must be attached for measurements.

Table 7.1

Frequency (MHz)	D (dB)	O/S (dB)
2	44	0
10	47	0
20	44	0
30	41	0
50	36	1
144	23	2

There is virtue in the modified bridge: Since a calibrated dial is not needed, it can be built with small trimmer pots with much better RF characteristics than encountered with pots with shafts. This will allow these traditional methods to be extended to higher frequency. (These experiments remain on our "to do" list at this writing.)

Fig 7.41 shows a return loss bridge (RLB,) a circuit with no adjustable elements. The signal coming from the detector port indicates the quality of the impedance match. Bridge use begins with a calibration, which places an open circuit at the unknown port. The detector level is carefully noted in dBm. Then the unknown termination is attached and the new detector level is recorded, again in dBm. The difference between the two in dB is called the return loss.

It is also interesting to observe voltage (rather than power) at the detector port. Assume we observe V_0 when the bridge is terminated in an open circuit and a smaller V_1 when loaded. The ratio V_1/V_0 is termed the voltage reflection coefficient, often signified with an upper case Greek Gamma, Γ. Return loss is related to Γ by $RL = -20 \log(\Gamma)$. Also, Γ is directly related to VSWR by $VSWR = (1+|\Gamma|)/(1-|\Gamma|)$. Hence, VSWR=2 corresponds to Return Loss = 9.54 dB and Γ=0.333.

One can use a short circuit instead of an open for calibration. In principle, the two responses will be identical.

There are two frequency dependent RLB characteristics that indicate performance. One is called directivity (D, dB), which is the indicated return loss when a good 50-Ω termination is attached to the unknown port. The other is the dB difference between an open circuit and a short circuit (O/S) at the unknown port. These parameters define the experiments we do when building a bridge.

Table 7.1 shows the results obtained with an experimental RLB. This represents the best transformer (at HF) we found after

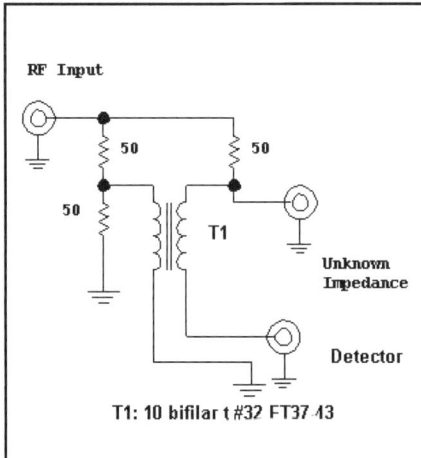

Fig 7.41—A return loss bridge is also known as a 6-dB hybrid. The detector impedance should be 50 Ω for accurate calibration.

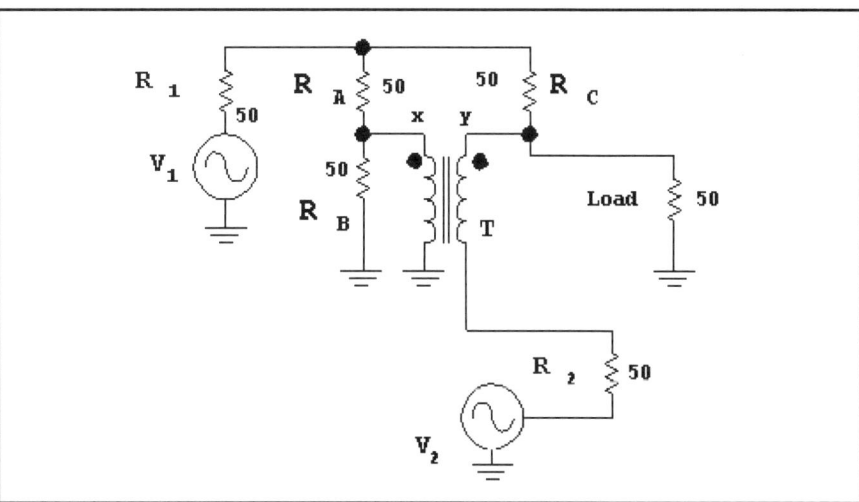

Fig 7.42—A RLB also finds use in combining two signal generators. The power delivered to the load is 1/4 of that available from each generator when the bridge is balanced.

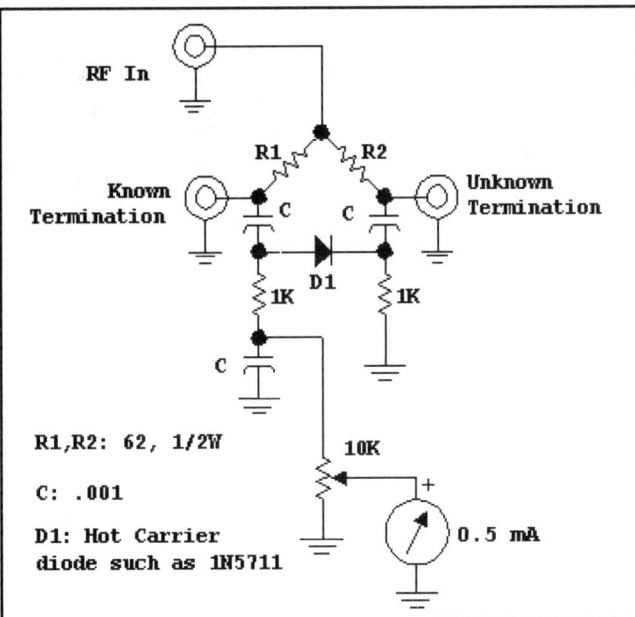

Fig 7.43—A bridge suitable for use through UHF. The symmetry of the schematic should be followed when building the instrument. Drive is 100 mW to 1 W. The "known" termination usually used is a 50 or 75-Ω coaxial terminator.

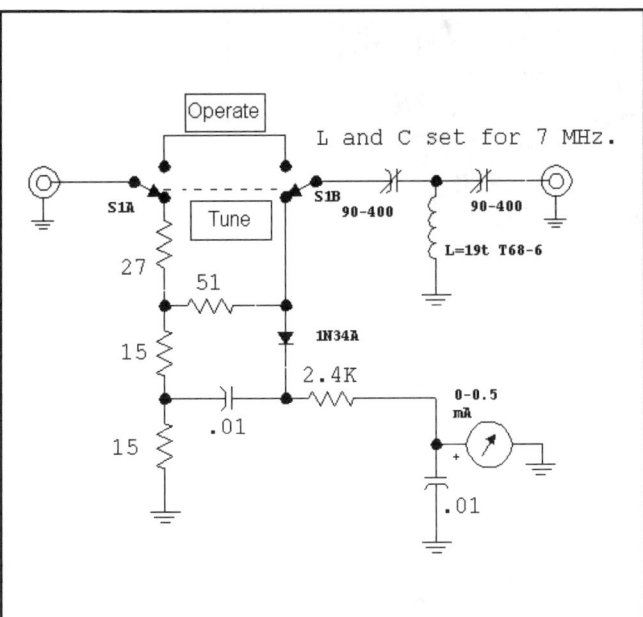

Fig 7.44—7 MHz transmatch and resistive bridge for portable operation. Variable capacitors are screwdriver adjusted, mica compression types. All resistors are 1/2 W. S1 is a DPDT slide or toggle switch. This design is suitable for transmitters up to about 3 W if the tuning is done quickly.

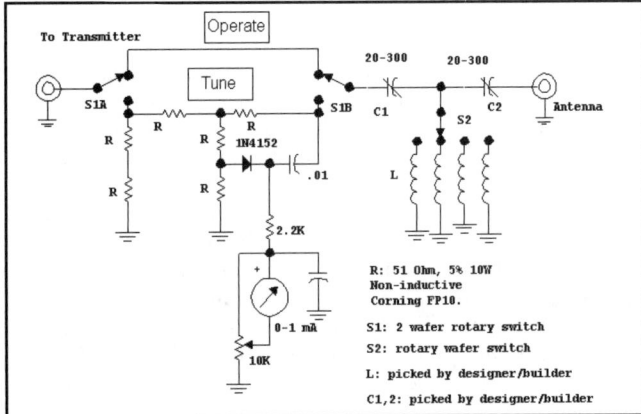

Fig 7.45—Higher power version of a transmatch with a resistive bridge. This unit is rated for 40 W or slightly more for short periods. The topology shown presents a 50-Ω load to the transmitter while attenuating the signal put on the air by 12 dB. If the resistors specified for R cannot be purchased, parallel combinations of 2-W resistors can be used.

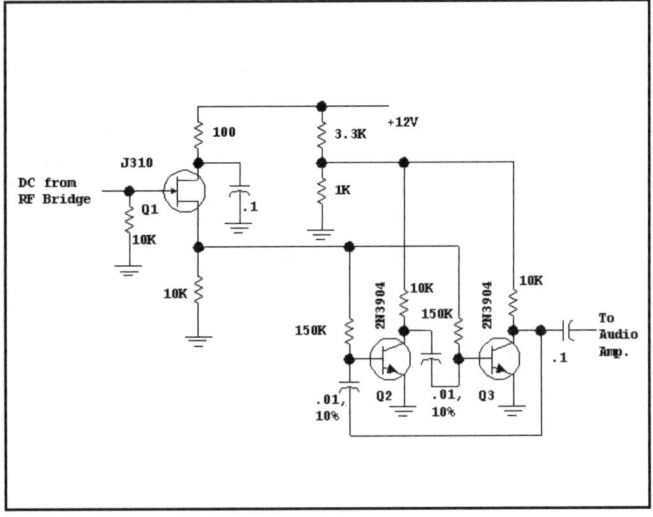

Fig 7.46—Audio meter replacement scheme for transmatch tuning. See text.

examining several. This bridge used 51-Ω, 1/10-W resistors and a transformer consisting of 10 bifilar turns of #28 on a FB73-2401. The high permeability core is preferred, providing an inductance of 175 µH for each winding.

A different transformer improved VHF performance at the cost of HF directivity. We saw 30-dB directivity at 144 MHz when the transformer used 5 of the 6 holes in a multi-hole bead, a FB43-5111. This configuration produces an inductance of 8.4 µH per winding.

The hybrid qualities of the return loss bridge are illustrated in **Fig 7.42**. Generator V_1 causes voltages x and y to be equal and in phase if the bridge is balanced. Hence, none of this power ends up in R_2, the impedance of the other source. But V_2 also sees a balanced bridge. The power delivered by V_2 forces the node with R_1 to be at signal ground, so none of the V_2 power ends up in R_1. These characteristics provide the isolation that we need when combining signals from two generators for IMD testing.

A conventional resistance bridge circuit with built-in detector is shown in **Fig 7.43**. This circuit functions into the UHF area, realized by small lead length and careful symmetry. A photo shows the inside of the circuit. This bridge works well at 144 and 432 MHz, as well as the HF spectrum.

A simple resistance bridge with included detector is often used for the adjustment of low power antenna tuners. This is often preferred over an in-line directional power meter, for the transmitter is always properly terminated during tuning. A circuit used with portable transceivers is shown in **Fig 7.44** where the components are appropriate for the 40-meter band. The slide or toggle switch is put into the "tune" position to adjust the circuit for best null. It is then returned to the "operate" position. A higher power home station version is shown in **Fig 7.45**. The low power variant uses a germanium diode while a silicon switching diode is used at higher power.

Some builders have used a light emitting diode to replace the meter indicating bridge balance. Performance is poor, especially for low power transmitters, for visual output is zero until about 1.6 V biases the LED. But meters are often heavy, difficult to find, and expensive. Some refined circuits use ferrite transformers for greater sensitivity.

An alternative scheme is shown in **Fig 7.46** where an audio oscillator replaces the visual output. The oscillator, a simple multi-vibrator using Q2 and Q3, is frequency modulated by the bridge signal with the *pitch* becoming higher with greater mismatch. The circuit is used by sending a string of dits into the transmatch. The pitch becomes identical for key up and key down when the match is perfect. The primary purpose of Q1, the JFET input, is to generate a dc offset from ground, so JFET type is extremely non-critical. An op-amp would also serve this function.

7.8 SPECTRUM ANALYSIS

What is a Spectrum Analyzer?

One of the most useful instruments the radio experimenter could have is the spectrum analyzer. Commercial versions are sophisticated and expensive, but excellent examples are beginning to appear on the surplus market. And there are now many available components that allow the enterprising experimenter to build his or her own spectrum analyzer.

The first question we must address is the most fundamental: What is a spectrum analyzer? In the general mathematical sense, the signals we encounter are generally collections of sine waves of the form:

$$A \cdot \sin(2 \cdot \pi \cdot f \cdot t)$$

...where A is an amplitude, f is frequency in Hz and t is time in seconds. We can regard this function as either one of time, t, or of frequency, f. In the most general sense, any function of time has a related spectrum or frequency domain representation. The two domains or viewpoints are related through a mathematical operation called the Fourier Transform.[12,13] Also see Chapter 10 of this volume.

Setting formalities aside, we look at electronic signals in the time domain with an oscilloscope or examine them against frequency with a radio frequency spectrum analyzer. We are already familiar with radio frequency spectra of several sorts, although they may not have been presented as such. A rudimentary spectrum analyzer, albeit un-calibrated, is shown in **Fig 7.47**. We have extracted our communications receiver from normal service and opened it to attach wires to the S-meter, a panel meter indicating the strength of received signals. This voltage is usually derived from the receiver AGC. The receiver is set to a band of interest and the tuning control is attached to a motor through a suitable pulley. The motor also drives a potentiometer that develops a voltage proportional to the frequency. The voltage from the pot indicating frequency is routed to the horizontal axis of an oscilloscope while the signal from the receiver's AGC, indicating signal amplitude, is applied to the 'scope vertical. The result is our spectrum analyzer.

The usual spectrum analyzer is calibrated in frequency, so we know the frequency representing the screen center. We also know the frequency span, the number of kHz or MHz associated with the dot as it sweeps from left to right.

The on-screen vertical position is also calibrated in the laboratory spectrum analyzer. While we obtained a *voltage* from the receiver to apply to the 'scope's vertical axis, we calibrate with regard to the *power* related to the signal that developed that voltage. The *top* of the screen is called the *reference level,* leaving the bottom with no special significance. When we see a signal on screen that just reaches the reference level, we know that it has a strength equal to that level. The usual spectrum analyzer displays signals logarithmically, so the calibration will be in terms of a num-

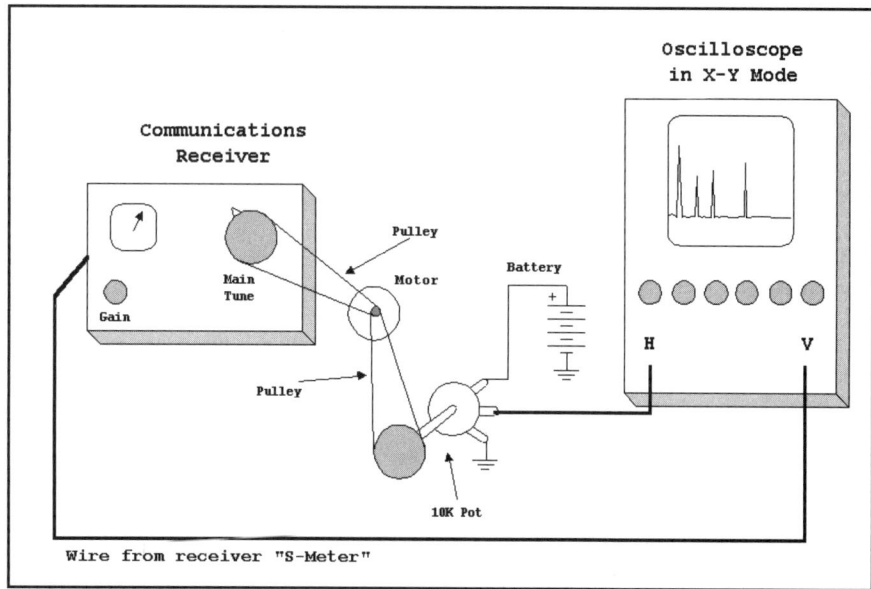

Fig 7.47—A rudimentary spectrum analyzer formed by applying motor drive to receiver tuning and to a pot that generates a voltage that indicates the tuned frequency. This voltage controls the X axis of an oscilloscope. The vertical Y axis is derived from the receiver S-meter circuitry. (Thanks to Bob Bales from Tektronix, Beaverton, OR who suggested this explanation.)

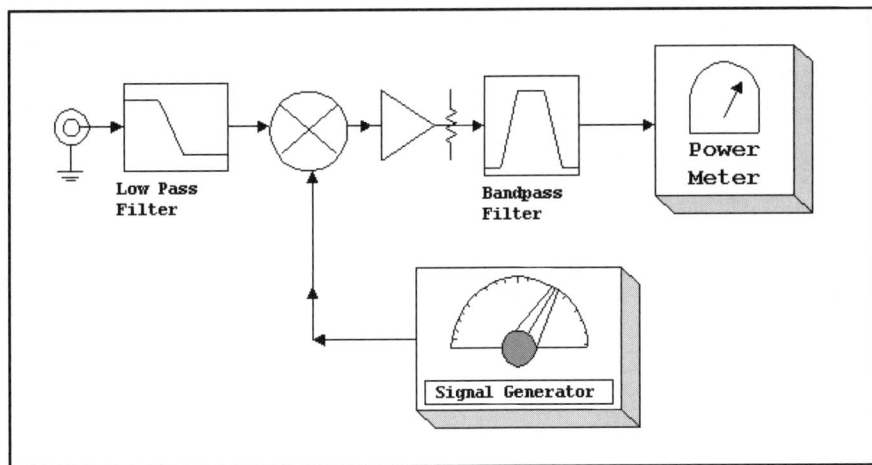

Fig 7.48—Measurement receiver allowing rudimentary spectrum analysis. Although this instrument is presented primarily to illustrate concepts, this unit could be built and would be useful. The amplifier could be a MAR-3 driving a MAV-11 (both from Mini-Circuits) with a 6-dB pad. The mixer might be a TUF-1 or similar part.

ber of dB per vertical division, for the decibel is also a log function. If we have our spectrum analyzer set up for 10 dB per major division, have a reference level of –30 dBm, and see a signal peak two divisions below the top, we conclude that the signal power is –50 dBm.

Spectrum analyzers come in many forms to cover many different frequency ranges. One that we will discuss in more detail tunes from 0 to 70 MHz. Instruments continuously sweeping and tuning from 0 to 2 or 3 GHz are common. Band switching units often tune from 0 to 21 GHz or even more.

The property of *selectivity* in a receiver becomes *resolution* in a spectrum analyzer. Resolution is the ability of an analyzer to resolve two signals that are close to each other in frequency. This is specified by the analyzer resolution bandwidth, RBW, usually equal to the 6-dB width of the filter in use. It is common for high performance spectrum analyzers to have resolution bandwidth selectable from 3 MHz down to 10 Hz. The extremely narrow bandwidth is useful for such tasks as examining 60 Hz sidebands on carriers or for digging way into the noise.

The typical analyzer is not a very sensitive instrument when compared with our receivers. A routine communications receiver might have a noise figure of 10 dB to yield an MDS of –137 dBm in a 500 Hz bandwidth. A typical NF might be 25 dB for an analyzer, resulting in an MDS of –119 dBm in a 1 kHz RBW. The analyzers are not lacking in dynamic range though. A typical analyzer will have a basic reference level of –30 dBm, but will include an input attenuator with a 60-dB range, allowing the reference level to be extended to +30 dBm, or one watt. A "proper" spectrum analyzer uses a front end that is strong enough to produce no internally generated third order IMD when all input signals are kept below the reference level, or "on screen."

Analyzers the Experimenter can Build

The equipment described above is not the ultimate, but merely the norm, representing what has been common within industry for the past 20 years or more. Equipment offering this performance is still rare in the basement lab of the typical experimenter. It would be a monumental task to duplicate a high performance laboratory instrument. But that is not our goal. Rather, all that we ask is to do some of the measurements, as needed for our experiments, with instruments that are simpler, but manageable. The concepts and some of the methods of the high end instruments will be applied to realize these goals.

Consider a very simple spectrum analysis receiver, shown in **Fig 7.48**. This is based upon a power meter that was described earlier in the chapter in Fig 7.13. The meter measured signals from approximately –80 to +10 dBm. We precede this meter with a 2 MHz wide bandpass filter at 110 MHz center frequency.[14] A remote signal generator is the local oscillator signal for a diode ring mixer followed by an amplifier and pad. The amplifier terminates the mixer and adds gain, allowing smaller signals to be seen. A low pass filter with a 70-MHz cutoff precedes the instrument, eliminating images.

Let's assume that we inject a 30-MHz signal from another generator into the input. We see no output until the *local* signal generator is tuned to 140 MHz when the input signal is converted to the 110-MHz IF. Changes in the input amplitude can easily be observed. We could use this instrument to tune a 30-MHz filter or amplifier. Tuning the local generator to 170 MHz allows 60 MHz to be received, allowing us to measure the second harmonic of the input signal. The 90-MHz third harmonic could be measured with the LO set to 200 MHz except that the 70-MHz input low pass filter would attenuate this response. (We could eliminate the input low pass filter from this instrument to produce an instrument that would allow the entire HF and VHF spectrum to be seen, although results would now be obscured by images.)

We now attach an antenna to the receiver and see considerable energy when tuned to the AM broadcast band around 1 MHz. However, we can't isolate one signal from the other because the 110 MHz bandpass filter is 2 MHz wide. The entire BC band fills the filter at once. This deficiency is altered with a 110 MHz filter with a narrower bandwidth. While crystal filters are possible at VHF, the more practical solution converts the signal to a second, lower IF.

A second problem occurs when we tune the analysis receiver to look at a low frequency: A spurious response is observed even with no applied input signal. This occurs because the LO is at 110 MHz, the intermediate frequency. This is a common characteristic of most swept front-end spectrum analyzers. Improved balance in the input mixer increases mixer LO to IF isolation to reduce the "zero spur" response.

Another subtlety becomes apparent when we actually build the analysis receiver of the figure: The balanced mixer must be reversed from the normal application. Most balanced diode ring mixers, such as the TUF-1 or SBL-1, have transformer coupled "LO" and "RF" ports with a dc coupled "IF" port. If low input frequencies are to be examined, the dc coupled port must be used as the RF input.

The instrument of Fig 7.48 is not a spectrum analyzer, for it lacks a graphic display. This is usually obtained by sweeping the frequency with time in unison with a sweep of the display. This begins by replacing the signal generator LO with a voltage controlled oscillator. The VCO is then swept with suitable circuitry. VCO design was discussed in Chapter 4.

A basic swept voltage generator is shown in **Fig 7.49**, beginning with the integrator circuit of part A. Starting with the capacitor discharged, apply a negative

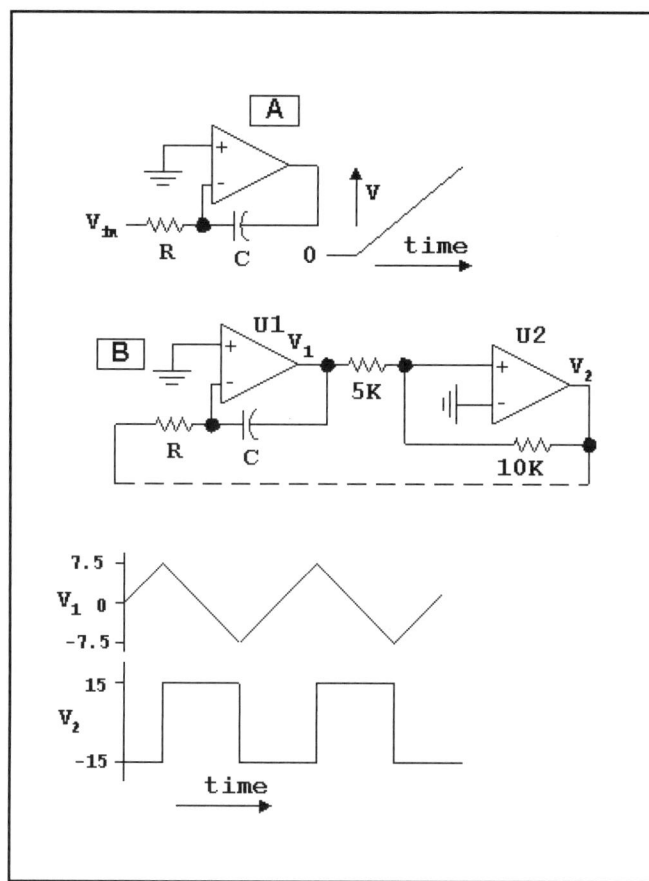

Fig 7.49—Part A shows an integrator circuit. This drives a level detector with hysteresis, U2 in part B. Feedback then creates a sawtooth generator. See text.

voltage to V_{in}. This is coupled to the inverting input, which causes U1's output to begin increasing. But this is coupled back to the inverting input through the capacitor. The equilibrium we require of a closed feedback loop in an op-amp is realized when the U1 output voltage ramps linearly upward. The current in the capacitor then equals that in the resistor, V_{in}/R. Had we applied a positive input we would generate a negative going ramp.

In part B of the figure, we drive the input of the next stage with the ramp. Assume U1 is ramping upward and that the output of U2 is negative against the −15 V power supply. The non-inverting input of U2 reaches 0 when U1's output is +7.5 V, a consequence of the voltage divider action. At this instant, the output of U2 changes state, now "slamming" against the +15 V power supply. If the U2 output becomes the driving source for the integrator input with the dotted connection, we obtain the sawtooth waveform shown for V1.

A practical sweep circuit grows slightly from that described. Diodes provide different slopes for the positive and negative going portions, for we use the left-to-right as the sweep and the other as a retrace. Potentiometers or switched resistors and/or capacitors are added to change sweep

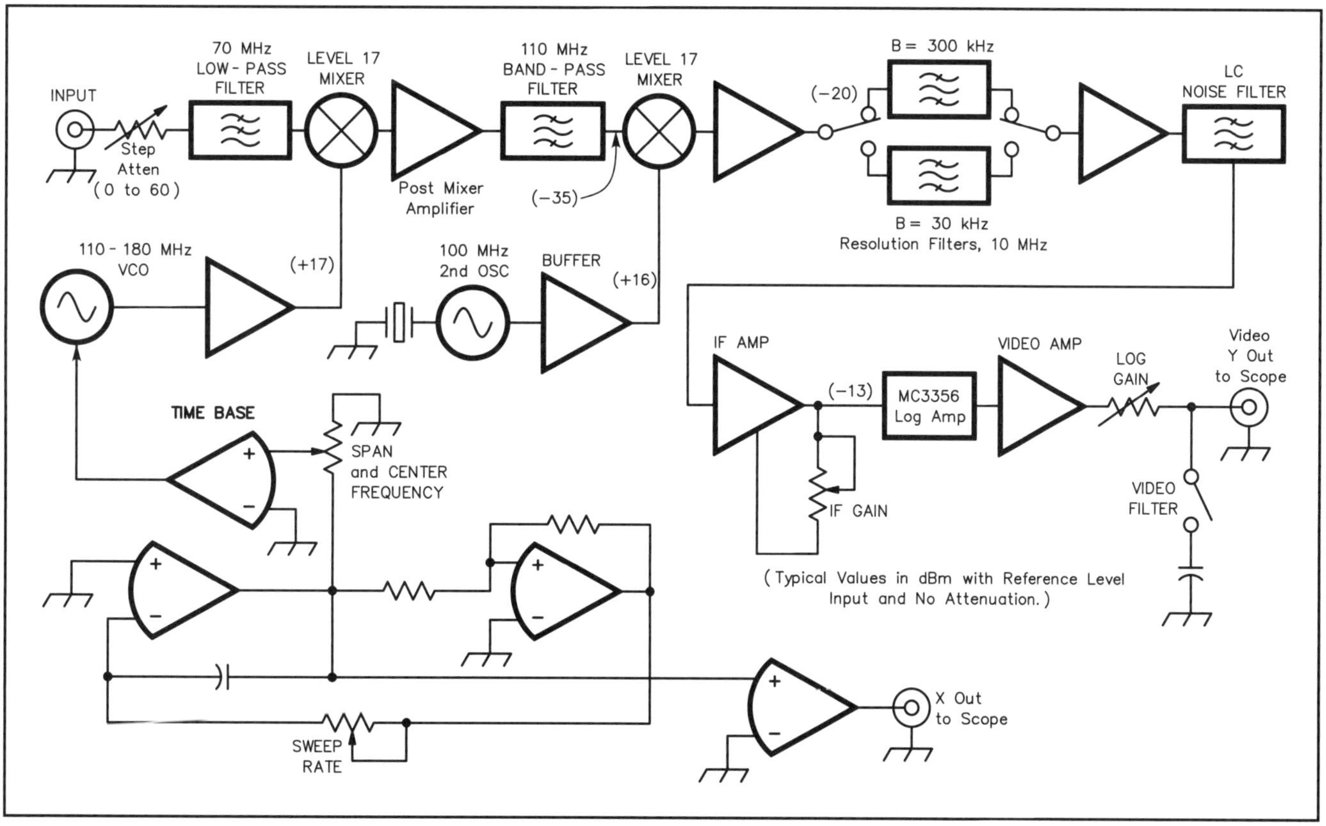

Fig 7.50—Block diagram of a spectrum analyzer the experimenter can build. A practical realization of this design is on the book CD. The 60-dB step attenuator can be an external accessory or built into the instrument.

Measurement Equipment 7.27

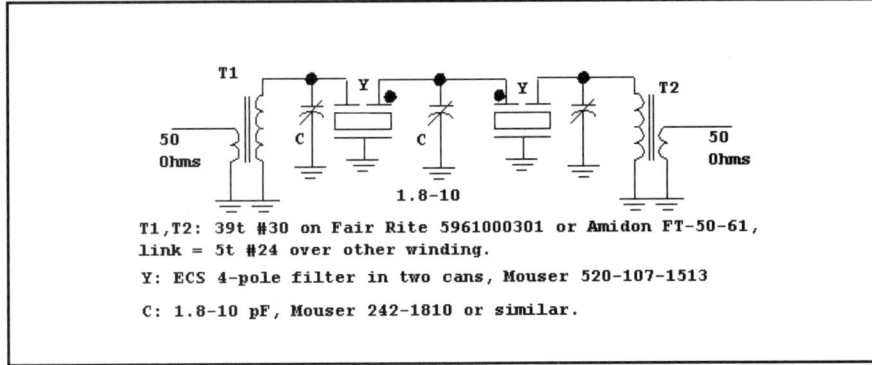

Fig 7.51—4th order monolithic crystal filter.

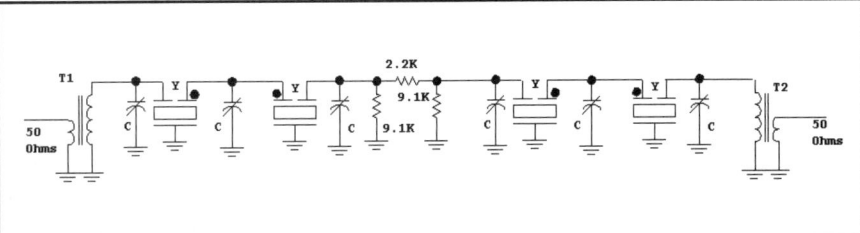

Fig 7.52—8th order crystal filter using two of the filters from Fig 7.51. Each filter block consists of a capacitor-filter element-capacitor-filter element-capacitor combination. These filters were the efforts of Jack Glandon, WB4RNO, and Fred Holler, W2EKB.

rates. V1 is ready to drive the X-axis of an oscilloscope while additional op-amps buffer the ramp and offset it as needed to drive the VCO.

An analyzer begins to emerge, shown in the block diagram of **Fig 7.50**. A commercially available varactor tuned VCO serves the LO function, with buffering to reach a level of +17 dBm. Dual conversion is employed to obtain a resolution bandwidth narrower than afforded by the VHF filter. High level mixers are used for reduced IMD.

This is a practical design that that has been widely duplicated.[15] Details are presented in the articles, which appear on the CD that accompanies this book. The rest of our discussion of spectrum analyzers is confined to general comments and thoughts for refinements of the *QST* design.

Two resolution bandwidths are available in the *QST* spectrum analyzer. One with a bandwidth of 300 kHz uses an LC filter while the other uses a commercial 30 kHz bandwidth crystal filter. Our 1st and 2nd IFs were 110.0 and 10.0 MHz, but 110.7 and 10.7 allow commercial crystal filter elements at 10.7 MHz to be used. These are ECS types X703ND and were purchased from Mouser or DigiKey.

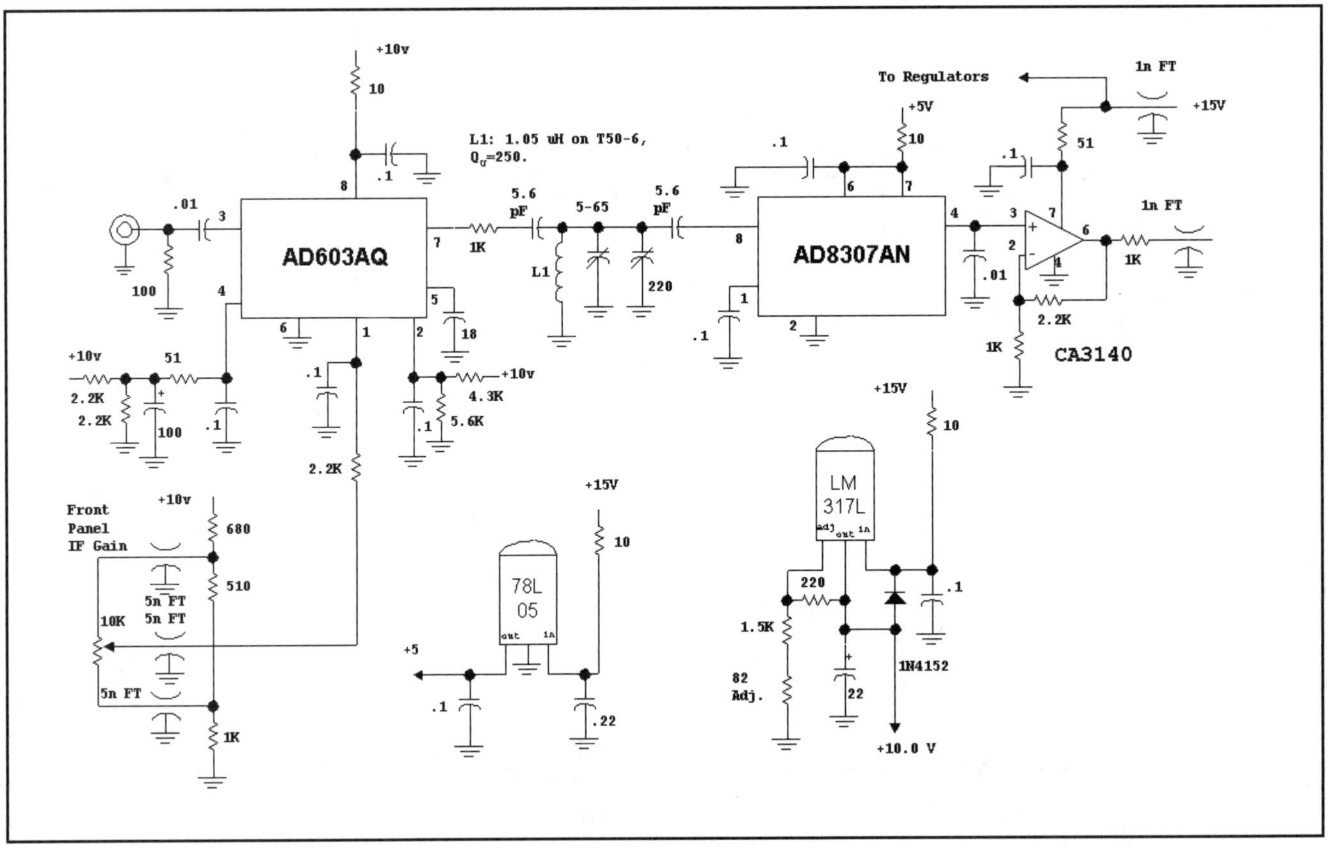

Fig 7.53—This IF and Log Amp section uses more accurate integrated circuits and replaces all circuitry of Fig 5 of the original article (see the book CD.) IF gain is variable from 10 to 50 dB. Resistors around the LM317L can be adjusted to set the 10 V level.

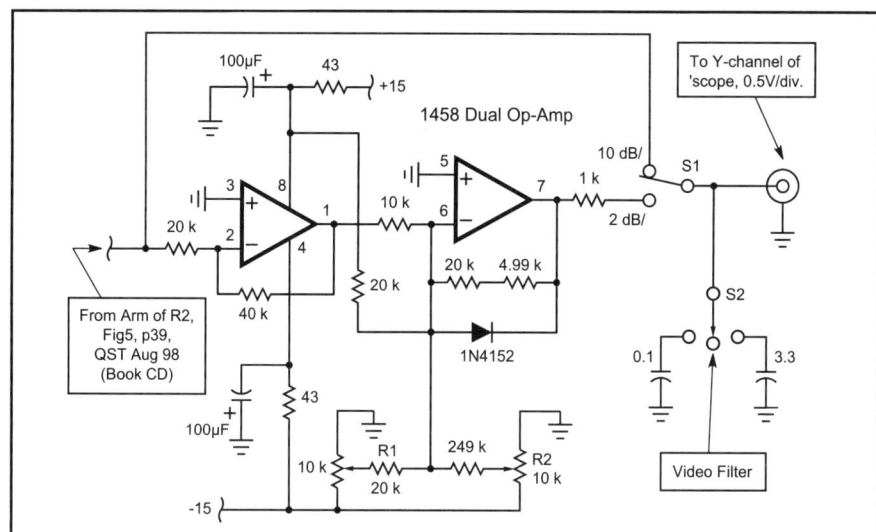

Fig 7.54—Circuit adding 2 dB per division to the spectrum analyzer. The video filter circuit is also included.

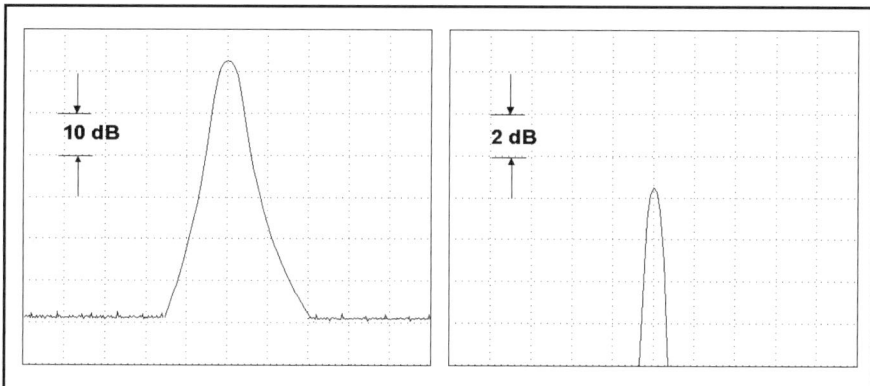

Fig 7.55—A 10 dB/div. signal at the left is adjusted to fill much of the screen. Switching to 2 dB/div. produces the display at the right. Adjusting the offset controls R1 and R2 allows moving the response anywhere on the CRT screen.

Fig 7.51 shows the schematic for a 4 pole filter using two packages. (One "product" from the catalogs includes two filter packages.) The termination for this filter is 3 kΩ at each end, realized with ferrite transformers. Owing to filter loss considerations, a Type 61 core is preferred over the higher permeability cores.

Although the performance was impressive, the stopband attenuation for the 4-pole filter was not adequate. Two stages of the circuitry of Fig 7.51 are cascaded to form an 8th order filter, shown in **Fig 7.52**. This filter has a stopband attenuation in excess of 90 dB, allowing a wide range of measurements. The filters are aligned for a compromise of rounded peak shape, low insertion loss, and stopband attenuation. Alignment can be done with the working analyzer and any convenient input signal.

IF filters for spectrum analyzer use are more critical than those used in a receiver. The analyzer operation essentially paints a picture of the filter shape over the complete dynamic range of the analyzer, so the filter should have a clean, spur-free response over this range.

The *QST* analyzer used the received signal strength indicator (RSSI) function from an early Motorola IC for the log amplifier. The parts were inexpensive and available at the time of publication. The AD8307 from Analog Devices is now commonly available and offers significantly better performance. The AD8307 has a wider dynamic range, improved accuracy, better temperature stability, and is the recommended part. However, it is not a pin-for-pin replacement, and it uses a different input power window, so the designer/builder will have to do some circuit development. The original system used discrete parts for the IF amplifier. An updated version that includes an AD603 as the IF amplifier, is shown in **Fig 7.53**. This circuit drives R2, the "log amp cal" pot, which then is routed to the added 2 dB per division board (described below) and then to the oscilloscope Y axis.

The analyzer contains a video filter, which consists of nothing more than a capacitor that is switched to parallel the video line to the oscilloscope Y-axis. This component, with the driving output resistance, serves to smooth the noise that otherwise creates a fuzzy line. The original video filter used a SPST toggle switch and a 0.1 µF capacitor. This has been replaced with a SPDT/Center-off toggle switch. Two capacitors of 0.1 and 3.3 µF are available, shown in **Fig 7.54**. The heavily filtered response is especially useful for noise measurements. Either filter may be useful in creating a trace that is more easily read on screen.

The spectrum analyzer user soon notices that the sweep rate must be changed with changes in filtering. This is usually a consequence of sweeping. The signal coming out of a filter can respond only as fast as the bandwidth of the filter allows. If, for example, our analyzer had a bandwidth of 1 MHz, we would expect to see output changes at the log amp commensurate with 1 µS. Any sweep rate available in the *QST* analyzer would be slow enough to keep up with such a bandwidth. But switching a 30-kHz filter into the system will cause the response shape to distort, never reaching the peak response seen with a slow sweep. Narrow video filtering does the same thing. Modern analyzers will automatically adjust sweep rates to match the selected resolution and video bandwidths.

Our spectrum analyzer is configured to produce 10 dB of change for every major division on the CRT screen, assuming an 8 division vertical range. This is in line with many traditional instruments. There are many situations when greater amplitude resolution is needed. One might be, for example, a measurement of resonator Q where one needs to accurately see a 3 dB change. This measurement is facilitated with the circuit of Fig 7.54. A front panel switch is added that allows the user to toggle between 10 and 2 dB per division.

The first op-amp of Fig 7.54 is set for an inverting voltage gain of 2 while the second has an inverting gain of 2.5 for a net of 5. The circuit can be offset by a large amount, which can be dialed in with R1 and R2. Any signal that appears on the screen in the 10 dB/div mode can be offset to appear anywhere on the screen with the 2 dB per division mode, illustrated in **Fig 7.55**.

A crystal oscillator presented earlier (Fig 7.29) is useful as a calibrator for the analyzer. It could be built in with a front panel BNC connector, or as a battery powered stand alone unit. The calibrator amplitude is adjusted with circuit component changes to deliver a level of –20 dBm while using a calibrated source as a "standard."

The calibrator or a signal generator can be used to calibrate the instrument. A signal of –20 dBm is applied to the analyzer input, which is usually run with at least 10 dB of input attenuation. The IF gain is set to generate a reference level response. The attenuator is then switched in 10 dB steps to move the response down the screen. If the signal does not line up on the major screen markers the log amp gain is changed and the process is repeated until reasonable log accuracy is realized. Analyzers using the AD8307 log amp are so accurate that the oscilloscope's vertical position control functions much like the IF gain control. There is no significance to the "screen bottom" setting in the 'scope in this application.

The AD8307 log amp accuracy is as good as or better than that of log amps in many spectrum analyzers found on the surplus market, allowing the builder/designer to realize outstanding performance with modest cost. Consumer communications ICs with built-in RSSI functions do not fare as well. But moderately accurate measurements are still possible by careful application of the step attenuator.

Consider a spurious response evaluation of a transmitter as a typical example of a measurement that asks for a dB ratio between two power levels. The transmitter is applied to the analyzer, taking care to keep all signals on screen. An extra attenuator or power tap may be needed to safeguard the analyzer from the high outputs available from a transmitter. The display level of the spur is carefully noted, perhaps by using the 2 dB/div mode for improved accuracy. The analyzer is tuned to the carrier signal and attenuation is added until the on-screen response equals that observed for the spur. This procedure is enhanced if 1 dB steps are available in the step attenuator. The spur level in dB with respect to the carrier (dBc) is then the amount of attenuation added. This measurement is as accurate as the step attenuator and has little to do with the analyzer characteristics. Harmonic distortion is a special case discussed later.

Shielding

One of the first questions asked when a designer embarks on the construction of a spectrum analyzer is "how much shielding is needed." While difficult to quantitatively answer, a little thought shows that shielding must be very good. The *QST* analyzer we have discussed has a minimum bandwidth of 30 kHz and a noise figure around 20 dB, so the minimum discernable signal is around –109 dBm. Yet we routinely use this instrument with 100 W transmitters. That power is +50 dBm, 159 dB above the analyzer MDS. This is the attenuation that must be provided in the overall measurement setup to be able to do good measurements. Part of this results from shielding and part comes from testing the transmitter with a non-radiating termination.

The popular boxes offered by Hammond, available in many catalogs, afford excellent shielding. These cast aluminum boxes have tight fitting bolt on lids and are easily drilled. A box is used for each major block in the RF chain, so one box contains the first mixer, post mixer amplifier, the VCO, and its buffer amplifier. The input low pass resides in a separate box with the 110 MHz first IF filter in another. The only "open" board in the analyzer contains the time base. Signals move into and out of the box on coaxial cable while dc bias and gain control lines are attached to feedthrough capacitors. The VCO tune line is on coax. Wires extending through rubber grommets in box walls are *not* suitable and should never be considered for RF application.

Use what is available for coaxial connectors. SMA or SMB are excellent, but expensive and not generally required for HF and VHF. BNC cables have become more affordable with the popularity of computer networks. A crimping tool is needed to take advantage of these parts. Inexpensive phono plugs and sockets (RCA) are suitable if carefully applied.

Application Hints

The spectrum analyzer is not merely an evaluation tool to test the rigs that are finished, although many folks treat it as such. Rather, the SA is used to measure things throughout the experimental experience. First and foremost, it is a sensitive meter used to examine signal levels, even when they are too weak to be seen with an oscilloscope. The sensitivity is the result of narrow bandwidth. Utility is maintained as a result of sweeping, eliminating the need to retune for various signal components.

The spectrum analyzer is almost always a tool for substitution measurements. As such, it is usually necessary to break a 50-Ω signal path and attach the spectrum analyzer. This is done in a breadboard by bolting a BNC connector to a ground lug and then soldering that lug to the ground foil near the circuit under test. The connector can be moved later, so it can be placed close enough to maintain short leads.

In other cases it is handy to attach a BNC chassis connector with ground lug to a short length of small coaxial cable (RG-174 or similar) with the other end of the cable soldered into the circuitry. The probing end should have a maximum ground length of perhaps one half to one inch with a similar length for the center conductor for HF and low VHF applications. The end of the center insulation is removed and soldered to a circuit board. It is vital to solder the cable ground to a circuit board ground close to the place where the measured signal currents flow. For example, if the output of a feedback amplifier was to be examined, you might "lift" a blocking capacitor from the output signal line. That capacitor can then be tack soldered to the cable center conductor. The ideal place for the cable ground is the board ground foil directly under the capacitor position. Removal of solder masking may be required in some cases. Alternatively, the ground connection for the bypass capacitor related to the feedback amplifier output could be used.

It is rarely valid to merely attach a cable ground at the edge of a board at, for example, a mounting hole. This procedure works well enough for high impedance probes from an oscilloscope while performing *in-situ* measurements. The feedback amplifier, in that case, still has the output currents flowing to a following stage. That termination was broken for our substitution measurement. Examine the complete loop starting and ending with the place where the center conductor and coax cable braid split. That loop should generally be small. If you are trying to evaluate the presence of spurious signals, you should not allow the loop to contain extra stages that might be carrying some of the contaminating signal.

Some applications are presented in the paper on the CD that accompanies this book.[16] The applications related to power meters, again on the CD, are also generally useful with spectrum analyzers.[17] Spectrum analyzer measurement of intermodulation distortion was discussed earlier in this chapter in the section on signal sources.

A common problem encountered when breadboarding a new circuit is a spurious oscillation. More often than not, this will occur at very high frequencies, often approaching the F_T of the offending transistor. A spectrum analyzer tuning only to 70 MHz will never see this directly, but the result is often still apparent on screen. This appears as a low level signal that moves in frequency as a hand or tool is placed close to the circuit. This is the result of mixing between the spurious oscil-

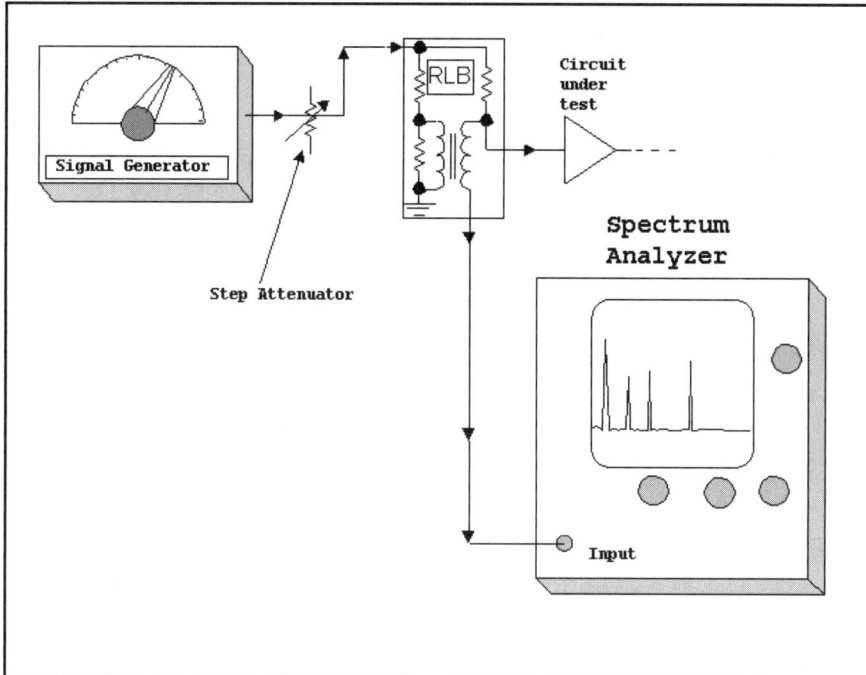

Fig 7.56—Return loss (VSWR) is easily measured during bench testing with a simple bridge.

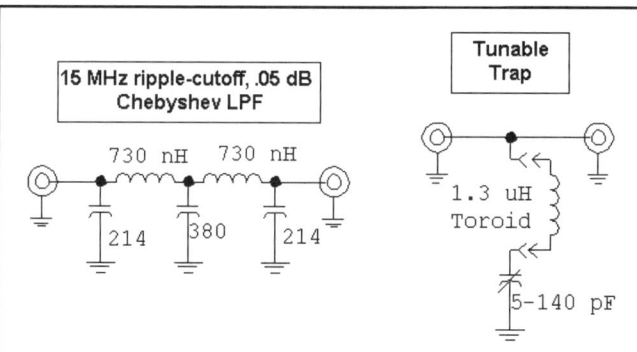

Fig 7.57—Low pass filter and tunable trap are used to evaluate harmonic distortion in the front end of an analyzer. These circuits were used to evaluate analyzer performance for measurement of 14-MHz harmonics from a transmitter.

lation and harmonics of signals that excite the circuit.

It is often useful to investigate the quality of impedance match, even with small signal amplifiers. A return loss bridge (discussed earlier in this chapter) is driven by a signal source and applied to a circuit under test. The generator power is turned down to a level that will not overdrive the amplifier under test. The return loss, which is directly related to VSWR, is then measured as shown in **Fig 7.56**.

Calibration During Measurements

A calibrator circuit was described earlier, a convenient means for checking analyzer amplitude and frequency calibration. But there is more to calibration for RF measurements.

Generally, the best procedure is to place no trust in the equipment that has not been earned. This applies especially to the homebrew spectrum analysis equipment described in this book, but is also important for the best laboratory instrumentation available.

Assume that we plan to measure the gain of an amplifier, and that we wish to get the most accurate number possible. The amplifier is set up with the appropriate power supply, a signal generator, and the spectrum analyzer or power meter. The set up is turned on and generally checked. The calibrations that have already been done for the analyzer are enough to get things started.

Once the system is working as expected, we now do a test set-up calibration. The amplifier is disconnected from the two coaxial cables and replaced with a through connector. This is a barrel or bulkhead connector in BNC cables or the equivalent in other cable types. It is important to use the same cables for the calibration as are used with the amplifier. The response is noted with the through connector. The amplifier is then inserted in its original position and the new response is noted. 2 dB per division is used for both measurements. The gain is then the difference between the two levels.

Newer commercial equipment is usually fairly accurate in the 1 or 2 dB per division ranges, so log errors are not major. However, when a homebrew analyzer based upon an IC RSSI function is used, the measurement should be done with a step attenuator rather than with numbers from the screen. This is a wise procedure with older commercial analyzers or with any measurements performed near the bottom of the log amplifier ranges, or with any measurements where noise levels are being compared.

Commercial spectrum analyzers feature highly refined frequency readouts. A cursor function can be activated that marks a trace on screen. The exact frequency is then displayed. Some instruments can be extremely accurate in this mode. The procedure is much more casual with the *QST* and other simple homebrew instruments. When we see a signal on screen with an unknown frequency, we carefully note the horizontal position, disconnect the input cable and attach a signal source adjusted for the same response, and read the frequency from a counter attached to the source.

The analyzer can be modified to incorporate a frequency counter. The frequency sweep would be stopped by opening the line from the center arm of the sweep rate pot.[18] There would still be horizontal motion on screen, but the amplitude would be fixed at that corresponding to screen center. This is called a "zero span" mode. The VCO could then be counted. Subtracting the first IF from this value gives a "center frequency."

Harmonic Distortion Measurements

Although common, this seemingly simple chore can be complicated by harmonics created within the spectrum analyzer. Measurements are meaningful only when we have confirmed the analyzer performance.

The evaluation can be done with several experiments. The first applies a signal to the analyzer from a generator and looks at the harmonic levels. The attenuation in the analyzer front end is changed. If both the fundamental and the indicated harmonic

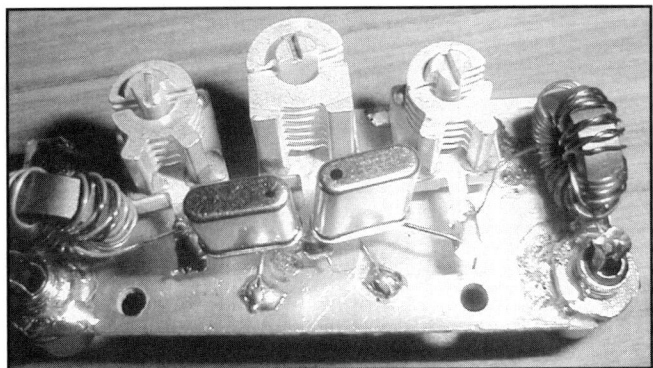

A close up photo of a 4th order filter built by WB4RNO. Any small trimmer capacitor with a suitably low minimum capacitance can be used.

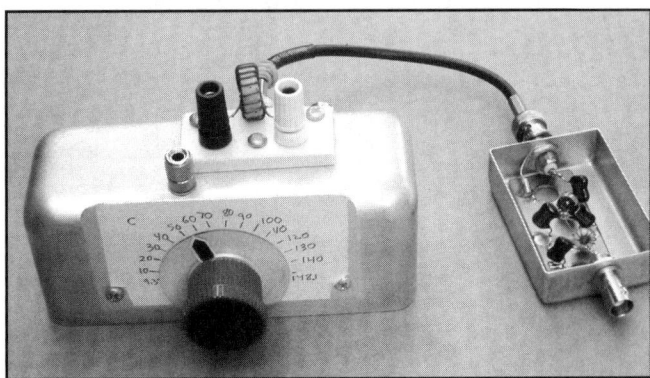

Clean-up gear to reduce the harmonic content of a signal source. This is used when evaluating a transmitter or other source for harmonic distortion.

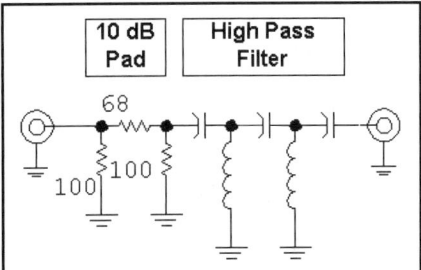

Fig 7.58—High-pass filter used for harmonic measurement. See text.

Fig 7.59—Front end for a triple conversion spectrum analyzer tuning to the low UHF spectrum. This analyzer has yet to be built, but is planned.

change in unison, the distortion is probably real and not an analyzer spur.

A second experiment places a low pass filter in the line from the generator to the analyzer. This will improve the generator performance, allowing the first experiment to be repeated with greater sensitivity. Again, identical tracking of fundamental and distortion tend to vindicate the analyzer, now at a level commensurate with the new harmonic attenuation level.

Traps can be used for further analysis. A tunable trap is shown in **Fig 7.57**. The trap is placed in the line between generator and analyzer and is tuned to attenuate the fundamental signal. If the trap is sharp, it can dramatically attenuate the fundamental with little impact on the harmonics. A 20 dB or greater attenuation of the fundamental without altering the harmonic guarantees the fidelity of the analyzer.

An analyzer can still be useful for analysis even when it is generating harmonics of its own. All that is required is to reduce the fundamental signal reaching the analyzer without altering the harmonic energy. This can be done with a high pass filter, shown in **Fig 7.58**. The high pass is preceded by a 10-dB pad, establishing a proper impedance environment for the generator (or transmitter) being evaluated for distortion. A measurement is performed without the trap to establish the fundamental power. The trap and pad are then inserted and the analyzer sensitivity is increased by the pad loss. The harmonic power is read to calculate a dBc value. If necessary, the trap can be cascaded with the high pass for further attenuation of the fundamental.

Expanding Performance

The *QST* spectrum analyzer tuned over a restricted range of 0 to 70 MHz with only two available resolution bandwidth positions. The VHF experimenter will want higher frequency performance.

Expanding the tuning range to higher frequency is easily realized, beginning with a review of the latest catalogs from Mini-Circuits and other vendors. A 100-200 MHz VCO was the basis for the *QST* design (Fig 7.50), but this could be replaced with other parts. One variation would use the POS-535 tuning from 300 to 525 MHz as the first LO. The first IF would become 300 MHz. A good choice for a second IF would then be 21.4 MHz where commercial monolithic crystal filters are available. A VHF 2nd LO will be needed, which could be free running or be multiplied up from a lower frequency crystal oscillator.

A triple conversion version of the analyzer is shown in the block diagram of **Fig 7.59**. This version tunes to 400 MHz with a first IF at 500 MHz. The second IF is then 110 MHz using the circuitry from the original design. This upgrade could be built as a supplement to the *QST* analyzer without disturbing the functionality of the original. This UHF extension uses only +7 dBm mixers, so the new design will not be as strong as the first with regard to distortion measurements. The 2nd LO could be homebrew or might use a second Mini-Circuits part.

The present analyzer can be supplemented with a block converter in much the same way that we add converters ahead of receivers for the higher HF or the VHF bands. A very simple block converter that we built uses a POS-200 (100-200 MHz) VCO driving a TUF-1 mixer. A 4 dB pad in the signal path sets the overall conversion gain at –10 dB. The 144 MHz amateur band is converted to 30 MHz when the LO is at either 114 or 174 MHz. Recall that the 3rd harmonic of an LO is generated within a diode ring mixer, often creating spurs, but also allowing third harmonic mixing. So setting the VCO to 157.3 MHz creates an effective LO of 472 MHz, which will convert 432 MHz to appear as 40 MHz. Mixer conversion gain is less with harmonic mixing and depends on the harmonic being used. The block converter output is filled with numerous spurious responses, but is nonetheless a useful and simple tool.

Figure 7.60 shows a narrow tuning range approach to spectrum analysis. This circuit was configured as a measurement receiver. It uses an outboard local oscillator to drive a diode ring mixer followed by a traditional post-mixer amplifier. The post-amp output is then applied to a narrow bandwidth 5 MHz crystal filter that then drives a log amp. There are two outputs. One is a built in meter while the other is a jack to drive a DVM. This instrument was originally configured to measure carrier and sideband suppression in single sideband transmitters, but has also found use in the pursuit of spurs from frequency synthesizers using direct digital synthesis. The instrument could also be configured for baseband measurements close to dc. It would then be useful for noise measure-

Crystal filter, log amp, and output driver for "Measurement Receiver."

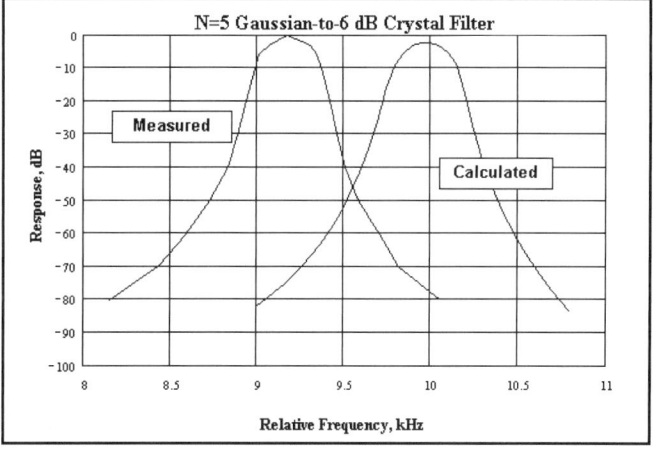

Fig 7.61—Crystal filter response for the circuit used in the measurement receiver. See text.

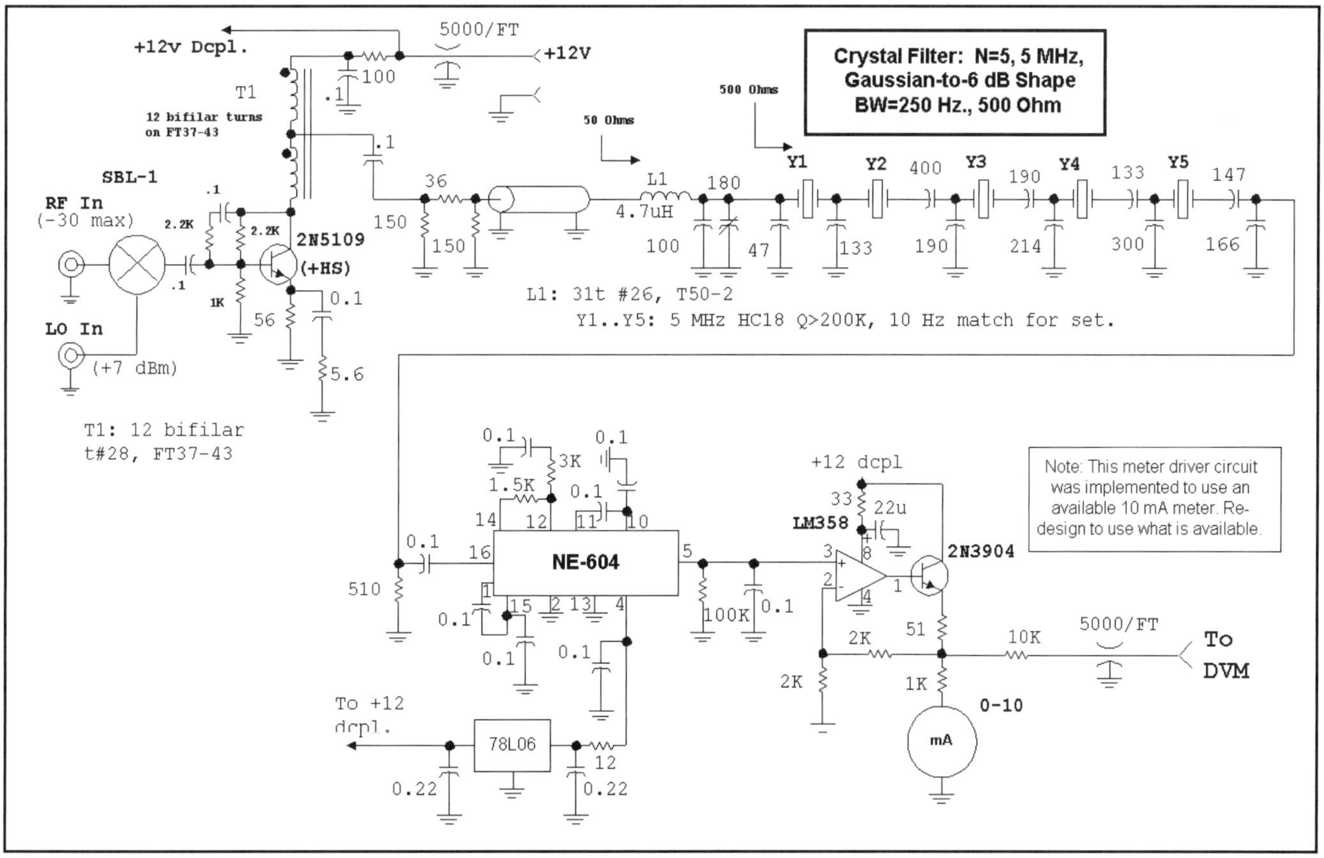

Fig 7.60—Measurement receiver for measurement of SSB transmitters. This unit used an available 10-mA meter movement with a high resolution scale, but can be adapted to available meters. This instrument can be adapted as a narrow tuning range spectrum analyzer, a refinement that we have yet to complete.

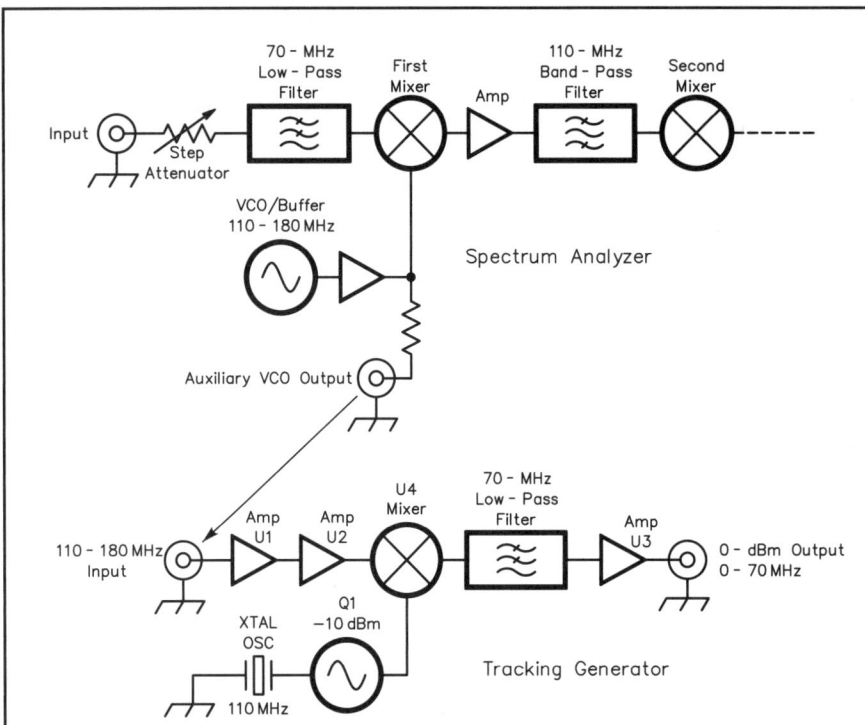

Fig 7.62—Functionality of a tracking generator and the mating spectrum analyzer front end. The complete design is included on the book CD.

Outside of measurement receiver.

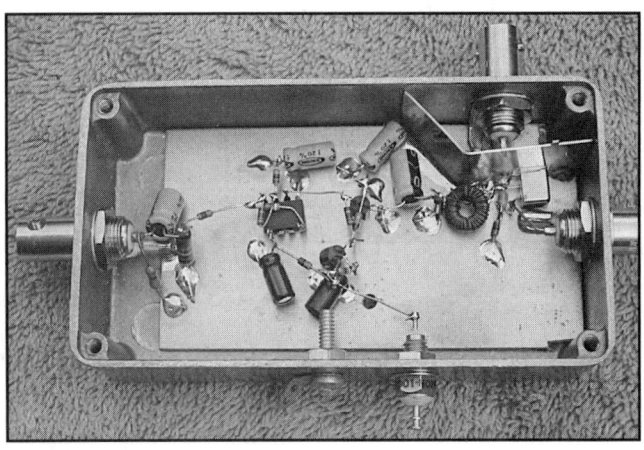

Converter for baseband spectrum analyzer on a PC. Used for evaluation of IMD in an HF transmitter.

ments in connection with oscillator phase noise evaluation.

The narrow crystal filter used in the measurement receiver is designed for a Gaussian-to-6 dB shape. Measured and calculated responses are shown in **Fig 7.61**. This filter shape is ideal for measurement applications, a consequence of the rounded, unambiguous peak with reasonable skirt response. The prospective builder is encouraged to design his or her own filter, for the component values will depend on crystal characteristics. The crystal used in this filter had a motional inductance of 98 mH and average unloaded Q over 200,000. The crystals were matched within 10 Hz. This response shape is generally very tolerant of component variations. Note that the traditional symmetry in component values is not present in this filter, even though the terminations are equal at 500 Ω at each end. Avoid narrow Chebyshev filters in analyzer applications.

This measurement receiver could be reconfigured as a spectrum analyzer with relative ease. A simple way to do this would be to modify the existing *QST* analyzer. Power supply and a sweep voltage from the analyzer could be brought to suitable connectors to drive the narrow bandwidth unit. The video output could be routed directly to the Y axis. The same sweep circuit and related panel controls would then control both spectrum analyzers.

A stand-alone swept VCO would be needed for the narrow bandwidth adapter. This, however, is not a difficult design task. It is wide bandwidth VCOs that offer greater challenge.

Tracking generators and filter measurements

Swept instruments are ideal for the alignment of filters of all types. Having a swept signal means that the entire frequency response can be displayed at one time. A tracking generator (TG) converts a spectrum analyzer to perform this task.

If we think of a spectrum analyzer as a special purpose receiver, a tracking generator is nothing more than a transmitter that transceives with the receiver. A block diagram is shown in **Fig 7.62**.

A sample of the swept first oscillator from the spectrum analyzer is required for the tracking generator. This signal is amplified and becomes the LO for a high level mixer, U4. The RF input for that mixer is a crystal controlled signal *exactly* at the spectrum analyzer first intermediate frequency. This frequency is easily measured by injecting a signal from a generator into the first IF with the spectrum analyzer set for the narrowest possible resolution bandwidth. This measurement needs to be done after the analyzer is finished and working, but prior to ordering a crystal for the TG.

This TG has an output of 0 dBm. This signal is a swept one that is always tuned to the same frequency that the analyzer sees. The great utility of a tracking generator over a simpler stand-alone swept oscillator is that the SA-TG combination allows observation in the narrow bandwidth of the analyzer. This results in a dramatic increase in measurement dynamic range. The evaluation of filter stopband attenuation details at levels well below the –100 dBc level are possible with a SA-TG combination. Full details of the TG are included on the CD that accompanies this book.

The extreme dynamic range comes with a price: The shielding of both the tracking generator and spectrum analyzer must be *very* good. As mentioned earlier, the SA-TG combination behaves like a transceiver. However, unlike the usual transceiver we might build for communications, the receiver and transmitter must both function *at the same time!* Signals that might leak from the TG to the SA will interfere with the intended one when testing filters. The observed result will often be a distorted filter shape with the edges of the filter skirts dipping into the analyzer noise floor. Another tell-tale indicator of these problems is a filter shape that changes with the position of some of the interconnecting coaxial cables.

As useful as the SA-TG combination can be, it presents a problem for the serious experimenter: Filters are so easily "tweaked" that builders may be tempted to ignore designing the filters in favor of empirical methods. Don't fall into this trap!

DFT Spectrum Analysis

The spectrum analyzers discussed so far have been of the *swept front end* type. The case where a block converter preceded a swept front end analyzer produced a *swept IF analyzer*. There is another popular analyzer that has become very common in recent times, the Fourier Transform Spectrum Analyzer. In this type, an incoming signal is converted to a digital stream of data with an analog to digital converter. The analog data feeding the converter is filtered with a low pass or bandpass filter to restrict the resulting digital data. The time domain representation is then subjected to mathematical calculations resulting in a frequency domain representation of the signal, a spectra. This is then graphically presented. The analysis used is a Discrete Fourier Transform, or DFT. The most popular DFT form is the so called Fast Fourier Transform, or FFT.[19]

The radio amateur is familiar with this method as a software technique. Audio signals are presented to the sound cards of personal computers. The resulting digital data is Fourier transformed in suitable software programs and displayed in one of several forms including the "waterfall" popular with digital communications modes.

DFT spectrum analyzers have two major advantages over swept tools: First, they

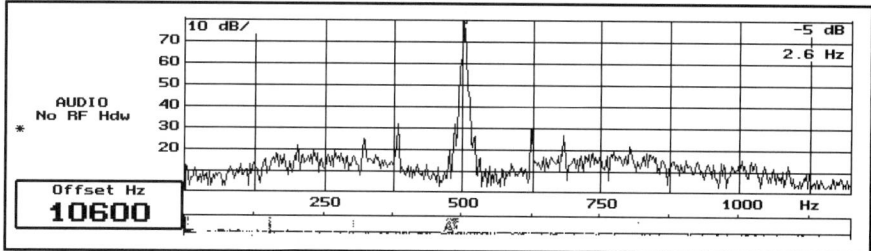

Fig 7.63—High resolution spectrum of a signal generator. The noise is phase noise on the generator. 120-Hz hum modulation is readily observed as well.

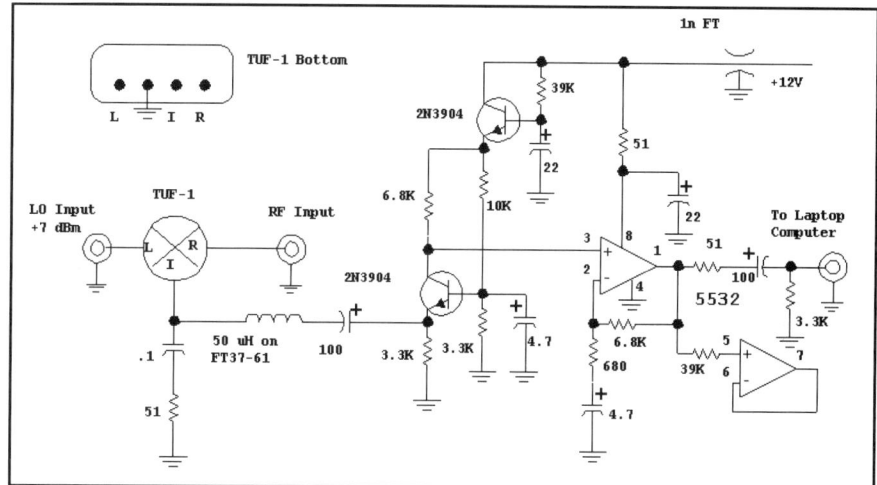

Fig 7.64—Block converter to heterodyne an RF signal to baseband where it can be observed with a spectrum analyzer running on a PC.

are capable of very high resolution (narrow bandwidth). Second, the spectrum shown represents the spectrum at one instant in time.

An FFT analyzer is very useful as a measurement tool. **Fig 7.63** shows an example where a signal generator was being investigated for phase noise. The noise shown in the figure is indeed noise, for a cleaner oscillator operating with the same analyzer parameters produced a similar spectrum, but without the noise. The resolution bandwidth for this example is 2.6 Hz! The hardware and software used for this example are discussed in much more detail in Chapters 10 and 11.

Although FFT methods often concern audio or "baseband," the concepts are capable of much more. So long as a signal can be sampled in time and converted to digital data, it can be transformed to the frequency domain. Many modern oscilloscopes are built with relatively low speed displays. But the incoming analog signal is anything but slow. The incoming data is amplified and/or attenuated and presented to a high speed "scan converter," essentially an A to D converter. Once the high speed signal is remembered, it can be read at a lower speed and displayed as a time signal. The data can also be presented to an FFT "engine," or computer to generate a corresponding spectra. While usually lacking the dynamic range of an analog spectrum analyzer, a spectra with a dynamic range of 50 dB or better is common with such oscilloscopes.

A block converter can be used to move part of an RF spectrum down to audio where it can be examined with an FFT type spectrum analyzer with an example shown in **Fig 7.64**. An external step attenuator and (optional) bandpass filter precede the converter. A diode ring mixer then moves the signal down. The rest of the circuitry is very much like that found in direct conversion receivers. This converter can be used ahead of the FFT analyzer implemented with the DSP hardware from Chapters 10 and 11. We have also used it with a personal computer sound card and modest cost software.[20] One must be careful with any of these schemes to avoid overdriving the A-to-D converter; overdrive can turn the entire screen to unrecognized gibberish! Sound card solutions seem less robust than the devoted DSP tools.

A block converter and a baseband FFT

analyzer are ideal for evaluation of SSB transmitter IMD. What had always been a difficult laboratory measurement is now available to almost all experimenters. A traditional two-tone audio generator was included earlier in this chapter.

The narrow resolution available from an FFT based analyzer will also allow the experimenter to measure in-band transmitter distortion. A tone spacing of around 100 Hz then becomes appropriate. In-band performance becomes important when an SSB transceiver is used to process narrow bandwidth information such as encountered in PSK31. Again, the availability of measurement tools provides the experimenter with great opportunity.

7.9 Q MEASUREMENT OF LC RESONATORS

Several schemes have been used for Q measurements over the years. They can all work well when carefully executed. Two schemes are presented here for LC tuned circuits. The first method measures the bandwidth of a tuned circuit configured as a symmetrically loaded bandpass filter with very high insertion loss. The schematic is shown in **Fig 7.65**.

The two coupling capacitors should be approximately equal. This prevents heavy loading by the input with weak output coupling which could create high insertion loss with a wider than minimum bandwidth. Equal values guarantee that the input and output each contribute equally to the loading. High insertion loss then ensures that the external loading is light so that bandwidth is determined only by resonator loss.

The measurement is done with a signal generator and sensitive detector such as a spectrum analyzer, a 50-Ω terminated oscilloscope, or one of the power meters described earlier. The generator is tuned for a peak response and the center frequency, f_0, is read with a counter attached to the generator. The output amplitude response is also noted. The signal generator drive is then increased by 3 dB, causing the output to increase by the same amount. The generator is then tuned first above, and then below the peak until the response is identical to the original amplitude. The frequencies of the upper and lower −3 dB points are noted and the difference is calculated as the BW. Then $Q_u = f_0/BW$ where both are measured in the same frequency units. If the insertion loss is 30 dB or more, the measured Q is very close to the unloaded value. See section 3.3. The measurement can be done with lower IL, but corrections will then be required to calculate Q_u from the measurement Q.

Another scheme for Q measurement uses resonator elements in a *trap* circuit, shown in **Fig 7.66**. Again, a tunable generator and a 50-Ω detector are used. However, instead of configuring the resonator as a lossy filter, we now configure it as a trap, a circuit that produces high attenuation at one frequency. The generator is tuned to find the null in the output response. The null depth, which can be very large, becomes a measure of the resonator Q.

Either a parallel connected series-tuned circuit or a series connected parallel-tuned circuit can be used as traps. There is usually little virtue of one type over the other. We generally prefer the series-tuned circuit because a grounded and calibrated variable capacitor can be used in the resonator. A photo shows a test fixture with a 140-pF variable capacitor and binding posts.

The generator is tuned to find the null response and the level is carefully noted. A spectrum analyzer is ideally used as the detector and should be in a 1 or 2 dB per

Fig 7.65—Measuring Q by determination of 3-dB bandwidth. The coupling capacitors, Cin and Cout, should be approximately equal and should be small enough that the insertion loss is 30 dB or more.

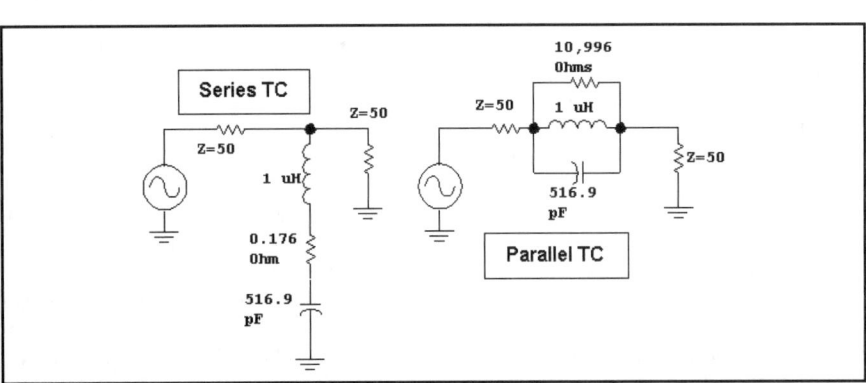

Fig 7.66—Measuring Q by determining the attenuation of a trap. A 7-MHz tuned circuit is used in this example with L=1 µH. The 0.176-Ω resistor in the series-tuned circuit and the almost 11-kΩ resistor in the parallel tuned circuit are models representing a 7-MHz Q of 250. The series-tuned circuit (STC) will have an attenuation of 43.1 dB while the PTC has 40.9 dB.

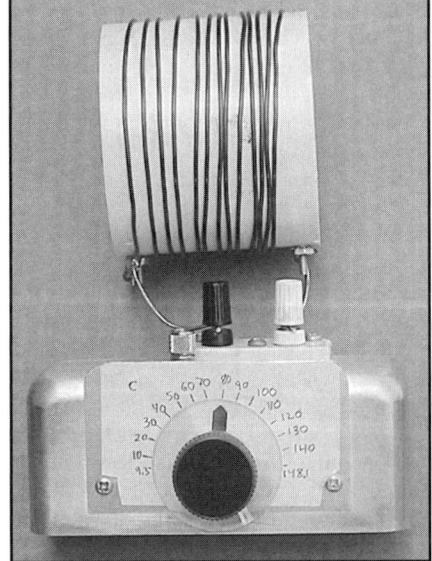

A test fixture simplifies Q measurement with the parallel connected series tuned trap method. The inductor shown was 13 turns of #14 enamel-covered wire wound on a 3.5-inch-diameter PVC pipe fitting. This coil had a measured Q of 371 at 7 MHz. The test fixture includes a grounded post allowing additional fixed capacitance to be added.

division sensitivity to provide amplitude resolution. The resonator is then disconnected and the generator is connected to the detector through a step attenuator. The attenuation is adjusted until the analyzer response is exactly the same as produced at the null. The attenuator value is then the null attenuation, A, in dB. Values of 60 dB or more are possible with some high Q tuned circuits.

This same measurement setup can be used to determine inductance if a calibrated capacitor is used. The unloaded Q is related to attenuation by

$$Q_S = \frac{4 \cdot \pi \cdot f \cdot L_u}{Z} \cdot \left(10^{\frac{A}{20}} - 1\right) \quad \text{Eq 7.4}$$

f, MHz; A, dB; L_u, uH; Z, Ohms

if the series tuned circuit form is used, or

$$Q_p = \frac{Z}{\pi \cdot f \cdot L_u} \cdot \left(10^{\frac{A}{20}} - 1\right) \quad \text{Eq 7.5}$$

f, MHz; A, dB; L_u, uH; Z, Ohms

...if the parallel tuned circuit is applied. Frequency is measured in MHz, A is in dB, and inductance is in µH for these equations. Z is the characteristic impedance of the measurement environment, usually 50 Ω.

It is useful to plot series resistance against attenuation for the parallel connected series impedance. This is shown in **Fig 7.67**. The experimenter may wish to build a similar curve for the series connected parallel impedance.

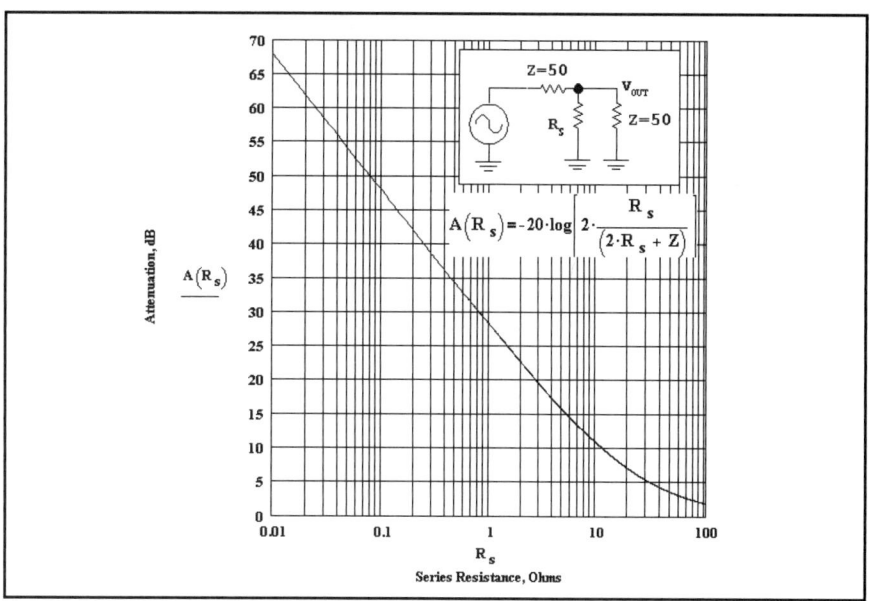

Fig 7.67—Attenuation vs R for the series impedance. See text.

It is important that a solid 50-Ω load and source impedance (Z in the equations) be used in this measurement. If the impedance is in question, use a 10 dB pad at both the generator and detector.

It is also important to prevent harmonics from confusing the results. This is guaranteed if you use a narrow bandwidth detector such as a spectrum analyzer. A wideband detector (a power meter or a 50 Ω terminated oscilloscope) will respond to harmonic energy that is not attenuated by the trap. The spectrum analyzer used for Q measurement could be very simple. Something as simple as a single tuned circuit preceding an oscilloscope would work so long as a pad was used to establish impedance. Alternatively, a very well low pass filtered signal generator could be used with any detector with adequate sensitivity.

The virtue of the trap scheme becomes apparent as soon as the two methods are compared. The traditional 3-dB bandwidth measurement depends on precisely establishing the 3-dB down level. A fraction of one dB error could still impact accuracy. In contrast, the depth of a null is often quite large for high Q resonators, and is easily measured with a step attenuator.

An accurate capacitance measurement tool such as the AADE or W7AAZ meters mentioned earlier is quite useful as a supplement to a Q measurement setup. With such a tool, accurate calibration of capacitors is ensured.

7.10 CRYSTAL MEASUREMENTS

A quartz crystal is modeled as a series RLC paralleled by a capacitance, **Fig 7.68**. Crystals are of special interest, for they are often used in construction of narrow filters. For this purpose, we need to know all of their parameters. Great precision is needed in knowing resonant frequency, for that strongly controls filter tuning. The knowledge of the other parameters is needed at an accuracy similar to that encountered in an LC filter.

There are numerous measurement schemes that will produce the four values. A 50-Ω measurement setup was presented in Chapter 3. Results from it are informative, especially if a batch of "junk box" crystals is encountered. However, more

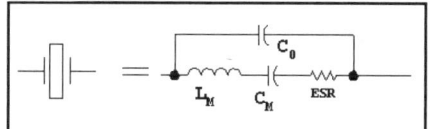

Fig 7.68—Model for a quartz crystal.

refined measurements are desired for filter design. An extremely useful, yet simple oscillator was also presented in Chapter 3 and is repeated here as **Fig 7.69**. A Colpitts oscillator with an emitter follower drives a frequency counter. A capacitor in series with the crystal, C_s, may be short circuited with a toggle switch. This produces a change in frequency that, when combined with the frequency and capacitor value, yield the motional capacitance, C_m. The motional inductance, L_m, is then calculated from series resonance, which is well approximated by the oscillator frequency when the switch is closed. The design equations are included in the figure. F is the frequency while DF is the frequency shift, both in Hz, when the switch is toggled; C_S and C_P, in Farads, are from the circuit. And as usual, $\omega = 2\pi F$.

If this test oscillator is built with Colpitts capacitors of C_p=470 pF and a series capacitor of C_s=33 pF, the circuit will function (fundamental mode) with crystals from 2 to 25 MHz. Simple equations are valid when C_p is more than $10 \times C_s$. It is

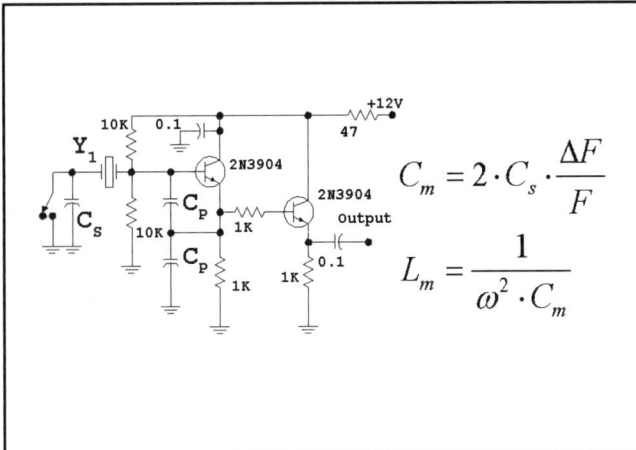

Fig 7.69—Colpitts oscillator for crystal testing, based on an insightful suggestion by G3UUR.

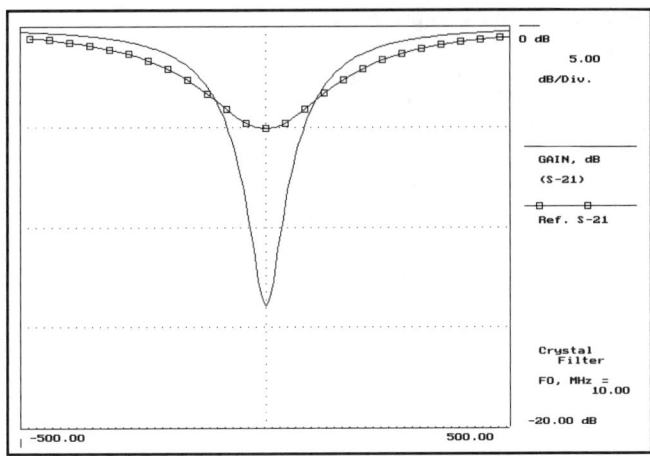

Fig 7.71—Sweeping two crystals while investigating their properties as traps. One has a Q of 40,000 while the one producing the deeper notch has a Q of 200,000. Notch depth is measured to determine Q.

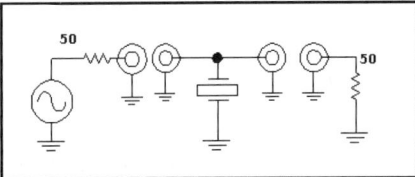

Fig 7.70—Using the trap nature of the crystal for a Q measurement.

also important that the C_s value be determined by measurements that include the switch. The 33 pF capacitor in our test set plus switch capacitance produced a net C_s=41 pF.

The crystal is essentially a series tuned circuit when operating near series resonance, so the series trap scheme described earlier for LC tuned circuits will also provide Q_U, as shown in **Fig 7.70**. Computer generated plots are shown for two different 10 MHz crystals in **Fig 7.71**. The shallow notch represents a low Q crystal with Q_u=40,000. The deeper and narrower notch corresponds to Q_u=200,000. The crystal Q relates to attenuation A in dB, motional L in Henry, frequency in Hz, and terminating resistance Z in Ω with...

$$Q = 4 \cdot \pi \cdot f \cdot L \cdot \frac{\left[10^{\left(\frac{1}{20} \cdot A\right)} - 1 \right]}{Z} \quad \text{Eq 7.6}$$

We performed an experiment with a crystal that had also been measured with earlier methods. The notch method for Q measurement yielded Q_U=202,000 with ESR=17.5 Ω. This was within a few percent of the earlier measurements. The ESR values for crystals are higher than we usually see with an LC resonator, so the notches are not as deep. This allows measurement with a power meter such as the AD8307 based design described earlier; a spectrum analyzer is not necessary. ESR can be 100 to 1000 Ω for very low frequency crystals, so the series connected parallel tuned circuit method might offer better measurements here.

Parallel capacitance, C_0, is easily measured with other tools such as the AADE or W7AAZ circuits. They are effective because those instruments operate at low frequency, around 1 MHz, well away from typical crystal resonance. With all four crystal parameters available, the designer/builder can proceed with the filter designs presented in Chapter 3.

The equipment described has also been used to evaluate HF ceramic resonators. In one measurement on an ECS type ZTA358MG (from Mouser) we saw L_M=761 µH, C_M=2.74 pF, C_0=31 pF, and Q_U=636. Series resonant frequency was well below the marked 3.58 MHz frequency at 3.38 MHz. The part is normally used in oscillators with a series capacitance.

7.11 NOISE AND NOISE SOURCES

Noise is generally the part of the response generated by our receivers that is undesired. However, we can also use noise as a measurement tool. By injecting noise into a communications system or component and examining the response, we can extract information about the system.

Figure 7.72 shows a simple noise source that is quite strong. This circuit delivers a noise output reaching –50 dBm at 10 MHz on a spectrum analyzer with a 300 kHz resolution bandwidth. This is more than 40 dB above the analyzer noise floor. If we apply this noise source to a filter, the signal appearing on screen is a picture of the filter response. While not nearly as useful as a tracking generator, it is still a simple and useful way to examine a filter. Gain stages can be added to the design to obtain even higher noise output.

The noise source of Fig 7.72 is not very flat with frequency. An improved source could be built with a Zener diode biased for a current of a few mA, with coupling into a high gain amplifier designed to have gain that is flat with frequency.

A noise source suitable for noise-figure measurement is shown in **Fig 7.73**. This circuit was designed by WØIYH and described in a paper included on the CD that accompanies this book.[21] The noise is generated by current flowing in D1 with S1 in the position shown in the figure. When the switch is toggled, current flows to forward bias the diode, preserving the source output impedance in the "off" state.

Paul Wade, W1GHZ, has also done some excellent work with noise generation, which is also included on the book CD.[22] Wade noted that an excellent noise source can be built with the emitter-base junction of a microwave transistor, using the diode as a Zener. Wade reports good results with the noise diodes operating as series elements.

The noise source of Fig 7.73 had an *excess noise ratio* (ENR) of 178 in the HF spectrum. This means that the noise power available from the source is 178 times (22.5 dB) stronger when the diode is biased into avalanche breakdown (Zener action) than when it is forward biased. If we were to attach this source to a perfect amplifier, one with no noise of its own, the resulting output noise would also change by 22.5 dB as the switch is toggled. An imperfect, real world amplifier will generate some noise of its own, so the output noise change will be *less* than 178 times when the diode is toggled. The output noise change is called the Y-factor and this measurement technique is called the Y-factor method. Noise factor is related to Y factor by

$$F = \frac{ENR}{Y - 1} \quad \text{Eq 7.7}$$

...where both ENR and Y are power ratios rather than dB values.

The noise sources are generally not difficult to build. However, calibration can be difficult. We borrowed a noise source to calibrate ours. See the two CD noise papers for more calibration information.

Noise figure for a receiver is measured with the test setup shown in **Fig 7.74**. The noise source is attached to a receiver antenna port with receiver AGC is turned off. The audio output is then applied to a true RMS reading voltmeter. We have used a surplus HP3400A and the Fluke Model 89 DVM. Alternatively, one can build an instrument using an Analog Devices AD636 that converts an arbitrary ac wave form to a dc signal proportional to the true RMS of that waveform. A paper describing this instrument is included on the book CD.[23] True RMS measurements are also done with relative ease with DSP software; see Chapter 11.

Consider an example: We toggle the switch to observe a 15.6 dB increase in audio output. This corresponds to a Y factor of 36.3. From **Eq 7.7**, the noise factor is then 5.04, which is a noise figure of 7 dB.

A practical detail complicates noise measurements when the bandwidth is narrow, such as the 500 Hz found in many CW receivers: The statistical variation with time of the noise from the receiver causes many meters to vary, making it difficult to obtain an accurate reading. The video filter of **Fig 7.75** averages the noise to reduce this problem. The dc output is applied to a high impedance voltmeter or oscilloscope.

The noise figure of amplifiers may be

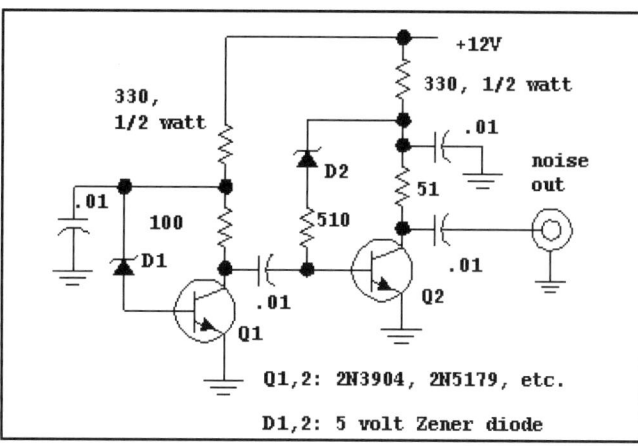

Fig 7.72—Noise in D1 is amplified in a two-stage amplifier, resulting in a strong noise source suitable for measurements. Virtually any diode or transistor types can be used in this source.

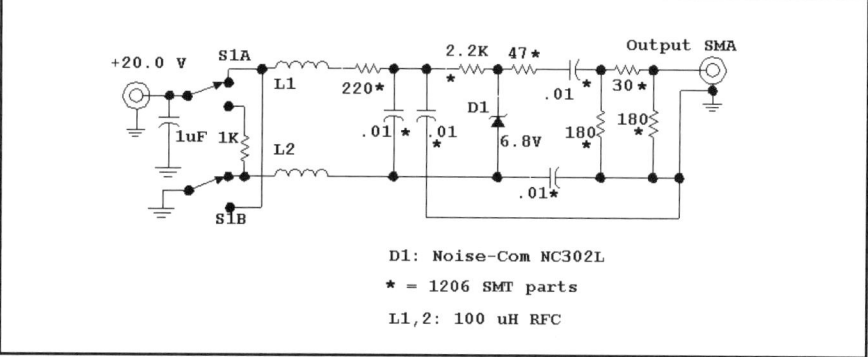

Fig 7.73—Noise source providing a flat frequency response over a wide bandwidth. Our source was built with surface-mounted components where possible. The diode was purchased from Noise Com, East 64 Midland Ave, Paramus, NJ 07652; tel 201-261-8797.

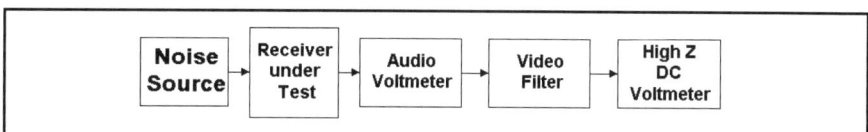

Fig 7.74—Test setup for noise figure measurement. The HP3400A is a true RMS audio voltmeter. This setup includes a video filter driving an oscilloscope, a refinement that may not be required. See text.

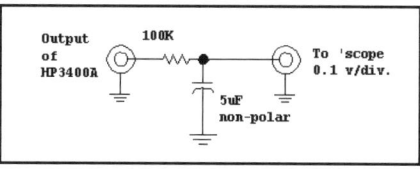

Fig 7.75—A simple video filter reduces meter-reading errors when working with narrow bandwidths.

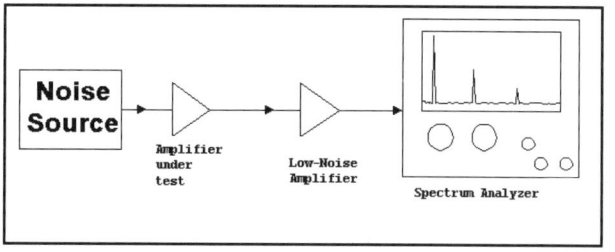

Fig 7.76—Test setup for noise figure measurement of an amplifier or other component.

evaluated with a spectrum analyzer in the test setup of **Fig 7.76**. The key element here is an auxiliary low noise amplifier (LNA) placed between the amplifier under test and the spectrum analyzer. This is needed because the noise figure of the typical analyzer is quite high. The LNA produces a cascade with a low combined noise figure not compromised by 2nd stage noise. See the discussion of noise figure in Chapter 2.

Begin a measurement with the noise source and both amplifiers off. Applying power first to the auxiliary LNA should produce an increase in output noise. Powering the amplifier under test should again increase the on-screen response. Switch the spectrum analyzer to a 1 or 2 dB per division vertical sensitivity and use extensive video filtering to replace trace "fuzz" with a smooth line representing averaged noise. Carefully note the on screen level of the noise. Then switch the noise source to the high noise position. Rather than reading a level from the screen, add attenuation in the analyzer front end until the trace is at the level seen earlier. An attenuator with 1 dB steps (or less) is preferred for this measurement. The amount of added attenuation is then the Y factor in dB. Converting this to a power ratio allows Eq 7.7 to be used.

The auxiliary low noise amplifier we used consists of a MRF544 followed by a Comm-Linear CLC425 operational amplifier.[24] Another suitable amplifier could be built with a cascade of MiniCircuits MAR-3 amplifiers, or similar parts, with a MAR-6 input stage, a configuration that should have a noise figure around 3.5 dB. Low noise figure designs were described in Chapter 6.

7.12 ASSORTED CIRCUITS

Testing AGC in receivers

The circuit shown in **Fig 7.77** is useful when observing the dynamics of a receiver AGC system with an oscilloscope. Named the "ditter," the circuit is an electronic switch with an off-to-on ratio of 80 dB at 14 MHz. The switching elements are inexpensive PIN diodes that are cascaded to obtain the desired off-to-on ratio. The circuit is balanced for the RF signal. However, the dc drive that turns the RF on and off is single ended. This prevents the control signal from creating a click that overwhelms the receiver. The topology was suggested by K7RO.

A slow pulse generator using a 555 timer drives the RF switch. Capacitor C1 controls the timing while the pot sets a duty cycle. A sample of the pulse provides a trigger signal for oscilloscope control. The signal biasing the diodes is filtered with C2 to prevent key clicks from an otherwise too fast rise and fall time. The circuit as shown has about a 1-mS rise, but a longer fall. Although the circuit was useful in studying some of the receivers in this book, a better timing circuit would be useful. One could use an external pulse generator or build a more refined one, probably using more than one timer.

The drive level should be confined to 0 dBm or less, which is adequate to overwhelm almost any receiver. Larger signals are partially rectified with the chosen PIN diodes.

An Experimenter's Receiving Converter

There are many situations where one wishes to receive signals at VHF to facilitate an experiment. A junk box crystal and diode ring mixer form the basis for the circuit of **Fig 7.78**. The crystal controlled oscillator at 25 MHz drives the diode ring at the standard +7 dBm level. Clearly, whatever crystal is available would be suitable.

In one application, we wished to check a 7-MHz transmitter for chirp, or slight change in frequency with keying. The best way to detect this is to listen to a harmonic. The receiver was attached to one of the mixer ports (either one is okay) and a 10X oscilloscope probe was attached to the other through a step attenuator. The transmitter, set for output at 7.04 MHz, was terminated in a load and the probe was attached to the termination. Third harmonic mixing was to be used, so we depend on a 75-MHz LO injection, a naturally strong response with a diode ring. The receiver was tuned to 2.44 MHz. This is the result of the 11th transmitter harmonic beating with the third LO harmonic. A chirp-free response was confirmed. A preselector filter can be used to reduce spurious responses for many applications.

Evaluating Noise in Local Oscillator Systems

The "critical path" for the construction of better communications equipment today is the local oscillator system in use.

Low distortion receiver front ends are becoming easy to build. Crystal filters with

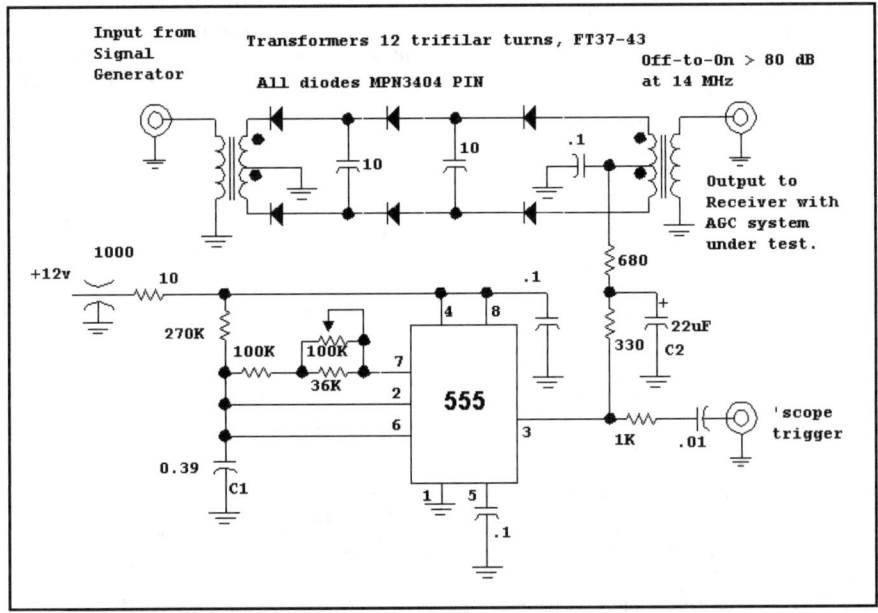

Fig 7.77—The *Ditter*, a circuit for generating keyed receiver input from a signal generator.

additional signal processing can provide outstanding selectivity, both close to a signal and well away from it. The various forms of frequency synthesis available to the builder all offer good frequency stability with the added bonus of electronic tuning. But the LO systems are compromised. Phase locked loop (PLL) systems tend to be plagued with phase noise. Synthesizers using direct digital synthesis (DDS) are often dominated by coherent spurious responses.

Although difficult problems to solve, the measurements are not that difficult. We illustrate the problem here with two measurement examples, the first with a commercial receiver using a synthesizer with both DDS and PLL. A crystal controlled oscillator (Fig 7.29) built with an internal battery, all housed in a well shielded box, was attached to the receiver input through a 10-dB pad and a step attenuator, initially set to 0 dB. The available input signal was confirmed to be –30 dBm at 7.018 MHz. The receiver, in CW mode, was tuned to this frequency with the setting stored in receiver memory. The receiver was then tuned downward while listening for responses with a well defined tone. AGC was on, for there is no provision to turn it off in the compromised receiver. A spur was found within a couple of kHz. The spur frequency was recorded in our notebook. The amplitude response was noted on an audio voltmeter attached to the receiver output. The tuning was then returned to the main signal and attenuation was inserted until the audio output equaled that seen with the spur. This occurred with 58-dB attenuation, so we infer the LO spurious response to be at 58 dB below the carrier, or at –58 dBc. This procedure was repeated as we found a large collection of spurious responses above and below the desired signal with results plotted in **Fig 7.79**.

There are difficulties encountered with this procedure. One must be sure the source is spur free. This was confirmed by repeating the experiment with a receiver using a traditional LC oscillator. You must also be sure that the signal from the source oscillator is not reaching the receiver by routes other than the antenna terminal. This can be confirmed by disconnecting the source from the attenuator to confirm that the signal disappears, or drops well below the level of the measured spurs.

Our second example evaluates phase noise with essentially the same procedure. Again start with a very strong signal, a –30 dBm input. Then tune away from the source frequency to a spacing of, for example, 10 kHz. Note the response in a true RMS reading audio voltmeter attached to the receiver output. Turn the source off momentarily to be sure that the noise decreases, for we wish to measure

The *Ditter* for AGC testing in a receiver. (Thanks for circuit suggestion from K7RO)

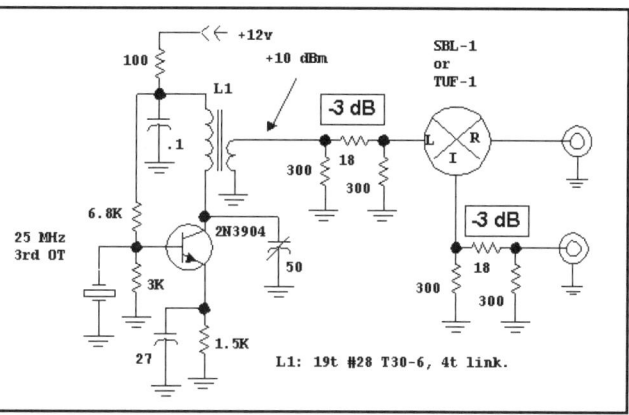

Fig 7.78— Receiving converter for experiments.

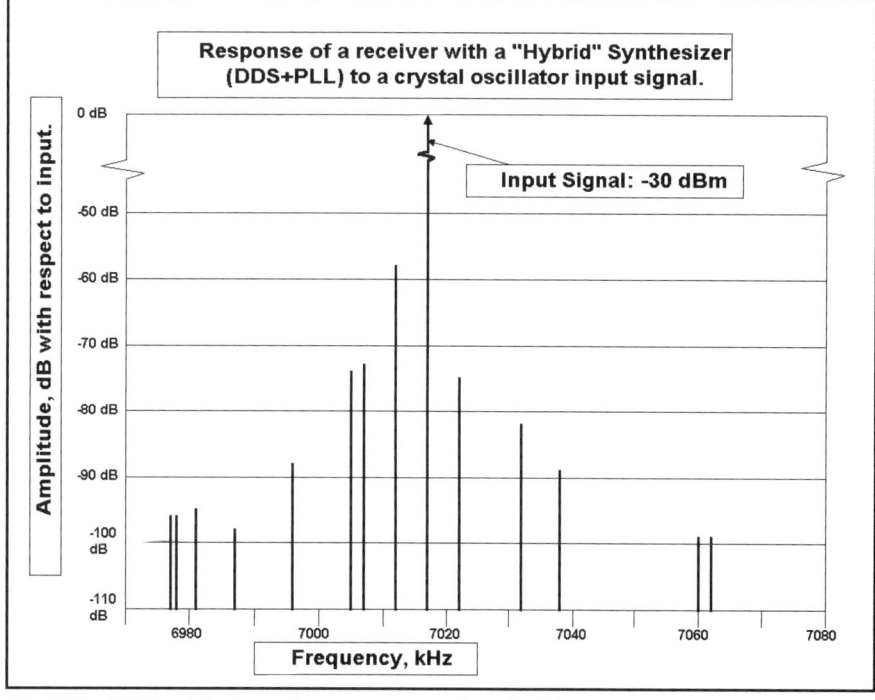

Fig 7.79—DDS-related spurious responses found with a commercial receiver.

Measurement Equipment **7.41**

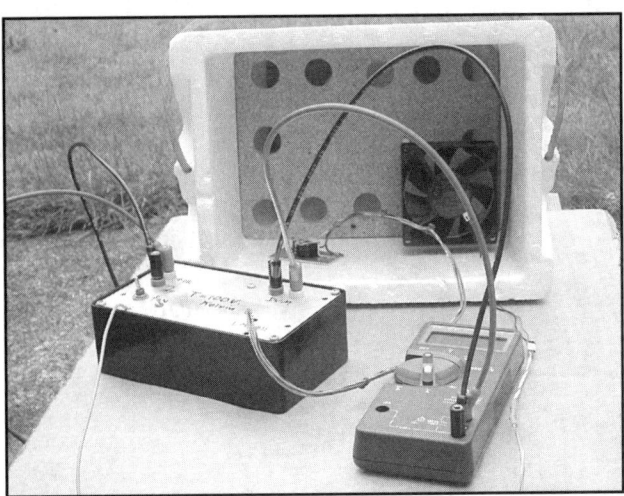

Control box, DVM, and "environmental chamber" for oscillator testing. The chamber has the lid removed so an oscillator can be placed inside. The lid is then placed on the box. A light bulb heater resides under the press wood base with holes. A 12-V fan moves the air within the box. Cables to the oscillator under test and the IC used for temperature measurement are routed under the lid edge.

the noise above the normal receiver background floor. Having recorded the response at 10 kHz offset, we return the tuning to the input signal. Attenuation is then added to bring the response down to the noise response level. In one measurement of this type reported in Chapter 4, we observed a noise response 110 dB down at a 5 kHz spacing. The receiver being measured had a 500 Hz noise bandwidth, so the spectral density of noise was 27 dB (10×Log[BW]) lower on a per Hz basis, or –137 dBc/Hz. It is necessary to normalize the response related to white (evenly distributed) noise, for that noise will change in proportion to bandwidth. In this example we attributed the observed noise to a VCO being tested, although it could have been the receiver LO. It was still a clean response compared with a typical DDS system like the one of Fig 7.79.

We often see equipment reviews where plots appear showing phase noise. Coherent spurs also appear in these plots. A per-Hz normalization is usually applied to the plot, for that is the most useful information form for pure noise. That normalization may or may not also be applied to the coherent spurs. The normalization, if applied, is not always stated in reviews. This problem disappears when you do your own measurements.

An Oven for Drift Compensation

A photograph shows an oven that we use for the evaluation and compensation of oscillators. The basic measurements were outlined in section 4.2. The "oven" is quite simple, starting with a Styrofoam box purchased at a local super market. The volume is approximately 600 cubic inches. The lower half of that space is occupied with a 60-W light bulb mounted in a ceramic socket attached to a wood strip. The cord for the bulb is run through a hole in the box.

A wood shelf with numerous 1-inch holes divides the box. The upper region contains a small dc fan that can be turned on to circulate the air and enough room for the oscillator module being tested and the temperature measuring circuitry. This oven measures temperature with a National Semiconductor LM3911 integrated circuit that is mounted in a small heat sink and then attached to a small circuit board. The LM3911 has been discontinued, replaced by a much better part from National, the LM45 that is supplied in a SOT-23 surface mount package. The part can be soldered to a small scrap of circuit board with a suitable bypass capacitor and the three wires needed to both power the device and to extract a signal. The output is read with a standard DVM with a sensitivity of 10 mV for each degree C change in temperature.

The oscillator under test is placed in the chamber and the lid is put in place. The oscillator is allowed to warm up while viewing output frequency on an external counter and initial temperature data is read. The light bulb is then turned on, allowing the temperature to climb. It's useful to cycle the bulb off and on, forcing the temperature to increase slowly. Once you have increased T by perhaps 20 degrees C, the fan is turned on for a short burst and the bulb is turned off, forcing the temperature to stabilize. If T seems fairly stable, new frequency data can be measured and **TCF** (**T**emperature **C**oefficient of **F**requency) can be calculated. It is not generally necessary to reach high temperatures, although an initial run up to perhaps 80C will serve to relieve stresses in the inductors resulting from the toroid winding. After a little data has been obtained, the lid can be removed, the bulb turned off, and the fan turned on. This will force the temperature to drop to room value in just a few minutes. The time is used for calculating the value of the temperature compensating capacitors needed.

The temperature compensation process is one that has left us with some very strong impressions:

1. An oscillator that we had regarded as being "pretty stable" with normal components drifts dramatically with the simple oven. This is *not* a minor, subtle effect, but dominant behavior.

2. Once we begin to apply compensation to the oscillator, just 2 or 3 runs will be enough to produce excellent stability.

3. A circuit that started as a "pretty stable" circuit is easily converted to "rock solid."

4. Circuits using really bad components regarding drift (such as varactor diodes) can still yield practical performance.

The whole process is an easy one. The one drawback is that it is somewhat time consuming, so we integrate it with other casual activities.

REFERENCES

1. W. Sabin, "A Series-Regulated 4.5- to 25-V, 2.5-A Power Supply," 2003 *ARRL Handbook*, Ch. 11 at 25-28.

2. W. Sabin, "Measuring SSB/CW Receiver Sensitivity", *QST*, October 1992, pp 30-34.

3. D. Bramwell, "Understanding Modern Oscilloscopes," *QST*, July 1976, pp 18-19.

4. J. Grebenkemper, "The Tandem Match —An Accurate Directional Wattmeter", *QST*, January 1987, pp 18-26.

5. R. Lewallen, "A Simple and Accurate QRP Directional Wattmeter", *QST*, February 1990, pp 19-23, 36.

6. W. Hayward and R. Larkin, "Simple RF Power Measurement", *QST*, June 2001, pp 38-43.

7. G. Daughters and W. Alexander, "Low Power Attenuators for the Amateur Bands," *73* Magazine, January 1967, pp 40-41.

8. D. Bramwell, "An RF Step Attenuator," *QST*, June 1995, pp 33-34.

9. R. Stone, "The UniCounter—A Multipurpose Frequency Counter/Electronic Dial", *QST*, December 2000, pp 33-37.

10. W. Carver, "The LC Tester", *Communications Quarterly*, Winter 1993, pp 19-27.

11. W. Hayward, *Introduction to Radio Frequency Design*, Prentice-Hall, 1982, and ARRL, 1994.

12. R. Bracewell, *The Fourier Trans-form and its Applications*, McGraw-Hill, 1969.

13. M. Engelson, *Modern Spectrum Analyzer Theory and Applications*, Artech House, 1984.

14. W. Hayward, "Extending the Double-Tuned Circuit to Three Resonators", *QEX*, March/April 1998, pp 41-46.

15. W. Hayward and T. White, "A Spectrum Analyzer for the Radio Amateur", *QST*, August and September 1998, pp 35-43 (Aug), 37-40.

16. Ibid.

17. W. Hayward and R. Larkin, "Simple RF Power Measurement".

18. W. Hayward and T. White, "A Spectrum Analyzer for the Radio Amateur".

19. R.W. Ramirez, *The FFT : Fundamentals and Concepts*, Prentice-Hall, 1985.

20. R.S. Horne, *Spectrogram*, Version 6.0.8, 2001, **www.n0ss.net**

21. W. Sabin, "A Calibrated Noise Source for Amateur Radio", *QST*, May 1994, pp 37-40.

22. P. Wade, "Noise Measurement and Generation", *QEX*, November 1996, pp 3-12.

23. W. Sabin, "Measuring SSB/CW Receiver Sensitivity".

24. S.O. Smith, "Build a 1-dB Noise Figure Amplifier for 50-ohm Systems", June 27, 1994 Analog Applications Issue, *Electronic Design*.

CHAPTER 8

Direct Conversion Receivers

8.1 A BRIEF HISTORY

In the early days of radio, signals were collected on a wire, converted from RF voltage and current to audio voltage and current with a crystal detector, and converted to acoustic energy with headphones (**Fig 8.1**). This worked well for spark and later AM broadcast signals, but with continuous waves, the output of the crystal detector was just a very weak dc voltage. A number of schemes were used to convert the CW to AM at the receiver, but the most sensitive method for detecting CW signals on a crystal detector required the use of an oscillator located near the receiver, as shown in **Fig 8.2**. When the oscillator was tuned close to the transmitted signal frequency, audible beats were produced by the crystal detector. The use of a "local oscillator" has been standard in receivers ever since.

The audible beat signal at the crystal detector is very weak. Early experimenters purchased the most sensitive headphones they could afford, and erected large antennas to collect as much signal as possible. Tuners included adjustments for both peaking the desired signal and achieving maximum power transfer between the antenna and detector. The technology for building highly sensitive headphones was already mature in the early days of radio, because the telephone system predated vacuum tube amplification by several decades. The first application of vacuum tubes in receiver circuits was for audio amplification. The "crystal detector" diode is considerably less sensitive as an envelope detector for AM than it would be with sufficient LO injection to serve as a product detector for CW, but early receiver lore involved using very low level LO injection. RF amplification was

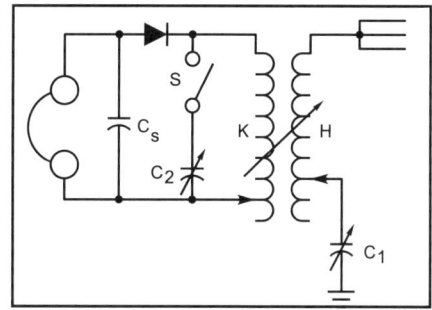

Fig 8.1—A fundamental crystal radio design.

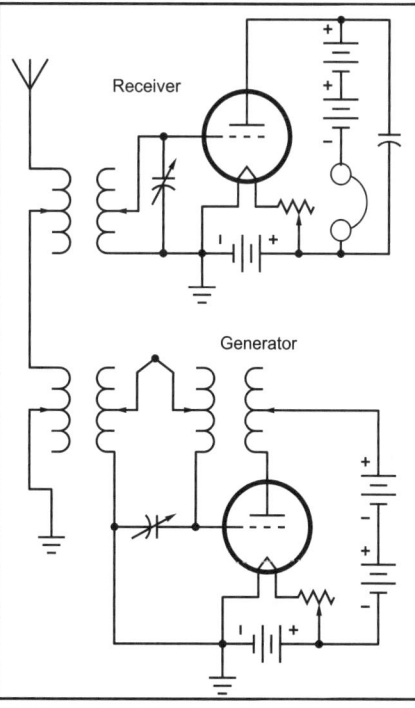

Fig 8.2—A classic radio enhanced with a local oscillator.

needed for AM, and (inevitably) early RF amplifiers using vacuum tubes were marginally stable, which lead directly to the discovery of regenerative receivers. Some RF amplifiers oscillated at two frequencies at once—which lead directly to the discovery of the superregenerative receiver. Cascading two regenerative detectors, one at HF and one at a superaudible frequency around 30 kHz, resulted in the superaudioheterodyne receiver, which was tricky to adjust and received every signal at two places on the dial.

Regenerative receivers were simple, inexpensive and worked well enough for amateur AM and CW work that receiver innovation stalled for more than a decade, until the bands became crowded enough that more selectivity was needed. The superheterodyne had been further developed for AM broadcasting, and by the mid 1930s, the transition to the superheterodyne for amateur high frequency work was nearly complete. High Frequency Regenerative receivers remained in *The ARRL Handbook* until the mid 1960s, and superregens are still widely used in toy walkie-talkies, radio controlled cars, and garage door openers.

Signal gain ahead of the detector is desirable if a diode is used to envelope detect AM, but for the linear modes, SSB and CW, the first stage of the receiver may be a lossy frequency converter, directly to audio. Such receivers are capable of outstanding performance at very high frequencies—something to think about the next time a State Patrolman recovers a weak echo from your speeding vehicle with a direct-conversion microwave receiver.

All of the technology—diodes, transformers, local oscillators and audio amplifiers— was available by 1920 to build high-performance direct conversion receivers for CW. There was little motivation for amateurs to develop such receivers at the time because regenerative receivers were adequate, simple and inexpensive. There was also a perception in that era that voice modes were the realm of experimenters and CW the realm of practical communicators. The situation is reversed today, with most technically advanced amateurs experimenting with non-voice modes, from minimalist HF CW stations through microwave systems for 1000-km tropospheric paths.

A radio experimenter is driven not by the desire to duplicate existing circuitry, but by the need to put a station on the air using whatever means are available, preferably without making expensive trips to the parts store. Marginal finances often unleash a wealth of ideas (the philosophy behind PhD programs and other monastic experiences). In the 1960s, when most HF stations operated at the 100 W level, the QRP Society embraced the philosophy of putting simple radio stations on the air and working DX using operator skill instead of transmitter power. Radio experimenters quickly expanded the QRP skill set to include radio design and construction, with an emphasis on elegant simplicity. With the disappearance of AM from the bands, and the emergence of CW as the experimenter's favored mode, the time was ripe for a reexamination of basic receiver circuitry. The '60s implementation of the direct conversion receiver was developed in parallel by a number of independent experimenters. All of the pieces were described in the mid '60s *ARRL Handbook*, but the editors clearly did not envision connecting them together into a receiver without an IF. Even the 1970s *ARRL Handbook* description of direct con-

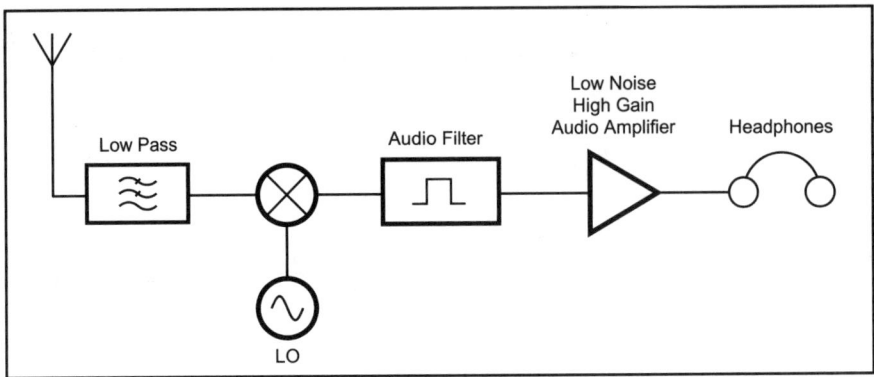

Fig 8.3—A block diagram of a basic direct-conversion receiver.

version receiver dynamic range and sensitivity exhibits gaps in understanding.

While the QRP Society provided the direct conversion receiver with a home, their fundamental philosophy also hampered its development. The QRP community embraces simplicity, and many of their designs are indeed simple and only just adequate. Examples of optimizing for simplicity are the numerous NE602 receiver circuits, which have surprising performance for so few parts. The usual first impression upon listening to a simple direct conversion receiver is that it sounds very good, but after making a few contacts most operators want something better. The something better is almost always a superhet. Wes Hayward correctly stated in *Solid State Design for the Radio Amateur*[1] that a direct conversion receiver with audio image rejection is at least as complicated as a simple superhet. This is even truer today, after another quarter century of superhet receiver evolution. The maturity of crystal ladder IF filter design has eliminated IF filter cost as a drawback for superhets, and easy-to-use ICs have reduced parts count below what was possible in the mid '70s.

A small group of experimenters stubbornly continued to develop the direct conversion receiver. Roy Lewallen's design[2] from 1980 is a timeless example of an optimized DSB design with CW filtering, and Gary Breed's 1988 design[3] nicely illustrates the practicality of eliminating the audio image. The KK7B designs published from 1992 through 1995[4-15] were originally intended to serve as VHF tunable IFs with microwave no-tune transverters, but were designed for broadband operation at any frequency from 25 kHz to 5 GHz. These designs have more components than the simplest superhets, but offer several performance advantages including freedom from birdies, ease of use throughout the radio spectrum, and superb in-channel audio fidelity.

By the year 2000, direct conversion receiver designs (**Fig 8.3**) pioneered by amateurs were making significant inroads into practical communications gear including family radio service transceivers, cordless phones, and cellular handsets. The number of papers on direct conversion presented at professional conferences has jumped from a few per decade to over a hundred in one year.

8.2 THE BASIC DIRECT CONVERSION BLOCK DIAGRAM

Fig 8.4 is the block diagram of a direct conversion receiver system for 40 meters. Unlike other figures in this text, the antenna and headphones are included in the diagram. The first block is the antenna. Its function is to collect as much of the desired signal, and as little noise and interference, as possible. While this seems obvious, few amateur or professional engineers actually think about the antenna when designing a receiver system. A 40-M dipole may provide a 1-mV rms noise floor in a 2-kHz bandwidth, during the evening, in the north central United States. Strong foreign broadcast stations may reach millivolt levels. Computer noise and "touch lamp" interference can reach 100-mV levels if the offending appliances are in the near field of the dipole. All of these signals are present at the downconverter.

Another important set of signals present at the downconverter input are FM broadcast stations. In urban areas, FM broadcast signals can produce signals of tens of millivolts in a few meters of wire. The 13th and 15th harmonics of 7 MHz are in the FM broadcast band, and most wideband mixers will downconvert signals near odd harmonics of the LO. The TUF-1 mixer recommended for several projects in this book has 34 dB more loss as a 13th or 15th harmonic mixer than as a fundamental mixer, when measured using a 7-MHz LO. A 1-mV signal at 91.5 MHz (easily obtained on a few meters of wire at KK7B,

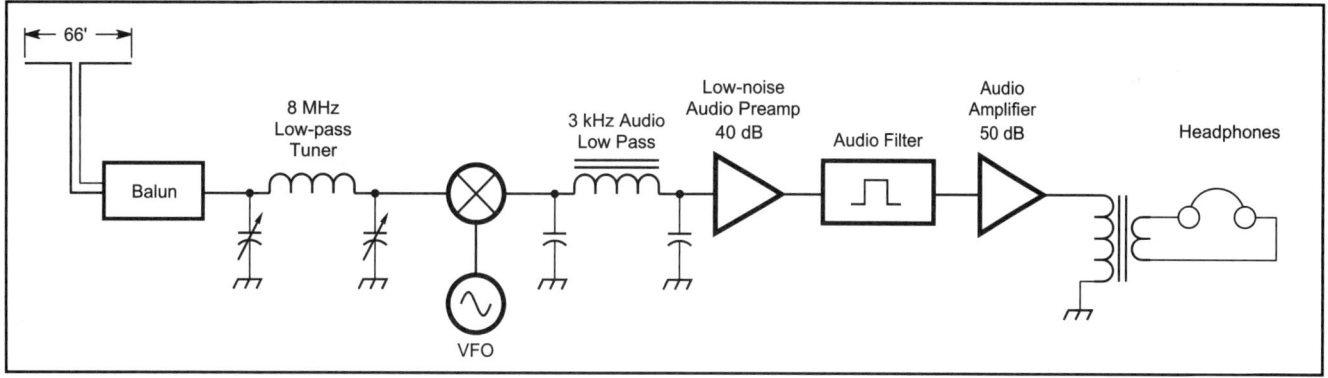

Fig 8.4—Block diagram of a 40-meter direct-conversion receiver.

Portland) is zero beat when the direct conversion receiver LO is tuned to 7.038 MHz, and the 34 dB of excess conversion loss reduces it to the equivalent of a 20-µV 40-meter signal at the antenna. It is easy to prevent these signals from arriving at the RF port of the mixer by using a low pass filter right at the mixer. They are VHF signals, so VHF construction techniques must be used. It is also important to prevent these FM broadcast signals from entering the receiver cabinet on power supply wires, speaker wires, headphone leads, CW key leads and microphone cords—all of which tend to be the right length to make efficient FM broadcast antennas.

The mixer itself can be any of several types, but the diode ring is a good choice for people who want simplicity, good performance, and understanding of how the mixer works. The details of the NE602 schematic are unpublished, and the bias controls to improve its performance are locked in place on the die.

Commonly used mixers have noise figures between 6 and 10 dB, and may have either conversion gain or loss. At first glance, conversion gain would seem to be an advantage. A receiver needs about 100 dB of gain between the antenna connector and headphones, and mixer gain makes the rest of the receiver easier to design. But there is a catch. Mixer gain occurs before any channel selectivity. The filter before the mixer in a direct conversion receiver passes an entire band, and the filtering after the mixer selects the desired signal. The mixer must linearly handle all of the strong and weak signals in the entire band, without distortion. If the mixer has gain, it amplifies all of the strong, undesired signals right along with the weak desired signal. High performance receivers, whether superhets or direct conversion, limit the amount of gain before the channel filter. Thus, minimum-parts-count casually designed receivers tend to have mixers with conversion gain, and more serious receivers have mixers with conversion loss.

Lossy mixers may be either the common diode ring and variations, or made up from transistors used as switches. A number of excellent passive FET mixers have been designed in the past few years, and they are now widely used in a variety of applications.

Mixer gain or loss does not affect receiver noise figure as much as might be suspected. Compare two receivers, each with a 2-dB noise figure, 13-dB gain RF preamplifier. Receiver #1 in **Fig 8.5** has a Mini-Circuits TUF-1 mixer with 5.7-dB loss and 7-dB noise figure, followed by an audio stage with 5-dB noise figure. Receiver #2 in **Fig 8.6** has the same RF preamplifier in front of a Gilbert Cell mixer with 8-dB noise figure and 10-dB gain, driving the same 5-dB noise figure audio amplifier. Using the cascaded noise figure formula presented elsewhere, Receiver #1 has a calculated 3-dB noise figure, and Receiver #2 has a 2.5-dB noise figure.

Now consider the fact that the Gilbert Cell receiver has 23-dB gain before any selectivity, and remember that short-wave Broadcast signals often reach millivolt levels. After the mixer downconverts the entire frequency spectrum present on the antenna and folds it in half around zero Hz, the circuitry connected to the IF port of the mixer selects a narrow portion of the spectrum and then amplifies it. Selectivity between the mixer and first audio amplifier is needed so that the first stage of audio does not have to linearly amplify the entire HF spectrum at once. A simple 10-kHz low-pass filter will narrow the frequency range to just 20 kHz centered around the LO frequency. Further band-limiting is normally included in the audio amplifier stages, but a wide-open direct conversion receiver sounds better on CW and SSB signals than any other receiver type, and should be experienced as a baseline for further receiver experimenting.

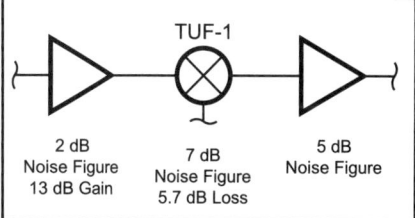

Fig 8.5—A preamp diode ring direct-conversion receiver.

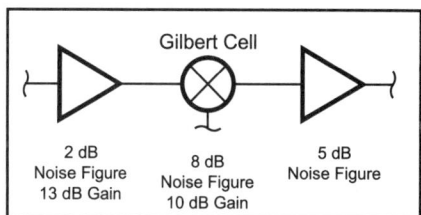

Fig 8.6—Block diagram of a preamp Gilbert direct-conversion receiver.

If the antenna provides 1 µV of noise floor and the headphones require 10 mV for comfortable listening, the receiver needs 80-dB gain. Very quiet locations may have a 0.1-µV 40-m noise floor, and low-sensitivity headphones might require 100 mV—which increases the gain requirement to 120 dB. Receivers without AGC require less gain than receivers with AGC, and also need a different listening style. A receiver described in the next chapter has more than 80 dB of undistorted headroom above the receiver noise floor. Some operators are accustomed to listening for weak signals with the receiver gain turned all the way up, and the receiver noise floor just below the pain threshold. If a click, pop or loud signal suddenly appears in the passband, the receiver is (theoretically) capable of providing an output that will break eardrums and melt headphones. Human ears have remarkable dynamic range. It is far more natural to lis-

The "ugly" MicroR1.

The MicroR1 built on a board.

ten to weak signals 60 dB below the pain threshold and match the receiver in-band dynamic range to the ear's capability.

In previous years the author has merely acknowledged that there are different listening styles, and some styles of listening require AGC more than others. However—two of our close friends (and strongest advocates of AGC), are nearing retirement with serious hearing loss. Both were licensed as novices in the early 1950s, and have spent half a century depending on receiver AGC to protect their ears. Setting receiver gain so that the noise floor from the antenna is well below the pain threshold and training the ears to listen is good hygiene. Weak signals will then be weak, strong signals will be strong, and only rarely will AGC be desired.

A Minimalist Direct Conversion Receiver

Not all direct conversion receivers have to be designed for high performance. Since the historical appeal of direct conversion is simplicity, it is appropriate to present a strict minimalist design. Simple NE602 based circuitry is presented elsewhere in the text. For this circuit, the use of specialized components is avoided. The receiver in **Fig 8.7** has each of the functional blocks from Fig 8.3. Q1 and its associated components is a simple Pierce oscillator. With the component values shown, it oscillates with every crystal tried from the author's junk box. The frequency may be trimmed a few kHz with a small (about 20 pF) trimmer capacitor in series with the crystal. Since both ends of the trimmer capacitor are floating, an insulated tuning tool or shaft should be used.

T1 is 10-trifilar turns of enameled wire on an FB 2410 ferrite bead. A transformer made of ten trifilar turns of plastic covered bell wire on a large ferrite RFI suppression core salvaged from a computer printer cable also works well. Diodes are 1N4148 or similar, and the three transistors are 2N3904 or similar small-signal NPNs.

The two stage audio amplifier has more than enough gain to bring the 40-m band US West Coast noise floor up to the audible level in portable CD player headphones. Coupling and feedback capacitors were selected by ear and back-of-the-envelope calculations from available values in the author's junk box. Gain is intentionally kept low for ear protection, and to eliminate the need for special construction techniques, a volume control, or shielding. The double tuned circuit on the RF input solves any harmonic mixing or AM broadcast detection problems, and the three adjustments may be tweaked to optimize signal power transfer from the antenna to the receiver. When signals are strong and shortwave broadcast interference is a problem, the coupling capacitor may be reduced and the input circuit optimized for desired signal-to-interference ratio rather than just maximum signal strength. The independent 9-V battery supply, balanced antenna and headphone connections, and no external ground connection eliminate ground loops and common mode problems. Current drain from the 9-V battery is about 8 mA.

This simple receiver is fun to listen to, particularly when it is open on the bench with all parts visible, and signals from 10,000 km away are rolling in. The accompanying photos show two different construction styles. Parts may be purchased new, or salvaged from old computer boards and transistor radios.

The receiver described in the preceding paragraphs is a nice illustration of how simple a "real" communications receiver can be. It also illustrates some of the challenges of simple receivers. Crystal control strictly limits tuning range, and limited selectivity requires skill in digging signals out of crowded bands. The challenges inherent in simple equipment are not necessarily disadvantages—it takes more skill to cross a harbor in a sailing dinghy than a motor boat. Copying signals from across the oceans with a three transistor circuit is similarly rewarding.

Just as sailors always want a bigger boat, radio experimenters always want to improve their receivers. The following paragraphs dig into the technical fundamentals needed to understand direct conversion receivers at a depth that allows performance to be pushed to superhet levels and beyond.

Direct Aversion

Before proceeding with the technical discussion, it is worthwhile to note that many otherwise rational human beings have an emotional aversion to direct conversion receivers. The basic block diagram is so simple and appealing than many unsuspecting designer-builders and engineering managers have fallen into the trap of believing that direct conversion is the "holy grail" of receivers, able to outperform the old, obsolete superheterodyne architecture at a fraction of the cost. Most attempts to build something cheaper and better than an existing, mature technology will fail. When the holy grail turns out to be a cracked clay cup, the designer involved may end up with a lingering bad taste in his mouth. Experienced professional and amateur technical writers tend to either love or hate direct conversion receivers, and this bias has often appeared in print.

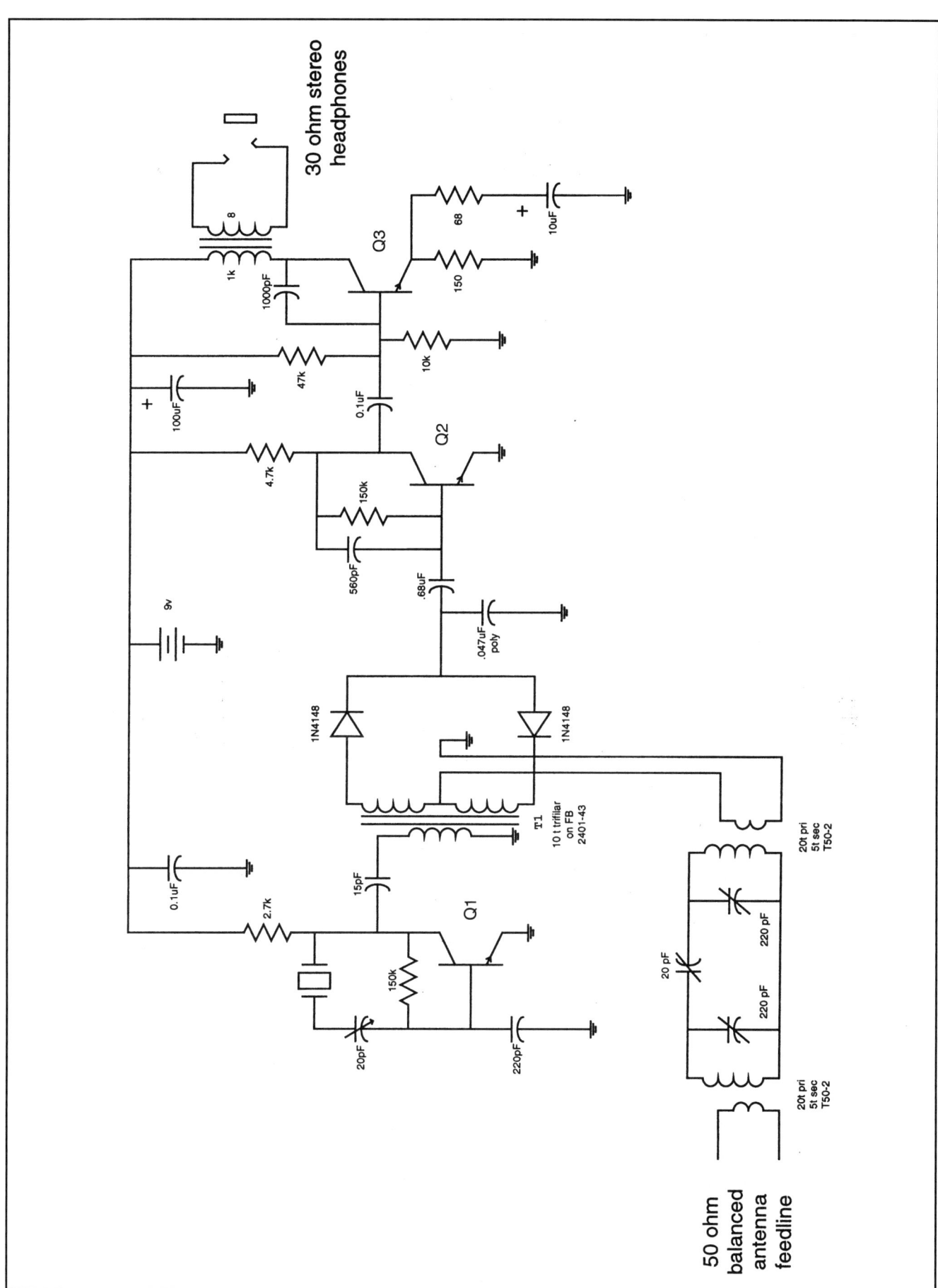

Fig 8.7—The schematic of the MicroR1.

8.3 PECULIARITIES OF DIRECT CONVERSION

The level of understanding represented in the preceding paragraphs is enough to build direct conversion receivers and use them to make contacts on the amateur bands, but they will exhibit some strange behavior that is not explained by conventional superhet thinking. Explaining the peculiarities of direct conversion receivers, and more importantly, designing and building a new generation that outperforms previous attempts, requires further study and a deeper understanding.

High Audio Gain

There are significant differences between the block diagrams and gain distributions of superhets and direct conversion receivers. Direct conversion peculiarities fall into two classes: problems from high audio gain and the effects of local oscillator radiation. AM demodulation, a common problem with direct conversion receivers, is a symptom of both high audio gain and LO radiation.

A typical direct conversion receiver has about 100 dB of gain from the mixer to the output. The output might be a 1-mA current flowing in a wire to the headphone jack. The ground wire coming back from the headphones also carries 1 mA. If the ground wire has 1 milliohm resistance, the voltage drop will be 1 µV, which is 100 times larger than the weakest audible signals. This sets up an ideal condition for audio oscillation or regeneration. Since it is impractical to reduce the resistance of all ground wires (#24 copper wire has about 2 milliohms per inch), it is very important that any ground return carrying output signals be separated from any input signal ground return. The easiest way to insure this is to use a separate ground wire for every component, and connect them all together at a single point. It is particularly important to treat the speaker or headphone jack as a component, and bring it's ground lead all the way back to the common ground connection rather than just grounding it to the radio case. This bears repeating: use two wires, a signal and a ground wire, to connect to the headphone jack or speaker, and do not ground the speaker or headphone jack to chassis ground. With a simple receiver, it is possible to actually connect the grounded leads of all components to the same point. **Fig 8.8** is a schematic showing how this can be done with the receiver in Fig 8.7. There are also magnetic and capacitive feedback mechanisms that become important at audio with 100 dB of gain. Often oscillations can be cured by moving around the wires carrying audio signals and power.

Inductors in the early stages of a direct conversion receiver should be of a self shielding type. Conventional Iron E core audio transformers are best avoided, although they have been successfully used on the input to high gain audio amplifiers in direct conversion receivers with several layers of magnetic shielding. The Toko 10 RB series of shielded inductors has been used for years, although the shielding is not perfect and they will pick up hum from nearby transformers. A small steel or mumetal enclosure around the audio pre-amp stages of a direct conversion receiver can reduce hum pickup by

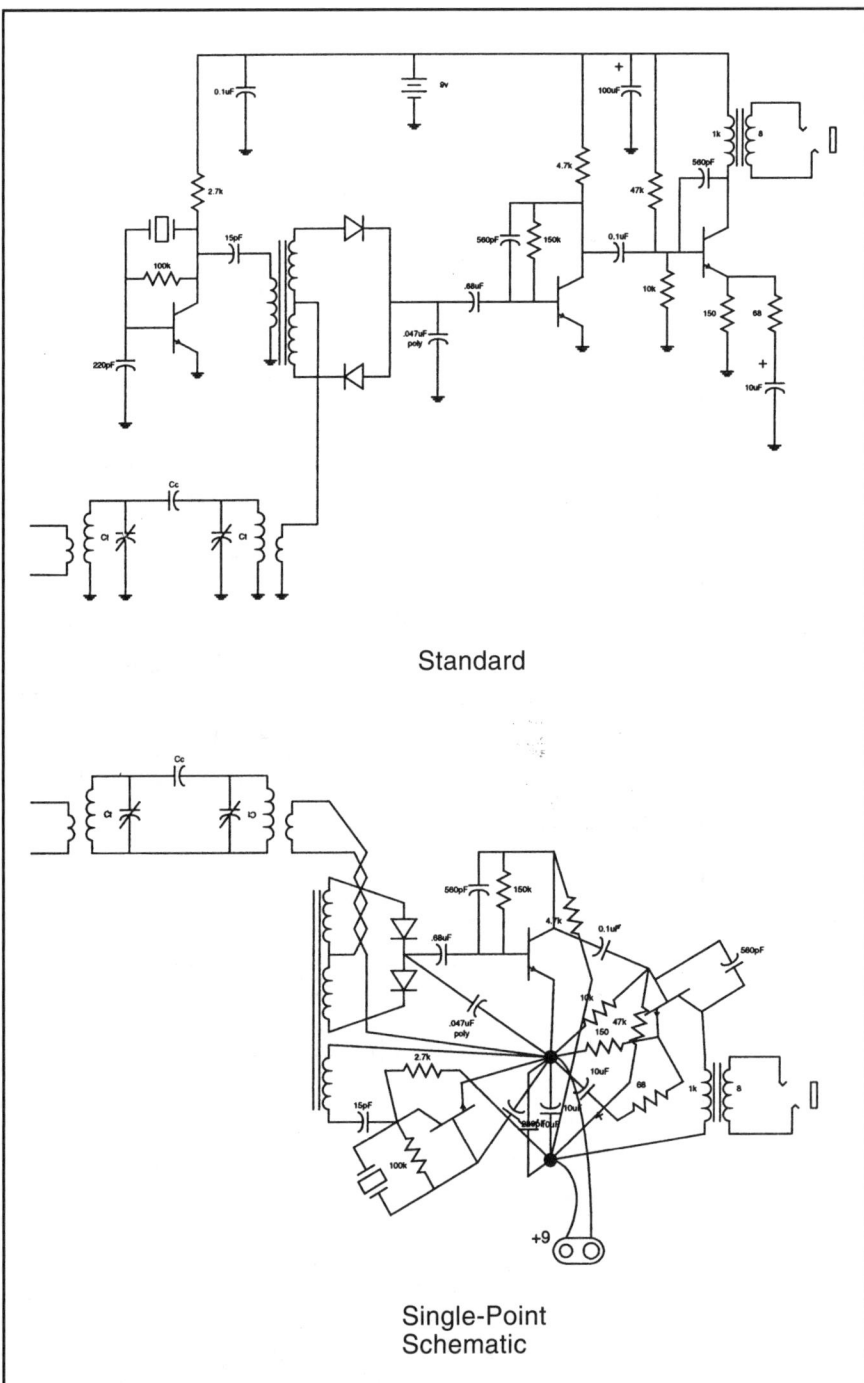

Fig 8.8—Compare the "standard" MicroR1 schematic above to the single-point schematic below.

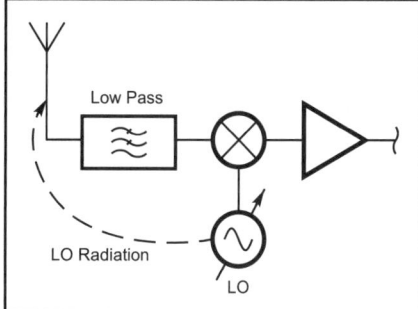

Fig 8.9—Local oscillator radiation.

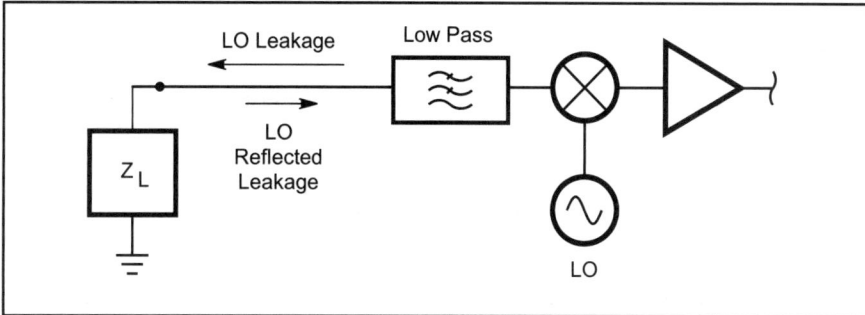

Fig 8.10—A mixer/LO with reflection coefficient.

many dB. Good direct conversion receivers tend to include high-pass filters in the audio chain, aggressively rolling off the audio response below about 300 Hz.

Microphonics, the loud clicks and pops when the receiver is bumped, are often blamed on high audio gain, but they are actually a symptom of Local Oscillator radiation, and can often be cured by improving receiver shielding.

Local Oscillator Radiation

Local oscillator radiation raises a whole new set of problems. **Fig 8.9** shows a simple direct-conversion receiver front end with local oscillator radiation arriving at the RF input port of the mixer. Since the LO is at the RF frequency, there is no possibility to use RF selectivity to reduce the level of LO at the mixer RF port (in a superhet, the LO and RF are separated by the IF, so the RF selectivity necessary for image rejection usually reduces the LO signal between the antenna and RF port of the mixer). At first glance, it appears that the LO signal at the mixer RF port will have no practical effect, because it is exactly zero beat. The mixer multiplies the RF port LO signal with the LO, and the output is pure dc:

low pass {a cos($2\pi f_o t + \phi$) cos($2\pi f_o t$)}
 = a/2 cos ϕ **Eq 8.1**

...where f_o is the LO frequency, a is the amplitude of the LO leakage, and ϕ is the phase difference between the LO and LO leakage.

DC at the IF will unbalance a balanced mixer, which causes it to radiate more LO. The additional LO radiation might be reflected by nearby objects or an imperfect antenna match. If the new term is in phase with the original radiated LO, this will further unbalance the mixer. Thus the amount of LO radiation is a function of the physical environment near the antenna.

This is not usually a problem at HF with large outdoor dipoles, but HF direct conversion receivers commonly exhibit disappointing performance with wire antennas connected directly to the back of the radio. A changing local electromagnetic environment around the antenna can be a particular problem at VHF and microwaves where antennas are small and good reflectors are numerous.

LO radiation and pickup by the antenna becomes more significant when either the amplitude or phase of the LO signal at the RF port of the mixer is time dependent. There are three major classes of time variation in the LO signal: transients, Doppler and modulated scatterers. Each of these will be treated separately.

Transients in LO radiation and reflection

One of the major annoyances with direct conversion receivers is microphonic clicks and pops when anything in the system experiences a mechanical change. **Figure 8.10** shows a mixer and LO system connected to a high-gain audio frequency IF amplifier and a load with some arbitrary reflection coefficient. As an example, suppose that the mixer is a Mini-Circuits TUF-1 and the LO is at 50 MHz. The data sheet shows 57 dB of LO to RF port isolation in this mixer at 50 MHz. With a +7 dBm LO, −50 dBm of LO power leaves the RF port of the mixer and is reflected from the load connected to the RF port. Let's pick an arbitrary reflection coefficient, say 0.2 at an angle of 45 degrees, for the load. The magnitude of the reflection coefficient will stay the same, but the angle will change as we vary the length of 50-Ω transmission line connecting the mixer to the load. −50 dBm in a 50-Ω system is 1 mV peak. The magnitude of the reflection is (0.2)x1 mV or 200 µV. The 200-µV signal reflected from the load arrives at the mixer, and with 6-dB conversion loss and the appropriate phase, becomes a 100-µV dc voltage at the IF port of the mixer and input to the audio amplifier. This voltage is too small to seriously unbalance the mixer, and is blocked from the following audio amplifier by the series input capacitor. However, if the connection to the load is broken, for example, by disconnecting the BNC connector, the reflection coefficient jumps from 0.2 at 45 degrees to 1.0 at some other angle. The signal at the RF port of the mixer jumps from 200 µV at some phase to 1 mV at some other phase. At the IF port, the signal jumps from 100 µV dc to 500 µV dc. The "before" and "after" voltages are both dc, but the jump between them is a transient, and is amplified by the audio amplifier. The output of the audio amplifier with a short transient into the input dc blocking capacitor is the impulse response of the amplifier. (If we recorded the shape of the amplifier output pulse on a digital oscilloscope, we could then perform an FFT and see the frequency response of the amplifier.) 400 µV is a big signal, and probably drives the amplifier into saturation. The output is a very loud pop in the headphones. The level of LO isolation in a direct conversion receiver can be quickly judged by simply disconnecting the antenna while listening. A loud pop indicates poor LO isolation.

As shown in equation **Eq 8.1**, the dc output of the mixer depends not only on the level of the LO signal at the RF port, but also on its phase ϕ. An abrupt change in phase with no change in reflection coefficient magnitude will also induce a pop in the headphones.

Mixer LO port to RF port isolation is only one way for LO to leak out of the system and return to the RF port. Any leakage from the LO compartment results in a signal that may be picked up by the antenna. Often a direct conversion receiver that works exceptionally well in the lab when connected to signal generators exhibits all manner of peculiar behavior when connected to an antenna. As long as the LO leakage is small and doesn't change with time, there will be no observable effects. If the LO leakage changes suddenly, however, there will be an audible response. A loose screw in a metal radio cabinet can cause a scratching sound when the radio is tuned, by changing the amount of LO that leaks out of the case and is picked up by the antenna. Direct conversion receivers that work well when first packaged in a shiny new aluminum enclosure often become microphonic as they age and the mating surfaces corrode. Direct conversion receivers soldered up in boxes made from copper-clad PC board age more gracefully.

Doppler Effects

Since direct conversion receivers can detect differences in the phase of a reflection, they are very sensitive to reflections from moving objects. Doppler becomes most important when the motion is fast enough that the Doppler modulation on the radiated LO signal is in the audio amplifier passband (**Fig 8.11**). The maximum Doppler shift for a signal radiated from point A, reflected from a moving object at point B, and received again back at point A is:

Doppler Frequency = $2 V_o/\lambda$ **Eq 8.2**

At 40 m, an airliner (**Fig 8.12**) passing

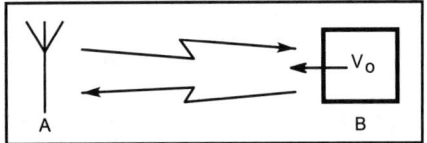

Fig 8.11—An illustration of RF Doppler.

directly overhead at 500 miles per hour (220 m/s) would induce a Doppler shift of 2×220/40 = 11 Hz. Airliners don't normally fly that fast when they are close to the ground, and 11 Hz is well below the audio range of interest, so we can ignore Doppler effects at HF. At 2 m, the Doppler shift from a 500 MPH airliner is 220 Hz, but airplanes flying that fast are normally a long way from the antenna. At microwaves, however, the story is entirely different. A 10368 MHz direct conversion CW receiver with LO leakage can detect all kinds of moving objects. With 3 cm wavelength, the Doppler shift from the airliner becomes 2 × 220/0.03 = 14.7 kHz which is at the top of the audible range. Cars at 50 MPH, however, have 1.47 kHz echoes, right in the middle of the audio passband for a conventional receiver. An audio phase-locked loop to recover the weak echo and an audio frequency counter can be used to remotely measure the speed of automobiles at ranges out to a mile or so, with very little radiated LO power. The direct conversion microwave receiver is sensitive not only to constant motion, but to vibration as well. Above 1 GHz extra care should be taken to make antennas for direct conversion receivers mechanically rigid. Some types of antennas, like horns, are less susceptible to reflecting surface vibrations than dish antennas, and Yagi antennas with mechanically resonant elements will induce spectral lines in the receiver audio output that can be seen using an audio FFT analyzer.

It is a useful exercise to estimate how far away objects can be and still produce Doppler effects in a receiver. Assume we have a 2-m receiver with very poor LO isolation, radiating 0 dBm from the antenna. Radiated power density (in watts per square meter) falls off as the surface of an expanding sphere:

$$\text{Power Density} \left(\text{watts/meter}^2\right) = \frac{P_o}{4 \pi R^2}$$

Eq 8.3

where P_o is the total radiated power and R is the distance between the source and the power detector

At 1 km, the power density is about 10^{-10} watts/m². Suppose this radiated LO energy bounces off of an airliner 1 km away with an effective radar cross section of 100 m². 10^{-8} watts will be bounced off the airliner. The spherically expanding scattered wave will have a power density of about 10^{-15} watts/m² after traveling the 1-km distance back to the receiver. A 2-m dipole has an effective capture area of about ½ m², so the signal bounced off the airliner is about 5 x 10^{-16} watts, or –123 dBm at the receiver antenna terminals. This is about 10 dB above the noise floor of a typical SSB receiver.

A more typical receiver will have much lower LO radiation, but moving objects within 10 meters of the antenna often result in a detectable output in the antenna. A half-wave dipole with a toggle switch in the middle is a useful VHF direct conversion receiver diagnostic tool. If you can hear the switch click in the headphones, you are detecting LO radiation.

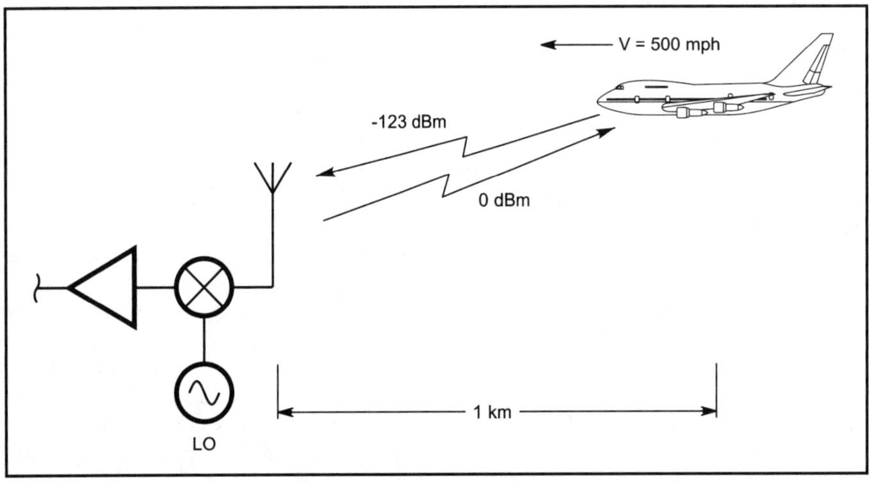

Tunable or Common Mode Hum

One of the direct conversion receiver peculiarities that puzzled early workers is the phenomenon of tunable hum. Receivers would have a particularly ragged sounding ac line noise hum that varied with changes in receiver tuning. This hum was particularly annoying in receivers that used a single high-Q tuned circuit at the RF port of the mixer—the common form of early direct conversion receiver. There were numerous theories for tunable hum— a few of them humorous in hindsight. In typical amateur fashion, lore developed that offered a set of fixes for tunable hum, including using an outdoor balanced antenna, using ferrite beads on the power supply leads, and using a battery power supply.

There is a difference between wisdom (don't eat raw pork) and understanding (Wow! Look what we see under the microscope!). Wisdom comes from experience, and understanding comes from study. For practical people like radio amateurs, wisdom usually comes long before complete understanding. Unfortunately, with the

Fig 8.12—2-m radiation from an airplane 1 km away.

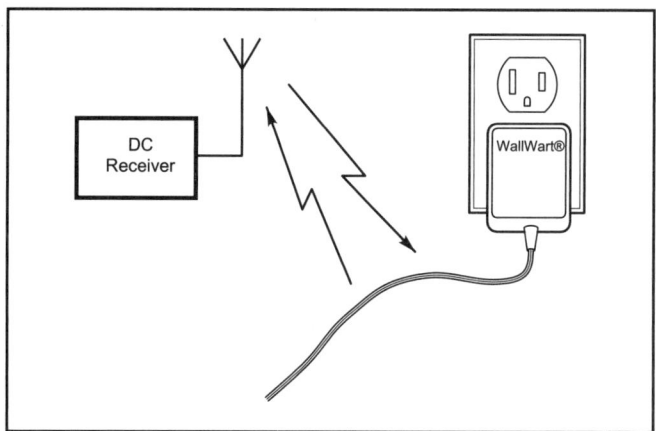

Fig 8.13—A tunable hum experiment.

Fig 8.15—The spectrum of a re-radiated LO.

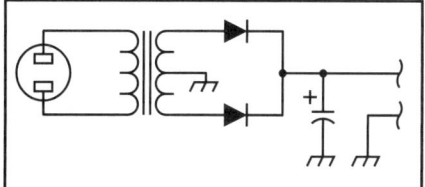

Fig 8.14—A power supply schematic.

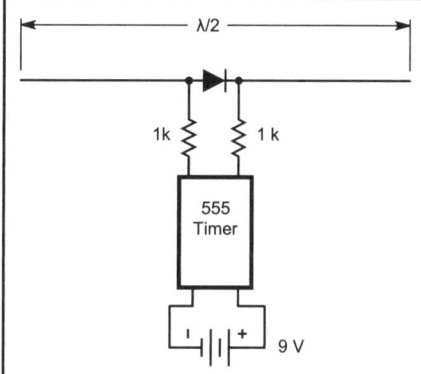

Fig 8.16—A hum probe.

proliferation of computer design, we are entering an age where folks are reluctant to do anything that can't be modeled mathematically and simulated. It is a good thing our ancestors weren't saddled with such nonsense, or they would have continued sticking their hands in the fire until medical science told them to stop. On the other hand, it is understanding that permits us to push the state of the art.

We now understand tunable hum well enough to dispense with the ferrite beads on battery power supplies and use indoor antennas on direct conversion receivers if we must, but much of old lore is still good. Battery supplies and a full-size outdoor antenna are recommended for reasons other than hum elimination.

Fig 8.13 shows a typical tunable hum experiment. The direct conversion receiver is connected to an antenna directly on the back panel. Right next to the antenna is a power cord going to a plug-in dc power supply. The power supply cord is a parasitic element of the antenna system. The power supply schematic is shown in **Fig 8.14**. Note that the power supply schematic is almost identical to the diode balanced modulator in the previous chapter. The modulating frequency is 120 Hz, due to the full-wave rectifier. The LO is picked up from the antenna wire, and then re-radiated with the 120-Hz sidebands. This wouldn't be much more than an annoyance, except that the 120-Hz modulating waveform is very rich in harmonics. The spectrum of a typical re-radiated LO signal is shown in **Fig 8.15**. The LO signal itself is at dc, and doesn't make it through the audio amplifier (although it may unbalance the mixer—increasing the strength of the radiated LO), but the sidebands are recovered by the mixer, and particularly the higher harmonics at 240 Hz, 300 Hz, 420 Hz etc. are subject to the full gain of the audio amplifier.

This explains the hum, and the harmonic content explains the raunchy sound, but why is it tunable? Refer again to equation **Eq 8.1**. The IF output of the mixer is a function not only of the amplitude of the signal at the RF port, but the phase ϕ. In fact, if the phase of the LO signal at the RF port is exactly 90° different from the LO drive, there is no detection of the sidebands at all. With a sharp single tuned circuit on the RF port, the phase varies more rapidly than the amplitude response as the tuning moves through resonance. At resonance, the phase shift through the tuned circuit will be zero, but off resonance the phase will smoothly tune from +90° to –90°. If there is some other phase shift path from the LO to the RF port of the mixer (there usually is), then at some point in the RF tuning, the hum will drop into the noise floor. Often the hum is eliminated at a point in the tuning where the sensitivity has been reduced to an unacceptable level.

It is interesting to observe that tunable hum is absent from *image-reject* direct conversion receivers. Common mode hum may still be present, but it is not tunable. An image-reject direct conversion receiver has two mixers with LO (or RF) ports 90° out of phase. After some baseband phase shifting, the IF outputs of these two mixers are added. If one mixer has zero common-mode hum, the other will have maximum hum. The sum will then have constant common-mode hum, regardless of any phase shifts in space or in the receiver RF path. Experimenters with image-reject direct conversion receivers who break the I and Q signal paths and listen to each channel separately often complain that "one channel has a lot of hum, but the other is fine" and try to eliminate the hum in the "bad channel" with improved bypassing and power supply decoupling, which is, of course, ineffective.

It is interesting to study receiver LO leakage with a "common-mode hum probe" consisting of an antenna, diode modulator, and modulating signal source. A modulating tone should be chosen that is not harmonically related to 60 Hz. At HF and VHF, a small loop antenna with a diode and a 555 timer works well. At microwaves, a dipole consisting of a diode and its leads serves well. **Fig 8.16** illustrates the circuit. If the probes are small enough, they may be used to find the LO leaks in a direct conversion system.

Eliminating LO Radiation Effects

Understanding common mode hum and

other LO radiation symptoms allows us to eliminate them. If we do not permit any LO signal to leak out into the RF environment around the antenna, then common mode hum cannot occur. There are several primary leaks that we must consider:

1. LO coupling through the mixer to the RF port and through the RF circuitry onto the antenna.

2. LO energy radiating from LO components on the circuit board.

3. LO energy on wires connected to the radio cabinet (**Fig 8.17**).

Reducing the amount of LO energy at the antenna connector involves mixer LO to RF port isolation, eliminating coupling from the LO components into the RF stages, and the reverse isolation of any amplifiers in the system. There are big differences in the LO to RF isolation of various mixers. Some unbalanced mixers have no LO to RF isolation at all. The mixers most suitable for direct conversion receivers are balanced. At 7 MHz, the LO to RF isolation of a TUF-1 mixer is more than 70 dB and the SBL-1 is around 65 dB. This is sufficient for acceptable direct conversion receiver performance with no RF amplifier. At 144 MHz, the TUF-1 LO to RF isolation has dropped to 50 dB and the SBL-1 has dropped to 45 dB. This is low enough to cause problems.

Additional isolation can be obtained by using an RF amplifier ahead of the mixer, as recommended in the excellent papers by Nick Hamilton.[16] This is good practice even at lower HF bands where an RF amplifier may not be needed for noise figure. It is important to note that reverse isolation varies widely between amplifier types. A Mini-Circuits MAR-2 with 12.5-dB gain has only 18-dB reverse isolation at 144 MHz, while a grounded gate U310 with 10-dB gain has 28-dB measured reverse isolation. A cascaded pair of grounded gate U310s on the input to a direct-conversion 2 m receiver can effectively eliminate LO energy coupled through the mixer through the RF amplifiers onto the antenna. At microwaves the differences can be even larger. The 12.5-dB gain MAR-2 has reverse isolation of 17 dB at 1296 MHz, while the 16-dB gain TriQuint 9132 has more than 45-dB reverse isolation.

Even if the mixer has good LO to RF port isolation and the RF amplifier has good reverse isolation, the LO can still couple onto the antenna connector if there is no shielding inside the radio case. The antenna connector should connect to the RF amplifier input with small coax, properly grounded at each end.

All of the components in the LO circuit can radiate LO energy. To gain some intuition for how effective components are as antennas, compare their size in wavelengths to the size of a mobile whip antenna on 80 meters. A typical mobile whip might be two meters tall, 0.025 wavelengths at 80 m. In a 40-m VFO, the individual components are very small in wavelengths, and would therefore make poor radiators. In a 2-m VFO, 0.025 wavelengths is only 0.05 meters, or about two inches. A two-inch long PC board trace could be as effective a radiator as an 80-meter mobile whip. Small magnetic antennas can be very effective. Think about the size in wavelengths of an AM radio ferrite loopstick. Small tuning coils and RF chokes are often the most significant sources of LO energy inside a radio cabinet. The use of shielded coils and toroids is recommended for all direct conversion applications. The most effective way to prevent LO radiation from components is to enclose the entire LO in a shielded enclosure. Small tin cans work well, and can be easily soldered in place. A PC board enclosure with soldered seams is superior to a machined aluminum box held together with screws.

It is meaningless to enclose the LO if there are holes in the enclosure with wires going in and out. The wire will pick up energy inside the box and conduct it outside, where it can be radiated or conducted onto other wiring. The LO signal itself should come out through coax or a coax connector, and dc wiring should use effective feedthrough capacitors and decoupling networks. The most careful VFO compartment shielding can be rendered useless if the VFO capacitor shaft goes through a hole in the compartment wall. Capacitor shafts can be significant radiators if they are not grounded to the wall near the entry hole (**Fig 8.18**). At VHF, a few inches of tuning control shaft through the radio panel can couple LO energy to the outside world. A grounded panel bearing is one option, but the common 1/4-inch sleeve types don't provide reliable grounding, and will result in common mode scratches as the radio is tuned. A better solution is to use a grounded sleeve bearing with a 1/4-inch non-metallic rod for the tuning shaft, and a shaft coupler to the capacitor shaft inside the sealed VFO compartment.

The same rules for keeping LO energy from radiating to the inside of the radio box and being picked up by the RF circuitry apply to keeping LO energy from radiating to the outside world on power supply,

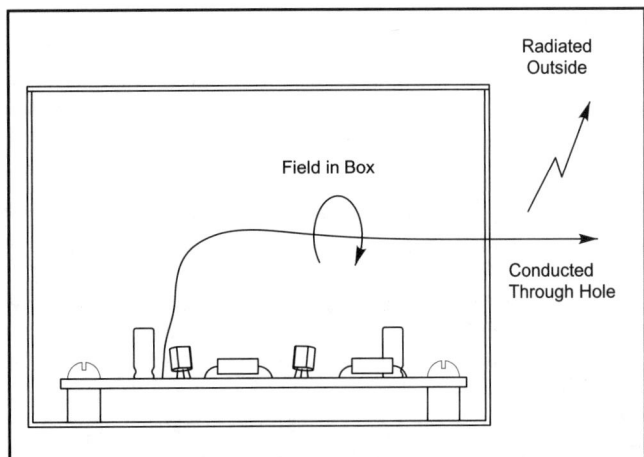

Fig 8.17—A wire pickup in an LO box.

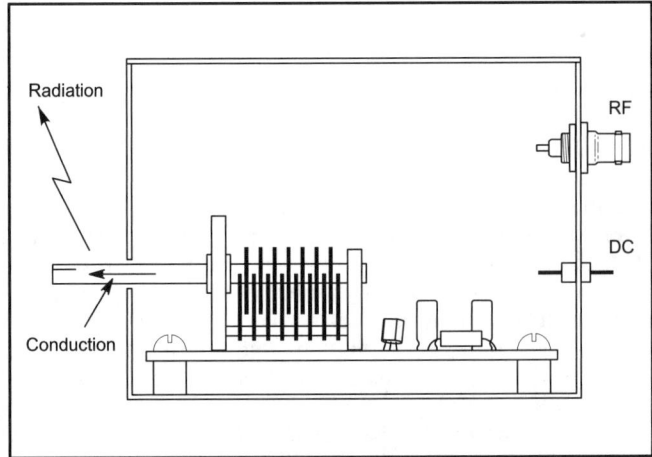

Fig 8.18—Capacitor shaft pickup in an LO box.

speaker, microphone and key leads. All dc and audio leads should be properly decoupled for RF. This can be a problem for speaker leads, since bypassing them to the chassis of a direct conversion receiver with high audio gain will introduce ground loop feedback. One way around the problem is to use a separate powered speaker, preferably with internal batteries, plugged into the headphone jack of the receiver.

A conservatively designed and built direct conversion receiver is double shielded, with internal enclosures around the VFO and RF circuitry, often a small steel or mumetal enclosure to reduce ac hum pickup around the audio preamp inductors, and an outer shielded enclosure. All RF connections are made using shielded connectors, preferably BNC at HF and SMA at VHF and up, and all dc and audio connections to the outside world properly bypassed. Care is also exercised so that mechanical connections like volume controls and the main tuning knob shaft do not conduct signals into or out of the receiver enclosure.

One technique that has been part of the lore for years is using a VFO followed by a frequency doubler. A balanced mixer is insensitive to energy at 1/2 or twice the LO frequency. The expression below shows multiplication of a low level 1/2 frequency signal with the LO. There is no output at dc.

$$a \cos [2\pi(2f_o)t + \phi] \cos 2\pi f_o t =$$
$$a/2 \cos [2\pi(3f_o)t + \phi] + a/2 \cos [2\pi f_o t + \phi]$$
Eq 8.4

Care must be taken to avoid radiating the frequency doubled signal, but a passive doubler right at the mixer port could be used. Then only the actual doubler circuitry must be shielded, and there are not even any dc power leads connected to stages carrying the on-frequency LO signal. In particular, the VFO shaft and capacitor body only have half-frequency energy, and may be left unshielded. The 40-m sleeping bag radio described later was built to test frequency doubling, and there is no separate shielding around the half-frequency VFO. As a fringe benefit, a CW transmitter using a frequency doubled VFO is much less susceptible to chirp than one with the VFO operating directly on frequency.

It might seem that it takes an awful lot of extra effort to build a good direct conversion receiver than to build a good superhet. This is not true. A *good* superhet requires exactly the same construction. Superhet

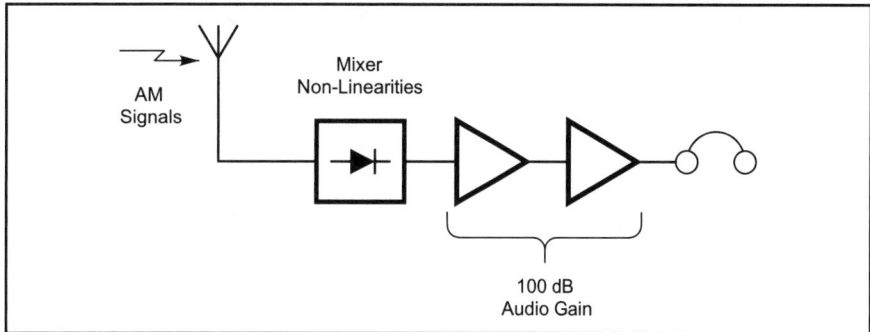

Fig 8.19—AM demodulator.

receivers with poor shielding have a different set of problems, like multiple internally generated spurious responses, poor image and IF rejection, and responses to strong out-of-band signals near harmonics of the oscillators. Good mechanical construction, shielding of individual stages, and proper bypassing and decoupling of power supply and audio leads makes a tremendous improvement in performance, whether the receiver is a conventional superhet, direct conversion, or a spectrum analyzer. Good mechanical construction is too expensive for mass produced or even kit radios, but is just a matter of planning, care, some worthwhile mechanical skills, and time for a designer-builder of a single radio. This is one area where a designer-builder can far exceed the mechanical quality and electrical integrity of a mass-produced receiver built under severe time and budget constraints, for example, a Collins 75S3C.

Adaptive Mixer Balance

Some balanced mixer types may be easily adjusted for LO radiation. The familiar "carrier balance" resistor adjustment in Gilbert Cell mixers is an example. It is possible, in concept at least, to measure the instantaneous LO level at the receiver antenna terminal, and vary a set of voltages in the mixer to force the LO leakage to zero. This technique permits eliminating not only stray LO energy from inside the mixer, but energy that arrives via other paths by canceling it with an equal-and-opposite mixer leakage signal. The mixer adjustment may be done once, during alignment or each time the radio is powered up, and then the balance adjustment locked in for normal operation.

There are sobering cautions that need to be mentioned. If the balance adjustment is done continually in real time, it must be recognized that adaptively nulling a signal by adding a sine-wave adjusted for precise amplitude and opposite phase is a form of phase-locked-loop. Since both phase and amplitude are variables, loop stability analysis becomes complicated. Designing an LO suppression loop that offers real benefit and remains stable over a wide range of operating conditions is an ambitious exercise. Another difficulty is that intentionally unbalancing the mixer to obtain a precise amplitude and phase carrier signal will null the LO at the expense of mixer 2nd order distortion performance.

AM Demodulation

A common problem with direct conversion receivers is demodulation of AM signals anywhere in the RF passband of the receiver. This is most often observed on 40 m when foreign broadcast signals are very strong. **Fig 8.19** illustrates the problem. Any mechanism in the mixer that produces a dc output at the mixer IF port from a signal at the RF port will result in the envelope of an AM signal appearing as weak audio, right at the input to a 100-dB gain audio amplifier. DC outputs occur when a mixer has second order distortion. Second order distortion is common when balanced mixers become unbalanced. Since the usual way that balanced mixers unbalance is the presence of LO signal at the mixer RF port, it is evident that AM demodulation is a symptom of both poor LO to RF isolation and high audio gain. Improving the shielding around the VFO, and LO to RF isolation often improve a receiver's immunity to AM demodulation. Receivers that use VFOs operating at half (or twice) the signal frequency usually have better AM rejection than receivers with fundamental VFOs, due to improved LO to RF isolation.

8.4 MIXERS FOR DIRECT CONVERSION RECEIVERS

The general properties of mixers are covered in a separate chapter, but the front-end of a direct conversion receiver is a unique application that puts some different demands on the mixer. To reduce LO radiation to an acceptable level, LO port to RF port isolation is needed. This usually requires a balanced mixer, but some other topologies are promising. The anti-parallel diode pair driven by a 1/2 frequency LO has been reported to work well, but has limited dynamic range and critical LO drive level requirements. Shunt FETs in switch mode have built-in LO to RF isolation. A number of experimenters have reported good success with different configurations of series FET switches using CMOS parts for several decades. The most common direct conversion mixers are Gilbert Cells like the NE602 and LM1496, and diode rings, both homebrew and commercial. Gilbert Cells have usually been used for low-cost-low-performance applications, but they should not be ruled out for higher performance receivers.

The important specifications for a direct conversion front-end mixer are noise figure (particularly 1/f noise figure when used with an audio IF), two-tone third-order dynamic range, 2nd order dynamic range, and LO to RF port isolation. Conversion gain or loss is less important, as it can be made up with gain elsewhere, and can not make up for poor noise figure.

Mixer recommendations

For the simplest direct conversion receivers, Gilbert Cells offer good performance at low current. The gain of a Gilbert Cell does not enhance receiver performance, since it occurs before any effective channel selectivity, but it does reduce the total receiver parts count. For some applications—carrying a rig into the mountains for a casual non-contest weekend backpacking trip, for example—the receiver is far less likely to fail from overload than from dead batteries. For such applications, "performance" takes on a different meaning, and a receiver that draws 5 mA outperforms one that draws 50 mA. For home station use or any kind of contest environment, a receiver with poor dynamic range can be as useless as one with dead batteries, and far more frustrating. For such applications, diode rings are recommended. For the designer builder, they have the advantage of a wealth of applications information and a published schematic.

Passive FET mixers in various configurations have dynamic range and noise advantages over both Gilbert Cells and diode rings. Considerably less has been published about passive FET mixers, although they are standard in cellular telephone handsets. This is an important area for amateur experimentation. Experiments are encouraged using both integrated quad analog switches and matched FETs on a single die in small multi-pin packages. Since the LO drive to a passive FET mixer goes to the high-impedance FET gate, little LO drive power is needed. The passive FET itself doesn't have a power supply. Thus passive FET mixers for direct conversion receivers offer the potential for the highest performance at the lowest operating current of any mixer type.

Direct Conversion Noise Figure

The noise figure of a direct conversion receiver mixer is generally different than the noise figure of the same mixer used in a superhet application, because of 1/f noise. Mixer noise figure does not have a neat and tidy definition, and mixer 1/f noise is even less well understood. Because of 1/f noise, diode ring mixers have noise figures in direct conversion receiver applications that range from within 1dB of their conversion loss to 15 or 20 dB worse. The increased noise figure is a result of excess noise at the IF port when the mixer is driven by the LO with the RF port terminated in a room temperature 50-Ω load. The noise spectrum is not necessarily a smooth 1/f curve, so merely observing the shape of the noise spectrum across a restricted audio passband is not enough to identify 1/f noise. Mixer noise figure is further complicated by the presence of noise on desired and image frequencies, noise in the bands around the harmonics of the LO, and the fact that the different contributions to mixer noise figure may be partially correlated. Rather than attempting to precisely define direct conversion mixer noise figure, this text will present a few measurements that provide some insight into noise in receiver systems, and will at least allow comparisons between different mixers and direct conversion receiver front-ends.

The first measurement is the noise figure of the audio amplifier itself. We have made this measurement with a hot-cold noise source. The audio amplifier is run at full gain in an environment with no hum or other noise pickup. The input to the audio amplifier is switched between two 50-Ω resistors, one at room temperature and the other at 77K. It is very important to measure the resistance of the cold resistor, to make sure it is still 50 Ω. Most resistors change value when the temperature drops that low. A series or parallel combination can be experimentally determined that provides a cold 50-Ω resistor. The output of the audio amplifier is connected to an averaging true RMS voltmeter reading in dB, and also a speaker or headphones. It is useful to listen while making the measurements, because the difference between hot and cold resistor noise can be heard in the headphones, and the measurements will be corrupted by any extraneous interference pickup, which can also be heard on the headphones. **Fig 8.20** gives noise figure as a function of the difference between the noise output from the hot and cold resistors in dB. The noise figure of the grounded base audio preamplifiers with diplexers in the receiver circuits in this text ranges from 5 to 7 dB.

The second step in the measurement process is to measure the conversion loss of the mixer. This can be done with a known RF signal at the RF port, a low-pass filter and 50-Ω termination on the IF port, and an RMS voltmeter across the 50-Ω resistor.

The last step in the measurement is to measure the excess IF noise when the mixer is connected to the audio amplifier and the LO is turned on. The input to the audio amplifier is switched between a room temperature resistor and the mixer, with LO drive and the RF port terminated in a room temperature 50-Ω load. At 14 MHz, a small sample of TUF-1 mixers produced between 1 and 6 dB more noise output than the 50-Ω room temperature termination. Two homebrew diode ring mixers using hand-wound toroids and 1N4148 diodes had less than 1-dB excess noise. A small sample of TUF-5 mixers operated at 1296 MHz and ADE-35 mixers at 2304 MHz had more than 10-dB excess noise. Special low-1/f noise diodes are used in 10-GHz direct conversion receivers for Doppler Radar applications.

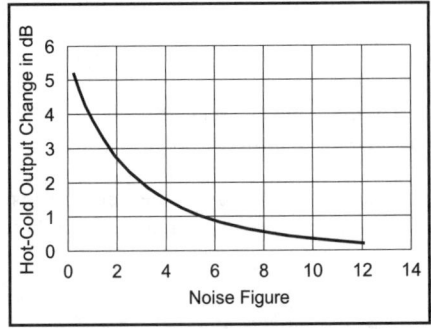

Fig 8.20—Hot-cold resistor noise figure differential.

This is a very small data set, and it is unwise to draw firm conclusions based on this limited information. More measurements are needed.

When the excess noise is low, a reasonable approximation to direct conversion receiver noise figure is just the baseband amplifier noise figure plus the mixer conversion loss. When excess mixer noise is present, the mixer loss and noise tend to dominate receiver noise figure, and baseband amplifier noise figure is less important. One experiment that may be done on the bench is to add attenuation between the mixer and baseband amplifier while observing receiver sensitivity. A 3-dB 50-Ω attenuator will drop the desired signals by about 3 dB, but it may also drop the receiver noise floor by about 3 dB, leaving the signal-to-noise ratio unchanged. Signals do not drop by precisely 3 dB, because the mixer impedance and the baseband amplifier input impedance are not exactly 50 Ω.

One way around the mixer excess noise uncertainty is to use a low-noise RF amplifier with enough gain to define the system noise figure. In this case it may be beneficial to include a resistive attenuator on the mixer output to optimize mixer dynamic range.

When used ahead of a DSB direct conversion receiver, a low-noise RF amplifier will have equal noise output on the desired and image bands. The image noise will reduce receiver output signal-to-noise ratio by 3 dB. Image noise may be suppressed by a narrow filter after the RF amplifier (practical for fixed-frequency applications), or by phasing, discussed in the following chapter.

Mixers with conversion gain, for example the Gilbert Cells used in LM1496 and NE602 integrated circuits, reduce the need for low-noise audio gain. The NE602 has low noise figure, which makes it attractive for simple receivers without RF amplification. The LM1496, biased for improved mixer linearity, is a better choice when an RF amplifier is used. In DSB direct conversion receiver applications with no provisions for suppressing image noise, each of these has the same 3-dB image noise penalty.

Based on these limited measurements and theory, a few guidelines for direct conversion receivers may be suggested. A homebrew diode ring with common 1N4148 silicon switching diodes, as used in Roy Lewallen's "Optimized QRP Transceiver"[17], with low-loss RF input circuitry and a grounded base audio amplifier, will provide an effective receiver noise figure around 10 dB, which is usually better than is needed at 7 MHz. Because the LO to RF isolation of homebrew mixers may not be as good as commercial packaged mixers using matched quads of Schottky diodes, the use of an RF amplifier ahead of the mixer is recommended. This will tend to negate any 1/f noise advantage of the homebrew switching diode mixer. In our HF designs, we tend to use small commercial packaged mixers, and about 10 dB of high reverse-isolation RF gain. This results in receivers that have noise figures in the 10-dB range, have very low LO radiation, and work well with common commercial packaged diode ring mixers. At VHF, we usually use about 20 dB of RF gain, and phasing to suppress image noise.

8.5 A MODULAR DIRECT CONVERSION RECEIVER

The "High Performance Direct Conversion Receiver" published in August 1992 QST[18] is a good benchmark. The ten-year-old design stands up well against more recent work, and the description is recommended reading. The circuitry presented here takes a slightly different approach, and takes advantage of a few improvements in our understanding during the past decade. A basic 40-m circuit is shown, but few changes are needed for operation on other bands.

The block diagram is shown in **Fig 8.21** and the schematic in **Fig 8.22**. The antenna is connected to a grounded-gate FET RF low-noise amplifier. The mixer is a Mini-Circuits TUF-3, with an audio diplexer and low-noise headphone amplifier. The VXO circuit provides clean sinewave +7 dBm drive to the mixer. For speaker output, a battery powered external speaker from RadioShack, or an amplified computer speaker is recommended.

RF Low-Noise Amplifier

The receiver gain distribution was designed for approximately 10-dB of RF gain ahead of the mixer. RF gain ahead of a diode ring mixer is not normally needed for receiver sensitivity below 10 MHz, but in a direct conversion application, there are other benefits to using an RF preamp. First, with RF gain up front, there is less need to design for low loss through the mixer IF termination and diplexer network. This permits the baseband circuitry to be optimized for selectivity and proper termination of both the mixer and diplexer network. Second, a grounded-gate FET amplifier typically has over 40 dB of reverse isolation, which adds directly to the LO to RF isolation of the mixer, and helps reduce the amount of LO radiation from the antenna. Third, with a buffer amplifier between the antenna connection and the mixer RF port, the mixer environment does not change when the antenna moves in the breeze. Fourth, Direct Conversion Receivers need good low-pass filters on the inputs, and the low-pass matching networks in and out of the FET provide all the attenuation needed. Finally, the simple mute switch turns the RF low-noise-amplifier into a strong 40-dB attenuator, which prevents any strong signals (for example from a companion transmitter) from arriving at the mixer diodes.

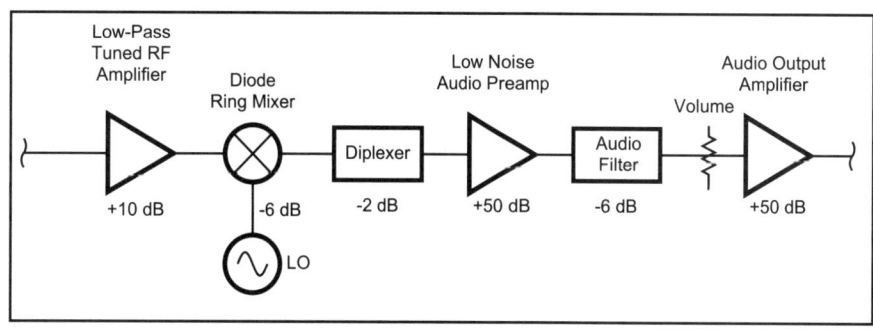

Fig 8.21—A modular receiver block diagram

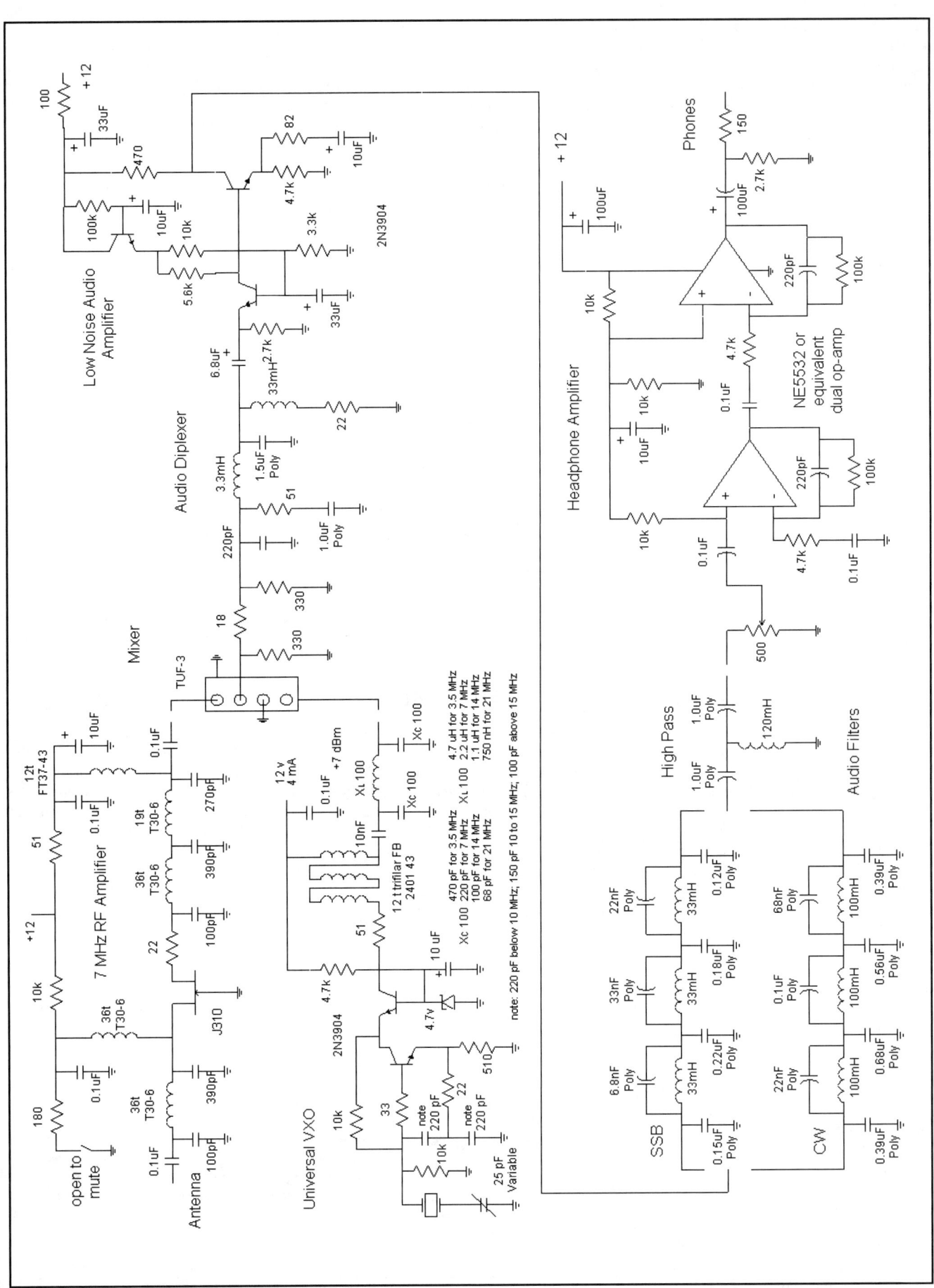

Fig 8.22—Modular receiver schematic.

Audio Diplexer

The diplexer network is designed to provide good selectivity before any wideband audio gain. This greatly improves the receiver close-in dynamic range, and permits the use of a grounded-base audio LNA operating at low current (0.5 mA) to set the impedance to 50 Ω. This audio diplexer is a little more selective than the ones described in the phasing chapter, because there is no need to precisely match amplitude and phase between two channels.

Audio low-noise amplifier

There are many audio low-noise amplifiers that will work in direct conversion receivers, but this one works well and has been widely duplicated for several decades. It has no flaws that impair performance in this application, so the design effort was focused elsewhere.

Filters

Passive audio filters work well, draw no current, and use inexpensive components available from several sources. The SSB and CW bandwidth filters shown are old favorites.

Headphone amplifier

The headphone amplifier provides audio gain to boost the signals from the low-levels in the signal processing components up to comfortable listening volume. This is a standard design, very similar to the headphone amplifier used in the Binaural Receiver[19] published in March '99 *QST*.

VXO

The VXO circuit is another old favorite, evolved over many years from a circuit published by Joe Reisert[20] as a frequency standard. There are a number of subtleties, including stiff regulation of the voltage on all three terminals of the oscillator transistor and the use of a Zener diode operated in the 4.7-V zero-temperature-coefficient sweet-spot. This VXO circuit tunes over about 5 kHz at 7 MHz, provides +7 dBm output and drifts a few Hz at turn-on.

Construction

The receiver was built on separate pieces of unetched copper-clad circuit board. The RF amplifier is on one piece, the VXO on a second piece, and the mixer and audio amplifier on a third piece. The audio filters are on separate pieces. There are a number of reasons for building the receiver on separate boards. The first is entirely practical—each piece is an evening project than can be built and tested as a stand-alone module. The second consideration is equally important: the RF amplifier is good for only one band; the VXO can be easily modified for different HF frequencies; and the mixer-audio board can be used on any frequency from 50 kHz through 250 MHz. By making the pieces separate, any of them may be replaced to put the receiver on a different frequency, or borrowed for a different project.

Generally speaking, receiver circuits built prototype-style on separate pieces of unetched copper-clad circuit board work better than PC board circuits. This is because the unetched copper-clad board permits both the short ground leads required by RF circuitry and the single-point grounding required by low-frequency high gain amplifiers. Receivers that must be mass-produced using PC boards often require many PC layout revisions to overcome the problems that arise when the prototype circuits are transferred to PC board construction.

The more components a receiver module has, the more practical it is to spend time developing a PC board design. For simple circuitry like the modules presented here, it is often more practical to use prototype construction, and avoid the headaches associated with PC board ground faults.

Applications

The modular high-performance direct conversion receiver presented here works equally well connected to an antenna, or as part of a superhet receiver. The well-defined near 50-Ω input impedance to the RF preamp provides a good termination for simple crystal filters, and the VXO circuit is a good BFO with enough tuning range to cover both sidebands.

8.6 DC RECEIVER ADVANTAGES

For much of their history, direct conversion receivers have been viewed as an adequate, simple substitute for more serious receivers. It is time to redefine direct conversion as an alternative architecture that poses a unique set of problems, but also offers significant advantages. Some of the important advantages are:

1. Simplicity
2. Few spurious responses
3. High spurious-free dynamic range
4. Very low distortion of the desired signal
5. Frequency range independence
6. Compatibility with DSP-based receiver architectures
7. Compatibility with adaptive receivers and antennas

Simplicity is best illustrated by the circuit in Fig 8.7. Build it ugly style in a few hours the Thursday evening before Field Day or the November CW Sweepstakes, string up a temporary 40-m dipole Friday evening, and spend a few hours over the weekend listening. Simplicity is appealing. Much of this text is devoted to pushing the performance envelope for designer built radio equipment. Spending two years building a receiver system that offers an incremental performance improvement that must be measured to be perceived is an interesting activity, but with a serious flaw. Suppose the number to be exceeded is the magic "100-dB SSB Bandwidth Two-Tone Third Order Dynamic Range." Magic to whom? Certainly not my teen-age daughter! But she will spend a few minutes politely listening to CW on headphones connected to a hand full of parts with a 9-V battery and some wires going out into the trees in the back yard—and when I ask her if she hears the kind of weak, warbly one and she says yes and then I tell her he's in St. Petersburg, Russia—her eyes light up. Now that's magic!

Superhets for SSB and CW have images, higher order undesired responses, and internally generated birdies. A direct conversion receiver with a low-pass filter between the antenna and mixer hears only signals within a few kHz of the LO. Period.

It is theoretically possible to design superhet receivers for arbitrarily good image and IF rejection, but in practice superhets must be designed to place images and IFs in parts of the spectrum with few strong signals. When image signals are 90-dB stronger than the desired signal, they will find a way into the receiver and cause problems. This severely constrains the choices of IF for frequency bands in heavily used portions of the spectrum. For example, what IF should be used

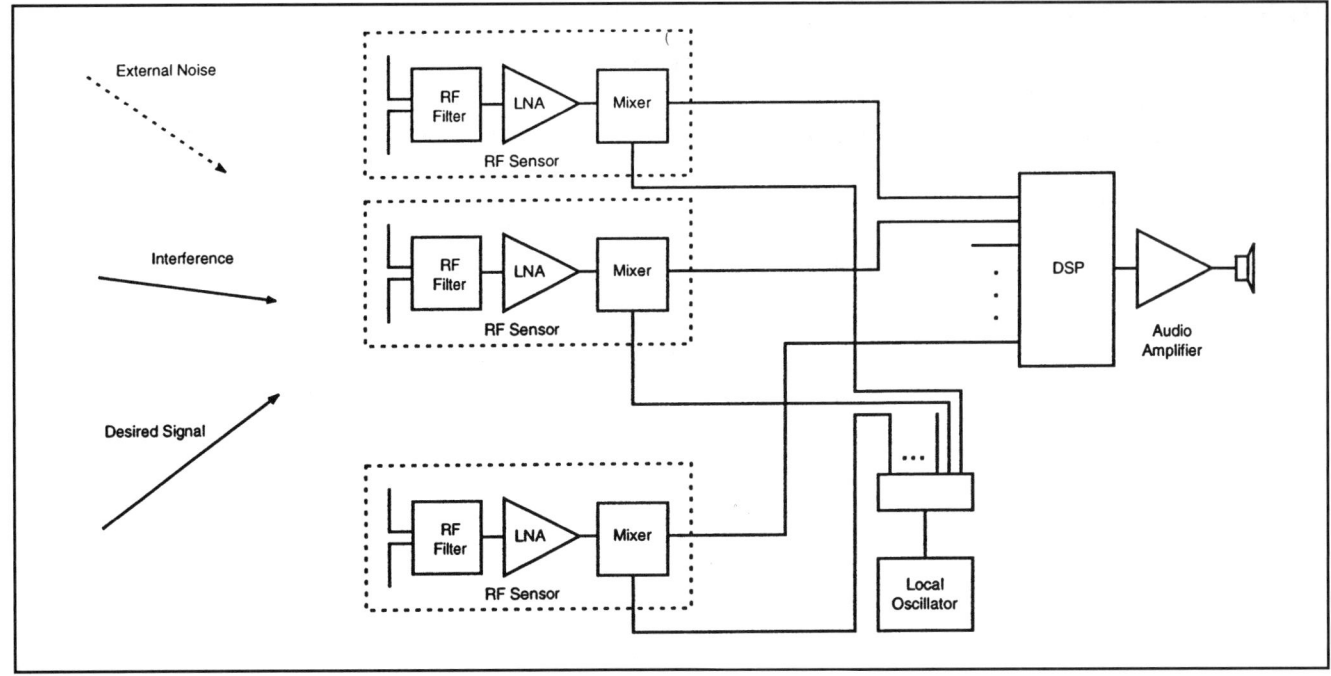

Fig 8.23—Advanced receiver architecture.

for a 144 to 148 MHz receiver? The industry standard IFs at 455 kHz, 10.7 MHz, and 21.4 MHz provide a selection of off-the-shelf filters. 455 kHz is too low for adequate image rejection. 10.7 is useful, but a bit low for providing good image rejection across a 4-MHz wide frequency range without retuning the RF amplifier. 21.4 MHz is attractive, except that with low-side injection, the image falls in the FM broadcast band, and with high side injection, the image is in TV channel 12. Direct conversion offers a technique for tuning across a wide frequency range and recovering 10-nV signals surrounded by 10-mV interfering signals. That is 120 dB of spurious-free dynamic range.

Because direct conversion receivers have only one frequency conversion stage, and it operates before significant receiver gain, mixer distortion does not significantly contribute to in-band intermodulation. The quality of the recovered audio is almost entirely determined by the distortion properties of the audio amplifier chain. Since audio engineers have spent decades reducing the distortion of high-gain audio amplifiers, simply following a diode-ring mixer with a low-noise preamplifier and high-fidelity audio amplifier will produce a receiver with significantly lower in-channel distortion than any commercial superhet. Audio engineers have also developed low-distortion gain control and gain compression techniques that operate strictly at audio, and that "audio AGC" technology is beginning to appear in amateur equipment.

The same block diagram works for direct conversion receivers whether the frequency of interest is 24 kHz or 24 GHz. A superhet designer will draw completely different block diagrams for a SSB receiver for those two frequencies. Furthermore, superhet frequency conversion plans must be designed with an understanding of the levels of all the potential sources of image, higher-order spurious responses, and birdies. A receiver optimized for 10 MHz might have a completely different frequency conversion plan than one optimized specifically for 14 MHz. For the amateur interested in the entire spectrum, the lessons learned and the time spent optimizing a 10-MHz direct conversion receiver apply just as well to a 2.4-GHz satellite receiver.

As DSP systems improve and become more widely used and understood, it becomes less and less attractive to compromise the signal with multiple frequency conversion, AGC, and crystal filter delay and ripple before it enters the DSP. Direct Conversion offers a way to simply translate a desired radio signal to the frequency range needed by the A to D converters ahead of a DSP engine (**Fig 8.23**). Soft-Radio advocates call this *Direct Sampling* and claim that there is no conventional radio at all—the computer is connected straight to the antenna. Such claims obscure the truth. Direct Sampling is just a different and convenient name for entirely conventional I and Q mixing, in the same sense that the term "wireless" allows people who have no understanding of radio to claim the title Wireless Expert. Such good natured competition between traditional radio designers and digital signal processing artists is a natural part of the evolution. Both camps need to realize that receivers of the future will use both skill sets. There is magic in simple radio circuits, but there is also magic in watching a signal below the noise level appear in a waterfall plot on a computer monitor.

Finally, in it's second hundred years, radio will experience significant changes. For six decades the usual way to collect and process HF and VHF signals has been a Yagi-Uda antenna with a single feed line connected to the back of a complex superhet receiver. Space diversity and adaptive antenna interference cancellation have been impractical because of the amount of hardware required and severe amplitude and phase matching constraints. The hardware problem is solved if each dipole antenna element has its own direct conversion down-converter, all of them driven by a single LO, and each connected to a separate input port of a computer sound card. The actual hardware is very simple, and with more than two dipoles, image-reject techniques can be combined with noise cancellation in the arrival-angle domain and adaptive CW interference cancellation in the frequency domain to produce an output signal-to-noise and interference ratio far better than the best conventional, single feed line system.

REFERENCES

1. W. Hayward and D. DeMaw, *Solid State Design for the Radio Amateur*, ARRL, 1986.
2. R. Lewallen, "An Optimized QRP Transceiver," *QST*, Aug, 1980, pp 14-19.
3. G. A. Breed, "A New Breed of Receiver," *QST*, Jan, 1988, pp 16-23.
4. R. Campbell, "Getting Started on the Microwave Bands," *QST*, Feb, 1992, pp 35-39.
5. R. Campbell, "No Tune Microwave Transceivers," *Proceedings of Microwave Update '92*, Rochester, NY, ARRL Publication Number 161, 1992, pp 41-54.
6. R. Campbell, "High Performance Direct Conversion Receivers," *QST*, Aug, 1992, pp 19-28.
7. R. Campbell, "High Performance Single-Signal Direct Conversion Receivers," *QST*, Jan, 1993, pp 32-40.
8. R. Campbell, "A Multimode Phasing Exciter for 1 to 500 MHz," *QST*, Apr, 1993, pp 27-31.
9. R. Campbell, "Single-Conversion Microwave SSB/CW Transceivers," *QST*, May, 1993, pp 29-34.
10. R. Campbell, "A Single Board No-Tune Transceiver for 1296 MHz," *Proceedings of Microwave Update '93*, Atlanta, GA, ARRL Publication Number 174, 1993, pp 17-38.
11. R. Campbell, "Subharmonic IF Receivers," reprinted from the *North Texas Microwave Society Feedpoint* in *Proceedings of Microwave Update '94*, Estes Park, CO, ARRL Publication Number 188, 1994, pp 225-232.
12. R. Campbell, "Simply Getting on the Air from DC to Daylight," *Proceedings of Microwave Update '94*, Estes Park, CO, ARRL Publication Number 188, 1994, pp 57-68.
13. R. Campbell, "A VHF SSB-CW Transceiver with VXO," *Proceedings of the 29th Conference of the Central States VHF Society*, Colorado Springs, CO, Jul, 1995, ARRL Publication Number 200, pp 94-106.
14. R. Campbell, "The Next Generation of No-Tune Transverters," *Proceedings of Microwave Update '95*, Arlington, TX, October, 1995, ARRL Publication Number 208, pp 1-22.
15. R. Campbell, "A Small High-Performance CW Transceiver," *QST*, Nov, 1995, pp 41-46.
16. N. Hamilton,"Improving Direct Conversion Receiver Design," *Radio Communications*, Apr, 1991.
17. R. Lewallen, "Optimized QRP Transceiver, *QST*, Aug, 1980, pp 14-19.
18. R. Campbell, "High Performance Direct Conversion Receiver," *QST*, Aug, 1992, pp 19-28.
19. R. Campbell, "A Binaural I-Q Receiver," *QST*, Mar, 1999, pp 44-48.
20. J. Reisert, "VHF/UHF Frequency Calibration," *Ham Radio*, Vol 17, Nr 10, Oct, 1984, pp 55-60.

CHAPTER 9

Phasing Receivers and Transmitters

9.1 BLOCK DIAGRAMS

The phasing method of single-sideband generation and reception has been discussed in the literature and incorporated in commercial products for over 50 years. The phasing method fell into disuse in amateur products from the late '60s through the '80s due to the popularity of transceivers built around a single IF crystal filter used for both sideband generation and receive selectivity. During this period, prices of old phasing transmitters dropped until they were only used on the air in modest stations scraped together on a budget, often by folks with no appreciation of the art of maintaining vintage radio gear. Sociology being what it is and amateurs being human, phasing transmitters were soon associated with poor signals, and their unfortunate operators were encouraged to upgrade or get off the air. Even scholarly authors during this period often used a little over-simplified mathematics to show that the phasing method was incapable of generating acceptable signals for the modern amateur bands.

How times have changed. During the '90s the vintage radio craze hit the amateur bands, and amateurs across the US began hearing signals from old Central Electronics transmitters, carefully restored, properly aligned, and conservatively operated. By comparison, the modern transceivers sounded thin and distorted. Modern radios have had to scramble to recapture the lost sound quality of the old rigs. Sociology still being what it is, there is now a market for low-distortion transmitters, and one amateur manufacturer has even introduced a full-sized transceiver with a Class A power amplifier. The lore has changed, and phasing transmitters and receivers now have the reputation for sounding better than conventional systems that use filters for opposite sideband suppression. As usual, careful study reveals that there is an element of truth in conventional wisdom, but that deeper understanding provides freedom from the bonds of lore.

Fig 9.1 is the block diagram of a conventional SSB exciter using a filter to remove the unwanted sideband. Since the filter passband frequency is fixed, the resulting SSB signal must be heterodyned to the desired final output frequency. Since it is difficult to build SSB bandwidth filters for frequencies above 50 MHz, there may need to be multiple frequency conversions to reach a microwave frequency. **Fig 9.2** is the block diagram of a phasing SSB exciter. The signal frequency networks all have considerable bandwidth, so operating the SSB modulator on the final output frequency is an option. Heterodyning the phasing exciter output to the desired output frequency also has merit, and was the method of choice in vintage gear. **Fig 9.3** shows a conventional superhet receiver with an SSB bandwidth IF filter to provide rejection of interference outside the desired bandpass, including rejection of the opposite sideband. **Fig 9.4** shows a superhet receiver with a phasing SSB demodulator at the IF. Note that the phasing system just rejects the opposite sideband—conventional selectivity is still

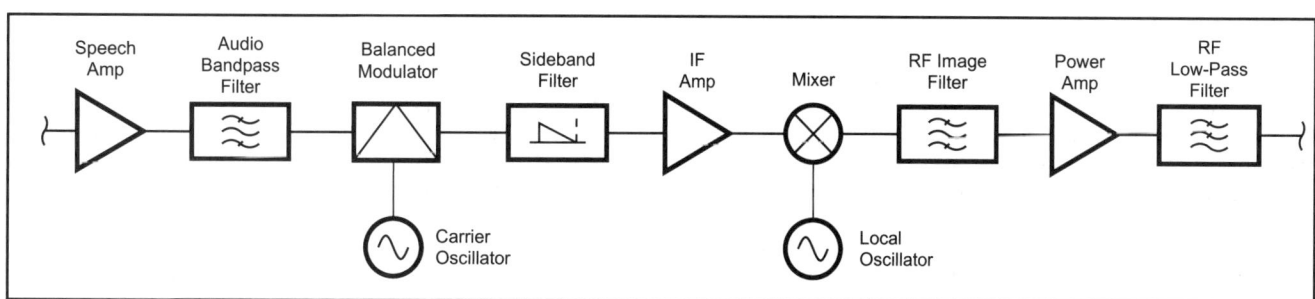

Fig 9.1—A block diagram of a conventional SSB exciter using a filter to remove the unwanted sideband.

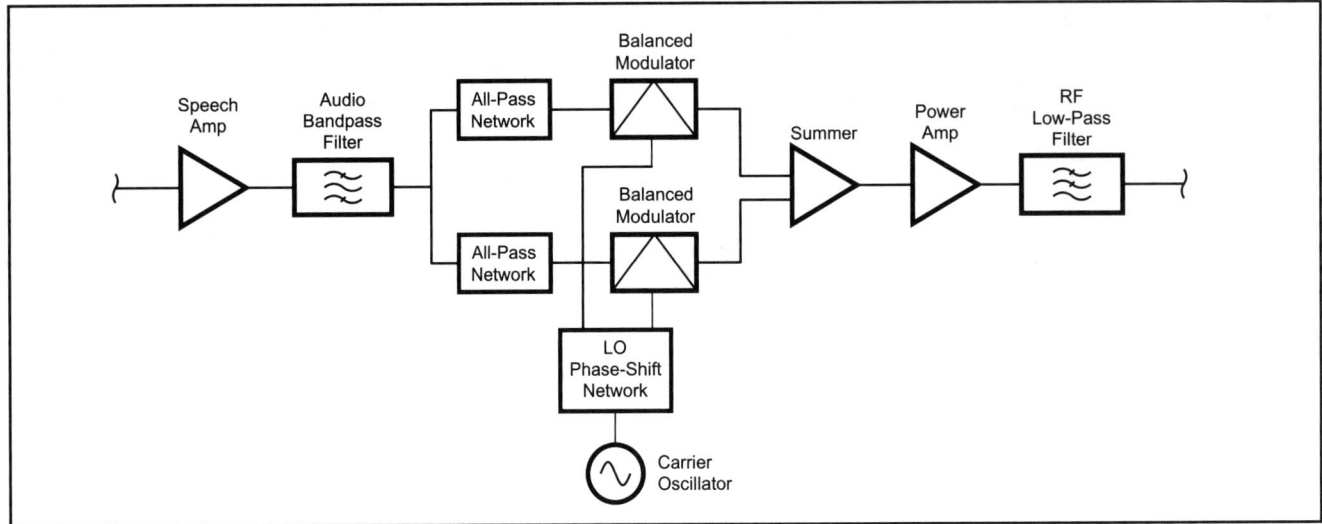

Fig 9.2—Block diagram of a phasing SSB exciter.

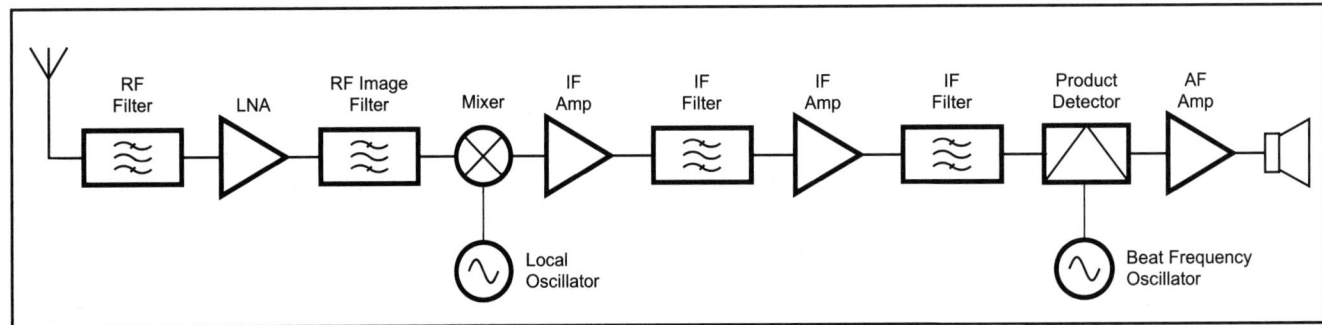

Fig 9.3—A conventional superhet receiver with an SSB bandwidth IF.

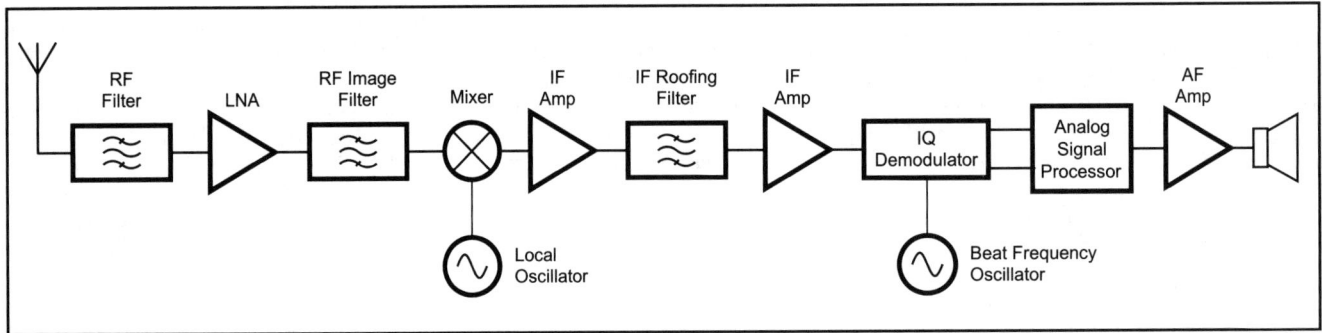

Fig 9.4—A superhet receiver with a phasing SSB demodulator at the IF.

needed to protect the receiver from interference at other frequencies. **Fig 9.5** is the block diagram of a phasing direct conversion receiver (high performance direct conversion receiver techniques are discussed in Chapter 8 of this book.) Phasing is added in **Fig 9.6** with baseband processing functions handled using a pair of analog-to-digital converters and a digital signal processor. Each of the systems shown in the block diagrams is optimum for certain applications, and a designer-builder needs to be familiar with the benefits and limitations of each before concluding that a particular radio architecture is best for a particular application.

Traditionally, phasing is presented as a transmit topic, with receivers tacked on as an "oh by the way, you can also..." This is fine until one wants to actually begin designing and building a receiver using phasing methods, at which point none of the math really makes sense, and signal levels, noise, and distortion terms that don't apply to transmitters become important. The treatment here will take the opposite tack, and discuss phasing direct conversion receivers in detail. There are several justifications for this. The first is that exploration of high performance phasing direct conversion receivers has been a major focus area for the author for over a decade, and many of the observations, much of the analysis, and the mathematical treatment have not been previously published—or at least not for a very

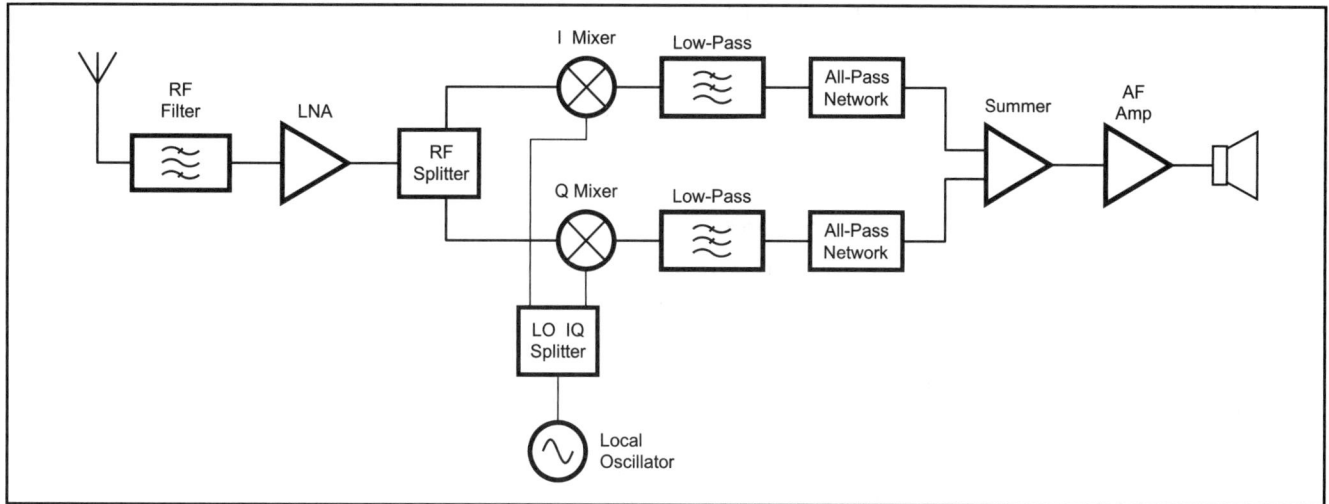

Fig 9.5—A block diagram of a phasing direct conversion receiver.

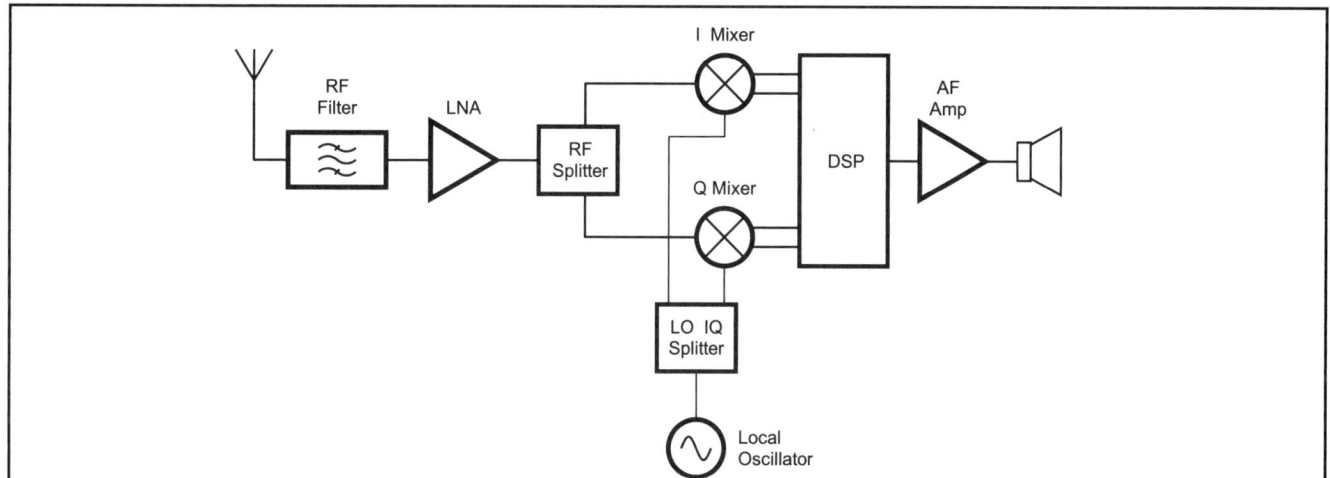

Fig 9.6—High performance direct conversion receiver technique with phasing added—with baseband processing in DSP.

long time. The second is that most of this decade of study has been a purely amateur activity, pursued because listening to that first phasing direct conversion receiver ten years ago was such a profound revelation. Phasing direct conversion receivers are an optimum choice for many applications, amateur and professional, whenever cost, distortion, spurious-free-dynamic range, frequency agility or adaptability to different bandwidths and modulation types are important. Furthermore, they are a rich field for experimentation and contribution to the amateur and professional literature. Finally, by describing the receiver mathematically using a general band-limited input signal $a_s(t)\cos[2\pi f_s t + \phi_s(t)]$, the discussion becomes independent of modulation type, and serious students of communications systems will have no difficulty converting to complex-envelope form, adding correlated and uncorrelated noise terms, and including the effects of various types of distortion.

The emphasis will be on direct conversion phasing receivers, rather than superhet receivers with phasing last-converters, because the direct conversion receiver generally presents a more difficult set of problems. However, it should be mentioned at this point that the ultimate receiver for weak CW and SSB signals in the presence of noise and strong-signal interference is most likely a hybrid superhet that includes a band-limiting filter followed by some IF gain and then a phasing product detector. This is certainly the approach being taken by makers of high-end amateur transceivers, and the technology will trickle down into the low end of the market, as it is less expensive than relying solely on mechanical, quartz crystal and ceramic filters for selectivity. The major difference between using the phasing system at the front-end of a direct conversion receiver or as the product detector for a hybrid superhet is in the gain, selectivity, and noise distributions in the receiver. These considerations will be discussed in detail in the R2pro design exercise.

9.2 INTRODUCTION TO THE MATH

Some mathematics is necessary for understanding how phasing receivers work. Fortunately, all of the necessary functions and identities may be found in a high school algebra and trigonometry textbook. That said, there is nothing trivial about the treatment that follows. It is deliberate and complete. It is also much less interesting than the pictures and schematics of the projects, and many of the subtleties were not appreciated by the author until long after the first signals began pouring out of the working receiver's speaker. Readers with an aversion to math in any form are invited to skip this section. Designer-builders who want to proceed directly to the R2pro design and projects section are encouraged to skim quickly through the math. Electrical Engineering graduate students should work slowly through the material step by step, because this stuff will be on the exam. Refer to Figs **9.7A-G** that appear after the equations.

The Basic Image-Reject Math From Receiver Point Of View

Any band-limited basic signal may be described as:

$$s_i(t) = a_s(t)\cos\left[2\pi f_s t + \varphi_s(t)\right] \quad \text{Eq 9.1}$$

…where f_s is the signal frequency; $a_s(t)$ is the time-varying signal envelope; and $\varphi_s(t)$ is the time-varying signal phase

Mixer

In an ideal mixer, a local oscillator multiplies this signal:

$$L_i(t) = 2\cos\left(2\pi f_o t + \varphi_o\right) \quad \text{Eq 9.2}$$

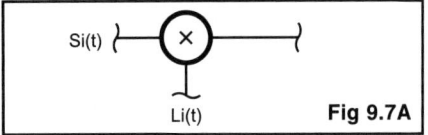

Fig 9.7A

…where the constant 2 simplifies later expressions, f_o is the LO frequency, and φ_o is the LO phase.

Multiplying the LO times the signal:

$$2\cos\left(2\pi f_o t + \varphi_o\right) a_s(t)\cos\left[2\pi f_s t + \varphi_s(t)\right]$$
$$= a_s(t)\cos\left[2\pi\left(f_o + f_s\right)t + \varphi_s(t) + \varphi_o\right]$$
$$+ a_s(t)\cos\left[2\pi\left(f_o - f_s\right)t - \varphi_s(t) + \varphi_o\right]$$

Eq 9.3

If the signal frequency f_s is lower than the LO frequency f_o, then the difference expression $(f_o - f_s)$ is a positive number.

Low-Pass Filter

In a receive downconverter application, the difference frequency expression is selected by a low-pass filter following the mixer, and the sum frequency $(f_o + f_s)$ expression is rejected. The downconverter output frequency range may extend from zero Hz up to the cutoff frequency of the low-pass filter, and this frequency range is referred to as "baseband." The baseband output is then:

$$b_i(t)$$
$$= a_s(t)\cos\left[2\pi\left(f_o - f_s\right)t - \varphi_s(t) + \varphi_o\right] \quad \text{Eq 9.4}$$

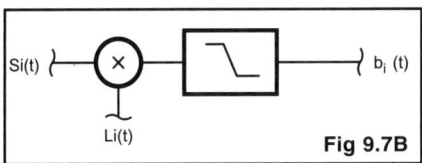

Fig 9.7B

Q Channel

In a phasing system, a second mixer multiplies the identical signal by an LO with $\pi/2$ phase delay. The two mixers and their signals are referred to as I for "in-phase" and Q for "quadrature." Since these expressions represent real signals and electronic components, they are not perfect. In particular, the amplitude of the signal at the Q mixer may not be identical to the I mixer amplitude, and the phase difference between the I and Q mixers may not be exactly $\pi/2$. We can incorporate these differences by introducing error terms into the signal and LO expressions.

$$s_q(t) = (1+\varepsilon)a_s(t)\cos\left[2\pi f_s t + \varphi_s(t)\right] \quad \text{Eq 9.5}$$

and

$$L_q(t) = 2\cos\left(2\pi f_o t + \varphi_o - \pi/2 + \delta\right) \quad \text{Eq 9.6}$$

…where ε is the amplitude difference between the I and Q signals and δ is the error in the $\pi/2$ phase delay. Note that the signal $s_q(t)$ at the input to the Q mixer is the same as $s_i(t)$ at the input to the I mixer except for the error ε.

Multiplying the phase-shifted LO and signal together in the Q mixer:

$$2\cos\left(2\pi f_o t + \varphi_o - \pi/2 + \delta\right)(1+\varepsilon)$$
$$a_s(t)\cos\left[2\pi f_s t + \varphi_s(t)\right]$$
$$= (1+\varepsilon)a_s(t)\cos$$
$$\left[2\pi\left(f_o + f_s\right)t + \varphi_s(t) + \varphi_o - \pi/2 + \delta\right]$$
$$+ (1+\varepsilon)a_s(t)\cos$$
$$\left[2\pi\left(f_o - f_s\right)t - \varphi_s(t) + \varphi_o - \pi/2 + \delta\right]$$

Eq 9.7

Once again, the low-pass filter rejects the sum frequency and passes the difference frequency, so we are left with:

$$b_q(t) = (1+\varepsilon)a_s(t)\cos$$
$$\left[2\pi\left(f_o - f_s\right)t - \varphi_s(t) + \varphi_o - \pi/2 + \delta\right]$$

Eq 9.8

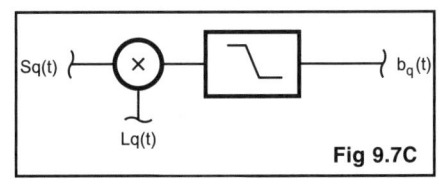

Fig 9.7C

Audio Phase-Shift Networks

In an image-reject receiver, the I and Q outputs of the mixers are then applied to the ports of a pair of all-pass networks that add an additional $\pi/2$ phase delay to the signal at the output of the Q mixer. An ideal all-pass network would introduce no additional amplitude or phase errors, but such errors occur in practice. In addition, the all-pass networks at baseband may have many octaves of bandwidth, and the amplitude and phase errors will vary across the baseband frequency range. We combine all of the amplitude errors into a single baseband frequency dependent error term $\varepsilon(f)$ and all of the phase errors into a single baseband frequency dependent phase error term $\delta(f)$. We also recognize that in practice the IQ all-pass network pair does not simply leave the I channel alone and add a constant $\pi/2$ phase delay to the Q channel, but introduces a frequency dependent phase shift to each channel, chosen so that the phase difference between the I and Q channels remains a (nearly) constant $\pi/2$. We combine this frequency dependent phase shift with the original LO phase φ_o and denote the result $\varphi_o(f)$. With the additional $\pi/2$ phase delay and all of the modified error and phase terms, the all-pass network Q baseband output becomes:

$$[1+\varepsilon(f)]a_s(t)\cos$$
$$\begin{bmatrix} 2\pi(f_o - f_s)t - \varphi_s(t) + \varphi_o(f) \\ -\pi/2 - \pi/2 + \delta(f) \end{bmatrix}$$
$$= [1+\varepsilon(f)]a_s(t)\cos$$
$$[2\pi(f_o - f_s)t - \varphi_s(t) + \varphi_o(f) - \pi + \delta(f)]$$

Eq 9.9

The I baseband output at the output of the all-pass network is:

$$b'_i(t) =$$
$$a_s(t)\cos[2\pi(f_o - f_s)t - \varphi_s(t) + \varphi_o(f)]$$

Eq 9.10

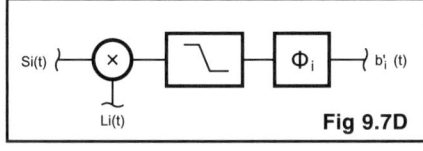

Fig 9.7D

Returning to the baseband Q term, use the trig identity:

$$\cos(a - \pi) = -\cos a \qquad \textbf{Eq 9.11}$$

to obtain

$$b'_q(t) = -[1+\varepsilon(f)]a_s(t)\cos$$
$$[2\pi(f_o - f_s)t - \varphi_s(t) + \varphi_o(f) + \delta(f)]$$

Eq 9.12

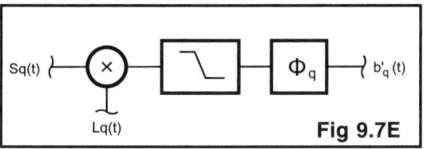

Fig 9.7E

Suppressing the Image

The baseband I and Q all-pass filter outputs are added to implement the image-reject function. To make the addition easier, the Q output may be broken down into separate terms:

$$-a_s(t)\cos$$
$$[2\pi(f_o - f_s)t - \varphi_s(t) + \varphi_o(f) + \delta(f)]$$
$$-\varepsilon(f)a_s(t)\cos$$
$$[2\pi(f_o - f_s)t - \varphi_s(t) + \varphi_o(f) + \delta(f)]$$

Eq 9.13

We may also separate the phase error $\delta(f)$ out using the trig identity:

$$\cos(a + b) = \cos a \cos b - \sin a \sin b$$

Eq 9.14

...where $2\pi(f_o - f_s)t - \varphi_s(t) + \varphi_o(f) = a$
and $\delta(f) = b$

$$b'_q(t) = -a_s(t)\cos$$
$$[2\pi(f_o - f_s)t - \varphi_s(t) + \varphi_o(f)]\cos[\delta(f)]$$
$$+a_s(t)\sin$$
$$[2\pi(f_o - f_s)t - \varphi_s(t) + \varphi_o(f)]\sin[\delta(f)]$$
$$-\varepsilon(f)a_s(t)\cos$$
$$[2\pi(f_o - f_s)t - \varphi_s(t) + \varphi_o(f)]\cos[\delta(f)]$$
$$+\varepsilon(f)a_s(t)\sin$$
$$[2\pi(f_o - f_s)t - \varphi_s(t) + \varphi_o(f)]\sin[\delta(f)]$$

Eq 9.15

At this point, it is convenient to make our first approximations. For phasing systems with opposite sideband suppression of more than 30 dB, the amplitude and phase error terms ε and δ must both be less than 0.1. Setting $\delta(f)$ to a maximum value of 0.1 and plugging it into the sine and cosine expressions:

$$\sin(0.1000) = 0.0998$$
$$\cos(0.1000) = 0.9950$$

...we may then use the "small angle" approximations:

$$\sin\varphi \approx \varphi \qquad \textbf{Eq 9.16}$$

$$\cos\varphi \approx 1 \qquad \textbf{Eq 9.17}$$

knowing that the approximation errors are very small in the range of interest. The approximation errors become vanishingly small when we reduce δ still further, to the limits needed for high performance systems.

Using the small angle approximations, the four Q terms at the output of the baseband all-pass network become:

$$b'_q(t) = -a_s(t)\cos$$
$$[2\pi(f_o - f_s)t - \varphi_s(t) + \varphi_o(f)]$$
$$+\delta(f)a_s(t)\sin$$
$$[2\pi(f_o - f_s)t - \varphi_s(t) + \varphi_o(f)]$$
$$-\varepsilon(f)a_s(t)\cos$$
$$[2\pi(f_o - f_s)t - \varphi_s(t) + \varphi_o(f)]$$
$$+\delta(f)\varepsilon(f)a_s(t)\sin$$
$$[2\pi(f_o - f_s)t - \varphi_s(t) + \varphi_o(f)]$$

Eq 9.18

We now use a second approximation. If ε and δ are less than 0.1, then their product must be less than $(0.1)^2 = 0.01$. Thus the last term above is always much less than the other three terms. Once again, the approximation error becomes vanishingly small for high performance systems. Discarding the last term, the Q signal at the output of the baseband all-pass network is:

$$b'_q(t) = -a_s(t)\cos$$
$$[2\pi(f_o - f_s)t - \varphi_s(t) + \varphi_o(f)]$$
$$+\delta(f)a_s(t)\sin$$
$$[2\pi(f_o - f_s)t - \varphi_s(t) + \varphi_o(f)]$$
$$-\varepsilon(f)a_s(t)\cos$$
$$[2\pi(f_o - f_s)t - \varphi_s(t) + \varphi_o(f)]$$

Eq 9.19

This signal is added to the I signal at the output of the baseband all-pass network:

$$b'_i(t) = a_s(t)\cos$$
$$[2\pi(f_o - f_s)t - \varphi_s(t) + \varphi_o(f)]$$

to obtain:

$$\text{LSB out} = +\delta(f)a_s(t)\sin$$
$$[2\pi(f_o - f_s)t - \varphi_s(t) + \varphi_o(f)]$$
$$-\varepsilon(f)a_s(t)\cos$$
$$[2\pi(f_o - f_s)t - \varphi_s(t) + \varphi_o(f)]$$

Eq 9.20

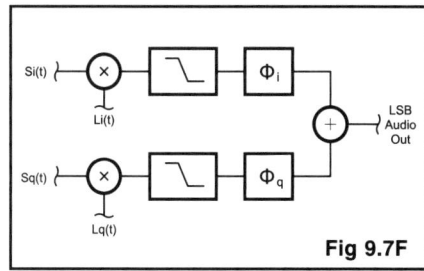

Fig 9.7F

Note that the equal and opposite signal components have added to zero, and only the error terms remain. Also note that the effect of a 0.1-radian phase error is identical to the effect of a 0.1 amplitude error. Finally, note that the two error terms are orthogonal (one is a sine, and the other is a cosine), so that each must be independently reduced to zero—an amplitude error will not cancel a phase error. The two error voltages add to make a resultant error signal, with magnitude:

$$\left\{[\delta(f)]^2 + [\varepsilon(f)]^2\right\}^{1/2} \quad \text{Eq 9.21}$$

Recovering the Desired Signal

Now examine the case of a signal frequency f_s greater than the LO frequency f_o.

The expression $(f_o - f_s)$ is now a negative number. The I baseband signal at the output of the I mixer is (as before):

$$b_i(t) = a_s(t)\cos\left[2\pi(f_o - f_s)t - \varphi_s(t) + \varphi_o\right]$$

...to make the frequency term $(f_o - f_s)$ positive, use:

$$(f_o - f_s) = -(f_s - f_o) \text{ to obtain}$$

$$a_s(t)\cos\left[-2\pi(f_s - f_o)t - \varphi_s(t) + \varphi_o\right] \quad \text{Eq 9.22}$$

Using the trig identity:

$$\cos a = \cos(-a) \quad \text{Eq 9.23}$$

...we obtain the I mixer baseband output:

$$b_i''(t) = a_s(t)\cos\left[2\pi(f_s - f_o)t + \varphi_s(t) - \varphi_o\right] \quad \text{Eq 9.24}$$

The Q mixer baseband output is (as before):

$$b_q(t) = (1+\varepsilon)a_s(t)\cos\left[2\pi(f_o - f_s)t - \varphi_s(t) + \varphi_o - \pi/2 + \delta\right]$$

Again using the $(f_o - f_s) = -(f_s - f_o)$ substitution and $\cos a = \cos(-a)$ identity, the Q mixer baseband output is:

$$(1+\varepsilon)a_s(t)\cos\left[2\pi(f_s - f_o)t + \varphi_s(t) - \varphi_o + \pi/2 - \delta\right] \quad \text{Eq 9.25}$$

At the output of the all-pass network, which adds $\pi/2$ phase delay and additional errors, the Q signal is:

$$b_q''(t) = (1+\varepsilon(f))a_s(t)\cos\left[2\pi(f_s - f_o)t + \varphi_s(t) - \varphi_o(f) - \delta'(f)\right] \quad \text{Eq 9.26}$$

Note that the combined phase error term $\delta'(f)$ is different than the previous case, because of the sign change on δ. Also note that the minus $\pi/2$ phase shift from the all-pass network has cancelled the plus $\pi/2$ phase shift from the LO.

Performing the same steps as before to reduce the signal at the Q baseband all-pass network output to separate components, we obtain:

$$a_s(t)\cos\left[2\pi(f_s - f_o)t + \varphi_s(t) + \varphi_o(f)\right]$$
$$+ \delta'(f)a_s(t)\sin\left[2\pi(f_s - f_o)t + \varphi_s(t) + \varphi_o(f)\right]$$
$$+ \varepsilon(f)a_s(t)\cos\left[2\pi(f_s - f_o)t + \varphi_s(t) + \varphi_o(f)\right]$$

Eq 9.27

Adding the I and Q outputs from the baseband all-pass network:

$$a_s(t)\cos\left[2\pi(f_s - f_o)t + \varphi_s(t) + \varphi_o(f)\right]$$
$$+ a_s(t)\cos\left[2\pi(f_s - f_o)t + \varphi_s(t) + \varphi_o(f)\right]$$
$$+ \delta'(f)a_s(t)\sin\left[2\pi(f_s - f_o)t + \varphi_s(t) + \varphi_o(f)\right]$$
$$+ \varepsilon(f)a_s(t)\cos\left[2\pi(f_s - f_o)t + \varphi_s(t) + \varphi_o(f)\right]$$
$$= 2a_s(t)\cos\left[2\pi(f_s - f_o)t + \varphi_s(t) + \varphi_o(f)\right]$$
$$+ \delta'(f)a_s(t)\sin\left[2\pi(f_s - f_o)t + \varphi_s(t) + \varphi_o(f)\right]$$
$$+ \varepsilon(f)a_s(t)\cos\left[2\pi(f_s - f_o)t + \varphi_s(t) + \varphi_o(f)\right]$$

Eq 9.28

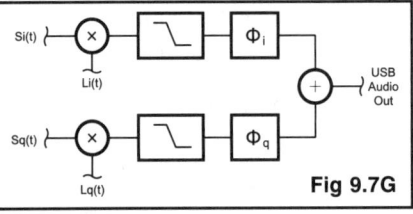

Fig 9.7G

...since $\delta'(f)$ and $\varepsilon(f)$ are both much less than 2, a reasonable approximation for the sum of the I and Q all-pass network outputs for an input signal with a frequency higher than the LO frequency is:

$$\text{USB out} = 2a_s(t)\cos\left[2\pi(f_s - f_o)t + \varphi_s(t) + \varphi_o(f)\right] \quad \text{Eq 9.29}$$

Once again, the accuracy of this expression becomes increasingly good as the amplitude and phase errors are reduced.

Sideband Suppression Expressions

In summary, it has been shown that signals at frequencies above the Local Oscillator frequency are downconverted and add at the output of the baseband all-pass network, while signals at frequencies below the Local Oscillator frequency are downconverted and subtract, leaving only the amplitude and phase error terms. It is a straightforward exercise, using the identical steps, to show that reversing the sign of either term, interchanging the LO phase shifts, interchanging the input ports of the all-pass network, or subtracting instead of adding the I and Q signals at the all-pass network output will result in adding the lower frequencies and canceling the higher frequencies.

Since the relative magnitude of the added signal is 2 and the magnitude of the error terms is:

$$\left\{[\delta(f)]^2 + [\varepsilon(f)]^2\right\}^{1/2}$$

...the familiar expression for opposite sideband suppression in dB for a given set of amplitude and phase errors is easily obtained:

Opposite sideband suppression in dB

$$= 20\log 1/2 \left\{[\delta(f)]^2 + [\varepsilon(f)]^2\right\}^{1/2} \quad \text{Eq 9.30}$$

For the effect of just an amplitude or phase error, the simpler expressions

$$20\log \varepsilon/2 \text{ (just amplitude error)} \quad \text{Eq 9.31}$$

$$20\log \delta/2 \text{ (just phase error)} \quad \text{Eq 9.32}$$

...may be used. The more complete expression above describes the opposite sideband suppression as a function of baseband frequency f for the case where the lower sideband is suppressed. These expressions may be used to obtain the common textbook plot of sideband suppression versus phase and amplitude errors. Plugging in a few numbers: if both the amplitude and phase have the maximum error of 0.1, the opposite sideband suppression is:

$$\left\{20\log\left[(0.1)^2 + (0.1)^2\right]^{1/2}/2\right\} = -23 \text{ dB} \quad \text{Eq 9.33}$$

Most textbooks quote amplitude and

phase errors in dB and degrees. To convert amplitude error ε to dB, use

$$20 \log [1 + \varepsilon] \quad \text{Eq 9.34}$$

...for ε = 0.1 in the example above, the amplitude error in dB is 20 log (1.1) = 0.83 dB.

To convert phase error in radians δ to error in degrees, multiply δ by 57.3 (degrees per radian). For the example above, the phase error in degrees is 5.73 degrees.

As an example going the opposite direction, suppose a phasing receiver system has 1-degree maximum phase error and 0.1 dB maximum amplitude error. What is the opposite sideband suppression? Converting 0.1 dB to ε:

$$\varepsilon = 10^{[(\text{error in dB})/20]} - 1 \quad \text{Eq 9.35}$$

$$= 10^{0.005} - 1 = 0.0116$$

...and converting the 1-degree phase error to radians

$$\delta = 1/57.3 = 0.0175$$

Using the expression for sideband suppression:

$$\{20 \log [(0.0116)^2 + (0.0175)^2]^{1/2} /2\}$$
$$= -39.6 \text{ dB} \quad \text{Eq 9.36}$$

This is an easy rule of thumb—to obtain 40 dB of opposite sideband suppression, the amplitude errors must be kept under 0.1 dB and the phase errors under 1 degree.

In the receive case analyzed here, summing the I and Q channel outputs suppresses the lower sideband. The upper sideband may be suppressed by first inverting the Q channel and then summing, which subtracts the I and Q channel outputs. Note that this is the reverse of what happens in a phasing SSB transmitter, where summing the I and Q channel RF outputs suppresses the upper sideband. This interesting result must be considered when designing phasing SSB transceivers.

9.3 FROM MATHEMATICS TO PRACTICE

It is tempting to believe that a good designer draws a perfectly analyzed block diagram, picks the circuit blocks out of a circuit catalog, connects them up, and has an operating receiver on the bench. If the performance is not perfect, then at least the flaws are perfectly understood and predictable. The truth is that the deeper one digs into receiver analysis, the more obvious the omissions and approximations in the mathematical treatment become. A diode ring mixer is not a perfect sine-wave multiplier, and the mathematics for the proper treatment of even simple amplifier distortion is beyond the scope of a practical text.

There are two very different approaches to receiver design and development. The first approach is to design each fundamental circuit block as carefully as possible using whatever analysis and measurement tools are available, and then connect the blocks together in a manner as close as possible to the way they were analyzed and measured. Because RF test and measurement equipment operates in a 50-Ω environment, all circuit blocks are designed and tested to interconnect using 50-Ω transmission lines. The basic rule is that connections between circuit blocks should carry sinusoidal voltages 50 times larger and in-phase with sinusoidal currents. If voltages are not sinusoidal, simple low-pass filters will remove harmonics, and if impedances are different from 50 Ω, transformers may be used. This technique results in receivers with very predictable performance, and many parts. A conservative frequency converter using this approach is shown in **Fig 9.8**.

The second approach is to decide what function needs to be accomplished, and design a circuit with as few components as possible that will perform the task. A minimum-parts-count frequency converter is shown in **Fig 9.9**. Clearly, the second circuit is simpler than the first. From the professional circuit design standpoint, the second circuit might even be called "better" because is uses fewer parts and less operating current to perform the same

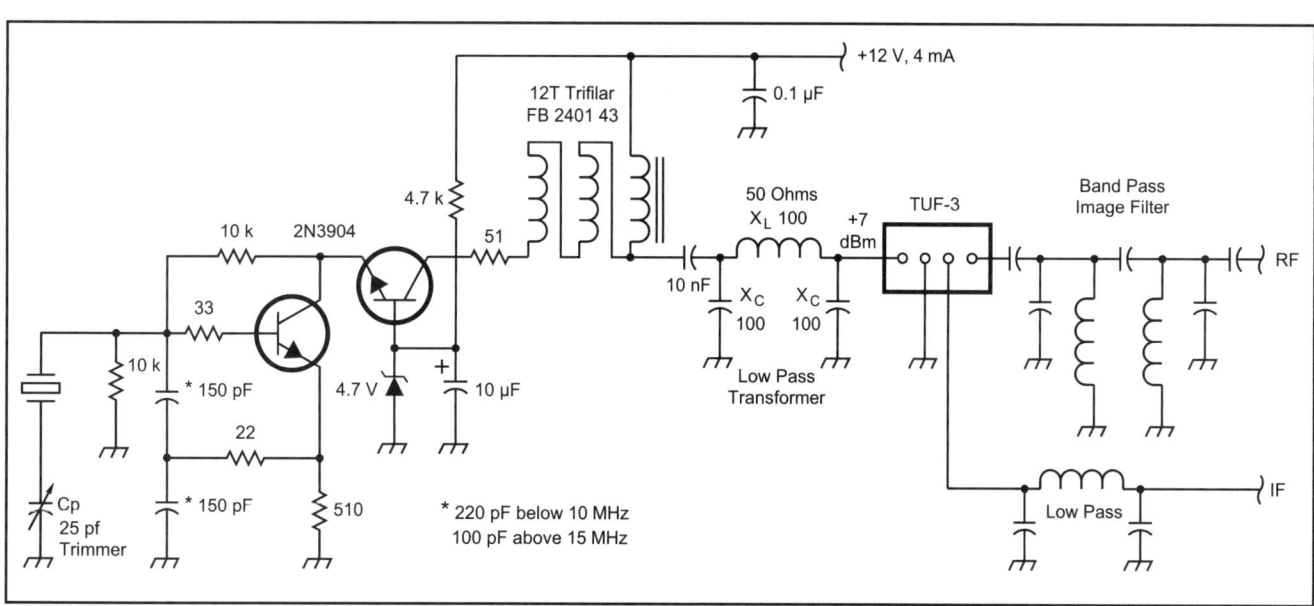

Fig 9.8—If voltages are not sinusoidal, simple low-pass filters will remove harmonics, and if impedances are different from 50 Ω, transformers may be used. This technique results in receivers with very predictable performance, and many parts. A conservative frequency converter using this approach is shown here.

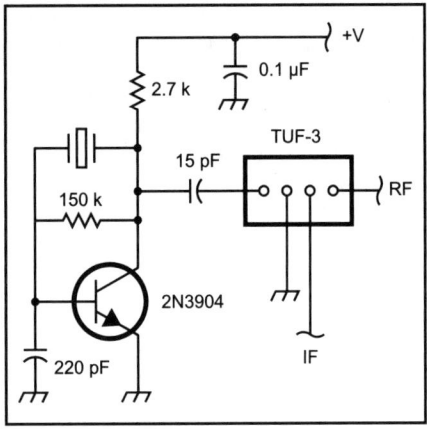

Fig 9.9—A minimum-parts-count frequency converter.

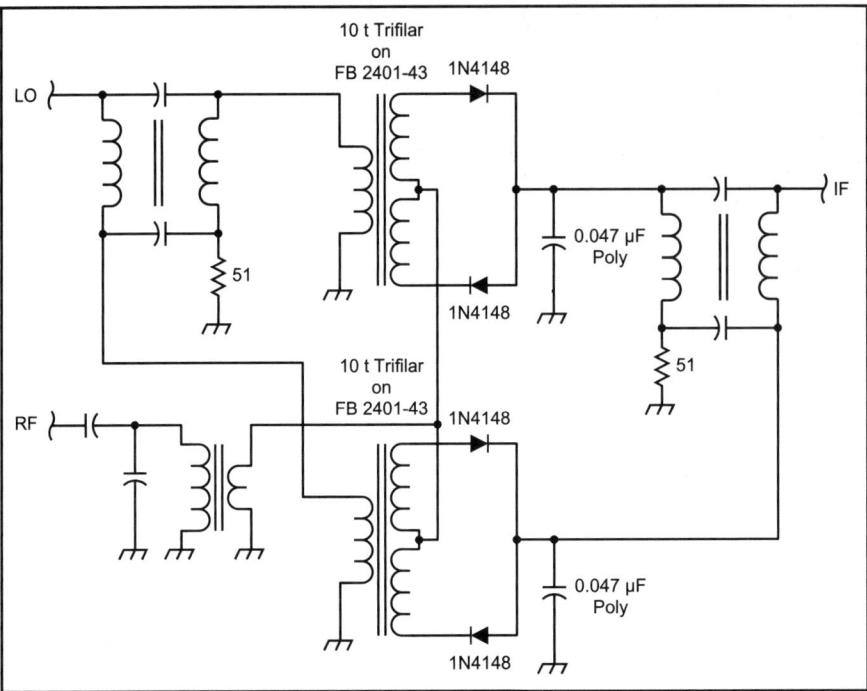

Fig 9.10—A minimum-parts-count image-reject detector that might be used in a simple CW receiver.

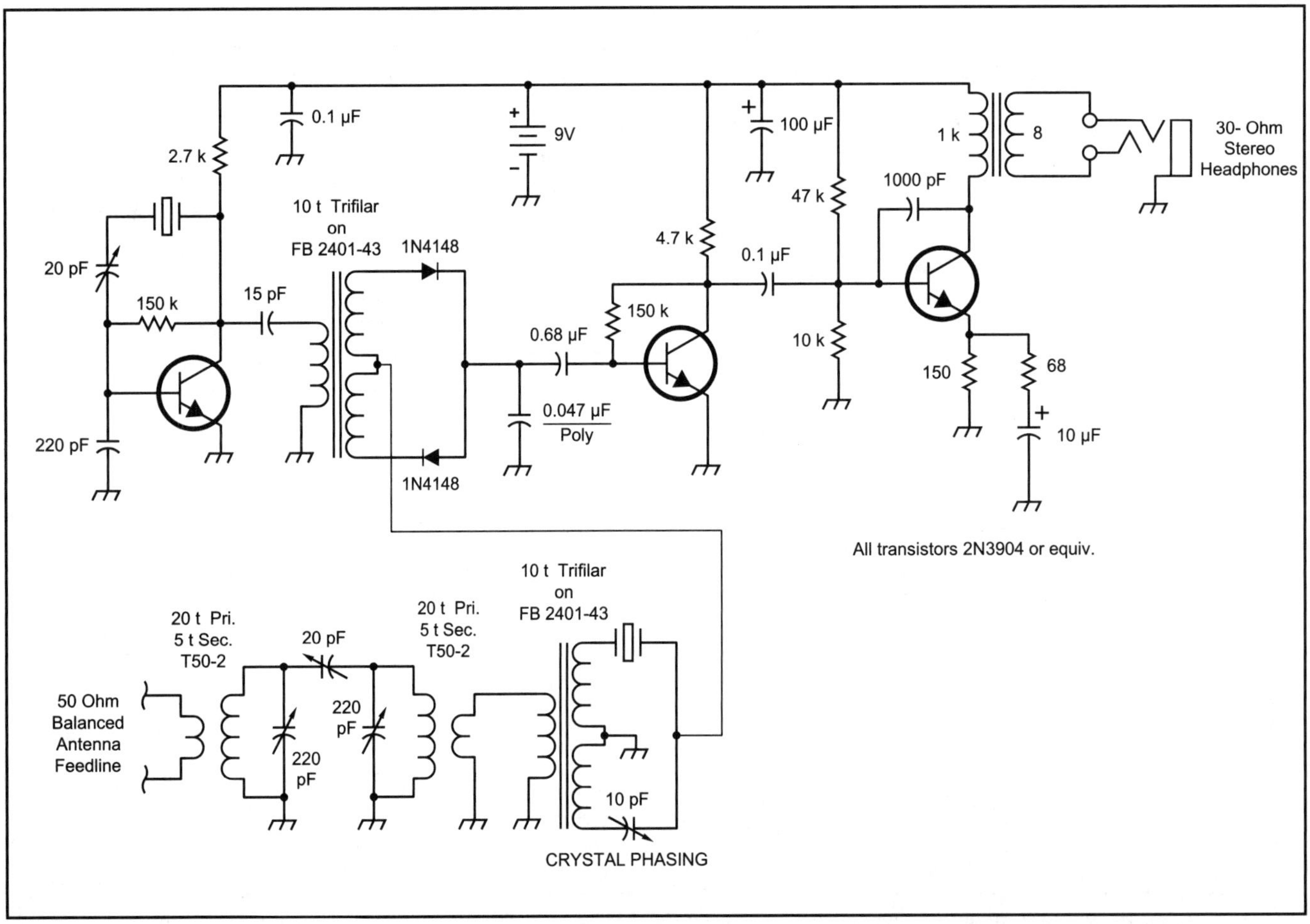

Fig 9.11—A simple fixed-frequency receiver using a single crystal filter. The two crystals are the same frequency, and the input circuit tunes from 3.5 to 7.5 MHz.

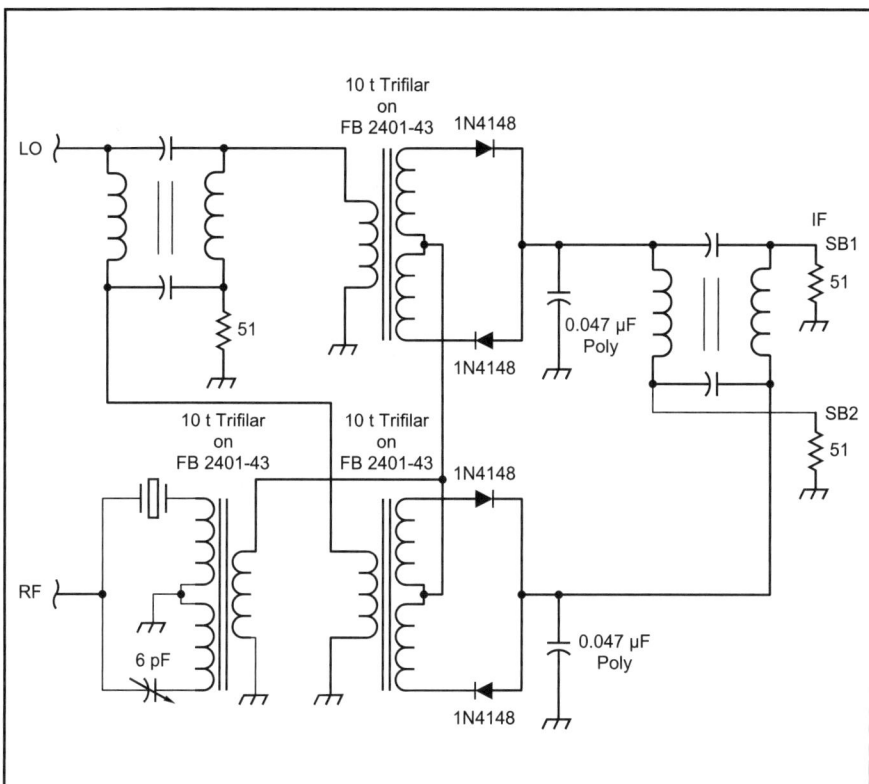

Fig 9.12—If the product detector is operated at a fixed frequency, crystal filter selectivity may be combined with a phasing product detector. This figure shows the basic circuit with a single-crystal CW filter connected directly to the product detector. It doesn't work as expected.

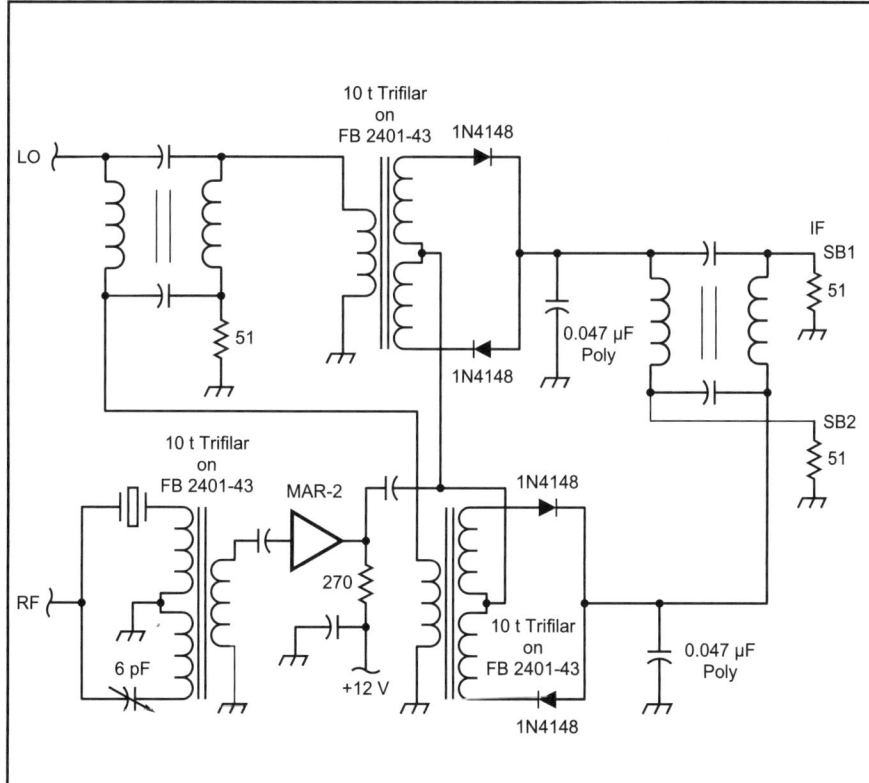

Fig 9.13—This circuit, with a buffer amplifier between the crystal filter and image-reject mixer, works as expected, with more than 40 dB of opposite sideband suppression at 1-kHz offset.

function. The difficulty arises when performance needs to be improved, or the circuit function is interconnected with other circuit blocks in a new and different way.

It is important to recognize that both approaches to RF circuit design are viable—the first offers higher performance from the outset, and a path to constant performance improvement by measuring and analyzing distortion and making incremental changes to the circuit blocks. The second approach involves more creativity and risk taking: attempts at new minimum-parts-count circuits often fail; and without 50 Ω ports, it is difficult to make diagnostic measurements without upsetting circuit behavior. Creative thinking, either in developing original circuits or pondering why they don't work as expected, is the delightful process designers use to solve problems.

There is a valid argument for both approaches to receiver projects—delightful simplicity is always a virtue—but there is a compelling argument for taking the methodical, analytical, 50-Ω approach to developing *phasing* receivers. A phasing receiver is a balanced system that depends on matching both amplitude and phase across significant bandwidths, through at least one frequency conversion, and with significant band limiting needed in both I and Q channels. Any deviation from perfect balance degrades opposite sideband suppression. Since amplitude and phase are both strong functions of termination impedances at mixer and amplifier ports, defining and controlling these impedances is the first step in building successful phasing receivers.

As an example of the problems that arise when impedance matching is neglected, let's look at a minimum-parts-count image-reject detector that might be used in a simple CW receiver. **Fig 9.10** illustrates the circuit. The RF ports of the two balanced diode mixers are simply tied together, and the LO and IF ports are quadrature split and combined using hybrid circuits. This circuit provides a useful reduction in opposite sideband interference. The selectivity curve is very similar to the classic receivers with single crystal filters and phasing controls.

The circuit in Fig 9.10 might be used as the product detector in a simple superhet receiver. For comparison, **Fig 9.11** is a simple fixed-frequency IF receiver using a single crystal filter. The image-reject product detector has a few more parts.

If the product detector is operated at a fixed frequency, crystal filter selectivity may be combined with a phasing product detector. **Fig 9.12** is the basic circuit with a single-crystal CW filter connected directly to the product detector. The crystal

Phasing Receivers and Transmitters 9.9

filter selectivity should add to the image-reject product detector circuitry, for very respectable performance. It does not work. The opposite sideband suppression is considerably less than expected.

The problem is that image-reject mixer behavior is strongly dependent on the impedances at the various mixer ports. By directly connecting the crystal filter, the mixer RF ports see an impedance that varies rapidly from one sideband to the other. The impedance in the desired band is resistive and reasonably well matched, but the impedance on the undesired sideband is almost perfectly reflective. A reflective mixer termination on one sideband and an absorptive termination on the other severely impacts image-reject mixer performance. In simulations, the opposite sideband suppression of the *filter* is maintained, but almost all of the opposite sideband suppression from the *image-reject mixer* circuitry is lost.

The circuit of **Fig 9.13**, with a buffer amplifier between the crystal filter and image-reject mixer, works as expected, with more than 40 dB of opposite sideband suppression at 1 kHz offset. A casual glance at this circuit would not hint that the added broadband components would significantly improve opposite sideband suppression.

The most termination-sensitive components in a phasing receiver or exciter are usually the mixers. Since providing wideband, resistive terminations to the mixer RF, LO and IF ports improves distortion performance in addition to opposite sideband suppression, it is simply good practice in phasing rigs. Paying attention to termination impedances usually adds components and complexity to circuits.

If adequate performance at minimum cost with few parts is the goal, it is unlikely that a phasing receiver or exciter can compete with a basic superhet. Making an intelligent choice about whether to use phasing techniques in a receiver involves weighing a number of factors. A strict phasing receiver can never achieve the opposite sideband selectivity of a good superhet with multiple crystal filters, and a superhet will always have more spurious responses and internally generated spurs than a direct conversion receiver. Nothing can compare with the sonic clarity of simple wide audio bandwidth direct conversion receivers. The choice of receiver architecture may not be made for purely practical reasons—an Amateur Radio designer-builder has the luxury of working on a technique purely for the joy of exploring new territory.

9.4 SIDEBAND SUPPRESSION DESIGN

The point of adding a phasing system to a receiver or exciter is to suppress one sideband. The first generation of amateur phasing circuits from the late 1940s into the 1950s were literally added on to conventional receivers and transmitters. Later commercial transmitters from Central Electronics, Hallicrafters, and others used conventional heterodyne methods, with phasing sideband selection and conventional tuned circuits at a fixed IF. Many recent improvements in performance have resulted from designing the entire radio system, from headphones and microphone to antenna, with phasing in mind. Before diving into more detailed system discussions, it is useful to discuss the amount of sideband suppression desired.

It is relatively easy to design and reproduce phasing circuitry to achieve undesired sideband suppression of more than 30 dB. With just a little more design care, and well-matched components, just over 40 dB of undesired sideband suppression may be routinely obtained. The receivers and exciter in the *QST* references from 1992 through 1995 all exhibit sideband suppression in the 41 to 43 dB range, when the circuit boards are used as intended. The receiver and exciter circuits shown at the end of this chapter consistently achieve sideband suppression in the mid-50 dB range, using 0.1% matched components and very careful alignment. It is worth emphasizing at this point that the level of sideband suppression depends on circuit design; precision components; and careful alignment. A 60 dB circuit can be designed, but component tolerances are unrealistically tight, alignment is difficult, and performance degrades as components age. A 40 dB circuit design works well with standard 1% components, and has quick and easy alignment that will hold for the life of the radio. 20 dB sideband suppression circuits work with junk box parts and no adjustments at all.

Before continuing with a further exploration of sideband suppression, a discussion of "how much is enough" is in order. As in most engineering questions, the answer begins with "that depends...." First of all, we should note that systems with no sideband suppression at all are entirely functional for some applications. A signal from a DSB transmitter is converted to SSB in the receiver, and once tuned in the operator can't tell the difference. Similarly, DSB receivers have been used for CW and SSB signals since the early days of radio.

DSB is attractive whenever simplicity is more important than spectral efficiency or interference rejection. A DSB transmitter may be paired with a direct conversion DSB receiver to build an ultra-simple rig. A disadvantage of such a radio is that it can not receive DSB very well, and it's transmitted signal must be received on a receiver with some provision for either suppressing one sideband or locking to the missing carrier frequency. Somehow a radio that cannot communicate with an identically equipped station seems incomplete.

Transmitters

A Single Sideband transmitter needs enough carrier suppression that the carrier is not evident when tuning in the signal, and enough opposite sideband suppression that the opposite sideband frequencies may be used for communications by other stations. 40 dB of carrier suppres-

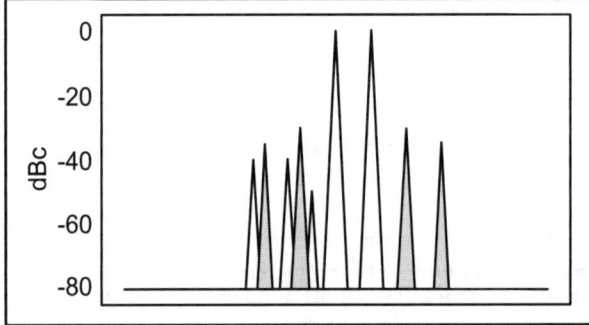

Fig 9.14—The spectrum of a typical SSB transmitter with two-tone modulation, with the carrier suppressed 50 dB, 40 dB opposite sideband suppression, and amplifier intermod products 30 dB (3rd) and 35 dB (5th) below either of the two desired output frequencies.

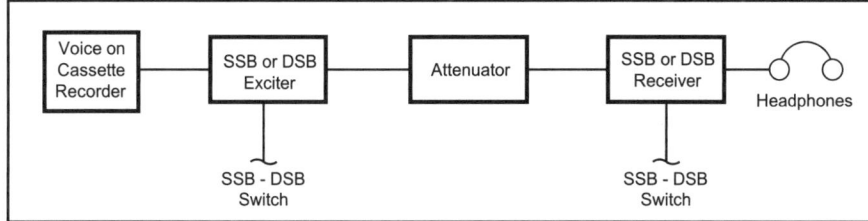

Fig 9.15—This test setup was used for a set of experiments to investigate the minimum sideband suppression needed for good SSB reception.

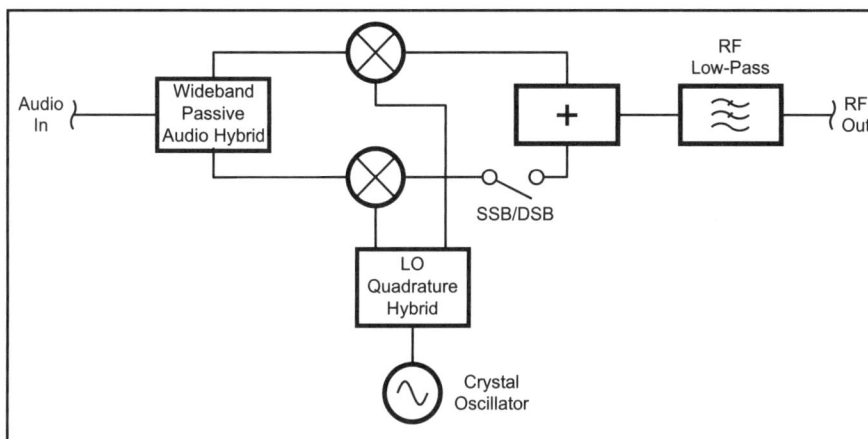

Fig 9.16—The exciter block diagram.

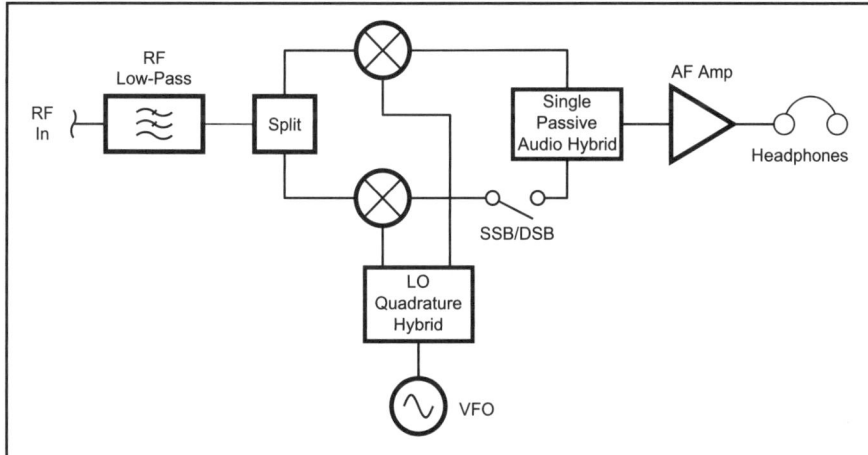

Fig 9.17—Receiver block diagram.

In the past few years, digital modes that use a computer sound card connected to the microphone input of an SSB transmitter have become popular. Transmitters for these modes benefit from having much lower distortion than SSB or keyed-carrier CW transmitters. Combining a phasing exciter with a crystal filter and very low distortion RF amplifier would make it possible to generate a PSK-31 signal that would be stunningly clean. PSK-31 operators display the whole spectrum of received in-channel distortion products on strong signals, so a clean signal is instantly recognizable on the air. Because PSK-31 stations operate in narrow bands, with tuning performed in baseband signal processing, a dedicated PSK-31 exciter and crystal filter can be built at the final output frequency, with no need for heterodyning.

Given that DSB transmitters are functional, and 40 dB of opposite sideband suppression is enough for SSB transmitter applications, are there any benefits to having less than 40 dB of opposite suppression but more than 0 dB? A set of experiments was performed to investigate the minimum sideband suppression needed for good SSB reception. **Fig 9.15** illustrates the test setup. **Fig 9.16** is the exciter block diagram, and **Fig 9.17** is the receiver block diagram. The exciter and receiver each have a switch to enable or disable the sideband suppression circuitry. The approximate sideband suppression available at the exciter, receiver, and the combined sideband suppression are shown in **Fig 9.18**.

Here are the comments copied from the lab notebook:

DSB-DSB Really hard to tune. Very poor sound.

DSB transmit, single-hybrid SSB Receive. Much better With the hybrid zero

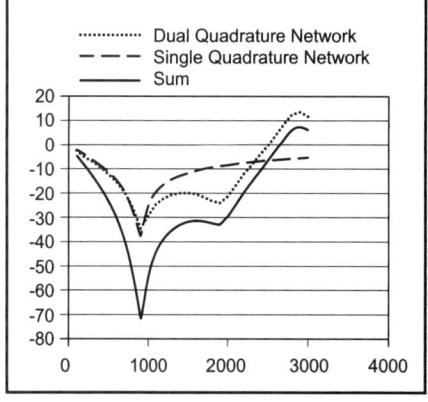

Fig 9.18—The approximate sideband suppression available at the exciter, receiver and the combined sideband suppression.

sion is generally considered sufficient, although at this level the carrier will often be noticeable to stations with good receivers and good ears. Opposite sideband suppression should be good enough that interference in the opposite sideband frequency band is dominated by amplifier intermod products, and not intelligible audio. **Fig 9.14** shows the spectrum of a typical SSB transmitter with two-tone modulation, with the carrier suppressed 50 dB, 40 dB opposite sideband suppression, and amplifier intermod products 30 dB (3rd) and 35 dB (5th) below either of the two desired output frequencies. This transmitter would sound very good on the air.

The intermodulation products are highlighted in gray. Clearly it is not necessary to suppress the opposite sideband in an SSB transmitter by much more than 40 dB, because the intermod products occupy the same frequencies and they are only about 30 dB below the desired sideband level in a well-designed transmitter. More carrier suppression is useful, however, because the carrier is present during breaks in speech.

near 1 kHz, the receiver has at least 10 dB sideband suppression over much of the 300 Hz to 3kHz speech range. The receive signal sounds like it has rapid QSB—identical to the familiar Airplane scatter QSB often experienced by VHF SSB operators.

Wideband passive hybrid SSB transmit, DSB receive. Better still—not bad at all. Probably quite acceptable for speech. The hybrid provides 15 to 20 dB of sideband suppression across most of the audio range. Rapid QSB can still be easily heard on music, but at nearly 20 dB down, the amount of QSB is only a few dB.

Wideband passive hybrid SSB transmit, single-hybrid receive. Very good. The combined sideband suppression of more than 25 dB across the audio range is good enough that it is hard to detect any effects from the inadequately suppressed sidebands.

Later experiments using a single-hybrid on both the receiver and exciter worked well for voice. **Fig 9.19** is a complete schematic of a simple voice exciter. Adding a low-noise, high-gain audio amplifier and switching results in a simple SSB transceiver, as shown in **Fig 9.20**.

The passive SSB modulator and demodulator with modest performance are significantly simpler than "serious" phasing receivers and exciters, and may be appropriate for some applications. **Fig 9.21** illustrates a modulator-demodulator circuit using a dual quadrature hybrid that provides 20 dB of opposite sideband suppression over a reasonable portion of the audio range. While over-simplified for most applications, its advantages are significant:

1. It is passive and bi-directional
2. There are no adjustments
3. Component values are not critical

The simple phasing systems described above do not provide the sideband suppression performance we have come to expect from conventional superheterodyne filter systems. We usually require better performance from the radios we design and build.

Good performance is available from pairs of 2nd-order networks, using common op-amp circuitry or RC networks like the classic B&W 2Q4. 2nd order networks are capable of providing sideband suppression of more than 30 dB across a voice bandwidth. Pairs of 3rd order networks

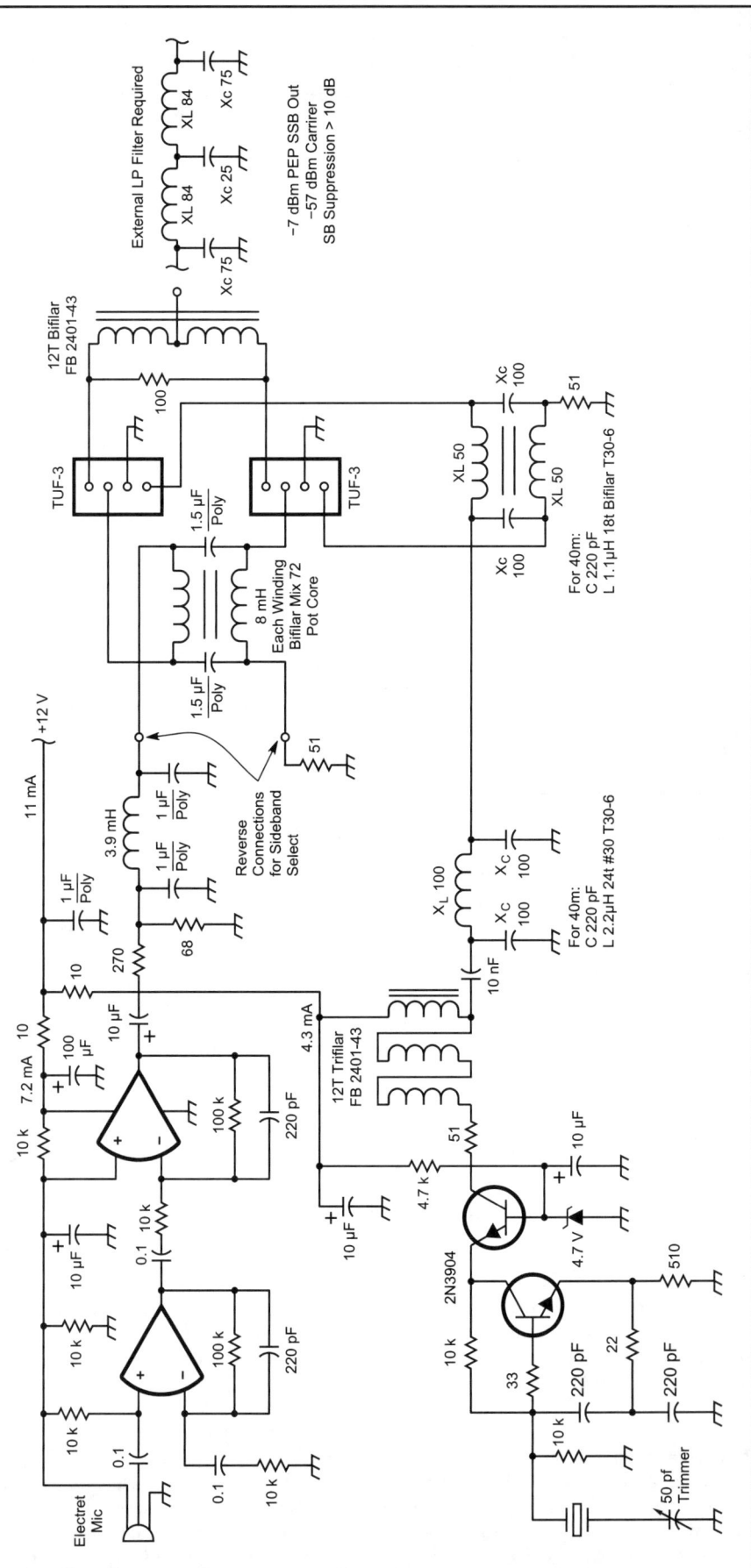

Fig 9.19—A complete schematic of a simple SSB exciter.

using op-amps easily provide more than 40 dB. Since op-amps, resistors and capacitors are all very inexpensive, the cost saving from relaxing the sideband suppression specification from 40 to 30 dB is seldom worthwhile. On the other hand, there is interest and value in revisiting classic circuitry, and a design using modern discrete components and a classic passive audio phase-shift network is appealing. As an aside—not every design should be built. There is tremendous value in notebook designs that work the problem without making it to the bench, and experiments on the bench that are never connected to the antenna.

Opposite Sideband Suppression in Receivers

For receivers, arguments can be made for almost any level of audio image suppression, from 100 dB to none at all. It is hard for a receiver with any degree of useful selectivity to compare with the sonic appeal of a wide-open direct conversion receiver or properly adjusted Regen. On the other hand, CW operators during a contest often try to copy weak signals at the noise floor in the presence of signals 90 dB stronger only a few kHz away. There is no easy "40 dB is enough" answer for receivers. Instead, there is a complex relationship between receiver topology, spectral purity, dynamic range, circuit complexity, expense, difficulty of adjustment, the need for AGC, operating habits, audio distortion, LO phase noise...the list is long enough that virtually every receiver experimenter will come up with a different requirement. There is, however, one piece of advice that has been distilled from several generations of SSB and CW receiver experimenters: time spent experimenting with a good, straight DSB direct conversion receiver connected to an antenna is part of your receiver education. You can't be a gourmet if you have never set foot in a kitchen, and there is a significant knowledge gap in your receiver background if you haven't performed the fundamental experiment of collecting radio signals on a wire, converting them to audio with a mixer and oscillator, amplifying them with a few transistors, and listening to them on headphones. This basic experience is the common ground shared by receiver experimenters.

Since there is no easy sideband suppression number, we will take a different approach to receiver opposite sideband suppression: how difficult it is to meet a particular spec. The simplest receivers have no provisions for reducing the opposite sideband, and they are so simple that the question "is additional selectivity desirable enough to warrant significant additional circuitry?" must always be asked. For many portable, emergency, and casual listening requirements, the answer is no. Furthermore, the simple receiver is such an important standard of comparison that it is useful to periodically design and build simple receivers for applications where relaxed selectivity requirements or better sounding audio are the goal.

Receivers Designed for Less than 20 dB Opposite Sideband Suppression

Having built and experimented with the "no selectivity" variant, a simple drop-in image-reject mixer can make a useful improvement in the performance of basic CW and SSB receivers. The circuit in **Fig 9.22** can replace the diode ring mixer in a 40-meter direct conversion rig. Opposite sideband suppression will be moderately good at a single frequency, near 800 Hz, and will degrade rapidly as the receiver is tuned away in either direction. The receiver response sounds very much like that of a 1940's classic receiver with a single crystal filter and front panel phasing control—with a single deep notch in the opposite sideband. The performance of this circuit is disappointing on the test bench, but it can sound very good on bands with few signals close to the noise level. It is primarily useful for CW, when combined with a narrow audio CW filter. Besides the obvious advantage of being a drop-in replacement for a diode ring mixer in a DSB receiver, this circuit is also attractive because it is entirely passive.

Receivers Designed for more than 30 dB Opposite Sideband Suppression

The next level of circuit complexity involves the use of a matched pair of product detectors and audio preamplifiers, driving a classic passive RC phase-shift network. This is appealing for historical reasons, particularly if discrete FETs are used to replace the standard vacuum tube functions. Simple direct conversion receiver circuits with good opposite sideband suppression—30 dB across an SSB bandwidth or 40 dB across a CW band—may be designed by optimizing for reduced parts count. Numerous examples of such receivers have appeared in European journals such as Sprat over the years. Once again, these receivers are appealing as design projects revisiting the classic homebrew projects of the past century. The drawback to these discrete transistor receivers is that they don't take advantage of the remarkable properties of operational amplifiers. Op-amps are little analog mathematical processors, and even if you skipped the math, it is important to remember that op-amps do math with fewer errors and approximations than discrete components.

Receivers Designed for more than 40 dB Opposite Sideband Suppression

If op-amps are to be used in a receiver, there is little point in restricting the audio phase-shift networks to 2nd order, and almost nothing to be gained by going to 4th order. Standard 3rd order networks can reliably provide more than 40 dB of opposite sideband suppression, the point at which limitations other than audio phase shift network phase and amplitude accuracy begin to dominate. The miniR2 block diagram shown in **Fig 9.23**, is an example of a good basic design for an image-reject direct conversion receiver. For a receiver without AGC, 40 dB of opposite sideband suppression sounds astonishingly good. CW signals simply disappear when a good phasing receiver is tuned through zero beat. This is a revelation to experimenters familiar with conventional superhet designs using SSB bandwidth filters, or simple CW crystal filters. The 40 dB opposite sideband range is the most practical realm for direct conversion phasing receivers. Receivers at this opposite sideband suppression level sound very good, can be reliably reproduced, provide more than enough selectivity for most HF and virtually all VHF applications, and will perform without adjustment indefinitely.

Receivers Designed for more than 50 dB Opposite Sideband Suppression

A well-designed 3rd order op-amp all-pass network built with selected components can provide more than 50 dB of opposite sideband suppression. 4th order networks can provide more than 70 dB of opposite sideband suppression, on paper. Large polyphase networks are capable of similar numbers. The difficulty is that very small differences in the phase-versus-audio frequency and amplitude-versus-audio frequency between the two channels puts a limit on sideband suppression. For 40 dB

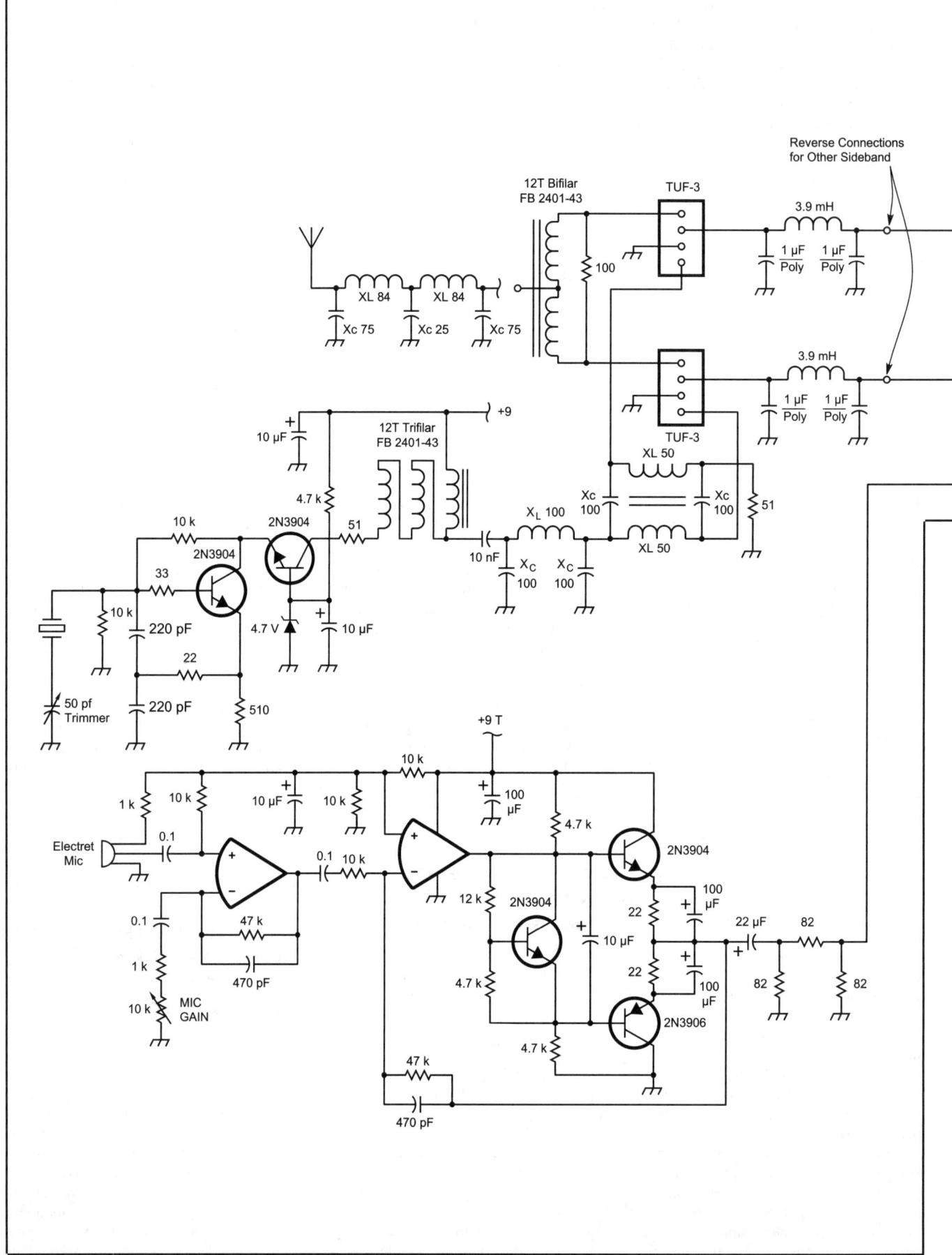

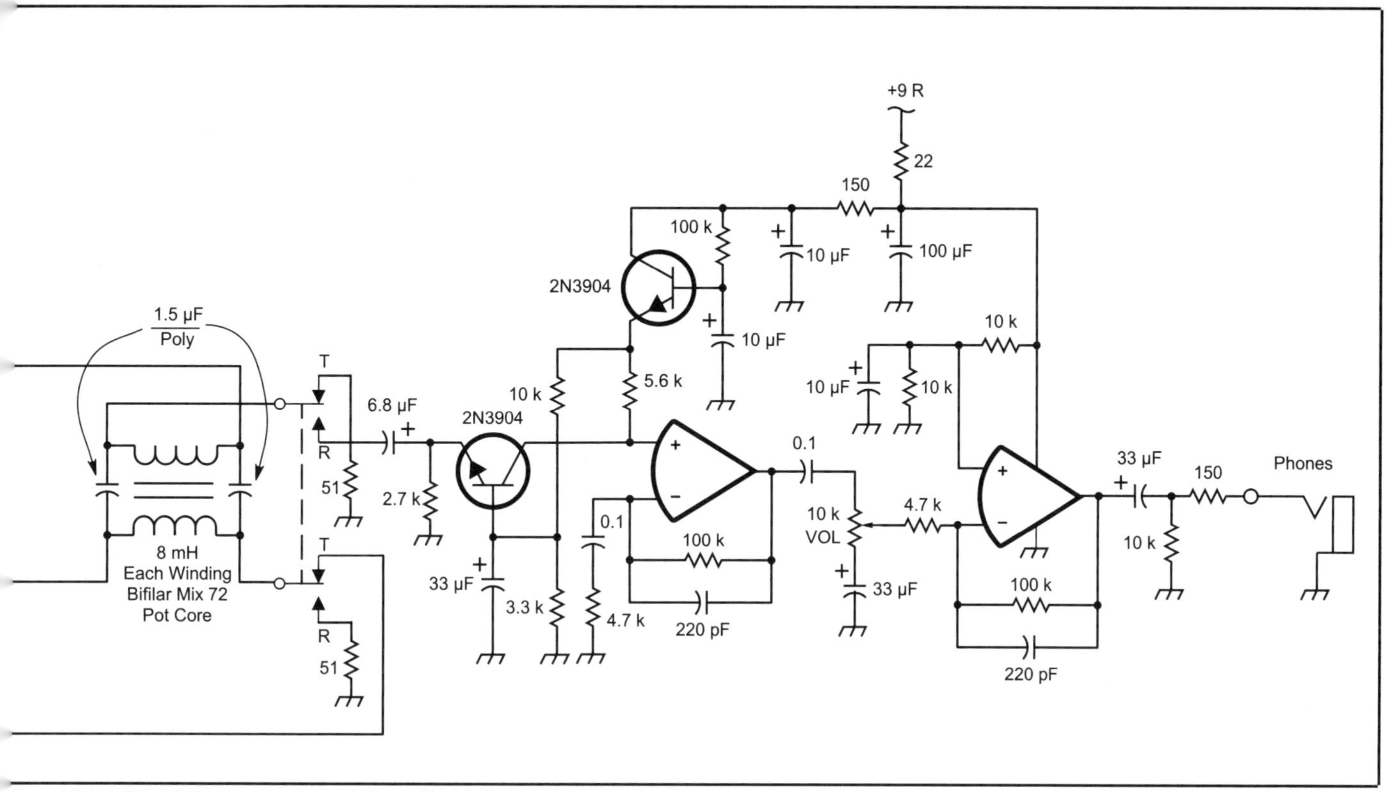

Fig 9.20—SSB transceiver schematic.

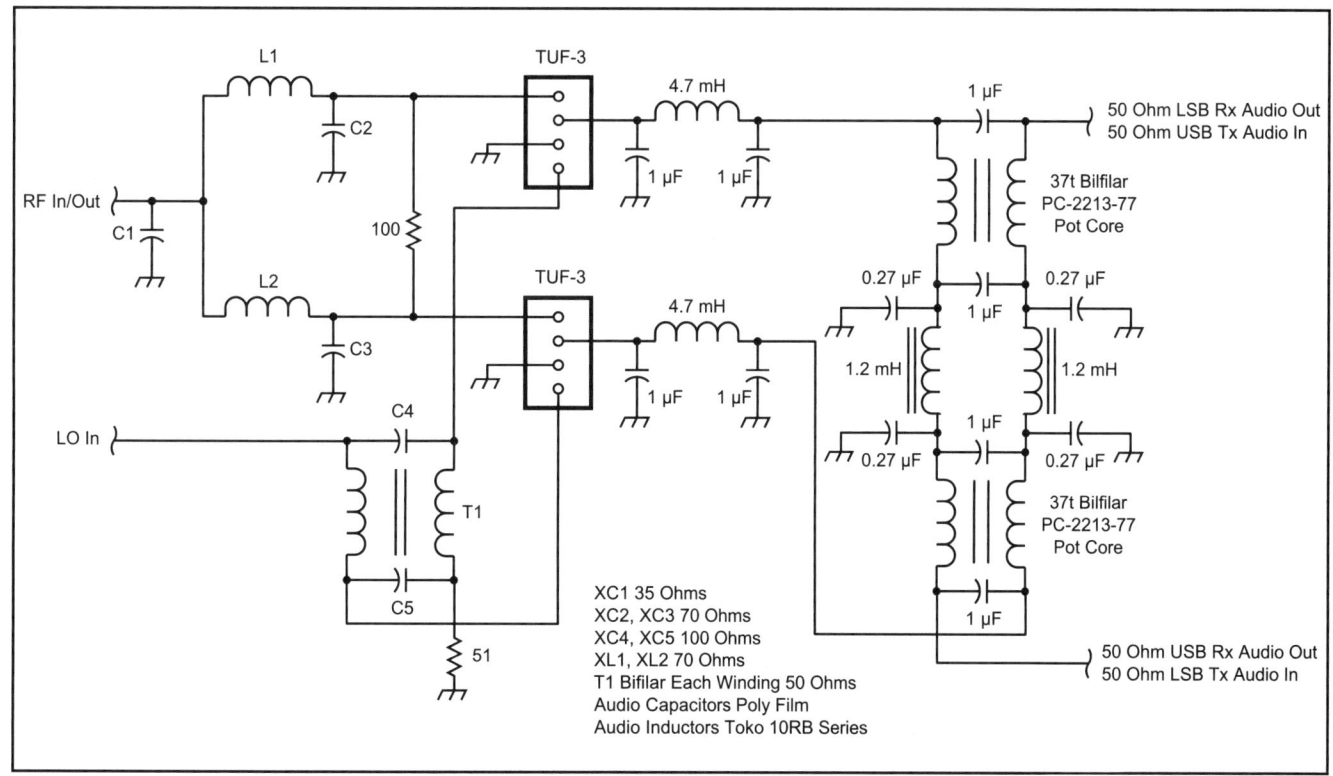

Fig 9.21—A modulator-demodulator circuit using a dual quadrature hybrid that provides 20 dB of opposite sideband suppression over a reasonable portion of the audio range.

Phasing Receivers and Transmitters   9.15

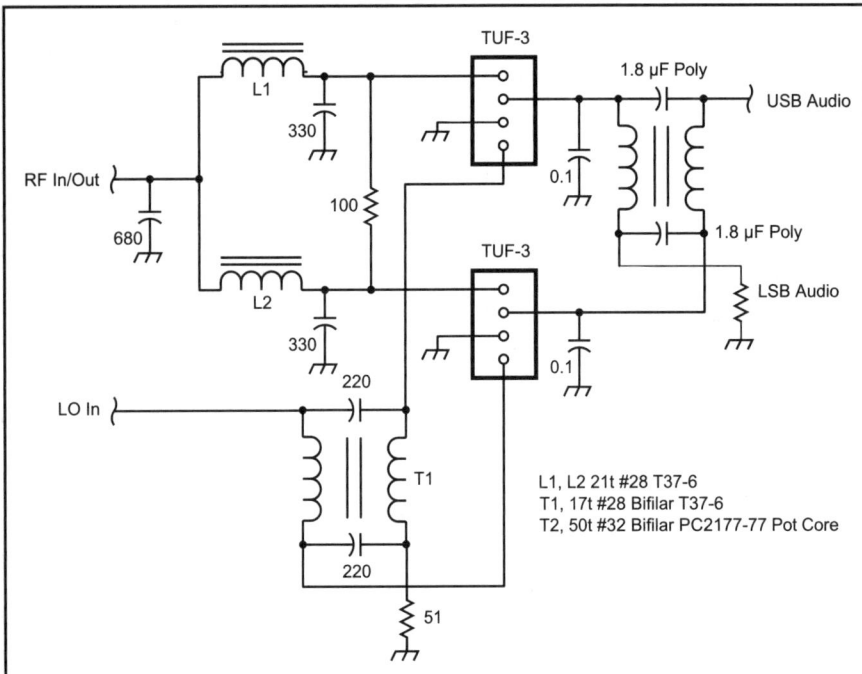

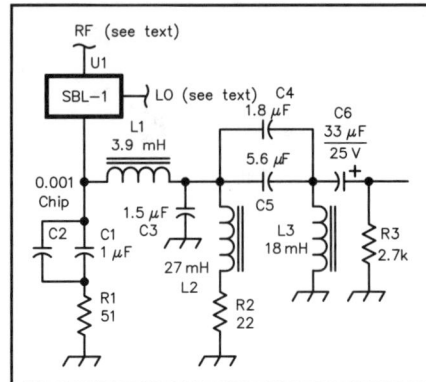

Fig 9.24—The complete schematic of the bandpass diplexer used in the R2.

Fig 9.22—A simple drop-in 40-m band image-reject mixer can make a useful improvement in the performance of basic CW and SSB receivers.

of suppression, differences must be less than one degree or 0.1 dB across the whole audio range. For 60 dB suppression, differences between channels must be less than 0.1 degree or 0.01 dB. The errors can occur anywhere in the system from the point where the I and Q channels split to the point where they are summed. Much attention has been given to the design of audio phase shift networks with arbitrarily small phase and amplitude errors, but the rest of the circuitry in the receiver I and Q channels needs to be perfect as well. Simply replacing the op-amp third order audio phase shift network in the January 1993 *QST* receiver (hereafter referred to as the "R2") with a nearly perfect DSP version does not significantly improve opposite sideband suppression.

If the R2 circuit is built using carefully matched (within 0.1%) components throughout, the opposite sideband suppression will be limited by differences in bandpass diplexer driving point impedance between the I and Q channels. **Fig 9.24** shows the complete schematic of the bandpass diplexer used in the R2. This is a doubly terminated network, intended for 50 Ω input and output terminations. The input termination is provided by the IF port impedance of the diode ring mixer. The output termination is provided by the input impedance of the grounded base amplifiers, which is determined primarily by the biasing. For 50 dB opposite sideband suppression, even the bias resistors must be matched to within 1%. The IF port impedance of a diode ring mixer varies with LO drive, which often changes across the receiver tuning range when using a quadrature hybrid in the LO signal path. The *PSPICE* simulation result in **Fig 9.25** shows the variation in phase across the audio passband when the driving impedance is 50, 75, and 100 Ω. For opposite sideband suppression of more than 40 dB across the 300 – 3000 Hz audio band the I and Q channel IF port impedances should differ by no more than 6 Ω. For 60 dB opposite sideband suppression, the I and Q port mixer IF impedances must be matched to within 0.6 Ω. This tight control of IF port impedance is more than we can expect from diode ring mixers.

Fig 9.26 shows the simplified diplexer networks used in the miniR2. Note that the 300 Hz High-Pass LC circuit has been eliminated, and the Low-Pass corner frequency has been moved up to 10 kHz. The miniR2 diplexer circuit is a little more tolerant of differences between mixer IF port impedances. **Fig 9.27** is a *PSPICE* simulation result showing miniR2 diplexer phase differences when the driving point impedance is 50, 75, and 100 Ω. This network is more tolerant of driving point impedance variations: plus or minus 9 Ω for 40 dB and 0.8 Ω for 60 dB opposite sideband suppres-

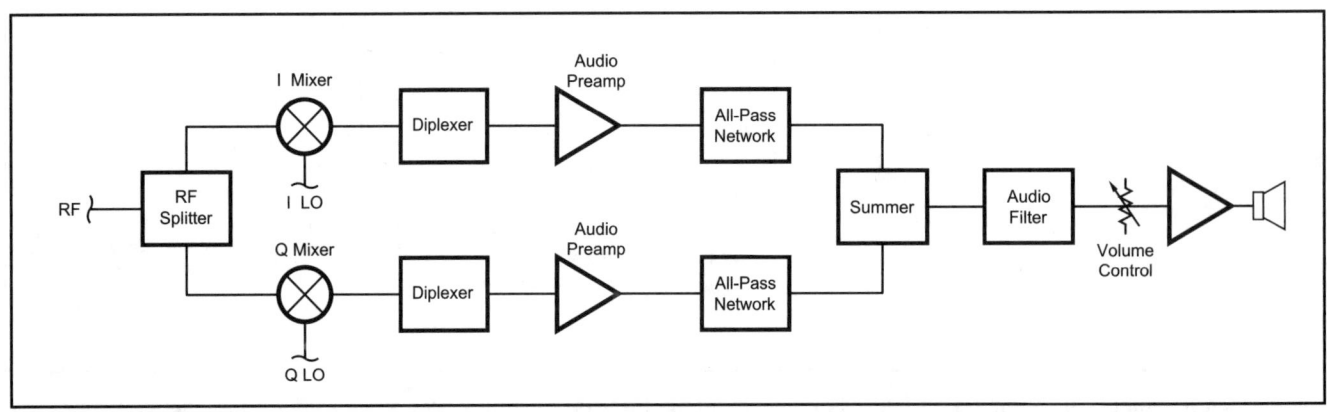

Fig 9.23—An example of a good basic design for an image-reject direct conversion receiver.

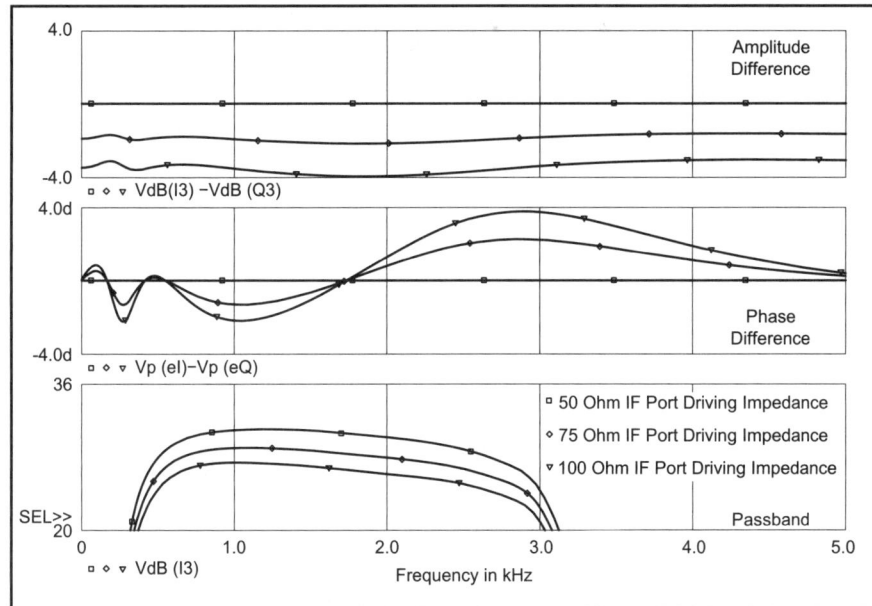

Fig 9.25—A *PSPICE* simulation shows the variation in phase and amplitude across the R2 audio passband when the driving impedance is 50, 75, and 100 Ω.

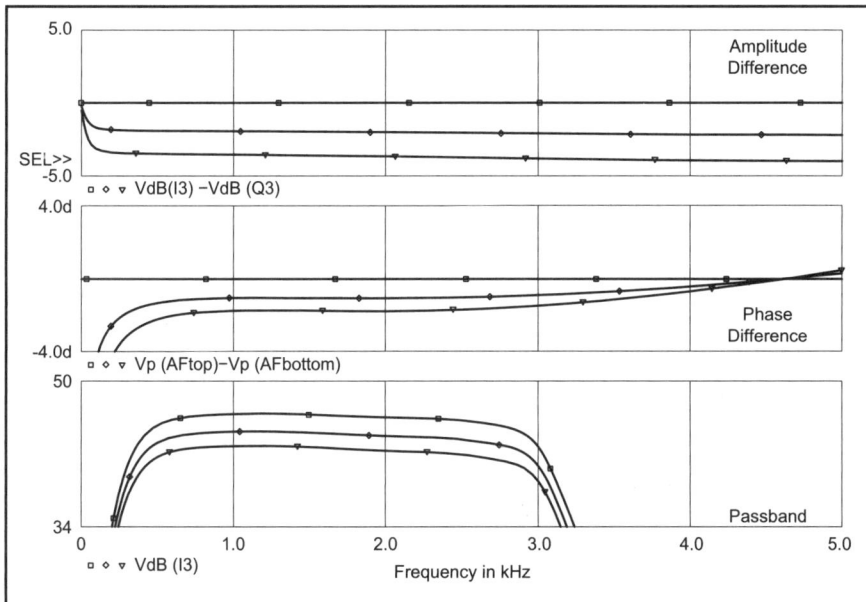

Fig 9.27—A *PSPICE* simulation result showing miniR2 diplexer amplitude and phase differences when the driving point impedance is 50, 75, and 100 Ω. This network is more tolerant of driving point impedance variations.

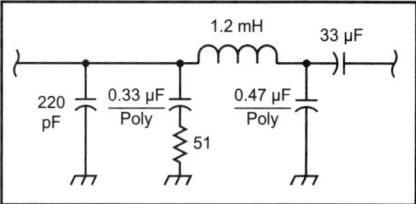

Fig 9.26—The simplified diplexer networks used in the miniR2.

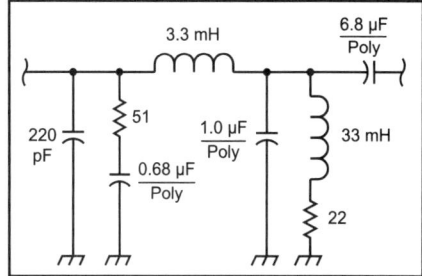

Fig 9.28—To reduce sensitivity to mixer IF port impedance and remove loose tolerance electrolytic capacitors from the I and Q signal paths, a new bandpass diplexer network was designed.

Diplexer Driving Point Impedance Measurements

An experimental receiver to study the effect of mixer IF impedance was built using the new diplexer circuit and all components matched to within 0.1%. LO drive at 14 MHz was provided by a Kanga Universal IQ VFO with the outputs carefully adjusted for equal amplitude and 90-degree phase shift. An independent phase trimmer was used on one mixer RF port. This receiver provided 43 dB of opposite sideband suppression across the 300 to 3000 Hz audio band. Then, the mixer IF ports were isolated from the diplexer inputs with 50-Ω 10-dB instrumentation attenuators. After readjusting the amplitude and phase trimmers, opposite sideband suppression improved to more than 50 dB across the 300 to 3000 Hz audio band. Switching from 10-dB to 20-dB attenuators and readjusting made a further small improvement—however at more than 53 dB opposite sideband suppression, all adjustments are an order of magnitude more critical than at the 40 dB level.

PSPICE simulations show that adding 6-dB pads between the mixer IF ports and diplexer inputs permits the experimental receiver circuit with carefully matched components to achieve 50 dB of opposite sideband suppression with a IF port impedance mismatch of up to 10 Ω. Following standard engineering practice, we

sion, if everything else in the receiver is perfect.

R2 receivers routinely exhibit 41 dB of opposite sideband suppression across the band, while miniR2 receivers typically are a few dB better. This indicates that sensitivity to mixer IF port impedance is well balanced with the errors obtained from using 1% tolerance components in the I and Q audio channels. Improving either just the phase shift network performance or just the IF port match will not significantly improve receiver opposite sideband suppression, because the other source of error will then limit performance.

To reduce sensitivity to mixer IF port impedance and remove loose tolerance electrolytic capacitors from the I and Q signal paths, a new bandpass diplexer network was designed. The new network is shown in **Fig 9.28**. It is simpler than the R2 network by 1 inductor, and the AC-coupled output eliminates the need for a blocking capacitor on the input to the audio preamp.

would avoid adding attenuation at this point, assuming that it would degrade receiver sensitivity, but standard practice is incorrect in this case. In fact the proper use of attenuation may permit us to redistribute receiver gain to improve both sensitivity and dynamic range.

Effect of Mixer IF Port Attenuation on Receiver Noise Figure

First we need to examine receiver noise figure. The techniques for measuring, calculating and even defining mixer noise figures are still evolving. A rigorous treatment is beyond the scope of this text. Standard practice calls for us to measure the audio amplifier noise figure (typically 5 to 6 dB for the R2 and miniR2 circuits) and add mixer conversion loss to obtain receiver noise figure. The resulting 12 dB noise figure is usually optimistic in practice, in part because mixers have excess noise when used with low frequency IFs. The excess noise has a 1/f character, but it is a mistake to assume that we should be able to observe a smooth 1/f spectrum in the noise output of a mixer. Low frequency diode noise mechanisms are not well understood, and the noise output varies widely between devices—even of the same part number and cut from the same semiconductor wafer. Furthermore, the noise output may have spectral peaks and dips that vary considerably from a smooth 1/f curve. Measurements of a small sample of TUF-1 mixers revealed excess baseband noise in an SSB bandwidth of between 1 dB and 7 dB. If a mixer has excess noise, attenuation after the mixer reduces the mixer noise along with the desired signal. Thus the signal-to-noise ratio changes by less than the attenuator value.

Adding an attenuator to the IF port of a diode ring mixer has an additional benefit. Mixer distortion is measured with perfect broadband 50 Ω terminations on all ports of the mixer. It is well known that the IF port termination can have a large effect on mixer dynamic range. By adding an attenuator to the mixer IF ports, dynamic range may be improved, and the expected mixer performance will be similar to the numbers in the Mini-Circuits data book. If mixer dynamic range is improved, then additional RF preamp gain may be added before the mixers. Careful selection of RF preamp gain, noise figure, and intercept performance may permit improved third-order performance and lower noise figure than the receiver without attenuators can provide.

Receivers Designed for more than 60 dB Opposite Sideband Suppression

Even if everything can be perfectly matched, baked in, trimmed, and then operated in a stable temperature controlled environment, it is still difficult to obtain more than 60 dB of opposite sideband suppression in a pure phasing receiver, because of distortion in the I and Q channel audio gain. A miniR2 has 50 dB of audio gain between the inputs to the I and Q preamps and the summing point. The gain control is after the summer, so this 50 dB gain is always in the system. With a 5 dB audio amplifier noise figure and no excess mixer noise, the noise floor at the summing point is:

-204 dBw/Hz + 5 dB Noise Figure + 34 dB SSB Bandwidth + 50 dB gain = –115 dBw

Using a 50-Ω reference voltage, this is an RMS noise voltage of 13 µV. For an HF application, it is common for the band noise to be 20 dB above the noise floor of the receiver. At VHF, at least 20 dB of LNA gain is likely to be used. In either case, the noise at the summing point would be about 100 µV RMS. Peaks could be much higher. A signal 60 dB above the band noise would be 0.1 V at the summing point. On the desired side of zero beat, this signal would be passed on to the volume control. On the other side of zero beat the 50 mV I channel signal would add to the –50 mV Q channel signal for a sum less than 100 µV. This means that the I and Q channels have to amplify signals 60 to 80 dB above the noise floor without distortion or compression. Harmonics and intermod products generated in the I and Q audio channels have different relative phase. It is also unreasonable to expect the two channels to have identical distortion characteristics. Distortion asymmetry is also an issue in phasing systems. Harmonic distortion is familiar to audio engineers. For harmonics more than 60 dB down, the total harmonic distortion (THD) specification is: THD < 0.1%. A receiver with THD 0.1% I and Q channels could handle an undesired signal 60 dB above the noise floor, but it would have no head room. As soon as a signal was strong enough to measure on the opposite sideband, distortion would begin to dominate. A better receiver would provide 60 dB of attenuation to a signal 80 dB above the noise floor, and no audible distortion products in the wrong sideband. Such signals are encountered on 20 meters during contests. This would require THD of 0.01% for undesired 1-V signals at the op-amp summing point. While this is possible using serious audio engineering techniques, it is clear that the quest for ever-higher opposite sideband suppression in phasing receivers has a practical limit. As in phasing transmitters, very low distortion is needed in the I and Q channels of a phasing receiver. The benefit for the user is that a carefully designed phasing receiver will sound exceptionally good. If the ultimate rejection to close-in interfering signals is desired, a different receiver architecture is needed. A superhet with a fixed IF and a carefully designed combination of crystal filters and/or phasing and/or DSP can provide over 100 dB of opposite sideband suppression across the entire 300 – 3000 Hz band.

Special considerations for CW

Many phasing direct conversion receivers have been built by dedicated CW operators who have no interest whatever in SSB bandwidths. Would such receivers benefit from redesigned audio phase shift networks? No. Remember that selectivity is improved by doing a better job of *rejecting* signals in the *stopband*, not passing signals in the passband. Thus it can be argued that the optimum phase shift network for a high performance CW receiver is exactly the same as the optimum network for SSB. In addition, a good CW receiver has several bandwidths, from narrow contest filters to wide open ones used for tuning around sparsely occupied bands. One major benefit of phasing receivers is the ease of making changes to the selectivity. It is easy to add filter options if frequencies from 200 Hz to 4000 Hz on the opposite side of zero beat are suppressed.

However, some receivers are optimized for simplicity, and there are other applications of simplified phasing method image-reject circuitry. If the audio band is limited to 300 – 1200 Hz, it is possible to obtain more than 50 dB of opposite sideband suppression with a pair of second order networks. An op-amp 2nd order network optimized for a CW-only receiver is shown later in this chapter. One application for CW bandwidth image-reject mixers is as the product detector following a simple CW filter. The combination of a crystal filter and image-reject product detector circuitry can provide better performance than either is capable of alone, as is demonstrated by the radios such as the Kenwood TS-570. By distributing the selectivity between a crystal filter before IF gain and a phasing product detector, the need for a "tail end" filter is generally avoided.

9.5 BINAURAL RECEIVERS

In a Binaural IQ receiver the I and Q channels are preserved all the way to the headphones. **Fig 9.29** (see next two pages) is a binaural receiver circuit from March 1999 *QST*. Sorting out the signals and interference is done using the ear-brain processor. As illustrated in the experiment described earlier, an outboard network built around an audio phase-shift network may be used to further process the I and Q channels. The network shown in **Fig 9.30** provides some sideband suppression and CW selectivity. The network in **Fig 9.31** provides ISB headphone output. Phasing circuitry and recombining are normally performed at low signal levels in receivers, to keep the amount of circuitry that must be precisely matched between the I and Q channels to a minimum. Binaural receivers built with standard tolerance components do not provide the I Q phase and amplitude precision needed to achieve high levels of opposite sideband suppression with outboard networks. Binaural receivers are a delightful way to listen, and also have many uses on the experimenter's bench. For example, a binaural receiver tuned across a CW signal from a crystal oscillator is a precise, low-distortion audio signal generator with matched I Q outputs—just the ticket for making circles on an X-Y oscilloscope.

Adjusting Phasing Rigs

One of the difficulties that renewed interest in phasing exciters and receivers has raised is that the lore of phasing rig adjustment has literally died out with the '40s and '50s generation of radio experimenters. Although modern components and modern component tolerances permit us to build phasing rigs that perform well beyond the capability of the classics, aligning them requires an unfamiliar set of skills. Some techniques, particularly those familiar to George Grammer at the ARRL, were well documented, but others are only preserved as vague recollections of observing the masters at work in their radio labs. Those of us who now experiment with phasing rigs have had to start from scratch and design new adjustment techniques, while remaining painfully aware that we are recreating a lost art.

In the math section, we found that we could combine all of the amplitude errors into a single term, and all of the phase errors into a single term. These two error terms are orthogonal—no amount of tweaking on the amplitude trimpot can correct a phase error, and vice versa. The situation is very much like shooting at a target. The

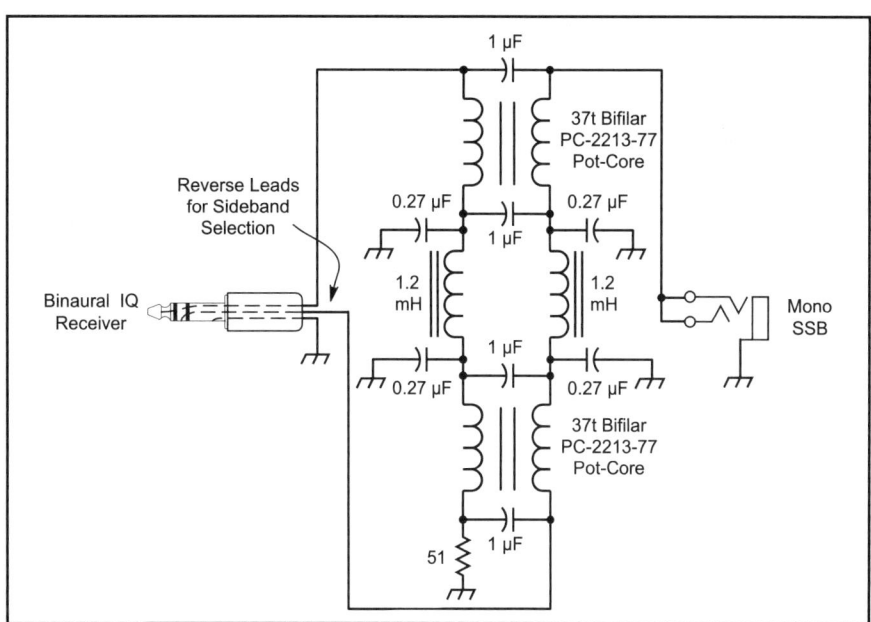

Fig 9.30—This outboard binaural network provides some sideband suppression and CW selectivity.

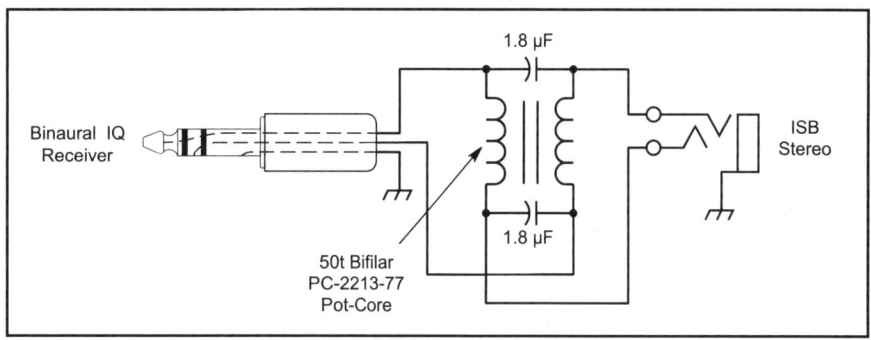

Fig 9.31—This outboard binaural network provides ISB headphone output.

sights have two adjustments: windage (or azimuth); and elevation. Both have to be properly adjusted to hit the center.

With two orthogonal error terms, a phasing rig needs two adjustments for opposite sideband suppression. This is a critically important point: no matter how many small amplitude errors we have in the system, we can tune them out with just a single amplitude balance adjustment. Similarly, all of the small phase errors in the system may be tuned out with just a single phase adjustment. We need precisely two adjustments in a phasing rig to null the opposite sideband.

The strategy for designing and building a successful phasing rig comes directly from the mathematics: design the system so that all the amplitude and phase errors are small; and include a single amplitude balance adjustment and a single phase trim adjustment to reduce the effect of the respective errors to zero.

Unlike most other tweaks in Amateur Radio, phasing adjustments cannot be tuned for maximum smoke. The easiest way to adjust a phasing receiver is to tune across a steady CW tone from an external signal generator, adjusting the phase and amplitude trimmers for minimum response on the undesired sideband. The signal generator must have adjustable output level so that the test signal can be kept between the receiver noise floor and distortion level on both the desired and undesired sidebands. The receiver and signal generator both need to be well shielded, to prevent signals from the generator leaking into the I or Q channel.

The easiest way to adjust a phasing ex-

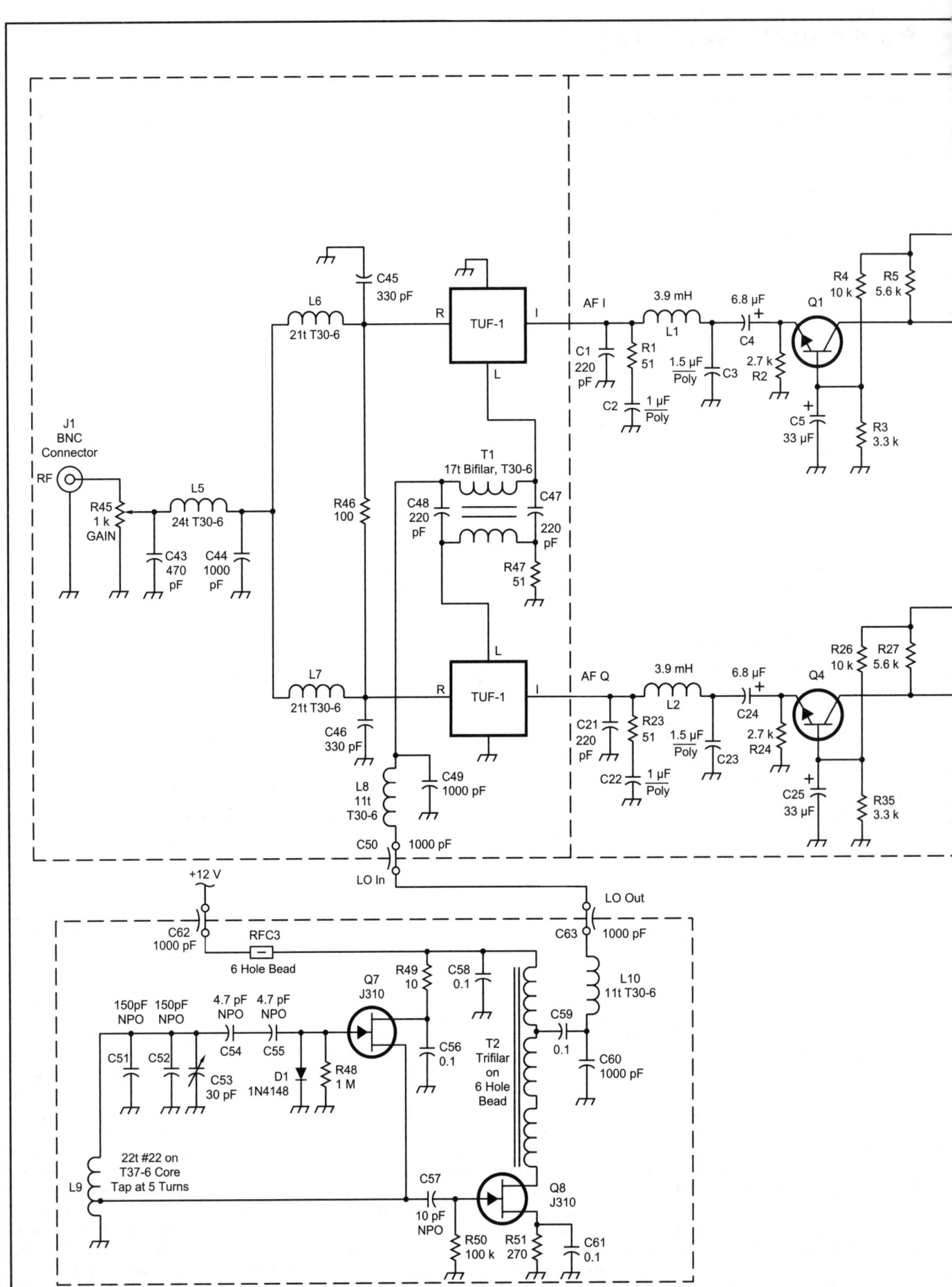

9.20 Chapter 9

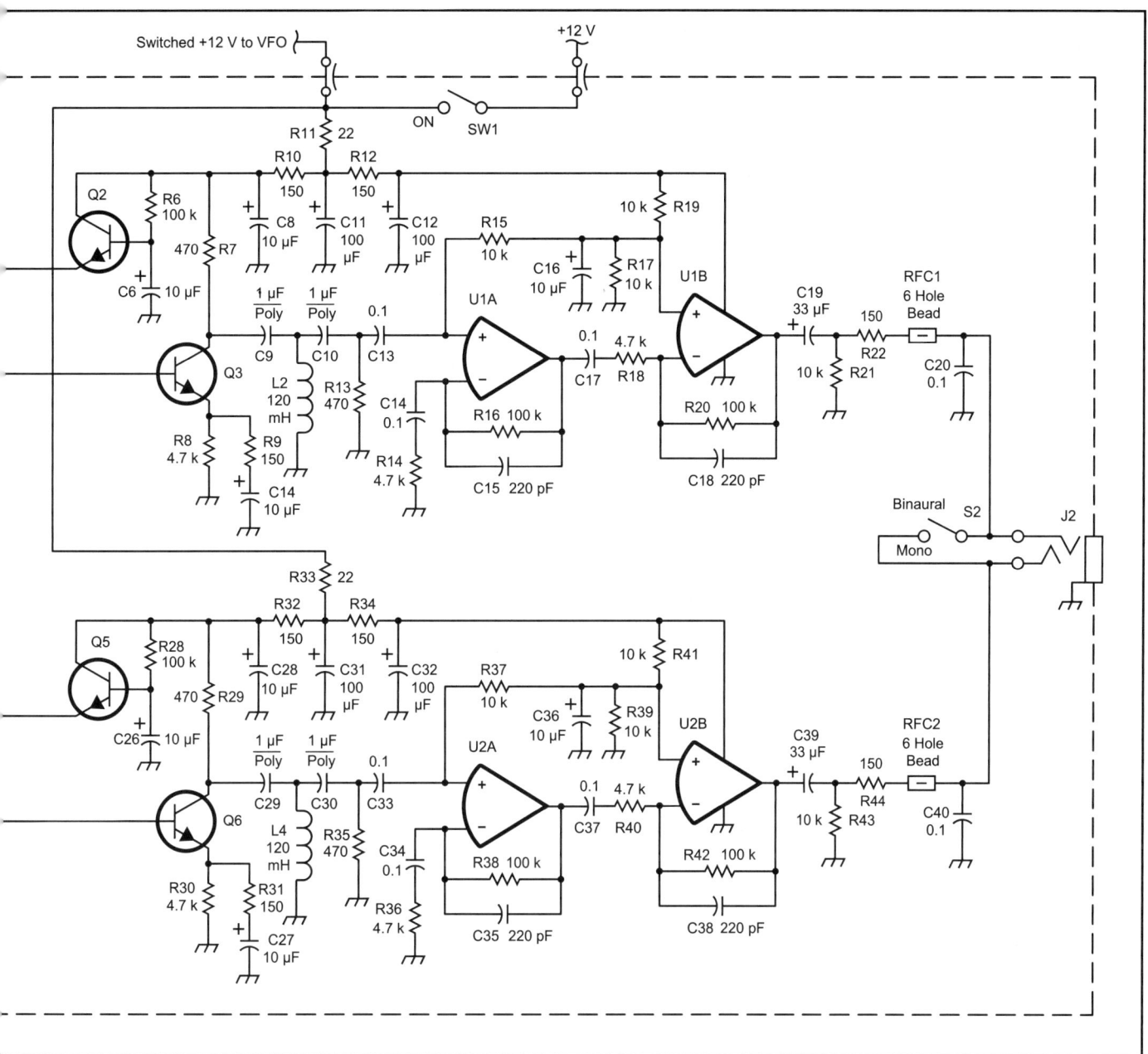

Fig 9.29—A binaural receiver circuit from March 1999 *QST*.

citer is to tune its low level output on a receiver with low distortion, very good selectivity, and selectable sidebands. Inject a pure sine wave audio tone into the microphone input and switch back and forth between the desired and undesired sidebands while adjusting the exciter phase and amplitude trimmers. Then sweep the audio tone frequency from 300 to 3000 Hz to verify that sideband suppression holds across the desired audio passband.

An SSB exciter with a pure sine wave audio tone into the microphone input generates a sine wave RF output. Residual carrier and opposite sideband energy amplitude modulates the desired sine wave RF output. The SSB exciter output may be observed on an oscilloscope, and phase and amplitude trimmers adjusted to reduce the audio amplitude variations in the output waveform. It is difficult to reduce spurious outputs by more than 40 dB while observing the exciter output on an oscilloscope, because the carrier, opposite sideband, distortion products, harmonics on the audio input tone, and power supply hum and noise all contribute to amplitude modulation of the desired sine wave RF output.

There is a clever old technique for adjusting opposite sideband suppression that does not require a good receiver or oscilloscope. The exciter output is connected through a suitable attenuator into a diode detector with headphones. With a low-level 1000-Hz sine wave tone injected into the microphone input, the SSB exciter will have a little output at the suppressed carrier frequency f_o; a desired sideband output 1000 Hz away; a suppressed opposite sideband 1000 Hz on the other side of the carrier frequency; and distortion products. The distortion products can be made arbitrarily small by reducing the audio tone level at the microphone input. The desired sideband, carrier, and opposite sideband all beat together in the diode detector, and the audible beats may be heard on the headphones. Imperfect carrier suppression results in a 1000-Hz audio tone, and poor opposite sideband suppression results in a 2000-Hz audio tone. The SSB exciter phase and amplitude trimmers may be adjusted for minimum 2000-Hz tone. If the exciter has carrier balance adjustments, they may be trimmed for minimum 1000-Hz tone.

Amplitude Balance Adjustment

The amplitude balance adjustment may be a variable gain element in either the I or Q channel anywhere from the point where the two paths separate to the point where they are summed together. It is usually easier to use a variable resistor at baseband, particularly if op-amp gain blocks are included in the system. A convenient amplitude trimmer for receivers is a ten-turn trim pot connecting the I and Q audio channels to the inverting input of the summing amplifier. If sideband switching is implemented by interchanging the I Q connections at the input to a precise audio phase shift network pair, balancing the amplitudes before the switch results in a system that has nearly equal sideband suppression on either sideband.

A significant point to watch for is that the variable gain element does not unbalance the drive or load impedance of bandpass or all-pass networks. An amplitude adjustment that behaves differently at low audio frequencies than at high audio frequencies, or that introduces phase errors across the audio frequency range, will make it impossible to obtain good opposite sideband suppression across the whole audio range. In exciters it is best to include a separate op-amp variable gain stage, to avoid upsetting either the mixer diplexer impedances or the op-amp all-pass network drive impedances. **Fig 9.32** shows possible locations for the amplitude balance adjustment in receivers, and **Fig 9.33** shows locations for exciters. Remember that only one amplitude adjustment is needed. The amplitude adjustments shown have no appreciable affect on phase. When DSP is used, it may be useful to do the

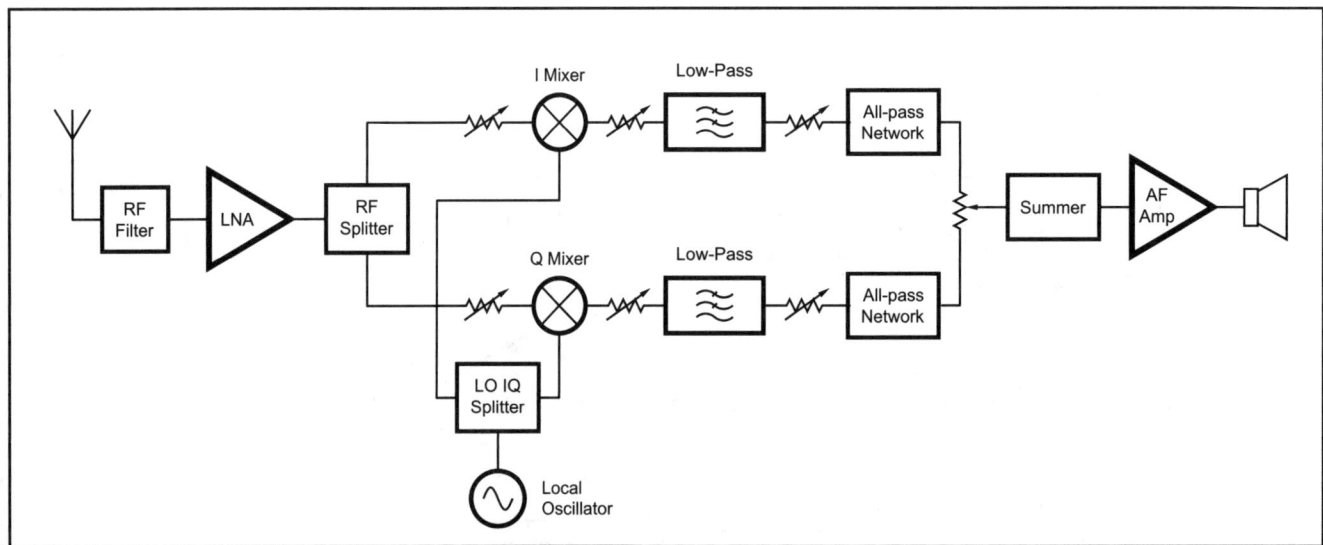

Fig 9.32—Possible locations for the amplitude balance adjustment in receivers.

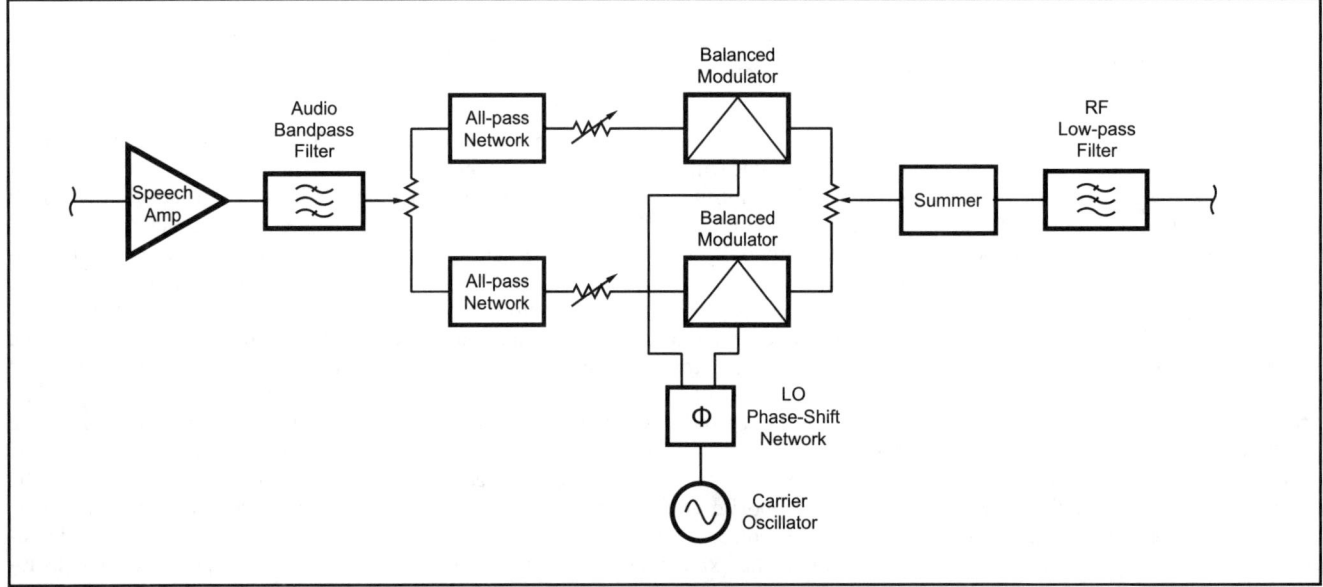

Fig 9.33—Possible locations for the amplitude balance adjustment in exciters.

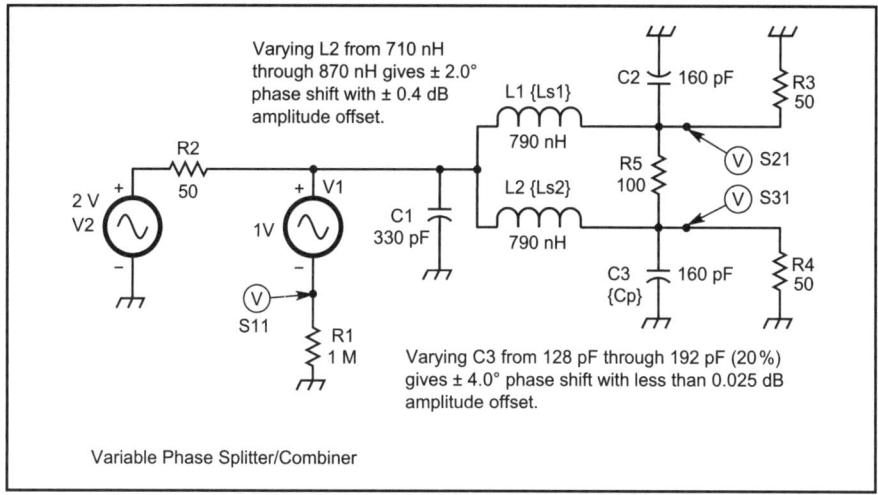

Fig 9.34—A variable phase splitter/combiner network for a 20-meter receiver or exciter. The PSPICE signal generators allow extraction of S11.

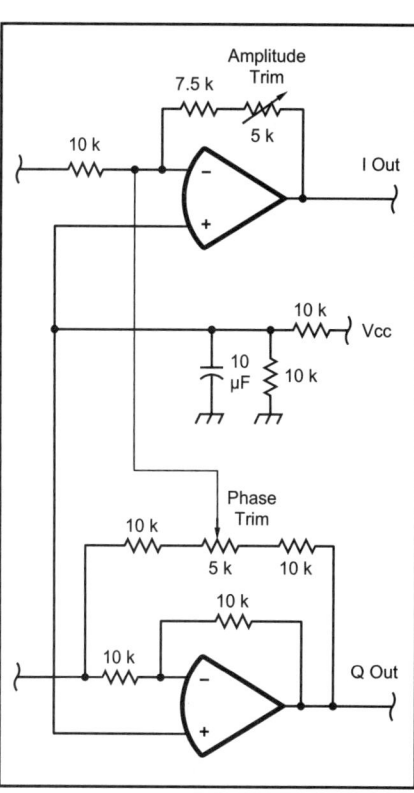

Fig 9.35—This op-amp circuit permits a small amount of Q-channel signal to be either added or subtracted to the I-channel signal.

amplitude balance trimming in software.

Phase Trim Adjustment

There are many possibilities for the location of the phase trimmer. Phase may be trimmed in the I and Q signal path anywhere after the audio phase-shift network in exciters and anywhere before the audio phase-shift network in receivers. LO Phase may also be trimmed at either the I or Q mixer LO port. As long as the phase errors in the system are small, only a single phase trim adjustment is needed, and it may be anywhere in the system. Some locations for the phase trim adjustment are better than others. The amplitude balance and phase balance in a phasing rig are mathematically independent, but it is not trivial to adjust phase without affecting amplitude as well. When mixers with saturating LO drive (for example, diode rings and Gilbert cells) are used, small changes in LO amplitude do not have a large effect on mixer performance. For this reason, including the phase trim adjustment in the mixer LO drive rather than in the RF or baseband path is good practice. On the other hand, low-pass filtering is needed at the output of phasing exciters and at the input to direct conversion receivers. A low-pass Wilkenson splitter is a useful RF splitter or combiner for a phasing rig, and using a variable capacitor for one element allows smooth adjustment of phase. **Fig 9.34** illustrates a network for a 20-meter receiver or exciter. The variable capacitor trims the phase over a plus or minus 4-degree range with 0.025 dB variation in amplitude.

It is possible to do the phase trimming at baseband, either in DSP or using op-amps. For complete suppression of the undesired

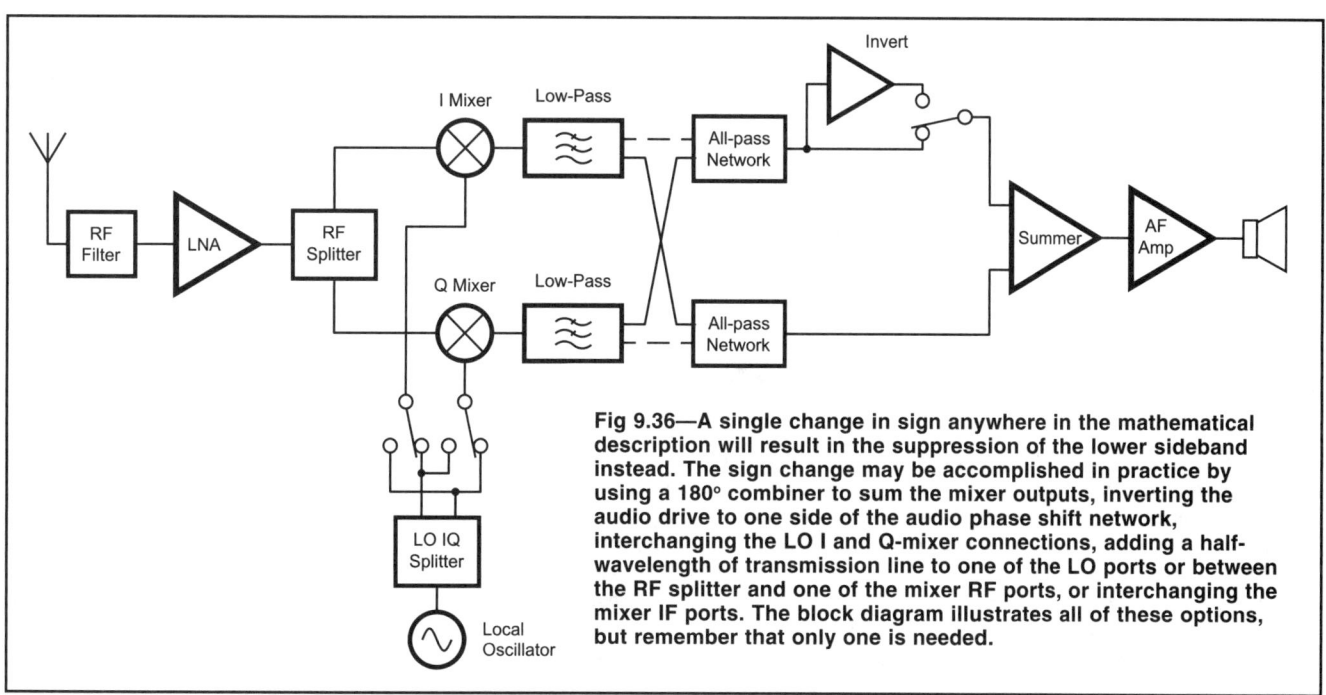

Fig 9.36—A single change in sign anywhere in the mathematical description will result in the suppression of the lower sideband instead. The sign change may be accomplished in practice by using a 180° combiner to sum the mixer outputs, inverting the audio drive to one side of the audio phase shift network, interchanging the LO I and Q-mixer connections, adding a half-wavelength of transmission line to one of the LO ports or between the RF splitter and one of the mixer RF ports, or interchanging the mixer IF ports. The block diagram illustrates all of these options, but remember that only one is needed.

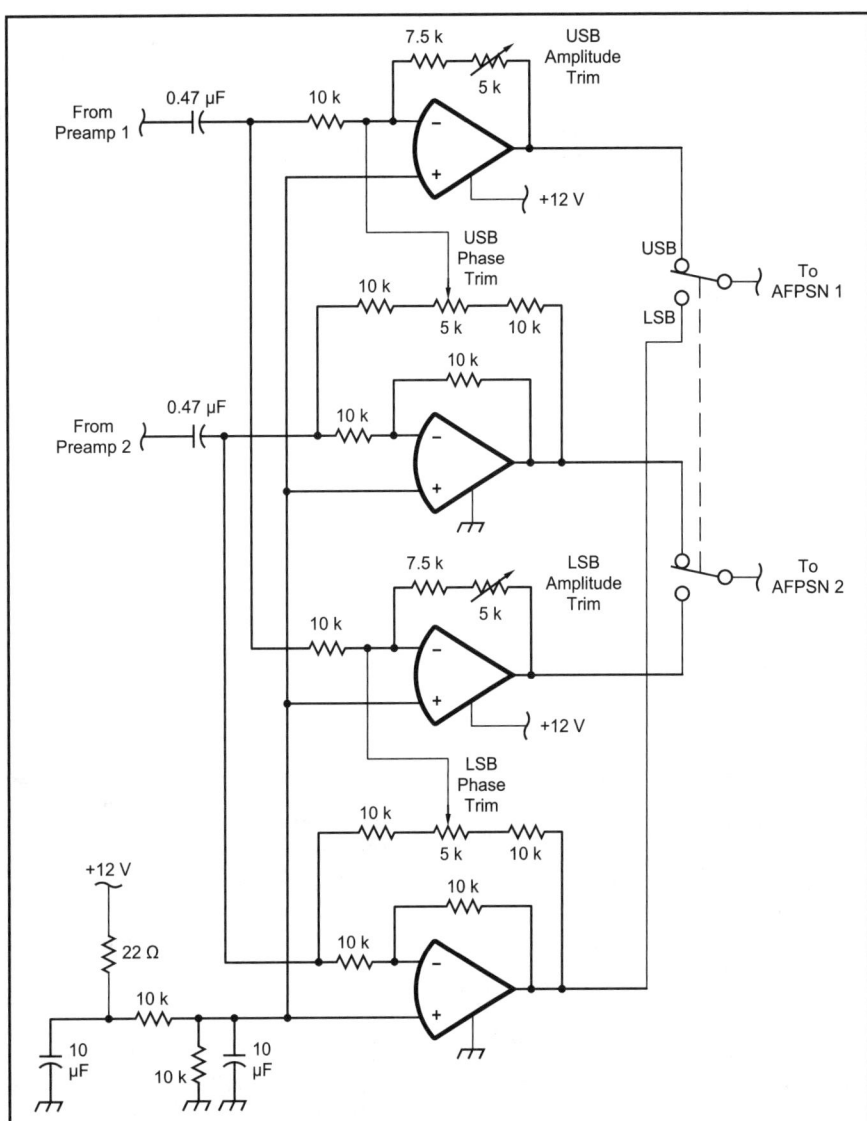

Fig 9.37—For systems that need to perform equally well on either sideband, the phase and amplitude adjustments may either be front panel mounted and adjusted every time the other sideband is selected, or an independent set of phase and amplitude adjustments may be used for each sideband.

tracted to the I channel signal. The same principle may be applied to receivers. It is necessary to do the phase trimming at a point in the audio circuitry where the signals in the two channels are 90 degrees apart, that is, between the mixers and the audio phase shift networks in both receivers and exciters.

Sideband Selection

In the mathematical description of a phasing receiver, the lower sideband is suppressed when the 90° shifted audio is multiplied with the 90° shifted LO, and the outputs of the two mixers are added. A single change in sign anywhere in the mathematical description will result in the suppression of the upper sideband instead. The sign change may be accomplished in practice by using a 180° combiner to sum the mixer outputs, inverting the audio drive to one side of the audio phase shift network, interchanging the LO I and Q mixer connections, adding a half-wavelength of transmission line to one of the LO ports or between the RF splitter and one of the mixer RF ports, or interchanging the mixer IF ports. The block diagram in **Fig 9.36** illustrates all of these options, but remember that only one is needed. Switching sidebands will generally introduce a different set of amplitude and phase errors. For systems that need to perform equally well on either sideband, the phase and amplitude adjustments may either be front panel mounted and adjusted every time the other sideband is selected, or an independent set of phase and amplitude adjustments may be used for each sideband. **Fig 9.37** shows one way this may be accomplished.

sideband, the I and Q channels after the audio phase-shift network in an exciter need to have the same signal, but 90 degrees out of phase. If there is a phase error, the angle between the I and Q channels will not be 90 degrees. It is possible to obtain exactly 90 degrees of phase shift by adding a small amount of the signal in the Q channel to the I channel. If the phase error is in the opposite direction, then a small amount of the Q channel signal can be subtracted to achieve exactly 90 degrees phase shift. The op-amp circuit in **Fig 9.35**, similar to one published by Blanchard, permits a small amount of Q channel signal to be either added or sub-

9.6 LO AND RF PHASE-SHIFT AND IN-PHASE SPLITTER-COMBINER NETWORKS

Numerous articles over the years have addressed the topic of LO phase shift networks for phasing rigs. The recent work by Blanchard is particularly recommended. In this section we will discuss the requirements and implications of different network selections, and present the networks that we have used extensively. Experimentation with other networks is highly recommended, as the ones presented here are not necessarily optimum, they are just familiar.

The first topic to address is the question of where to put the 90 degree phase shift: in the RF path or the LO path. There is an easy answer to this question that is usually correct. The RF path contains signals that must be precisely matched in amplitude and phase between the I and Q RF channels. The LO path has a pair of sine waves with precisely defined phase, but we are usually not too concerned with LO amplitude, and we never need it to be matched to hundredths of a dB. Simple phase shift networks provide precise 90° phase shift over a wide bandwidth, but the amplitude is only balanced at a single frequency.

Equal amplitude I and Q LO may be obtained by following such a network with a limiter. Phase shift networks using splitters and lengths of transmission line, either actual coax or lumped element equivalents, have well matched amplitude over a wide frequency range, but 90° phase shift at only one frequency. It is difficult to build a passive network that provides both precise amplitude balance and a 90° output pair over a wide RF bandwidth. With wideband op-amps, we can use the same circuitry from 3 to 30 MHz that we use from 300 to 3000 Hz, but we wouldn't want to use a wideband unity-gain op-amp circuit as the RF input stage of a receiver. On the other hand, there are many simple in-phase splitters that provide good phase and amplitude accuracy over a wide bandwidth. For this reason, we almost always put the 90° phase shift network in the LO path and an in-phase splitter in the RF path.

One reason that we might choose to use in-phase LO and quadrature RF is that the RF ports of diode-ring mixers are often better behaved than the LO ports. Experimenters who build their first phasing rigs are often amazed at how much different an LO phase shift pair works when connected to mixers than when it is observed with 50-Ω loads on an oscilloscope. It is common for the phase adjustment range to be too small, and additional capacitors often need to be tacked on the bottom of the circuit board at one mixer LO port or the other. In many applications, the phasing receiver or exciter only needs to operate at a single frequency or over a very narrow band—for example, when following a

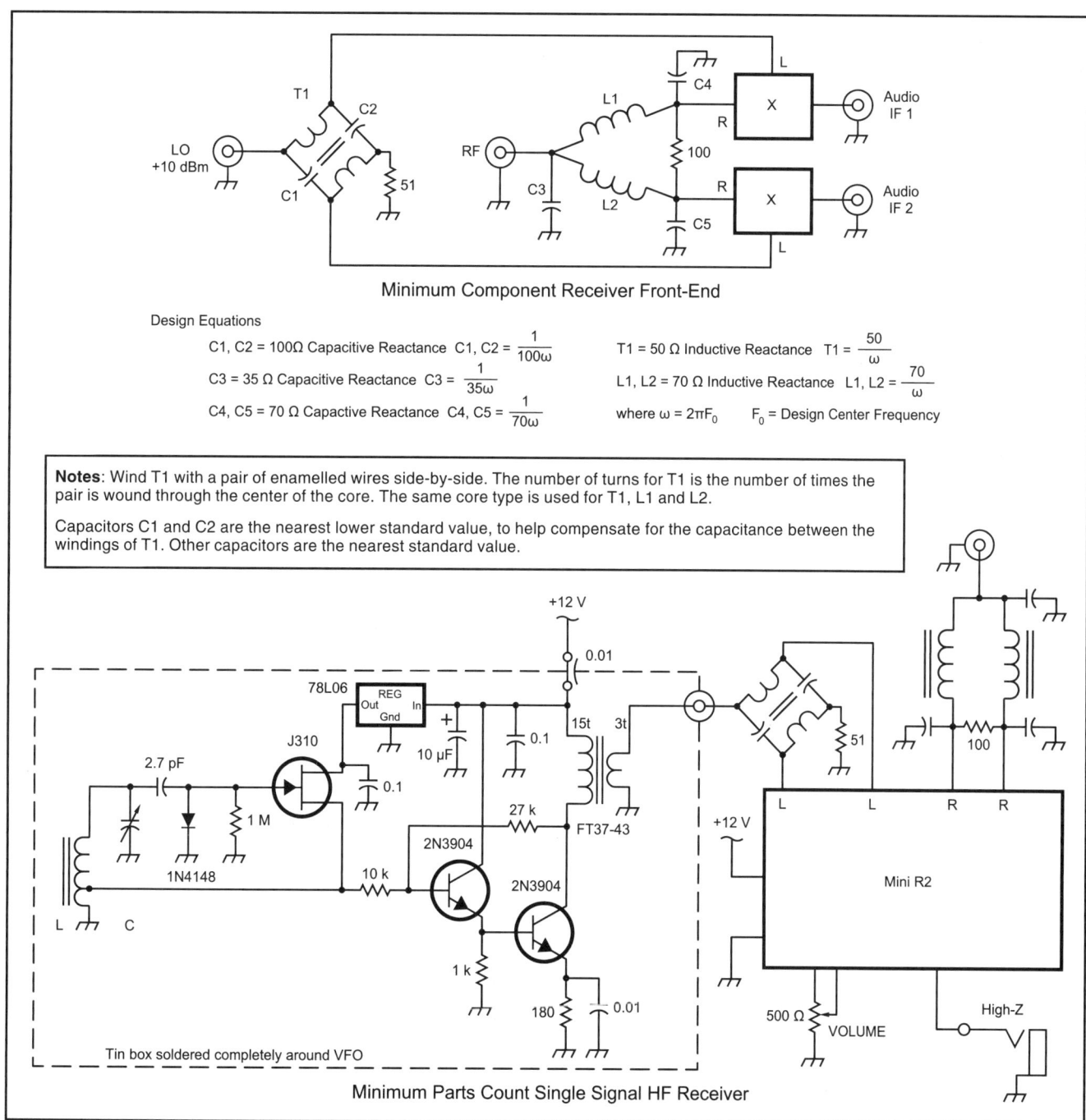

Fig 9.38— A good combination of LO quadrature network and RF splitter for HF and low VHF single-band receivers and exciters.

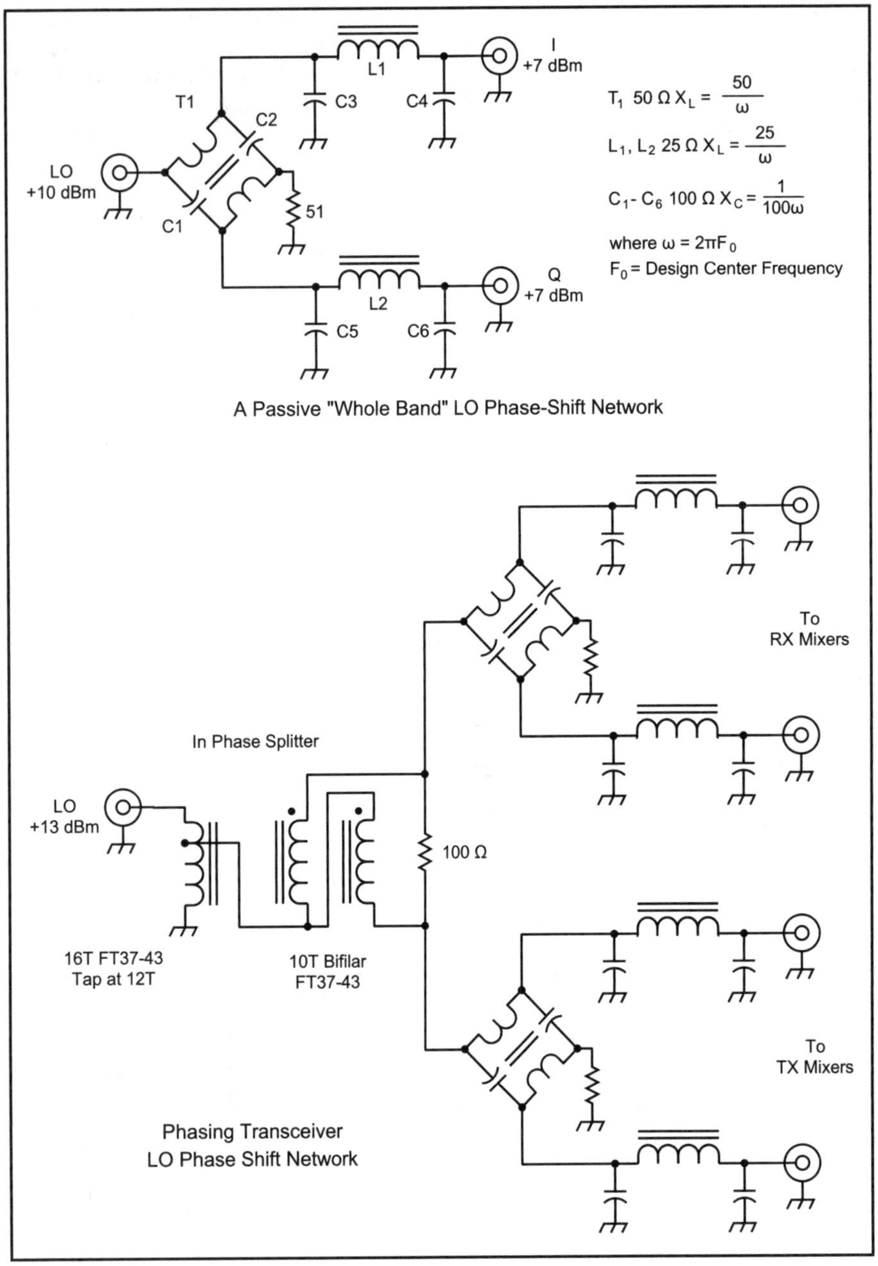

Fig 9.39—The LO quadrature network has wideband phase balance and acceptable amplitude balance over any amateur band, and the combined low-pass filter-splitter for RF provides a natural and well-behaved phase adjustment point. Here we move the phase adjustment to the LO path.

crystal filter or when used with a VXO as a tunable IF for microwaves. In this case the benefits of connecting the quadrature network to the RF ports instead of the LO mixer ports and using in-phase LO splitting may outweigh the bandwidth penalty.

A good combination of LO quadrature network and RF splitter for HF and low VHF single-band receivers and exciters is shown in **Fig 9.38**. The LO quadrature network has wideband phase balance and acceptable amplitude balance over any amateur band, and the combined low-pass filter-splitter for RF provides a natural and well-behaved phase adjustment point. **Fig 9.39** moves the phase adjustment to the LO path. This arrangement has been used extensively in amateur phasing exciters and receivers, and is attractive for band switched applications.

The bifilar toroid quadrature hybrid described in the reference by Fisher may be converted to a broadband structure by connecting a second network through a pair of transmission lines. The transmission lines are usually lumped element equivalents at frequencies below 50 MHz. The network shown in **Fig 9.40** is used in a receiver that covers 6.8 to 11 MHz without band switching. Front panel phase and amplitude trimmers are appropriate in such a receiver.

At VHF, a pair of transmission lines may be used, either with an in-phase splitter or just soldered together. It will be necessary to trim the length for maximum opposite sideband suppression. This is a tedious process, more so if connectors have to be unsoldered and resoldered every time the line length is trimmed. If anything in the system changes—any other transmission line length or the VSWR at any port—the line length will have to be readjusted. This brings up an interesting point: it is generally not appropriate to use modular construction and connectors between the

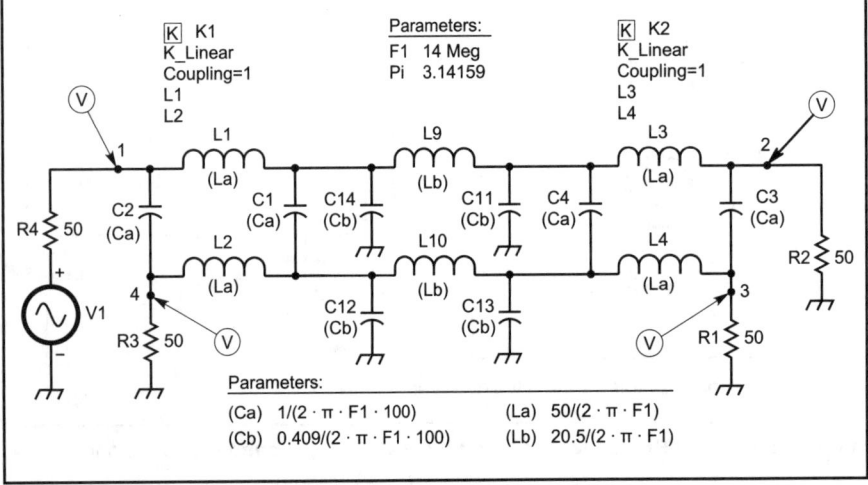

Fig 9.40—The bifilar toroid quadrature hybrid described by Fisher may be converted to a broadband structure by connecting a second network through a pair of transmission lines. The transmission lines are usually lumped element equivalents at frequencies below 50 MHz. This network is used in a receiver that covers 6.8 to 11 MHz without band switching.

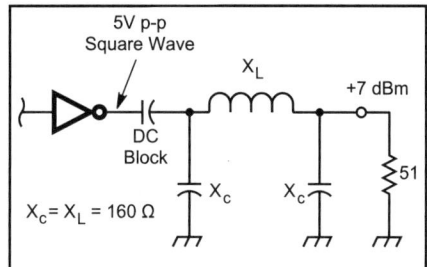

Fig 9.41—CMOS logic with a 5 V supply can drive +7 dBm into diode mixers using this circuit. The pi network converts the high-impedance IC square wave output into a sine wave and transforms the impedance down to drive a 50-Ω load.

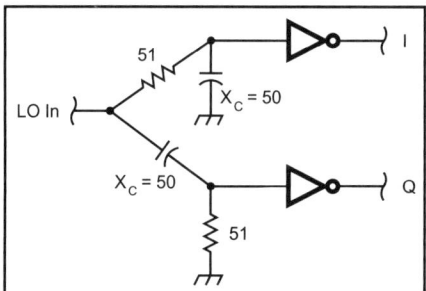

Fig 9.42—A simple logic LO phase-shift network.

stages of a phasing rig. It is much better to adjust it once, solder everything in place, and then leave it alone. If the rig has cables with connectors, they will eventually be needed for other projects and borrowed. Then new cables will have to be made up to get the phasing rig running again, and at 2 meters a few tenths of an inch makes a difference. Three of the most reliable rigs at KK7B use phase shift networks that were adjusted by squeezing turns on a toroid, and then the turns were locked in place with nail polish. All three still provide more than 40 dB of opposite sideband suppression after years of portable operation and world travel.

Digital ICs configured as frequency dividers can provide accurate 90° phase shift, and have often appeared in print. They have been used less often, partly because logic levels are not the appropriate drive for any of the more common mixers used in receivers and exciters, and partly because many more people have written about phasing rigs than have actually designed and built them. There may be parts of the brain that, once used to grasp fundamental digital concepts, are no longer capable of understanding basic RF. If so then the reverse is also probably true. Experiments with logic phase shift networks and commutating mixers are highly encouraged. CMOS logic with a 5 V supply can drive +7 dBm into diode mixers using the circuit in **Fig 9.41**. The pi network converts the high-impedance IC square wave output into a sine wave and transforms the impedance down to drive the 50-Ω load. The pi network output capacitor is a convenient point to trim the phase. A simple logic LO phase-shift network is shown in **Fig 9.42**. Instead of a frequency divider to obtain the 90° output pair, an RC network is used. The inverters following the RC network act as hard limiters, and the networks on the output provide +7 dBm into 50 Ω and a convenient phase trim.

Some DDS ICs provide I and Q outputs. These may be used with a broadband RF splitter and switched RF low-pass filters to build simple general coverage phasing rigs, and experiments along these lines are encouraged. For wideband rigs, it is convenient to do both the amplitude and phase trimming at baseband, using the op-amp circuitry shown earlier. The phase noise performance of wide range DDS based Local Oscillators

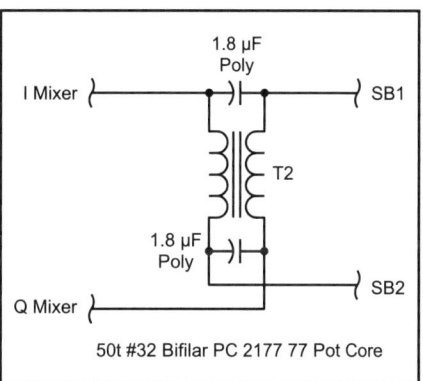

Fig 9.43—This simple quadrature hybrid circuit has good performance at only one audio frequency, but it is truly elegant in its simplicity and provides trivial sideband switching, draws no current, and offers the possibility of binaural independent sideband listening.

is often not in the "high-performance receiver" category, and the miniR2 circuit provides more than enough signal processing performance. The extra design and construction time and expense to use the R2 and R2pro circuitry is wasted if receiver system performance is limited by the LO.

Digital LO generation, and LO buffer amplifier distortion generate LO signals that may be very rich in harmonics. Harmonics are important in phasing systems, because a phase shift in a harmonic will shift the phase of the composite waveform. Even if the I Q LO provides a perfect pair of sine waves, harmonics are generated in the mixer. A conservative approach to control of harmonic phase is to drive the mixer LO ports with wideband buffer amplifiers and resistive attenuators. For most applications, a more practical approach is to have a wide range available on the phase trim adjustment to compensate for harmonic phase effects.

If the phase trim adjustment does not have enough range, a common technique is to tack a small value (start with a few pF) capacitor from one mixer LO port to ground. If the opposite sideband suppression improves, leave the chip capacitor in place and readjust. If opposite sideband suppression degrades, move the capacitor to the other mixer. Add enough capacitance that the phase trim adjustment range permits the opposite sideband suppression to be nulled. It may be necessary to add a surprisingly large value capacitor before the phase is equalized. 100 pF will shift a 40-meter signal in a 50-Ω system about 10 degrees.

Audio Phase Shift Networks

A collection of audio phase-shift networks is shown in the next set of figures. The simple quadrature hybrid circuit in **Fig 9.43** has good performance at only one

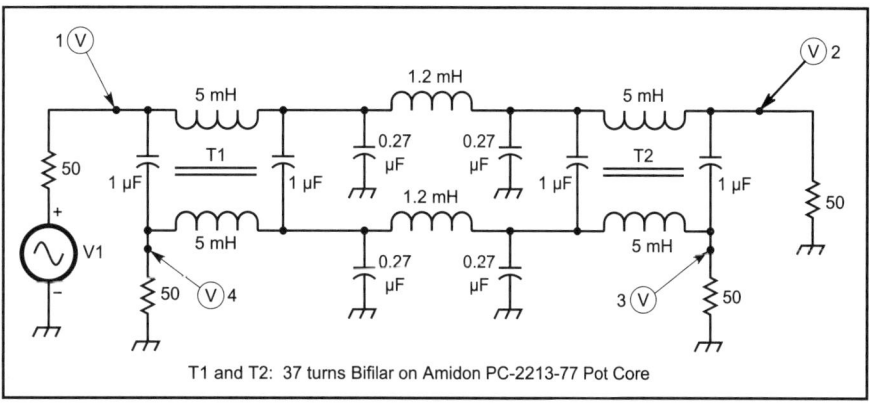

Fig 9.44—A broadband version of the circuit in Fig 9.43 provides marginal performance over a wider bandwidth, but good performance nowhere.

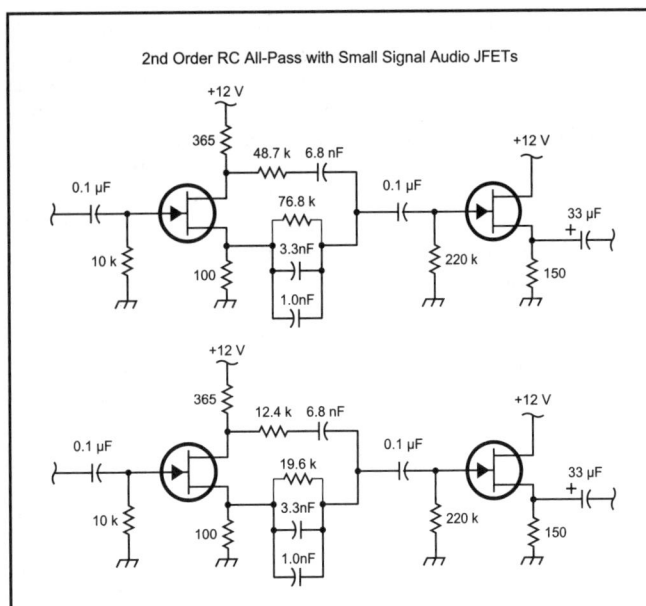

Fig 9.45—FET drive and load circuits for using classic second-order RC networks in exciters and receivers.

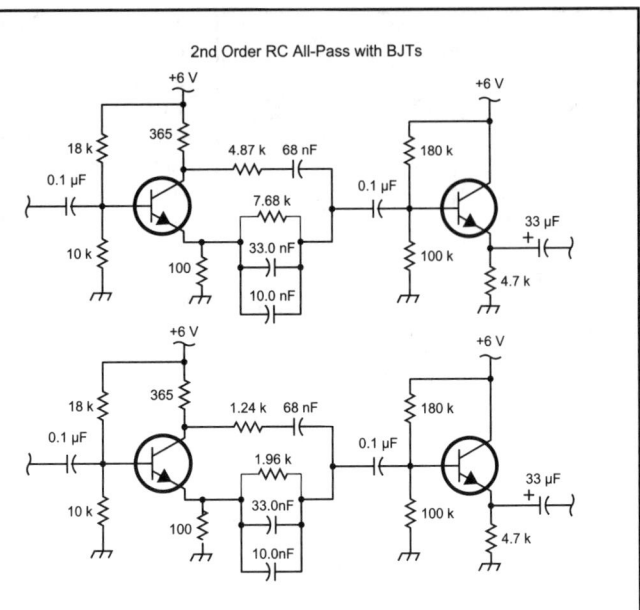

Fig 9.46—BJT drive and load circuits for using classic second-order RC networks in exciters and receivers.

audio frequency, but it is truly elegant in its simplicity and provides trivial sideband switching, draws no current, and offers the possibility of binaural independent sideband listening. It offers a real performance improvement over the simplest DSB direct conversion and regenerative receivers. The broadband version in **Fig 9.44** provides marginal performance over a wider bandwidth, but good performance nowhere. One difficulty with passive LC audio quadrature hybrid networks using pot-core inductors is maintaining inductor tolerances. The inductance can vary over a wide range depending on the tightness of screw holding the pot core halves together, and a mechanical jolt can result in a big inductance shift.

Second-order RC audio phase-shift networks were used in the classic homebrew and commercial rigs of the '50s. They are capable of good performance in both exciter and receiver applications, but will not provide the same level of performance as the common third-order op-amp networks or polyphase RC networks. Since networks with better performance are no more difficult to build, there is no obvious technical reason to use the classic circuitry in a rig with modern parts. There is, however, an appeal to simple circuitry, and even the solid-state circuits of the '60s are now old enough to be included in the classic category. A complete phasing transmitter using point-to-point wiring and only two and three terminal devices (no ICs) could be part of a '60s vintage homebrew station, and more importantly, could sound exceptionally good on the air. It is critical to remember that the drive and load impedances, and the relative drive levels, are part of the network. **Figs 9.45** and **9.46** show several different drive and load circuits for using classic second-order RC networks in exciters and receivers. Components are standard 1% resistors and matched capacitors.

Fig 9.47 is a single stage op-amp all-pass network. This is such a common circuit in phasing rigs that it is useful to examine its behavior. At DC, C1 is an open circuit. The gain from Vi through the non-inverting input is +2. The gain from Vi through the inverting input is –1. These two add together for a net gain of +1 at DC. At high frequency, C1 effectively shorts the inverting input to ground. Then the gain from Vi through the non-inverting input is 0, and the gain from Vi through the inverting input is still –1. The sum is –1. The frequency f_o occurs when $XC1 = R1$. The voltage at the non-inverting input at f_o is $0.5(1-j)$. The gain from Vi through the non-inverting input at f_o is $1-j$. The sum of the outputs from Vi through the inverting and non-inverting inputs is $-1 + (1-j) = -j$. Thus, the all-pass op-amp circuit has unity gain all the way from dc to high frequencies, and a phase shift of $-90°$ at f_o. A phasing rig with just one op-amp all-pass network could have perfect opposite sideband suppression at one frequency f_o. By adding a second all-pass network in the other channel with a different frequency f_o, a phase difference of approximately 90° can be maintained over a small bandwidth. This might be useful for a simple CW receiver or an SSB transmitter with very relaxed (20 dB) opposite sideband suppression requirements. **Fig 9.48** is a pair of all-pass networks with the 90° frequencies chosen for good suppression over an audio band from 470 to 900 Hz, and **Fig 9.50** is a pair that provides at least 21 dB suppression from 360 Hz to 2050 Hz. **Figs 9.49** and **9.51** show the phase errors from 0 to 4 kHz. The errors may be reduced by adding more sections and re-calculating the all-pass network frequencies. Adding a second pair of op-amps allows us to achieve better opposite sideband suppression performance over wider bandwidths. **Fig 9.52** illustrates a

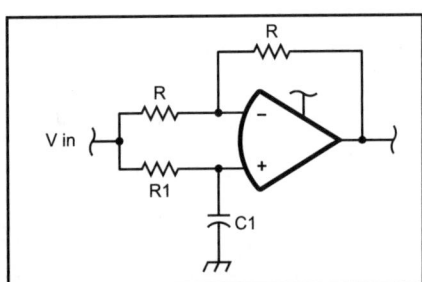

Fig 9.47—A single-stage op-amp all-pass network.

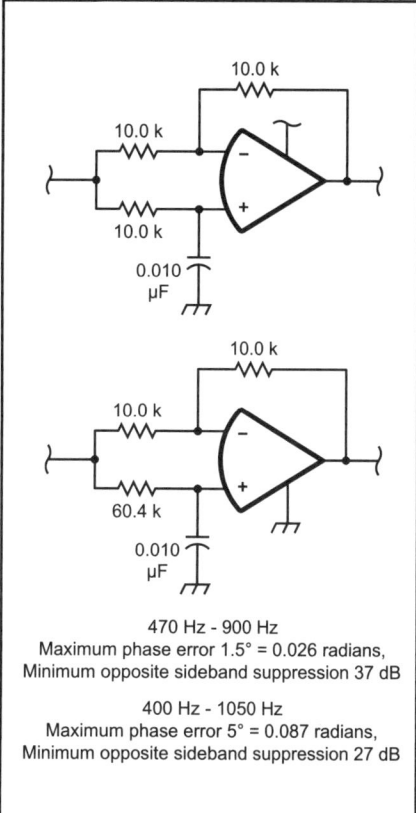

Fig 9.48—A pair of all-pass networks with the 90° frequencies chosen for good suppression over an audio band from 470 to 900 Hz.

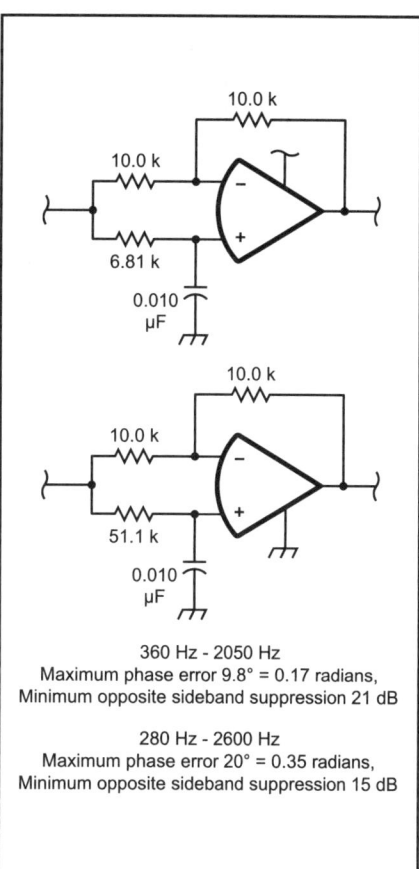

Fig 9.50—A pair of all-pass networks that provide at least 21 dB suppression from 360 to 2050 Hz.

second-order all-pass network pair for CW receivers that provides more than 50 dB of opposite sideband suppression from 300 Hz to 1120 Hz, and **Fig 9.54** is one that provides more than 36 dB of opposite sideband suppression from 250 Hz to 3650 Hz for SSB operation. **Figures 9.53 and 9.55** show the phase errors for these two networks.

Adding a third pair of op-amps allows us to build a network with small enough amplitude and phase errors that we could achieve almost 60 dB of sideband suppression from 270 Hz through 3600 Hz, if the rest of the receiver were perfect. With this network, other receiver considerations will set the practical limit for sideband suppression. For most applications, the third-order all-pass network pair shown in Fig 9.56 is recommended. **Fig 9.57** shows the phase errors. Op-amps, resistors and capacitors are inexpensive, and this network has been widely duplicated. Note that one resistor value, 1.52 kΩ, is not a standard 1% component. A 1.50-kΩ and a 20-Ω resistor in series will stand side-by-side on the PC board.

Phase Shift Network Component Tolerances

With 1% tolerance resistors and capacitors right out of the bag, the network in Fig 9.56 will reliably provide more than 40 dB opposite sideband suppression. **Fig 9.58** is a simulation of the phase error when component values vary by 1% or less. Selecting the resistors and capacitors by hand using an accurate ohm and farad meter will improve performance. **Fig 9.59** is a simulation with 0.5% errors, and **Fig 9.60** is a simulation with 0.2% errors. More precise matching beyond 0.1% does not provide any practical benefit with 3rd order networks, because the design errors in the network are then larger than the component tolerance errors, as shown in **Fig 9.61**. Note that the capacitors and 10.0 k resistors all have the same value, and may be matched to each other, rather than an absolute standard. 1% resistors are cheap—it may be easiest to just measure a bunch of the 6 values needed and select those that are closest to the design value.

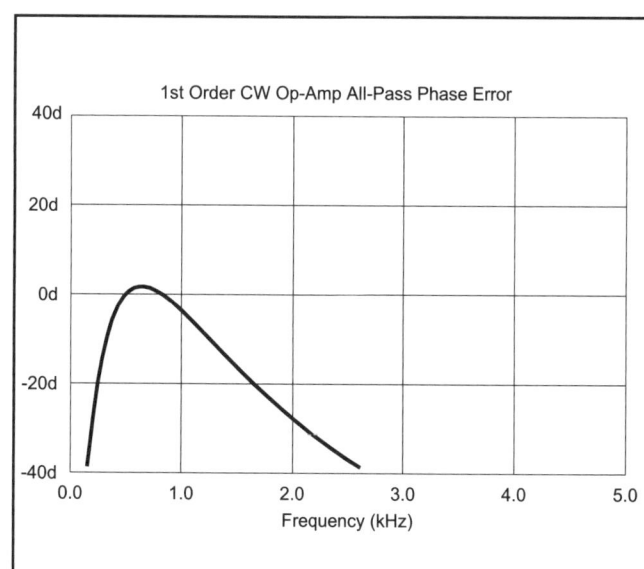

Fig 9.49–Phase errors of the Fig 9.48 network pair.

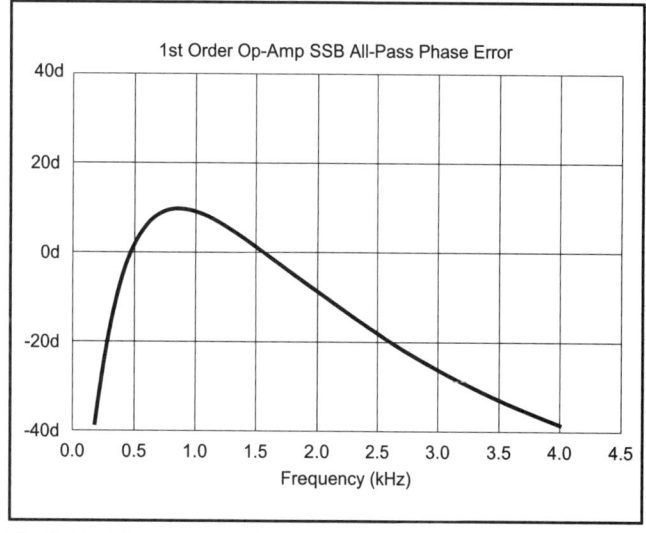

Fig 9.51—Phase errors from 0 to 4 kHz. The errors may be reduced by adding more sections and recalculating the all-pass network frequencies.

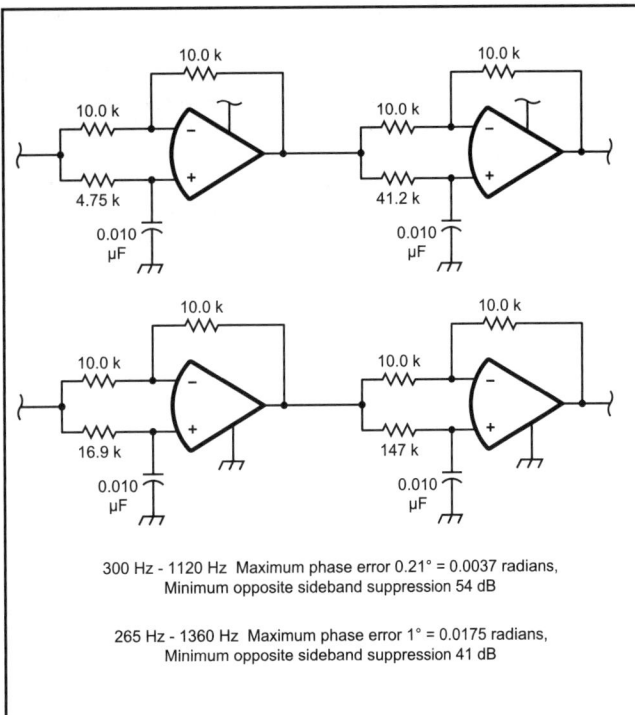

Fig 9.52—A second-order all-pass network pair for CW receivers that provides more than 50 dB of opposite sideband suppression from 300 Hz to 1120 Hz.

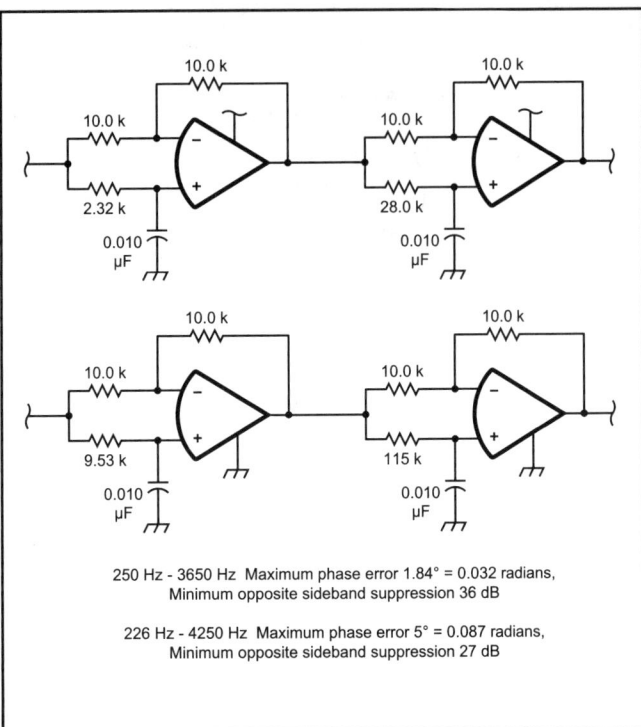

Fig 9.54—This second-order all-pass network provides more than 36 dB of opposite sideband suppression from 250 Hz to 3650 Hz for SSB operation.

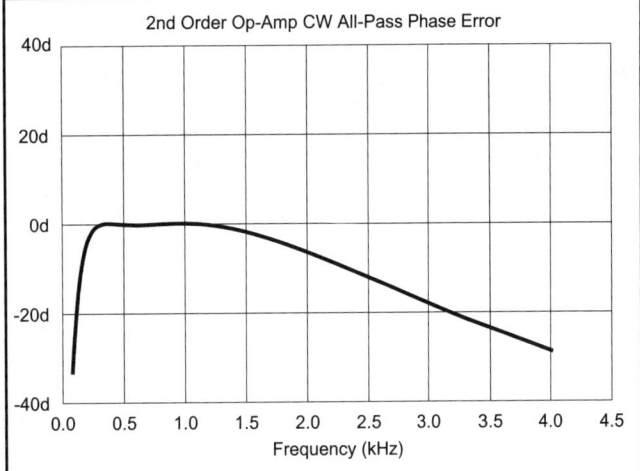

Fig 9.53—Phase errors in the second-order all-pass network shown in Fig 9.52.

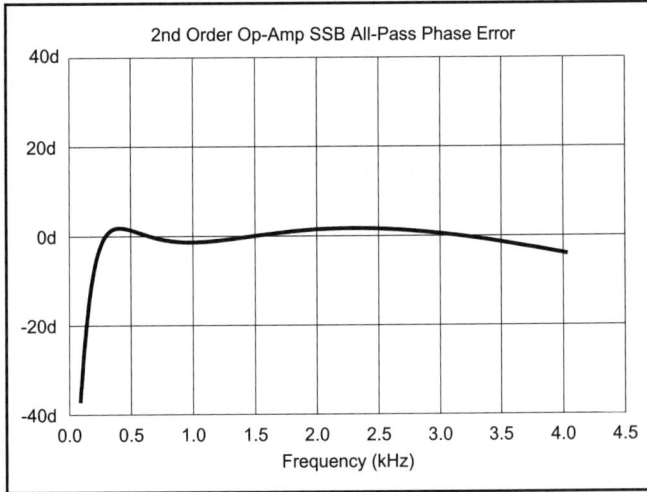

Fig 9.55—Phase errors in the second-order all-pass network shown in Fig 9.54.

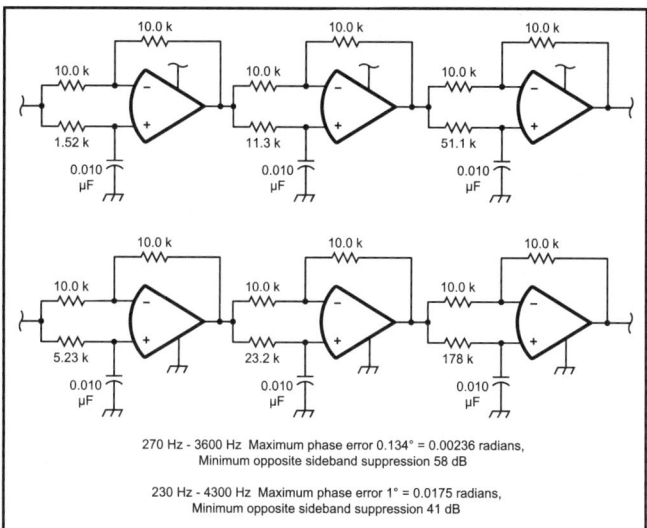

Fig 9.56—Adding a third pair of op-amps allows us to build a network with small enough amplitude and phase errors that we could achieve almost 60 dB of sideband suppression from 270 Hz through 3600 Hz, if the rest of the receiver were perfect.

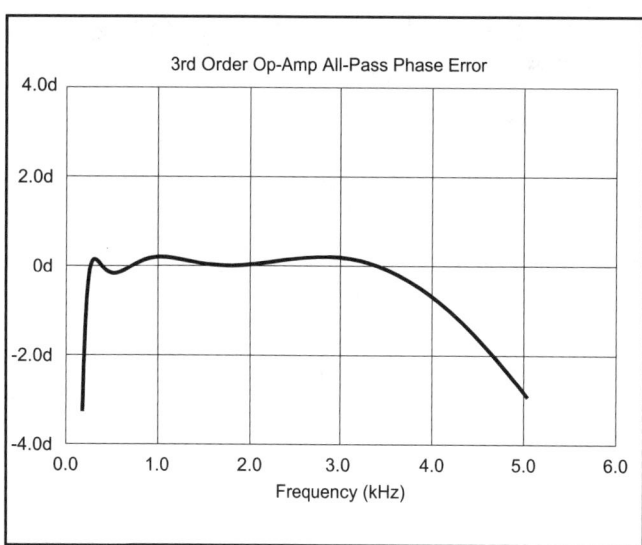

Fig 9.57—Phase errors in the network shown in Fig 9.56. Note the change in scale.

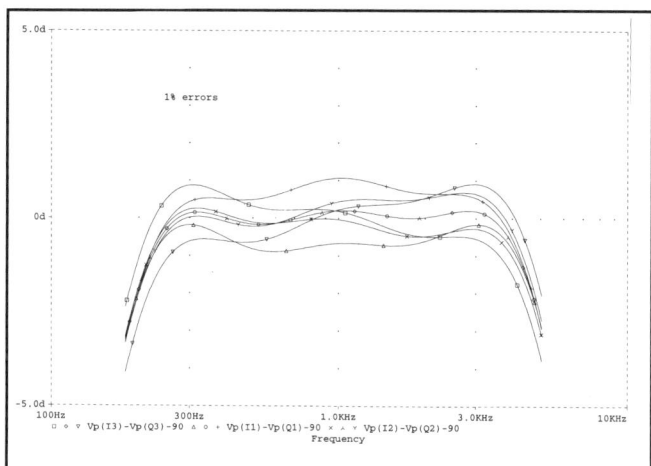

Fig 9.58—A simulation of the phase error when component values vary by 1% or less. Selecting the resistors and capacitors by hand using an accurate ohm and farad meter will improve performance.

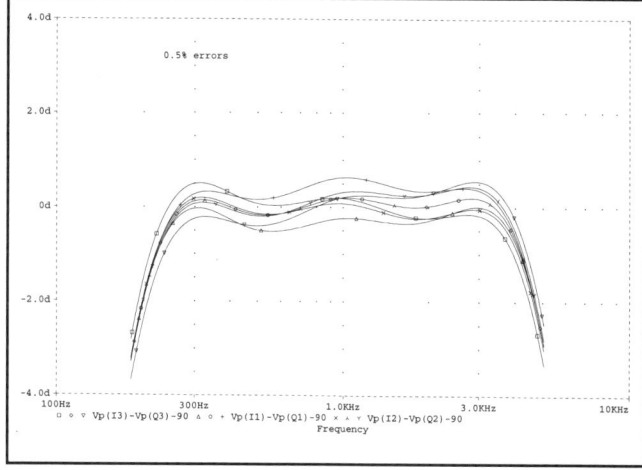

Fig 9.59—A simulation with 0.5% errors.

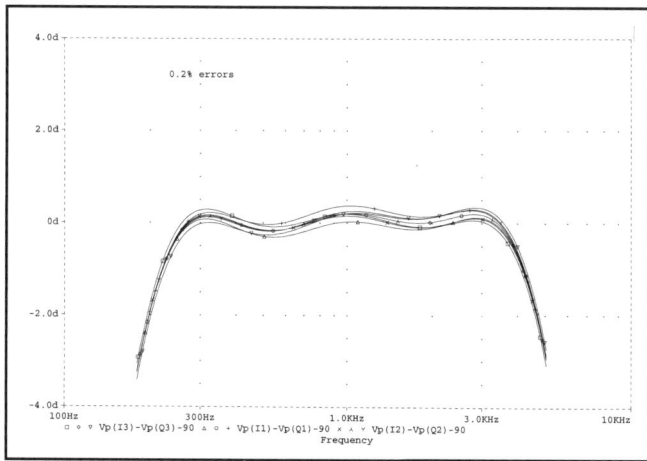

Fig 9.60—A simulation with 0.2% errors.

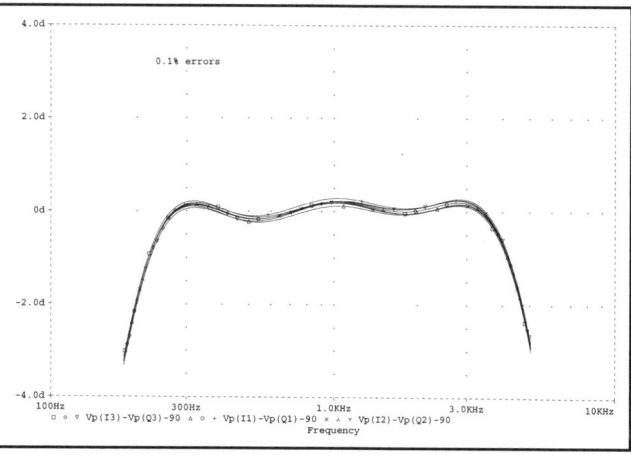

Fig 9.61—More precise matching beyond 0.1% does not provide any practical benefit with 3rd order networks, because the design errors in the network are then larger than the component tolerance errors.

9.7 OTHER OP-AMP TOPOLOGIES, POLYPHASE NETWORKS AND DSP PHASE SHIFTERS

Many other passive and active audio phase shift networks are possible, and have been described in the literature. The ones described above are ones that we have used and recommend. There are several other op-amp all-pass networks that have been used in phasing exciters and receivers. Many others are possible. Polyphase networks, described in the *ARRL Handbook*, may also be used in receiver and exciter applications. They are capable of excellent phase and amplitude balance across the passband. There are a few subtleties to consider in deciding between an op-amp all-pass network and a polyphase network. Polyphase networks are lossy, so more gain needs to be used ahead of them in receiver applications. The sideband cancellation actually occurs in the network, so no summing amplifier is needed afterward. This eliminates the possibility of trimming the summing amplifier for amplitude balance, requires that sideband selection be performed by reversing the LO drive to the mixers or inverting the output of one of the audio preamps, and requires duplicating the phase shift network for ISB applications.

Polyphase networks are 4 phase networks, and in I Q systems two of the phases are neglected. There are advantages to 4-phase receivers and exciters, however. Four-phase exciters have inherent carrier balance, as long as the four mixers are identical. This may be useful at VHF and microwaves, where it is difficult to obtain adequate carrier suppression with an I Q mixer pair. Polyphase network performance degrades rapidly outside the design passband, so it is useful to design the network for a significantly wider bandwidth than will actually be used. The biggest advantage of polyphase networks is that they are symmetrical, and therefore have self-correcting properties. Phase errors in the input section of the network are corrected by later sections. This allows reduced tolerance components to be used in part of the network. Good examples of rigs using polyphase networks are in the literature.

DSP may also be used to generate an I Q pair. This option is discussed in much more detail in the DSP chapters.

Some workers have included phase trim resistors in the audio phase-shift networks of phasing rigs. This is discouraged for several reasons. First of all, it is unnecessary. A network that can support 50 dB of opposite sideband suppression can be built just by measuring the parts before construction. At this level, other errors in the system will begin to dominate. Secondly, all of the RC combinations in an op-amp all-pass network interact. The only reasonable method of tweaking the individual RC time constants involves a special phase-shift test procedure, and the adjustments might not be correct once the network is removed from the test fixture and inserted into a real receiver or exciter. Finally, it is possible to have too many adjustments. Imagine a car with a V-8 engine, and separate timing for the spark to each cylinder brought back to the dashboard and under control of the driver. Some things are better done correctly the first time, and then left alone. A notable exception to this is systems employing DSP. When the phase shift network is under software control, it is possible to optimize a large number of variables during a self-test routine.

9.8 INTELLIGENT SELECTIVITY

A final philosophical comment regarding the optimization of opposite sideband suppression is in order. The first 20 dB of opposite sideband suppression provides a real improvement in signal-to-noise level for SSB and CW signals, by removing the image noise contribution from the unused sideband. Once image noise is 20 dB down, it is hard to measure any further improvement in signal-to-noise ratio by suppressing it further. Additional opposite sideband suppression is needed to suppress interfering signals in the unused sideband, which may be much stronger than the desired signal. In a receiver with "intelligent selectivity," the available resources can be optimized to suppress the interference, rather than to improve the opposite sideband suppression spec across the audio passband. This is significant, because the impulse response of a receiver with good selectivity in the traditional sense is significantly different than one with a wide response and a few deep nulls. Also, interference can take many forms, and it has long been recognized that optimizing the receiver to suppress nearby strong CW interference makes the receiver less robust to impulse type interference. Spending a few hours with a binaural IQ receiver is useful in understanding the implications of selectivity and interference rejection.

9.9 A NEXT-GENERATION R2 SINGLE-SIGNAL DIRECT CONVERSION RECEIVER

The R2pro is an image-reject direct conversion receiver subsystem consisting of several circuit boards. It is intended for applications where a performance improvement over the basic miniR2 circuit is desired, or for experimental applications where access to signals throughout the system is needed. For most applications, the miniR2 circuit provides excellent performance using off-the-shelf parts. The R2pro requires hand-matched components and careful measurements during construction. It is intended to be used with RF gain, and its design flexibility requires that some engineering decisions be made by the builder.

Review of Previous Work

The phasing receiver described in January 1993 *QST* was developed in parallel with the "High Performance Direct Conversion Receiver," described in the August 1992 issue. All of the basic circuitry from the straight DSB receiver was duplicated onto the phasing receiver circuit boards, with appropriate additions for eliminating the undesired sideband. The audio quality of the August 1992 DSB direct conversion receiver remains a benchmark for amateur receivers. The phasing version sounds good, but summing two channels with different time delays (as required by the image-reject circuitry) modifies the impulse response of the channel, and the receiver loses some of its presence. This is exactly the same effect one encounters with an SSB bandwidth crystal filter in a conventional superhet.

After several hundred R2 receivers had been built, the second-generation miniR2 circuit was developed. The miniR2 circuit board is half the size of the original R2, and has only headphone output. MiniR2 circuitry is simplified and has improved tolerance of component variations, so that good performance may be obtained without hand-matching the audio diplexer components. The audio filter component count was reduced to fit all of the parts on the small circuit board, but audio quality was not compromised. The miniR2 is suitable for use with headphones or an external audio power amplifier. The complete schematic for the miniR2 circuit board is in **Fig 9.62**. There is only one modification from the original *QST* article circuit—the 0.1µF capacitor in series with the inverting input to the summing amplifier. This capacitor eliminates sensitivity to dc power supply voltage variations.

Many experimenters have used the basic R2 and miniR2 circuitry as the foundation for experiments using DDS frequency synthesizers and DSP audio signal processing, as suggested in the original *QST* articles. We have built a dozen different R2 and miniR2 receivers and transceivers for a wide variety of fixed and portable applications—often with outstanding results, and sometimes immediately indicating directions for further work.

After all this learning experience, it was natural to update the original high-performance phasing receiver circuit. A number of revised versions have been built—but the requirement that the new version work better than the original is tough. The original circuitry, and the circuit board layout, were optimized over a period of more than a year of continuous activity.

Updating the R2

The first task in updating the R2 circuit was to determine what needed to change. The following list was formulated:

- Replace the SBL-1 mixers with the TUF-3 package.
- Replace the LM 387 audio IC with a modern low-noise dual op-amp
- Revise the audio diplexers for better tolerance to component variation
- Improve opposite sideband suppression
- Improve receiver system noise figure
- Improve audio stability
- Make it easier to build advanced experimental receivers
- Design a receiver circuit that rewards component selection with performance
- Eliminate distortion from the muting circuit
- Improve LO reverse isolation

The new receiver was named the R2pro. The philosophy is that the R2pro trades more expensive construction, more expensive components, component matching, design flexibility, and a higher level of builder knowledge and experience for slightly improved performance over the miniR2. The miniR2 circuit is a better choice for most applications, particularly when small size or battery operation is desired. The R2pro is for designer-builders who want to go to the extra effort and expense required to push a receiver to the limits of the direct conversion architecture.

Multiple Circuit Boards

There is a significant problem with direct conversion receivers built on a single circuit board. *RF* grounding and shielding techniques are very different than the grounding and shielding techniques needed for high-gain *audio* amplifier circuitry. If the low-level RF signals, high-level LO signal, all the mixer conversion products, and high-gain audio amplifier are all on the same circuit board, there must be compromises in grounding and shielding. These compromises were handled on the R1, R2 and miniR2 boards by designing the ground traces such that the audio stages saw an approximate single-point-ground and the area around the mixers was an unbroken ground plane. Any of these single-board receivers can be made to oscillate by connecting the power-supply or speaker ground wire to the wrong point on the circuit board ground, even though all of the grounds are connected together. For a review of audio grounding techniques, see Horowitz and Hill, *The Art of Electronics*.

The conflicting requirement for an RF tight enclosure and a single-point audio ground makes it difficult to package single board direct conversion receivers. Early versions of the R1 and R2 direct conversion receivers pictured in *QST* were enclosed in soldered-up copper-clad PC board enclosures. Other packages, particularly those made of aluminum pieces held together with screws—are prone to intermittent audio oscillations and microphonics. Breaking up the receiver into separate functional blocks—each with its own circuit board—provides more grounding flexibility. Then the PC board with the mixers can be completely shielded, and the PC board with the audio output amplifier can have a single point ground. By optimizing the gain partitioning and packaging of the receiver, hum and microphonics can be eliminated and the placement of ground connections becomes much less critical. As a fringe benefit, breaking up the PC board makes it easier to build experimental versions using DSP, different mixers, audio processors and power amplifiers etc.

Block by Block R2pro Circuit Description

The R2pro block diagram is shown in **Fig 9.63**. Note that the R2pro system design includes an RF preamp, and that the audio output stage is a completely separate block.

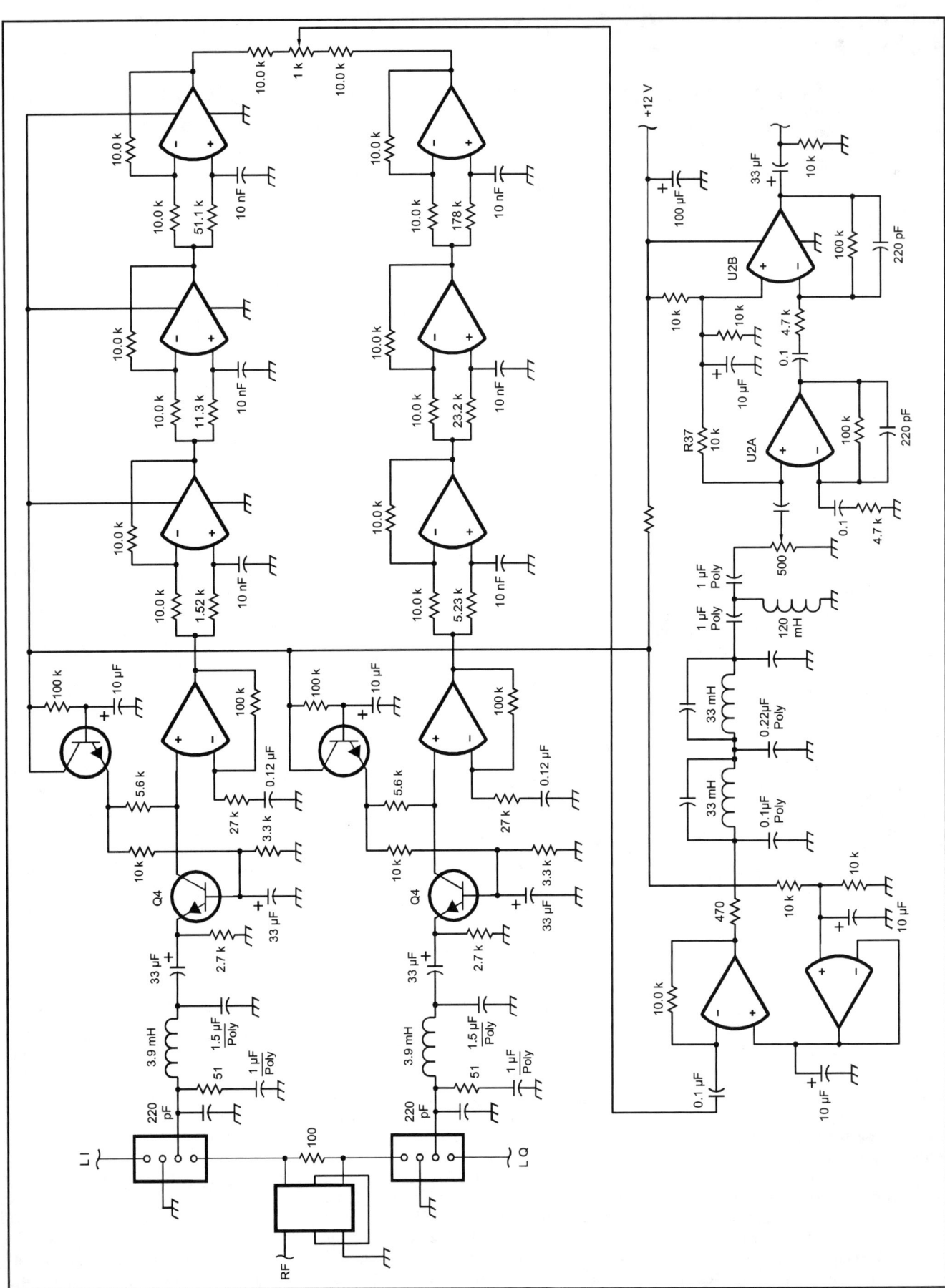

Fig 9.62—This simplified version of the mini R2 uses some different parts values and requires matching of the diplexer components.

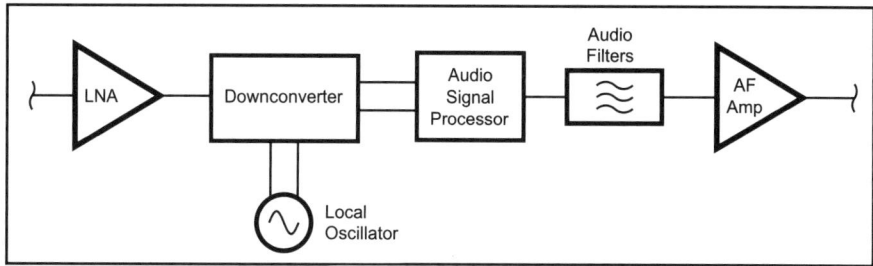

Fig 9.63—The R2pro block diagram.

RF PREAMP

The first block in the R2pro receiver subsystem is the RF preamp. The use of a preamp permits additional mixer loss in the design for improved dynamic range, improved phase and amplitude balance over the baseband frequency range, constant impedance at the downconverter RF port, and lower LO radiation from the receiver RF port. The basic design shown in **Fig 9.64** is highly recommended, but any low-noise, moderate-gain 50-Ω bandpass amplifier with high reverse isolation (S12) may be used. Because direct conversion receivers are sensitive to signals near the odd harmonics of the desired signal, it is necessary to provide significant attenuation to signals above the band of interest. This is particularly important in metropolitan areas with many FM broadcast signals. A separate RF-tight enclosure is appropriate.

The grounded gate circuit in Fig 9.64 was designed specifically to use in front of direct conversion receivers at MF through VHF. Low-pass filtering in the input and output match to the transistor provides the necessary attenuation of signals near odd harmonics of the LO. The bias switch is part of the receiver mute circuit, and switches the amplifier gain between +13 dB and –40 dB. The grounded gate topology is a strong 40-dB attenuator when it is reverse biased, and can be switched in as a front-end attenuator when very strong signals are present, without introducing front-end distortion. It is common for direct conversion receivers to experience audible pops during full break-in CW operation. One source of these pops is the dc shift at the mixer IF port when the strong TX signal appears at the mixer RF port. One solution is to switch in a large attenuator between the antenna switch and mixer RF port. The "sleeping bag radio" described in Chapter 12 uses a similar preamp circuit in front of a miniR2 board, and has absolutely clean transmit/receive switching at all volume levels. **Fig 9.65** shows the swept frequency response for several different bands. The typical input intercept of +13 dBm is a good match for receivers with standard level diode ring mixers.

The amplifier noise figure of approximately 4 dB and the relatively low gain of

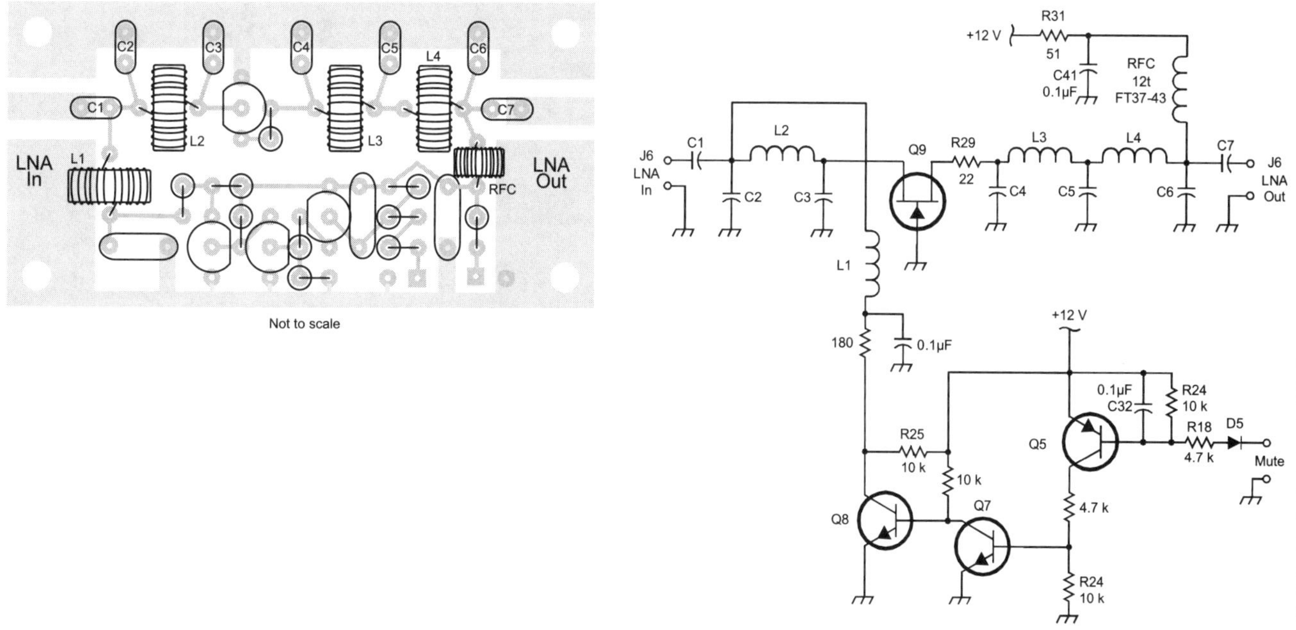

Band	C1	L1	C2	L2	C3	C4	L3	C5	L4	C6	C7
3-4	820p	1.3μ	1800p	4.0μ	820p	100p	20μ	680p	3.8μ	470p	2200p
6-8	470p	680n	820p	2.0μ	470p	56p	10μ	390p	1.9μ	220p	1000p
9-11	330p	450n	680p	1.5μ	330p	39p	6.8μ	270p	1.4μ	180p	1000p
13-15	220p	330n	470p	1.0μ	220p	27p	4.7μ	180p	1.0μ	120p	1000p
18-22	180p	240n	270p	760n	120p	18p	3.5μ	120p	760n	100p	1000p
24-30	150p	160n	220p	560n	100p	12p	2.7μ	82p	540n	56p	680p

Fig 9.64—The use of a preamp permits additional mixer loss in the design for improved dynamic range, improved phase and amplitude balance over the baseband frequency range, constant impedance at the downconverter RF port, and lower LO radiation from the receiver RF port. The basic design shown here is highly recommended, but any low-noise, moderate-gain 50-Ω bandpass amplifier with high reverse isolation (S12) may be used.

the preamp stage have the effect of reducing the receiver noise figure without severely impacting two-tone third-order dynamic range. Third-order dynamic range near 100 dB is possible with standard level diode-ring mixers and a narrow CW bandwidth. High-level mixers permit better dynamic range numbers, if the LO system is quiet enough.

The direct conversion receivers described by the author in *QST* in 1992-1995 were all developed using a full-sized elevated 40-m dipole in a quiet lakeside location in the Upper Peninsula of Michigan. At this location, signals from all over the US and Canada were quite strong, and the antenna noise power was always high enough that a 15-dB noise figure was always adequate. There are other locations that can benefit from quieter receivers, even on 80 meters. In the mountains of the Pacific Northwest, band noise levels on 40 meters are commonly well below the

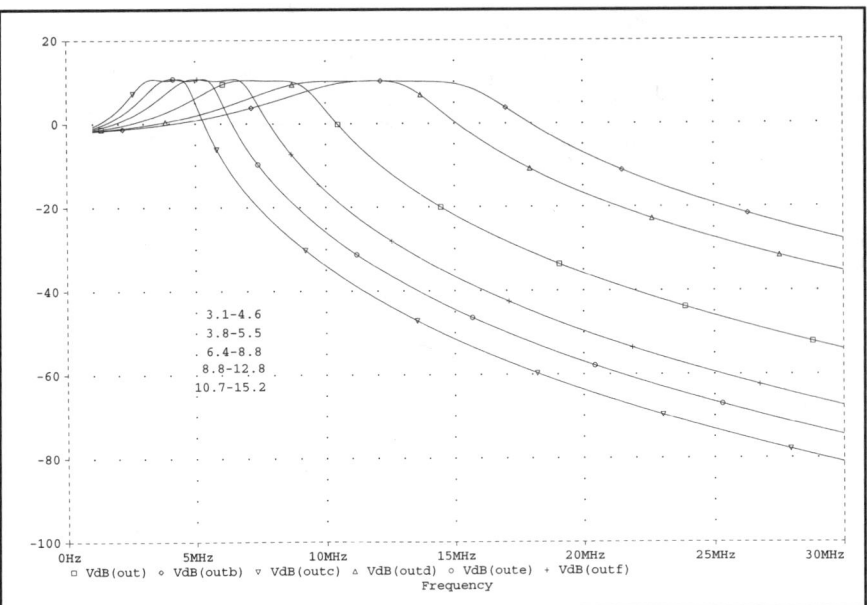

Fig 9.65—LNA swept frequency plot.

Fig 9.66—Downconverter schematic diagram.

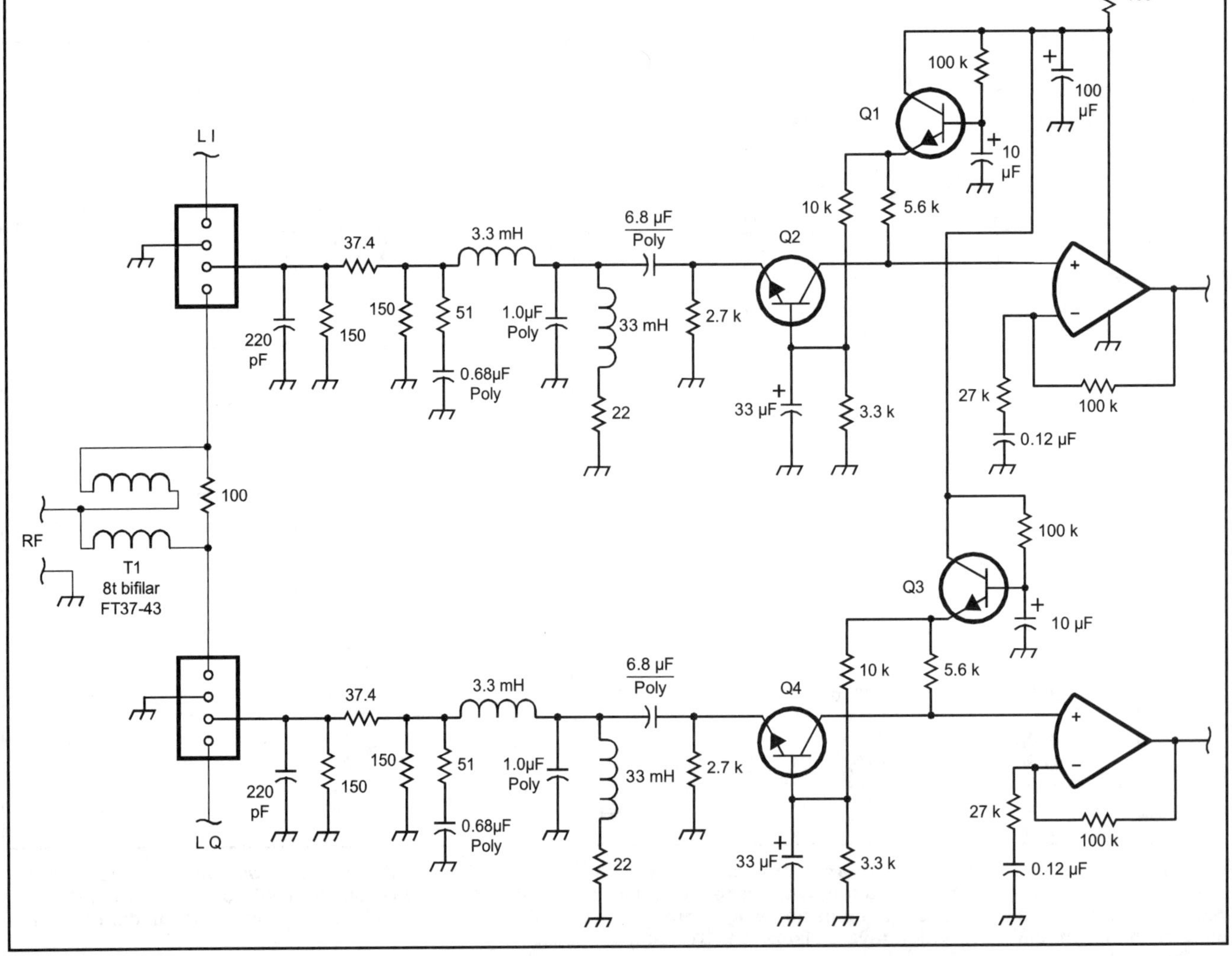

accepted numbers in the amateur and professional literature. For mountain portable operation, receiver noise figures should be below 10 dB for all HF bands, and much better noise figures may be useful above 20 meters, particularly when using directive antennas.

For wide-band systems, a broadband impedance transformer can replace the tuned low-pass output on the RF preamp. This will permit coverage of multiple bands, but the low-pass function is still important and must be included somewhere in the receiver RF path. When a lower noise figure is desired, a two stage grounded-gate RF preamp is a good choice. Two of the Fig 9.64 circuits packaged separately with coax connectors is a high-performance construction option.

In summary, here are a few good reasons to include RF gain in any direct conversion receiver:

1. Improved Noise Figure.
2. Electronic front-end gain switching
3. Reverse isolation to eliminate LO radiation
4. Improved receiver gain distribution

For phasing direct conversion receivers there are additional advantages:

1. Providing mixer RF port impedance that doesn't change with antenna tuning
2. Option to use attenuators on all mixer ports

DOWNCONVERTER

After the preamplifier is the down-converter block, shown in **Fig 9.66**. (A layout and photo are shown in **Figs 9.67** and **9.68**.) The downconverter includes an RF in-phase splitter, two mixers, IF port attenuators, a matched pair of diplexer networks, and a matched pair of audio LNAs. All of the resistors in the downconverter board should be 1% metal film. The input splitter is somewhat different than earlier versions. Rather than attempting to match to 50 Ω, the splitter shown matches the mixer inputs to a lower impedance—but achieves nearly perfect amplitude balance and very low loss over a very wide frequency range. The upper frequency limit is reached when the winding on T1 approaches a quarter wavelength. At the lower frequency limit, amplitude balance is still perfect, but isolation is poor. If operation down to 50 kHz is desired, more turns on a type 71 core could be used. At 144 MHz and above, a few bifilar turns through a small ferrite bead work well. The mixers are type TUF-3, which offer better port-to-port isolation and lower conversion loss than TUF-1 mixers from 150 kHz through 225 MHz, the usual operating range of R2 type systems. TUF packaged mixers are available for direct conversion applications at frequencies up to 2500 MHz. The small sample of microwave diode mixers we have measured have higher 1/f noise than we have seen with TUF-1, TUF-3 and SBL-1 mixers. Microwave Doppler Radar systems use special low-1/f noise diodes.

After the mixers are a pair of matched attenuators. The 6 dB attenuators shown in the schematic should be used for most applications. If more gain is available before the mixers, more attenuation may be used. These attenuators serve three very useful purposes: they ensure textbook termination of the mixer IF ports; they attenuate mixer 1/f noise; and they provide a well defined source impedance to drive the matched diplexer networks. Mixer IF termination has been widely discussed in the literature. Mixer 1/f noise degrades receiver noise figure. Different mixers, even matched TUF-3s with the same date code, have widely varying amounts of 1/f noise. Attenuation between the mixer and the audio preamp can't improve receiver noise figure, but it can reduce the effect of mixer 1/f noise. Advanced receiver artists are encouraged to study this. The R2pro circuit balances preamp gain and post-mixer attenuation to set the receiver noise figure and dynamic range, so that receiver performance is relatively independent of mixer 1/f noise.

The third very important function of the post-mixer attenuators is to set the driving point impedance to the matched diplexer networks. In the original R2, the diplexers are connected directly to the mixer IF port impedance, which varies with LO drive level. If one mixer has more LO drive than the other (a common condition) the phase and amplitude response of one diplexer network will be slightly different than the other. These differences are typically enough that the ultimate opposite sideband suppression of R2 systems across an SSB bandwidth is about 41 dB—even with perfect audio phase-shift networks. By contrast, the miniR2 with off-the-shelf components often exhibits nearly 50 dB of opposite sideband suppression.

The diplexer networks are slightly simplified from the original R2 networks. The R2 networks provided rapid roll off both above and below the 300 to 4000 Hz audio band. The roll off below the audio range does not contribute much to useable receiver dynamic range, but it does introduce rapid phase shifts in the critical 300 to 600 Hz frequency range. When R2 receivers are optimized for SSB operation, the suppression of the opposite sideband in the 300 to 600 Hz range is often right at the 40 dB spec. If the receiver is optimized for CW operation, sideband suppression usually falls off at higher audio frequencies. The miniR2 and R2pro eliminate the rapid roll off at the low end of the audio range, which permits good performance through the CW range when the receiver is optimized for SSB. Another change from the R2 and miniR2 circuits is the elimination of the electrolytic capacitors from the critical audio signal paths. The R2pro has only matched polypropylene capacitors in the audio path prior to the summing network.

The roll off above the audio range is retained from the R2, with slight changes to make the receiver less sensitive to component tolerance. For good performance, it is necessary to match the diplexer components in R2pro to within 1%, just as in the original R2. If this is not done, opposite sideband suppression is likely to be poor across the audio band. By contrast, the diplexers in the miniR2 were designed to be used with standard tolerance components. The benefit of using the R2pro

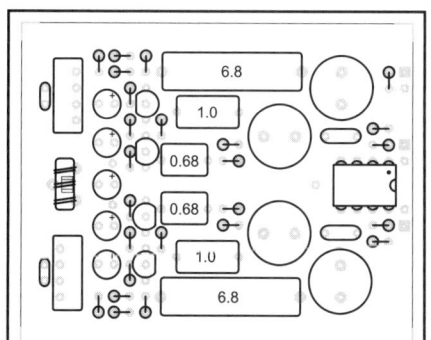

Fig 9.67—The downconverter board layout.

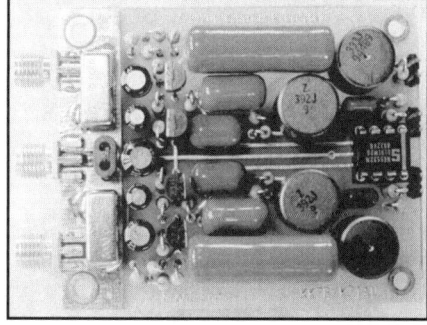

Fig 9.68—A view of the downconverter board.

diplexers with matched components is that the close-in dynamic range is good. R2pro two-tone measurements may be made at tone spacings of 10 kHz and 5 kHz.

The usual grounded-base audio preamp stages are used following the diplexer networks. There are other audio preamps that will work, but the grounded base stages have the advantage of having an input impedance that is set by the current through the transistors, which may be set up precisely using 1% resistors. The grounded base stages drive the non-inverting inputs of a low-noise dual op-amp, which provides low impedance drive to the following stages. Note that the outputs are not dc blocked. This is so that the low impedance drive from the dual op-amp can directly drive the audio phase-shift network. Because these outputs carry dc, there is the potential to short them and damage the dual op-amp. IC sockets are appropriate.

It is critical that everything in the I and Q channels of the downconverter block be well matched. In most cases, it is the I Q downconverter block, and not the audio phase-shift network, that sets the ultimate limitation on receiver opposite sideband suppression. The baseband LNA pair is nearly identical to the version used in the miniR2, with the exception that 1% resistors are used in all locations and transistor pairs Q1—Q3 and Q2—Q4 should be matched. This may be done by comparing the dc voltages on the I and Q outputs of the downconverter block using a digital voltmeter. First insert a temporary jumper between the emitter and collector holes for transistors Q2 and Q4. Then select a pair of devices for Q1 and Q3 that results in equal output voltages. The voltages should be matched to within 2%. Then solder in Q1 and Q3, remove the jumpers, and select a second pair of devices for Q2 and Q4 that results in equal dc voltages at the I and Q outputs. Since the gain and input impedance for these common base bipolar amplifiers are set by the quiescent currents, and the currents result in voltage drops across the 1% resistors, setting the dc voltages equal results in well-matched gain and input impedance for the baseband LNA pair.

The noise figure of the receiver is determined by the performance of the early stages. It is necessary to have enough gain in the early receiver stages to over-ride the noise of the later stages of the receiver. The analog signal processor block has a relatively high noise figure, resulting from the cascade of unity-gain op-amp phase

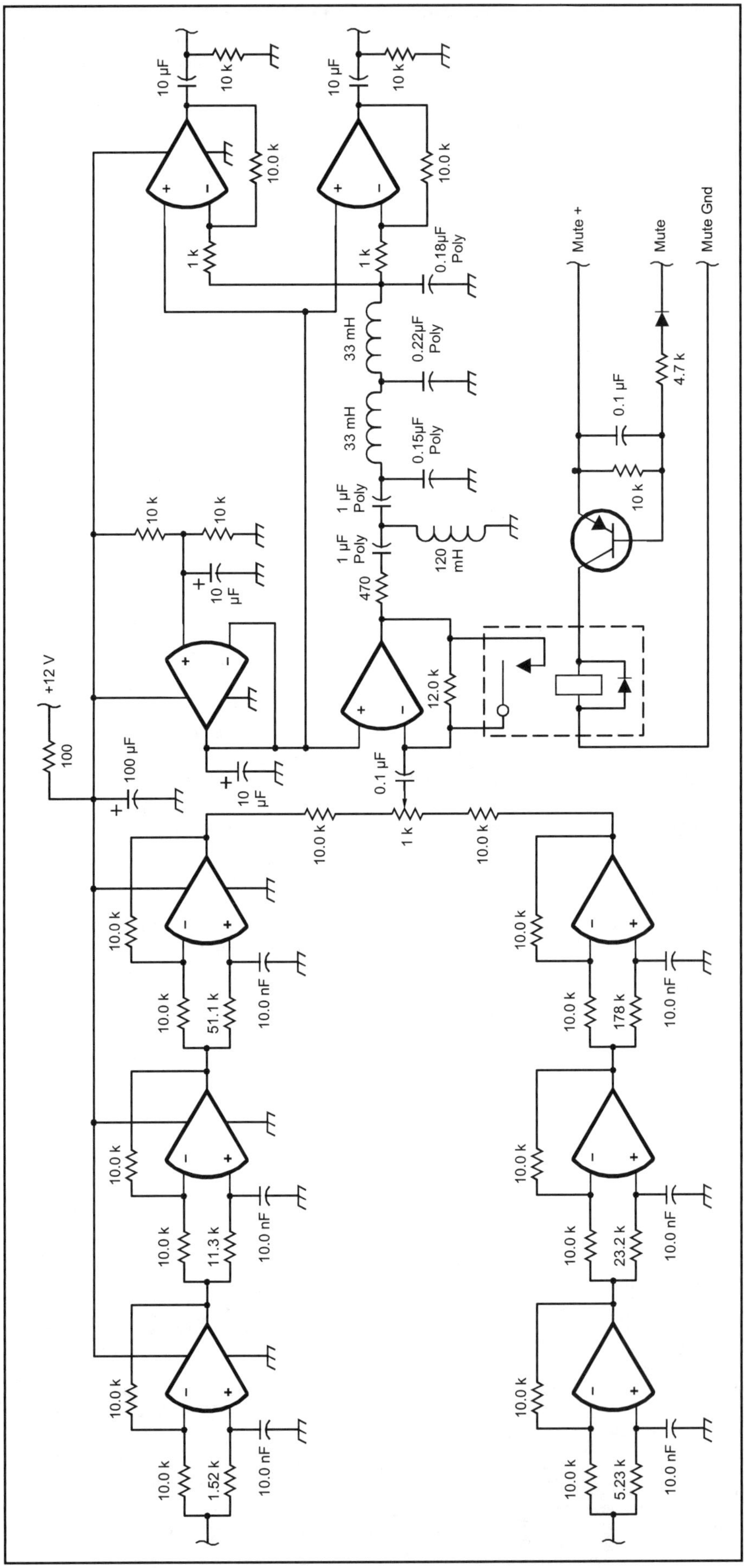

Fig 9.69—ASP schematic.

shift networks and the lossy bandpass filtering. The downconverter PC board gain is set by the ratio of the op-amp series and feedback resistors to a value that overrides the noise of the analog signal processor but that does not severely compromise in-band dynamic range. With the component values shown, the mixer loss is approximately 6 dB, there are 6-dB pads following each mixer, the bandpass diplexers have just under 2-dB loss, the grounded-base LNA stages have a noise figure of about 5 dB and approximately 40-dB gain, and the op-amp LNAs have 11-dB gain. Thus the total gain for the downconverter stage is about 37 dB and the noise figure at the downconverter RF input is approximately 19 dB. With all components matched to within 1%, the amplitude and phase errors in the I and Q outputs should be less than 0.1 degree and 0.02 dB across the baseband output range from 200 Hz to 4000 Hz.

Since the downconverter block contains both RF and low-noise audio signals, it must be constructed using good RF and audio practice. Audio signal levels are low and the gain is moderate so conventional RF grounding and shielding practices may be used for the downconverter block. With LO signals floating around on the same frequency as the desired input signal, shielding is very important. The circuit board is designed to fit inside a Hammond 1590B die-cast aluminum box. An enclosure soldered up from tin sheet or PC board scraps is even better. The RF and LO inputs should enter through coax connectors. Type BNC, SMA and RCA phono are all acceptable. The audio outputs should leave through either coax connectors or matched 1nF feedthrough capacitors. The audio output signals include dc bias for the Op-Amps in the analog signal processor. For connection to the high impedance inputs of a DSP processor or oscilloscope, dc blocking capacitors may be used. The dc power supply lead should be connected using a feedthrough capacitor and external series resistor.

ANALOG SIGNAL PROCESSOR

The third block in the R2pro system is the analog signal processor (ASP) shown in **Fig 9.69**. (A board layout and photo is shown in **Figs 9.70** and **9.71** respectively.) This board contains the audio phase-shift network, the summer, and a wideband passive audio filter. The audio gain is low, but the signal levels are also low, so this board should not be located where it can pick up power supply or computer noise. There are no RF signals present, so audio grounding rules apply. The single audio ground rail runs up the middle of the PC board between the ICs. The power supply line is decoupled by the 100 µF capacitor and 100 Ω series resistor. Do not bypass the hot end of the 100 Ω resistor to ground. The dc bias to the non-inverting inputs to the analog signal processor comes from the previous stage.

There is only one change in the audio phase-shift network from the version used in the miniR2. 1.52 kΩ is not a standard value in the 1% series. It is obtained by connecting a 1.50-kΩ and 20-Ω resistor in series. With the audio phase-shift network components (resistors and capacitors) selected to within 0.1% of their marked value, more than 60 dB of opposite sideband suppression could be obtained—if the rest of the receiver were perfect. By selecting these components, the builder can be assured that the audio phase shift network is not limiting receiver performance. The image-reject mixer provides an attenuation band that covers the entire opposite sideband from 200 Hz to over 4000 Hz. This attenuation band is ideal for CW or SSB receivers, and provides very good selectivity when combined with audio channel filters.

Following the audio phase-shift network is a summing amplifier. The amplitude balance adjustment is conveniently located at the input to the summing amplifier. The summing amplifier drives a 250 Hz to 4000 Hz bandpass filter. This filter serves as a roofing filter, and provides optimum performance from optional external digital and analog filters that may be added to the output of the analog signal processor block. Roofing filter performance is good enough that it can serve as the only bandpass filtering in the receiver for high-fidelity listening. The output of the roofing filter drives a second gain block that provides an ideal filter termination for textbook bandpass response. The gain of the output gain block is set by the feedback resistor. With the values shown, the gain of the analog signal processor block is approximately 13 dB. It is possible to increase the gain of the output gain block to directly drive medium impedance headphones. The analog signal processor block also contains a mute circuit. Grounding the mute terminal drops the gain of the summing amplifier to zero. The mute circuit uses a reed relay with completely independent power, ground and control circuit. This permits the relay to be controlled by front panel switches and TR switching logic without corrupting the

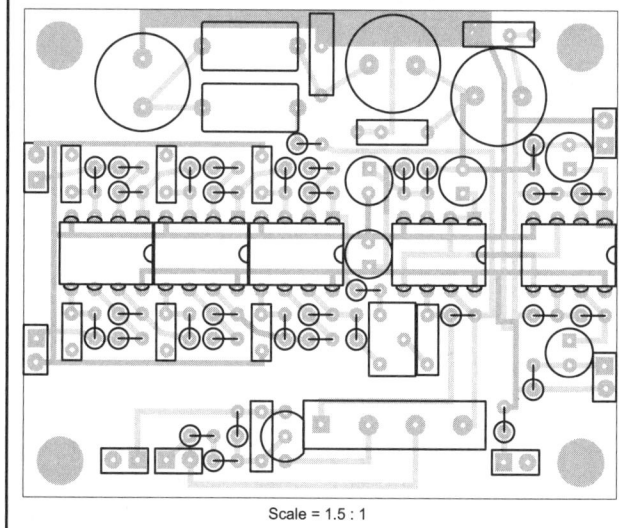

Fig 9.70—ASP layout.

Fig 9.71—The analog signal processor.

Analog Signal Processor signal ground and power supply lines. Use of a relay also eliminates the low level distortion introduced by an FET switch. The sealed reed relay switching time of a few milliseconds is quick enough for full break-in operation on fast CW or digital modes.

The analog signal processor board has two isolated, independent outputs. The first output is normally connected through optional filters and the volume control to the audio output circuit board. The second output may be used to drive a signal level meter or audio derived gain control system. This is the ideal take-off point for DSP filters, FFT analyzers, home audio system stereo amplifiers, outboard audio filters or the computer sound card. Output levels may be independently selected by changing the output stage feedback resistors. The 1-kΩ input resistors should not be changed, as they provide the termination impedance for the roofing filter.

For construction hints on mounting and connecting to the ASP board, take the cover off a stereo receiver or amplifier and look at the circuitry around the magnetic phono cartridge inputs. Don't expect to find RF shielding, but a well defined single ground connection, shielded wire or twisted pair with the ground connected only at one end, and power connections directly to the big power supply capacitor are common. This PC board should be mounted on nylon standoffs with a single wire to ground at the power supply.

OPTIONAL FILTERS

The low output impedance of the analog processor with a series 470-Ω resistor, and the 500-Ω volume control provide proper terminations for a wide variety of passive filters. **Fig 9.72** is a pair of useful audio 500-Ω filters using standard value inductors and capacitors that have been used in a number of our radios. Also see the photo in **Fig 9.73**.

Signal levels are high enough at this point that open PC board construction is acceptable. If Wide SSB, Narrow SSB and CW options are all installed, it is useful to add attenuation to the SSB filters so that either gain or receiver output noise remain constant as filters are switched. 500-Ω attenuators are easy to construct. Use the resistor values from the *ARRL Handbook* tables, and multiply all resistor values by 10. For example, a 500-Ω 6-dB pi-network pad has a 390-Ω series resistor and 1.5-kΩ shunt resistors.

Signal channel selectivity is distributed through the baseband gain path. The bandpass diplexers pass a 300 Hz to 4000 Hz channel with smooth rolloff outside the passband to enhance phase-shift network performance and provide graceful impulse response. The baseband LNA and gain block have wide bandwidth, to preserve amplitude and phase balance between the I and Q channels. After the summing amplifier, the 3rd order Butterworth High-Pass filter and 5th order Butterworth Low-Pass filter provide a flat passband with good impulse response at the medium frequencies. This roofing filter provides all the band-limiting needed for a high-fidelity SSB or CW receiver—and it is recommended that the receiver be put into operation with no additional filtering before adding narrow bandwidths. Some of the most skilled and avid CW operators are now using very wide bandwidth receivers when band conditions permit, because such receivers preserve the quality of

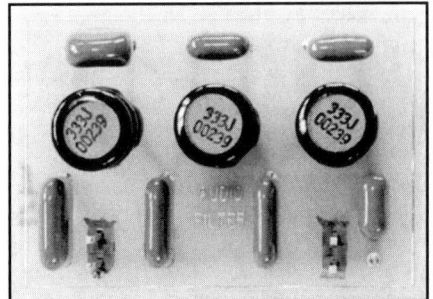

Fig 9.73—SSB and CW filters.

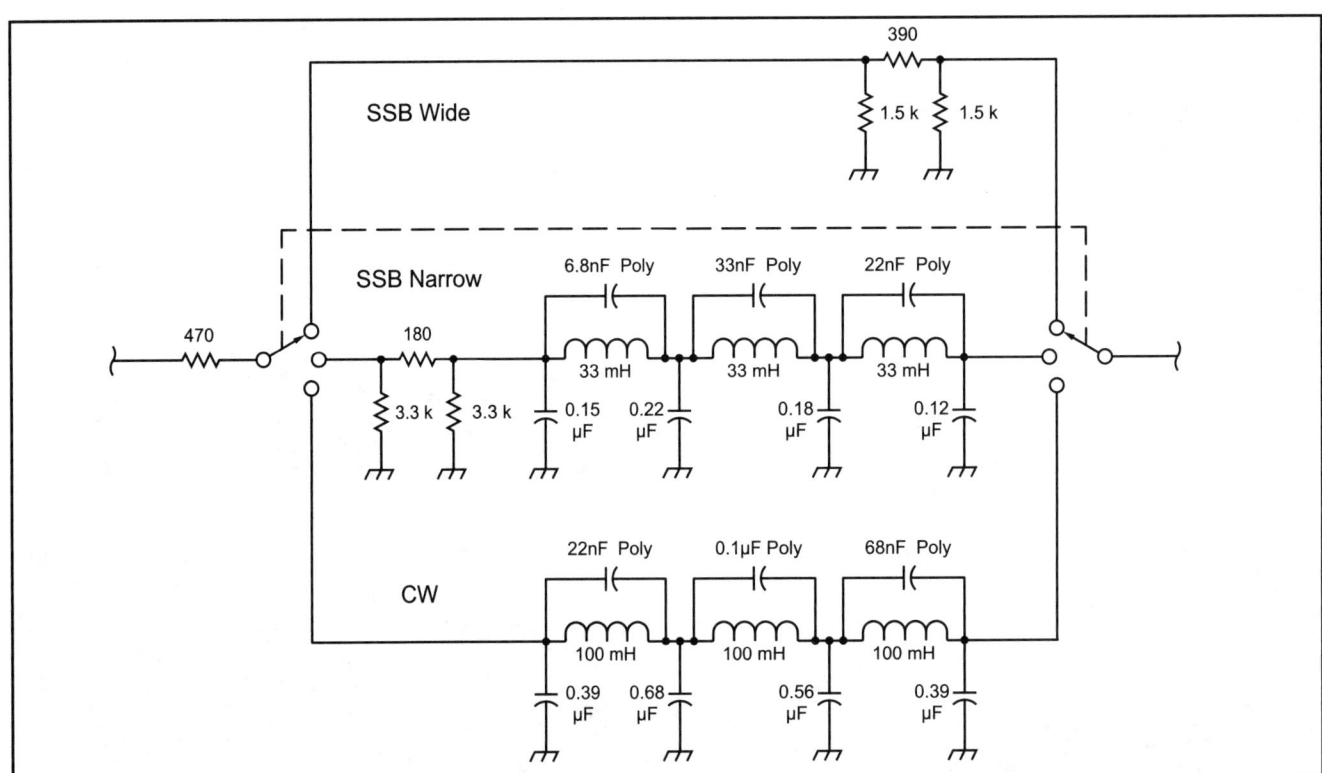

Fig 9.72—A pair of useful audio 500-Ω SSB and CW filters using standard value inductors and capacitors that have been used in a number of our radios.

transmitted signals and allow a much better perception of the texture of the band. Interestingly, low-audio-frequency impulse response is dominated by the effectively very steep skirts of the receiver response due to the high-pass filtering and the operation of the phase-shift image-reject circuitry.

Switched-capacitor and DSP filters may also be used at this point in the circuit. It is necessary to observe appropriate input signal levels, and bear in mind that the dynamic range and noise figure of the DSP may limit receiver performance. At the output of the analog processor, the receiver has an in-channel two-tone dynamic range of well over 60 dB and total harmonic distortion lower than 0.1%. By this point in the receiver, the noise floor, dynamic range and in-channel distortion have been set. DSP at this point can not improve these numbers—it can only provide wonderfully flexible filtering and additional whistles and bells. When the digital signal processing is carefully designed, it can add to the utility of the receiver without corrupting basic performance. If the DSP system has too few bits, if the A-to-D converters have a high noise figure, or if the signal levels are set up improperly so that the available DSP dynamic range is not used—a poor receiver with wonderfully flexible filtering will result. The audio recording industry has pushed the state-of-the-art in DSP well beyond the needs of this receiver. In particular, noise-free digital delay offers the possibility of intelligent audio AGC systems that go well beyond the best commercially available amateur receiver systems.

The R2pro is set up so that sophisticated laboratory instrumentation may be used to observe the distortion at all points in the signal path. The ear can often detect distortion that is difficult to measure, and the ear-brain quickly learns to recognize different distortion and noise mechanisms. The acid test is to set up the receiver with a switch that completely bypasses the DSP, and equal gain in the DSP and non-DSP modes. When the DSP is set for wide bandwidth, and switching between modes is completely transparent, the operator can be confident that the DSP system is not corrupting receiver performance.

AUDIO POWER AMPLIFIER

An audio power amplifier circuit is shown in **Fig 9.74** (also see the board layout in **Fig 9.75** and the photo in **Fig 9.76**.) Any audio amplifier with enough gain may be used at this point, but it is a shame to connect a low distortion receiver to an inexpensive IC amplifier with questionable fidelity. The version in Fig 9.74 has a gain of 46 dB, with the volume control arrangement shown. Since the audio power amplifier has high gain and is capable of medium power operation, signal currents flow in the power supply wires. It is critical that the power amplifier use appropriate audio amplifier construction practice. In particular, both speaker wires must connect to the appropriate points in the circuit. Do not use the chassis as the negative speaker lead connection or as the negative power supply lead to the audio output amplifier. The circuit board layout works well when connected directly to the speaker, and to the power supply

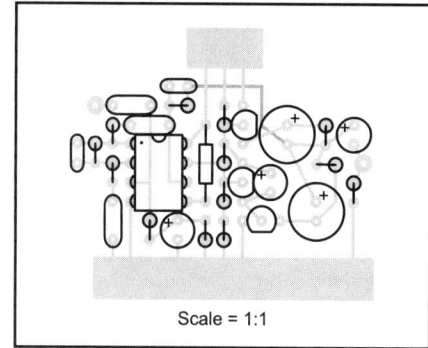

Fig 9.75—Board layout for the audio power amplifier.

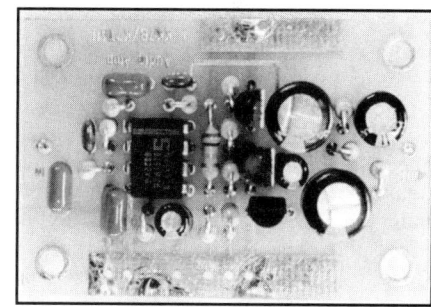

Fig 9.76—The audio power amplifier.

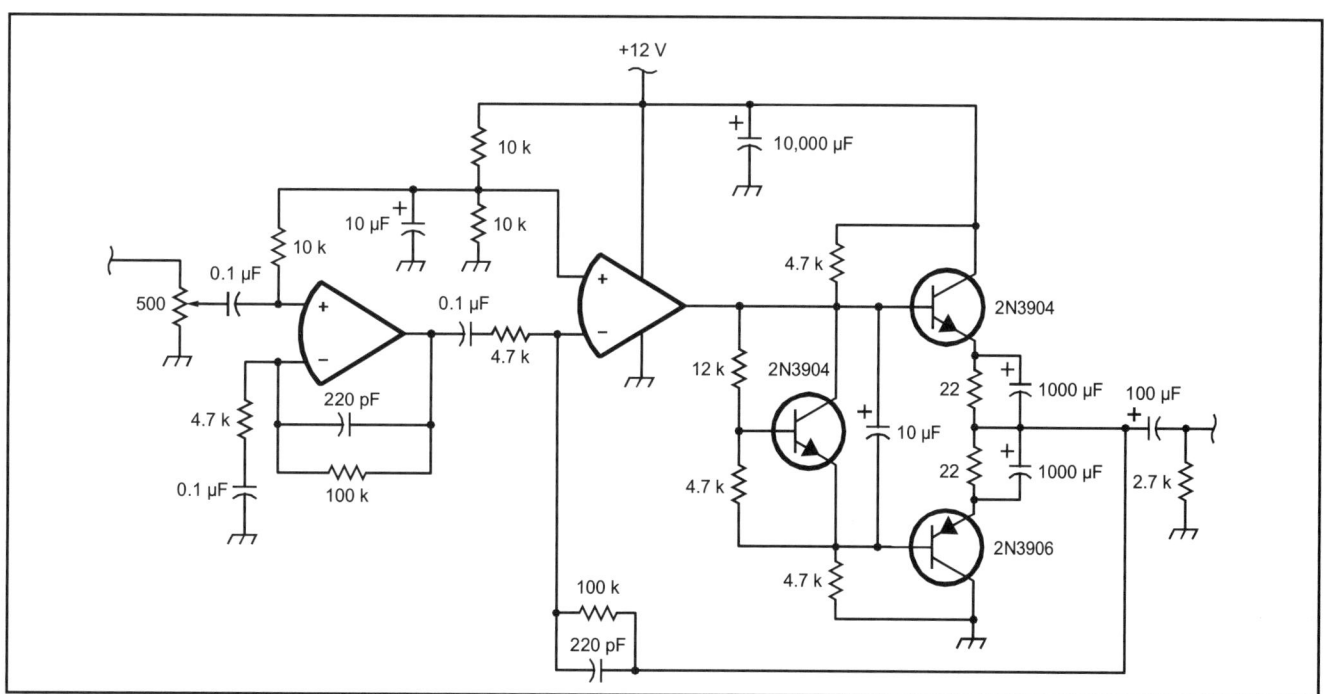

Fig 9.74—An audio power amplifier circuit.

capacitor with #18 wires. Feedback problems (howling) in direct conversion receivers can often be cured by using a separate battery power supply for the audio power amplifier. While this is not always attractive for normal operation, temporarily operating the audio power amplifier circuit board from a separate battery supply can serve as a very useful troubleshooting tool when trying to figure out which ground wire needs to be cut to eliminate the offending ground loop.

This audio power amplifier provides reasonable output with headphones or a small speaker in a quiet room. For more volume, an external power amplifier should be used. Some external sound card amplification systems for computers are quite good. Others are quite inexpensive. Each has its merits.

LOCAL OSCILLATOR

A local oscillator is not included in the R2pro receiver system, but the choice of LO in large part determines the success of the finished project. Two local oscillators that have been used to build excellent direct-conversion receivers are a well-shielded JFET Hartley and a moderately well-shielded JFET Hartley driving a balanced frequency doubler. When the diode doubler is used in a circuit with toroid inductors, open PC board construction is acceptable. The Kanga UVFO circuit in Chapter 12 works well and provides additional useful features such as CW offset and a keyed auxiliary output. Because of differences in the way even and odd harmonics add, direct conversion receivers that use odd harmonic frequency multipliers must be very well shielded.

While analog local oscillators represent mature technology and simple elegance, the state of the synthesizer art continues to progress rapidly. The best hybrid DDS—PLL synthesizers are very, very good, and continue to improve. The R2pro circuit blocks provide a convenient platform for experiments with different types of synthesizers.

Sideband Switching, Binaural, and ISB modes

It is not trivial to set up a switched-sideband phasing image-reject receiver system with equal sideband suppression on either sideband. This is particularly the case for the R2pro, with available sideband suppression of over 50 dB. The reason for the difficulty is subtle. In a phasing system, all the cumulative amplitude errors throughout the system may be compensated with a single amplitude trimming adjustment. Similarly, all of the cumulative phase errors may be trimmed out with a single phase trim. When the sideband switch is thrown, the receiver configuration changes, and the distribution of amplitude and phase errors is likely to change. Our R2 and miniR2 receiver trimmed for more than 40 dB opposite sideband suppression on one sideband typically exhibit less than 30 dB opposite sideband suppression when connections to the analog signal processor are reversed. Readers fluent in image-reject concepts can investigate options for sideband switching that preserve the distribution of amplitude and phase errors when switching sidebands. A good strategy is to trim the errors before the audio phase shift network, so that at the input to the nearly ideal analog signal processor the I and Q channels have precisely equal amplitude and 90º phase shifts. Reversing connections at this point will then switch sidebands without redistributing the errors.

One viable method to provide good sideband suppression in a switched-sideband receiver is to make the amplitude and phase trim adjustments front-panel controls. This is particularly attractive for receivers that cover a wide frequency range, as phase shifts will likely need to be tweaked when changing bands. Judging from the front panels of many high-end radios, there is no penalty for providing additional operator control over receiver functions. A well-shielded external crystal calibrator with variable output is a useful accessory for a receiver with front-panel phase and amplitude trims. It is important that the test signal enter the receiver on the antenna connector, and that all leakage paths into the I and Q RF circuitry are 60 or 70 dB down.

For single-band switched-sideband receivers, there are other options. From the basic theory, four trimming adjustments (one amplitude and one phase trim for each sideband) are needed to optimize suppression of either sideband. A very conservative option is to use two independent down-converter and analog signal processor PC boards, with switched (or split) LO and RF inputs. An independent LO (or RF) phase trim can then be implemented for each downconverter, and one analog signal processor can be set up for upper sideband and the other for lower sideband. The desired sideband may then be selected by switching between analog processor outputs. Of course, an additional audio power amplifier could also be added for full Independent Sideband operation. The trimming adjustments for suppression of opposite sidebands are completely independent in this implementation.

Binaural operation is simple to add to an ISB receiver with two identical audio channels. Binaural ISB, with one sideband in each ear, just requires additional switching. For Binaural IQ, as described in March 1999 *QST*, the I and Q outputs of the downconverter board are amplified by a stereo amplifier. A number of experimenters have noted that Binaural IQ receivers sound best with very little audio filtering. A versatile receiver might have a switch that provides wide open Binaural IQ for tuning around the band and then a number of narrow band options for communicating with individual stations.

Some of the receiver circuitry in the previous paragraphs adds many parts to achieve a very tenuous performance advantage. Philosophically, minimum parts considerations should not apply to high-performance phasing direct-conversion receivers. Also philosophically, front-panel amplitude and phase trim adjustments are an elegant solution, and are really cool to play with. The philosophy behind each receiver is different, however—which may be the whole point of this entire book.

Trimming

Finally, here are a few words on the actual process of trimming a phasing receiver for best opposite sideband suppression. A "target" analogy is a useful way to think about trimming a phasing receiver. The undesired

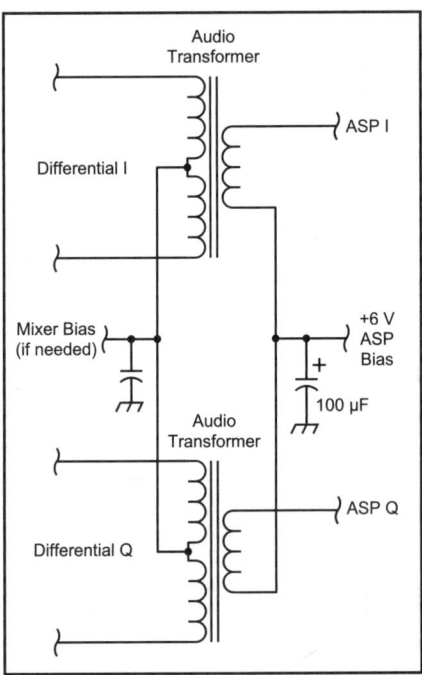

Fig 9.77—A circuit for connecting an I Q balanced mixer output pair into the I and Q inputs of the R2pro analog signal processor board.

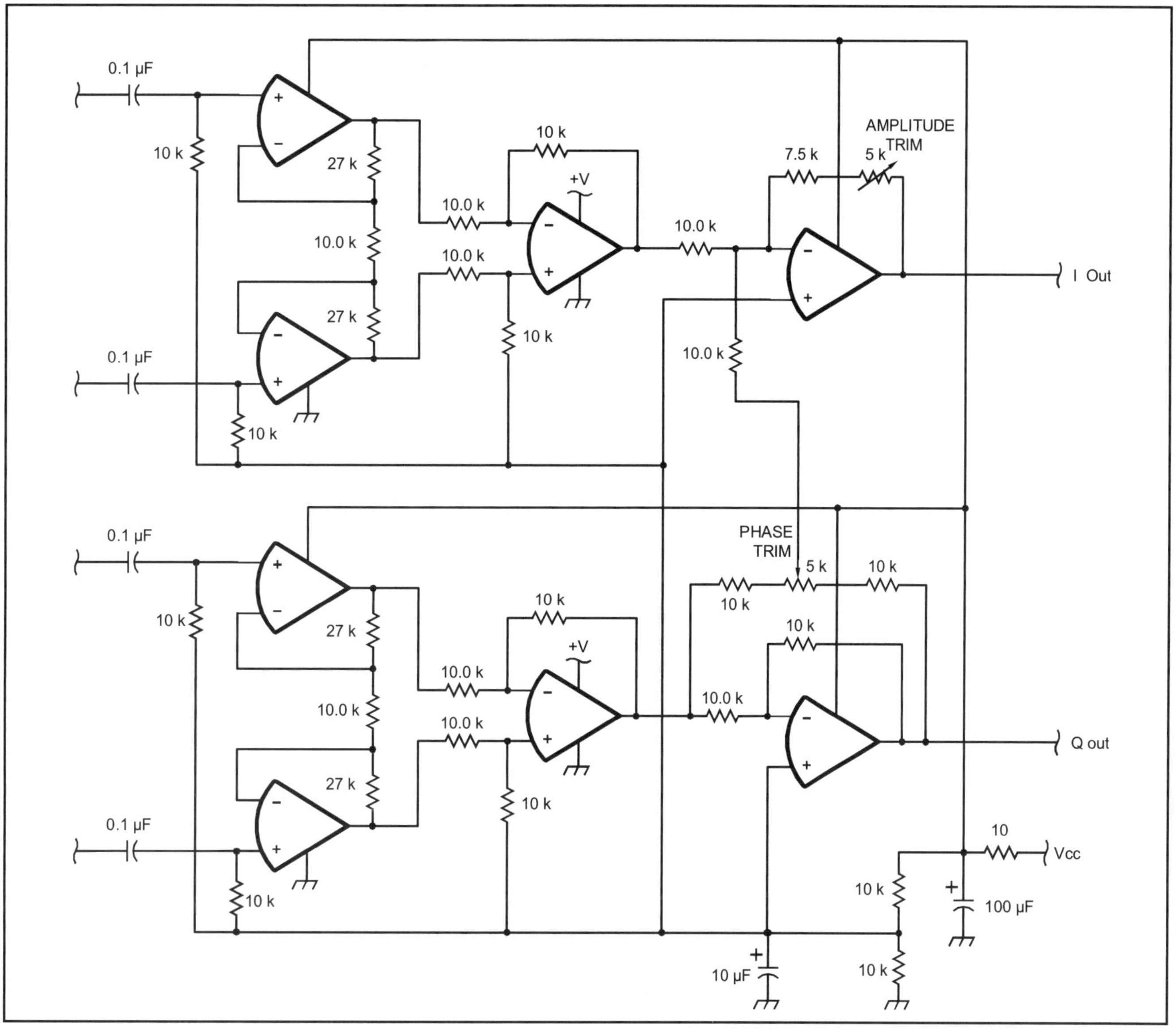

Fig 9.78—Connecting the I Q balanced mixer output pair into the I and Q inputs of the R2pro analog signal processor using a pair of differential op-amp circuits.

opposite sideband level is the distance from the center of the target. The two adjustments, amplitude and phase, are like the windage and elevation adjustments on a gun sight. If one adjustment is way off, adjusting the other one will have little effect on distance from the center of the target. Once one adjustment is perfect, the other adjustment will have a very large effect.

In a phasing receiver, the output we hear when tuned to the wrong sideband is the level of the undesired signal, which represents distance from the target center. There is no indication whether amplitude, phase, or both need to be adjusted. If neither adjustment has much effect, then both are way off. Adjust first one, then the other, while listening to the undesired signal level. As the adjustments approach the optimum values, they become more criti-

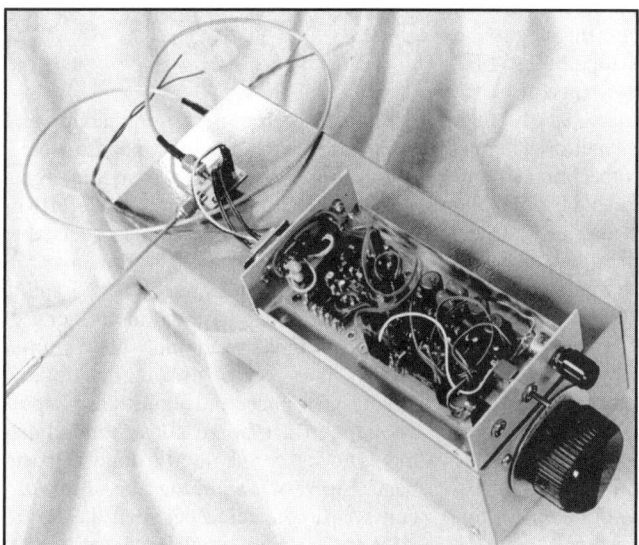

Fig 9.79—The interface circuit board connected between the R2pro ASP and a commercial IQ mixer operating at 2.3 GHz.

Phasing Receivers and Transmitters 9.43

cal. It should be possible to reduce any sine wave frequency in the audio passband down below the noise level. If the signal is strong, it will be possible to reduce the fundamental below the noise while hearing the distortion products. It is important to listen while adjusting, because a meter can't tell the difference between the signal being suppressed, the desired channel noise floor, and distortion products. Once a single-frequency tone is suppressed below the noise floor, tune the receiver slowly to change the tone frequency and observe its suppression. In a properly adjusted R2pro, the suppression will be more than 50 dB over the entire audio frequency range. If it is not, re-optimize the receiver using a different tone frequency. Frequencies near the middle of the receiver audio passband are most useful.

A phasing receiver will always have some opposite sideband suppression. If it does not, then one of the two channels is not working. If the signal has equal strength on either side of zero beat, don't touch the amplitude and phase trimmers, fix the broken I or Q channel first.

Once a phasing receiver using modern components is optimized, the phase and amplitude adjustments hold very well. The prototype miniR2 on 20 meters still exhibits 43 dB opposite sideband suppression from 300 to 3000 Hz after six years, a circumnavigation, numerous camping trips, and a number of disassemblies to display the circuitry.

Interface Circuitry For Other Mixer Types

Much of our work in the amateur bands uses diode ring mixers. Diode rings work well, are available in small quantities in many different varieties, and offer good performance in familiar, mature circuits. Much of our work in our professional lives has been in the development of passive FET mixers of various topologies. FET mixers offer a number of performance trade-offs with diode rings, and often the passive FET mixers are superior. There is also a wide variety of other mixer types including active mixers using Bipolar and CMOS transistors that may be the best choice for some applications. Classic vacuum tube beam deflection mixers, and future optical mixers offer interesting experiment possibilities. This paragraph presents a few interface circuits that have been developed to interconnect passive FET balanced and I Q mixers to the baseband circuitry developed for the R2pro. Much of this work is in the microwave bands, and outside the scope of this text.

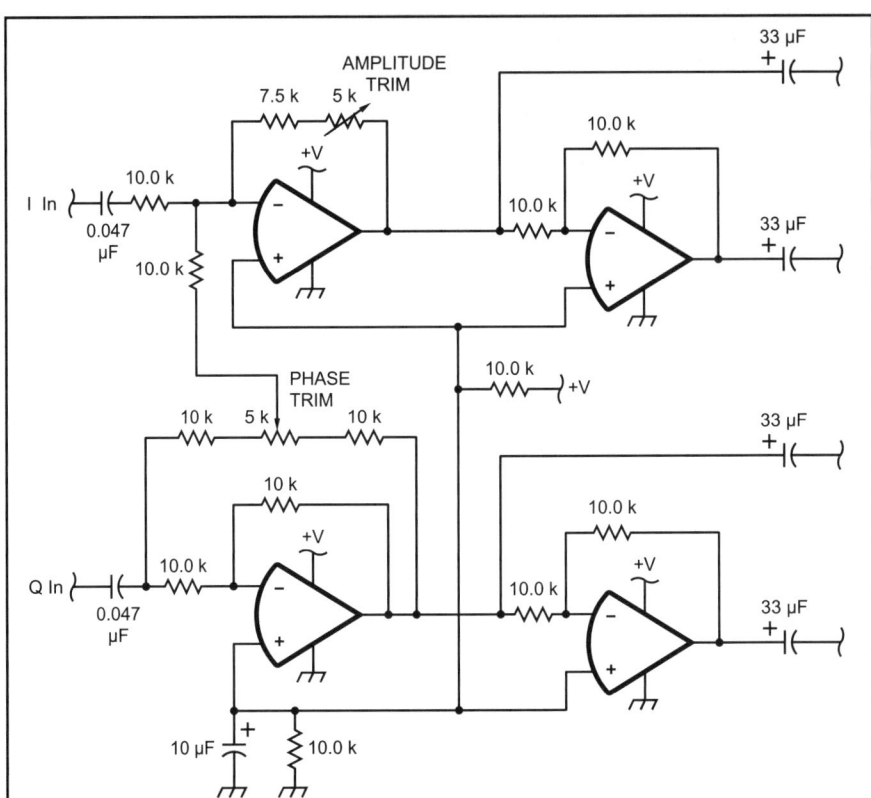

Fig 9.80—A circuit that provides dc-isolated balanced I and balanced Q drive to the inputs of an I Q upconverter.

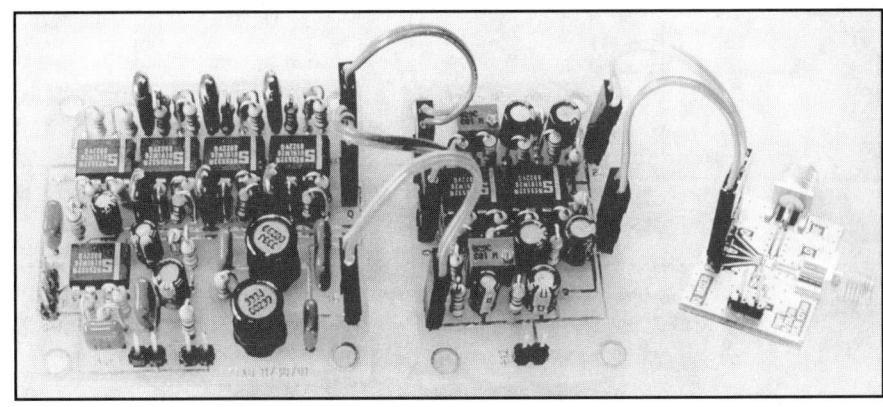

Fig 9.81—A prototype microwave SSB exciter connected to a commercial passive IQ FET mixer at 2.3 GHz.

Fig 9.77 is a circuit for connecting an I Q balanced mixer output pair into the I and Q inputs of the R2pro analog signal processor board. The center-tapped floating transformer primaries may be used to provide operating bias to the mixer if needed, and 6 V bias to the ASP I and Q inputs is provided by the transformer secondaries. **Fig 9.78** accomplishes a similar task using a pair of differential op-amp circuits. The phase and amplitude trimpots on the interface board allow both adjustments to be conveniently done at baseband. **Fig 9.79** is a photograph of this circuit board connected between the R2pro ASP and a commercial IQ mixer operating at 2.3 GHz.

Passive FET mixers are also used as upconverters, and **Fig 9.80** is a circuit that provides dc isolated balanced I and balanced Q drive to the inputs of an I Q upconverter. **Fig 9.81** is a photograph of a prototype microwave SSB exciter connected to a commercial passive FET mixer at 2.3 GHz.

Alternative mixer types are a rich field for amateur experimentation, and there is much progress to be made in this area. Between the 50-Ω interface circuitry described for diode rings and the balanced circuitry presented here, an experimenter should have the tools needed for experiments with many different mixer types.

9.10 A HIGH PERFORMANCE PHASING SSB EXCITER

After completing the R2pro design, it was natural to take a similar approach to the basic phasing exciter. The design of the resulting circuit is described here. In block diagram form, and even in simple circuit implementations, a phasing SSB exciter and SSB receiver have much in common, but as circuitry is optimized for each application, significant differences become apparent. A few differences are:

1. The audio drive signals at the exciter diode ring IF port are only about 10 dB below the LO drive. The diode ring thus contributes significant distortion, and its IF port impedance will vary dynamically with drive.

2. The overall gain from microphone input to exciter output is much lower than the gain in a receiver. Curing unwanted audio feedback and oscillations in an exciter are not significant design tasks.

3. Carrier suppression is an issue, and can not be helped by RF amplifier reverse isolation.

4. RF feedback from the antenna back in to the modulator or LO tuned circuit causes FM

5. There are significant differences in the handling of SSB and CW

6. There are significantly different grounding considerations.

Since there are so many different requirements between optimized receiver and exciter circuitry, each exciter circuit block was redesigned, borrowing sub-circuits from the receiver and previous designs where performance met the exciter requirements.

Microphone Amplifier

The microphone amplifier input is the connection point for a dynamic or electret mike element. It needs to interface to a wide variety of signal sources without changing its gain or passband characteristics. The microphone amplifier defines the noise floor inside the channel during pauses between words, or when using an external digital signal source connected to the exciter audio input. Typical inexpensive electret elements with integral FET amplifiers have an output voltage of about 20 mV and a signal to noise ratio of more than 60 dB. The mike amplifier needs to

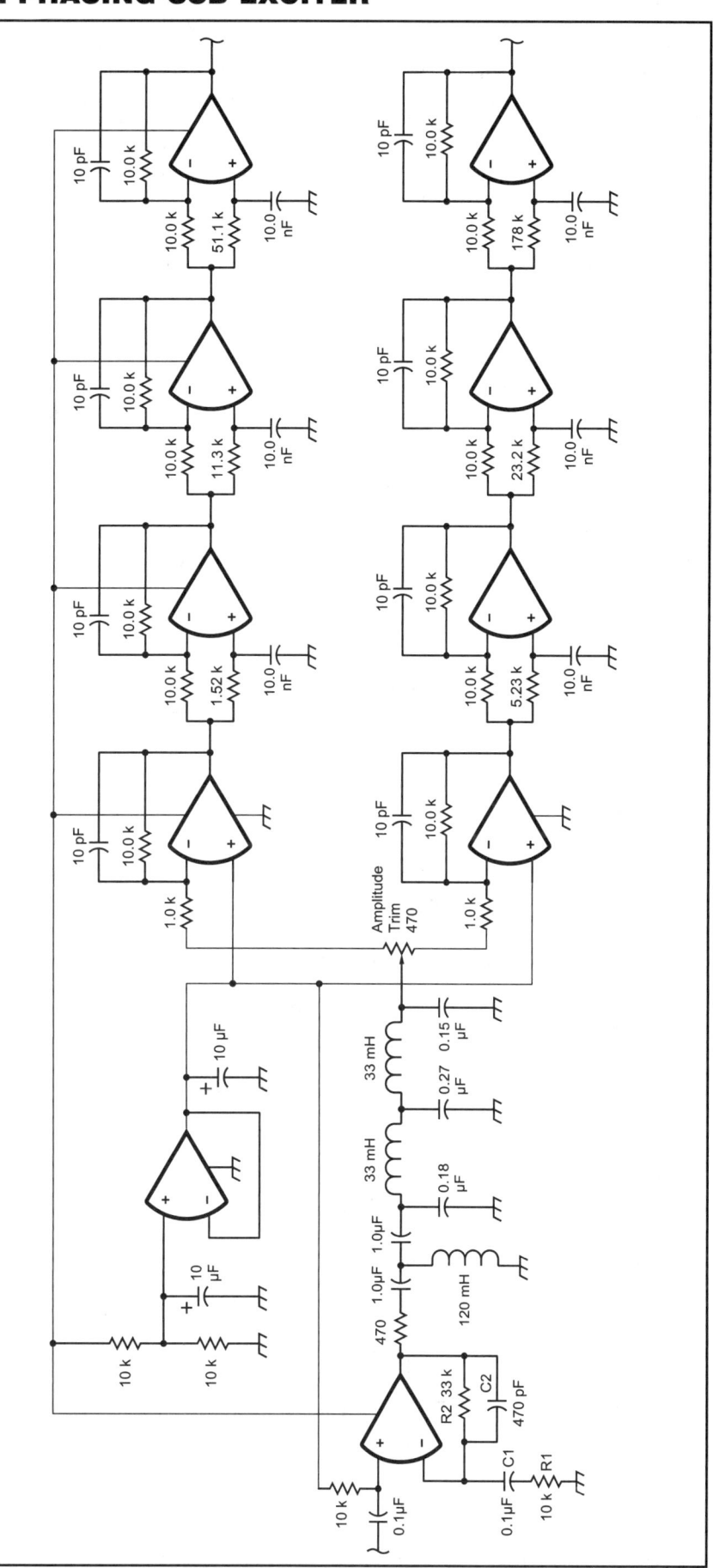

Fig 9.82—This schematic is a speech amplifier and analog signal processor. The I and Q audio outputs may be directly connected to either the modulator circuit shown in Fig 9.83 or the balanced output circuit in Fig 9.80.

have input noise much less than 20 µV across the speech passband to ensure that the exciter noise is below the microphone noise. Typical low-noise Op-Amps have input noise voltage of less than $10 nV/Hz^{1/2}$. Thus the equivalent input noise from the op-amp in a 4-kHz bandwidth is about 630 nV—90 dB below the microphone output. This is good enough for any microphone likely to be used in amateur service.

It is useful to calculate the output noise floor of the exciter when the microphone is disconnected. If the rms input noise of the mike amplifier is 630 nV across the speech bandwidth and the transmitter linearly amplifies a 20-mV signal up to, for example, 10 W (22.4 V rms) into a 50-Ω load, then the transmitter has a total of 61 dB linear gain from the microphone input to the antenna. The output noise voltage is 61 dB stronger than 630 nV, or 700 µV rms. The noise power at the output is 10 nW—low power even by QRP standards. When the inexpensive electret microphone is connected, the noise output increases by 30 dB, up to about 10 µW. This is strong enough to easily hear in nearby receivers on the quiet VHF bands.

The microphone amplifier circuit in **Fig 9.82** has an input impedance of 10 kΩ, 10 dB gain, a high-pass characteristic defined by R1 and C1 and a low-pass provided by R2, C2. For maximum fidelity and flexibility in tailoring the microphone response, the mike amplifier passband is flat from 150 Hz to 4 kHz, with very graceful roll-off above and below. The output impedance of the Op-Amp is raised to about 500 Ω with the series resistor, to drive the LC speech filter.

High Fidelity Speech Filter

The speech filter is designed for high quality speech and rapid roll-off above the desired passband. A 1-dB ripple Chebyshev low-pass prototype was scaled to 500 Ω and 4 kHz to provide the high frequency filter edge, and a single series capacitor provides one high-pass pole at 100 Hz. The filter output is terminated in the 470-Ω input resistor to the inverting input of the output op-amp.

The gain distribution through the exciter audio is designed to minimize off-channel noise and the impact of component tolerances on opposite sideband suppression. Most of the audio gain is before the LC speech filter, so that the filter will have maximum effect on off-channel amplifier noise. The 1-dB ripple Chebyshev speech filter has rapid phase and amplitude variations near the upper passband edge, so this filter is placed before the audio channel is split into I and Q paths. A matched pair of such filters could be used at the output of the I and Q phase shift circuitry to suppress the op-amp phase-shift network noise, but then the component tolerances would have to be unreasonably tight. Instead, a pair of simplified 50-Ω LC low-pass filters is used after the I and Q audio power amplifier stages to remove the

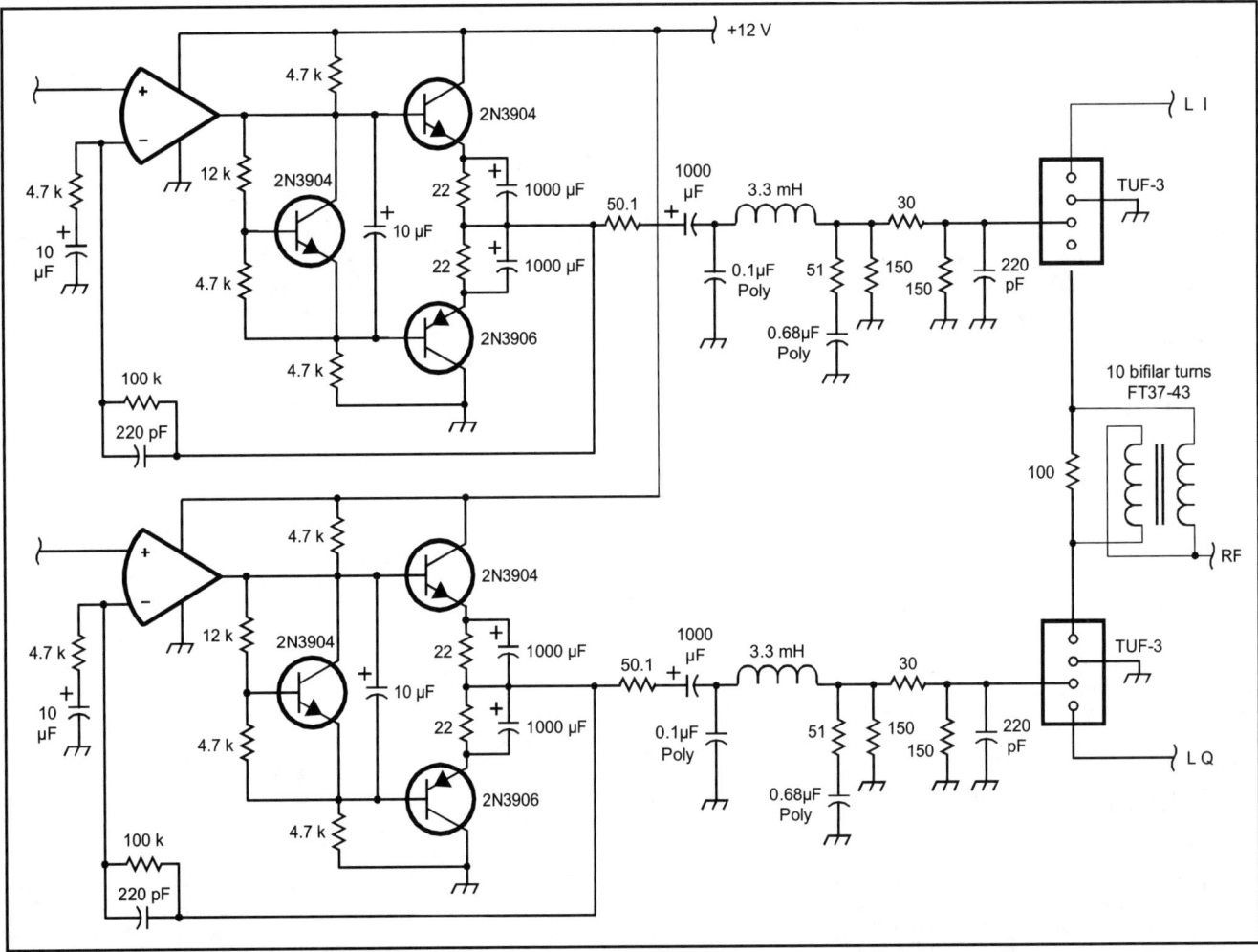

Fig 9.83—The modulator circuitry shown here is connected directly to the output of the audio phase-shift network.

broadband noise from the active phase-shift network and I and Q power amplifiers. These 50 Ω LC low-pass filters were designed for amplitude and phase errors small enough for more than 50 dB of opposite sideband suppression when built with 1% matched components.

Buffer Amplifiers

The LC speech filter termination drives a pair of buffer amplifiers through the amplitude balance pot. These buffer amplifiers provide low impedance drive to the audio phase-shift network. This is a change from the April 1993 *QST* circuit that drove the phase shift network directly from the amplitude balance pot. The original circuit could be adjusted for more than 40 dB of opposite sideband suppression, but both the amplitude and phase needed significant re-adjustment when switching sidebands. The new circuit may be adjusted for almost 50 dB of opposite sideband suppression with very little trimming needed when switching sidebands.

Audio Phase Shift Network

The audio phase shift networks are copied directly from the R2pro circuit. There is no need to change component values. There is some degradation of sideband suppression at audio frequencies below 200 Hz, but less than one would experience with a filter exciter. Using the values derived for the receiver provides maximum suppression of adjacent-channel interference. Dual op-amps are used instead of the quad op-amps specified in the earlier *QST* circuit to ease board layout and reduce the number of parts that need to be kept in stock. With parts selected to 0.1% tolerance, this phase shift network pair will provide more than 50 dB of opposite sideband suppression from 300 to 3500 Hz.

Mixer IF Port Driver Amplifiers

The modulator circuitry shown in **Fig 9.83** is connected directly to the output of the audio phase-shift network. As in the R2pro circuitry, this connection is dc coupled and carries the 6 V bias for the modulator op-amps. The I and Q output audio amplifiers are changed significantly from the earlier design. One issue is that diode ring IF port impedance is a function of both LO drive level, and for modulator service, IF drive level. Since the diode ring IF port is the termination for the LC noise filter, any change in impedance will create phase and amplitude errors between the two channels. Not only do such errors limit the amount of sideband suppression that may be obtained, they will change when tuning across the band, and require readjusting the exciter when switching sidebands. A significant reduction in phase and amplitude errors caused by diode ring IF port impedance variations may be made by adding a 6-dB 50-Ω attenuator between the LC filter and the diode ring IF port. This attenuator may also improve diode ring intermod distortion performance.

The input termination to the I Q LC filter pair is provided by the low impedance output of the audio power amplifier circuitry with a 50-Ω series resistor and 1000 μF dc blocking capacitor. The dc blocking cap could have been used to shape the channel, but then it would have had to be a precision component. Since 10 μF capacitors with the necessary tolerance are both expensive and very large, the capacitor value was increased to the point where a standard tolerance electrolytic could be used. A 1000 μF capacitor with a 50-Ω load has a high-pass pole at 3.2 Hz. A +50% capacitance error from 1000 μF to 1500 μF in just the I channel introduces less than 0.1 degree of differential phase error in the low end of the audio passband.

The appropriate drive level for the diode rings is determined by the desired amount of third order distortion. There is a trade-off between third-order distortion, carrier level, and exciter noise. Exciter third order distortion may be reduced to an arbitrary low level by driving the IF port at low level, but then the RF output is low relative to the diode-ring LO output, and more noisy gain must be used to reach the desired RF output level. With +7 dBm LO drive and two 0 dBm tones on the IF ports of a TUF-1 mixer, the RF third-order products are only 15 dB down from the −9.0 dBm desired outputs. This might be acceptable for some simple VHF or microwave applications where the mixer is connected directly to the antenna—but it is hardly in keeping with a high-performance phasing exciter.

Of particular importance is the fact that mixer intermod products do not have the same phase relationships between the I and Q channels as the desired signals that produced them. The largest signals in the opposite sideband of a phasing exciter are usually intermod products, not the suppressed sideband. Thus it is meaningless to build a phasing exciter with phase and amplitude accuracy to provide 50 dB of opposite sideband suppression, and then over-drive the I and Q mixers so that the intermod products are only 30 dB down.

Measurements

A TUF-1 mixer was measured with two −10 dBm IF tones and a 22 MHz, +7 dBm LO. The desired outputs dropped to −15.3 dBm, and the 3rd order intermod products dropped to 47.5 dB below each desired tone. −15.3 dBm outputs from −10 dBm inputs indicates a conversion loss of only 5.3 dB. The 22 MHz carrier feedthrough is at −63.3 dBm, or 48.0 dB below either tone of the two-tone output. At 7 MHz the carrier suppression improves to 49.9 dB below either of the two tones.

From these experiments with −10 dBm two-tone drive into a single mixer, the carrier and intermod products are both more than 47 dB below either tone. This puts them −53 dB below the PEP output. Combining a pair of these mixers as an SSB modulator makes a further improvement. The carriers from the two mixers are 90 degrees out of phase, so the resultant voltage is 1.414 times the voltage of each carrier. The desired sideband adds in phase, so the resultant voltage is 2.0 times the voltage for either mixer output. A passive combiner involves an impedance transformation, so the resultant voltages are reduced by 0.707 into a 50-Ω load. The final output tones are then 3 dB stronger than the tones from a single mixer, but the combined carrier outputs are the same as for a single mixer.

The situation is more complicated for intermod products. Some of them add in phase, some cancel, and some add with 90 degree phase shift. The worst case is when the intermod products add in phase, exactly the same as the desired sideband.

An SSB modulator built with two TUF-1 mixers operating at a carrier frequency of 22 MHz, with two −10 dBm tones into each mixer IF port, will have desired sideband output tones of −12.3 dBm (−15.3 dBm + 3 dB), a carrier 51 dB below either tone, and intermod products at least 47 dB below each tone. This performance is a good fit with a precise phase shift SSB system that provides 50 dB of opposite sideband suppression.

The IF amplifier driver amplifiers are also potential sources of distortion. With a 6-dB pad between each LC low-pass filter mixer IF port, filter loss, and the 6-dB loss through the 50-Ω series termination resistor, the total loss between the driver amplifier and mixer IF port is about 14 dB. Two −10 dBm tones is −4 dBm PEP, so the driver amplifier must supply a two-tone +10 dBm with distortion products well below the level produced by the mixer. Fortunately, a suitable amplifier was designed as the audio output stage for the R2pro. At the +10 dBm PEP output level, distortion products are all more than 60 dB below each of the desired tones.

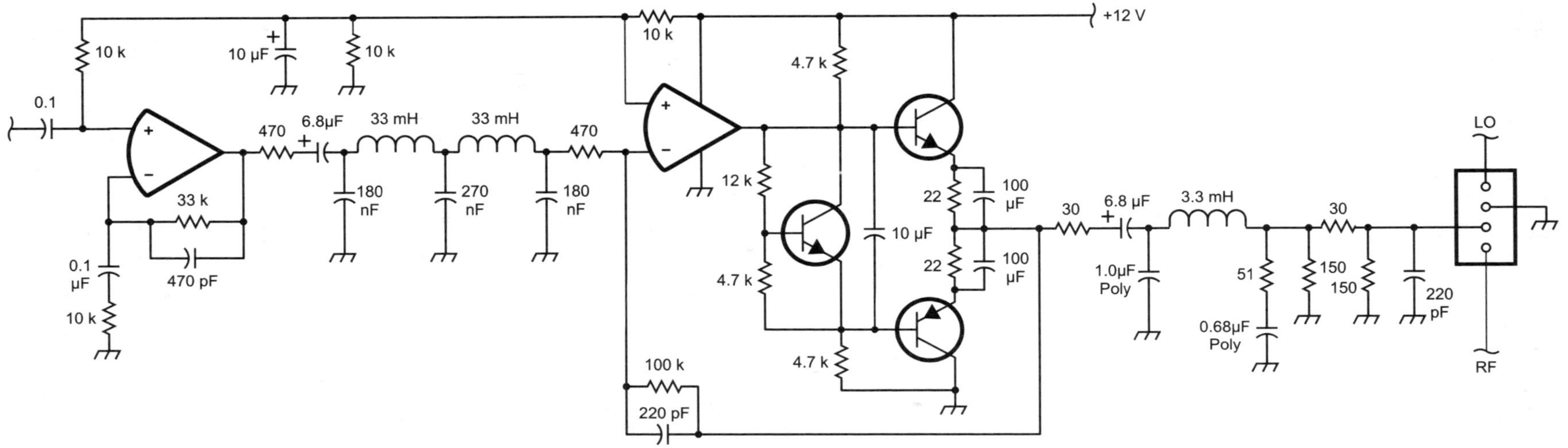

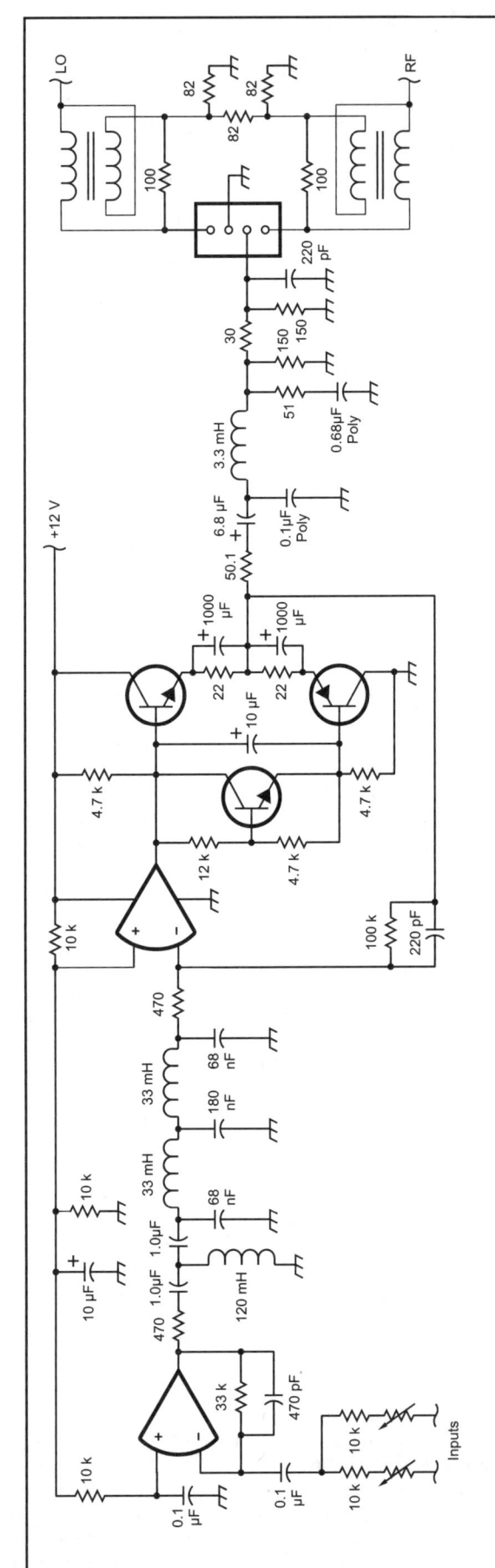

Fig 9.84—A complete low-distortion DSB modulator with 50-Ω output.

Fig 9.85—An AM exciter that generates a DSB signal and then adds the correct amount of carrier to obtain 100% modulated AM at very low distortion.

Mixer Environment

To obtain 50 dB opposite sideband suppression, amplitude errors between the I and Q channels across the entire speech passband must be held to less than about 0.03 dB, and phase errors must be held to less than 0.007 radians (0.4 degrees). Since mixer port terminations affect both conversion loss and the phase behavior of any LC networks connected to the ports, it is important for the mixers to operate in as ideal an environment as possible. Good 50-Ω terminations on all three mixer ports, constant LO drive level, and good isolation between the RF ports of the I and Q mixers are all necessary to maintain sideband suppression. Isolation between the I and Q mixer RF ports is needed because the LO leakage from one mixer is 90 degrees out of phase with the LO drive to the other mixer. This is precisely the phase that results in maximum sensitivity to recovery of phase noise or other fluctuations on either mixer.

On each mixer port, 6-dB resistive pads will generally improve opposite sideband suppression across the audio and RF passband. In transmit applications, the noise figure penalty is less of a concern, so the use of a 6-dB pad on each IF port, and a 6 dB increase in audio drive level, is good practice. Pads on the LO ports of the mixer help maintain opposite sideband suppression when LO connections are changed (or cables are flexed). LO port pads should be used if sufficient LO drive level is available. Above 20 MHz, the Mini-Circuits MAV-11 provides a simple way of obtaining +17 dBm of LO drive. After a twisted-wire hybrid splitter, the I and Q LO levels will both be +14 dBm. 6 dB pads (and a little circuit loss) will drop this to the appropriate drive level for standard level diode ring mixers. A 6 dB pad on the RF port helps maintain constant mixer behavior across a wide RF band. An alternative to a resistive pad on the RF port is an amplifier with a good, broadband, resistive input match and high reverse isolation. The reverse isolation prevents changes in the amplifier output load from appearing at the mixer summer.

Sideband Selection

There are a number of options for sideband selection. Reversing the LO connections to the mixers, reversing the I and Q audio drive connections to the modulator drivers, or introducing a 180 degree phase shift in either the I or Q audio drive will all work. One advantage of taking great care to operate the mixers in a 50-Ω environment and making the audio phase shift network as accurate as possible is that the amplitude and phase trim adjustments are likely to need very little trimming when switching sidebands. The sideband selection method chosen depends to a large extent on whether the exciter is to be used at a single frequency, or will be required to cover a multi-octave range, and whether the I and Q audio drive is obtained from a DSP chip or an analog IC chain.

A DSB Modulator

The same basic circuits that are used to build up a phasing exciter may be used to build up a DSB or filter-type SSB exciter. **Fig 9.84** is a complete low-distortion DSB modulator with 50-Ω output. The microphone gain should be set up so that the output level at each sideband is –15 dBm.

DSB with Carrier

There are applications for a very low distortion AM exciter. **Fig 9.85** is an AM exciter that generates a DSB signal and then adds the correct amount of carrier to obtain 100% modulated AM at very low distortion. Two inputs are provided, so that the exciter may be connected directly to the stereo output of a CD player. With a +10 dBm LO in the 1 MHz range, this exciter may be used to play collections of vintage radio programs over lovingly restored AM broadcast radios. Use low-pass Pi networks to connect to the 25-Ω RF and LO ports.

9.11 A FEW NOTES ON BUILDING PHASING RIGS

Some of our phasing rigs have been learning experiences, and some are fine radios that have displaced all the commercial equipment in the author's home and portable stations. The most successful radios have a few features in common.

1. Separate receiver and exciter circuitry. The individual components in phasing rigs are inexpensive, and it is false economy to include complex switching networks so that a circuit block used in the receiver may also be used in the exciter. Complicated switching schemes to re-use receiver components in the SSB exciter is an obsolete concept that became popular in the 1960's to save money on expensive crystal filters, and to reduce the number of vacuum tubes and filament current drain.

2. A common VFO for full transceive operation, but independent LO phase shift networks. A conservative approach is to distribute low level LO signals on 50 Ω lines to buffer amplifiers and LO phase-shift networks in the exciter and receiver modules. This eliminates interaction between the receiver and exciter adjustments.

3. Buffered RF ports on both the receiver and exciter. A receiver LNA with good reverse isolation and a relatively broadband, near 50-Ω RF output should be hard-wired to the RF input of the image-reject mixer. The exciter image-reject mixer should be hard-wired to a broadband, 50 Ω low-level amplifier input. The LNA and exciter low-level output amplifier should be built into the receiver and exciter modules.

4. Good RF filtering and a very clean LO. Phasing circuitry does a fine job of eliminating the opposite sideband, but it does nothing to reduce strong off-channel and out-of-band signals that can cause interference through various distortion mechanisms.

5. Modular construction using feedthrough capacitors and mechanically solid RF-tight enclosures. Not only are individual modules easier to test and align, they hold their alignment when interconnected, and greatly reduce spurious responses and outputs. Modular construction with 50 Ω interconnecting signal cables and bypassed dc connections should be used whenever performance is more important than construction time.

The philosophy behind our phasing rigs is also worth noting. Early amateur work, and much of the professional use of phasing techniques, has been motivated by the desire to cut costs. In contrast, our work has been primarily directed toward improved performance compared with the usual inexpensive narrow-IF-filter superheterodyne approaches. It is an interesting exercise to build and communicate with a radio having only a few parts, but that is a different experience from using a system designed for smooth operation and high performance. For minimum parts count projects, simple *DSB* direct conversion receivers and simple superhets are often the best choice.

9.12 CONCLUSION

In the 25 years since publication of *Solid State Design for the Radio Amateur*, much has changed. Some of the most simple, light-weight mountain rigs include microprocessor frequency control and superhet receivers with crystal filters carefully designed for optimum CW intelligibility. Rack-mount direct conversion receivers are used in high-end weak-signal tropospheric scatter UHF SSB and CW stations. EME contacts have been made using a few watts of transmit power and truly awesome receiver signal processing power.

At the end of this chapter it is useful to explore some of the advantages of phasing receivers and exciters.

1. Phasing techniques work at any frequency. This can be used to eliminate frequency conversions in heterodyne receiver and transmitter system, which makes it easier to avoid internal and external spurious responses and achieve spectral purity. The same baseband processor may be used with simple RF circuitry on any amateur band from 170 kHz through millimeter waves.

2. Phasing receivers and exciters require low distortion mixers and audio amplifiers. While it is *possible* for a conventional superhet receiver or exciter to sound good, most published designs and commercial products do not. High fidelity is *necessary* for a phasing rig. Now that there are many published receiver and exciter phasing circuits to duplicate, the designer-builder can confidently construct a very fine sounding radio system.

3. The emphasis on low distortion all the way through the RF to audio chain means that there is no penalty for using audio filtering for selectivity. High-performance audio filters may be realized using conventional L C networks or digital signal processing systems.

4. Phasing rigs inevitably have lower in-channel distortion than conventional superhets using narrow filters. Low in-channel distortion provides a significant performance improvement on any mode that injects a baseband signal into the SSB microphone input and recovers the signal from the receiver audio output. This includes conventional SSB and all of the present and future modes using Computer Sound Cards interconnected with the radio.

5. The basic phasing rig block diagram has many components that may be replaced by DSP and DDS systems. DDS and DSP are two areas in which the state of the art is rapidly advancing. Phasing receivers and exciters provide the radio experimenter with the interface between antennas and the latest advances in signal processing technology.

6. The final advantage to phasing systems is philosophical. A basic superhet receiver with a crystal filter is fairly easy to explain and understand. It is also straightforward to build, and alignment is simple. When badly constructed and poorly adjusted, it still provides adequate performance. A phasing receiver is no more complicated than a superhet, but its underlying principles are more subtle. Care in construction pays off, and listening while playing with the phasing adjustments is really very cool. An amateur who has built up a phasing receiver, looked at the I and Q channel signals on a dual-trace oscilloscope, and tweaked the phase and amplitude adjustments while listing to an opposite-sideband signal drop into the noise acquires a depth of understanding far beyond that of most wireless graduate students and many of their professors. The best part is that understanding of phasing systems comes from experimenting with simple circuits and thinking—the tinkering comes first—then the understanding. In this area the amateur with his simple workbench; primitive test equipment; and time to contemplate, has a profound advantage over both the engineering student with a computerized bench and exam next week, and the professional engineer with a million-dollar lab and a technician to run it.

REFERENCES

1. R. Campbell, "LO Phase Noise Measurement in Amateur Receiver Systems", *Proceedings / Microwave Update '99*, Plano, TX, October 1999, ARRL Publication number 253, Newington, CT, 1999, ISBN: 0-87259-772-5, pp 1-12.

2. R. Campbell, "A Binaural IQ Receiver", *QST*, March 1999, pp 44-48.

3. R. Campbell, "Medium Power Diode Frequency Doublers", *Proceedings / Microwave Update '99*, Plano, TX, October 1999, ARRL Publication number 253, Newington, CT, 1999, ISBN: 0-87259-772-5, pp 397-406.

4. R. Campbell, "Microwave Downconverter and Upconverter Update", *Proceedings / Microwave Update '98*, Estes Park, CO, October 1998, ARRL Publication number 241, Newington, CT, 1998, ISBN: 0-87259-703-2, pp 34-49.

5. A. Ward, "Noise Figure Measurements", *Proceedings / Microwave Update '97*, Sandusky, OH, October 1997, ARRL Publication number 231, Newington, CT, 1997, ISBN: 0-87259-638-9, pp 265-272.

6. R. Campbell, "Direct Conversion Receiver Noise Figure", *QST*, February 1996, pp 82-85.

7. R. Campbell, "Binaural Presen-tation of SSB and CW Signals Received on a Pair of Antennas", *Proceedings / 18th Annual Conference of the Central States VHF Society*, Cedar Rapids, IA, July 1984.

8. W. Hayward and J. Lawson, "A Progressive Communications Receiver", *QST*, November 1981, pp 11-21.

9. S. Bedrosian, "Normalized Design of 90° Phase Difference Networks", *IRE Transactions on Circuit Theory*, June 1960, pp 128-136.

10. R. Fisher, "Broad-Band Twisted-Wire Quadrature Hybrids", *Transactions on Microwave Theory and Techniques*, May 1973, pp 355-357.

11. R. Harrison, "A Review of SSB Phasing Techniques", *Ham Radio*, Vol. 11, No. 1, January 1978, pp 52-63.

12. J. Reisert, "VHF/UHF Frequency Calibration", *Ham Radio*, Vol. 17, No. 10, October 1984, pp 55-60.

13. B. Blanchard, "RF Phase Shifters for Phasing-Type SSB Rigs", *QEX*, January/February 1998, p 34.

CHAPTER 10

DSP Components

The basic concepts of performing signal processing functions in a computer go back many years. Much of this processing was performed on relatively slow computers, where signals were treated as a series of numbers. But, Digital Signal Processing, or DSP, as applied to communications systems is more: It refers to the conversion of conventional analog signals into digital words, then processing these words for some useful purpose and the conversion back to analog signals. In addition, all of this must occur fast enough to keep up with the incoming signal. That is to say, the computation is "in real time."

The increased speed of digital computing hardware along with improvements in low-cost converters for input and output devices has brought DSP to many everyday products. This has made possible some functions that were difficult to perform in analog hardware. In addition, there are reduced production costs associated with using DSP, all of which is attractive to equipment manufacturers and homebuilders alike. Not surprisingly, there are also limitations in using DSP to replace analog functions. These lie primarily in the areas of speed and dynamic range.

Figure 10.1 illustrates the implementation of a bandpass filter first as a conventional LC design and then as a DSP element. The LC design is obviously simple in only requiring 6 components. It can be built over a wide range of frequencies and consumes no power. However, in order to achieve high Q in the inductors it may occupy a fair volume and, particularly at lower frequencies, may become heavy.

In contrast, the DSP version has much greater hardware complexity. Most of this is hidden away inside integrated circuits, but even the interconnect wires (PC board traces) will count in the tens or hundreds for most implementations. The DSP implementation might consume a few watts of power, as well. However, once the filter program is written, it is precisely duplicated by any number of builders. Once the signal has been converted to digital form it is often easy to add other functions, such as AGC, or to increase the performance of the filter considerably beyond that which is practical for the analog filter. For this reason, it would be unusual to see a DSP based circuit that was as simple as just a band-pass filter. The DSP implementation is limited in the upper frequency that can be used and is most often seen for frequencies in the 10's of kHz. The increasing processing rates of DSP devices can be

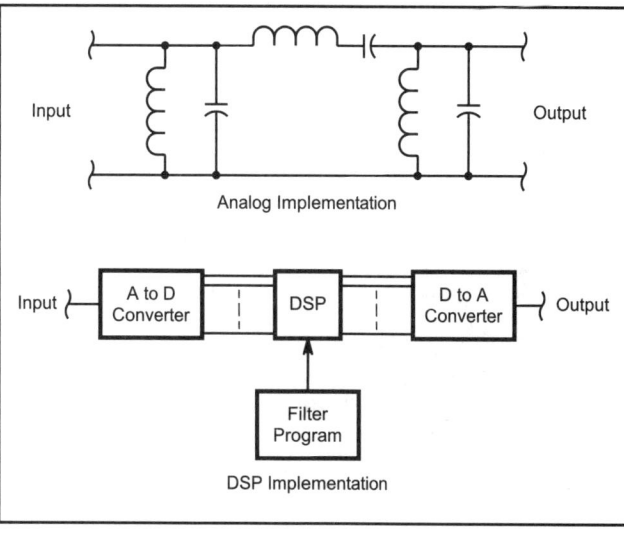

Fig 10.1—Alternate analog and DSP implementations of a band-pass filter.

expected to push these frequencies up in the future.

In this chapter, we will explore the types of DSP building blocks that can replace or supplement analog circuitry. Where possible, comparisons with similar analog functions will be made. This will help to give a rational basis for mixing DSP functions into communications gear in the places where it "makes sense." Examples of mixed analog and digital circuitry will show how these building blocks can be used for both audio and IF applications.

This chapter will attempt to provide enough detail to allow construction or modification of working "DSP components." In the case of hardware construction, this usually requires that the builder is able to write down a schematic diagram complete with component values. For our software case, there is no direct equivalent of the schematic diagram. Many have tried to use various forms of "flow diagram" to communicate the contents of programs. For logic decisions, this can be a useful tool. However, for a computational algorithm, such as a digital filter, the flow diagram does not add clarity over communicating directly with a well-commented computer program, written in a reasonably clear language. This approach will be applied here.

This chapter places emphasis on working DSP components. The background mathematics is not emphasized. However, there are other texts, such as that by Doug Smith[1], KF6DX, which should be consulted to add this perspective.

10.1 THE EZ-KIT LITE

One of the interesting parts of circuit design is the selection of components. For instance, we might need a basic NPN transistor to operate at low signal levels and since the "junk-box" has a supply of 2N2222 we will use them. These devices are readily available from a number of sources, inexpensive and chosen for those reasons, as much as technical ones. However, as the complexity of the circuit function increases, the devices become more specialized and the number of sources diminishes. For instance, most integrated RF amplifiers, even at low power levels, are available from only one or two sources. When we get to DSP devices it is a case of each manufacturer having a separate processor that not only doesn't substitute for any other, but that have different internal structures requiring different programming languages.

For these reasons, it is necessary to pick a specific language and a specific processor family when describing the operation of a DSP function. If this is not done, the description becomes quite mathematical and remote from an actual working program. The Analog Devices ADSP-2100 family and specifically the ADSP-2181 are used in this chapter to describe the DSP functions. This choice was made for several reasons:

1 – The assembly language is easy to follow

2 – Good support manuals are available
3 – The EZ-Kit Lite makes getting started simple.

This, however, is not to say that the Analog Devices ADSP-21xx series is the best solution for a particular problem. However, this is a good all-around processor and provides a consistent language to illustrate the examples that follow.

Fig 10.2 is a block diagram of the EZ-Kit Lite board. The processor is an ADSP-2181 that has both 16K on-chip words of 16 bit data memory and 16K on-chip words of 24-bit program memory. This is more than adequate for any likely amateur project. When the board is powered down, programs can be stored in a 27C080, or in a smaller EPROM. The firmware procedure for loading from this 8-bit EPROM storage to the 24-bit program memory is part of the DSP hardware. The EPROM is not used after program loading is completed. The EZ-Kit Lite executes 33 million instructions per second.

Communications with a PC through a serial port requires a software UART (Universal Asynchronous Receiver/Transmitter) to be run in the EZ-Kit, but the hardware to change to RS232 levels is part of the board.

Analog input and output takes place through a dual (stereo) set of converters in an AD1847 CODEC.* The sampling rate of the CODEC is programmable up to 48 kHz and supports an analog bandwidth of about 20 kHz.

Other digital lines are available for con-

---

*The term CODEC stands for Coder/Decoder and refers to the combination of Analog-to-Digital and Digital-to-Analog conversions, along with dynamic-range compression algorithms. For the applications in this book, no compression algorithms are used, but we will still refer to the conversion package by its common nickname CODEC.

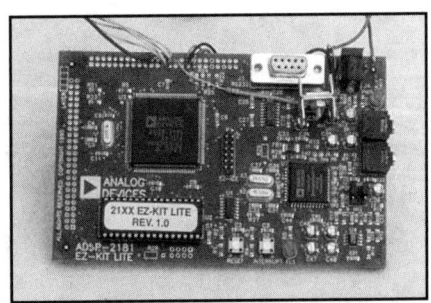

The EZ-Kit Lite.

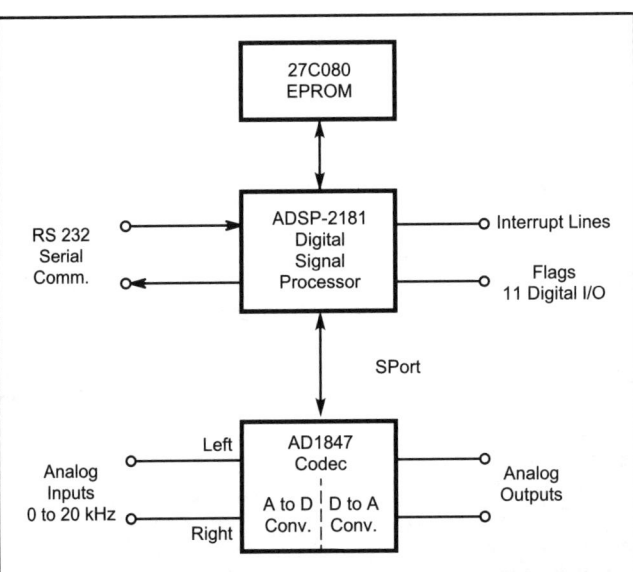

Fig 10.2—Block diagram of the EZ-Kit Lite from Analog Devices. The CODEC has dual A/D and D/A converters. Memory in the ADSP-2181 can be loaded from the EPROM.

trol purposes and connections are supplied for adding almost any kind of memory or I/O device.

Mixed-Modes

All real-life signals are analog in their nature. This means that a signal level is not constrained to a fixed set of levels, but rather may take on any level as time passes. Even the outputs of digital logic circuits are not just "0" or "1" but instead consist of waveforms that have rise-times, ringing and other variations. All of the RF, IF, and audio signals used in radio systems are, more obviously, analog.

DSP provides an alternate way to deal with these analog signals. This involves approximating the analog signal with a series of digital numbers, processing these numbers with some sort of computer and then creating a processed analog signal that again only approximates the desired result. It is important to keep in mind that the signal of real interest is the analog one. The digital calculations are only a means to obtain the processed signal. In order to maintain an adequate approximation of the analog signal, one must examine the computer routines and in some cases take special precautions. The human ear is often the final judge of DSP distortion. Most people cannot hear digitized distortion when 7 or 8 bits are used in the representation. Even with a 16-bit processor, care must be taken to ensure that this number of bits is retained accurately.

Why DSP?

Traditionally signal generation and processing has used analog components. Most of this book involves these techniques. A transistor oscillator can create a signal of good spectral purity. Inductors and capacitors make fine signal filters. Combined with a few transformers and diodes, one has a mixer capable of handling a very wide range of signal levels. The simplicity of this approach has great appeal and for many projects, it is clearly the proper approach. The arguments for putting some portion of the equipment into a DSP process generally are:

• Increased performance in networks such as filters, 90-degree phase-shift networks and banks of filters.
• Better precision in operations such as SSB generation.
• Simpler reproduction of software, relative to hardware.
• The availability of functions that are difficult to implement in hardware, such as adaptive filters.
• The DSP processor likely will have extra time available for conventional control functions, such as displays or switches.

From a manufacturer's point of view, where a commercial product is involved, much of this can result in lower production costs at high volumes. For the experimenter, producing a project for himself, this can simplify the project as well, assuming that much of the project can be based on existing programs. However, if one must develop the entire program, it may well turn out that the time required is considerably above that of similar hardware.

Arguments in favor of using analog components generally center about the following considerations:

The A/D and D/A conversion processes tend to restrict the dynamic range of the process.
• The bandwidth of the process is too great for a DSP.
• The basic complexity of the DSP is not justified,
• The power consumption is higher than the analog counterparts.
• Programs and debugging of programs requires new skills.

As with any other technology, one must weigh the various considerations and decide if DSP is the best approach to a particular application.

Dynamic Range

In any communications system the lowest level of a signal that can be handled is limited by noise, and some form of overload sets the highest level. The ratio of these two levels, usually expressed in dB is the dynamic range of the system. Systems using DSP have dynamic range limitations, as do analog systems, but the form of noise and overload effects can be quite different. In well-designed systems, the limitations on dynamic range normally come from the conversions to or from analog signals. Internally, the DSP can handle a wide range of signals, because of the resolution of data words and by the use of level shifting algorithms, such as AGC.

For both A/D and D/A converters, noise is introduced by the minimum resolution of the converters. In addition, as will be seen below, some converters may have higher levels of noise associated with the conversion process itself. As converters get faster, they tend to have fewer bits per word with a larger least-significant bit and this represents more noise. This is not always a problem, since a faster converter spreads the noise over a wider frequency range. The noise in a single communications channel may actually be less with the wider bandwidth converter. This is due to the noise, from the A/D encoding process, being spread over a wider frequency bandwidth and a smaller percentage of this noise hitting within the communications channel.

The EZ-Kit Lite uses the AD1847 CODEC for both the A/D and D/A conversions. This is of the *sigma-delta** type[2] that is commonly used in DSP applications. The internally generated noise for this conversion process can be considerably greater than that associated with a least-significant bit. **Figure 10.3** is an oscilloscope picture of the noise associated with the A/D converter running with a 48-kHz sample rate and no input signal. The levels were measured by using the DSP to multiply the A/D noise by 100, making it of sufficient level to cover the D/A noise. The RMS A/D noise can be seen to be 153 µV, or about 8 times the level attributable to the least-significant bit. This effectively limits the useful bits to 16–3 or 13.

The corresponding D/A noise, shown in **Fig 10.4**, has an rms level of about 200 µV, which is slightly greater than the A/D noise. It is more difficult to quantify this since the bandwidth of the noise on the output of the D/A converter is much wider than half the sample rate. The level given

---

*Sigma-delta A/D converters use low-resolution conversions (usually 1 bit), operating at very high conversion rates. The very high digitizing noise is reduced by digital filtering, which accepts only a small part of the noise frequency spectrum. Further noise reduction comes from feedback loops that are able to shape the noise spectrum to move much of the noise energy to high frequencies allowing it to be removed by the digital filters. Similar processes are used to reduce the noise in the sigma-delta D/A converters.

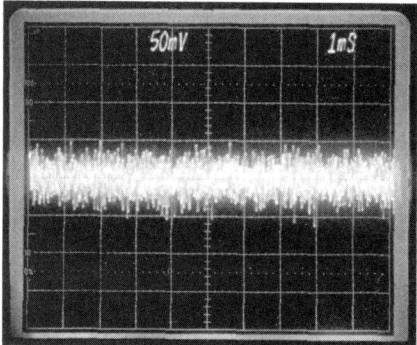

Fig 10.3—Oscilloscope trace of the A/D converter noise in the EZ-Kit Lite. There was no input signal to the converter and the DSP was used to amplify the noise by 100. This was then applied to the D/A converter to produce the trace shown. Each vertical division is 50 millivolts and each horizontal division is 1 millisecond.

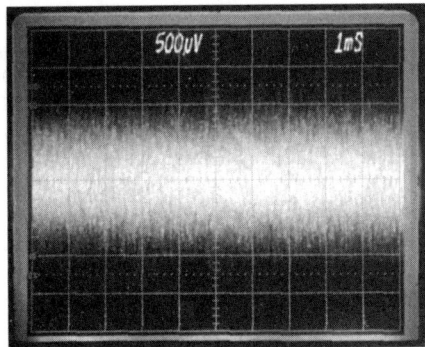

Fig 10.4—Oscilloscope trace of the D/A converter noise in the EZ-Kit Lite. No signal was driving the converter and the oscilloscope bandwidth had been limited to 30 kHz. Each vertical division is 500 µV and each horizontal division is 1 mS.

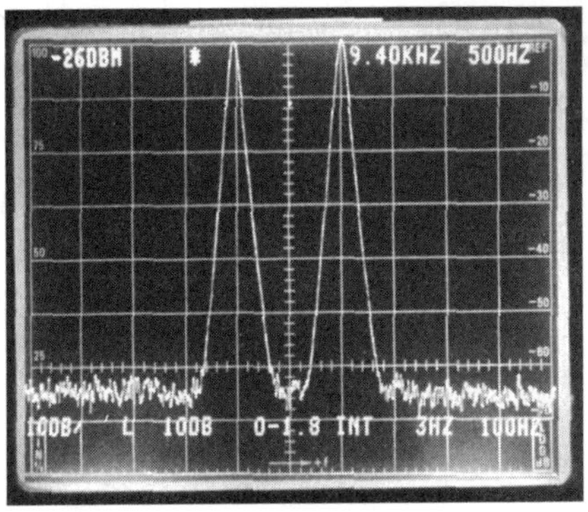

Fig 10.5—D/A output spectrum for two sine waves at 8.9 and 9.9 kHz. Each signal was 2.0 V p-p so that the peak level for both sine waves was 4.0 V p-p, which is full scale for the D/A converter. The noise floor, which is about 65 dB below each of the sine waves, is mainly from the spectrum analyzer.

input signal by quantizing it into a series of small steps. On a detailed scale, these input/output characteristics do not appear at all linear. However, as long as the input signals are within the range of the digital words, the process, on a large scale, is often very linear. This results in the small step non-linearities dominating and the resulting intermodulation distortion being spread over a very large number of products, in a noise-like fashion. The term intermodulation ceases to be a good descriptor. As an example, **Fig 10.5** shows the spectrum of two sine waves produced by DSP computation and converted to analog signals by the AD1847 CODEC. No conventional intermodulation products are observable, although the sine waves are using the full available range of the D/A converter. Although mostly obscured by the spectrum-analyzer noise floor, if it could be seen, the distortion product from the two signals would appear to be similar noise.

In contrast to analog circuit distortions, the overload point of the digital signal is abrupt and creates severe distortions. Depending on the nature of the computation, either the signal output will reach a maximum value and not go any further, or even worse, it may wrap around between the greatest positive and the most negative values. In DSP processors, such as the ADSP2181, this choice of overload responses is programmable. Never-the-less consideration must be taken to avoid problems from operating in these signal regions.

above was estimated by placing an RC low-pass filter, down 3 dB at 30 kHz, on the output of the converter. This limited the noise to roughly the band of interest (24 kHz for a 48-kHz sample rate).

It is often desirable that the noise associated with the analog processes prior to the digital hardware be amplified until it is somewhat stronger than this "digital" noise. However, doing this reduces the total dynamic range. These are the same tradeoffs between overload prevention and signal sensitivity that have always existed in analog signal design.

The number of bits of the A/D converter limits the top end of dynamic range. Depending on the type of converter, this may result in abrupt compression or it may generate erroneous values. Although this latter form of distortion can obliterate the ability to receive a signal, either effect is a severe form of distortion

Intermodulation distortion in analog equipment is usually dominated by the third and fifth order products (see Chapter 2). This is due to the gradual nature of the non-linearities of analog components. In contrast, the digital process distorts an

10.2 A PROGRAM SHELL

We now need to digress from the signal processing subject to gain a general understanding of the process of programming a DSP microprocessor. The details shown here are specific to the EZ-Kit, but all DSP microprocessor environments have a corresponding process.

The EZ-Kit Lite requires sizeable amounts of programming before it can be used for even the most trivial DSP function. Much of this is associated with programming the CODEC that provides the A/D and D/A conversions. An example of this is setting the sample rate to 48 kHz as is used in the example programs. It is important that these hardware initialization chores be performed correctly, but most often the DSP programmer need not be concerned about the details involved.

For this reason, the EZ-Kit manufacturer provides a program shell. This is a computer program that does almost no useful work other than to pass data through unchanged. It provides a place where a DSP function can be placed to create a useful program.

Fig 10.6 shows the overall flow of the shell, which is the same for any of the programs in this book. When first started, the program initializes the parameters of the hardware and software. This is only done once, although the program may continue to operate for days, months or longer. Following initialization, the program goes into a continual loop. In the figure, this loop is referred to as a background process.

The operations in the background process loop can range from no process to a complicated mathematical computation, such as a Fast Fourier Transform. As much processing as possible should be put here. The only requirement for being part of the background is that the processing does not require periodic computations at precise time intervals. Examples of background processes would be the reading of a switch or the outputting of data to a controlling PC. These operations need to be done quite often, but the exact times are not critical.

Computations that must be done periodically are handled by interrupts. The interrupt is a signal sent to the DSP to request special processing. In our case, the reason for the interrupt is that another 1/48,000 second (about 20.8 µs) has elapsed. The specific hardware that generates the interrupt is the CODEC. Typical of

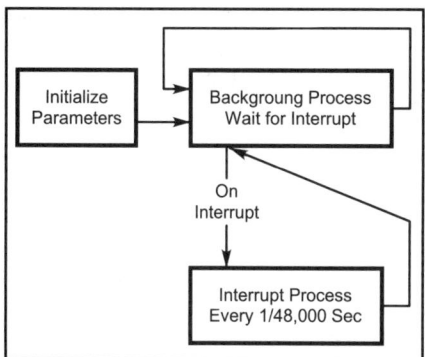

Fig 10.6—Main flow of the DSP programs. To give some feel for the numbers involved, the interrupt rate is shown as 48,000 per second. Depending on the application, this rate might range from 6,000 to 100,000 interrupts per second.

the types of process that must be done periodically are the reading of the A/D data, the computational update of a digital filter, or the outputting of data to the D/A converter. If any of these events do not occur on their precise, periodic schedule, there will be considerable distortion in the signal waveforms coming from the processor.

When the processor receives an interrupt, the background program instruction in progress is completed and the program then "jumps" to the location assigned for processing the interrupt. After the interrupt processing is completed, the program jumps back to the next place in the background process and continues with the background computations. This leaves a maximum amount of time for background processing, while still guaranteeing that the periodic needs will always be met. Recall that the basic processor can execute 33 million instructions per second, much faster than the 48-kHz rate of jumping to an interrupt routine.*

Several things can go wrong when the program is jumping to different places in the program at seemingly random times, however. The interrupt process could take longer than 20.8 microseconds, in which case the next interrupt would arrive before the first processing was complete. Called an *interrupt overrun*, this results in only partial completion of the interrupt process

*The ratio of the instruction rate and the interrupt rate determines the maximum number of instructions allowed in the interrupt routine. For our case, this is 33,000,000/48,000 or 687 instructions. Of course, if the interrupt routine always used this maximum number, there would be no time left for the background process. The balance between the two processes is part of the design process.

with very detrimental results. The program must be designed to keep all processing sufficiently short to prevent this. In addition, the background will generally be using a variety of computational registers. If the interrupt routine changes these registers, there will be errors in the resultant data in the background process. The interrupt routines must make sure that any register that it uses is restored before the background process resumes. In the case of the Analog Devices ADSP-2100 series of processors, this is very easily done for one interrupt. All of the computational registers are duplicated and they can be changed by the single instruction ena sec_reg or dis sec_reg. As one might surmise from the instructions, the two register banks are referred to as primary and secondary.

Programming within the Shell

No attempt will be made here to go through all the details of the shell program. A copy is included on the CD-ROM as *C1SHELL.DSP*. Comments have been added to the original Analog Devices program which explain most of the operation. Although it is not necessary to know all the details of this code, it is instructive to see a few lines of the program to understand the overall structure of a DSP program.

For those that have not yet written a DSP program, this programming information may seem mysterious and difficult to follow. It may be useful for the reader to skim through this section and the following one on "autobuffering", with the idea of returning when it is time to actually put a program together. The concepts here are important for making the DSP program, but not necessary for seeing how DSP fits into the "bag of tricks" for improving our communications circuitry.

When the DSP program first runs, a number of hardware and software parameters are initialized. In the program this looks like:

start: imask=0; { Turn off all interrupts }
      call init0; { Instructions that simulate easily }
      call init1; { And those that do not }

The first instruction is to prevent an interrupt from occurring in the program operation, before the initialization is complete. The two subroutine calls, "call init0" and "call init1" do the initialization. Two calls are used as a convenience when testing the programs using the emulator program provided with the EZ-Kit Lite.

Certain items, such as hardware interrupts, require extra effort for simulation but can be omitted for much program testing. When this is the case, the call to init1 can be "commented" out of the program.

For our shell program the background process is particularly simple:

again: { We have no background process. If we did, it would go here.}
jump again; { Go round and round forever }

This starts with a label "again:" that is not an instruction, but merely a name for the location in memory where the actual instruction jump again is located. The net result of this is that the instruction is executed repeatedly. This does nothing useful, but does allow the program to wait for an interrupt to occur. When this happens, the operation of the program is transferred to the interrupt routine. The return from the interrupt routine will once again go back to the "jump again" loop.

The interrupt routine, often called an "ISR" for interrupt service routine, is again simple:

input_samples:
ena sec_reg;   { use secondary register bank }

mr0=dm(rx_buf+1);   { Get left audio from A/D }
mr1=dm(rx_buf+2);   { Right }

 { This shell does no processing to the signals, other than to pass them through. Processing would go here. }

dm(tx_buf+1) = mr0; { Send left audio to D/A }
dm(tx_buf+2) = mr1; { Right audio}
dis sec_reg;   { Back to primary registerbank }
rti;   { This undoes the interrupt }

The first instruction switches all computational registers to the secondary set. All computation will be performed using the values in the secondary register set, while the primary register set is fully preserved for future use. The next instruction, mr0=dm(rx_buf+1), uses the computational register, mr0 as temporary storage for the number that was in memory at the address rx_buf+1. This is the data from the A/D for the left channel signal. Then, mr1 is loaded with the data from the A/D for the right channel signal.

To make a more useful program, we could now perform some signal processing action on one or both of these signals. However, since this is only an "empty" shell we will just send the data to the D/A converters for both the left and right signals. Putting the numbers back in memory at the addresses tx_buf+1 and tx_buf+2 does this. The primary registers are then brought back as the active computational registers and the processing is restored to the background process by the rti instruction.

Autobuffering

A potentially puzzling question is "who put the data into memory at dm(rx_buf+1) and who is taking it back out from dm(tx_buf+1)?" There is specialized hardware, called *autobuffering*, built into the processor that is able to exchange data between a serial port and data memory. The address in memory where this occurs is set up as part of the initialization process. These memory address were given the symbolic names rx_buf for incoming data and tx_buf for outgoing data. Left channel data is located 1 address location past the start of the data areas, referred to as rx_buf + 1 and the right channel data is 2 address locations past the start of the data area. The transfer of the data takes place without any processor instructions being required.

Every 1/48,000 second the CODEC, which includes the A/D, initiates a serial data transfer that is handled through the autobuffering. The completion of this transfer causes an interrupt in the DSP. This, in turn, causes the background activity to be stopped and our interrupt processing to begin.

The interrupt routine is in program memory at the symbolic address input_samples. This address is jumped to at the time of the interrupt as the result of a table of instructions that is placed in the first 48 instructions of program memory. These mini-programs are each 4 instructions long and the one used for the serial port used with the CODEC looks like:

```
jump input_samples {14: SPORT0 rx }
rti;        { Three filler instructions }
rti;        { so that there are a total of 4 }
rti;
```

The jump instruction is all that is needed for our shell program and so the remaining three instructions are filled out with do-nothing instructions, in this case they are rti, or return-from-interrupt instructions. The particular instruction is not important. The use of rti is often intended to prevent problems in case of accidental interrupts, but the utility of this is questionable and the real reason is to comply with a convention!

There are always 11 more interrupt mini-programs, most of which are not used. As can be seen from the full program listing, each serves a particular interrupt, if the interrupt mask enables it. Each of these has a specific address in memory. Our serial-port program is at address 14 hex (20 decimal.)

10.3 DSP COMPONENTS

When a piece of electronic equipment is assembled in a traditional way, a number of components are soldered together. These components can be fundamental ones, such as a resistor or a diode. In some cases, though they will be complex building blocks, such as a phase-locked loop built in an integrated circuit. In the same manner, one can look at DSP functions as components that can replace, or add to the analog components. In the following pages we will explore some of these DSP components, and see how they fit into radio designs.

Amplifiers and Attenuators

As DSP components, amplifiers and attenuators consist of multiplying the signal by a constant. If the constant is greater than 1.0 we have an amplifier and if it is less than 1.0 we have an attenuator. For instance, a 4-dB attenuator could consist of a signed multiplication:

```
my0=20675;  { -4 dB as a fraction of
              32768 }
mr=mr1*my0 (ss); { The signal is in
                   mr1 already }
```

It is assumed that the input signal has already been placed in the mr1. The instruction my0=20675 places a constant into one of the multiplier input registers, called my0. The output is called mr and for the ADSP-2100 series of processors this is a 40-bit register divided into three parts, called mr2, mr1 and mr0. For our case of the multiplication of two 1.15 format signed numbers,* the 16-bit signed result is in the mr1 register.**

The attenuation value in my0 is the 1.15 format fraction corresponding to the voltage ratio for –4 dB. In equation form this is:

$$my0 = (\text{int})\left(32768 \cdot 10.0^{\frac{-A}{20.0}}\right)$$

Eq 10.1

where A is the attenuation value in dB, which in our case is 4.0. The (int) operator

*See the sidebar "Decimal numbers in a fixed-point DSP" for a description of the number formats.

**The mr0 register contains the least-significant 16 bits that are used if we want to work with more than 16 bits. The high 8 bits in the mr2 register are available for functions that use "multiply and accumulate." This allows one to multiply two numbers together and add the product to a previous result. This is common operation in DSP.

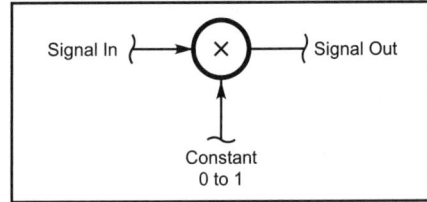

Fig 10.7—DSP attenuator using a multiplier. This multiplication operation occurs for every input signal sample.

indicates that we will use the closest integer to the calculated value. Fig 10.7 shows this attenuator in block diagram form.

This simple arrangement does not work for amplifiers. In 1.15 format, the largest number is 32767/32768, which is slightly less than 1.0. This can be overcome by the use of shifting. For instance, a "voltage" gain of 4.0 (as a ratio), or 12.04 dB, is achieved by shifting the binary number for the signal level to the left by two bits, as illustrated in **Fig 10.8**. In general, we need better control of gain than can be obtained with powers of 2 and this is achieved by cascading the shifting operation with the attenuation operation. As a more general example, a gain of 3.5, or 10.88 dB, is illustrated in **Fig 10.9**. In program form this would look like:

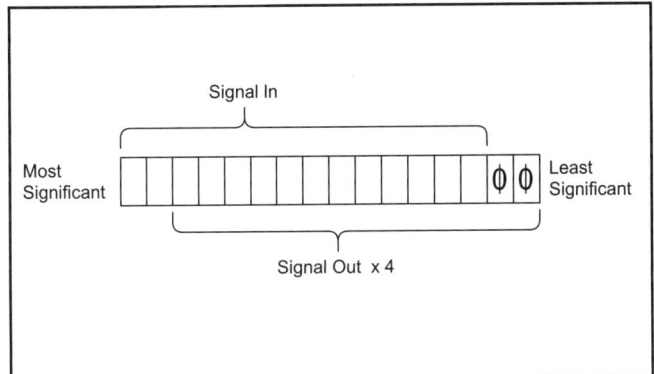

Fig 10.8—DSP gain of 4 using a shift register. The shift operation allows any amount of shifting, either up or down, in a single operation.

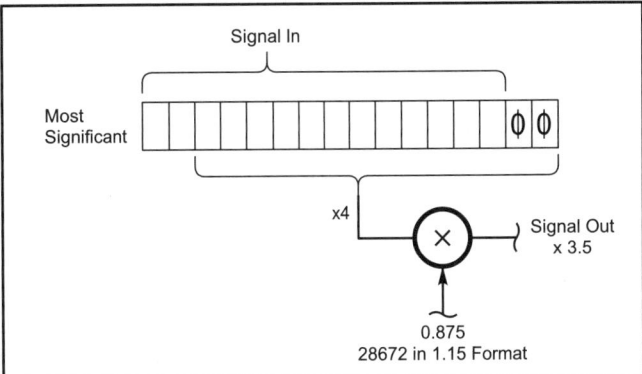

Fig 10.9—DSP gain of 3.5 using a shift register and a multiplier. A gain of 4 is first applied by the shift register, as was done in **Fig 10.8**. Following the shifter, an attenuation of 0.875 is applied, using the multiplier of **Fig 10.7**. This brings the net gain to 3.5.

```
sr=ashift mr1 by 2 (hi);{ The signal is in
             mr1; shift 2 bits }
my0=28672;   { 0.875 in 1.15 format }
mr=sr1*my0 (ss); { Multiply the shifted
             signal by my0 }
```

with the result again in the mr1 register.

The examples shown here are for constant values of attenuation. In many instances, it is necessary to have the gain the result of some calculation. The si register is useful for this case, allowing the number of bits of shift to depend on a register value. One should take care that the number of bits of shift is not more than necessary. If a large amount of shift is followed by a large amount of attenuation, there will be a loss of accuracy (dynamic range). The attenuation constant in my0 should be between 0.5 and 1.0.

10.4 SIGNAL GENERATION

Generation of signals using DSP is easily done. The primary advantages are the accuracy of the waveform and its stability over time. DSP signal generators tend to be limited to frequencies in the low MHz range, or less, due primarily to the computational load. Two examples of signal generation, the sine wave and random noise, are shown here.

Sine Wave Generator

One basic component that is needed for many DSP programs is a sine-wave generator. Digital generators can be implemented either as lookup tables or as calculated functions.

Lookup tables consist of a large block of data in memory that has every sine-wave value stored according to the phase angle. In its pure form this could require 65K words of storage for 16 bit phase angles. This is the fastest implementation, but obviously is impractical for many applications, because of the memory needs.

Various schemes allow the reduction of memory usage.[3] The most obvious is to use the symmetry of the sine wave and only compute values for a 90-degree segment from 0 to 90 degrees. This reduces the table to a fourth of the original size in exchange for a few computer instructions.

Other methods reduce the table size further by approximating the output waveform. This can be done as a series of steps where the output does not change, although the input phase does; this has very little computational overhead. More exact results are obtained by approximating the sine wave with a series of straight lines connecting the lookup-table values, but with higher computational overhead.

At the other extreme is direct calculation of the function.[4] This uses very little memory, but each data point requires, for our example, about 27 DSP machine cycles. This is quite acceptable for many applications. In terms of computing time, each data point takes 27 × .03 = 0.81 microseconds on the ADSP-2181.

The method again starts by dividing the sine wave into four regions of 90 degrees each as shown in **Fig 10.10**. For any point between 0 and 90 degrees, the sine wave is approximated by the following polynomial equation.[5]

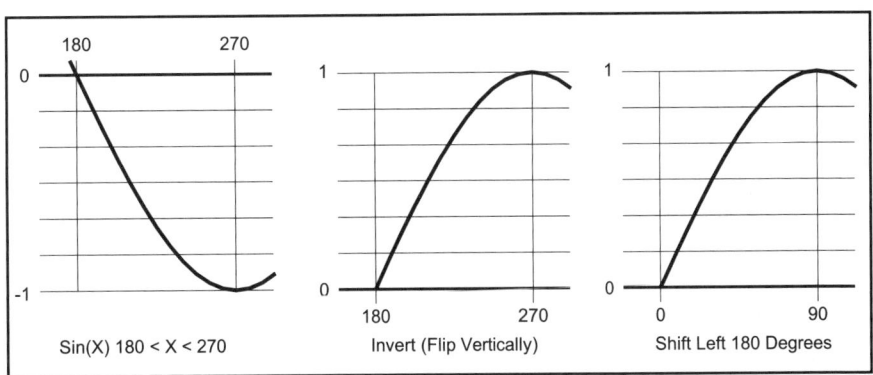

Fig 10.10—The values of sin(x) between 180 and 270 degrees are seen to be the same as those from 0 to 90 degrees, after the curve has flipped vertically and shifted 180 degrees. This symmetry allows the values from 0 to 90 degrees to be the only ones that need be calculated.

$$\sin(x) = 3.140625x + 0.02026367x^2$$
$$- 5.325196x^3 + 0.5446778x^4$$
$$+ 1.800293x^5$$

Eq 10.2

where x is the angle in degrees divided by 180. In the fixed point processing of the DSP (see the sidebar), the equation requires integer coefficients and takes the form

$$\sin(X) = 12864X + 83X^2 - 21812X^3$$
$$+ 2231X^4 + 7374X^5 \quad \textbf{Eq 10.3}$$

Two items are being dealt with in creating this equation. First, the coefficients have been scaled up to be 16-digit integers. But, in addition, they have been scaled back by a factor of 8 to insure that overload does not occur when the DSP calculation is only partially completed.

The calculation of the sine-wave value by these equations is valid only for 0 to 90 degrees. In fixed-point values this corresponds to 0 to 65536/4, or 0 to 16384. To deal with all possible angles from 0 to 360 degrees, the values are corrected according to the symmetry rules, such as those given above.

The five coefficients for the calculation of the polynomial are kept in a program-memory table called sin_coeff. Access to this table is discussed below, and is initialized in the first two lines of the sin routine. The next four lines are to divide the input data into four 90-degree segments. Note that the program constants are given as hexadecimal numbers. This requires a bit of translation to the more familiar decimal numbers. Many hand-held calculators have this translation, making the task simpler. In the program instruction my1=ar, both of these computational registers will have a value that is somewhere between 0 and 16383 decimal, or 0000 to 3FFF hexadecimal. This is the input value to the polynomial calculation.

The instruction mf=ar*my1 (RND), mx1=pm(i4,m4); indicates that the mf register will hold the results of the rounded multiplication of the ar and my1 registers, and that the mx1 register will be loaded with the first polynomial coefficient that was in program memory (pm.) The comma shows that both halves of this computation occur simultaneously, i.e., this is a single instruction. Not all instructions can be combined this way, but when it is possible, there is quite a bit of savings in processor time. The register mf now contains the input value squared.

Next mr=mx1*my1 (SS), mx1=pm(i4, m4); multiplies the first coefficient in mx1 by the input value in my1, leaving the product in mr, and also loads the second coefficient into mx1 register, overwriting the first coefficient.

The remainder of the polynomial calculation continues in a similar fashion. For efficiency in program size, the middle three multiplications are put into a loop. The register cntr controls the loop and it is automatically decreased with every loop. Loop initialization is performed by the instruction do approx until ce;.

After the polynomial is calculated, the value is adjusted according to the 90-degree segment of the input. Finally rts; is a subroutine return.

Using The Sine Wave Routine

Incorporation of this routine into the program shell takes only a few instructions. First, we need to initialize the frequency of the sine wave to some value, which for this example will be 1000 Hz. A number called "dphase" is set up in memory:

.var/dm dphase; { For generation of sine wave }

and this is initialized to the nearest integer value to the phase shift that occurs during 1/48,000 second, given by 1000*65536/48000 = 1365.33. This is put into data memory by:

ax0=1365; { 1000.24 Hz }
dm(dphase)=ax0;

The sine wave calculation consists of adding this phase change to the last phase value and using this in our sine wave routine. The program segment that goes into the middle of the ISR looks like:

ax1 = dm(dphase); { Phase increment for oscillator }
ay1 = dm(phase); { Last phase }
ar = ax1 + ay1; { New phase }
ax0 = ar; { The phase input to sin is reg ax0 }
dm(phase) = ar; { Save for next data point }
call sin; { Phase in ax0, Sin returned in ar }

Finally the sine wave is sent to both the left and right D/A:

dm(tx_buf+1) = ar; { Send sine wave to Left D/A (Codec) }
dm(tx_buf+2) = ar; { Right D/A }

A sin Routine

The routine for the EZ-KIT Lite looks like the following:

```
sin:    m4=1;  l4=0;                              { Use i4,m4 index registers to }
        i4=^sin_coeff;                            {   point to polynomial coeffs }
        ay0=H#4000;                               { This is 90 degrees }
        ar=ax0, af=ax0 and ay0;                   { Check 2nd or 4th quad. }
        if ne ar=-ax0;                            { If yes, negate input }
        ay0=H#7FFF;                               { This is a mask to replicate data, }
        ar=ar and ay0;                            {   while removing the sign bit }
        my1=ar;
        mf=ar*my1 (RND), mx1=pm(i4,m4);           { mf = input**2 }
        mr=mx1*my1 (SS), mx1=pm(i4,m4);           { Start polynomial calculation }
        cntr=3;                                   { Loop for 3 of 5 coefficients }
        do approx until ce;
        mr=mr+mx1*mf (ss);                        { More polynomial calculation }
approx: mf=ar*mf (rnd), mx1=pm(i4,m4);            { Power increase; get next coef}
        mr=mr+mx1*mf (ss);                        { Do last polynomial calculation }
        sr=ashift mr1 by 3 (hi);                  { Mult *8 (shift left 3) }
        sr=sr or lshift mr0 by 3 (lo);            { Convert to 1.15 format }
        ar=pass sr1;                              { See if result >=1.0 }
        if lt ar=pass ay0;                        { If so, saturate, i.e. set to 0x7FFF }
        af=pass ax0;                              { See if input was negative }
        if lt ar=-ar;                             { If so, negate output}
        rts;
```

Index Registers

The sin program uses *index registers,* in particular **i4**, along with the modifying registers **m4** and **l4**. These allow access to sequential addresses in memory without having to spend DSP computational time.

In the sine wave calculation, **m4=1** indicates that after the index register **i4** is used, we want to move sequentially to the next higher address. **l4=0** indicates that there is never a wrap-around in the addresses that are generated by adding on the **m4** value. And **i4=^sin_coeff** sets index register 4 to the address of **sin_coeff**, a table in program memory that was loaded with five polynomial coefficients by the assembler directives:

```
.var/pm  sin_coeff[5];
.init    sin_coeff: H#324000, H#005300, H#AACC00,
                    H#08B700, H#1CCE00;
```

This usage of the index registers is illustrated by the instruction **mx1=pm(i4,m4)**; indicating that the computational register **mx1** will be loaded with the contents of program memory at address **i4**, and then **i4** will have the value **m4** (one) added to it, for use next time. Other values of **m4** can be used, including negative ones, to allow stepping through tables in any equal arrangement.

The ADSP-2100 series of DSP have 8 index registers, named **i0** to **i7**. The **m0** to **m7** modify registers are used to change the address of the index registers after they are used. With some restrictions, the number of the index register need not be the same as that of the modify register. For instance, **i0** can be modified by **m0**, **m1**, **m2** or **m3**. The length registers always correspond to a particular index register and can be a value such as **l0 = 10** which means that the buffer that starts with the address in **i0** has a length **10**. When the 10th value is either read or written, the address in **i0** will not be incremented again by **m0**. Instead the address will be taken back to the initial value given to **i0**. This is the meaning of a *circular buffer.* If **l0** had been given a value of 0 the DSP would interpret this as a special case with **i0** indexing into a conventional non-circular buffer.

Program memory is 24 bits per instruction. Tables are often stored in program memory, but most often only 16 bits worth of data is used, since this corresponds to the size of most computations and of the data memory words. To make the data line up properly, 8 zero bits must be appended to each table entry stored in program memory. As an example, the first **sin_coeff** entry is the hex number 324000. The last two zeros are the extra 8 bits. Removing these we have the hex number 3240, which converts to a decimal value of 12864, which is the first coefficient of the sine calculation.

We should remember that we have only calculated a series of numbers that represent the sine wave at specific points, as shown in **Fig 10.11**. Before this is a "clean" sine wave it is necessary that this be converted to a continuous curve. In the case of the EZ-Kit, the low-pass filter to accomplish this is included in the D/A converter of the CODEC. The need for this conversion becomes more obvious as the frequency of the sine wave increases and fewer points are calculated per cycle.* **Fig 10.12** illustrates this for an 8500-Hz sine wave with a 48-kHz sample rate. To a good approximation, this collection of sample points will be converted to a continuous sine wave by the application of a low-pass filter at half the sample rate, on the output of the D/A converter. If one studies the apparently random collection of data points, it will become apparent that they are indeed sample points along a sine wave with about 48/8.5=5.6 data points per cycle.

*See chapter 4, section 4.7, for further discussion of hardware DDS computations. The process is identical, except that in the DSP case, one may need to use the sine-wave for internal functions such as driving a software mixer instead of always driving a D/A converter to produce an analog output signal.

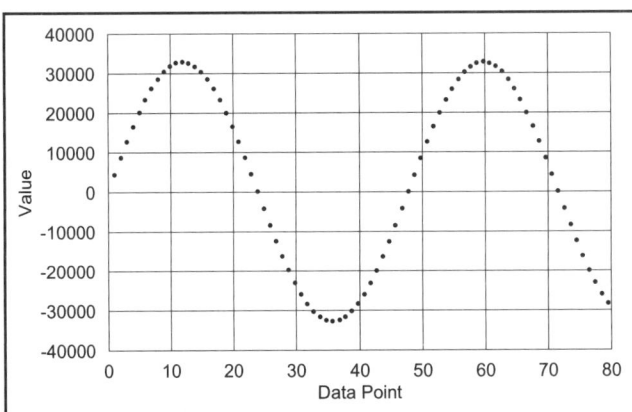

Fig 10.11—Calculated points for a 1000-Hz sine wave sampled at 48 kHz. The ability of these points to be smoothed to a continuous sine-wave curve is readily apparent.

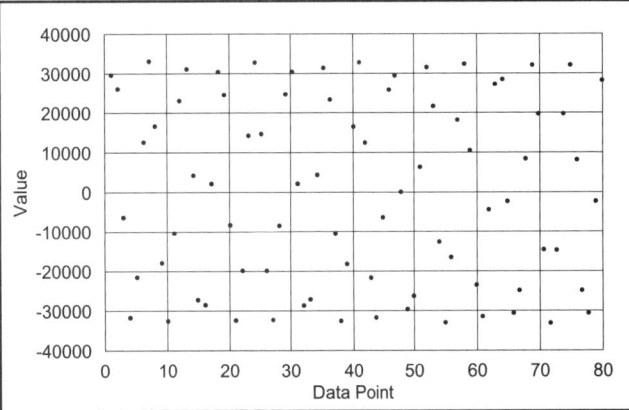

Fig 10.12—Calculated points for an 8500 Hz sine wave. The sample rate is identical with that of **Fig 10.11**. Careful study will show that these are indeed sample points on a sine wave. The ability of the low-pass filter to connect these points into a smooth curve is not so obvious, yet the resulting sine wave is exact.

10.5 RANDOM NOISE GENERATION

For the testing of transmitters and receivers it is often useful to have a noise-like signal. In the area of modulation and coding, interesting experiments can be performed by using a controlled noise source. A simple example is to add Morse code to the noise and test various filters and signal processors for the accuracy of copy by an operator.

One feature of a digital computer is the total predictability of computational results. This seems inconsistent with generating noise, and in a philosophical sense, it is! However, in a practical sense, the noise generator can be made to have a repetition period long enough that it is functionally random. For instance, the noise generator that will be described here repeats its pattern in about 25 hours running in the EZ-Kit Lite. Within that period, the output seems "noise-like" by most measures, although each successive output is totally determined by the previous output.

One algorithm, called the linear congruence method,[6,7] produces most of the computer-generated random numbers of the world. Three constants must be selected for this method, and large amounts of study have gone into the rules for selecting

---

Decimal Numbers in a Fixed Point DSP

The fixed point DSP use an arithmetic system called *2's Complement*.* In this system, positive numbers start at zero, represented by all binary bits being zeros, and progress to larger values by adding 1 to the next lower number. This progresses until all of the bits are 1, except for the farthest left bit that is always a zero for positive numbers. In the simple case of a three-bit system, the positive values would be

    011 binary   3 decimal
    010 binary   2 decimal
    001 binary   1 decimal
    000 binary   0 decimal

The 2's complement negative numbers are created by interchanging all binary values, bit-by-bit, and then adding 1 while saving the right-hand three bits. For instance, the decimal value +2 is 010 and if we interchange the binary values, we have 101. Adding 1 to this yields 110, which represents the decimal value -2. The same two operations will also bring us back to +2 indicating consistency. Applying this rule to the four values above produces the following table for the negative values:

    000 binary   -0 decimal
    111 binary   -1 decimal
    110 binary   -2 decimal
    101 binary   -3 decimal

The values for –0 and +0 are the same, which fits our idea of "nothing!" And the three true negative values all have a leading one, which is consistent with the positive values having a leading zero. However, the binary value of 100 does not appear in either table. Since it has a leading one, indicating a negative number, and it fits in the binary sequence either below –3 or above +3, it will be assigned the decimal value of –4. It does not follow the 2's complement rules for negation, since it produces the same 100 value. The last table entry is thus:

    100 binary   -4 decimal

Now, the operations of addition can be performed by following the same rules that we have in the decimal system, except that a *carry* will be generated when the result exceeds 1 instead of when it exceeds 9. For the binary system this occurs when we add 1+1. That is:

    0 + 0 = 0 No Carry
    0 + 1 = 1 No Carry
    1 + 1 = 0 Carry Generated

When there are multiple places in addition, the carry is added as a 1 in to the next position to the left.

So, for our 3-bit example, decimal values 1 plus 2 is
    001
    +010
    011

or decimal 3. This applies equally well to negative numbers and extends to subtraction, which starts to explain the wide use of 2's complement arithmetic systems in binary computers!

Our 3-bit example shows the operation of the number system, but it does not convey a feel for working with numbers in a 16-bit DSP system. The following table shows a few of the decimal values, and their binary representations for the larger number system:

Largest positive number
    0111 1111 1111 1111 binary +32767 decimal
    - - - - - - - - - -
    0000 0000 0000 0111 binary +7 decimal
    - - - - - - - - - -
    0000 0000 0000 0010 binary +2 decimal
    0000 0000 0000 0001 binary +1 decimal
    0000 0000 0000 0000 binary +0 decimal
    1111 1111 1111 1111 binary –1 decimal
    1111 1111 1111 1110 binary –2 decimal
    - - - - - - - - - -
    1111 1111 1111 1001 binary –7 decimal
    - - - - - - - - - -
    1000 0000 0000 0000 binary –32768 decimal

In *fixed-point arithmetic*, the standard way to use this arithmetic system to represent decimal numbers is to divide the number value by some power of 2. For instance, if all the values are divided by 32768 (2 to the 15th power) the table looks like: (see top of next page)

In this case, the last column is the *fractional representation* of these same 2's complement numbers. The

---

* Processors, such as the ADSP-2181 allow for either "Unsigned" arithmetic, or for "Signed 2's complement arithmetic." Because of it's greater generality, only the latter type is considered here. See Reference 4 for details of unsigned arithmetic.

these constants, as can be read about in the references. From the point-of-view of the noise-generator user, it is usually sufficient to borrow upon others study of these constants and apply them. This generator comes from the formula

v(n+1) = (a×v(n) + c) mod m
where
  v(n+1) = current generator output
  v(n) = last generator output

a, c, m are constants
mod m means dividing by m and taking only the remainder.

The constants are carefully chosen not only to produce good random numbers, but also to simplify the computation using our fixed-point processor. One good set is

a = 1664525
c = 32767

$m = 2^{32} = 4,294,967,296$

The length of time before the random noise repeats is determined by m. The value used here is the largest that can be used with a 32-bit word size. This requires double precision calculations, but if we restricted our calculation to 16 bits, the result would repeat $2^{16} = 65536$ times faster, or about every 1.36 seconds. For some purposes, this could cause strange results.

Largest positive number	0111 1111 1111 1111 binary	Fractional 32767 / 32768 = 0.99997
	0000 0000 0000 0111 binary	7 / 32768 = 0.00021
	0000 0000 0000 0000 binary	0 / 32768 = 0.0
	1111 1111 1111 1111 binary	(65535-65536) / 32768 = -0.00003
	1111 1111 1111 1001 binary	(65529-65536) / 32768 = -0.00021
Most negative value	1000 0000 0000 0000 binary	(32768-65536) / 32768 = -1.00000

total range is from −1.0 to almost 1.0. With 16 bits available, the step size (the fractional value of the least-significant bit) is 1/32768 or about 0.00003.

Sometimes the range of numbers being represented do not lie between −1 and +1. This is handled by dividing the binary representations by some other power of 2 than 32768. If the numbers were between −8.0 and 8.0 the divisor would be 4096 (2 to the 12th power.) The price paid for this is the resolution step size is now 1 / 4096 or about 0.00024.

Note that the divisors such as 32768 or 4096 are only implied, and not carried in any way with the 2's complement numbers. When writing a DSP program it is necessary to keep track of the number form. If a subroutine is expecting numbers in one format and they arrive in a different one, erroneous results will occur. Comments in the DSP program should carry the format information.

The notation describing the divisor value is not consistent in all literature. Often times a divisor of 32768 is called Q15 notation, since there are 15 bits to the right of the implied decimal point. The divisor of 4096 would be Q12. In their literature, Analog Devices uses the terminology 1.15 for Q15, 4.12 for Q12 and so forth. In this book we will continue this notation.

Addition is the operation for which 2's complement arithmetic fits perfectly. So long as the implied decimal points are the same for two numbers, they can be added without regard for their sign. As long as there are enough bits for the result, it will be correct. However, if there is not sufficient room for the result, bad things happen. For instance if we add the decimal representations of 15,000 and 20,000 together, one would expect to get 35,000. However, this is larger than can be represented with 15 bits, which is 32767. This will result in generating a carry bit that hits, of all places, in the sign bit. If we proceed blindly ahead we will have the erroneous negative value 35000-65536=−30536. This is called *wrap around*.

DSP program writers must take steps to prevent wrap around from occurring. In many cases, the DSP microprocessor can cause the results of computations to go to maximum positive or negative values in the case of overflow, preventing wrap around. In other cases, a formal check of the numerical values is required with appropriate adjustment of the data.

Multiplication of numbers occurs frequently in DSP programs. The sign bit adds an extra complexity to this operation. For instance, 3 times 2 would *seem* to produce the following, in binary signed 1.3 format numbers:

```
    0010    Signed 2
  x 0011    Signed 3
    0010
    0010
    0000
    0000
  0000110   Signed 6
```

But this is not what is found if one operates a DSP microprocessor. Instead, the result will be shifted one bit to the left and the result, in binary, is 00001100 that would seem to be 12 in decimal. The DSP signed multiplier has been built to acknowledge that each number being multiplied has a sign bit, but the result doesn't need two sign bits. Thus all results of signed multiplies are shifted left.

This all sounds somewhat arbitrary until it is seen that if there is an implied decimal point in the numbers, it will move one position to the right with each multiply, unless the shifting of one bit occurs. Dividing the numbers in the previous example by 8 turns them into Q1.3 format numbers. Doing the example again with Q1.3 format and the decimal point shown results in:

```
     0.010  or Signed 2/8
   x 0.011  or Signed 3/8
     0010
     0010
     0000
     0000
  0.000110  or Signed 6/64
```

Notice that only 6 places are needed to the right of the decimal point. Along with a single sign bit, 7 bits are required.

The generator, in DSP code is:

```
my1=25;                      { Upper half of a (1664525/65536) }
my0=26125;                   { Lower half of a, the remainder }
mr=sr0*my1(uu);              { 32 bit multiply: a(hi)*v(lo) }
mr=mr+sr1*my0(uu);           {   and a(hi)*v(lo)+a(lo)*v(hi) }
si=mr1;                      { Temp storage to free mr1 }
mr1=mr0;                     { LS Word of a*v(mid) }
mr2=si;                      { 8 bits of
mr0=h#fffe;                  { c=32767, left-shifted by 1 }
mr=mr+sr0*my0(uu);           { (above) + a(lo)*v(lo)+c }
sr=ashift mr2 by 15 (hi);
sr=sr or lshift mr1 by -1 (hi);  { Right-shift by 1 }
sr=sr or lshift mr0 by -1 (lo);  { Now have uniform rn in sr1
```

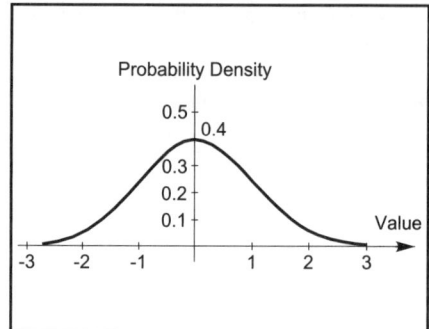

Fig 10.13—Gaussian noise probability curve, showing relative probability of being in the vicinity of any value. The curve extends forever on either side of the graph, but the probability of achieving these values rapidly becomes insignificant.

This program from the Analog Devices library[8] is an example of a routine that is carefully tuned for a particular application. In order to make the repeat period very long, the random number is generated as a 32-bit unsigned number. The constant multiplier, a, is 21 bits long and so the product can be up to 32+21=53 bits. The final operation of the algorithm, as shown above, is to divide by 2³² and then take the 32-bit remainder. At this point the top 32 bits will be discarded. The program does this, in part, by never generating that part of the product at all. If one examines the construction of a 64-bit product from two 32-bit numbers (using a 16 bit processor) it is seen that there are four terms to be added together. The product of the high-order 16 bits of v, with the high-order 16 bits, need never be produced.

The choice of m as a power of 2 is a common trick to avoid explicit division. A right shift of the data equal to the value of the exponent is all that is needed.

Selecting the desired words does a shift of 32 bits. This makes the three shifts at the end of the listing a surprise, at first. These three shifts are really only a shift of 1 bit corresponding to a division by 2. It is needed to correct for the shift in the multiplier result for unsigned multiplies, as discussed in the Decimal Number sidebar.

The resulting random numbers, left in the sr1 register, are equally likely to be anywhere between 0 and 65535, the full range of a 16-bit number. This is referred to as a Uniform Random Number.

Gaussian Random Numbers

What we have from the Uniform random number generator is not quite the noise that occurs in receivers, called Gaussian noise. Gaussian noise can take any value, but with decreasing probability as the magnitude of the value gets greater, as illustrated in **Fig 10.13**. There are a several ways to convert our random numbers into Gaussian noise, all of which must be approximations. There is always some overload point in real hardware, and Gaussian noise does not allow this! Fortunately, the probability of achieving these levels is very small, and as a practical matter can generally be ignored.

One simple way to generate Gaussian noise is to simply add several of the outputs of our uniform random number generator together. This is well founded on a mathematical principle known as the Central Limit Theorem.[9] The more numbers we add together, the better the approximation becomes. This is done in DSP by a loop (**see box at bottom of page**).

Most of the instructions in the loop are to free up the shift register for the division by 8. The division is needed to prevent overflow when 8 numbers are added together. One subtle operation is the use of an arithmetic shift (rather than a logical shift) to divide by 8. Doing this implies that the random number that ranged between 0 and 65536 is now being treated as a signed number ranging between -32768 and 32767. In fractional, 1.15 format this corresponds to numbers between -1.0 and 0.99997.

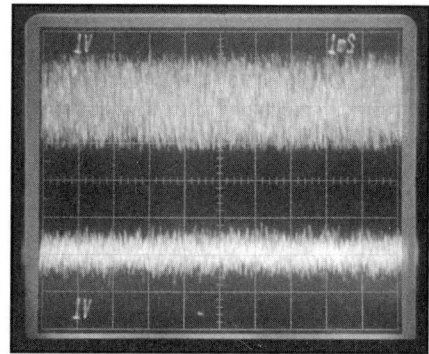

Fig 10.14—Oscilloscope picture of random noise as generated by the listings in the text. The upper trace is *uniform* random noise and the lower trace is *Gaussian*.

Program For Generating Random Gaussian Noise From 8 Uniform Noise Samples

```
getrnd:
    my1=25;                  { Upper half of a (1664525/65536) }
    my0=26125;               { Lower half of a, the remainder }
    af=pass 0;               { Clear the arithmetic accumulator }
    cntr=8;                  { The number of uniform rn added }
    { Now loop 8 times to generate a noise sample: }
    do randloop until ce;    { Decrease cntr until 0 }
        sr1=dm(seed_msw);    { Get the 32 bit seed from last }
        sr0=dm(seed_lsw);    { call to this fcn or last loop }
        { The Random Number Generator, shown above, goes here,
          leaving the result in the sr0 and sr1 registers }
        dm(seed_msw)=sr1;    { Save new seed, high 16 bits }
        dm(seed_lsw)=sr0;    { and low 16 }
        { Uniform random number still in sr1. Add to accumulator: }
        sr=ashift sr1 by -3 (hi);  { Divide by 8, ie, shift right 3 }
randloop: af=sr1+af;         { Accumulate 8 uniform rn }
    rts;                     { Random 16-bit value in af }
```

10.12   Chapter 10

One of the advantages of the DSP approach of noise generation is the ability to know the noise power precisely.* This is found by considering the process used to generate the noise samples:
• 1/3 is the average power for −1.0 to +1.0 uniform random numbers.
• This is diminished in power by $(1/8)^2 =$ 1/64 for the shift by 3 bits.
• This is increased by 8 for adding the 8 numbers together.
• The final result is a total noise power of $1/(8 \times 3) = 0.04167$ W.

The process of combining the 8 uniform random numbers has reduced the power from 0.333 to 0.04167, but the maximum possible values have been kept at −1 and +1. We are increasing the peak-to-average ratio, a necessary operation if a Gaussian approximation is to result.

The generation of each Gaussian noise value by this method requires 134 instruction cycles, or about 4 microseconds of EZ-Kit Lite processor time.

Fig 10.14 is an oscilloscope plot showing both the uniform random numbers before scaling (top) and the Gaussian noise, both to the same scale. It can be seen that the Gaussian noise clusters about the center value, much more than the uniform generator. It is not so obvious that the attainable peak values are the same for both plots. The Gaussian generator produces these peak values very infrequently!

*The normalized values of numbers range from −1.0 to 0.99997, which can be thought of as voltages. In order to think about power in the DSP computation we must square the voltage and divide by the "resistance." For simplicity, the resistance value is chosen to be 1 Ω and the power is just the normalized value squared.

10.6 FILTERING COMPONENTS

After A/D encoding of an analog waveform, such as an audio or an IF signal, we can then apply frequency selective filtering to the waveform. Such filters, called *digital filters* can be implemented in DSP with all the conventional passband shapes such as Low-Pass, High-Pass and Band-Pass. The input to the filter consists of a sequence of numbers representing successive samples of a voltage. Each sample period the filter performs some calculations on the new sample. This involves values that were previous samples and in some cases the results of the previous calculations. By carefully designing this calculation it is possible to make its output level very sensitive to the frequency of the input, which is what we mean by frequency domain filtering.

There are two basic ways to implement a digital filter, called IIR and FIR filters. The distinction in the arrangement of the calculation is not great. The IIR filters involve the results of previous calculations and FIR filters do not. Nevertheless, this small difference has major influences on both the design and the operation of the filter.

IIR Filters

IIR stands for Infinite Impulse Response and refers to the fact that, in principle, the output of the filter continues forever after an input has been removed. In actuality it does not, of course, since the output of a properly designed filter will get smaller with time and eventually become smaller than the smallest number our processor can recognize. The simplest IIR filter is the analog of the RC low-pass filter shown in **Fig 10.15**. The digital implementation consists of adding a small fraction of the new input to a fraction of the last filter output. If we call the filter input sample x_i and the filter output sample y_i then our filter consists of the single calculation:

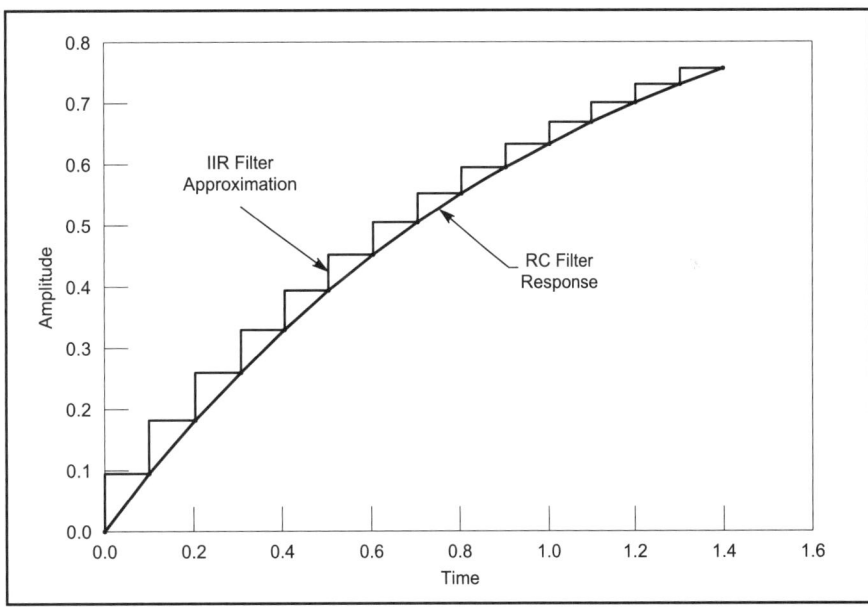

Fig 10.16—The charging response for the RC filter and the IIR filter approximation.

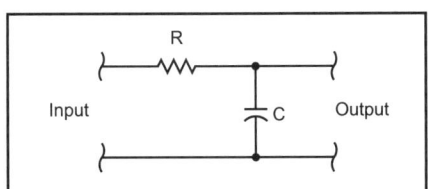

Fig 10.15—Simple RC low pass filter in analog form.

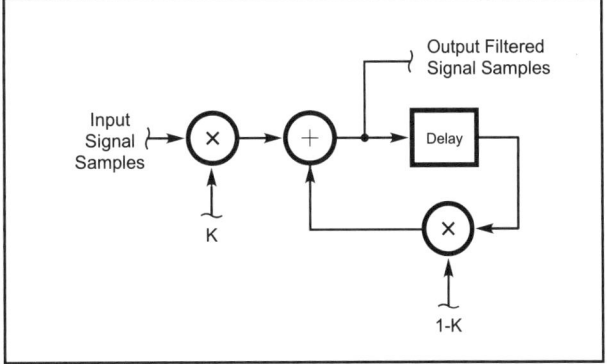

Fig 10.17—Block diagram of the simple IIR filter that has the response of an analog RC low-pass filter. The output signal is delayed by a sample period and a fraction of this is fed back to be summed. This use of feedback is characteristic of IIR filters.

$y_i = K x_i + (1-K) y_{i-1}$ **Eq 10.4**

where K is between 0 and 1, typically 0.001 or less. **Figure 10.17** is a block diagram of this filter. Operation of this simple filter can be calculated for the first few terms while the input rises from 0 to 1. We assume that the output is 0 when we start and that K=0.1 (this big value for K makes things happen faster for our example):

New Input, x_i	$K x_i$	$(1-K) y_{i-1}$	New Output, y_i
0.0	0.0	0.0	0.0
1.0	0.1	0.0	0.1
1.0	0.1	0.09	0.19
1.0	0.1	0.171	0.271
1.0	0.1	0.2439	0.3439

It can be seen that the output is growing towards 1.0, but with smaller steps with each new input. This is the same exponential growth that we associate with the RC filter. **Fig 10.16** shows both the charging characteristics of the RC filter and our digital equivalent. If we allow the process to continue for a very long time, the output will achieve a value of essentially 1.0. At that point the response is as follows:

New Input, x_i	$K x_i$	$(1-K) y_{i-1}$	New Output, y_i
1.0	0.1	0.9	1.0
1.0	0.1	0.9	1.0

Notice that if the input and output are the same there is no change in the output, as would be expected for the RC filter. RC filters are characterized by their time constant, T, in seconds that is equal to the product of the resistance and the capacitance. This is the time for the capacitor to charge to 63% of its final value. Design of the equivalent digital filter involves choosing the value K according to:

$K = 1 / [0.5 + (T / T_s)]$ **Eq 10.5**

where T_s is the time between successive input samples.

The RC IIR filter, implemented in DSP assembly language, is shown in the box to the right.

Notice that in conventional 16-bit representation of signed decimal numbers the value 32768 (or 2^15) would be 1.0 if it was not the wrap-around point and therefore 32768 represents –1.0. This is why it is used for the calculation of 1.0-K. For example, if the value for K in the DSP program is 5 representing a decimal value of 5/32768 or 0.0001526, then 32768 – 5 would be 32763 representing 32763/32768 or 0.99985.

One limitation of our routine is the smallest value for K being 1/32768 or 0.00003. This means the longest possible time constant is 32767.5 times the period between samples. To circumvent this problem we would need to use more than 16 bits in our arithmetic. This is available as standard arithmetic in some processors. For 16 bit processors it is implemented through multiple precision arithmetic. The price is slower processing. The routine given here computes a new filter output in 0.18 microseconds on the ADSP-2181 whereas a double precision version would be roughly twice as long.

The simple IIR filter has limited performance and a frequency response that drops off at only 6-dB per octave. Although slow in rolling off the frequency response, this is adequate for many applications. Improved performance comes from using not only the current input sample but also one or more of the previous input samples. Additionally, one or more of the previous output values can be used along with the current output. Each of these inputs and outputs has a different K value by which it is multiplied. This provides high filtering performance for the small computational complexity involved. As with most things, there are some drawbacks. Determination of the K values for a particular filter response involves some complexity. Narrow-band IIR filters often involve small K values that end up requiring multiple precision arithmetic. This can end up negating the simplicity arguments. There can be numerical stability* problems associated with computational accuracy as well as detrimental effects from the phase

---

* *Numerical stability* here refers to the inherent errors in the calculations causing the algorithm to produce errors of major proportion. This most often happens when subtracting two numbers that are almost the same value. For these occasions, special care may be required, such as the use of multiple precision arithmetic, using 32 or more bits in a data word, in place of the normal 16 bits.

---

response of the IIR filters such as unnecessary ringing. Never the less, the IIR filter has many applications where its computational efficiency makes it the filter type of choice. However, because of the drawbacks listed, we will concentrate on the alternate category, the FIR filter.

FIR Filters

For filters of higher complexity it is often desirable to use the FIR filter, standing for Finite Impulse Response. These filters never use the previous outputs of the filter computation, but do use the current input along with many of the previous inputs. Analog circuit designers have used the corresponding circuit called a transversal filter as was described in Chapter 3.

DSP construction of the FIR filter is very simple, as shown in the block diagram of **Figure 10.18**. The signal is already available in sampled form from the A/D converter. A delay line consists of places in memory for some quantity of previous samples. Each time a new sample arrives we put it into the beginning of the delay-line memory. Multiplying all the samples by constant numbers and then adding them together form new outputs. The constant multiplier numbers are referred to as the FIR coefficients, or tap weights. The filter design consists of choosing the coefficients to suit the particular application. As with analog filters, there are tradeoffs between the complexity (number of coefficients), pass-band ripple and the out-of-band rejection.

The FIR structure can be used to form filters that are highly selective to the frequency of a sine wave input signal. All of the response characteristics of L-C filters, such as Butterworth and Chebyshev are possible with the FIR filter.

The actual implementation of the FIR filter will be shown in DSP assembly language. This is not hard to follow and allows us to see the type of optimization that has been done to the DSP hardware to make these calculations particularly efficient. For a 10-coeffi-

---

Program for IIR Filter

{ The new sample is in register mx0, the previous output of the filter is in RAM at the location save_y and K is a constant defined at the top of the program by #define K=5;}

```
my0 = K;              { Load register my0 with charging constant }
mr = mx0 * my0 (ss);  { Multiply the sample by K, both signed integers }
mx0 = dm(save_y);     { Get the last output }
my0 = (32768 – K);    { Let the assembler figure out 1.0 – K }
mr = mr + mx0*my0 (ss); { Diminish last output and add new contribution }
dm(save_y) = mr1;     { Get ready for next time, output left in mr1 }
```

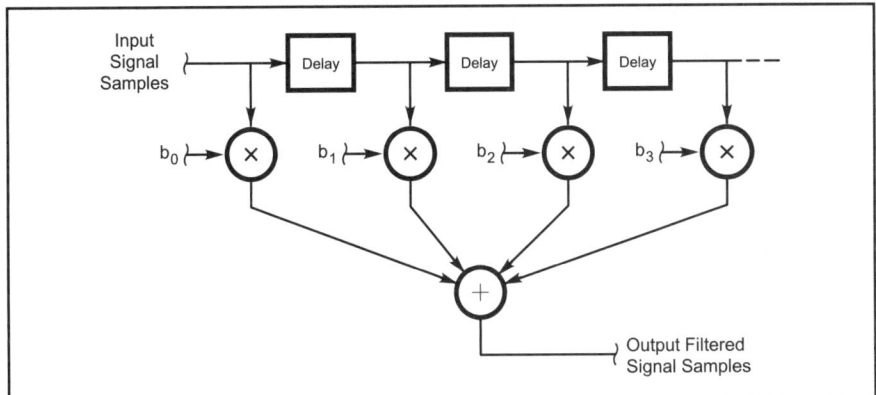

Fig 10.18—Block diagram of the software operations for the FIR filter. The input signal samples are delayed by multiples of the sample period. After multiplication by the filter coefficients, shown here as b_i, the results are summed to produce the filtered output sample. The output values are not brought back into the calculation as was done with IIR filters. The filter can be extended to the right to increase the performance. Filters with more than 100 coefficients are common.

cient filter we start with the initialization shown in Box 1.

These three instructions are part of initialization of the program and are executed only once, when the program is first run. The first instruction again uses index registers that were described on page 10.8. All three instructions set up the registers for the indexed access to the input data delay line. The 'hat' symbol seen in i0 = ^circ_data_buffer should be read as "the address of" and creates a constant that can be automatically determined when the program is assembled and linked.

The remainder of the instructions for the FIR filter are executed periodically when new data points are available. The new signal value arrives in the ax0 register as shown in Box 2.

The filtered output is in the multiplier accumulator register, mr1. The instruction dm(i0, m0) = ax0; uses the index registers to place the new data point into our buffer and, importantly, to increment i0 to the next location in the buffer. Since the buffer is circular the new data point will replace the oldest data in the buffer and leave the address in i0 pointing to the next oldest data point.

Next are three instructions to setting up the index register, i4, which is the address of a series of constants that are our FIR filter coefficients. These registers could have been set up at initialization time by making l4 = 10, but are shown this way to emphasize that the FIR filter calculation always start with the same coefficient. The coefficients are, interestingly, stored in program memory, pm(i4, m4). This is a convenience for speeding up the calculation as will be seen below.

Proceeding in the program, we encounter mr = 0, mx0 = dm(i0, m0), my0 = pm(i4, m4); which is a multifunction operation executed entirely within one instruction cycle. This clears the multiply accumulator, mr which is a 40-bit register consisting of mr0 for the least significant 16 bits, mr1 for the middle 16 bits, and an 8-bit overflow register mr2. In addition two multiply input registers mx0 and my0 are loaded with data from the delay line, dm(i0, m0) and a coefficient pm(i4, m4). Here is where the efficiency of storing the coefficients in program memory occurs. Separate hardware exists inside the DSP microprocessor for accessing data and program memory. This allows the loading of mx0 and my0 to occur simultaneously.

The do-loop counter, cntr, is loaded with 9, the number of coefficients, less 1. Do firloop until ce; is an instruction that does housekeeping chores necessary to do repeating calculations and prepares us for the FIR filter.

With everything in place we are ready to do the actual FIR filter calculation:

Firloop: mr = mr + mx0 * my0 (ss), mx0 = dm(i0, m0), my0 = pm(i4, m4); is another multifunction operation that executes in a single instruction cycle. It multiplies the contents of registers mx0 and my0, adds these onto the contents of mr and then reloads mx0 and my0 with new values from data and program memory. The designation (ss) indicates that both mx0 and my0 are to be treated as 2's complement signed numbers. The label 'Firloop:' indicates that this is the end of our do-loop. In this case, the loop is only one instruction long, and so this multiply and accumulate operation is repeated 9 times.

After the multiply and accumulate operations, we fall through to one last multiply and accumulate. This one uses the (rnd) designator that still treats the inputs as signed numbers, but also rounds the mr1 register (the output) according to whether mr0 is more or less than a half. Rounding is done on only the last accumulate. Note that at this point we have used all 10 coefficients.

Box 1 – DSP program for FIR filter initialization

```
i0 = ^circ_data_buffer;   { Points to a circular buffer, i.e., a delay line }
l0 = 10;                  { i0 points to a circular buffer of length 10 }
m0=1;                     { Increment i0 by m0=1 after use }
```

Box 2 – DSP program for FIR filter computation

```
dm(i0, m0) = ax0;                              { Enter the new data point into delay line }
i4 = ^fir_coeffs;                              { Points to start of a table of 10 constants }
l4 = 0;                                        { This buffer need not be circular }
m4 = 1;                                        { Increment i4 by 1 after use }
mr = 0, mx0 = dm(i0, m0), my0 = pm(i4,m4);     { Initial data load }
cntr = 9;                                      { This sets the number of 'do' loops }
do firloop until ce;                           { Loop 9 times, ie, until counter empty (ce) }
Firloop:   mr = mr+mx0*my0 (ss), mx0=dm(i0,m0), my0=pm(i4,m4);
mr = mr + mx0 * my0 (rnd);                     { This is the tenth calculation }
```

Table 10.1

List of operations for 10 coefficient FIR filter showing memory locations

```
dm(3)=New data value
mr=0
mr=mr+dm(4)*pm(1)
mr=mr+dm(5)*pm(2)
mr=mr+dm(6)*pm(3)
mr=mr+dm(7)*pm(4)
mr=mr+dm(8)*pm(5)
mr=mr+dm(9)*pm(6)
mr=mr+dm(10)*pm(7)
mr=mr+dm(1)*pm(8)
mr=mr+dm(2)*pm(9)
{ End of loop }
mr=mr+dm(3)*pm(10)
```

Table 10.1 shows what is happening. Here we have used the shorthand terminology of dm(i) being the ith memory location in our circular buffer. Likewise pm(j) is the jth coefficient in the program memory table. We assume that we came upon this calculation at a time when dm(2) had just been read and we next need to use dm(3). This is where we put the new data point. The multiply and accumulates can be seen to occur 10 times. At the eighth of these we have reached dm(11), which is outside our buffer, so we "wrap around" to the start of the circular buffer at dm(1).

Observe that we have incremented the i0 value 11 times for our 10 coefficients. This causes the operation to start one location further around in the circular buffer next time a data point is processed. This is equivalent to pushing the data through a delay line, but requires no actual movement of data, only the pointer to the data, i0.

The FIR filter calculation can be seen to be straightforward. In the ADSP-2181 it requires about $10+N_f$ instruction cycles for a filter with N_f coefficients. A complex, high performance filter of 200 coefficients would need 210 instruction cycles. If this was repeated at an 8-kHz rate we would be using 8000×210=1,680,000 cycles out of a possible 33.3 million, or only about 5% of the available processing time.

So far we have a way to compute the filter output if we could find out what coefficients to use. The next section shows a way to find them.

FIR Filter Design by the Window Method

The relationship between the frequency response of a FIR filter and the coefficient values is a mathematical formula called the discrete Fourier transform.[10] The details of the transform will not be dealt with here since for most purposes it is not necessary to actually evaluate it. Instead, one can start with a general transform of an ideal rectangular frequency response. For instance, if we wish to pass 400 to 800 Hz the ideal frequency response would be 1.0 within that frequency band and 0 elsewhere. The Fourier Transform of this simple response shape has been done for us, and all we need to do is to plug in the values corresponding to 400 and 800 Hz. Since this is a sampled data operation the sample frequency, say 8000 Hz, is involved as well. In equation form the coefficients are:

$$c_k = \frac{\sin\left(2\pi k \frac{f_H}{f_S}\right)}{\pi k} - \frac{\sin\left(2\pi k \frac{f_L}{f_S}\right)}{\pi k}$$

Eq 10.6

for k=0 to $N_f/2-1$, and N_f is the number (an even number*) of coefficients to be found. A special case is k=0:

$$c_0 = 2 \cdot \left(\frac{f_H - f_L}{f_S}\right)$$

Eq 10.7

* The formulas are shown here for an even number of coefficients. The form for an odd number is slightly different and although not covered here, is included in the design program.

f_L and f_H are the lower and upper band-pass cutoff frequencies, and f_S is the sample rate, all in Hz. Only half of the coefficients are calculated since they divide into halves that are symmetric, as shown in **Fig 10.19**. This same formula applies equally well to low-pass and high-pass filter design by setting $f_L=0$ or $f_H=(f_S/2)$ respectively.

Unfortunately, filters designed by this formula have several flaws. The response curve of **Fig 10.20** is the result of analyzing our filter. The pass-band is not flat, the sides of the filter are not vertical and probably worst of all, the out-of-band response is only 20 to 30 dB below that of the pass-band. What went wrong? Well, we have tried to describe the filter response with too few elements. Our sampled data cannot describe the extremely fast transitions such as occur at the edges of the pass-band. This design approach compromises the out-of-band attenuation in favor of small transition bands.

Fortunately, it is possible to easily cure the poor out-of-band attenuation. By systematically adjusting the c_k coefficient values, it is possible to push down the out-of-band response. The process for doing this is called windowing. The price that we pay for improved out-of-band rejection is a more gradual transition between the pass-band and the stop-band. This is usually an acceptable tradeoff.

Most FIR filter design descriptions include a variety of windowing methods. Here we will only show one method, the Kaiser window. This is a particularly useful technique:

• It provides an adjustable method for trading off maximum out-of-band response, in dB, for cutoff rate at the pass-band edge.

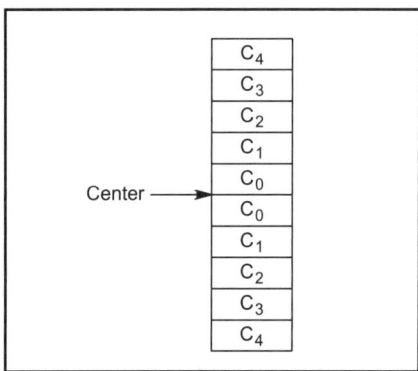

Fig 10.19—Table of FIR filter coefficients for N_f=10. Only half of the coefficients are calculated and are placed in the second half of the table. The first half of the table is arranged symmetrically as shown. The design program performs these operations automatically. If the number of coefficients is odd, the symmetry remains about the middle coefficient, which must then be doubled in value, since it only occurs once.

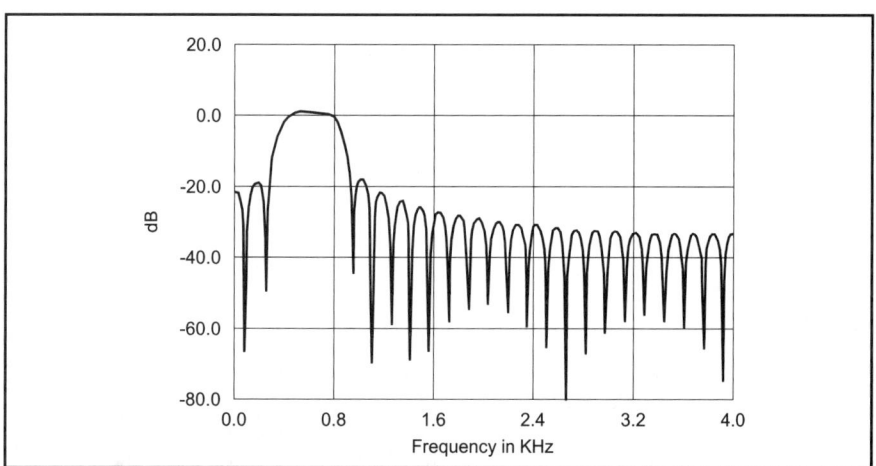

Fig 10.20—Response Curve for a 50-coefficient FIR filter designed to pass 400 to 800 Hz with an 8-kHz sample rate. No windowing function was used with a resulting high out-of-band response.

- The out-of-band response drops rapidly as one moves away from the passband edge. Typically, close-in responses are not as troublesome as those far out.
- The design process, though not trivial, involves a computation not a great deal more complicated than other standard windowing methods.

Implementation of a Kaiser window involves choosing a dB level for the maximum out-of-band attenuation response, Kdb. This would typically be a number in the 30 to 80 dB range. A BASIC program[11] can be used for determining the Kaiser window as well as the coefficient values for the FIR filter. The results of using this program to apply a 30-dB Kaiser window to our band-pass filter can be seen in **Fig 10.21**.

To better understand the design of a FIR filter using the Basic program, we will show the details for a simple 10 coefficient low-pass filter. Keep in mind that our performance will not be particularly good and most FIR filters use more coefficients, perhaps 30 to 300. Assuming our sampling rate is 8 kHz and we want the low-pass to cutoff at 2.5kHz, we run the program as follows:

FIR Filter Design, Low-pass, Band-pass or High-pass
Number of FIR coefficients? 10
Sample rate, Hz? 8000
Lower Cutoff Frequency, Hz, between 0 and half of sample rate? 0
Upper Cutoff Frequency, Hz, between 0 and half of sample rate? 2500
Stop-band Attenuation, dB (e.g. 55.0)? 30
Coefficient 1 = .0158115
Coefficient 2 = .0304284
Coefficient 3 = −.0976571
Coefficient 4 = .0379926
Coefficient 5 = .5243738
Coefficient 6 = .5243738
Coefficient 7 = .0379926
Coefficient 8 = −.0976571
Coefficient 9 = .0304284
Coefficient 10 = .0158115

The coefficients are decimal numbers and not the integers required by many DSP functions. Conversion to integers is accomplished by the following part of a Basic program that could be attached onto our FIR design program:

```
FOR j = 1 TO nf
b(j) = INT(32768 * b(j))
IF b(j) < 1 THEN b(j) = b(j) + 1
PRINT "Coefficient "; j; " = "; b(j)
NEXT j
```

This works for 16-bit integer arithmetic. For 24 bit integer arithmetic we replace the 32768 which is 2^15 by 8388608 which is 2^23. Here is what we get from running this program on our 10-coefficient filter (because of the symmetry we will only show the first 5 coefficients):

Coefficient 1 = 518
Coefficient 2 = 997
Coefficient 3 = −3200
Coefficient 4 = 1245
Coefficient 5 = 17183

FIR filter coefficients will normally be placed into program memory (PM) for the Analog Devices ADSP-2100 series of DSP. The assembler for the Analog Devices EZ-Kit requires that this data be presented in 24-bit format, left justified and right padded with zeros. This is most easily handled in hexadecimal since the right zeros appear as '00' on the end, each corresponding to four binary bits each equal to zero. A Basic program to convert the original decimal b(j) coefficients would be:

```
DIM H$(301)
FOR j = 1 TO nf
H$(j) = HEX$(b(j))
IF b(j) >= 0 THEN GOSUB POSH ELSE GOSUB MINH
PRINT H$(j)
NEXT I%
STOP

POSH:
G$ = H$(I%)
IF LEN(G$) = 1 THEN G$ = "000" + G$ + "00"
IF LEN(G$) = 2 THEN G$ = "00" + G$ + "00"
IF LEN(G$) = 3 THEN G$ = "0" + G$ + "00"
IF LEN(G$) = 4 THEN G$ = G$ + "00"
H$(I%) = G$
RETURN

MINH:
H$(I%) = RIGHT$(H$(I%), 4) + "00"
RETURN
```

Again the resulting hex output for the first 5 coefficients is:

020600H
03E500H
F38000H
04DD00H
431F00H

These coefficients would normally be placed into a separate data file, rather than cluttering up the assembly listing.

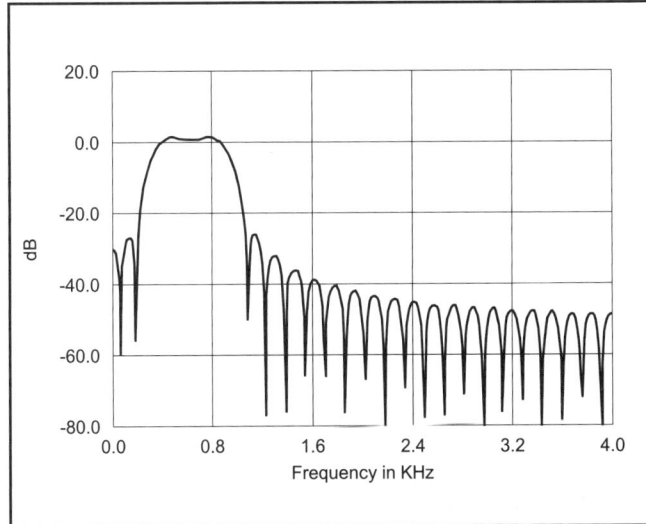

Fig 10.21—Response Curve for the 50-coefficient FIR filter of Fig 10.20 when using a 30-dB Kaiser windowing function to reduce the out-of-band response.

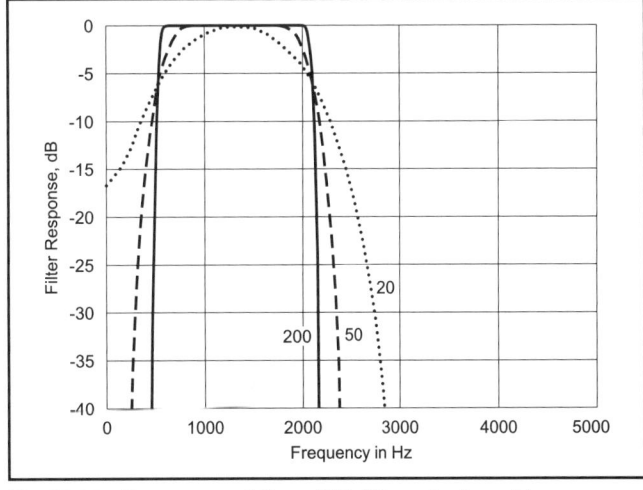

Fig 10.22—Response of three FIR filters designed to cover 500 to 2000 Hz at 6 dB points. The number of coefficients has been set to 20, 50 and 200. The sampling rate for the system was 9600 Hz. The sharpness of the filter is seen to be strongly dependent on the number of coefficients.

FIR-Filter Performance

In Chapter 3, it was shown that passive filters designed from LC components, or active filters using op-amp circuitry, all become sharper in response as there complexity increased. Not surprisingly, this follows for FIR filters as well where the complexity is measured in terms of the number of coefficients.

Fig 10.22 shows the response curves for three FIR filters using 20, 50 and 200 coefficients. All filters were designed to cover 500 to 2000 Hz at −6 dB relative response. With 200 coefficients, the response drops to −40 dB in about 80 Hz, whereas with 20 coefficients the same amount of attenuation occurs over about 680 Hz. This change in performance is very much like that seen in Chapter 3 as the number of resonators was changed.

It might also be noted from the figure that the responses at the high and low cut-off frequencies are nearly mirror images of one another. The rate of cutoff of the filter depends on the number of coefficients, the side lobe levels and the sampling rate of the system, but not on the width of the filter. This can be seen further in **Fig 10.23**, where the bandwidth of the filter was changed, but the number of coefficients was kept at 200. The frequency scale has been narrowed to show the response details better. Note that the cutoff shape is very similar for the different bandwidths. An interesting characteristic is that the very narrow filters start showing insertion loss, as can be seen with the 100-Hz bandwidth. This happens when the top portions of the response curve from the high and low frequency sides start to overlap.

Figure 10.24 shows the details of the out-of-band response for the 500-Hz filter of Fig 10.23. The design value for the side lobes was −50 dB. As is characteristic of the Kaiser-window FIR filters, the first out-of-band side lobe is at the −50 dB level, but as the frequency gets farther from the pass band, the side lobes continue to drop. For many receiver applications, this is a reasonable response. Interfering transmitter spectrums tend to be

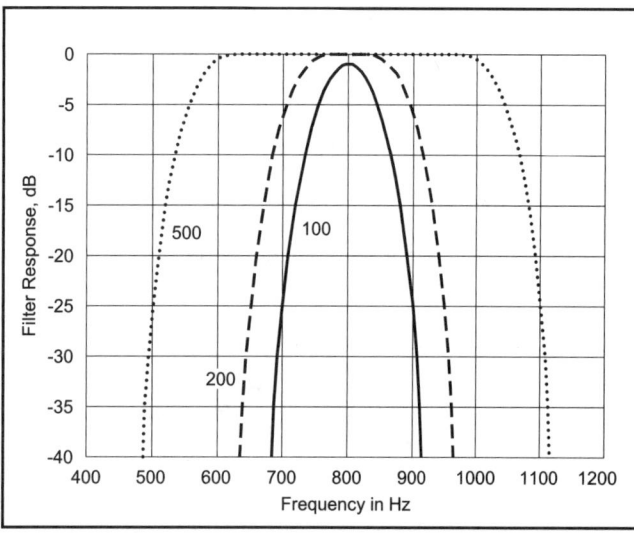

Fig 10.23—Response of three FIR filters designed for a center frequency of 800 Hz, using 200 coefficients and a sampling rate of 9600 Hz. The −6 dB bandwidth was designed to be 100, 200 and 500 Hz.

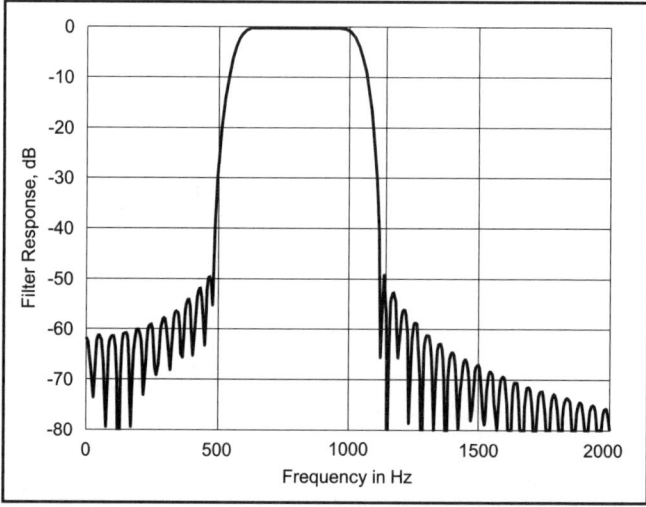

Fig 10.24—The out-of-band response for the 500-Hz filter of Fig 10.23. The design value for the side lobes was −50 dB.

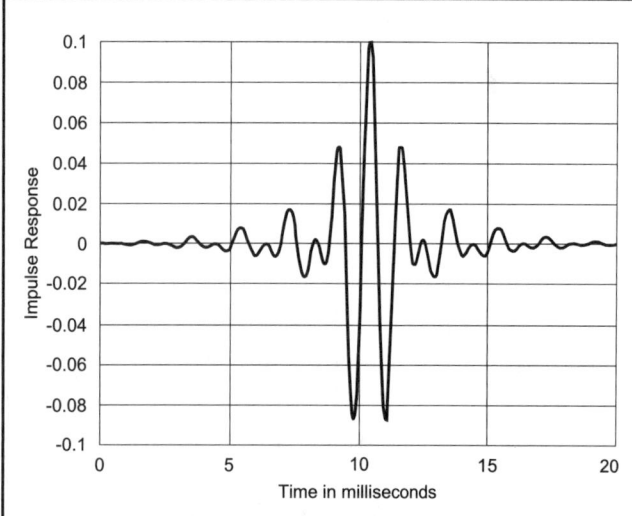

Fig 10.25—Impulse response of a Kaiser-window FIR filter designed for a center frequency of 800 Hz, using 200 coefficients and a sampling rate of 9600 Hz. The −6 dB-bandwidth was designed to be 500 Hz.

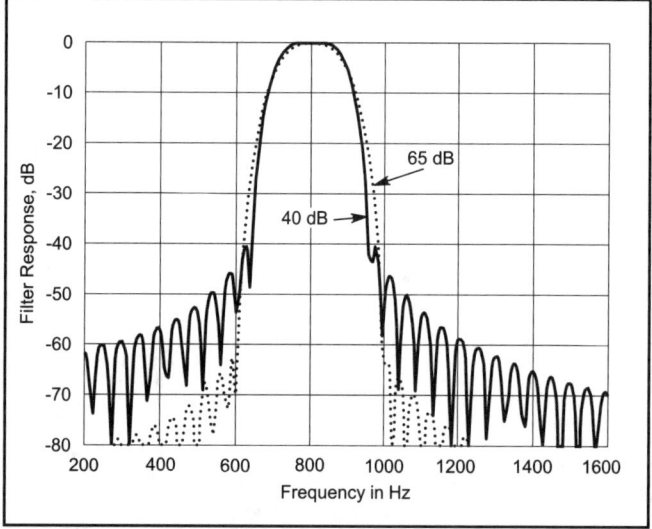

Fig 10.26—Response of a Kaiser-window FIR filter designed for a center frequency of 800 Hz, using 200 coefficients and a sampling rate of 9600 Hz. The −6 dB-bandwidth was designed to be 200 Hz. The two response curves correspond to design side-lobe levels of 40 and 65 dB.

Alternate DSP Devices

The examples in Chapters 10 and 11 are all built around a single DSP processor, the Analog Devices ADSP-2181. This makes the programs easier to follow since the language is not changing from example-to-example. However, it obscures the fact that a number of excellent alternate devices are available from several manufacturers. For specific applications, a particular device may excel over others.

At 33 MHz, the ADSP-2181 does not represent the fastest available processor, either. For audio applications, this is often not important. With a little care in programming, it is usually possible to pack the last IF and audio functions of a communications receiver and transmitter into a device such as this. Examples of this are in Chapter 11 of this book.

Bread-boarding of fast processors such as used for DSP is not simple. Multi-layer PC boards are of major benefit and the IC packages most often use a large number of fine-pitch pins, making connections unsuitable for wires. For these reasons, the use of a "demo board" makes experimentation much easier. Most manufacturers offer demo boards for their DSP devices, often bundled with some collection of support software. Before selecting a particular DSP device for a project, it is best to determine the current offerings of these boards. The prices vary widely, often reflecting the bundled software.

Representative families of low-cost DSP processors are reflected in the table below. These are not the high-end products from the various manufacturers, since these often represent unneeded expense as well as higher power consumption. The changing nature of these processor families suggests that one should check the manufacturers Web sites for the current data.

In addition to specialized DSP processors, it is quite practical to use a PC directly. High-end Intel, AMD or Motorola processors are able to provide performance levels comparable to the better dedicated DSP device. A sound board provides the CODEC functions. This is not as compact a solution as the dedicated DSP board and thus can't easily be regarded as a "component." The programming environment is complicated by the general-purpose operating systems in use.

An example of an alternate demo-board is the "TMS320C3x Starter Kit" from Texas Instruments. The hardware consists of a 3.5 by 5.0 inch PC board with a TMS320C31 32-bit floating-point processor and a TLC32040 A/D and D/A converter. It is bundled with an assembler and an emulator type of debugger. An interface is provided to control the board from a PC.

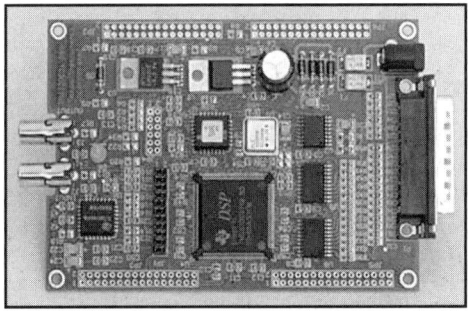

The TMS320C3x Starter Kit from Texas Instruments.

Manufacturer	DSP Processor	Number of Bits	Floating Point	Processor Rate, MIPS
Texas Instruments	TMS320VC5416	16	No	160
Texas Instruments	TMS320C31-50	32	Yes	25
Motorola	DSP56309	24	No	100
Analog Devices	ADSP2181	16	No	33*
Analog Devices	ADSP2191	16	No	160
Analog Devices	ADSP21065	32	Yes	40

*This is the ADSP2181 as used in the EZKIT Lite, put here for comparison purposes. Versions are available that operate at 50 MIPS.

---

strongest close to their center, and are therefore not filterable when close to the receiver pass band. When there is greater separation between the interfering transmitter and the receiver pass band, where filtering is more effective, the attenuation of the Kaiser-window filter is greater.

In Chapter 3, it was noted that LC filters tend to have added group delay near the edges of the pass band. This is associated with undesirable "ringing" for the filters. FIR filters are usually designed with coefficients that are symmetrical about their center values. This produces a group-delay response that is exactly flat with frequency. The amount of delay is half the number of filter coefficients, multiplied by the sampling period. The response of the filter to a very short impulse is easy to find as it is just the values of the filter coefficients. **Fig 10.25** shows the impulse response for the 500-Hz bandwidth filter of Fig 10.24. The vertical scale shows the coefficient values for a filter with a gain of 1.0 and should be examined here for relative values. The horizontal axis has been scaled in time to correspond to the 9600-Hz sampling rate, i.e. a sampling period of 1/9600=0.1042 milliseconds. The figure shows a considerable amount of ringing still exists, although the group delay is flat. This ringing is a fundamental consequence of the fast cut off characteristic of the filter. Other filter designs can have less ringing, but only by sacrificing the sharp frequency response.*

---

* An example of a non-ringing filter is given by C. R. MacCluer, W8MQW, "A Matched Filter for EME," *Proceedings of the Central States VHF Society, 1995*, p 24. These filters have a frequency response, at frequency f, of $\sin[2*pi*(f-f_o)*T]/[2*pi*(f-f_o)*T]$, where f_o is the center frequency and T is the length, in seconds, of the sine-wave burst (CW dot). This "sin(x)/x" response creates a slow fall-off with frequency, but the peak signal-to-noise ratio of a CW dot is maximized. The non-ringing characteristic produces an interesting and pleasant "sound" when used in the audio path of a receiver. Because of the spectral side lobes, it can be difficult to tune in a signal by ear. However, when on-frequency, the filter provides excellent CW copy. Another example of this filter implementation is included with the DSP-10 transceiver software that is part of the *Experimental Methods in RF Design* CD.

A further parameter that is available to the Kaiser-window FIR filter designer is the side lobe level. **Figure 10.26** shows the frequency response of filters designed to 40 and 65 dB levels. These filters both have the same nominal 200-Hz bandwidth at -6 dB points. The most obvious feature is the side lobe response far from the pass band, which is about 20 dB lower for the 65 dB case. In addition, it can be seen that the design with the lower out-of-band response is also less sharp around the pass band. The response at 40 dB below the peak is 296 Hz wide for the 40-dB filter and 344 Hz for the 65-dB filter. Thus the penalty for having the lower out-of-band side lobes is poorer passband shape.

Hilbert Transforms

One of several specialized applications for FIR filters is the Hilbert 90-degree transform. These are a close counterpart to the broadband 90-degree phase-shift networks discussed in Chapter 9. They are characterized by a constant 90-degree phase shift and an amplitude response that covers a wide frequency range. The flatness of the frequency response as well as the bandwidth that can be covered depend on the size of the FIR filter, i.e., the number of coefficients.

The Hilbert transform has a fixed delay in addition to the 90-degree phase shift. In order to produce two signals differing in phase by exactly 90-degrees, it is necessary to place a fixed delay in the second path. A DSP implementation of the fixed delay requires only a few instructions.

The interested reader should study the 18-MHz transceiver in Chapter 11, which uses one of the Hilbert transforms in the SSB generation and detection.

10.7 DSP IF

Computers, and specifically DSP microprocessors, are limited in their processing speed. The instruction set for the DSP makes it faster for signal processing, but DSP is still best suited for signals in the 10's of kHz or less.* Audio processing is easily in this range and not surprisingly, has been a major application for DSP in radio systems. Interesting applications are possible by use of a low frequency IF, however.

Fig 10.27 is a block diagram of a radio receiver, implemented with the last IF in a DSP at 7.5 kHz. One would prefer an IF as low as possible, which is often quite practical. For instance, if the analog IF has a bandwidth of 5 kHz, the 60-dB points for a reasonable crystal filter might be 15 kHz apart. This will allow the use of an IF as low as 2.5 to 7.5 kHz with the image rejection being always greater than 60 dB (see **Fig 10.28**). With the proper A/D converter, this would be supported by a sampling rate of about 20 kHz.

Fine Tuning

A major advantage of the DSP IF is the simplicity of fine frequency control. We have already seen that we can easily generate a sine wave in software with good frequency resolution. This is ideal for use as the oscillator for frequency conversion. This can be a shift in the IF, or more often,

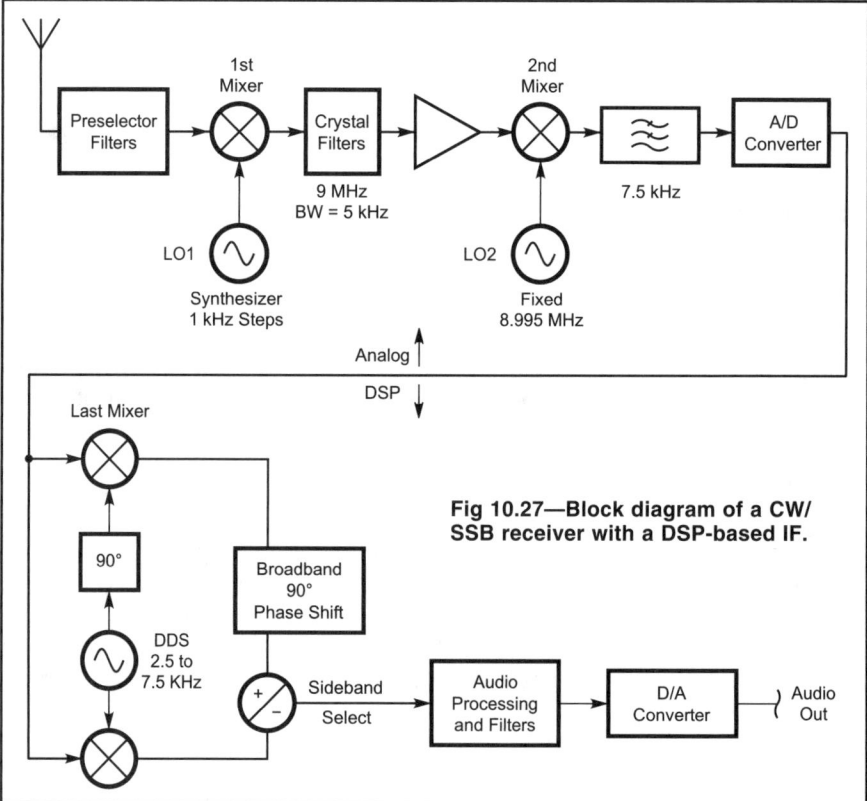

Fig 10.27—Block diagram of a CW/SSB receiver with a DSP-based IF.

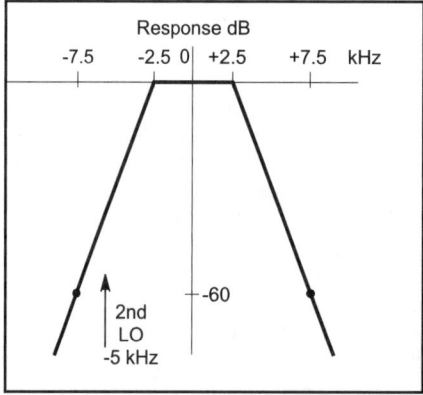

it is the final conversion often called the BFO. As we will see, the input and output frequencies of the conversion process can overlap and so there is considerable freedom in choosing the IF.

---

* The ADSP-2181 in the EZ-Kit Lite that has been used for the examples executes 33 instructions per microsecond. Each instruction can be a simple operation, such as adding of two numbers, or it can be a multiple part instruction that multiplies two numbers together, adds these to an existing sum, fetches two different values from memory and updates a loop counter. This latter type of instruction is an example of the specialized instructions that allow high computation rates in a DSP microprocessor.

Fig 10.28—The required response curve for the crystal filter used in the receiver of Fig 10.27. The frequencies shown are relative to the IF center. Image responses are limited by having 60 or more dB of rejection at 5 kHz from the band edge.

10.8 DSP MIXING

The double-balanced mixer of Chapter 5 has wide application as an analog component. The simplicity of a DSP implemented mixer can be surprising at first introduction:

mr=mx0*my0 (ss);

That is, only a simple signed multiply is required. If mx0 and my0 registers represent sine waves, then mr will represent a signal containing only the sum and difference frequencies. The rejection of signals passing from the inputs (mx0, or my0) to the output (mr), called port-to-port isolation in conventional mixer descriptions, is for practical purposes perfect.

This very high isolation allows the input and output frequencies to be in overlapping bands. Additional processing is needed since one usually only desires the sum or the difference frequencies. An example of this is a Hilbert Re-tuner described by Forrer.[12] This process corresponds to the Phasing method of SSB detection, described in Chapter 9.

10.9 OTHER DSP COMPONENTS

There are many functions that lend themselves to DSP implementation in a radio. We only touch upon many of them here. The following should be thought of as a starting point for further exploration!

Automatic Gain Control (AGC)

Figure 10.29 is a block diagram of a DSP implementation of a classical AGC feedback loop. The control point for the loop, shown in the figure, is the IF signal after A/D conversion. The function of the loop is to keep the control-point amplitude close to constant. A detector is used to measure the envelope of the IF signal. This is low pass filtered and adjusted in level by the AGC Filter. The filter output goes back through a D/A converter to control the gain of an IF amplifier. In addition, the AGC controls a digital gain multiplier that is within the loop.

The analog gain control is used to ensure that the A/D converter is operated well into its operating range, while still preventing overload. The digital part of the loop keeps the total signal level near a constant level at the output.

The response of the filter going to the analog IF amplifier, referred to in the figure as the slow loop must cutoff at a low enough frequency to allow stability, including the delay effects of the A/D converter. The converter delay is often many hundreds of microseconds resulting in a maximum AGC bandwidth in the tens of Hertz. This is too slow to provide adequate attack response on a rising strong signal, and requires that the A/D converter not be set to operate too close to it's overload point. This is usually possible to arrange in the design.

Improvement comes from the internal DSP fast loop in Fig 10.29. This feedback loop does not include the A/D converter and is limited only by the sample rate of the data. The signal levels should be set so that this loop is the gain controlling function for normal operation.

One of the big advantages of a feedback AGC system is its ability to work with highly inaccurate gain control functions. In the case of the DSP, however, this is not needed. Gain can be controlled by either multiplication, or multiplication along with a binary shift. Either of these functions are accurate to a fraction of a dB and can be used with open loop control. The general scheme for this AGC system is **Fig 10.30**. The analog feedback slow loop is maintained for very strong signals, but the DSP gain control is placed after the detector. This allows a delay to be placed in the signal path, so that the signal levels are well known when the control is applied. That is, the gain is reduced in a "circuit" before the signals arrive at that point. This feed forward approach is capable of very good sounding AGC, since the accuracy of the control and the response time have been made independent. Methods of this sort have been in use for several years in DSP based transceivers offered by Rohde and Schwarz.[13]

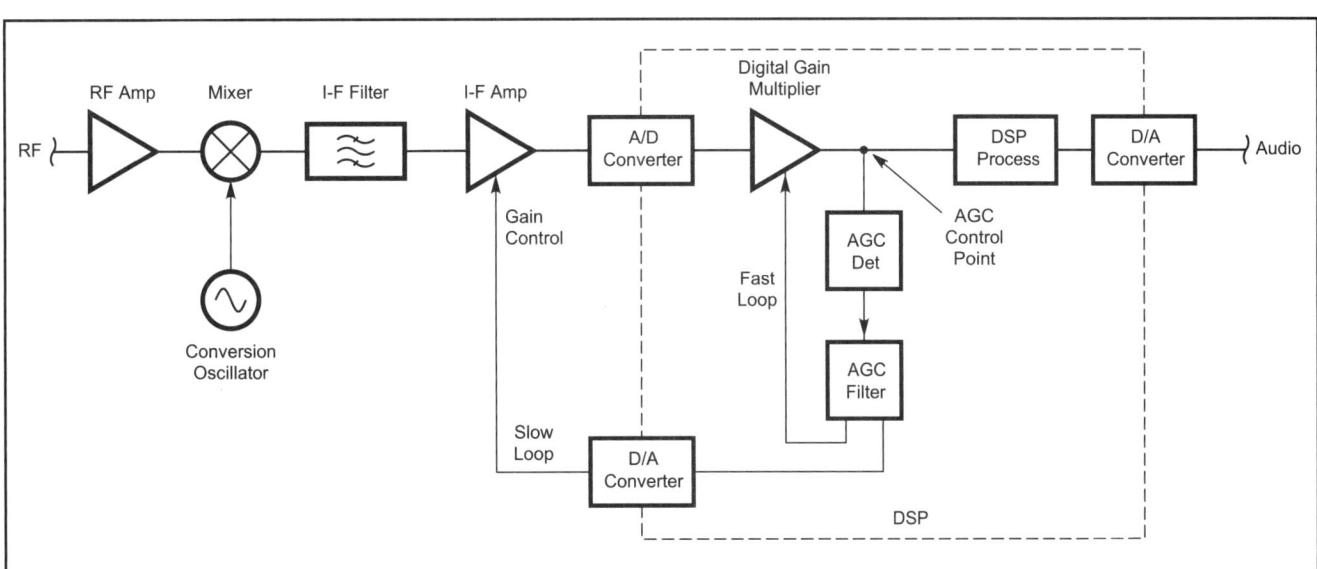

Fig 10.29—DSP-based feedback type of AGC showing a combination of analog and digital gain-control points.

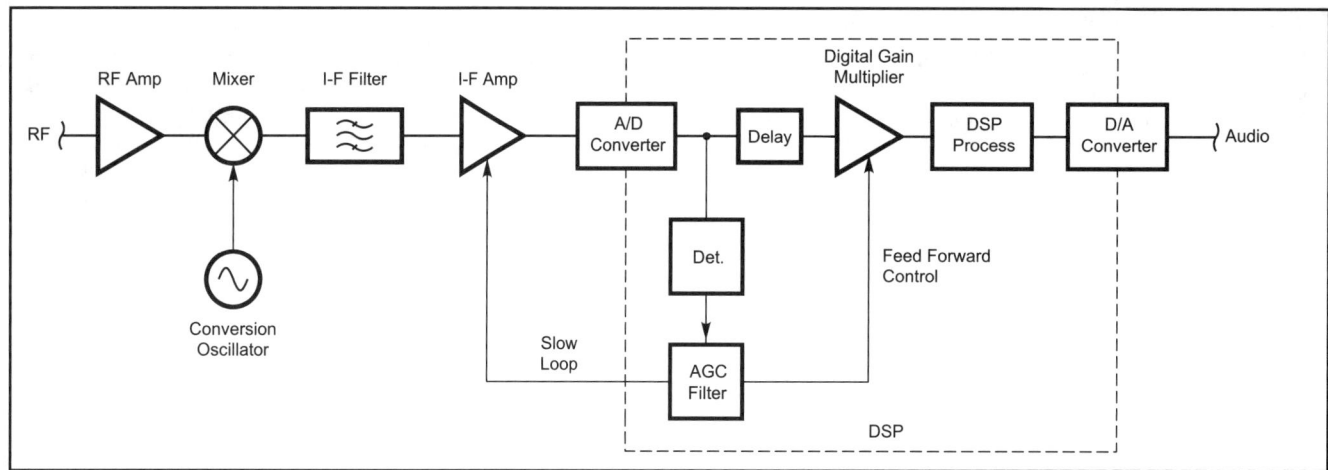

Fig 10.30—DSP-based AGC with analog feedback and digital feed forward control.

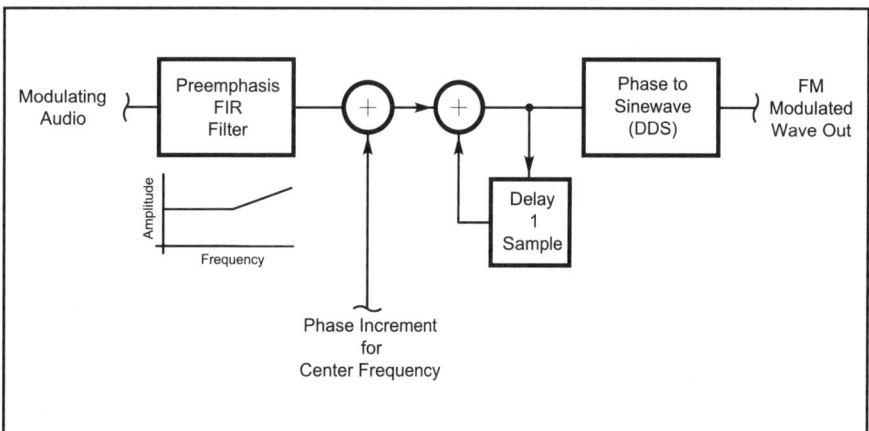

Fig 10.31—Direct generation of FM signal.

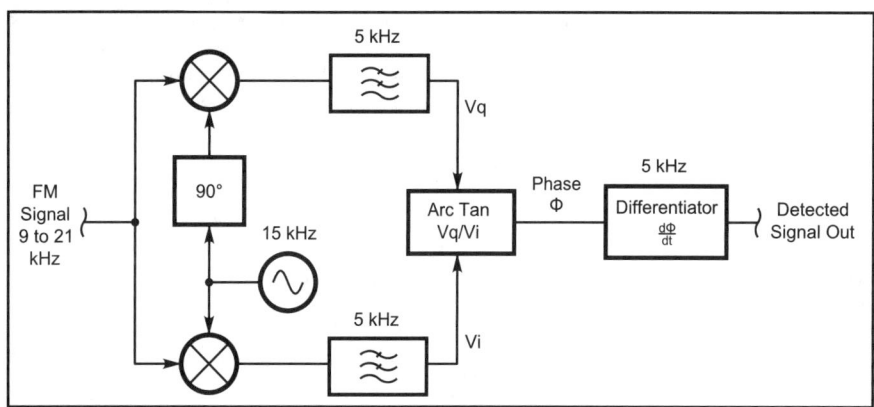

Fig 10.32—An FM detector built using an arctangent phase detector and a differentiator.

FM Reception

As is the case for analog Frequency Modulated (FM) discriminators,[14] a number of methods exist for the DSP-based detection of an FM signal. FM is a special case of phase modulation and one of the best FM detectors starts with a phase detector, as shown in **Fig 10.32**. The FM signal at IF, shown here as 9 to 21 kHz is mixed with a pair of constant frequency signals at mid-band (15 kHz). These two mid-band signals differ in phase by 90 degrees and, with DSP, can be generated as two separate signals. Low pass filters, in this case at 5 kHz, remove the signals at the sum frequency, leaving just the difference signals. Since these two signals were derived from the 90-degree mixing process they are called quadrature signals (see Chapter 9) and can be shown to retain all of the information that was originally in the IF signal.

The phase angle of the input signal, relative to the 15-kHz center sine wave, can be determined from the two quadrature signals, v_i and v_q by:

$$\varphi = \tan^{-1} \frac{v_q}{v_i} \qquad \textbf{Eq 10.8}$$

Arc tangent functions can be computed by polynomial approximations, in a fashion very similar to that used to compute a sine wave earlier in this chapter.[15]

Frequency is defined as the rate-of-change of phase. The mathematical term for this operator is the derivative and the functional block for finding it is the differentiator. When reduced to a DSP program, all that is required is to subtract the current phase value from the previous value. In general it is necessary to watch the phase value where it passes through 360 degrees, since that point and 0 degrees are the same. If the phase value has been scaled so that

FM Transmission

Earlier in this chapter the DDS method of generating sine waves was described that was based on incrementing a phase value by a constant amount called a phase increment. The frequency of the sine wave is proportional to the phase increment. FM modulation can be accomplished by varying the phase increment in accordance with the modulation waveform. This is inherently of very low distortion. Most FM systems employ some preemphasis for the higher modulation frequencies that can be accomplished by placing a FIR or IIR filter ahead of the modulator. **Fig 10.31** shows the overall arrangement.

360 degrees is the entire range of the 2's complement arithmetic (0 to 65535 for 16-bit arithmetic) then the rollover at 360/0 degrees is automatically treated correctly for either direction of rollover.

Thus the output of the difference operation is the FM demodulated signal. In general, it is necessary to place this through an appropriate de-emphasis filter to reduce the high frequency boost introduced at transmission time. This could be the simple RC IIR filter described earlier.

10.10 DISCRETE FOURIER TRANSFORM

In Chapter 7 we explored using Spectrum Analyzers to observe the content of signals in the frequency domain. They consisted of a detector for measuring signal amplitude coming from a receiver along with a local oscillator for tuning the receiver. The local oscillator was made voltage tunable so that it could be swept across a range of frequencies. When combined with an oscilloscope for displaying the signal amplitude, analysis of the signal spectrum was possible.

An alternate DSP implementation of the Spectrum Analyzer, using the Discrete Fourier Transform (DFT), has some attractive features. The swept local oscillator and associated mixer are not needed in

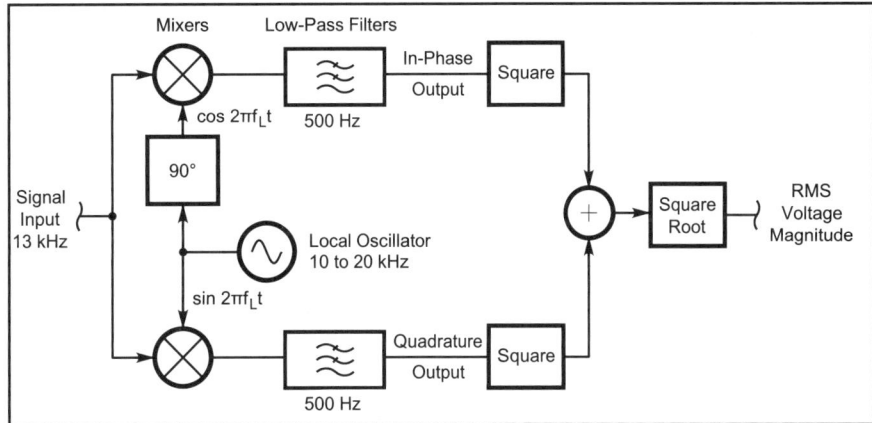

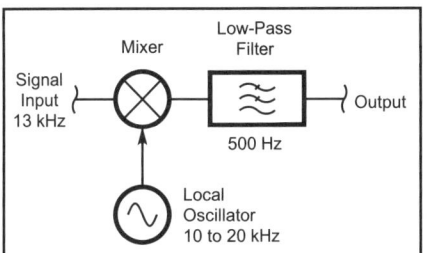

Fig 10.33—A first implementation of a circuit to measure signals in the 10- to 20-kHz frequency range. The output of this circuit is sensitive to both the frequency of the input signal and its phase, relative to the local oscillator.

Fig 10.34—An improved implementation of the circuit of Fig 10.33. The in-phase and quadrature outputs will never be zero simultaneously, regardless of the input phase relative to the local oscillator. Blocks have been added to square the in-phase and quadrature outputs, add these together and then take the square root. This produces the RMS voltage of the signal input at the frequency of the local oscillator.

Mathematics of the Discrete Fourier Transform

Mathematical formulations of the Fourier transform are given in many books. In general, the DFT has inputs and outputs consisting of complex numbers described as $v_{Rk} + j\, v_{Ik}$ where v_{Rk} and v_{Ik} are called the "real" and "imaginary" parts of the complex number. This use of complex numbers has considerable convenience in writing and evaluating equations. However, the mystical sound of "imaginary" numbers and associated use of $j=\sqrt{-1}$ can be removed if an alternate description of the complex number as "an ordered pair of real numbers" is used. This illustrates that each input to the DFT is a pair of real numbers that are treated by a specific set of rules (equations) to produce a set of ordered pairs of real numbers at the output. Ordered pairs merely means that the first number (real) is not to be interchanged with the second number of the pair (imaginary).

With this in mind, we can examine the kth outputs of the DFT with a complex input:

$$X_{Rk} = \sum_{n=0}^{N-1} V_{Rn} \cdot \cos(2\pi k n / N) - \sum_{n=0}^{N-1} V_{In} \cdot \sin(2\pi k n / N)$$

and...

$$X_{Ik} = \sum_{n=0}^{N-1} V_{Rn} \cdot \sin(2\pi k n / N) + \sum_{n=0}^{N-1} V_{In} \cdot \cos(2\pi k n / N)$$

Here we have separated the real and imaginary inputs, V_{Rn} and V_{In} as well as having separate equations for the real and imaginary output parts, X_{Rk} and X_{Ik}. Notice that all mention of j disappears and the real and imaginary parts are kept separate by placing a subscript R or I on the variable.

We show the kth output pair, but there are a total of N of these output pairs corresponding to k values from 0 to N-1.

If the inputs have zero imaginary parts, such as is the case for a time waveform, the second sum in each equation will become zero and the DFT outputs simplify to:

$$X_{Rk} = \sum_{n=0}^{N-1} V_{Rn} \cdot \cos(2\pi k n / N)$$

and...

$$X_{Ik} = \sum_{n=0}^{N-1} V_{Rn} \cdot \sin(2\pi k n / N)$$

These are the versions that are described by circuit analogs in the text.

hardware form. The output spectrum is being constantly generated instead of waiting for the tuning to sweep by, providing higher sensitivity and faster update rates. However, the DFT is limited, by both A/D encoding and computing rates, in the frequency range that can be covered.

The operation of the DFT can be understood by a thought implementation of an analogous traditional hardware circuit. This starts by assuming we wish to examine the output of a receiver IF in the 10- to 20-kHz range. Initially, assume that the only signal present sits at 13kHz.

We wish to find out what signals are in this IF band and what their strength might be. We begin with a balanced mixer capable of operation at these low frequencies, as shown in **Fig 10.33**. We drive the mixer with a suitable local oscillator, capable of covering 10 to 20 kHz and run the output through a very narrow low-pass filter. As we tune the LO, we see no output until we get close to 13 kHz, due to the low-pass filter. Then we start to see low frequency outputs. When the LO is exactly at 13 kHz, the output is a dc signal that we can measure with a voltmeter.

We might be tempted to note the dc level coming from the mixer and use this to infer the strength of the incoming 13-kHz signal. However, this would generally produce an error, for we know nothing of the phase of the LO with respect to the signal we are trying to measure. Recall the phase detector characteristic investigated in Chapter 4, section 7 shows that the mixer output depends on the phase angle between the RF and LO signals. For 90-degree phase differences this output will be zero, clearly the wrong answer! This dilemma can be solved by replacing the single mixer with a pair of identical mixers, both driven from a common signal or RF port, but driven with a pair of LO signals with 90-degrees phase difference. This is illustrated by the block diagram of **Fig 10.34**, where we have simplified the circuit by using a single oscillator and a 90-degree phase shifter. Now, as the phase of the input is varied, we will see the output of one mixer go to zero while the other peaks. The true (RMS) output voltage magnitude is obtained by squaring each of the two mixer output voltages, adding, and taking the square root.*

Clearly, we can replace the hardware mixers with a DSP version. The 10-to-20-kHz signal is applied to an A-to-D converter to produce a time-sampled version of the signal. This is applied to a pair of

*If one only wants the power of the signal as an output, the square-root block can be omitted.

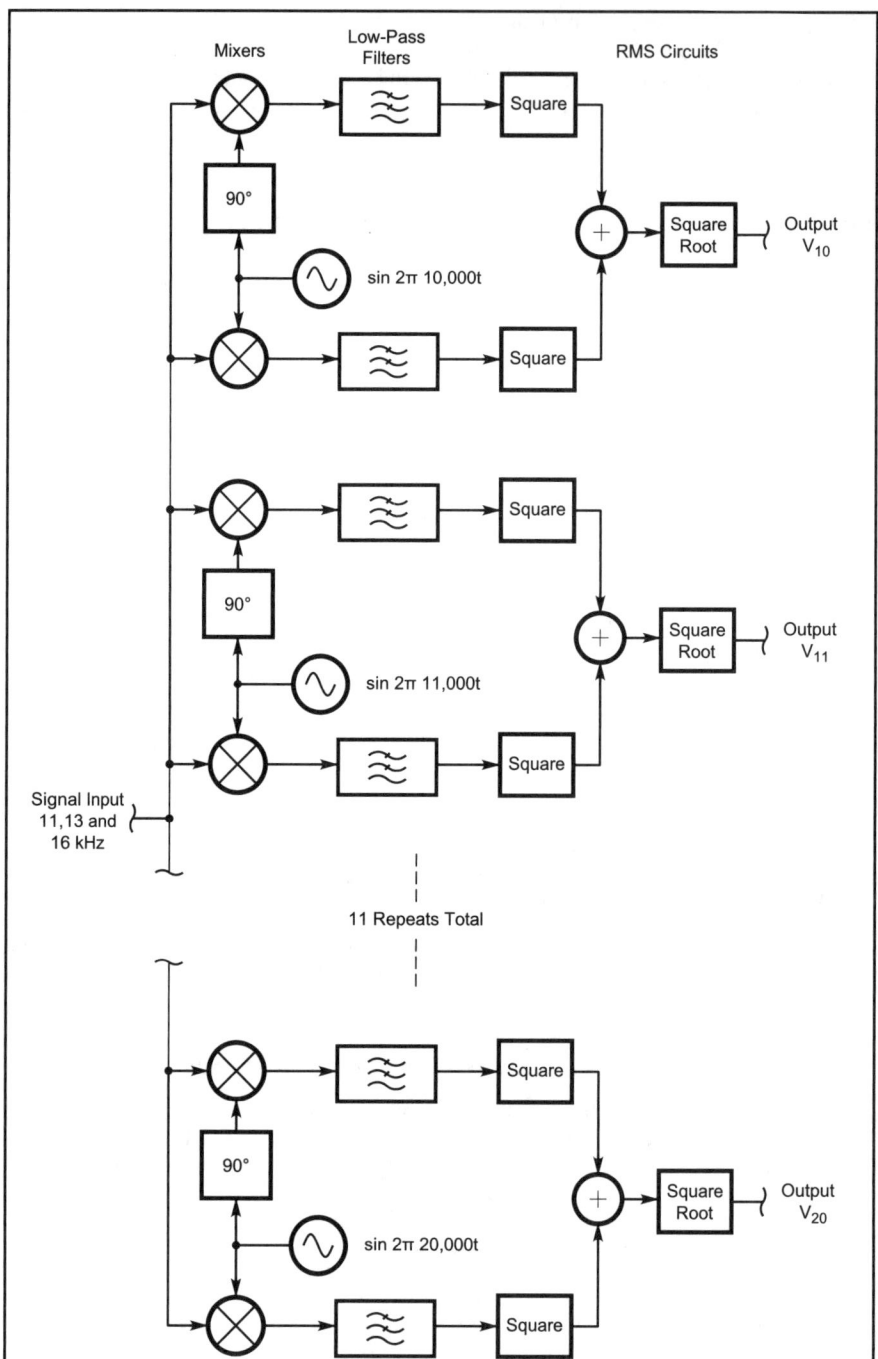

Fig 10.35—A filter bank type of Spectrum Analyzer, built from multiples of the in-phase/quadrature filters of Fig 10.34. As discussed in the text, this structure is equivalent to a Discrete Fourier Transform, followed by the RMS squaring and square-root circuits.

DSP mixers, one driven with a $\cos(2\pi f_L t)$ signal while the other is driven in quadrature by $\sin(2\pi f_L t)$. The outputs are low pass filtered to eliminate any sum terms, leaving only the base-band outputs. These can be used to calculate the output voltage, just as we did with the hardware mixer. This is just a phasing method receiver as discussed in Chapter 9.

Let's continue our thought implementation by adding more signals in the 10- to 20-kHz band. The original 13-kHz signal is supplemented with a weaker one at 11 kHz, and perhaps another at 16 kHz. One way to estimate the overall spectra would be to add two more mixer pairs with a pair driven at each of the new input frequencies. However, let's get even more general. Instead of adding just two more mixer pairs, we will assemble a collection of 11 of these circuits with a quadrature pair at each 1-kHz increment from 10 to 20 kHz.

Fig 10.36—A detailed block diagram of the DFT with only "real" input data, such as from samples of a time waveform. The multiplying (mixing) signals are calculated sine and cosine values with frequencies spaced every f_s/N Hz, where f_s is the sampling rate for the data. The resulting outputs are referred to here as "In-phase" and "Quadrature" data.

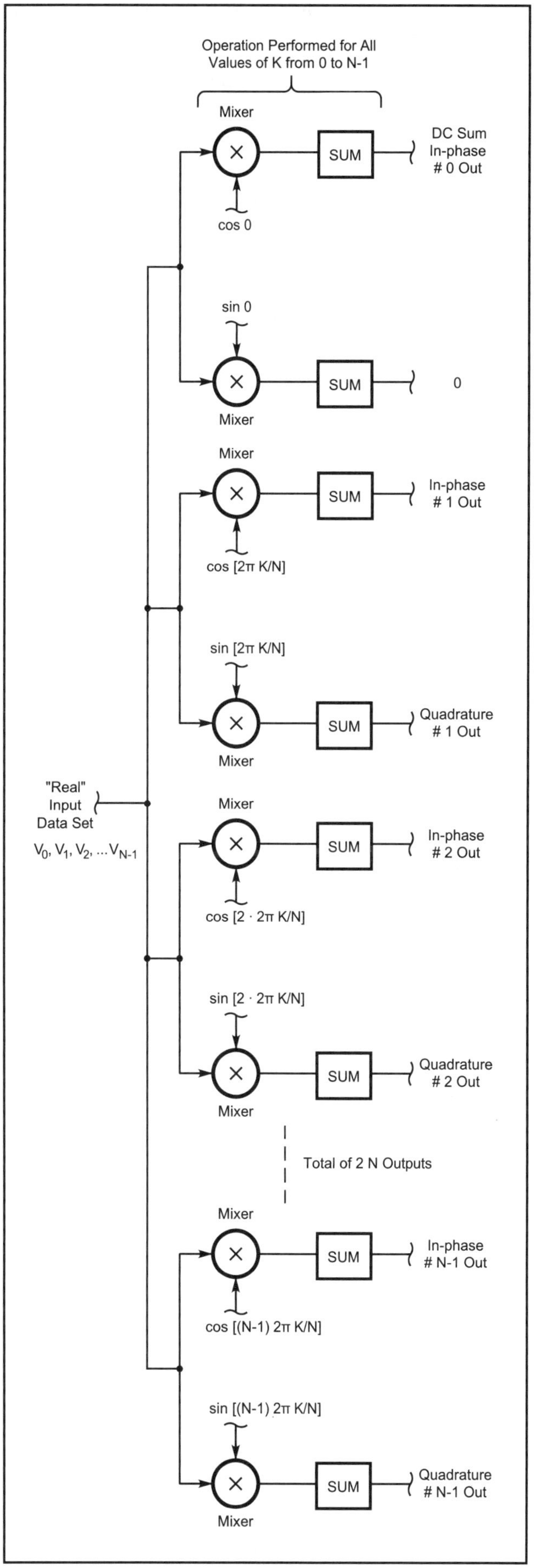

Figure 10.35 shows a block diagram of our growing collection of thought hardware. Most outputs will be close to zero, but we will see substantial outputs corresponding to 11, 13 and 16 kHz.

We now have a "bank of filters" type spectrum analyzer. We could have achieved the desired result by actually building 11 band-pass filters, each followed by a suitable detector. Instead, we have achieved the same result with mixers driven by quadrature-local-oscillator signals.

These systems are fundamentally different than the usual "swept front-end" spectrum analyzer. If we were to build one of those for this example, we might use a swept local oscillator that tuned from, for example, 60 to 70 kHz. A single mixer would heterodyne the input up to a narrow band-pass filter at 50 kHz, followed by a suitable detector. As the oscillator sweeps the input frequency from 10 to 20 kHz, the signal-amplitude output for the incremental kHz points will be virtually the same as we obtained from the banks of mixer pairs. However, while the swept system provides information for one frequency at a time the filter bank provides all outputs simultaneously.

Banks of oscillators, mixers and low-pass filters become unwieldy if built from hardware. But we can build up their equivalent DSP components as is shown in **Fig 10.36**. As shown in Fig 10.37, oscillators are replaced by quadrature sine and cosine wave computations. Numerical multipliers replace the mixers. The low-pass filters are replaced by summing several multiplier outputs. This needs to be repeated for each of the frequencies of interest, such as our integral frequencies from 10 to 20 kHz. Put into this mathematical form, we have recreated the DFT algorithm.* Those inclined towards mathematical descriptions can also see this from the equations in the sidebar, "Mathematics of the Discrete Fourier Transform." Most implementations of the DFT would compute the spectral outputs from 0 to 9 kHz as well as the 10- to 20-kHz outputs shown, but this is not required to be a DFT.

*As will be discussed, the full DFT is more general and allows the input to be a complex number. Here, we are dealing with a simplified case where the "imaginary part" of the input is zero.

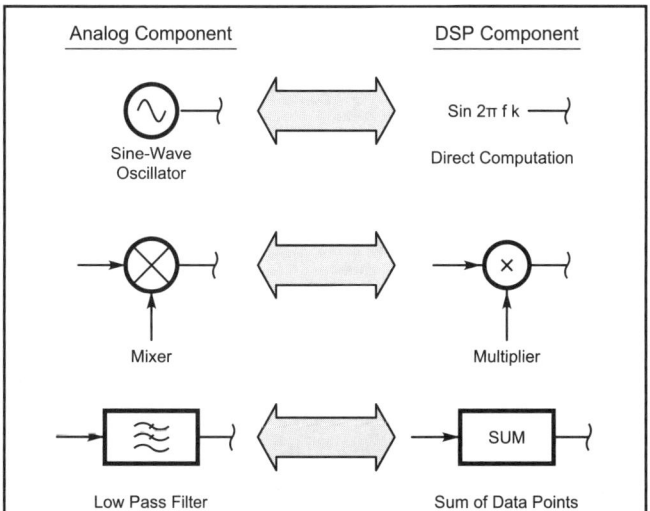

Fig 10.37—Equivalent analog and DSP components that are used to create an "equivelent circuit" for the discrete Fourier transform (DFT).

Terminology for the DFT differs from that used for hardware. Our block diagram of Fig 10.35 is in the latter term. Restructured in conventional DFT terminology, Fig 10.36 shows the same filter bank implementation. The RMS voltage blocks have been removed to show only the DFT.

Implementing the DFT

The "discrete" in DFT tells us that the system is only using data samples, as we would get an A/D converter. The Nyquist criteria requires the sample rate to be at least twice the highest frequency of interest. This would require a sample rate greater than 2×20 kHz for the thought implementation above.

The more points in the sample, the greater resolution we can achieve in estimating the related spectrum. This can be put into the formula:

$B = f_s / N$ **Eq 10.9**

where B is the frequency spacing between adjacent spectrum samples (filter-bank centers), f_s is the sample rate, and N is the number of sample points being averaged. One divided by B gives the length of time over which samples were collected. The frequency spacing B can easily be made quite small. For example, if the sampling rate is 10 kHz and there are 1024 samples in the DFT, the resolution B will be 10,000/1024 or 9.77 Hz. By selecting suitable f_s and N it is practical to have resolutions of less than 1 Hz.

The streamlined class of algorithms most often used to compute the DFT is called the Fast Fourier Transform (FFT).[16] These algorithms eliminate the redundant calculations that occur when N equals 2 raised to an integer power. The efficiency of the FFT allows large numbers of points to be included in a DFT computation. N values of 64 to 4096 are common. The details of the FFT require some study to follow, but for most applications this need not be done since prewritten subroutines can be used.[17] Rather than focusing on the details of the FFT, the important element is to understand the general nature of the DFT and the meaning of the resulting data.

FFT implementations usually compute N quadrature pairs of outputs. If only a few outputs are needed, it is often simpler to implement a band-pass filter bank. An efficient implementation of this is the Goertzel algorithm.[18]

The Ins and Outs of the DFT

When one uses the DFT, interpretation of the input and output data can be confusing. To understand how these data are used, we will examine finding the frequency spectrum of a time waveform.

The DFT algorithm operates on a block of N input-data points, each of which is a sample of a time waveform, such as an IF or AF signal. The DFT is expecting N complex input numbers that are divided into two groups, the "real" and the "imaginary" values. These are historic names used with complex numbers and should be thought of as merely a way to keep the groups separate. For our case, the N real values will be the waveform time samples and the imaginary group will all be zero.*

After the DFT calculation is completed, there will be non-zero values in each of the real and imaginary groups. These represent the zero-degree and 90-degree amplitude components of the frequency spectrum, referenced to a sine wave at the center frequency of each of the output frequencies.

The spacing between spectral data points is $B = f_s / N$. If we have N outputs from the DFT these will seem to extend from 0 frequency to $N \times f_s / N$ or f_s which is the sampling frequency. This is inconsistent with the Nyquist sampling theorem, which says the highest frequency for which we can extract unambiguous information is half of the sampling frequency. This is resolved when we look at the DFT output. It will be seen that each output point appears twice. The first N/2 data points apply for frequencies up to half the sampling frequency and the second half are their mirror image. The practical result is that one merely discards the redundant data to the right and uses the left data.

An example of this is in **Fig 10.38** showing a time waveform with N=16 and the resulting spectral power from a DFT. The output power values to the right of center are seen to be mirror images of those to the left.

Fig 10.39 illustrates this operation of the

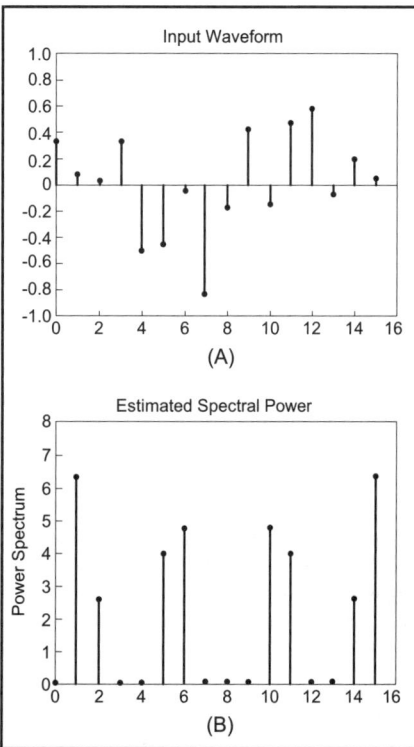

*Operating the DFT with half the inputs set to zero suggests wastefulness! It is possible to place a second time waveform in place of the zeroed imaginary group. The output values then contain co-mingled spectral data that can be sorted out with simple additions and subtractions. This can be a major computational saving for some applications, but with some possibility of added noise for fixed-point DSP.

Fig 10.38—This diagram shows (A) 16-member time waveform and the power for the DFT output. To emphasize the discrete nature of the data involved, the values are shown as dots with attached vertical lines. Note that the spectral power is symmetrical about the 8th output.

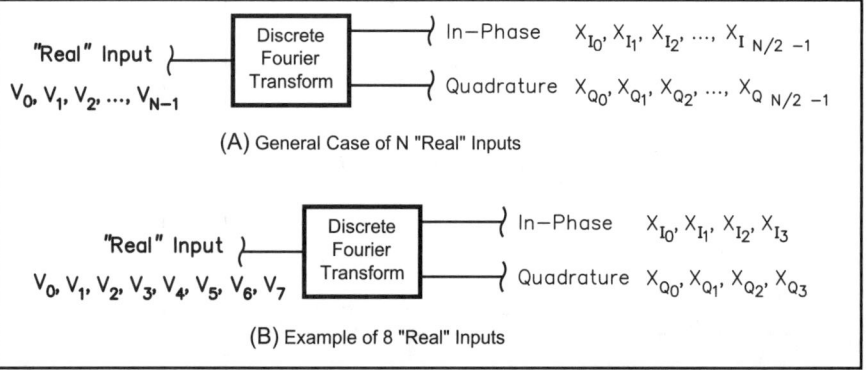

Fig 10.39—Block diagram of the Discrete Fourier Transform with a time waveform input. The output information is referred to here as "In-phase" and Quadrature." For this case of all "real" inputs, the number of output pairs is half the number of input samples. The upper figure applies to any number of sample data points. The lower figure is specific to 8 input sample data points.

DFT on a real time series in block diagram form. This is shown with a "real" input since the imaginary input was set to zero. To make their role more obvious, the outputs are now called "in-phase" and "quadrature." N inputs numbered 0 to N-1 will produce pairs of outputs numbered 0 to (N/2)-1. The lower figure shows this for the specific case of N=8. There are 8 inputs, numbered 0 to 7 and 4 pairs of outputs numbered 0 to 3.

DFT Spectral Frequency Response

Since the DFT of a time waveform is equivalent to a bank of band-pass filters, they must have a frequency response. We can use the mixer/low-pass filter (LPF) analogy to find this response. **Fig 10.40** shows the response of an LPF constructed by adding 16 points together, just as is done for a 16-point DFT. The data sample rate was set at 1000 Hz producing a frequency bin spacing of:

$$B = f_s / N = 1000/16 = 62.5 \text{ Hz} \quad \text{Eq 10.10}$$

The 3-dB point on the response curve is at 27.8 Hz. The mixer input signal that produces this LPF input can be on either side of the LO. Thus the overall 3-dB bandwidth is twice the LPF response or 55.6 Hz, or 89% of the bin spacing.

At the bin spacing the response is down 3.92 dB. The fall-off rate of this low-pass filter response is not particularly fast, with the first side-lobe response down only about 13 dB. This means that the output of the DFT will tend to respond to signals far from the associated LO frequency. The use of "windowing" functions to improve this off-frequency response is discussed below.

Power from the DFT

Often it is desirable to estimate the power associated with each of the output frequencies of the DFT. The in-phase and quadrature outputs correspond to the sides of a right triangle and the power to the hypotenuse squared:

$$P_i = V_{I\,i}^2 + V_{Q\,i}^2 \quad \text{Eq 10.11}$$

An example of a spectrum analyzer built using the power outputs from the DFT is the DSP-10 2-M radio, originally described in *QST*.[19] The narrow bandwidths that are achieved with the DFT are useful for detection and observation of weak signals. **Fig 10.41** is the Spectrum Analyzer display from that radio while receiving weak carriers. Signals below about –150 dBm are too weak to be heard by the ear, but narrow bandwidths of the DFT make these easily

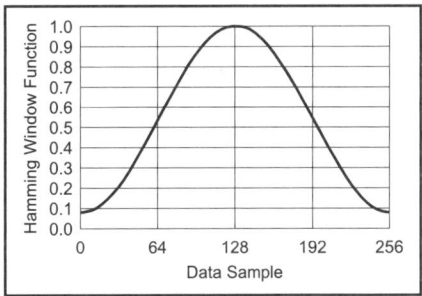

Fig 10.42—The Hamming window function, used to weight data sets to reduce spectral spreading. The data point values are multiplied by the corresponding window function to taper the values to small levels at the beginning and end of the data set.

seen on the display. The DSP-10 also uses the DFT outputs to provide weak signal communications modes. This is illustrated by examples in Chapter 12.

Other DFT Applications for Signal Processing

The spectral power data is useful for understanding the nature of signals being received. There are characteristic signatures or "looks" for particular modulation forms. CW, SSB, FM and data signals can be identified by their spectrum, without knowing any details of the information content. In addition, the DFT can be used to provide data for other functions, such as FM squelch, noise blankers and a transmitter predistorter that is discussed below. In the case of the FM squelch, the presence of a signal causes a reduction in the high frequency noise from the FM detector. By examining the power in various DFT outputs it is possible to sense the presence of a signal. In a similar way, comparing the various outputs of the DFT can sense the broadband nature of impulsive noise.

Windowing of DFT Data

A DFT operates on a fixed number of data points, collected at a uniform rate. The DFT behaves as though the signal went on forever, but with the assumption that the next set of samples will look exactly like the set we measured. And the next, as well... This is all fine except that it is highly

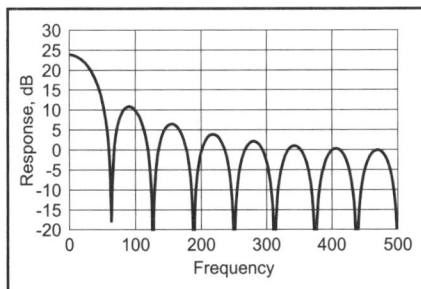

Fig 10.40—Response of a Low-Pass filter constructed by summing 16 data samples together, as occurs in the DFT. The data was samples at 1000 per second.

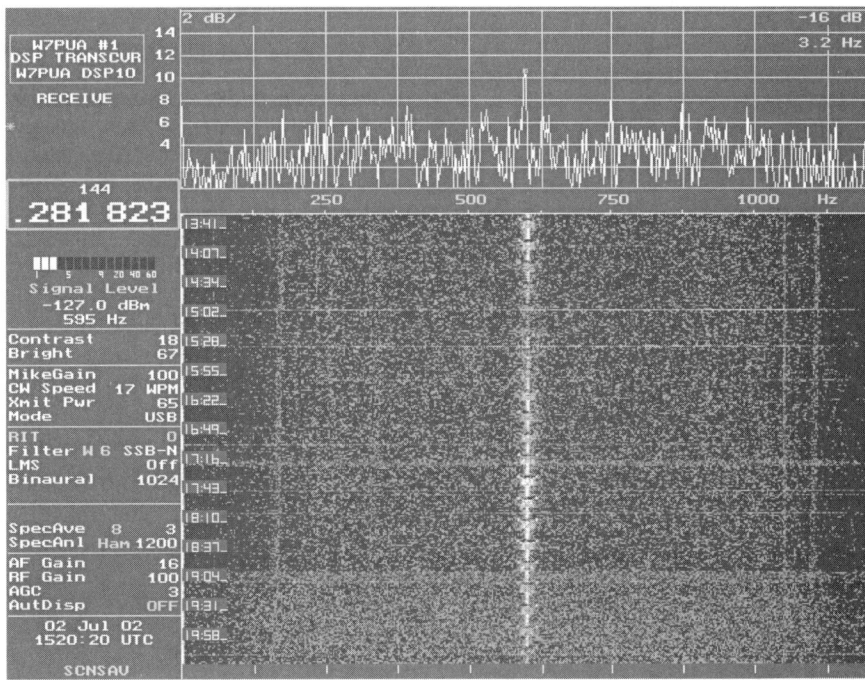

Fig 10.41—A Spectrum Analyzer display while receiving weak signals with the DSP-10. Signals below about –150 dBm are too weak to be heard by the ear, but the narrow bandwidths of the DFT make these easily seen on the display.

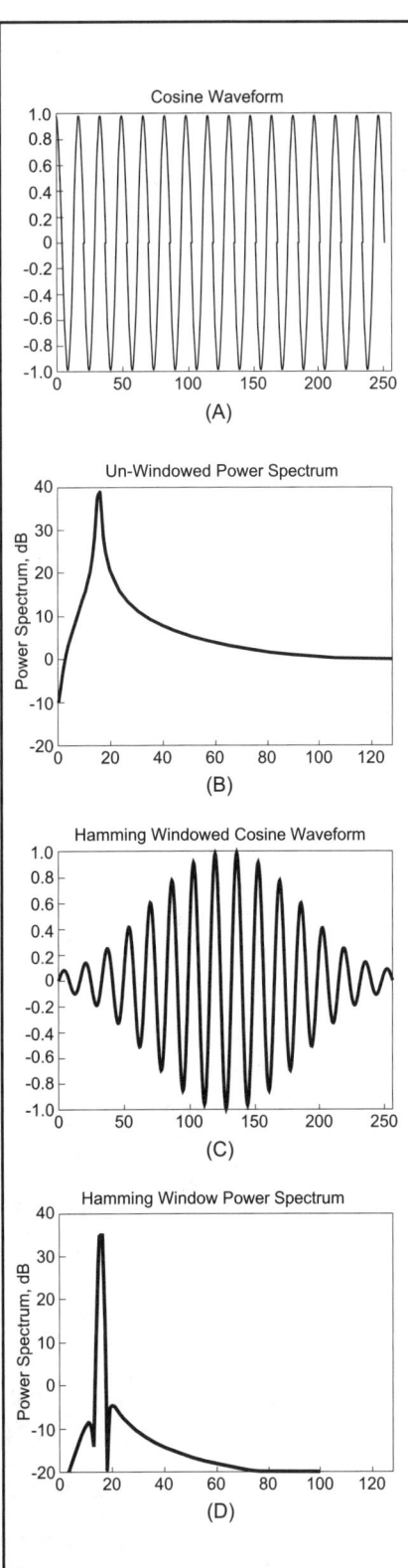

Fig 10.43—Illustrating the use of windowing to minimize spectral leakage, the figures show (a) a cosine waveform, chosen to not meet up at the endpoints, (b) the resulting un-windowed DFT power spectrum, (c) the same cosine waveform with a Hamming window applied, and finally, the much narrowed DFT power spectrum from the windowed waveform (d).

unlikely that the last point of the data set will end on the same value as it started with, or with the same slope, and the same curvature as it started. As a result, there is almost always a major jump (discontinuity) when passing between the end points. The spectral energy of this jump is spread over all frequencies and tends to be strong enough to overwhelm a low-level signal near the frequency of a strong one. The jump causes a "sidelobe structure" that drops off very slowly in frequency. The term "leakage" is often used, as the signal at one frequency appears to leak to other frequencies. This makes for a measurement of limited utility.

The best solution to this jump problem is to taper the data towards zero in the region near the edges of the sample period. If the data at the edges is zero, then the jump will also be zero. There are endless ways to taper the data and they are called *windowing functions*. A classic curve, shown in **Fig 10.42**, is the Hamming window. It has a first sidelobe down 43 dB. Many alternative windowing functions have been devised with an excellent summary in the book by deFatta, et.al.[20]

Experimentation is involved in selecting a windowing function. Each of these functions represents a distortion of the input data and a tradeoff must be made between distorting the data and the spreading of the spectrum from leakage. The usual data distortion makes spectral widths appear wider than they are; this is often quite an acceptable compromise.

Figure 10.43 shows the DFT of a cosine wave, with and without a Hamming window. The waveform without windowing (a) has been chosen to not have the last data point line up with the first one. This results in the wide and poorly defined power spectrum in (b). Application of the Hamming window results in the tapering of the data as seen in (c). The improvement in the associated power spectrum is seen in (d). Several imperfections remain. The spectral width is not a single narrow line, but overlaps 2 bins at the top and more down the sides of the spectral estimate. In addition, once 40 dB below the peak of the spectrum, the width gets quite broad. To some extent, these imperfections are part of having only a sample of the waveform and therefore making only an "estimate." However, by changing the windowing function, one can trade off the areas where a compromise is made.

10.11 AUTOMATIC NOISE BLANKERS

Noise blankers attempt to determine when a broadband noise pulse is present and during that period to "turn off" the receiver processing. Both of these functions can be performed in DSP. Two general problems exist in the operation of this type of noise blanker. Signals can be interpreted as noise, causing cross modulation onto the desired signal from the interfering signal, and the blanking process may introduce unwanted signals that resemble the interference. The design must attempt to minimize these problems, but to some degree noise blankers will have these characteristics.

Most noise blankers attempt to use the bandwidth of the interfering noise as an identifying criteria. Impulsive types of noise tend to have short duration, and to be quite strong in a wider-band receiver. This type of signal produces a rapidly rising pulse, limited by the bandwidth of the measurement. For instance, an IF bandwidth of 10 kHz can pass an impulse noise signal with a rise time of about 70 microseconds. The fastest rise time for a 3-kHz SSB signal is over 200 microseconds. A satisfactory blanker can result if one is able to provide the wider bandwidth and identify the strong signals with fast rise times. Often DSP IF bandwidths may not be as wide as desired and this can be a limitation of the noise blanker operation.

The blanking operation is ideal for DSP implementation. As was discussed in mixer operation, the simple act of multiplying two signals together is "double balanced" and neither input signal is fed through to the output. When the blanking operation is in an off state, the signal can be completely removed. Alternatively, a substitute signal can be created that is the prediction of the desired signal, based on its past characteristics. For a simple example, if the input signal was a CW tone, it would be logical to continue the last tone that was not blanked. Some delay is needed to give time for the blanking decision to be made. This delay can be implemented in DSP in a few processor instructions. More general predictors are also possible for cases such as noise input or an SSB signal.

Fig 10.44 shows a block diagram of a DSP implementation of a noise blanker. The envelope detector determines the maximum amplitude of the IF signal. It would look at both the positive and negative excursions of the signal in order to respond, as quickly as possible, to any rapidly rising noise burst. A 2500-Hz low-pass filters extracts the signal envelope. In a similar fashion, the output of a 12-kHz filter responds to all signals present in the pass band. If only the desired signal is present, the outputs of the two filters would be very similar. However, a noise burst would produce a greater response from the wider-band filter. This difference can be sensed by taking the ratio* of the two outputs. A comparator can sense if the noise response is over a threshold and then produce a blanking signal.

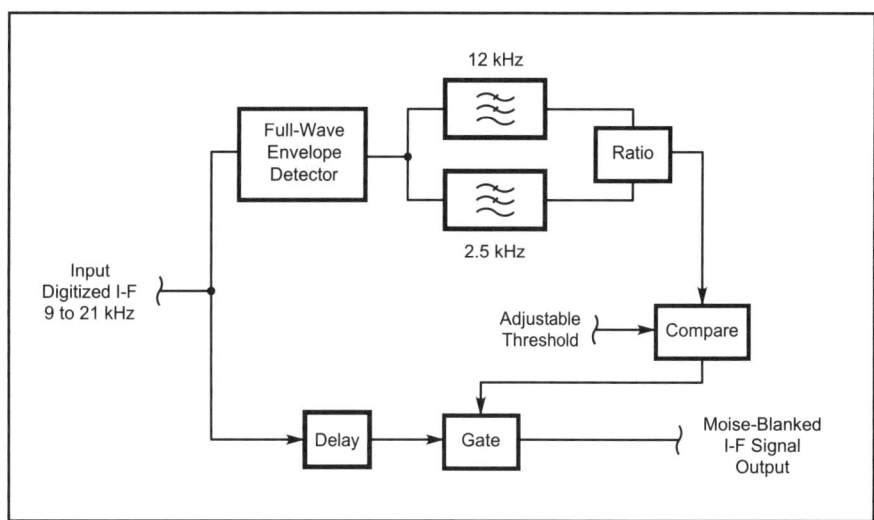

Fig 10.44—Block diagram of a noise-blanker suitable for implementation as a DSP function. An envelope detector follows the amplitude of the wide-band (12 kHz) signal. Two low-pass filters are used to determine the presence of a noise burst, which then gates the received signal. A signal delay allows time for the decision making.

10.12 CW SIGNAL GENERATION

We have discussed the generation of a sine wave and gating this on and off can generate crude CW signals. It is well known that spectral broadening (key clicks) will result from sudden on/off transitions. The keying can be made to have much better transitions by treating the process as amplitude modulation as shown in **Fig 10.45**. Here the logical signal from the keying device is placed through a low-pass filter to convert it to an analog signal of limited bandwidth. The process of amplitude modulation then produces a spectrum that is twice as wide as the limited bandwidth.

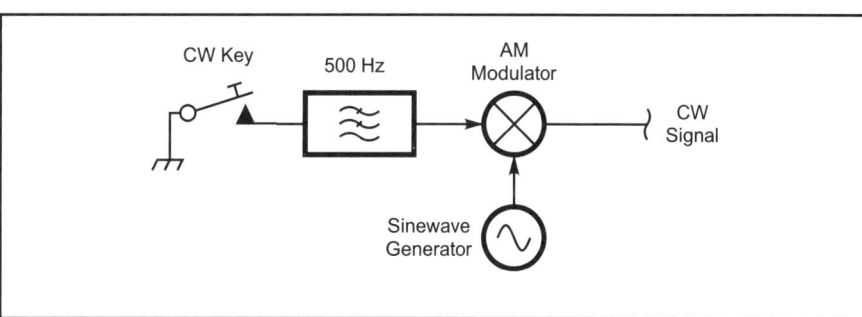

Fig 10.45—Block diagram of a CW generator using pulse shaping and an amplitude modulator. This limits the spectrum of the keyed waveform. The AM modulator in its DSP implementation is a multiplication of the two signals.

10.13 SSB SIGNAL GENERATION

All of the techniques for SSB generation shown for analog equipment in Chapter 6 can be implemented in DSP. Often the most attractive approach is the phasing method as was discussed in Chapter 9. The challenges of tight component tolerances and component drift are not problems in the software implementation and high carrier and opposite sideband rejections are easily achieved.

As an alternative to the phasing method, it is practical to implement a filter type of SSB generator. Typically this would utilize an IF in the 5- to 25-kHz range and analog mixing to convert the results to the operating frequency.

The FIR filters, mixers and sine wave generators shown above can be combined to implement a DSP IF sideband generator.

Alternatively, it is practical to have a hybrid analog/digital approach where the two quadrature audio signals are generated in the DSP and the mixers and conversion oscillator are conventional analog components. This approach lends itself to error compensation for the analog components. An example of this approach is the 18-MHz transceiver of Chapter 11.

Predistorter Distortion Reduction

SSB signals are raised in power level by amplifiers that often have intermodulation distortion products only 25 to 35 dB below the peak transmitted level. These distortion products are spread in frequency and can cause interference in adjacent channels. One can limit these product levels by reducing the output level of the amplifier or operating the amplifier in Class A; doing this results in poor dc-to-RF power efficiency for the amplifier.

---

*Division is not usually a fast operation in a fixed point DSP microprocessor. It is often desirable to find the logarithm of two values and subtract them. For applications such as the noise blanker, the logarithm function does not need high accuracy and can be implemented as a series of straight lines. This can be a relatively fast process.

One alternate solution that allows the efficiency to remain high while reducing distortion is called predistortion. For example, if the only amplifier distortion was gain compression, as shown in **Fig 10.46**, one can imagine that the distortion could be removed, if a gain-expanding pre-distorter was placed ahead of the amplifier. The predistorter would have the opposite gain characteristic to the amplifier, as shown in the upper part of the figure. For an analog implementation, it might be possible to use some diodes arranged as shown in **Fig 10.47**. If we were fortunate, the diodes would provide the proper amount of gain expansion to remove the inherent gain compression of the amplifier, at least over a restricted operating range.

A more elaborate gain expander can be built using the computational ability of a DSP device. It is presented here to indicate the potential for DSP components to improve the distortion performance as well as to suggest some possible directions that could be explored. This is not an implementation of a predistorter, but rather a conceptual treatment. The ambitious experimenter is encouraged to pursue this area since the potential benefits are substantial.

An example of such an implementation is shown in **Fig 10.48**. A polynomial is shown as the gain expansion curve. Within broad restrictions, it is possible to approximate a gain expansion curve to any precision by using enough terms in the polynomial. Results from a simulation of an amplifier and predistorter are shown in **Figs 10.49** through **10.52**. In this example, the amplifier is modeled as a linear amplifier of gain 1.0 (0 dB) along with a cubic distortion term, which is often the dominant distortion for amplifiers. For those inclined to describe this mathematically, the output voltage, v_o, in terms of the input voltage v_a is:

$$v_o = v_a - 0.1 v_a^3 \qquad \text{Eq 10.12}$$

where the 0.1 multiplier is chosen to be convenient as an example. If two sine waves of equal 1.54-V peak-to-peak input are applied to the amplifier without predistortion, the resulting intermodulation spectrum will be that shown in Fig 10.49. Here the intermodulation products are about 31 dB below the peak output; this is probably typical of the levels found in linear power amplifiers.

Next a mathematical predistorter was placed ahead of the amplifier. It is a simple polynomial device that has an output/input relationship:

$$V_a = V_i \times (1.0147 - 0.0409 v_i^2 + 0.1930 v_i^4) \qquad \text{Eq 10.13}$$

The coefficients for this predistorter were found by curve fitting with a spreadsheet program to be close to the inverse of amplifier distortion. The squared and fourth power terms treat the positive and negative waveform values in an identical manner, which is a computational convenience. This is only an example of a predistorter polynomial. The selection of the polynomial complexity, or choosing a different form of predistorter, is all part of the design process.

Fig 10.50 shows the input waveform for

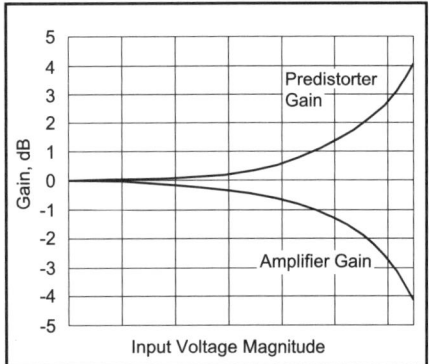

Fig 10.46—Amplifier (lower graph) and predistorter gain characteristics. The two devices are cascaded to result in a net gain that is always 0 dB. The gain of the devices is shown as 0 dB for low-levels, which is not usually the case and these, should be thought of as relative gains.

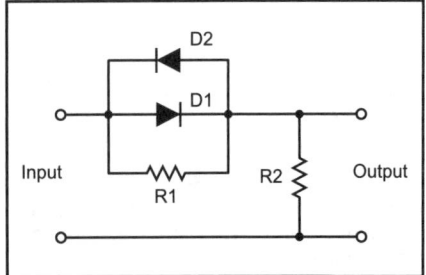

Fig 10.47—Schematic diagram of a simple gain expanding predistorter. This analog circuit is constrained by available diode types, but does provide a general gain characteristic that is opposite to that of amplifier gain compression.

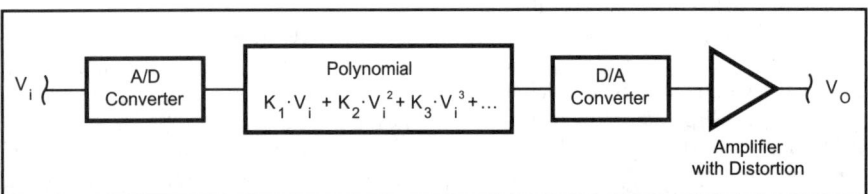

Fig 10.48—Block diagram of a gain expander that could be implemented in a DSP system. The A/D and D/A converters are shown to emphasize the points where the signal has a digital form. In general, it would be combined with other digital blocks. As the complexity of the polynomial gets greater, the potential for reducing distortion improves.

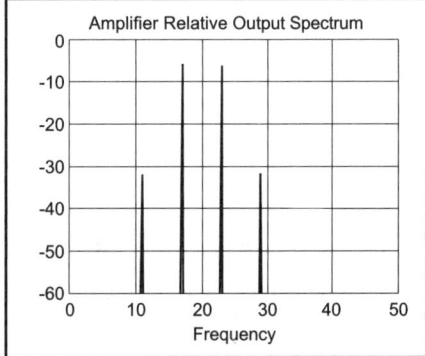

Fig 10.49—Amplifier output spectrum showing the two desired signals at frequencies of 17 and 23 and the third-order intermodulation products at frequencies of 12 and 29. These frequencies were chosen to be easy to simulate, but the results apply generally to any two-tone test frequencies. There are no intermodulation products of order higher than three, for the amplifier as it was modeled.

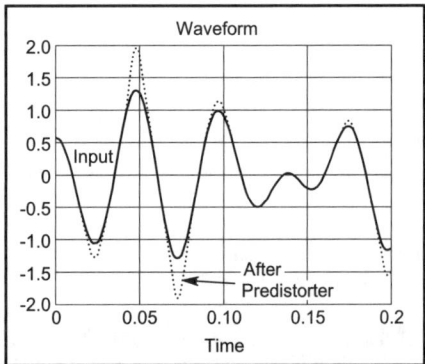

Fig 10.50—Waveforms before and after the predistorter. Only the extreme voltages are increased by the predistorter. This increases the drive to the amplifier to overcome the amplitude compression in the amplifier.

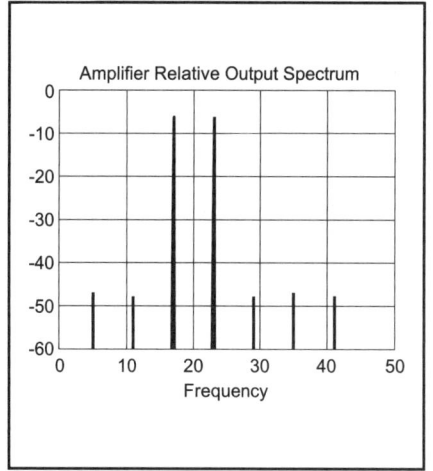

Fig 10.51—Output spectrum for the same amplifier as used in Fig 10.49, except with the predistorter ahead of the amplifier. The third-order products have been reduced by about 17 dB. Fifth and seventh order products can be seen on either side of the third-order products. The predistorter and its interaction with the amplifier characteristics introduced these.

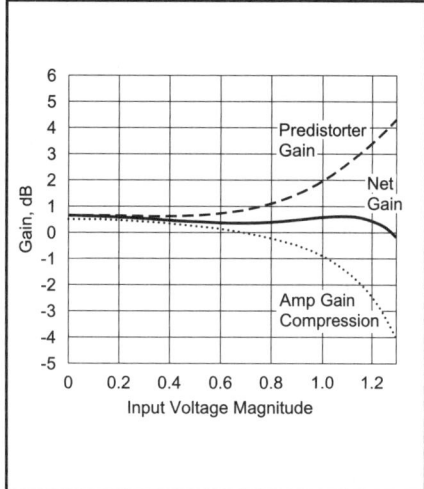

Fig 10.52—Simulated amplifier and predistorter gain characteristics. The predistorter has been designed to minimize the error in the net gain for voltages from 0 to 1.25. All voltages are referenced to the input to the predisorter, and the input to the amplifier can be greater due to the predistorter gain expansion.

the simulated amplifier, both with and without the predistortion. For small signals the predistorter has no effect on the waveform. This seems reasonable, since small signals tend to have very little amplifier distortion. As the signal levels exceed 0.5 V the effect of the predistorter becomes significant. The drive level is increased considerably on the waveform peaks. As the amplifier output tries to compress, the predistorter drives it harder to bring it back to linearity. **Fig 10.51** is a plot of the resulting amplifier spectrum when the two desired outputs have the same level as for Fig 10.49. The third order intermodulation products are now about 48 dB below the peak output, an improvement of 17 dB.

The gain characteristics for this example are shown in **Fig 10.52**. The amplifier gain is down about 2.6 dB for an input level of 1.20 V. For this same level, the predistorter has a gain increase of 2.6 dB and the net gain is about 0 dB, representing no distortion. Below this level, the correction is not perfect, but stays within about 0.2 dB of 0 dB.

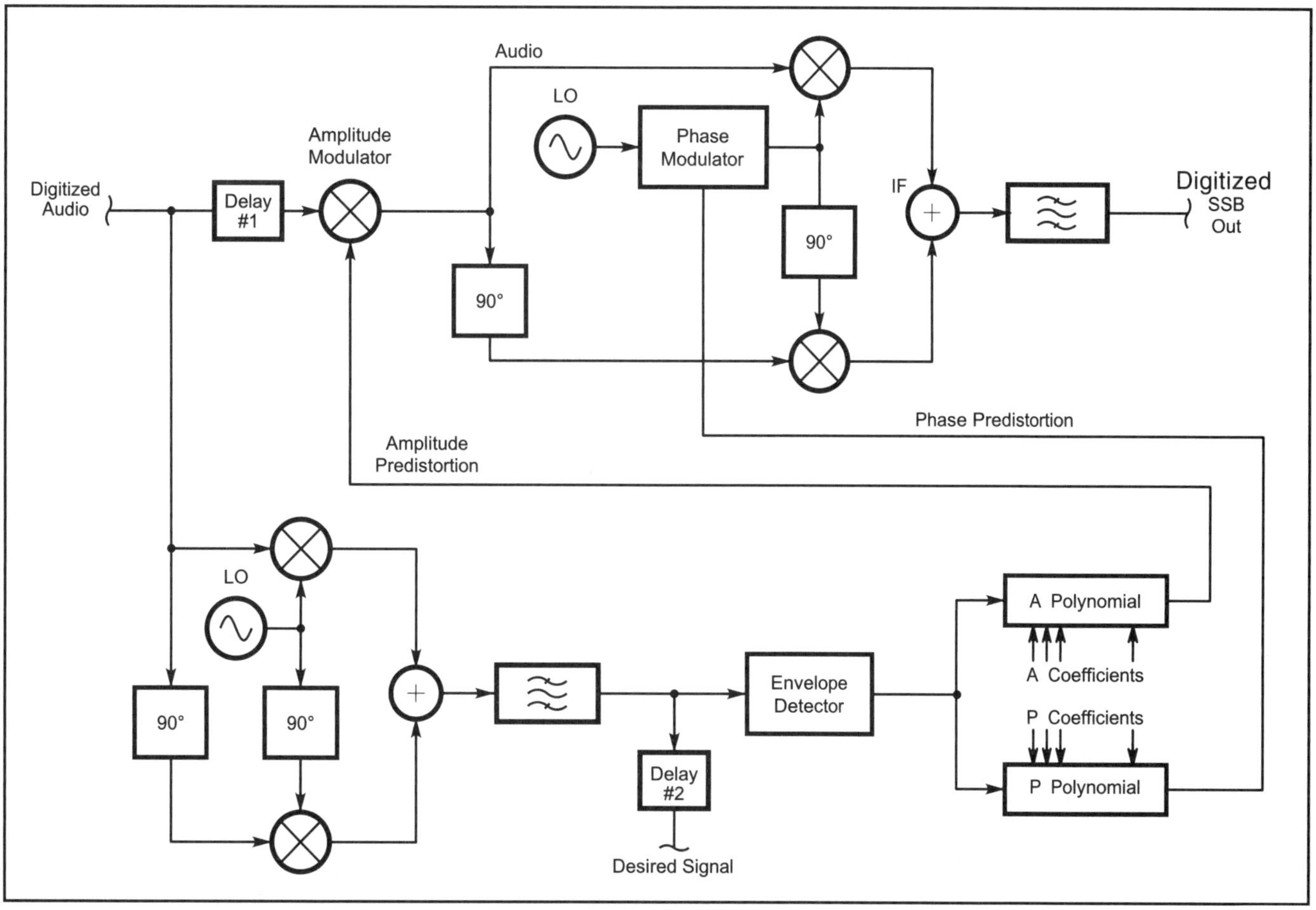

Fig 10.53—Block diagram of an SSB transmitter with predistortion in both amplitude and phase. The lower portion of the diagram is conventional phasing type of SSB generator that serves to determine the desired envelope amplitude, which determines the polynomial predistortion. All components shown are implemented in DSP.

If this predistorter was applied to a real amplifier, the results would be disappointing. This is because we have built a paper amplifier that has no phase distortion at large signal levels. Transistor amplifiers are not this simplistic and require correction for phase as well as for amplitude. However, the technique shown above works equally well for phase corrections. A polynomial of the input voltage can be used to determine the needed phase predistortion. **Fig 10.53** is a block diagram of an SSB transmitter with both amplitude and phase corrections being applied. It is necessary to know the envelope of the desired signal and the lower SSB generator in the figure serves this purpose. Amplitude and phase modulation for the predistortion can be applied to a second SSB generator as shown. All local oscillators (LO) are at the frequency of the (suppressed) transmit I-F carrier.

In general, it is not satisfactory to use a fixed set of coefficients for the polynomials. Time, temperature, load impedance and other factors will change these. This suggests a feedback process that can observe the success of the predistorter and attempt to improve this by changes in the coefficients. The first step in such a process is to make a measurement of the amplifier output distortion. This could be a spectral analysis of the output spectrum, since we desire to not have any power outside a particular frequency band. The spectral analysis can be done by converting the frequency of the amplifier output back to a low frequency and applying a DFT to the signal, using DSP. Alternatively, one could take the converted signal and compare it with the desired signal in Fig 10.52, attempting to make the amplifier output a multiplied replica of the drive signal. This again is straightforward in a DSP implementation, but one must allow for delays and constant phase shifts that occur in the amplifier.

Next, a process for changing the predistortion polynomial coefficients must be designed. This can proceed at a slow rate relative to the changes in the transmitted signal. It is only necessary to follow temperature or other long-term affects. A number of sophisticated procedures exist for determining the coefficients.[21] But, it is possible to get good performance from operations as simple as trial-and-error. This, easy-to-follow procedure changes one of the coefficients by a small amount and then observes the amplifier output. If the distortion is reduced, the change is left. If not, a trial in the opposite direction is made. A lack of improvement at this point means that the original coefficient was satisfactory. Then the procedure repeats the steps for the next coefficient. So long as the starting coefficients are not totally unreasonable, this will normally progress to the optimum set of coefficients.

Fig 10.51 shows that 5th and 7th order intermodulation products have been introduced by the predistorter. These high-order products are potentially more harmful than the original, but larger, 3rd order product. The high order products are controllable in amplitude by a combination of the operating level and the predistorter design. Care should be taken to evaluate these effects.

Predistortion systems can be seen to have some complexity in their operation. But the rewards are quite great. Not only does the amplifier distortion reduction mitigate "spectrum pollution," but the efficiency of the amplifier is effectively improved.

REFERENCES

1. D. Smith, *Digital Signal Processing Technology*, ARRL, 2001.

2. P. Horowitz and W. Hill, *The Art of Electronics*, Cambridge University Press, 1989, Chapter 9. This is a discussion of A/D converters including sigma-delta.

3. D. Garcia, "Precision Digital Sine-Wave Generation with the TMS32010," paper #8 in *Applications Manual, Digital Signal Processing with the TMS320 Family, Theory, Algorithms and Implementations, Volume 1*, Texas Instruments, 1986. This gives a good discussion of the approximation tradeoffs associated with lookup tables. Program listings are specific to the TMS32010, but the discussion is quite general.

4. *Digital Signal Processing Applications Using the ADSP-2100 Family, Volume 1*, Prentice-Hall, 1992.

5. D. J. DeFatta, J. G. Lucas, W. S. Hodgkiss, *Digital Signal Processing: A System Design Approach*, John Wiley, 1988. This is a great book, if you are comfortable with some college level math, but it is not a math book like some DSP books!

6. W. H. Press, S. A. Teukolsky, W. T. Vetterling, B. P. Flannery, *Numerical Recipes in C*, Cambridge University Press, 1992. This book discusses the background, implementation and limitations of the method, as well as a large number of computer methods for numerical calculations.

7. P. Horowitz and W. Hill, *The Art of Electronics*.

8. See Reference 4.

9. W. Davenport and W. Root, *An Introduction to the Theory of Random Signals and Noise*, McGraw-Hill, 1958, Ch. 5. The Central-Limit Theorm of statistics states that under some very general conditions, the sum of a number of random variables approaches the Gaussian distribution as the number gets large. Most college level statistics books cover this theorm as well as signal analysis books such as this one.

10. *The ARRL Handbook for Radio Amateurs*, ARRL, 2002. Chapter 18 contains an introduction to the Fourier transform.

11. The FIR filter design program is included on the CD-ROM for this book as FIRDES1.BAS. The Basic program will run on most Basic interpreters such as have been included with DOS and Windows™ operating systems up through Windows98™.

12. J. Forrer, "A DSP-Based Audio Signal Processor," *QEX*, September, 1996, pp 8-13.

13. U. Rohde, personal correspondence with Wes Hayward, 1997.

14. *The ARRL Handbook*, reference 10 above, includes examples of several types of FM detectors.

15. Reference 4, Chapter 4 includes an Arctangent routine that could be used as the basis for an FM detector.

16. E. O. Brigham, *The Fast Fourier Transform*, Prentice-Hall, 1974. For those comfortable with the concepts of calculus, this is a wonderful reference book. The Discrete Fourier Transform properties and the "fast" implementations are both well covered. Similar material is covered in R. W. Ramirez, *The FFT Fundamentals and Concept*, Prentice-Hall, 1985. In addition, there is a summary of the DFT in the *ARRL Handbook*, Reference 10 above.

17. Chapter 6 of Reference 4 contains a variety of FFT routines.

18. Section 14.5 of Reference 4 contains an implementation of the Goertzel algorithm for DTMF decoding.

19. R. Larkin, "The DSP-10: An All-Mode 2-M Transceiver Using a DSP IF and PC-Controlled Front Panel," *QST*, in three parts, Sep 1999, pp 33-41; Oct 1999, pp 34-40; Nov 1999, pp 42-45.

20. See Reference 5.

21. T. R. Cuthbert, Jr., *Optimization Using Personal Computers With Applications to Electrical Networks*, John Wiley & Sons, 1987. This book covers the mathematical side of *optimization* and is good for those wanting to spend some time on the subject. Knowledge of Calculus and Linear Algebra is required to fully use the material, but BASIC programs and examples are provided for those who wish to approach the subject experimentally.

CHAPTER 11

DSP Applications in Communications

In Chapter 10 a number of DSP building blocks, such as oscillators, filters and modulators were explored. In many cases the blocks were alternatives to traditional analog functions, while in other cases, such as the discrete Fourier transform, we are introducing functionality that was not previously practical. In this chapter, we will explore methods for combining several blocks to produce a piece of communications equipment. We will be integrating three types of functions:

• Traditional analog components, such as RF amplifiers and RF mixers.
• DSP components, such as were covered in Chapter 10.
• Controls for both of these types of components. Most often this is associated with operator interaction, involving both displays and interface controls.

The control of the communications equipment can usually be improved by some sort of computer, which is often a dedicated microprocessor. This may be a good approach, depending on the complexity of the devices. An alternative, however, is to use the same DSP device that is processing signals to do the control functions. This approach will be used several times in this chapter, with the result of needing less total hardware and only a single computer program.

The journey of an experimenter who decides to investigate these DSP projects will begin with the EZ-KIT Lite from Analog Devices. The first things that might be done with this DSP board are simple demonstrations such as audio filters, which are well described in the manuals supplied with the board. Several of these can be tied into an existing receiver and used directly for on-the-air experiments.

This chapter focuses on the processing of signals, but before getting to that we need to look at some basic control techniques. The first issue we will address is that of computer interrupts, which are fundamental to having the DSP programs operate in synchronism with the attached hardware.

All the DSP programs needed to bring life to these projects are included on the CD-ROM with the book and are not repeated in the text. Shown in this chapter are a few fragments of the programs to illustrate a number of detailed operations. It is recommended that the reader look at the complete program, on occasion. This gives a "big picture" view of combining fragments into a working DSP program.

11.1 PROGRAM STRUCTURE

All computer programs have some form of overall structure, ranging from trivial to excessively complex. Often times the structure is largely determined by a group of programs, collectively referred to as an *operating system*. For a PC, this constrains all programs to certain conventions while allowing multiple programs to share resources, such as memory or processor time. To the person writing a program this can be both a convenience as well as a source of anxiety. Having a set of subroutines available to handle standard operations can speed up program writing. However, if there are multiple users of resources, there may be no guarantee that a particular program will finish its task when needed. "Real-time" programming becomes problematic under these circumstances.

For simple DSP programs, it is often possible to operate with no real-time operating system. All resources are allocated when the program is designed. The overhead of the operating system is avoided and the programs are guaranteed to complete their tasks on time. All the programs in this chapter will use this approach and have the same structure. This consists of a background program that processes all data that has no time deadlines, and a single *Interrupt Service Routine* (ISR) that includes all routines that must be completed on a periodic basis.

Interrupts

As discussed in Chapter 10, data processing devices require some method to change the program operation, based on some electrical input. Called interrupts, these methods involve some internal dedicated hardware to make changes to the processor state. Normally the minimum operation is a change in the address of the program being executed. The programmer must have placed appropriate instructions at the interrupt-altered address.

A complication for interrupt programming is the potential for multiple interrupts. For example, in a DSP program, these might be an operation to output data to a D/A converter and a need to output serial data to a serial port. The first interrupt might come from the D/A converter and the second from a hardware timer that is often built on the same IC as the DSP device. The programmer must ensure that these two interrupts will be processed correctly, regardless of when the interrupts occur, including the case of one interrupt occurring while a second interrupt is being processed. For our example, the data to the D/A converter must be processed before the next D/A request is received. If this is not done, there will be large amounts of signal distortion associated with a missing data output.

A simple plan that ensures a minimum amount of time will be available for interrupt processing is to use only one interrupt that occurs on a periodic basis. Although this may require some planning to accommodate all processes, the simplicity of this scheme opens additional interrupt processing time in two ways:

• There is no possibility of two interrupts occurring at the same time and therefore no worst-case timing constraints to allow all processes to be finished

• No communication is required between processes about tasks that need to be performed. That is, the operating system is built-in to the program at design time.

If there is only one interrupt, all interrupt processing should be completed in one period, leaving the system free at the time of the next interrupt. This is the way that communication between tasks is minimized. This processing should include everything that needs to be completed before the next interrupt.

Additionally, any process that does not need to be completed before the next interrupt should be placed into the background process. Examples of this are the updating of data for a display or the reading of a knob or a switch. Again, these processes can be arranged in a sequential order with no need to monitor the time increment needed. So long as the interrupt process leaves any time at all, the background will be processed. Determining whether this is happening at a fast enough rate can be done at design time. It will only happen more slowly if it is being monitored by some part of the process.

Fig 10.6 in the previous chapter illustrates the single timed interrupt structure used for all of the projects in this chapter. Even more elaborate processes, such as the DSP-10 transceiver (only outlined in this chapter but included on the book CD), will continue to use the same structure.

11.2 USING A DSP DEVICE AS A CONTROLLER

The "S" in *DSP* is for *signal*, and one usually thinks of such a microprocessor as being for signal handling functions. However, applications usually need some form of control functions, in addition to processing signals. As will be seen it works quite well to use the same processor for control purposes, resulting in an overall reduction in hardware and software complexity by eliminating the need for a separate controller and the associated interfacing. All of the control program can be implemented as a background activity that essentially runs on a "time available" basis. This way the time critical functions such as signal generation or filtering are not affected. The following discussions of the rotary encoder and an LCD panel are examples of using the DSP device as a general-purpose controller.

Rotary Encoder

Simple control functions can use push buttons to communicate our desires to the DSP. But if a numerical value is to be transmitted push buttons can be awkward, and we must look to either a keyboard or a rotary knob as a control device.

A knob is often easier to use for applications such as changing a frequency. Reading the position of a knob is commonly done with a *rotary optical encoder*.[1] This operates by shining an LED light source through an encoding pattern onto a pair of optical sensors. The encoding pattern rotates with the knob. After conversion to logic level signals, the outputs of the sensors take on the pattern shown in **Fig 11.1**. The sequencing of the two outputs, A and B, prevent their changing at the same time. The logic that determines the direction of turning proceeds as follows. If output A and output B are both low, the next change will be to high on output B if the motion is clockwise. If instead, the next change is to high on output A, it would indicate counter-clockwise rotation. For all four combinations of high and low, we can make a similar determination by examining the figure.

Once the direction of rotation is determined, a counter can be increased or decreased at each transition. Implementing this counter with digital hardware is a possibility, but the example here uses a DSP software implementation. The counter output can control the frequency of an oscillator or other such functions.

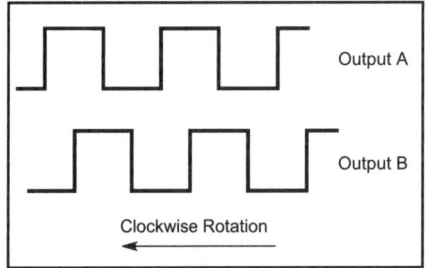

Fig 11.1—This diagram shows the logic levels that occur at the two rotary encoder outputs, as it rotates. At no time do both of the outputs change levels simultaneously.

Three-Wire Serial Interfaces

Serial hardware interfaces are common for communicating between devices. This simple interface is often implemented using three wires, a **data** wire, a **clock** wire to tell when the data is valid and a **latch** wire to tell when the new serial data should be used. This is compatible with shift registers used as receiving devices. Since the data is never used until a latch signal is applied, it is possible to share data and clock lines, as will be seen below. In addition, serial devices are often built to be cascaded allowing multiple devices to be talked to with a single set of wires.

An example of expanding the serial interface to multiple devices is **Fig A** which uses two cascaded shift registers to double the number of parallel outputs to 16. The QH' output is intended for cascading the shift registers. The number of outputs can be increased this way without limit other than the increase in time required to make a change in the outputs.

Many standard functions, in integrated-circuit form, are available with a serial interface. Examples are frequency synthesizers, A/D converters and D/A converters. Often it is possible to cascade

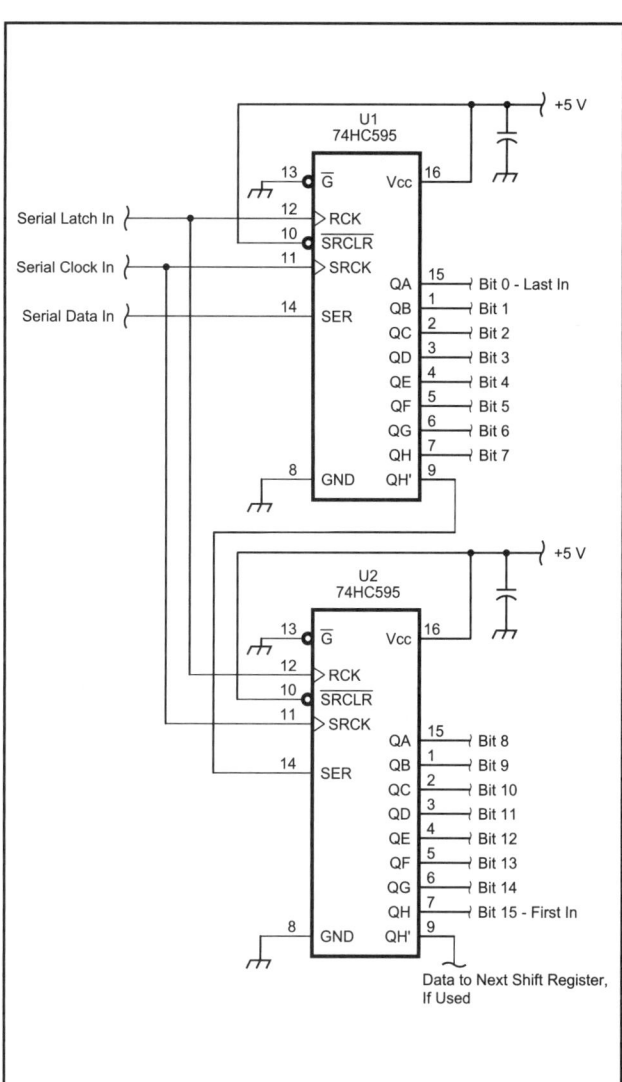

Fig A—Schematic diagram of two cascaded serial-in/parallel-out shift registers providing 16 logic level outputs.

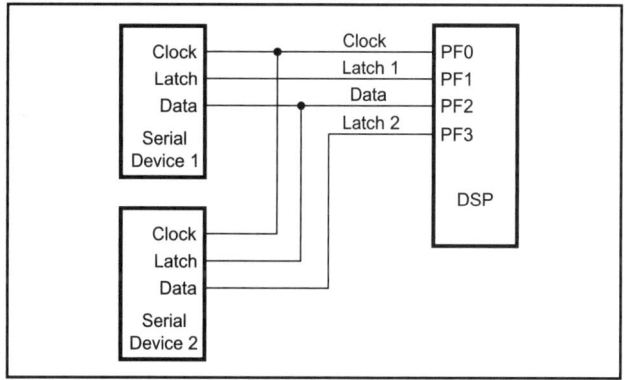

Fig B—Schematic diagram of two cascaded serially programmed devices requiring only three wires from the controller.

Fig C—Schematic diagram of two serially programmed devices sharing data and clock wires, but having individual latch lines.

serial devices using a common set of three serial programming lines. This requires more clocking events per program, but the time for this activity is often available.

For example, **Fig B** shows a serially programmed National LMX1501A frequency synthesizer cascaded with an 8-bit shift register. The shift-register arrangement is identical with that of Fig A, except that the cascading output QH' is used to send data on to the frequency synthesizer IC. The data passes through the shift register and on to the internal shift registers of the synthesizer. Common clock and latch lines are used for both devices. We need to be careful that all timing constraints for the various devices are met. An example of such a constraint is the RC network on the data line going into the synthesizer. This provides a delay of about a half microsecond, guaranteeing that the synthesizer has clocked in the data from QH' before it changes due to the clock signal. Some devices may clock fast enough for the network to not be needed, but this must be examined on an individual basis.

Sometimes the time required to program a very long serial stream is excessive, or the serially programmed device may not have an output to support cascading. For these cases, it is possible to share data and clock wires, but to have separate latch wires as is shown in **Fig C**. The data is clocked into both devices at the same time, but only the device receiving a latch signal will act on the data.

The three-wire interface is quite flexible in its usage. In many cases it is the only form for which a particular device may be available. However, in some sense it transfers the simplicity of the interface back to the software that provides the drive. This generally is a satisfactory result since wiring up parallel interfaces with 8, 16 or possibly more wires is very repetitious and not as challenging as software!

The particular encoder used here was a Clarostat 600EN-128 with a resolution of 256 changes per rotation. A variety of encoders are available most of which can be adapted to this application, as well as the possibility of a home-built encoder as described in Reference 1.

Many possibilities exist for connecting the rotary encoder to the processor. **Fig 11.2** illustrates one of the simplest ways to accomplish this. Here the two encoder outputs are connected to *Programmable Flag* inputs, PF0 and PF1. These inputs are part of a set of 8 pins that are dedicated to input and output of digital data (I/O). Within the processor these pins can be defined as either inputs or outputs by writing to a memory-mapped register. Once this is done the pin logic levels can be read from a second memory-mapped register. The only constraint on this implementation is the limited number of pins available.

Expansion of the number of digital I/O lines can be accomplished by connecting flip-flops to what is referred to as *I/O Space*. This allows 16 bits to be read (or written) at a time and requires minimal support hardware. An alternative is to continue using the Programmable Flags, but adding serial-to-parallel conversion hardware (shift registers) as is illustrated in **Fig 11.3**. A major advantage of this scheme is its compatibility with multitudes of serially programmed devices (see sidebar "Three-Wire Serial Interfaces"). Referring to Fig 11.3, there are three lines, *data, clock* and *latch,* to transmit the serial data from the processor to the shift register. **Fig 11.4** shows the timing diagram for producing 8 bits of parallel data from the shift register. The data line sets the value of the individual bits. After the data line has achieved a well-defined value, the clock makes a zero-to one transition that loads the current data value into the shift register. This is repeated a total of 8 times, at which point the entire 8-bit byte has been loaded into the shift register. The order of the shift register is such that the most significant bit (Qh) is the first bit in, and the least significant bit (Qa) is the last bit into the shift register.

To this point, we have converted serial data from the processor into parallel data lines. If we are to read the logic levels of a multiplicity of external lines, it will easily use up the free programmable flag lines. One simple interface that is particularly suited to occasional reading of lines is the digital multiplexer. Figure 11.3 shows the 8-input multiplexer using a 74HC151 IC. The particular line that is to be read by the processor is selected by the 3-bit address coming from Qa, Qb and Qc of the shift

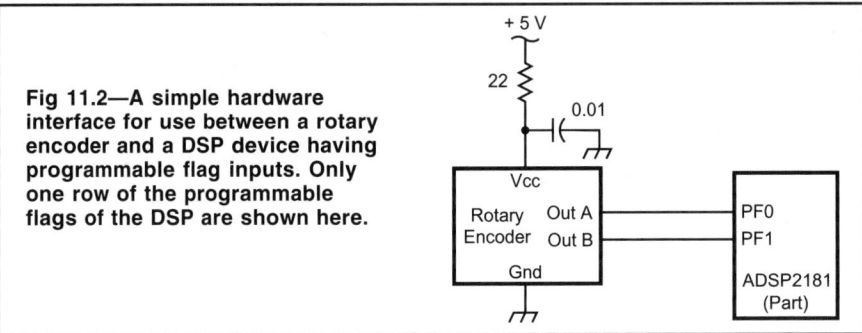

Fig 11.2—A simple hardware interface for use between a rotary encoder and a DSP device having programmable flag inputs. Only one row of the programmable flags of the DSP are shown here.

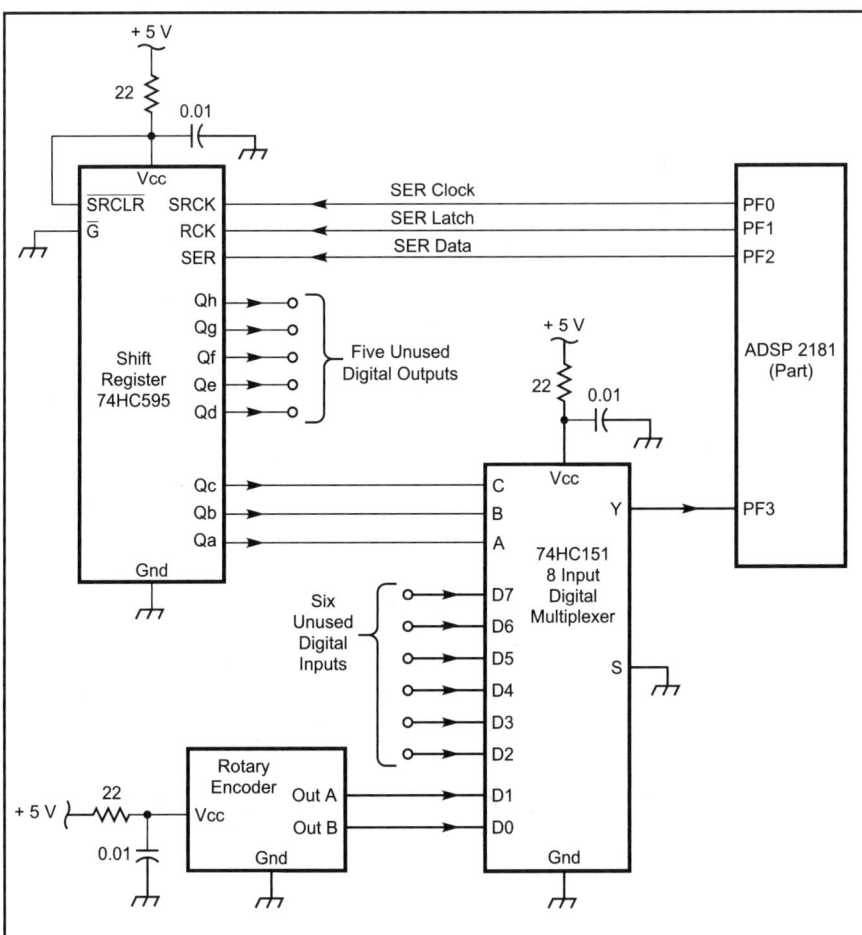

Fig 11.3—An alternative approach to expansion of the number of digital I/O lines is the addition of serial-to-parallel conversion hardware as shown here.

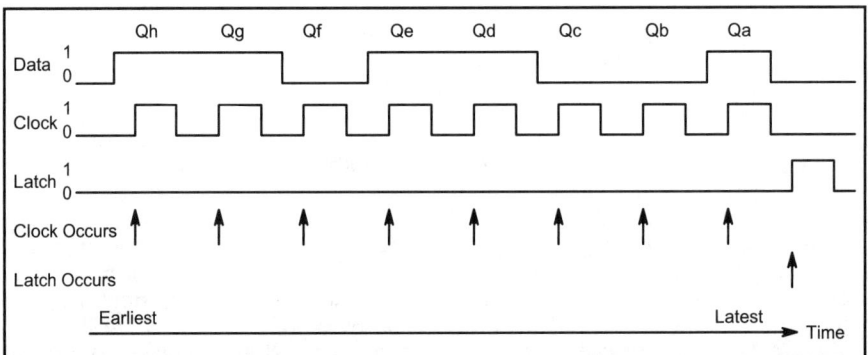

Fig 11.4—Timing diagram for loading the eight-bit 74HC595 shift register with an example binary value of 11011001. Both clocking and latching occur when the signals go from logic 0 to logic 1.

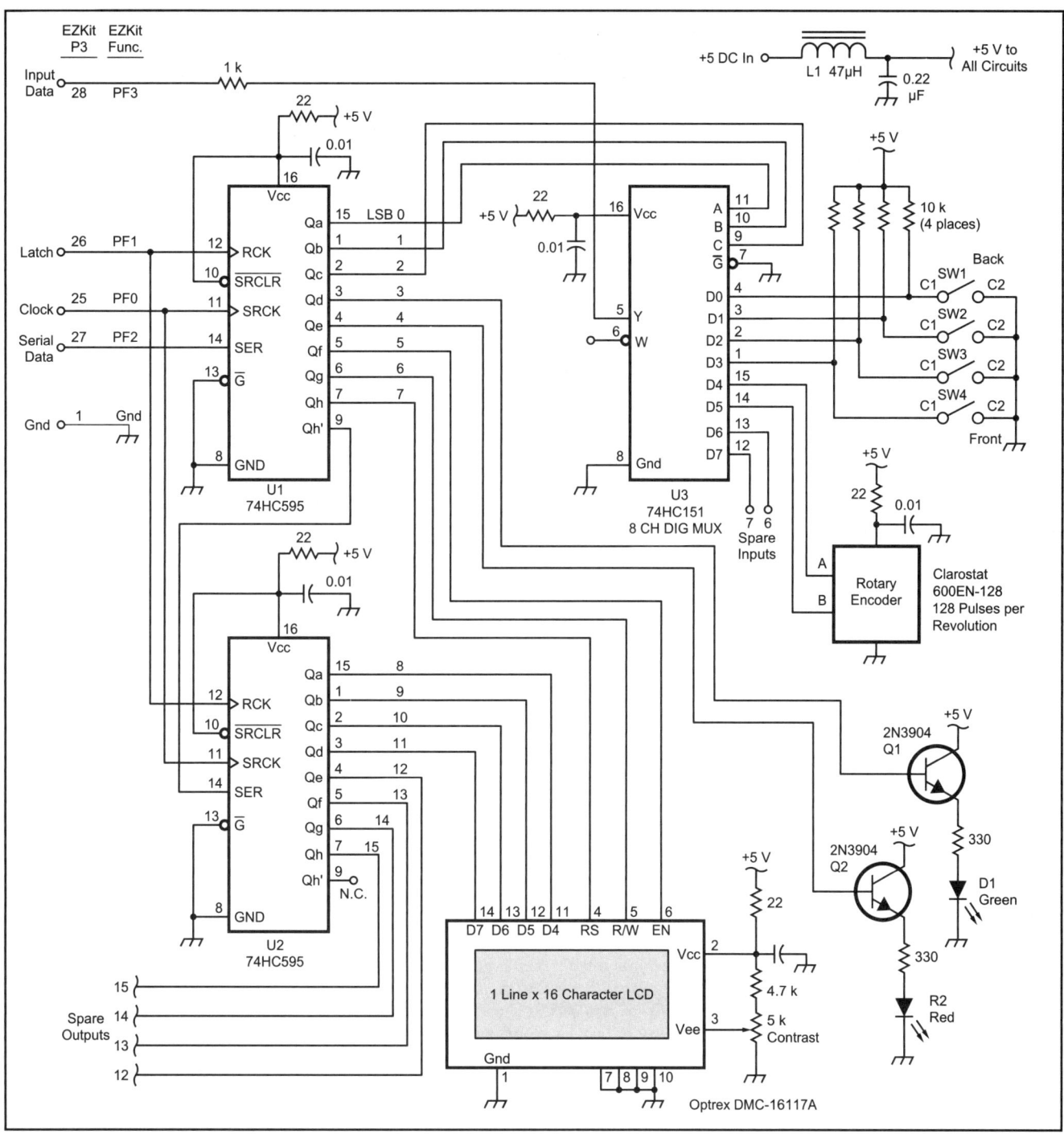

Fig 11.5—Schematic diagram of the hardware interface between a DSP device and multiple control devices, including a rotary knob, four push buttons, two LED indicators and an LCD display.

register. The output of the multiplexer goes to the processor pin PF3. This is programmed to be an input pin during the initialization of the processor.

As a final step in the evolution of control box schematics, **Fig 11.5** shows a complete interface including the rotary encoder for the knob, four push buttons, two LED indicators and a 16-character LCD panel. Four of the parallel inputs are used to read the state of the push buttons. The two LED indicators are driven by simple emitter followers, Q1 and Q2, from two of the parallel outputs.

The LCD panel has several options for an interface. Rather simple is the seven-wire arrangement shown in Fig 11.5. Four wires are for data that can be sent a half-byte at a time and the other three wires control the reading of the data by the LCD. All seven wires come from the parallel output interface produced by the shift registers U1 and U2. The control of the LCD panel will be discussed further below when we look at the methods for using the DSP as a control device.

Programming the Rotary Encoder

A complete example program for the rotary encoder is *C11KNOB.DSP*, included on the book CD. The software is centered on a routine, *knob*. This routine compares the two bits that describe the current knob state with those for the previ-

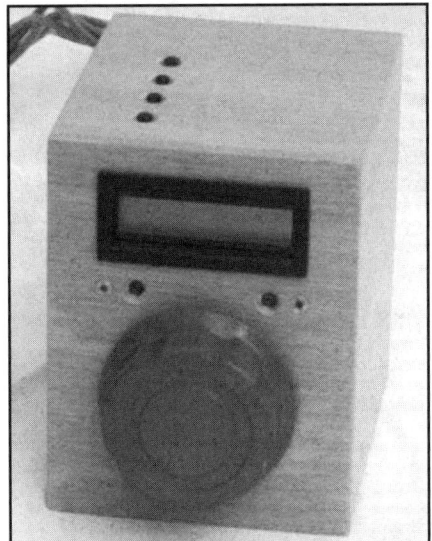

The knob box was built from thin plywood. An inner box made from scrap circuit board material contains the logic circuitry shown in Fig 11.5. The four push buttons are placed on the top of the box as a convenience in using the box. It is light enough that it wants to move when the buttons are pushed! The LCD display is above the knob. A plastic bezel trims off the display.

Box 1 - DSP routine to determine knob rotation using a lookup table. The output in ax0 is –1, 0, or 1 for counter-clockwise movement, no movement or clockwise movement.

```
knob:
     ay0 = 4;  call inbit;        { LSB of knob state, in ax0 }
     mr1 = 0;                     { In case bit 3 of ax0 = 0 }
     ar = tstbit 3 of ax0;        { Find out }
     if eq jump kn1;              { Yes, it is = 0 }
     mr1 = 1;                     { The other case, bit 3 of ax0=1 }
kn1: ay0 = 5;  call inbit;        { Similar stuff for next to LSB }
     ar = tstbit 3 of ax0;
     if eq jump kn2;
     ar = setbit 1 of mr1;
     mr1 = ar;
kn2:
     ar = dm(knob_st);            { Here with new state in mr1 }
     sr = lshift ar by 2 (hi);    { Knob state at last measurement }
     ay0 = sr1;                   { Move left 2 bits }
     ar = mr1 or ay0;             { 4 bit state }
     dm(knob_st) = mr1;           { Current state for next time }
     ay0 = ^encoder;              { The lookup table address }
     ar = ar+ay0;                 { Get location in the table }
     i4 = ar;  m4 = 0;  l4 = 0;   { The i4 index register gives the }
     ay0 = pm(i4, m4);            { easy way to get a table entry }
     none=pass ay0;               { Set flags, based on table entry }
     rts;                         { With -1, 0, or +1 in ay0 }
```

Box 2 - Lookup table for determining knob rotation

```
.var/pm  encoder[16];       { Rel Adr=Last state*4+New state }
.init encoder:
    0, H#FFFF00, H#000100, 0,
    H#000100, 0, 0, H#FFFF00,
    H#FFFF00, 0, 0, H#000100,
    0, H#000100, H#FFFF00, 0;
```

Box 3 - Program to modify a program variable, amult, using the routine *knob*.

```
call knob;                { See if knob has moved (in ay0) }
ar=dm(amult);             { Alter by either 0, -1 or +1 }
ar=ar+ay0;                { We add, but ay0 may be + or - 1 }
dm(amult)=ar;             { For next time & use by others }
```

ous state and makes one of three choices:
• No Change
• Knob moved counter-clockwise, one count
• Knob moved clockwise, one count

This occurs in the following manner. The inputs come from another routine *inbit* that returns, in register **ay0**, the logic levels of the hardware input lines connected to the 74HC151 digital multiplexer of Fig 11.5. Bits 4 and 5 of **ay0** contain the multiplexer inputs D4 and D5, which are the A and B outputs of the rotary encoder. The previously measured values for these lines are stored in a data memory location dm(knob_st). By comparing the old and the new measurements, it is possible to deduce the knob movement, if any (See sidebar "Using a Table Lookup to Determine Knob Motion"). The implied movement is stored in a 16-member lookup table. This is certainly not the only way to deduce the knob movement, but it has the appeal of being easy to understand. In general, solutions that use a little more memory, but are easy to understand, have much appeal! The entry point to the lookup table is constructed from the old and new knob states by shifting the old state left to bits 2 and 3 and putting the new state in bits 0 and 1. This creates a 4-bit binary number that ranges in value from 0 to 15. All combinations of old and new state are included. The lookup table returns a value of -1, 0 or +1, as shown in **Box 1**.

The lookup table is entered into the program as part of program memory as shown in the snippet in **Box 2**. The encoder table is stored as 24-bit data in pm, but used as 16-bit data in the DSP. The 00s on the end of the hex values are 8 bits, set to 0, that are never used, but are very necessary to make the bits line up when read as 16 bit values.

It is now possible to alter a value, such as the amplitude multiplier for a signal by calling the *knob* routine. As an illustration, we can modify a memory "gain" value called *amult*, as shown in **Box 3**.

More elaborate programming would allow different changes to be made depending on the knob rotation. This could be used for operations such as changing a filter or a frequency band.

LCD Panel

The *liquid-crystal display* (LCD) is convenient for displaying data from our DSP device. These displays range from the

Inside the knob box is a second box for the digital electronics. Pigtail wires run to the EZ-KIT Lite. For this box, a plug was placed on the pigtail wires to allow the same EZ-KIT Lite to be used for other projects. Any type of plug would be suitable.

simple character display to a large matrix with colors. We will only deal with the least complex of these, but the principles required to extend the complexity will be the same. The display shown here has 16 characters, arranged in a single row. Any of the alphanumeric characters and a variety of symbols can be displayed. The particular display used here is the Optrex DMC-16117A, but a variety of products are available from Optrex and other manufacturers. The programming of many of these displays is similar to that shown here. Check the manufacturer's data sheets for the particular panel for details.

Programming the LCD panel through the serial-hardware lines is straightforward, but will appear to be somewhat laborious. The panel requires a sequence of commands be sent to initialize the controller. Once this is done, the individual characters of the display can be set by two byte commands. The emphasis here will be on the general nature of using the DSP as a controller, rather than on the specific procedures for this display. The details of this example are included with the programs for the "Knob Box," along with an application using the box, the two sine wave plus noise generator. Both of these projects are shown later in this chapter.

When a character is sent to the LCD, it is displayed at the left edge, and all existing data on the display are pushed a character to the right. If one wants to write any new character, it is necessary to write all 16 positions in sequential order. For an example, we will display a 16-bit number in decimal form. This will include a leading negative sign if appropriate, or a leading blank if the number is zero or positive. These numbers, in decimal form, can range from –32768 to 32767. Including the minus sign, up to six characters are needed. To simplify the display arrangement, we will always leave room for six characters. We could write a long program routine to convert the number into numeric characters and to load these into the LCD display. Doing this can make a program difficult to follow and prevents reuse of any of the program pieces for other purposes. Writing the program as a collection of subroutines minimizes these problems.

We will now look at some of the details of these five subroutines. For selected portions of the routines, the detailed program instructions are shown. The fully commented source programs are included on the *Experimental Methods in RF Design* CD as part of the program C11KNOB.DSP.

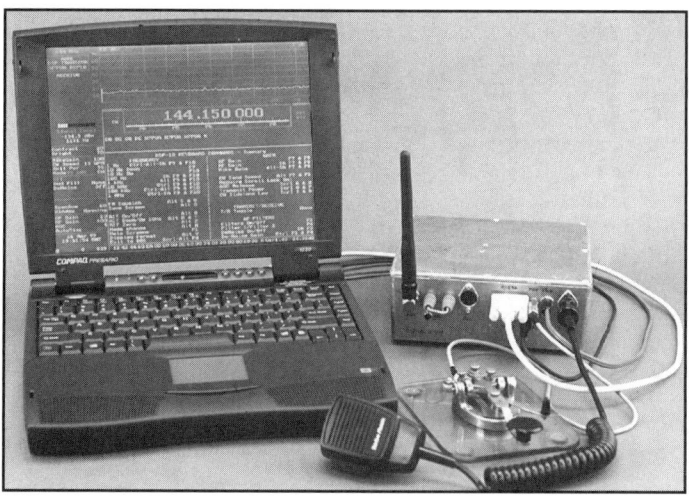

A complete QRP rig for 2-meters, the DSP-10, is built around a minimal amount of hardware and the software running in the laptop PC. Along with the RF hardware in the die-cast box is an Analog Devices EZ-KIT Lite that serves as the last IF and audio portions of the transceiver. See page 11.27 for more information.

Using A Table Lookup To Determine Knob Motion

The table that is stored at the program memory table "encoder" is reconstructed here with the table address offset in binary and the table entries as decimal numbers:

4-Bit Address Offset	Entry
0000	0
0001	−1
0010	1
0011	0
0100	1
0101	0
0110	0
0111	−1
1000	−1
1001	0
1010	0
1011	1
1100	0
1101	1
1110	−1
1111	0

The address offset is shown as a binary number, corresponding to decimal equivalent numbers of 0 to 15. The binary values are the encoder-output logic levels for the last measurement followed by those for the current measurement. All 16 possible combinations are in the table. Relating these to the knob encoder, the binary numbers are B'A'BA where the primed values refer to the last measurements and B and A are the two logic outputs from the encoder.

Some of the address offsets, such as 0101 or 1111, have the same old and new values and correspond to no motion of the knob. All four of this type can be found in the table to have an entry value of 0 indicating "no change."

Next are address offsets such as 0001. Here the B output has remained logic-level 0, but the A output has changed from 0 to 1. Referring back to the encoder logic of **Fig 11.1** it can be seen that only if the knob has counter-clockwise motion is this possible. This results in an entry of -1. In a similar fashion, an offset of 0010 can only occur for clockwise rotation and an entry value of 1 results. If the knob is controlling a value, such as frequency, the new value can result from adding the table entry to the old frequency.

Note that there are four address offsets, such as 0011 or 1001 that should never occur. These correspond to both A and B outputs of the encoder changing at the same time. Fig 11.1 would suggest that this cannot occur. However, if the knob is rotated so fast that a state is skipped over, the 0011 combination may be encountered. This combination tells us that the encoder has changed by two positions, but there is no clue as to the direction. For this reason, the table entry must be zero, meaning that no change will be made.

Converting a Binary Number to Individual ASCII Digits

Fig 11.6 illustrates the programming of the LCD to display a 16-bit signed integer. The subroutine *n2bcd* converts the 16 bit number into six ASCII characters* that are left in a six position array in data memory. Each character is broken into four-bit halves, called nibbles, ready to be sent to the display by the subroutine *outch*. The routine *lcd4* supports *outch* by moving four bits into the shift register using multiple calls of the subroutine *load16*. This subroutine handles the pulsing of hardware lines to move data into the shift register. Completing the needed subroutines is *delay*, slowing the DSP process to ensure that the waveforms going to the shift registers have sufficient time to be correctly formed.

Changing the 16-bit number to 6 ASCII

*Most computer users are familiar with the ASCII character code as the language of text files or serial ports, where 128 different symbols are encoded into 7-bit binary numbers. *The ARRL Handbook* includes the details.

characters was seen to be the function of the subroutine *n2bcd*. This is done by considering each character position in order. If the number is negative, the first position is loaded with an ASCII minus sign. Otherwise it is loaded with a space or "blank" character. The number is then negated if it was negative.

The numeric value to be placed in each character position is determined by repeated subtractions. For instance, for the digit following the sign, we subtract 10,000 (decimal) from it. If this produces a negative result the number must be less than 10,000 and we will put a '0' character in the second table position and move to the 1000s digit. Otherwise we put a one in the second table position and repeat the 10,000 subtraction. This continues through '3', which is the largest value possible for the 10,000s digit, at which point the subtraction must have a negative result. **Fig 11.7** is a flow chart that illustrates this process for the 10,000 digit, and the program fragment in **Box 4** shows these same steps in assembly language.

The second instruction loads the *ay1* register with the ASCII value for the character zero, which is 30 hex or 48 decimal.

This is simpler than counting the number of subtractions and then adding 30 hex to it. Since all of the characters from '0' to '9' are in sequence in ASCII, the results are the same.

The subroutine repeats the same series of subtractions for the 1000s digit, except that here the number of subtractions possible may be as high as nine. This continues through the unit digit, after which all of the six character positions will hold the proper ASCII character. When we humans write a two-digit number in a six-digit space, we leave blanks in the four leading zero spots. These could be converted, but we will keep things simple by leaving these in place since it is not wrong.

This routine demonstrates the complexity occurring when converting a number built on powers of two to one built on powers of 10. For each power of 10, like 10000, 1000, 100,..., subtraction must be used to successively remove the powers of 10. The routine could be shortened by building it out of loops, but generally with the ADSP-2181 program memory is not in short supply. In-line routines, such as used here are often easier to debug and can execute faster than their looped equivalents.

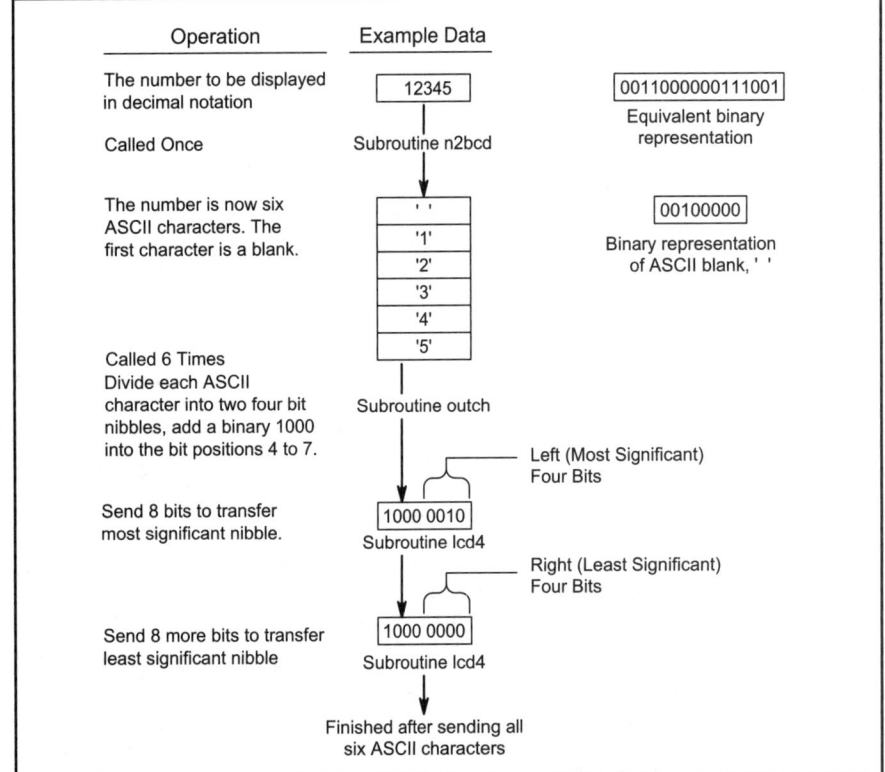

Fig 11.6—Data structures used in converting a 16-bit signed number into a form for sending to the LCD display. Three subroutines are used to break the number into characters, prepare a character for transmission and to send a four-bit nibble as required by the LCD display.

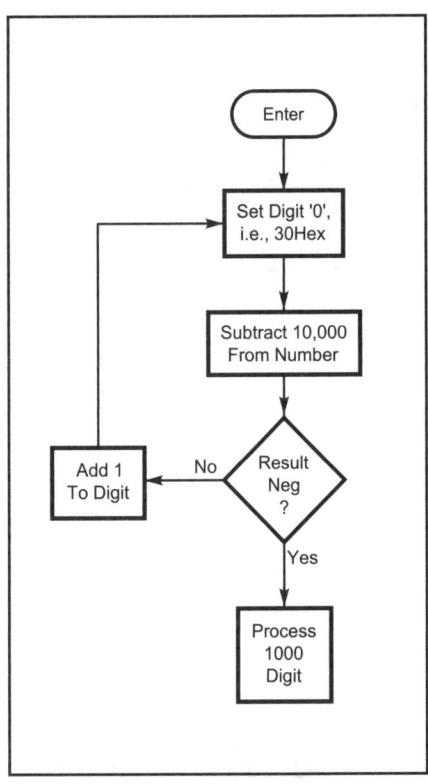

Fig 11.7—Flow diagram of a portion of the n2bcd subroutine, showing the extraction of the 10,000's digit. The digit is converted to ASCII by adding the value 30 hex.

Box 4 - DSP program to determine the ASCII value corresponding to the 10,000's digit.

```
{ The number to be converted to BCD is in data memory dm(temp1) }
      ay0 = 10000;              { Find the 10,000s digit }
      ay1 = h#30;               { '0' to count the subtractions }
n2a:  ar = dm(temp1);           { Test the current reduced number }
      af = ar - ay0;
      if lt jump n2b;           { Done for this digit }
      ar = ar - ay0;            { Not done, reduce working number }
      dm(temp1) = ar;
      ar = ay1 + 1;             { Increase current digit }
      ay1 = ar;                 { This is where it is kept }
      jump n2a;                 { Continue subtractions }
n2b:  dm(digit + 1) = ay1;      { store the ASCII value in memory }
```

We now have six characters in a memory array ready to be sent to the display. This is transmitted to the LCD as nibbles, each containing four-bits of the character. To indicate that this information is display data, a binary one is placed in the left-hand position of the eight. All of this is handled by a subroutine, called *out_ch*.

Going back to the schematic of the display in Fig 11.5, of the 16 bits of shift-register output lines, only seven go to the LCD. So, we need to be careful that sending data to the LCD does not change the other outputs. This is accomplished by using a logical OR instruction with a copy of all the outputs kept in data memory as dm(data16). Other data manipulation steps are needed to be consistent with the requirements of the LCD hardware. The subroutine *lcd4* performs these operations for both nibbles. **Fig 11.8** shows the flow of this subroutine.

The only missing operation now is a method to load the 74HC595 shift registers with serial data (see the sidebar on page 11.2, "Three-Wire Serial Interfaces"). This is accomplished by use of a subroutine *load16*, outlined in **Fig 11.9**. One advantage of this modular subroutine structure is the ability to use this same routine for any operation that requires altering the outputs of the shift registers. The figure and the commented listing on the *Experimental Methods in Radio Frequency Design* CD-ROM can be examined to see the detailed operation. However, one recurring element is to send a pulse on a hardware line. In assembly language sending a positive going pulse typically looks like **Box 5**.

The routine "delay3" does nothing for 3 microseconds. This allows plenty of time for the feed-through filters coming from the PF leads to achieve their full rise. The delay routine could have been written as a loop, such as

```
delay3:
    cntr=97;
    do dly3a until ce;
dly3a:      nop;
    rts;
```

but this has a drawback. There are only four places on the counter stack. Every time a new value is loaded into the "cntr" register, the current value is placed on the counter stack. There is only room for four values on this stack and a fifth attempt will result in counter data being lost. To leave room for other routines, the delay routine uses extra space in program memory to save space on the counter stack. It looks like:

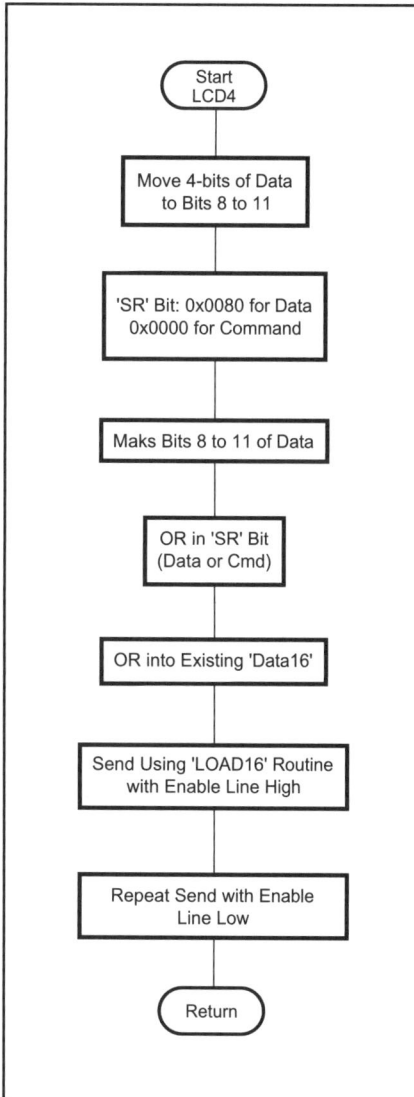

Fig 11.8—Flow diagram for the subroutine *lcd4* that transmits 4 bits of data or command to the LCD panel, while not changing the other outputs of the hardware shift register.

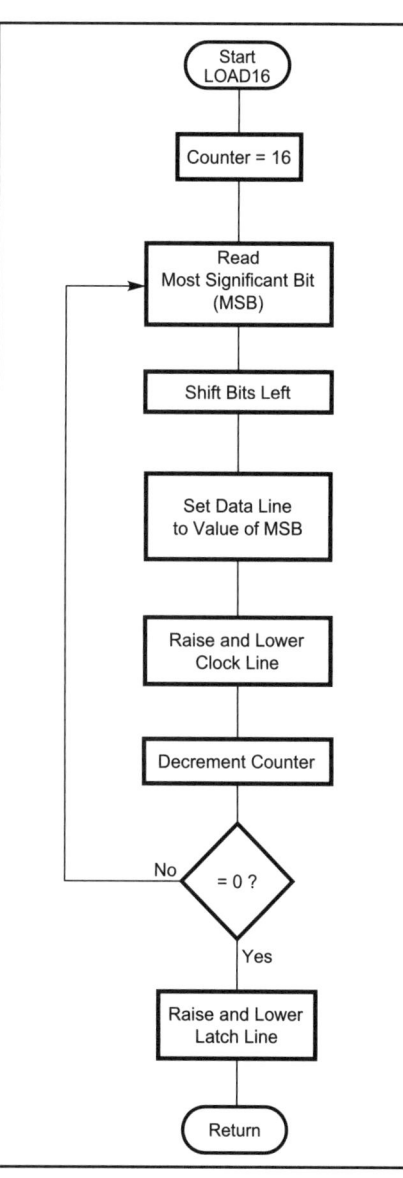

Fig 11.9—Flow diagram of the subroutine *load16*. This transfers 16 bits of data to the hardware shift registers.

> **Box 5 - DSP assembly language to create a 3 microsecond pulse on the hardware line, PF1.**
>
> ```
> { Latch the data with a pulse on bit 1 }
>     ax0 = dm(PFDATA);      { Get the current PF data }
>     ar = setbit 1 of ax0;  { Make bit 1 a 1, it was 0 }
>     dm(PFDATA) = ar;       { Send to hardware, via dm }
>     call delay3;           { Pulse is 1, Wait 3 microseconds }
>     ax0 = dm(PFDATA);      { Get the PF data again}
>     ar = clrbit 1 of ax0;  { Bring hardware line to 0 }
>     dm(PFDATA) = ar;       { Again send to hardware, via dm }
> ```

```
delay3:

    nop; nop; nop; nop; nop;
    nop; nop; nop; nop; nop;

    { ... And 8 more lines of
    NOPs here ... }

    nop; nop; nop; nop; nop;
    nop; nop; nop; nop; nop;

    rts;
```

Either routine performs no function during its execution. If an interrupt occurs during the delay routine, it will only increase the delay time, which will not be harmful.

Returning to the *load16* routine, the memory location dm(PFDATA) is one of a number of dedicated memory locations that are treated as registers.[2] The lower 8 bits of PFDATA correspond to the 8 pins of *Programmable Flag* called *PF0* to *PF7* in hardware terms. These pins can be programmed to be either inputs or outputs. If they are outputs, as we need for the shift register data, clock and strobe, writing to the location dm(PFDATA) will change the pins to the new value. Reading from dm(PFDATA) tells the program the current setting of all pins while writing will set the levels.

The *load16* routine proceeds through all 16 bits by finding from dm(data16) the desired bit value, putting this onto bit 2, and then moving the clock line, bit 0, from 0 to 1 and back. Delays are inserted at each point to make sure that the data arrives before the clock pulse and that all pulses are long enough to reach their full extreme values. Finally the strobe line, bit 1, is moved from 0 to 1 and back, latching the 74HC595 shift-register data by moving it to the output pins.

11.3 AN AUDIO GENERATOR TEST BOX

A device using the capabilities of the Knob Box is the Audio Generator. This provides an output signal from the EZ-Kit consisting of two sine waves and a random noise. This is useful for transmitter testing using either one or two tones. The noise signal can be useful for transmitter testing or for simulating the reception of signals in noise. Each sine wave can have its frequency set to any value from 1 Hz to 20 kHz, and the RMS amplitude can be varied in 0.1-mV (100-microvolt) steps. The noise is always Gaussian and flat with frequency. The noise RMS amplitude can also be varied in 0.1-mV steps.

This audio generator also illustrates the building block assemblage that we are using. The sine wave and noise generators come from Chapter 10 routines, and the knob and LCD hardware and software are those that have just been discussed. In the following section, we will tie these together into a handy test box.

All signals from the generator have great relative-amplitude accuracy. The absolute accuracy of the D/A converter output is only about 10%. This is a scaling error only and can be removed by calibration of the particular converter. Even without an absolute calibration, the signal-to-noise ratio or the ratio of two signal voltages can be set very accurately, typically better than 0.1 dB.

The distortion in the generator output is very low at about 0.025 per cent. Distortion is a much more important parameter for this type of application.

The four button switches on the knob box control the various functions. Button 1 scrolls through the display controlling which of the three waveforms is being controlled:
  Sine wave 1
  Sine wave 2
  Noise
Button 2 selects the knob function:
  Amplitude
  Frequency
Button 3 is left unused to allow for future additions, and Button 4 toggles all outputs between on and off. The red LED indicates the on/off state.

The display has 16 characters, adequate to indicate the generator state. For instance, if Button 1 selects the first sine-wave generator, the display would be

"1 fffffHz vvv.vmV"

where the first 1 means that the data applies to generator 1, fffff is the frequency in Hz and vvv.v is the RMS output level in millivolts.

Fig 11.10 is a block diagram of the soft-

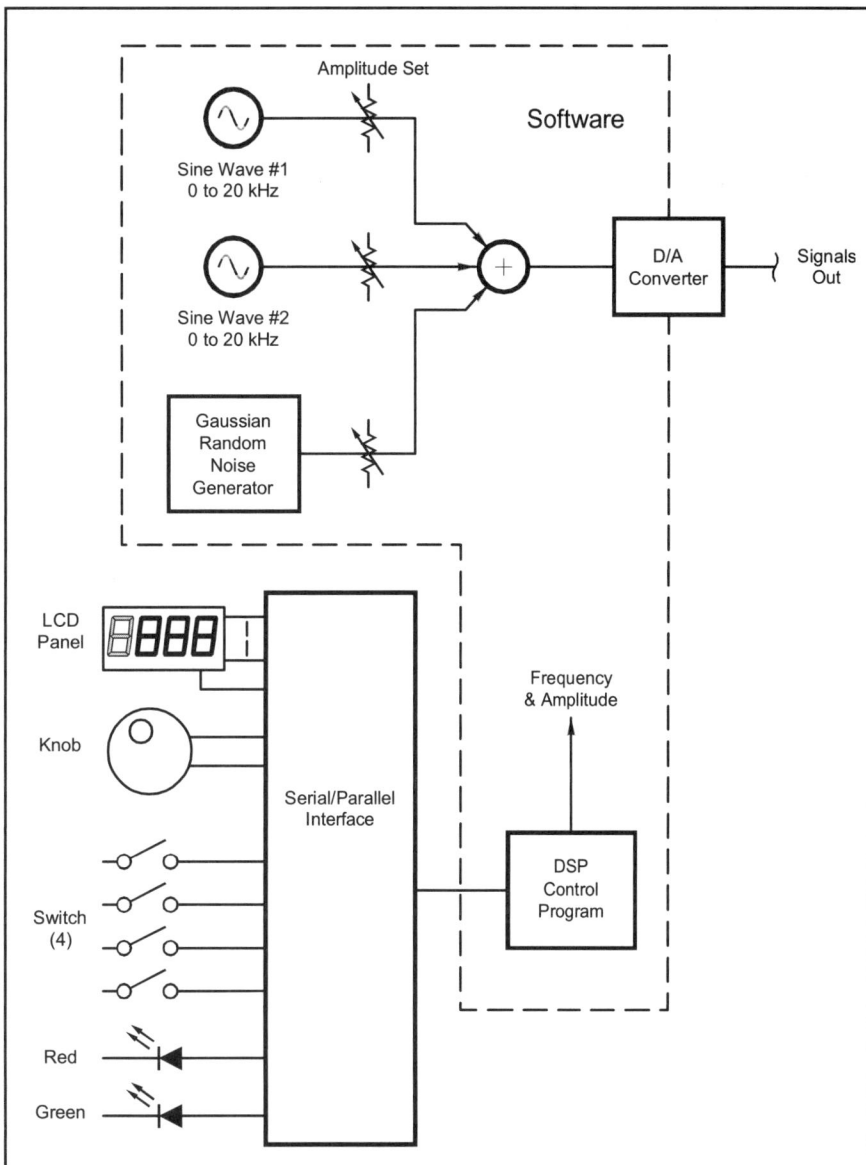

Fig 11.10—Overall block diagram of the tone and noise generator. The knob controls both the frequency of the sine-wave generators and the amplitudes of the three signals. The function of the knob is determined by the push buttons. The 16-character display is also driven by the interface circuitry controlled by the DSP software.

Box 6 - DSP routine to set phase increment for sine-wave generator.

```
{ Frequency in Hz in the ar register. To convert to a phase increment
  we need to multiply by 65536/48000. But in the 1.15 arithmetic, the
  biggest value is 1.0. So, we multiply by FR2PH=0.5*65536/48000=0.6827 and
  then shift left 1 bit, the same as multiplying by 2. }
     .const    FR2PH=0X5762;   { Hex for 0.6827 in 1.15 format }
{ And the code in the main body of the program: }
my0=FR2PH;
mr=ar*my0 (ss);            { The fractional multiply, and }
sr=ashift mr1 by 1 (hi);   {   the multiply by 2, which is }
sr=sr or lshift mr0 by 1 (lo);  { in two parts to get LS bit }
```

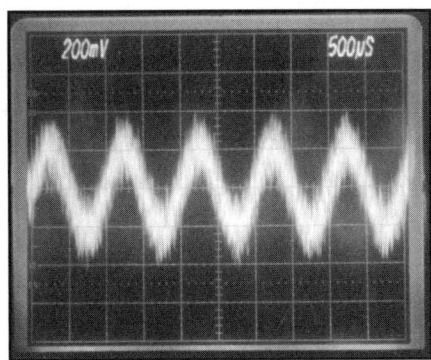

Fig 11.11—Oscilloscope trace of the Audio Generator output. One sine wave is set to 150-mV RMS and the other to zero. The noise level is 50-mV RMS making the S/N 9.5 dB (20*log(3)). The sine-wave frequency is 1000 Hz.

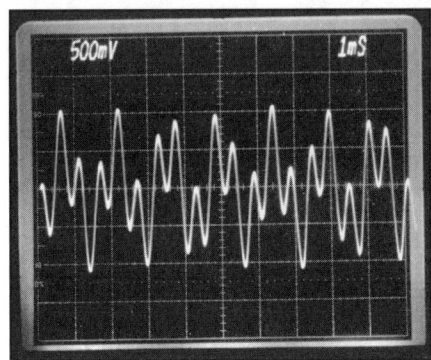

Fig 11.12—Oscilloscope trace of the Audio Generator output. The sine-waves are of equal amplitude and the frequencies are 700 and 1900 Hz. The noise is set to zero.

can be seen in the full listing that is available in the program *c11tbox.dsp* on the CD-ROM that accompanies this book. The more interesting areas are the details that must be handled to make the signal generator operate properly.

For instance, the display for frequency is in integer Hz, from 1 to 20,000. The sine-wave generator has a resolution of about 0.73 Hz. The knob could be used to change frequency in either in steps of 1 Hz or 0.73 Hz. Either way, a conversion must be made to the other resolution step. The method used was to always change the desired frequency by 1 Hz, and then to convert this to a phase increment corresponding to the 0.73 Hz step. This results in the knob always producing a visible frequency change on the display, but about 1/3 of the possible generator frequencies are not used. The conversion from a frequency in the **AR** register to a phase increment in the **SR1** register is as follows in **Box 6**.

Figs 11.11 and **11.12** are example waveform outputs from the Audio Generator. Output levels and frequencies are shown in the captions.

ware and hardware functions involved. The individual functions, such as sine-wave generation, knob control and LCD display have all be covered earlier and will not be repeated here. The details of the integration of these program components

If the D/A converter is operated below its overload point the distortion, including intermodulation, can be expected to be very small. The principle drawback to this approach is the limited frequency range. For the hardware used here it is not practical to operate much above 20 kHz.

11.4 AN 18 MHZ TRANSCEIVER

This CW/SSB transceiver operates in the 17-meter amateur band from 18.068 to 18.168 MHz. Direct conversion, as discussed in Chapters 8 and 9, is used for both the receiver and transmitter. All RF functions are built with conventional hardware, but the audio functions are DSP based. In addition, control functions were delegated to the DSP, to the extent possible.

The general arrangement of the transceiver is shown in the block diagram, **Fig 11.13**. The receiver begins with a single tuned circuit and an RF amplifier. The considerations for signal-to-noise ratio, dynamic range and LO radiation were discussed in Chapter 8 and apply here. In

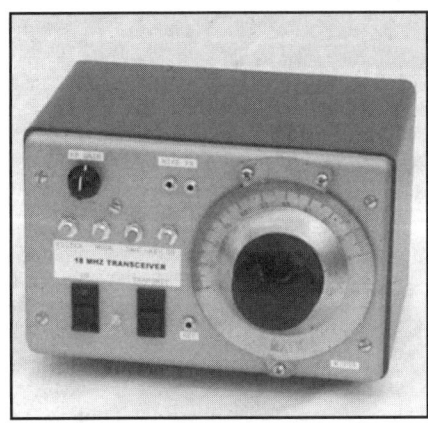

The 18-MHz Transceiver.

order to use the same filters and mixers on both receive and transmit, there is a PIN diode switch following the RF amplifier. For reception, this switch also provides a simple method for manually controlling the RF gain, as the PIN diode can also be used as an adjustable resistor.

Two mixers are connected to the RF circuits through a power divider. A 90-degree power divider supplies the conversion oscillator for the two mixers. In reception, this creates the 'In-phase' and 'Quadrature' or I and Q signals at audio. After low-pass filtering, an A/D converter that is part of the DSP board, digitizes the two signals.

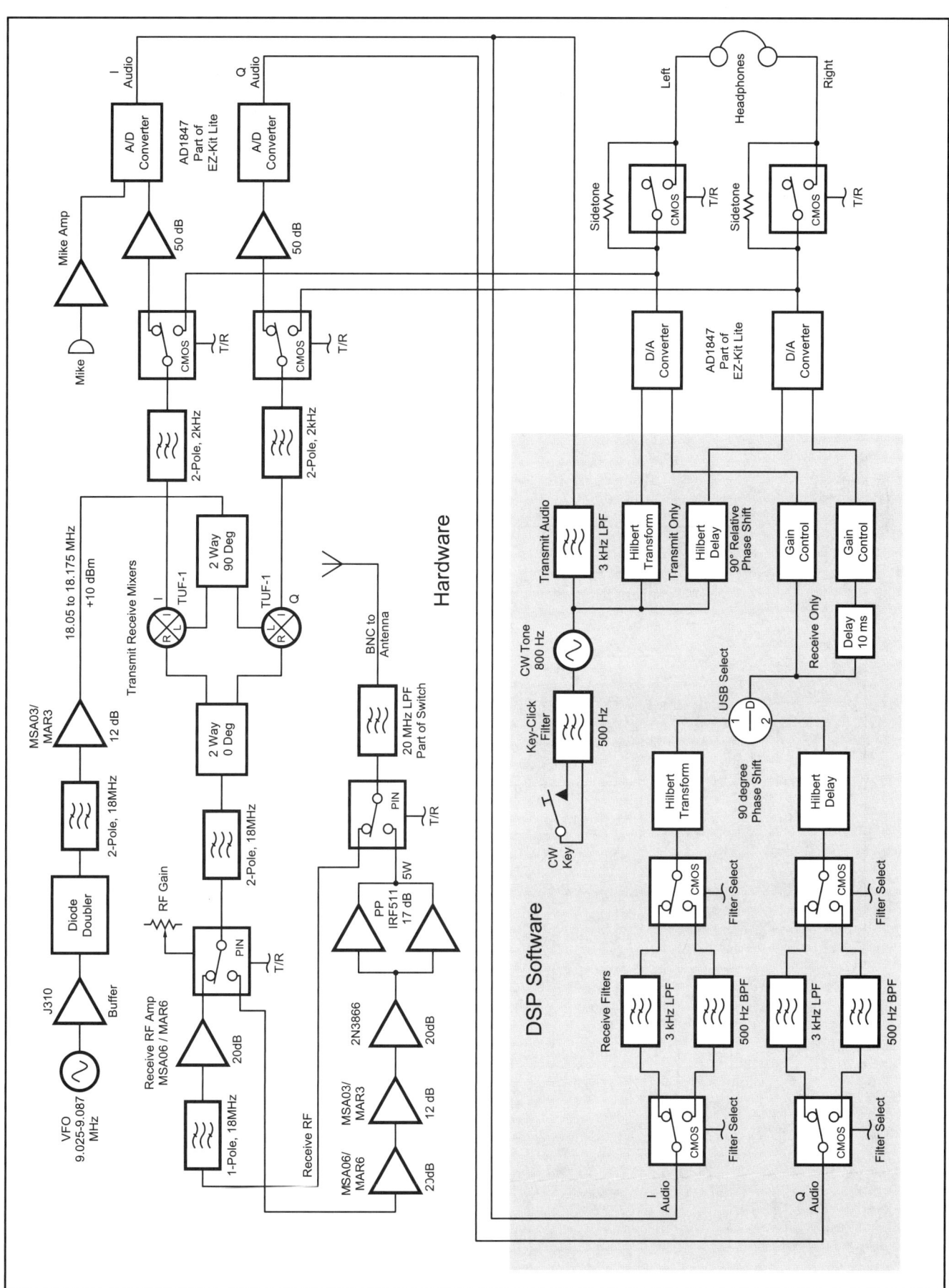

Fig 11.13—Block diagram of the 18-MHz transceiver showing the division of the functions between conventional hardware and DSP software.

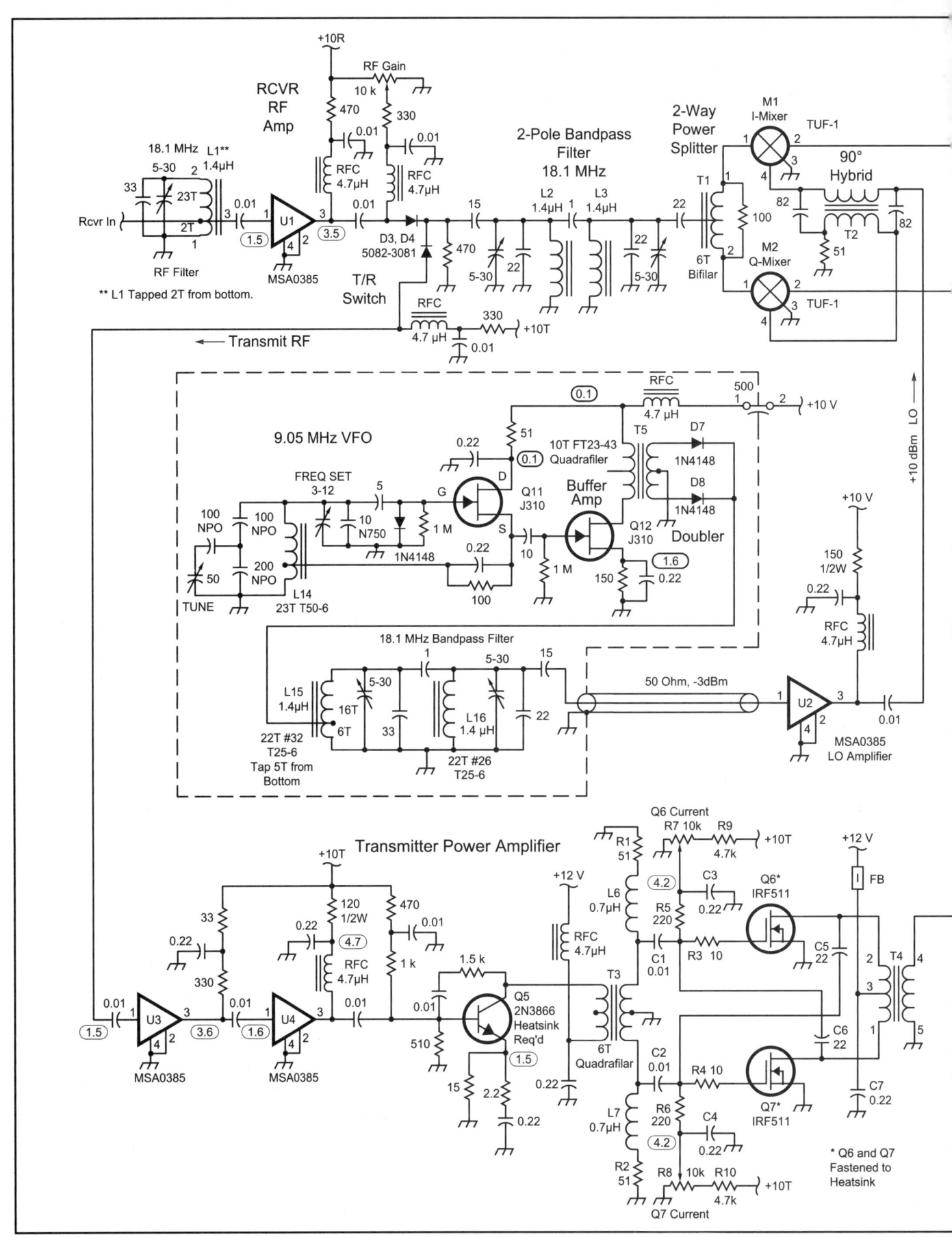

Fig 11.14—Schematic diagram of the hardware used with the 18-MHz transceiver (continued on next two pages).

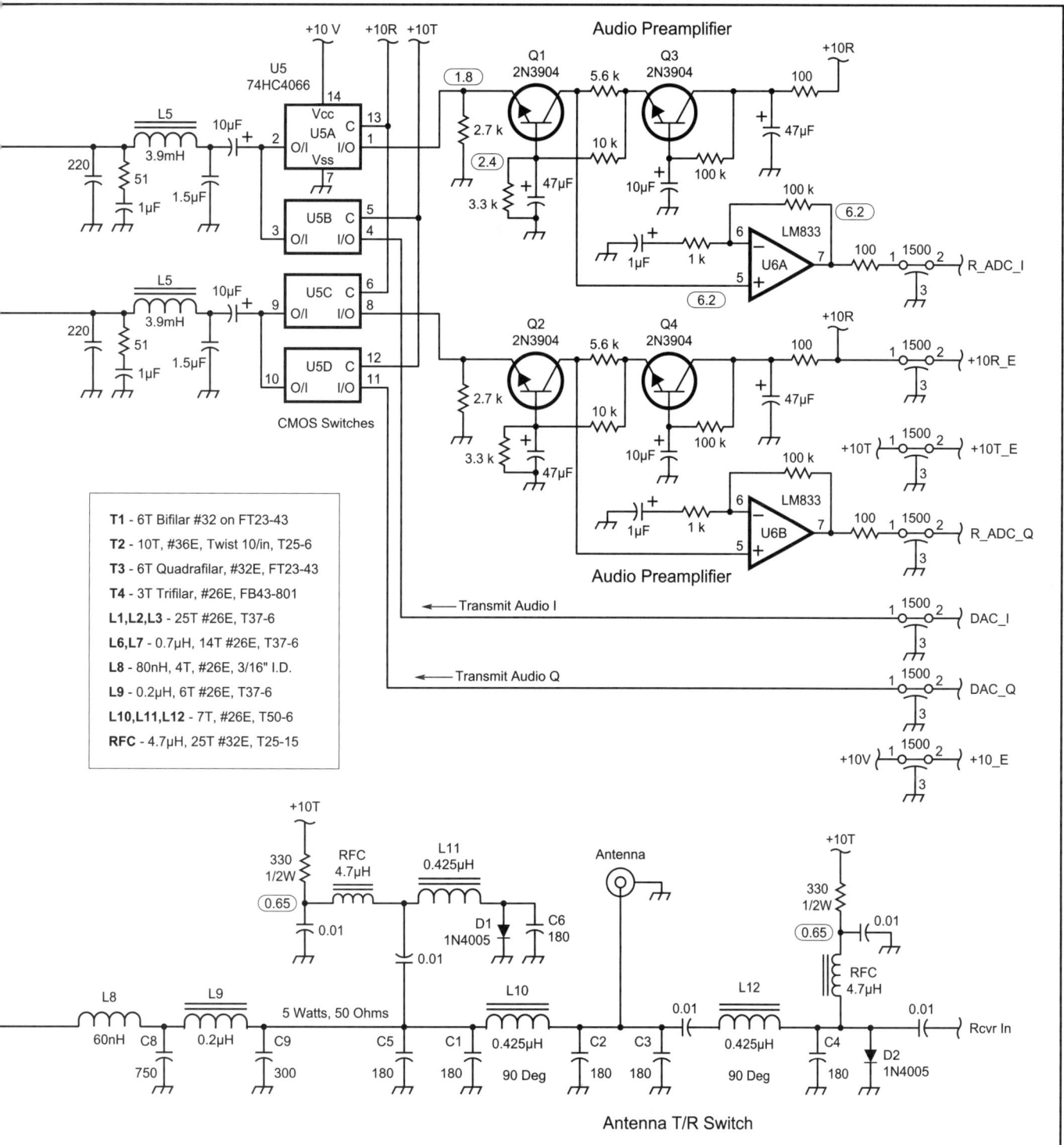

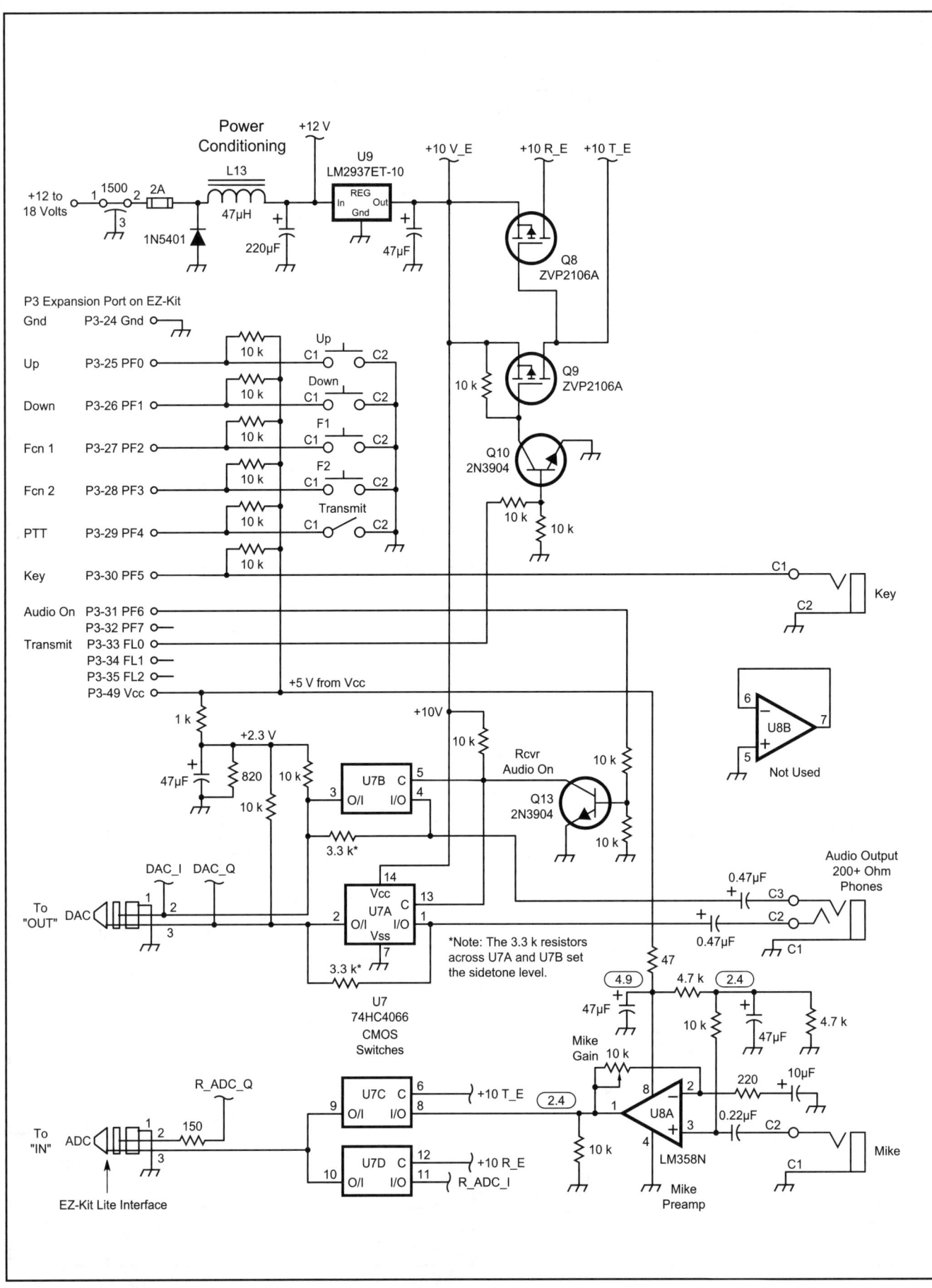

Fig 11.14 continued.

The I and Q audio signals are put through individual audio filters in the DSP. Two filter bandwidths are provided, a 3-kHz low pass filter and a 500-Hz filter, suitable only for CW. Due to the DSP implementation, the I and Q filters are identical in their response. In order to have single-sideband reception, a broadband 90-degree phase difference must be applied to the two audio signals. This is done with a DSP filtering technique called the Hilbert transform. The received upper-sideband signal can then be formed with a simple subtraction of the audio signals. Dividing the audio signal into left and right channels and applying a delay to one of these provide binaural reception. A D/A converter then converts the audio back to analog form, ready to go to headphones.

Transmission reverses most of the signal paths from those of reception. For SSB, a microphone preamp provides some voltage gain ahead of the A/D converter. Low-pass DSP audio filtering restricts the transmitted bandwidth, remembering that we have no I-F filtering to do this. Hilbert transforms produce the 90-degree phase difference needed for the suppression of the lower sideband. The transmitter signal is converted to analog form in the same D/A converter that was used in the audio output of the receiver. After going back through the I-Q mixers, the RF signal is quite low in amplitude. Four stages of amplification raise this to about 5-W SSB PEP or CW amplitude.

For CW transmission, the on-off key signal goes through a 500-Hz LPF to restrict key-clicks. The filtered signal amplitude modulates a pair of 800-Hz tones. These tones are generated in the DSP to differ in phase by 90 degrees, again ready to be converted to analog signals for the I-Q mixers. We again used a method that works well because of the accuracy of DSP, but is considered poor practice in hardware form.

The VFO is quite conventional. A frequency doubler increases the isolation between the 9-MHz VFO and the 18-MHz RF signals.

RF Hardware Details

To simplify the hardware, a number of silicon MMICs are used as amplifiers. As shown in the RF schematic, **Fig 11.14**, the receiver RF amplifier, U1, is a broadband device with a gain of about 20 dB. This is an Agilent (HP) MSA0685, or equivalently, the Mini-Circuits MAR-6. These devices have input and output impedances that are close to 50 Ω, broadband gain and reasonable output powers and inter-modulation levels. These devices are available in a number of different gain and power levels. They require external blocking capacitors, dc power feed RFCs and current limiting resistors. Probably the biggest drawback to the use of these devices is their power consumption. Their efficiency is about half of that achievable with a well designed transistor amplifier, due mainly to the power lost in the current limiting resistor.

Preceding the RF amplifier is a single tuned circuit built around the inductor L1. This restricts the signals that are seen by U1. It is particularly important to reduce the level of inputs at half frequency, or about 9 MHz. Otherwise, these signals are prone to being doubled in the amplifier, making the 17-meter band come to life at times it is not! Two more tuned circuits, built around L2 and L3 provide most of the RF selectivity. This filter uses a configuration of S. B. Cohn[3,4] using capacitive coupling on the ends to match impedance levels. The 15 pF on the input matches to 50 Ω while the 22 pF on the output side matches to 25 Ω, suitable for connecting to the two 50-Ω mixers.

Between the RF amplifier and the filter is a PIN diode switch controlled by the transmit receive (T/R) voltages. For transmit, this connects the filter to the transmit RF amplifier. In the receive case, it serves this same switching function but, also the current through the diode can be varied by the RF gain control. This allows about 40 dB of control range, and is of considerable value when working strong local stations.

A two-way isolated power splitter, T1, applies the received signal to the two mixers. Usually these splitters include a transformer to change the impedance level from 50 to 25 Ω. As was discussed above, this impedance transformation is part of the RF filter.

The mixers are double-balanced TUF-1 types from Mini-Circuits. These provide excellent isolation between the LO and RF ports; this is the transmit carrier rejection. The LO drive differs in phase by about 90 degrees for the two mixers providing one of the necessary elements for the "phasing method" of SSB detection and generation.

The RF phase-shift network (see the discussion in Chapter 9) consisting of a tightly coupled inductor, T2, the two 82-pF capacitors and the 51-Ω terminating resistor. This network has rather sophisticated operation, considering its simplicity. The LO signal is divided into two equal mixer drive signals with a 90-degree phase difference. In addition, there is isolation between the two outputs that go to the mixers. Ideally, no power is transferred to the 51-Ω resistor. It serves to provide isolation when one signal is applied at just one of the mixers.

The drawback of this phase-shift network is that it only works over a narrow band of frequencies. The power division is equal only at the center frequency, and the isolation deteriorates out-of-band as well. This causes the harmonic energy generated in the mixer diodes, due to the LO drive, to redistribute itself in strange ways, as can be observed on an oscilloscope. However, the important equal power and 90-degree relationship is preserved at the fundamental frequency. Because of this, the circuit generates outputs of the correct amplitudes and phase.

AF Circuitry

The receive path signals are generally too weak for the A/D converter without amplification. Full scale for the A/D converter is about ±2 V or a 4 V swing. About 14 bits are above the A/D noise level within an audio bandwidth. This sets the minimum input-signal requirements at about $4/2^{14}=4/16384=244$ microvolts. Bringing a 0.1-microvolt signal up to this level requires about 67 dB of audio gain. This is provided by grounded-base transistor Q1 (or Q2) and a low-noise op-amp, U6A (or U6B). Further details of this circuit can be found in Chapter 8.

The receive audio path to the A/D converter has switches, U7C and U7D, allowing the microphone audio to be connected to the A/D converter during transmit. These are 74HC4066 CMOS types, which show an "On" resistance of 35 Ω, typically. For reception this can have an effect on the noise figure. One simple method of minimizing this affect is to parallel two or more switches by mechanically stacking them and soldering the pins together. Alternatively, four MOSFET devices, such as the 2N7000, could be substituted for the CMOS switches.

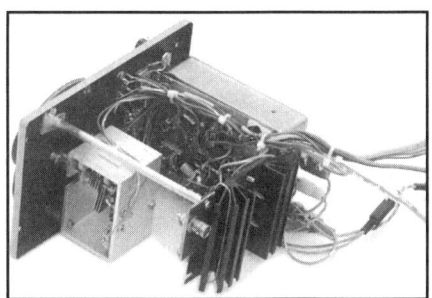

General inside view of the 18-MHz transceiver.

VFO

FET Q11 is a conventional Hartley VFO shown in Fig 11.14, operating at half of the output frequency. The tuning capacitor was capacitively tapped down on the tuned circuit to make the tuning range just over 100 kHz. Q12 buffers the output of the VFO. Diodes D7 and D8 are a balanced doubler that is reasonably efficient at producing even harmonics and suppresses the fundamental frequency and odd harmonics. This reduces the filtering needs on the output of the doubler; the double-tuned circuit built around L15 and L16 produces a clean spectrum, as was illustrated in Chapter 5.

In the interest of good mechanical stability, the VFO was built in a surplus aluminum box with relatively thick walls. The coils were all fastened in place with dabs of silicone sealant. Multiple aluminum spacers hold the VFO to the steel front panel. Almost no microphonics can be sensed when the case is tapped with a hard object. This is often a problem with VFOs built for higher frequencies.

Considerable experimentation was done to make the VFO temperature stable. The procedure was straight from Hayward.[5] After about 7 or 8 tries, a simple compensation consisting of a 10-pF N750 parallel capacitor was found to make the temperature drift of the 18-MHz frequency only 25-Hz per degree C. There is probably good fortune involved in getting the compensation that good, as an apparently identical 10 pF produced a drift of about 50-Hz per degree. Either way, it is worth the effort to do the experiments and compensate the VFO, since the uncompensated stability was measured at –470-Hz per degree C.

Power Amplifier

A single low cost IRF511 MOSFET was tried as an output amplifier. It produced about 3 W of power at 13.6 V. Higher supply voltages produced much more output, but battery operation was one of the goals for this rig. To produce a 5-W output, two of the MOSFETs were placed in the push-pull configuration shown in the schematic. Ferrite cores were used in the input and output transformers.

As is usually the case for these devices (see Chapter 2), HF stability required some extra components. The major culprit in degrading the stability is the 30-pF feedback capacity from the drain to the gate. Good stability and gain at 18 MHz could be achieved by applying some cross neutralization from the two 22-pF capacitors. It was found, however, that there was a tendency toward oscillation in the 2 to 4-MHz region. This is associated with the cut-off phase-shift of the input and output

18-MHz transceiver shielded box circuit detail showing extensive use of the "ugly" construction method.

transformers. Two steps were taken to keep this from being a problem. First, the amount of neutralization was limited to the 22-pF value instead of using the full 30-pF value. Second, a low-frequency input-loading network was added to each device, consisting of L6 and L7, along with the associated 51-Ω resistors. The resulting amplifier is measured to be unconditionally stable for all input and output impedances, throughout the HF spectrum.

A low-pass filter/matching network was placed on the amplifier output. L8 and L9 and the associated capacitors limit the harmonics and also step the 7-Ω output impedance up to 50 Ω. This network limits the frequencies for which this amplifier can be used. Other portions of the amplifier are useful from 1.8 to 30 MHz.

Antenna Switching

Low cost rectifier diodes (see Chapter 6) switch the antenna between the transmitter power amplifier output and the receiver input. A simpler, series-tuned approach, as was also used in Chapters 6 and 12, would probably have worked at this power level. However, this is an example of a solid-state RF switch that can be applied at quite high power levels. The use of impedance inverters for fast antenna switching has roots at least as far back as the early days of radar where it was implemented in waveguide.[6] The following discussion shows how these concepts were applied to this transceiver.

Pi-networks, consisting of L10, L11 and L12 along with their associated 180-pF shunt capacitors, act as 90 degree phase shifters at 18 MHz. Just like their counterparts, the "quarter-wave transformer," these networks serve as impedance inverters. This means that if one end has a low impedance placed across it, the impedance seen looking into the other end will be high. The opposite is true as well; if a high impedance is placed across one end, the other end will show a very low impedance.

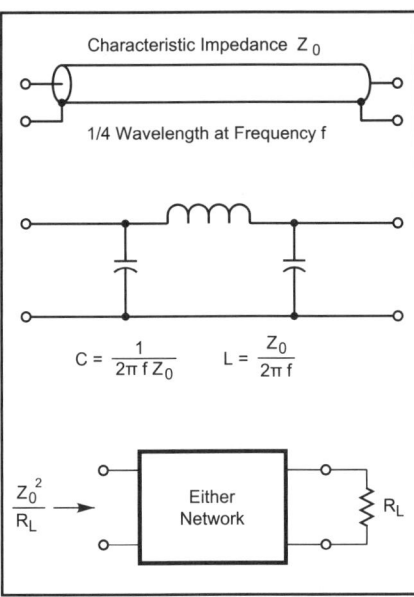

Fig 11.15—Schematics and design equations for impedance inverters built from transmission lines and lumped capacitors and inductors. At the single frequency, f, the two circuits have identical behavior.

Fig 11.15 shows the design for this network. Both the capacitors and the inductor are chosen to have the same reactance at the center frequency. This reactance has the same role as the characteristic impedance of the quarter-wave transformer.

In the antenna T/R switch of Fig 11.14, the inverting network consisting of L12, C3 and C4 acts as low-pass filters during receive, with the signal passing without attenuation. In transmit, diode D2 is conducting and its low impedance shorts out the receiver input. The inverting network uses this low impedance to cause a very high impedance to appear across C3.

The same effect occurs at the transmitter output, due to diode D1 and the inverting network consisting of L11, C5 and C6. During transmit, when D1 is conducting, the impedance seen at the transmitter output, across C5, is very high. Here we also exploit the reverse effect. During receive D1 is not conducting and therefore presents a high impedance, primarily the diode capacity of a few pF. This is transformed through the inverting network to produce a low impedance at the transmitter output, disconnecting any effects of the MOSFET amplifier. The next inverting network, L10, C1 and C2 transform this back to a very high impedance at the antenna connection point.

A single value of capacity, 180 pF, was used for all the networks, for convenience. If they are available, the parallel 180-pF capacitors can be replaced with a single 360 pF.

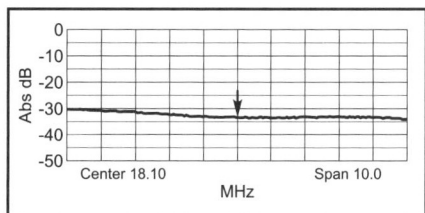

Fig 11.16—Measured isolation of the antenna T/R switch between the transmitter and the receiver.

The inverting networks are relatively non-critical. Any tuning that might be needed can come from squeezing or spreading the turns on the coils.

The Antenna T/R switch was tested as a component by breaking the leads going to the transmitter output and the receiver input. The 18-MHz insertion loss from the antenna connector was 0.33 dB to the transmitter (in transmit) and 0.25 dB to the receiver (in receive). Receiver isolation is a measure of the amount of power going from the transmitter to the receiver input, when the switch is in transmit. As can be seen in **Fig 11.16** this was measured to be about 33 dB at 18 MHz. For a 5-W transmitter this keeps the power at the receiver input below 4 dBm, well below the maximum safe input level for the RF amplifier.

DSP Circuits

For this rig we have chosen to move much of the circuitry into DSP. This is an alternative to conventional analog circuits. In some cases we can improve upon the performance that could be expected from the analog equivalent, but in most cases it comes down to what is easiest. The DSP is again done with a demo board. In some sense, the demo board is a component that is generally easier to install than the parts that it replaces.

One might argue that it takes more time to write the DSP software than building hardware. This is almost certainly true for the first time with a circuit block. However, seldom do we need to write software the "first time." In many cases, we can borrow from previous work or find suitable beginnings in reference books. The material presented here falls in this category. However, this is not to discourage anyone from taking the code apart and trying there own ideas and algorithms. There can be great fascination with writing a program and seeing it produce useful results, such as a DX QSO!

The DSP program for the 18-MHz transceiver not only processes the audio signals for the transmitter and receiver, but controls the simple functions such as transmit and receive switching, reading the panel button switches and lighting the transmit LED. Instead of laboriously describing all of the DSP programs, the following will describe the most important elements of the program. Much of what will be left out is repetitious or is obvious, once one understands the basics of the program writing.

The full DSP program listing for the transceiver is available on the book CD-ROM as *TR18.DSP*.

Reception

The basic reception scheme, shown in **Fig 11.17**, is the direct-conversion I-Q (phasing) method. The basic principles have been around for a long time and have been implemented in analog circuits, as was shown in Chapter 9, and DSP as was done by Rob Frohne, KL7NA.[7] The logical juncture between the RF circuits and the DSP is at the outputs of the mixers. The first of the low-pass filtering is done in hardware. This limits the level of out-of-band signals that are seen by the A/D converter.

Almost all of the band-pass shaping is done in the DSP. Two identical filters are used, one in the I channel and another one in the Q channel. If the signal that we are receiving is of a single frequency, such as a CW signal, the I and Q channels will be a single-frequency audio signal. The frequency will be the difference between the 18.1-MHz LO and the incoming signal. Ideally the amplitudes will be identical and they will be 90 degrees out-of-phase. The actual phase difference will track that of the LOs applied to the mixers.

Applying a 90-degree phase-shift across the audio spectrum and either adding or subtracting the resulting two signals accomplishes SSB reception. The 90-degree shift will bring the two audio signals together so that they are either in-phase, or 180 degrees out-of-phase. Addition, or subtraction, then makes the two signals either add to double amplitude, or to cancel to zero. The choice of sign determines whether upper or lower sideband reception is being used.

Regardless of how it is implemented this "phasing method" has two standard problems. First, producing a constant amplitude, constant 90-degree phase shift over a wide band of frequencies is always an approximation. Second, the mixers, LOs and analog filters all introduce small phase and amplitude errors. Both of these factors, explored in some detail in Chapter 9, serve to limit the ability to eliminate the undesired sideband, referred to as opposite side-band rejection. A DSP

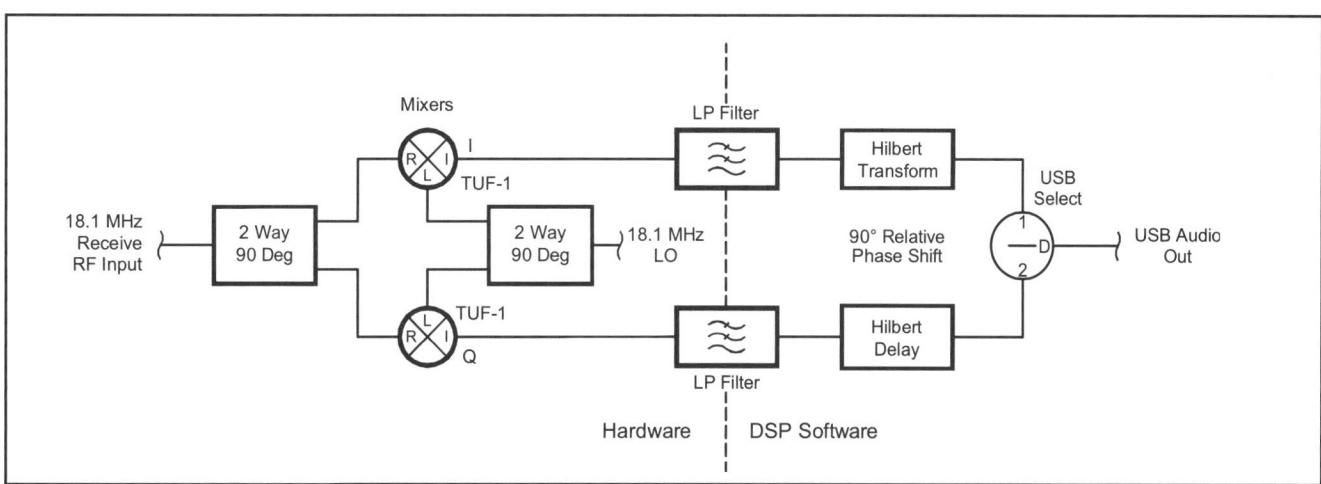

Fig 11.17—Simplified block diagram showing the phasing method of reception used in the 18-MHz transceiver. The circle at the right with a minus sign subtracts input signal 2 from input signal 1.

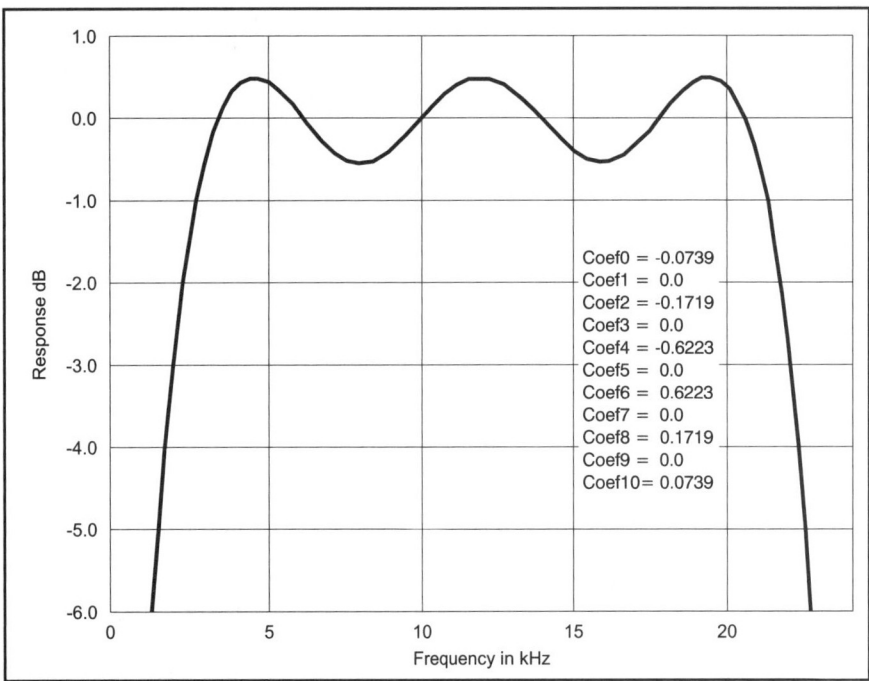

Fig 11.18—Coefficients and amplitude response for a very simple 11-tap Hilbert transform. This is shown to illustrate the method, as one would never use a transform with only 11 taps for SSB generation.

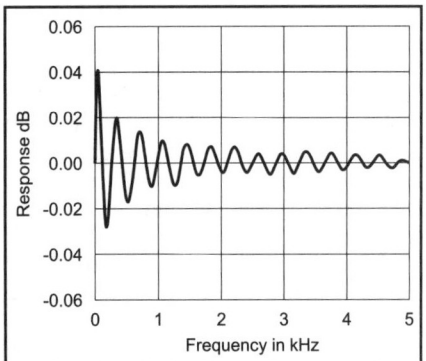

Fig 11.19—The amplitude response of a Hilbert transform using 247 taps and a sample rate of 48 kHz.

implementation of the phasing method does not inherently provide a higher level of unwanted-sideband rejection relative to analog methods. Rather, it should be looked at as an alternative implementation that is potentially easier to implement. This is particularly true if the DSP hardware is being used for other purposes anyway and the only addition is in the software area.

In Chapter 9, the reasons for needing a wideband audio 90-degree (relative) phase shift network were explored. An analog method was used in that chapter to achieve that response, typically using 6 op-amps and precision RC networks. In DSP implementations, the same function is accomplished by a "Hilbert transform." The implementation of this transform has a structure identical to the FIR filter that was discussed earlier. The distinguishing characteristic is the particular choice of FIR coefficients. The coefficients and frequency response for a simple 11-tap Hilbert transform are shown in **Fig 11.18**. The response of this transform is exceedingly far from flat. It cuts off below about 3 kHz and has about a half dB ripple above this frequency However, it does allow us to examine several important characteristics of this transform:

• Every other coefficient is zero
• The second half of the coefficients is the negative of the first half
• The amplitude response varies across the passband
• The phase shift is not shown, as it is 90 degrees, plus a constant delay, at all frequencies. If the number of taps is an odd number, the constant delay is an integral number of sample periods, and easily compensated for. The difference between the Hilbert transform output and a constant delay leaves a very accurate 90-degree differential phase shift.

The amplitude response of the Hilbert transform is never completely flat with frequency. As we saw, with only 11 taps, performance is so poor that one would not consider it in a transceiver. As the number of taps is increased, it is possible to not only cover a wider frequency range, but to also diminish the ripple in the pass-band.

Further, our 48-kHz sample frequency is high, as is discussed below. The frequency response of the FIR filters scales with sample frequency. For the 18-MHz transceiver, we can allow the 48 kHz to remain, if the number of taps is raised. A value of 247 was selected. The computational load is only about half of what it seems, since every other coefficient is zero and does not need to be computed. **Fig 11.19** shows the resulting response, which is typically flat within about 0.01 dB and always within 0.04 dB. Going back to the phasing analysis of Chapter 9, this contributes a typical opposite sideband response of 20 log e/2 which for the 0.01 dB error (voltage error e=.00115), results in an opposite sideband suppression of –20 log(0.00115/2) or about 65 dB. Repeating this calculation for the worst case 0.04-dB error, the opposite sideband is –53 dB.

The DSP program snippet in **Box 7** is the Hilbert transform and compensating delay. The structure is so similar to the conventional FIR filter described in Chapter 10, that only the Hilbert transform specific portions will be discussed.

The zero coefficient values are not entered at all in the table hilbert_coeff, cutting the table size almost in half. To see how zero multiplies occur, it is useful to remember that the data are arranged in a circular buffer. The second time the instruction, dm(i0, m1)=mr1 occurs, the new data are placed at the location in the buffer pointed to by i0 and the pointer is increased by m1, which has a value of one. Within the FIR multiply-and-accumulate loop, mx0 is loaded with data from mx0=dm(i0, m0) where m0 has a value of two. This causes the pointer, i0, to be incremented by the value two after the data are fetched from memory, skipping every other data point. When the counter reaches zero, the loop is broken and after the last computation, i0 is left pointing to the oldest point in the buffer. The next time through the do_hilbert routine, placing data into the buffer causes an increment of one and the FIR computation moves up by one. This brings us to the first of the data points that were passed over in the last FIR computation cycle. And the process continues, moving up one point in the buffer each cycle.

The block diagram of **Fig 11.20** illustrates this same Hilbert transform operation. The top 'I' path is a simple delay to compensate for the flat delay of the transform. The bottom 'Q' path is a FIR filter in structure, but only the even numbered coefficients are used since the multiplications for the coefficients of zero value are omitted.

As is normally the case with broadband

Box 7 - DSP program for computing a 90° differential phase shift using the Hilbert transform.

{The following are constant and memory declarations placed at the top of the overall program:}

```
.const    H3=247;            { Num taps in Hilbert
                               FIR filt }
.const    H3P1ON2=124;       { This is  (H3+1)/2  }
.const    H3M1ON2=123;       { ...and (H3-1)/2  }
.const    H3M3ON2=122;       { ...and (H3-3)/2  }
```

{The Hilbert coefficients are stored in program memory(pm) so they can be fetched at the same time as data is brought in from data memory (dm). The values are read from a file hil_3_48.dat where the values are given as 24-bit hex numbers. The values are left justified 16-bit numbers and padded on the right with two hex zeros. A sample of coefficients would look like:}

```
          021E00
          01F500
          01D000
          01AF00}
.var/pm/circ    hilbert3_coeff[H3P1ON2];
.init                       hilbert3_coeff: <hil_3_48.dat>;
```

{ Each data memory location for the Hilbert transform is declared as follows: }

```
.var/dm/circ    h3delay[H3M1ON2];    { Delay line }
.var/dm/circ    h3data[H3];          { Buffer for data }
.var/dm         m1_sav;              { Allows reuse of m1 }
.var/dm         h3delay_i0_sav;      { Allows reuse of i0 }
.var/dm         h3data_i0_sav;       { Allows reuse of i0 }
```
{———————————————————————}

{ Initialization of the Hilbert transform takes place once at the beginning of the program operation. Zeroing of arrays is useful for simulation, but is not needed for transceiver operation, and is not done here. }

```
          i0=^h3delay;            { Address of delay line
                                    memory }
          dm(h3delay_i0_sav)=i0;
          i0=^h3data;
          dm(h3data_i0_sav)=i0;
```
{———————————————————————}

{ This is the Hilbert transform subroutine. It is called during the 48-kHz rate interrupt to generate a 90-degree phase shift between the I and Q channels. Hilbert has independent inputs and outputs for delayed and phase shifted paths. Uses HIL_3_48.DAT running at 48 KHz in order to get response down to 300 Hz.}

```
          Delayed path:   ar in, ax1 out.
          90 deg path:    mr1 in, mr1 out.}
          do_hilbert:                 { 48 KHz Hilbert for receiving }
             dm(m1_sav) = m1;
             m1 = 1;
```

{ First the delayed path to compensate for the Hilbert delay }

```
          i0 = dm(h3delay_i0_sav);   m0=0;  l0=%h3delay;
          ax1 = dm(i0, m0);          { get ax1, the delayed output }
          dm(i0, m1) = ar;           { Put new data in, update ptr }
          dm(h3delay_i0_sav) = i0;   { Save pointer }

             { Next the actual Hilbert transform:    }
          i0=dm(h3data_i0_sav); m0=2; l0=%h3data;  { i0
                                             points to data }
          i4=^hilbert3_coeff; m4=1; l4=%hilbert3_coeff;
          dm(i0, m1)=mr1;            { Enter new data and bump ptr 1 }

          mr=0, mx0=dm(i0, m0), my0=pm(i4, m4);
          { FIR multiply and Accumulate loop: }
          cntr=H3M3ON2;
          do hil_loop until ce;
          hil_loop:    mr=mr+mx0*my0(SS), mx0=dm(i0, m0),
my0=pm(i4, m4);
             { Process the last point: }
          mr=mr+mx0*my0(SS), mx0=dm(i0, m1), my0=pm(i4, m4);
          mr=mr+mx0*my0(RND);    { mr1 = phase shifted
                                            output }
          if mv sat mr;
          dm(h3data_i0_sav)=i0;
          m1 = dm(m1_sav);
          rts;
```

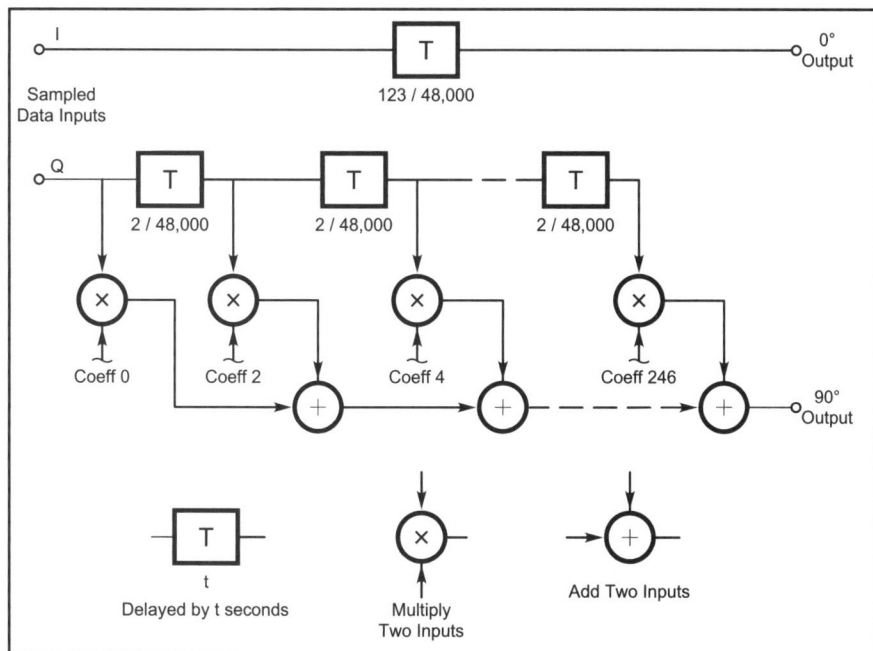

Fig 11.20—Block diagram of the Hilbert transform with 247 taps. The blocks marked 'T' are delays of multiples of sample periods, as indicated on the diagram. Each sample period is 1/48,000 second.

phase shift networks, there is a fixed delay that is much greater than the delay associated with the 90-degree phase shift. For the 247-tap Hilbert transform, and our 48-kHz sample rate, this delay is 0.5* (247-1)/48,000 or 0.0025625 seconds (about 2.6 ms). Other than the need to compensate for this delay, there are no operational problems for an SSB or CW radio.

The second problem in our phasing method of SSB reception was phase and amplitude errors between the two channels. These errors are associated with the mixers and LO hardware and will most likely stay relatively constant over time. If we knew what the errors were we could add in an "anti-error" and have perfect opposite side-band rejection. The degree to which this can be accomplished in practice results in typically 20 dB in improved side-band rejection. Temperature extremes will not allow this to be kept with a simple correction, but the results can be surprisingly good. The problem of knowing what the error is can be solved by merely adjusting the correction until the opposite sideband disappears.

To understand this process one should think of the error between the desired I (or Q) signal as both an amplitude and a phase shift. This is referred to as an "error vector" and is illustrated in **Fig 11.21**. In the example, not only is the actual signal longer (bigger amplitude) than the desired signal, but there is a phase shift between the two. To correct the signal, we must subtract the error vector from the actual signal. To do this, we take advantage of the fact that the actual I and Q signals are roughly 90-degrees apart in phase shift. By taking a fraction of the I signal and adding it to a fraction of the Q signal, it is possible to create the negative of the error signal—just what we need to suppress the opposite sideband. **Fig 11.22** shows an implementation for our correction of the sideband suppression. The constants, I_Gain, Q_Gain and QI_Cross_Gain are all numbers between –1 and 1. Both I_Gain and Q_Gain would not be needed if we allowed gains greater than 1. But it is a convenience to not do this and it is relatively easy to provide the two gain values. Therefore, one of those gains will be set to 1.0, which, in fractional integer arithmetic, is the fraction 32767/32768, entered as a hex value of H#7FFF (see the discussion of fixed-point arithmetic in Chapter 10).

The other gain of the I_Gain/Q_Gain pair can then be set to a value close to 1.0, as determined experimentally. The cross-gain value should be small, but it can be either plus or minus. A value such as +0.05 might be typical and is represented as the fraction 0.05*32768/32768 or 1638/32768 and entered into the program as 1638.

Listing TR18D shows the USB reception routines, including the vector correction. The usual declarations of constants and memory, by name, are at the top of the program.

The three constants that are needed to

```
Listing TR18D

Phasing method receiver including error correction. The inputs are I and Q
signals that have been low-pass filtered.


{The following are constant declarations, placed
   at the top of the overall program:}

.const     RGAIN_I=32400;      { Adjust value to suit }
.const     RGAIN_Q=32767;      { Adjust value to suit }
.const     RGAIN_IQ=2060;      { Adjust value to suit }

{———————————————————————————————————————————————}

{The I data is at memory location 'save_i' and the Q data is in sr0}

      ar = dm(save_i);              { Move the I signal data to ar }
      my1 = RGAIN_I;                { I Gain correction factor }
      mr = ar * my1 (SS);           { I signal * correction }
      dm(save_i) = mr1;             { Temporary storage }
      my1 = RGAIN_IQ;               { Generate the IQ cross }
      mr = ar * my1 (SS);           {   correction factor }
      ay0 = mr1;                    { Save cross-correction factor }
      my1 = RGAIN_Q;                { Q chan gain correction }
      mr = sr0 * my1 (SS);          { Q signal * correction }
      ar = mr1 + ay0;               { Add in cross-correction }

      mr1=dm(save_i);               { For hilbert }
      call do_hilbert;              { 90 deg; ar, mr1 in; ax1, mr1 out }
      ay1 = mr1;                    { Get ready to subtract Q out }
      ar = ax1 - ay1;               { - = usb }
                                    { USB audio output is now in register ar }
```

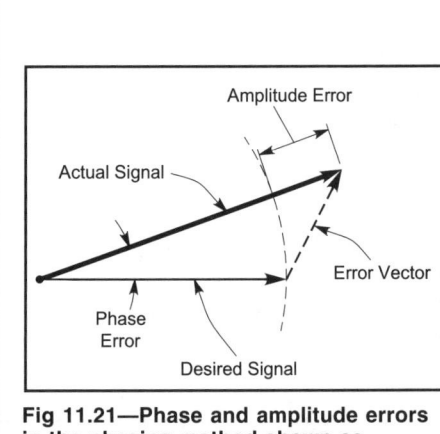

Fig 11.21—Phase and amplitude errors in the phasing method shown as vectors.

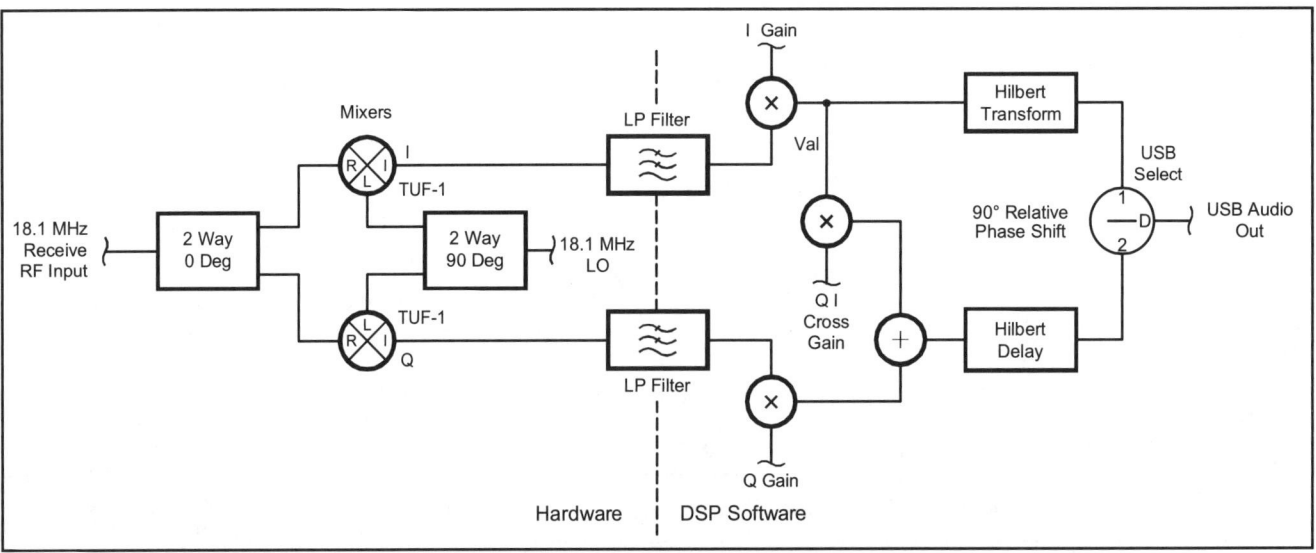

Fig 11.22—Block diagram of a phasing method receiver with DSP software error correction. The cross gain is shown going from the I channel to the Q channel. It will work equally well going in the reverse direction, but both directions are never needed.

suppress the opposite sideband are entered as constants. This is a very simple system, but requires re-assembly of the program to null the sideband. Experience has shown this a reasonable approach, since the settings do not normally need to be changed often. Multiplication by both RGAIN_I and RGAIN_Q occurs each time through the routine, even though one of these constants will have the value of 1.0. This simplifies the adjustment of the constants since we don't know which will have the 1.0 value.

The Hilbert transform, discussed above, is a subroutine invoked by 'call do_hilbert.' This applies the differential phase shift so that the USB can be formed with simple subtraction 'ar=ax1-ay1.'

Also shown in the listing is the audio gain control. One of the conveniences of a DSP implementation is having gain control steps in constant dB amounts. For analog gain controls, this is approximated with what are called "log" potentiometers. Our DSP implementation starts with the binary shifter as a basic component. If the signal word is shifted left by one bit, the result is an increase in level of 6.0 dB. Shifts to the right decrease the audio level by the same amount. This has the desired equal dB amounts per step, as well as great simplicity. The drawback is that the steps are too big. Experience suggests that 1-dB steps seem too small, but 1.5 to 2-dB steps allow one to choose a comfortable audio level with a reasonable number of button pushes.

We implement 1.5-dB steps by having a table of four entries corresponding to gains of 0, –1.5, –3.0, and –4.5 dB. This table, stored in program memory, is called 'aud_gain' and provides multipliers that can be used between the 6.0-dB steps. As an example, a gain of –1.5 dB is a voltage ratio of $10^{(-1.5/20)} = 0.8414$. In fractional arithmetic this is a value of $0.8414*32768 = 27571$, which in hexadecimal form is H#6BB3. The program memory words are 24-bits wide, but only 16 bits of this are available when used as data. The bits will be properly aligned if the hex values are padded on the right with '00.' Thus, the –1.5-dB entry in hexadecimal is H#6BB300.

The button control parts of the program have setup two values for the audio gain control, 'af_gain' which contains one of the 1.5-dB step multipliers, and 'af_shift', which is the number of 6-dB steps. These shifts can be either plus or minus.

Audio Filtering

The general nature of FIR filters has already been covered. Here we apply these principles with two receive filters: a 3-kHz low-pass filter suitable for all modes and a 500-Hz wide band-pass filter for CW use.

The index register pointer, i0, of the DSP is used to find the data points for the FIR filter. Initialization of this register is critical. Omitting this can cause hours of grief in getting the DSP program to operate. The program may function at times and fail at others, depending on the random initialization. The program instructions for this initialization are:

i0=^idata;

dm(fir1i_i0_sav)=i0;

When the FIR filter is called, the pointer i0 is loaded by the instruction

i0=dm(fir1i_i0_sav);

all of which allows i0 to be reused in many routines.

Binaural Delay

This feature is always in operation for the transceiver. The addition of a delay of about 10 milliseconds in the sound heard by one ear, relative to the other has interesting effects, very closely related to the I-Q binaural effects used in Chapter 9. The noise heard by the two ears loses correlation and allows the mind to better distinguish between a CW tone and the noise. On CW, the tone takes on the effect of having a spatial position that depends on the tone frequency. The noise position is, in effect, always moving around "inside your head."

As a signal is tuned, the phase relationships between the tones heard by the ears changes for the delay system. For the I-Q binaural, it is a constant 90 degrees while the phase shift for the delay binaural increases with frequency. For the 10 millisecond delay the phase shift is 90 degrees at $1/(4*0.01)=25$ Hz and changes quite rapidly with tuning. Thus, the two systems do not have the same sound when tuned. In either system the noise is uncorrelated and the sound is similar, not unlike an FM stereo radio without an antenna. Probably the biggest difference is that the I-Q binaural system receives both sidebands, whereas the delay binaural is compatible with SSB. The delay binaural is in the final audio path and is compatible with any mode.

Implementation of a binaural delay requires some memory for storing the signal, but very little computation is needed. **Listing TR18E** is the portion of the DSP program required.

Operation of this delay line is closely related to the address generators used by the ADSP-2181 DSP.

A segment of memory, such as our 'delay [DELAY_SIZE] can be designated as circular by the key word 'circ.' DELAY_SIZE is the same as the constant 512 and so this many words of data memory are set aside. Each word is 16 bits, adequate to store one sample of the

---

Listing TR18E

DSP program snippets for delay binaural sound.

```
{The following are constant and memory declarations, placed
  at the top of the overall program:}
.const          DELAY_SIZE=512;
.var/dm/circ    delay[DELAY_SIZE];      { The delay line, binaural }
.var/dm         del_i0_sav;             { Storage when not in interrupt }
{————————————————————————————————————————}
{ This part of the program is executed at startup to initialize the
  pointer to the delay line, delay[]. }
    ax0 = ^delay;                       { Get the address of delay line }
    dm(del_i0_sav) = ax0;               { The pointer is saved here }
{————————————————————————————————————————}
{ This program snippet is executed at each 48 kHz interrupt to put the
  left channel data into the delay line, and to take the delayed data
  out for the right channel. Left data is in register sr1:}
    i0=dm(del_i0_sav);                  { Load i0 pointer }
    m0=0;                               { Do not adjust the pointer, now }
    l0=DELAY_SIZE;                      { The length of the circular line }
    mr1=dm(i0, m0);                     { Remove the delayed signal }
    m0=1;                               { Now increment pointer on write }
    dm(i0, m0)=sr1;                     { Put the new signal in the line }
    dm(del_i0_sav)=i0;                  { Save the pointer for next time }
    dm(tx_buf+2) = mr1;                 { Send audio data to right D/A }
```

audio waveform. This is illustrated in **Fig 11.23**. There are 8 address generators, and the binaural delay uses only one of these, generator zero. Three parameters control the generator, i0, m0 and l0. i0 is a pointer, meaning that it is an address in memory. m0 is an increment amount that tells the generator to add the value of m0 to i0 after doing either a read or write operation. l0 applies if the buffer is circular, and tells the address generator to not point to memory locations past the base location plus l0, but instead to wrap around to the beginning. Note that m0 can be zero or negative. Negative values mean that the progress through the circular buffer is in the reverse direction.

Returning to the listing, the value of the pointer is restored to i0 with 'i0=dm(del_i0_sav)' and m0 is set to zero, meaning that no change will occur to i0 when the delay line is accessed. l0 is set to the length of the circular buffer, DELAY_SIZE. The right audio channel signal sample is next removed from the line with 'mr1=dm(i0, m0)' and left temporarily in register mr1. The increment register, m0, is now changed to one. When we put the new audio data into the delay line with dm(i0, m0)=sr1, the pointer, i0, will now have one added to it following the memory write. What this does is to move i0 to the location of the now oldest data point. After 512 applications of this routine, the pointer will be again pointing to the data point that was just entered. This delays the data by 512/48,000 of a second, or about 10.7 milliseconds.

SSB Transmission

The phasing method for SSB reception that was described above is reversible for transmission. The audio signal is placed through a Hilbert transform to produce a signal with 90-degrees phase shift, relative to a signal with a simple delay. Both signals can be then be passed through D/A converters and applied to a pair of mixers. The mixers have 0 and 90-degree LO signals, just as in reception. The sum or difference of the two mixer outputs at RF is now the desired SSB signal, ready for amplification.

The opposite side-band suppression that can be achieved depends on the care taken in matching the mixers and in achieving exactly 90-degree phase differences for the LO signals. But, as was done in reception, it is possible to apply software corrections to the audio signals to improve the cancellation at the mixer outputs. This is illustrated in **Fig 11.24**. The DSP implementation is parallel to that used for reception. A separate set of constants are needed, as there are differences in the audio paths, due primarily to the differences introduced by T/R switching.

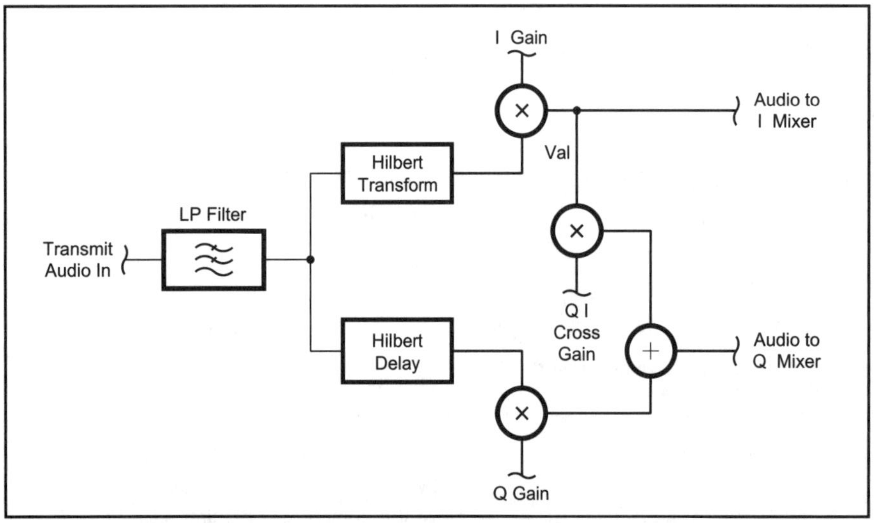

CW Transmission

This mode requires that the frequency of the transmitted signal and the received "zero-beat" signal be offset by a tone frequency, such as 800 Hz. Some sort of TR activated switching device can be used in the VFO to provide the offset as was seen in Chapter 6. Alternatively, an audio tone can be generated and passed through the SSB generator. The VFO never changes frequency and the offset can be precise. Unfortunately, there often are two undesired signals accompanying the CW signal. The VFO output must be suppressed by the quality of the mixer balance. Mixers, such as the Mini-Circuit TUF-1 can have 50 to 60 dB of inherent L-R balance. It is often possible to increase this by 10 dB or more by adding a very small gimmick capacitor between the LO signals and the mixer output. **Fig 11.25** illustrates a general approach for increasing the mixer balance in this way.

The second undesired signal is the opposite side-band. This, however, is the same problem that was solved for SSB with the I-Q vector correction. This suggests a method for adjustment of the correction constants. If we transmit a CW tone and receive the unwanted side-band on a local communications receiver, the S-meter can be used to find a null. The correction constants are those that make the signal disappear.

Fig 11.26 shows the screen of a spectrum analyzer attached to the output of the 18-MHz transceiver with the key down. The VFO is in the center of the screen at 18.100 MHz Each division is 500 Hz and the tone frequency is 850 Hz. With USB being used, the transmitted signal is above the carrier frequency. Suppression of the carrier is 48 dB and the opposite side-band, 850 Hz below the center, is 63 dB below the transmitted signal. An additional signal can be seen 1700 Hz above the VFO frequency. This is due to the modulation of the second harmonic of the 850-Hz tone. This undesired output is suppressed 50 dB. One would always want all spurious signals to be undetectable, but in the real-world way of such things, these levels are acceptable. This level of spurious signal will, in general, be covered by the key-clicks in almost any CW transmitter.

Key-click suppression is normally dealt with by limiting the rise-time of the keying waveform. It can be shown that this will cause the key-click spectrum to fall off much faster as one tunes off the CW signal. It is possible to increase the rate even more if, not only the rise-time is limited,

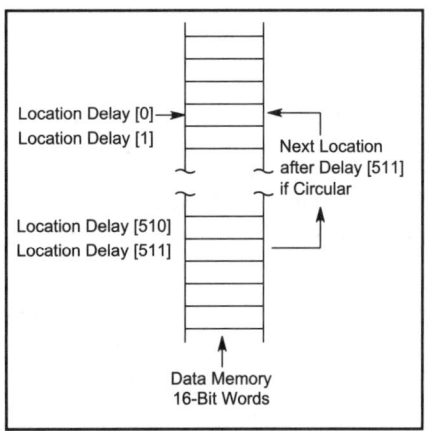

Fig 11.23—Circular data buffer used to implement a 512 point delay line as is used for delay binaural operation.

Fig 11.24—Simplified block diagram showing the I and Q corrections to improve the unwanted sideband rejection for transmit.

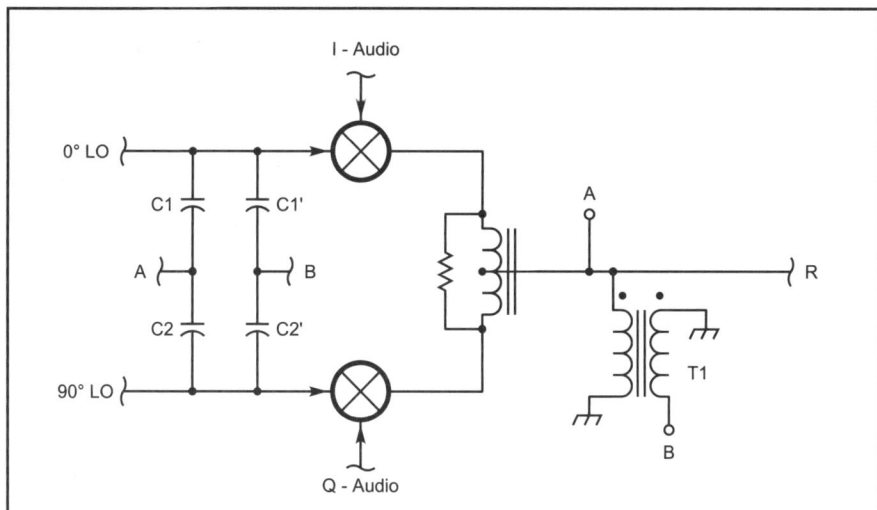

Fig 11.25—Schematic diagram a circuit for increasing L-R isolation of a balanced mixer. In order to minimize the capacitance values, one should never use both C1 and C1' or C2 and C2', as this would only increase the size of both capacitors. All capacitors are a fraction of a pF, made from gimmick wires, which are merely two enamel covered wires twisted together. The transformer, T1, is 5 turns of #26 bifilar wire on a small ferrite core, such as Amidon FT-23-43.

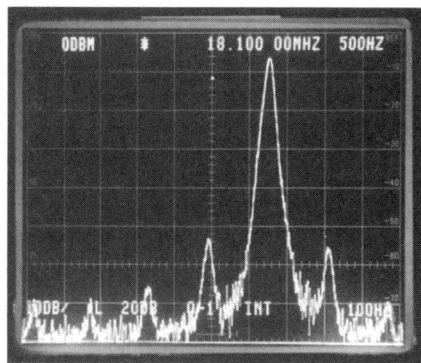

Fig 11.26—Output spectrum of 18-MHz transceiver in CW mode. The carrier is at the center of the screen. The transmitted signal is the large response 1.7 divisions to the right. The small response the same distance to the right is the unwanted sideband. Measurements were done with a Tektronix 494 analyzer.

but the keying waveform is made to have rounded corners at turn-on and turn-off. A direct way to insure that this happens is to pass the keying waveform through a low-pass filter and then use the resulting waveform to amplitude modulate the RF signal. In our case, the modulation can be applied to the 800-Hz tone, before it goes to the Hilbert transform and then to the mixers. As an added benefit, the 800 Hz is available for use as a transmitter side tone, ensuring that a station is tuned in correctly when the received tone is the same as the side tone.

The filter used here is a 500-Hz LPF. The 48-kHz sampling rate requires about 200 taps on the FIR filter, but the DSP is not busy during CW transmission, so this is not a problem. As shown in **Listing TR18F**, amplitude modulation in the DSP is accomplished by generating a sine-wave at the CW offset (800 Hz) and multiplying this by the output of the key-click LPF. This is repeated for a 90-degree phase shifted Q signal by generating a cosine wave and repeating the modulation. The output of the key-click low-pass filter has overshoot that is slightly greater than the input. This is a necessary part of limiting the transmit spectrum. To ensure that this is not saturated by the low-pass FIR filter, the input to the filter is reduced in amplitude by a factor of 0.9l, as shown.

The I and Q corrections for improving the side-band suppression uses the constant values GAIN_I, GAIN_Q and GAIN_IQ. As was the case for reception,

---

Listing TR18F

DSP routines used to generate a CW transmit signal

```
    xi_cw:                          { If key is down, put a .9 (29491) into CW fir
filt.
                                    Modulate fir output onto carrier. This scheme
                                    allows top space for overshoot in the fir. }
    ax0 = dm(key);                  { Get hardware CW key data }
    none = pass ax0;
    ar = 0;                         { CW off }
    if ne jump xi_cw1;              { CW key is up }
    ar = 29491;                     { 0.9 to key click filter }
xi_cw1: call fir_xmt_cw;            { Input in ar, output in mr1 }
    my0 = mr1;
    ax0 = dm(cw_dphase);            { Phase increment for lo }
    ay0 = dm(cw_phase);             { Last phase }
    ar = ax0 + ay0;                 { New phase }
    dm(cw_phase) = ar;              { For next time }

    ax0 = dm(cw_phase);
    call sin;                       { ax0=Phase, Sin returned in AR }
    mr=ar*my0(SS);                  { CW Gate }
    ar = -mr1;                      { Make USB}

    my1 = GAIN_I;                   { Gain correction factor }
    mr = ar * my1 (SS);             { Keyed sine wave * correction }
    ar = mr1;                       { Corrected I signal }

    my1 = GAIN_IQ;
    mr = ar * my1 (SS);
    dm(t1) = mr1;                   { Cross-correction for Q }
    dm(tx_buf + 1) = ar;            { In-phase transmit i-f sig out }

                                    { That takes care of I, now Q: }
    ax0 = dm(cw_phase);             { The phase used for I chan }
    ay0 = 16384;                    { 90 degrees for quadrature lo }
    ar = ax0 + ay0;                 { Q chan phase }
    ax0 = ar;
    call sin;                       { Cos lo sig, sin() preserves my0 }
    mr = ar * my0(SS);              { CW Gate for Q signal }
    my1 = GAIN_Q;                   { Q chan gain correction }
    mr = mr1 * my1 (SS);
    ay0 = dm(t1);                   { Now add in cross-correction }
    ar = mr1 + ay0;

    dm(tx_buf + 2) = ar;            { Quadrature transmit sig out }
```

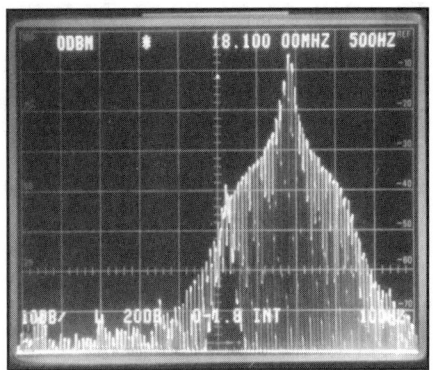

Fig 11.27—Measured spectrum of the transceiver in CW, when being keyed on and off at 10 dots/sec. The horizontal scale is 500 Hz/div and the vertical scale is 10 dB/div.

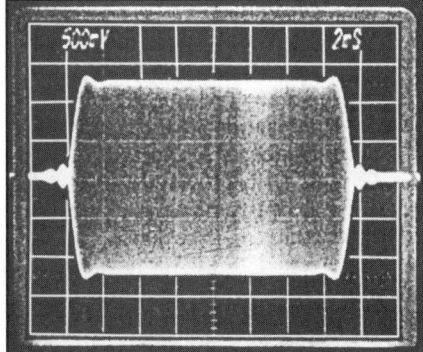

Fig 11.28—RF waveform that results from the keying low-pass filter. The small ripples at the ends of the waveform are a result of the key-click reduction. This waveform was measured on the DSP-10 transceiver, outlined later in this chapter, that uses the same keying system as the 18-MHz transceiver.

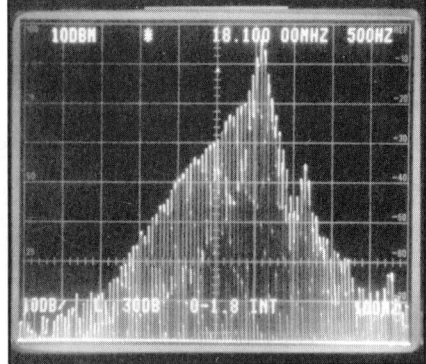

Fig 11.29—Measured spectrum of a commercial transceiver in CW operation, when being keyed on and off at 10 dots/sec. The horizontal scale is 500 Hz/div and the vertical scale is 10 dB/div. This spectrum is typical of signals on the air with their key-click spectrums limited by rise and fall times. It is shown here for comparison with the DSP derived spectrum of Fig 11.27.

either GAIN_I or GAIN_Q should be kept at a value of +1 (32767 integer.)

The resulting key-click spectrum (see **Fig 11.27**) is cleaner than many commercial transmitters and sounds very good on the air. The spectrum is down about 30 dB at an offset of 500 Hz. **Fig 11.28** is the keying waveform at the output of the key-click low-pass FIR filter. The small ripples that both precedes and follows the main keying transitions are characteristic of a frequency constrained waveform. These ripples are not heard by the ear when receiving the signal. If they were not present, the ear would hear the well known key-click sound. For comparison, **Fig 11.29** is representative of the key click spectrum for transmitters that shape the keying by limiting the rise and fall times. This was measured on a commercial transmitter of 1990 vintage. The far-out spectrum tends to fall off more slowly than the DSP shaped system produces.

Control Functions

Four push-button switches are used to communicate data into the DSP for the 18 MHz transceiver:
Button 1 - Turn the audio gain up 1.5 dB.
Button 2 - Turn the audio gain down 1.5 dB.
Button 3 - Alternate between Upper Sideband and CW modes.
Button 4 - Alternate between a wide-band SSB filter and a narrow-band CW filter.

Operation of all four push-buttons is the same.

Push-button switches are prone to having multiple on/off states when they are first pushed, referred to as "contact bounce." The effects of this can be eliminated with hardware de-bounce circuits, or in our case this can be done in the DSP. A software counter, bcounti is used to measure how long the ith switch has been depressed. The counter is initially set for a value of 100, meaning that the switch has not been pushed. The interrupts occur every 1/48,000 second at which time the switch state is read. If the switch has been pushed, the counter is diminished by one, but not allowed to go less than zero. If the switch has not been pushed, the counter is increased by one, but not allowed to go above 100.

In the background portions of the program, the counters are examined. If any of them are at zero, they are considered to have been pushed, that is, the button has been down at least 100/48,000=2.083 milliseconds and is now "de-bounced." So, the appropriate action for the switch, such as turning up the audio gain is performed. Next, a flag is set so that further indications of the switch having been pushed will be ignored until the counter has returned to 100. This is saying that each push of the button must be followed by a release. There are no extra repeated actions for holding the buttons down.

The details of this de-bouncing and button interpretation are covered with comments in the program *TR18A.DSP* on the book CD for those wanting to see an example.

Sampling Rates For The 18-MHz Transceiver

The A/D and D/A converters for the transceiver operate at a 48-kHz rate. This provides an audio response to at least 20 kHz. In the case of the transmitter, it is totally inappropriate to transmit signals with such bandwidths, and low pass filtering is provided to prevent this. In the case of the receiver, it is interesting to be able to have wider bandwidths than the conventional SSB filters give. Typically, in the interest of QRM rejection, these filters cut-off in the 2.5 to 3.0 kHz region. Some people find the narrower filters create a muffled sound to the audio. A high sampling frequency gives ample opportunity to experiment with this.

Another example of an algorithm that benefits from a high sampling rate is a noise-blanker. Signals are easily stored in a delay line while decisions to blank are made. As discussed in Chapter 10, if sufficient bandwidth is available, the presence of noise could be determined by the nature of the wide-band signal relative to the desired signal being received.

It is challenging to maintain high opposite side-band rejection with an analog I-Q phase-shift network. In the case of the Hilbert transform approach in DSP, the only difficult part is keeping good amplitude response at low frequencies (around 300 Hz.) The high frequency side of the Hilbert response continues up to within a few hundred Hz of half the sampling frequency. Thus the opposite sideband rejection bandwidth can be very wide.

One of the interesting effects from using a wide bandwidth for SSB reception is a new view of transmitter splatter. One hears the transmitter splatter, not as off-frequency hash, but rather as a distortion to the voice. It is possible to make judgments of transmitter cleanliness by tuning the signal in and listening for the distortion. The excellent linearity of the A/D converters makes the receiver an insignificant contributor to the distortion being heard.

As usual, there are some negative features of using a high sampling frequency. The most obvious is the increased load on the processor. With a sampling frequency of 48 kHz, there is a maximum time of 1/48000=20.833 microseconds to process the interrupt. The ADSP-2181 processor completes 33 instructions per microsecond and so there are a maximum of 20.833x33=687 instructions per interrupt. During reception these are allocated roughly as:

FIR Filter I	143
FIR Filter Q	143
Hilbert Transform	148
I-Q vector correct	10
Audio gain control	4
Binaural delay	7
Other receiver jobs	62
Buttons	56
Total	573

This uses about 84% of the available time, but leaves adequate time for the background processing. Background tasks are chosen because they have neither deadlines, nor rates of occurrence that they must achieve.

A second drawback to a high sampling rate is the response of FIR filters. These can have fast rates of cut-off outside of the pass-band, but the filter shape still scales with sampling frequency. We get satisfactory response for the 18-MHz transceiver using a 48-kHz sampling rate. But if we needed greater selectivity, there would be two approaches possible. We could run a lower sampling rate. A rate of 10 to 15 kHz would still support excellent audio response for communications.

A second way that allows the FIR filters to have a low sampling rate and also have a wide-band system available is to use multiple rates. This approach, called decimation is illustrated in **Fig 11.30**. The basic process is to limit the bandwidth to

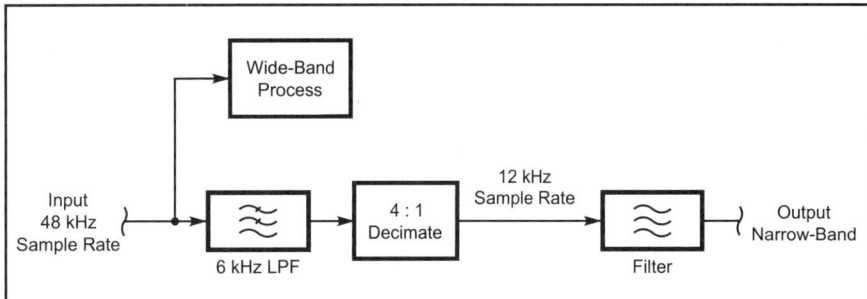

Fig 11.30—Use of decimation to improve filter response and to reduce computational load. The process of decimation first low-pass filters the data and then discards a fraction of it that is no longer needed to satisfy the Nyquist sampling criteria.

a fraction of the total bandwidth using a low-pass filter. In this example, the filter cuts off all significant signals above 6 kHz. Next 3 out of every 4 samples are discarded. The Nyquist sampling criteria is met since the new sampling rate of 12 kHz is at least twice the frequency of any signal that we are processing. The selectivity of all filters, low-pass, band-pass or any other, will be improved by a factor of four. Alternatively, the selectivity can be maintained, but the numbers of taps in the FIR filters can be reduced.

The gains of decimation are great. Not only can the number of FIR taps be reduced, but the processing load is also reduced because the sampling rate is down.

Analog vs Digital

One may already have noticed some strong resemblances between the R2. receiver and the mixer/I-F circuits of this 18-MHz rig. It is interesting to compare the two circuits to see where the use of DSP has changed the implementation.

The I-F filter/diplexer, built around L4 and L5 is identical. Switches, U5A-U5D are added to allow T/R switching and so are needed with either implementation. The R2 uses audio filtering that is in the DSP for this rig. Audio amplification is needed for both implementations since the signal levels coming from the mixers can be of sub-microvolt levels. For the DSP implementation, RF filtering, consisting of 1500-pF feed-through filters, is needed to keep noise from the DSP processor from getting back into the RF circuits. And, of course, the biggest difference is that the DSP implementation requires A/D and D/A converters plus the processor.

The overall complexity and power consumption of the DSP implementation are both greater than that of their analog counterparts. The compensating feature is the performance of functions such as filtering and sideband suppression, along with the ability to make changes and add features without hardware changes.

11.5 DSP-10 2-METER TRANSCEIVER

As the complexity of an electronic project grows, the amount of time and technical skill required for successfully completion easily exceeds the allowable bounds for "weekend experimenters." Much of the material in this book emphasizes ways to have success with a project by using simple approaches and limiting the features. QRP amateur construction and operation has thrived on this approach. This view can be modified somewhat when the project has significant portions implemented in software. An example is the DSP-10 all-mode 2-meter transceiver using a DSP-based last I-F and audio sections with a computerized front panel.

The details for the DSP-10, including the *QST* article and all of the computer programs, are included on the *Experimental Methods in RF Design* CD. The following material is an overview of the project that shows the overall scope and content. Most of the DSP programs involve routines that have been discussed in Chapters 10 and 11. A major distinction is that the control program, written in the language 'C', runs on a personal computer (PC) and communicates with the DSP through a 9600-baud serial connection.

Fig 11.31 is an overall block diagram of both the hardware and DSP software for the transceiver. Dual conversion is used in the mixing process to convert between the 144-MHz RF frequency and the 10-to-20-kHz DSP I-F. Coarse tuning with 5-kHz steps is done at the 126-MHz first conversion synthesizer. Fine tuning to less than 1-Hz steps is provided by the DSP software. PIN-diode and CMOS switches select the direction of signal flow in the RF hardware.

All signal generation and detection is DSP based in the general style of the 18-MHz transceiver described previously in this chapter. At the 10-to-20-kHz I-F, two software I-Q mixers are driven by software generated sine waves at 0 and 90-degree relative phase shifts. This forms the basis for precise SSB conversion to audio. For FM, an *arc tangent detector* is used as outlined in section 10.9, FM Reception. Audio processing starts with an AGC followed by FIR filters for either band-pass filtering or *LMS denoise* filtering.

An FFT spectrum analyzer operates continuously, providing a spectral display on the PC. The spectral data is sent to the PC via a serial port operating at 9600 baud. The UART*for the serial port is in the DSP software, again simplifying the needed hardware. A continuous display of the data is very useful for determining the usage of the spectrum as well as for detecting the presence of signals that are too weak to be heard by ear. The DSP-10 also has provision for very weak signal (but slow) communications using this data.** More is said about this in Chapter 12.

DSP-10 Front Panel

In order to provide an adequate human interface for a transceiver of this complexity, the control comes from a PC. Even at that it represents a rudimentary approach to a "front panel" in that only keyboard commands are used and the program runs under DOS. Control settings, such as **FREQUENCY** and **AUDIO GAIN** are displayed along the left side of the panel. On the right side, the topmost portion of the screen is keyboard driven transmit data that will be sent in Morse code. Following down the right side is a spectrum analyzer display that represents the current receiver audio.

Below the spectrum analyzer display is a large block containing a long-term presentation of spectral signal strength, called a *waterfall display*. Brighter colors represent greater signal strength as illustrated in **Fig 11.32**. Each time that the spectrum is calculated for the upper display, a new row of pixels is added to the waterfall display. Eventually the display area is fully used and the display must scroll up to show only the newest data. This general type of spectral display has been widely used to look for patterns representing "coherent" signals. The human capability for pattern recognition operates well here.

Finally along the bottom of the screen is a status line that can be used for a variety of purposes ranging from diagnostic status information to the current position of the Moon or Sun.

Additional DSP-10 Features

Also available through the software are:
- Eight audio filters of varying characteristics

---

*Universal Asynchronous Receiver-Transmitter (UART) interprets and transmits the serial data at a serial port for communications with a computer. Devices performing these functions are available as integrated circuits, but can be implemented in software where adequate computing time is available.

**Two specialized weak-signal modes, called LHL-7 and PUA43 are fully described on the book CD. At VHF and microwave frequencies, these techniques have been used to communicate at signal levels more than 20-dB below the levels possible with conventional CW.

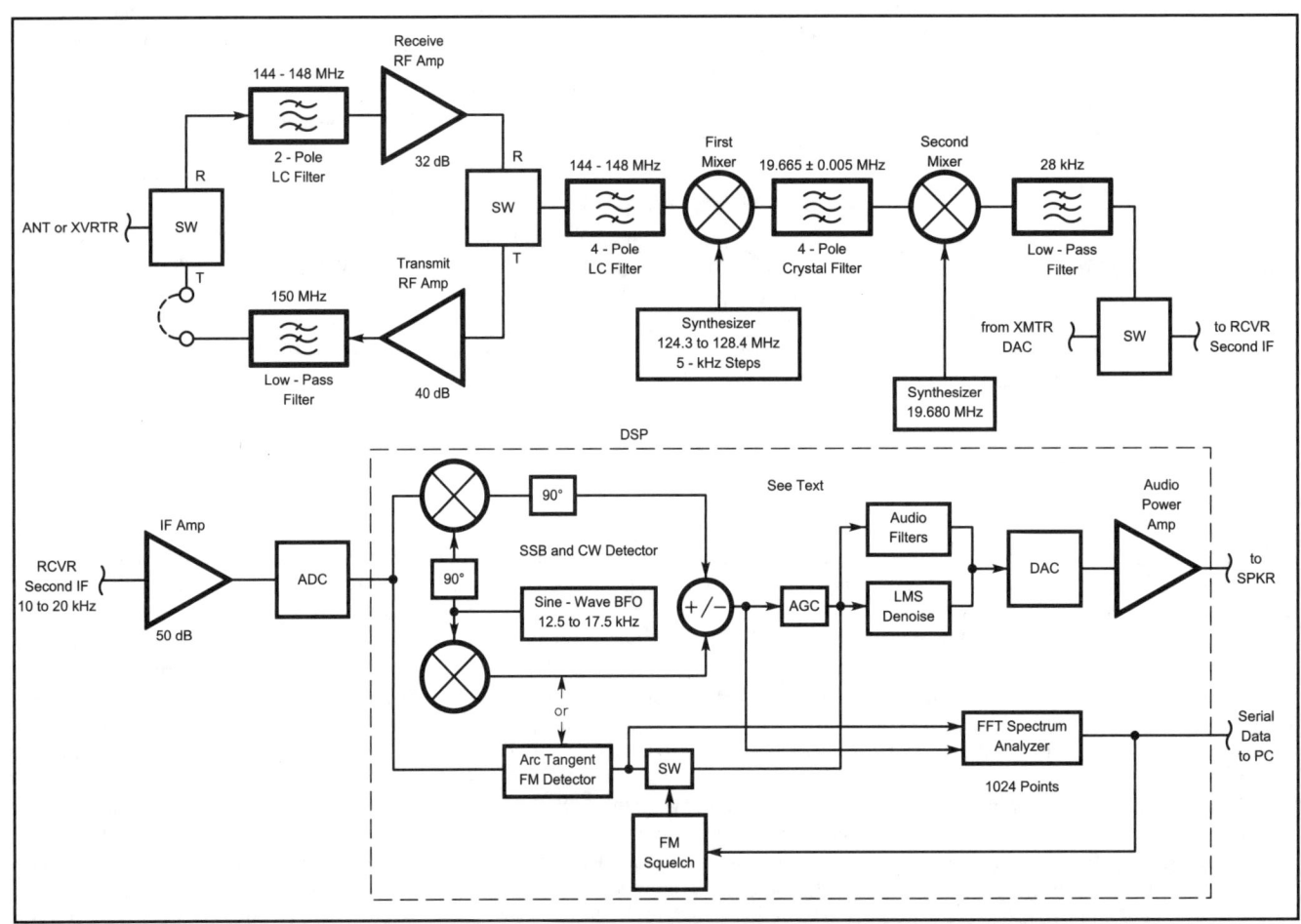

Fig 11.31—Overall block diagram for the DSP-10 2-meter transceiver. The portion inside the dashed lines is implemented as a DSP program. Not shown here are the control and display functions that are implemented in a PC.

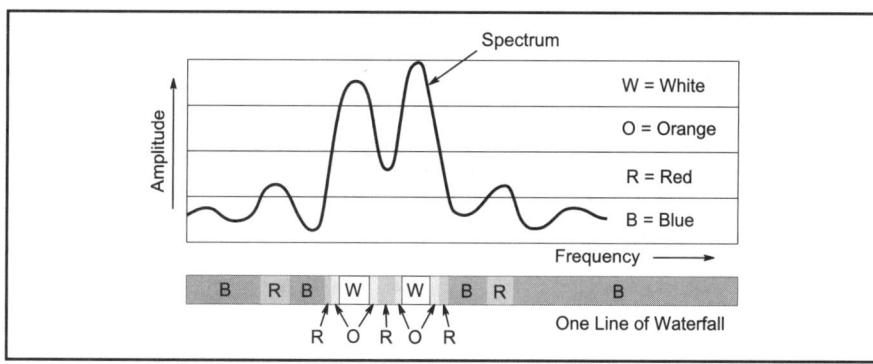

Fig 11.32—Diagram showing how the upper spectral is "sliced" into colors to form the one line of the waterfall display. While this simplified diagram has only four colors, the waterfalls usually have 16 colors or more. Added colors improve the ability to see weak signals against a noise background.

- One audio filter that can be customized
- Auto-Notching of tones
- Automatic correction of receiver frequencies for EME* operation[8]
- A variety of long-term averaging methods
- Frequency corrections for external transverters**
- Accurate S-meter reading displayed in dBm
- Saving of spectral data in computer files

This summary of the features illustrates the potential of adding sophistication to the radios operation through software. The initial radio can be quite primitive with the features growing with time. New features are added to existing radios by loading the new software. This process lends itself to group activities, where the final product can be shared by software distribution.

An additional characteristic of the software-based radio is the ability to change its "personality" by the loading of different software. Often, new modes of operation and control of the radios operation may be added as software is written. However, the hardware design process is challenged to anticipate future applications. Adding a little more control, such as a gain adjustment, to the hardware may allow considerable growth in capability by future software changes. However, adding control of enough functions and having adequate bandwidth for future needs may instead add considerably to the cost and complexity of the hardware. Which brings back the point made for all-hardware radios, that the price of trying to make a software radio totally flexible may well be an unfinished project!

DSP-10 Multi-Rate Processing

As discussed earlier, the only hardware interrupt occurs at a 48-kHz rate. Certain processes, such as the audio filtering and serial data transmission, do not require this high rate of processing. To minimize the processing time required, much of the processing is performed at 1/5 rate, or 9600 Hz. Since this is a sub-multiple of the basic rate, only the one interrupt routine is needed.

Within the interrupt routine, a software divide-by-five is used to determine which of the 9600 rate routines are to be processed. Even though the processing load will generally not be evenly divided between the five 9600 rate routines, all of the remaining time is still available for the background routines. The key design parameter is the longest running of the five routines. This must not exceed the 1/48000 second (20.833 microseconds) that is available between interrupts.

Provision is made for using a triggered oscilloscope to measure the amount of time spent in the interrupt routines. At the start of each interrupt routine, a hardware logic level output is set high. Returning from the interrupt routine sets the line low. This allows an oscilloscope to see each of the five routines and their running times. Most triggered oscilloscopes have a variable "time/div" which needs to be set to just cover the 5x20.833=104.2 microseconds. Usually it undesirable for the oscilloscope to trigger for the next 104.2 microseconds. If there is a "Hold-Off" adjustment on the oscilloscope, this is easily handled. Otherwise, some care in setting the trigger level will normally result in a consistent trigger point.

DSP-Based Audio Processor

The DSP-10 radio uses an I-F of 10 to 20 kHz with a digital sampling rate of 48 kHz. However, nothing restricts using the I-F portion of the radio without RF hardware by extending the input frequency range down into the audio range. When the "BFO" gets to zero Hz, one has an audio processor. What this means is that the same EZ-KIT Lite DSP board used for the other projects in this chapter becomes a full-featured audio processor, suitable

---

*EME refers to the Earth-Moon-Earth path of signal reflection. Due to the Earth's rotation and the non-circularity of the Moon's orbit, there is a Doppler shift in the returned signal. This shift is up to about ±400 Hz at 2-meters and proportionally more at higher frequencies.

**Operation at frequencies other than 2-meters is possible by using *transverters* to produce external frequency mixing of both the transmitted and received signals.

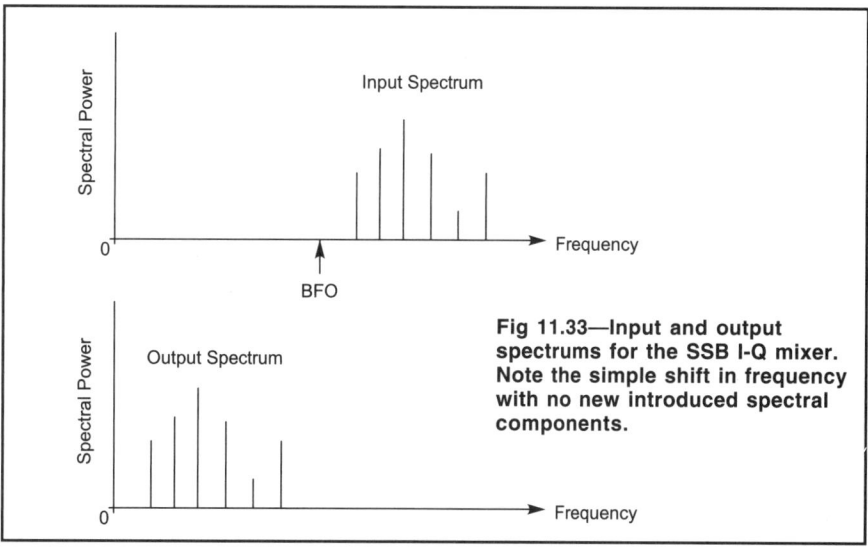

Fig 11.33—Input and output spectrums for the SSB I-Q mixer. Note the simple shift in frequency with no new introduced spectral components.

for use with any transceiver. Only two elements are *not* fully achieved without adding the DSP-10 RF hardware:
- Accurate RF frequency control under an external 10-MHz reference
- Tight integration of the control functions, such as frequency display and transmit/receive sequencing.

The overall block diagram of the audio processor is the DSP portion of Fig 11.31 that is inside the dashed lines. Modes such as FM make little sense when the input is the audio coming from a receiver, but they remain available waiting for an application! Since the SSB mixing structure remains on the input to the audio processor, it is possible to provide a frequency offset,[9] as shown in **Fig 11.33**. The I-Q mixing removes the lower sideband that would appear as a mirror image of the input spectrum, folded about the BFO frequency. The very high balance of the DSP multiplier mixers then allows the input and output spectrums to overlap without interference. The frequency display is modified for the audio processor and displays a *Frequency Offset* in Hertz in place of the radio frequency.

The DSP-10 audio processor can be used as a 0 to 20-kHz spectrum analyzer. At any time the frequency band being observed can be 1200, 2400 or 4800-Hz wide with resolution bandwidths of about 3, 6 or 12 Hz respectively. The vertical display can be set to 1, 2, 5 or 10 dB/div and unlimited video averaging is available through the PC software.

The DSP and PC programs that are used for the DSP-10 RF operation also support the audio processor. The executable programs, along with all source code are available on the *Experimental Methods in RF Design* CD. The general requirements for the audio processor are:

- An EZ-KIT Lite to run the DSP program.
- A PC to run the control program. This runs under DOS and uses 640x480 VGA 16-color graphics. A serial port is needed for communicating with the EZ-KIT. The computer need not be fast; a 486 level is adequate. This is a great application for the old computers that are collecting dust somewhere.
- An audio cable connecting between the receiver audio output and the EZKIT input.
- If transmit functions are to be used, an audio cable and possibly level adjustment circuitry is needed between the EZKIT output and the transceiver microphone jack.
- If a parallel port is available there are optional T/R controls from the PC program. These come from the parallel port as TTL levels and usually need some level conversion.

With these minor adaptations the audio processor is compatible with most of the other projects in this book.

Fig 11.34 shows the audio processor screen with a CW DX pileup. This is interesting to observe, but there was no magic as far as copying the stations! However, there is utility in using this type of spectral display for choosing a frequency on which to operate.

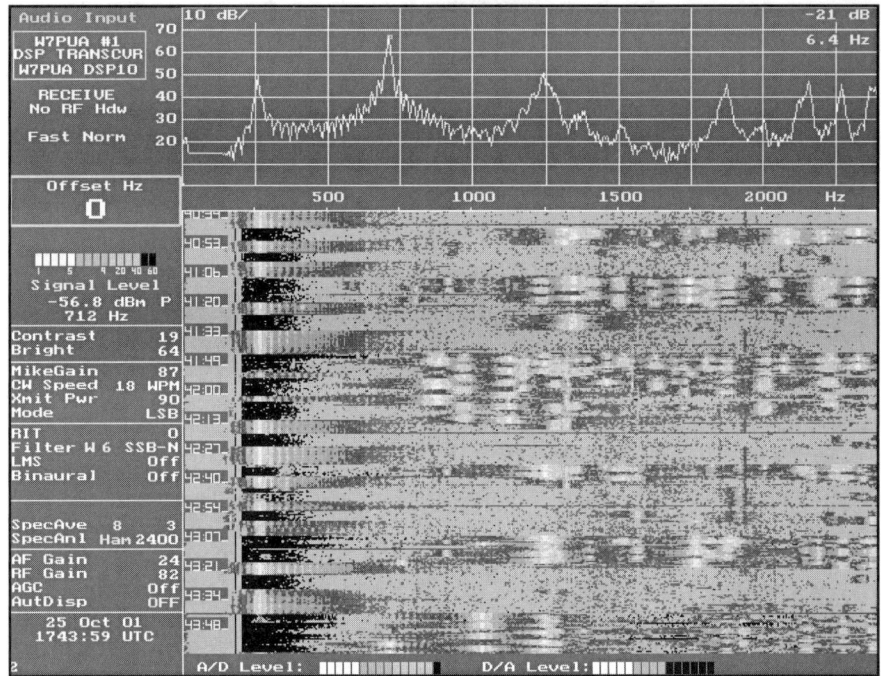

Fig 11.34—Audio processor display with operation on 10-meter CW. The top graph is the latest measured audio spectrum, which is updated every 0.6 seconds. Each of the approximately ten peaks are CW stations. The lower waterfall display shows the signal strength for each frequency plotted downward as time progresses. The time in minutes and seconds is shown at the left edge of the waterfall display. As explained in the text, brighter colors on the waterfall represent stronger signals.

The station at about 250 Hz is the DX station. He has asked stations calling him to operate at higher frequencies. The multiplicity of stations desiring a QSO and responding to the request are to the right at offsets up to at least 2400 Hz.

The bandwidth occupied by each station is mainly set by the rise and fall waveforms of the CW keying (key clicks) as was discussed for the 18-MHz transceiver.

Extensions

The features of the DSP-10 and the associated audio processor happen to be associated with weak-signal communications. Since all of the source files are available, it can be a good place to begin a project for very different uses. This might be a data communications mode, a propagation monitor or a radio astronomy project. Or it might be some only slightly related area such as ornithology research. It is often easier to modify a software project that is working than to bring up a new one from an "empty file." Either way, though, the software approach allows a different kind of flexibility than can be achieved in hardware modification. The opportunities for exploration are endless!

REFERENCES AND NOTES

1. D. D. Rasmussen, "A Tuning Control for Digital Frequency Synthesizers," QST, Jun, 1974, pp 29-32. This article on the inner workings of the rotary optical encoder has all of the information needed to construct an encoder instead of purchasing a manufactured version.

2. There are many registers that control functions or select options. Those that are selected through data memory mapped locations must not also be used for other data storage. More information on these registers is available from "EZ-KIT Lite Reference Manual," Analog Devices that is supplied as part of the EZ-KIT.

3. S. B. Cohn, "Direct-Coupled-Resonator Filters," *Proc. IRE*, Vol 45, Feb, 1957, pp 187-196.

4. G. L. Matthaei, L. Young, E. M. T. Jones, *Microwave Filters, Impedance-Matching Networks, and Coupling Structures*, McGraw-Hill, 1964. Reprinted in 1980 by Artech House, Inc., Dedham, MA. Section 8.11 covers the direct-coupled resonator filters. The remainder of this book is a wealth of RF and microwave design information.

5. W. Hayward, "Measuring and Compensating Oscillator Drift," *QST*, Dec, 1993, pp 37-41.

6. G. C. Southworth, *Principles and Applications of Waveguide Transmission*, Van Nostrand, 1950, p 606.

7. R. Frohne, "A High-Performance, Single Signal, Direct-Conversion Receiver with DSP," *QST*, Apr, 1998, pp 40-45.

8. An excellent discussion of the general characteristics of EME communications is Chapter 10, "Earth-Moon-Earth (EME) Communications" by D. Turin and A. Katz, from the book *The ARRL UHF/Microwave Manual, Antennas, Components and Design*, ARRL, 1990.

9. J. Forrer, "A DSP-Based Audio Signal Processor," *QEX*, Sep, 1996. This article provides background information on several of the basic routines as well as a set of routines that can be run on the EZ-Kit Lite. This material is contained on the book CD.

CHAPTER 12

Field Operation, Portable Gear and Integrated Stations

This book is perhaps more personal than its predecessor with the individual chapters written by easily identifiable individuals. But there is also a strong common thread of interests among us: we all enjoy a wide sampling of frequency bands, ranging from VLF through microwaves; we all have equipment that we have built that we take to unusual locations, ranging from the hills of Michigan's Northern Peninsula to the mountains of the Pacific West to the coastal waters of Oregon; We all operate stations from home, with virtually all of that operation using, or relating to equipment we have built; Although QRP is a frequent pursuit, we all use higher power at times, and we all integrate experimental activity with station operation.

This chapter illustrates some of that activity, both from the field and at home. A variety of rigs are described, showing one or more of our interests. The equipment is presented not for exact duplication, but mainly as encouragement for other experimenters. None of the equipment we have built will include the features that another designer/builder will want. But, the tools of the other chapters can be evoked for the design of whatever you might need.

12.1 SIMPLE EQUIPMENT FOR PORTABLE OPERATION

A longtime favorite activity at W7ZOI has been portable operation, predominantly from the mountains of the western United States. Many of our mountain rigs are simple (non-phasing) direct conversion CW designs. While not optimum for contests (such as Field Day), they are otherwise adequate. These are the rigs that are thrown into the pack when we just want to make a few enjoyable backcountry contacts. They also provide a link to the outside world when we hike alone. The several rigs described here are not presented for exact duplication, but as a source of ideas for the designer/builder.

Batteries and Power Sources

A wide variety of batteries offer portable power for the experimenter. Rechargeable Nickel-Cadmium (NiCd) or Nickel Metal Hydride (NiMH) batteries are ideal for radio applications, for they are capable of high current output, reasonable total capacity, and are easily charged. They also feature rather stable output voltage.

In spite of these virtues, the ubiquitous alkaline flashlight cell remains the most popular energy source. The reason is simple: the total energy per pound contained is far in excess of that available from popular rechargeable cells. A 1.2-V NiCd AA cell has a typical capacity of 500 mA-hours with the ability to be recharged for up to 1000 cycles. An alkaline cell, used only once, weighs about the same amount with a rated capacity of 2800 mA-Hours. The cell voltage can vary from 1.5 V at the beginning of use to 0.8 V for complete discharge. Data is available on the Web at **data.energizer.com/** and **www.duracell.com/OEM/Primary/Alkaline/**. Some emerging but more expensive battery technologies are also of interest.

The experimenter may wish to measure battery performance. Single cell testing is adequate, but the test should emulate the expected duty cycle, for total energy available from batteries may depend upon the way it is extracted. Accordingly, we graphically examined a typical CW transmission. A dash length is three times that of a dot while a space following either is one dot length. The pause after a letter is three dot lengths while the pause after a word is five. Our sample "transmission" then produced a duty cycle of just over 50%. A similar receiving period accompanies this during a contact, reducing operation to a 25% key down duty cycle. Most of us spend at least as much time listening as we do making contacts. So, we estimated a typical key down use as being $1/8$, or 12.5%, increasing to 25% during contests.

A circuit that will test batteries with a 12.5% duty cycle is shown in **Fig 12.1**. A 7555 timer IC oscillating at an audio rate is divided with a chain of 14 divide-by-2 elements within a 74HC4060 IC. Q13 and Q14 outputs are decoded to produce a 25% duty cycle. These are then combined with the Q6 output to create a "string of dits" with net duty cycle of 12.5%. A 74HC138 one-of-eight decoder extracts the keying signal, which then controls a power MOSFET switch.

Resistance RRX sets the load during receive periods with RTX switched in during "key down" intervals. A 1-Ω MOSFET on-resistance is part of the transmit load. The resistor values can be changed to accommodate other conditions

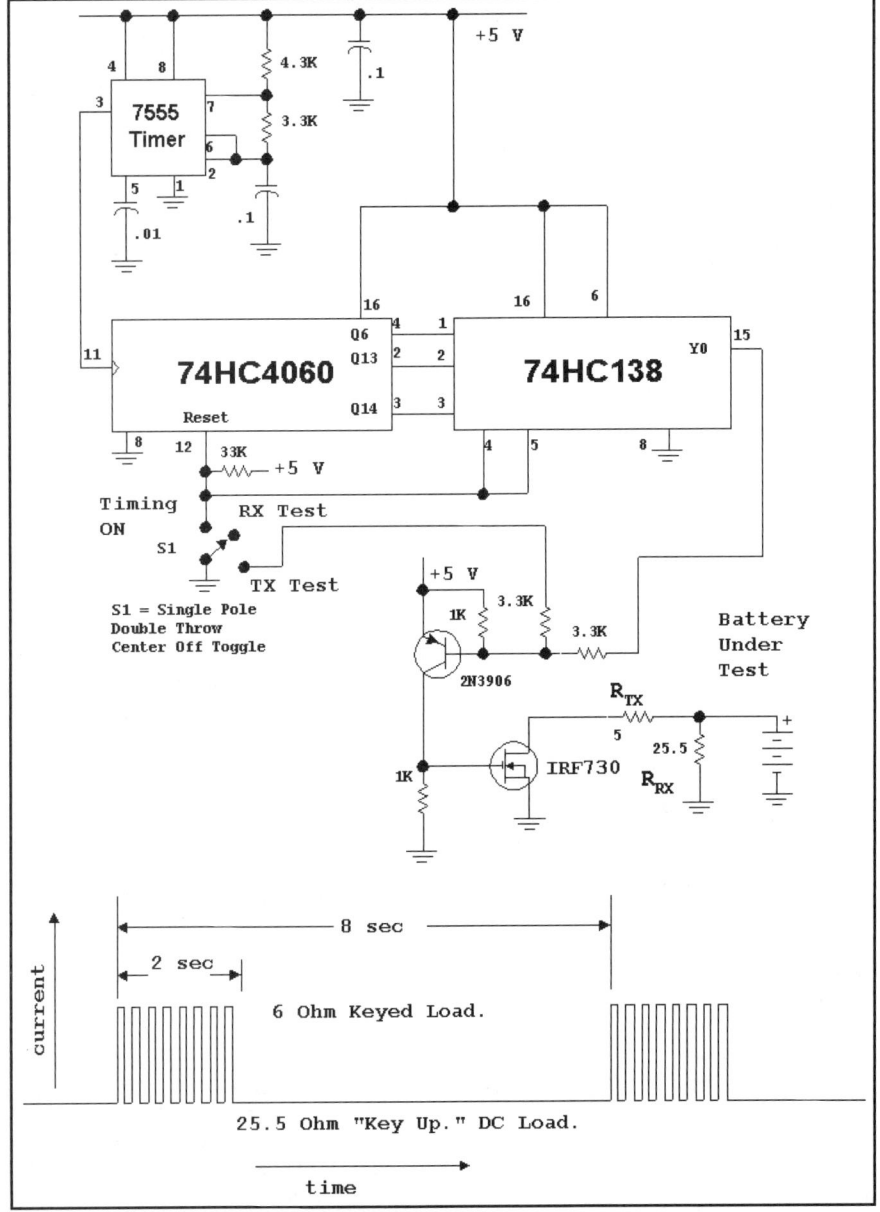

Fig 12.1—Timing circuit for testing a single cell battery at 50-mA receive and 300-mA TX current. RTX and RRX will change with a different transceiver.

switched between 8 and 10 cells. Clearly, there are numerous opportunities available for the experimenter.

Portable Antennas

Choosing a backcountry antenna presents interesting problems. The stay-at-home radio amateur generates numerous exciting ideas when first considering field operation. Thoughts of exotic beams hanging between the trees or other available structures are common. But these grand plans often change after the first trip when the complications of getting lines into available trees are encountered. Also, the impact of long runs of coaxial cable is greater when they must be carried over a few miles of trail.

Our main antenna is an inverted-V dipole. The inverted form is preferred over a flat dipole because only one support is needed. We usually carry three 50-ft pieces of 1/8 inch nylon cord. Two pieces are tied together and attached to a rock that is launched into a tree. This line supports the dipole center and the feedline. Once in the tree, only one line is needed to support the center. The remaining two pieces then support the dipole ends. If suitable rocks are not found, a cloth bag filled with smaller rocks, sand, or even snow can be used.[1] Some back-country radio amateurs will tie antenna ends to a cord that is then tied to a rock. The rock is flung into the tree where it remains suspended during operation. This is a poor practice if there is the slightest chance that the knot will become undone in the wind and drop the rock on a passing hiker!

Dipole center insulators are easily fabricated from hardware store plastic water pipe fittings. Plastic insulated wire is usually used for portable antennas, with the ends secured with nylon cord or rope, so end insulators are never needed.

The height of a dipole impacts performance. More often than not, we are satisfied with an antenna that is only 25 or 30 feet above ground, high enough for effective daylight 7-MHz operation. A higher antenna will do as well during the day, and will develop the low angle radiation needed for longer distance nighttime operation. But it will also require that more rope and feedline be packed up the trail. A simple transmatch (shown later) is usually used, even with dipoles.

End fed wire antennas are especially useful in the field, featuring a complete lack of feedline. A halfwave wire (67 feet at 7 MHz) is easily hauled into a tree with a single line. The polarization is usually a mixture of vertical and horizontal. The wire end near camp is fixed in place with

or batteries. While the scheme is certainly not a standard, it approximates actual use with a repeatable experiment. This scheme tests the battery with a pulsed constant resistance load. The manufacturers also show battery behavior with constant current. Switch S1 allows the circuit to be switched off to read the receive voltage or toggled to a "key down" mode to measure transmit current. Manual measurements are done with a DVM.

Fig 12.2 is typical of the data we obtained, based upon the load presented by the "Western Mountaineer" transceiver described later. There was about a 0.1-V difference between R and T loading over the entire battery life. This is the result of internal battery resistance of about 0.33 Ω. The perturbation at 360 minutes showed the result when the test was terminated in the evening, but restarted the next morning.

The battery life exceeds 1000 minutes for an AA cell for a key down voltage of 1.1 at "end of life." This constrains our equipment design if we wish to obtain maximum battery life. The AA cell is probably suitable for higher transmit current, limited by internal resistance.

We have modified one transceiver (below) to include a voltage measurement circuit and use a battery pack that can be

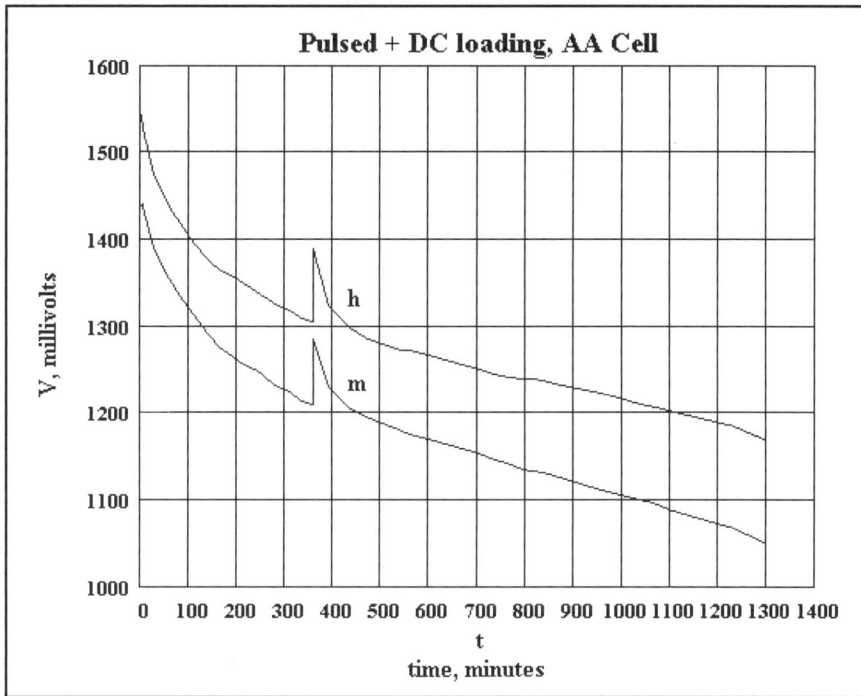

Fig 12.2—Battery voltage during pulsed testing for a single AA cell. See text for conditions.

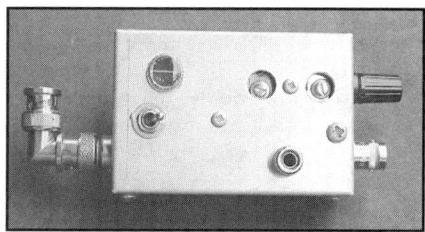

Portable transmatch using screwdriver adjustments.

Center of inverted-V dipole. The rope supports both the antenna and the transmission line.

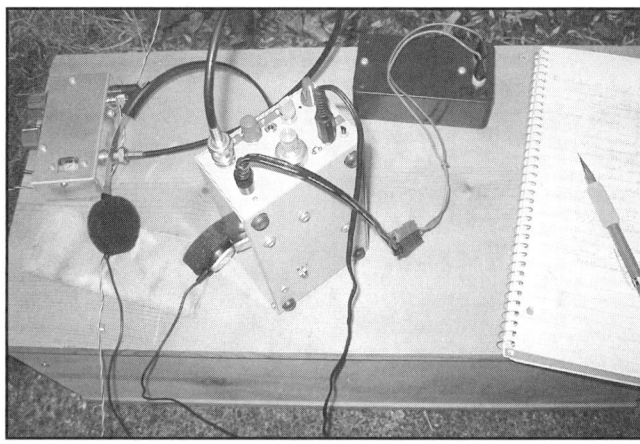

Back yard experiments should include some listening to be sure the antenna is really functioning.

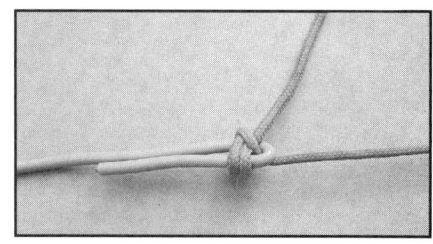

The end of a portable antenna requires no insulator. A tie-off cord or rope with the insulation on the wire is sufficient.

A dipole insulator is fabricated from an end cap of PVC pipe. This cap is 1.25-inch OD.

a short piece of rope and fed with a transmatch. One or two quarter wavelength pieces of wire are laid on the ground to form a reference for the transmatch. A transmatch that differs from that used with dipoles is usually required, for typical Z is around 3000 Ω. Measurements on back yard systems show that while an end fed wire without a reference radial or two can sometimes function, the match is then susceptible to hand capacitance effects. These problems disappear with even one radial. Slip the radials into the brush where they won't be under foot.

An end-fed full wavelength wire also enjoys a lack of feedline, and can be configured to generate a dominant horizontally polarized signal with lower angle components. The full wave wire can also be configured as a loop.

An antenna support is a problem when operating above timberline. We have carried a 12-foot telescoping whip (14 inches collapsed) to support a dipole. The whip base is lashed to a rock, ice ax, or ski pole. Fishing poles of various sorts are popular among QRP enthusiasts, some coming in lengths of 20 ft or more. Sometimes no support at all is needed; a dipole on snow or dry rocks can still function, although experimentation is required.

VHF antennas present a different challenge. Our standard portable mast uses 0.625-OD aluminum tubing in the form of two 5-foot tent poles, each in three sections. A ten-foot length is formed with a connecting piece of 0.75-inch OD tubing. A slip ring provides a guy point at the 5-foot level. The usual antennas used at 144 and 432 MHz are coax fed Yagis.[2]

Bands and Modes

The dominant band we use in the mountains remains 7-MHz CW. Other operators have different preferences. A good friend and hiking companion, WA7MLH, has done a great deal of winter camping from snowshoes or skis. Jeff has found both 80-meter CW and 75-meter SSB to be effective. Unfortunately, 80-meter CW often lacks people with whom to converse. The higher bands can be great fun when working other QRP stations. The antennas are usually a bit easier at 14 MHz and above. Simplicity remains the best guideline. Some simple beams are useful for Field Day and other committed radio events, but are not recommended for routine backpacking where the radio gear is a secondary goal.

The Trail-Friendly Radio

The term "Trail Friendly Radio," or TFR was introduced in 1996 by members of the "Adventure Radio Society (ARS)," an informal group of QRP enthusiasts who regularly take radio gear beyond the limits of motorized travel.[3] A TFR need not look like the usual home bound transceivers that must sit on tables or shelves. Some of the following equipment is in the TFR category. Also see the "Sleeping Bag Radio" described elsewhere in this chapter.

Equipment for backpacking or other field use should be lightweight, compact, and should be easy to operate. A minimum of controls is desirable, and they should be capable of use even when the operator wears gloves or mittens. Temperature testing prior to use is vital.

The Adventure Radio Society sponsors an informal, monthly contest called the "Spartan Sprint" that emphasizes these ideals. The scoring for this contest is essentially the number of contacts divided by the total station weight, including key, headphones, and batteries. It is common to encounter several stations in the contest with total station weight under a pound, with some around 0.1 pound! This is realized only with meticulous attention to details such as small circuit boards with less than normal thickness, screwdriver tuning (with very light-weight tools), rigs without cabinets, Lithium batteries, and absolute minimum power. While most "winners" are operating from a home environment, some are taking this minimalist equipment into the field. More is to be found on the ARS Web site.

Alternative Power

Many of the folks participating in QRP and in backpacking radio are also intrigued with alternative energy. The most common form is solar power, although the present "wind-up" broadcast receivers suggest many mechanical sources, including wind and waterpower.

Some simple circuits for use with solar panels are shown in **Fig 12.3**. With some solar cells and rechargeable batteries, it is permissible to merely connect the panel to the battery, perhaps with a diode to prevent leakage into the panel. The current from the panel should be less than the maximum allowed charging current for the battery. Current is confined with a series current limiter, shown in Fig 12.3A. With the components shown, current is limited at either 50 or 130 mA from the charger. This circuit should not be used without a rechargeable battery, for that would allow voltages greater than 15 to be applied to the transceiver. Figure 12.3B uses a shunt regulator with current limiting to either charge a battery or to supply a voltage regulated output. The latter occurs when

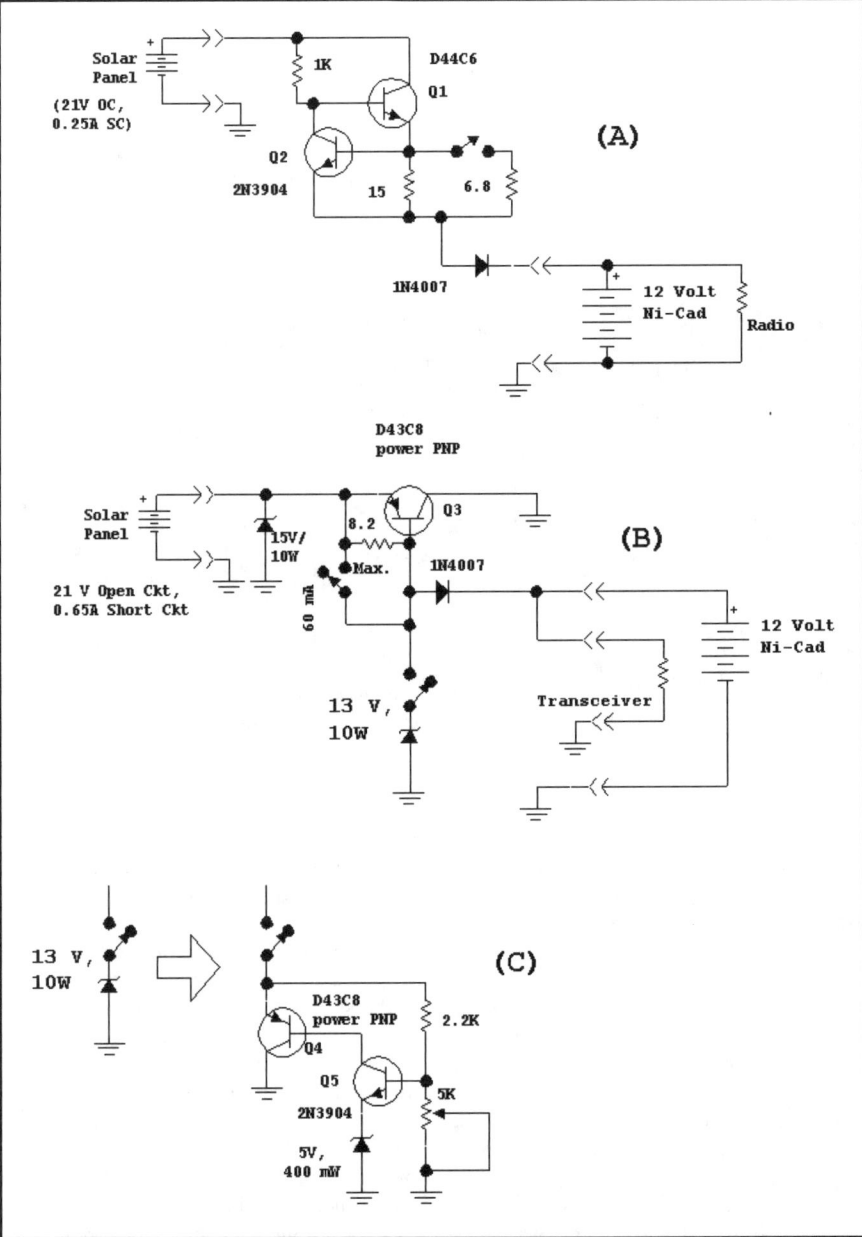

Fig 12.3—Some circuits for handling solar panels. See text for discussion. A TIP-32 may be used for the power PNP, replacing the D43C8.

A solar panel provides energy to keep batteries "topped off" during a 1993 Field Day operation. The operator is sitting in the tent to escape a light rain.

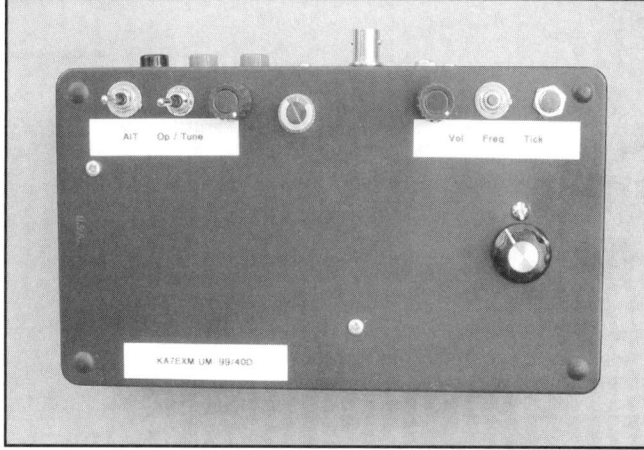

Front panel of Portable CW transceiver. The station weight, including batteries, earphones, keyer paddle, transmatch, and an end fed antenna, is about 2 pounds. The transceiver includes a bridge and VSWR indicating meter, so the transmatch consists of nothing more than the matching network.

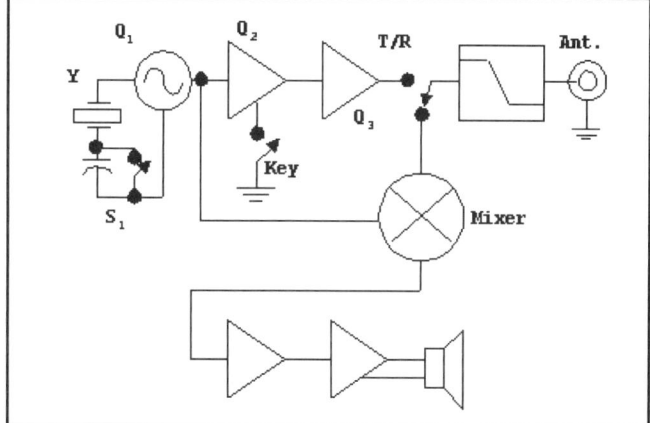

Fig 12.4—Block diagram for a simple direct conversion transceiver. A single crystal oscillator serves a dual function.

the 13-V Zener diode is switched into the circuit and is useful when making contacts with the solar panel being the only energy source. The 15-V Zener diode protects the transceiver against excessive voltage. Q3 can dissipate the full energy capability of the panel, so a heat sink should be used. Solar panels are capable of short circuit operation without damage. Power Zener diodes are expensive and are best replaced with the adjustable shunt regulator circuit shown in Fig 12.3C with Q4 also attached to a heat sink. The designer/builder should investigate modern battery management integrated circuits from Maxim and other vendors.

Micro-Mountaineer-Class Transceivers

A simple transceiver can be built with a single crystal controlled oscillator serving a dual function: The oscillator is the frequency control for a simple two or three stage transmitter; the oscillator is also the LO for a direct conversion receiver. A block diagram is shown in **Fig 12.4**.

This transceiver topology is the result of current operating practices where operators calling CQ will rarely look for an answer more than a kHz away from their transmitter frequency. With such a practice, there is little value in receiving on frequencies other that those where your transmitter can function.

Some sort of offset capability is required for the crystal oscillator in such a transceiver, needed to produce a beat note that can be heard when a station is zero beat with your transmitter. This can be an inductor or capacitor in series with the crystal. The extra element can be switched in or out automatically with the keying, or can be manually activated by the operator. These differences are all details that the experimenters can individually implement.

A simple Micromountaineer transceiver results from combining the "Beginner's Transmitter" of Chapter 1 with the Micro-R1 basic direct conversion receiver of Chapter 8. The sidetone oscillator and transmit-receive switch included with the transmitter complete the station.

A contemporary version of the Micromountaineer was presented in *QST* for July, 2000 with the article included on the book CD. That version featured 2N3904 transistors throughout the RF portion of the design with a NE602-LM386 combination as the receiver. (See the beginner's receiver in Chapter 1.) MOSFET switches are used in the T/R system for a rig that can be built for any band from 1.8 to 50 MHz. The 28-MHz version has been used for contacts all over North America and Japan.

The July 2000 version lends itself well

A Micromountaineer class transceiver uses internal crystals, but accepts an external VFO.

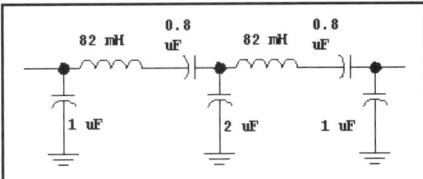

Fig 12.5—An audio bandpass filter for use with the *QST* July 2000 Micromountaineer.

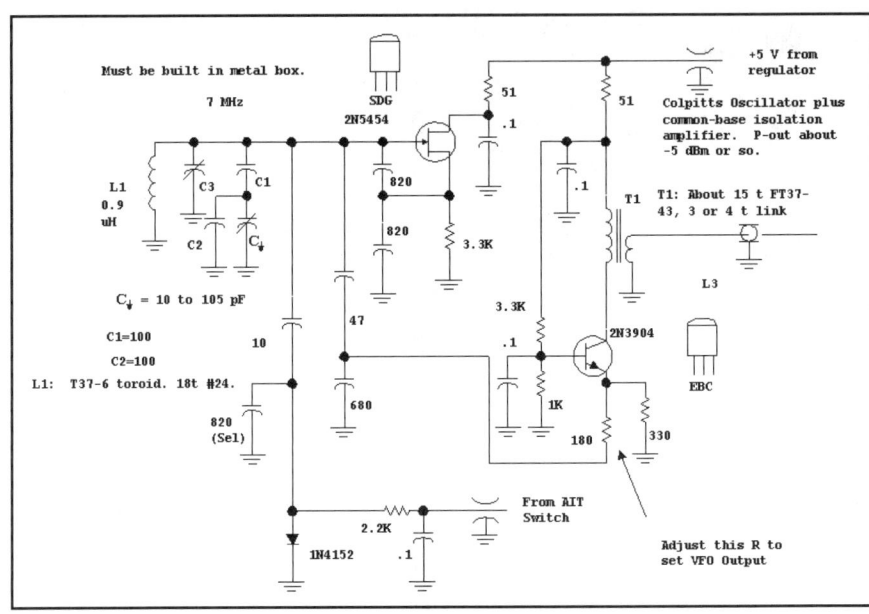

Fig 12.6—7-MHz VFO for use with the July 2000 *QST* transceiver.

to modifications. **Fig 12.5** shows a passive LC audio filter that can be added in the headphone lead to substantially improve selectivity. Ed Kessler, AA3SJ, built this circuit.

A variable frequency oscillator is easily added to Micromountaineer class portable rigs. **Fig 12.6** shows one that was added to the *QST* July 2000 version built by Roger Hayward, KA7EXM. The VFO operates at the 7-MHz output frequency, so it is vital that the oscillator be shielded from the rest of the circuitry. If the oscillator frequency was reduced to 3.5 MHz and was followed by a frequency doubler, no shielding would be needed. This transceiver is shown in the photographs. **Fig 12.7** shows the modifications used within the transceiver. The previously tuned output at Q2 was replaced with a ferrite transformer. The VFO signal is then injected at the base of that stage. The gain is set with the addition of Q2 emitter components while a dc signal from the AIT switch is routed to the feed-through capacitor feeding the 1N4152 diode. The capacitor marked "sel" in the VFO may be selected to set the offset with the value show producing about 800 Hz in the KA7EXM transceiver.

A 1-kΩ resistor is added to the transceiver to feed a sample of the oscillator signal to a frequency counter. KA7EXM used a Frequency Mite from Small Wonder Labs for this function. See the discussion of counters in Chapter 4. The interface from the main transceiver board to the counter should be coaxial cable or a twisted wire pair.

This transceiver also includes a built in electronic keyer. Both the keyer and frequency counter provide sidetone outputs that are routed to the audio system. The modification to the audio on the trans-

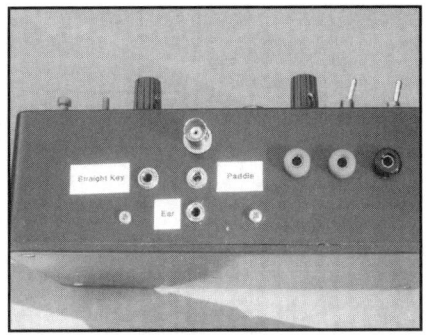

End view of the KA7EXM 7-MHz Micromountaineer.

The external "plug-on" VFO for use with the hand held rig.

ceiver is shown in **Fig 12.8**. The user may wish to disable the sidetone oscillator included on the original *QST* design.

The KA7EXM version of the *QST* transceiver was built as a Trail Friendly Radio as described above. It was put in a plastic box (approximately 3 × 5 × 9 inches) with internal shielding of the VFO, shown in the photographs. Controls are on the larger surface with all interface attachments to one end. While this may not be optimum for a classic home station environment with table and chair, it worked well when used on backpacking trips in Oregon's Cascade Mountains.

Earphones, rather than speakers, should always be used with portable transceivers. This is a courtesy to other back-country travelers.

There are clearly numerous modifications and variations that can be applied to this project with new bands being of special interest. Versions with the VFO operating at the output frequency would work well at 1.8, 3.5 and 10.1 MHz. Variations using a frequency doubler following the VFO would be preferred at 7 MHz and higher with a heterodyne VFO offering better performance at 21 MHz and higher.

A photograph shows a diode ring product detector based variation that we built and used in the mid 1980s time frame. Crystal control was included with a pair of internal crystals. However, an outboard VFO could also be attached when desired. Banana plugs and jacks provided a convenient mechanical interface. Coaxial cable provides a VFO output connection and a power supply interface between units. The offset control to the VFO was multiplexed

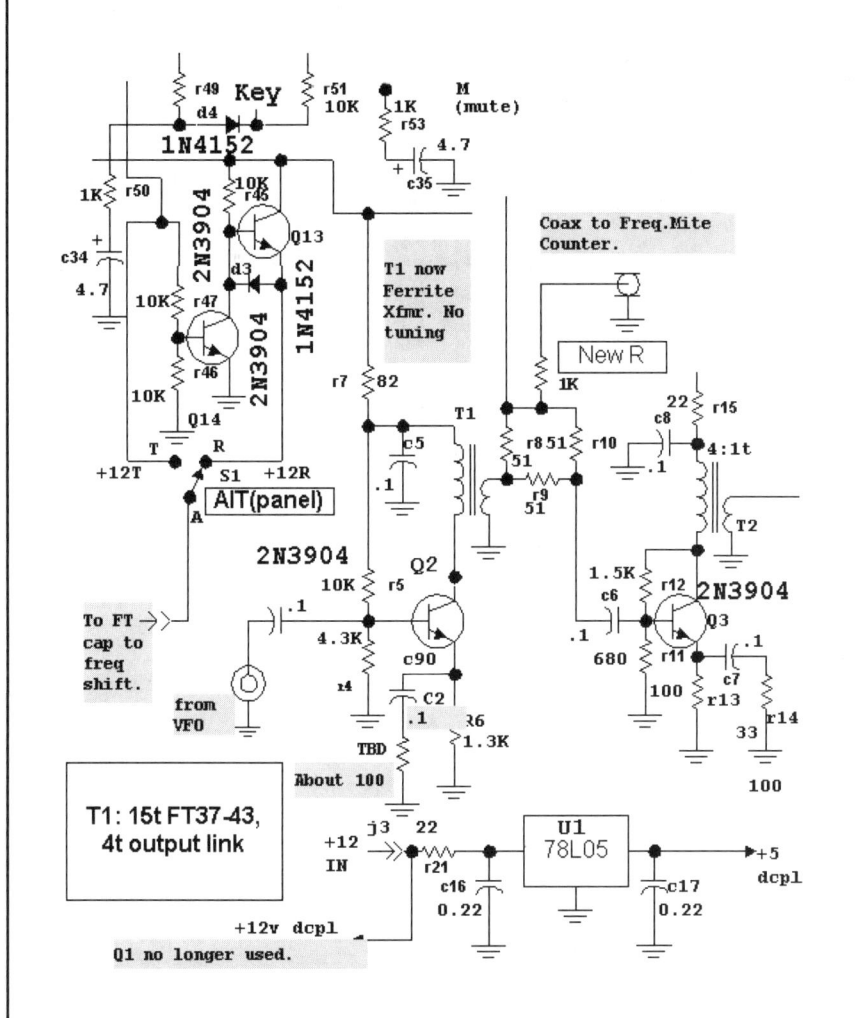

Fig 12.7—Modifications applied to the July 2000 *QST* transceiver when a VFO was added. See text.

All controls and I/O lines attach to the end of the "Western Mountaineer," allowing it to reside in a small camera case inside a parka. The turns counting dial is on a 10 turn pot to control a temperature compensated VCO. The knob in the upper right corner allows the supply voltage to be "measured."

with the RF line. This transceiver has seen nearly two decades of 40-meter CW use. The VFO is usually included, but is left at home or in a base camp during summit climbs where weight must be minimal.

The "Western Mountaineer"

This rig is a simple direct conversion transceiver based upon the popular Phillips NE-602 Gilbert Cell mixer. The name was chosen because the rig was designed for use in the mountains of the western USA where strong international broadcast signals are rarely a problem. Builders in the eastern USA or in Europe will find this circuit unsuitable and should consider a diode ring based design such as the still excellent W7EL transceiver.[4]

The VFO and transmitter, shown in **Fig 12.9**, begins with a high-C Colpitts oscillator tuned with a varactor diode, D2. This circuit is temperature compensated with two methods. Part of C2 consists of polystyrene elements with most capacitance built from NP0 parts. The tuning diode is then compensated with D1, a second silicon diode. This oscillator was discussed in Chapter 4. R1 is selected to determine the diode current. It is vital that a thermal chamber be used to adjust the temperature compensation. Details are presented in the book CD[5] and in Chapters 4 and 7.

The VFO operates directly at the 7-MHz transmitter output frequency, making oscillator shielding vital. The shield was built from tin sheet stock. A wall was built around the part of the circuit board containing the oscillator and soldered directly to the ground foil. A lid was attached, leaving access to C1. Compensation diode D1

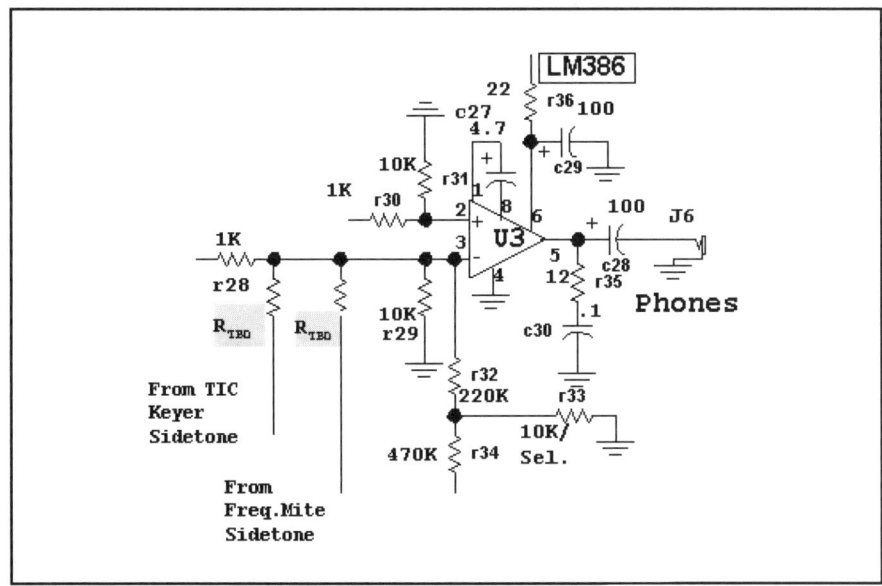

Fig 12.8—Sidetone signals from the counter and the keyer may be injected as shown. Removing R23 will disable the original sidetone from the output.

Interior of KA7EXM transceiver. The original plan called for internal batteries, but they didn't quite fit.

is enclosed in the same compartment.

The VFO is tuned with R2, a pot controlling a current pulled from the summing node of op-amp U3A. A CW offset of about 800 Hz is provided with Q13. This is configured for the Almost Incremental Tuning scheme outlined in Chapter 6. RIT could be implemented if desired; see Chapter 4.

The VFO output is buffered and amplified in several stages, eventually driving a power amplifier, Q5 and Q6, consisting of a pair of 2N3904 transistors with an output of 0.6 W. An output low pass filter provides impedance matching to the PA and harmonic attenuation.

The receiver, shown in **Fig 12.10**, begins with the NE-602 product detector, U2. The detector output is then dc coupled to U4, which then drives U6, an RC active peaked low pass filter. An interesting subtlety was discovered when this topology was first built: although the bias was as expected with about 4 V dc through the chain of U4 and U6, the voltage changed

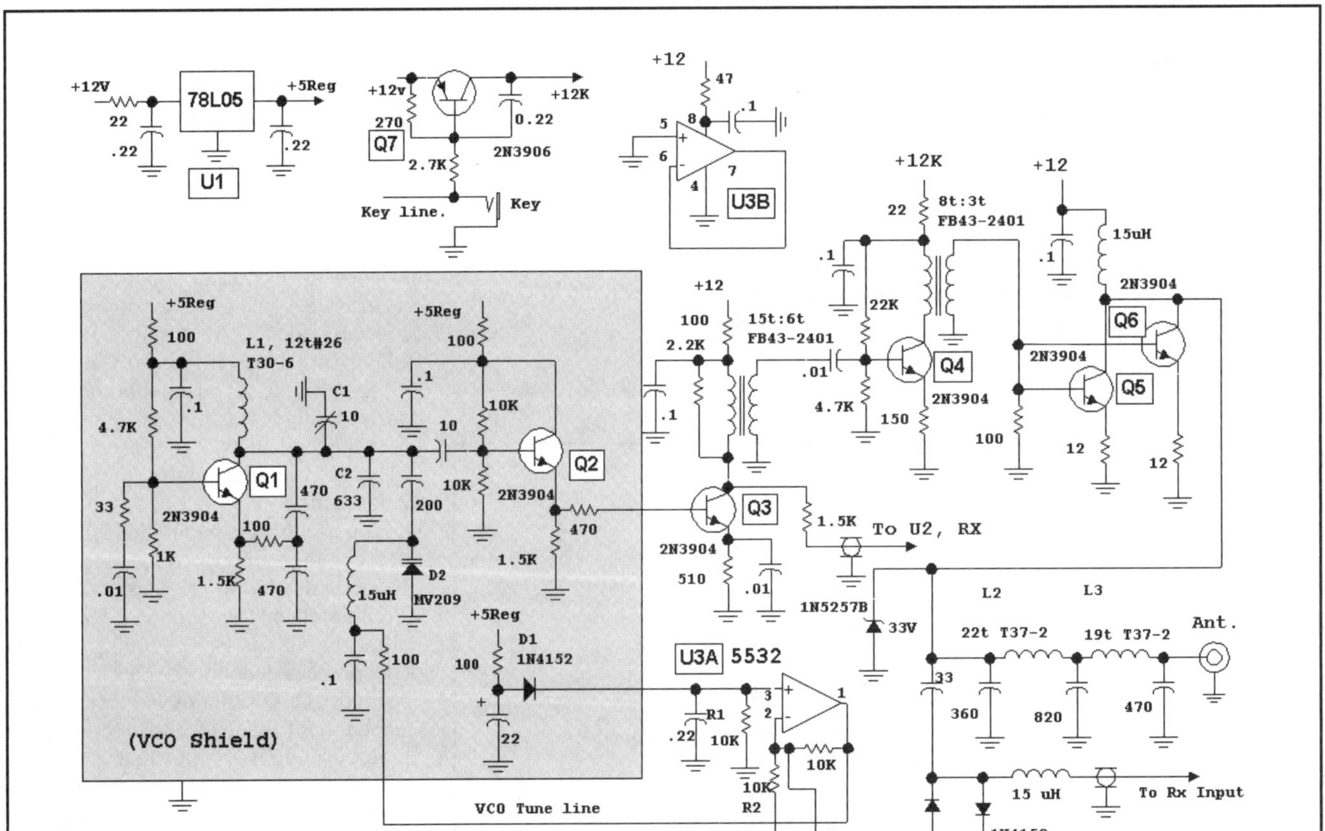

Fig 12.9—The VFO and transmitter portion of the "Western Mountaineer" direct conversion transceiver.

by several volts when the LO was attached to U2, pin 6. This was the result of unbalance in the input circuitry driving pins 1 and 2. Changing to a fully balanced topology at T1 eliminated the problem. If the circuit was duplicated today, we would use ac coupling between U2 and U4.

The receiver is muted with two FETs during transmit intervals. Q12 was usually adequate. Initially a pair of back-to-back diodes was used across U5A, but they distorted on loud signals. Complete muting was not possible after diode removal, so Q14 was added. Q12 could probably be eliminated.

The receiver schematic includes a voltage comparator using U7A. This circuit is driven by a front panel mounted potentiometer, R4. As R4 is varied, the voltage on the non-inverting input of U7A also changes. The reference voltage at the inverting input is merely the 5 V regulated supply. The output of U7A changes state when the two op-amp inputs are equal, which toggles the sidetone (Q9 and Q10)

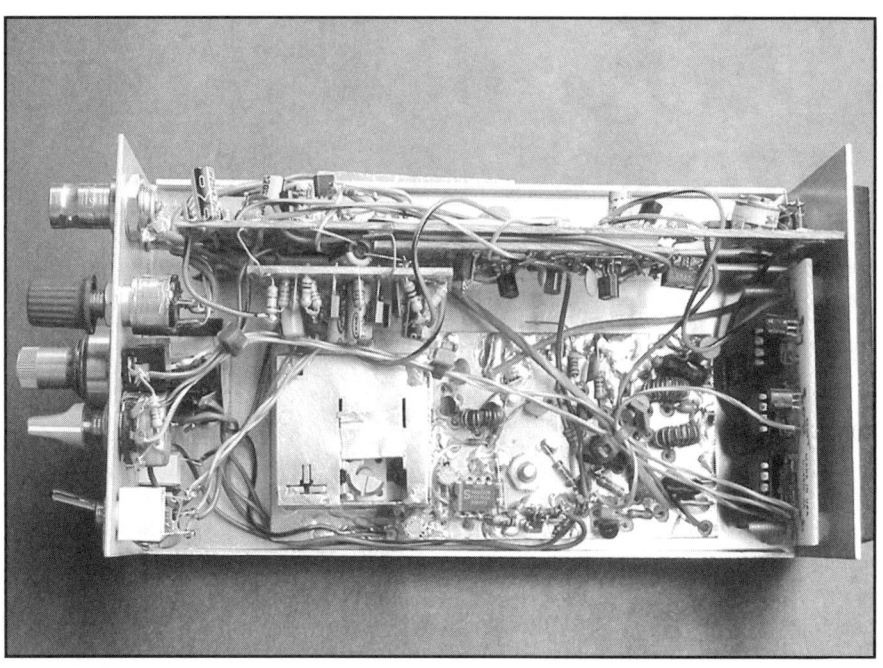

Inside shot of the Western Mountaineer showing the VFO and transmitter, excluding voltage-measuring circuitry.

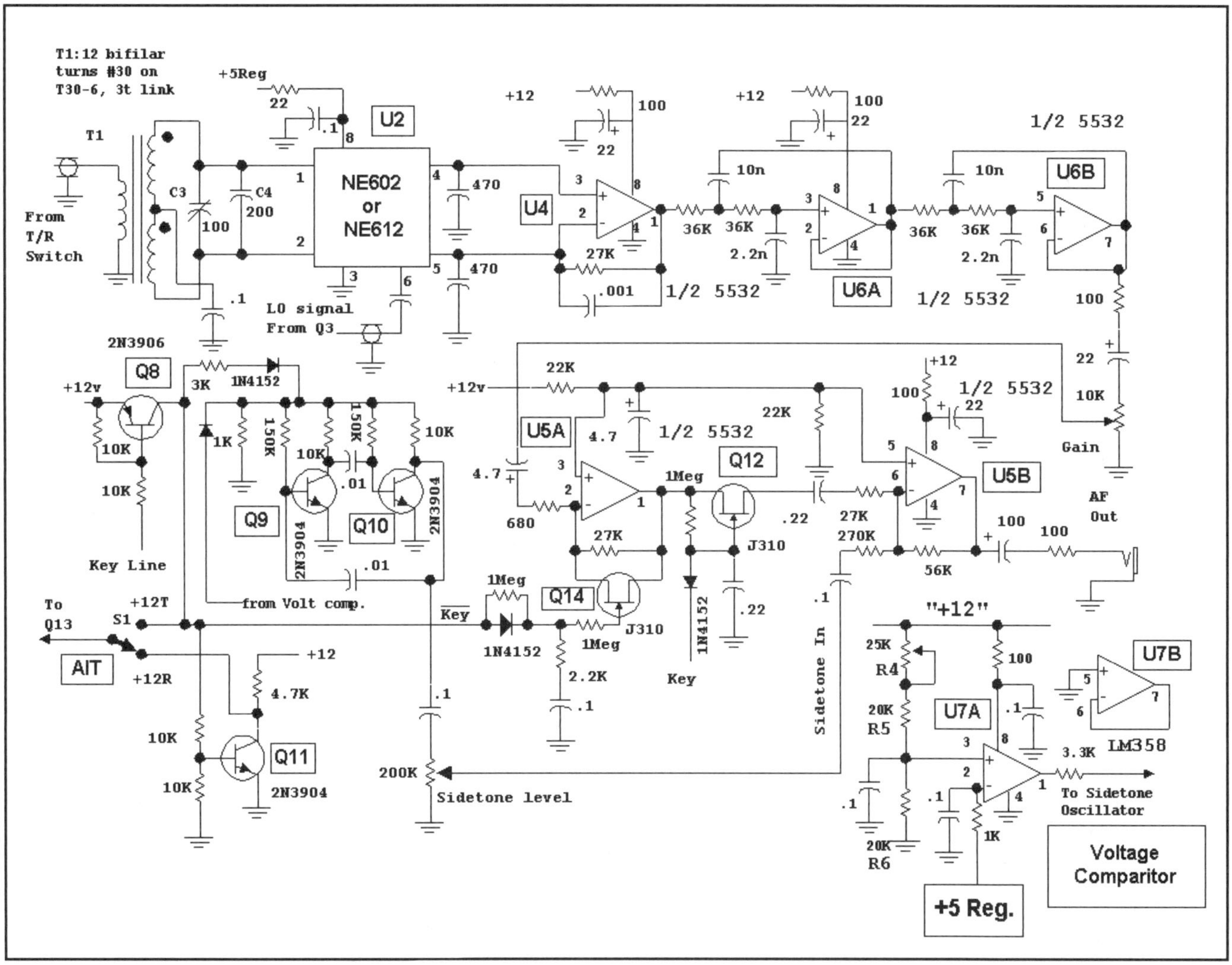

Fig 12.10—Receiver portion of the "Western Mountaineer" transceiver.

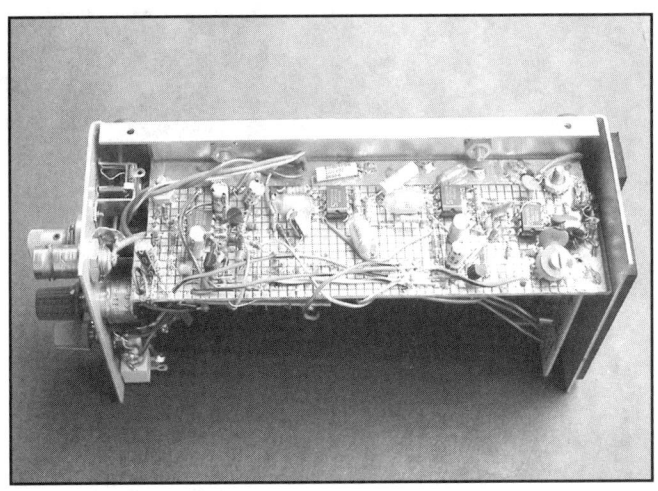

Inside shot of the Western Mountaineer showing the receiver board.

Shot of the Western Mountaineer installed in a protective case, including battery pack.

A Micromountaineer-Class transceiver in use for Field Day. The rig is in use here on the 9500-foot summit of Oregon's Mt McLoughlin.

Here W7ZOI tries to get in just a few more Field Day contacts before the rain becomes more intense. *KK7B photo.*

oscillator on or off. This serves as a method for measuring the battery voltage without a voltmeter. R4 is a 25-kΩ pot, a small part that was on hand. The designer/builder may wish to use other values. The same results will be obtained if R5 and R6 are scaled with R4. The pot is normally set to rest in a position that inhibits oscillation.

The transceiver was examined for output power and key down current consumption as supply voltage changed. This is vital information for equipment that will operate from a power source that may change as it is consumed. The results are shown in **Fig 12.11**. The receive current is nearly constant at 50 mA for this transceiver, the result of having used a large number of 5532 op-amps. The designer/builder may wish to find substitutes that consume less power while still offering low noise. U4 and U6A should use fairly low noise parts while the rest of the op-amps are less critical.

The transceiver is breadboarded on PC board material containing a matrix of islands where components are mounted. The TX board had components on the ground foil side while the RX used a surface mount like scheme with standard leaded components. The rig has most input and output cables attached to the small end of a $2 \times 3.5 \times 6$ inch box, shown in photos. This allows it to reside in a small camera bag that also includes a battery pack. The rig can even be operated from inside a down parka during winter

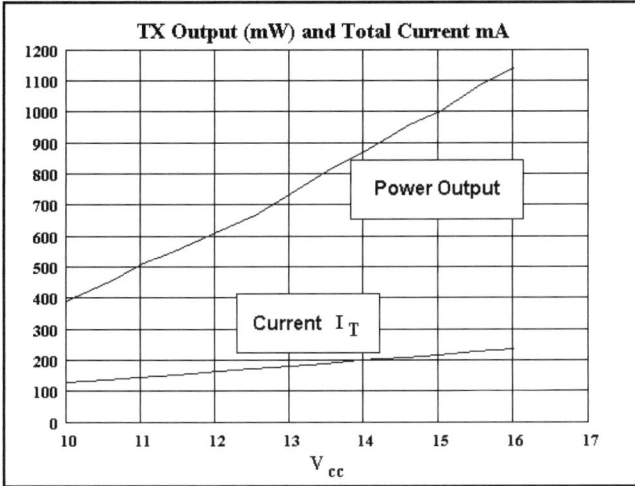

Fig 12.11—Power output and key down current consumption for the transceiver for voltages from 10 to 16 V.

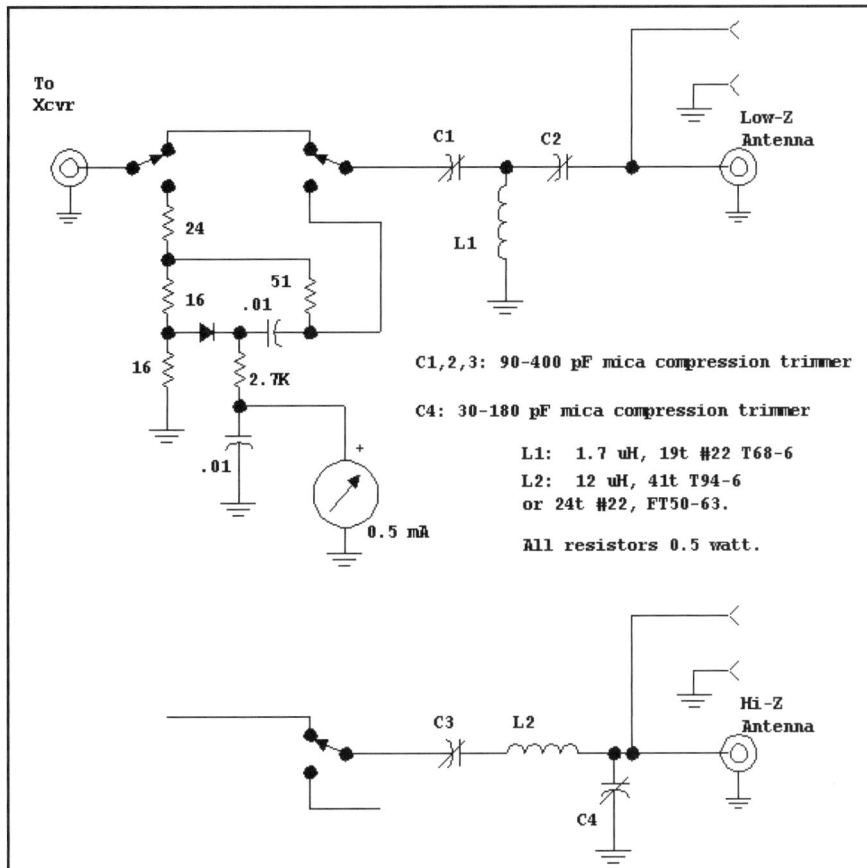

excursions. A keyer is built into the rig.

A portable transmatch is shown in two forms in **Fig 12.12**. This circuit uses screwdriver adjusted trimmer capacitors. While less convenient than capacitors with knobs, the compact and lightweight features are useful for backpacking applications.

Single Signal Systems

While the work reported here uses direct conversion for portable rigs, there is certainly nothing to preclude the use of super-heterodyne equipment. The "Unfinished" transceiver described next has been used for a number of Field Day events, always with good performance. The ultimate portable rig might well be a single signal design (superhet or phasing) optimized for low current. An excellent beginning design is a transceiver described by Benson.[6] This design has been extended in numerous kits built by QRP clubs world wide including the popular NorCal-40. Additional information is presented in the ARRL compendium, *QRP Power*, ARRL, 1996.[7]

Fig 12.12—A small transmatch suitable for portable use. The bridge is switched into the signal path only when tuning. A small screwdriver is included for tuning. The upper circuit is suitable for coax lines while the lower one is intended for end fed wires. Component values are set for 7-MHz antennas.

12.2 THE "UNFINISHED," A 7 MHZ CW TRANSCEIVER

This transceiver (single conversion super-heterodyne, 5-MHz IF with 2-MHz LO) has earned the name "Unfinished," for it is an ongoing effort that has been in a state of transition for over a decade. It has been a perpetual design platform to try new circuit ideas as they are generated. A homebrew crystal filter provides selectivity. This is intended here to be a source of ideas rather than a construction project.

Fig 12.13 shows the LO and RIT system, which tunes from 2 to 2.1 MHz, producing coverage of the bottom 100 kHz of the band. A JFET, Q7, serves as an oscillator with a bipolar buffer, Q6. Temperature was compensated with a polystyrene capacitor, adjusted with an experimental oven. (See Chapter 7) Q8 and a Zener diode provide a stable voltage for the system, although an IC regulator would serve as well. The output is low pass filtered and routed to a diode ring receiver mixer. A low power tap is extracted for use with an IC transmit mixer. A pair of varactor diodes are used as part of an RIT system.

The 2-MHz LO is built in an aluminum box, approximately $2 \times 2 \times 5.5$ inches. No lid is used, for isolation requirements are minimal.

The receiver front end is shown in **Fig 12.14**. A diode ring mixer, U1, is preselected with a double tuned circuit and followed by a bipolar post-mixer amplifier, Q1. A 2N5109 or equivalent is used, although a 2N3904 could also be applied. This transceiver is sometimes used for portable applications, so post-amp current is modest. A pad and a homebrew crystal filter follow the amplifier. The circuit shown here has a bandwidth of 250 Hz with 500-Ω terminations. The filter is designed for a Gaussian-to-6 dB shape, which has minimal ringing, even with the narrow bandwidth. The rounded peak shape is selective enough to be extremely effective, yet the low number of crystals produces a response that maintains a receiver "brightness" rarely experienced with narrow, multi-resonator filters. Impedance match is carefully controlled at 500 Ω around the crystal filter.

The Gaussian-to-6 dB filter shape is an especially good one for the experimenter, for it is very tolerant of changes in crystal characteristics or filter capacitors. Altering crystal motional L from the design value of 0.1 Henry by +/− 30%, or dropping Q_U from 200,000 to 50,000 still produced useful filters.

The receiver has a noise figure of about 17 dB with an input intercept of around +15 dBm for a two-tone DR of 97 dB. High-level mixers and a higher current

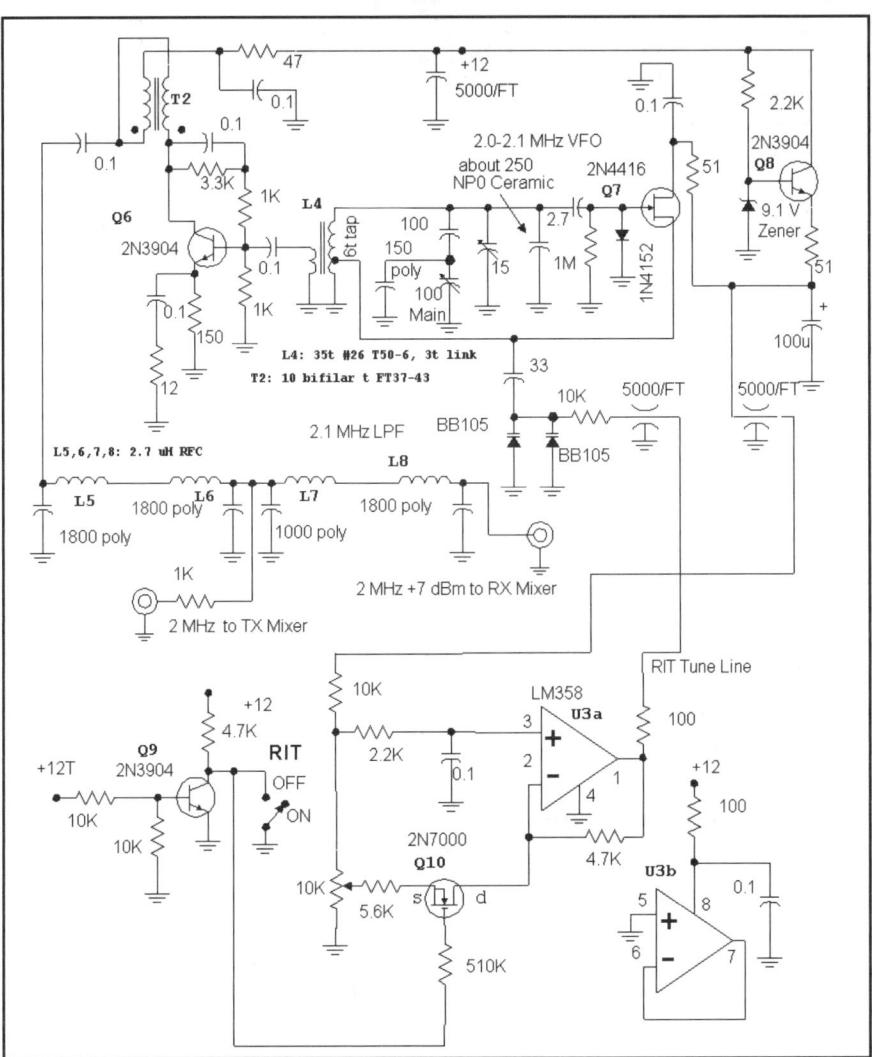

Fig 12.13—VFO and RIT for the Unfinished.

Audio circuitry for the "Unfinished-7" Transceiver. The rectangular cutout locates a crystal filter from an earlier version.

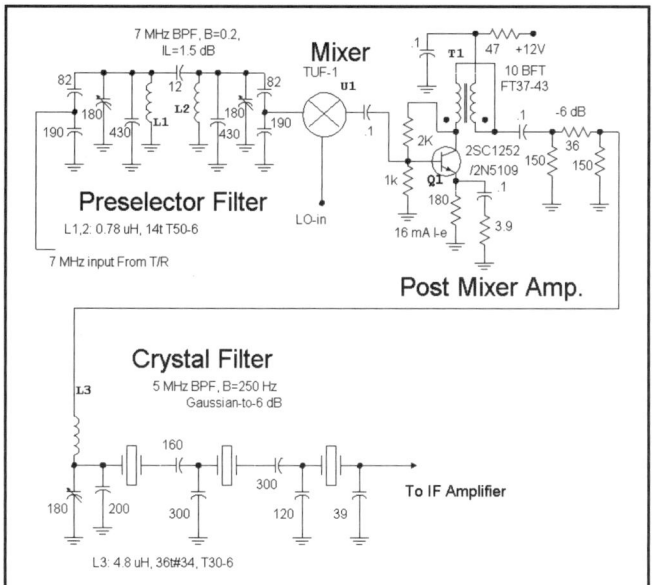

Fig 12.14—Receiver front end for the Unfinished. A Gaussian-to-6 dB shaped crystal filter is included. The double-tuned circuit is not symmetric, because an adjustment was made to compensate for interaction with the tuned circuit in the T/R system.

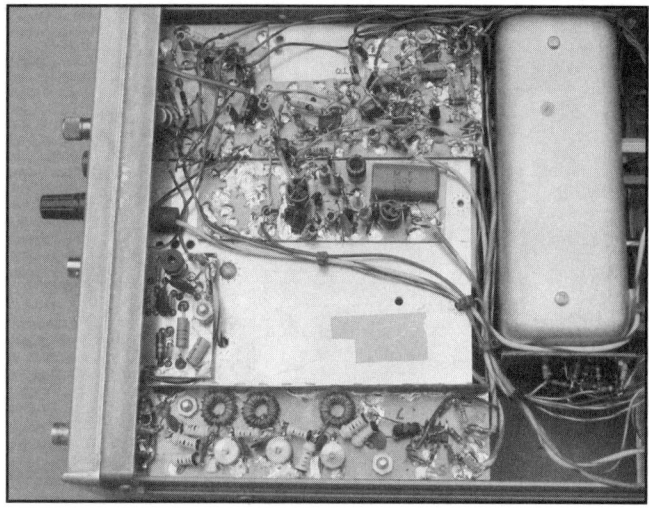

The bottom inside view of the "Unfinished-7" Transceiver. The upper circuitry includes audio, regulators, and sidetone. The board along the lower edge is the transmit mixer and triple tuned bandpass filter. The transmitter driver is the small board above the bandpass. The VFO module is at the right.

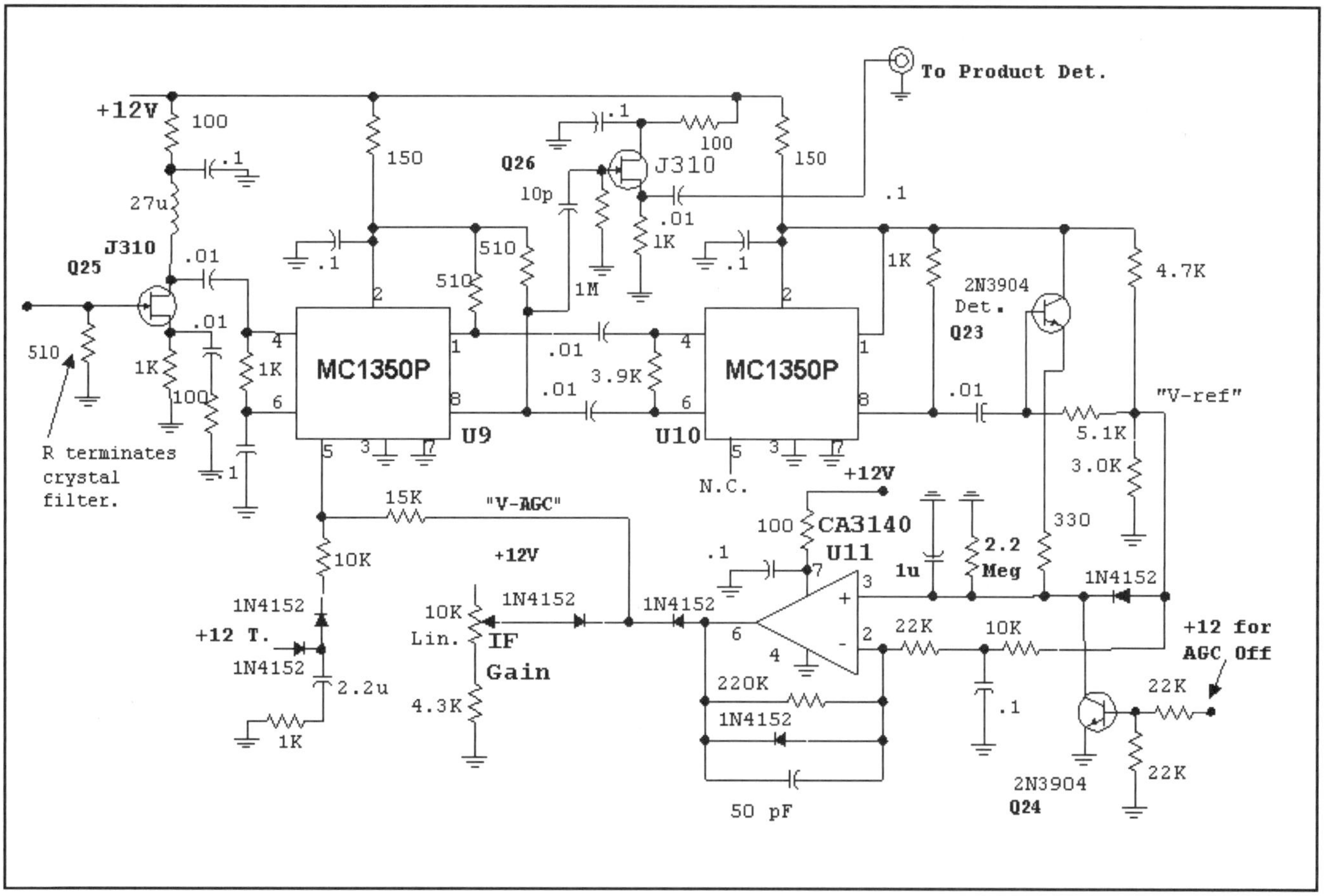

Fig 12.15—IF Amplifier. See text for details.

Field Operation, Portable Gear and Integrated Stations

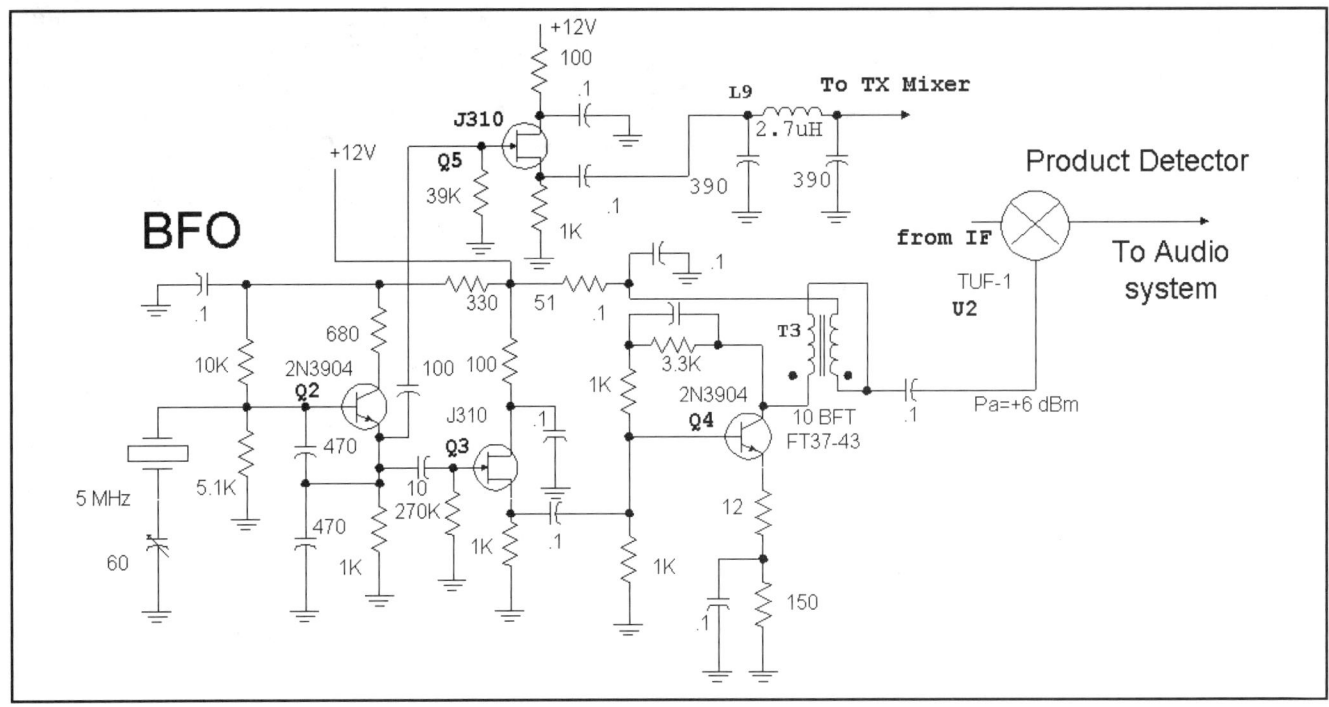

Fig 12.16—BFO and product detector for the Unfinished.

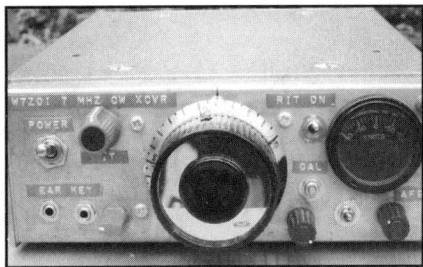

The "Unfinished-7" Transceiver front panel.

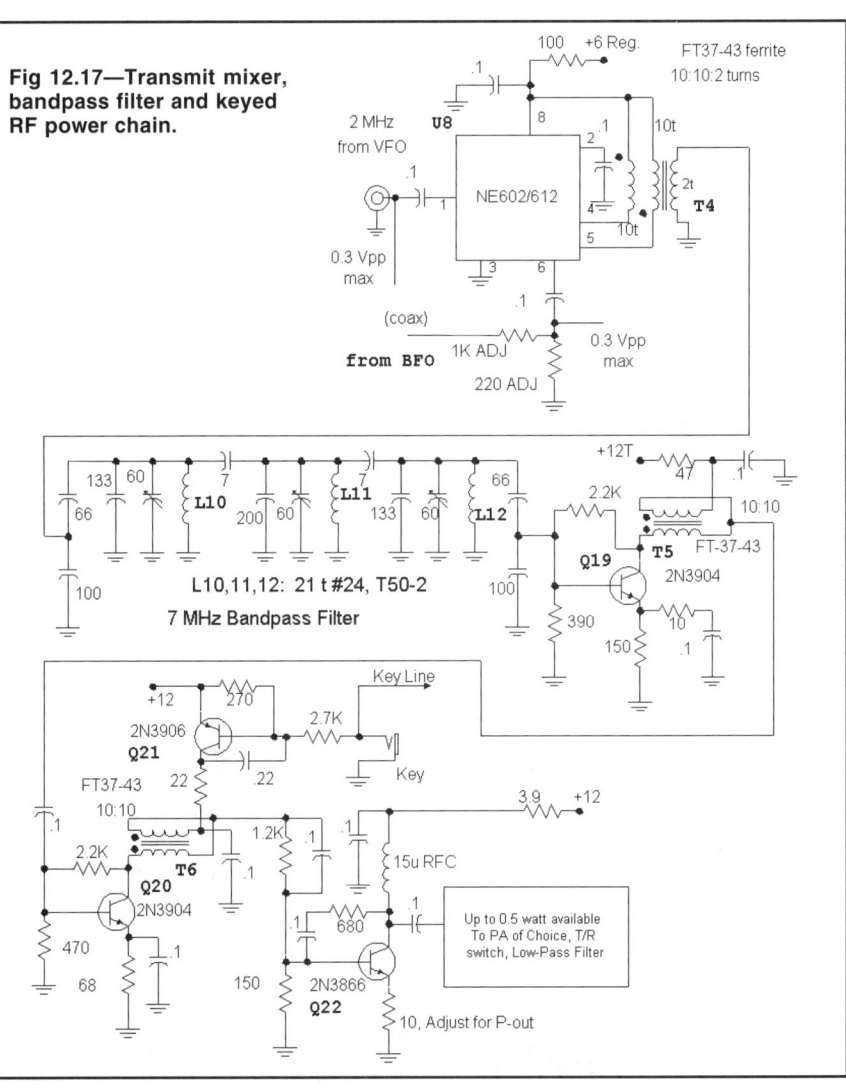

Fig 12.17—Transmit mixer, bandpass filter and keyed RF power chain.

post mixer amplifier should easily extend this well past 100 dB.

The IF amplifier, shown in **Fig 12.15**, is effective, but is probably the weak point in the design. A lower noise IF would extend the overall receiver two-tone DR slightly, as discussed earlier in this chapter. This system uses a pair of MC1350P integrated circuits, but only one has AGC applied. The output of the second is detected with transistor Q23, producing a dc signal that is applied to op-amp U11 that feeds the AGC signal to the first MC-1350P. A JFET follower, Q26, provides output to the detector. A JFET follower, Q25, precedes the first MC1350. However, the impedance is only 500 Ω, set by an input resistor. A higher impedance at this point would drop the IF noise figure. This AGC system is strictly an "ear-saver," with a threshold set high to preserve a clean response. This is a choice available to the designer/builder.

The product detector and BFO are

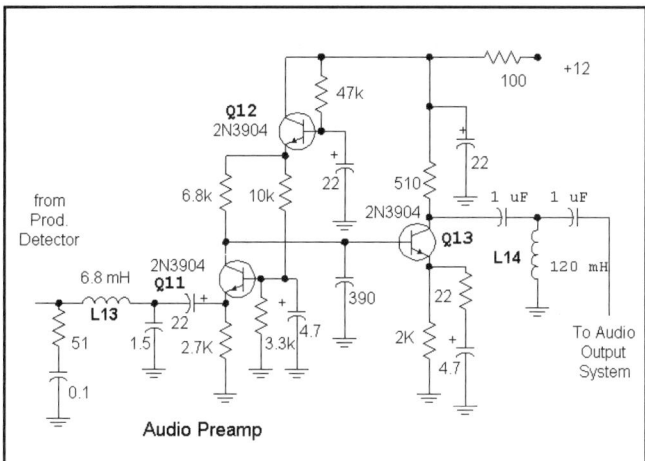

Fig 12.18—Audio preamplifier for the Unfinished.

Top inside view of the "Unfinished-7" Transceiver. The VFO module is at the right. The board parallel to the VFO is the double tuned front-end filter, mixer, and post-mixer amplifier. The third order Gaussian-to-6-dB crystal filter and IF amplifier are along the bottom with the BFO and product detector just above. The crystal calibrator and transmitter output amplifier are toward the upper left.

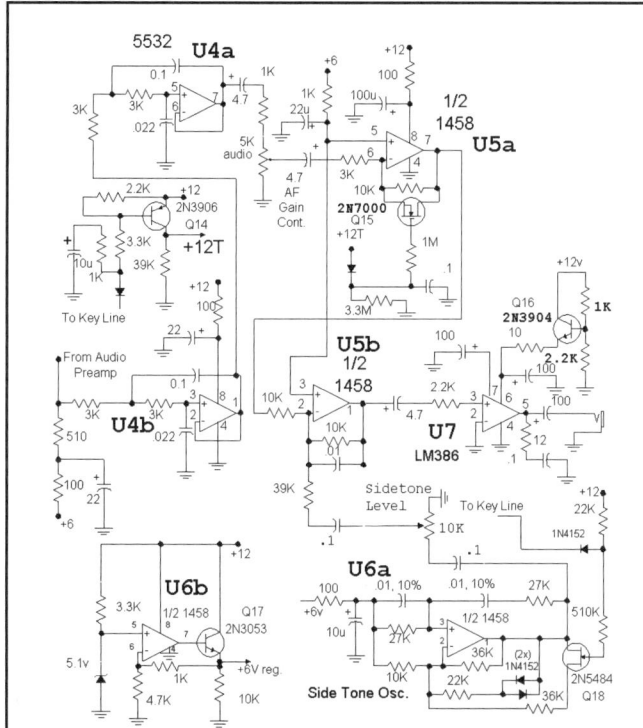

Fig 12.19—Audio output system. An RC active low pass filter, sidetone oscillator, 6-V regulator and T/R control are included.

The transmit mixer and triple tuned bandpass filter. Shield strip along one side of the board helps to confine ground currents.

shown in **Fig 12.16**. A bipolar transistor oscillator is followed by a pair of FET followers. One drives a bipolar power amplifier that then drives a diode ring product detector while the other routes signal to the transmit mixer. A separate keyed carrier oscillator was originally used. However, this produced a slight chirp. Any detectable chirp was deemed intolerable, so the design was altered. The RIT is always activated during use, with the "center" position providing a zero offset situation. A simple crystal calibrator (not shown) allows calibration in the field.

The transmit mixer, 7-MHz bandpass filter, and RF power chain are shown in **Fig 12.17**. A modest NE602, U8, works well as the transmit mixer. The BFO and VFO signals are both confined to 0.3 V peak-to-peak at the IC. This is a place where measurement is important, for "more" is not better. A triple tuned bandpass filter terminated in an un-keyed amplifier, Q19, follows the mixer. The circuitry from U8 through Q19 is built on a separate board with a long narrow shape with little shielding. The signal from Q19 is routed to a keyed amplifier, Q20 and Q22 with output up to 0.5 W. The Q22 emitter resistor is adjusted for the desired drive to the PA in use. No power amplifier design is shown, allowing the designer/builder to use what he or she needs. Spectral purity was measured with a high efficiency 8-W PA in place (See the W7EL "Brickett" described in Chapter 2). Two non-harmonic output spurs were found close to the 7-MHz carrier at the –60 and –63 dBc levels.

Field Operation, Portable Gear and Integrated Stations 12.15

The product detector drives an audio preamp, shown in **Fig 12.18**. An input LC low pass filter drives a familiar common base stage, followed by a common emitter amplifier driving a high pass LC filter.

The rest of the audio system is shown in **Fig 12.19**. U4a and b form a 4-pole RC active low pass filter with a peak at 850 Hz, -6 dB cutoff at 1.3 kHz, and a –40 dB response at 3.3 kHz. This low pass is a wonderful supplement to the minimal, but carefully designed IF crystal filter. U5 provides additional audio gain and a convenient place for receiver muting. U7, an ubiquitous LM386, provides audio output. Bypassing of pin 7 improved power supply rejection problems that produced a thumping sound with strong CW signals.

U6a and Q18 form perhaps the best side tone oscillator we have used. The op-amp is a Wienbridge oscillator with back-to-back limiting diodes. The circuit is close to oscillation with an open key. Circuit gain is changed by FET switch Q18 when the key is pressed. Q18 was picked for low pinchoff of –1.5 V. The relatively small gain shift produces a sidetone output that is free of clicks. Output is extracted from a point that does not change dc level when keyed.

U6B with Q17 form a 6-V regulated supply. This is used in the audio system as well as in the transmit mixer. Q14 provides a switched +12 V in transmit. The transceiver is breadboarded with no printed circuits, allowing frequent and convenient changes.

Although this rig is featured here as an experimental vehicle, it has done well in extended operation for several years of home use as well as several backpacked Field Day efforts.

12.3 THE S7C, A SIMPLE 7-MHZ SUPER-HETERODYNE RECEIVER

This receiver began with a long list of goals. It was to be a super-heterodyne design, offering the basic selectivity, sensitivity, and stability of the classic topology. The design was to use generic devices, avoiding the market driven whims of the semiconductor manufacturers. An adaptable circuit was desired, something that could be altered for other bands and modes. Low power consumption was a goal, allowing the circuit to function for an extended period with a handful of AA cells. And, above all else, it was to be a simple design, suitable for both the beginner and the seasoned designer/builder. The resulting superhet example shown is for the 7-MHz CW band, generating the S7C designator.

A block diagram for the receiver is shown in **Fig 12.20**. A cascode JFET mixer front end is driven by a VXO. While the tuning range is restricted, the stability is excellent. The restricted range simplifies construction, for no dial drive mechanism is required. The mixer then drives a two-crystal filter embedded between two bipolar transistor amplifiers. The output is routed to a product detector, audio amplifier, and headphones.

The circuit, shown in **Fig 12.21**, began with the elements of the "Micro-R1" Minimalist Direct Conversion Receiver presented in Chapter 8. Q1 is an audio output amplifier driven by audio preamplifier, Q2. A crystal controlled BFO, Q3, provides the needed injection for a two-diode product detector. The only changes of significance are the addition of an audio gain control and a few component value changes. The most significant of these is C4, which is larger than the value used in the original direct conversion receiver. This component was increased to provide greater flexibility in setting the BFO, Q3, to the proper frequency.

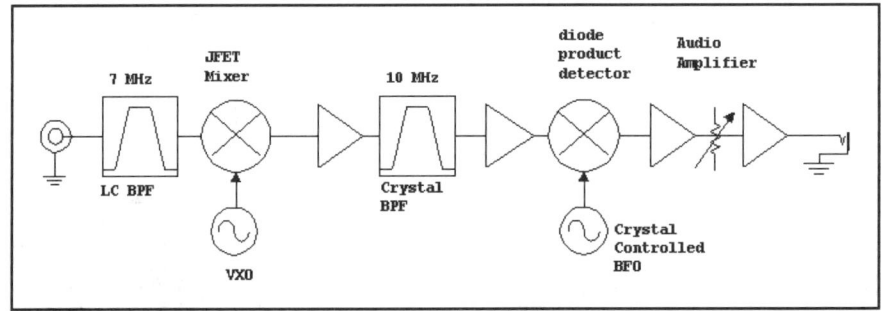

Fig 12.20—Block diagram for the S7C.

Top inside view of the simple superhet receiver. The left board houses the front-end mixer and the VXO. The board at the right contains the IF and crystal filter. The power connector uses a quick disconnect normally used with audio speaker cables.

The audio and BFO sections were breadboarded on a scrap of circuit board material, a 7-MHz crystal was dropped in at Y2, and the receiver was tested as a direct conversion circuit. The original Micro-R1 of Chapter 8 was driven by a link coupled double tuned circuit. The low impedance of the link provided the low audio impedance needed for proper detector operation. We added a radio frequency choke (value not critical) to the circuit to obtain the required gain. C1, a 5-65-pF trimmer capacitor in the BFO allowed some tuning around the crystal frequency. We eventually substituted a fixed capacitor in the circuit for C1, saving the trimmer for yet another project. The builder should review the discussion in Chapter 8.

The IF amplifier was built next. This design obtains selectivity from a double tuned circuit using two crystals. The filter is placed between two feedback amplifiers, each followed by a 6-dB pad. Each amplifier is biased for a 3-mA emitter current with a 9-V supply. The amplifiers and pads are designed for a characteristic impedance of 150 Ω, a departure from the more common 50-Ω designs. The product detector works well when driven from this

12.16   Chapter 12

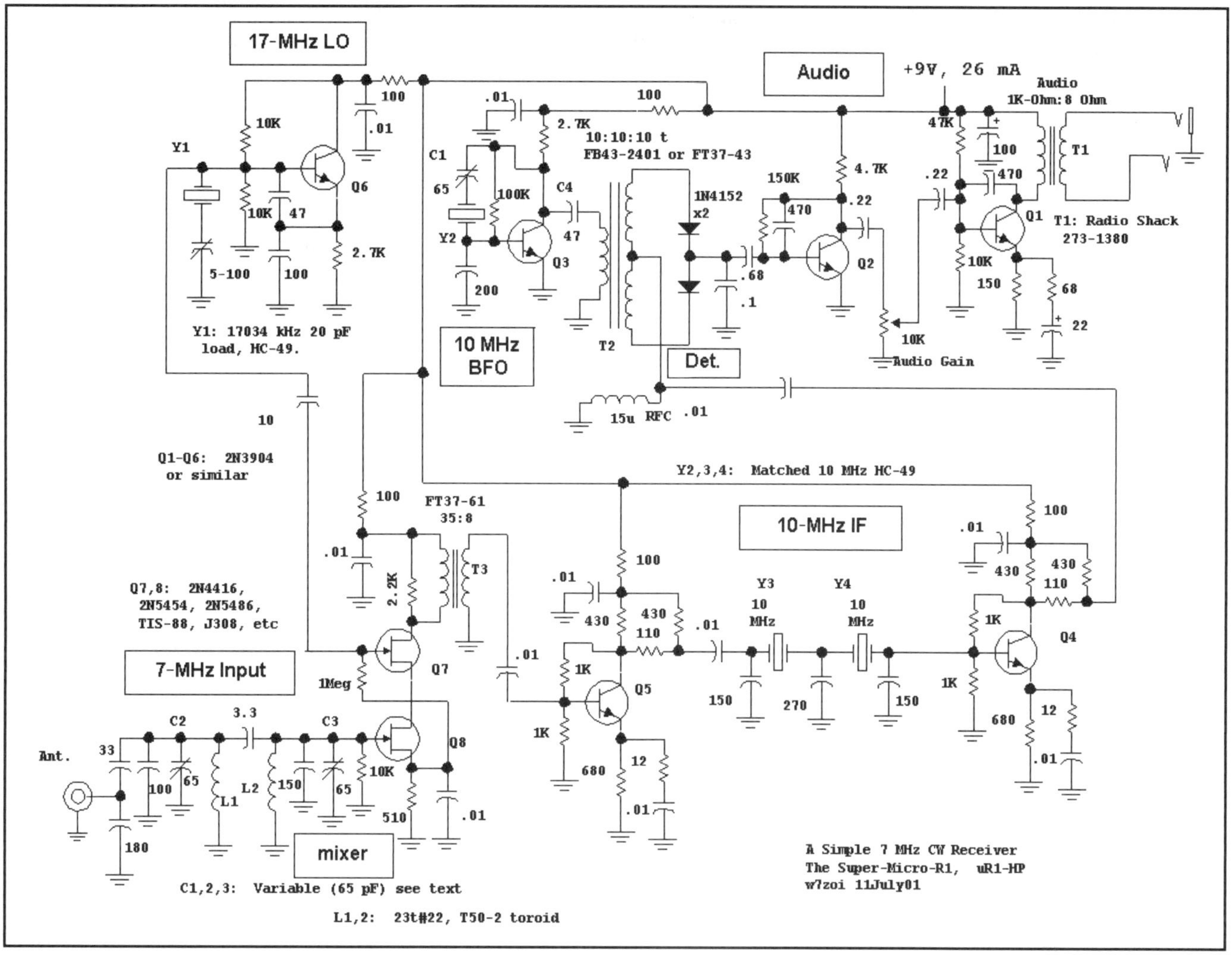

Fig 12.21—Schematic for the 7-MHz super-heterodyne.

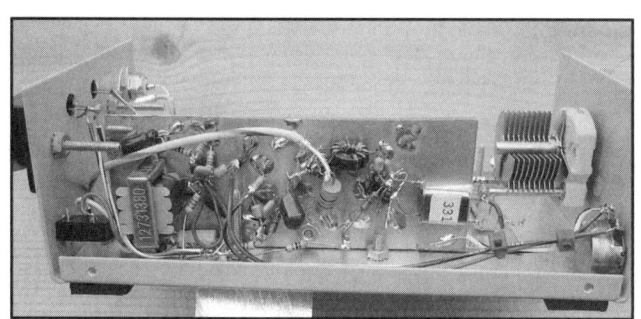

The audio and product detector board for the simple superhet receiver.

impedance. The crystal filter was also designed for 150-Ω terminations at each end.

The builder should purchase a few inexpensive HC-49 crystals from one of the popular mail order sources (Mouser, Digi-Key, etc.) The crystals are then matched with an oscillator circuit and a frequency counter. The BFO (Q3) could even be used as the test oscillator if you don't wish to build a separate test circuit. Y3 and Y4 should be within about 50 Hz of each other. Y2, the BFO crystal, is much less critical, for that frequency will be adjusted with C1. See Chapter 3 for information on crystal filters.

We used a 10-MHz IF in this example, for crystals were available in our junk box. This presented a problem, for 10-MHz signals from WWV and/or WWVH leaked through the front end and could be heard. This emphasizes the need for front-end selectivity. We'll discuss this later.

The IF system was breadboarded on a small scrap of PC board material and tested with the product detector. While detailed evaluation of the IF filter would happen later, we used a signal generator to confirm the functionality of the circuit. The single signal response was dramatic, considering the circuit simplicity.

The next part that was built was the 17-MHz VXO, Q6. This circuit used a crystal that had been specially ordered for the desired frequency, although the crystal is not otherwise special. We wished to have the tuning approximately centered at 7.040 MHz, the gathering spot for North American QRP operators. Our IF turned out to be centered at 9.9989 MHz, just over one kHz below 10 MHz. The sum of these frequencies is 17.039 MHz. VXOs tend to tune upward with much greater ease than they do downward, so we picked a frequency of 17.034 and ordered an HC-49

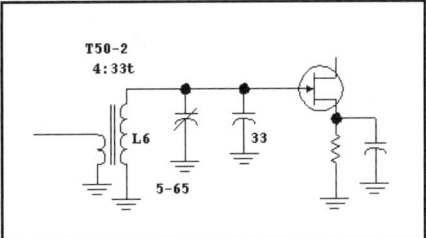

Fig 12.22—Single tuned mixer input circuit.

Front panel view of the simple superhet.

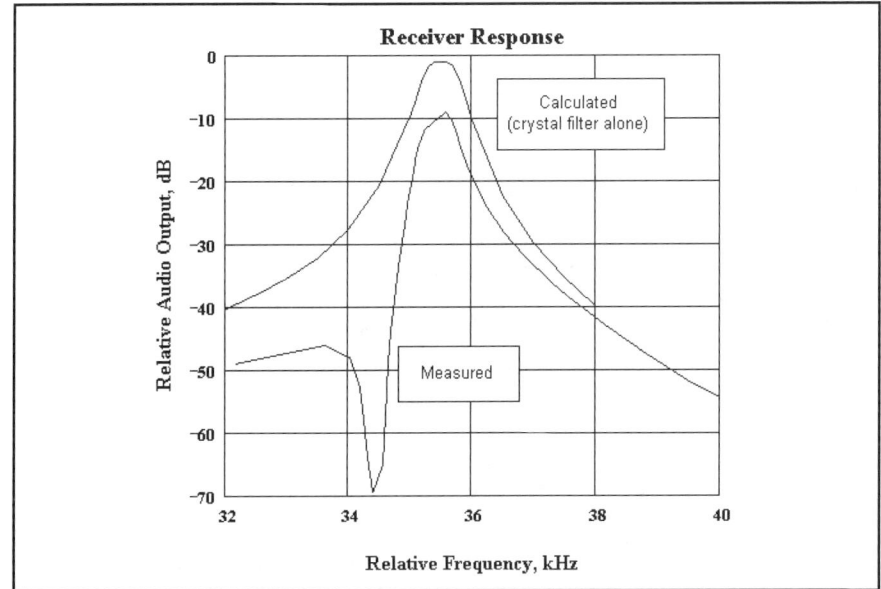

Fig 12.23—Measured audio output as a signal generator is tuned through the receiver. The calculated response of the crystal filter alone is superimposed for comparison. The BFO was set up for a 1-kHz beat note for this measurement.

cased fundamental mode crystal, specified for a 20-pF load capacitance. The final tuning range for our receiver was from 7030 to 7045 kHz. (The crystal was measured using equipment described in Chapter 7, resulting in Lm = 3.72 mH and C0 = 6 pF.) The builder will need to pick a different crystal frequency for compatibility with an alternative IF or target frequency. The VXO was built on yet another scrap of circuit board, and was eventually moved to the breadboard containing the mixer.

The receiver is completed with a front-end mixer. Several circuits were tried, producing the cascode of two JFETs, Q7 and Q8. This mixer has no balance, so it will function as an amplifier, allowing input RF signals to appear at the output. This is the route of the 10-MHz feed-through problem mentioned earlier. The mixer can also become an oscillator operating at the frequency of the input tank. This oscillation was easily suppressed with the 2.2-kΩ resistor in the Q7 drain circuit. If you encounter a problem here, reduce the value of this resistor. A tuned circuit at T3 on a powdered iron toroid would be a preferred solution.

This mixer has some strong virtues. First, it is quiet: We measured a 10-dB noise figure with this circuit. The current is low at about 3 mA. Very little LO power is required, allowing drive from simple oscillators. We found that the performance is best with a signal at the gate of Q7 of about 5 V peak-to-peak. This circuit is similar to the popular dual gate MOSFET mixers that were common in receivers in the 1970 to 1990 timeframe. We measured IIP3 of +5 dBm for this mixer, making it suitable for wide dynamic range applications.

The mixer is also breadboarded on scraps of PC board material. The ferrite output transformer, T3, is wound on a low loss –61 core material, offering better gain than a more common –43 core. The FET type used was a 2N5454, again a choice dictated by the junk box. These parts had I_{DSS} = 10mA and V_P = –3 V. However, there is nothing special about this FET. Virtually any of the common JFETs will work well. If a higher I_{DSS} part is used it may be worthwhile to experiment with the bias resistor.

The mixer in our receiver used a double tuned input circuit. The front-end selectivity eliminated all traces of the feed-through from WWV. Initial experiments used a single tuned input, shown in **Fig 12.22**. An external low pass filter (7th order 7.5-MHz cutoff Chebyshev, see Chapter 1) was then effective in eliminating WWV feed-through. A 10-MHz trap (LC or crystal) could also suppress the spurious response.

Results and Variations

This receiver is a joy to use. The first experiment that is always performed with a new receiver is a session of listening. The narrow bandwidth is effective on a moderately crowded band, yet the use of just two crystals produces a bright and lively sound not compromised by excess filtering. The constrained gain, modest selectivity, and lack of AGC make the receiver especially useful when the 40-meter band is dominated by the thunderstorms of late summer.

After a period of listening, we measured the receiver and experimented with some alternative circuits. A 7-MHz signal generator was applied to the receiver to determine the selectivity, shown in **Fig 12.23**. The single-signal character is clear. The response null occurs as the generator is tuned through zero beat, a result of the audio characteristics.

We measured an MDS of –138 dBm with this receiver, consistent with the NF measurement and an overall bandwidth slightly narrower than the 500 Hz of the crystal filter.

The stability of the VXO was excellent, but left us wondering what was happening just a few kHz down the band. So, we temporarily replaced the VXO with a 3-MHz signal generator, which worked well. A simple single transistor oscillator would serve in this application.

Some users will want more selectivity. The crystal filter could be redesigned to use more crystals. A simple alternative would add another crystal filter just like the first one. The impedance at the output of T3 and the input impedance of Q5 are both 150 Ω, so the filter would be properly terminated in this position. The additional two crystals should be frequency matched to Y3 and Y4.

Ed Kessler, AA3SJ, built a similar receiver with inexpensive off the shelve crystals for the IF and the VXO. In his version, he used 4.0 MHz for the IF with an LO at 11.046 MHz. The LO used a "super VXO" with two parallel crystals, a topology discussed in Chapter 4.

12.4 A DUAL BAND QRP CW TRANSCEIVER

This transceiver began as an experiment to investigate electronic band switching methods, but evolved into an enjoyable QRP rig. The super heterodyne design, **Fig 12.24**, covers the 14- and 21-MHz CW bands with an output of two watts. An available junkbox 9-MHz crystal filter provided receiver IF selectivity. This circuit is described to illustrate ideas rather than for duplication.

Band selection begins with a mechanical switch in the transmitter portion of the circuit. The three-section switch selects the two ends of the transmitter low pass filters and establishes dc lines that route throughout the transceiver for frequency control. For example, a line labeled "+12(21)" provides +12 V only when the rig operates in the 21 MHz band.

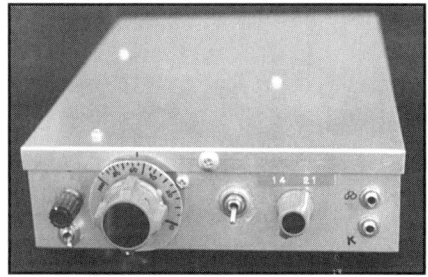

Front panel view of dual band transceiver.

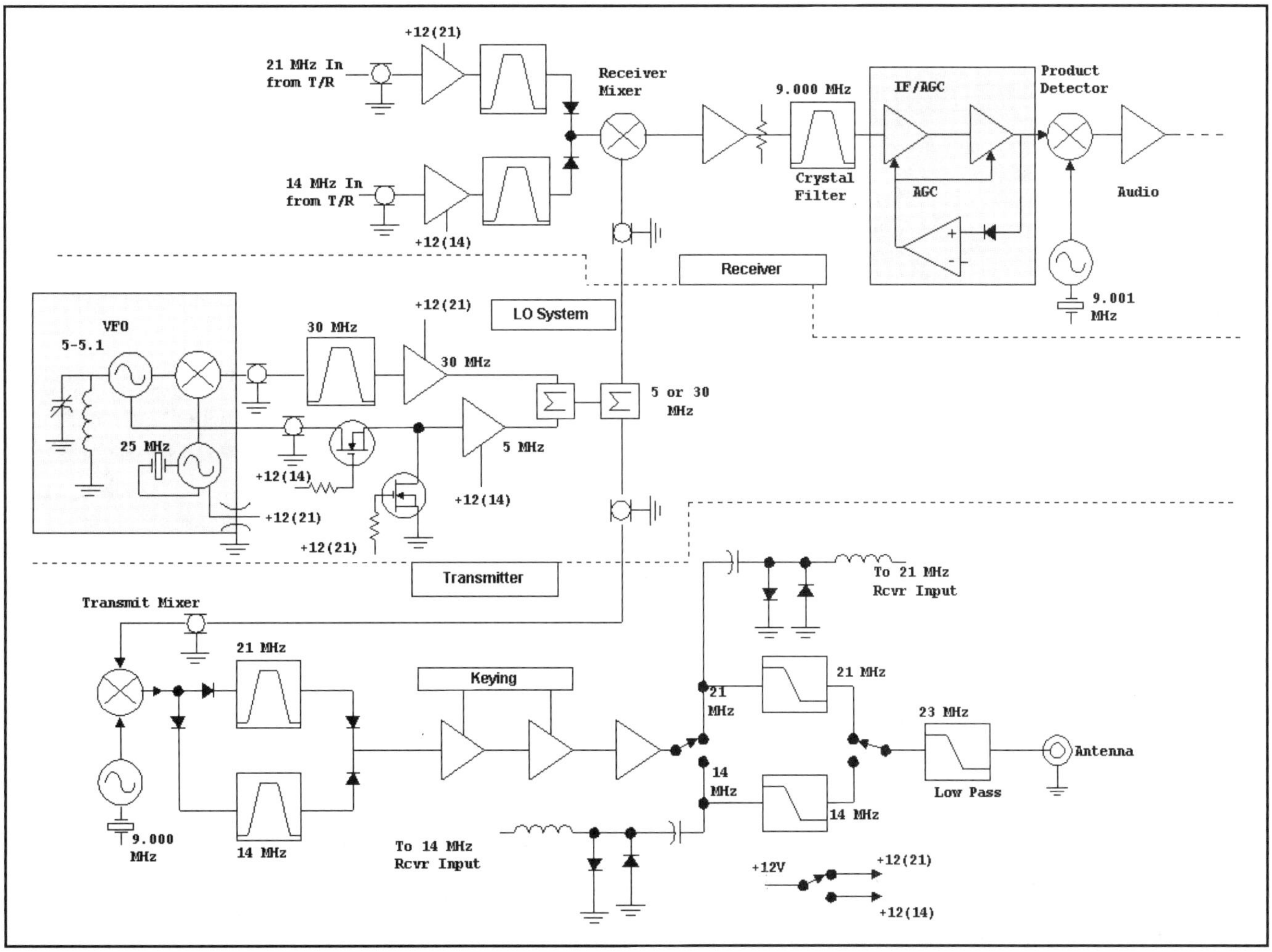

Fig 12.24—Block diagram for the dual band transceiver. The upper region is the receiver with the transmitter at the bottom of the page. LO details appear in the middle of the block.

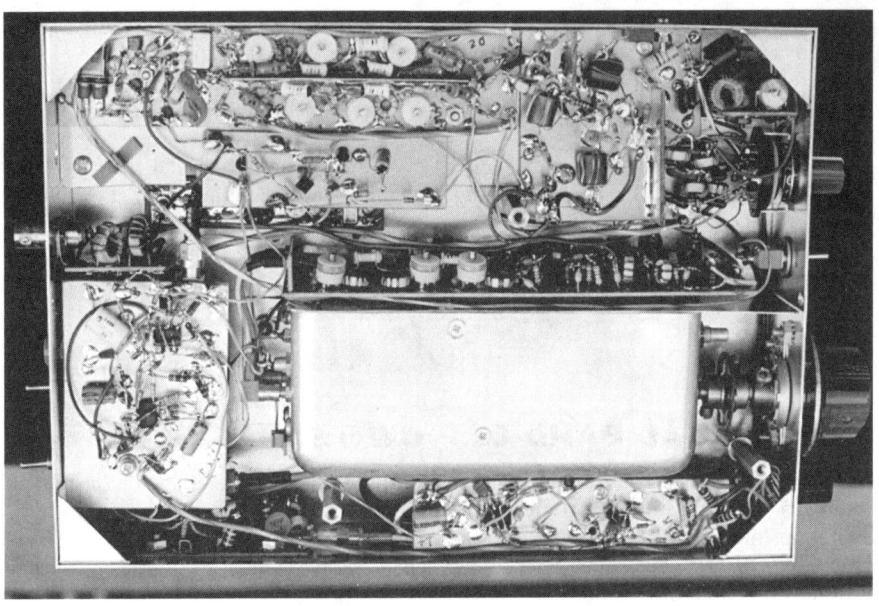

Inside view of dual band transceiver. Mounted below the VFO enclosure are the LO chain bandpass filters. The PA is bolted to the side of the box near the bandswitch. The triple tuned transmitter bandpass filters are along the lower edge of the photo. Most receiver front-end circuitry is hidden below the transmitter chain. Audio, product detector, and BFO circuitry are along the upper edge of the photo. The IF amplifier is between the VFO and the rear apron with the crystal filter under the board.

Local Oscillator System

The LO uses a 5-MHz LC oscillator, a mixer, and a 25-MHz crystal controlled oscillator, shown in **Fig 12.25**. This portion of the LO resides in a shielded box. A signal is extracted from the VFO resonator to drive a common base buffer, Q2. The output is applied to a resistive power splitter with one output available at a coaxial connector. The other output is filtered and applied to a diode ring mixer, U2. The "LO" for that ring mixer is the 25-MHz crystal controlled oscillator which is active only when the 15-meter band is selected. Signal levels are stabilized with an 8-V regulator. Powers are measured and carefully established before the module is sealed, ideally with a spectrum analyzer. The mixer output is attached to coaxial cable with short leads and then to an output connector with the desired 30-MHz signal and a 20-MHz image.

RF outputs from the oscillator module are applied to a filter board, shown in **Fig 12.26**. The 30-MHz signal drives a three-section bandpass filter. Feedback amplifiers Q5 and Q6 increase the 30-MHz level to +11 dBm after low pass filtering.

The 5-MHz signal from the VFO module is attenuated in a 6-dB pad and then applied to a series MOSFET switch, Q9. This switch is "on" only in 14-MHz operation. The output is then increased in cascaded feedback amplifiers, Q7 and Q8, and low pass filtered, generating an available power of +12 dBm for use with 14-MHz operation. The gain is slightly lower in Q7/Q8 than in Q5/Q6. Only one of the two outputs is available at a time, for only one

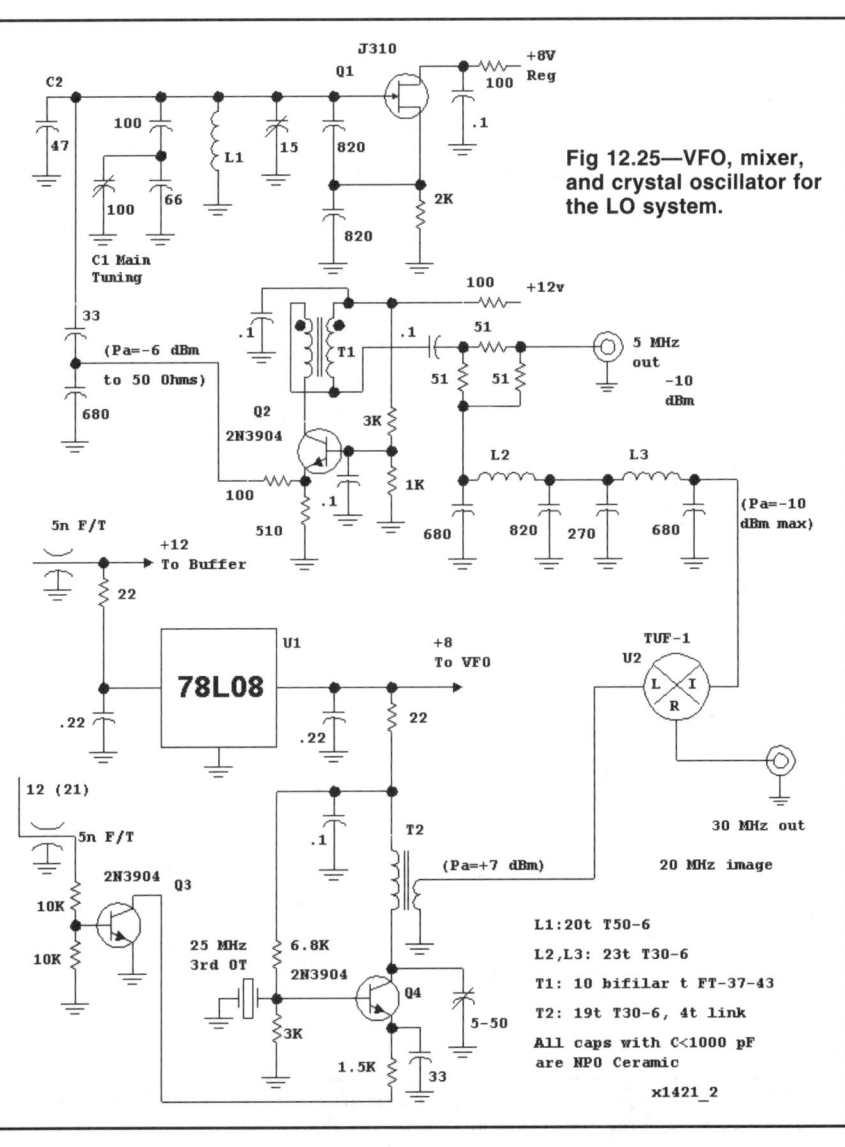

Fig 12.25—VFO, mixer, and crystal oscillator for the LO system.

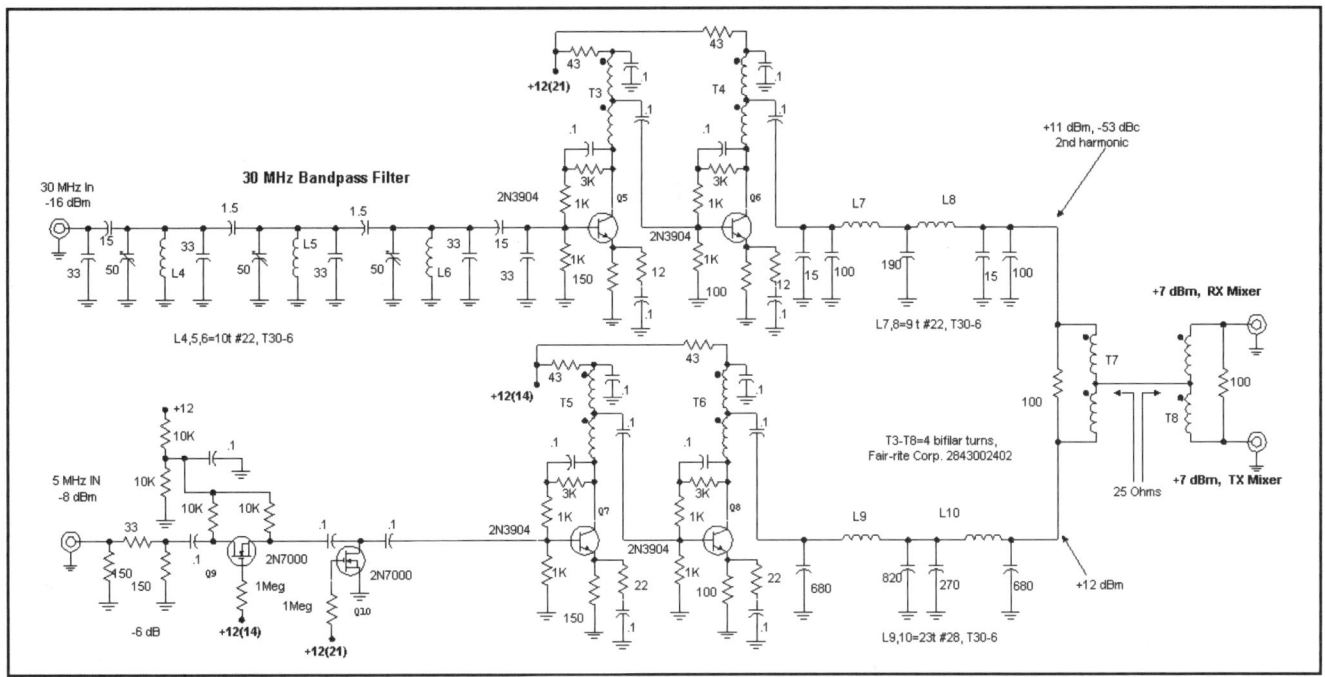

Fig 12.26—The LO signals are processed in this board. The 30-MHz signal is bandpass filtered, amplified, and low pass filtered. The 5-MHz signal is amplified and low pass filtered. Outputs are combined with a 0-degree hybrid. Another hybrid splits the signals, providing +7 dBm for both the transmit and receive mixers.

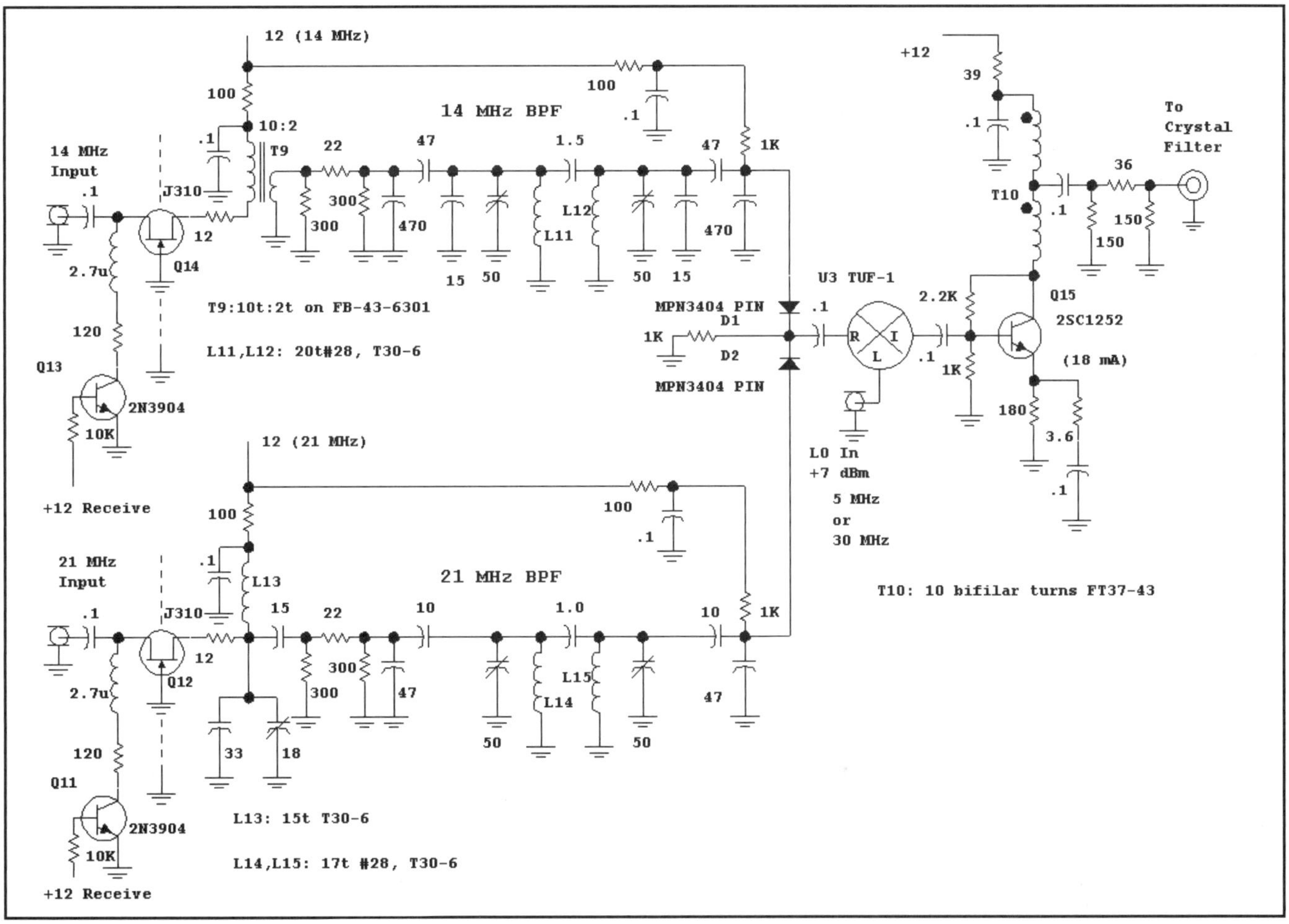

Fig 12.27—The receiver front end for the dual band transceiver. PIN diode switching is used to select the bandpass filter output appropriate to the band in use.

Field Operation, Portable Gear and Integrated Stations

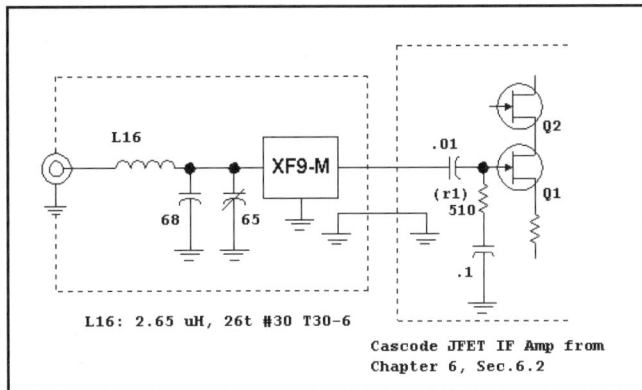

Fig 12.28—Input section of the crystal filter and IF amplifier for the transceiver. See text.

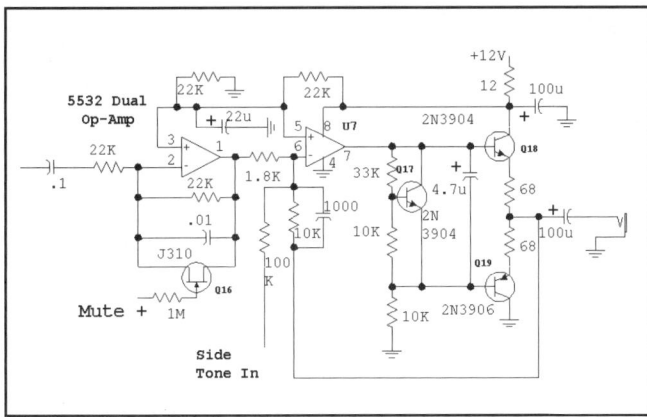

Fig 12.30—An audio output amplifier for the receiver.

bank of amplifiers is biased on. Suppression of the 5-MHz component during 21-MHz operation is improved with a shunt MOSFET switch, Q10. The two outputs are combined without switching in a 0-degree hybrid built from T7. The output would contain both signals if both were on at the same time. The resulting output is split into two equal, but isolated components with another hybrid, T8. The result is a pair of +7-dBm signals for the two diode ring mixers in the receiver and transmitter.

The harmonics are more than 50 dB below the desired LO outputs, and images are difficult to find. Before the shunt FET switch, Q10, was added, some 5-MHz energy could be seen when the 30-MHz component was dominant. However, adding the switch pushed the 5-MHz component to the –80 dBc level. This is more extreme than needed, but instructive.

Receiver Circuits

The receiver is much like others we have described. A low-gain, moderately low-noise RF amplifier drives a diode ring mixer. The RF amplifiers were designed for good input match rather than lowest noise. A post mixer amplifier, Q15, provides signals to a crystal filter. A JFET based IF amplifier adds gain and provides a convenient place for AGC. Another diode ring serves as the product detector with a conventional audio chain.

The front end, the only place where band switching is needed, is shown in **Fig 12.27**. Each of the RF amplifiers, Q12 and Q14, is powered only when the respective band is selected. Transistor switches remove current from the RF amplifiers during transmit intervals.

MPN3404 PIN diodes are used for band selection. There are slight differences in the two RF amplifiers. That for the 14-MHz band uses a ferrite transformer while the output in the 21-MHz circuit is tuned. A pad (just over 3 dB) drops the gain a bit and helps to fix the impedance for the following double tuned bandpass filters.

The diode ring mixer is followed by a post mixer amplifier with modest current of 18 mA. This then drives the crystal filter and IF circuit, shown in the abbreviated circuit of **Fig 12.28**. The input 50 Ω is transformed up to 500 Ω with the L-network shown. A variety of IF amplifiers have been used in this circuit, most with low gain. The one presently in use is that from Chapter 6 using cascode connected J310 JFETs. The original circuit was modified by changing the input resistor to 510 Ω to properly terminate the German (KVG XF9-M) crystal filter we used. The

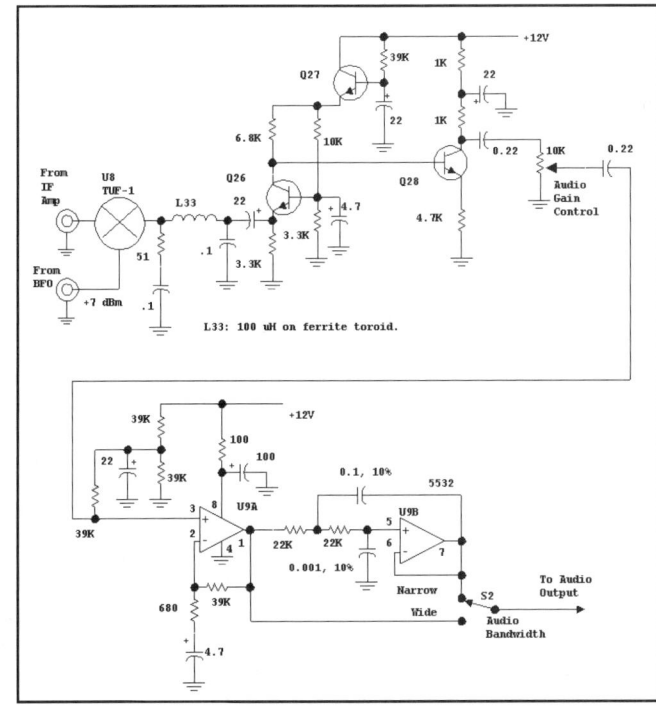

Fig 12.29—Product detector and audio amplifier. The emitter of Q28 may be bypassed for gain higher than needed here.

designer/builder may wish to add a transformer to match between the crystal filter and the 2.2 kΩ originally in place; the higher impedance will allow greater gain, lower noise figure, and greater flexibility in AGC threshold adjustment.

An early version of this receiver used nothing more than a single JFET as the IF amplifier. Only manual IF gain control was used; most of the overall gain was obtained at audio. Performance was excellent for use in working other QRP stations. However, we found it lacking for general use when stronger signals were routine. The present system includes AGC with an adjustable threshold.

The detector and audio system, shown in **Fig 12.29**, is the "standard" used throughout the book for direct conversion

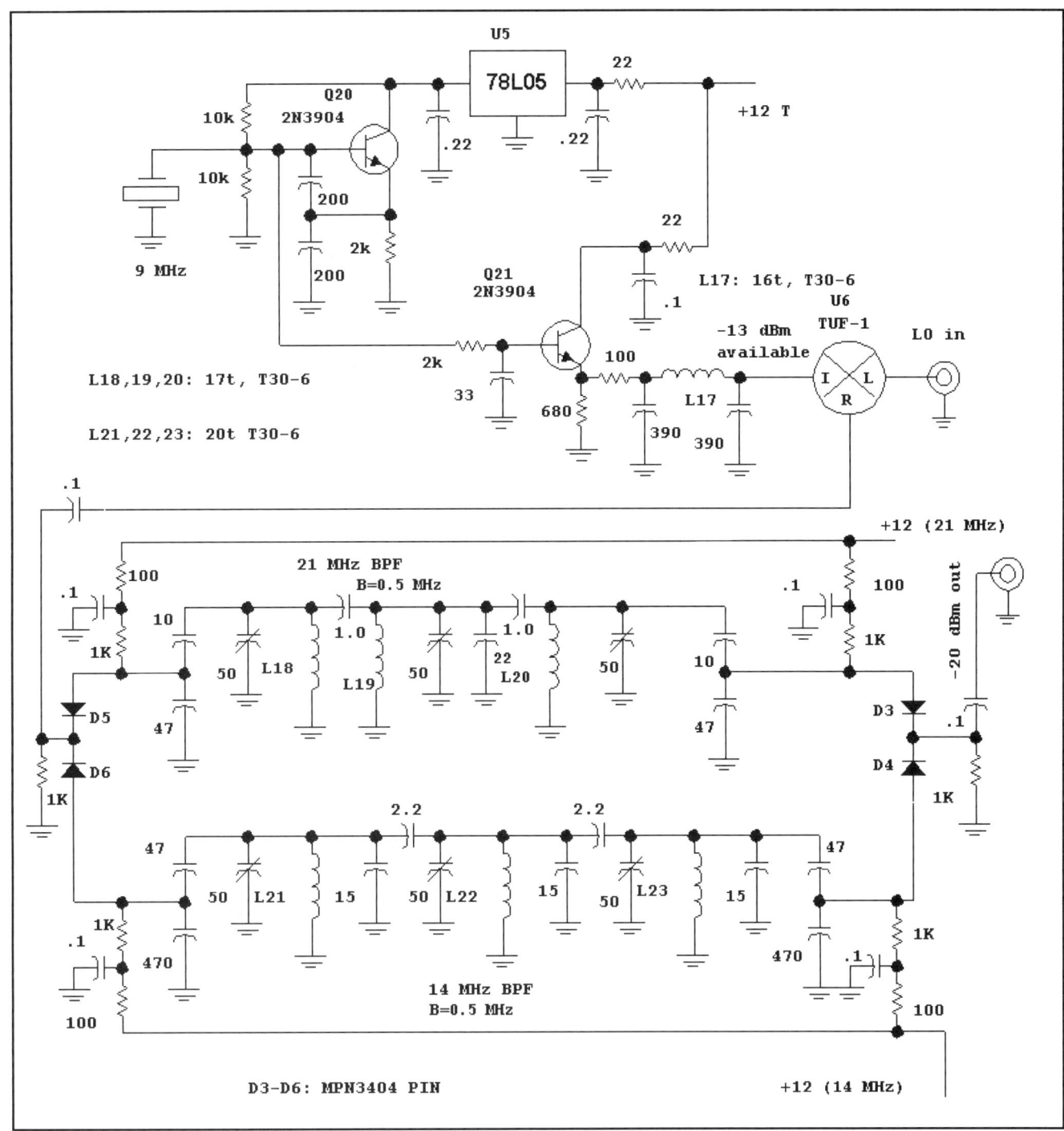

Fig 12.31—Transmit mixer with PIN diode switched bandpass filters. See text for details.

systems and simple superhets. A TUF-1 diode ring product detector drives a common-base amplifier. The second audio stage operates at a gain of about 0.2, but it could be increased as needed. After the audio gain control, an op-amp provides voltage gain, followed by a switchable peaked low pass filter with a Q of 5.

The circuit shown in **Fig 12.30** using plastic transistors and an op-amp will drive a small speaker. The high open loop gain of the op-amp keeps distortion low. This circuit, with only 10 mA in Q18 and Q19, would benefit from increased standing current, reducing clipping that occurs with high output.

The rest of the receiver is routine and is not repeated here. The crystal controlled BFO and sidetone oscillator are not shown. This receiver measured NF=11 dB, IIP3=+3 dBm, for DR=93 dB with a 500 Hz bandwidth. The receiver AGC is degraded by BFO energy reaching the IF system. BFO and IF shielding would both improve performance.

Transmitter Details

A simple heterodyne process generates the output signals for the transmitter, shown in **Fig 12.31**. A 9-MHz crystal oscillator is applied as the RF signal to a diode ring mixer. The larger drive at 5 or 30 MHz comes from the LO chain. The

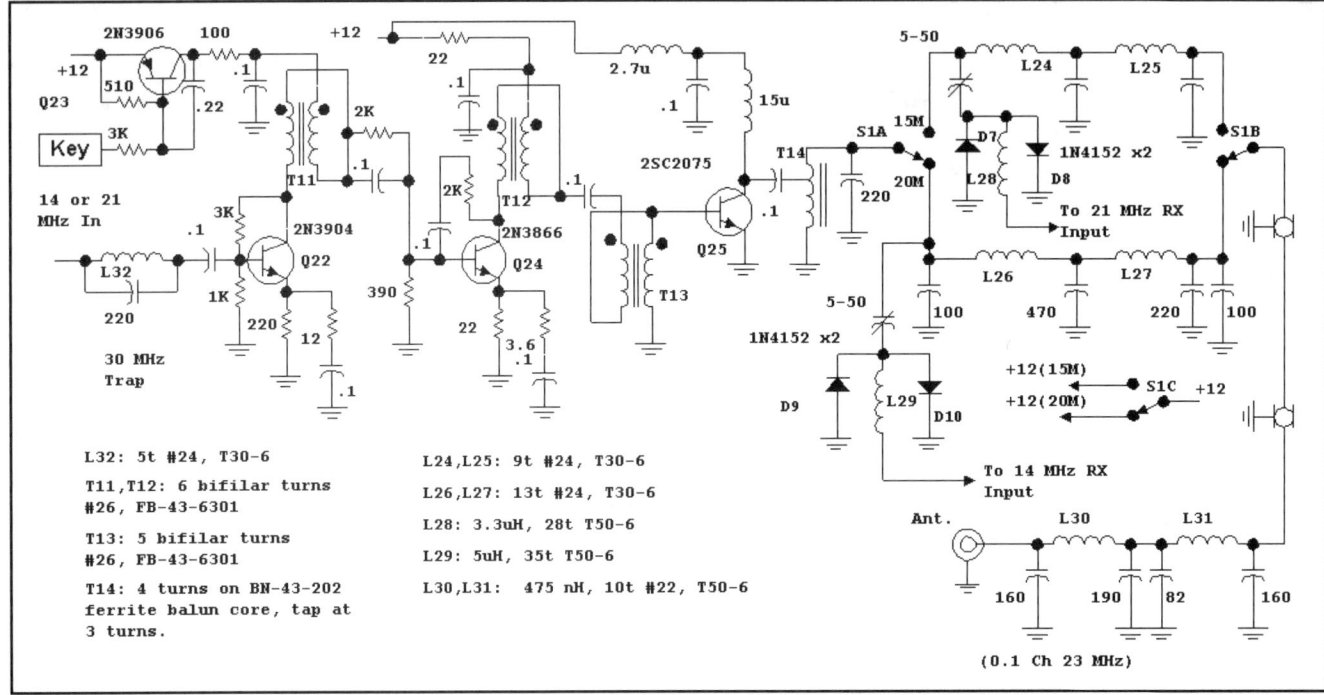

Fig 12.32—RF power chain for the transceiver.

mixer output is then filtered in one of two PIN diode switched bandpass filters.

The initial transmit mixer system used double tuned circuits for both bands and had no 9-MHz low pass filter. The results were interesting. Although the 21-MHz observed output was clean, there were spurious outputs related to the 14-MHz band. These occurred at 13 and 16 MHz at –52 and –56 dBc. The 13-MHz spur was a 1:2 spur that could be solved with reduced harmonics in the 9-MHz drive. The higher frequency spur was related to a 5:1 product. (A N:M spurious output frequency results from $Nxf_{LO} +/- Mxf_{RF}$; See Chapter 5.) The third order low pass filter was added to the 9-MHz RF, pushing the first spur to the –72-dBc level with no change in the other.

The 14-MHz double-tuned circuit was changed to a triple tuned filter with a bandwidth of 0.5 MHz. The higher frequency spur was now suppressed to –75 dBc and the lower one was lost in the noise. We later found some 30-MHz energy in the 21-MHz output, which prompted a change to a triple tuned filter for that band as well. None of these results would ever have been observed without the use of spectrum analyzer for the experiment. But the result is a justification for using a triple tuned bandpass over a simpler double tuned circuit when one seeks improved spectral purity. While triple tuning uses more components, it is no more difficult to design or tune at HF than one with two resonators.

The transmitter power chain, shown in **Fig 12.32**, begins with a 30-MHz trap, tuned by compressing turns on L32. A two-stage driver amplifier then provides the bulk of the gain and adequate drive power for Q25, the 2SC2075 output stage. A wideband transformer, T14, reflects a load of about 28 Ω to the PA collector. Both driver stages are keyed to produce a backwave below –70 dBc. Low pass filters for both bands are selected with the mechanical band switch. A final 23-MHz low pass is then added to the output.

We were still able to find two spurs in the output for each band. They were, however, all at –62 dBc or less. The worst harmonic was the 2nd when operating at 14 MHz at –63 dBc. With the exception of the VFO, only incidental shielding is used.

12.5 WEAK-SIGNAL COMMUNICATIONS USING THE DSP-10

Chapter 11 contained an overview of the DSP-10 DSP-based 2-meter transceiver and the associated audio processor. The published material on this project is on the CD-ROM and has the details necessary to build and modify this radio. An interesting application of the DSP-10 is the processing of signals to allow detection of stations too weak to hear with the ear, and to allow communication with these stations at very slow data rates. This is an example of what is practical to achieve using the programmable aspects of the radio. As was discussed in the overview, there are many other possible applications. In addition to the following summary of weak signal operation, detailed material is available on the CD-ROM that accompanies this book.[8]

Additive Noise

The expression *weak signals* is a relative term. Normally, the signal is referenced to the received noise level. Of course, the nature of this noise changes with frequency and conditions. Interfering signals and static from lightning can provide a complex noise environment that is most challenging to the weak-signal enthusiast. Simplifying matters for our consideration here, the primary noise source considered is the well-behaved thermal noise, also known as white Gaussian noise (WGN). The "white" refers to the flatness with frequency and "Gaussian" refers to the probability distribution, also called normal or bell-shaped. WGN dominates the VHF and higher frequencies, but this source extends down into the HF bands as well.

This WGN is added to the signals received at the antenna terminals. This is a result of our receiver being linear. As was discussed in Chapter 2, filtering can reduce this additive noise, since it is flat with frequency. This gives us a way to remove noise from signals, so long as the bandwidth of the signal is less than the filter bandwidth.

Signals and Multiplicative Noise

The signals being transmitted for weak-signal work can generally be chosen to occupy reasonable bandwidths.[9] Simple modulation methods are the most easily dealt with and can generally be used. An example is a single frequency tone, transmitted for a predetermined amount of time. This signal can be extended to two or more tones in order to convey information as frequency shift keying. This idea will be explored further below, but here it is important to observe that the received signal is not generally an attenuated version of that transmitted. Instead, as the signal passes through the transmission media (atmosphere, ionosphere, Moon reflection, etc.) modulation is applied to the signal. This is akin to the modulated signals described in Chapter 6.

As the signal passes through the transmission media the amplitude varies—in amateur lingo, this is QSB. Typically, this variation is random in nature. What we have is a signal with amplitude modulation (AM). Frequency sidebands will appear on either side of the transmitted carrier as with all AM signals. The frequency offset of the sidebands depends on the speed with which the amplitude varies. Faster changes produce sidebands farther from the carrier. In addition, the length of the transmission path will vary, again often randomly. Movement of the refractive and reflective layers causes this. In this case, we have phase modulation (PM), again producing sidebands on either side of the signal.

It is possible for the AM and PM sidebands to add and cancel in different ways for those above the carrier than for those below. Consequently, the modulation placed on the signal by the transmission media is not symmetrical about the carrier frequency, and may not look like a typical modulation spectrum. As a modulation it is multiplicative noise and different from the additive noise just discussed. We do not have the option of removing this noise by filtering, since lowering the bandwidth removes the signal along with the noise. The propagation media places a lower limit on the filter bandwidth usable with a narrow-band signal.

A General Approach

A wonderful paper by K3NIO[10] outlines this weak-signal communications problem and proposes a practical solution that he and K8DKC demonstrated on 20 meters. Poor's model for signal and noise were the ones we have above and his communications system, built around RTTY and FSK, applied these principles:

• Maximize the transmitter average power by having it on continuously
• Minimize the receiver (pre-detection) bandwidth, consistent with the signal and propagation path modulation
• Use detectors to estimate the signal amplitude at each frequency
• Trade off time and sensitivity by following the detectors with low-pass filter(s) to provide averaging (integration) of the signal amplitude.

The performance of the system was limited by the availability of low-pass filters (RC networks), suitable for very long integration times. But as Poor points out, so long as one can build a low enough cut-off to the filter, the ultimate sensitivity of this approach is limited only by our patience for the answers to appear.

Going to more than two frequencies was not part of the 1965 system, but was known to offer improvement for communications systems.[11] Today the multi-tone filtering can be performed by discrete Fourier transforms (see Chapter 10). Long-term integration is easily done in a digital computer. The following two examples, taken from the DSP-10, show how these ideas can be applied using DSP techniques.

Example 1 - EME-2 for Moon-bounce Echoes

THE GOAL

A 2-meter station, with the antenna

The K3NIO Experiments

The 1965 experiment reported by K3NIO represented an early attempt at signal processing to receive beyond the limits of the human ear. K3NIO and his collaborator in this effort, K8DKC, were RTTY enthusiasts and had frequency shift keying (FSK) equipment available. They did their 14-MHz experiments in the late evening hours when the band was essentially dead. The transmitters were set up for narrow FSK and keyed with standard CW, driven with an automatic keyer set for a typical speed of 3 words per minute. The two stations were separated by 500 miles, used three element Yagi antennas and 1-kW transmitters. The stations were crystal controlled to provide stability that was not common in 1965.

Their receiving system is shown in the block diagram below. The normal 14-MHz receiver had improved selectivity, provided with an audio bandpass filter. The audio signal was applied to a limiter, and then to a frequency discriminator. The output from that circuit is a dc level indicating the frequency of a tone moving through the system. The dc was filtered, or averaged with an RC active low pass filter with a 1-Hz cutoff. The resulting dc then drove a comparator and a strip chart recorder, allowing visual copy of CW.

The results were dramatic. Essentially, they found it possible to make slow speed contacts, even when they could not detect the presence of any signal when listening to the receiver operating in the normal mode.

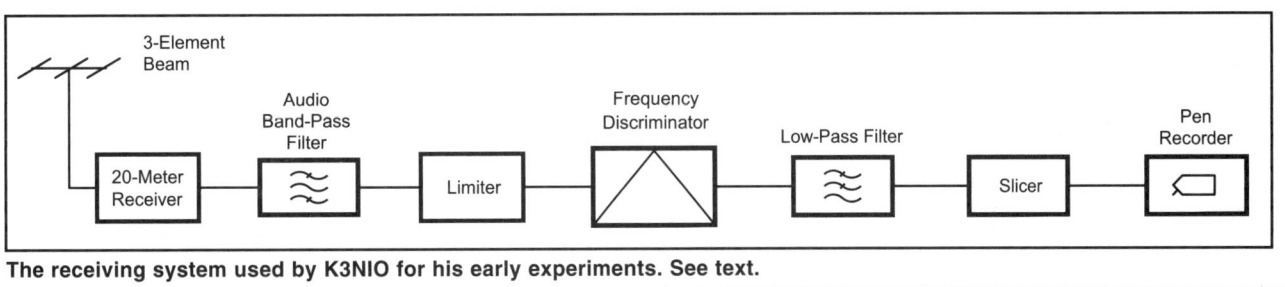

The receiving system used by K3NIO for his early experiments. See text.

pointed at the Moon, can transmit a pulse for roughly two seconds and then receive the resulting echo. This comes back 2.6 seconds after it was transmitted, as shown in **Figure 12.33**. Adding to the challenge, if this "Moon-bounce" station is of modest proportions, the received signals will be extremely weak. For instance, a station with two 12-element Yagis and 500 W of transmitter power can expect to see an average power return of about –160 dBm. For the noise levels encountered on this band the resulting signal-to-noise ratio might be about –5 dB in a 50-Hz bandwidth, which is totally inaudible. Regardless, the goal of this example is to be able to measure this and much weaker echoes coming back from the Moon. The value, in addition to satisfying a general curiosity, is allowing the measurement of the system performance of the station and the propagation path.

As a reference point, we should examine just how well this "marginal" Moonbounce station can hear his echoes. Helping the situation, the signal strength fades above and below the average return. This is due to the irregular surface of the Moon and the shifting nature of the path. With some patience, the signal will appear for a second or so at, perhaps, 6 dB higher level or 1 dB S/N. Additionally, if the antenna is along the Earth's surface the signal reflected from the ground will sometimes add to that coming in directly, adding as much as 6 dB more to the signal. Now we are up to about 7 dB S/N. At this level, a perceptive operator will sense by ear the presence of a Moon-bounce echo. However, if the station is located where ground reflections are poor, such as at the edge of the forest, the echoes may never be heard.

Looking for a way to use DSP to enhance the detectability of the echo, one should explore the elements outlined above. First, we narrow the pre-detection bandwidth to the limit set by the modulation of the propagation path. On 2-meters this is generally 1 Hz or less. Next, any amount of improvement is possible by post-detection averaging that we call long-term integration. This resulted in a mode called EME-2 that was implemented in the DSP-10 software, as will be described below. However, before exploring these receiver concepts, it is worth considering the transmitter side to see if we might do better there as well.

TRANSMITTER WAVEFORMS

In the discussion above, we decided it was desirable to increase the average power of our transmitter by having it on as much as possible. Holding the key down for two seconds and listening for about 3 is only on 40% of the time. It might be possible to transmit on one frequency for 2 seconds and then move a MHz higher and transmit for seconds two through four. If the transmitter and receiver could be separated sufficiently, either in a geographical sense or by use of filtering, such as that of FM repeaters, this might be a preferred method of operation. But for most stations, the simplicity of merely sharing a single antenna by means of an antenna relay is an overwhelming consideration. The loss of average power can still be made up for by more integration.

The waveform considered here is a constant-frequency sine wave, keyed on and then off two seconds later, generating a pulse. One might hope that a more elaborate modulation would be helpful for identifying the returned signal. Radar designers have considered this problem for many years. In terms of detectability, the theory offers no encouragement in this area. The key factors are the power in the transmitted pulse and the care with which the receiver pre-detection filter is "matched" to the received waveform.[12] Thus, we might as well work with the simple approach and that is a keyed sine wave.

PRE-DETECTION FILTERING

The one-Hz filter for our system is a major challenge for LC construction, but is easily accomplished with the discrete Fourier transform (DFT) of Chapter 10. There are other possible DSP implementations, but the DFT provides a bank of filters that is useful for estimating the noise level and for the case that the signal is not received on frequency for some reason. The filter response of the DFT may not be the exact matched filter, but the bandwidth is close to proper and the losses for improper shape are not large.

The DSP-10 implementation of the DFT has several bandwidths available, in steps of two, with the narrowest being about 2.3 Hz. This is not a fundamental restriction, but neither does it provide optimal performance. Those with an interest in this area might explore using narrower bandwidths by increasing the sampling time interval.

LONG-TERM INTEGRATION

At each filter bin of the DFT the power can be calculated as the square of the received envelope (see Chapter 10). This power can be added up for a number of bins near that where the signal should be received. The bins on either side are estimates of the noise power and the center bin is signal-plus-noise power. From these two quantities an estimate of the signal strength alone can be made, using only subtraction.

A complication in continuing the integration process for extended periods is the changing Doppler shift of the return signal.[13] In the DSP-10 implementation of this process, the Doppler calculation is quite elaborate and accurate to better than 1 Hz at 2-meters. This allows the integra-

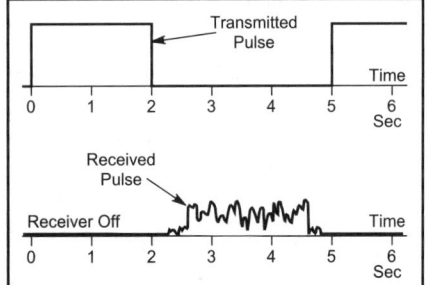

Fig 12.33—Timing diagram showing the two-second pulse being transmitted and the delay before the reception of the weak echo. This timing is repeated every five seconds for the EME-2 measurement mode.

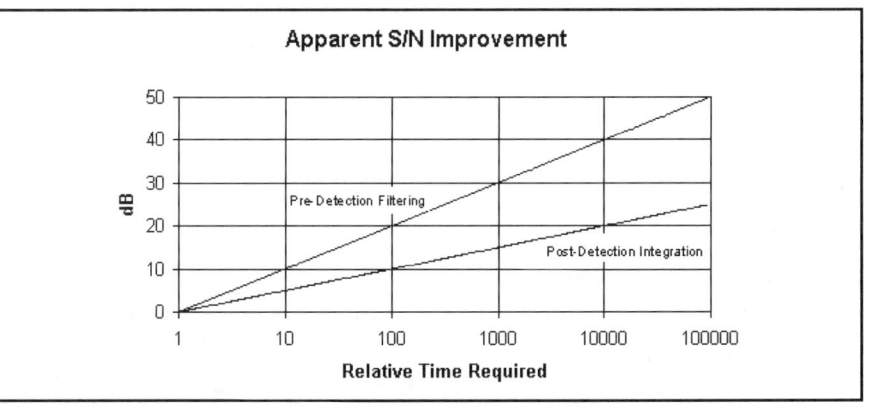

Fig 12.34—This is a comparison of the improvement in apparent signal-to-noise ratio for the pre-detection filtering and long-term post-detection integration.

tion to continue as long as the Moon is within view.

The remaining element is a means of displaying the return value. Two systems have proven of value for EME-2. A simple table of the signal-plus-noise estimates, expressed in dB, for 21 bins, centered on the return frequency provides most of the data. Along with this is the number of power values that have been integrated. A graphical plot of this same data also allows one to easily digest the results of a test and is always available.

A comparison of the improvement in apparent signal-to-noise ratio for the pre-detection filtering and long-term post-detection integration is shown in **Fig 12.34**. For either method, the parameter describing the amount of improvement is time. Expressed in dB, the rate of improvement is twice as great for the pre-detection filtering. This obviously only applies to the extent that multiplicative noise from the modulation path is not a limiting factor.

A SAMPLE OF EME-2

A number of tests have been made using EME-2 in the DSP-10. These have verified the concept that the amount of integration determines the sensitivity and there is no obvious lower limit to the process.[14] One of these test results is shown in **Fig 12.35**, where a reasonably modest 100 W was used by W7PUA with a single Yagi, having a 34-foot boom. This antenna, a commercial M[2] 2MXP28 product used with a home-built combining hybrid, has circular polarization to minimize the degradations from Faraday rotation.[15]

The lower trace is the result of one two-second-pulse return. Because the bandwidth of each DFT is wider than that of the pulse, there are nine DFT's involved in generating this trace. The amplitude of the signal-plus-noise shown here is about 6 dB over the average noise and somewhat stronger than average. The upper trace is the result of averaging 71 two-second pulse returns together, requiring about six minutes. The noise averages to its power at all frequencies while the signal-plus-noise at the 323 Hz line is about 2.4 dB greater. After this many pulses, the signal return on the upper trace becomes very well defined and the level of the return can be measured quite accurately. This signal echo was never heard by ear.

Example 2 - PUA43 for Weak Signal Communications

The work of K3NIO suggests the possibility of using the EME-2 approach with frequency-shift keying as a modulation method for weak-signal communication. Looking at the spectral plot for EME-2 certainly supports the idea that one might communicate by lining up multiple frequencies, each somehow corresponding to a portion of a message. The reference by Murray Greenman, ZL1BPU, points out the advantage of using more frequencies than the two used by Poor. With an eye towards pushing the limits of slow, weak-signal communication, a modulation and coding system was implemented in the DSP-10 that applied these principles. This used 43-tone modulation, where each tone represented a different symbol such as an alphabetic character. At the time a number of different schemes were being tried, and this particular one was nicknamed *PUA43*.

PUA43 sends the same message repeatedly, once or twice during each minute. It is quite structured. The message length can only be either 28 or 14 symbols long, each corresponding to specific two-second time periods. The number of minutes that the message is sent is determined by the users, giving flexibility for improving weak-signal copy by using many repeats of the same message.

Power received for each of the symbols is added over multiple repeats, just as was

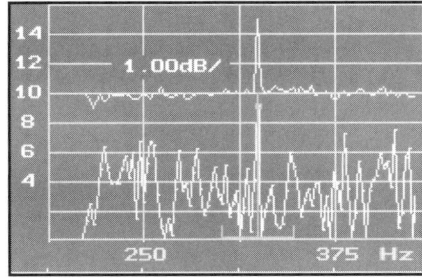

Fig 12.35—This portion of a DSP-10 screen shot shows the graphical output with the EME-2 mode Moonbounce echo. Some editing has been done to remove uninteresting parts of the display. The vertical scale is relative power in dB and the horizontal scale is audio pitch in Hertz. The bottom trace is the power average of one return. The upper trace results from averaging 71 of the lower traces together. The return signal has had its frequency adjusted for Doppler shift and always lines up with the vertical line at 323 Hz. The scale is different for the two traces, with 2 dB per division for the lower trace and 1 dB per division for the top averaged trace. At 144 MHz, the transmitter power was 100 W and the antenna was a single 34-foot Yagi.

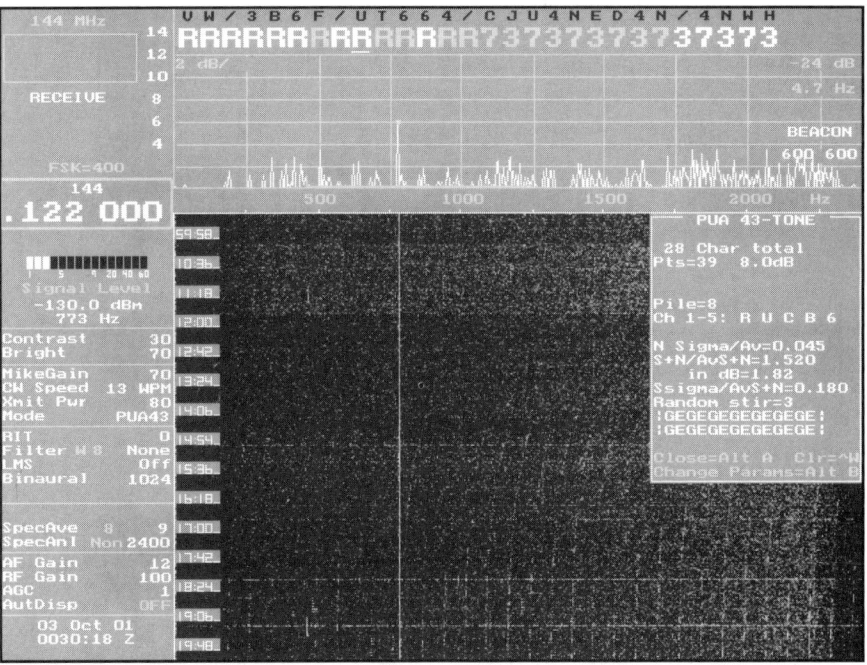

Fig 12.36—Screen shot from DSP-10 showing the reception of a PUA43 message by W7LHL. The signal-plus-noise to noise ratio of this plot is similar to that of the EME-2 reception of the previous example. The frequency band for the 43 frequencies in use extends from 450 to 1238 Hz, corresponding to the DFT bin spacing of 4.3 Hz that was being used. The large characters at the top of the screen are the most likely possibilities. The smaller characters above them are the second most likely. Various informational items relative to both transmission and reception are in the box on the right side of the screen. The straight line down the waterfall is a local interfering signal that is being ignored by means of frequency randomization.

done for each frequency in EME-2. Examining the power corresponding to the 43 possible symbols generates the display of the 14 or 28 characters. The most likely (highest power) and second-most likely symbols are displayed. The display color depends on the confidence of the particular character being correct, based on the measured noise characteristics.

An example of signal reception is in **Fig 12.36**, again on 144 MHz. The waterfall display (see Chapter 11) shows very little evidence of any signal being present, other than an interfering signal that is coming straight down the waterfall at about 770 Hz. The copy of the message, seen in large letters at the top of the screen is the result of integration of power for 39 minutes.

Several provisions of the PUA43 mode enhance the copy of signals. Every minute the frequency corresponding to a particular symbol changes by a positive offset that is the same for all symbols. The frequencies outside the frequency band being used for the 43 symbols are wrapped around to the bottom part of this band. This randomization, called stirring, causes coherent interfering signals (birdies) to get moved around to various symbols, rather than appearing as a false symbol. Additionally, there are unused frequencies between the 43 symbol frequencies. These are for noise estimation and serve two purposes. Knowing the noise levels across the band allows any variations in gain to be corrected so that they do not bias the symbol selection toward particular frequencies. Also, knowing the signal-to-noise ratio allows the confidence in a particular character being correct to be found, enhancing the data presented to the operator.

A characteristic of most weak-signal schemes is a need for accurate frequency control at the transmitter and receiver. This mode works best when the frequency can be controlled within a few Hz. As was done for EME-2, the PUA43 type of modes can be used for Moon reflections with the Doppler corrections that are available in the DSP-10. This adds a slight complication in needing to know the latitude and longitude of both stations.

The performance of this type of mode can be very good. A signal-to-noise ratio of −10 dB in a 50-Hz bandwidth will allow good copy of a message in about 6 minutes. As noted above, CW copy by ear might need 16 to 18-dB higher levels. Additional time allows even lower signal-to-noise ratios, but quadrupling the time used only has the effect of doubling the transmitter power. Though most people will not have interest in using extremely long times for a transmission, even a few minutes of transmission will provide a major improvement relative to audible copy. A number of terrestrial and EME contacts have been made using the PUA43 mode. Perhaps one of the more interesting early EME contacts is that done Feb 25, 2001, by Ernie Manly, W7LHL, and Larry Liljeqvist, W7SZ, on 1296 MHz using only 5 W on each end. The antennas were ordinary surplus TVRO dishes of 10 and 13-foot diameter.

Further Directions

The DSP enhanced copy of weak signals provides an alternative to bigger antennas and higher power. One can expect that various schemes will be developed to use this capability. These should improve on the examples that are shown here.

Other avenues exist that emphasize different elements of signal propagation. One example of this is the work of Joe Taylor, K1JT with the *WSJT* program.[16] This uses a multiple frequency modulation and coding scheme, called FSK441, that is optimized to use bursts of signal, such as occur with meteor scatter. This contrasts strongly with the approach of the PUA43 mode that must grind out signal copy, based only on the average power being received. Each propagation situation needs to be considered as a strong determining factor in the system to be used.

12.6 A 28 MHZ QRP MODULE

One approach to adding new bands to an existing low power station is to build an add-on module where a stand-alone transmitter is combined with a receiving converter. This example interfaces with a home station CW receiver (Chapter 6) with a 4-MHz input. This module uses a 28 to 4-MHz receiving converter and a VXO based 28-MHz CW transmitter. The power output is purposefully confined to 1 W, adding sport to an already exciting band. A single crystal provides a transmitter tuning range of over 60 kHz.

The Transmitter

The transmitter shown in **Fig 12.37** begins with a VXO operating at 18.7 MHz. This free running oscillator is eventually frequency divided by 2, creating a square wave. The third harmonic of that signal, at 28 MHz, is selected with a bandpass filter, amplified, and keyed to form the transmitter. The VXO circuit with oscillator Q1 was originally like others shown in Chapter 4, providing about a 40-kHz tuning range at 28 MHz.

Inside view of the 10-meter module with the VXO and triple tuned bandpass filter in the center. The receiver RF amplifier board is at the bottom of the photo.

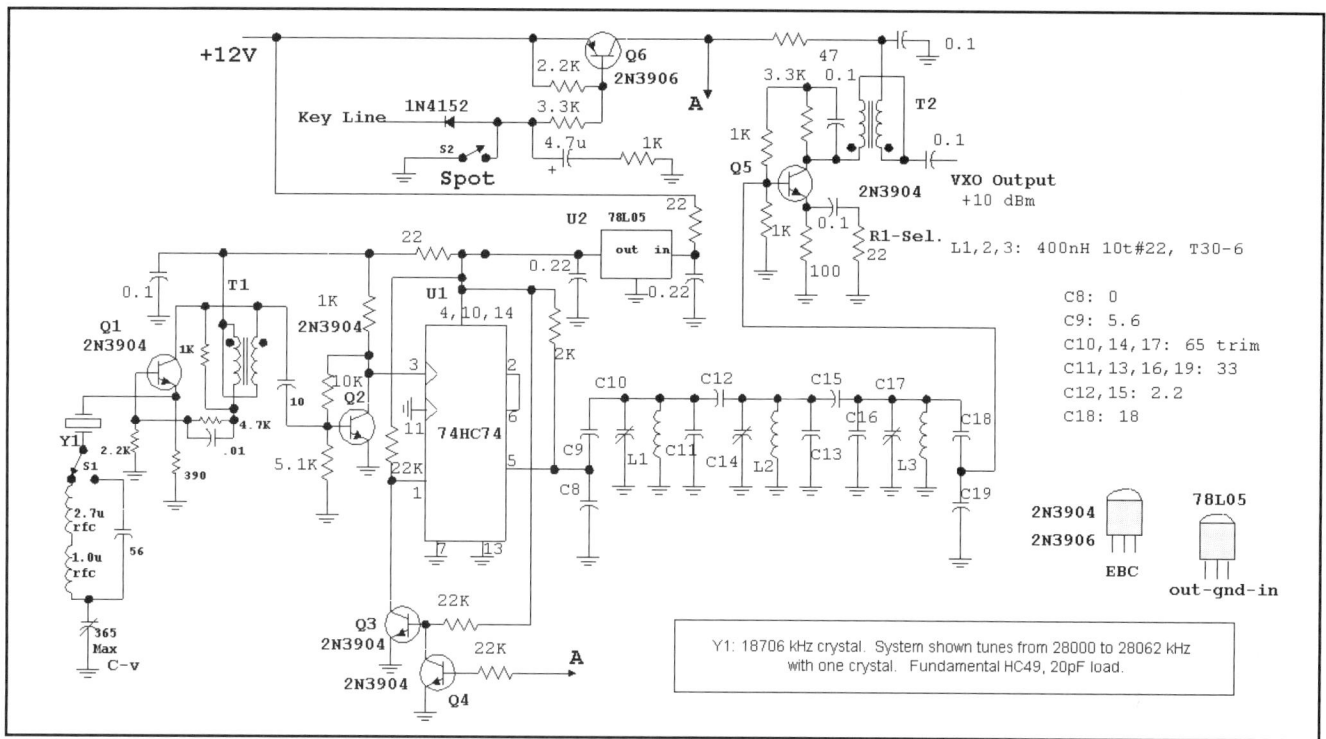

Fig 12.37—An 18.7-MHz VXO (Q1) is frequency divided by 2 with U1 to form a square wave. The third harmonic is selected with the bandpass filter and amplified to a 10-milliwatt output level. T1 and T2 are 10 bifilar turns on FT-37-43 cores. S1 is a wafer switch with low capacitance. A toggle switch should *not* be used here.

Inside view of the 10-meter module. The VXO board is below the board containing the rest of the transmitter. The output low pass filter and T/R relay are on the small board at the upper right. The delay control is on the side panel.

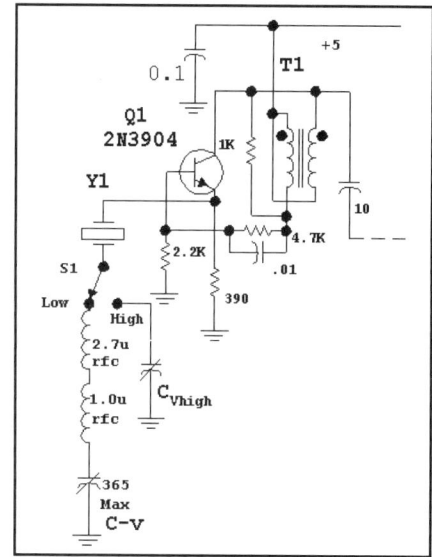

Fig 12.38—An even larger tuning range is available with a separate tuning control for each range. C_{VHigh} is selected from the junk box to have a low minimum capacitance.

The circuit was modified to use two ranges and now tunes from 28.000 to 28.062 MHz with the available components. The low end of the band is tuned when S1 inserts a series inductance in the circuit. Experiments showed an even larger upward range was available if a separate tuning capacitor was used for each range. This variation is shown in **Fig 12.38**. Experimentation is almost always useful with VXO circuits. (We measured our crystal as having $L_m = 3.01$ mH and $C_0 = 6$ pF.)

The transmitter continues in **Fig 12.39** with a driver using a parallel pair of 2N3904 transistors. The power amplifier is a 2N3866 with a 1-Ω emitter degeneration resistance. A 7-element low pass filter follows the transmitter, suppressing harmonics and other spurious responses. The only harmonic observed was the second at –69 dBc. The 18-MHz output is present in the output, but at the –73 dBc level.

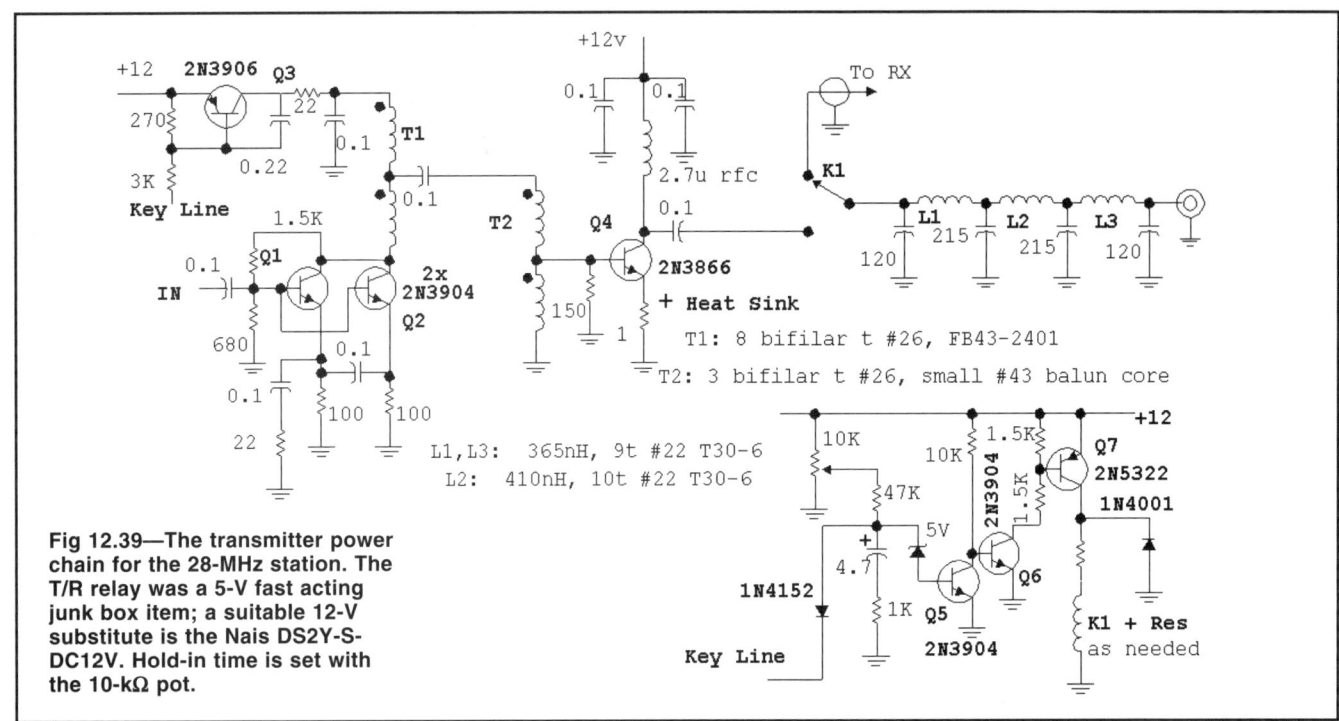

Fig 12.39—The transmitter power chain for the 28-MHz station. The T/R relay was a 5-V fast acting junk box item; a suitable 12-V substitute is the Nais DS2Y-S-DC12V. Hold-in time is set with the 10-kΩ pot.

Receiving Converter

A diode ring mixer is the basis of the receiving converter, driven from a crystal-controlled oscillator using a 32-MHz third-overtone oscillator. The post mixer amplifier is a common gate JFET with a drain current of about 13 mA. A narrow bandwidth 4-MHz output feeds a wide bandwidth bandpass filter. The mixer is preselected with a double tuned circuit.

An RF amplifier is included in the receiver. We used a circuit left from an earlier effort employing a dual gate MOSFET. A common gate JFET, described in Chapter 6, would be ideal, offering low noise figure with less gain.

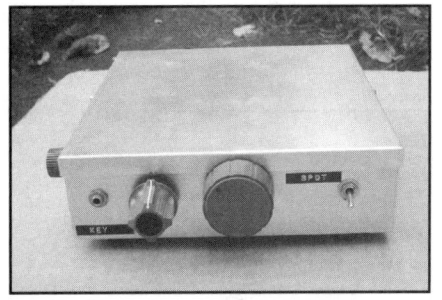

Front panel view of the 10-meter module.

12.7 A GENERAL PURPOSE RECEIVER MODULE

This module is essentially the heart of a direct conversion receiver. A TUF-3 diode ring was chosen for improved performance at lower frequency, although the TUF-1 will fit the board. The mixer is followed by an LC low pass filter and an audio amplifier chain using a mixture of bipolar transistors and op-amps. Muting circuitry, an RC active low pass filter, an audio attenuator, and a sidetone oscillator are included on the single board.

The module works very well as a direct conversion receiver. Careful attention to grounding in the early audio stages has eliminated many of the traditional problems encountered, which were described in Chapter 8. The board is sized to fit in a Hammond 1590B box with feed through capacitors and coax connectors, effectively reducing spurious responses from local VHF signals.

The schematic is shown in **Fig 12.40**. A low pass filter using a ferrite toroid inductor follows the ring mixer. The one we used was a pre-wound 55-μH part from the junk box, but would ideally use higher inductance with a larger core. An increase in the value of C2 would then improve the low pass filtering. The toroid form is preferred, for it is less susceptible to hum pickup than the other inductors often used. A resistor, R1, provides a termination for sum products exiting the ring mixer.

The audio amplifier begins with a common base stage offering a 50-Ω impedance to the mixer. A degenerated common emitter amplifier, Q3, follows this. At this point the user could exit the board to drive a volume control and/or LC filter. This option is shown in **Fig 12.41**. The filter is a three element high pass configured to suppress frequencies below 300 Hz. A low pass could be cascaded if desired. We have used the board without this filter. Ideally, the signal after the high pass filter, if used, would exit the enclosure on a feedthrough capacitor. The rest of the circuitry (described below) would then be built on a separate board without shielding.

The first op-amp stage includes an FET switch for receiver muting. An RC active low pass filter, U1b, follows this. This circuit is programmable by the designer/builder. The response of the filter alone

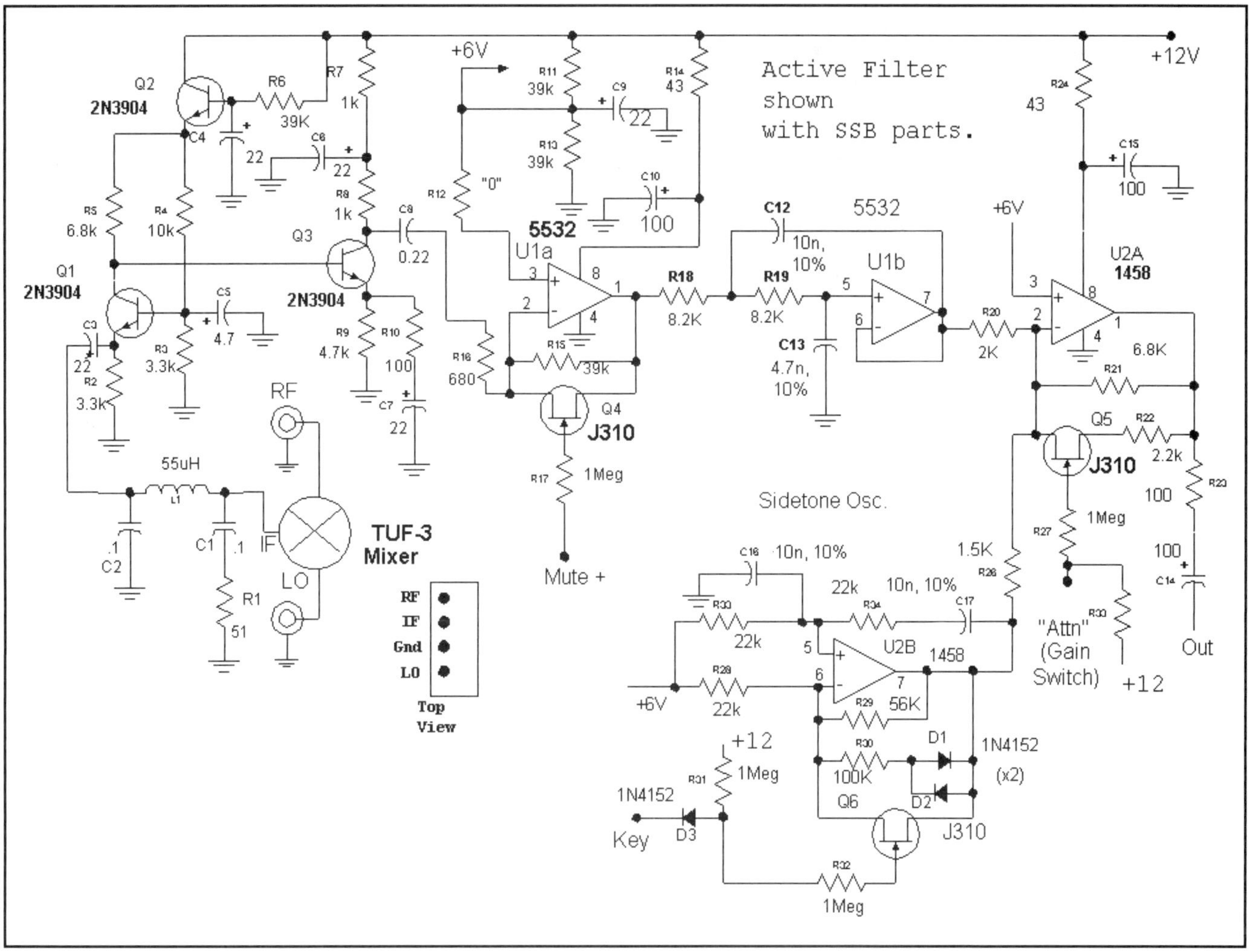

Fig 12.40—General-purpose direct-conversion receiver.

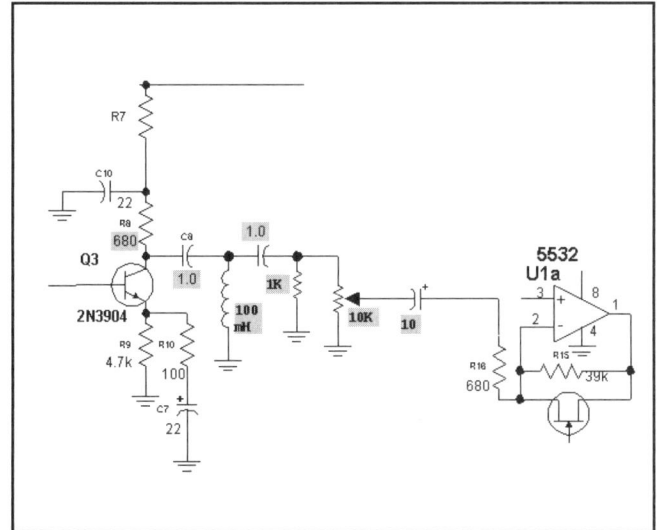

Fig 12.41—Option with an added audio gain control. Also shown is an LC high pass filter. The altered or added components are highlighted.

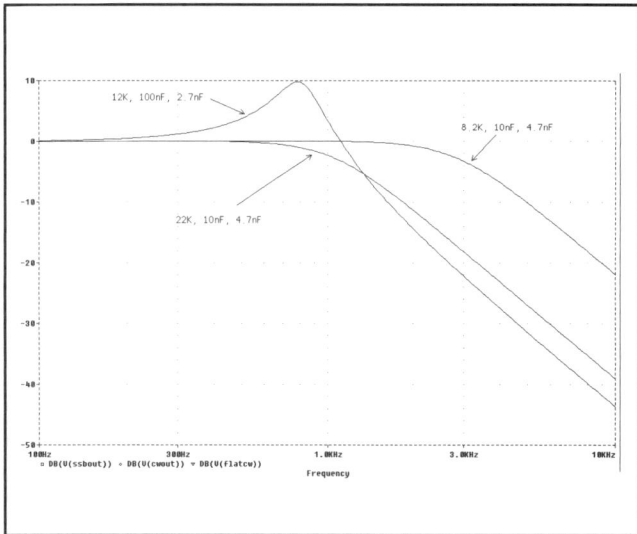

Fig 12.42—Calculated response for low pass filter with three different component value sets.

Field Operation, Portable Gear and Integrated Stations 12.31

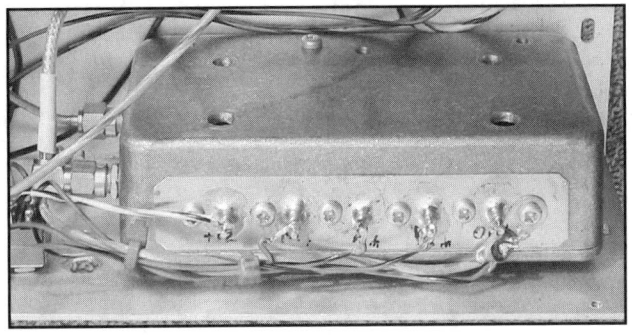

A shot of the module installed in shielded enclosure. A box built from circuit board would also work well.

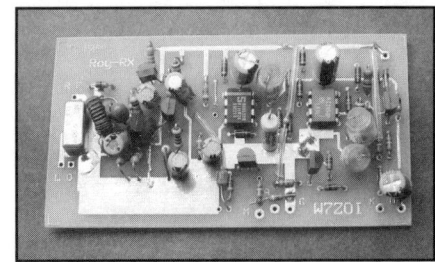

General purpose direct conversion module contains a diode ring mixer, audio amplifier, active audio filter, gain programmable active filter, and sidetone oscillator. This board is normally mounted inside a shielded box with coax connectors and feed-through capacitors for all interfaces. Two boards can be used for a binaural receiver.

Table 12.1
General-Purpose Receiver Module—Components for the Low Pass Filter

Bandwidth and Shape	R18 and R19	C12	C13
3 kHz flat	8.2 kΩ	10 nF	4.7 nF
1 kHz flat	22 kΩ	10 nF	4.7 nF
Peak at 700, Q=3	12 kΩ	100 nF	2.7 nF

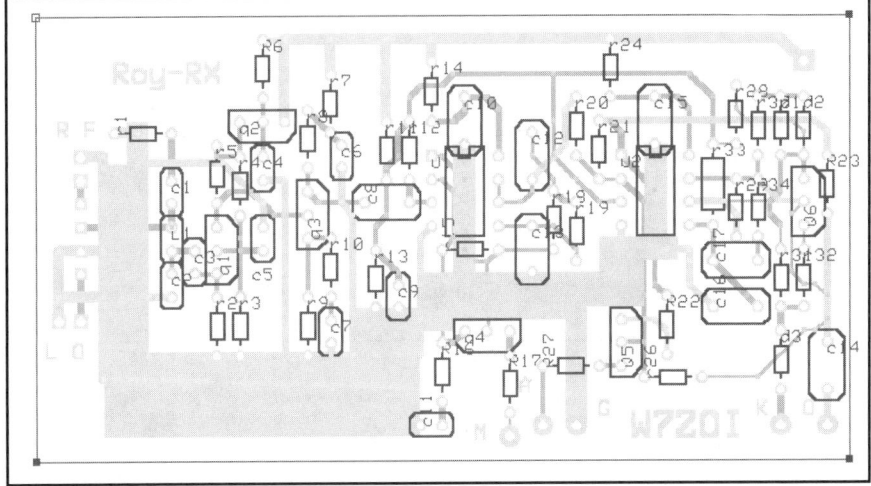

Fig 12.43—View of the component side of the circuit board. Copper runs on both sides of the circuit board are shown. The board layout is double sided, through-hole plated, and was done with the program *Express PCB Version 2.1.1* found at www.expresspcb.com.

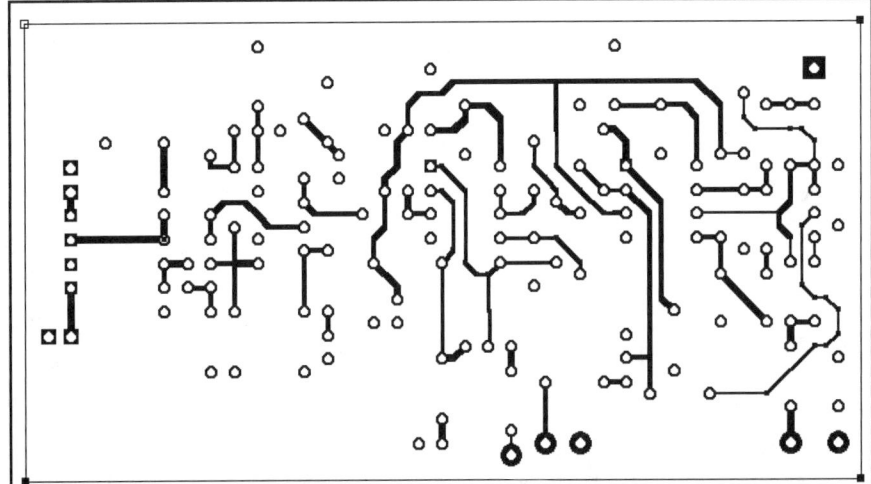

Fig 12.44—This view is identical to that of **Fig 12.43**, but shows only the runs on the opposite side of the board.

(without the rest of the receiver) is shown in **Fig 12.42** for three component value sets summarized in **Table 12.1**.

An inverting amplifier, U2A, with a gain that can be switched with an external signal, follows the active low pass filter. A 12-dB gain step is available with the components shown. This op-amp has enough output to drive low impedance headphones.

The remaining half of U2 serves as a sidetone oscillator. This Weinbridge topology was used in the "Unfinished" transceiver discussed elsewhere.

There is considerable flexibility available in this design. If a simpler receiver is needed, U1b is capable of driving headphones, allowing U2 to be eliminated. Gain can be programmed in the second audio stage with changes in R10, in U1A through R15 and R16, and in U2A.

We have used these modules in three different receiver types. The first is a simple direct conversion receiver where the circuitry and performance are very much like that of the W7EL classic so long as the board is well shielded and used with a well isolated LO. Second, we have used a pair of these as a binaural receiver.[17] Finally, the board has been a handy "tail end" for several superhet rigs. A pair of the boards could be used to build a phasing receiver, although there is probably too much selective circuitry in the version shown, encouraging a redesign using the guidelines of Chapter 9. The PC board layout used is shown in **Figs 12.43** and **12.44**. Repeated building of the same design justifies a printed board. The name on the board, "Roy-Rx," indicates that this is a variation of the Roy Lewallen design from *QST*, August, 1980.[18]

12.8 DIRECT CONVERSION TRANSCEIVERS FOR 144 MHZ SSB AND CW

These transceivers were built using prototype circuit boards during the development of the line of products sold by Kanga US. They illustrate different packaging techniques, and also some of the effort that goes into moving from prototype or ugly construction to a commercially available production circuit board. Both transceivers use identical circuitry, and the basic design is intended as a tunable IF for microwave transverters. A wooden box was chosen to investigate the problems that result from having no shielding at all around the circuit boards. The radio works well as a tunable IF, but is subject to hum and noise pickup when directly connected to a nearby, non-directional 2-meter antenna. It works fine on the 2-meter band, however, with a small Yagi 10 meters away, and pointed away from the transceiver. The version built in the gray steel chassis has no shielding between PC boards, but is well shielded from the outside world. It works with a whip antenna, but has some microphonics that are not present in direct conversion rigs with more extensive shielding.

The circuitry is all on three printed circuit boards. The block diagram is shown

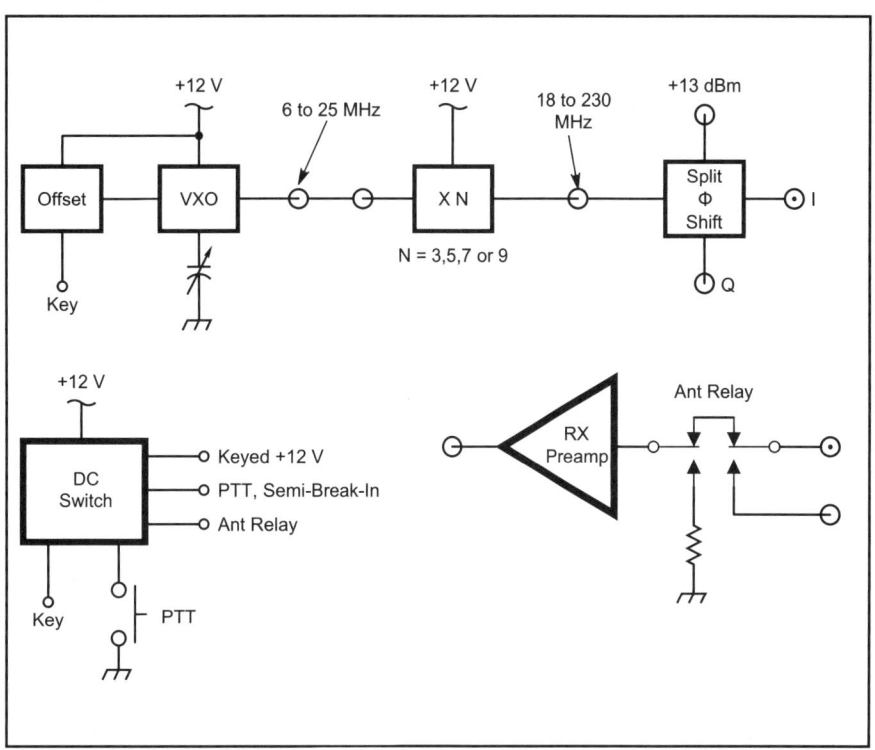

Fig 12.46—Block diagram of LM2 PC board, which contains the VXO, LNA and TR switching circuits.

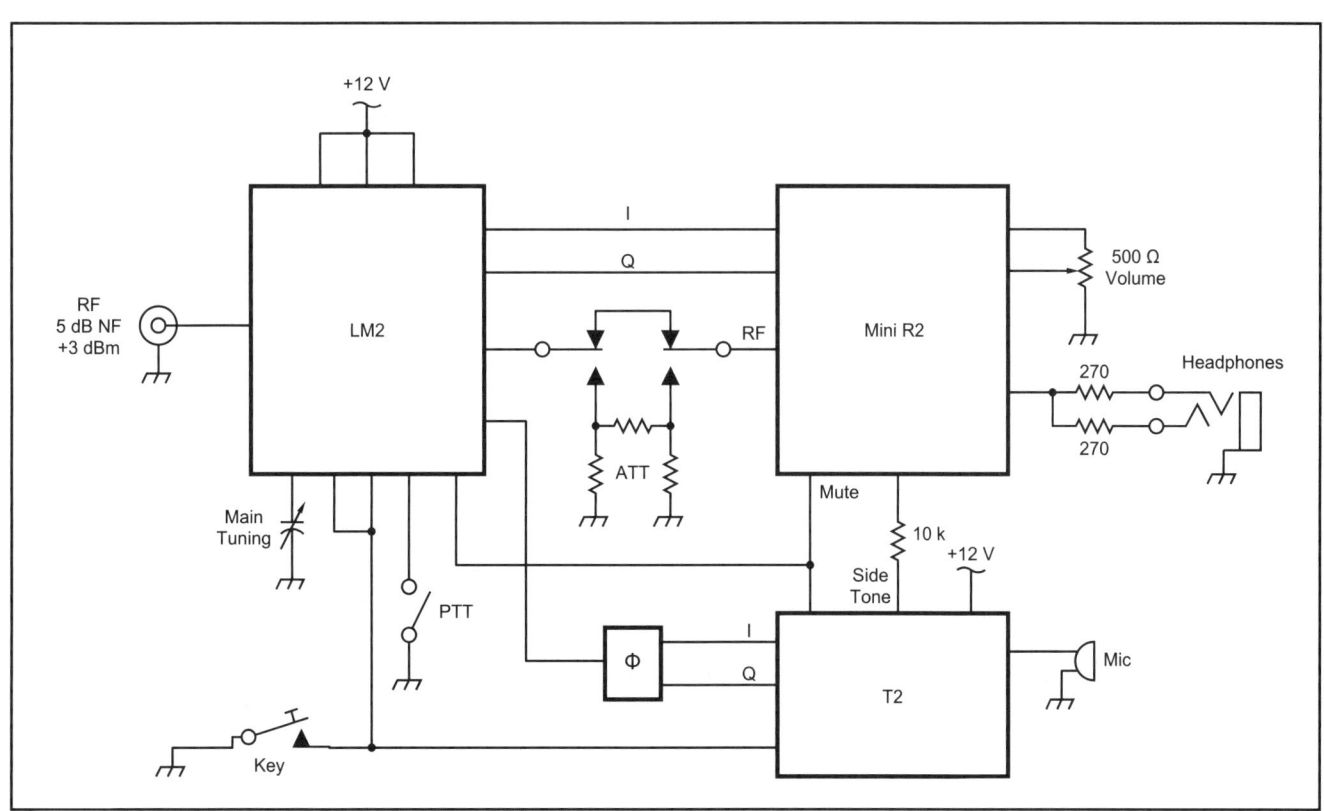

Fig 12.45—Block diagram of direct-conversion 144-MHz SSB/CW transceiver.

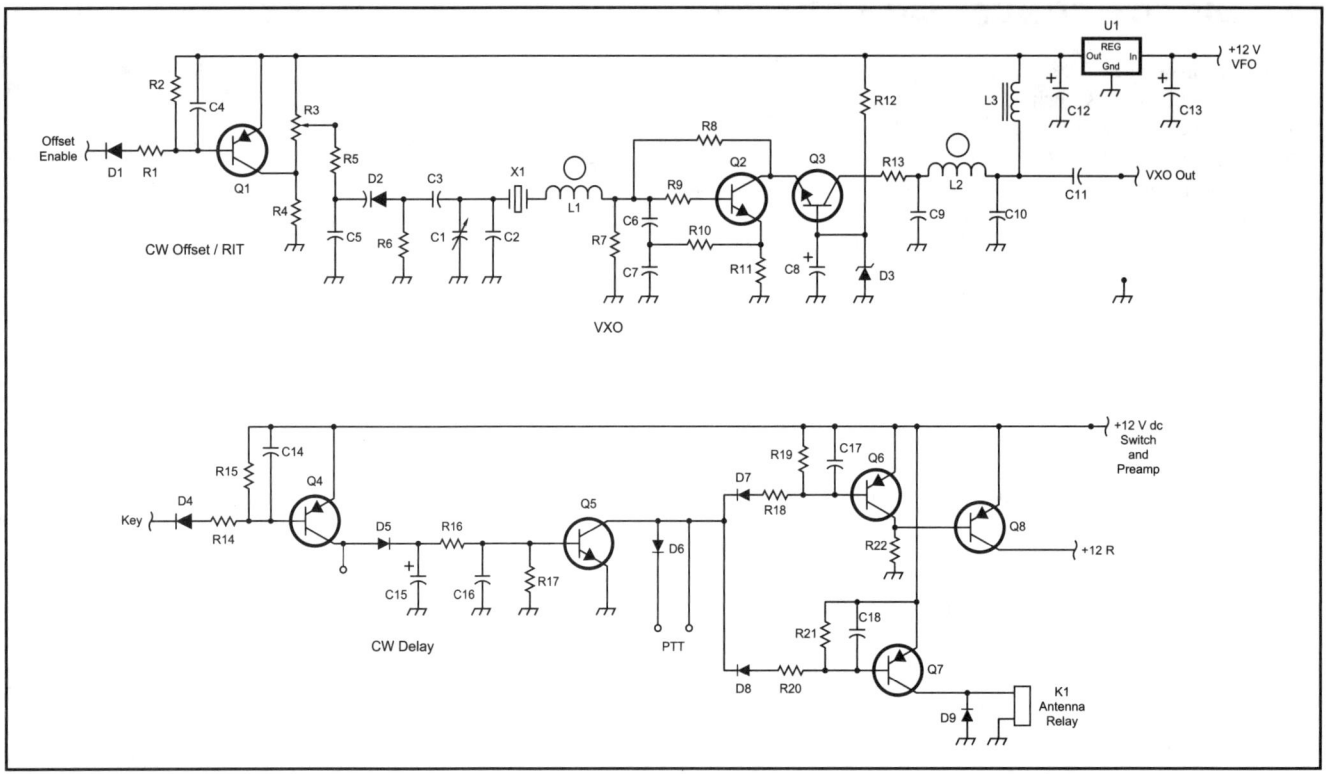

Fig 12.47—LM2 schematic 1.

Fig 12.48—LM2 schematic #2 and parts list.
R1 4.7 kΩ
R2 10 kΩ
R3 50 kΩ Trimpot Panasonic 36C series
R4 47 kΩ
R5 100 kΩ
R6 1 MΩ
R7 10 kΩ
R8 10 kΩ
R9 33 Ω
R10 22 Ω
R11 510 Ω
R12 3.9 kΩ
R13 51 Ω
R14 4.7 kΩ
R15 10 kΩ
R16 4.7 kΩ
R17 10 kΩ
R18 4.7 kΩ
R19 10 kΩ
R20 4.7 kΩ
R21 10 kΩ
R22 10 kΩ
R23 1 MΩ chip
R24 120 Ω 1/2 W
R25 100 Ω chip
R26 100 Ω chip
R27 51 Ω chip
R28 510 Ω
C1 Approx 40 pF variable Main Tuning. See Text.
C2 Upper frequency limit or temperature comp. See Text.
C3 RIT range set. See Text.
C4 0.1 μF Panasonic V series
C5 0.01 μF disk ceramic
C6 See Table 12.3
C7 See Table 12.3
C8 10 μF electrolytic
C9 See Table 12.3
C10 See Table 12.3
C11 0.01 μF disk ceramic
C12 4.7 μF tantalum
C13 10 μF electrolytic
C14 0.1 μF Panasonic V series
C15 22 μF tantalum CW semi-break-in delay
C16 0.1 μF Panasonic V series
C17 0.1 μF Panasonic V series
C18 0.1 μF Panasonic V series
C19 22 pF chip
C20 0.01 μF chip
C21 10 μF electrolytic
C22 See Table 12.2
C23 See Table 12.2
C24 See Table 12.2
C25 See Table 12.2
C26 See Table 12.2
C27 See Table 12.2
C28 See Table 12.2
C29 0.01 μF chip
C30 0.01 μF chip
C31 See Table 12.2
C32 See Table 12.2
C33 See Table 12.2
C34 See Table 12.2
C35 See Table 12.2
C36 See Table 12.2
C37 See Table 12.2
C38 See Table 12.2
C39 See Table 12.2
C40 See Table 12.2
C41 See Table 12.2
C42 See Table 12.2
C43 0.01 μF chip
C44 See Table 12.2
C45 See Table 12.2
C46 See Table 12.2
C47 See Table 12.2
C48 See Table 12.2
L1 VXO range inductor, 33t T37-2 toroid. See Text.
L2 See Table 12.3
L3 See Table 12.3
L4 See Table 12.2
L5 See Table 12.2
L6 See Table 12.2
L7 6 turns FT 25-43 ferrite toroid
L8 See Table 12.2
L9 See Table 12.2
L10 See Table 12.2
L11 See Table 12.2
L12 See Table 12.2
L13 See Table 12.2
L14 See Table 12.2
D1 1N4148
D2 MV2107 or similar tuning diode
D3 4.7-V Zener
D4 1N4148
D5 1N4148
D6 1N4148
D7 1N4148
D8 1N4148
D9 1N4148
Q1 2N3906
Q2 2N3904 or PN5179
Q3 2N3904 or PN5179
Q4 2N3906
Q5 2N3904
Q6 2N3906
Q7 2N3906
Q8 2N3906
U1 78L09
U2 78L06
U3 74AC04
U4 MAV-11 or MAB-4. See Text.
U5 Toko splitter
U6 Toko splitter
U7 MAR-6
K1 OMRON 65V-2-H
X1 Crystal See Text

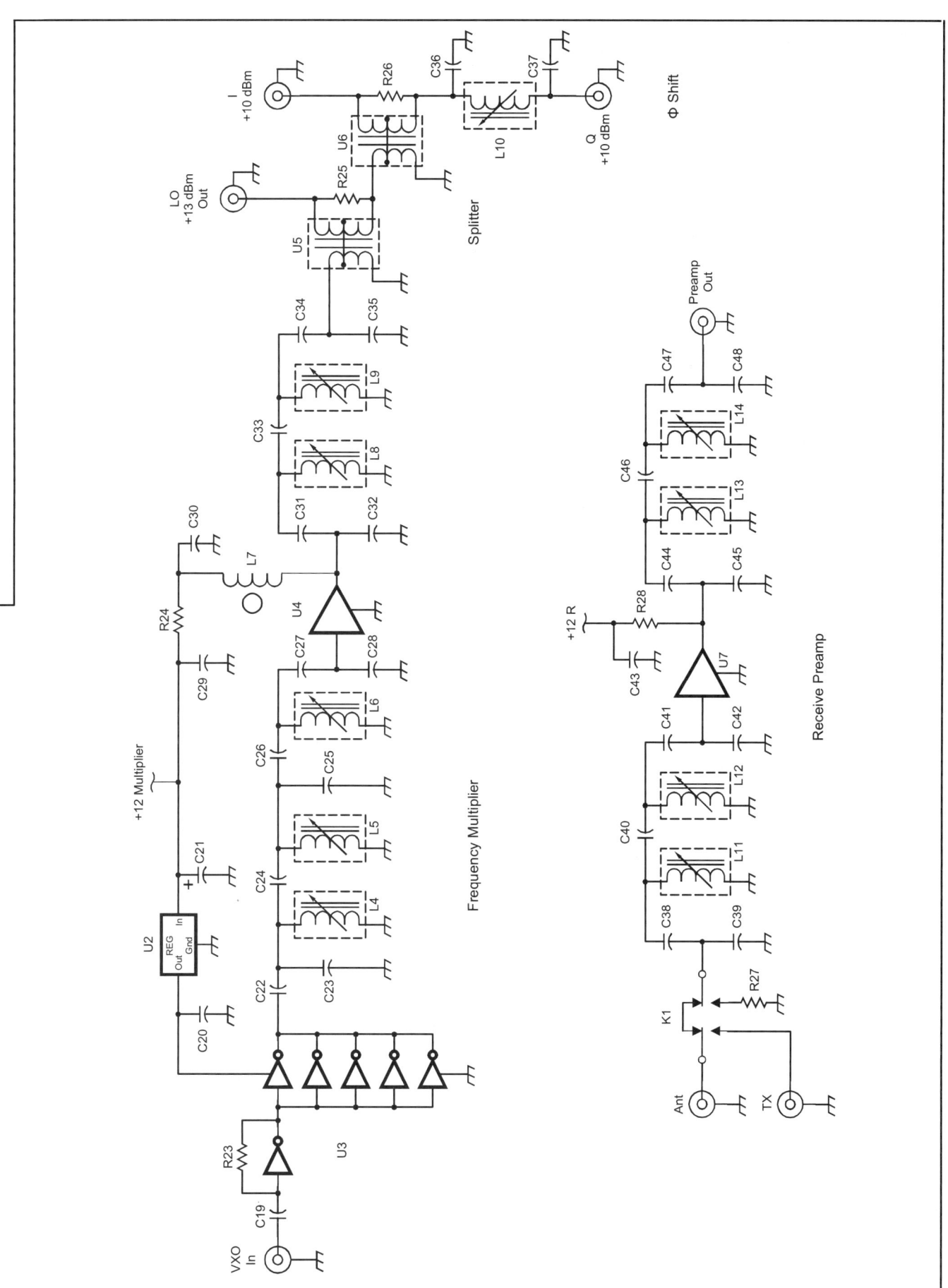

Field Operation, Portable Gear and Integrated Stations

Table 12.2
Filter and Phase Shift Compnents

All chip capacitor values are in pF, 1206- or 0805-series Panasonic. All inductor values in nH, MC122- or MC134-series Toko with case.

Component	Frequency (MHz)						
	18	21	24	28	50	144	222
C22	56	56	39	33	20	3.9	3.9
C23	68	68	47	47	22	5.6	3.9
C24, C26, C33, C40, C46	10	10	10	10	5	1	1
C25	120	120	76	68	39	9.1	6.8
C27, C31, C34, C38, C41, C44, C47 C28, C32, C35, C39, C42, C45	180	180	120	120	56	12	8.2
C48	390	390	270	270	150	47	27
C36, C37	180	150	120	120	68	22	15
L4, L5, L6, L8, L9, L11, L12, L13, L14	422	422	422	350	226	108	53
L10	422	383	350	291	159	53	32

in **Fig 12.45**. The miniR2 and T2 PC boards have been previously described in *QST*.[19,20] The LM2 PC board contains the VXO, LNA and TR switching circuits. The LM2 block diagram is shown in **Fig 12.46**. **Figs 12.47** and **12.48** are the LM2 schematics. In **Figs 12.49** and **12.50** you'll see the wood-boxed transceiver, and **Figs 12.51** and **12.52** are the version in the metal chassis.

Table 12.3
VXO Components

All capacitor values are in pF, Panasonic 100 V C0G, monolithic ceramic. L2 values represent the suggested number of turns on a T37-2 toroid core. Adjust for maximum output across 50Ω. L3 values are in µH using a JW Miller epoxy conformal coated iron core.

Component	Frequency Range (MHz)				
	6-8	8-10	10-15	15-20	20-26
C6, C7	220	220	150	100	82
C9	150	120	82	68	56
C10	680	560	390	330	220
L2	24	21	19	17	16
L3	18	15	12	8.2	6.8

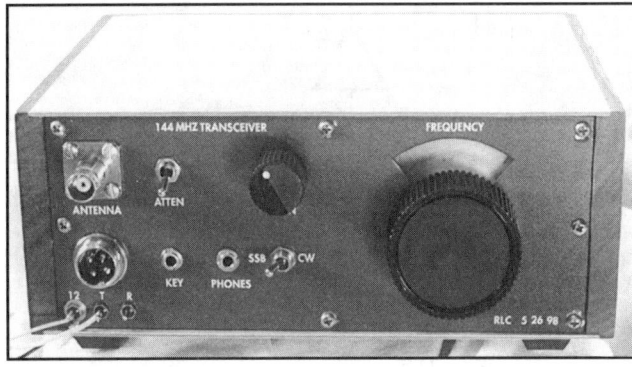

Fig 12.49—Wood Box 144-MHz transceiver.

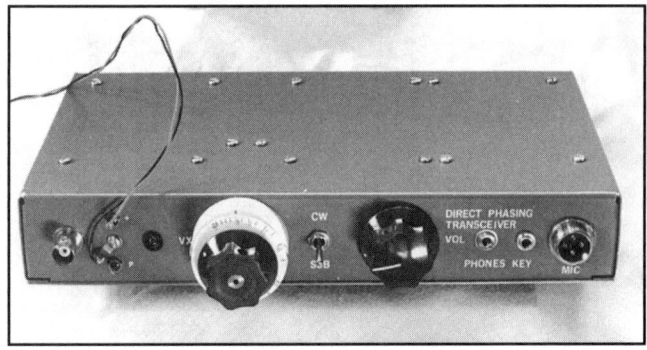

Fig 12.51—The Metal Box 144-MHz transceiver.

Fig 12.50—An interior view of the Wood Box 144-MHz transceiver.

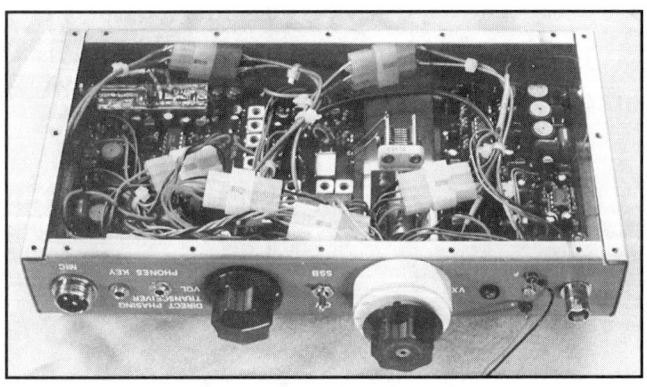

Fig 12.52—An inside look at the Metal Box 144-MHz transceiver.

12.9 A 52 MHZ TUNABLE IF FOR VHF AND UHF TRANSVERTERS

This transceiver was designed and built to serve as the base station tunable IF for weak signal SSB and CW DXing on the bands from 222 through 2304 MHz. It is mounted in a large rack-mount box, and is connected to a set of rack mount transverters. A front-panel switch selects the desired transverter. The transverters provide 100-W output on 222 and 432 MHz, 10 W on 903 MHz, 15 W on 1296 MHz and 4 W on 2304 MHz, with less than 2-dB noise figure on each band. 52 MHz was chosen for the IF because it is not harmonically related to any of the desired band segments, and there is no CW or SSB activity near 52 MHz to cause IF breakthrough problems.

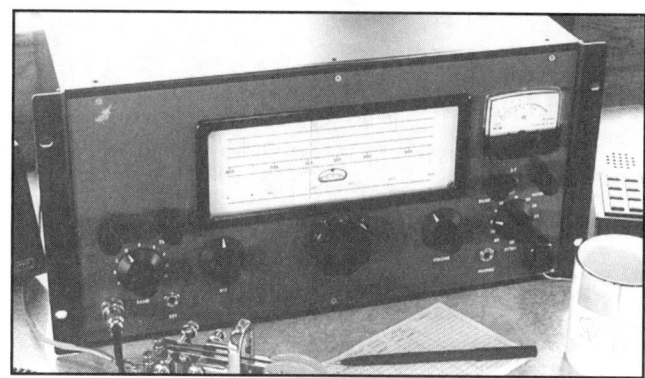

Fig 12.53—The 52-MHz IF transceiver in operation.

Fig 12.53 is a photograph of the IF transceiver in operation, and the block diagram is shown in **Fig 12.54**. Modular construction is used, and each module is mounted in a shield box. The T2 exciter and LO modules are build in boxes soldered up from PC board material; the R2 receiver is in a steel chassis. The filters and preamp are in aluminum boxes with screw-on covers. The receiver and exciter each has its own independent phase-shift network with an air-variable phase trim capacitor, hardwired directly to the receiver or exciter circuit board.

The LO phase shift adjustments and amplitude trimmer adjustments are accessible on top of the shielded enclosures, but after initial alignment they have remained untouched during the 6 years (and a move half-way across the country) that the rig has been in service. Detailed schematics

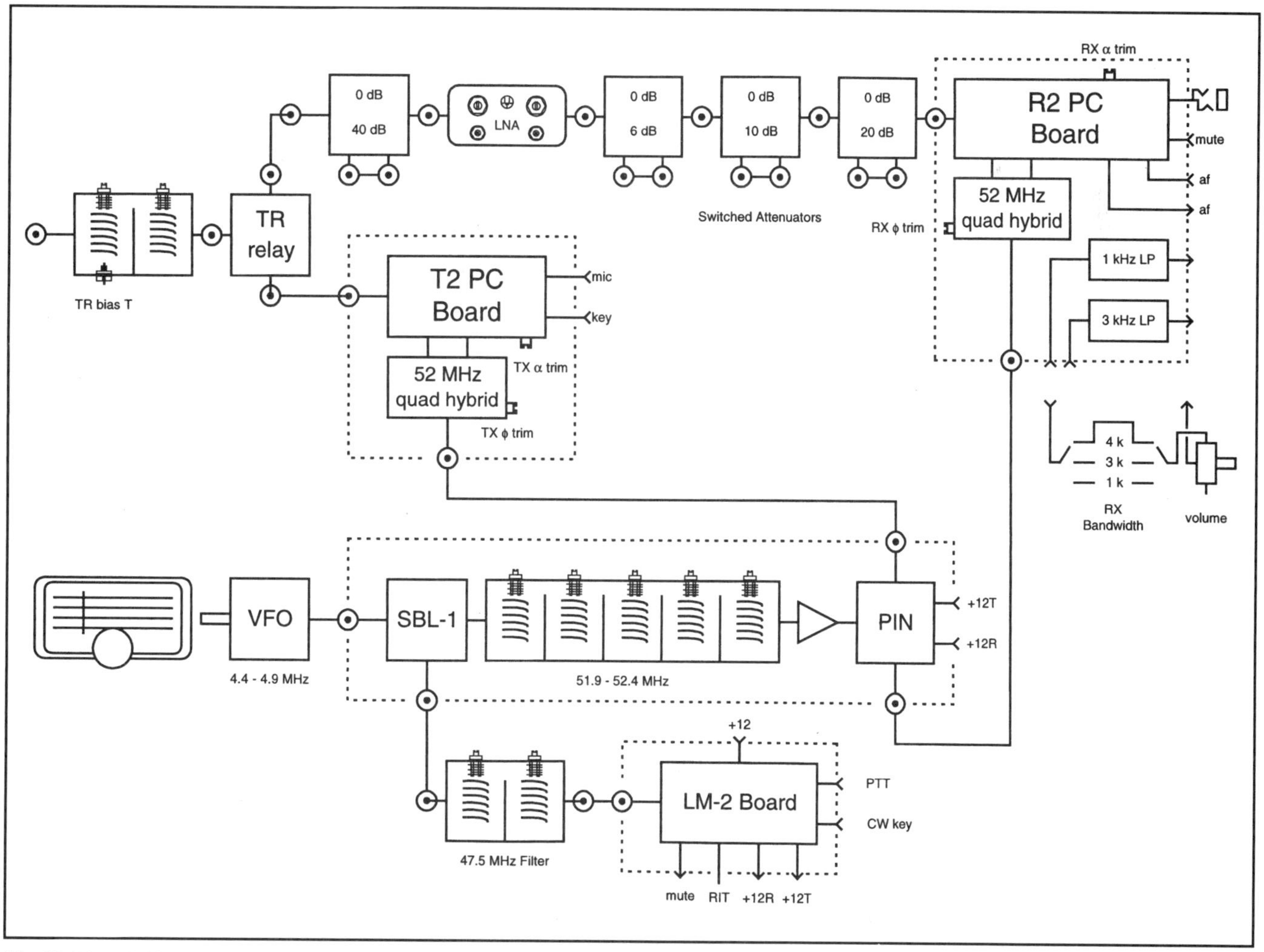

Fig 12.54—52-MHz IF transceiver block diagram.

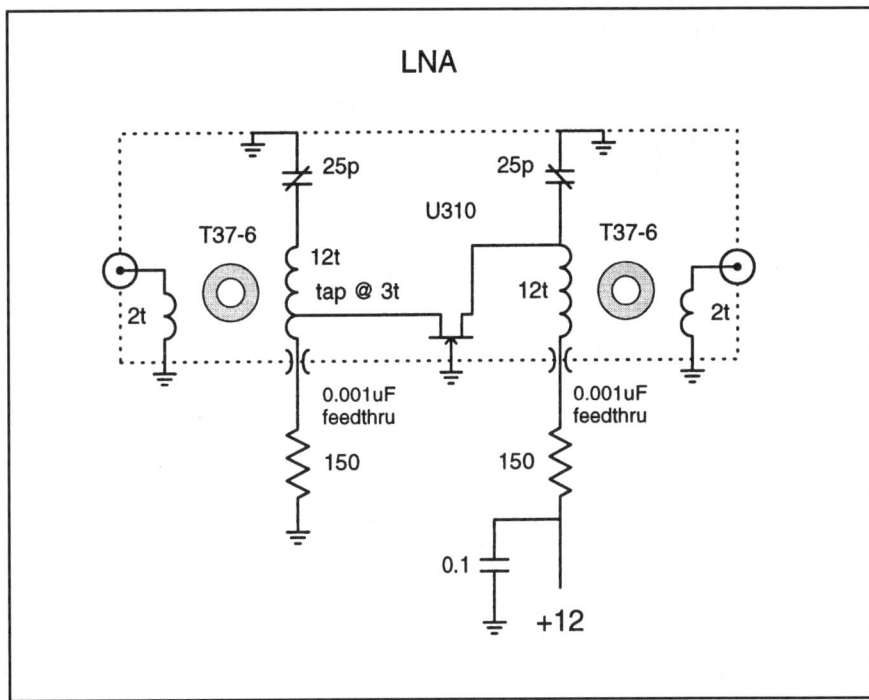

Fig 12.55—LNA schematic.

of each of the circuit blocks are shown in **Figs 12.55** through **12.61**. **Figure 12.62** is a close-up of one of the LO phase-shift networks, illustrating the mechanical and electrical symmetry and connection of the phase-trim capacitor. **Figure 12.63** is a view of the 52-MHz filter. **Figure 12.64** is a look under the hood, and **Fig 12.65** is a bottom view, showing much of the circuitry.

The Local Oscillator system is premixed from the 4-MHz range up to 52 MHz. A 5-section helical resonator filter selects the 52-MHz product, rejects the 44-MHz image, and provides additional attenuation of the 48-MHz premix oscillator. The output tunes from 51.9 MHz to 52.4 MHz, and the vintage Eddystone Dial provides a smooth, slow tuning rate and may be reset to within 1 kHz.

This IF transceiver was built to replace a commercial 6-meter rig being used as a tunable IF in a competitive VHF contest station. The commercial rig had a few spurs and birdies, and synthesizer noise burbles that sounded like weak signals

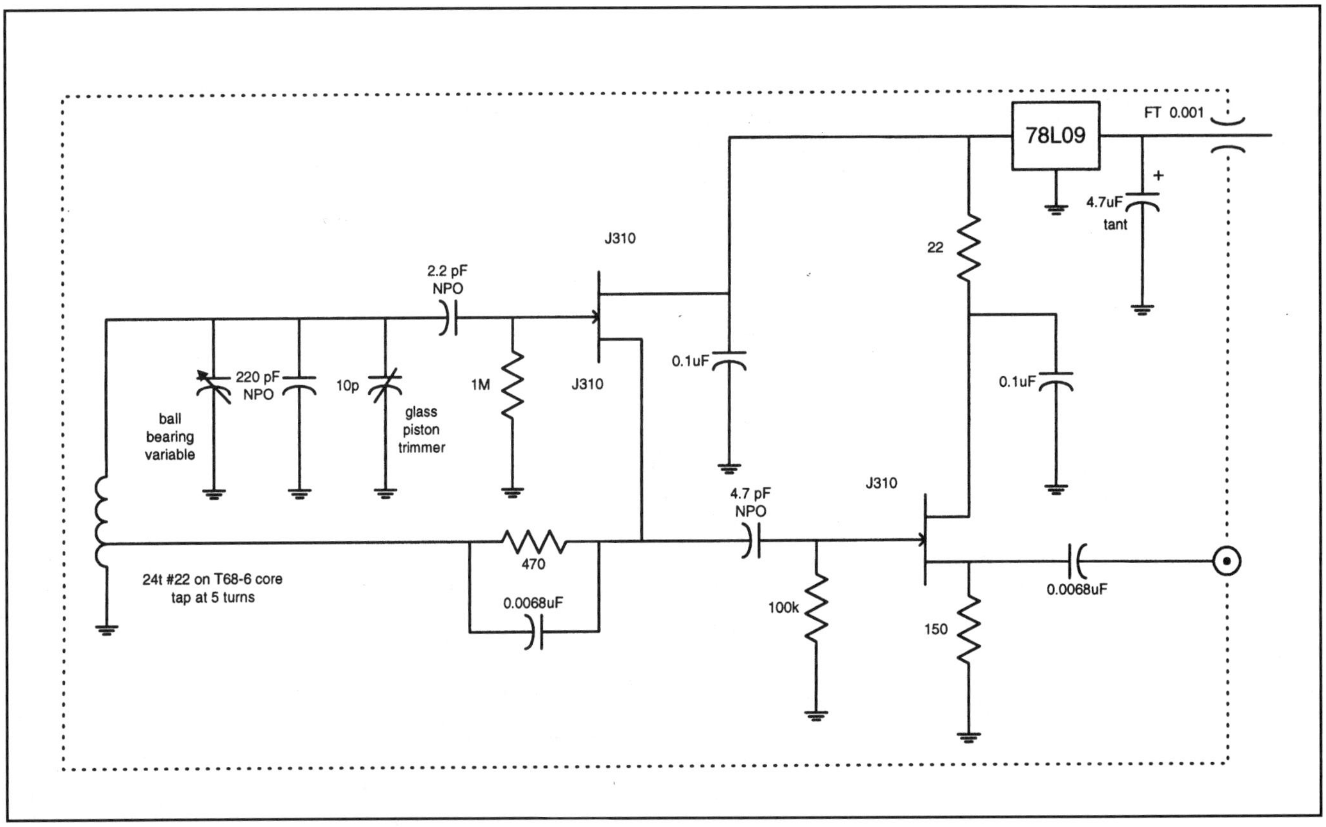

Fig 12.56—The 4.4-4.9 MHz VFO schematic

12.38   Chapter 12

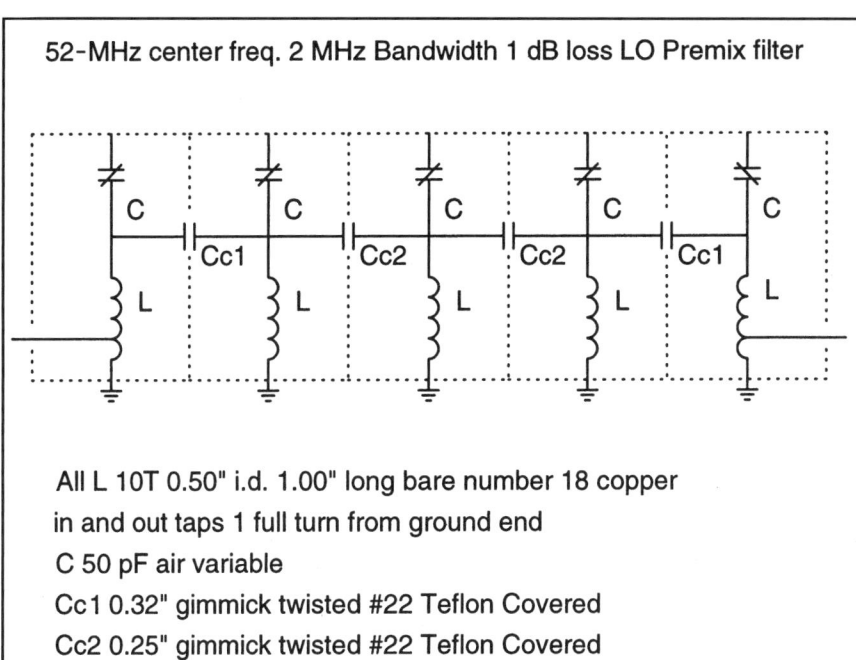

Fig 12.57—Schematic of the 52-MHz premix filter.

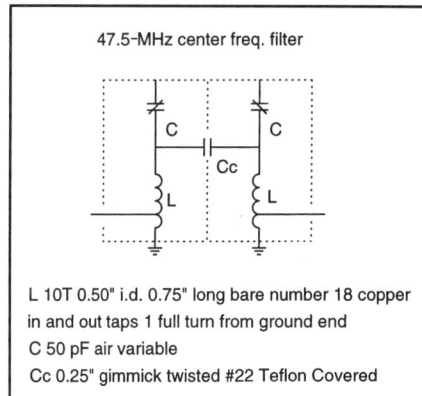

Fig 12.58—The 47.5-MHz premix oscillator filter.

when tuning for UHF DX. In addition, the audio distortion of the commercial radio contributed to operator fatigue over the course of a weekend contest. The homebrew 52-MHz transceiver has no spurious responses or birdies, and all undesired outputs are more than 70 dB below the desired output.

Modular construction with individual shielded modules, and a spacious cabinet, contributes to a very large piece of radio equipment with fine performance.

This 52-MHz tunable IF is a "work in progress," with unfinished audio gain control, metering, and mode selection functions. It has been in service for 6 years, and every year or so a function will be added. There is ample room inside for additional circuit modules, and room on the front panel for additional controls.

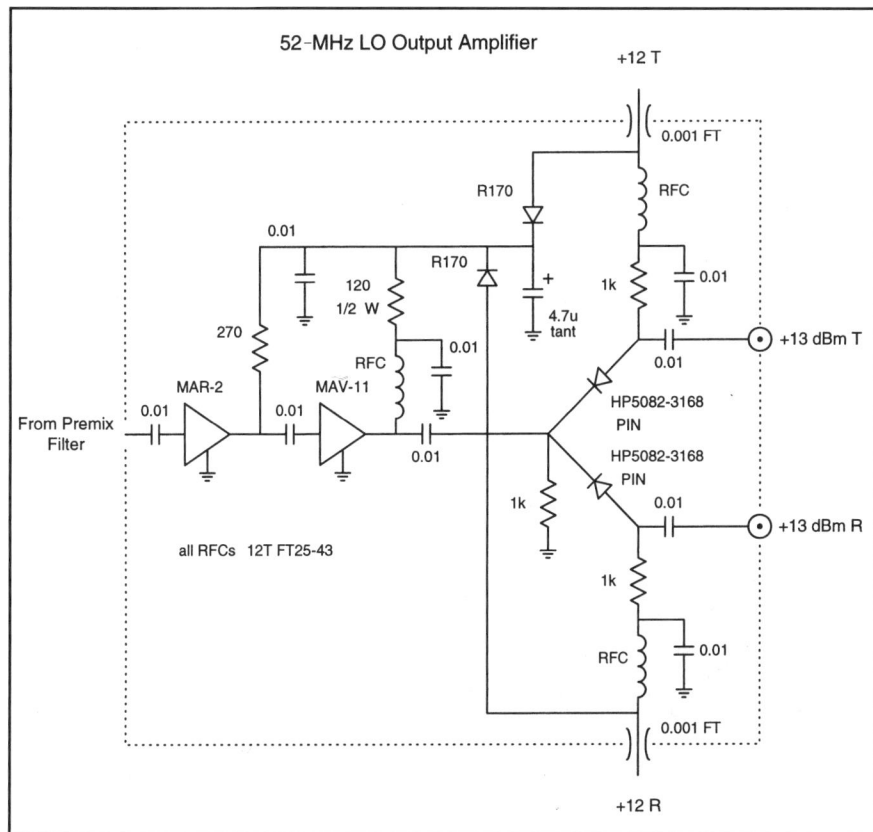

Fig 12.59—The 52-MHz premix LO output amplifier.

Field Operation, Portable Gear and Integrated Stations

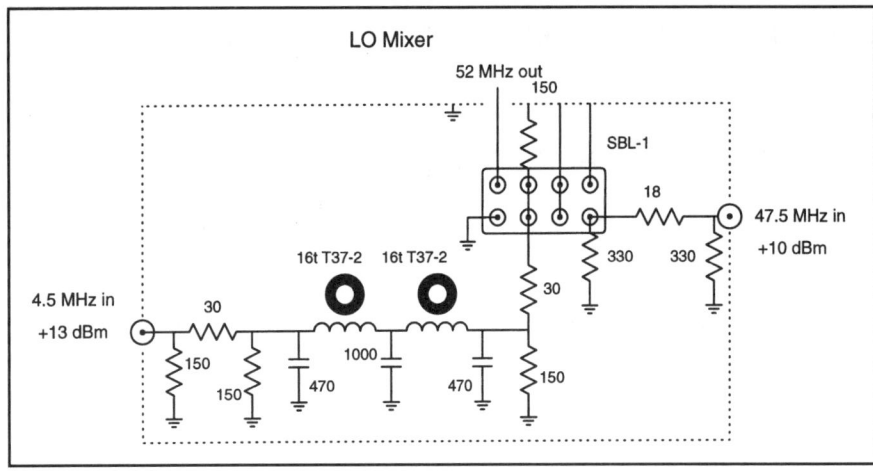

Fig 12.60—Schematic of the premix LO mixer.

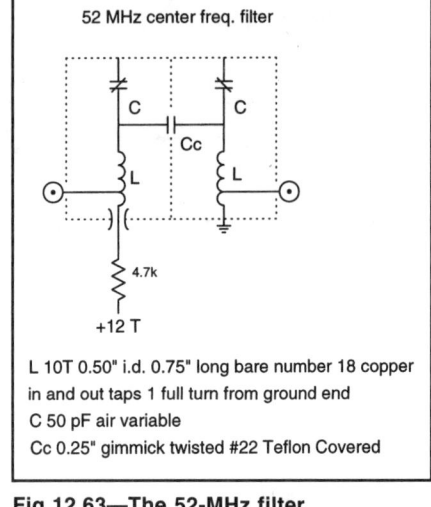

Fig 12.63—The 52-MHz filter.

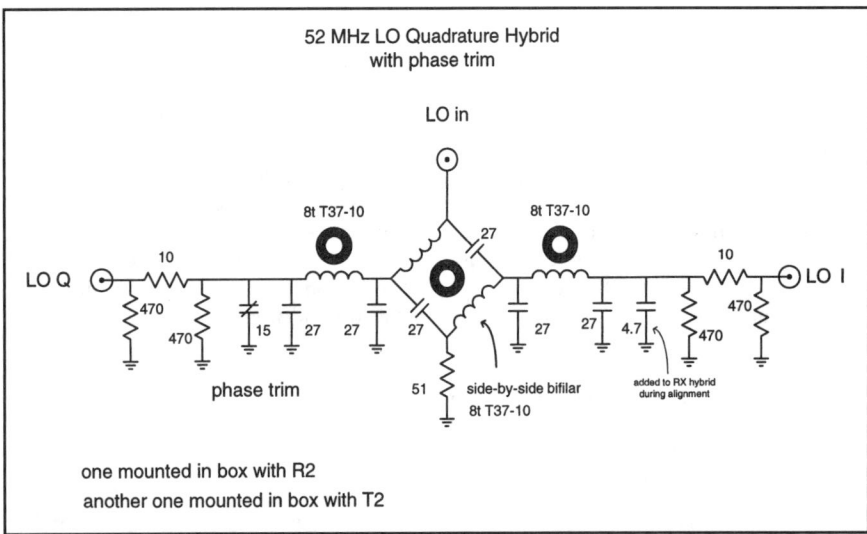

Fig 12.61—Schematic of the 52-MHz LO quadrature hybrid.

Fig 12.62—Close-up of LO quad-rature hybrid.

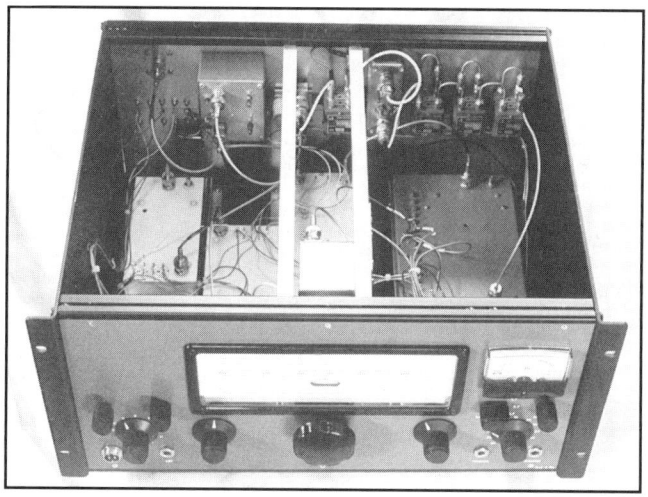

Fig 12.64—A peek at the inside top of the 52-MHz transceiver.

Fig 12.65—The inside bottom of the 52-MHz transceiver.

12.10 SLEEPING BAG RADIO

On winter camping trips in the Northwest and Michigan's Upper Peninsula, radio operation typically occurs at night, while snuggled deep inside a warm sleeping bag. This is a different environment that completely changes the usual ergonomics of a radio. This 40-meter CW transceiver is designed to sit on either its back or bottom, with all connections and controls on the front/top. It is stable in either position. The controls are kept to a minimum, with a large, stiff tuning knob, a volume control, and RIT. CW is full break-in, and the use of a keyed receiver LNA along with conventional receiver muting eliminates any receiver thumps during keying. The radio is built in two die-cast boxes screwed together, with feedthrough capacitors to carry the signals and power into the back compartment. The back compartment contains an interchangeable receiver circuit board, which may be either an R1 direct conversion receiver, a mini-R2 receiver, or a binaural receiver. This radio has a solid feel to it, and is heavy enough that it is unwelcome on a weeklong summer trek through the backcountry—but for a short overnight jaunt on snowshoes it is ideal. The tuning knob is large enough to tune with mittens, and stiff enough that it doesn't move when bumped.

The four photographs in **Figs 12.66** through **12.69** illustrate the construction. **Figure 12.70** illustrates how the receiver compartment is double shielded from the outside world. All connections into the receiver compartment are made using 0.001 µF feedthrough capacitors into the VFO/PA compartment. **Figure 12.71** is a block diagram. The VFO/frequency doubler is shown in **Fig 12.72**, the PA, using a high-gain differential amplifier driving a 5-W CB power transistor is

Fig 12.66—The Sleeping Bag Radio.

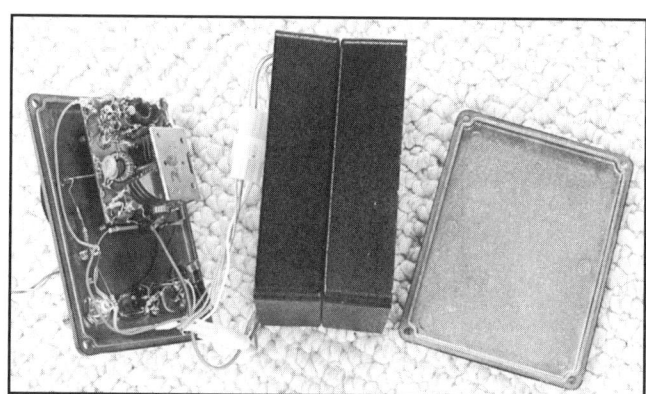

Fig 12.67—Sleeping Bag Radio VFO.

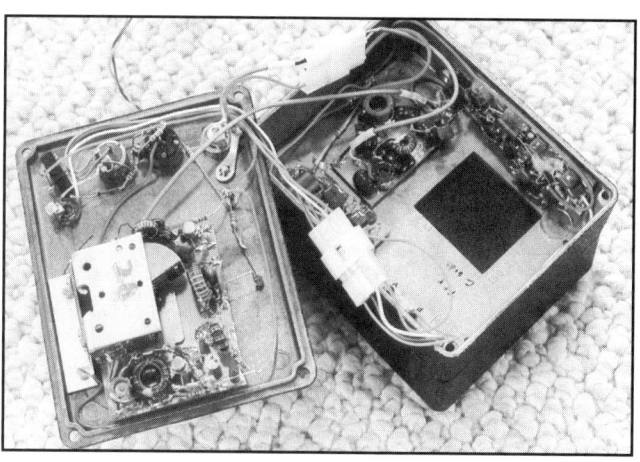

Fig 12.68—The PA compartment.

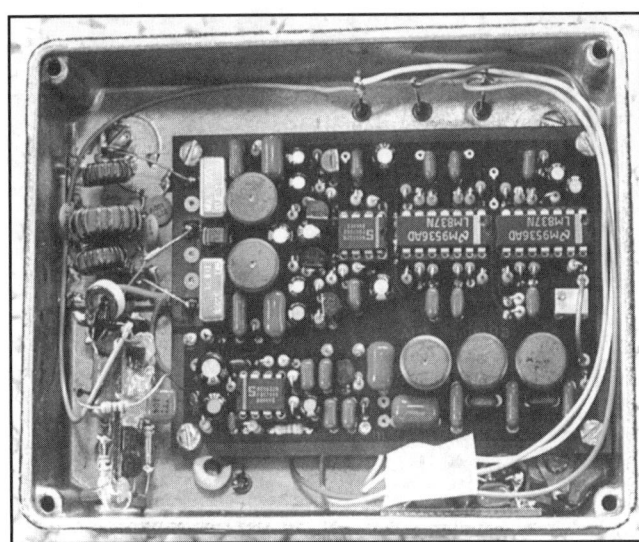

Fig 12.69—The Sleeping Bag Radio receiver compartment.

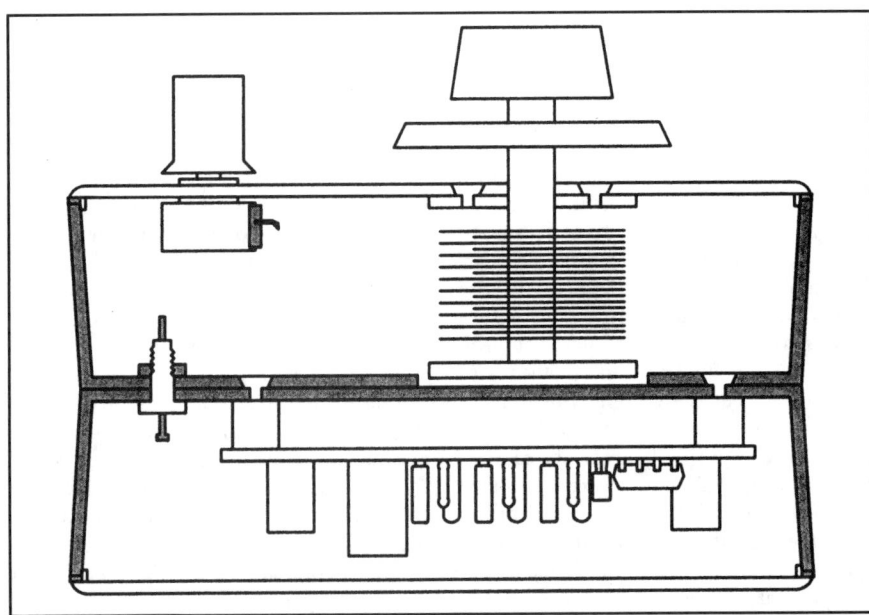

Fig 12.70—A Sleeping Bag Radio construction sketch.

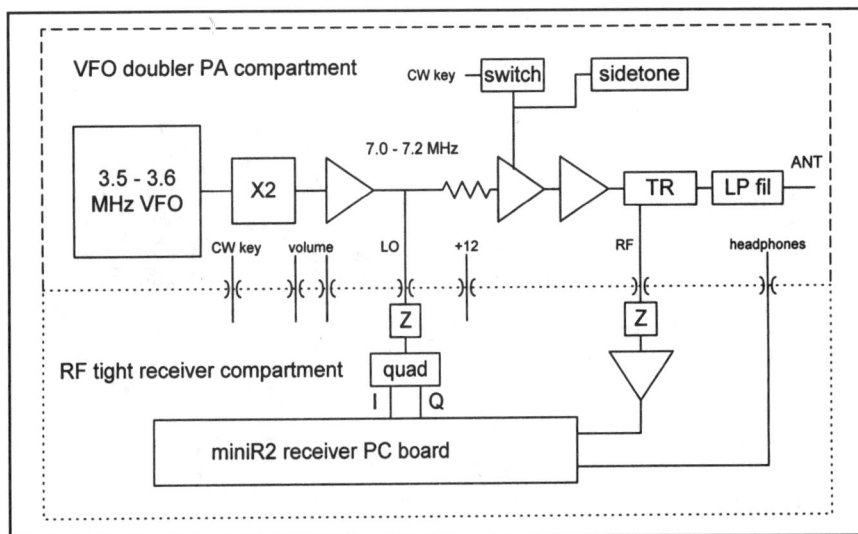

Fig 12.71—Block diagram of the Sleeping Bag Radio.

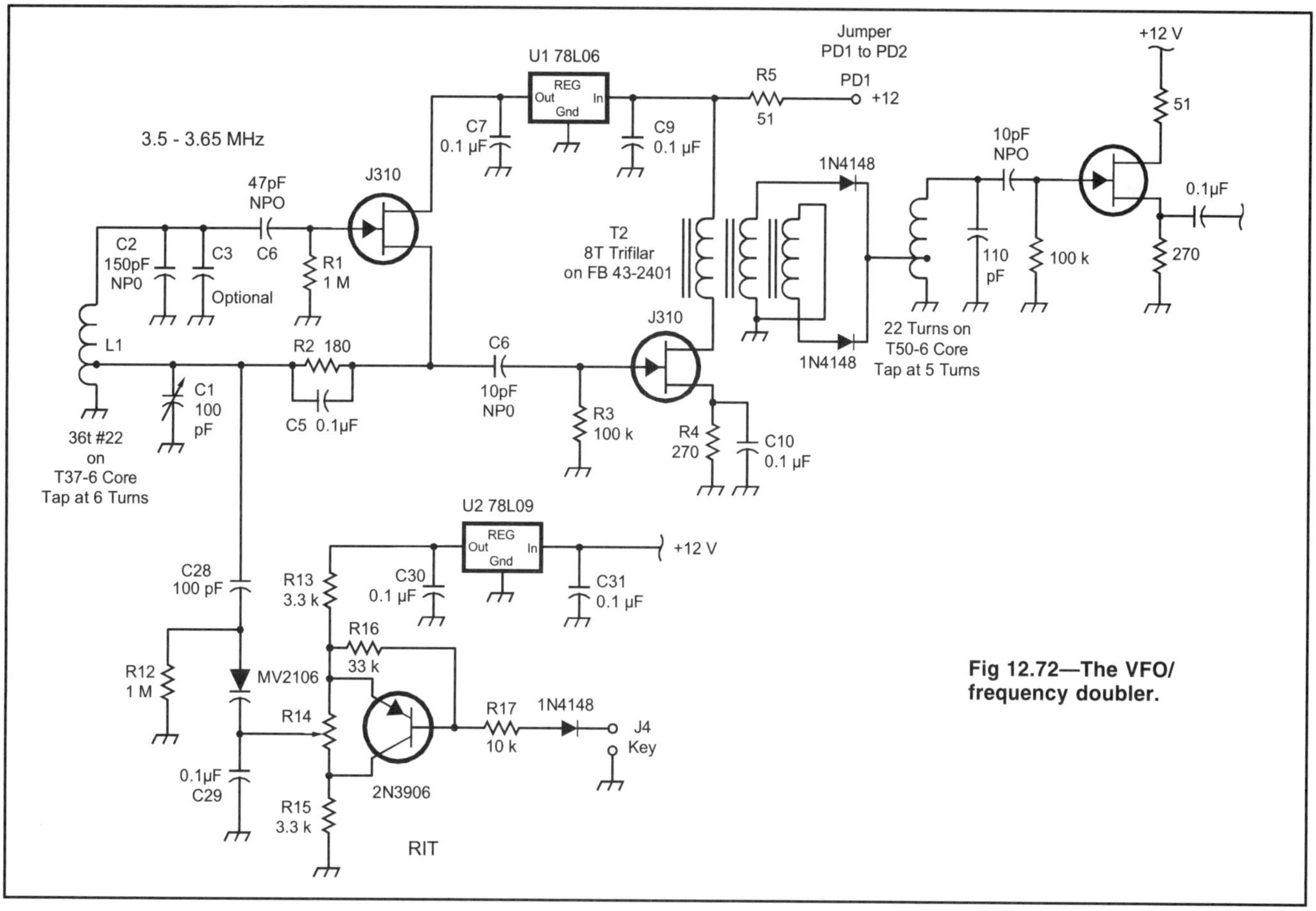

Fig 12.72—The VFO/frequency doubler.

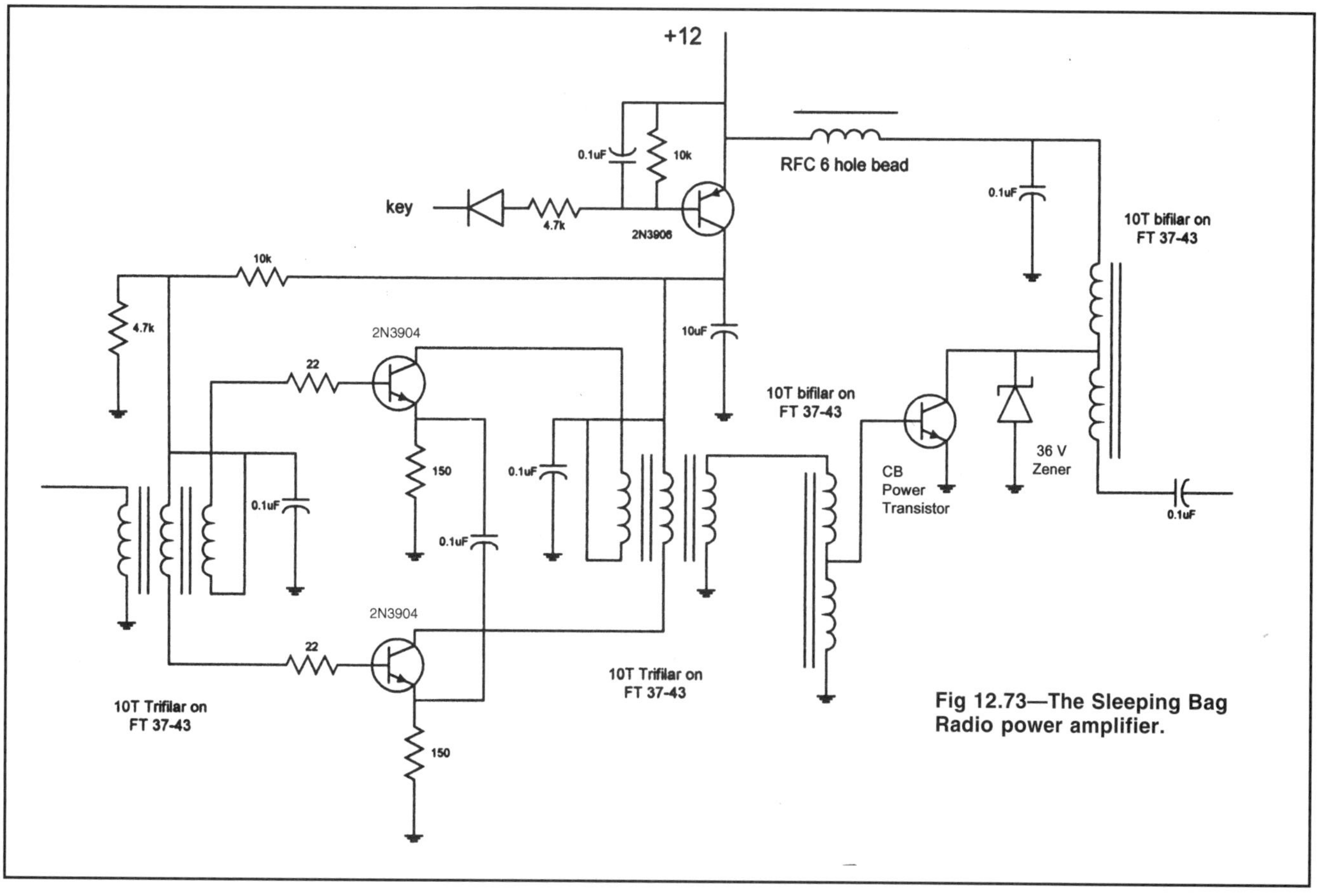

Fig 12.73—The Sleeping Bag Radio power amplifier.

Field Operation, Portable Gear and Integrated Stations

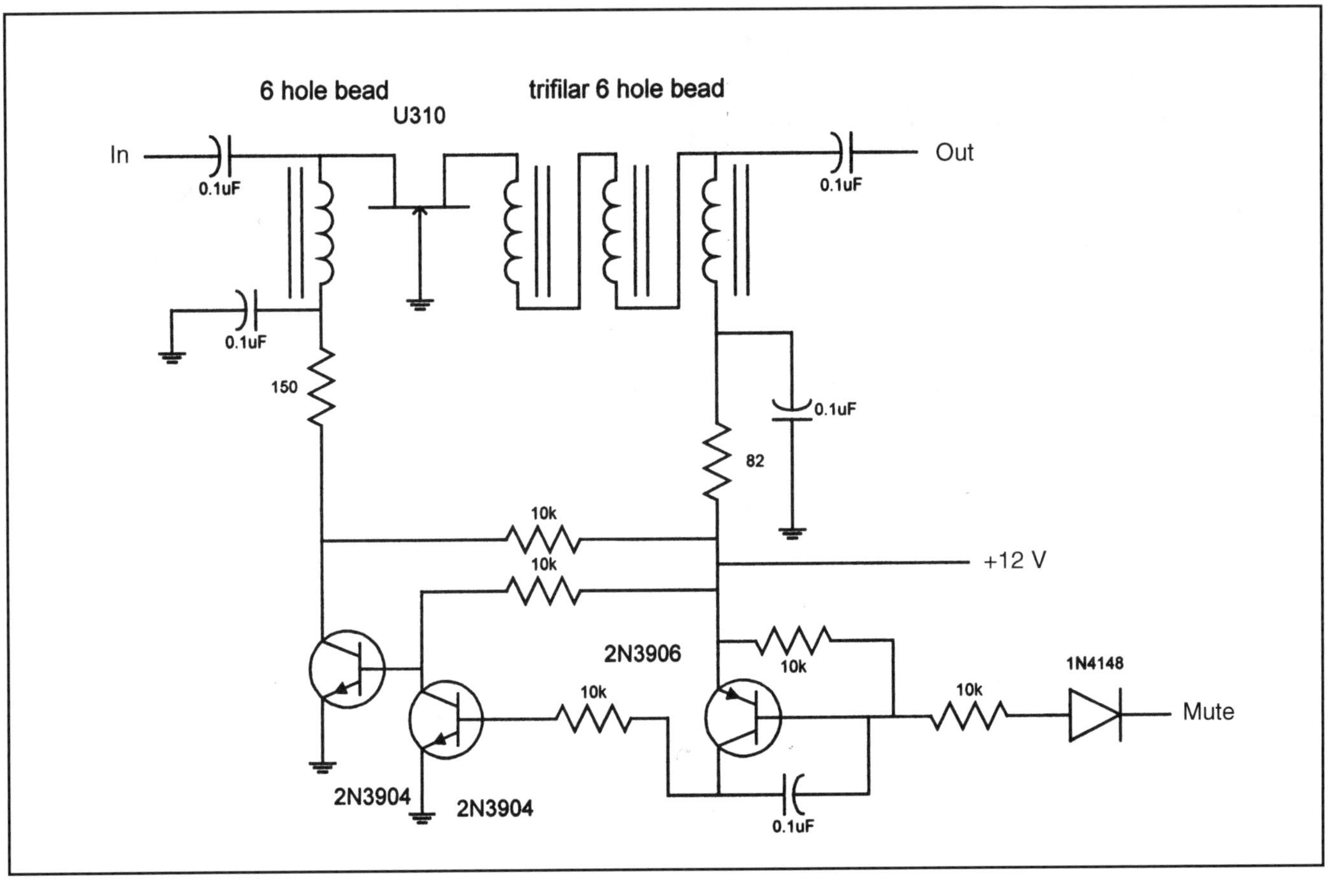

Fig 12.74—The LNA/attenuator.

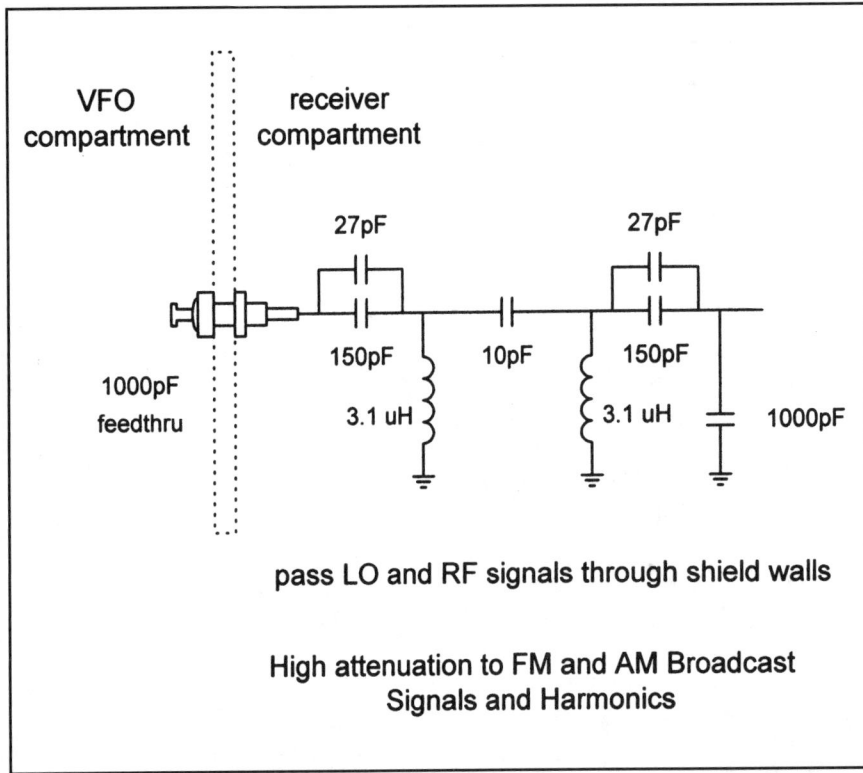

pass LO and RF signals through shield walls

High attenuation to FM and AM Broadcast Signals and Harmonics

shown in **Fig 12.73**, and the LNA/attenuator is shown in **Fig 12.74**. The 7-MHz RF and LO signals are routed through the shield walls on the feedthrough capacitors using the bandpass networks shown in **Fig 12.75**. This is the best CW transceiver I have ever used.

Fig 12.75—The 7-MHz bandpass feedthrough filter used in the Sleeping Bag Radio.

12.11 A 14 MHZ CW RECEIVER

This is a simple home station receiver for the CW portion of the 20-meter band. It uses R2pro circuit boards and a Kanga UVFO universal VFO board, along with lightweight aluminum chassis construction. **Fig 12.76** is a construction sketch, and **Fig 12.77** is a block diagram. The R2pro receiver circuit boards are described in detail in Chapter 9. **Fig 12.78** is a schematic of the UVFO board. **Figs 12.79**, **12.80** and **12.81** illustrate the construction. There are two selectable bandwidths and front-panel muting for use with a small QRP transmitter or vintage 40-W tube transmitter. Appearance and controls are basic. Performance is uncompromising, with over 50 dB of opposite sideband suppression, 9-dB noise figure, a slow tuning rate, 80 dB between the receiver noise floor and onset of audio clipping, 92-dB SSB bandwidth two-tone third-order dynamic range, and absolutely no spurious responses or synthesizer noise.

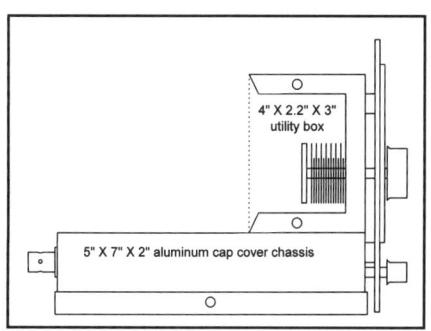

Fig 12.76—A construction sketch of the 14-MHz R2pro.

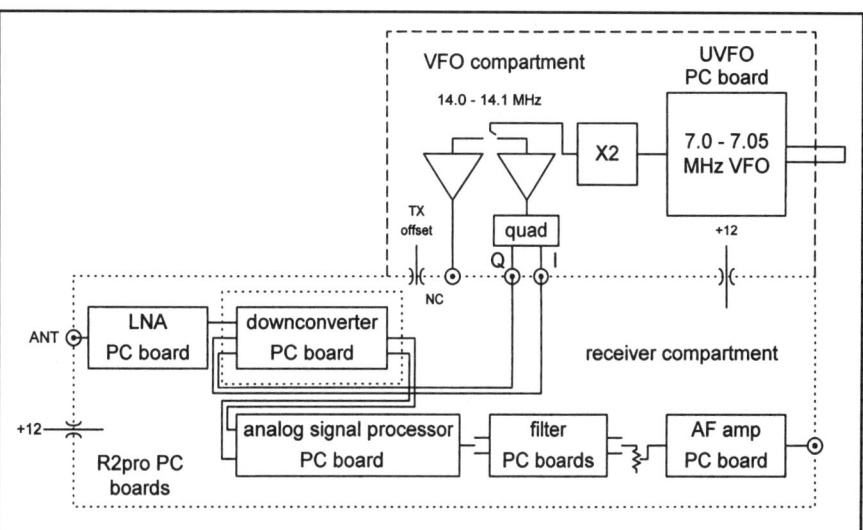

Fig 12.77—The 14-MHz R2pro block diagram.

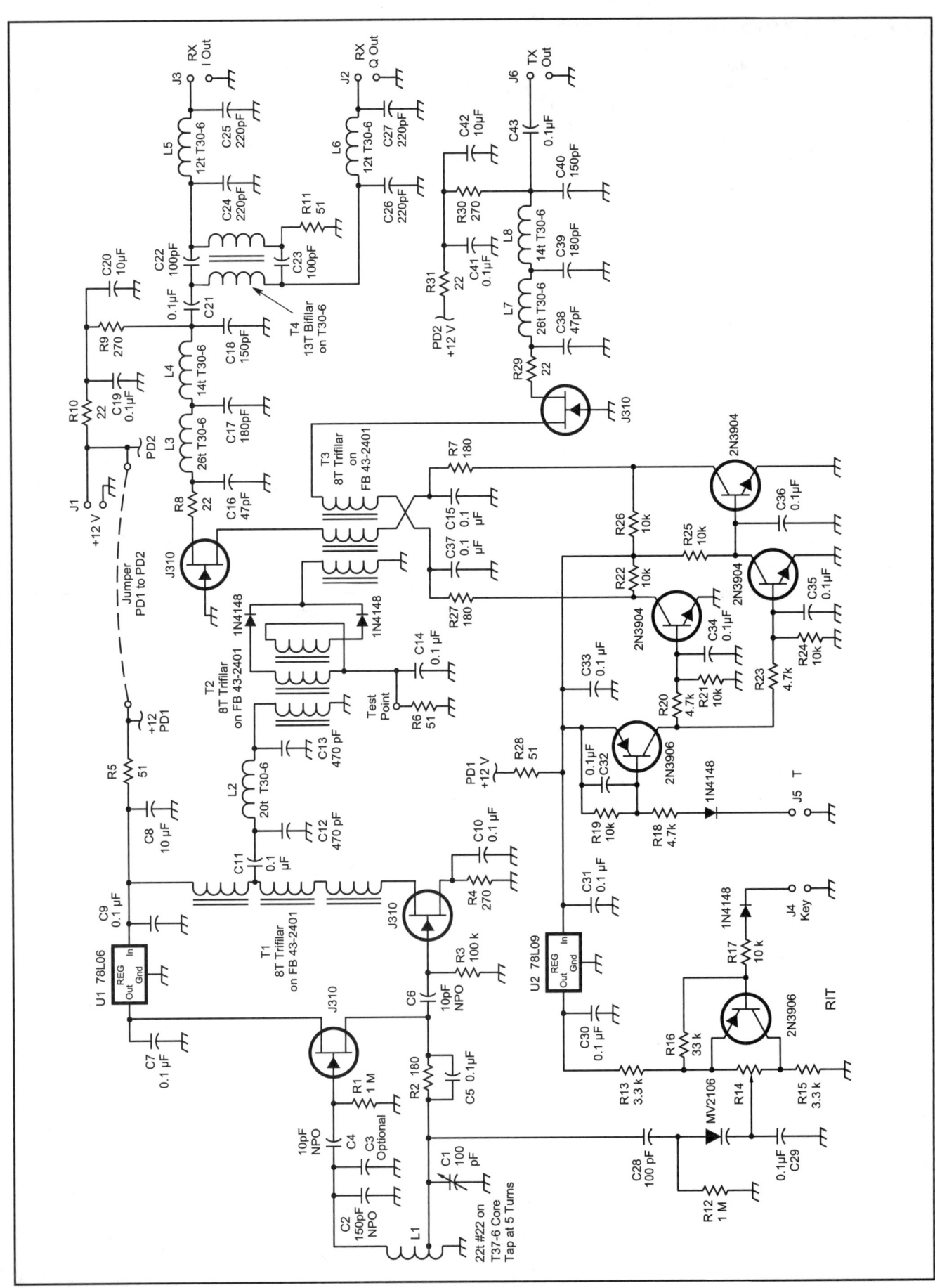

Fig 12.78—14 MHz UVFO schematic.

Fig 12.79—14 MHz R2pro front view.

Fig 12.80—The UVFO.

Fig 12.81—The R2pro circuit boards.

REFERENCES AND NOTES

1. "How to Frustrate a Bear," *Back-packer Magazine*, Oct, 2001, p 86.
2. Britain, "Some Really Cheap Antennas", *CQ VHF*, Aug, 1998 and Oct, 1998.
3. "From Our Vantage Point,"*The Sojourner*, on-line travel magazine of the Adventure Radio Society (ARS), May, 1998, www.natworld.com/ars/.
4. R. Lewallen, "An Optimized QRP Transceiver," *QST*, Aug, 1980, pp 14-19.
5. W. Hayward, "Measuring and Compensating Oscillator Frequency Drift," *QST*, Dec, 1993, pp 37-41.
6. D. Benson, "A Single-Board Super-het QRP Transceiver for 40 or 30 Meters," *QST*, Nov, 1994, pp 37-41.
7. J. Kleinman and Z. Lau, *QRP Power*, ARRL, 1996.
8. Detailed operation of the various weak-signal modes is described in the file README20.TXT. The source code, in 'C', for these modes is primarily in the files U_CODE.C, UMATRIX.C and MOONSUN.C. The specification for the 'PUA43' code is in the file PUA43_02.ZIP. All of these files are included on the CD-ROM.
9. Different countries have different restrictions on the amateur use of data modes. For US amateurs, a short summary of the interpretation of FCC regulations on these matters is the sidebar by Paul Rinaldo, "Is Hellschreiber Permissible Under Part 97?," *QST*, Jan, 2000, p 54. Before using any mode on the air, it is important to determine the legality of its usage and the frequencies that are allowable.
10. V. Poor, "R9/S1," *QST*, Oct, 1965, pp 33-37. This was not the introduction of these ideas, but it is a good summary of the amateur experimenter art of the time.
11. The advantages of multi-tone keying, along with historic background is in the article by M. Greenman, "MFSK for the

New Millennium," *QST*, Jan, 2001, pp 33-36.

12. Interested readers might start their exploration for further information with the "Matched Filter" topic in books such as D. K. Barton, *Radar System Analysis*, Prentice-Hall, Englewood Cliffs, NJ, 1964.

13. D. Turrin, and A. Katz, "Earth-Moon-Earth (EME) Communications," *The ARRL UHF/Microwave Experimenters Manual*, ARRL, 1990, Chapter 10.

14. Urban dwellers might quarrel with this statement, since coherent "birdies" coming from the all pervasive electronic gadgetry in people's houses will make extended integration times frustrating! EME-2 includes provisions for randomizing the transmitting frequency effectively to shift the interfering signals around, making them noise-like. This prevents the interference from adding in any particular bin but does not remove the equivalent noise power that is added.

15. See reference 13.

16. J. Taylor, "WSJT: New Software for VHF Meteor-Scatter Communication," *QST*, Dec, 2001, pp 36-41.

17. R. Campbell, "A Binaural I-Q Receiver," *QST*, Mar, 1999, p 44.

18. R. Lewallen, "An Optimized QRP Transceiver," *QST*, Aug, 1980, pp 14-19.

19. R. Campbell, "High-Performance, Single-Signal Direct-Conversion Receivers," *QST*, Jan, 1993, pp 32-40.

20. R. Campbell, "A Multimode Phasing Exciter for 1 to 500 MHz," QST, Apr, 1993, pp 27-31.

CD-ROM Contents

The material contained on the CD-ROM packaged on the inside back cover of this book contains articles, reference material, and software. This material is organized in the following directories:
   \software
   \articles
   \dsp

The **\dsp** directory contains specific lists of material for the DSP programs in Chapters 10 and 11 and the DSP-10 2-meter transceiver project.

ARTICLES AND REFERENCES

All of the following articles and references are on the CD-ROM in Adobe *Acrobat* PDF format. Double-click **articles.pdf** to access a summary of these materials. Alternatively, open any PDF document in the **\articles** directory to access that specific article. The article filename on the CD-ROM is shown after each reference listing.

If you do not have the Adobe *Acrobat* Reader on your computer, it is available for free downloading at **http://get.adobe.com/reader**

1. D. Benson, "Freq-Mite — A programmable Morse Code Frequency Readout," *QST*, Dec, 1998, pp 34-36. **qst199812.pdf**
2. D. Bramwell, "Understanding Modern Oscilloscopes," *QST*, Jul, 1976, pp 18-19. **qst197607.pdf**
3. D. Bramwell, "An RF Step Attenuator," *QST*, Jun, 1995, pp 33-34. **qst199506.pdf**
4. G. A. Breed, "A New Breed of Receiver," *QST*, Jan, 1988, pp 16-23. **qst198801.pdf**
5. R. Campbell, "Binaural Presentation of SSB and CW Signals Received on a Pair of Antennas," *Proceedings of the 18th Annual Conference of the Central States VHF Society*, Cedar Rapids, IA, Jul, 1984. **pmu1984.pdf**
6. R. Campbell, "Getting Started on the Microwave Bands," *QST*, Feb, 1992, pp 35-39. **qst199202.pdf**
7. R. Campbell, "High Performance Direct Conversion Receivers," *QST*, Aug, 1992, pp19-28. **qst199208.pdf**
8. R. Campbell, "No Tune Microwave Transceivers," *Proceedings of Microwave Update '92*, Rochester, NY, Oct, 1992, ARRL Publication number 161, pp 41-54. **pmu1992.pdf**
9. R. Campbell, "High Performance Single-Signal Direct Conversion Receivers," *QST*, Jan, 1993, pp 32-40. **qst199301.pdf**
10. R. Campbell, "A Multimode Phasing Exciter for 1 to 500 MHz," *QST*, Apr, 1993, pp 27-31. **qst199304.pdf**
11. R. Campbell, "Single-Conversion Microwave SSB/CW Transceivers," *QST*, May, 1993, pp 29-34. **qst199305.pdf**
12. R. Campbell, "A Single Board No-Tune Transceiver for 1296," *Proceedings of Microwave Update '93*, Atlanta, GA, Sep, 1993, ARRL Publication number 174, pp 17-38. **pmu1993.pdf**
13. R. Campbell, "Simply Getting on the Air from DC to Daylight," *Proceedings of Microwave Update '94, Estes Park, CO*, Sep, 1994, ARRL Publication number 188, pp 57-68. **pmu1994a.pdf**
14. R. Campbell, "Subharmonic IF Receivers," reprinted from the *North Texas Microwave Society Feedpoint* in *Proceedings of Microwave Update '94*, Estes Park, CO, Sep, 1994, ARRL Publication number 188, pp 225-232. **pmu1994b.pdf**
15. R. Campbell, "A VHF SSB-CW Transceiver with VXO," *Proceedings of the 29th Conference of the Central States VHF Society*, Colorado Springs, CO, Jul, 1995, ARRL Publication number 200, pp 94-106. **pmu1995b.pdf**
16. R. Campbell, "The Next Generation of No-Tune Transverters," *Proceedings of Microwave Update '95*, Arlington, TX, Oct, 1995, ARRL Publication number 208, pp 1-22. **pmu1995a.pdf**
17. R. Campbell, "A Small High-Performance CW Transceiver," *QST*, Nov, 1995, pp 41-46. **qst199511.pdf**
18. R. Campbell, "Direct Conversion Receiver Noise Figure," *QST*, Technical Correspondence, Feb 1996, pp 82-85. **qst199602.pdf**
19. R. Campbell, "Microwave Downconverter and Upconverter Update," *Proceedings of Microwave Update '98*, Estes Park, CO, Oct, 1998, ARRL Publication number 241, pp 34-49. **pmu1998.pdf**
20. R. Campbell, "A Binaural IQ Receiver," *QST*, Mar, 1999, pp. 44-48. **qst199903.pdf**
21. R. Campbell, "LO Phase Noise Management in Amateur Receiver Systems," *Proceedings of Microwave Update '99*, Plano, TX, Oct, 1999, ARRL, Publication number 253, pp 1-12. **pmu1999a.pdf**
22. R. Campbell, "Medium Power Diode Frequency Doublers," *Proceedings of Microwave Update '99*, Plano, TX, Oct, 1999, ARRL, Publication number 253, pp 397-406. **pmu1999b.pdf**
23. B. Carver, "High Performance Crystal Filter Design," *Communications Quarterly*, Winter, 1993, pp 11-18. **cq199301a.pdf**
24. B. Carver, "The LC Tester," *Communications Quarterly*, Winter, 1993, pp 19-27. **cq199301b.pdf**
25. B. Carver, "A High Performance AGC/IF Subsystem," *QST*, May, 1996, pp 39-44. **qst199605.pdf**
26. R. Fisher, "Twisted-Wire Quadrature Hybrid Directional Couplers," *QST*, Jan, 1978, pp 21-23. **qst197801.pdf**
27. J. Grebenkemper, "The Tandem Match – An Accurate Directional Wattmeter," *QST*, Jan, 1987, pp 18-26. **qst198701.pdf**
28. R. Hayward, "The Ugly Weekender II, Adding a Junk Box Receiver," *QST*, Jun, 1992, pp 27-30. **qst199206.pdf**
29. W. Hayward and R. Bingham, "Direct Conversion; A Neglected Technique." *QST*, Nov, 1968, pp 15-17. **qst196811.pdf**
30. W. Hayward and J. Lawson, "A Progressive Communications Receiver," *QST*, Nov, 1981, pp 11-21. **qst198111.pdf**

31. W. Hayward and R. Hayward, "The Ugly Weekender," *QST*, Aug, 1981, pp 18-21. **qst198108.pdf**
32. W. Hayward, "The Double Tuned Circuit: An Experimenter's Tutorial", *QST*, Dec, 1991, pp 29-34. **qst199112.pdf**
33. W. Hayward, "Reflections on the Reflection Coefficient: An Intuitive Examination," *QEX*, Jan, 1993, pp 10-23. **qex199301.pdf**
34. W. Hayward, "Measuring and Compensating Oscillator Frequency Drift," *QST*, Dec, 1993, pp 37-41. **qst199312.pdf**
35. W. Hayward, "Electronic T/R Switching," *QEX*, May, 1995, pp 3-7. **qex199505.pdf**
36. W. Hayward, "Refinements in Crystal Ladder Filter Design," *QEX*, Jun, 1995, pp 16-21. **qex199506.pdf**
37. W. Hayward, "Extending the Double Tuned Circuit to Three Resonators," *QEX*, Mar/Apr, 1998, pp 41-46. **qex199803.pdf**
38. W. Hayward and T. White, "A Tracking Signal Generator for Use with a Spectrum Analyzer," *QST*, Nov, 1999, pp 50-52. **qst199911b.pdf**
39. W. Hayward and T. White, "A Spectrum Analyzer for the Radio Amateur," *QST*, Aug and Sep, 1998, pp 35-43. **qst199808.pdf, qst199809.pdf**
40. W. Hayward and T. White, "The Micromountaineer Revisited," *QST*, Jul, 2000, pp 28-33. **qst200007.pdf**
41. W. Hayward and R. Larkin, "Simple RF-Power Measurement", *QST*, Jun, 2001, pp 38-43. **qst200106.pdf**
42. N. Heckt, "A PIC-Based Digital Frequency Display," *QST*, May, 1997, pp 36-38. **qst199705b.pdf**
43. H. Johnson, "Helical Resonator Oscillators." **w4zcb.pdf**
44. R. Larkin, "The DSP-10: An All-Mode 2-Meter Transceiver Using a DSP IF and PC-Controlled Front Panel," *QST*, Sep, 1999, pp 33-41; Oct, 1999, pp 34-40; Nov, 1999, pp 42-45. **qst199909.pdf, qst199910.pdf, qst199911.pdf**
45. R. Larkin, "An 8-Watt, 2-Meter Brickette," *QST*, Jun, 2000, pp 43-47. **qst200006.pdf**
46. R. Lewallen, "An Optimized QRP Transceiver," *QST*, Aug, 1980, pp 14-19. **qst198008.pdf**
47. R. Lewallen, "A Simple and Accurate QRP Directional Wattmeter," *QST*, Feb, 1990, pp 19-23. **qst199002.pdf**
48. J. Makhinson, "A Drift-Free VFO," *QST*, Dec, 1996, pp 32-36. **qst199612.pdf**
49. J. Makhinson, "DEMPHANO, A device for measuring phase noise," *Communications Quarterly*, Spring, 1999, pp 9-17. **cq199904.pdf**
50. J. Reisert, "VHF/UHF Frequency Calibration," *Ham Radio*, Oct, 1984, pp 55-60. **hr198410.pdf**
51. D. Rutledge, et al, "High-Efficiency Class-E Power Amplifiers," *QST*, May, 1997, Part I, pp 39-42, and Jun, 1997, Part II, pp 39-42. **qst199705a.pdf, qst199706.pdf**
52. W. Sabin, "Measuring SSB/CW Receiver Sensitivity," *QST*, Oct, 1992, pp 30-34. **qst199210.pdf**
53. W. Sabin, "A Calibrated Noise Source for Amateur Radio," *QST*, May, 1994, pp 37-40. **qst199405.pdf**
54. W. Sabin, "Diplexer Filters for an HF MOSFET Power Amplifier," *QEX*, Jul/Aug, 1999, pp 20-26. **qex199907.pdf**
55. W. Sabin, "A 100-W MOSFET HF Amplifier," *QEX*, Nov/Dec, 1999, pp 31-40 **qex199911.pdf**
56. B. Shriner and P. Pagel, "A Step Attenuator You Can Build," *QST*, Sep, 1982, pp 11-13. **qst198209.pdf**
57. K. Spaargaren, "Frequency Stabilization of LC Oscillators," *QEX*, Feb, 1996, pp 19-23. **qex199602.pdf**
58. J. Stephensen, "Reducing IMD in High-Level Mixers," *QEX*, May/Jun, 2001, pp 45-50. **qex200105.pdf**
59. P. Wade, "Noise Measurement and Generation," *QEX*, Nov, 1996, pp 3-12. **qex199611.pdf**
60. A. Ward, "Noise Figure Measurements," *Proceedings of Microwave Update '97*, Sandusky, OH, Oct, 1997, ARRL Publication number 231, pp 265-272. **pmu1997.pdf**

SOFTWARE

- LADPAC-2008, Design programs for Windows. Run **setup.exe** and follow the on-screen directions to install the software.
- Analysis of mixing with a JFET (Mathcad file **mixer_jfet1.mcd**, Adobe Acrobat file **mixer_jfet1.pdf**). See Chapter 5, section 1. Using **mixer_jfet1.mcd** requires Mathsoft *Mathcad* version x.x or higher. **Mixer_jfet1.pdf** is compiled from screenshots showing the equations used in the *Mathcad* file, useful the those who don't have *Mathcad*.

DSP (DIGITAL SIGNAL PROCESSING) Programs for Chapters 10 and 11

The programs for Chapters 10 and 11 are in the directories CHAP10 and CHAP11. For each **c1xxx.dsp** file there is also a **c1xxx.exe** file created by the **ld21** linker as described in **read.txt**. The contents of the two directories are:

CHAPTER 10

c1shell.dsp Basic DSP structure for EZKIT- Lite
c1shell.exe
c1sin.dsp Generates single sine wave at 1000 Hz
c1sin.exe
c1sin2.dsp Generates 2 sine waves at 700 and 1900 Hz
c1sin2.exe
c1spn.dsp Generates 1000 Hz sine wave plus Gaussian noise
c1spn.exe
c1fir.dsp FIR filter coefficients
c1fir.exe
fir200bp.dat Part of **c1fir.dsp** - Band pass FIR filter coefficients
firdsn3.bas A QBASIC program for calculating FIR filters using the Kaiser window method.

CHAPTER 11

c1knob.dsp Interaction with a rotary knob, switches, LCD display
c1knob.exe
c1tbox.dsp Uses the c1knob to generate 2 sine waves plus noise
c1tbox.exe
c18.dsp An 18 MHz I-Q transceiver for CW and USB
c18.exe
lp2_8.dat Part of **C18.dsp** - Low pass FIR filter coefficients
lp_5_48.dat Part of **C18.dsp** - Low pass FIR filter coefficients
bpcw1.dat Part of **C18.dsp** - CW audio FIR filter coefficients
hil_3_48.dat Part of **C18.dsp** - Hilbert transform for 90 degree phase shift. These are coefficients for a specialized FIR filter.

All of the **c1xxx.exe** programs can be put into EPROM for loading when the EZKIT-Lite starts operation. See the Analog Devices PROM Splitter for details.

Documentation for the DSP-10 2-Meter Transceiver

Included in five directories is a complete set of documentation for the DSP-10 2-meter transceiver. All **.TXT** files are simple ASCII text with embedded end-of-lines. All **.HTM** files can be read on a Web browser.

This documentation is up-to-date as of March 2002. Further data may be available on the internet. The URL currently is **http://www.proaxis.com/~boblark/dsp10.htm**. If the Web page location is changed it will still include the word **ABCDSP10ABCD** that may be helpful for locating it with a search engine! See the **.txt** files listed below for more information.

Here is a quick summary of the contents to help in finding files.

ARTICLES
Contains the three *QST* articles from Sept-Nov 1999 in **.PDF** format.
1. R. Larkin, "The DSP-10: An All-Mode 2-Meter Transceiver Using a DSP IF and PC-Controlled Front Panel," *QST*, Sep, 1999, pp 33-41; Oct, 1999, pp 34-40; Nov, 1999, pp 42-45.

HARDWARE
dsp10hdw.txt - General notes, corrections and improvements.
dsp10n45.txt - Assembly notes for the project
dsp10pd2.txt - Assembly part-by-part list, with locations on the PCB
dsp10ph5.htm - Part list for purchasing parts
u15_mod.htm - Improvement information referenced by **dsp10hdw.txt**
u15mod1.gif - A sketch required for **u15_mod.htm**.
f10.gif - A corrected figure 10 for the *QST* articles.
f11.gif - A corrected figure 11 for the *QST* articles.

EXECUTABLE
Uhfa.exe - DOS Executable front panel program
Uhf3.exe - Machine language program (NOT A DOS **.EXE** file)
Egavga.bgi - Borland graphics drivers for PC
Gnugpl.txt - User license (Please Read)
Uhfa_43a.rnd - Random number list for several of the weak signal modes.
Readme16.txt - Software user information for basic modes
Readme20.txt - Additional user information, including weak-signal modes.
Wat_exe.txt - A reminder that **UHF3.EXE** is NOT a DOS **.exe** file.

SOURCE CODE AND MISCELLANEOUS
CSRC - Source code for the PC program, in Borland C: 28 files.
DSPSRC - Source code for the EZKit program: 33 files.

Included in the last two directories are two batch files, **U.BAT**, that assembles and links the program from the various modules. The file, **U3.BAT**, serves the same function for the DSP program.

The file **Pc_dsp2.txt** in the directory **CSRC** has the details of the communication between the PC and the DSP.

INDEX

Editor's Note: Except for commonly used phrases and abbreviations, topics are indexed by their noun names. Many topics are also cross-indexed, especially when noun modifiers appear (such as "Modulator, Balanced" and "Balanced, Modulator"). The letters "ff" after a page number indicate coverage of the indexed topic on succeeding pages.

18-MHz
    Schematic diagram: 11.14ff
    Transceiver: 11.12ff
        DSP circuits used: 11.19ff
    Transceiver output (CW) spectrum: 11.25
    Transceiver, sampling rates: 11.26–11.27
2-m
    Transceiver (DSP-10): 11.27ff
28-MHz QRP module: 12.28
40-m
    D-C receiver block diagram: 8.3
7-MHz portable transmatch: 7.24

A
AA3X: 2.38
AD8307: 7.8
Adaptive
    Mixer Balance: 8.11
Adjustment
    Amplitude balance: 9.22–9.23
    Phase trim: 9.23–9.24
Advanced Power Technology: 2.37
Adventure Radio Society (ARS): 12.4
    Spartan Sprint: 12.4
AGC (Automatic gain control)
    Amplifier: 6.20
    Audio derived: 6.22
    Hang system: 6.25
    Intermediate frequency (IF) amplifier: 6.15ff
    Pop: 6.22
    Testing of, in receivers: 7.40
    Threshold: 6.19
Almost incremental tuning (AIT): 6.67
AM: 3.17, 6.1
    Demodulation: 8.11
    Exciter, low-distortion: 9.48–9.49
Amateur Radio: 1.1ff
Amidon Inc.: 3.32
Amplifier: 2.1
    Audio power: 9.41
    Automatic gain control (AGC): 6.16

Bidirectional: 6.60
Bipolar transistor: 6.16
Buffer: 9.47
Circuits: 2.1
Classes of amplifier operation: 2.31ff
    Class A: 2.11ff, 6.55
    Class AB: 2.31–2.32, 6.56
    Class AB1: 2.31
    Class B: 2.31
    Class C: 1.18ff, 2.31ff
    Class D: 2.31–2.32
    Class E: 2.31ff
Common source JFET: 6.33
Differential amplifier (diff-amp): 2.16ff
General-purpose IF: 6.20
High-performance post-mixer: 6.47
Intermediate frequency (IF) and AGC: 6.15ff
Junction field effect transistor (JFET)
    Bidirectional: 6.62
    Cascode pair: 6.18
    Common gate, RF: 6.12
    Common source: 6.13
    Gain of: 6.33
Keying of transmitter stage: 6.63ff
Large signal amplifiers: 2.1
Lichen transceiver power chain: 6.79
Limiting, using digital IC: 5.18
Linear power: 6.54
Low noise (LNA)
    Swept frequency plot: 9.36
Low-noise RF: 8.13
Metal oxide silicon field effect transistor (MOSFET)
    IF: 6.17, 6.24
    RF: 6.13
Microphone: 9.45–9.46
Mixer IF-port driver: 9.47ff
Monoband SSB/CW transceiver power chain: 6.86
Noise: 2.19
    Thermal: 2.19
Operational: 2.16ff, 3.25ff
    1458: 3.26–3.27

5532:	3.26
741:	3.26–3.27
LM-324:	3.25
LM-358:	3.25
Topologies:	9.32
Oscillation:	2.31
Post mixer, with JFET:	5.14
Power for 50 MHz:	6.86
Power, with IRF511 MOSFETs:	11.18
Radio frequency (RF):	6.12
Roofing filter:	6.50
Small signal:	2.1
Speech, analog signal processor:	9.45–9.46
SSB (linear amplifiers):	2.37
Transparency:	2.26
VXO transmitter, with digital frequency multiplier:	5.20

Amplitude and phase
    Errors, with phasing method: 11.22
Amplitude balance adjustment: 9.22–9.23
Amplitude modulation (AM): 6.1–6.2
    Double-sideband, full-carrier: 6.7
Analog
    vs. digital: 11.27
Analog Devices: 10.2
    AD1847 CODEC: 10.2ff
    ADSP-2100 family: 10.2, 10.5
    ADSP-2181: 10.2, 10.4
    Analog Devices 9831: 4.26
    EZ-Kit Lite: 10.2, 11.1
Analog signal processor (ASP): 9.39–9.40
Analog to digital (A/D) converters: 10.3
    A/D noise: 10.3
    D/A noise: 10.3
    Dynamic range, limits of: 10.4
    Sample rate: 10.4
    Sigma-delta A/D converters: 10.3
Angle, Chip, N6CA: 6.12
Antenna
    Transmit/receive (T/R), switching: 6.68, 11.18
Appliance: 1.4
Applications
    Of spectrum analyzers (hints for use): 7.30
ARRL Field Day: 12.4
ARRL Handbook. (See *The ARRL Handbook*)
ASCII
    Digits from binary number (converting): 11.8
AT cut (*See* Crystal, quartz)
Attenuators: 7.10–7.11
    10-dB pad: 6.14
    Continuously variable: 7.10
    Fixed: 7.11
    Pi (π) and Tee: 7.10
    PIN Diode: 6.18
    Power Pi (π): 7.10
    Radio frequency (RF): 6.12
    Schematics and design equations for: 7.9
    Step: 7.11
Audio: 6.1
    Amplifier: 1.12
    Derived automatic gain control (AGC): 6.23
    Filter, SSB and CW: 9.40
    Filtering, DSP in: 11.23–11.24

Gain
    High, in D-C receivers: 8.6–8.7
    Generator: 11.11–11.12
    Lichen transceiver, receive: 6.78
    Phase shift network (PSN): 9.47
    Power amplifier: 9.41
    Processor, DSP-based: 11.29
PSN
    Modulator circuitry: 9.46
    Signal sources: 7.13
Auto-transformer: 2.36
Available noise: 2.20
Available power: 2.14, 2.19

B
Backwave: 2.32, 6.64
Balanced
    Mixer: 5.5, 5.7
        Better L-R isolation of: 11.25
    Modulator: 6.2, 6.56
Band-spread tuning: 1.10
Bandpass diplexer network: 9.17
Bandpass filter (*See also* Filter)
    14-MHz for VXO transmitter: 5.20
    21-MHz for VXO transmitter: 5.21
    Lichen transceiver: 6.76-6.77
    Monoband SSB/CW transceiver: 6.83
Bandwidth: 6.11
    Resolution: 7.26
Bartlett's bisection theorem: 3.6
Baseband: 6.1, 6.3
Beat-frequency oscillator (BFO): 6.6, 6.85
Bell Labs: 4.2
Bells and whistles: 1.4
Berlin, Howard: 3.27
Beta cutoff: 2.9
Bidirectional amplifier: 6.61
    JFET: 6.62
Bifilar windings: 3.33ff
Binary
    Number conversion to ASCII digits: 11.8
Binaural
    Delay: 11.23–11.24
    Mode: 9.42
Binaural receiver: 9.19ff
BJT (base-junction transistor) model: 2.10
Bleeder resistor: 1.15
Block converter: 7.35
Block diagrams: 1.4, 1.6, 6.4
    14-MHz R2pro CW receiver: 12.46
    14-MHz receiver: 6.27
    18-MHz transceiver: 11.13
    2-m (DSP-10) transceiver: 11.28
    40-m D-C receiver: 8.3
    52-MHz IF transceiver: 12.38
    Basic D-C receiver: 8.2
    CW transmitter: 6.5
    Direct conversion 144-MHz transceiver: 12.33
        LM2 PC board: 12.33
    Direct-conversion (D-C) receiver: 6.6
    Double-sideband transmitter: 6.7
    Dual-band QRP CW transceiver: 12.19
    Elements of: 1.6

Filter-type SSB exciter: 9.1
General-purpose receiver front end: 6.32ff
High performance D-C receiver
High-dynamic-range receiver: 6.44
Hilbert transform, 247-tap: 11.21
I and Q corrections
    Better sideband rejection using: 11.24
Image-rejecting D-C receiver: 9.16
Lichen transceiver: 6.71
Mixer: 5.1
Mixer/LO with reflection coeff.: 8.7
Modern front end: 6.46
Modular receiver: 8.13
Monoband SSB/CW transceiver: 6.83
Phasing D-C receiver: 9.3
Phasing receiver
    DSP error correction: 11.22
Phasing-type SSB exciter: 9.2
Preamp diode ring D-C receiver: 8.3
Preamp, Gilbert D-C receiver: 8.3
R2pro: 9.35
Receiver front end: 6.11
Single-conversion superheterodyne receiver: 6.6
Single-sideband (SSB) transceiver: 6.9, 6.61
Single-sideband (SSB) transmitter: 6.7
Sleeping Bag Radio: 12.43
Superheterodyne receiver
    with a phasing SSB demodulator: 9.2
    with a SSB IF bandwidth: 9.2
Superheterodyne single-sideband (SSB) receiver: 6.8
The S7C superhet receiver: 12.16
Tone and noise generator: 11.11
VXO transmitter with digital frequency multiplier: 5.19
Blocking elements: 2.31
    Blocking capacitor: 2.31
Bolometer: 2.13
Boltzmann's constant: 2.2ff, 6.10
Bottom, Virgil: 3.17
Boulouard, Andre: 3.36
Breadboard circuits: 1.2
    Breadboard: 1.2
    Low inductance grounding: 1.3
    Manhattan breadboarding: 1.3
    Quasi-Printed boards: 1.3
    Ugly construction: 1.2ff
Bridge
    Impedance measurement using: 7.21ff
    Rectifier: 1.14
    Return loss (RLB): 7.22
    RF impedance: 7.23
    RF resistance: 7.22
    Suitable for UHF: 7.24
    Wheatstone: 7.21
    Wien: 7.13
Buffer amplifier: 1.17–1.18, 9.47
Butterworth filter (See Low-pass filter)
Bypassing and decoupling: 2.28ff
    Grounded points: 2.28
    Parasitic inductance: 2.28
    Problems of: 2.30
    Signal grounded: 2.28
    Tantalum electrolytic capacitors: 2.30

C

Calibration
    during measurements: 7.31
Capacitance
    Measurement: 7.11–7.12
Capacitor
    Phasing: 3.17
    Small numeric value: 3.15
Capital Advanced Technologies: 1.2
Carrier: 2.14, 2.21, 4.10, 6.1
    CW, generation: 6.58
    Oscillator, for monoband SSB/CW transceiver: 6.85
Carrier to noise ratio (CNR): 4.10, 4.12
Carver, Bill, W7AAZ: 2.28, 3.24ff, 6.24ff
Cathode ray tube (CRT): 7.3
Central limit theorem: 10.12
Chamber testing
    Of oscillators: 7.42
Chebyshev filter (See low-pass filter)
Circuit boards
    Multiple, in D-C receivers: 9.33
Clapp oscillator (See Oscillator)
Clarke and Hess: 3.34
Classes of amplifier operation (See Amplifier)
Clean equipment (signals): 1.5
Clock wire: 11.2–11.3
CODEC (coder/decoder): 10.2ff
Cohn, S. B.: 3.10, 3.21
Color burst crystal: 6.90
Colpitts oscillator (See also oscillator): 1.13, 7.37–7.38
Common base amplifier (CB): 2.8
    Current gain: 2.8
    Voltage gain: 2.8
Common-collector amplifier (CC): 2.7
Common-emitter amplifier (CE): 1.13, 2.7
Common-mode
    Choke: 2.16, 3.34
    Drive: 2.16
    Hum: 8.8–8.9
    Input range: 2.18
Common-source JFET amplifier: 6.33
Communications
    DSP applications in: 11.1ff
    Weak signal
        Using the DSP-10: 12.24
Communications Concepts, Inc.: 2.38
Compact Software
    Super Spice: 3.25
Compensation
    Of oscillator drift: 7.42
    Temperature, process of: 7.42
Component Testing
    Setup for: 7.20
Computer programs
    ARRL Radio Designer: 3.4
    GPLA: 3.4
    Structure of: 11.1–11.2
Controller
    DSP device as: 11.2
Conversion gain
    Mixer: 5.6
Conversion loss
    Mixer: 5.6

Conversion oscillator: 5.1
Converter: 6.41
   An experimenter's receiving: 7.40
   Block: 7.35
   D/A: 11.1
   For baseband spectrum analyzer: 7.34
   Frequency
      A minimum-parts-count: 9.8
      RF to TTL/CMOS: 7.12
Converting
   Binary number to ASCII digits: 11.8
Coupling coefficient: 3.33
Creeping features: 1.4
Cross modulation: 6.28
Crystal
   Color burst: 6.90
   Filter: 3.17, 6.48
      4th order monolithic: 7.28
      8th order (ref. WB4RNO and W2EKB): 7.28
      Bidirectional: 6.62
      Response: 6.27, 6.84
   Measurement of: 7.37–7.38
   Oscillator: 1.11, 1.17, 4.14, 6.65
   Quartz: 3.17, 4.14
      AT cut: 3.17
      Equivalent series resistance (ESR): 3.17
      Model: 7.37
      Motional parameters: 3.18
      Piezo-electric effect: 3.17
      Resonant frequency: 3.17
      Surface effects: 3.17
   Testing of, using Colpitts oscillator: 7.38
   Variable oscillator (VXO): 6.91
Current controlled device: 2.3
Current gain (b): 2.3
Current source: 2.7
CW: 1.2
   Carrier generation: 6.58
   Considerations, of phasing D-C receivers: 9.18
   Monoband transceiver: 6.83
   Receiver: 6.6
   Receiver, 14-MHz: 12.46
   Transceiver, portable: 12.5
   Transmission with DSP: 11.24ff
   Transmitter: 6.4ff
      IF amplifier: 6.58

D
D-C receiver
   A minimalist: 8.4–8.5
D/A converter: 11.1
Darlington configuration: 2.27
Data wire: 11.2–11.3
dBm: 7.6
dBW: 7.6
DC measurements: 7.2
Dead bug style: 4.30
Decibel (dB): 2.14
   Arithmetic: 7.6
   Ratio: 2.14
Decoupling resistor: 1.18
DeFatta, D. J. et al: 10.28
Degeneration: 2.25–2.26
   Resistance: 2.25
DeMaw, Doug, W1FB: 1.1
Demodulation
   AM: 8.11
Denormalization equations: 3.4
Design
   Receiver: 9.7ff
Detector: 1.10, 6.19, 6.23
   Peak: 7.5
   Phase: 4.19ff
   Product: 5.1
DFT (Discrete Fourier Transform): 7.35
Diagram
   Shift-register timing: 11.4
Diagrams, block. (See Block diagrams)
Differential amplifier (*See* Amplifier)
Differential-mode drive: 2.16
Digi-Key: 12.17
   Catalog: 1.2
Digital
   vs. analog: 11.27
Digital noise: 10.4
Digital signal processing (DSP): 10.1ff
   Alternate DSP devices: 10.29
   Audio processor: 11.29
   Automatic noise blankers: 10.28–10.29
   Building blocks: 10.2
   CODEC (coder/decoder): 10.2ff
   Components: 10.2, 10.6
      Amplifiers: 10.6
      Attenuators: 10.6
      Automatic gain control (AGC): 10.21–10.22
      Discrete Fourier transform (DFT): 10.23ff
         Ins and outs of: 10.26
         Spectrum analyzer: 10.24
      FM reception: 10.22–10.23
      FM transmission: 10.22
      Multiplier: 10.7
      Shift register: 10.7
   Device, as a controller: 11.2
   Digital filter: 10.2, 10.13
      DSP IF: 10.20
      Finite impulse response (FIR) filter
         Computation: 10.15
         Hilbert transform: 10.20
         Kaiser window: 10.16ff
         Performance: 10.18ff
      Infinite impulse response (IIR) filter: 10.13–10.14
DSP program
   Autobuffering: 10.6
   Gaussian random numbers: 10.12
      Gaussian noise: 10.9
   Index registers: 10.8
      Polynomial coefficients: 10.8
      Sequential addresses: 10.8
   Instruction: 10.5
   Interrupt overrun: 10.5
   Interrupt service routine (ISR): 10.5–10.6
   Jump instruction: 10.6
   Primary register set: 10.5
   Secondary register set: 10.5
Dynamic range: 10.3
Fast Fourier Transform (FFT): 10.4

In communications: 11.1ff
Phase shifters: 9.32
Phase-locked loop: 10.6
Process: 10.3
   Adaptive filters: 10.3
   SSB generation: 10.3
Program shell (also Shell program): 10.4–10.5
Signal generators: 10.7
   Integer coefficients: 10.7
   Sine wave: 10.7
      Calculated functions: 10.7
      Lookup tables: 10.7
SSB signal generation
   Gain expander: 10.30
   Predistorter: 10.30ff
   Predistorter distortion reduction: 10.29ff
   Predistortion: 10.30
   Predistortion polynomial coefficients: 10.32
   Transmitter: 10.31–10.32
Why DSP?: 10.3
Digital voltmeter (DVM): 4.5, 7.2
Diode: 2.1
   Equation: 2.2
   Frequency doubler: 5.16
   Frequency tripler: 5.17
   Ideal: 2.1, 2.4
   Junction: 2.1
   Mixer: 5.3
      Ring: 5.13
      Ring, commutating balanced: 5.8
   Offset voltage: 2.1
   PIN: 6.16ff
   Polarity dependent properties: 2.1
   Saturation current: 2.2
   Small signal model: 2.2
   Switching: 6.62
   Varactor: 4.17, 6.67
      Motorola MV209: 4.17
   Zener: 2.34–2.35, 4.4
Diode ring
   Preamp, D-C receiver: 8.3
Dip meter: 7.12
Diplexer: 2.40ff, 3.36–3.37
   Low-pass output filter: 2.42
Direct digital synthesis (DDS): 4.18, 4.26
   Spurious responses related to: 7.41
Direct-conversion (D-C) receiver: 1.6ff, 6.6, 6.10, 8.1ff
   Block diagram of, basic: 8.2
   Modular: 8.13ff
   Noise figure: 8.12
   Peculiarities: 8.6–8.7
   Single-sideband (SSB): 6.7
Directional coupler: 3.16, 3.36
Discrete Fourier transform (DFT)
   Spectral frequency response: 10.27
   Spectrum analyzer display: 10.27
   Windowing functions: 10.28
      Hamming: 10.27–10.28
Dishal Method: 3.9
Display
   Waterfall: 11.28–11.29
Distortion: 2.10, 2.12
Ditter (The *Ditter*): 7.40

Dobbs, George, G3RJV: 1.9, 1.11
Domain: 3.1
   Frequency: 3.1, 6.1ff
   Time: 3.1, 6.2ff
Doppler
   Effects: 8.8
   RF, Illustration of: 8.8
Double sideband (DSB)/CW 50 MHz station: ... 6.90ff
Double-sideband AM: 6.7
   Transmitter: 6.7
Double-tuned circuit (*See* Filter)
Doubly-terminated filter (*See* low-pass filter)
Downconverter: 9.37ff
Drift
   Compensating for oscillators with: 4.4ff, 7.42
Drive, common-mode: 2.16
Dropout: 1.14–1.15
DSB
   Modulator, low-distortion: 9.47ff
   With carrier: 9.49
DSP-10 2-m radio: 10.27, 12.24ff
Dual-gate MOSFET mixer: 5.12
Dummy load: 1.16, 2.33, 7.8
   50-ohm termination: 1.16
Dynamic range (DR): 6.29ff, 7.20
   Compression algorithms: 10.2
   Receiver with enhanced: 6.44ff

E

Easy-90 receiver: 6.34
Ebers-Moll equations: 2.10–2.11
Electronic T/R system: 2.33
EME-2 moon-bounce mode: 12.25ff
   Faraday rotation: 12.27
   Pre-detection filtering: 12.26
   Transmitter waveforms: 12.26
Emitter bypassing: 2.31
Emitter degeneration: 1.13, 2.7ff
Emitter follower: 2.7ff
   Input resistance: 2.7
   Output impedance: 2.7
   Voltage gain: 2.7
Emitter resistance: 2.8
Emitter saturation current: 2.10
Encoder
   Rotary optical: 11.2
   Rotary, programming of: 11.5–11.6
Engelbrecht, R. S.: 3.38, 6.47
ENR (Excess noise ratio): 2.21, 7.39
Environmental chamber
   For oscillator testing: 7.42
Epiphyte transceiver: 6.71
EPROM: 10.2
Equations
   Calculating power from oscilloscope readings: 7.9
Equivalent series resistance (ESR)
   Value in crystals: 7.38
Error, Phase and amplitude
   Phasing method: 11.22
Excess noise ratio (ENR): 2.21, 7.39
Excessive miniaturization: 1.4
Exciter
   AM, low-distortion: 9.48–9.49

Experiments
    Tunable hum: 8.9
Express PCB, Version 2.1.1: 12.32
EZ-Kit Lite: 10.2ff, 11.1

F
Fair-Rite (Amidon) cores: 2.31, 3.34
Faraday rotation: 12.27
Faraday's Law: 3.33
Fast Fourier Transform (FFT): 7.35, 10.4
FCC: 1.5
Feedback: 2.19
    Amplifier: 2.24ff
    Negative: 4.1ff
    Positive: 4.1ff
Ferrite balun core: 2.36
    Binocular type: 2.36
Ferrite bead: 1.17
Ferrite transformer: 3.17, 3.33
    Magnetic field: 3.33
Ferrite transmission-line transformers: 3.34
FFT (Fast Fourier Transform): 7.35, 10.4
Field Day (ARRL): 12.11
Field effect transistor (*See* Transistor, field effect)
Filter
    Active: 3.24
        Selectivity from audio filtering: 3.24
        Voltage controlled voltage source (VCVS): 3.24
    All pass: 3.1ff
    At VHF and higher: 3.11
    Audio, SSB and CW: 9.40
    Band reject: 3.1
    Bandpass: 3.1ff, 5.4, 12.14ff
        14-MHz, for VXO transmitter: 5.20
        21-MHz, for VXO transmitter: 5.21
        Active: 3.26
        Coupling: 3.9
        Finite impulse response (FIR): 3.28
        Infinite gain multiple feedback (IGMFB): .. 3.26–3.27
        LC: 3.8
        Lichen transceiver: 6.76
        Losses in: 3.8
        Monoband SSB/CW transceiver: 6.85, 6.88
        Multiple resonator: 3.9
        Stopband attenuation: 3.12ff
        Transmission line resonators: 3.11
        Triple tuned: 3.11, 12.13
    Bandwidth: 3.2
    Crystal: 3.1ff, 12.13
        4th order monolithic: 7.28ff
        8th order: 7.28
        Bandwidth: 3.20
        Bidirectional: 6.62
        Butterworth design: 3.23
        Group delay: 3.23–3.24
        KVG XF9-M (German): 12.22
        Linear phase: 3.24
        Lower sideband ladder topology: 3.19
        Mesh: 3.19–3.20
        Min-loss (Cohn filter): 3.21ff
        Response: 6.28, 6.84
        Using 3.58-MHz TV color burst: 3.20–3.21
    Crystal, 4th order monolithic: 7.28ff

    Crystal, 8th order: 7.28
    Double tuned circuit (DTC): 3.10
        Design: 3.14
        Top-coupled: 3.10
        Transmission line: 3.15
    DSP: 3.1
        Audio filter: 3.28
    Finite impulse response (FIR): 3.1ff
        Taps: 3.28
    Frequency domain response: 3.1
    Hairpin: 3.16
    High fidelity speech: 9.46–9.47
    High-pass, for harmonic evaluation: 7.32
    Impedance matching networks: 3.29, 3.32
        Directional impedances: 3.29
        L-network: 3.30
        π-network: 3.30
        T-network: 3.30–3.31
    Infinite impulse response (IIR): 3.1
    Input impedance match as performance measure: 3.2
    Insertion loss (IL): 3.1ff
    LC: 3.1
    Loop: 4.18
    Lossless: 3.1
    Low-pass
        In harmonic evaluation: 7.31
        Lichen transceiver: 6.82
    Measurements, and tracking generators: 7.34
    Optional, for phasing receivers: 9.40–9.41
    Passband: 3.1ff
        Ripple: 3.1–3.2
    Passive: 3.1
    Preselector: 6.44
    Quartz crystal: 3.3ff
    RC active: 3.1
    Real: 3.1
    Receiver
        Crisp sound: 3.23
    Resonator: 3.9ff
        Acoustic: 3.8
        Electric: 3.8
        Microwave: 3.9
        UHF helical: 3.9
        VHF helical: 3.9
    Response improvement with decimation: 11.27
    Roofing: 6.46
    Shape: 3.9
    Simple video: 7.39
    Spectrum analyzer IF: 7.29
    Stopband: 3.1ff
    Time delay: 3.1
    Transfer properties: 3.1
        Voltage transfer function: 3.1ff
Filter (*See also* High-pass filter)
Filter (*See also* Low-pass filter)
Filtering
    Audio, DSP in: 11.23–11.24
    LMS denoise: 11.28
Fisher, Reed, W2CQH: 3.36ff
Flag
    Programmable: 11.4, 11.10
Formulas
    Power density: 8.8

Forward bias: 2.1
Fourier Transform: 7.25
    Discrete Fourier transform (DFT): 10.23ff, 12.25
    Fast Fourier transform (FFT): 10.26
Frequency
    Carrier: 4.2
    Counter: 1.11, 4.5, 4.29, 7.11
        Accuracy: 4.31
    Domain: 2.10, 6.1ff
        Mixer output: 5.5, 5.12
    Doubler: 5.16
    Incremental tuning: 6.66
    Intermediate (IF): 5.1, 6.6
    Measurement: 7.11–7.12
    Multiplier: 5.1, 5.16, 6.91
    Normalized rate of change of (TCF): 4.5
    Offset: 6.66
    Shift: 6.65
    Synthesis: 4.18
    Synthesizer: 4.1, 4.31
    Tripler: 5.17
Frequency converter
    A minimum-parts-count: 9.8
Front-end design, receiver: 6.27, 6.30
    General-purpose: 6.32
    Modern: 6.46
FSK441: 12.28

G
G3UUR method: 3.19
G3UUR oscillator: 3.19
Gain
    High audio, in D-C receivers: 8.6–8.7
Gain compression: 2.21–2.22, 6.28
    Mixer: 5.6
General-purpose receiver front end: 6.32
Generator
    Audio: 11.11–11.12
    Swept voltage: 7.26–7.27
    Tracking: 7.34
Generators and sources: 7.13ff
GI3XZM: 1.11
Gilbert cell: 1.7, 1.9, 4.20
    Balanced modulator: 6.57
    Bipolar junction transistor mixer: 5.11
    Mixer: 5.10, 6.54, 6.62, 12.7
Gilbert D-C receiver
    Preamp: 8.3
Gilbert, Barrie: 5.10
Golden screwdriver: 1.4
Greenman, Murray, ZL1BPU: 12.27
Gumm, Linley, K7HFD: 4.12–4.13

H
H-mode mixer: 5.15, 6.48ff
Hairpin circuit: 3.15
Hairpin filter: 3.16
Ham radio: 1.1, 1.11
Hamilton, Nick, G4TXG: 3.36
Hamming window function: 10.27
Hang automatic gain control (AGC) system: 6.23
Harmonic: 2.21
    Distortion: 1.19, 2.14ff, 6.28

Distortion, measurements of: 7.31–7.32
    Suppression: 1.19
Harmonics: 2.10
Hartley oscillator (*See also* Oscillator): 1.9
Hawker, Pat: 5.15
Hayward, Roger, KA7EXM: 12.6
Hayward, Wes, W7ZOI (author): 12.1, 12.10
Helical resonator: 3.16
Hexfet amplifiers: 2.37
High fidelity
    Speech filter: 9.46–9.47
High frequency effects: 2.9
High level FET mixer: 5.15
High-level mixer: 6.47
High-pass filter: 1.10, 3.1ff
    Bandstop: 3.8
    For harmonic evaluation: 7.32
    Transfer functions: 3.26
    Voltage Controlled Voltage Source (VCVS): 3.25
High-performance post-mixer amplifier: 6.47
Hilbert transform
    247 taps/48-kHz sampling: 11.20
    247-tap, block diagram: 11.21
Homebrewing: 1.1
Horrabin, Colin, G3SBI: 5.15, 6.47–6.48
HP-8970 Noise Figure test set: 2.21
HP3400A true-RMS voltmeter: 4.17
Huff 'n Puff scheme: 4.6ff
Hum
    Probe: 8.9
    Tunable or common mode: 8.8–8.9
Hybrid: 3.35
Hybrid-π model: 2.9

I
I-V characteristic: 2.1
Ideal diode (*See* diode)
Ideal element: 2.1
Ideal transformer: 3.32
IF (Intermediate frequency)
    Filters, for spectrum analyzers: 7.29
IIP3 (receiver input intercept): 7.18
    Test setup to determine: 7.19
Image
    Response: 5.4
    Signal: 5.4, 6.6
    Suppression: 5.4
Image-rejection detector
    A minimum-parts-count: 9.8
IMD testing: 7.17
Impedance match/missmatch measurement: 2.15
Impedance transformation circuits: 2.33
Inductance
    Common mode: 3.35
    Measurement: 7.11–7.12
Inductor
    Self-shielding type: 8.6
Injection locking: 4.20
Input intercept: 6.30
    Mixer: 5.6
Insertion power gain: 2.14
Instruments: 2.14
    Power meters: 2.14

RF detection: 2.14
   Spectrum analyzers: 2.14
   Wideband oscilloscopes: 2.14
   Wideband voltmeters: 2.14
Integrator: 1.20
Intercept point: 6.30
Interface
   Circuitry for other mixer types: 9.44
   Three-wire serial: 11.2–11.3
Intermediate frequency (IF): 5.1, 6.6, 6.15
   Amplifier and AGC: 6.15
   Field effect transistor (FET) system examples: 6.23
   General-purpose amplifier: 6.20
   Speech processor: 6.59
   Systems: 6.18
Intermodulation distortion (IMD): 2.21ff, 6.28
   Mixer: 5.6
   Order: 2.21
   Ratio: 2.22
   Testing: 7.17
International Rectifier (Hexfets): 2.37
Interrupt service routine (ISR): 11.1
Interrupts: 10.4, 11.1–11.2
Introduction to Radio Frequency Design: .. 2.8ff, 3.9, 4.33
Inverting input: 2.18
ISB
   Mode: 9.42
Isolation
   Mixer: 5.4

J
JAØAS: 4.16
JFET (*See* Transistor, field effect)
JH1FCZ: 4.16
Johnson, D. E.: 3.25
Johnson, Harold, W4ZCB: 6.48, 6.52
Junction diode (*See* diode)

K
K3BT: 2.39
K3NIO: 12.25, 12.27
K3NIO Experiments (The): 12.25
K4XU: 2.37
K8DKC: 12.25
Kanga US: 12.33
Keep It Simple, Stupid (KISS): 1.4
Kessler, Ed, AA3SJ: 12.19
Keying
   Transmitter: 6.64
   Waveform: 6.64
Kitchin, Charles, N1TEV: 1.9
Koren, V.: 2.28
Kurokawa, K.: 3.38

L
L-leakage: 3.34
L-network: 1.18, 3.38
Large scale integration chips (LSI): 4.25
Large signal amplifiers: 2.1, 2.10
Latch wire: 11.2–11.3
LC Tester by Bill Carver, W7AAZ: 7.12
Learn by doing: 1.5
Leeson, D. B.: 4.11

Lewallen, Roy, W7EL (*See* W7EL)
Lichen transceiver: 4.18, 6.71ff
   Carrier oscillator: 6.77
   Local oscillator: 6.77
   Low-pass filter: 6.82
   Main board: 6.73
   Mixer injection switching: 6.76
   Receive audio: 6.78
   RF power chain: 6.79
   Transmit bandpass filter: 6.76
Liebenrood, John, K7RO: 6.62
Liljeqvist, Larry, W7SZ: 12.28
Limiting amplifier: 5.18
Linear power amplifier: 6.54
Liquid-crystal display (LCD)
   With DSP data device: 11.6–11.7
LM317 voltage regulator: 1.15
LM386 audio amplifier: 1.7ff
LMS denoise filtering: 11.28
LO to RF isolation: 1.9
Local oscillator (LO): 5.1, 6.2, 6.41, 9.42
   Eliminating radiation effects: 8.9ff
   Mixer drive level: 5.6
   Monoband SSB/CW transceiver: 6.83
   Radiation and reflection
     Transients: 8.7–8.8
Loop filter: 4.18ff
Lore: 1.4, 2.29
Low frequency
   Resolution: 7.11
Low-noise amplifier (LNA)
   Swept frequency plot of: 9.36
Low-noise RF amplifier: 8.13
Low-pass filter: 1.10ff, 2.33, 3.1ff, 10.9ff, 12.30
   3rd-order: 3.3
   Bessel: 3.3
   Butterworth: 3.3ff, 10.14
   Cauer-Chebyshev (elliptic): 3.7, 3.16
   Chebyshev: 1.20, 2.33, 3.3ff, 10.14, 12.18
   Cutoff frequency: 3.1, 3.3
   Doubly-terminated: 3.2
   For harmonic distortion evaluation: 7.31
   Lichen transceiver: 6.82
   Odd-order Pi: 3.3
   Passband: 3.1
   RC active: 3.25
   Stopband: 3.1
   Transfer function: 3.4ff
   Transformation: 3.8
   Trap frequencies: 3.7
   Ultra-spherical: 3.5

M
Maas, Steve: 5.4
MacCluer, C. R., W8MQW: 10.29
Makhinson, Jacob, N6NWP: 6.47
Manhattan breadboarding (*See* Breadboard circuits)
Manhattan construction (*See* Breadboard circuits)
Manly, Ernie, W7LHL: 12.27–12.28
Master oscillator, power amplifier (MOPA): 6.5
Matched (source to load): 2.14
MathCad 7.0: 4.33

Mathcad file
   On book CD: 5.2
Mathematical analysis: 1.6
Mathematics
   Audio phase-shift network: 9.4
   Image-rejection: 9.4
   Low-pass filter: 9.4
   Mixer: 9.4
   of image suppression: 9.5
   of recovering the desired signal: 9.6
   Q-channel: 9.4
   Sideband suppression: 9.6
MAX038 (Maxim): 7.13
Maximum smoke: 1.4
MDS measurement: 7.18
Measurement: 2.14
   Calibration during: 7.31
   DC: 7.2
   Impedance
      Bridge use in: 7.21ff
   Impedance, of diplexer driving point: 9.17–9.18
   In situ (in-place): 2.14, 7.1
   Mixer noise figure: 5.6
   Noise figure, test setup for: 7.39
   Of crystals: 7.37–7.38
   Of frequency, inductance and capacitance: 7.11–7.12
   Of harmonic distortion: 7.31–7.32
   Of IIP3: 7.18
   Of MDS: 7.18
   Of Q, in LC resonators: 7.36–7.37
   Receiver, for SSB transmitters: 7.33–7.34
   RF power: 7.5ff
   Substitution: 2.14
   Test equipment for: 7.1ff
   Using substitution in: 7.1
Measurement receiver: 7.26
Mechanical displacement: 3.17
Metcalf, Mike, W7UDM: 6.61
Meter, S: 6.21
Micro-Mountaineer Transceivers: 12.5–12.6
   Western Mountaineer: 12.7ff
Micro-R1: 12.16
Micro-strip: 3.36
   Transmission line: 3.15
Micrometals, Inc.: 3.14ff, 4.6
   T30-6, a common toroid core: 3.31
   Toroid numbering scheme, copyright: 3.32
Microphone
   Amplifier: 9.45–9.46
Microphonics: 8.7
Microwatt meter circuits: 7.7
Microwave
   SSB exciter prototype: 9.44
Mini-Circuits
   MAR-2: 2.27
   Mixer: 5.15
   POS-110 VCO: 4.21
   SBL-1 mixer: 4.19
Minimum detectable (or *discernable*) signal (MDS):
.... 6.11, 6.29
Mixer: 2.19, 5.1, 6.5ff
   Amplifier, post: 5.14
   Balance: 5.5

Balance, adaptive: 8.11
Balanced
   Increased L-R isolation of: 11.25
Bipolar transistor: 5.3
Conversion gain: 5.6
Conversion loss: 5.6
Diode: 5.3
   Ring: 5.13
   Ring, commutating balanced: 5.8
Dual gate MOSFET: 5.12
Environment: 9.49
FET: 6.47
For D-C receivers: 8.12
Gain compression: 5.6
Gilbert cell: 5.10, 6.54, 6.62
H-mode: 5.15, 6.48ff
High-level: 5.15, 6.48
IF-port driver amplifiers: 9.47ff
Injection switching: 6.76
Input intercept: 5.6
Intermodulation distortion (IMD): 5.6
Isolation: 5.4
JFET with LO: 5.1
Local oscillator (LO) drive level: 5.6
Measurement: 5.4
Mini-Circuits: 5.15
MOSFET
   Dual gate: 5.12, 12.18
MOSFET ring: 5.9
NE-602: 5.10
Noise figure: 5.6
Other types, interfaces for: 9.44
Output: 5.5
Recommendations: 8.12
Specification: 5.4
Switching-mode: 5.4, 6.47
   Commutating, with FET: 5.8
Mixer/LO
   Block diagram, with reflection coeff.: 8.7
Mixing product detector: 1.13
MMICs: 7.8
Moda, Giancarlo, I7SWX: 6.48
Mode
   Binaural: 9.42
   ISB: 9.42
Model: 2.1
   Current generator: 2.11
   Field effect transistor (FET): 5.1
   Of a quartz crystal: 7.37
Modeling: 2.1
   Model: 2.1
   Model current generator: 2.11
   Process: 2.11
Modular equipment: 1.4
Modulation
   Amplitude: 6.1–6.2
   Cross: 6.28
Modulator
   Balanced: 6.2, 6.56
   Circuitry used with audio PSN: 9.46
   DSB: 9.49
   Low-distortion DSB: 9.47ff
Monoband SSB/CW transceiver: 6.83
   BFO/carrier oscillator: 6.85

Control circuits: 6.86, 6.90
Local oscillator: 6.84
Power chain: 6.86
Receiver circuits: 6.90
SSB generator: 6.85
MOSFET (*See* Transistor, field effect (FET))
Mouser Electronics: 12.17
Multiple-port networks: 3.35
Splitter/Combiner: 3.35
Multiplier
Frequency: 5.1, 5.16
MWS Wire Industries: 3.33
Multifilar® parallel banded magnet wire: 3.33

N
NE-602 Integrated circuit: 1.7ff
Mixer: 5.10
Negative feedback: 1.12, 2.4ff
Network
All-pass pair: 9.29
All-pass, second-order: 9.30
Audio phase-shift (PSN): 9.27ff
Bandpass diplexer: 9.17
Bifilar toroid quadrature hybrid: 9.26
In-phase splitter-combiner: 9.24ff
LO and RF phase-shift: 9.24ff
LO quadrature: 9.26
Op-amp, all-pass, single-stage: 9.28
Phase-shift
Component tolerances for: 9.29ff
Polyphase: 9.32
Simple logic LO phase-shift: 9.27
Noise: 7.38ff
Additive: 12.24
Bandwidth: 6.29
Evaluating, in local oscillators: 7.40
Figure: 2.20–2.21, 6.10ff
Direct conversion: 8.12
Measurement: 2.21, 2.27
Measurement of mixer: 5.6
Mixer: 5.6
Receiver
Effect of mixer IF-port attenuation: 9.18
Test setup to measure: 7.39
Figure differential
Hot-cold resistor: 8.12
Gaussian, white (WGN): 12.24
Power: 10.13
Signals and multiplicative: 12.25
Sources: 7.38ff
Temperature: 6.11
Noise factor (*See* Noise, Figure)
Noise gain: 2.20
Noise power: 2.20
Non-inverting input: 2.18
Nonlinear device: 5.3
Normalized resistance: 2.14
Norton, D.: 2.27–2.28
Notes
On phasing rig construction: 9.49
NPO (*See* oscillator, drift, compensating for)
Nyquist criteria: 10.26

O
Ohm's Law: 2.1ff
Open loop gain: 2.19
Operating system (OS): 11.1
Operational amplifier (*See* amplifier)
Optical (Rotary) encoder: 11.2
Optrex DMC-16117A display: 11.7
Oscillator
Beat-frequency (BFO): 6.6, 6.85
Butler: 4.15
Carrier: 6.85
Circuits: 6.65
Clapp: 4.2, 4.14
Colpitts: 1.13, 4.1ff, 7.37–7.38
VHF: 4.9
Conversion: 5.1
Crystal controlled: 4.1, 6.65, 7.16ff
Crystal controlled, for 7 and 50 MHz: 7.17
Crystal controlled, for receiver MDS: 7.18
Crystal, for receiver input intercept (IIP3): 7.18
Drift, compensating for: 4.3, 7.42
Negative positive zero (NP0): 4.3ff
Hartley: 1.9, 4.1ff, 7.15
LC: 4.1, 6.66, 7.12
Lichen transceiver, carrier: 6.77
Local: 4.1ff, 5.1, 6.2, 6.41, 9.42
Evaluating noise in: 7.40
Lichen transceiver: 6.77
Monoband SSB/CW transceiver: 6.84
Negative resistance: 4.1
Noise: 4.10
Spectrum of: 4.11
Wideband: 4.11
Permeability-tuned: 4.17
Pierce: 4.14
Seiler: 4.2ff
Synthesized: 4.6
Testing of, in environmental chamber: 7.42
Vackar: 4.2ff
Variable-frequency (VFO): 6.65, 6.84
Voltage-controlled (VCO): 4.17, 6.52
Wide-range tuning: 7.15
Oscilloscope: 2.14–2.15, 7.3ff
10X probe: 7.4
Block diagram (partial): 7.4
RF power measurement using: 7.8ff
Traditional measurements (K7OWJ reference): 7.5
Trigger level: 7.4
Output impedance transformation: 2.12
Output intercept (OIP3): 6.30, 7.20
Output power: 2.7
Oven
For evaluating oscillator drift: 7.42
Oxner, Ed: 5.8, 5.15

P
Pi-type matching network: 1.19, 2.25
Parasitic inductance (*See* bypassing and decoupling)
Parts list
Easy-90 receiver: 6.35
Peak detector: 7.5

Phase
   and amplitude
      Errors, with phasing method: 11.22
      Shifters, DSP: 9.32
Phase detector: 4.19ff
Phase locked loop (PLL): 4.18ff, 7.41, 10.6
   Diode ring phase detector: 4.20
   Loop filter: 4.21, 4.24
   Pull-in range: 4.20
   Synthesizer: 4.25
   Tracking filter: 4.22
Phase noise: 4.1ff
   Blocking: 6.28
   Measurement: 6.52
Phase trim adjustment: 9.23–9.24
Phasing
   Receiver trimming: 9.42ff
   Receivers and exciters
      Adjusting: 9.19ff
   Receivers and transmitters: 9.1ff
   Rig construction
      Notes: 9.49
   SSB exciter, high-performance: 9.45
Phasing capacitor: 3.17
Phasing mathematics: 9.4ff
Phasing method: 1.6–1.7
PIN diode: 6.16ff
   Attenuator: 6.18
   Transmit/receive (T/R) switch: 6.69
Pinch-off voltage: 2.5, 2.9
PLL (Phase-locked loop): 4.18ff, 7.41
Polyphase networks: 9.32
Portable operation: 12.1
   Battery-voltage testing: 12.3
   Batteries and power sources: 12.1
      Alkaline flashlight cell: 12.1
      Nickel Metal Hydride (NiMH): 12.1
      Nickel-Cadmium (NiCd): 12.1
   DSB/CW 50 MHz station: 6.90ff
   Portable antennas: 12.2
      Inverted-V dipole: 12.2
   Portable transmatch: 12.3
   Sleeping bag radio: 12.4
Power amplifiers: 2.31ff
   50 MHz: 6.86ff
   A CW-QRP Rig: 2.33ff
   Audio: 9.41
   Class-A: 6.55
   Class-AB: 6.56
   Classes of operation: 2.31
   Using IRF511 MOSFETs: 11.18
Power available: 2.14
Power density formula: 8.8
Power gain: 2.7, 2.14
   Transducer: 3.1
Power measurement: 2.13
Power meter
   Logarithmic: 7.7
   Low-level: 7.6
   QRP (Lewallen reference): 7.7
   W7EL design: 7.6
Power pad: 7.6
Power resistors
   At RF: 7.10

Power supply: 1.14
   Schematic: 8.9
Power tap: 7.8
Power termination: 7.6
Preamplifier
   Use, permitting mixer loss: 9.35
Preselector filter: 6.44, 6.51
Primitive explanations: 1.1
Printed circuit boards (PCB): 1.2
   Etchant: 1.2
      Ferric chloride: 1.2
   Insulating material: 1.2
      Epoxy-fiberglass: 1.2
   Photo-resist material: 1.2
   Printed metal runs: 1.2
   Surface mount technology (SMT): 1.2ff, 2.29
      Surfboards: 1.2
Probe
   Hum: 8.9
Processing
   Multi-rate, in DSP-10: 11.29
Processor
   DSP-based audio: 11.29
Product detector: 5.1
Programmable divider: 4.25
Programmable flag: 11.4, 11.10
Programming the rotary encoder: 11.5–11.6
PSPICE
   Simulations of phase and amplitude variations: 9.17
PUA43, Weak signal communications mode: 12.27–12.28

Q

   and filter losses: 3.8
   Determination of, via bandwidth measurement: 7.36
   Loaded: 4.12
   Loaded tank: 4.10
   Measurement of, in LC resonators: 3.9, 7.36–7.37
   Measurement, test fixture for: 7.36
   Quartz crystal: 3.17
QEX: 3.21
QRP: 1.4
   Complete rig for 2m (DSP-10): 11.7
   Power meter: 7.7
   Transceivers: 1.4
QRP Power: 12.11
QRP Quarterly: 3.33
QST: 1.2, 2.28ff, 12.6
Quadrature coupler: 3.36–3.37
   Twisted-wire hybrid directional: 3.36
Quarter wavelength line, synthetic: 3.32
Quartz crystal: 3.17

R

R2 Receiver
   A next-generation, single-signal conv.: 9.33
   Updating: 9.33
R2pro receiver: 9.33ff
Radiation
   Eliminating in an LO: 8.9ff
Radio frequency (RF)
   Amplifier: 6.12
   Attenuator: 6.12
Ramp: 7.3

Ratios
    Power: 2.7
    Voltage: 2.7
Receiver: 6.1
    14-MHz: 6.34
    AGC, testing of: 7.40
    Binaural: 9.19ff
    Converter: 6.91ff
    Design and development: 9.7ff
    Design of 20- to 60-dB sideband suppression: 9.13ff
    Detector: 1.9ff
    Direct-conversion (D-C): 6.6, 6.10, 8.1ff, 12.31
        Single-sideband (SSB): 6.7
    Dynamic range (DR): 6.29
        Enhanced: 6.44
    Easy-90: 6.34
    Factor: 6.30
    Front-end
        Cross modulation: 6.28
        Design: 6.27, 6.30
        Gain compression: 6.28
        General-purpose: 6.32
        Harmonic distortion: 6.28
        Intermodulation distortion (IMD): 6.28
        Phase-noise blocking: 6.28
        Reciprocal mixing: 6.28
    Fundamentals: 6.9
    High performance D-C: 9.3
    Incremental tuning (RIT): 6.66
        Direct-conversion (D-C) transceiver: 6.67
        Superheterodyne: 6.66
    Input intercept: 6.30
    Modular D-C: 8.13ff
    Modular, block diagram: 8.13
    Noise figure
        Effect of mixer IF-port attenuation on: 9.18
    Phasing: 9.1ff
    Phasing D-C: 9.3
    R2pro: 9.33ff
    Regeneration: 1.10, 1.11
    Regeneration control: 1.10, 1.11
    Regenerative: 1.9ff
    Schematic of a modular: 8.14
    Schematic of binaural from Mar. '99 *QST:* 9.20–9.21
    Simple fixed-frequency: 9.8
    Single-signal superheterodyne: 6.6
    Superheterodyne: 6.15
    The Triad: 6.48, 6.52
    Tickler coil: 1.9
Receiver module, general purpose: 12.30
Reciprocal mixing: 6.28
References: 5.21, 6.94
Resistors
    Hot-cold noise figure differential: 8.12
    Power, at RF: 7.10
Resolution
    In a spectrum analyzer: 7.26
    Low-frequency: 7.11
Resolution bandwidth (RBW): 7.26
Resonator: 1.10, 3.8ff
    Helical: 3.16
    Tank: 1.10
    Transmission line: 3.15

Return loss (VSWR): 2.16, 7.31
Return loss bridge (RLB): 2.16, 7.22
    Directivity: 2.16
Reverse biased: 2.1–2.2
RF
    Low-noise amplifier: 8.13
RF amplifier: 1.10
    Lichen transceiver: 6.79
RF Doppler
    Illustration of: 8.8
RF impedance bridge: 7.23
RF load: 7.7
RF power measurement: 1.15, 7.5ff
RF probe: 1.16–1.17
RF resistance: 7.10
RF resistance bridge: 7.22
RF signal generator
    3-45 MHz: 7.15
    Lab-quality: 7.13
    Traditional, gen. purpose servicing: 7.13
RF sources
    General purpose: 7.14
Rhode and Schwartz: 10.21
Rhode, U.: 2.28, 4.13
Ripple: 1.14
Roofing filter: 6.46, 6.50
    Amplifier: 6.50
Rotary optical encoder: 11.2
RSGB Radio Communications Handbook: 4.10
Ruthroff: 3.34

S
S meter: 6.21
Sabin, William, WØIYH: 2.43, 6.56, 7.38
Sampling Rate
    For 18-MHz transceiver: 11.26–11.27
Saturation current (*See* diode)
Saturation region: 2.5
Saw tooth waveform: 7.3
Schematic diagrams: 1.6
    10.1-MHz converter: 5.13
    14-MHz CW receiver
        Universal VFO: 12.47
    18-MHz transceiver: 11.14ff
    28-MHz QRP module
        Transmitter power chain: 12.30
        VXO & frequency divider module: 12.29
            Modified tuning range: 12.29
    52-MHz tunable IF
        4.4-4.9-MHz VFO: 12.39
        47.5-MHz premix oscillator filter: 12.40
        52-MHz filter: 12.41
        52-MHz LO quadrature hybrid: 12.41
        52-MHz premix filter: 12.40
        52-MHz premix LO output amplifier: 12.40
        LNA: 12.39
        Premix LO mixer: 12.41
    Analog signal processor (ASP): 9.38
    Audio power amplifier: 9.41
    Bandpass diplexer: 9.16
    Basic miniR2: 9.34
    Better L-R isolation of balanced mixer: 11.25
    Bidirectional amplifier: 6.60

Binaural receiver, Mar. '99 *QST*: 9.20–9.21
Broadband quadrature hybrid: 9.28
Carrier-oscillator for CW: 6.60
CW/SSB IF amplifier: 6.58
Downconverter: 9.36
Drive and load designs: 9.28
DSB/CW 50-MHz station
    Receive converter: 6.93
    Transmitter: 6.92
    VFO: 6.93
    VXO and frequency multiplier: 6.90–6.91
Dual-band QRP CW transceiver
    Audio output amplifier: 12.22
    IF amplifier & filter section: 12.22
    LO signal processor board: 12.21
    Product detector & audio amplifier: 12.22
    Receiver front end: 12.21
    RF power amplifier chain: 12.24
    Transmit mixer & PIN diode filters: 12.23
    VFO, mixer & crystal oscillator: 12.20
Easy-90 receiver: 6.34ff
Frequency multiplier: 5.18
Frequency tripler: 5.17
Gen purpose, direct conversion receiver: 12.31
    Option for audio gain control & filter: 12.31
General-purpose receiver front end: 6.32
Gilbert cell mixer with discrete transistors: 5.11
H-mode mixer: 6.48
Hardware interface
    DSP to multiple control devices: 11.5
High-performance post-mixer amplifier: 6.47
IF speech processor: 6.59
Image-rejection mixer for 40m: 9.16
Image-stripping preselector filter: 6.45
Keying shape of amplifier stage: 6.63
LC oscillator: 6.66
Lichen transceiver
    Audio system and AGC detector: 6.79
    Bandpass filter: 6.78
    Carrier oscillator: 6.77
    Local oscillator: 6.77
    Main board: 6.73
    RF driver: 6.80
Limiting amplifier: 5.18
LM2 transceiver, 144-MHz SSB & CW
    LM2 schematic 1: 12.34
    LM2 schematic 2: 12.35
Micromountaineer transceiver
    7-MHz VFO: 12.6
    Circuitry to inject sidetone signals: 12.7
    Rig modifications to add VFO: 12.7
MicroR1: 8.5
Modular receiver: 8.14
Modulator-demodulator: 9.15
Monoband SSB/CW transceiver
    BFO and carrier generator: 6.85
    Control circuits: 6.90
    Local oscillator amplifier: 6.85
    Power amplifier for 50 MHz: 6.89
    QRP amplifier: 6.89
    SSB generator: 6.87
    Transmitter power chain: 6.86
MOSFET mixer: 5.12
PIN diode transmit/receive (T/R) switch: 6.69
Post-mixer amplifier: 5.14
Power amplifier for 50 MHz: 6.86
Power supply: 8.9
Receiver incremental tuning (RIT): 6.66
S7C superhet receiver
    Single-tuned mixer input: 12.18
Simple quadrature hybrid: 9.27
Simple SSB exciter: 9.12
Sleeping Bag Radio
    Bandpass feedthru filter: 12.45
    LNA/attenuator: 12.45
    Power amplifier: 12.44
    VFO, doubler: 12.44
Solar panel interfaces: 12.2
SSB Transceiver: 9.14–9.15
Timing circuit for battery testing: 12.2
Transmit/receive (T/R) antenna switching: 6.68
Unfinished transceiver
    Audio output & control system: 12.15
    Audio preamplifier: 12.15
    BFO and product detector: 12.14
    IF amplifier: 12.13
    Receiver front end: 12.13
    Transmit mixer, filter, keyed RF amplifier: 12.14
    VFO and RIT: 12.12
VXO transmitter with digital frequency multiplier: 5.19
    21-MHz bandpass filter: 5.21
    Power amplifier: 5.20
Western Mountaineer transceiver
    Receiver: 12.9
    Transmatch: 12.11
    VFO and transmitter circuitry: 12.8
Second-order intermodulation distortion (IMD): 6.28
Seiler oscillator (*See* Oscillator)
Selectivity: 9.32
Serial three-wire interface: 11.2–11.3
Servo loop: 4.19
Shielding
    Of spectrum analyzers: 7.30
Sideband: 5.4
    Inversion: 5.4
    Selection: 9.49
    Suppression, in transmitters: 9.10ff
    Switching: 9.42
Sidetone oscillator: 1.21–1.22
Signal analysis: 6.2
Signal generator: 1.11, 2.15
Signal generator extender: 7.16
Signal grounded (*See* bypassing and decoupling)
Signal processing: 6.1
Signetics: 1.7
Siliconix
    Commutating double-balanced mixer: 5.8
Silverman, Hal, W3HWC: 3.31
Sine wave: 6.1
Single-sideband (SSB) (*See also* SSB)
    Gen. board for monoband SSB/CW transceiver: 6.87
    Monoband transceiver: 6.83
    Receiver
        Direct-conversion (D-C): 6.7
    Signal: 6.4
    Transmission with DSP: 11.24

Transmitter: 6.7
    IF amplifier: 6.58
Single-signal superheterodyne receiver: 6.6
Sleeping bag radio: 12.42ff
Small-signal amplifiers: 2.1
Small-signal bipolar transistor model (*See* Transistor)
Small-signal diode model (*See* Diode)
Smith chart: 2.29ff, 3.31
Smith, Doug, KF6DX: 10.2
Solar panel: 12.5
Solid State Design for the Radio Amateur: 1.1, 1.4
Solid State Radio Engineering: 2.32
Source resistor method: 2.5
Sources
    Noise: 7.38ff
Sources and generators: 7.13ff
Spectral power density: 2.20
Spectral purity: 1.18, 2.41
Spectral voltage density: 2.20
Spectrum
    18-MHz CW transceiver output: 11.25
    Of a re-radiated LO: 8.9
    Of a typical SSB transmitter: 9.11
Spectrum analysis: 7.25ff
Spectrum analyzer: 1.5
    Application hints: 7.30
    Converter, for baseband: 7.34
    DFT use in: 7.35
    Experimenters, block diagram of: 7.27
    IF filters for use in: 7.29
    Lichen transceiver two-tone test: 6.81
    Output: 4.11–4.12
    Reference level on screen of: 7.25
    Resolution: 7.26
    Rudimentary: 7.25–7.26
    Shielding: 7.30
    Triple conversion: 7.32
Speech processor
    Intermediate frequency (IF): 6.59
Spittle, Derry, VE7QK: 6.71
SPOT switch: 6.67
SPRAT: 1.11
Spurious
    Emissions: 1.5
    Responses (DDS-related): 7.41
    Responses (Mixer): 5.5
Square-law device: 1.9
Squeeging: 4.4
SSB (*See* also Single-sideband): 1.2, 3.17
    Exciter prototype
        Microwave: 9.44
        Gear: 1.4
    Phasing exciter, high performance: 9.45
SSB transmitter
    Measurement of: 7.33–7.34
Structure
    Of computer programs: 11.1–11.2
Summing node: 2.19
Super-Star Professional, Eagle Software: 3.27
Superheterodyne: 1.6–1.7, 12.16
    Receiver: 6.15
        Single-signal: 6.6

Suppression
    of opposite sideband in receivers: 9.13
    of sideband
        Design, in transmitters: 9.10ff
Surface mount technology (SMT). (*See* printed circuit boards (PCB))
Swept voltage generator: 7.26–7.27
Switching
    Antenna: 11.18–11.19
    Diode: 6.62
    Mode mixer: 5.4, 6.47
    of sidebands: 9.42

T
Table
    Lookup, to determine knob motion: 11.7
    Output power of JFET mixer, 5.1: 5.2
Tantalum electrolytic capacitors (*See* bypassing and decoupling)
Taylor, Joe, K1JT: 12.28
TCF (Temperature coefficient of frequency): 4.5ff, 7.42
Tee network: 2.36
    L-C-C type: 2.36
Tektronix 494A: 4.26
Temperature
    Coefficient of frequency (TCF): 4.5ff, 7.42
    Coefficient of inductance (TCL): 4.5
    Compensation: 4.4, 7.42
    Compensation process: 7.42
    Kelvin (K): 6.10
Terminator: 7.8
Test
    Notes from min. sideband supp. experiments: 9.11–9.12
    Setup for component testing: 7.20
    Setup for minimum sideband suppression: 9.11
    Setup for noise-figure measurement: 7.39
    Setup for receiver dynamic-range measurement: 6.29
    Setup to evaluate NE-602 mixer: 5.11
Test equipment: 1.5, 7.1ff
    Audio generator: 7.13
    Dip meter: 7.12
    DVM (Digital voltmeter): 7.2
    LC tester by Bill Carver, W7AAZ: 7.12
    Logarithmic power meter: 7.7
    Oscilloscope: 7.3ff
    QRP power meter (Lewallen ref.): 7.7
    RF measurement: 7.1ff
    RF signal generators: 7.13ff
    Spectrum analyzers: 7.25ff
    True RMS voltmeter: 7.2
    Two-tone audio generator: 7.13
    VTVM (Vacuum Tube Voltmeter): 7.2
    W7EL power meter: 7.6
Test fixture
    For Q measurement: 7.36
The ARRL Handbook: 1.2, 2.23
The Art of Electronics: 2.8
Third-order intercept point: 2.22
    Third-order input intercept: 2.22
    Third-order output intercept: 2.22
Third-order intermodulation distortion (IMD): 6.28
Three-terminal devices: 2.8
Tickler coil: 1.9, 4.12
Time domain: 2.10

Time domain waveform: 6.2ff
    Diode ring commutation mixer: 5.9
Timing diagram
    Shift register: 11.4
Tolerance
    Component, in phase-shift networks: 9.29ff
Toroid: 1.10, 3.31ff, 4.5
    Ferrite inductor: 12.30
    Powdered iron: 3.31
Tracking filter: 4.22
Tracking generator: 7.34
Trail-friendly radio (TFR): 12.4, 12.6
Transceiver
    18-MHz
        DSP circuits: 11.19ff
    52-MHz tunable IF for VHF/UHF transverters: . 12.37ff
    An 18-MHz: 11.12ff
    CW/SSB: 11.12ff
    Design: 6.53
    Direct-conversion (D-C): 6.65
        For 144-MHz SSB and CW: 12.33
            Metal box version: 12.36
            Wood box version: 12.36
        Frequency offset: 6.67
        Receiver incremental tuning (RIT): 6.67
    DSP-10 (2-m): 11.27ff
    Epiphyte: 6.71
    Lichen: 6.71ff
        Carrier oscillator: 6.77
        Low-pass filter: 6.82
        Main board: 6.73
        Mixer injection switching: 6.76
        Receive audio: 6.78
        RF power chain: 6.79
        Transmit bandpass filter: 6.76
    Monoband SSB/CW: 6.83
    Single-sideband (SSB): 6.9
    Superheterodyne: 6.66
        Receiver incremental tuning (RIT): 6.66
Transconductance (g_m): 2.3ff
    Hexfets: 2.37
Transducer power gain: 2.7ff, 3.1
Transform
    Fourier: 7.25
    Hilbert, 247 taps/48-kHz sample: 11.20
Transformer
    Bifilar windings: 3.33
        Multifilar® parallel-banded magnet wire: 3.33
    Ferrite: 3.17, 3.33
    Ideal: 3.32, 3.34
    Wideband: 3.35
Transients
    In LO radiation and reflection: 8.7, 8.8
Transistor: 2.1
    Beta (β): 2.3ff
    Bipolar junction transistor (BJT): 2.1ff
        Bidirectional amplifier: 6.61
        Gilbert cell mixer: 5.11
        Mixer: 5.3
    Bipolar transistor biasing: 2.4
    Bipolar, amplifier: 6.16
    Field effect (FET): 2.1ff, 4.3ff
        Channel: 2.9

Common drain (source follower): 2.8, 2.9
Common gate: 2.8
Common source: 2.8, 6.33
GaAs MOSFET: 2.9, 4.12
HEXFET: 2.37ff
High-speed CMOS: 4.29
Junction (JFET): 2.5–2.6, 4.12
    Amplifier: 6.33
    Balanced mixer: 5.7
    Bidirectional amplifier: 6.62
    Cascode pair amplifier: 6.18
    Common gate RF amplifier: 6.12
    Common source amplifier: 6.13
    IF amplifier: 6.17
    Mixer with LO: 5.1
    Post mixer amplifier: 5.14
Metal oxide silicon (MOSFET): 2.5ff, 4.12, 4.23
    Availability: 6.14
    IF amplifier: 6.17, 6.27
    RF amplifier: 6.13
Mixer
    Commutating, switching mode: 5.8
    High level: 5.15
    Modeling: 5.1
    Passive mixer: 6.47
Small signal, bipolar model: 2.3
Transmatch: 2.33ff
    Portable, for 7 MHz: 7.24
Transmission
    Of CW/SSB using DSP: 11.24ff
Transmission line
    Microstrip: 3.15, 3.31
    Transformer: 3.31, 3.34
        Balun: 3.34
        Current balun: 3.34
        Isolation transformer: 3.34–3.35
        Q-section (Quarter-wave line): 3.31, 3.34
        Synthetic: 3.31
Transmit-receive system (T/R): 1.20, 2.41
    Antenna switching: 6.68
    PIN diode: 6.69
Transmitter: 6.1
    CW: 6.4ff
    Design: 6.53
    Double-sideband AM: 6.7
    Intermediate frequency (IF) systems: 6.57
    Phasing: 9.1ff
    Sideband suppression design: 9.10ff
    Single-sideband (SSB): 6.7
    VXO, with digital frequency multiplier: 5.19
Trask, C.: 2.28, 3.34
Triad receiver: 6.48, 6.52
Trigger level: 7.4
Trigonometric identities: 6.2–6.3
Trimming
    a phasing receiver: 9.42ff
Triple conversion
    Spectrum analyzer: 7.32
Tunable hum: 8.8–8.9
TVRO dish: 12.28
Two-tone dynamic range: 6.29
Two-tone generator: 7.13–7.14

Two-tone test
    Lichen transceiver: 6.81

U
UART: 10.2
Ugly construction (*See* Breadboard circuits)
Ugly Weekender: 4.27–4.28
UHF
    Bridge suitable for: 7.24
Unfinished, The (aka *The Unfinished-7*): 12.12ff
Uniform random noise: 10.12

V
Vackar oscillator (*See* Oscillator)
Vacuum Tube Voltmeter (VTVM): 7.2
Varactor diode: 6.67
Variable crystal oscillator (VXO): 4.15–4.16, 6.91
    Super: 4.16
Variable-frequency oscillator (VFO): 6.65, 6.84
Video
    Simple filter for: 7.39
Voltage
    Controlled oscillator (VCO): 6.52
Voltage-driven component: 2.3
Voltage drop: 2.1
Voltage follower: 2.18
Voltage gain: 2.3ff
Voltage limiting: 2.13
Voltage reflection coefficient (Γ): 2.15
Voltage regulator: 1.15
    LM317: 1.15
    Switching-mode regulator: 1.15
Voltage standing wave ratio (VSWR): 2.15
Voltage-variable resistor: 2.9
VXO (*See also* Variable crystal oscillator): 12.28–12.29
VXO extender: 4.33
VXO transmitter
    Digital frequency multiplier: 5.19

W
W7AAZ: 3.19
W7EL: 2.27ff, 3.36, 4.6, 12.7, 12.32
    Optimized QRP transceiver: 1.3
    Power meter: 7.6
    The "Brickette": 2.37, 2.40
W7LHL (*See* Manly)
W7PUA: 12.27
W7ZOI: 12.1, 12.10
WA3RNC: 1.9
WA7MLH: 1.4, 12.4
WA7TZY: 4.17
Wade, Paul, W1GHZ: 7.38
Walkman®: 1.11–1.12
Waterfall display: 11.28–11.29
Waveform
    Frequency domain: 6.1ff
        Mixer output: 5.5, 5.12
    Keying: 6.64
    Saw tooth: 7.3
    Time domain: 6.2ff
        Diode ring commutation mixer: 5.9
        Diode switching-mode mixer: 5.3
Waveforms: 2.11
Waveforms, Class C amplifier: 2.34
Wenzel, Charles: 5.17
Western Mountaineer: 4.17, 12.2ff
Wheatstone bridge: 7.21
Wien bridge circuit: 7.13
Wilson, Robert, KL7ISA: 3.31
Wireless technology: 1.2
WSJT program: 12.28
WWV, WWVH: 12.17

X
XR-2206 (Exar): 7.13

Z
Zener diode: 2.33ff
Zverev: 3.11

FEEDBACK

Please use this form to give us your comments on this book and what you'd like to see in future editions, or e-mail us at **pubsfdbk@arrl.org** (publications feedback). If you use e-mail, please include your name, call, e-mail address and the book title, edition and printing in the body of your message. Also indicate whether or not you are an ARRL member.

Where did you purchase this book?
☐ From ARRL directly ☐ From an ARRL dealer

Is there a dealer who carries ARRL publications within:
☐ 5 miles ☐ 15 miles ☐ 30 miles of your location? ☐ Not sure.

License class:
☐ Novice ☐ Technician ☐ Technician with code ☐ General ☐ Advanced ☐ Amateur Extra

---

Name _____ ARRL member? ☐ Life ☐ Annual ☐ No

Call Sign _____

Daytime Phone ( ) _____ Age _____

Address _____

City, State/Province, ZIP/Postal Code _____

If licensed, how long? _____ e-mail address: _____

Other hobbies _____

Occupation _____

For ARRL use only	EXP METHODS
Edition	1 2 3 4 5 6 7 8 9 10 11
Printing	1 2 3 4 5 6 7 8 9 10 11

From _____
_____
_____

Please affix postage. Post Office will not deliver without postage.

EDITOR, EXPERIMENTAL METHODS IN RF DESIGN
ARRL—THE NATIONAL ASSOCIATION FOR AMATEUR RADIO
225 MAIN STREET
NEWINGTON CT 06111-1494

— — — — — — — — — — — — — — please fold and tape — — — — — — — — — — — — — —